Studies in Surface Science and Catalysis 159

NEW DEVELOPMENTS AND APPLICATION IN CHEMICAL REACTION ENGINEERING

Studies in Surface Science and Catalysis

Advisory Editors: B. Delmon and J.T. Yates
Series Editor: G. Centi

Vol. 159

NEW DEVELOPMENTS AND APPLICATION IN CHEMICAL REACTION ENGINEERING

Proceedings of the 4th Asia-Pacific Chemical Reaction Engineering Symposium (APCRE '05), Gyeongju, Korea, June 12-15, 2005

Edited by

Hyun-Ku Rhee

Seoul National University, Korea

In-Sik Nam
Jong Moon Park

Pohang University of Science and Technology, Korea

ELSEVIER

Amsterdam – Boston – Heidelberg – London – New York – Oxford – Paris
San Diego – San Francisco – Singapore – Sydney – Tokyo

Elsevier
Radarweg 29, PO Box 211, 1000 AE Amsterdam, The Netherlands
The Boulevard, Langford Lane, Kidlington, Oxford OX5 1GB, UK

First edition 2006

Library of Congress Cataloging-in-Publication Data
A catalog record for this book is available from the Library of Congress

British Library Cataloguing in Publication Data
A catalogue record for this book is available from the British Library

ISBN-13: 978-0-444-51733-3
ISBN-10: 0-444-51733-2
ISSN (Series): 0167-2991

For information on all Elsevier publications
visit our website at books.elsevier.com

Printed and bound in The Netherlands

06 07 08 09 10   10 9 8 7 6 5 4 3 2 1

Working together to grow
libraries in developing countries

www.elsevier.com | www.bookaid.org | www.sabre.org

ELSEVIER   BOOK AID
             International   Sabre Foundation

CONTENTS

BIOLOGICAL AND BIOCHEMICAL REACTION ENGINEERING

CATALYSIS AND CATALYTIC REACTION ENGINEERING

CHEMICAL REACTION ENGINEERING IN MICROELECTRONICS

ENVIRONMENTAL REACTION ENGINEERING

FLUIDIZED BED AND MULTIPHASE REACTORS

FUEL CELLS AND ELECTROCHEMICAL REACTION ENGINEERING

MICRO-REACTION TECHNOLOGY

MODELING, SIMULATION AND CONTROL OF CHEMICAL REACTORS

NANO MATERIALS SYNTHESIS AND APPLICATION

NOVEL REACTORS AND PROCESSES

POLYMER REACTION ENGINEERING

Preface

This Proceedings of APCRE'05 contains the articles that were presented at the 4th Asia-Pacific Chemical Reaction Engineering Symposium (APCRE'05), held at Gyeongju, Korea between June 12 and June 15, 2005, with a theme of "New Opportunities of Chemical Reaction Engineering in Asia-Pacific Region". The authors were invited to submit their manuscripts during the APCRE'05 at Gyeongju. It was understood that every manuscript should be reviewed by two experts in the corresponding area and, if accepted, the manuscripts will be published in the book series "Studies in Surface Science and Catalysis" by Elsevier BV early in 2006.

Following the tradition of APCRE Symposia and ISCRE, the scientific program encompassed a wide spectrum of topics, including not only the traditional areas but also the emerging fields of chemical reaction engineering into which the chemical reaction engineers have successfully spearheaded and made significant contributions in recent years. Indeed, APCRE'05 focused on the following areas and naturally the articles are classified into the same areas as in the APCRE'05:

Biological and Biochemical Reaction Engineering
Catalysis and Catalytic Reaction Engineering
Chemical Reaction Engineering in Microelectronics
Environmental Reaction Engineering
Fluidized Bed and Multiphase Reactors
Fuel Cells and Electrochemical Reaction Engineering
Micro-reaction Technology
Modeling, Simulation and Control of Chemical Reaction Systems
Nano Materials Synthesis and Application
Novel Reactors and Processes
Polymer Reaction Engineering

Out of 284 papers presented at the APCRE'05, 190 papers have been accepted after a cautious review process and included in this Volume. In addition, six plenary lectures and 11 invited lectures are placed in two separate chapters in the front. One author was allowed to have no more than four articles in the Proceedings. The articles were contributed by chemical reaction engineers and scholars from academia and R&D sector as well as from industrial sector, representing 16 countries not only from the Asia-Pacific region but from the western hemisphere.

We would like to express our sincere gratitude to all the authors for their valuable contribution and to the members of the Organizing Committee for sparing their valuable time and efforts to carry out the review process so successfully. We hope that this Proceedings may serve as noticeable references for the scientific and industrial communities in the years to come, contribute to make an overview of new developments and application in chemical reaction engineering, and allow chemical reaction engineering to make its full contribution to further advancement of respective countries in the Asia-Pacific region and other parts of the world.

Hyun-Ku Rhee
In-Sik Nam
Jong Moon Park

December 20, 2005

ORGANIZING COMMITTEE MEMBERS

Prof. Hyun-Ku Rhee (Chairman)
Seoul National University

Prof. In-Sik Nam (Secretary General)
Pohang University of Science & Technology

Prof. Yoon Bong Hahn
Chonbuk National University

Prof. Yong Kang
Chungnam National University

Prof. Jong-Ho Kim
Chonnam National University

Prof. Kwan-Young Lee
Korea University

Prof. Dong-Keun Lee
Gyeongsang National University

Prof. Tae-Jin Lee
Yeungnam University

Prof. Won Mook Lee
Hanbat National University

Dr. Suk Woo Nam
Korea Institute of Science and Technology

Prof. Young Woo Nam
Soongsil University

Prof. Jong Moon Park
Pohang University of Science and Technology

Prof. Seung Bin Park
Korea Advanced Institute of Science and Engineering

Prof. Young-Koo Yeo
Hanyang University

INTERNATIONAL ADVISORY COMMITTEE MEMBERS
(APCRE Working Party)

PLENARY LECTURES

Studies in Surface Science and Catalysis, volume 159

Hyun-Ku Rhee, In-Sik Nam and Jong Moon Park (Editors)

Challenging reaction engineering problems

Dan Luss

Chemical Engineering Department
University of Houston, Houston, TX 77204 USA

1. INTRODUCTION

We are living in a rapidly changing world. Demographic changes due to the aging population in Europe, US and Japan and the population explosion in Asia are bound to generate major economic changes. The pace of globalization is accelerating and we all face issues of food, health, safety, comfort, energy and environment. The accelerating world wide economic, political, scientific, technological and social changes affect all of us and the chemical industry.

Past success of the chemical industry was accomplished by discovery and commercialization of novel products and development of manufacturing processes. Recently the research and development activities in the US have recently been drastically reduced as the management of many companies believes that the probable success of development of novel profitable chemicals is rather small. Moreover, the difference in the price of key raw materials, such as natural gas, between the US and other countries stopped the expansion of various processes (such as polyethylene production). The realization that various US plants will eventually have to be closed, has further decreased the research activity. At present many companies concentrate on incremental advances in order to engineer me-too or me-slightly-better or slightly-cheaper products, not on major developments. I believe that evolving world-changing scientific innovations will offer the chemical industry a great opportunity to contribute to enabling technologies and manufacture many new products. Our challenge is to benefit from these opportunities.

It is very difficult to predict the potential impact of any innovation. The steam engine was developed in England to remove water from coal mines. Yet, other applications of this invention significantly contributed to the build up of the economic power of England. World-changing scientific innovations are the products of long term research. The current accelerating rate of change is likely to lead to the simultaneous evolution of several scientific innovations. These will lead to the development of enabling technologies and novel products. We have a great opportunity to become important contributors to these enabling technologies and to the manufacturing of the novel products. A biotech revolution has been initiated by the scientific advances in molecular biochemistry, genomics, proteomics and engineering of metabolic pathways. Among others it will lead to development of new drugs that change the way we live and significantly improve the prevention, diagnosis, and treatment of diseases. This revolution is in its infancy and its full impact will become evident in about 10-15 years. Recent advances in nanoscience will most likely create another technological revolution that will impact many sectors including engineering, medicine and energy. These innovations as well as advances in molecular simulation, high-throughput screening methods and the ever increasing computational power will lead to major changes in the nature and diversity of the products and processes of the chemical industry [1]. The chemical industry needs to participate in the development of all the related enabling technologies from an early stage to fully benefit from being involved in the manufacturing of these products.

2. THE BIOTECHNOLOGY OPPORTUNITIES

The recent advances in the biological sciences will lead to the development of many new products, such as novel pharmaceuticals, food additives and vitamins, biochemicals, biopolymers, and biofuels and will increase agriculture productivity. These products will enable us to be healthier, live longer, eat improved foods, develop new energy sources and products and treat our waste. Diagnostics will enable pinpointing medical problems. The medicines will cure and prevent many clinical problems. They will break down brain blood clots that lead to strokes, prevent heart disease by minimizing plaque formation and cure various cancers and AIDS. Drugs will be developed that regenerate nerve tissue in order to overcome memory loss with age and to treat dementia. Others will block absorption of glucose and sucrose benefiting type-2 diabetes patients and enable reduction of the calories gained from meals. Future developments will include more selective delivery to the organ or tissue where they are needed, development of semi-synthetic tissues and organs, gene therapy. Embryonic stem cells are likely to cure many sickness. The ongoing advances in genomics and proteomics and the increased computational power are likely to enable future development of patient specific drugs. The demand for drugs that cure and prevent health problems such as strokes, heart attacks, cancer, memory loss or even just obesity is expected to be huge. It is essential to produce them at prices affordable by society as the demand for them may lead to economic and ethical dilemmas about the high expense at the "end of life". To benefit from the demand for these novel products we need to gain some knowledge of biology, immunology, physiology and medicine.

The acetone-butanol process developed by Chaim Weizmann was the first production of industrial chemicals in fermentors . The application of ChE technology to the production of penicillin, was the first demonstration of its power to significantly increase the productivity of large biochemical reactors. While biological reactions are conducted in cells, our task is to design large scale reactors and their operating procedures. Since most biological reactions are conducted in dilute aqueous solutions, we need to develop novel, economical separation processes such as filtration, centrifugation, crystallization and membrane and chromatographic separation.

At present many small-molecule natural products are made from microbes in large (>100,000 L) fermentors followed by a series of separation and purification operations. Recent developments of bioprocesses include the production of high-purity therapeutic proteins, using genetically engineered cells and the large-scale production of plant-based medicines, such as the anticancer drug Taxol, from rare and difficult-to-grow plant species.

It is important that chemical engineers master an understanding of metabolic engineering, which uses genetically modified or selected organisms to manipulate the biochemical pathways in a cell to produce a new product, to eliminate unwanted reactions, or to increase the yield of a desired product. Mathematical models have the potential to enable major advances in metabolic control. An excellent example of industrial application of metabolic engineering is the DuPont process for the conversion of corn sugar into   1,3-propanediol, which is used in the synthesis of polypropylene terephthalate (Sorona$^{\circledR}$) by the reaction

This biological process replaces the chemical route of producing it from propylene and has several important advantages, namely

- Lower manufacturing cost.
- Lower capital investment.
- Smaller environmental impact.

The development of additional commercial processes utilizing metabolic engineering is a major challenge and opportunity. These processes will enable a transition from oil and natural gas feedstocks to renewable feedstocks such as; corn, soybeans, switch grass and biomass (straw, wood, agriculture and municipal waste). The success of the transition to renewable feedstocks will have in addition to its commercial benefits, a major impact on the image of the industry and our ability to attract bright students. The scale-up and design of these processes will require gaining sufficient background in life sciences to interact with biologists.

The many recent advances in tissue-engineered products include cartilage and artificial skin, biocompatible materials for organ replacements and artificial bones and teeth. The aging of the population will certainly increase the demand for these products. ChEs have an opportunity to participate in the development of large-scale processes for the production and packaging of these products that combine living cells and polymers in a sterile environment. CRE methodology will be also very useful in developing pharmacokinetic models of the human body that are needed to develop therapeutic strategies for the delivery of chemotherapeutic drugs and in assessing risk from exposure to toxins and pollutants.

3. NOVEL MATERIALS OPPORTUNITIES

Nanoparticles consist of clusters (of atoms or molecules) small enough to have material properties very different from the bulk. Most of the particle atoms or molecules are near the surface. The particle properties vary with the nanoparticle size. Many important scientific questions about their properties still need to be resolved. Nanoscience is likely to create a revolution in materials technology. Nanoparticles are expected to have applications in various technologies involving colloids, emulsions, polymers, paints, ceramic and semiconductor particles, and metallic alloys. When nanotechnology advances from laboratory demonstrations to widespread fabrication and manufacturing, it is likely to enable development of entirely novel electronic, fluidic, photonic, or mechanical nanodevices with critical dimensions smaller than 20 nm. Nanoparticles may enable production of tiny components of much faster computers and development of a variety of medical and imaging applications. A major challenge in commercializing nanodevices is the need to drastically reduce the high production price of these particles, and to improve the control of the particle size distribution. The methods employed for production of microelectronics are unlikely to provide inexpensive access to nanostructures. In contrast, ChE technologies may enable a more economic production of nanostructures with specified composition, size and properties.

A major growth opportunity of ChE is the synthesis, processing and manufacturing of high quality novel electronic, polymeric and ceramic materials with desirable properties. The properties of many materials can be improved by incorporating several materials into a composite to gain beneficial properties. The potential growth area includes components of fuel cells, voltaic cells for solar energy storage, improved fiber optics, etc. One such challenging opportunity with a very large potential market is ceramic automobile engines. The high temperature at which ceramic materials can operate will enable a significant increase in the fuel efficiency of the engine. The fragility of current ceramic materials and the difficulty of machining them are preventing this application. An important challenge is to overcome this fragility by incorporating ceramics and resins into composites, that are stable at high temperatures and easily machined.

Polymers and copolymers are among the most beneficial materials produced by synthetic chemistry. The invention and commercialization of new polymeric materials with radical new properties provides an opportunity to monopolize the market and justify the expense involved in the research and development. The commercialization of new polymers or copolymers always presents scale-up and design challenges. Scientists have recently developed new polymeric materials whose commercial impact has yet to be realized. Examples are semiconductive and conductive polymers and amphiphilic dendritic block copolymers. Other promising materials, such as polymers for (targeted) drug delivery and tissue engineering, have the potential to benefit the biomedical field, but are still in a relatively early stage of commercial development. The three-dimensional structure of macromolecules is another important synthetic variable. New materials with controlled branching sequences or stereo regularity provide tremendous opportunity for manipulating material properties. New polymerization catalysts and initiators for conducting controlled free-radical polymerization enable the synthesis and production of many new polymers. Metallocene catalysts enable production of polymers and copolymers with desired stereochemistry and properties.

Experience with chlorofluorocarbons (CFCs), insecticides such as DDT, herbicides and fertilizers has taught us that extended stability of these products may lead to unexpected harmful results. An important challenge will be to develop novel products that have a limited stable life and then decompose so that they do not persist in this environment. Examples are the development of plastic packages that decompose and degrade with time and of agricultural chemicals that do not harm unintended targets and are not overly persistent.

The development of new processes always generates challenging design and operation problems and provides engineers with an opportunity to make important contributions. For example, industrial experience shows that gas-phase olefin polymerization in fluidized bed reactors using metallocene catalysts can lead to local overheating and formation of polymer sheets that have require reactor shut-down. This severe unexpected operation problem required Exxon to declare *force major* on its production of linear low-density polyethylene in May of 1997 [2]. This engineering problem has still not been completely solved and is an example of the need of creative engineering to identify and solve the problems associated with the commercialization of new processes. Another important challenge is the design of a reactor that will produce a uniform blend of two polymers.

Novel catalysts enable synthesis of novel products, increase the yield and selectivity of the desired products, use less expensive reactants, minimize the formation of by-products and decrease energy consumption. Industry is willing to conduct catalyst research and development as it is much less expensive than developing a new process. While significant know-how exists about the development and engineering the performance of homogeneous catalysts, the traditional development of new heterogeneous catalysts is more empirical and depends on the nature of the desired reaction(s) pathway. The discovery of new heterogeneous catalysts has been accomplished either by pure serendipity, trial-and-error approaches, or design by analogy to existing catalysts. These empirical methodologies will still be used in future development of new catalysts for producing specific materials.

Future hetrogeneous catalyst design may benefit from know-how about the design of homogeneous catalyst design. Supported metallocene catalysts are the first heterogeneous catalysts that enable rational catalyst design so that it will produce a polymer with desired stereochemistry and properties. These novel catalysts will enable the development of many new products. An important advance will be the development of man-made catalysts that imitate enzymes, which have an extraordinary high activity and selectivity. The understanding of the mechanism by which enzymes operate has enabled the development of several medical drugs. However, the lack of an understanding of how the enzymes attain their high selectivity has prevented the development of analog man-made catalysts.

Important advances occurred in recent years in our ability to develop novel catalysts for specific applications. Combinatorial methods have been successfully used for many yeas by biochemists to develop various drugs. There has been in recent years a drive to apply this high-throughput approach to expedite and increase the efficiency of screening of candidate catalysts. There are by now several commercial companies which provide equipment and software for conducting high-throughput screening and selection of catalysts, which is significantly more efficient and economical than the traditional experimental techniques. The combinatorial screening is especially advantageous when the catalyst is not sensitive to (or requires) pretreatment and does not decay rapidly during its sojourn in the reactor. Among others, this screening has already enabled a more efficient development of several novel metallocene catalysts for the production of polymers and copolymers with desired stereochemistry and properties. The combinatorial screening is expected to become an important tool in the arsenal of researchers trying to develop new catalysts. Recent theoretical developments and advances in computational and visualization capabilities suggest that a molecular scale approach, including molecular modeling, may provide useful insight to future screening of potential catalysts.

4. OTHER CHALLENGES

A major challenge will be to develop new processes or step-up technologies that increase the yield and/or selectivity, use cheaper raw materials, decrease energy consumption, minimize the product separation and purification needs and lower capital investment. Innovative step-out technologies can still have a major impact on existing processes. An excellent example of such an accomplishment is the reactive distillation process developed by Eastman Chemicals for production of methyl acetate by via the reaction [2]

$$HOAc + MeOH \rightleftharpoons MeOAc + H_2O$$

The conventional process consists of a reactor followed by eight distillation columns, one liquid-liquid extractor and a decanter. The reactive distillation process consists of one column that produces high-purity methyl acetate that does not require additional purification and there is no need to recover unconverted reactant. The reactive distillation process costs one fifth of the conventional process and consumes only one fifth of the energy.

Another important challenge is to enhance the reliability of the design and scale up of multi-phase reactors, such as fluidized bed reactors and bubble-columns. The design uncertainty caused by the complex flow in these reactors has often led to the choice of a reactor configuration that is more reliable but less efficient. An example is Mobil use a packed-bed reactor for the methanol to gasoline process in New Zealand, even though a fluidized bed is expected to provide superior performance. A reliable scale-up procedure will enable conducting many exothermic reactions in fluidized bed reactors rather than the much more expensive multi-tube reactors.

Society requires that future production of chemicals be safe and environmentally benign. Thus, chemical reaction engineers will have to develop novel processes that are less polluting than the existing ones, or modify existing processes. The reduction of chemical emissions may require use of alternative raw materials and treating of the waste generated within the plant. CRE will have to develop reactors that destruct pollutants formed upstream by another process, such as the NO_x produced in organic oxidation processes. These reactors will have to operate under transient feed conditions (changes in feed composition and flow rates) and lead to a high conversion over a wide range of residence times and feed compositions. These requirements may lead to the development of novel reactor configurations and mode of operation, such as a reverse-flow reactor that can destroy volatile organic compounds or NO_x at a higher efficiency than traditional packed- bed reactors. The need to eliminate emission of pollutants will motivate development of many novel reactor configurations. One such potential development is a decanting two-phase reactor in which an organic reactant is completely converted in the reactor [4]. In addition to minimizing plant emissions, ChE will be involved outside the plants in clean up of toxic waste dumps, radioactive waste, contaminated soil and groundwater. Moreover, CRE expertise will be needed to eliminate formation of smog, acid rain and pollution of river and lakes. Another important activity will be the efficient sequestration of carbon dioxide.

Significant improvements have been made in the development of catalytic mufflers that reduce and destroy the effluent carbon monoxide, hydrocarbons and NO_x from passenger automobiles and large trucks exhausts. There is still a need to develop efficient catalytic converters for motorcycles and scooters, which are a major source of pollution in many cities in the Far East. Another important task will be the development of efficient converters that minimize the emission of NO_x and particulates from diesel engines. This, in turn, will enable a wider application of this engine, the efficiency of which exceeds that of gasoline automobile engines. While air pollution is the subject of the media concern, water pollution is most likely to become a major problem in the future. It will be a much more difficult to solve problems associated with water contamination, as the time constant of remediation of polluted water resources is much larger than that of the atmosphere. It is essential that reaction engineers will tackle these important problems, which by default are handled at present mainly by civil engineers.

ChE will be challenged to develop creative process for recycling of waste materials, rather than burning or discarding them. One potential process is bioremediation, using genetically modified microorganisms to decompose the waste. This approach has already been applied to the treatment of oil spills.

An important task will be to develop a more efficient and clean use of fossil fuels - petroleum, natural gas, and coal and the generation of alternative clean energy resources.

8

The increase in the price of oil and natural gas motivates the chemical industry to develop processes that use alternative raw materials and to develop efficient and economical processes for liquid fuels synthesis from coal and natural gas. An innovative promising approach for producing gasoline from methane is presented in [5]. Other important tasks are development of efficient methods for producing liquid fuels from unconventional sources such as oil shale, tar sands, and deep-sea methane hydrates.

An important challenge will be participation in the development of alternative fuels. One activity will be development of efficient and economical fuel cells. I conjecture that both large solid oxide fuel cells and miniature ones for use in accessories, such as laptops, cameras and cellular phones are likely to find commercial applications ahead of the use of hydrogen fuel cells in automobile. ChE will be involved in the transformation of solar energy to electricity. One option is to accomplish this by photovoltaic cells. This will require manufacturing photocells that are efficient, long lasting and inexpensive. Anther option is use of solar pools to capture the solar energy. Nuclear reactors are expected to gain in the near future public acceptance and be built again. Their operation will require development of efficient ways to properly concentrate and process the radioactive waste from these reactors. Biomass is another alternative renewable fuel source

Another important task will be to provide society with professional guidance and advice about the feasibility and merit of proposed technological solutions. It is our duty to point out when proposed processes create unrealistic expectations. An example of such guidance is the recent article by Shinnar [6] that motivated a discussion of the feasibility of use of hydrogen fuel cells in automobiles.

The chemical industry has missed in the past several opportunities to become the leader in the manufacturing of several new products and materials, such as semi-conductors. We cannot afford and should not repeat such mistakes. Many challenging engineering problems will have to be solved before the upcoming world-changing scientific innovations can be exploited. The key to benefiting from these is to identify these problems, to figure out how we can solve them using CRE methodology and tools and to generate the resources needed for their development and commercialization. If we do this and let the upcoming innovations change the chemical industry we shall be like the legendary phoenix, that when old burns itself to ashes, and then rises from them youthfully to begin a new life.

REFERENCES
[1] Beyond the molecular frontier: Challenges for chemists and chemical engineers, National Academies Press, Washington, DC, 2003.
[2] D. Rotman, *Chem. Week*, 159, (May 21,1997) 23.
[3] J.J. Siirola, Advances in Chemical Engineering, 23, 1-62 (1996).
[4] J. Khinast, D. Luss, T.M. Leib, and M.P. Harold, *AIChE J.*, 44 (1998) 1868.
[5] http://www.grt-inc.com
[6] R. Shinnar, Technology in Society, 25 (2003) 455: CEP, 5 (Nov, 2004) 5.

Studies in Surface Science and Catalysis, volume 159
Hyun-Ku Rhee, In-Sik Nam and Jong Moon Park (Editors)

Combinatorial computational chemistry: first principles quantum methods as a tool for industrial innovations

E. Broclawik[a,b], A. Govindasamy[c], C. H. Jung[c], C. Lv[c], R. Raharintsalama[c], H. Tsuboi[c], M. Koyama[c], M. Kubo[c,d] and A. Miyamoto[a,c]

[a] New Industry Creation Hatchery Centre, Tohoku University, 6-6-10 Aoba, Aramaki, Aoba-ku, Sendai 980-8579, Japan

[b] Institute of Catalysis, Polish Academy of Sciences, ul. Niezapominajek 8, 30-239 Kraków Poland

[c] Department of Applied Chemistry, Graduate School of Engineering, Tohoku University, 6-6-07 Aoba, Aramaki, Aoba-ku, Sendai 980-8579, Japan

[d] PRESTO, Japan Science and Technology Agency, 4-1-8 Honcho, Kawaguchi, Saitama 332-0012, Japan

1. INTRODUCTION

Combinatorial computational chemistry has recently made spectacular progress and tremendous impact on the development of a variety of engineering materials. It has also been extensively used to identify, optimize, and rationally select new materials. Studies at atomic and electronic level are expected to play a key role in predicting new materials with unusual characteristics. Combinatorial computational chemistry approach is particularly desirable when experimental information at atomic scale is scarce. Initially, computational chemistry was used to elucidate the physico-chemical properties of standard molecules, however recently it contributes much to the design of new materials and catalysts as well as to predict the unexplored properties by the significant advancement of the theory. One of the most anxiously awaited outcomes that theory can provide to aid the experiment is the understanding of chemical reaction mechanisms on electronic and atomic level, which is essential for the materials and catalysts design. Advanced areas of research require computationally demanding first principles calculations of chemical accuracy and they are indispensable for deep understanding the intricate reaction mechanisms in modern catalysis or biological sciences. Last but not least accurate calculations provide an excellent calibration for the development of novel, simplified methodologies derived from *ab initio* quantum chemistry, massively applicable for real systems and paving the direct road from theory to modern technology.

It is thus obvious that among numerous computational methods, first principles quantum chemical approach is indispensable. However, initially first principles quantum chemical calculations required the use of models consisting of a few atoms (clusters) and the range of properties was limited. Since the advent of modern computing resources, as well the models could be extended to cover larger variety of structures as the methodology has been greatly advanced to cover wider variety of properties. Among the most important theoretical developments was a break-through in traditional quantum chemistry by introducing Density functional theory (DFT) by Hochenberg, Kohn and Sham in 1960-ties [1]. DFT offered the possibility to include electron correlation by addressing the electron density itself. At the same time, DFT preserved simple conceptual orbital picture of the electronic structure and calculation modesty becoming thus a powerful and computationally robust tool in hands of chemists and physicists. In this way the spectrum of systems treated with sufficient accuracy could be extended beyond applications to organic chemistry as to include transition metal and rare earth chemistry. This provided the access to wide area of homogeneous (transition metal complexes) and heterogeneous catalysis [2-9]. DFT paved also the pathway to periodic electronic structure methods covering thus solids and surfaces and opening the door for materials science applications in modern electronics, energy resources and storage [10-20]. On top of pushing methodological concepts to more advanced level, the extension to the excited states and their properties should be mentioned. Time dependent DFT (TDDFT) was introduced to practical implementations starting from 1990-ties [21,22]. Applications to optical properties of materials and to photochemical reactions immediately followed [23-27].

Variety of software systems has been developed to serve in modern research. The selected examples presented in this review will be based on such programs as DMol^3 and Castep included in Accelrys machinery [28,29], ADF suite [30], DFT has also been incorporated into the Gaussian [31] and Jaguar [32] in the Maestro system. Therefore theoretical chemistry definitely entered the field of modern applications in up-to date technological areas. In this junction a novel methodology for real systems, a tight-binding quantum chemical molecular dynamics (TB-QCMD) program "Colors" developed recently in our laboratory should be mentioned, based on DFT-derived tight-binding approximation carefully parameterized from accurate DFT results [33]. In the following review we will illustrate the interface between computational first principles quantum chemistry and modern science and technology, with the highlight on the most advanced methodological developments: explicit searches for catalytic reaction mechanisms in ground and excited states.

2. TRANSFORMATIONS OF ORGANO COMPOUNDS TRIGGERED BY THE INTERACTION WITH TRANSITION METAL AND RARE EARTH SYSTEMS

2.1. Palladium hydrido-complexes

The expanding field of surface organometallic chemistry offers new possibilities to homogeneous and heterogeneous catalysis. Preserving catalytic activities, immobilization of catalyst species on solid surfaces is important for industrial applications. For this purpose, fundamental understanding of reaction mechanisms catalyzed by organometallic complexes is essential. β-Hydrogen elimination of transition metal alkyls is one of the most important elementary processes in organometallic chemistry and catalytic reactions. Pd-catalyzed reactions have been reported in many papers, including catalytic reactions such as oxidation reactions, allylic alkylation reactions, telomerization reactions, water gas shift reaction, reduction of various substrates and others. In this junction, DFT studies were carried out to explore the role of cationic species in the decomposition mechanisms of ethyl-palladium complexes to form ethene and hydrido-palladium complexes [2,4]. The Material Studio DMol^3 program by Accelrys was used to perform DFT calculations, the methodological novelty consisted in including solvent effects by both explicit solvent molecules and COSMO method implemented in DMol^3.

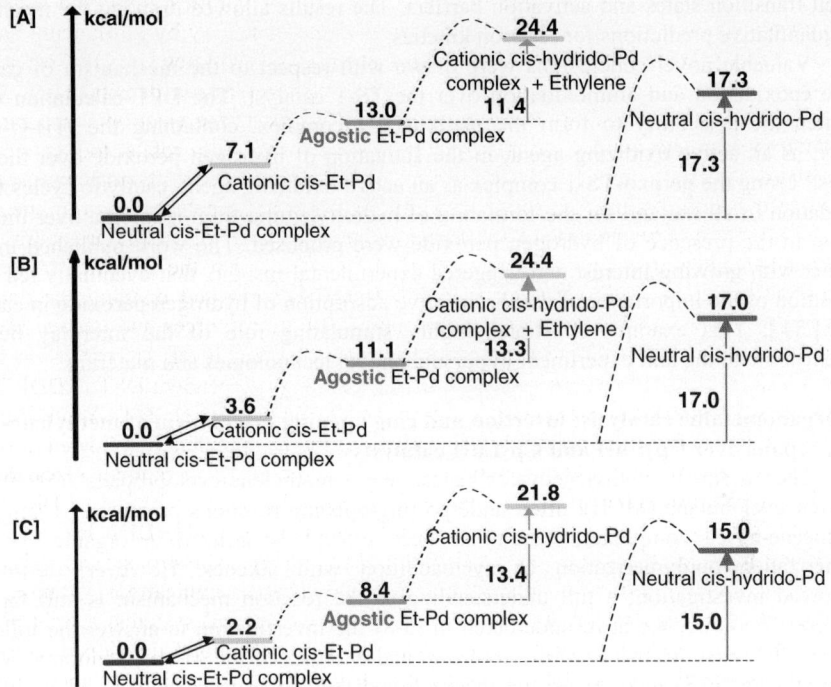

Fig. 1 Relative energy comparison between neutral and solvent coordinated cationic species, associated with counter anion Cl⁻ [A], Br⁻ [B], and I⁻ [C].

Taking selected models for intermediates and final states we followed the mechanism and energetics of the decomposition reaction of monoethyl-palladium complexes. Sample results are shown in Fig. 1 for Cl, Br and I, respectively. The results showed that the relative energy levels (in kcal/mol) of the solvated cationic agostic complexes with respect to those of solvated neutral *cis*-ethyl-Pd complexes were Cl (13.0) > Br (11.1) > I (8.4). For the cationic reaction pathways, the energy differences between the solvated cationic agostic *cis*- ethyl-palladium complexes and the corresponding final states (ethene and the solvated cationic hydrido-palladium complexes) were Cl (11.4) < Br (13.3) <I (13.4). On the other hand, for the neutral reaction pathways, analogous energy differences from the solvated neutral *cis*-ethyl-palladium complexes to the final states were Cl (17.3) > Br (17.0) > I (15.0). Thus the reaction pathway via cationic complexes seemed to be preferable over the pathway via neutral complexes from thermodynamic point of view, which is in agreement with the experiment.

2.2. Hydrogen peroxide activation on titanosilicalite catalyst

In the same spirit DFT studies on peroxo-complexes in titanosilicalite-1 catalyst were performed [3]. This topic was selected since Ti-containing porous silicates exhibited excellent catalytic activities in the oxidation of various organic compounds in the presence of hydrogen peroxide under mild conditions. Catalytic reactions include epoxidation of alkenes, oxidation of alkanes, alcohols, amines, hydroxylation of aromatics, and ammoximation of ketones. The studies comprised detailed analysis of the activated adsorption of hydrogen peroxide with explicit transition states and activation barriers. The results allowed drawing the mechanism with quantitative predictions for reaction kinetics.

Valuable novel conclusions were drawn with respect to the mechanism of catalytic ethene epoxidation and ammoxidation over the TS-1 catalyst. The DFT calculation results exhibited the possibility to form the peroxo-TS-1 complex, containing the (Ti)-O-O-(Si) moiety, as an active oxidizing agent, in the activation of hydrogen peroxide over the TS-1 catalyst. Using the peroxo-TS-1 complex as an active oxidizing agent, catalytic cycles for the epoxidation of ethene and for the formation of hydroxylamine from ammonia over the TS-1 catalyst in the presence of hydrogen peroxide were proposed. The work published in 2001 was met with growing interest and triggered experimental insights that eventually led to the recognition of the importance of the dissociative adsorption of hydrogen peroxide in catalytic cycle [34]. This example illustrates highly stimulating role of the interplay between theoretical modeling and experiment in pursuing novel technologies and materials.

2.3. Organometallic catalysis: insertion and ring opening mechanism of methylene-cyclopropane over Cp_2LaH and Cp_2LuH catalysts

The strained methylenecycloalkanes, e.g., methylene-cyclopropane (MCP) and methylenecyclobutane (MCB), often undergo ring-opening reactions promoted by single-site metallocene-based complexes, which has been effectively utilized in organic synthesis, polymerization/copolymerization or cycloaddition with alkenes. However, despite the widespread investigation, a full understanding of the reaction mechanism is still far from complete. Therefore, we have undertaken in [5-8] the investigation to answer the following questions: (i) what are the possible structures and energies of the assumed minima (M) and transition states (TS) involved in the insertion and the subsequent cleavage of proximal or distal bonds? (ii) Are there any differences in the ring opening mechanism over Cp_2LaH and Cp_2LuH catalysts as they give rise to different polymerization products?

The insertion and the ring-opening pathway of MCP over Cp_2LaH and Cp_2LuH catalysts were addressed using DFT method. Lanthanide-based systems are highly complicated due to relativistic effects thus the latter were carefully accounted for in the calculations. All stationary points along the reaction cycle were defined and frequency calculations were carried out to determine their nature, as well as to estimate the zero-point energy correction. Figure 2 depicts the geometry of stationary points along with the corresponding relative energies of various structures. In this energy profile, the sum of the energies of Cp_2LnH and MCP in isolated state is considered as the reference energy. It can be noticed that the coordination geometries of the various minima and transition states for both Cp_2LaH-MCP and Cp_2LuH-MCP systems are similar suggesting that the reaction may follow an identical pathway at the initial stages although they give very different products.

For the various processes; a lower energy barrier is noted for Cp_2LaH than Cp_2LuH suggesting that the insertion and ring opening is more facile in the case of the former. Finally, it may be concluded that the activation of MCP initiated by Cp_2LnH proceeds initially *via* the formation of Cp_2LnH-MCP complex followed by 1,2-insertion with the deformed MCP structure. Subsequently, the ring opening occurs at the proximal bond with a simultaneous hydrogen transfer followed by the establishment of Ln-C4 bond. With additional MCP molecules, the M3 may proceed to a smooth polymerization or a dimerization.

The work discussed in the preceding paragraphs apart from being a valuable study clarifying novel aspects of metalloorganic catalysis by lanthanide complexes has the additional aspect upgrading its importance. It served as the calibration data for TB-QCMD simulation of the ring opening of MCP assisted by Cp_2LaH using "Colors" program [35].

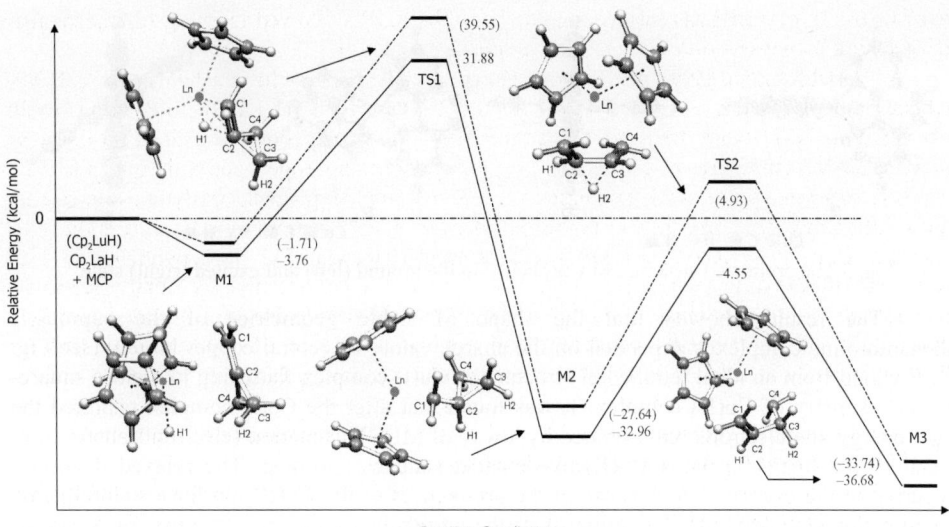

Fig. 2. Potential energy profile of 1,2-insertion and ring opening processes of Cp₂LnH-MCP systems obtained at RPBE level

The ring-opening mechanism was well supported by the snapshots and the overlap bond population obtained from TB-QCMD simulations, where the formation of new C–H and La–C bonds and the dissociation of La–H and proximal C–C bonds could be tracked. The obtained dynamic ring opening mechanism was similar to the static mechanism, however, a novel transition state was also proposed for insertion reaction of alkenes, with tetrahedral h4-coordination. This example perfectly illustrates the importance of mutual interplay between high-level first principle methodologies and simplified methodologies derived from *ab initio* quantum chemistry, massively applicable for real systems.

3. ECXITED STATES AND SPECTRSCOPIC PROPERTIES OF ORGANO-METALLIC AND POLYMER MATERIALS

3.1. Ligand to metal charge transfer complexes

The researches on photophysics and photochemistry of organometallic complexes have received intensive attention due to the great applications of these complexes in many fields such as photocatalysis, optical reaction, material design, and solar energy conversion. One of the most intriguing spectroscopic characteristics for the organometallic complexes is an intramolecular electron transfer during the excitations. At this junction we have undertaken DFT studies to determine the structural changes and charge transfer processes that occur in the complex of cuprous(I) bis-phenanthroline (Cu(phen)$_2^+$) upon oxidation to copper(II), Cu(phen)$_2^{2+}$, and relate these changes to the MLCT excited state of Cu(phen)$_2^+$ [24,25]. All calculations in this investigation have been performed using DFT by using Gaussian98 (G98) and Amsterdam density functional program package (ADF). Here the relativistic terms were calibrated by a combined scalar relativistic zero order regular approximation (ZORA). The vertical excitation energies and oscillator strengths of low-lying singlet and singlet excitation states were derived by TDDFT.

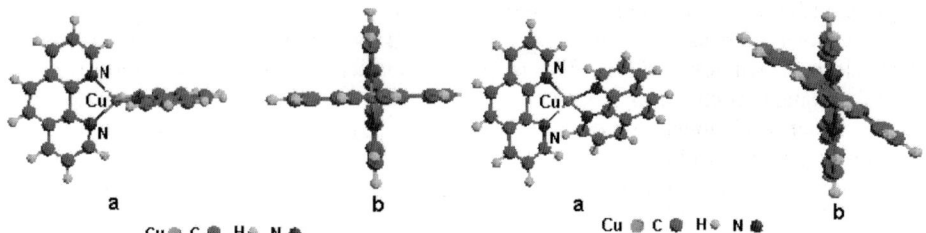

Fig. 3 The optimized structure of Cu(phen)$_2^+$ in the ground (left) and excited (right) state.

The results showed that the shape of stable geometries of the cuprous(I) phenanthroline complexes depended on the charge values of central copper largely (see Fig. 3). It varied from an ideal tetrahedral structure of Cu(I) complex flattening to a close square-planer geometry of Cu(II) complex. It was found that after the Cu(I) complex captured the light energy, the electrons redistributed by a way of MLCT excitation (electron transfer from metal to the ligands), and a Cu(II)-like excited state was formed. The relaxed distortion occurred in the excited state because of the preferences of the Cu(II) ion for a square-planar coordination environment. The similarity of the properties for the excited state of Cu(phen)$_2^+$ and Cu(phen)$_2^{2+}$ decreased the steric energy barrier of the evolution from Cu(I) to Cu(II) complexes through the intermediate of excited state. The large structural distortion between ground state and excited state of Cu(phen)$_2^+$ reduced the lifetime of excited state, and thus the modification in structure to limit the structural distortion between ground state and excited state are necessary to increase the efficiency of solar energy conversion.

In similar spirit we have investigated the excitation of the MLCT in the UV- vis spectra of the complexes Cp$_2$MCl$_2$ (Cp- C$_5$H$_5$, M-Ti, Zr, Hf) by DFT [26,27]. The nature of the main spectral features has been interpreted on the basis of the electronic structure of the complexes. These complexes represented an iso-structural series with a pseudotetrahedral geometry. With the transition from ground state to excited state, the electron transferred from Cp ligands to the central metals, accompanying with an increase in the bond length of M-Cl. The two lowest adsorption bands are predominantly corresponding to the excitation of Cp ligand to the metals charge transfer character. The energies and oscillator strengths of the lowest excitations depended strongly on the central metals. The overall effect was thus to increase the energy gap between the highest occupied molecular orbitals (HOMOs) and the lowest unoccupied molecular orbitals (LUMOs) and to increase the excitation energies along this series. The results of triplet states have been used to simulate the phosphorescence spectrum. The predicted level patterns of the lowest triplet excited states are in excellent agreement well with the phosphorescence data available.

3.2. Electronic and molecular properties of ground and excited states for conducting polymers

The conjugated polymer, poly (3,4-ethylenedioxythiophene) (PEDOT) is a low band-gap polymer with high conductivity and good thermal and chemical stability. PEDOT chosen for this study is the most widely used, industrially important electrically conducting polymer. Its major advantage is high stability in the doped state, with conducting properties that remain almost unaltered under aging in environmental conditions. Unfortunately, PEDOT is an insoluble polymer. This drawback can be circumvented by polymerization in combination with a water-dispersible polyelectrolyte such as poly(4-styrenesulfonate) (PSS). PEDOT doped with PSS has good transparency in the visible spectrum, high conductivity and very good film forming properties. Oxidized PEDOT, doped with PSS, has been widely used both in light-emitting devices as hole transport buffer layer and in photovoltaic devices as hole-collecting layer such as the electrodes of capacitors and photo-diodes, antistatic coatings, electro-chromic windows, field effect transistors, and hole transport material. We have undertaken quantum chemical calculations to understand the electronic nature the properties of excited states of models for these low band-gap polymers, which may be related to the light-emitting behavior [23].

Theoretical calculations were performed using G98, ADF, and DMol^3 programs. In the case of G98, Hatree-Fork method was used for the optimization of ground state and adiabatic transition energies were obtained by optimizing the first excited state (S_1) using CIS method. The low-lying singlet and triplet excitation energies and oscillator strength were derived by TDDFT. The geometry and the electronic properties of singlet and triplet excited state of ethylenedioxythiophene (EDOT) and styrenesulphonic acid was studied using ADF program. The periodic structure and band-gap of PEDOT with several repeating units were obtained by using Dmol^3 program.

The analysis of the excitation orbital character showed that the highly intense vertical transition in EDOT is mainly due to the electron transfer from the HOMO to the LUMO. During this transition the molecule is excited from its ground state (S_0) to the first excited state (S_1). In styrene-sulphonic acid the intense transition is from S_0 to the second excited state (S_2) and it is due to the electron transfer from HOMO to LUMO+1. This shows that not only the transition energy but also its intensity is required to predict experimental spectrum. The aromatic C-C bonds are elongated both in styrene-sulphonic acid on excitation, which clearly indicates that the transition is of π-π* type transition. However, the planarity of the thiophene ring in EDOT is destroyed during excitation and thus the degeneracy of π and π* is lifted and the state becomes stabilized. This shows that the planar geometry of EDOT may be easily distorted either by the excitation or by donating or accepting electrons. This in turn may influence the electronic band gap. The band gap energy for PEDOT is evaluated by periodic DFT method and it decreases with the increments in the number of repeating unit in the periodic unit cell. The calculated energy gap for PEDOT with tetramer repeating-unit agrees reasonably well with the experimental value. This example illustrates the need of interplay between different quantum chemical methodologies in order to obtain properties of advanced materials from multidisciplinary perspective.

4. SUMMARY AND CONCLUSION

In this brief review we illustrated on selected examples how combinatorial computational chemistry based on first principles quantum theory has made tremendous impact on the development of a variety of new materials including catalysts, semiconductors, ceramics, polymers, functional materials, etc. Since the advent of modern computing resources, first principles calculations were employed to clarify the properties of homogeneous catalysts, bulk solids and surfaces, molecular, cluster or periodic models of active sites. Via dynamic mutual interplay between theory and advanced applications both areas profit and develop towards industrial innovations. Thus combinatorial chemistry and modern technology are inevitably interconnected in the new era opened by entering 21st century and new millennium.

[1] P. Hohenberg and W. Kohn, Phys. Rev. B, 136 (1964) 864; W. Kohn and L. J. Sham, Phys. Rev. A, 140(1965) 1133.
[2] R. Raharintsalama, H. Munakata, M. Koyama, M. Kubo, A. Miyamoto, Appl. Surf. Sci., 244 (2005) 631-635.

16

[3] H. Munakata, Y. Oumi, A. Miyamoto, J. Phys. Chem. B 105 (2001) 3493-3501.
[4] R. Raharintsalama, H. Munakata, A. Endou, M. Kubo, A. Miyamoto, Trans. Mater. Res. Soc. Jpn. 29 (2004) 3751-3754.
[5]Y. Luo, P. Selvam, A. Endou, M. Kubo, A. Miyamoto, J. Am. Chem. Soc., 125 (2003) 16210-16212.
[6]Y. Luo, P. Selvam, M. Koyama, M. Kubo, A. Miyamoto, Chem. Lett., 33 (2004) 780-785.
[7] Y. Luo, P. Selvam, M. Koyama, M. Kubo, A. Miyamoto, Inorg. Chem. Commun., 7 (2004) 566-568.
[8] Y. Luo, P. Selvam, Y. Ito, M. Kubo, A. Miyamoto, Inorg. Chem. Commun., 6 (2003) 1243-1245.
[9] A. Endou, T. W. Little, A. Yamada, K. Teraishi, M. Kubo, S. S. C. Ammal, A. Miyamoto, M. Kitajima, F. S. Ohuchi, Surf. Sci., 445 (2000) 243-248.
[10] K. Nishitani, S. Takagi, M. Kanoh, T. Yokosuka, K. Sasata, T. Kusagaya, S. Takami, M. Kubo, A. Miyamoto, Jpn. J. Appl. Phys., 42 (2003) 5751-5752.
[11] M. Elanany, M. Koyama, M. Kubo, P. Selvam, A. Miyamoto, Micropor. Mesopor. Mat., 71 (2004) 51-56.
[12]C. H. Jung, Y. Ito, A. Endou, M. Kubo, A. Imamura, P. Selvam, A. Miyamoto, Catal. Todday, 87 (2003) 43-50.
[13]Y. Luo, X. H. Wan, Y. Ito, S.Takami, M. Kubo, A. Miyamoto, Appl. Surf. Sci., 202 (2002) 283-288.
[14]Y. Luo, X. H. Wan, Y. Ito, S. Takami, M. Kubo, A. Miyamoto, Chem. Phys., 282 (2002) 197-206.
[15] X. H. Wan, X. J. Wang, Y. Luo, S. Takami, M. Kubo, A. Miyamoto, Organometallics, 21 (2002) 3703-3708.
[16]R. V. Belosludov, S. Takami, M. Kubo, A. Miyamoto, Y. Kawazoe, Mat. Trans., 42 (2001) 2180-2183.
[17]H. Tamura, H. Zhou, S. Takami, M. Kubo, A. Miyamoto, J. Chem. Phys., 115 (2001) 5284-5291.
[18] H. Zhou, Y. Yokoi, H. Tamura, S. Takami, M. Kubo, A. Miyamoto, M. N. Gamo, T. Ando, Jpn. J. Appl. Phys., Part 1, 40 (2001) 2830-2832.
[19] X. Wan, K. Yoshizawa, N. Ohashi, A. Endou, S. Takami, M. Kubo, A. Miyamoto, A. Imamura, Scripta Mater., 44 (2001) 1919-1923.
[20] N. Ohashi, K. Yoshizawa, A. Endou, S. Takami, M. Kubo, A. Miyamoto, Appl. Surf. Sci., 177 (2001) 180-188.
[21] E. K. U. Gross, J. F. Dobson, M. Petersilka, Density Functional Theory of Time-dependent Phenomena, Springer: New York, Herdelberg, 1996.
[22] M. E. Casida, In Recent Developments and Applications of Modern Density Functional Theory; J. M. Seminario, Ed.; Elsevier: Amsterdam, 1996; p 391.
[23] A. Govindasamy, C. Lv, X. J. Wang, M. Koyama, M. Kubo, A. Miyamoto, Appl. Surf. Sci., 244 (2005) 195-198.
[24] X. J. Wang, C. Lv, M. Koyama, M. Kubo, A. Miyamoto, J. Organomet. Chem., 690 (2005) 187-192.
[25] X. J. Wang, C. Lv, A. Endou, M. Kubo, A. Miyamoto, J. Organomet. Chem., 678 (2003) 156-165.
[26]X. J. Wang, S. Takami, M. Kubo, A. Miyamoto, Chem. Phys., 279 (2002) 7-14.
[27] X. J. Wang, X. H. Wan, H. Zhou, S. Takami, M. Kubo, A. Miyamoto, J. Mol. Struct. Theochem, 579 (2002) 221-227.
[28] B. Delley, J. Chem. Phys., 92 (1990) 508-517; ibid., 113 (2000) 7756-7764.
[29] M. D. Segall, P. L. D. Lindan, M. J. Probert, C. J. Pickard, P. J. Hasnip, S. J. Clark, M. C. Payne, J. Phys.-Condens. Mat., 14 (2002) 2717-2744.
[30] G. de Velde, F. M. Bickelhaupt, E. J. Baerends, C. Fonseca Guerra, S. J. A. van Gisbergen, J. G. Snijders, T. Ziegler, J. Comput. Chem., 22 (2001) 931-967.
[31]M. J. Frisch et al. Gaussian 03, Revision B.03, Gaussian Inc., Pittsburgh, PA, 2003.
[32]Schrodinger, Inc., Portland, Oregon "JAGUAR 4.0", 2000.
[33]P. Selvam, H. Tsuboi, M. Koyama, M. Kubo, A. Miyamoto, Catal. Today, 100 (2005) 11-25.
[34]F. Bonino, A. Damin, G. Ricchiardi, M. Ricci, G. Spano, R. D'Aloisio, A. Zecchina, C. Lamberti, C. Prestipino, S. Bordiga, J. Phys. Chem. B. 108 (2004) 3573-3583.
[35] Y. Luo, Ph.D. Thesis, Tohoku University, 2003.

Studies in Surface Science and Catalysis, volume 159
Hyun-Ku Rhee, In-Sik Nam and Jong Moon Park (Editors)

Automotive applications of chemical reaction engineering and future research needs

Se H. Oh and Edward J. Bissett

General Motors Research & Development Center, Warren, Michigan 48090, USA

The stricter future emission standards require further improvement in emission control performance through the optimization of catalytic converter properties and its operating conditions. The "single-channel" 1-D monolith model has been used extensively at GM to guide the development and implementation of emission control systems with improved cold-start emission performance while reducing the noble metal usage and hardware development time/iterations. This paper discusses some examples of model applications to three-way catalyst systems as well as future research needs and directions in emission control system modeling, including the development of elementary reaction step-based kinetic models.

1. INTRODUCTION

Catalytic converters, which have been in widespread use since the introduction in the U.S. in the fall of 1974, have proven to be effective at reducing automobile exhaust emissions of carbon monoxide, hydrocarbons and nitrogen oxides. In the face of the ever-tightening exhaust emission standards over the next decade or so, it is of critical importance to further improve the performance and durability of a catalytic converter while reducing its noble metal usage requirements. Since the performance of a catalytic converter is a complex function of its design and operating parameters, mathematical modeling (rather than an empirical approach) promises to be helpful in accelerating the development and vehicle implementation of optimum emission control systems (i.e., those with high emission performance and low noble metal content) with reduced resources requirements.

Most of the monolith catalytic converter modeling studies reported in the literature [1-3] have focused on the "single-channel" 1-D model to describe the behavior of adiabatic (i.e., completely insulated) monoliths exposed to a uniform flow distribution at the front face. The single-channel 1-D model was subsequently extended to include the effects of flow maldistribution and heat loss to the surroundings [4, 5]. Monolith modeling with these effects accounted for would require simultaneous consideration of the entire array of monolith channels (3-D model) because of the channel-to-channel variations in conversion and thermal behavior, and interactions among neighboring channels through radial heat conduction.

In this paper, we first briefly describe both the "single-channel" 1-D model and the more comprehensive 3-D model, with particular emphasis on the comparison of the features included and their capabilities/limitations. We then discuss some examples of model applications to illustrate how the monolith models can be used to provide guidance in emission control system design and implementation. This will be followed by brief discussion of future research needs and directions in catalytic converter modeling, including the development of elementary reaction step-based kinetic models.

2. CATALYTIC CONVERTER MODELS

At the heart of an automotive catalytic converter is a catalyzed monolith which consists of a large number of parallel channels in the flow direction whose walls are coated with a thin layer of catalyzed washcoat. The monolith catalyst brick is wrapped with mat, steel shell and insulation to minimize exhaust gas bypassing and heat loss to the surroundings.

2.1. Single-Channel (1-D) Model

The single-channel model is a transient, 1-D monolith model with the spatial variable x only (coordinate along the flow direction), which is derived from fundamental principles of heat and mass conservation under the assumptions of a uniform flow distribution at the monolith face and adiabatic operation of the converter. The major physical/chemical features included in the model are (1) convective heat and mass transport in the bulk gas phase, (2) gas-solid heat and mass transfer, (3) axial conduction and accumulation of heat in the substrate, and (4) catalytic reactions and attendant heat release in the catalyzed washcoat layer deposited on the monolith substrate. Pore diffusion resistances within the thin (typically ~30 μm) washcoat layer were neglected. Also, the accumulation of mass and heat in the gas phase was neglected since their time constants are typically much smaller than that of the solid-phase thermal response. With this quasi-static approximation invoked, the resulting model equations consist of a time-dependent partial differential equation (PDE), which describes the solid-phase energy balance (transient, heat conduction equation with a heat source term), coupled with a system of differential-algebraic equations (DAE). Detailed description of the model formulation and the basic governing equations is given in our earlier papers [2, 6].

Broadly speaking, this model seeks to predict temperature and species concentrations, in both the gas and solid phases, as a function of time and axial position along the monolith length. The numerical solution method employed involves a uniform-mesh spatial discretization and subsequent time-integration for the PDE using a standard, robust software (such as LSODI found in ODEPACK), and x-integration by LSODI for the DAE system [6].

The model considers the noble-metal catalyzed oxidation reactions of CO, two hydrocarbons of differing reactivities and H_2, and the reaction kinetics was described by the global rate expressions of the dual-site Langmuir-Hinshelwood type [2].

2.2. Three-Dimensional Model

The three-dimensional model has been developed by extending the single-channel 1-D model described above to include the effects of flow maldistribution at the monolith front face and heat loss from the converter skin to the surroundings, which give rise to channel-to-channel variations in the temperature (as well as concentration) profile along the monolith length. This makes it necessary to include in the model three-dimensional heat conduction in the monolith substrate and through the surrounding materials (mat, steel shell and insulation) to account for interactions among neighboring channels through heat conduction.

The basic scheme for the numerical solution is the same as that used for the 1-D model, except that in this case the solid temperature field used to solve the DAE system for each monolith channel must be calculated from the *three-dimensional* solid-phase energy balance equation. The three-dimensional energy balance equation can be solved by a nonlinear finite element solver (such as ABAQUS) for the solid-phase temperature field while a nonlinear finite difference solver for the DAE system calculates the gas-phase temperature and concentrations as well as the solid-phase concentrations. Additional details on the numerical methods can be found elsewhere [5].

2.3. Comparison of 1-D and 3-D Models

The predicted conversion and thermal behavior of the 1-D and 3-D monolith models was compared under step-change conditions where a cold catalytic monolith is suddenly exposed to a step flow of stabilized exhaust gas at an elevated temperature [5]. It was shown that both models predict substantially different thermal response characteristics established within the monolith during the converter warm-up period. The 3-D model predicts that the center portion of the monolith (where the exhaust flow is higher) is heated up preferentially, resulting in steep temperature gradients in both the axial and radial directions. Obviously, such detailed information on the thermal behavior, which provides a basis for the analysis of thermal stress and fatigue in the monolith converter assembly, cannot be predicted by the 1-D model. However, the lightoff curves in the conversion-time domain predicted by the two models have been shown to be very similar under the step-change conditions [5]. These similarities in emission predictions can be attributed to the fact that the preferential heating near the center of a monolith tends to be counterbalanced by the shorter residence time of the gas in the region resulting from the higher flow rate there. It should also be noted that similar conversion predictions of the 1-D and 3-D models seem to carry over to simulations over actual driving cycles, such as the European Driving Cycle [7]. Furthermore, predictions from the 1-D model of converter-bed temperatures and converter-out mass emissions have been shown to agree well with those measured during the cold-start portion of actual vehicle emission tests [6]. This indicates that the 1-D model, though not capable of predicting the radial temperature profiles and the severity of the temperature gradients, is adequate in describing the conversion performance of a monolith during its warm-up, thus providing a useful design tool for emission control system development.

3. COLD-START EMISSION CALCULATIONS

For late-model gasoline vehicles, the vast majority (typically > 80%) of tailpipe hydrocarbon emissions occurs during the cold-start period of the FTP (EPA-prescribed driving schedule used for vehicle emission testing). The single-channel (1-D) monolith model with global rate expressions has been extensively used at GM to provide guidance in the design, optimization and implementation of emission control systems with improved cold-start emission control performance. Some examples of model applications include: (1) Determine the noble metal loading and catalyst volume required for a given emission performance target, (2) Quantify effects of substrate properties, (3) Quantify impact of changes in exhaust system architecture and engine management strategy, (4) Quantify impact of poisoning and thermal aging, and (5) Optimize the design and operation of electrically heated converters.

In this section we will first discuss the mode of converter warmup/lightoff following a cold start, and then the results of cold-start emission calculations for two different vehicle emission control systems to illustrate some of the model applications mentioned above.

3.1. Converter Warmup/Lightoff Behavior with and without Poisoning

Our earlier converter modeling study [3] has shown that during the cold-start period (when a cold monolith converter is suddenly exposed to hot exhaust gas), the upstream section of the monolith is first heated up to the reaction temperatures by the hot exhaust, leading to converter lightoff, and that the reaction is confined to a small fraction of the total monolith length (typically 1 to 1.5 inches) near the inlet throughout the warmup period. (The downstream section of the monolith remains cold for much of the time, contributing very little to converter lightoff.)   In addition, as the monolith converter is poisoned during vehicle use, the reaction zone tends to be shifted downstream because the upstream section is preferentially deactivated as a result of the deposition of engine oil-derived poisons (e.g., P and Zn) onto the catalytically active surface.   Such poison accumulation has been shown to delay converter lightoff [8]; the poisoned section must be heated up first before the catalytically active downstream section reaches the reaction temperatures.

3.2. Parametric Analysis on Dual-Brick Converter Performance

Here we describe the results of cold-start emission calculations for a GM vehicle equipped with a 2.2-L engine.   Cumulative tailpipe mass emissions during the first 2 minutes of the FTP (referred to as "cycle 1 emissions") will be used here as a measure of cold-start emission performance.   The *baseline* emission control system for this vehicle consists of an underfloor converter aged on an engine dynamometer to 50,000 miles equivalent, which contains 2 monolith bricks (87.7 cm^2 frontal area, 400 square cells/in^2): 3 in.-long front brick containing 225 g/ft^3 Pd and 6 in.-long rear brick containing 90 g/ft^3 Pt/Rh.   The measured cycle 1 conversions (i.e., conversions averaged over the first two minutes of the FTP) were 36.4% for HC and 23.0% for CO.   It was observed that the measured cycle 1 HC conversion remained virtually unchanged when increasing the noble metal loading of the rear brick from 90 to 120 g/ft^3 (while holding all the front brick properties constant), suggesting that the rear brick contributed very little to the cold-start emission control performance for this specific vehicle emission control system of interest here.   So our discussion below will focus on the results of parametric sensitivity analysis on the front brick only.   Also, the predicted cycle 1 conversions of 36.2% for HC and 26.1% for CO compare favorably with the measured cycle 1 conversions mentioned above, demonstrating the capability of the 1-D model to predict cold-start emission performance reasonably well.

Figure 1 shows how the cycle 1 HC emission is affected by the length (volume) of the front brick for various poison penetration depths.   In the calculations, the frontal area as well as the noble metal (Pd) concentration in the front brick was held constant.   The model predicts that the cold-start emission performance initially improves with increasing the front brick volume, but there is a critical brick size beyond which the emission benefit starts to diminish, consistent with the discussion in Section 3.1.   This critical brick length depends strongly on the poison penetration depth into the monolith: ~2 in. with 1 in. poison penetration, and ~3 in. with 2 in. poison penetration.   Accurate prediction of the critical brick size is important because using a front brick much larger than the critical size would be a waste of expensive noble metals without any significant cold-start emission benefit.   Notice also that the asymptotic emission levels achieved at sufficiently long front brick lengths increased substantially with catalyst poisoning (~30% increase in emission upon 2 in. poison accumulation).   As discussed above, this poison-induced deterioration of cold-start emission performance predicted here is a direct consequence of the poisoned upstream section of the monolith acting as a heat sink during the converter warmup process.

Note that in Fig. 1, the Pd concentration in the front brick was held constant at 225 g/ft^3, so that the amount of Pd in the front brick is directly proportional to its length (volume). Another interesting converter design question is "What is the optimum front brick size when a fixed amount of noble metal is allowed to be used?"   Figure 2 shows the results of such calculations for the noble metal content of 5.3 g Pd in the front brick.   Interestingly, the predicted cold-start HC emission goes through a minimum as the front brick length is varied, and the optimum brick size at which the emission is the lowest depends again on the poison penetration depth.   The cold-start emission performance suffers when the brick size is too small as a result of insufficient contact time between the exhaust gas and the catalyst.   If the brick size is too large, on the other hand, the amount of Pd per unit monolith volume in the upstream section (which is primarily responsible for converter lightoff) becomes too small, resulting in poor cold-start emission performance.

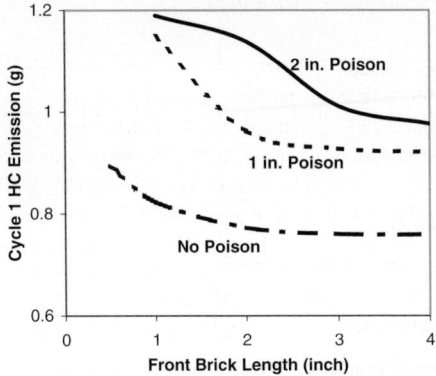 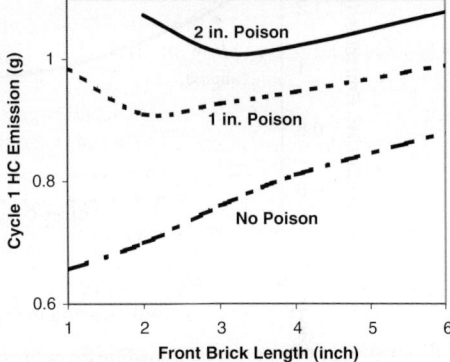

Fig. 1. Effects of front brick volume on cold-start emissions at various poison penetration depths (underfloor converter location; front brick with 225 g/ft^3 Pd loading + 6 in. rear brick with 90 g/ft^3 noble metal loading).

Fig. 2. Effects of front brick volume on cold-start emissions at various poison penetration depths (underfloor converter location; front brick containing 5.3 g Pd + 6 in. rear brick with 90 g/ft^3 noble metal loading).

Figure 3 shows the impact of the Pd loading in the front brick (g Pd per ft^3 monolith volume) on cycle 1 HC emissions at two different converter locations for the case of a 3 in.-long front brick with 1 in. poison penetration.   At the underfloor converter location, the model predicts the expected result that the cold-start emission level generally decreases with increasing Pd loading; however, the emission benefit of increasing Pd loading diminishes at high loadings.   The results of such calculations can be used to determine the Pd loading required to achieve a given emission performance target and to quantify the sensitivity of emission performance to variations in noble metal loading, providing some sense of "bang for the buck" for a specific vehicle emission control system of interest.   In addition, the closeness between the post-front-brick and tailpipe HC emissions in Fig. 3 indicates that the rear brick contributes very little to the cold-start emission performance except at very low Pd loadings (< 50 g/ft^3) in the front brick, in agreement with our emission test data discussed earlier.   Also shown in Fig. 3 is the prediction that the cold-start HC emission is drastically reduced when moving the converter closer to the engine.   (The inlet exhaust gas temperature profile to the close-coupled converter during the FTP was increased by as much as 150 to 200°C compared to the converter at the underfloor location included in the baseline emission control system.)   It is also interesting to note that at the close-coupled converter location, the predicted cold-start emission performance is little affected by the Pd loading in the front brick as long as the Pd loading is higher than 50 g/ft^3.   For example, a 6-fold increase in Pd loading from 50 to 300 g/ft^3 is predicted to reduce the cycle 1 HC emission only by 0.035 g.

Additional simulations show that the cold-start emission performance of a close-coupled converter is also much less sensitive to poisoning than that of an underfloor converter (results not shown). These results suggest that optimizing the converter location provides an effective means of improving cold-start emission performance at lower noble metal loadings.

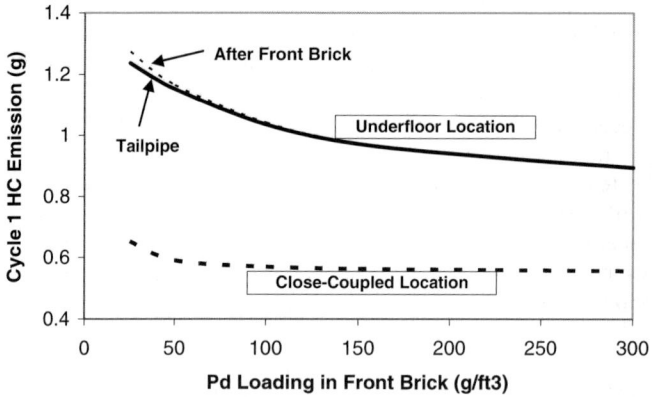

Fig. 3. Effects of front brick Pd loading on cold-start emissions at two converter locations (1 in. poison penetration; 3 in. front brick + 6 in. rear brick with 90 g/ft^3 noble metal loading).

The results of Fig. 3 clearly illustrate the importance of optimizing not only converter design parameters but also its operating environment in developing effective emission control systems. In fact, our experience (as well as the simulation results presented here) has shown that the emission performance of an exhaust aftertreatment system is often affected to a greater extent by converter operating environment. (In addition to close-coupled converter location, air injection near the exhaust valve in combination with fuel-rich engine operation has also been shown to increase the exhaust temperature as a result of the exothermic heat generated by the gas-phase oxidation reactions occurring in the exhaust port region, providing substantial improvements in cold-start emission performance [9].) Since the operating environment of a converter is influenced by the engine management system as well as the exhaust system architecture, we cannot consider the catalytic converter in isolation, but rather the entire engine/exhaust aftertreatment system needs to be considered for the design, optimization and implementation of vehicle emission control systems. Such a systems modeling approach would require development, validation and integration of all the necessary component models (e.g., engine-out emission model, exhaust port/manifold model, pipe heat transfer model, converter performance model, catalyst aging model).

The prediction in Fig. 3 that the cycle 1 emission level becomes insensitive to Pd loading variation in the regime of high Pd loadings can be explained by the fact that the benefits of improving the catalyst activity (by increasing the *active* metal surface area and/or the *specific* reaction rate) tend to be tempered by the material balance requirement that the reactant consumption rate cannot exceed the external mass transfer rate at the surface concentration equal to zero (i.e., mass transfer-controlled condition). Thus, in emission control system development, catalyst activity improvement must be combined with optimization of converter operating environment (e.g., increasing the converter inlet gas temperature) in order to maximize the emission performance and/or minimize the noble metal usage requirements.

It is worth mentioning that the results discussed in Sections 3.1 and 3.2 provide a rationale behind the recent trends toward a close-coupled converter location as well as dual-brick converter design (i.e., a relatively small-volume catalyst brick with high Pd loadings followed by a larger Pt/Rh catalyst brick) in order to improve converter lightoff performance without excessive noble metal usage.

3.3 Electrically Heated Converter (EHC) Design

In the discussions above, the engine exhaust was considered to be the only energy source for heating the catalyst to the required reaction temperatures. One cold-start emission control strategy that received considerable attention in the early to mid 1990's is to supply an external energy input through the use of an electrically heated converter (EHC; metal-substrate monolith heated resistively) placed upstream of a main catalytic converter. Early EHC systems consist of a catalyzed electric heater of a relatively large size (150-500 cm^3) to provide an external energy input as well as additional catalytic conversion during the cold-start period. However, such large-volume electric heaters, though providing substantial catalytic conversion over itself once warmed-up, were found to be rather ineffective at reducing cold-start tailpipe HC emissions. As will be illustrated below, our modeling studies help explain why large electric heaters are ineffective and point ways toward improved EHC system design and operation. For the discussion of simulation results here, we consider a GM vehicle equipped with a 3.8-L V6 engine whose emission control system consists of a metallic catalyzed electric heater of varying sizes placed immediately upstream of a ceramic-substrate main converter containing a Pt/Rh-impregnated monolith brick (87.7 cm^2 frontal area x 28 cm long with a noble metal loading of 32 g/ft^3).

Figure 4 shows how variations in the electric heater size affect the relative contributions of the heated element and the main (unheated) converter to the overall HC conversion performance of the EHC system for the case of 20 s heating at 2500 W. As expected, the HC conversion over the heater increases with increasing heater volume. With a large electric heater, however, the conversion performance of the main converter (as given by the difference between the dashed and solid curves in Fig. 4) is predicted to be substantially lower than that with a small-volume heater, so that the best overall conversion performance (i.e., lowest tailpipe emissions) can be obtained in the regime of small heater volumes.

The greater conversion enhancement of the main converter by a small-volume electric heater is a direct consequence of its effectiveness in heating the main converter located downstream. This aspect is clearly illustrated in Fig. 5, which compares the predicted solid-phase temperatures at a location 2.54 cm downstream from the front face of the main converter for three different heater volumes (2500 W power supply for 20 s in all cases). It can be seen that the main converter bed temperature rises more rapidly with small heater volumes (compare 6 cm-long vs 0.8 or 0.4 cm-long heater). This prediction is reasonable because as the heated volume decreases, the amount of energy accumulated within the heater would decrease, and thus a larger fraction of the supplied energy (exhaust sensible heat, electrical energy, and reaction exotherm) would become available to heat up the main converter. However, there is a critical heater size below which no significant improvement in converter heat-up characteristics (and thus no emission benefit) is obtained. For example, decreasing the heater length from 0.8 to 0.4 cm does not significantly improve the main converter heat-up characteristics, in accord with the predicted insensitivity of tailpipe emissions to heater size variation for heater lengths shorter than 1 cm in Fig. 4. Furthermore, the main converter bed temperature trace with the 0.4 cm-long uncatalyzed heater (see dashed curve in Fig. 5) follows closely that predicted with the catalyzed heater of the same size.

This suggests that the primary function of the small-volume electric heater in the EHC system is to transfer the supplied electrical energy downstream for rapid lightoff of the main converter rather than to provide additional catalytic conversion. In fact, consistent with this argument, computer simulations for the EHC system with a 0.4-cm-long heater predicted very similar tailpipe HC emissions regardless of whether or not the electric heater is catalyzed (see Fig. 4).

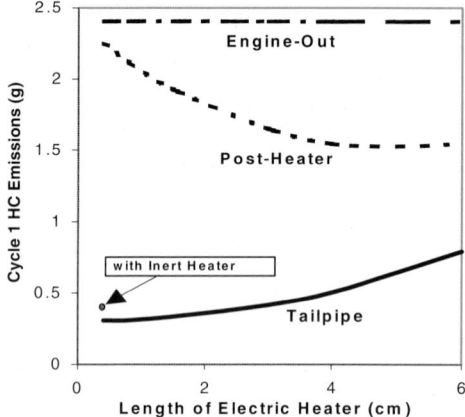

Fig. 4. Effects of electric heater volume on the post-heater and tailpipe HC emissions (20 s heating at 2500 W). Also shown is the tailpipe HC emission predicted with the 0.4 cm-long inert electric heater.

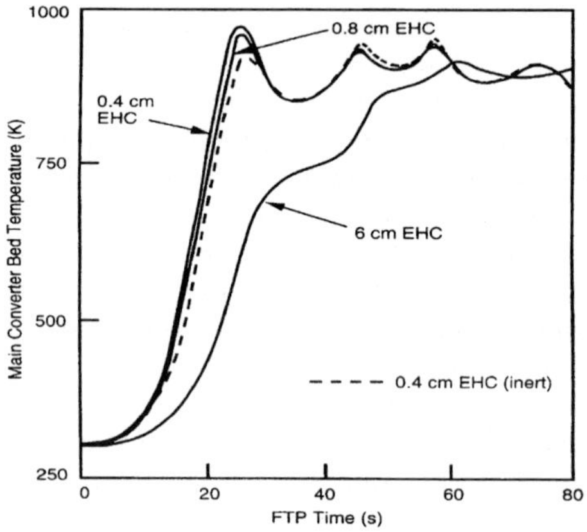

Fig. 5. Time variations of the main converter bed temperature predicted with electric heater of different volumes (20 s heating at 2500 W). Also shown is the bed temperature trace with the 0.4 cm-long inert electric heater.

In view of their effectiveness in reducing cold-start emissions, we are particularly interested in the behavior of small-volume electric heaters. For an uncatalyzed (inert) heater of sufficiently small size, the following explicit, analytical asymptotic solution can be obtained [10]:

$$T_g = T_{gi}(t) + \frac{xP}{F(t)}(1 - e^{-t/\gamma})$$

$$T_s = \frac{P}{\mu}(1 - e^{-t/\gamma}) + \frac{1}{\gamma}e^{-t/\gamma}\int_0^t \left[T_{gi}(\tau)e^{\tau/\gamma} + \frac{xP}{F(\tau)}(e^{\tau/\gamma} - 1) \right] d\tau$$

where constant electrical power P is assumed to be supplied to a heater with the dimensionless volume μ (normalized to the characteristic volume which is proportional to the exhaust flow rate and inversely proportional to the product of the interphase heat transfer coefficient within the heated element and its geometric surface area per unit volume), and γ represents the ratio between the characteristic volumes for the interphase heat transfer and solid-phase heat accumulation. Note that the above equations allow for the transient nature of the inlet gas temperature T_{gi} and flow rate F encountered during vehicle emission tests. The asymptotic solution for the solid-phase temperature T_s contains a term inversely proportional to μ whereas the gas-phase temperature T_g is independent of μ. The key feature of the asymptotic solution, then, is that decreasing the heater volume μ will increase the solid temperature of the electric heater but the gas temperature will not exhibit corresponding increases in the regime of small heater volumes ($\mu \ll 1$). This insensitivity of the gas temperature to variation in μ can be attributed to the fact that the higher solid temperatures developed within the smaller heater tend to be compensated by poor solid-to-gas heat transfer (required for gas heating with the electrical energy supplied to the metal substrate) resulting from its insufficient geometric surface area available, so that there is no change in exit gas temperature from the heater and thus in tailpipe emissions. The quantity of particular interest here is T_g at the heater exit ($x = 1$) since the electric heater in EHC systems is most effective when its exit gas temperature is raised as high and as quickly as possible. The above formula allows one to easily calculate the outlet gas temperature from a sufficiently small electric heater as a function of heating time for the general cases of time-varying inlet gas temperature and flow rate. This information on $T_g(x = 1)$ vs t can be used to determine the time required to reach a typical catalyst lightoff temperature for a given electrical power level. Alternatively, the power level required to achieve a predetermined catalyst lightoff time can also be estimated using the same formula.

The asymptotic techniques used here do not allow the precise determination of the "best" heater size (i.e., heater size small enough to give minimum tailpipe emissions, but not so small as to generate excessive solid temperatures) but the "best" heater size regime should be expected in the transition between $\mu = O(1)$ and $\mu \ll 1$. This suggests that the "best" physical heater volume is the one that provides just enough geometric surface area for effective heat transfer from the solid phase (resistively heated) to the exhaust gas, so that proper gas heating using the supplied electrical energy can be accomplished.

The results discussed in Sections 3.1 and 3.2 suggest that the function of a small-volume heater as a heat transfer device, designed to supply heat to the main converter for its rapid activation, can be augmented by coupling it with a highly active lightoff catalyst located downstream. This concept has been validated both experimentally and computationally, and provides a scientific basis for the evolution of EHC design toward "cascade" configurations which include a small-volume electric heater followed by an unheated "lightoff" catalyst containing high concentrations of noble metals [11, 12].

4. CONCLUSIONS AND FUTURE RESEARCH NEEDS

It is reasonable to conclude from this study that the single-channel 1-D monolith model with global rate expressions is capable of predicting the cold-start emission performance of a 3-way catalytic converter reasonably well, thus providing a useful tool for the optimization of vehicle emission control system design and implementation. However, "tuning" of some of the reaction rate constants in the global rate expressions is often required to accurately simulate the particular catalytic converter system under consideration, and such model calibration needs to be minimized by improving the global kinetics through further theoretical and experimental studies focusing on such topics as exhaust hydrocarbon species grouping strategy; promotional role of H_2 and its reactivity; kinetic effects and oxygen storage/release characteristics of base metal additives (e.g., Ce); reaction mechanism and kinetics for NO reduction; and impact of catalyst aging on reaction kinetics.

In the long run, our modeling efforts should be directed toward developing converter models with more accurate and detailed kinetics which is based on elementary reaction steps occurring on the catalyst surface. The elementary reaction step-based modeling approach is well suited to handling a situation where rate-determining steps are likely to change, allows one to extend/adapt the model more easily to different catalyst formulations or more complex reacting mixtures, and can be extrapolated with more confidence to outside the ranges where experimental data exists. However, the application of such elementary reaction step-based kinetic models to the actual exhaust environment results in a large number of reaction steps [e.g., 13, 14], and our limited mechanistic understanding of some of the relevant reactions (e.g., those involving NO and HC's) makes the reliable estimation of all the rate constants difficult. Consequently, a significant number of the kinetic parameters are often determined by regression analysis of the experimental data that the model is supposed to predict. In order to maximize the benefits of the microkinetics-based modeling approach, it is necessary to minimize the number of "adjustable" parameters by determining as many kinetic parameters as possible from independent experiments and/or first principles calculations [15, 16]. It is also important to perform a thermodynamic consistency check on the estimated kinetic parameters to ensure that they are physically reasonable [17, 18].

The results presented in this paper clearly demonstrate the need/benefit to simulate the entire engine/exhaust aftertreatment system for the design, optimization and implementation of vehicle emission control systems. Such a systems modeling approach would require development, validation and integration of all the necessary component models (e.g., engine-out emission model, exhaust port/manifold model, pipe model, converter performance model, catalyst aging model). Among these component models, prediction capabilities for engine-out emissions under cold-start and transient conditions as well as those for catalyst aging are not adequate at this point and need further development. Also, there are some significant numerical issues related to the integration of the individual component models, such as the numerical efficiency, accuracy and robustness of the entire emission control system model, given that each of the component models has a different characteristic response time and governing equations of different structures. Other challenges include potential incompatibility of the various numerical approaches in use with the component models, and difficulties associated with numerical coupling and information sharing between them.

REFERENCES

[1] R.H. Heck, J. Wei, and J.R. Katzer, AIChE J., 22 (1976) 477.
[2] S.H. Oh and J.C. Cavendish, Ind. Eng. Chem. Prod. Res. Dev., 21 (1982) 29.
[3] S.H. Oh and J.C. Cavendish, Ind. Eng. Chem. Prod. Res. Dev., 22 (1983) 509.
[4] K. Zygourakis, Chem. Eng. Sci., 44 (1989) 2075.
[5] D.K.S. Chen, S.H. Oh, E.J. Bissett, and D.L. Van Ostrom, SAE Paper No. 880282 (1988).
[6] S.H. Oh, E.J. Bissett, and P.A. Battiston, Ind. Eng. Chem. Research, 32 (1993) 1560
[7] G.C. Koltsakis, P.A. Konstantinidis, and A.M. Stamatelos, Appl. Catal. B: Env., 12 (1997) 161.
[8] S.T. Darr, R.A. Choksi, C.P. Hubbard, M.D. Johnson, and R.W. McCabe, SAE Paper No. 2000-01-1881 (2000).
[9] K. Kollmann, J. Abthoff, W. Zahn, H. Bischof, and J. Göhre, SAE Paper No. 940472 (1994).
[10] E.J. Bissett and S.H. Oh, Chem. Eng. Sci., 54 (1999) 3957.
[11] G. Brunson, J.E. Kubsh, and W.A. Whittenberger, SAE Paper No. 932722 (1993).
[12] L.S.Jr. Socha, D.F. Thompson, and P.A. Weber, SAE Paper No. 940468 (1994).
[13] D. Chatterjee, O. Deutschmann, and J. Warnatz, Faraday Discuss., 119 (2001) 371.
[14] J.H.B.J. Hoebink, J.M.A. Harmsen, M. Balenovic, A.C.P.M. Backx, and J.C. Schouten, Topics in Catal., 16/17 (2001) 319.
[15] J.A. Dumesic, D.F. Rudd, L.M. Aparicio, J.E. Rekoske, and A.A. Trevino, *The Microkinetics of Heterogeneous Catalysis*, American Chemical Society, Washington, DC (1993).
[16] R.A. Van Santen and M. Neurock, Catal. Rev. - Sci. Eng., 37 (1995) 557.
[17] S. Sriramulu, P.D. Moore, J.P. Mello, and R.S. Weber, SAE Paper No. 2001-01-0936 (2001).
[18] A.B. Mhadeshwar, H. Wang, and D.G. Vlachos, J. Phys. Chem, B, 107 (2003) 12721.

Studies in Surface Science and Catalysis, volume 159
Hyun-Ku Rhee, In-Sik Nam and Jong Moon Park (Editors)

Enzymatic Processes for Fine Chemicals and Pharmaceuticals: Kinetic Simulation for Optimal R-Phenylacetylcarbinol Production

N.Leksawasdi[a,b], B.Rosche[a] and P.Rogers[a]

[a]School of Biotechnology and Biomolecular Sciences
University of New South Wales, Sydney, NSW 2052

[b]Department of Food Engineering, Faculty of Agro-Industry,
Chiang Mai University, Chiang Mai 50100, Thailand

Abstract

An optimal feeding strategy with a 1:1.2 molar ratio of benzaldehyde: pyruvate has been developed to maximize the enzymatic production of the pharmaceutical intermediate R-phenylacetylcarbinol (PAC) based on a previously published model for this intermediate. The results of the simulation indicate with pyruvate decarboxylase (PDC) at an initial carboligase activity of 4 U ml^{-1} that up to 730 mM PAC would be produced in 81 h. However, the experimental results show an appreciably lower maximum level of PAC (viz 300 mM) produced after only 54 h, although enzyme activity was maintained at similar or higher values than in the simulation results. It is possible that the increasing PAC concentrations and associated by-products (acetoin and acetaldehyde) have resulted in significant inhibition of PDC during the course of the biotransformation and future model development will need to include one or more product inhibition terms in its structure.

Nomenclature

Nomenclature

A	pyruvate concentration (mM)	K_b	intrinsic or microscopic binding constant
B	benzaldehyde concentration (mM)		for benzaldehyde (mM^{1-h})
E	PDC activity at time t (U ml^{-1} carboligase)	K_{ma}	affinity constant for pyruvate (mM)
		K_{mb}	affinity constant for benzaldehyde (mM)
E_o	initial concentration of PDC enzyme	P	PAC concentration (mM)
EQ	binary enzyme complex between PDC and 'active acetaldehyde'	Q	acetaldehyde concentration (mM)
		R	acetoin concentration (mM)
h	exponent with similar functionality to Hill coefficient for benzaldehyde (no unit)	t	time (h)
		t_{lag}	lag time (h)
		U	carboligase enzyme activity unit
i	iteration loop identifier of each species to be used in numerical integration	V_p	overall rate constant for the formation of PAC (μmol h^{-1} U^{-1})
k_{d1}	first order reaction time deactivation Constant (h^{-1})	V_q	overall rate constant for the formation of acetaldehyde (ml h^{-1} U^{-1})
k_{d2}	first order benzaldehyde deactivation coefficient (mM^{-1} h^{-1})	V_r	overall rate constant for the formation of acetoin (l^2 h^{-1} U^{-1} mol^{-1})

Introduction

The growth of industrial scale biotransformation processes has increased significantly over recent years with more than 130 reported in 2002 [1]. Major activity is focussed on the production of chiral pharmaceuticals although a number of large scale industrial processes have been developed for the food and agricultural industries.

The advantages of such biotransformation processes are: (1) the relatively high yields which can be achieved with specific enzymes, (2) the formation of chiral compounds suitable for biopharmaceuticals, and (3) the relatively mild reaction conditions. Key issues in industrial–scale process development are achieving high product concentrations, yields and productivities by maintaining enzyme activity and stability under reaction conditions while reducing enzyme production costs.

Ephedrine and pseudoephedrine are a vasodilator and decongestant respectively used widely in the treatment of asthma and the symptoms of colds and influenza. These pharmaceuticals were derived originally from the plant *Ephedra sinica* and used in traditional Chinese medicinal preparations. Although some are still produced from such sources, the major production is via a fermentation process followed by a chemical catalytic reaction. As shown in Figure 1, the intermediate *R*-phenylacetylcarbinol (PAC) is produced by decarboxylation of pyruvate followed by ligation to benzaldehyde.

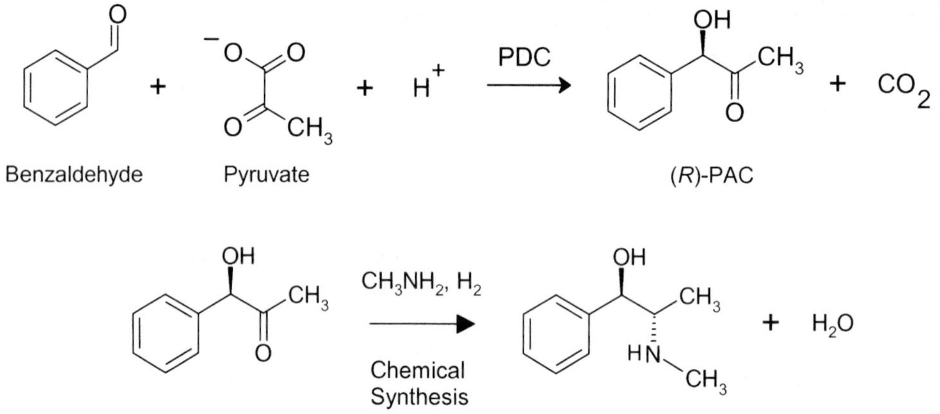

Figure 1.
Formation of PAC from benzaldehyde and pyruvate catalysed by PDC and reductive amination of *R*-PAC to produce the chiral biopharmaceutical product ephedrine.

In current industrial practice, benzaldehyde is added to fermenting baker's yeast (*Saccharomyces cerevisiae*) with resultant PAC production occurring from the yeast-derived pyruvate. Typically PAC concentrations of 12-15 g l^{-1} are produced at yields of 65-70% theoretical in a 10-12 h biotransformation process. [2]. Appreciable concentrations of benzyl alcohol are produced as by-product due to oxidoreductase activity in the fermentative yeast.

An enzymatic process using partially purified pyruvate decarboxylase (PDC) with added pyruvate overcomes the problems of benzyl alcohol formation and limiting availability of pyruvate [3]. As a result increased concentrations, yields and productivities of PAC were achieved with concentrations of PAC in excess of 50 g l^{-1} (330 mM) in 28 h and yields on benzaldehyde above 95% theoretical [4-6]. Screening of a wide range of bacteria, yeasts and other fungi as potential sources of stable, high activity PDC for production of PAC confirmed a strain of the yeast *Candida utilis* as the most suitable source of PDC [7].

The most limiting factor for enzymatic PAC production is the inactivation of PDC by the toxic substrate benzaldehyde. The rate of PDC deactivation follows a first order dependency on benzaldehyde concentration and reaction time [8]. Various strategies have been developed to minimize PDC exposure to benzaldehyde including fed-batch operation, immobilization of PDC for continuous operation and more recently an enzymatic aqueous/octanol two-phase process [5,9,10] in which benzaldehyde is continuously fed from the octanol to the enzyme in the aqueous phase. The present study aims at optimal feeding of benzaldehyde in an aqueous batch system.

Mathematical Model for Enzymatic PAC Production

A model developed by Leksawasdi et al. [11,12] for the enzymatic production of PAC (P) from benzaldehyde (B) and pyruvate (A) in an aqueous phase system is based on equations given in Figure 2. The model also includes the production of by-products acetaldehyde (Q) and acetoin (R). The rate of deactivation of PDC (E) was shown to exhibit a first order dependency on benzaldehyde concentration and exposure time as well as an initial time lag [8]. Following detailed kinetic studies, the model including the equation for enzyme deactivation was shown to provide acceptable fitting of the kinetic data for the ranges 50-150 mM benzaldehyde, 60-180 mM pyruvate and 1.1-3.4 U ml^{-1} PDC carboligase activity [10].

Product

$$\frac{d[P]}{dt}\bigg|_i = V_p \left[\frac{K_b [B_i]^h}{1+K_b [B_i]^h} \right] \left[\frac{[A_i]}{K_{ma} + [A_i]} \right] [E_i]$$

Reactants

$$\frac{d[B]}{dt}\bigg|_i = -\frac{d[P]}{dt}\bigg|_i \qquad \frac{d[A]}{dt}\bigg|_i = -\frac{d[P]}{dt}\bigg|_i - \frac{d[Q]}{dt}\bigg|_i - 2\frac{d[R]}{dt}\bigg|_i$$

By-products

$$\frac{d[Q]}{dt}\bigg|_i = V_q [A_i][E_i] - V_r [Q_i][A_i][E_i] \qquad \frac{d[R]}{dt}\bigg|_i = V_r [Q_i][A_i][E_i]$$

Figure 2 .
Mathematical model employed in modeling the biotransformation for PAC [11].

$$\left.\frac{dE}{dt}\right|_i = \begin{cases} 0 & ; \quad t \le t_{lag} \\ -\left(k_{d1} + k_{d2}.b\right)E_i & ; \quad t > t_{lag} \end{cases}$$

Equation for PDC deactivation [8].

Values of kinetic constants which provided the 'best fit' of the four data sets within the above substrate ranges were determined (Table 1) and used in the subsequent study to determine optimal feeding rates to maximize PAC concentration.

Table 1
Kinetic parameters used in the construction of feeding profile.

Constants	Unit	Values
V_p	μmol h^{-1} U^{-1}	24.8
K_b	mM$^{-1.34}$	1.00×10^{-4}
h	no unit	2.34
K_{ma}	mM	10.6
V_q	ml h^{-1} U^{-1}	1.56×10^{-2}
V_r	l^2 h^{-1} U^{-1} mol^{-1}	2.51×10^{-3}
k_{d1}	h^{-1}	2.64×10^{-3}
k_{d2}	mM^{-1} h^{-1}	1.98×10^{-4}
t_{lag}	h	5.23

Optimization of PAC Production

(a) Simulation Profiles

Based on the model described in the previous section, an optimization strategy has been developed for substrate feeding with maximum PAC production as its objective function. A molar ratio of pyruvate: benzaldehyde of 1.2:1 was maintained in the feed as some of the pyruvate is converted also to by-products acetaldehyde and acetoin.

The optimal feeding profile based on the model is shown in Figure 3 and the simulation profiles are shown in Figure 4 for initial substrate concentrations of 90 mM benzaldehyde and 108 mM sodium pyruvate, and initial PDC activity of 4.0 U ml^{-1} carboligase. Feeding was programmed at hourly intervals and the initial reaction volume would increase by 50% by the end of the simulated biotransformation.

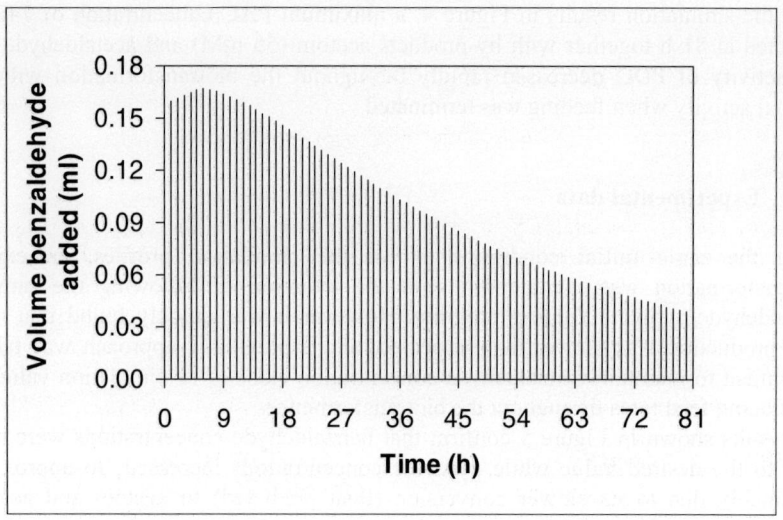

Figure 3.
Prediction of benzaldehyde pulse feeding profile to optimize the production of PAC in fed batch PAC biotransformation process at 6°C. A molar ratio of 1.2:1 pyruvate to benzaldehyde was used, with AR grade benzaldehyde and a 1.4 M solution of pyruvate used in the simulated feeding.

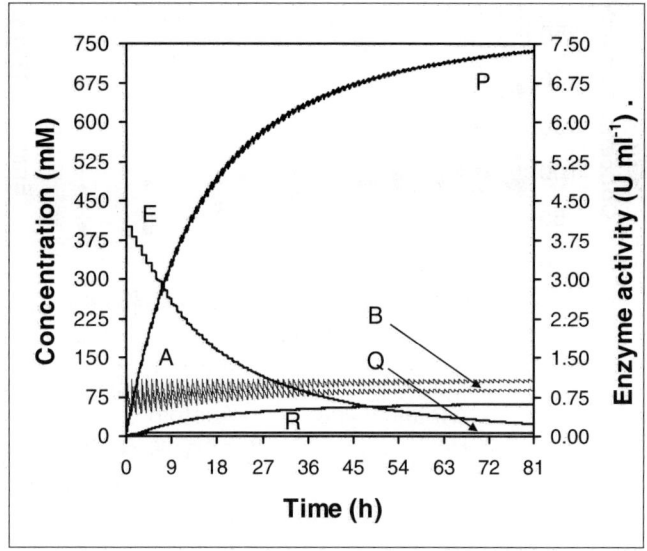

Figure 4.
Simulation profile of fed batch PAC biotransformation kinetics at 6°C with initial PDC activity of 4.0 U carboligase ml^{-1}, 90 mM benzaldehyde and 108 mM sodium pyruvate. Feeding was performed hourly as illustrated in Fig. 3 and the initial reaction volume of 30 ml (which would be used experimentally) increased to 45 ml at the end of reaction.

From the simulation results in Figure 4, a maximum PAC concentration of 740 mM was predicted at 81 h together with by-products acetoin (55 mM) and acetaldehyde (10 mM). The activity of PDC decreased rapidly throughout the biotransformation with only 5% residual activity when feeding was terminated.

(b)   Experimental data

Using the same initial conditions as for the simulation profiles, an experimental biotransformation was commenced with the intention of following the same optimal benzaldehyde/pyruvate feeding program. However it was quickly found that the rate of PAC production was slower than expected, and a pragmatic approach was taken in the experiment to maintain benzaldehyde concentration close to its simulation value (90 mM) by reducing feed rates throughout the biotransformation.

The results shown in Figure 5 confirm that benzaldehyde concentrations were maintained close to the desired value while pyruvate concentrations increased   to approx. 200 mM presumably due to its slower conversion (than predicted) to acetoin and acetaldehyde. After 55 h a maximum PAC concentration of approx. 300 mM was reached, while acetoin and acetaldehyde concentrations were 20 mM and 5 mM, respectively. A residual PDC activity of 35% remained at this time.

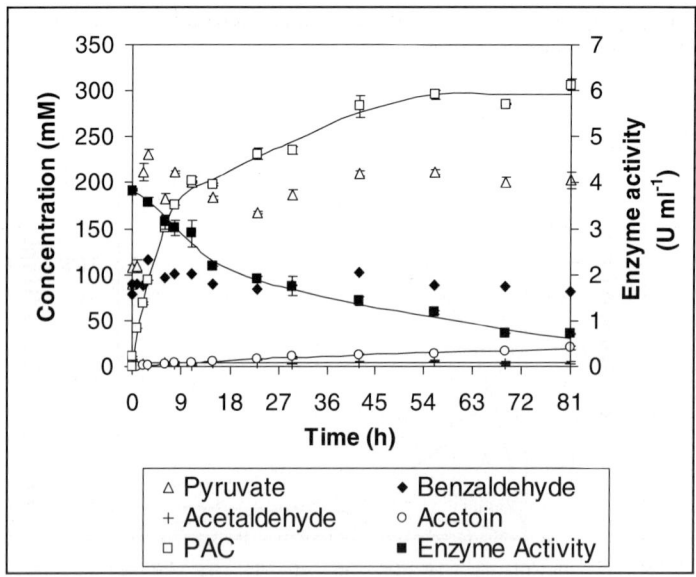

Figure 5.
Fed- batch PAC biotransformation kinetics at 6°C with optimal feeding program. Initial and final reaction volumes after 81 h were 30 and 45 ml, respectively. The biotransformation was carried out in 2.5 M MOPS, 1 mM MgSO$_4$, 1 mM TPP with initial pH of 6.5 and initial carboligase activity of 3.82 U ml^{-1}. The pH adjustment was performed manually using 30% (v/v) H$_2$SO$_4$.

Discussion and Conclusions

Comparison of experimental data from the optimal feeding program with simulation results has demonstrated much lower PAC concentrations than those predicted by the model (e. g., 300 mM compared to 700 mM, respectively at 54 h) with the comparative difference confirming the need for further model development.

It is relevant also to compare the results in Fig.5 with previously published data for PAC production under similar environmental conditions, where with higher concentrations of initial benzaldehyde (600 mM), pyruvate (400 mM) and PDC activity (8.4 U ml^{-1}) a similar maximum concentration of PAC of 330 mM was produced [6]. PDC stability was similar in both processes with half life values of approximately 27h. However, PAC production was much faster in the benzaldehyde emulsion system, presumably due to higher initial enzyme concentration.

The conclusion from this comparison is that the optimal feeding program with its lower benzaldehyde concentrations did not result in any increase in PDC stability or final PAC concentration. This suggests that a component(s) other than benzaldehyde (with its limited solubility of 90-100 mM in this system containing 2.5 M MOPS buffer) is more critical in achieving increased PAC concentrations and productivities.

It is possible that product inhibition by PAC or by-product inhibition by acetoin or acetaldehyde may play a more important role than benzaldehyde in influencing PAC production in this aqueous phase system.

From analysis of the data in Figure 5 it is clear that the rate of PAC formation is fairly constant for the first 9 h to approx. 180 mM PAC and subsequently declines after this to reach a maximum value of 300 mM in 54 h. Unpublished studies by our group have shown no significant inhibition effects of PAC up to concentrations of 154 mM [13]. However, it is possible that appreciable inhibition could occur at higher PAC concentrations. By-products acetoin and acetaldehyde may also be inhibitory in the latter stages of the biotransformation although both concentrations were below 5 mM at 9 h.

As a result of the difference between the experimental and simulation results future development will focus on inclusion in the model of a PAC product inhibition term with a thresh-hold concentration value of approx 180 mM.

In a further approach, an organic/aqueous two-phase system has been designed in which benzaldehyde, PAC and acetaldehyde partition strongly into the organic phase. Acetoin partitioned preferentially into the aqueous phase while pyruvate showed a total preference for this latter phase. In the various two-phase processes PAC concentration in the enzyme-containing aqueous phase reached a maximum value of 165 mM [5] which is below the thresh-hold value estimated in the present study. A wide range of solvents was screened for use in the two-phase process and octanol was selected for the organic phase in view of its relatively high partition coefficient for benzaldehyde and low water solubility and toxicity. Several modes of operation of the aqueous/octanol two-phase system were evaluated and concentrations up to 165 g l^{-1} PAC (octanol phase) / 25 g l^{-1} PAC (aqueous phase) were attained in a slowly stirred phase separated process [5,9,10]. The results with this system illustrate that high yields and productivities, as well as significantly increased PAC concentrations can be achieved in a two-phase process.

References

[1] A. Strathof, S. Panke, S. and A. Schmid, Current Opinion in Biotech., 13 (2002) 548.

[2] P.Rogers, H.S.Shin and B.Wang, Adv Biochem Eng.56 (1997) 33.

[3] H.S.Shin and P. Rogers, Biotech. Bioeng., 49 (1996) 52.

[4] B.Rosche, N.Leksawasdi, V.Sandford, M.Breuer, B.Hauer and P.Rogers,
Appl.Microbiol.Biotechnol. ,60 (2002) 94.

[5] B.Rosche, V.Sandford, M. Breuer, B. Hauer and P.Rogers, J.Mol.Cat.B:Enzymatic, 19-20
(2002)109.

[6] B.Rosche, M.Breuer, B.Hauer and P.Rogers, Biotech. Letts., 25 (2003) 847.

[7] B.Rosche,M. Breuer, B.Hauer and P. Rogers, Biotech. Letts., 25 (2003) 841.

[8] N.Leksawasdi, M.Breuer, B.Hauer, B.Rosche and P.Rogers, Biocat Biotransf., 21 (2003) 315.

[9] V.Sandford, M.Breuer,B. Hauer, P.Rogers and B.Rosche, B. Biotech.Bioeng.,(2005) (in press).

[10] B.Rosche,M. Breuer, B.Hauer and P. Rogers, Biotech. Letts., (2005) (in press).

[11] N.Leksawasdi, Y.Chow, M.Breuer, B.Hauer, B.Rosche and P.Rogers, J.Biotech.,111
(2004) 179.

[12] N.Leksawasdi, B.Rosche and P.Rogers, Biochem. Eng.J.,23 (2005) 211.

[13] V.Sandford. Enzymatic Process Development for R-PAC: An Intermediate for Ephedrine and
Pseudoephedrine Production. PhD Thesis UNSW (2002).

Studies in Surface Science and Catalysis, volume 159
Hyun-Ku Rhee, In-Sik Nam and Jong Moon Park (Editors)

Industrial and real-life applications of micro-reactor process engineering for fine and functional chemistry

Volker Hessel[a,b], P. Löb[a] and H. Löwe[a,c*]

[a] *IMM Institut für Mikrotechnik Mainz GmbH, 55129 Mainz, Germany*
[b] *Eindhoven University of Technology, Department of Chemical Engineering and Chemistry, 5600 MB Eindhoven, The Netherlands*
[c] *Johannes Gutenberg University Mainz, Institute of Organic Chemistry, 55128 Mainz, Germany*
*   *Corresponding author: loewe@imm-mainz.de*

1. CHEMICAL MICRO PROCESS ENGINEERING

Chemical micro processing can both improve current chemical processes [1,2] and act as an enabling technology towards novel chemistry [3]. By using microstructured reactors "game-changing" optimization of existing processes can be achieved by means of process intensification, based on increased mass and heat transfer as well as better residence-time control and regular-patterned fluid dynamics [1,2]. As a result, there is considerable evidence meanwhile how microstructured reactors can improve existing chemical processes, in particular concerning the fields of fine-chemical synthesis, particle formation, and energy generation via fuel processing [1-9]. With respect to fine-chemical synthesis, for example, selectivity improvements were shown for a large number of organic reactions, including most of the well-known name reactions such as the Michael addition, Knoevenagel condensation, Hantzsch pyrazole synthesis, Wittig reaction, and many more [1,3]. Often overlooked so far, however, is the real potential of this technology, which is to perform novel ways of chemistry up to the so-called "dream reactions".

Recently, microstructured reactors have stepped into chemical production [4] and thus *micro-reactor process and plant design*, including economic incentives, is the issue at this time. For this purpose, large-capacity microstructured apparatus is needed ('micro inside, fist- to shoebox size outside') and plant concepts have to be proposed which include all process steps.

2. INDUSTRIAL CASE STUDIES – FOR PROCESS DEVELOPMENT

Most convincing for testing of the new tool microstructured reactor and the new type of processing, named chemical micro process engineering, is to cite real-life applications. In the following some new examples are given, either with IMM involved as research entity or with IMM tools being used. Also industrial process development studies are presented. In the next chapter, the first examples for industrial case studies, either with IMM or other suppliers' tools, which go for chemical production are given. There are certainly more such examples, some of which are known to the authors; however, these are kept confidential so far, and some may be made public in the next time. The following chapter presents IMM in-house process developments which were made to be launched to the clients.

2.1. Phenyl boronic acid synthesis (Clariant /Frankfurt) [10]

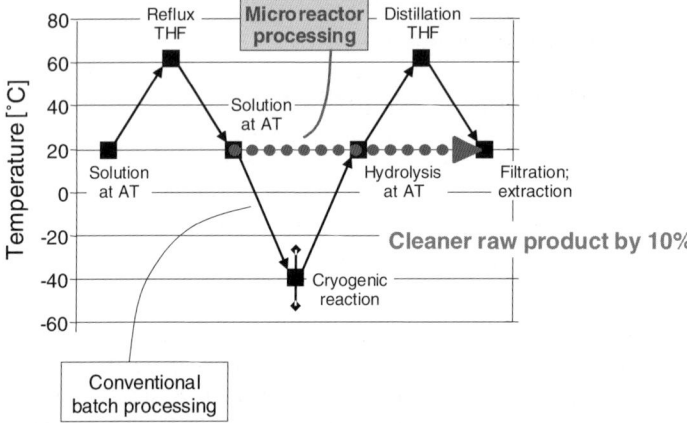

Mixing sensitivity is particularly pronounced for the class of organometallic reactions. Often these reactions are carried out under cryogenic conditions to get acceptable yields. This can be changed when using microstructured reactors. In this way, the phenyl boronic acid synthesis from phenyl magnesium bromide could be performed at high selectivity even at room temperature [10]. The yield was raised by about 25% as compared to the industrial batch production process. Energy savings are both given by shifting the former cryogenic process to room temperature and by achieving a highly pure crude product, thereby rendering the former energy-consumptive distillation step unnecessary (see Fig. 1). Thus, having higher selectivity did affect not only the reaction itself, but also downstream purification.

Fig. 1   Temperature profile of the phenyl boronic acid synthesis along the major steps of the process flow scheme. The difference in the temperatures of the conventional batch and the micro-reactor processes stand for the reduction in energy consumption and respective heat-transfer equipment when using the latter [10]

2.2. Azo pigment Yellow 12 manufacture (Trustchem/Hangzhou) [11]

By the use of microstructured mixers, pigment and other particulate syntheses can be improved. In this way, finer particles with more uniform size distribution were yielded for the commercial azo pigment Yellow 12 (see Fig. 2) [11]. The particles formed in the microstructured mixer have better optical properties such as the glossiness or transparency at similar tinctorial power. Since the micro-mixer made pigments have more intense colour, lower contents of the costly raw material in the commercial dye products can now be employed which increases the profitability of the pigment manufacture.

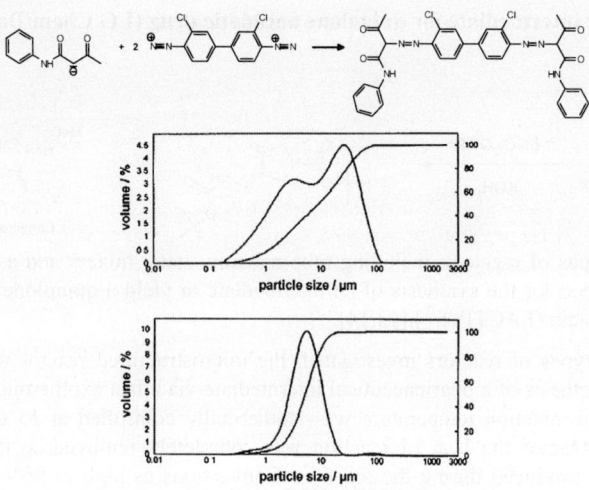

Fig. 2   Particle size distributions, (by volume) of the impeller-batch (top) and the micro-mixer-continuous-flow (bottom) processes when manufacturing the pigment Yellow 12. The cumulative distributions are given as well [11]

2.3. (S)-2-Acetyl tetrahydrofuran synthesis (SK Corporation/Daejeon) [12]

In the (S)-2-acetyl tetrahydrofuran (ATHF) synthesis, the Grignard reagent MeMgCl is very reactive and not easy to handle in large scale [12]. The Grignard reaction can not only cause safety and hazardous problems at industrial scale, but there are also issues of chirality conservation. The α-hydrogen of the starting material is unstable under basic conditions, and consequently, racemization may occur. The optical purity of the micro-reactor product was 98.4% as compared to 97.9% at batch level. Further, there are selectivity issues, i.e. an over-alkylation to tertiary alcohol must be avoided. Also, the individual impurity level must be less than 0.2%. The micro-reactor impurity was 0.18% by minimization of back mixing, while the batch impurity was 1.56% (see **Table 1**). Accordingly, with fine thermal and flow control, the productivity and economics of this process are increased.

Table 1
Selectivity (individual impurity) and optical purity of the batch and micro-reaction technology (MRT) processes for (S)-2-acetyl tetrahydrofuran synthesis

	Individual impurity	Optical purity
Batch	1.56%	97.7%
MRT	0.18%	98.4%

2.4. Synthesis of intermediate for quinolone antibiotic drug (LG Chem/Daejeon) [13,14]

Gemifloxacin (FACTIVE ™)

Five different types of reactors, including tube reactors, static mixers and a microstructured reactor, were tested for the synthesis of an intermediate to yield a quinolone antibiotic drug, named Gemifloxacin (FACTIVE®) [13,14].

Among several types of reactors investigated, the microstructured reactor was successfully applied to the synthesis of a pharmaceutical intermediate via a fast exothermic Boc protecting reaction step. The reaction temperature was isothermally controlled at 15°C. By using the microstructured reactor the heat of reaction was completely removed so that virtually no byproducts were produced during the reaction. Conversions as high as 96% were achieved. The micro-reactor operation can be compared with other reactors, however, which need to be operated at 0°C or –20°C to avoid side reactions.

3. INDUSTRIAL PRODUCTION

3.1. Nitro glycerine production plant for acute cardiac infarction (Xi'an Huian Industrial Group/Xi'an) [4]

A nitroglycerin pilot plant has been installed at Xi'an site in China and is currently started up [4]. The manufactured nitroglycerin will be used as medicine for acute cardiac infarction. Therefore, the product quality must be on highest grade, and initial tests already revealed higher selectivity and purity. The plant is foreseen to operate safely and fully automated. As a second step, a plant for downstream purification, of notably larger size and complexity as the reactor plant, is going to be developed. Environmental pollution should be excluded by advanced waste water treatment and a closed water cycle. Thus, the purification plant will have parts for washing and drying of the nitroglycerin and, in a final stage, also encompass formulation and packaging.

3.2. Production-oriented development of plants for polycondensation of LED-materials by Suzuki coupling (COVION works; IMM involved) [15]

In a government funded project (POKOMI, 16SV1981, funding agency: VDI/VDE/IT und BMBF), started in spring 2005, a micro-reactor engineered plant is developed by IMM and mikroglas chemtec Company, which is equipped with process control and on-line analysis for the synthesis of polymeric, light emitting semi-conductors for use in displays [15]. At the example of the Suzuki-coupling reaction as common synthesis route for polymeric semiconductors the economic and ecologic improvements for chemical industry are going to be demonstrated by the COVION company, Frankfurt/Hoechst site. Target is the transfer of micro-reactor engineering into chemical production. Generic trends to reliability, operation times, and controllability of micro-reactor plants are going to be deduced and, in particular, for the integration of microstructured reactors into existing plants ('plant upgrading'; 'multi-scale approach'). This involves also considerations on cost analysis with rentability and amortisation time and analysing corresponding process development time and process safety.

3.3. Pilotplant (10 t/a) with 3 micro-reactor blocks for MMA manufacture at Idemitsu site (MCPT works) [16]

Fig. 3 Industrial pilot plant (10 t a^{-1}) with three micro-reactor blocks for radical polymerisation at the Idemitsu site in Japan [16].

Iwasaki and Yoshida [16] designed and constructed an industrially-suited pilot plant for the free radical polymerisation of methylmethacrylate (MMA) with a capacity of 10 t a^{-1}, which was installed at the industrial site at the Idemitsu Company in Japan. Eight micro-reactor blocks are arranged in a parallel manner (see Fig. 3). Each of these blocks contains three micro-tube reactors (500 mm internal diameter, 2 m length) in series. The performance of this pilot plant was similar (polydispersity index, yield and average number-based molecular weight) to that of a single micro-reactor tube. This demonstrates that external numbering-up concepts can be successfully applied to industrial use.

3.4. Production operation (1700 kg h^{-1}) of a microstructured mixer-reactor for high-value polymer intermediate product (FZK works) [17]

At the Institut für Mikroverfahrenstechnik (IMVT) in the Forschungszentrum Karlsruhe, the group of Schubert et al. developed a customised microstructured reactor of about shoebox-size for use at the company DSM Fine Chemicals GmbH in Linz/Austria. During a ten-weeks lasting production campaign over 300 tons of a high-value product were manufactured for plastics industry [17]. The central element of this new production plant is a microstructured reactor, made from a special Nickel alloy, being 65 cm long and 290 kg heavy. This device is operated at a throughput of 1,700 kg liquid chemicals per hour (see Fig. 4). The pre-term 'micro' thus relates to the interior of the microstructured reactor; in the micro-mixer unit, the reactants are fed via several ten thousands of micro channels and thereafter react in such micro-flow patterned fluidic environment. In this way, a central reaction route at DSM is replaced, which was conducted before in a very large reactor tank encasing several thousands of explosive and corrosive chemicals. The micro-reactor yield could be increased thereby as compared to the former route; the process safety for handling the corrosive chemicals was still enhanced when using the micro-reactor process. The use of raw materials and the waste streams were reduced, improving the cost analysis and eco-efficiency of the process.

Fig. 4 Production-type microstructured reactor for throughput at 1,700 kg h^{-1} and transfer of a power of 100 kW. This apparatus was used for manufacture of a high-value product for plastics industry at DSM in Linz/Austria [17]

4. CURRENT MICRO-REACTOR PROCESS DEVELOPMENT AT REAL-LIFE EXAMPLES FOR FUTURE INDUSTRIAL USE

4.1. Kolbe-Schmitt Synthesis (own works) [18]

Liquid-phase reactions are typically limited from cryogenic temperatures (> -70°C) to about 150°C, in rare cases to 200°C or more. Even at the latter highest temperature level, many organic reactions remain much too slow for an efficient use in microstructured reactors, which are favourably operated from milliseconds to seconds. To overcome this, meso-scaled tubing reactors (1-2 mm ID), encompassed by micro mixers and heat exchangers on demand, with back-pressure regulation allow for a facile high p,T operation (*p* being pressure, *T* being temperature) beyond 200°C and 50 bar [18].

In this way, the operational range of the Kolbe-Schmitt synthesis using resorcinol with water as solvent to give 2,4-dihydroxy benzoic acid was extended by about 120°C to 220°C, as compared to a standard batch protocol under reflux conditions (100°C) [18]. The yields were at best close to 40% (160°C; 40 bar; 500 ml h^{-1}; 56 s) at full conversion, which approaches good practice in a laboratory-scale flask. Compared to the latter, the 120°C-higher micro-reactor operation results in a 130-fold decrease in reaction time and a 440-fold increase in space-time yield. The use of still higher temperatures, however, is limited by the increasing decarboxylation of the product, which was monitored at various residence times (τ).

| Aqueous Kolbe-Schmitt synthesis from resorcinol to 2,4 dihydroxy carboxylic acid | Further Kolbe-Schmitt products from: |

In less than one minute, half of the 2,4-dihydroxy benzoic acid is decomposed already at 160°C in the micro-reactor setup [18]. Thus, a study was conducted to find optimal process parameters for T and τ to achieve efficient high p,T operation via discrimination between the desired electrophilic substitution and undesired decarboxylation routes. The best operation point was at (200°C; 40 bar; 2000 ml h^{-1}; 16 s).

There is also substrate selectivity; e.g., the isomeric product 2,6-dihydroxy benzoic acid exhibits lower decarboxylation rates [19]. Thus, the use of different phenolic substrates provides different reaction scenarios with different rates of substitution relative to decarboxylation. Accordingly, substrates such as 1,4-dihydroxy benzene (hydroquinone) and 1,3,5-trihydroxy benzene (phloroglucinol) were converted to their respective carboxylic acids to reveal the impact of substitution isomerism and electron-richness of the core [19]. The synthesis of hydroquinone gives only a few per cent of the desired Kolbe-Schmitt product at high conversion, while the main product is benzoquinone, obviously via oxidation as predominant side reaction route. On the contrary, for the phloroglucinol a very efficient Kolbe-Schmitt synthesis route could be established using the micro reactor with yields up to 65%. Even more pronounced as for the resorcinol case, the experiment revealed decarboxylation as a side reaction route; here already being noticeable at temperatures of 140°C, while this was only the case above 200°C for resorcinol. Thus, a detailed process parameter variation was undergone to find optimum selectivity at reasonable conversion.

4.2. Hydrogenation of nitrobenzene (done with UCL, London) [20,21]

Nitro aromatics owe their great importance in organic synthesis for being intermediates for the generation of the respective anilines by hydrogenation [22]. For instance, pharmaceuticals are produced via that route [22]. The hydrogenations of nitro aromatics have high intrinsic reaction rates, which however cannot be exploited by conventional reactors as they are unable to cope with the large heat releases due to the large reaction enthalpies (500 – 550 kJ mol^{-1}) [20,23]. For this reason, the hydrogen supply is restricted, thereby controlling reaction rate. Otherwise, decomposition of the nitro aromatics or of partially hydrogenated intermediates can occur [20]. As model reactions, the hydrogenations of nitro benzene and p-nitro toluene over supported noble metal catalysts were investigated in a micrstructured falling film microreactor.

For nitrobenzene hydrogenation, the overall mass transfer coefficient k_La was conservatively estimated (based on the film thickness in the middle of the channels) to be in the range 3-8 s^{-1} [21]. By reducing the flow rate, and therefore the liquid film thickness, one would obviously increase the k_La values. However, the need to avoid liquid film dry-out imposes a minimum practical flow rate. It is worth noting that decreasing the flow rate, results also to an increase of residence time and specific interfacial area. Despite the large mass transfer coefficient, analysis of the results indicated that the system was operating in between mass transfer and kinetic control regimes. As a comparison, for intensified gas liquid contactors k_La can reach 3 s^{-1}, but for bubble columns and agitated tanks it does not exceed 0.2 s^{-1} [24].

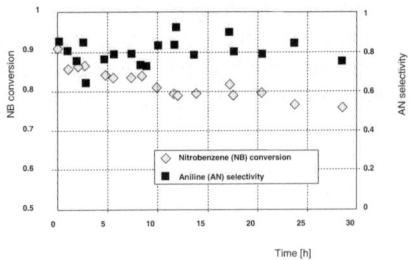

Fig. 5 Comparison of nitrobenzene conversion and aniline selectivity as a function of reaction time for the incipient wetness catalyst [20]

For the hydrogenation of nitrobenzene by Yeong et al. a wide variation of preparation procedures for the palladium catalyst was tested [20]. A sputtered palladium catalyst exhibited low conversion and large deactivation of the catalyst (60°C; 4 bar). The corresponding selectivity was also low. A slightly better performance was obtained after an oxidation / reduction cycle. Following a steep initial deactivation, the catalyst activity stabilised at 2 - 4% conversion and at about 60% selectivity. After reactivation, selectivity approached initially 100%. As side products, all intermediates except phenylhydroxylamine were identified. For a UV-decomposed palladium catalyst, a conversion was found slightly higher than for the sputtered one. A similar spectrum of side products as for the sputtered catalyst was given. For an impregnated palladium catalyst, complete conversion was achieved and maintained for six hours. Selectivity decreased with time, but remained still at a high level. The best performance of all catalysts investigated was found for an incipient-wetness palladium catalyst. Having initially more than 90% conversion, a 75%-conversion at selectivity of 80% was reached for long times on stream (see Fig. 5).

The catalyst life-time for the four types of catalysts, prepared by different preparation routes, depends on the catalyst loading which is related to the preparation route [20]. The larger the loading, the longer the catalysts could be used before reactivation. The four catalysts had the following sequence of life-time and activity:

wet-impregnation > incipient wetness > UV-decomposition of precursors > sputtering.

Several reactivation routes of the used catalyst were tested such as dissolution of organic residues by dichloromethane or burning of them by heating in air. In this way, initial activity was recovered, thus regaining complete conversion.

4.3. Brominations of aromatics and alkylaromatics (own works) [25,26]

The bromination of meta-nitrotoluene is an example for a high-temperature, high-pressure (high p,T) side-chain bromination of alkylaromatics [25].

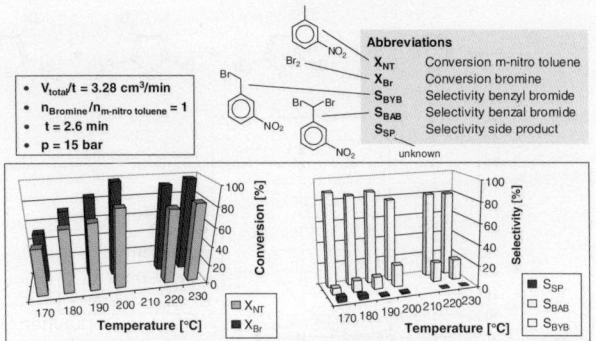

Fig. 6 High p,T operation for the radical side-chain bromination of m-nitro toluene in a micro-mixer-reactor setup. The large increase in operational temperature increases conversion at good selectivities, which tend to decline slightly with temperature. The two-fold substituted product, m-nitro toluene benzal bromide, is formed in larger amounts at temperatures above 200°C (IMM, unpublished results)

The transformation from batch to continuous processing, the safe operation with bromine at temperatures over 170°C and the decrease of reaction time, i.e. increase of space–time yields, were drivers for the development here.

Molar ratios of bromine to m-nitrotoluene ranging from 0.25 to 1.00 were applied. The reactants were contacted in an interdigital micro mixer followed by a capillary reactor. At temperatures of about 200°C nearly complete conversion is achieved (see Fig. 6). The selectivity to the target product benzyl bromide is reasonably high (at best being 85%; at 200°C and higher being 80%). The main sideproduct formed is the nitro-substituted benzal bromide, i.e. the two-fold brominated side-chain product.

A continuous bromination reaction of thiophene in a micro reactor rig was developed, using pure bromine and aromatic (solvent-free) without any catalyst [26]. Favourable room temperature processing could be established. This micro-reactor operation led to high selectivities (at about complete conversion) and yields up to 86%, which is better than for batch processing (77% yield) and literature data (50% yield) (see Fig. 7). In addition, the reaction time could be reduced from about 2 hours to less than 1 second. Correspondingly, the space-time yields were by order of magnitude higher for the continuous micro-reactor process as compared to batch operation. The micro reactor allowed a fast parametric study and in this way to reliably find optimal operating conditions. In this way, optimal process parameters with regard to temperature and the bromine-to-thiophene molar ratio could be determined.

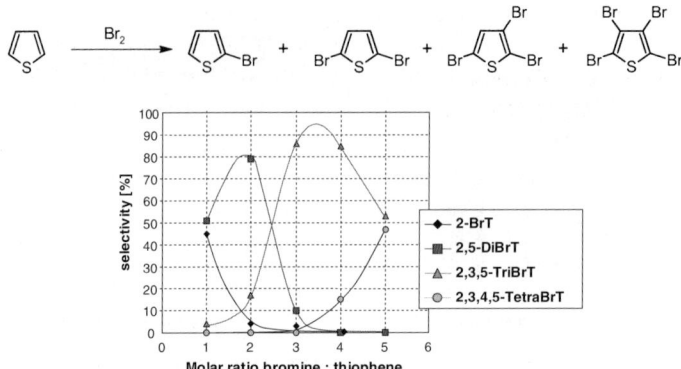

Fig. 7 Selectivity to mono-, di-, tri-, and tetra-substituted thiophenes as a function of the molar ratio of
bromine to thiophene [26]

5. FUTURE DIRECTIONS – ESTABLISHING A NOVEL CHEMISTRY BY ENABLING FUNCTION

5.1. Traditional chemistry – Processes follow limitations of reactors

We should tailor future chemical processes much better to the needs of micro reactors and, in this way, reveal the real potentials of these processes themselves. Using traditional ways of batch processing, the organic chemist has voluntarily chosen to restrict his syntheses to slow reactions or, if being fast, to artificially slow them down – which is some sort of *"subduing chemistry"* [3]. With micro reactors, such limitations are not given anymore and this paves the ground to a *novel chemistry*. to the notable start-up and shut-down times and the preference of sequential over combined processing, the organic chemist has voluntarily chosen to restrict his syntheses to slow reactions or, if being fast, to artificially slow them down – which is some sort of "subduing chemistry".

5.2. Novel chemistry – Novel, better protocols due to better performing reactors

High p,T processing

The temperature level of many organic routes is often simply defined by the boiling point of the solvent, since work-up is done under reflux conditions [18]. Many processes are quite slow at such temperatures. Basically, there is no kinetically or mechanistically derived reason why one should not work at temperatures up to 100°C higher than the solvents' boiling points. With microstructured reactors, such limitations are not given anymore. In order to maintain the mono-phase liquid operation, the routes have to be conducted at elevated pressures, typically in the range of 50 bar. This has been termed high-p,T processing.

Contacting 'all-at-once'

If aggressive reactants, formerly added drop-wise, are contacted 'all at once', explosions may happen in a typical batch; at least, the selectivity will be unacceptable [18]. Not so when using a microstructured reactor, as e.g. found for the hydrolysis of benzal chlorides. Here, a high selectivity and conversion was found for micro-reactor processing, although the same operation would result in unstable, explosive conditions with conventional processing.

Use of hazardous elements - direct routes

"Dream reactions" can be performed using chemical micro process engineering, e.g., via direct routes from hazardous elements [18]. The direct fluorination starting from elemental fluorine was performed both on aromatics and aliphatics, avoiding the circuitous Anthraquinone process. While the direct fluorination needs hours in a laboratory bubble column, it is completed within seconds or even milliseconds when using a miniature bubble column. Conversions with the volatile and explosive diazomethane, commonly used for methylation, have been conducted safely as well with micro-reactors in a continuous mode.

Routes in the explosive regime

Several examples were reported for conducting routes in the explosive regime [18]. Among them and most prominent was the detonating-gas reaction, using pure hydrogen and oxygen mixtures.

Simplified protocols

In a micro reactor, there is much more surface available than in standard reactors [18]. Thus, surface-chemistry routes may dominate bulk-chemistry routes. In this context, it was found sometimes micro-reactor routes can omit the addition of costly homogeneous catalysts, since the surface now undertakes the action of the catalyst. This was demonstrated both at the examples of the Suzuki coupling and the esterification of pyrenyl-alkyl acids.

6. SUMMARY

In summary, chemical micro process engineering, i.e. the use of microstructured reactors for chemistry and chemical engineering, offers promising perspectives for tomorrow's processing. This will not really change the large world-scale plants in a mid-term view, but certainly will affect industrial scale-up involving respective pilot and laboratory plants. In addition, laboratory plants for research synthesis and analysis will change to an extent. Actually, the two latter processes have started already [4].

REFERENCES

[1]   V. Hessel, S. Hardt and H. Löwe; Chemical Micro Process Engineering - Fundamentals, Modelling and Reactions, Wiley-VCH, Weinheim (2004).
[2]   V. Hessel, H. Löwe, A. Müller and G. Kolb; Chemical Micro Process Engineering - Processing and Plants, Wiley-VCH, Weinheim (2005).
[3]   V. Hessel, P. Löb and H. Löwe; Current Organic Chemistry 9, 8 (2005) 765.
[4]   A. M. Thayer; Chem. Eng. News 83, 22 (2005) 43.
[5]   H. Pennemann, P. Watts, S. Haswell, V. Hessel and H. Löwe; Org. Proc. Res. Dev. 8, 3 (2004) 422.
[6]   P. Pennemann, V. Hessel and H. Löwe; Chem. Eng. Sci. 59, (2005) 4789.
[7]   V. Hessel, H. Löwe and F. Schönfeld; Chem. Eng. Sci. 60, (2005) 2479.
[8]   G. Kolb and V. Hessel; Chem. Eng. J. 98, 1-2 (2004) 1.
[9]   V. Hessel and H. Löwe; Chem. Eng. Technol. 26, 1 (2003) 13.
[10]   V. Hessel, C. Hofmann, H. Löwe, A. Meudt, S. Scherer, F. Schönfeld and B. Werner; Org. Proc. Res. Dev. 8, 3 (2004) 511.

[11]   H. Pennemann, V. Hessel, H. Löwe, S. Forster and J. Kinkel; Org. Proc. Res. Dev. 9, 2 (2005) 188.

[12]   J. Kim, J.-K. Park and B.-S. Kwak; Proceedings of the "4th Asia-Pacific Chemical Reaction Engineering Symposium, APCRE05", pp. 441; (12 - 15 June 2005); Gyeongju, Korea.

[13]   J. Choe, K.-H. Song and Y. Kwon; Proceedings of the "4th Asia-Pacific Chemical Reaction Engineering Symposium, APCRE05", pp. 435; (12 - 15 June 2005); Gyeongju, Korea.

[14]   J. Choe, Y. Kwon, H.-S. Song and K.-H. Song; Korean J. Chem. Eng. 20, 2 (2003).

[15]   www.mstonline.de/foerderung/projektliste/detail_html?vb_nr=V3MVT016, "VDI/VDE/IT and BMBF web information, mst online", 2005.

[16]   T. Iwasaki and J.-I. Yoshida; Proceedings of the "8th International Conference on Microreaction Technology, IMRET 8", (10 - 14 April 2005); Atlanta, USA.

[17]   www.fzk.de/fzk/idcplg?IdcService=FZK&node=2374&document=ID_050927, "Forschungszentrum Karlsruhe", FZK Press release 13/2005, 06 July 2005.

[18]   V. Hessel, C. Hofmann, P. Löb, J. Löhndorf, H. Löwe and A. Ziogas; Org. Proc. Res. Dev. 9, 4 (2005) 479.

[19]   V. Hessel, C. Hofmann, P. Löb, H. Löwe, M. Parals and A. Ziogas; Org. Proc. Res. Dev. (2005) submitted for publication.

[20]   K. K. Yeong, A. Gavriilidis, R. Zapf and V. Hessel; Catal. Today 81, 4 (2003) 641.

[21]   K. K. Yeong, A. Gavriilidis, R. Zapf and V. Hessel; Chem. Eng. Sci. 59, (2004) 3491.

[22]   R. Födisch, D. Hönicke, Y. Xu and B. Platzer; in M. Matlosz, W. Ehrfeld and J. P. Baselt (Eds.) Microreaction Technology - IMRET 5: Proc. of the 5th International Conference on Microreaction Technology, pp. 470, Springer-Verlag, Berlin, (2001).

[23]   M. W. Losey, M. A. Schmidt and K. F. Jensen; in W. Ehrfeld (Ed.) Microreaction Technology: 3rd International Conference on Microreaction Technology, Proc. of IMRET 3, pp. 277, Springer-Verlag, Berlin, (2000).

[24]   S.-Y. Lee and Y. P. Tsui; Chem. Eng. Progress 7 (1999) 23.

[25]   P. Löb, V. Hessel, H. Klefenz, H. Löwe and K. Mazanek; Lett. Org. Chem. (2005) in print.

[26]   P. Löb, H. Löwe and V. Hessel; J. Fluorine Chem. 125, 11 (2004) 1677.

Studies in Surface Science and Catalysis, volume 159
Hyun-Ku Rhee, In-Sik Nam and Jong Moon Park (Editors)

Synthesis and catalytic applications of uniform-sized nanocrystals

Jongnam Park, Jin Joo, Youngjin Jang, and Taeghwan Hyeon*

National Creative Research Initiative Center for Oxide Nanocrystalline Materials and School of Chemical Engineering, Seoul National University, San 56-1, Shilim-dong, Kwanak-gu, Seoul 151-744, South Korea, Fax: 82-2-888-1604, thyeon@plaza.snu.ac.kr

We developed a new generalized synthetic procedure to produce monodisperse nanocrystals of many transition metals, metal oxides, and metal sulfides without a size selection process. Highly-crystalline and monodisperse nanocrystals were synthesized from the thermal decomposition of metal-surfactant complexes. We synthesized monodisperse spherical nanocrystals of metals (Fe, Cr, Cu, Ni, and Pd), metal oxides (γ-Fe_2O_3, Fe_3O_4, $CoFe_2O_4$, $MnFe_2O_4$, NiO, and MnO), and metal sulfides (CdS, ZnS, PbS, and MnS). We reported the ultra-large-scale synthesis of monodisperse nanocrystals by the thermolysis of metal-oleate complexes. We synthesized as much as 40 grams of monodisperse magnetite nanocrystals using 1 L reactor. By controlling the nucleation and growth processes, we were able to synthesize monodisperse magnetite nanoparticles with particle sizes of 4, 6, 7, 8, 9, 10, 11, 12, 13, 14, and 16 nm. Multi-gram scale synthesis of CdS, ZnS, PbS, and MnS were achieved from the thermolysis of metal-surfactant complexes in the presence of sulfur.

We synthesized uniform Cu_2O coated Cu nanoparticles from the thermal decomposition of copper acetylacetonate, followed by air oxidation. We successfully used these nanoparticles for the catalysts for Ullmann type amination coupling reactions of aryl chlorides. We synthesized core/shell-like Ni/Pd bimetallic nanoparticles from the consecutive thermal decomposition of metal-surfactant complexes. The nanoparticle catalyst was atom-economically applied for various Sonogashira coupling reactions.

1. INTRODUCTION

The synthesis of nanoparticles has been intensively pursued not only for their fundamental scientific interest, but also for many technological applications [1]. For many of these applications, the synthesis of monodisperse nanoparticles (standard deviations $\sigma \leq 5\%$) with controlled particle sizes is of key importance, because the electrical, optical, and

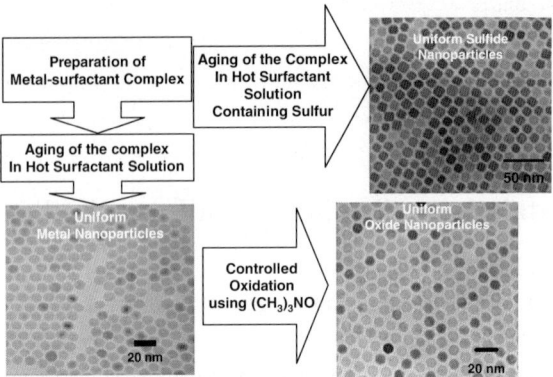

Scheme 1. Synthesis of uniform sized nanocrystals of metals, metal oxides, and metal sulfides from thermal decomposition of metal-surfactant complexes.

magnetic properties of these nanoparticles depend strongly on their dimension. For example, the color sharpness of semiconductor nanocrystal (quantum dot)-based optical devices is strongly dependent on the uniformity of the nanocrystals [2], and monodisperse magnetic nanocrystals are critical for the next-generation multi-terabit (Tbit/in^2) magnetic storage media [3]. Recently, several colloidal chemical synthetic procedures (or mechanisms) have been developed which produce monodisperse nanoparticles of various materials. Herein, we would like to summarize our recent progress on the synthesis of uniform-sized nanocrystals from the thermal decomposition of metal-surfactant complexes. We discuss on the synthesis of uniform-sized oxide nanocrystals via non-hydrolytic sol-gel reactions. The catalytic applications of core/shell nanoparticles of Ni/Pd and Cu/CuO will be also presented.

2. RESULTS AND DISCUSSION

2.1 Synthesis of uniform-sized nanocrystals from thermal decomposition of metal-surfactant complexes

2.1.1 Nanocrystals of ferrites

Our research group synthesized monodisperse iron nanoparticles from the high temperature (300 °C) aging of iron-oleic acid metal complex, which was prepared by the thermal decomposition of iron pentacarbonyl in the presence of oleic acid at 100 °C [4]. We were able to synthesize monodisperse nanoparticles with sizes ranging from 4 nm to 20 nm without using any size selection process. Initially, the iron oleate complex was prepared by reacting Fe(CO)$_5$ and oleic acid at 100 °C. Iron nanoparticles were then generated by aging the iron complex at 300 °C. The TEM image of the iron nanoparticles revealed that the nanoparticles were monodisperse and the electron diffraction pattern showed that the nanoparticles were nearly amorphous. The XRD pattern of the sample after being heat-treated in an argon atmosphere at 500 °C revealed a bcc α-iron structure. Particle size could be controlled from 4 nm to 11 nm by using different molar ratio of iron pentacarbonyl to oleic acid. Iron nanoparticles with particle sizes of 7 nm and 11 nm were prepared using 1:2 and 1:3 molar ratio of $Fe(CO)_5$:oleic acid. In order to produce nanoparticles larger than 11 nm, a reaction mixture with a 1:4 molar ratio of $Fe(CO)_5$ and oleic acid was used during the synthesis. However, the particle size of the resulting nanoparticles was still around 11 nm. We could produce iron nanoparticles with particle sizes bigger than 11 nm by adding more iron oleate complex to previously prepared 11 nm iron nanoparticles, and then aging at 300 °C. Using this synthetic procedure, we were able to tune the particle sizes of the nanoparticles from 11 nm to 20 nm.

The synthesized monodisperse iron nanoparticles were then transformed to monodisperse magnetic iron oxide (maghemite and magnetite) nanocrystals by controlled oxidation using trimethylamine oxide $((CH_3)_3NO)$, as a mild oxidant. In the current synthetic process, the size uniformity was determined during the synthesis of metallic nanoparticles. Transmission electron microscopic images of the particles showed 2-dimensional and 3-dimensional assembly of particles, demonstrating the uniformity of these nanoparticles. Electron diffraction, X-ray diffraction, and high-resolution transmission electron microscopy proved the highly crystalline nature of the nanoparticles. Particle size could be varied from 4 nm to 20 nm by altering the experimental parameters. The dominating size controlling factor was the molar ratio of iron pentacarbonyl to oleic acid. As described in the synthesis of iron nanoparticles, nanoparticles larger than 11 nm were synthesized by seeded growth process, employing addition of more iron-oleate complex into 11 nm sized particles followed by aging at 350 °C. The resulting larger iron nanoparticles were then transformed to iron oxide nanocrystals by oxidizing with trimethylamine oxide. X-ray absorption spectroscopy (XAS) and X-ray magnetic circular dichroism spectroscopy (XMCD) studies of the iron oxide nanocrystals revealed that γ-Fe_2O_3 (maghemite) phase was the dominant phase of the small 4 nm sized iron oxide nanocrystals, whereas the proportion of the Fe_3O_4 (magnetite) component gradually increases with increasing particle size.

Using a similar procedure, based on the thermal decomposition of a metal-surfactant complex followed by mild oxidation, we synthesized highly crystalline and monodisperse nanocrystals of cobalt ferrite ($CoFe_2O_4$), manganese ferrite ($MnFe_2O_4$) MnO, and Ni [5].

In December 2004, our group reported the ultra-large scale synthesis of monodisperse nanocrystals using inexpensive and non-toxic metal salts as reactants [6]. Using this synthetic procedure, we were able to synthesize as large as 40 grams of monodisperse nanocrystals in a single reaction, without going through a size sorting process. Moreover, the particle size can be controlled simply by varying the experimental conditions. The current synthetic procedure

is very general and nanocrystals of many transition metal oxides were successfully synthesized using a very similar procedure. Instead of using toxic and expensive organometallic compounds such as iron pentacarbonyl, we prepared the metal-oleate complex by reacting inexpensive and environmentally-friendly compounds, namely metal chlorides and sodium oleate. The metal-oleate complexes were prepared by reacting metal chlorides and sodium oleate

Fig. 1 TEM image of 12 nm sized monodisperse magnetite nanocrystals

in a mixed solvent composed of water, ethanol, and hexane with a volume ratio of 1:1:2 at 70 °C. The following is a typical synthetic procedure for monodisperse iron oxide

(magnetite) nanocrystals with a particle size of 12 nm. 36 g (40 mmol) of purified Fe(III)

oleate, which was synthesized from the reaction of iron chloride (FeCl$_3$) and sodium oleate, and 5.7 g (20 mmol, Aldrich 90%) of oleic acid were dissolved in 200 g of 1-octadecene at room temperature. The reaction mixture was heated to 320 °C with a heating rate of 3.3 °C/min, and then kept at that temperature for 30 min. When the reaction temperature reached 320 °C, a severe reaction occurred and the initial transparent solution became turbid and brownish black. The resulting solution containing the nanocrystals was then cooled to room temperature, and sufficient amount of ethanol was

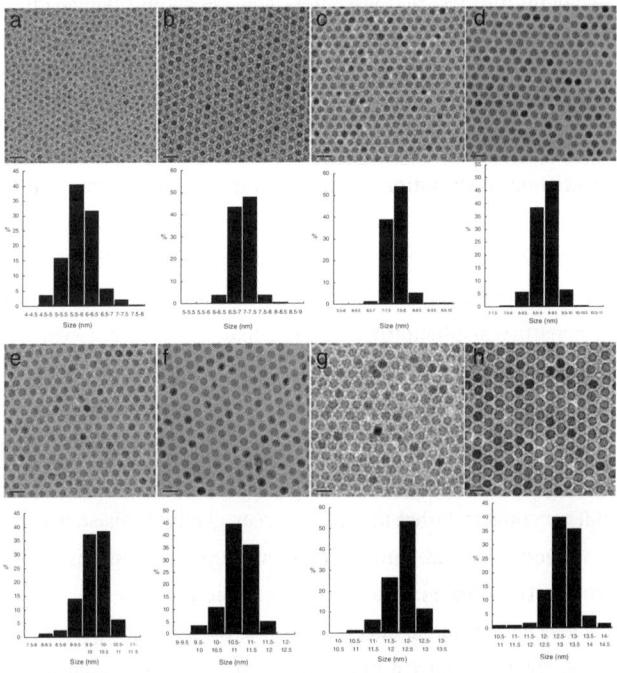

Fig. 2. TEM images and the corresponding particle size distribution histograms of (a) 6 nm, (b) 7 nm, (c) 8 nm, (d) 9 nm, (e) 10 nm, (f) 11 nm, (g) 12 nm, and (h) 13 nm sized iron nanoparticles showing the one nanometer level increments in diameter. The scale bars at the bottom of the TEM images indicate 20 nm

added to the solution to precipitate the nanocrystals. The nanocrystals were separated by centrifugation. The amount of the separated nanocrystals was as large as 40 grams with a yield of > 95%. The nanocrystals were easily able to be re-dispersed in various organic solvents including hexane and toluene. Figure 1 shows the TEM image of the 12 nm sized monodisperse magnetite nanocrystals produced as much as 40 grams using 500 mL of solvent in 1 L reaction vessel.

We synthesized monodisperse iron oxide nanoparticles with a continuous size spectrum of 6, 7, 8, 9, 10, 11, 12, and 13 nm [7]. The synthesis of the iron nanoparticles with particle sizes of 6, 7, 9, 10, 12, 13, and 15 nm was achieved by the controlled additional growth of the previously synthesized monodisperse iron nanoparticles, and the overall synthetic procedure is similar to seed-mediated growth. These monodisperse nanoparticles were obtained directly without a size-selection process, and the synthetic procedure is highly reproducible. The subsequent chemical oxidation of these nanoparticles produced monodisperse and highly-crystalline iron oxide nanocrystals. This concept of continuous growth without additional nucleation would be applicable to other materials for the 1-nm incremental size controlled synthesis of monodisperse nanoparticles. The transmission electron microscopic (TEM) images of these air-oxidized iron oxide nanoparticles of 6, 7, 8, 9, 10, 11, 12, and 13 nm are shown in Figure 2. All of the nanoparticles, except 6 nm sized particles ($\sigma = 8.5\%$), exhibited particle size distributions with standard deviations (σ) less than 5%.

2.1.2 Metal sulfides

We synthesized semiconductor nanocrystals of PbS, ZnS, CdS, and MnS through a facile and inexpensive synthetic process [8]. Metal-oleylamine complexes, which were obtained from the reaction of metal chloride and oleylamine, were mixed with sulfur. The reaction mixture was heated under appropriate experimental conditions to produce metal sulfide nanocrystals. Uniform cube-shaped PbS nanocrystals with particle sizes of 6, 8, 9, and 13 nm were synthesized. The particle size was controlled by changing the relative amount of $PbCl_2$ and sulfur. Uniform 11 nm sized spherical ZnS nanocrystals were synthesized from the reaction of zinc chloride and sulfur, followed by one-cycle of size selective precipitation. CdS nanocrystals that consist of rods, bipods and tripods were synthesized using a reaction mixture containing 1:6 molar ratio of cadmium to sulfur. 5.1 nm sized spherical CdS nanocrystals were obtained using a reaction mixture with a cadmium to sulfur molar ratio of 2:1. MnS nanocrystals with various sizes and shapes were synthesized from the reaction of $MnCl_2$ and sulfur in oleylamine. Rod-shaped MnS nanocrystals with an average size of 20 nm (thickness) × 37 nm (length) were synthesized using 1:1 molar ratio of $MnCl_2$ and sulfur at 240 °C. Novel bullet-shaped MnS nanocrystals with an average size of 17 nm (thickness) × 44 nm (length) that were synthesized from the reaction of 4 mmol of $MnCl_2$ and 2 mmol of sulfur at 280 °C for 2 hours. Shorter bullet-shaped MnS nanocrystals were synthesized using a 3:1 molar ratio of $MnCl_2$ and sulfur. Hexagon-shaped MnS nanocrystals were also obtained. All of the synthesized nanocrystals were highly crystalline. Figure 3 shows TEM images of uniform spherical shaped ZnS nanocrystals with a particle size of 8 nm (Figure 3a), CdS nanocrystals that consist of bipods and tripods with an average thickness of 5.4 nm (Figure 3b), uniform 13 nm sized cube-shaped PbS nanocrystals (Figure 3c), and hexagon-shaped MnS nanocrystals (Figure 3d).

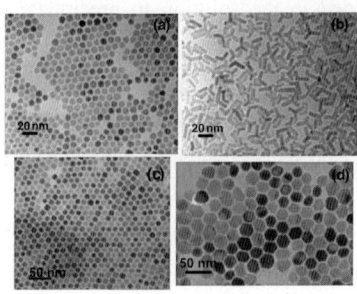

Fig. 3. TEM images of (a) ZnS nanocrystals, (b) CdS nanocrystals, (c) cube-shaped PbS nanocrystals, (d) hexagon-shaped MnS nanocrystals.

2.2. Catalytic applications of core/shell nanoparticles of Ni/Pd and Cu/Cuo

2.2.1 Synthesis of Ni/Pd bimetallic core/shell nanoparticles and their applications to Sonogashira coupling reactions

We synthesized core/shell-like Ni/Pd bimetallic nanoparticles from the consecutive thermal decomposition of metal-surfactant complexes [9]. Our strategy is as follows: Ni-TOP complex was decomposed at relatively low temperature of 205 °C, where Pd-TOP complex was rarely decomposed. After aging at 205 °C for 30 min to decompose Ni-TOP complex completely, the temperature was slowly increased to 235 °C to decompose Pd-TOP complex, generating Pd shell on the top of Ni core. The core/shell structure of the nanoparticles was characterized by collecting TEM images and EDX data of the aliquots of the reaction mixture and using a field-emission Auger electron spectroscopy (FE-AES) of the finally synthesized nanoparticles. These data clearly demonstrated that bimetallic Pd-Ni nanoparticles having Ni-rich core and Pd-rich shell were synthesized.

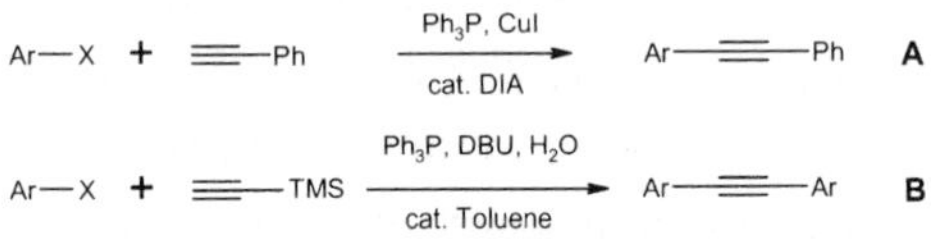

Ar—X + ≡—Ph —(Ph$_3$P, CuI / cat. DIA)→ Ar—≡—Ph   **A**

Ar—X + ≡—TMS —(Ph$_3$P, DBU, H$_2$O / cat. Toluene)→ Ar—≡—Ar   **B**

Scheme 2. (a) Sonogashira coupling reaction (b) sila-Sonogashira coupling reaction

We tested the catalytic activity of the Ni/Pd nanoparticles and similar sized Pd nanoparticles for the following Sonogashira coupling reactions using equal amount of palladium in the reaction mixtures. As expected, the Ni/Pd nanoparticles showed much better catalytic activity than Pd nanoparticles, which resulted from the larger number of nanoparticles derived from the core/shell structure. The nanoparticle catalyst can be recycled and reused at least 5 times without losing the catalytic activity, demonstrating the semi-heterogeneous characteristics.

2.2.2. Synthesis of Cu₂O coated Cu nanoparticles and their successful applications to Ullmann-type amination coupling reactions of aryl chlorides

We synthesized uniform Cu_2O coated Cu nanoparticles from the thermal decomposition of Cu-surfactant complexes in hot surfactant solution [10]. The detailed synthetic procedure is as follows. A solution containing 0.1 g of $Cu(acac)_2$ and 10 mL of oleyamine was slowly heated to 230 °C. And then, the solution was kept at this temperature for 6 hours, producing a red-colored colloidal solution. Transmission electron microscopic (TEM) image showed that uniform 15 nm sized copper nanoparticles were produced. When 0.5 g of precursor was used in the synthesis, smaller 12 nm sized copper nanoparticles were synthesized. When the synthesized Cu nanoparticles were exposed to air, the color of the nanoparticle solution was changed to blue, which shows that copper oxide, Cu_2O, was formed. TEM image revealed that the particle size and shape of the nanoparticles are kept nearly unchanged after the air oxidation and HRTEM image showed a polycrystalline nature of the nanoparticle. X-ray diffraction pattern revealed that Cu_2O shell was formed from sacrificing copper core.

We investigated the catalytic performance of the Cu_2O coated copper nanoparticles for Ullmann coupling reactions. When the coupling reactions using aryl bromides such as 2-bromopyridine and 4-bromodacetophenone were conducted in DMSO at 150 °C, the reactions proceeded completely. Even though the coupling reactions using copper-based catalysts have been intensively studied and

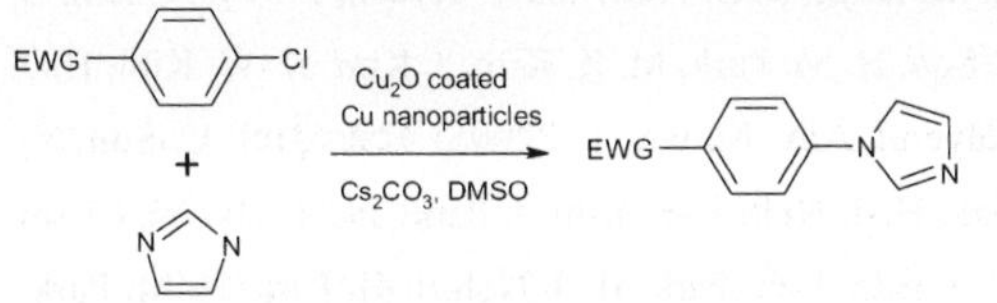

Scheme 3. Ullmann-type amination coupling reaction

applied in many industrial processes for more than 100 years, most of the reported coupling reactions using copper-based catalysts, however, have used expensive aryl iodides and aryl bromides as reactants. Recently, in the coupling reactions using palladium-based catalysts, employing cheap aryl chlorides instead of expensive aryl bromides or aryl iodides is a very important issue. Surprisingly enough, when we conducted Ullmann-type amination reactions of various aryl chlorides with electron withdrawing groups using the Cu_2O coated copper nanoparticles as catalysts, the reactions proceeded very well. This high catalytic activity of the Cu_2O coated Cu nanoparticles seems to result from high surface area derived from the nanoparticles. In addition, the coordination of oleylamine on the nanoparticles might affect the catalytic activity.

ACKNOWLEDGMENT

We would like to thank the Korean Ministry of Science and Technology for the financial support through the National Creative Research Initiative Program.

REFERENCES

[1] (a) G. Schmid, Nanoparticles: From Theory to Application, Wiley-VCH: Weinheim, 2004. (b) K. J. Klabunde, Nanoscale Materials in Chemistry, Wiley-Interscience: New York, 2001. (c) J. H. Fendler, Nanoparticles and Nanostructured Films, Wiley-VCH: Weinheim, 1998. (d) A. P. Alivisatos, Science, 271 (1996) 933. (e) C. Pacholski, A. Kornowsli, and H. Weller, Angew. Chem. Int. Ed., 41 (2002) 1188. (f) T. Hyeon, Chem. Commun., (2003) 927. (g) S. -W. Kim, S. U. Son, S. S. Lee, T. Hyeon, and Y. K. Chung, Chem. Commun., (2001) 2212.

[2] (a) M. Nirmal, and L. Brus, Acc. Chem. Res., 32 (1999) 407. (b) C. B. Murray, C. R. Kagan, and M. G. Bawendi, Annu. Rev. Mater. Sci., 30 (2000) 545. (c) A. L. Rogach, D. V. Talapin, E. V. Shevchenko, A. Kornowski, M. Haase, and H. Weller, Adv. Func. Mater., 12 (2002) 653.

[3] (a) S. Sun, C. B. Murray, D. Weller, L. Folks, and A. Moser, Science, 287 (2000) 1989. (b) D. E. Speliotis, J. Magn. Magn. Mater., 193 (1999) 29. (c) R. C. O'Handley, Modern Magnetic Materials, Wiley: New York, 1999.

[4] T. Hyeon, S. S. Lee, J. Park, Y. Chung, and H. B. Na, J. Am. Chem. Soc., 123 (2001) 12798.

[5] (a) T. Hyeon, Y. Chung, J. Park, S. S. Lee, Y.-W. Kim, and B. H. Park, J. Phys. Chem. B., 106 (2002) 6831. (b) E. Kang, J. Park, Y. Hwang, M. Kang, J.-G. Park, and T. Hyeon, J. Phys. Chem. B, 108 (2004) 13932. (c) J. Park, E. Kang, S. U. Son, H. M. Park, M. K. Kim, J. Kim, K. W. Kim, H.-J. Noh, J.-H. Park, C. Bae, J.-G. Park, and T. Hyeon, Adv. Mater., 17 (2005) 429. (d) S. U. Son, Y. Jang, K. Y. Yoon, C. An, Y. Hwang, J.-G. Park, H.-J. Noh, J.-Y. Kim, J. Park, and T. Hyeon, Chem. Commun., (2005) 86. (e) J. Park, E. Kang, C. J. Bae, J.-G. Park, H.-J. Noh, J.-H. Park, H. M. Park, and T. Hyeon, J. Phys. Chem. B, 108 (2004) 13598.

[6] J. Park, K. An, Y. Hwang, J.-G. Park, H.-J. Noh, J.-Y. Kim, J.-H. Park, N.-M. Hwang, and T. Hyeon, Nature Mater., 3 (2004) 891.

[7] J. Park, E. Lee, N.-M. Hwang, M. Kang, Y. Hwang, J.-G. Park, H.-J. Noh, J.-Y. Kim, J.-H. Park, and T. Hyeon, Angew. Chem. Int. Ed., 44 (2005) 2872.

[8] J. Joo, T. Yu, H. B. Na, and T. Hyeon, J. Am. Chem. Soc., 125 (2003) 11100.

[9] S. U. Son, Y. Jang, J. Park, H. B. Na, T. Hyeon, H. M. Park, H. J. Yun, and J. H. Lee, J. Am. Chem. Soc., 126 (2004) 5026.

[10] S. U. Son, I. K. Park, J. Park, and T. Hyeon, Chem. Commun., (2004) 778.

INVITED LECTURES

Studies in Surface Science and Catalysis, volume 159
Hyun-Ku Rhee, In-Sik Nam and Jong Moon Park (Editors)

Single Event MicroKinetics (SEMK) as a tool for catalyst and process design

J. W. Thybaut, C. S. Laxmi Narasimhan and G. B. Marin

Laboratorium voor Petrochemische Techniek, Ghent University,
Krijgslaan 281 – S5, B-9000 Gent, Belgium.

ABSTRACT

Increased computational resources allow the widespread application of fundamental kinetic models. Relumped single-event microkinetics constitute a subtle methodology matching present day's analytical techniques with the computational resources. The single-event kinetic parameters are feedstock invariant. Current efforts are aimed at mapping catalyst properties such as acidity and shape selectivity. The use of fundamental kinetic models increases the reliability of extrapolations from laboratory or pilot plant data to industrial reactor simulation.

1. FUNDAMENTAL KINETIC MODELLING

Simulation models are essential tools for reactor design and optimization. A general simulation model consists of a reactor and a reaction model [1]. The reactor model accounts for the reactor type and for the flow pattern in the reactor, while the reaction or kinetic model describes the kinetics of the chemical reactions occurring.

Describing the kinetics of complex feedstock processing requires a compromise between accuracy, analytical capabilities and computational resources [2,3]. Generally stated, the more explicitly a kinetic model accounts for the individual reactions which are occurring, the better it is in describing the simulated process but the higher the CPU time required. The better describing capabilities include, e.g., the feedstock independence of the kinetic parameters and a high accuracy over an extended range of operating conditions. Less explicit models consider various components as one reacting pool of components, a so-called lump, the properties of which are an average over the components included in that lump [4,5]. Today's increased computational resources allow the use of more detailed microkinetic models for industrial simulations and, hence, a lot of effort is put into the development of fundamental kinetic models [6,7].

2. SINGLE-EVENT MICROKINETICS FOR HYDROCRACKING

Innumerable reactions occur in acid catalyzed hydrocarbon conversion processes. These reactions can be classified into a limited number of reaction families such as (de)-protonation, alkyl shift, β-scission,... Within such a reaction family, the rate coefficient is assumed to depend on the type, n or m,, cfr. Eq. (1), of the carbenium ions involved as reactant and/or product, secondary or tertiary. The only other structural feature of the reactive moiety which needs to be accounted for is the symmetry number. The ratio of the symmetry number of the reactant, e.g., $\sigma_{O_{i,j}}$, to that of the transition state, e.g., $\sigma_{\neq_{ikqr}}$, equals the number of so-called single events [2,3,8].

Hydrocracking is a bifunctional process requiring metal as well as acid sites. Saturated reactants are dehydrogenated $\left(K_{deh,P_i;O_{i,j}} \right)$ on the metal sites yielding dehydrogenated species which are easily protonated $\left(\tilde{K}^0_{prot.O_{ref};m} \right)$ on the acid sites. The carbenium ions resulting from this protonation are susceptible to skeletal rearrangement reactions such as alkyl shifts $\left(\tilde{k}^{AS} \right)$ and PCP (protonated cyclopropane) branching reactions $\left(\tilde{k}^{PCP} \right)$ and to cracking via β-scission $\left(\tilde{k}^{\beta} \right)$. Prior to these chemical elementary steps a physisorption $\left(K_{L,P_r} \right)$ step occurs which leads to an enrichment of the reactant in the catalyst pores.

Reaction rates for the acid catalyzed elementary steps in hydrocracking can be expressed as follows when the metal catalyzed (de)-hydrogenation reactions are in quasi equilibrium:

$$r^{AS/PCP/\beta}_{m_{i,k};n_{q,r}} = \frac{C_t \dfrac{\sigma_{O_{i,j}}}{\sigma_{\neq_{ikqr}}} \tilde{k}^{AS/PCP/\beta}(m;n) \tilde{K}_{prot,O_{ref};m} \tilde{K}_{isom,O_{i,j};O_{ref}} K_{deh,P_i;O_{i,j}} C_{sat,P_i} K_{L,P_i} \dfrac{p_{P_i}}{p_{H_2}}}{1 + \sum\limits_{j=1}^{n} K_{L,P_j} p_{P_j}} \quad (1)$$

It was assumed in the derivation of this rate that the carbenium ion concentrations were negligible. At lower total pressures this assumption is not always holding. In such a situation it suffices to replace the denominator in this equation by an expression accounting for the carbenium ion concentrations [9].

3. ACCOUNTING FOR THE CATALYST PROPERTIES

The single-event microkinetic concept ensures the feedstock independence of the kinetic parameters [8]. Present challenges in microkinetic modelling are the identification of catalyst descriptors accounting for catalyst properties such as acidity [10,11] and shape selectivity [12,13].

3.1. Catalyst Acidity

The catalyst acidity is determined by the number of acid sites and their acid strength. The total concentration of acid sites, C_t, can be obtained from independent TPD measurements. The average acid strength of the sites is characterized by the alkene standard protonation enthalpy, ΔH^0_{prot}, and is typically determined by regression using reference component hydrocracking data (see Fig. 1). Variations in acid strength affect the carbenium ion stability and, hence, only the elementary steps in which a net production or consumption of carbenium ions occurs, i.e., protonation and deprotonation in hydrocracking, depend on the strength of the acid sites. The other elementary steps such as alkyl shifts, PCP-branching reactions and β-scissions, without net production or consumption of carbenium ions, are not affected by variations in acid strength [11].

Although more recent quantum chemical calculations involving long-range and environmental effects inside zeolite crystals are resulting in similar stabilities for carbenium ions and alkoxy species [14,15], it has previously been suggested from quantumchemical calculations [16,17] and spectroscopic studies [18] that the protonated intermediates on acid

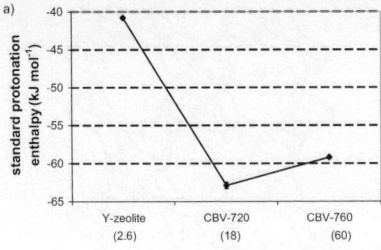 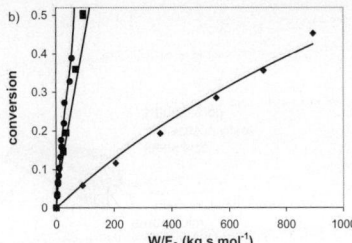

Fig. 1.   a) Standard protonation enthalpy in secondary carbenium ion formation on H-(US)Y-zeolites with a varying Si/Al ratio. b) Effect of the average acid strength for a series of H-(US)Y zeolites: experimental (symbols) versus calculated results based on the parameter values obtained in [11] (lines) for n-nonane conversion as a function of the space time at 506 K, 0.45 MPa, H_2/HC = 13.13 (Si/Al-ratios: ♦: 2.6, ●: 18, ■: 60)

catalysts are alkoxides and that the transition states exhibit a carbenium ion like character. As a result a charge separation between the hydrocarbon and the catalyst lattice occurs in the transition state. Variations in acid strength were found to only have minor effects on the alkoxide stability but do affect the transition state formation. The higher the charge separation in the transition state, the more pronounced this effect. Hence, the activation energy of the elementary steps belonging to reaction families requiring different degrees of charge separation in the transition state could be affected to a different extent by changes in acid strength [19]. The variation in charge separation within the reaction families that determine the selectivities in hydrocracking is relatively limited and, hence, no effect of the catalyst acidity is observed on the hydrocracking selectivities in the range of operating conditions typically investigated [11,20]. In catalytic cracking, however, differences in selectivity are observed when using catalysts with a different acidity [21]. This is most probably due to the higher number of reaction families involved in catalytic cracking and more pronounced differences in charge separation among these families.

3.2.   Shape selectivity

Because the pore dimensions in narrow pore zeolites such as ZSM-22 are of molecular order, hydrocarbon conversion on such zeolites is affected by the geometry of the pores and the hydrocarbons. Acid sites can be situated at different locations in the zeolite framework, each with their specific shape-selective effects. On ZSM-22 bridge, pore mouth and micropore acid sites occur (see Fig. 2). The shape-selective effects observed on ZSM-22 are mainly caused by conversion at the pore mouth sites. These effects are accounted for in the hydrocracking kinetics in the physisorption, protonation and transition state formation [12].

Alkane physisorption on ZSM-22 can be described using an additivity method accounting for the number of carbon atoms inside and outside the ZSM-22 micropores [22]. Linear alkanes can fully enter the micropores while branched alkanes can only enter the pore mouths. Multiple physisorption modes exist at the pore mouths where branched alkanes can enter the pore mouth with each of their 'straight ends'.

Alkene protonation at pore mouths can exclusively lead to secondary carbenium ions. In addition, the alkene standard protonation enthalpies increase with the number of carbon atoms inside the micropore because charge dispersive effects are supposed to be more effective on carbon atoms inside the micropores.

Transition state formation is sterically hindered at ZSM-22 pore mouths if the elementary reaction requires the ionic centre to move too far away from the deprotonated acid

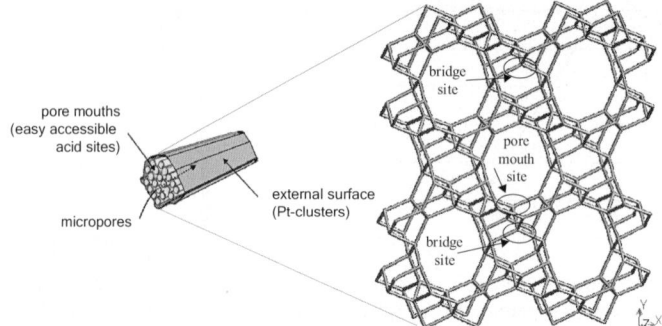

Fig. 2.   Isometric view of ZSM-22 crystallite : location of pore mouth, bridge acid and micropore acid sites.

site. This is, e.g., the case in alkyl shifts leading to a positional shift of the ionic centre and the branch. For the same reason geminally dibranched hydrocarbon formation is also forbidden at pore mouths [12].

4. SEMK BASED PROCESS DESIGN

4.1.   Relumped Single-Event MicroKinetics
Single-event microkinetics describe the hydrocarbon conversion at molecular level. Present day analytical techniques do not allow an identification of industrial feedstocks in such detail. In addition current computational resources are not sufficient to perform simulations at molecular level for industrial feedstock conversion. These issues are addressed using the relumping methodology.

Also in relumped form, single-event microkinetics account for all reactions at molecular level [2,3,13]. This requires a molecular composition of the lumps considered. The definition of the lumps in hydrocracking is such that thermodynamic equilibrium can be assumed within the lumps. Per carbon number 12 lumps are considered, i.e., normal, mono-, di- and tribranched alkanes, mono-, di-, tri- and tetracycloalkanes and mono-, di-, tri- and tetra-aromatic components.

4.2.   Two-stage hydrocracking
A typical two-stage hydrocracking process scheme is presented in Fig. 3. The first stage is typically focussed on hetero-atom removal and aromatic hydrogenation. A high-pressure separation is performed between the two stages to remove the light, hetero-atom containing components formed in the first-stage reactor. The second-stage reactor is working at low sulfur and nitrogen conditions and, hence, high conversions can be achieved in this reactor. Subsequently a product fractionation is performed and the heaviest fraction is recycled to the second stage reactor. The catalyst used in the simulations was NiMo on amorphous silica alumina. The lower acid strength of the silica alumina was accounted for using a standard protonation enthalpy 30 kJ mol^{-1} less negative than on USY (see Section 3.1).

Aromatic hydrogenation is accounted for using earlier developed rate equations [23]. The lower hydrogenation activity of NiMo compared to Pt is accounted for by increasing the activation energy by 23 kJ mol^{-1}. The inhibition by hetero atom containing components on the catalyst activity is twofold: the effect on the acid catalyzed reactions is accounted for via an extra adsorption term for nitrogen containing compounds, while the inhibition on the metal

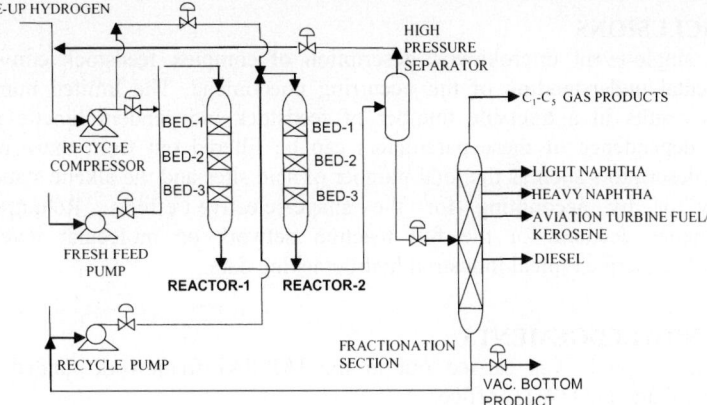

Fig. 3.   Schematic representation of a typical two-stage hydroconversion unit. The reactors contain multiple catalyst beds and quench zones. The second stage reactor is a recycle reactor.

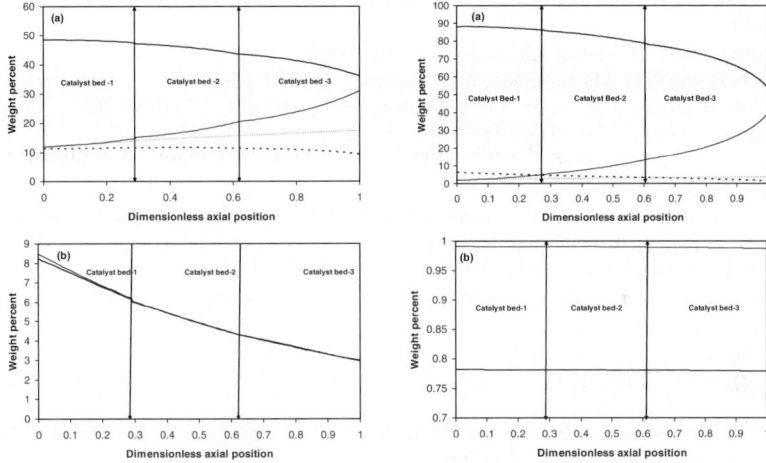

Fig. 4.   Weight percent profiles through first-stage (left) and second stage reactor of a) alkanes (full lines) and cycloalkanes (dashed lines) and b) aromatic components. Thick lines correspond to C_{23}^+ fractions, thin lines to C_{23}^- fractions. Operating conditions: p_t: 17.5 MPa; LHSV: 1.67 m_L^3 $(m_r^3$ h$)^{-1}$; molar H_2/HC: 18; T_{inlet}: 661 K (reactor 1) 622 K (reactor 2). Catalyst: NiMo on amorphous silica-alumina.

catalyzed reactions is expressed by including the hydrogen sulfide and ammonia partial pressure with a negative partial reaction order.

Weight percent profiles of the main fractions in the hydrocracking reactors illustrate the main differences in stage 1 and stage 2 hydrocracking. The aromatic content is significantly reduced in the first-stage reactor, while the thermodynamic equilibrium is practically achieved in the second. The conversion of the heavy fractions is much higher in the second stage than in the first stage reactor. The weight percent of C_{23}^+ alkanes decreases from 90 to 50 wt% in the second stage reactor, whereas in the first stage reactor the C_{23}^+ fraction is only reduced from 50 to 35 wt%. The highly paraffinic nature of the product stream of the second stage is caused by the aromatic hydrogenation and the relatively easy dealkylation of the corresponding cyclo-alkanes.

5. CONCLUSIONS

A single-event microkinetic description of complex feedstock conversion allows a fundamental understanding of the occurring phenomena. The limited number of reaction families results in a tractable number of feedstock independent kinetic parameters. The catalyst dependence of these parameters can be filtered out from these parameters using catalyst descriptors such as the total number of acid sites and the alkene standard protonation enthalpy or by accounting for the shape-selective effects. Relumped single-event microkinetics account for the full reaction network on molecular level and allow to adequately describe typical industrial hydrocracking data.

6. ACKNOWLEDGEMENTS

This research was carried out in the IAP-PAI framework funded by the Belgian government and the DWTC office.

REFERENCES

[1]   G.F. Froment and K.B. Bischoff, Chemical Reactor Analysis and Design 2nd ed., J. Wiley, New York, 1990.
[2]   G.G. Martens and G.B. Marin, AIChE J., 47 (2001) 1607.
[3]   J. W. Thybaut and G.B. Marin, Chem. Eng. Technol., 26 (2003) 509.
[4]   S. M. Jacob, B. Gross, S. R. Voltz, and V. W. Weekman, AIChE J., 22 (1976) 701.
[5]   S.A. Quader, S. Singh, W.H. Wiser and G.R. Hill, J Inst. Petrol., 56 (1970) 187.
[6]   K.C. Waugh, Chem. Eng. Sci, 51 (1996) 1533.
[7]   P. Stoltze, Progr. Surf. Sci., 56 (2000) 65.
[8]   Vynckier, E., and Froment, G. F., "Modeling of the kinetics of complex processes based upon elementary steps", in: Kinetic and Thermodynamic Lumping of Multicomponent Mixtures (G. Astarita and S. I. Sandler, Eds.) Elsevier, Amsterdam (1991) 131-161.
[9]   J. W. Thybaut, C. S. Laxminarasimhan, P. A. Jacobs, J. A. Martens, J. F. M. Denayer, G. V. Baron and G. B. Marin, Catal. Lett. 94 (2004) 81.
[10] B. A. Watson, M. T. Klein and R. H. Harding, Ind. Eng. Chem. Res., 35 (1996) 1506.
[11] J. W. Thybaut, G. B. Marin, G. V. Baron, P. A. Jacobs and J. A. Martens, J. Catal., 202 (2001) 324.
[12] C. S. Laxmi Narasimhan, J. W. Thybaut, G. B. Marin, P. A. Jacobs, J. A. Martens, J. F. Denayer and G. V. Baron, J. Catal., 220 (2003) 399.
[13] C. S. Laxmi Narasimhan, J. W. Thybaut, G. B. Marin, J. F. Denayer, G. V. Baron, J. A. Martens, P. A. Jacobs, Chem. Eng. Sci., 59 (2004) 4765.
[14] X. Rozanska, R. A. van Santen, T. Demuth, F. Hutschka and J. Hafner, J. Phys. Chem. B 107 (2003) 1309.
[15] L. A. Clark, M. Sierka and J. Sauer, J. Am. Chem. Soc., 125 (2003) 2136.
[16] V. B. Kazansky, M. V. Frash and R. A., van Santen, Appl. Catal. A Gen., 146 (1996) 225.
[17] A. M. Rigby, G. J. Kramer, and R. A. van Santen, J. Catal., 170 (1997) 1.
[18] N. C. Ramani, D. L. Sullivan, J. G. Ekerdt, J. Catal., 173 (1998) 105.
[19] M. A. Natal-Santiago, R. Alcala and J. A. Dumesic, J. Catal., 181 (1999) 124.
[20] J. F. Denayer, G. V. Baron, G. Vanbutsele, P. A. Jacobs and J. A. Martens, J. Catal., 190 (2000) 469.
[21] M.-F. Reyniers, Y. Tang and G. B. Marin, Appl. Catal. A Gen., 202 (2000) 65.
[22] C. S. Laxmi Narasimhan, J. W. Thybaut, G. B. Marin, J. A. Martens, J. F. Denayer and G. V. Baron, J. Catal., 218 (2003) 135.
[23] J. W. Thybaut, M. Saeys and G. B. Marin, Chem. Eng. J., 90 (2002) 117.

Studies in Surface Science and Catalysis, volume 159
Hyun-Ku Rhee, In-Sik Nam and Jong Moon Park (Editors)

61

Selective oxidation of p-xylene to terephthaldehyde (TPAL) on W-Sb oxides

W.-H. Lee*, S. W. Lee, K. H. Kim, Y. S. Lim, J. H. Chae, H. K. Yoon, D. I. Lee

LG Chem, Ltd./Research Park, 104-1 Moonji-dong, Yuseong-gu, Daejon, Korea

ABSTRACT

Selective oxidation of p-xylene to terephthaldehyde (TPAL) on W-Sb oxide catalysts was studied. While WO_3 was active in p-xylene conversion but non-selective for TPAL formation, addition of Sb decreased the activity in p-xylene conversion but increased TPAL selectivity significantly. Structure change was also induced by Sb addition. Evidences from various characterization techniques and theoretical calculation suggest that Sb may exist as various forms, which have different p-xylene adsorption property, reactivity toward p-xylene and TPAL selectivity. Relative population of each species depends on Sb content.

1. INTRODUCTION

The current commercial process for TPAL production involves handling of Cl_2 and HCl, which is not environmentally friendly. Much effort has been made to develop a heterogeneous catalytic process, that is, the selective oxidation of p-xylene to TPAL [1- 4]. Yoo et. al. [3] had discovered that Fe/Mo/DBH were selective in oxidizing para-xylene to TPAL and para-tolualdehyde in high yields. Yet, the yield and the catalyst selectivity were not high enough for commercialization. Recent patent [4], however, shows the promising results using W-Sb-Fe Oxides. In this paper, our findings on the effects of Sb in W-Sb oxide catalysts, which we have studied systematically, are discussed.

2. EXPERIMENTAL

W-Sb oxides were prepared by mixing aqueous solutions of ammonium tungstate and antimony tartarate, followed by drying in an air-circulated oven at 120°C and then calcination at 650°C.

Selective oxidation of p-xylene was carried out over the temperature range of 450-590°C at an atmospheric pressure using an 8-channel parallel tubular reactor system made in-house. Each reactor was charged with 0.25 g of catalyst. The feed composition was: p-xylene/O_2/N_2 = 0.25/1/98.75 (%). The total flow rate was 200ml/min. Reactants and products were analyzed with an on-line GC. The products identified with GC were terephthaldehyde (TPAL), p-tolualdehyde (PTAL), CO_2, CO, and benzotolualdehyde.

Catalysts were characterized using SEM (Hitachi S-4800, operated at 15 keV for secondary electron imaging and energy dispersive spectroscopy (EDS)), XRD (Bruker D4 Endeavor with Cu K radiation operated at 40 kV and 40 mA), TEM (Tecnai S-20, operated at 200 keV) and temperature-programmed reduction (TPR). Table 1 lists BET surface area for the selected catalysts.

Table 1
BET surface area of $W_{12}Sb_xO_y$ catalysts

Catalysts	W	$W_{12}Sb_{0.2}$	$W_{12}Sb_{0.6}$	$W_{12}Sb_2$
BET surface area (m^2/g)	2.1	2.4	4.4	1.6

3. RESULTS

Fig. 1 shows SEM micrographs of W-Sb oxides with various Sb contents. WO_3 consists of particles of irregular shape whose sizes are around 100 ~ 300 nm. As Sb content increases up to 0.6, the average size of particles having spherical shapes decreases to ~ 100 nm. The plate-like particles are also observed in W-Sb oxides and the size increases with Sb content. EDS analysis revealed that they are Sb-rich oxide particles. At higher Sb content ($x \geq 1$), the morphology of the catalyst particles changes significantly. At $x = 1$, the catalyst consists of polyhedral particles with various shapes. At $x = 3$, the catalyst consists of bar-shaped large particles.

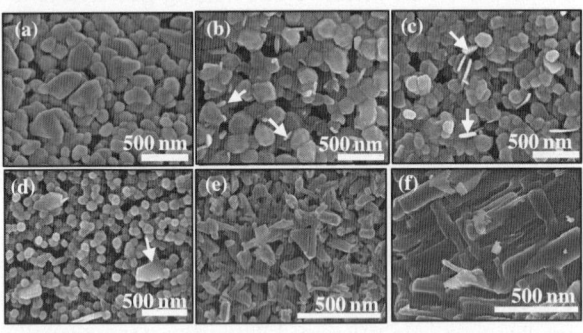

Fig. 1. SEM micrographs of $W_{12}Sb_xO_y$, where x = (a) 0 (b) 0.2 (c) 0.4 (d) 0.6 (e) 1.0 (f) 3.0

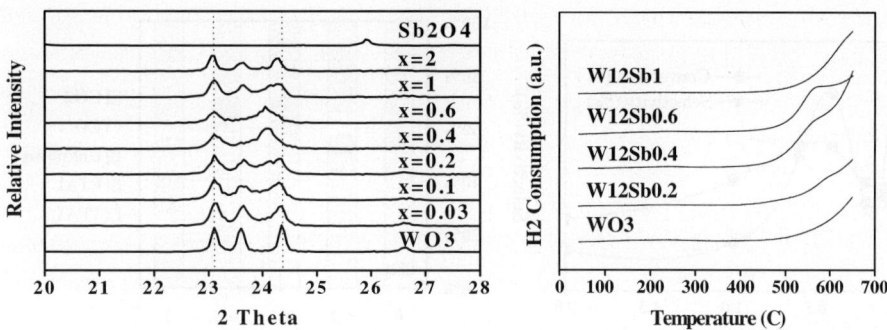

Fig. 2. (Left) XRD patterns for WO_3, $W_{12}Sb_xO_y$ and Sb_2O_4; (Right) H_2 TPR for WO_3 and $W_{12}Sb_xO_y$.

Fig. 2 shows XRD patterns and H_2 TPR results for W-Sb-O with various Sb contents. The triclinic structure of WO_3 shows three distinct peaks between $2\theta = 23°$ and $25°$. As Sb content increases up to 0.6, the last two peaks diminish gradually with a new peak appearing at about $2\theta = 24°$. As Sb content increases to 1 or higher, the XRD pattern abruptly changes again, which can be explained from SEM micrographs.

H_2 TPR results (Flow rate of 5% $H_2 = 20cc/min$, Ramp rate= $10°C/min$) show that Sb modifies the catalyst surface reducibility . The hydrogen consumption on WO_3 takes place above 500 °C. It is evident from TPR results that new surface species were formed by Sb addition. The population of such species, that is represented as the area under the new peak, increases with increasing Sb content with the maximum at Sb=0.4 and then decreases with further increasing Sb content.

Fig. 3 shows the effects of Sb in W-Sb oxides on the p-xylene conversion and TPAL selectivity as well as product distribution during the selective oxidation of p-xylene at 550 °C on the catalysts. Although not presented in this paper, the p-xylene conversion increased with increasing reaction temperature. The p-xylene conversion data in Fig. 3 show three distinct regions depending on Sb content. In the first region, the activity decreases drastically with increasing Sb content, reaching the minimum at Sb=0.2. Then, it increases with increasing Sb content until the maximum activity was achieved at Sb = 0.4. The p-xylene conversion decreases again significantly with further increase in Sb content. The TPAL yield was the highest at Sb = 0.4. The same trend is valid for all reaction temperatures employed in this study. Fig.3 shows that WO_3 is not selective for TPAL formation but TPAL selectivity at 550 °C increases significantly with Sb content up to 0.6 and then decreases with further increasing Sb content. Although data are not presented in this paper, TPAL selectivity increases with reaction temperature while CO_2 and CO selectivities decrease, indicating that the reaction routes to TPAL and carbon oxides may be different and the activation energy for the former is greater. It is evident from the product distribution in Fig. 3 that Sb addition enhances TPAL

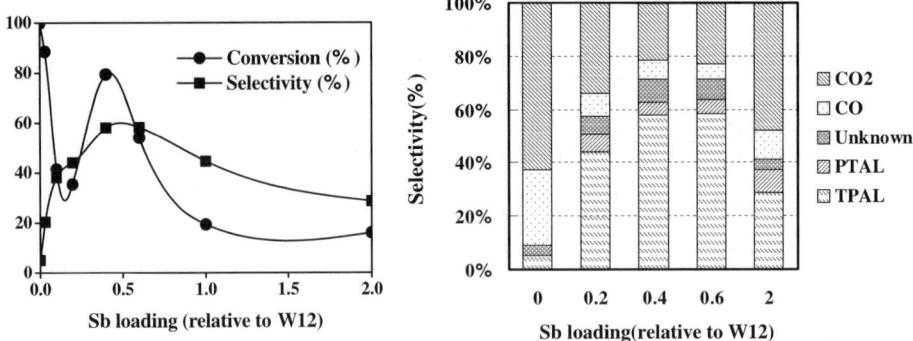

Fig. 3. (Left) p-Xylene conversion and TPAL selectivity at 550 °C as a function of Sb content; (Right) Product distribution at 550 °C as a function of Sb content.

formation and reduces CO_2 and CO formation. PTAL formation is less influenced by Sb content.

We have performed DFT calculations using the DMol3 program package [5] to gain insight into the doping effects of Sb on the WO_3 catalyst. CASTEP [6], a plane-wave pseudo-potential total energy package, is used to determine the relative stabilities of W-Sb oxide surfaces. Perdew-Wang [7] functional together with ultra-soft pseudo-potentials [8] and plane-wave basis set was used. Structure optimization was performed using the BFGS minimization technique with the third and fourth layers constrained. Calculation showed that W ions replaced with Sb ions in the first layer (see Fig. 4) were found to be more stable than those in the second layer by 0.65 eV, implying that the substituted Sb would be mainly located in the surface' rather than at the bulk.

Assuming that substituted Sb at the surface may work as catalytic active site as well as W, First-principles density functional theory (DFT) calculations were performed with Becke-Perdew [7, 9] functional to evaluate the binding energy between p-xylene and catalyst. Scalar relativistic effects were treated with the energy-consistent pseudo-potentials for W and Sb. However, the binding strength with p-xylene is much weaker for Sb (0.6 eV) than for W (2.4 eV), as shown in Fig. 4.

4. DISCUSSION

Molecular modeling showed that WO_3 surface has a structure where p-xylene may adsorb well, explaining the high activity of WO_3 in p-xylene conversion. Experimental data in this study show that Sb addition modifies significantly the activity and selectivity properties of WO_3 as well as the catalyst structure. Fig. 3 shows the minimum and maximum p-xylene

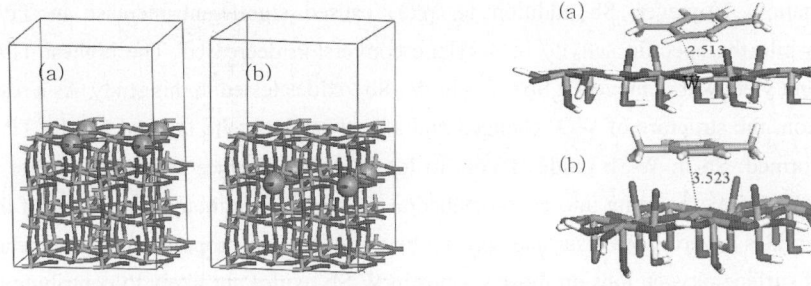

Fig. 4. (Left) Sb location in $W_{12}Sb_xO_y$ catalysts (a) in 1st layer (b) in 2nd layer; (Right) Effect of Sb on p-Xylene adsorption over (a) WO_3 catalyst surface and (b) W-Sb oxide catalyst surface.

conversion depending on Sb content. However, if BET surface area is taken into consideration, the specific activity in p-xylene conversion would decrease with increasing Sb content.

TPR results indicate that a new surface phase form on W-Sb oxides and the reactivity of its oxygen ions toward hydrogen is higher than that of WO_3 surface. Thin layer of Sb oxides, not easily detected in XRD measurements, may form as a result of Sb addition and cover WO_3 surface partially, which would reduce the number of p-xylene adsorption site and lower the activity. Molecular modeling also indicates that Sb favor to locate on the surface substituting some of W ions and that the binding strength with *p*-xylene is much weaker for Sb (0.6 eV) than for W (2.4 eV). In that case, the sticking probability of p-xylene to reaction sites will decrease significantly and as a result the activity also will decrease. However, once adsorbed, the weak adsorption may favor the generation of gaseous TPAL before further oxidation to CO or CO_2. Furthermore, the replacement of W (6-valence) with Sb (5-valence) causes the decrease of reactive surface oxygen O$^-$, which favors the selective oxidation of p-xylene to TPAL.

XRD pattern also shows that phase transformation occurs from triclinic WO_3 to tetragonal structure of W-Sb oxides as the Sb content increases. Electron diffraction analysis suggests that there occurs incorporation of Sb ions into some of cubo-octahedral sites in WO_3 leading to such structure change. It is not clear how these Sb ions may affect the properties of adsorption of p-xylene onto WO_3. They may weaken the binding energies of neighboring oxygen ions for TPAL formation as well as p-xylene onto W-Sb oxides resulting in the high TPAL selectivity. It is likely that different species described above may coexist and their relative population may depend on the Sb content.

5. CONCLUSION

WO_3 was active in selective oxidation of p-xylene conversion but was not selective for

TPAL formation. However, Sb addition to WO_3 caused significant increase in TPAL selectivity while the specific activity in p-xylene conversion decreased. The highest TPAL selectivity and yield was achieved at Sb = 0.4 in $W_{12}Sb_x$ oxides tested in this study. As a result of Sb addition, the structure of WO_3 changed and a new catalytic site favored for the TPAL formation formed. Sb in W-Sb oxides favors to locate on the surface as Sb ions either by substituting W or incorporating into cubo-octahedral sites in WO_3 although formation of thin layer of Sb oxides covering WO_3 surface may not be excluded. The properties of p-xylene and reactivity of surface oxygen ions on those species in W-Sb oxides are likely to contribute the enhanced TPAL selectivity

6. REFERENCES

[1] Simmons, Kenneth E., Williams, James E., US Patent No. 4017547 (1977)

[2] J.S. Yoo, J.A. Donohue, M.S. Kleefisch, P.S. Lin, S.D. Elfline, *Appl. Catal.A ,***105** (1993) 83.

[3] J.S. Yoo, M.S. Kleefisch, J.A. Donohue, US Patent No. 5324702 (1994)

[4] N. Kishimoto, I. Nakamura, Y. Nagamura, A. Nakajima, M. Hashimoto, K. Takahashi, US Patent No. 6458737 (2002).

[5] B. Delley, J. Chem. Phys. 113 (2000) 7756

[6] V. Milman, B. Winkler, J. A. White, C. J. Pickard, M. C. Payne, E. V. Akhmatskaya, and R. H. Nobes, Int. J. Quantum Chem. 77 (2000) 895

[7] J.P. Perdew and Y. Yang, Phys. Rev. B 45 (1992) 13244

[8] D. Vanderbilt, Phys. Rev. B 41 (1990) 7892

[9] A.D. Becke, Phys. Rev. A 88 (1988) 3098

Studies in Surface Science and Catalysis, volume 159
Hyun-Ku Rhee, In-Sik Nam and Jong Moon Park (Editors)

Catalytic Reaction Engineering for Environmentally Benign Chemical Processes

Akira Igarashi

Dept. of Environmental Chemical Engineering, Kogakuin University, 2665-1 Nakano-machi, Hachioji-shi, Tokyo 192-0015, Japan

1. DEVELOPMENT OF NEW CATALYTIC REACTION ENGINEERING

Green Chemistry (GC) is a globally utilized, environmentally friendly, chemical technology. GC emphasizes precaution (prevention) over diagnosis and treatment, and attempts to minimize the impacts of chemical processes and products on the environment and human health by considering their entire life cycle at the design stage, thus promoting R&D [1]. Integrating GC principals into existing businesses, such as the chemical process industry, is of primary importance.

Since the first synthesis of ammonia, catalyst development and chemical reaction engineering have been instrumental in the creation of the chemical process industry. As a result, catalytic processes have contributed much to the realization of prosperous civilization. In the future, catalytic processes are expected to fulfill important roles in petroleum refining, chemical processing, and environmental preservation. However, at present, many catalytic processes discharge large amounts of byproducts and consume large amounts of auxiliary raw materials.

Catalytic processing currently faces a range of challenges. In particular, environmental objectives call for processes with much higher conversion efficiency and better selectivity to minimize the release or removal of products. Such objectives would not be cost effective without economic incentives. Future sustainable growth relies on the development of environmentally benign catalytic processes, as well as on improved disposal and recycling schemes. There is a need for "green" catalytic processes with clean waste streams and the generation of raw materials from waste products.

The need for environmentally benign catalytic reaction engineering translates into four key areas of research [2]:

1. Treatment of waste streams: removal of contaminants such as NO, SO_2, H_2S, NO_3^{2-}, NH_3 from gas or water;
2. Clean and efficient processes: production of organic chemicals without salt, replacement of HF or H_2SO_4, catalytic combustion, high-temperature selective oxidation fuel cells;
3. New products: stereo-selective production of pharmaceuticals and agrochemicals, new oil products and energy carriers, replacement of CFCs;
4. Recycling: cracking of polymers, degradation of chlorine-containing organics.

Solutions to these problems require improved catalyst formulations and the development of alternative processes. However, most reactions satisfying these objectives are very difficult to achieve

68

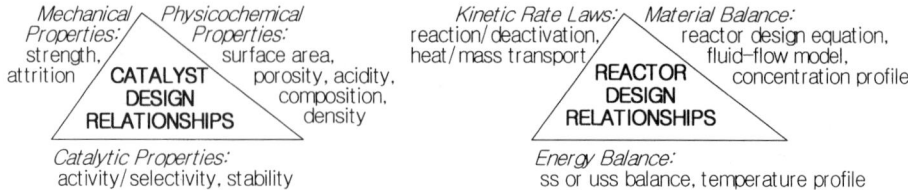

Fig. 1. Triangle concept for catalyst and reactor design

at present. Conventional development strategies, in which the catalyst is developed independently of reactor design, will be unable to provide a rapid solution, as the necessary type of reaction is typically very difficult to achieve. There is a need to carry out catalyst and reactor development simultaneously, and to improve the integration of these two aspects of catalytic reaction engineering. Thus, a new field of science combining catalytic chemistry and reaction engineering is necessary.

Currently, there is a large gap between nano-scale catalytic chemistry, which focuses on the analysis of chemical phenomena based on an understanding at the atomic or molecular level, and macroscopic reaction engineering, which involves computer-aided modeling of transport phenomena. The conventional concept of catalyst design is an optimized combination of interdependent mechanical, physicochemical and catalytic properties [3], while reactor design is conventionally based on simultaneous solution of mass balance, the rate law and energy balance, involving quantitative optimization of the mathematical relationships for reaction/transport kinetics, material balance and energy balance [4]. Fig. 1 shows the triangle concept for catalyst and reactor design [3, 4]. To fill this gap, the concept of optimization needs to be introduced into the catalytic chemistry design process, and an understanding of chemical phenomena needs to be incorporated into reaction engineering. This can be achieved by establishing a new hybrid discipline that integrates chemical phenomena and transport phenomena, namely catalytic reaction engineering, as shown in Fig. 2 [5]. This new discipline should be included as an integral part of any rational catalyst and process design methodology. It could be said that the keywords of catalytic reaction engineering are "nano" and "macro", and that control of the reaction field is achieved by combining "nano" and "macro" processes.

2. THE "CATALYTIC REACTION ENGINEERING TOWARD GREEN CHEMICAL PROCESSES" PROJECT IN JAPAN

Based on the above background, the "Catalytic Reaction Engineering toward Green Chemical

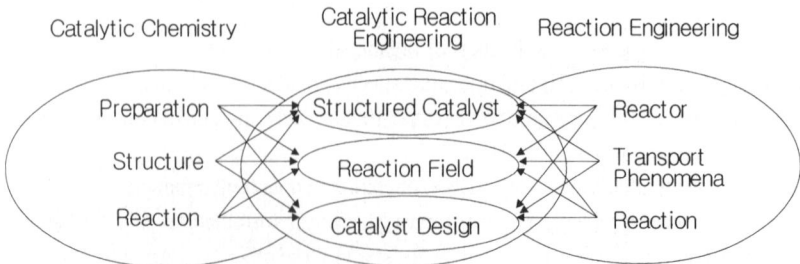

Fig. 2. Catalytic reaction engineering as a new hybrid discipline

Processes" project was undertaken between 2001 and 2003, funded under the Scientific Research in Priority Areas program of the Ministry of Education, Culture, Sports, Science and Technology of Japan [5]. The project representative is Prof. T. Hattori of Nagoya University. The project was supported by 21 research groups, with each division working together toward achieving the overall results of catalytic reaction engineering. Research was divided into three areas: creation of reaction fields using structured catalysts; control of catalytic functions using reaction fields; and the design of catalytic reaction fields toward green chemical processes, for which details are shown below.

2.1. Creation of reaction fields using structured catalysts

The research group led by the author, focused on the development of structured catalytic reactors for a catalyst/reactor system, which proposed new reaction fields to replace the conventional packed-bed catalytic reactor. This group aimed to improve heat transfer, reaction separation, and active species control in the reaction field by integrating reaction fields into a structured catalytic reactor. The group was divided into three subgroups, working towards the "creation of reaction fields considering heat transfer using a plate-type catalytic reactor"; "creation of reaction fields considering separation using selective permeable membranes"; and, "creation of a selective oxygen species formed on electrochemical membrane cells".

The authors developed a multi-layered microreactor system with a methanol reformer to supply hydrogen for a small proton exchange membrane fuel cell (PEMFC) to be used as a power source for portable electronic devices [6]. The microreactor consists of four units (a methanol reformer with catalytic combustor, a carbon monoxide remover, and two vaporizers), and was designed using thermal simulations to establish the appropriate temperature distribution for each reaction, as shown in Fig. 3.

The microreactor was constructed from thirteen microchanneled glass plates stacked with anodic bonding and placed in a vacuum package for thermal isolation. The appropriate catalyst for each reaction, namely the high-performance $Cu/ZnO/Al_2O_3$ catalyst developed in our previous study for methanol reforming [7], a 5wt.%-Pt/Al_2O_3 catalyst prepared by the impregnation method for catalytic combustion of methanol, and a commercial PROX catalyst for CO removal, was deposited on the microchannel of each reactor. When the microreactor was heated by applying voltage to a thin film heater attached to one side of the reformer, the temperature distribution observed for each unit approximated the simulated results. Finally, methanol reforming was achieved in the microreactor using heat supplied from the internal catalytic combustor. The reforming temperature of the

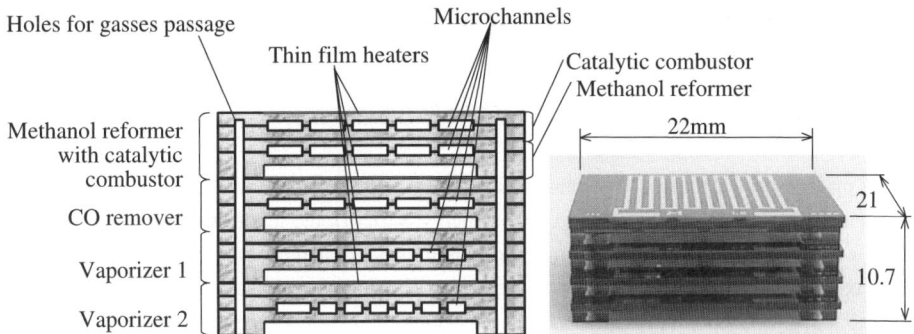

Fig. 3. Multi-layered microreactor system with methanol reformer

microreactor could be maintained at 280 °C without a supply of electrical power. A hydrogen production rate sufficient to generate 2.5 W of electrical power was obtained.

2.2. Control of catalytic functions using reaction fields

The selectivity of the main product, as well as catalytic activity and the rate of catalyst fouling, change with reaction conditions such as reaction temperature and reaction media. For example, the life of a solid acid catalyst in the isomerization of aliphatic hydrocarbons is strongly dependent on the reaction media, with very short lifetimes in inert gas, and long lifetimes in hydrogen. The purpose of the research group led by Prof. T. Takahashi was to clarify the effect of the reaction field with respect to optimization of catalytic functions. The following reaction fields were considered as those affecting catalytic functions: a reaction field using dilution gas or dilution solvent (a new type of reaction field), a reaction field for achieving the multiple functionality of the catalyst, and a reaction field using energy other than heat to drive catalytic reactions. This group was divided into three subgroups, namely "Effects of reaction fields on catalytic reactions", "Catalyst design with multiple reaction fields", and "Formation of reaction field by the addition of energy other than heat".

2.3. Design of catalytic reaction fields toward green chemical processes

Since the controlling factors of catalysis often depend on the reaction field, the elucidation of correlations between the reaction field and the catalysis is a major subject of the optimization of catalytic reaction systems. In this group, led by Prof. T. Hattori, catalysis was studied at the most microscopic level. The purpose of this group was to clarify the role of catalytic chemistry in catalytic reaction engineering in terms of the effect of the reaction field on catalytic reactions considering both reaction engineering and chemistry. To date, such research has only been carried out from a chemical perspective. This research group was divided into three subgroups, namely "Molecular reaction engineering for catalytic reaction fields", "Catalytic reaction field design for liquid-phase reactions", and "Chemical approaches to engineering problems".

2.4. Publication of results

The project made it possible to establish the systematization base of catalytic reaction engineering as a new hybrid discipline. This project produced many useful and progressive results, as presented at the 10th Asian Pacific Confederation of Chemical Engineering (APCChE 2004), which was held in Kitakyushu, Japan in October, 2004. Representative papers, except for previously published papers, are published in the "Journal of Chemical Engineering of Japan," issued by the "Japan Society of Chemical Engineering, Japan" in 2005.

3. HYDROGEN PRODUCTION BY STEAM REFORMING

A major aspect of the new environmental requirements is creating a method of energy conversion. For example, hydrogen is anticipated to be the energy carrier of the future, and hydrogen utilized for energy can be produced by steam reforming of natural gas or naphtha (e.g., in Japan and Korea), by partial oxidation of fuel oil, and by coal gasification. The natural gas steam reforming process is the most attractive production technique from the viewpoint of energy consumption and capital cost, and has thus become the dominant method employed through the world.

Steam reforming of hydrocarbons is a highly endothermic reaction, requiring temperatures of approximately 800 °C to proceed. These severe conditions require thermally stable catalysts and reaction tubes. Large amounts of hydrocarbons need to be burned to maintain the high reaction temperature, and the excess steam is used to prevent carbon deposition on the catalyst surface. Hydrogen production by steam reforming consumes large amounts of energy and produces vast amounts of carbon dioxide as a combustion product up to 0.9 kg per 1 m^3 hydrogen, in addition to carbon dioxide as a reaction product, in the case of natural gas. Thus, while hydrogen has been promoted as a form of clean energy, it is ironic that hydrogen is currently produced from fossil resources with profound environmental impact.

Advances in catalyst, desulphurization and reaction tube technology have lead to increases in the scale and reaction pressure of the steam reforming process. While further technological innovation is not expected, the recent emphasis on environmental accountability has lead to an explosion of research on inexpensive, efficient and low-waste hydrogen production processes, and the demand for hydrogen for technologies such as fuel cells has increased. Thus, there is a burgeoning need for improved mechanical strength and load-responsiveness of the catalyst with respect to daily start-up and shut down (DSS) operation.

Currently, coupling catalysis with separation through membranes, namely membrane reactors, is becoming feasible for practical use. The application of hydrogen-permeable membrane reactors to the steam reforming brings about the drastic lowering of the reaction temperature. Other beneficial effects include savings on energy consumption, and the prolonging of reactor materials and catalyst life. Kikuchi et al. applied the Pd/ceramic composite membrane, which consists of a thin palladium layer deposited by electron-less plating on the outer surface of ceramics, to the methane steam reforming in temperature ranging from 350 to 500 °C [8].

A commercially-supported nickel catalyst was also shown to work well for the production of hydrogen. As hydrogen is selectively removed from the reaction system, the thermodynamic positions of the reactions are shifted to the product side, and 100% conversion of methane to hydrogen and carbon dioxide can be attained even at temperatures as low as 500 °C. The product hydrogen is essentially free of carbon monoxide, and can be applied to a PEMFC. A similar steam reformer was tested by Tokyo Gas Co. and Mitsubishi Heavy Industries on a larger scale to develop a practical apparatus and to confirm the applicability to a fuel processor for a PEMFC [8]. The reformer was found to operate as a membrane reactor and achieved a high level of conversion of city gas to carbon dioxide and hydrogen, which could be supplied to a PEMFC, thereby generating electric power without any decline in performance over time. Fig. 4 shows the structure of the steam reformer equipped with the palladium membrane.

4. CONCLUSIONS

In the future, catalyst reaction engineering will play a more major role in environmentally benign chemical processes than at present. Catalytic reaction engineering should be an integral part of any rational catalyst and reactor design. Miniaturization of catalytic reactors using structured catalysts and micro-structured reactors is a new research direction. These technologies, which include membrane reactors to overcome equilibrium constraints, in conjunction with the development of hydrogen processors, will become very important in the environmentally friendly production of chemicals.

72

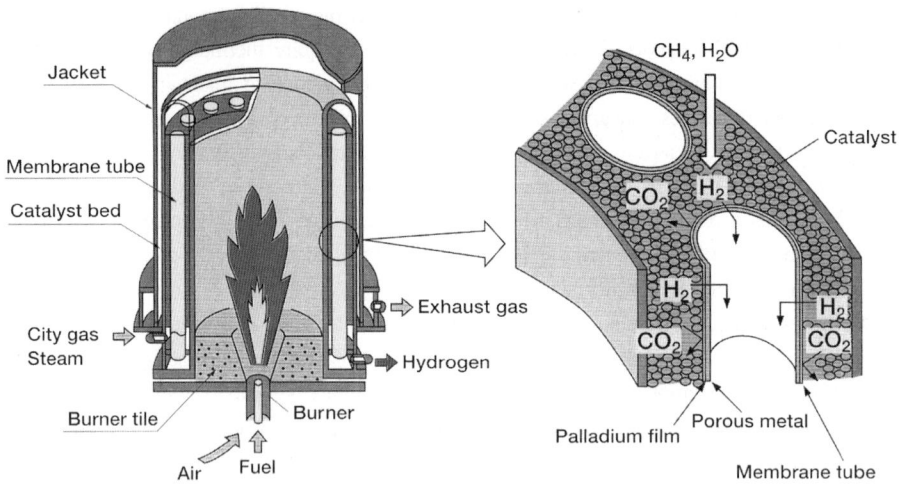

Fig. 4. Structure of stream reformer equipped with palladium membranes

REFERENCES

[1] M. Misono, J. Ind. Eng. Chem., (Korea), 7 (2004) 1126.

[2] R. A. van Santen, Chem. Eng. Sci., 50 (1995) 3027.

[3] J. T. Richardson, Principles of Catalyst Design, Plenum Press, New York, 1989.

[4] R. J. Farrauto and C. H. Bartholomew, Fundamentals of the Industrial Catalytic Process, Blackie Academic & Professional, London, 1997.

[5] A. Igarashi, APCChE 2004, 1D01-2, Kitakyushu, Japan (October, 2004); J. Chem. Eng. Japan, in press.

[6] Y. Kawamura, N. Ogura, T. Yahata, K. Yamamoto, T. Terazaki, T. Yamamoto, and A. Igarashi, APCChE 2004, 2D03, Kitakyushu, Japan (October, 2004); J. Chem. Eng. Japan, in press.

[7] Y. Kawamura, K. Yamamoto, N. Ogura, T. Katsumata, and A. Igarashi, J. Power sources, in press.

[8] E. Kikuchi, CATTEC, 1 (1997) 67.

Studies in Surface Science and Catalysis, volume 159
Hyun-Ku Rhee, In-Sik Nam and Jong Moon Park (Editors)
© 2006 Elsevier B.V. All rights reserved

Chemically Assisted Formation of Nanocrystals for Micro-electronics Application

Zerlinda Tan, Rohit Gupta, S. K. Samanta, Sungjoo Lee, and Won Jong Yoo

Silicon Nano Device Laboratory, Department of Electrical and Computer Engineering, National University. of Singapore, Singapore, email: eleyoowj@nus.edu.sg

1. INTRODUCTION

Nanocrystals are receiving significant attention for nano-electronics application for the development of future nonvolatile, high density and low power memory devices [1-3]. In nanocrystal complementary metal oxide semiconductor (CMOS) memories, an isolated semiconductor island of nanometer size is coupled to the channel of a MOS field effect transistor (MOSFET) so that the charge trapped in the island modulates the threshold voltage of the transistor (Fig. 1).

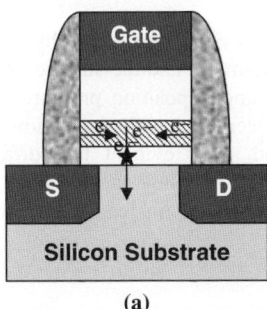
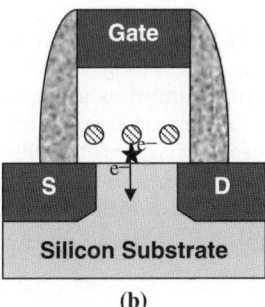

(a)   (b)

Fig. 1. Schematic of (a) conventional continuous floating gate CMOS memory and (b) nanocrystal CMOS memory. Discrete charge storage in nanocrystal memories reduces the possibility of charge loss through defects in the underlying tunnel dielectric. ▨▨ = floating gate ★ = defect in tunnel oxide

The use of floating gate composed of isolated nanocrystals overcomes the problems of charge loss encountered in conventional flash memories, allowing more aggressive scaling of the tunnel dielectric to achieve smaller operating voltages, better endurance, and faster program/erase speeds. The electrical performance of such devices depends on quality of the dielectrics and nanocrystals formed in the devices. Until now, semiconductor nanocrystal based nonvolatile memories have been widely reported. However, metal nanocrystals with high density of state and smaller energy perturbation due to carrier confinement can be strong contenders for nanocrystal based memories. The growth control of nanometer sized dots with good surface coverage is important for memory application. In this paper, we wish to demonstrate the chemically assisted formation of SiGe and Ni nanocrystals for non-volatile memory applications, and the electrical performance of devices fabricated using these nanocrystals.

74

2. FORMATION OF SILICON GERMANIUM (SiGe) NANOCRYSTALS

SiGe nanocrystals were formed *in situ* on HfO_2 and SiO_2 using a multi-module CVD system at the temperature of 500°C. SiGe nanocrystals were deposited using equal flow rates of SiH_4 and GeH_4, in the pressure range of 0.5 – 5Torr on 5nm thick thermally grown SiO_2 or metal organic CVD HfO_2. The effect of deposition time, pressure and substrate material on the size and density of SiGe nanocrystals was studied.

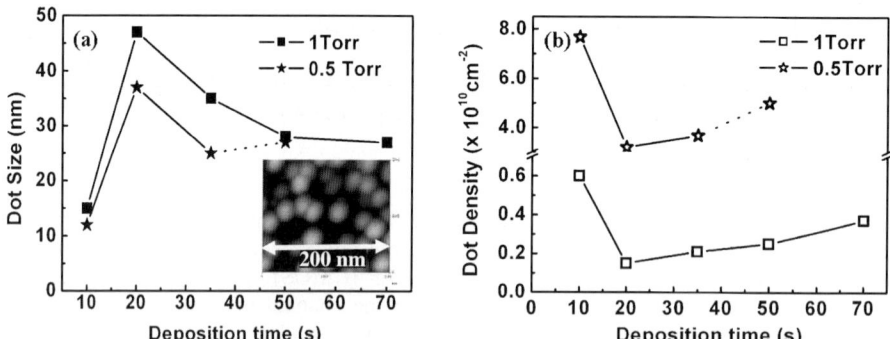

Fig. 2. Effect of deposition time and deposition pressure on (a) dot size (inset shows atomic force micrograph of dots formed at 0.5Torr) (b) dot density.

The formation of SiGe nanocrystals on SiO_2 at 1Torr, 10s was clearly observed by atomic force microscopy (inset of Fig. 2(a)). Fig. 2 shows the mean diameter and the surface density of the nanocrystals formed as a function of deposition time and deposition pressure. The mean diameter of the nanocrystals initially increases then decreases with deposition time whereas the nanocrystal density follows the opposite trend. It is evident that different mechanisms dominate in shorter and longer deposition times. According to Kim *et al*, the formation of SiGe on a dielectric surface preferentially occurs on nucleated Si through impingement [4].

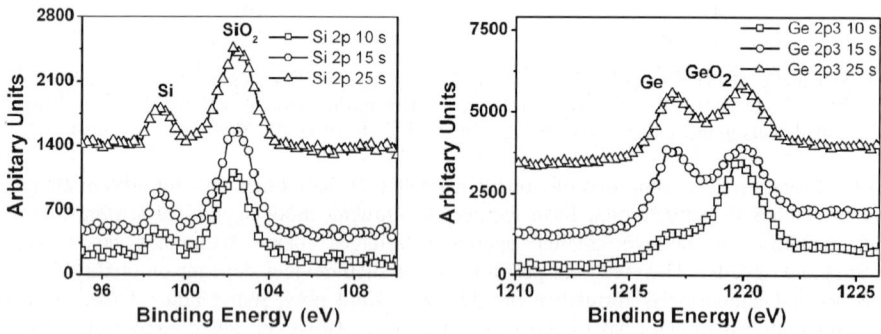

Fig. 3. XPS of SiGe nanocrystals on HfO_2 (*ex situ* measurements) [5].

Fig. 3. shows the X-ray photoelectron spectroscopy (XPS) results on the HfO_2. The HfO_2 surface with SiGe nanocrystals showed the presence of Ge (1217eV), GeO_2 (1220eV), Si (98.5eV) and SiO_2 (102.5eV). The component due to Ge^{4+} and Si^{4+} are present at all deposition times due to the oxidation of elemental Ge and Si upon exposure to air. It can be observed that the component due to elemental Ge increases steadily with deposition time. Auger electron spectroscopy detected 18.4 and 12.3 atomic % of Ge on SiO_2 and HfO_2, respectively, at the deposition time of 5s, deposition pressure of 5Torr and deposition temperature of 500°C. From these results, it appears that Ge nucleation takes place preferentially on pre-nucleated Si sites on the dielectric and the direct nucleation on vacant dielectric sites is less efficient. The existence of more pre-nucleated Si sites on SiO_2 compared to HfO_2 enhances the nucleation of Ge on SiO_2 compared to HfO_2. This clearly suggests different nucleation mechanism of Ge on insulators as compared to that of Si. In our previous work [5], we suggested a possible SiGe nanocrystal growth mechanism as shown below.

$SiH_{4(gas)} + vacant\ site \leftrightarrow SiH_{4(adsorbed)} \leftrightarrow SiH_{2(adsorbed)} + H_{2(gas)}$

$SiH_{2(adsorbed)} \leftrightarrow Si_{(solid)} + H_{2(gas)}$

$GeH_{4(gas)} + Si_{(solid)} \leftrightarrow SiGe_{(solid)} + 2H_{2(gas)}$

Nucleation of Ge is difficult because Ge tends to etch the oxygen species that is present in the oxide substrate. A significant amount of the Ge supplied to the SiO_2 surface above 500K is converted to volatile GeO upon arrival hindering the accumulation of Ge adatoms [6]. Although the nucleation of Ge is difficult, it is still possible when the deposition time is long enough for sufficient Ge adatoms to gather on the dielectric surface. This means that when the deposition time is long enough such that a critical amount of Ge is accumulated on the substrate, the formation of Ge nucleus seed for further growth of SiGe nanocrystals is still possible. Our results suggest that different growth mechanisms exist during shorter and longer time depositions: during shorter deposition times, the SiGe nanocrystal formation mechanism is dominated by the nucleation of Ge on pre-nucleated Si sites on the dielectric; at longer deposition times, many small SiGe nanocrystals can begin to grow, seeded by Ge nuclei which have a longer incubation time.

It was found that the nanocrystal density can be effectively controlled by the deposition pressure. For deposition times between 10 to 70 s at 1 Torr, the density of the nanocrystals formed was of the order of 10^9 cm^{-2}. By reducing the deposition pressure to 0.5 Torr, nanocrystal density near to 10^{11} cm^{-2} can be consistently achieved for deposition times between 10 to 50s (Fig. 2b). This is because the nucleation density of Si is strongly dependant on the presence of Si-OH bonding [7]. At pressure as high as 1 Torr, the diffusion loss of atomic hydrogen is responsible for the suppression in the formation of Si nuclei which are needed as a seed layer for the formation of SiGe nanocrystals. Due to the high density of nanocrystals at 0.5Torr, visible agglomeration was observed at 50s (Fig. 2b). To obtain small and high density nanocrystals, it is important to quench the deposition just after nucleation. By careful process optimization a mean nanocrystal size of about 10nm and density of about $10^{11}cm^{-2}$ can be achieved. In this study, the smallest nanocrystal was obtained at 500°C and ~14sec, after the onset of nucleation at 5 – 10s.

76

3. FORMATION OF NICKEL (Ni) NANOCRYSTALS

Further enhancement in programming efficiency and retention may be achieved by integrating metal nanocrystals with physically thick high-k materials with lower tunneling barriers such as HfO_2 (conduction band offset = 1.5eV, k-value ~ 20) compared to conventional SiO_2 (conduction band offset = 3.2eV, k-value ~ 3.9). The high work function of Ni (4.9eV) enables the creation of a deep potential well for the trapping of charge carriers. In addition, a high density of states is available for the storage of many electrons in metal nanocrystals allowing more effective coupling of the nanocrystal floating gate to the conduction channel. Furthermore, the nanocrystal formation temperature of Ni (~600°C) is lower compared with other high work function metals such as Pt (~900°C). The low temperature of formation is important to maintain the quality of the underlying HfO_2 dielectric [8]. After pre-gate cleaning and surface nitridation, 5nm thick HfO_2 was prepared by metal organic CVD at 400°C, followed by post deposition annealing for densification. A thin Ni layer was sputtered on HfO_2 and annealed to form Ni nanocrystals in N_2 ambient.

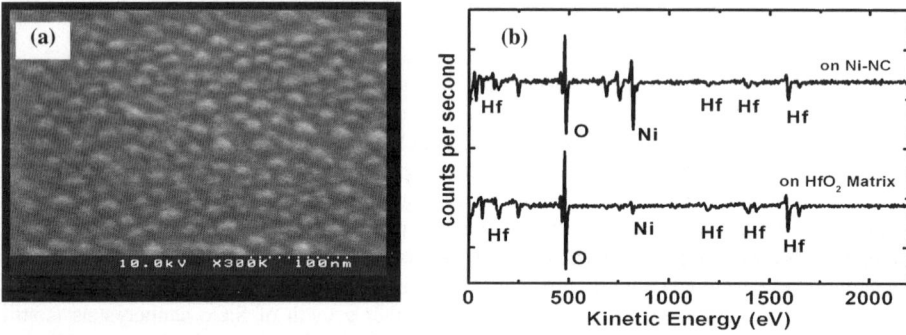

Fig. 4. (a) Ni nanocrystals formed on HfO_2 after sputtering followed by annealing and (b) AES analysis of the elemental composition of the islands and matrix suggested well separated metal nanocrystals.

Ni nanocrystal formation was observed on the uncapped samples by scanning electron microscopy (SEM). Fig. 4(a), shows the Ni nanocrystals formed on HfO_2 after sputtering a thin Ni film (1-5nm) followed by rapid thermal annealing. Atomic force microscopy (AFM) reveals that the initial HfO_2 surface is relatively smooth (rms = 0.168nm). After depositing a thin layer of Ni, some thickness perturbation was observed (rms = 0.739nm), however no island formation was observed yet. Upon annealing, well separated nanocrystals were observed on the substrate. Auger electron spectroscopy (AES) distinguishes the Ni nanocrystals clearly from the surrounding matrix with strong Hf signals and suppressed Ni signals in the HfO_2 matrix compared to the Ni islands (Fig. 4(b)). This confirms the separation between the Ni islands. The undulations observed in the as deposited Ni thin film may be attributed to hillock formation caused by the high stress built up in the thin metal film due to its high surface to volume ratio. It is believed that annealing helps to relieve the stress in the thin Ni film by giving the Ni atoms enough surface mobility to rupture forming islands on the dielectric surface. Island formation allows the highly stressed Ni film to arrive at a more stable state by minimizing surface energy [3]. The dewetting of thin Ni films can also be attributed to the reactivity of the metal absorbates to the oxygen in the substrate. The smaller bond enthalpy of Ni-O (382 kJmol^{-1}) compared to Hf-O (802 kJmol^{-1}), not only drives the three dimensional agglomeration of the Ni nanocrystals but also helps to ensure the chemical stability of Ni on HfO$_2$.

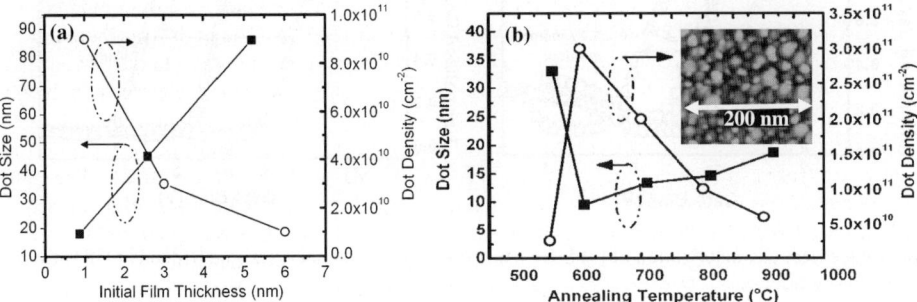

Fig. 5. Size and density of Ni nanocrystals as a function of (a) initial Ni film thickness and (b) annealing temperature (inset shows the AFM micrograph of Oswalt ripening at high temperatures).

The formation of Ni nanocrystals is strongly dependant on the initial Ni film thickness and the annealing temperature (Fig. 5.). Although the deformation process in nanometer thick metal films is not fully understood, it is believed that deformation processes governed by abundant grain boundaries may become important in ultra thin films [9]. Since the grain size and grain density of Ni thin films is limited by the film thickness (thinner Ni films tend to have smaller grain size and higher grain density) [10], faults which nucleate from the grain boundaries leading to the formation of islands may cause the observed smaller dot size in thinner films. According to the AFM results in Fig. 5(b), the optimum temperature to obtain minimum nanocrystal size and highest nanocrystal density is about 600°C. At higher temperatures, nanocrystal size is increased whereas density is reduced. It is understood that the critical radius of Ni nanocrystals is attained at 600°C where the Ni adatom chemical potential is balanced by the chemical potential of the nucleus. At temperatures greater than 600°C, the chemical potential of the nucleus is lower than that of the adatom, as a result, the nucleus grows [11]. After annealing above 800°C, Oswalt ripening takes place causing a wide distribution of nanocrystal size (inset of Fig. 5(b)). By optimizing the process parameters, small nanocrystals with mean size of 9nm and estimated density of about 3 x 10^{11} cm^{-2} were obtained (inset of Fig. 5(b)). Even smaller nanocrystals may be obtained by further reducing the initial film thickness and increasing the ramping rate of rapid thermal annealing.

4. ELECTRICAL CHARACTERIZATION

MOS capacitors with SiGe and Ni nanocrystals were fabricated by capping the samples with HfO$_2$ dielectric (12nm) and TaN metal gate (150nm) by reactive sputtering. The fabrication was completed with gate patterning, etching and forming gas anneal at 420°C for 30 min. Memory effect of the SiGe and Ni nanocrystals was evaluated by capacitance-voltage (C-V) measurements. From C-V characteristics (Fig. 6.), the memory window of the Ni nanocrystal memory device (2.0V) is larger than that of the SiGe nanocrystal memory device (1.7V) even though the nanocrystal density is both ~10^{11}cm^{-2}. This may be attributed to the large charge storage capacity of metal nanocrystals due to its high density of states.

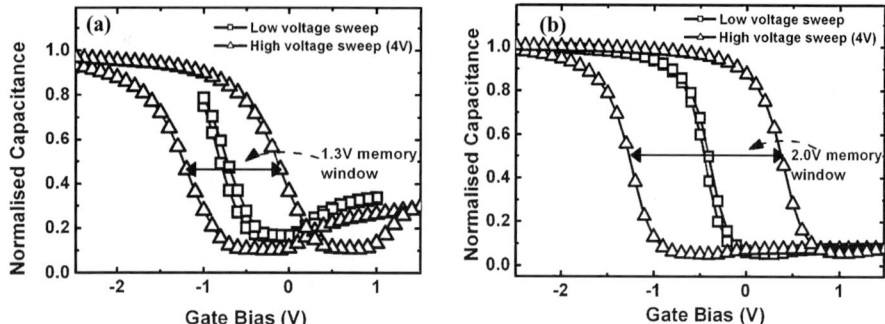

Fig. 6. Electrical characterization of (a) SiGe nanocrystals (b) Ni nanocrystals embedded in HfO_2 from MOS capacitors.

4. CONCLUSION

In this work, SiGe nanocrystals with density up to $10^{11}cm^{-2}$ were formed by the *in situ* CVD process using SiH_4 and GeH_4, with the help of pre-nucleated Si seeds on HfO_2 and SiO_2. Also, we demonstrated the self-assembly of Ni nanocrystals on HfO_2 by sputtering and rapid thermal annealing. Promising memory effects were observed from MOS structures utilizing SiGe and Ni nanocrystals embedded in HfO_2.

ACKNOWLEDGEMENT

This work was supported by the grant R-263-000-266-305 from the Agency for Science, Technology, and Research in Singapore.

REFERENCES

[1] R. Muralidhar, R. F. Steimle, M. Sadd, R. Rao, C. T. Swift, E. J. Prinz, J. Yater, L. Grieve, K. Harber, B. Hradsky, S. Straub, B. Acred, W. Paulson, W. Chen, L. Parker, S. G. H. Anderson, M. Rossow, T. Merchant, M. Paransky, T. Huynh, D. Hadad, K-M. Chang and B. E. White Jr., IEDM Tech. Dig., (2003) 601.
[2] D.-W. Kim, T. Kim and S. K. Banerjee, IEEE Trans. on Electron Devices, 50 (2003) 1823.
[3] Z. Liu, C. Lee, V. Narayanan, G. Pei and E. C. Kan, IEEE Trans. on Electron Devices, 49(2002) 1614.
[4] D. W. Kim, S. Hwang, T. F. Edgar and S. Banerjee, J. Electrochem. Soc., 150 (2003) G240.
[5] R. Gupta, W. J. Yoo, Y. Wang, Z. Tan, G. Samudra, S. Lee, D. S. H. Chan and K. P. Loh, Appl. Phys. Lett., 84 (2004) 4331.
[6] S. K. Stanley, S. S. Coffee and J. G. Ekerdt, Appl. Surf. Sci., in press (2005).
[7] K. Makihara, H. Deki, H. Murakami, S. Higashi and S. Miyazaki, Appl. Surf. Sci., 244 (2005) 75.
[8] C. H. Choi, S. J. Rhee, T. S. Jeon, N. Lu, J. H. Sim, R. Clark, M. Niwa and D. L. Kwong, IEDM Tech. Dig., (2002) 857.
[9] E. Ma, Science, (2004) 623.
[10] D. Wang, U. Geyer, S. Schneider and G. von Minnigerode, Thin Solid Films, 292 (1997) 184.
[11] R. M. Tromp and J. B. Hannon, Surf. Rev. and Lett., 9 (2002) 1565.

Studies in Surface Science and Catalysis, volume 159
Hyun-Ku Rhee, In-Sik Nam and Jong Moon Park (Editors)

Application of sol-gel techniques in fabrication of fuel cells

Seong-Ahn Hong[1] and Suk Woo Nam[2]

[1]National RD&D Organization for Hydrogen and Fuel Cells
[2]Fuel Cell Research Center
 Korea Institute of Science and Technology, Seoul 136-791, Korea

ABSTRACT

Sol-gel techniques have been widely used to prepare ceramic or glass materials with controlled microstructures. Applications of the sol-gel method in fabrication of high-temperature fuel cells are steadily reported. Modification of electrodes, electrolytes or electrolyte/electrode interface of the fuel cell has been also performed to produce components with improved microstructures. Recently, the sol-gel method has expanded into inorganic-organic hybrid membranes for low-temperature fuel cells. This paper presents an overview concerning current applications of sol-gel techniques in fabrication of fuel cell components.

1. INTRODUCTION

A fuel cell is a membrane reactor in which two coupled electrochemical reactions are taking place on both sides of the membrane. A typical fuel cell couples electrochemical hydrogen oxidation with oxygen reduction reaction to produce electricity and heat. Since energy conversion process in the fuel cell does not involve intermediate combustion cycle, energy conversion efficiency from fossil fuels to electricity is much higher than conventional methods of power generation. Higher conversion efficiency without direct combustion of fuel yields less amount of CO_2 and other emissions, making a fuel cell ideal for future power generation.

A fuel cell consists of an ion-conducting membrane (electrolyte) and two porous catalyst layers (electrodes) in contact with the membrane on either side. The hydrogen oxidation reaction at the anode of the fuel cell yields electrons, which are transported through an external circuit to reach the cathode. At the cathode, electrons are consumed in the oxygen reduction reaction. The circuit is completed by permeation of ions through the membrane.

Table 1. Applications of sol-gel methods in fabrication of fuel cell components.

Fuel Cell	Components	Sol-gel application
SOFC	Anode, cathode, electrolyte	Powder synthesis
	Electrolyte	Thin film coating
	Cathode	Multilayer structure formation
	Electrode/electrolyte interface	Interface modification
MCFC	Cathode	Protective coatings to decrease dissolution rate
	Anode	Modification of surface wetting characteristics
PEMFC	Electrolyte	Synthesis of organic/inorganic hybrid membranes
	Electrode	Active catalyst coating

Porous electrodes are commonly used in fuel cells to achieve high surface area which significantly increases the number of reaction sites. A critical part of most fuel cells is often referred to as the triple phase boundary (TPB). These mostly microscopic regions, in which the actual electrochemical reactions take place, are found where reactant gas, electrolyte and electrode meet each other. For a site or area to be active, it must be exposed to the reactant, be in electrical contact with the electrode, be in ionic contact with the electrolyte, and contain sufficient electro-catalyst for the reaction to proceed at a desired rate. The density of these regions and the microstructure of these interfaces play a critical role in the electrochemical performance of the fuel cells [1].

Sol-gel techniques have been applied to fabricate ceramic or glass materials in a wide variety of forms: ultra-fine powders, thin film coatings, fibers, monoliths, microporous membranes or highly porous aerogels. Advantages of the sol-gel processes include excellent control of microstructure, ease of compositional modification at relatively low temperature by using simple and inexpensive equipments [2]. Applications of the sol-gel processing in fabrication of solid oxide fuel cell (SOFC), molten carbonate fuel cell (MCFC) and polymer electrolyte membrane fuel cell (PEMFC) are summarized in Table 1. The sol-gel techniques have been not only used for powder synthesis or thin film coating, but also applied to modify the electrode or the electrode/electrolyte interface. In the case of powder synthesis, sol-gel method has been used to keep compositional homogeneity of the materials for fuel cell components [3-5]. Thin film coating or modification of fuel cell components is carried out by sol-gel process to form improved microstructure so that electrochemical properties of the components or reaction rates can be enhanced. Examples of sol-gel coating to form fuel cell components are as follows.

2. SOFC

2.1 Thin electrolyte films

Thin film coating is one of the well-known applications of sol-gel technology. Thin ceramic electrolyte films have been deposited by sol-gel coating on a porous electrode to form an electrode-supported SOFC. Since internal resistance of the cell decreases with decreasing electrolyte thickness, the SOFC with thin electrolyte films can be operated at reduced temperature of about $700^{\circ}C$. Very thin films (0.5-10μm) of yttria-stabilized zirconia (YSZ) have been deposited onto porous electrodes using the sol-gel technique [6]. The sol-gel chemistry is also modified to permit direct film deposition on porous substrates in a single step. Another proposed sol-gel method uses spin coating while heating the substrate [7]. Sol-gel coating has been also tried to remove defects produced after the slurry coating of YSZ on a porous anode [8]. Sol–gel coating of YSZ layers was also performed to protect a doped ceria electrolyte against reduction in the reducing atmospheres [9].

Either particulate sol or polymeric sol has been used for thin film coatings. The polymeric sol was fabricated by partial hydrolysis of corresponding metal alkoxide. If the rate of hydrolysis or condensation is very fast, then some kinds of organic acids, beta-dicarbonyls, and alkanolamines have been used as chelating agent in sol–gel processes to control the extent and direction of the hydrolysis-condensation reaction by forming a strong complex with alkoxide. [2].

2.2 Multilayer cathode

Considerable efforts have been made to improve the SOFC performance by using a composite cathode. The cathode composed of $La_{1-x}Sr_xMnO_{3+\delta}$ (LSM) and YSZ has been extensively studied to reduce the cathode polarization. It was reported that addition of YSZ to the LSM cathode enlarged the TPB area, extending the reaction site to the electrode and significantly reducing the cathode polarization. Preparation of this composite cathode includes solid-state mixing of LSM and YSZ powders, followed by high temperature sintering above 1200 °C in order to facilitate the adhesion between the cathode and the YSZ electrolyte. The high processing temperature, however, is not desirable since interfacial reaction between the LSM cathode and the YSZ electrolyte would take place to form highly resistant products, such as $La_2Zr_2O_7$, which eventually reduce the cell performance.

By using a sol-gel process the microstructure and composition of the cathode can be controlled more easily at lower processing temperature. Therefore, layers of LSM cathode were deposited on YSZ electrolyte by sol-gel and dip coating method [10]. The layers consisted of the thin and functionally cathodic interlayer deposited on the electrolyte and thicker but porous cover layer for current collection. The porosity and composition were adjusted by controlling sol composition. Similar microstructure was obtained for $La_{0.8}Sr_{0.2}Mn_{1-y}Fe_yO_{3+\delta}$ [11], $La_{2-x}NiMnO_{4+\delta}$ [12] and $Sr_{0.5}Sm_xCoO_3$ [13] materials. Functionally graded 4-layer cathode was prepared by sol-gel/slurry coating for honeycomb SOFC [14]. The interlayer in contact with the electrolyte consisted of 50% LSM and 50% gadolinia-doped ceria (GDC) and the composition was changed from catalytically active layer to a current collection layer. The particulate sol of GDC was prepared and then the sol was mixed with LSM or other current collecting materials to produce the slurry for multilayer coating. By using the sol-gel method, calcination temperature can be reduced, avoiding undesirable reaction between YSZ electrolyte and cathode materials.

2.3 Modified electrode/electrolyte interface

In solid electrolyte fuel cells, the challenge is to engineer a large number of catalyst sites into the interface that are electrically and ionically connected to the electrode and the electrolyte, respectively, and that is efficiently exposed to the reactant gases. In most successful solid electrolyte fuel cells, a high-performance interface requires the use of an electrode which, in the zone near the catalyst, has mixed conductivity (i.e. it conducts both electrons and ions). Otherwise, some part of the electrolyte has to be contained in the pores of electrode [1].

Sol-gel technique has been used to deposit solid electrolyte layers within the LSM cathode. The layer deposited near the cathode/electrolyte interface can provide ionic path for oxide ions, spreading reaction sites into the electrode. Deposition of YSZ or samaria-doped ceria (SDC, $Sm_{0.2}Ce_{0.8}O_2$) films in the pore surface of the cathode increased the area of TPB, resulting in a decrease of cathode polarization and increase of cell performance [15].

Sol-gel technique has also been applied to modify the anode/electrolyte interface for SOFC running on hydrocarbon fuel [16]. A Ni/YSZ cermet anode was modified by coating with SDC sol within the pores of the anode. The surface modification of Ni/YSZ anode resulted in an increase of structural stability and enlargement of the TPB area, which can serve as a catalytic reaction site for oxidation of carbon or carbon monoxide. Consequently, the SDC coating on the pores of anode leads to higher stability of the cell in long-term operation due to the reduction of carbon deposition and nickel sintering.

3. MCFC

3.1 Protective coatings for cathode

Dissolution of the state-of-the-art NiO cathode into the electrolyte and subsequent precipitation of Ni in the matrix is one of major factors limiting the lifetime of MCFC. The growth of Ni precipitates in the matrix results eventually in electric shorting between cathode and anode. Recently, a number of attempts were made to cover the pore surface of NiO cathodes with layers of stable materials to alleviate NiO dissolution [17]. The stable materials coated on NiO surface encompass mostly $LiCoO_2$ or $LiCo_{1-x}Ni_xO_2$. The coated layers were expected to reduce the surface area of NiO exposed to the carbonate melts, thereby reducing the rate of NiO dissolution and improving the cell life. A number of stabilized NiO cathodes have been prepared by sol-gel coating method with $LiCoO_2$ [18], Co/Mg [19], and TiO_2 [20]. According to the results obtained from single cell tests under pressurized conditions, the $LiCoO_2$-coated cathodes did suppress the NiO dissolution by about 50% at 1 and 3 atm. However, the suppression of the NiO dissolution was much less effective, only by 15%, at 5 atm. The NiO dissolution was relatively independent of the amount of $LiCoO_2$ incorporated into the NiO cathode [21].

3.2 Modification of anode

Electrolyte loss occurring in long-term operation of MCFC is another problem to be solved for practical application of MCFC. For commercialization, the MCFC should show stable performance over 40,000 hours. Electrolyte loss in MCFC is caused by various factors, e.g., corrosion of components, creepage, reaction with cell components and direct evaporation. These factors result in poor electrolyte–electrode contact, a small electrochemical reaction area, high ohmic resistance, gas cross-over, and gas leaking. Excessive injection of electrolyte into a cell without the electrode modification caused rapid reduction of the cell performance due to the flooding of the cathode. On account of the poor electrolyte wettability on the Ni-based anode compared with the electrolyte-philic nature of the cathode, the electrolyte distribution is easily localized on the latter electrode.

Sol-gel coating has been used to modify wetting characteristics of MCFC anode. The surface of the anode is modified with bohemite sol by using a dip-coating method. The coated film changes into lithium aluminate particles and produces micropores in the anode during cell operation. Consequently, the surface modification leads to increase in the electrolyte filling contents in the anode pores without any significant degradation in cell performance. In addition, the modification by the sol–gel coating technique induces an anti-sintering ability due to the lithium aluminate particles coated between the nickel particles. The results suggest that the surface modification can increase the structural stability of the anode, as well as the operation time, due to the additional electrolyte in the cell [22].

4. PEMFC

For last few years, extensive studies have been carried out on proton conducting inorganic/organic hybrid membranes prepared by sol–gel process for PEMFC operating with either hydrogen or methanol as a fuel [23]. A major motivation for this intense interest on hybrid membranes is high cost, limitation in cell operation temperature, and methanol crossover of plain perfluorosulfonic acid membranes such as Nafion® [24]. The flexibility in sol–gel approach offers the potential for molecular engineering of composition and properties of a diverse range of materials. The presence of organic phase makes the hybrid materials more flexible while their thermal stability has greatly been enhanced by inorganic part. The sol-gel applications in formation of hybrid membrane can be classified into modification of existing Nafion® membranes and synthesis of new hybrid membranes.

Nafion® membranes have been modified by sol–gel processing to incorporate hydrophilic metal oxides. The hybrid membranes were made mostly by either infiltration of Nafion® with sol–gel solutions or mixing sol–gel and Nafion® solutions [25]. A variety of metal oxides such as SiO_2, ZrO_2, and TiO_2 have been formed by *in-situ* sol-gel reaction inside the Nafion® membranes [26]. The SiO_2-incorperated Nafion® membranes were further impregnated with Pt to form self-humidifying membranes [27]. The sol-gel reaction at elevated temperature of 70°C produces stable SiO_2 nanoparticles uniformly distributed in the electrolyte. The hybrid membranes composed of Nafion® and SiO_2 or TiO_2 were fabricated by mixing Nafion® ionomer solution with sol-gel derived metal oxide nanoparticles and recasting the mixed solution [28-30]. The nanoparticles of metal oxide were formed by either external sol-gel process or *in-situ* process. Phosphotungstic acid has been added to Nafion®/SiO_2 membranes to improve proton conductivity [31].

Concerning hybrid membranes which are not based on Nafion®, a number of polymers have been applied with sol-gel processing to form inorganic/organic membranes, and recent studies are well summarized in reference [23]. Other than hybrid membranes, inorganic membranes such as nanoporous TiO_2 and P_2O_5- TiO_2-SiO_2 glasses have been prepared by sol-gel methods to produce proton-conducting layers [32,33].

5. Concluding Remarks

Sol-gel techniques have been successfully applied to form fuel cell components with enhanced microstructures for high-temperature fuel cells. The applications were recently extended to synthesis of hybrid electrolyte for PEMFC. Although the results look promising, the sol-gel processing needs further development to deposit micro-structured materials in a selective area such as the triple-phase boundary of a fuel cell. That is, in the case of PEMFC, the sol-gel techniques need to be expanded to form membrane-electrode-assembly with improved microstructures in addition to the synthesis of hybrid membranes to get higher fuel cell performance.

REFERENCES

[1] Fuel Cell Hand Book (7th ed.), U.S. Department of Energy (2004).
[2] C.J. Brinker and G.W. Scherer, Sol-Gel Science, Academic Press, Boston (1990).
[3] M. Gaudon, C. Laberty-Robert, F. Ansart, P. Stevens, A. Rousset, Solid State Sciences, 4 (2002) 125.
[4] F.J. Lepe, J. Fernandez-Urban, L. Mestres, M.L. Martinez-Sarrion, Journal of Power Sources, 151 (2005) 74.
[5] F.-Y. Wang, S. Chena, Q. Wanga, S. Yua, S. Cheng, Catalysis Today, 97 (2004) 189.
[6] N. Q. Minh and T. Takahashi, Science and Technology of Ceramic Fuel Cells, Elsevier, Amsterdam (1995)..

[7] N.X.P. Vo, S.P Yoon, S.W. Nam, J. Han, T.-H. Lim, S.-A. Hong, Key Engineering Materials,. 277-279 (2005) 455.

[8] S.D. Kim, S.H. Hyun, J. Moon, J.-H. Kim, R.H. Song, Journal of Power Sources, 139 (2005) 67.

[9] S.-G. Kim, S.P. Yoon, S.W. Nam, S.H. Hyun, S.-A. Hong, Journal of Power Sources, 110 (2002) 222.

[10] M. Gaudona, C. Laberty-Robert, Florence Ansart, L. Dessemond, P. Stevens, Journal of Power Sources, 133 (2004) 214.

[11] P. Lenormand, S.Castillo, J.-R.Gonzalez, C. Laberty-Robert, F. Ansart, Solid State Sciences, 7 (2005) 159.

[12] M.L. Fontaine ., C. Laberty-Robert, F. Ansart, P. Tailhades, Journal of Power Sources, in press.

[13] Z, Tang, Y, Xie, H, Hawthorne, D, Ghosh, Journal of Power Sources, in press.

[14] S. Zha, Y. Zhang, M. Liu, Solid State Ionics, 176 (2005) 25.

[15] S.P. Yoon, J. Han, S.W. Nam, T.-H. Lim, I.-H. Oh, S.-A. Hong, Y.-S. Yoo, H.C. Lim, Journal of Power Sources, 106 (2002) 160.

[16] S.P. Yoon, J. Han, S.W. Nam, T.-H. Lim, S.-A. Hong, Journal of Power Sources 136 (2004) 30.

[17] S.-G. Kim, S.P. Yoon, J. Han, S.W. Nam, T.-H. Lim, S.-A. Hong, H.C. Lim, Journal of Power Sources, 112 (2002) 109.

[18] S.T. Kuk, Y.S. Song, K. Kim, Journal of Power Sources, 83 (1999) 50.

[19] E. Park, M.Z. Hong, H. Lee, M. Kim, and K. Kim, Journal of Power Sources, 143 (2005) 84.

[20] M.Z. Hong,, H.S. Lee, M.H. Kim, E.J. Park, H.W. Ha, and K. Kim, Journal of Power Sources, in press.

[21] J Han, S.-G. Kim, S.P. Yoon, S.W. Nam, T.-H. Lim, I.-H. Oh, S.-A. Hong and H.C. Lim, Journal of Power Sources, 106 (2002) 153.

[22] J.Y. Youn, S.P. Yoon, J. Han, S.W. Nam, T.-H. Lim, S.-A. Hong, K.Y. Lee, Journal of Power Sources, in press.

[23] D.J. Jones and J. Roziere, in Handbook of Fuel Cells (W. Vielstich, A. Lamm, H.A. Gasteiger eds.), Vol 3, Wiley (2005) 447.

[24] R. Thangamuthu and C.W. Lin, Journal of Power Sources 150 (2005) 48.

[25] L.C. Kleina, Y. Daiko, M. Aparicio, F. Damay, Polymer, 46 (2005) 4504.

[26] N.H. Jalani, K. Dunn, R. Datta. Electrochimica Acta, 51 (2005) 553.

[27] H. Hagihara, H. Uchida, M. Watanabe, Electrochimica Acta, in press.

[28] A. Sacc, A. Carbone, E. Passalacqua, A. D'Epifanio, S. Licoccia, E. Traversa, E. Sala, F. Traini, R. Ornelas, Journal of Power Sources, 152 (2005) 16.

[29] V. Baglio, A.S. Arico, A. Di Blasi, V. Antonucci, P.L. Antonucci, S. Licoccia, E. Traversa, F. Serraino Fiory, Electrochimica Acta, 50 (2005) 1241.

[30] R. Jiang, H.R. Kunz, J.M. Fenton, Journal of Membrane Science, in press.

[31] W. Xua, T. Lua, C. Liua, W. Xing, Electrochimica Acta, 50 (2005) 3280.

[32] M.T. Colomer, Journal of the European Ceramic Society, in press.

[33] T. Uma., S. Izuhara, M. Nogami, Journal of the European Ceramic Society, in press.

Studies in Surface Science and Catalysis, volume 159
Hyun-Ku Rhee, In-Sik Nam and Jong Moon Park (Editors)

Advances in multiphase reactors for the fuel industry

Tiefeng Wang, Fei Wei, Jinfu Wang, Fei Ren, Yi Cheng and Yong Jin

Department of Chemical Engineering, Tsinghua University, Beijing 100084, P.R. China

1. INTRODUCTION

Fuel industry is of increasing importance because of the rapidly growing energy needs worldwide. Many processes in fuel industry, e.g. fluidized catalytic cracking (FCC) [1], pyrolysis and hydrogenation of heavy oils [2], Fischer-Tropsch (FT) synthesis [3,4], methanol and dimethyl ether (DME) synthesis [5,6], are all carried out in multiphase reactors. The reactors for these processes are very large in scale. Unfortunately, they are complicated in design and their scale-up is very difficult. Therefore, more and more attention has been paid to this field. The above mentioned chemical reactors, in which we are especially involved like deep catalytic pyrolysis and one-step synthesis of dimethyl ether, are focused on in this paper.

2. GAS-SOLID FLUIDIZED BED REACTOR

In the last two decades, considerable progress has been made in developing and applying riser reactors for FCC, deep catalytic cracking (DCC) and pyrolysis [7]. On one side riser reactors offer significant advantages over conventional bubbling fluidized bed reactors, such as high gas-solid contact efficiency, high gas and solids throughput and the ability to handle cohesive particles. But on another side they suffer from severe solids backmixing due to non-uniform gas and solids flow. However in residual oil catalytic cracking, it would be advantageous to have short residence time (less than 1.0 s) with narrow residence time distributions (RTD). Downer reactors can largely overcome the disadvantages of riser reactors.

2.1. Comparison of riser and downer

The hydrodynamics in risers and downers are compared in Figs. 1(a-c). Figure 1a shows that the solids fraction distribution in the downer is much more uniform, with the difference in concentration between the core and the annulus much smaller than in the riser. Only an annular region of relatively high solids fraction exists near the wall in downers. Figure 1b shows that the radial profiles of vertical particle velocity of downers are more uniform than in risers. Figure 1c shows that the RTD of solid particles is rather narrow in the downer than in the riser. The wide tail of the solids RTD in the riser indicates strong solids backmixing, showing that downer reactors operate much closer to plug flow than riser reactors [9,10]. Therefore, the riser reactor has the advantages in providing shorter contact time and being operated in higher operating temperature that are necessary for a pyrolysis reaction.

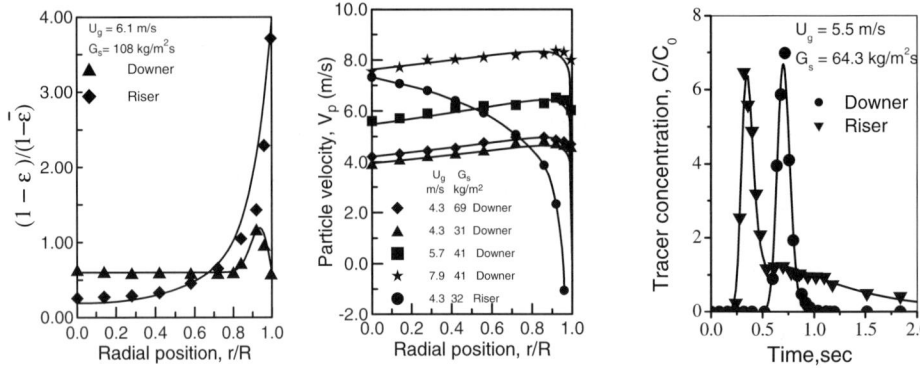

(a) Profile of solid fraction [8]   (b) Profile of particle velocity [8]   (c) RTD of solid particles [9]

Fig. 1. Comparison of riser and downer

2.2. Coupled CFB reactor

While having some advantages over riser reactors, downer reactors also suffer from some serious shortcomings, such as a low solids holdup in the bed, difficulty in even distribution of injected residual on the catalysts, and a high sensitivity to the structure of the inlet [11,12]. Therefore, the development of a new coupled CFB reactor that can fully utilize the advantages of the riser and the downer is of interest.

2.2.1. Riser-downer coupled CFB

To overcome the shortages of the downer, a novel riser-downer coupled CFB reactor was proposed for the FCC process [13], as shown in Fig. 2a. This riser-downer coupled CFB reactor has several promising characteristics. First, the solid fraction is high with intensive gas-solid mixing in the top of the riser, which can ensure high contact efficiency between injected residual and catalysts in the initial stage. The subsequent downer reactor possesses a uniform radial flow structure and a low axial back-mixing that can efficiently suppress the occurrence of the secondary reactions, which may be significant especially at the high feed conversion in series reactions. Second, the length ratio of the riser and the downer can be changed, and this changes the residence time and its distribution. This allows adjustment of the reaction to suit market requirements. A mini-commercial unit of FCC process has been operated and the expected results are shown in Table 1.

Table 1
Comparison of the mini-commercial riser-downer unit with the industrial riser

	T, °C	dry gas %	coke %	liquid yield, %	olefin in gasoline, %	octane number
Industrial riser	500	5.10	9.03	80.6	45	90
Riser-downer unit	540	3.27	9.97	82.17	28	94.8

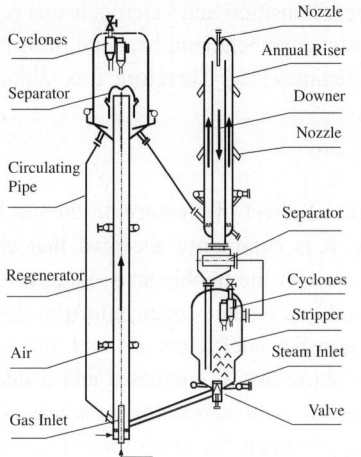

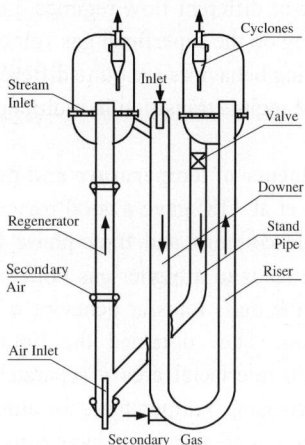

Fig.2a. Schematic of the riser-downer coupled reactor

Fig.2b. Schematic of the downer-riser coupled reactor

2.2.2. Downer-riser coupled CFB

The CFB catalytic cracking reactor plays an important role in the petroleum industry because of its better gas-solids contact and narrow residence time distribution, but its non-uniform radial flow structure and the extensive backmixing of gas and solids lead to a lower conversion rate and poorer selectivity to desired intermediate products [14].

For fluidized catalytic cracking of heavy residual oil and reduced olefin content in product, a coupled downer-riser CFB system can be used [15]. In this process, there are two distinct stages: the first stage is the fast cracking reaction of the hydrocarbon which requires high temperature and short residence time with less backmixing; the second stage is the hydrogen-reforming reaction which requires lower temperature and longer residence time and is not sensitive to backmixing. Therefore, a downer-riser CFB can combine the advantages of the plug flow and short residence time in the downer and a high solids holdup in the riser.

3. GAS-LIQUID-SOLID REACTOR

The heterogeneous, strongly exothermic catalytic reaction and high reaction temperature and pressure are the common characteristics of several important fuel industry processes. Due to good heat control performance to avoid the axial and radial temperature gradient, gas-liquid-solid slurry airlift loop reactors are preferred for processes such as Fischer-Tropsch synthesis, hydrogenation of heavy oil, liquid-phase synthesis of DME and methanol. In the following, the flow regime and the influences of pressure and temperature in bubble columns and airlift loop reactors are discussed firstly. Then we focus on the three-phase slurry airlift reactors (ALR) for its attractive advantages and promising wide applications.

3.1. Flow regime

The hydrodynamics in a gas-liquid or gas-liquid-solid system is characterized by the presence of different flow regimes, i.e., homogeneous, transition and heterogeneous regimes, depending on the superficial gas velocity. The hydrodynamic behavior, heat and mass transfer, and mixing behaviors are quite different in different regimes [16]. Therefore, it is important to study the regime transition in multiphase flows [17].

3.2. Influence of temperature and pressure

Fan et al. [18] gave a good review on the studies of effect of pressure on the gas holdup in bubble columns and three-phase fluidized beds. It is commonly accepted that elevated pressure leads to a higher gas holdup due to a decrease in the bubble size. Yang et al. [19] studied the mass transfer behavior of a three-phase slurry system under industrial operating conditions. They obtained the liquid-side mass transfer coefficient k_L and the specific gas-liquid interfacial area a separately. The results show that k_L increases and a decreases with increasing temperature; k_L almost keeps constant with increasing pressure, while a increases remarkably with increasing pressure. They obtained the experimental correlations for k_L and a at high temperature and pressure.

3.3. Hydrodynamic behavior of ALR

Gas holdup and liquid circulation velocity are the most important parameters to determinate the conversion and selectivity of airlift reactors. Most of the reported works are focused on the global hydrodynamic behavior, while studies on the measurements of local parameters are much more limited [20]. In recent years, studies on the hydrodynamic behavior in ALRs have focused on local behaviors [20-23], such as the gas holdup, bubble size and bubble rise velocity. These studies give us a much better understanding on ALRs.

3.3.1. Influence of gas distributor
The bubble size distribution is closely related to the hydrodynamics and mass transfer behavior. Therefore, the gas distributor should be properly designed to give a good performance of distributing gas bubbles. Lin et al. [21] studied the influence of different gas distributor, i.e., porous sinter-plate (case 1) and perforated plate (case 2) in an external-loop ALR. Figure 3 compares the bubble sizes in the two cases. The bubble sizes are much smaller in case 1 than in case 2, indicating a better distribution performance of the porous sinter-plate. Their results also show the radial profile of the gas holdup in case 1 is much flatter than that in case 2 at the superficial gas velocities in their work.

3.3.2. Influence of internals
At a given superficial gas velocity, the gas holdup decreases with increasing liquid circulation velocity. In a large scale ALR, internals are needed to regulate the liquid circulation velocity, break bubbles and improve the hydrodynamic behavior. Figure 4 shows the influence of the internal (*Perforated Plate*) on the overall gas holdup [22]. The gas holdup in the riser with the perforated plate (case a) is about twofold of that without a perforated plate (case b). Zhang et al. [23] showed that the specially designed internal (*Bubble Scraper*) also has remarkable effect of breaking bubbles, and makes the radial profiles of the gas holdup, bubble rise velocity and liquid velocity more uniform.

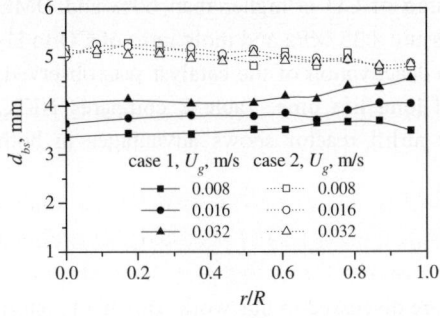

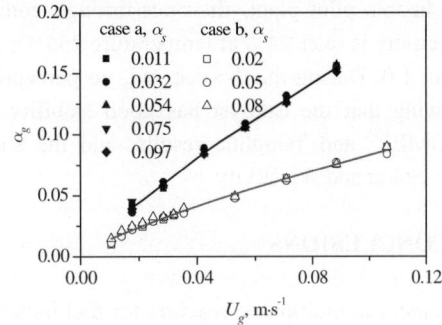

Fig. 3. Radial profiles of the bubble size with different gas distributors (air-water system) [22].

Fig. 4. Influence of internals on the gas holdup (air-water-solid slurry system) [23].

3.4. Spherical reactor

Most processes in the fuel industry are carried out under high pressure conditions with large scale production. However, the manufacture of large (such as 6-10 m in diameter) cylindrical reactors to be operated under high pressure such as 10 MPa is very difficult. A novel spherical reactor is proposed for the relative thin-wall reactor design in high pressure processes. In that case the wall thickness can be decreased by half at the same operating pressure. The hydrodynamic behavior of the reactor consisting of serials of spherical units with novel internals has been studied. The results show that not only the distributions of local parameters can be even in redial direction and the back-mixing of liquid flow also could be restricted. The spherical reactor has economical feasibility and great potential for large scale production in the fuel industry.

3.5. One-step DME synthesis

The external-loop slurry airlift reactor was used in a pilot plant (3000 t/a) for one-step synthesis of dimethyl ether (DME) from syngas. Specially designed internals were used to intensify mass transfer and heat removal. This new technology is highly efficient and easy to scale-up to industrial.

Table 2
Comparison of the pilot-plant results

Conditions	NKK	LPDME^{TM}	Tsinghua Univ.
H_2/CO ratio	1	0.7	1
Catalyst		Cu-Zn-Al+γ-Al$_2$O$_3$	LP201+TH16
CO conversion, %	40	22	63
Selectivity, %	90	40-90	94
Reactor type	Slurry bubble column	Slurry bubble column	Slurry airlift reactor
Design scale, t/d	5	10	10

In this pilot plant, the once-through conversion of CO is higher than 60% and DME selectivity is over 94% at temperature 255 ℃, pressure 4.35 MPa and mole ratio of CO to H_2 about 1.0. During the 3-week run, no perceptible deactivation of the catalyst was observed, showing that the catalyst has good stability and long life time. Table 2 compares NKK, LPDME[TM] and Tsinghua results, and the slurry airlift reactor shows advantages in both conversion and selectivity.

4. CONCLUSIONS

Advances in multiphase reactors for fuel industry are discussed in this work. Downer reactors have some advantages over riser reactors, but suffer from some serious shortcomings. The coupled reactors can fully utilize the advantages of the riser and the downer. For fuel industry that involves gas-liquid-solid system, slurry bed reactors especially airlift reactors are preferred due to their performance of excellent heat control and ease of scale up. For high-pressure processes, the spherical reactor is promising due to its special characteristics.

REFERENCES

1. L. Reh, Chem. Eng. Tech., 18 (1995) 75.
2. F. M. Dautzenberg and J. C. De Deken, Catalysis Reviews-Sci. & Eng., 26 (1984) 421.
3. H. P. Jr. Withers, K. F. Eliezer and J. W. Mitchell, Ind. Eng. Chem. Res., 29 (1990) 1807. (F-T)
4. R. Krishna and S. T. Sie, Fuel Proc. Tech., 64 (2000) 73.
5. S. Ledakowicz, M. Stelmachowski and A. Chacuk, Chem. Eng. Proc., 31 (1992) 213.
6. S. Lee, M. R. Gogate and C. J. Kulik, Chem. Eng. Sci., 47 (1992) 3769.
7. Y. Jin, Y. Zheng and F. Wei, CFBirculating Fluidized Bed Technology VII, J.R. Grace, J.-X Zhu and H. de Lasa (eds.), 2002, 40.
8. D.R. Bai, Y. Jin, Z.Q. Yu and N.J. Gan, Circulating Fluidized Bed Technology III, P. Basu, M. Horio and M. Hasatani, M. (eds.), Pergamon Press, Toronto, 1991, 51.
9. F. Wei and N.J. Gan, J. Chem. Ind. & Eng. China (Chinese), 45 (1994) 230.
10. F. Wei, Z. Wang, Y. Jin, Z. Q. Yu, W. Chen, Powder Technol., 81 (1994) 25.
11. R. Deng, F. Wei, Y. Jin, Q. Zhang, Y. Jin, Ind. Eng. Chem. Res., 41 (2002) 6015.
12. P. M. Johnston, H. I. de. Lasa, J. X. Zhu, Chem. Eng. Sci., 54 (1999) 2161.
13. H. Liu, R.S. Deng, L. Gao, Wei, and Y., Ind. Eng. Chem. Res., 44 (2005) 733.
14. F. Wei, X. Wan, Y. Hu, Z. Wang, Y. Yang and Y. Jin, Chem. Eng. Sci., 56 (2001) 613
15. Z.Q. Li, C.N. Wu, F. Wei and Y. Jin, Powder Tech., 139 (2004) 214.
16. J. Zahradník and M. Fialová, M., Chem. Eng. Sci., 51(1996) 2491.
17. T.F. Wang, J.F. Wang and Y. Jin, Chem. Eng. Sci., 2005 accepted.
18. L.S. Fan, G.Q. Yang, D.J. Lee, K. Tsuchiya1 and X. Luo, Chem. Eng. Sci., 54 (1999) 4681.
19. W.G. Yang, J.F. Wang and Y. Jin, Chem. Eng. & Tech., 24 (2001) 651.
20. M. Utiger, F. Stuber, A.M. Duquenne, H. Delmas and C. Guy, Can. J. Chem. Eng. 77 (1999) 375.
21. J. Lin, M.H. Han, T.F. Wang and J.F. Wang, Chem. Eng. J., 102 (2004) 51.
22. T.F. Wang, J.F. Wang, B. Zhao, F. Ren and Y. Jin, Chem. Eng. Commn., 191 (2004) 1024.
23. T.W. Zhang, J.F. Wang, T.F. Wang and Y. Jin, Chem. Eng. Process., 44 (2005) 81.

Studies in Surface Science and Catalysis, volume 159
Hyun-Ku Rhee, In-Sik Nam and Jong Moon Park (Editors)

The Impact of Interdisciplinary Study on Biochemical Reaction Engineering

Jian-Jiang Zhong

State Key Laboratory of Bioreactor Engineering, East China University of Science and Technology, 130 Meilong Road, Shanghai 200237, China   e-mail: jjzhong@ecust.edu.cn

1. INTRODUCTION

Biochemical reaction engineering (BRE) plays an important role in the biotechnology world, especially in the industrial application of biotechnology/bioscience research achievements. With the rapid development of biotechnology, there is an accelerated trend in interdisciplinary research between BRE and other fields. Under the circumstances, novel BRE concepts/methodologies are developed to handle new processes and new targets, which also contributes to the advancement of biotechnology and other disciplines. In this article, by taking cell cultures for production of valuable secondary metabolites as a typical case, several examples are demonstrated regarding the impact of interdisciplinary study on BRE.

2. BIOREACTOR DESIGN FOR SHEAR SENSITIVE CELL CULTURES

A stirred bioreactor is easy to maintain homogeneous conditions by mechanical agitation even at a high cell density. It also possesses several other advantages such as existing industrial capacity and reliability. Therefore, modified stirred bioreactors with low shear stress have received great interest for their application to shear-sensitive plant cell cultures [1]. However, until now, there is a lack of reports on scale-up of modified stirred bioreactors for plant cell cultures. In addition, although there are many reports on plant cell cultures including large-scale processes [1-3], very few reliable scale-up strategies have been presented in spite of a great need of such information for industrial application of the cell culture technology.

The application of chemical engineering principles to bioprocesses successfully led to the development of a novel low-shear cell culture bioreactor, called centrifugal impeller bioreactor (CIB) [4,5]. The novel bioreactor has advantages of more efficient mixing, reduced surface liquid turbulence, lower shear stress and higher oxygen supply capacity over a cell-lift bioreactor, which is widely used for animal cell cultures. In cell cultures of *Panax notoginseng* for production of ginseng saponin and polysaccharide, successful application and laboratory scale-up of the CIB were realized towards efficient bioprocessing of useful metabolites in high-density cell cultivations [6].

Because of low shear stress generated by CIB, *P. notoginseng* cells could be cultivated in a relatively wide range of agitation speeds (80-210 rpm) in 3-L CIB without detrimental effects on cell growth and product formation. Initial k_La level, which had significant effects on *P. notoginseng* cell cultures, was identified to be a key factor for CIB scale-up, and a high cell density and hyperproduction of ginseng saponin and polysaccharide were successfully obtained in the 30-L CIB [6]. Sucrose feeding based on specific oxygen uptake rate (SOUR) and residual sugar concentration further enhanced the production of cell mass, ginseng saponin and polysaccharide in both 3-L and 30-L CIB. Furthermore, by adopting a fed-batch cultivation strategy, a maximum DW and concentrations of total saponin and polysaccharide in 30-L CIB were enhanced to 30.3 ± 1.0, 2.1 ± 0.1 and 3.5 ± 0.2 g/L with their corresponding productivity of 1467 ± 87, 102 ± 13 and 179 ± 18 mg/(L·d), respectively (Fig. 1). The success of CIB scale-up in high-density *P. notoginseng* cell cultures both in batch and fed-batch modes suggests that this novel bioreactor can have great potential in high-density cultivation of other plant cells to achieve a highly productive process [6]; and it may be used as an interesting tool to solve some related BRE problems, especially for shear sensitive biological systems including higher fungi fermentations [7].

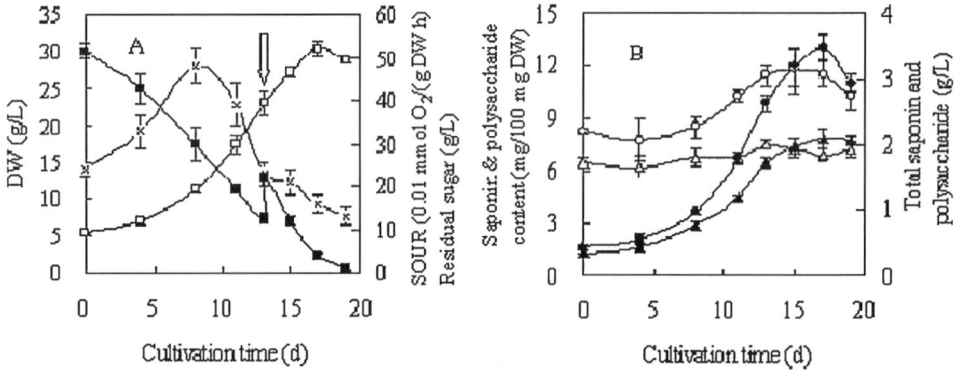

Fig. 1. Fed-batch cultivation of *P. notoginseng* cells in 30-L CIB. Symbols: DW (□), SOUR (×), residual sugar (■), saponin content (△) and production (▲), and polysaccharide content (○) and production (●).

3. NEW SYNTHETIC INDUCERS FOR PLANT SECONDARY METABOLISM

Although plant secondary metabolites are not involved in the basic metabolic processes of living cells, it is well known that they are indispensable for the survival of plants against various external stimuli. In the past decade, there have been numerous reports on the use of various biotic or abiotic elicitors for enhancement of plant secondary metabolite biosynthesis. There have been very few studies on the use of synthetic chemicals to enhance the production of plant secondary metabolites. Recently, we have been searching for novel jasmonate elicitors that can significantly promote secondary metabolite biosynthesis by plant cells. Integration of chemical technology with BRE resulted in the innovation of chemically synthesized very powerful inducers, which can act as a driving force to achieve an industrially interesting highly productive bioprocess for useful plant secondary metabolite production [8-11].

In this work, a suspension culture of *Taxus chinensis*, which produces a bioactive taxoid, taxuyunnanine C (Tc), was taken as a model plant cell system. Experiments on the timing of jasmonates addition and dose response indicated that day 7 and 100 μM was the optimal elicitation time and concentration for both cell growth and Tc accumulation [8]. The Tc accumulation was increased more in the presence of novel hydroxyl-containing jasmonates compared to that with methyl jasmonate (MJA) addition. For example, addition of 100 μM

2,3-dihydroxypropyl jasmonate (DHPJA) on day 7 led to a very high Tc content of 47.2±0.5 mg/g (at day 21), while the Tc content was 29.2±0.6 mg/g (on the same day) with addition of 100 μM MJA (Fig. 2). Quantitative structure-activity analysis of various jasmonates suggests that the optimal lipophilicity and the number of hydroxyl group may be two important factors affecting their elicitation activity. In addition, the jasmonate elicitors were found to induce plant defense responses, including oxidative burst and activation of L-phenylalanine ammonia lyase (PAL). Interestingly, a higher level of H_2O_2 production and PAL activity was detected with elicitation by the synthesized jasmonates compared with that by MJA [8], which well corresponded to the superior stimulating activity for the former case.

Since the synthetic steps are simple and other chemical reagents are not so expensive compared with the starting material (MJA), it is considered that those new jasmonate analogues may have promising applications to other cell culture systems as well as for large scale production of valuable plant secondary metabolites. The work is also considered to have a high pharmacological and industrial impact due to the efficient stimulation of taxoids and ginsenosides biosyntheses.

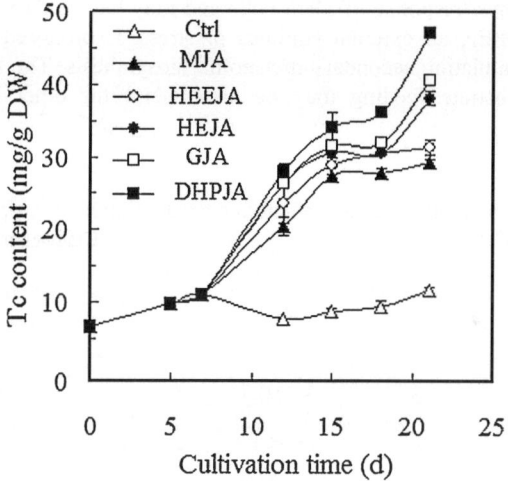

Fig. 2. Time profiles of Tc content in *T. chinensis* suspension cultures elicited with methyl jasmonate (MJA), 2-Hydroxyethoxyethyl jasmonate (HEEJA), 2-hydroxyethyl jasmonate (HEJA), D-Glucosyl jasmonate (GJA) and 2,3-dihydroxypropyl jasmonate (DHPJA). One hundred μM of each elicitor was added to the cultures in 1μL of ethanol per 1mL of culture medium on day 7 of cultivation. Data are the means of three flasks and vertical bars show standard deviations.

4. PULSED ELECTRIC FIELD AS A NEW ABIOTIC ELICITOR

In the interdisciplinary field of biophysics and biotechnology, the bioeffects of electric field have received considerable interest for both fundamental studies on these interaction mechanisms and potential application. However, the effects of pulsed electric field (PEF) on secondary metabolism in plant cell cultures and fermentation processes have been unknown. Therefore, it would be very interesting to find out whether PEF could be used as a new tool for stimulating secondary metabolism in plant cell cultures for potential application to the value-added plant-specific secondary metabolite production. Furthermore, if the PEF permeabilization and elicitation are discovered in a cell culture system, the combination of these effects may lead to highly efficient production of intracellular secondary metabolites with their continuous biosynthesis and release into medium, which is beneficial to industrial application.

In our collaboration between Biochemical Engineering Laboratory and a group from Physics Department, the PEF effects on the growth and secondary metabolite production by plant cell culture were investigated by taking suspension cultures of *T. chinensis* as a model system [12]. The cultured cells in different growth phases were exposed to a PEF (50 Hz, 10 V/m) for various periods of time. A significant increase in intracellular accumulation of taxuyunnanine C (Tc), a bioactive secondary metabolite, was observed by exposing the cells in early exponential growth phase to a 30 min PEF. The Tc content (*i.e.* the specific production based on dry cell weight) was increased by 30% after exposure to PEF, without loss of biomass compared with that of the control. Combination of PEF treatment with sucrose feeding was proven useful in improving the secondary metabolite formation. Production of reactive oxygen species, extracellular Tc and phenolics (Table 1) was all increased, while cell capacitance was decreased with PEF treatment. The results show that PEF induced the defense response of plant cells and may have altered the cell/membrane's dielectric properties. PEF, an external stimulus or stress, is proposed as a promising new abiotic elicitor for stimulating secondary metabolite biosynthesis. The PEF approach and its combination with substrate feeding may be also useful for other biochemical reaction processes.

Table 1
Effects of PEF on extracellular Tc and phenolics accumulation in suspension cultures of *T. chinensis*

Extracellular Accumulation (mg/L)	Time after elicitation (d)	Control	PEF	Stimulating activity*
	0	3.0±0.2	3.0±0.2	~
Tc	3	3.3±0.3	5.3±0.5	60%
	0	18.9±1.9	18.9±1.9	~
Phenolics	3	19.8±0.5	25.1±1.1	30%

*The stimulating activity is the ratio of average accumulation titer under elicitation to that of control.

5. CALCIUM SENSORS AND SIGNAL TRANSDUCTION

Calcium is considered as the most versatile intracellular messenger, able to couple a wide range of extracellular signals to specific responses. In recent years, evidences suggest that extracellular Ca^{2+} affects plant secondary metabolite production. However, there have been no reports on the cellular signal responses to quantitative external levels of Ca^{2+}, in spite of the potential significance of such information for biotechnology application by regulation of cellular physiology and metabolism. Information regarding the effect of external Ca^{2+} signal on enzyme activities of plant secondary metabolism is also not available.

Use of biochemical and biological information for bioprocesses is also significant to the advancement of BRE. Here, the information on the signal transduction from external Ca^{2+} was utilized for regulation of ginsenoside biosynthetic pathway of cultured cells of *P. notoginseng*. A quantitative study on the effects of external calcium and calcium sensors was conducted to manipulate the ginsenoside heterogeneity in the plant cells, which is a new and very interesting topic [13-15].

In our work, by comparing the chemical structures of ginsenoside Rd and Rb_1, we hypothesized that there might exist an enzyme (called as UDPG:ginsenoside Rd glucosyltransferase, UGRdGT), which catalyzes the synthesis of ginsenoside Rb_1 from Rd. Experimental work was done in finding and characterizing this new glucosyltransferase [14], which is important to the various ginsenosides biosyntheses and their heterogeneity.

The synthesis of intracellular ginsenoside Rb_1 by *P. notoginseng* cells in 3-day incubation was found to be dependent on medium Ca^{2+} concentration [13]. At an optimal Ca^{2+} concentration of 8 mM, a maximal ginsenoside Rb_1 content of 1.88 ± 0.03 mg g^{-1} dry weight was reached, which was about 60% and 25% higher than that at Ca^{2+} concentration of 0 and 3 mM, respectively. Ca^{2+} feeding experiments confirmed the Ca^{2+}-concentration dependent Rb_1 biosynthesis. To understand the mechanism of the signal transduction from external Ca^{2+} to ginsenoside biosynthesis, intracellular content of calcium and calmodulin (CaM), activities of calcium/calmodulin dependent NAD kinase (CCDNK), calcium-dependent protein kinase (CDPK) and UGRdGT in the cultured cells were analyzed [13]. The intracellular calcium content and CCDNK activity were increased with an increase of external Ca^{2+} concentration within 0-13 mM. In contrast, the CaM content and activities of CDPK and UGRdGT reached their highest levels at 8 mM of initial Ca^{2+} concentration, which well coincided with the behavior of ginsenoside Rb_1 synthesis. In Ca^{2+} feeding experiments, it was confirmed that the intracellular content of calcium and CaM, and activities of CCDNK, CDPK and UGRdGT showed a similar Ca^{2+}-concentration dependency compared to non-feeding experiments.

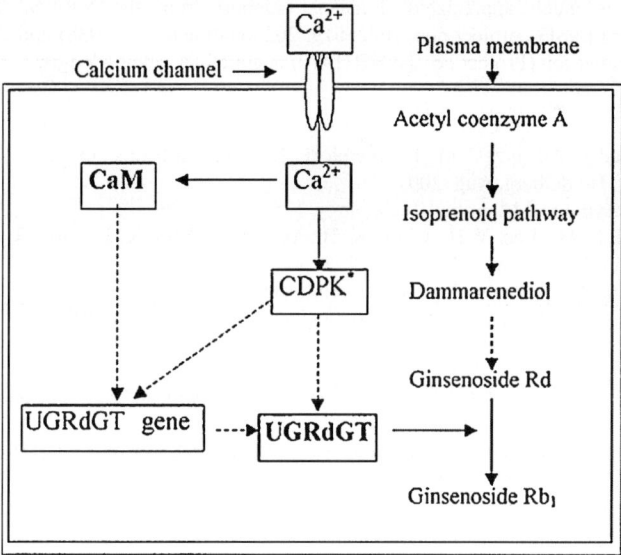

Fig. 3. A proposed signal transduction pathway regarding the external Ca^{2+} effect on ginsenoside Rb_1 synthesis by *P. notoginseng* cells. Ca^{2+} signal changes are triggered by external Ca^{2+} concentrations. The calcium signatures are decoded by calcium sensors, CaM and CDPK. UGRdGT is possibly modulated by the sensors in a direct or indirect (dashed lines) way. Changes of CDPK activity may result from increased synthesis or posttranslational modification of the enzyme (shown as CDPK*).

Based on our results and literature information, a model on the effects of external calcium signal on ginsenoside Rb₁ biosynthesis via the signal transduction pathway of CaM, CDPK and UGRdGT is proposed (Fig. 3) [13, 15]. It is considered that regulation of external calcium concentration will be a new useful strategy for manipulating ginsenoside Rb₁ synthesis of ginseng cells by BRE approach in a large-scale cultivation process. The knowledge and information obtained may be also helpful to other plant cell culture work.

6. CONCLUDING REMARKS

By taking plant secondary metabolite production as an example, this article addresses the impact of interdisciplinary study on the advancement of biochemical engineering. The chemical engineering principle is critical to the novel bioreactor design; chemistry and chemical technology apparently plays an important part in the development of new powerful synthetic elicitors; the finding of pulsed electric field as a new abiotic inducer is due to the physics contribution; while the discovery of a new enzyme and the interesting role of Ca^{2+} signals mostly owes to the information from biochemistry and plant physiology. As a trend, such interdisciplinary studies are starting to become a major player in pushing the biochemical engineering science forward.

ACKNOWLEDGEMENTS

The author wishes to thank Profs. Xuhong Qian, Shu-De Chen and Yufang Xu for their collaboration on the related work. The experimental investigations performed by Dr. Xian-Bing Mao, Mr. Zhi-Gang Qian, Dr. Si-Jing Wang, Dr. Wei Wang, Ms. Hong Ye, Dr. Cai-Jun Yue, Mr. Zhan-Ying Zhang and Dr. Zhen-Jiang Zhao is much appreciated. Financial support from the National Natural Science Foundation of China (NSFC project nos. 20225619, 20236040 and 30270038) and Shanghai Science & Technology Commission (Project no. 04QMH1410) is gratefully acknowledged.

REFERENCES

1. Zhong, J.J. (ed.) Advances in Biochemical Engineering/Biotechnology, 72: Plant Cells. Springer-Verlag, Heidelberg, Aug. 2001.
2. Scragg, A. H. Enzyme and Microbial Technology, 12: 82-85, 1990.
3. Son, S. H.; Choi, S. M.; Lee, Y. H.; Choi,, K. B.; Yun, S. R.; Kim, K. J.; Park, H. J.; Kwon, O. W.; Noh, E. W.; Seon, J. H.; Park, Y. G. Plant Cell Reports, 19: 628-633, 2000.
4. Wang, S.J., Zhong, J.J. Biotechnology and Bioengineering, 51: 511-519, 1996.
5. Wang, S.J., Zhong, J.J. Biotechnology and Bioengineering, 51: 520-527, 1996.
6. Zhang, Z.Y., Zhong, J.J. Biotechnology Progress, 20: 1076-1081, 2004.
7. Mao, X.B., Zhong, J.J. Biotechnology Progress, 20: 1408-1413, 2004.
8. Qian, Z.G., Zhao, Z.J., Xu, Y.F, Qian, X.H., Zhong, J.J. Biotechnology and Bioengineering, 86: 809-816, 2004.
9. Qian, Z.G., Zhao, Z.J., Tian, W.H., Xu, Y.F., Zhong, J.J., Qian, X.H. Biotechnology and Bioengineering, 86: 595-599, 2004.
10. Zhao, Z., Xu, Y., Qian, Z., Tian, W., Qian, X., Zhong, J.J. Bioorganic & Medicinal Chemistry Letters, 14: 4755-4758, 2004.
11. Wang, W., Zhao, Z.J., Xu, Y.F., Qian, X.H., Zhong, J.J. Biotechnology and Bioprocess Engineering 10: 162-165, 2005.
12. Ye, H., Huang, L.L., Chen, S.D., Zhong, J.J. Biotechnology and Bioengineering 88: 788-795, 2004.
13. Yue, C.J., Zhong, J.J. Biotechnology and Bioengineering, 89: 444-452, 2005.
14. Yue, C.J., Zhong, J.J. Process Biochemistry, 40: (in press), 2005.
15. Zhong, J.J., Yue, C.J. Advances in Biochemical Engineering/Biotechnology, 100: (in press), 2005.

Studies in Surface Science and Catalysis, volume 159
Hyun-Ku Rhee, In-Sik Nam and Jong Moon Park (Editors)

Some studies on materials and methane catalysis for solid oxide fuel cells

Ta-Jen Huang

Department of Chemical Engineering, National Tsing Hua University,
Hsinchu, Taiwan, R.O.C.

1. INTRODUCTION

This presentation reports some studies on the materials and catalysis for solid oxide fuel cell (SOFC) in the author's laboratory and tries to offer some thoughts on related problems. The basic materials of SOFC are cathode, electrolyte, and anode materials, which are composed to form the membrane-electrode assembly, which then forms the unit cell for test. The cathode material is most important in the sense that most polarization is within the cathode layer. The electrolyte membrane should be as thin as possible and also posses as high an oxygen-ion conductivity as possible. The anode material should be able to deal with the carbon deposition problem especially when methane is used as the fuel.

2. CATHODE METERIALS

At present, perovskite-type mixed conductors such as $LaSrMO_3$ (M = Co, Fe, Mn) are the most widely used and studied materials for SOFC cathodes. An important advantage of using mixed-conductor electrodes in SOFCs is to spread the triple-phase boundaries, as demonstrated by using either copper oxide (CuO) or $Y_1Ba_2Cu_3O_{7-\delta}$ as the cathode materials in our earlier investigations [1, 2]. It was found that CuO itself posseses enough oxygen vacancies needed for the role of a mixed-conductor [3]. Nevertheless, the properties of these materials may be modified with dopants.

It was found [1] that the activity of oxygen reduction on a CuO electrode is closely related to the electronic conductivity and the oxygen ion vacancy density in the surface layer of the electrode. Doping Li into CuO to create the oxygen-ion vacancies and adding Ag to modify the electrical conductivity provide new ways of enhancing the activity of an oxide electrode for oxygen reduction and thus improving the cathode performance. However, an optimal doping content of Ag exists and is around 50 mol% Ag [3]. It was also found [2] that a layer of strong oxygen-adsorption catalyst such as Pt or Ag coated on the $Y_1Ba_2Cu_3O_{7-\delta}$ electrode is able to largely enhance the activity of oxygen reduction by improving the ability of oxygen to be adsorbed on the electrode surface.

As shown in Fig. 1, the behavior of the I-V curve for the electrode with a CuO under-layer has an abrupt shift (jump) when the overpotential reaches a value around -200 mV. This behavior is attributed to a dramatic change of the path of oxygen-ion transport from that of Fig. 2a to that of Fig. 2b. This is due to part of the oxygen species migrating directly through the CuO under-layer in the form of ions. Therefore, an enormous amount of oxygen vacancies should have been created in CuO at the overpotential around -200 mV to cause the dramatic change of the cathode behavior. This mechanism of the creation of the oxygen vacancies is further supported by the results of doubling the thickness of the CuO under-layer,

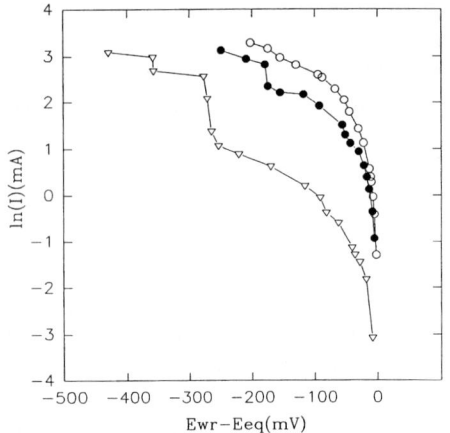

Fig. 1. Current-overpotential curves of 50 mole% Ag-YSZ electrodes with an under-layer of CuO. (○): Ag-YSZ; (●): Ag-YSZ/CuO with CuO = 0.004 g; (▽): Ag-YSZ/CuO with CuO = 0.008 g. Operating condition: 800 °C and P_{O2} = 0.21 atm [3].

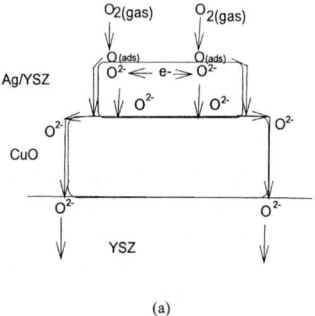

(a)

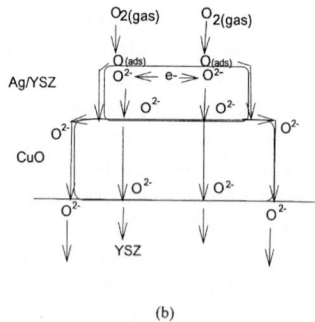

(b)

Fig. 2. Models for oxygen-ion transport mechanism over Ag-YSZ/CuO electrodes: (a) at low overpotential, (b) at high overpotential [3].

as also shown in Fig. 1, where an increase of the thickness of the CuO under-layer cause the dramatic change (shift) of the cathode behavior to occur at a higher overpotential due to the increase of the energy of the lattice-oxygen migration in this thicker CuO layer [3].

For $Y_{1-x}Sr_xMnO_3$ ($0\leq x \leq0.2$), it has been proposed [4] that the conductivity were affected by both Sr doping and oxygen interstitial. For $Y_{0.6}Sr_{0.4}Mn_{1-y}Co_yO_3$ ($0\leq y \leq0.4$), the effect of Co substitution for Mn on the crystal structure, electrical conductivity and thermal expansion properties were investigated [5]. By X-ray powder diffraction, the crystal structure was found to change from hexagonal symmetry of $Y_{0.8}Sr_{0.2}MnO_3$ to orthorhombic of $Y_{0.6}Sr_{0.4}Mn_{1-y}Co_yO_3$. The differences in the structure of the unsubstituted $Y_{1-x}Sr_xMnO_3$ ($0.2\leq x \leq0.4$) are attributed to the average ionic radii of the cations and the amounts of Mn^{4+} present. With Co substitution, the activation energy increases and thus the electrical conductivity decreases. In addition, the relative densities of the materials reached ~94% with sintering at 1350°C for 12 h and had higher concentration of the available lattice sites, thus showing higher conductivity, than that with sintering at 1300 ^{o}C for 6 h, which achieved ~70% relative density. It was also found [5] that the thermal expansion coefficient (TEC) increases as the Sr and Co content of $Y_{1-x}Sr_xMn_{1-y}Co_yO_3$ increases and those with Co content of y = 0.2 exhibit TEC compatibility with YSZ, the commonly-used electrolyte material.

3. ELECTROLYTE MATERIALS

Doped zirconia, especially 8 mole% yttria-stabilized zirconia (YSZ), has been most widely used as the electrolyte materials for SOFC. The typical operating temperature for the state-of-the-art zirconia-based SOFC is 1000°C, although thin film approaches on laboratory scales have been successful in lowering the cell operating temperature below 800 ^{o}C. For the intermediate-temperature SOFCs with the operating temperature below 800 ^{o}C, new materials have been researched extensively, with doped ceria as a promising candidate. A modified sol-gel process has been proposed [6] for preparing samaria-doped ceria powders as the electrolyte materials. It involves treating the gel with high-carbon (long chain, high boiling point) alcohol. It yields near-completely soft-agglomerated nanocrystalline powders which can be easily sintered in air to yield near-fully relative density at 1300 ^{o}C, which is significantly lower than temperatures of 1400-1500°C required by the doped-ceria powder

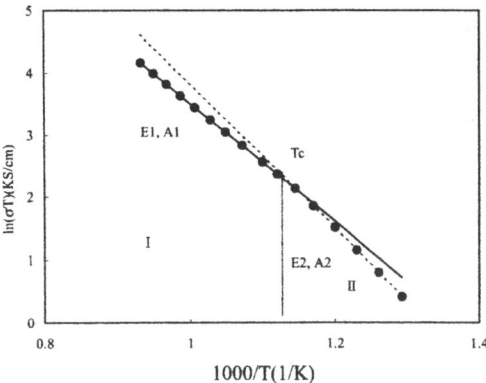

Fig. 3. Illustration of the curvature separated by T_c into two regions I and II [7].

100

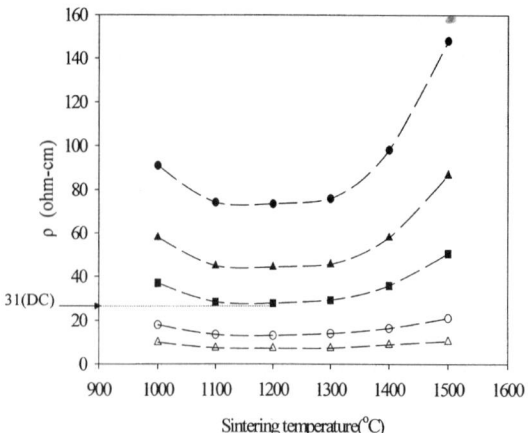

Fig. 4. Overall resistivity vs. sintering temperature at various measurement temperatures: (●) 600 °C, (▲) 650 °C, (■) 700 °C, (○) 800 °C, (△) 900 °C. 31(DC) denotes the value of grain resistivity from the DC measurement at 700 °C [8].

prepared via the previous processes of the sol-gel method.

The conductivity of the electrolyte materials was found to be completely due to oxygen vacancy conduction [7]. For samaria-doped ceria, the conductivity increases with increasing samaria doping and reaches a maximum for $(CeO_2)_{0.8}(SmO_{1.5})_{0.2}$, which has a conductivity of 5.6×10^{-1} S/cm at 800 °C. A curvature at $T = T_c$, the critical temperature, has been observed in the Arrhenius plot, as shown in Fig. 3. This phenomenon may be explained by a model which proposed that, below T_c, nucleation of mobile oxygen vacancies into ordered clusters occurs, and, above T_c, all oxygen vacancies appear to be mobile without interaction with dopant cation. In addition, the composition dependences of both the critical temperature and the trapping energy are consistent with that of the activation energy.

It was also found [8] that the sintering conditions have significant effects on the resistivity of the $Sm_{0.2}Ce_{0.8}O_{1.9}$ material. As shown in Fig. 4, the overall resistivity decreases with lower sintering temperature and attains a minimum at the sintering temperature of 1100-1200 °C, which is about 31 ohm-cm at 700 °C measurement. This makes the $Sm_{0.2}Ce_{0.8}O_{1.9}$ material capable of working as SOFC's electrolyte at temperatures lower than 700 °C to avoid possible reduction of cerium (4+) and thus suitable for intermediate-temperature SOFC.

4. ANODE MATERIALS

In the search of high-performance SOFC anode, doped ceria have been evaluated as possible anode materials [9, 10]. Comparing Ni-samaria-doped ceria (SDC) with Ni-YSZ, the Ni-SDC anode exhibits higher open-circuit voltages and a lower degree of polarization with either methanol as the fuel, as shown in Fig. 5, or methane as the fuel, as shown in Fig. 6. It was found that the depolarization ability of the anode is associated with the catalytic activity, the electrical conductivity, and the oxygen ionic conductivity of the anode materials [9]. It was also found that the anodic polarization and electro-catalytic activity strongly depend on the Ni content in the anode, and the optimum result for the Ni-SDC anode is achieved with 60

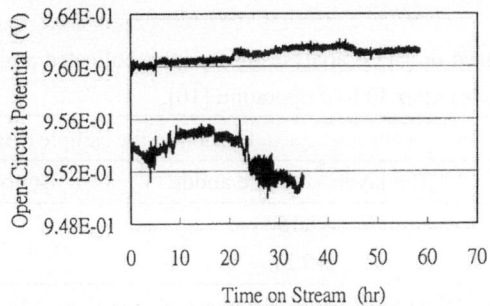

Fig. 5. Change of the open-circuit potential with time for steam reforming of methanol over the 30 wt% Ni-SDC and 30 wt% Ni-YSZ electrode-catalyst. Upper: Ni-SDC; lower: Ni-YSZ. Operating conditions: 800 °C, 1 atm, $H_2O/CH_3OH = 2$, space time = 0.37 s [9].

wt% Ni, as also shown in Fig. 6. The good performance of this Ni-SDC anode appears to be due to its microstructure [10].

5. METHANE CATALYSIS

Methane decomposition is the most important reaction step, especially for high-temperature operations. Thus, carbon deposition occurs commonly and is a major problem, especially with the Ni-based anode. However, carbon deposition may not deactivate the anode [10, 11]. In some cases, the anode activity increases due to carbon deposition which increases the electrical conductivity of the low-Ni-content anode [11].

For direct oxidation of methane, it is shown [12] that the mechanism is:

$$CH_4 + * \rightarrow CH_x* + \tfrac{1}{2}(4-x) H_2 \tag{1}$$

$$CH_x* + O* \rightarrow CO + x/2 H_2 + 2 * \tag{2}$$

where * denotes an active site over Ni/SDC anode, and O* denotes an adsorbed oxygen species over this site, which has been transported from the cathode in the form of oxygen ion.

The CH_x species produced in reaction (1) is the source of carbon deposition. However,

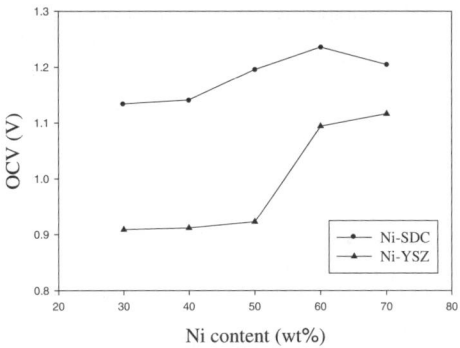

Fig. 6. Variations of the open-circuit voltage (OCV) with nickel content of Ni-SDC and Ni-YSZ anodes. Operating conditions: 600 °C, 1 atm; feed composition CH_4:Ar = 10:90; total flow rate: 100 ml/min [10].

102

Table 1

Carbon formation based on TPO* analyses of samples collected at the top and bottom layers over the Ni-SDC anodes after 30 h of operation [10].

	Carbon in the sample (mg)	
	Top layers over the anode	Bottom layers over the anode
30wt% Ni-SDC	17.15	4.28
60 wt% Ni-SDC	19.21	2.57

* Operating conditions of TPO (temperature-programmed oxidation): 50% O_2 in Ar, 1 atm, 10 $^{\circ}$C/min from 25 to 900 $^{\circ}$C, total flow rate: 30 ml/min; sample weight: 20 mg.

the microstructure of Ni-SDC plays an important role in the performance and stability of the anodic electro-catalyst. The 60 wt% Ni-SDC anode exhibits the best performance and stability, as indicated by the highest OCV as shown in Fig. 6. This is considered to be due to optimization of the anode microstructure at this Ni content, in which optimum distribution and connection between nickel and SDC are obtained. Thus, carbonaceous deposits merely grew on the surfaces of the anodic electro-catalyst, with the least possibility of growing within and, hence, damaging the internal structures of this anode. This is verified by quantifying the amount of coke formed over the Ni-SDC anodes, as shown in Table 1. Note that two layers of coke were formed over the anode: one layer of coke can be removed with supersonic vibration in a water bath and is called the top layer; the other is very tightly adhered to the anode and can be removed only with forced cut-off, and this is called the bottom layer. Also note that additional carbonaceous product was formed but can be blown away and is not counted as coke here. Indeed, it is seen that the bottom layer of coke formed over the 60 wt% Ni-SDC anode was less than that over 30 wt% Ni-SDC. Note also that the bottom layer of coke would deactivate the anode but the top layer may not. As a result, the 60 wt% Ni-SDC anode has improved activity and stability [10, 11].

Over the anode, the hydrogen and CO produced via reactions (1) and (2) are then oxidized at the anode by reacting with the oxygen species transported from the cathode. The catalysis of the fuel such as methane at the anode and oxygen at the cathode becomes increasingly important with demanding catalytic activity as the SOFC operation temperature decreases, which is the aim under intensive research efforts.

REFERENCES

[1] C.L. Chang, T.C. Lee and T.J. Huang, J. Appl. Electrochem., 26 (1996) 311.
[2] C.L. Chang, T.C. Lee and T.J. Huang, J. Solid State Electrochem., 2 (1998) 291.
[3] C.L. Chang, C.C. Hsu and T.J. Huang, J. Solid State Electrochem., 7 (2003) 125.
[4] T.J. Huang and Y.S. Huang, Mater. Sci. & Eng. B, 103 (2003) 207.
[5] C.Y. Huang and T.J. Huang, J. Mater. Sci., 37 (2002) 4581.
[6] G.B. Jung, T.J. Huang, M.H. Huang and C.L. Chang, J. Mater. Sci., 36 (2001) 5839.
[7] G.B. Jung, T.J. Huang and C.L. Chang, J. Solid State Electrochem., 6 (2002) 225.
[8] G.B. Jung and T.J. Huang, J. Mater. Sci., 38 (2003) 2461.
[9] L.K. Tseng and T.J. Huang, J. Chin. Inst. Chem. Eng., 31 (2000) 493.
[10] J. B. Wang, J.C. Jang and T.J. Huang, J. Power Sources, 122 (2003) 122.
[11] J.J. Chen, M.S. Thesis, National Tsing Hua University, Hsinchu, Taiwan, 2003.
[12] T.J. Huang and T.C. Yu, Catal. Lett., in press (2005).

Studies in Surface Science and Catalysis, volume 159
Hyun-Ku Rhee, In-Sik Nam and Jong Moon Park (Editors)
103

Hydrodynamics, Heat and Mass Transfer in Inverse and Circulating Three-Phase Fluidized-Bed Reactors for WasteWater Treatment

Sang Done Kim[a]* **and Yong Kang**[b]

[a]Department of Chemical and Biomolecular Engineering, Korea Advanced Institute of Science and Technology, Daejeon 305-701, Korea(* kimsd@kaist.ac.kr)
[b]School of Chemical Engineering, Chungnam National University, Daejoen 305-764, Korea

1. INTRODUCTION

Recent research development of hydrodynamics and heat and mass transfer in inverse and circulating three-phase fluidized beds for waste water treatment is summarized. The three-phase (gas-liquid-solid) fluidized bed can be utilized for catalytic and photo-catalytic gas-liquid reactions such as chemical, biochemical, biofilm and electrode reactions. For the more effective treatment of wastewater, recently, new processing modes such as the inverse and circulation fluidization have been developed and adopted to circumvent the conventional three-phase fluidized bed reactors [1-6].

In wastewater treatment processes, solid materials including catalyst supporter, bio-media, food particles, adsorbent or absorption media are normally porous, small and light particles. These relatively low density solids are easily floating in the continuous liquid medium due to the buoyance force of solid particles. To fluidize these low density particles, the inverse fluidized bed reactor has been employed in which the liquid phase flows downward against the buoyance force acting on the floating particles and the upward gas flow. This countercurrent gas/liquid flow mode can increase the gas holdup. With increasing gas holdup, the performance of reactors for wastewater treatment could be improved, since the bubbles can provide the microorganism with oxygen and the liquid with gas reactants [5-9].

In spite of attractive advantages of conventional three-phase fluidized bed reactors, the range of U_L has to be extremely low and narrow, when the small or porous particles are fluidized by viscous liquid medium, which is often encountered in the processes of wastewater treatment. To overcome this limitation and increase the treatment efficiency, the three-phase circulating fluidized-bed reactor has been devised in order to recycle solid particles from the reactor through a downcomer to the reactor. The circulation mode can minimize the dead zone by increasing the contacting efficiency among three (gas-liquid-solid) phases with sufficiently high U_L. In addition, the deactivated solid media can be regenerated continuously by means of the circulation mode of particles [2-4, 10,11].

To provide the prerequisite knowledge for designing the three-phase fluidized-bed reactors with new modes, the hydrodynamics such as phase holdup, mixing and bubble properties and heat and mass transfer characteristics in the reactors have to be determined. Thus, in this study, the hydrodynamics and heat and mass transfer characteristics in the inverse and circulating three-phase fluidized-bed reactors for wastewater treatment in the present and previous studies have been summarized. Correlations for the hydrodynamics as well as mass and heat transfer coefficients are proposed. The areas wherein future research should be undertaken to improve the state of the present knowledge have been defined with recommendations.

2. THREE-PHASE INVERSE FLUIDIZED-BED REACTORS

2.1. Hydrodynamics

The inverse fluidization of low density particles in wastewater treatment reactors can be divided into two modes ; the one is gas-liquid countercurrent flow mode comprising a continuous flow system and the other is gas only flow mode in a batch system, since the floated low density particles can be easily fluidized by the upward gas flow. For the wastewater treatment in bioreactors, the biomass is usually growing on the surface of the fluidized particles to form a biofilm. In a batch system, the collision frequency and pressure on the particles increase with increasing gas velocity (U_G) but show their maxima with increasing particle holdup. The effects of U_G are similar to those of other studies [9, 12, 13], but the effects of solid holdup are not consistent with the results of other investigations in which the particle fluctuation frequency increases with increasing solid holdup up to 0.6. It is interesting to note that the increase trend of fluctuation frequency and dispersion coefficient of particles with increasing U_G in a continuous inverse beds [14] is very similar to that of particle collision frequency and pressure in the batch system. However, the unified correlations to predict the frequency of collision or fluctuations of fluidized particles have not been proposed up to now due to the limited research works.

It has been reported that the gas phase holdup or bubble size(L_V) increases with increasing U_G, U_L or liquid viscosity (μ_L); its rising velocity(U_B) increases with increasing U_G or μ_L, but decreases with increasing U_L; the frequency(F_B) of rising bubbles increases with increasing U_G or U_L but decreases with increasing μ_L. The gas (or bubble) holdup in the beds of relatively heavier particles is higher than that in the beds of relatively lighter particles, since the bubble size in case of the former is smaller than that of the latter. The bubble properties and liquid axial dispersion coefficient (D_z) have been correlated with the experimental variables as [8, 15-17]:

$$L_V = 1.18 U_G^{0.235} U_L^{0.335} \mu_L^{0.391} \left(\frac{\rho_S}{\rho_L} \right)^{3.61} \quad (1) ; \qquad U_B = 0.420 U_G^{0.114} U_L^{-0.173} \mu_L^{0.132} \left(\frac{\rho_S}{\rho_L} \right)^{2.13} \quad (2)$$

$$F_B = 12.4 U_G^{0.052} U_L^{0.356} \mu_L^{-0.056} \left(\frac{\rho_S}{\rho_L} \right)^{-5.51} \quad (3) ; \qquad D_z = 0.157 U_G^{0.263} U_L^{0.225} \mu_L^{-0.018} \left(\frac{\rho_S}{\rho_L} \right)^{2.26} \quad (4)$$

with correlation coefficients of 0.92, 0.95, 0.94 and 0.92 for Eqs. (1)-(4), respectively.

2.2 Heat and Mass Transfer

Very limited data on the heat and mass transfer in three-phase inverse fluidization systems is available up to now. For the wastewater treatment, the reactor temperature should be controlled and maintained within a certain level to optimize the reactor performance, since the temperature of reactor or process can provide the microorganisms with favorable circumances.

The heat transfer coefficient (h) has been determined by measuring the temperature difference between the immersed heater and the bed. The h value increases with increasing U_G (Fig. 1(a)), but exhibits a maximum value with increasing U_L (Fig. 2(a)). The effects of U_G on h is dominant, since the bubbling phenomena become more vigorous due to the increase of gas holdup and turbulence intensity with increasing U_G. The reason why the h value exhibits a maximum with U_L can be due to the considerable decrease in the solid holdup (ε_s) in the higher range of U_L. As in the conventional beds, in the lower range of U_L, the h

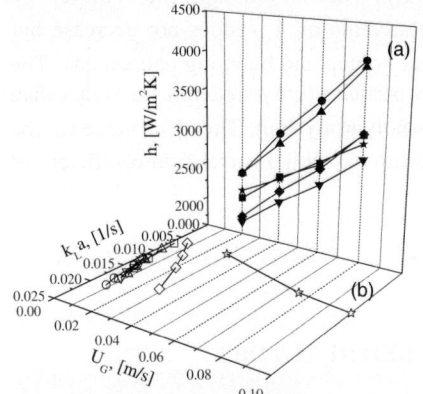

d_p (mm)	ρ_s (kg/m³)	U_L (m/s)	Liquid	Solid	Author
■ 4	966.6	0.01	Water	PE	
▲ 4	966.6	0.03	Water	PE	
● 4	966.6	0.05	Water	PE	Cho
▼ 4	877.3	0.01	Water	PP	et al. [6]
◆ 4	877.3	0.03	Water	PP	
★ 4	877.3	0.05	Water	PP	
□ 4	966.6	0.01	Water	PE	
○ 4	966.6	0.03	Water	PE	Kim
△ 4	877.3	0.02	Water	PP	et al. [18]
▽ 4	877.3	0.03	Water	PP	
◇ 3.5	650	0.0283	Water	PS	Nikolov et al. [5]
☆ 3.5		0.0289			

Fig. 1. Effect of U_G on h (a) and k_La (b) in three-phase inverse fluidized beds.

value increases with increasing U_L due to the increases of turbulence intensity of fluid element as well as mobility of fluidized particles, however, the turbulence and particle mobility would decrease with a further increase in U_L due to the considerable decrease of ε_S, compensating for the increase of gas and liquid holdups. Relatively heavier particles would be more effective for the heat transfer due to their potential for bubble breaking and consequent increase in contacting frequency with heater surface. The h value can be predicted from Eq. (5), with a correlation coefficient of 0.95[6].

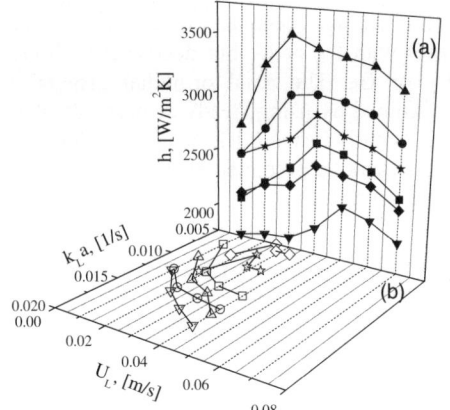

d_p (mm)	ρ_s (kg/m³)	U_G (m/s)	Liquid	Solid	Author
■ 4	966.6	0.002	Water	PE	
● 4	966.6	0.004	Water	PE	
▲ 4	966.6	0.006	Water	PE	Cho
▼ 4	877.3	0.002	Water	PP	et al. [6]
◆ 4	877.3	0.004	Water	PP	
★ 4	877.3	0.006	Water	PP	
□ 4	966.6	0.002	Water	PE	
○ 4	966.6	0.004	Water	PE	Kim
△ 4	877.3	0.004	Water	PP	et al. [18]
▽ 4	877.3	0.006	Water	PP	
◇ 3.5	650	0.007	Water	PS	Nikolov et al. [5]
☆ 3.5	650	0.014	Water	PS	

Fig. 2. Effect of U_L on h (a) and k_La (b) in three-phase inverse fluidized bed.

$$Nu = \left(\frac{hd_p(1-\varepsilon_s)}{k_L\varepsilon_s}\right) = 0.050\left(\frac{C_p\mu_L}{k_L}\right)\left[\frac{d_p\rho_L(U_G+U_L)}{\mu_L\varepsilon_s}\right]^{0.810} \tag{5}$$

The volumetric gas-liquid mass transfer coefficient (k_La) has been obtained by fitting the concentration profile of dissolved oxygen to the axial dispersion model [8, 18]. The value of

$k_L a$ increases with increasing U_G(Fig. 1(b)). With increasing U_L, the $k_L a$ value increases initially but approaches to an asymptotic value with a further increase in U_L(Fig. 2(b)). This trend of $k_L a$ is very similar to that of gas holdup; the gas holdup increases initially and approaches to an asymptotic value with increasing U_L. That is, although the value of ε_S decreases considerably in the higher range of U_L, the value of $k_L a$ does not decrease but maintains an asymptotic value due to the effects of gas holdup and bubbling phenomena. The values of $k_L a$ in the beds of relatively higher density particles (polyethylene) are higher than those in the beds of relatively lower density particles(polypropylene). The $k_L a$ value can be correlated based on the isotropic turbulence theory as Eq. (6) with a correlation coefficient of 0.95.

$$St = \left(\frac{k_L aL}{U_L}\right) = 0.719\left(\frac{U_L}{U_G + U_L}\right)^{-3.10}\left(\frac{\rho_s}{\rho_L}\right)^{1.74} \tag{6}$$

3. THREE-PHASE CIRCULATING FLUIDIZED-BED REACTORS

3.1. Hydrodynamics

Comparing with the conventional three-phase beds, the axial solid holdup distribution is much more uniform and the radial distribution of gas holdup (ε_G) is much flatter in circulating beds, due to the relatively high U_L and solid circulation. The values of ε_G and bed porosity can be predicted by Eqs. (7) and (8) with a correlation coefficient of 0.94 and 0.95, respectively.

$$\varepsilon_G = 0.07U_G^{0.492}U_L^{-0.023}G_S^{-0.047} \tag{7}; \quad \varepsilon_G + \varepsilon_L = 1.23U_G^{-0.003}U_L^{0.279}G_S^{-0.023} \tag{8}$$

Bubble size in the circulating beds increases with U_G, but decreases with U_L or solid circulation rate (G_S); bubble rising velocity increases with U_G or U_L but decreases with G_S; the frequency of bubbles increases with U_G, U_L or G_S. The axial or radial dispersion coefficient of liquid phase (D_Z or D_r) has been determined by using steady or unsteady state dispersion model. The values of D_Z and D_r increase with increasing U_G or G_S, but decrease (slightly) with increasing U_L. The values of D_Z and D_r can be predicted by Eqs.(9) and (10) with a correlation coefficient of 0.93 and 0.95, respectively[10].

$$Pe_{z,cir}\left(=\frac{d_p U_L}{D_{z,cir}}\right) = 5.38\left(\frac{d_p}{D}\right)^{0.77}\left(\frac{U_L(\varepsilon_G + \varepsilon_L)}{U_G + U_L}\right)^{2.54} \tag{9}; \quad Pe_{r,cir}\left(=\frac{d_p G_S}{D_{r,cir}\rho_S}\right) = 0.086\left(\frac{d_p}{D_c}\right)^{0.58}\left(\frac{U_L(\varepsilon_G + \varepsilon_L)}{U_G + U_L}\right)^{0.69} \tag{10}$$

3.2. Heat and Mass Transfer

The available data on the heat and mass transfer coefficients in three-phase circulating fluidized-bed reactors are comparatively sparse. The heat transfer coefficient(h,cir) has been measured in the immersed heater-to-bed system. The value of h_{cir} increases with increasing U_G or G_S but exhibits a local maximum with increasing U_L(Figs. 3(a)-5(a)). The mass transfer coefficient ($k_L a,cir$) has been recovered by employing the similar dispersion model as in the case of inverse or conventional fluidized beds[18]. The $k_L a,cir$ value increases with increasing U_G, G_S or d_p, but does not change considerably with increasing U_L. The effects of G_S on $k_L a,cir$ are not consistent; Yang et al.[2] reported that the value of $k_L a,cir$ showed a maximum with increasing G_S, but Kim et al.[18] pointed out that the value increased gradually with G_S.

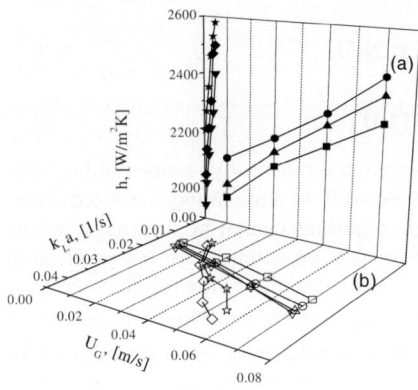

d_p (mm)	ρ_s (kg/m³)	G_s (kg/m²s)	U_L (m/s)	Liquid	Solid	Author
■ 2.1	2500	2	0.27	Water	Glass bead	
▲ 2.1	2500	4	0.27	Water	Glass bead	Cho et al. [4]
● 2.1	2500	6	0.27	Water	Glass bead	
▼ 0.401	1130	0.05(ε_s)	0.02	Waste water	Polymer resin	
◆ 0.401	1130	0.1(ε_s)	0.02	Waste water	Polymer resin	Kang et al. [21]
★ 0.401	1130	0.15(ε_s)	0.02	Waste water	Polymer resin	
□ 1	2500	2	0.17	Water	Glass bead	
○ 1.7	2500	2	0.22	Water	Glass bead	Kim et al. [22]
△ 2.1	2500	2	0.26	Water	Glass bead	
▽ 3	2500	2	0.33	Water	Glass bead	
◇ 0.4	2460		0.0812	Water	Glass bead	Yang et al. [2]
☆ 0.4	2460		0.0632	Water	Glass bead	

Fig. 3. Effects of U_G on h (a) and k_La (b) in three-phase circulating fluidized bed.

d_p (mm)	ρ_s (kg/m³)	G_s (kg/m²s)	U_G (m/s)	Liquid	Solid	Author
■ 2.1	2500	2	0.01	Water	Glass bead	
● 2.1	2500	2	0.03	Water	Glass bead	Cho et al. [4]
▲ 2.1	2500	2	0.05	Water	Glass bead	
▼ 0.401	1130	0.05(ε_s)	0.003	Waste water	Polymer resin	
◆ 0.401	1130	0.1(ε_s)	0.003	Waste water	Polymer resin	Kang et al. [21]
★ 0.401	1130	0.15(ε_s)	0.003	Waste water	Polymer resin	
□ 1	2500	2	0.05	Water	Glass bead	
○ 1.7	2500	2	0.05	Water	Glass bead	Kim et al. [22]
△ 2.1	2500	2	0.01	Water	Glass bead	
▽ 3	2500	2	0.01	Water	Glass bead	
◇ 0.4	2460		0.0361	Water	Glass bead	Yang et al. [2]
☆ 0.4	2460		0.0542	Water	Glass bead	

Fig. 4. Effect of U_L on h (a) and k_La (b) in three-phase circulating fluidized bed

d_p (mm)	ρ_s (kg/m³)	U_L (m/s)	U_G (m/s)	Liquid	Solid	Author
■ 2.1	2500	0.25	0.03	Water	Glass bead	
● 2.1	2500	0.27	0.03	Water	Glass bead	Cho et al. [4]
▲ 2.1	2500	0.29	0.03	Water	Glass bead	
▼ 2.1	2500	0.31	0.03	Water	Glass bead	
□ 1	2500	0.21	0.03	Water	Glass bead	
○ 1.7	2500	0.24	0.03	Water	Glass bead	Kim et al. [22]
△ 2.1	2500	0.28	0.07	Water	Glass bead	
▽ 3	2500	0.40	0.07	Water	Glass bead	
◇ 0.4	2460	0.0722	0.0542	Water	Glass bead	Yang et al. [2]
☆ 0.4	2460	0.0722	0.0361	Water	Glass bead	

Fig. 5. Effects of G_S on h (a) and k_La (b) in three-phase circulating fluidized bed.

Although the values of k_La_{cir} in the literature are reasonable and comparable each other, the different trend mentioned above may be due to the different operating conditions. The gas-liquid interfacial area(a) and liquid side mass transfer coefficient(k_l) have been determined from the knowledge of measured values of gas holdup and k_La_{cir} [11]. The values of a and k_l increase almost linearly with increasing U_G or U_L. The values of h_{cir} and k_La_{cir} in circulating beds can be predicted by Eqs.(11) and (12) with a correlation coefficient of 0.92 and 0.93, respectively[19].

$$h_{cir} = 2776U_G^{0.080}U_L^{0.032}G_S^{0.065} \quad (11) \; ; \qquad k_L a_{,cir} = 0.263U_G^{0.275}U_L^{0.017}G_S^{0.155}d_P^{0.097} \quad (12)$$

4. RECOMMENDATIONS FOR THE FUTURE STUDY

Although some correlations have been proposed to predict the hydrodynamics and heat and mass transfer characteristics in three-phase inverse as well as circulating fluidized beds, unified correlations covering wide range of operating conditions cannot be drawn due to the extremely limited data reported in the literature. With increasing potential for the application of these reactors in the fields of wastewater treatment processes, the following features have to be examined in the future study. Effects of solid (d_p, shape, ρ_S etc.) and liquid properties (μ_L, σ_L, ρ_L, etc.) on the hydrodynamics and heat and mass transfer have to be determined for designing the reactors. Unified correlations to predict the hydrodynamics and heat and mass transfer characteristics are essentially needed in the future study. For modeling of these reactors, the model study with reasonable and practical assumptions is also required. Since the reactors are composed of multiphase-dynamic system, a stochastic approach in addition to the deterministic one has to be considered to analyze, diagnose the fault and conduct on-line control of the system. In addition, various kinds of modifications of three-phase inverse as well as circulating fluidized beds have to be investigated for the development of more effective and economic wastewater treatment reactors.

REFERENCES

1. L. S. Fan, Gas-Liquid-Solid Fluidization Engineering, Butterworth, Boston, MA(1989) 368.
2. W. G. Yang, J. F. Wang, W. Chem and Y. Jin, Chem. Eng. Sci., 54(1999)5293.
3. J. Zhu, Y. Zheng, D. G. Karamanev and A. S. Bassi; Can. J. Chem. Eng., 78(2000)82.
4. Y. J. Cho, P. S. Song, S. H. Kim, Y. Kang and S. D. Kim, J. Chem. Eng. Japan, 34(2001)254.
5. V. Nikolov, I. Farag, I. Nikov, Bioprocess Bioeng., 23(2000)427.
6. Y. J. Cho, H. Y. Park, S. W. Kim, Y. Kang and S. D. Kim, Ind. Eng. Chem. Res., 41(2002)2058.
7. P. Buffire and R. Moletta, Chem. Eng. Sci., 54(1999)1233.
8. S. W. Kim, H. T. Kim, P. S. Song, Y. Kang and S. D. Kim, Can. J. Chem. Eng., 81(2003)621.
9. P. Buffiere and R. Moletta, Chem. Eng. Sci., 55(2000)5555.
10. Y. Kang, Y. J. Cho, C. G. Lee, P. S. Song and S. D. Kim, Can. J. Chem. Eng., 81(2003)1130.
11. S. Han, J. Zhou, Y. Jin, K. C. Loh and Z. Wang, Chem. Eng. J., 70(1998)9.
12. M. P. Comte, D. Bastoul, G. Hebrard, M. Roustan and and V. Lazarova, Chem. Eng. Sci., 52(1997)3971.
13. C. L. Brieno, Y. A. A. Ibrahim, A. Margaritis and M. A. Bergougnou, Chem. Eng. Sci., 54(1999)4975.
14. S. M. Son, S. H. Kang, Y. Kang and S. D. Kim, J. KIChE, 42(2004)332.
15. S. M. Son, S. H. Jung, S. W. Kim, Y. Kang and S. D. Kim, Prof. 9th Asian Conf. on Fluidized-Bed and Three-Phase Reactors(2004)265.
16. H. T. Kim, P. S. Song, S. W. Kim and Y. Kang, J. KIEC, 13(2002) 691.
17. S. M. Son, P. S. Song, C. G. Lee, S. H. Kang, Y. Kang and K. Kusakabe, J. Chem. Eng. Japan, 37(2004)990.
18. S. W. Kim, P. S. Song, H. T. Kim, Y. Kang and S. D. Kim, J. KIChE, 40(2002)482.
19. Y. J. Cho, S. J. Kim, S. H. Nam, Y. Kang and S. D. Kim, Chem. Eng. Sci, 56(2001)6107.
20. H. T. Kim, P. S. Song, S. W. Kim and Y. Kang, J. KIEC, 13(2002) 691.
21. T. Kang, P. Song, G. Choi, Y. Cho, Y. Kang, H. Choi and S. D. Kim, J. KIChE, 40(2002)640.
22. S. J. Kim, J. S. Kim, C. K. Lee, Y. Kang and S. D. Kim, Proc. 14th symp. on Chem. Eng. (2001)187.

Studies in Surface Science and Catalysis, volume 159
Hyun-Ku Rhee, In-Sik Nam and Jong Moon Park (Editors)

New Developments in Polymer Reaction Engineering

Kyu Yong Choi

Department of Chemical and Biomolecular Engineering, University of Maryland, College Park, MD 20742, U.S.A.

1. OVERVIEW

In recent years, the polymer industry has become very competitive: product quality specifications have tightened, profit margins have declined, and a timely introduction of new products to existing and new markets has become critical for staying in business. Due to the modern changes in life-styles and rapid advances in nano-, bio-, and information technology, the polymer industry is making key adjustments in R&D and business strategies to develop high value-added and/or target-oriented specialty polymers for emerging markets. However, identifying a promising future market for novel materials developed in research laboratories is neither easy nor straightforward.

Recent developments in new polymerization catalysts and chemistry, such as single site olefin polymerization catalysts, ring-opening metathesis and ring-opening polymerization techniques, and living free radical polymerization methods, offer new and exciting opportunities for molecular-level designing of polymer architecture (e.g., dendritic polymers, star polymers, block copolymers, gradient copolymers, single-chain organic nanoparticles, etc.). Polymers with precisely controlled structures and functionalities can be used as the active component, or they can be used in the manufacture of advanced microelectronics and magnetic storage media. The advances in chemistry have allowed polymer R&D to become increasingly interdisciplinary; polymer scientists and engineers collaborate in projects with biologists, physicists, chemists, medical doctors and scientists, biochemical engineers, electronics engineers, mechanical engineers, and materials scientists to develop novel materials and the means for fabrication and manufacturing. Excellent communication skills, learning proficiencies, and intellectual abilities are essential for a successful interdisciplinary collaboration.

The polymer industry is also globalizing: manufacturing bases, as well as R&D functions, are frequently relocated to other countries that offer a more profitable business environment; technology outsourcing is becoming a favored form of low-cost technology acquisition; mergers and technical alliances between the companies have become so frequent that sometimes it is hard to identify who is who in the industry. Strict domestic and international regulations on environmental emissions also require new process technology, new raw materials, or even new polymerization chemistry. High-speed internet and information technology of global computing and data exchange network is also changing the culture of modern scientific and engineering research.

In such a dynamic technology and business environment, it is crucial for the polymer industry to secure technical capability and superiority in producing new advanced materials that are needed in highly sophisticated and specialized applications such as electronics, optics, coatings, energy, and cosmetics in the most cost effective and timely manner. In this context, current industrial polymer research seems to be highly directed towards product development rather than process research. It is interesting to note that the university is emerging as a powerful source of new technology. In fact, many universities operate technology transfer offices, and they are becoming more aggressive in licensing inventions and technologies resulting from discoveries by university researchers. Sometimes, a conflict of interest in the ownership of intellectual properties (IP) becomes a huddle in university-industry collaborative research.

Significant changes are also occurring in the ways polymeric materials are manufactured: when manufacturing bulk commodity polymers that are typically of low cost, process competitiveness is pursued through debottlenecking or process intensification. Advanced model-based polymer reactor controls are designed and implemented to reduce reaction time and to minimize off-specification products; polymerization conditions are then modified and optimized to develop novel product grade slates or to simplify the product portfolio to reduce manufacturing and inventory cost.

Sometimes, new values are added not only to the polymer itself, but also to the shape or physical state of the processed polymers to maximize the profit opportunity. For example, when a company develops a novel polymeric material and its manufacturing technology, the company may prefer to make their novel polymers available to customers in the form of intermediate consumer products, such as high performance films or fibers, rather than manufacturing and selling bulk resins to industrial customers. To do so, the company should have a line of technical capabilities from polymer synthesis to consumer product manufacturing.

New discovery or success in a laboratory or bench scale synthesis of novel materials may not always be commercialized at a larger scale. This occurs because the physical changes, material transport effects, non-ideal mixing and control of local concentration gradients that were previously insignificant issues in the laboratory setting may have become the major issues in commercialization [1]. Many of the polymer's structural parameters are difficult to measure on-line or in-line, which makes direct feedback control of polymer properties very difficult or infeasible. Thus, 'recipe'-driven reactor control techniques are frequently used and inherently such techniques result in product quality inconsistency and low productivity. Polymer reaction engineering has dealt with these process-related problems in the past two decades. It is expected that these principles and methodologies are strongly needed for solving from molecular-scale to reactor-scale problems when manufacturing sophisticated and precisely controlled polymer materials. In what follows, a few examples are discussed to illustrate some recent developments in polymer reaction engineering at the author's own discretion.

2. POLYMERIZATION PROCESS INTENSIFICATION

In conventional chemical manufacturing processes, profitability increases as reactor capacity increases (e.g., polyolefins, poly(vinyl chloride)). However, there are some new and important developments that are growing in other directions. Process intensification concept is an example. The process intensification refers to technologies that replace large, expensive, and energy-intensive process equipment with smaller, less costly, and high-performance process equipment in a simplified and more flexible process configuration [2,3]. Originally, the concept was limited only to the equipment and engineering methods. However, the development of new reaction chemistry and their applications to process design are changing the concept of process intensification because these novel developments in 'soft' side technology may require different types of process hardware.

In the polymer industry, post-reaction product treatment processes such as liquid-solid separation, drying, precipitation, particle size control, and polymer purification are very complex and costly. Future polymer plants should be designed such that process equipment can be easily and quickly converted to making new products at minimal cost and with minimal or zero off-specification products. Radically different new reactor design concepts and more efficient and intense high shear mixing devices need to be integrated with reaction process for overall polymerization process intensification. For example, high solid polymerization or solvent-free polymerization processes in spray towers or in kneaders offer the possibility to reduce the reactor size [2]. Alarcia et al [4] show that employing a continuous emulsion polymerization reactor with model-based design techniques can intensify batch and semi-batch operations for the production of acrylic waterborne adhesives. The need for process intensification in the polymer industry can be quite strong. The lifetime of many polymeric consumer products and the materials used in them are expected to decrease; the lifetime of a new polymer plant will also be shorter than old plants. Therefore, new plant design and operational concepts need to be developed to most effectively adapt a polymer manufacturing process technology to a rapidly changing market environment.

3. MODELING, DESIGN AND CONTROL OF POLYMER MICROSTRUCTURE

Developing an ability to precisely control the polymer microstructure is one of the most important objectives in the manufacture of industrial polymers because a polymer's physical and thermodynamic properties are strongly dependent on the polymer microstructure represented by chain length distribution, copolymer composition and monomer sequence length distributions, long-chain and short-chain branching, stereoregularity, crosslinking, etc. The reaction chemistry, catalysts, and monomers determine many of these important molecular architectural properties that are unalterable once the polymerization is complete. Therefore, it is crucial to understand how to control the polymer microstructure during polymerization. Obviously, the first step in designing the structure control technique is to develop an understanding of reaction chemistry of polymer chain formation. In heterogeneous polymerization processes, physical transport processes (mass and heat transfer limitations) often mask the polymerization chemistry and kinetics.

3.1 Controlling polymer microstructure through novel catalysis and chemistry

In the past three decades, industrial polymerization research and development aimed at controlling average polymer properties such as molecular weight averages, melt flow index and copolymer composition. These properties were modeled using either first principle models or empirical models represented by differential equations or statistical model equations. However, recent advances in polymerization chemistry, polymerization catalysis, polymer characterization techniques, and computational tools are making the molecular level design and control of polymer microstructure a reality.

Perhaps the most notable and exciting events of polymerization science in the past decade are the development of single site catalysts for olefin polymerization and the living/controlled free radical polymerization techniques. The single site catalyst technology permits the tailoring of polymer chain length distribution. The living free radical polymerization techniques offer the convenience and versatility of free radical polymerization with living polymerization capabilities that were believed to be feasible only through ionic reactions that have stringent monomer/solvent purity requirements. These new living polymerization techniques include atom transfer radical polymerization (ATRP), reversible addition-fragmentation transfer (RAFT), and stable free radical polymerization (SFRP) [5-13]. Although there are some technical problems to be resolved in each of these methods (e.g., colour and odor, terminal site unsaturation, slow reaction rate, control of stereochemistry), there is no doubt that the potential applications of these techniques to bio and nanotechnology are enormous. These techniques allow researchers to synthesize a variety of polymeric materials with precise control over shape, size, and functional group placement that were previously impossible using traditional polymerization chemistry. For example, customized polymer microstructures represented by block copolymers, gradient polymers, dendritic polymers, and hyperbranched polymers can be obtained by living free radical polymerization techniques. One of the major challenges with these techniques includes simplified, yet effective, synthetic routes for mass production, which are critical to fully exploit the physical properties of the novel materials that are currently obtainable only in very small quantities.

3.2 Multi-scale modeling problems in heterogeneously catalyzed α-olefin polymerization

α-olefin polymers represent perhaps the least expensive, yet most versatile thermoplastic polymers. Modern high performance catalysts (metallocenes and Zieger-Natta type catalysts) offer unprecedented capability of tailoring polymer microstructure that impacts the polymer's ultimate end-use properties. Using a class of single site metallocene catalysts, one can control the polymer microstructure represented by long-chain branching (LCB), short-chain branching (SCB), monomer sequence length distribution, stereoregularity, copolymer composition distribution, chain length distribution, etc. In propylene polymerization, new chain structures (e.g., isotactic stereoblock, hemiisotactic polypropylenes, gradient copolymers, etc.) that did not exist in the past can now be designed. The metallocene technology has been extended to the synthesis of vinyl polymers such as syndiotactic polystyrene.

Olefin polymerization processes are also optimally adjusted or customized to fully take advantage of high performance catalysts in a commercial process. A catalyst that works perfectly well in a laboratory reactor system often causes unexpected operational problems in a large-scale continuous polymerization reactor (e.g., reactor fouling, electrostatic effect, particle overheating, fines and agglomeration). The exact causes of such discrepancies are not always understood and hence costly scale-up tests are necessary before the catalyst is accepted for commercial scale reactors. To address these problems, a profound knowledge and understanding of reaction and process characteristics is required.

Advanced computational models are also developed to understand the formation of polymer microstructure and polymer morphology. Nonuniform compositional distribution in olefin copolymers can affect the chain solubility of highly crystalline polymers. When such compositional nonuniformity is present, hydrodynamic volume distribution measured by size exclusion chromatography does not match the exact copolymer molecular weight distribution. Therefore, it is necessary to calculate the hydrodynamic volume distribution from a copolymer kinetic model and to relate it to the copolymer molecular weight distribution. The finite molecular weight moment techniques that were developed for free radical homo- and co-polymerization processes can be used for such calculations [1, 14,15].

A certain class of long-chain branched polyethylenes synthesized over metallocene catalysts exhibits significantly different rheological behaviors from those without branching. For a linear binary olefin copolymerization process, Soares and coworkers [16-19] developed a model to calculate the instantaneous chain length and copolymer composition using the modified Stockmayer bivariate distribution function. With the advancement in the temperature rising elution fractionation (TREF) technique, long-chain and short-chain branched polymers can be characterized quantitatively.

Recently, Teymour and coworkers developed an interesting computational technique called the digital encoding for copolymerization compositional modeling [20,21]. Their method uses symbolic binary arithmetic to represent the architecture of a copolymer chain. Here, each binary number describes the exact monomer sequence on a specific polymer chain, and its decimal equivalent is a unique identifier for this chain. Teymour et al. claim that the proposed modeling technique enables the description of polymeric chains not only at the chain length and composition levels, but down to the distinction of the specific sequencing on these chains.

Modeling of polymer morphology is an important issue in olefin polymerization. The intrinsic polymerization kinetics of heterogeneous olefin polymerization is often masked by intraparticle and/or interfacial mass and heat transfer limitations. In the past two decades, a large number of papers have been published on the modeling of polymer particle growth and morphology, mostly based on the multigrain model proposed by Ray and coworkers [22-25]. Understanding the mechanism of catalyst/polymer particle fragmentation and the subsequent development of particle morphology is important in optimizing catalyst design and polymer morphology. Recently, Kosek and coworkers [26] developed a novel computational technique using a multi-scale approach to couple the kinetics of polymer particle growth and mophogenesis. Through microscopic imaging, the structure of a catalyst particle is characterized with a phase function and a three-dimensional pore structure is reconstructed.

3.3 Monomer sequence length distribution and penultimate effect in ethylene-cycloolefin copolymers synthesized over homogeneous metallocene catalysts

Cyclic olefin copolymer is a thermoplastic polymer of ethylene and cyclic olefin that can be synthesized over homogeneous metallocene catalyst by an addition polymerization mechanism without ring opening. The copolymer's glass transition temperature (Tg) can be varied from about 20 to 260°C by varying the cyclic olefin comonomer content. The polymer microstructure modeling work by Park and Choi [27-28] illustrates that a well-known terminal model fails to describe the copolymerization kinetics when a bulky comonomer, such as norbornene, is incorporated into a polymer chain. They show that a penultimate model that accounts for the strong steric effect exerted by a bulky comonomer at an active transition metal site yields accurate model predictions. The monomer sequence length distribution analysis also indicates that the ethylene block length becomes smaller than 60 monomer units as the norbornene content in the copolymer exceeds 12-14 mol-%, making the copolymer completely amorphous and transparent. This computational result is in agreement with experimentally measured crystallinit limit by Kaminsky and Knoll [29].

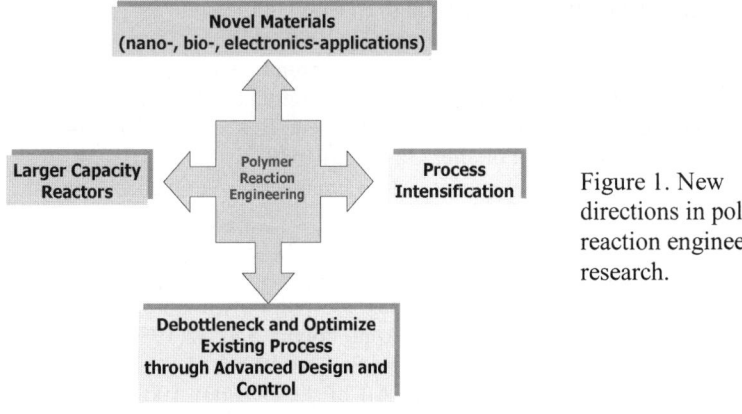

Figure 1. New directions in polymer reaction engineering research.

114

4. CONCLUDING REMARKS

Novel laboratory discoveries and findings can bear fruits only when they are materialized through innovative engineering. Advances in nano-, bio- and information technology, and powerful modeling and computational tools offer exciting new opportunities for polymer reaction engineers. Figure 1 illustrates the four major directions in polymer reaction engineering research. For polymer reaction engineers, participation in interdisciplinary collaborative projects with scientists in other disciplines will become crucial to adopt new science and principles, and also to expand new research potentials. It is also expected that advanced computing techniques and new polymer chemistry will impact the intensification of polymerization process operations through innovative equipment design. Finally, it should be noted that one thing that never changes is the importance of fundamental understanding of reaction phenomena from the molecular scale to the reactor scale.

REFERENCES

[1] W.J. Yoon, Y.S. Kim, I.S. Kim, and K.Y. Choi, Korean J. Chem. Eng., 21(2004), 147.
[2] K.D. Hungenberg, DECHEMA Monographs, 5 (2004), 9.
[3] C. Tsouris and J.V. Porcelli, Chem. Eng. Prog., Oct. (2003), 50.
[4] F. Alarcia, J.C. de la Cal, and J.M. Asua, DECHEMA Monographs, 138 (2004), 79.
[5] M.K. Georges, R.P.N. Veregin, P.M. Kazmaier, and G.K. Hamer, Macromolecules, 26(1993), 2987.
[6] K. Matyjaszewski, T.E. Patten, and J. Xia, J. Am. Chem. Soc., 119(1997), 674.
[7] M. Hölderle, M. Baumert, and R. Mülhaupt, Macromolecules, 30(1997), 3420.
[8] S.A.F. Bon, M. Bosveld, B. Klumperman, and A.L. German, Macromolecules, 30(1997), 324.
[9] H. de Brouser, J.G. Tsavalas, F.J. Schork, and M.J. Monteiro, Macromolecules, 33(2000), 9239.
[10] A. Butté, G. Storti, and M. Morbidelli, Chem. Eng. Sci., 54(2000), 3485.
[11] M.F. Cunningham, Prog. Polym. Sci., 27(2002), 1039.
[12] M. Zhang and W.H. Ray, J. Appl. Polym. Sci., 86(2002), 1630.
[13] K. Matyjaszewski, DECHEMA Monographs, 138(2004), 107.
[14] T. Crowley and K.Y. Choi, Polym. React. Eng., 7 (1999), 43.
[15] T. Crowley and K.Y. Choi, Comp. Chem. Eng., 23 (1999), 1153.
[16] J.B.P. Soares and A.E. Hamielec, Macromol. Theory Simul., 5(1996) 547.
[17] J.B.P. Soares and A.E. Hamielec, Macromol. Theory Simul., 6(1997) 591.
[18] J.B.P. Soares, J.D. Kim, and G. Rempel, Ind. Eng. Chem. Res., 36 (1997) 1144.
[19] J.B.P. Soares and A.E. Hamielec, Polymer, 36 (1995) 1639.
[20] F. Teymour, Paper presented at Polymer Reaction Engineering V, Quebec, Canada (2003).
[21] R. Tabash and F. Teymour, Chem. Eng. Sci., 59(2004), 5129.
[22] S. Floyd, K.Y. Choi, T.W. Taylor, and W.H. Ray, J. Appl. Polym. Sci., 32(1986), 2935.
[23] S. Floyd, K.Y. Choi, T.W. Taylor, and W.H. Ray, J. Appl. Polym. Sci., 31(1986), 2231.
[24] S. Floyd, R.A. Hutchinson, and W.H. Ray, J. Appl. Polym. Sci., 32(1986), 5451.
[25] R.A. Hutchinson and W.H. Ray, J. Appl. Polym. Sci., 41(1990), 51.
[26] Z. Grof, J. Kosek, M. Marek, and P.M. Adler, AIChE J., 49(2003), 1002.
[27] S.Y. Park, K.Y. Choi, K.H. Song, and B.G. Jeong, Macromolecules, 36 (2003) 4216.
[28] S.Y. Park and K.Y. Choi, Macromol. Mater. Eng., 290 (2005) 353.
[29] W. Kaminsky and A. Knoll, Polym. Bull., 31(1993), 175.

Studies in Surface Science and Catalysis, volume 159
Hyun-Ku Rhee, In-Sik Nam and Jong Moon Park (Editors)
© 2006 Elsevier B.V. All rights reserved

CO2 and energy: Strategy for our future

T. Kojima

Dept. of Materials and Life Science, Faculty of Science and Technology, Seikei University, 3-3-1 Kichijojikitamachi, Musashino-shi, Tokyo 180-8633, Japan

1. INTRODUCTION

1.1. COP3

In December of 1997, COP3 (the third session of the Conference of the Parties to the Framework Convention on Climate Change) was held in Kyoto. Outline of the agreement is as follows;

a) by the period of 2008 to 2012 compared with 1990 levels (compared with 1995 levels for HFC, PFC, SF6),

b) reduction in greenhouse gas emissions at least 5% totally for developed countries; e.g., -8% for EU, -7% for USA, -6% for Japan,

c) in terms of CO_2-equivalent emissions; x21 for methane, x310 for N_2O, x1300 for HFC, x6,500 for PFC and x23,900 for SF6,

d) inclusion of the effects of land-use changes; CO_2 absorption by forests established after 1990, and

e) introduction of trading of emission reductions between countries.

The Protocol has not been ratified by USA or Australia but the ratification by Russia in 2004 made it coming into effect in 2005.

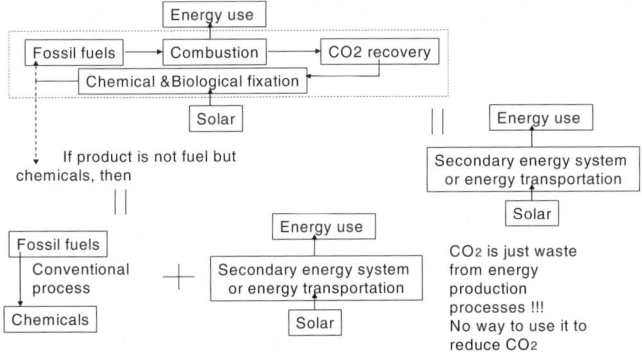

Fig. 1. Evaluation of chemical and biological fixation of CO2

116

1.2. Environmental issues

In the present paper I will discuss CO_2 and energy issues together with the environment and resources issues because other environmental issues and resources problems are the most important issues for the survival of the human beings, comparable to the global population problem and north/south issues [1]. The research activities concerning the problems in the laboratory of the present author are also introduced only by the references.

Up to now, various environmental issues called "pollution" problems were focused in 1970's in Japan. Most of the problems have been solved so far, however still we have serious problems of soil pollutions together with the waste treatment [2, 3].

2. CLASSIFICATION OF MEASURES

The measures for the carbon dioxide problem are classified as follows [4].

a) Primary energy: changes from fossil fuel resources to renewable or nuclear energies and from higher carbon content fossil fuels to lower ones.

b) Secondary energy system: energy conservation, material and energy recycling, change in life style, and improvement in efficiencies of energy use.

c) CO_2 from other than energy.

d) CO_2 recovery and storage: separation system, ocean disposal, subterranean storage, and biological and chemical fixation.

e) CO_2 absorption from atmosphere: afforestation, rock weathering, and ocean absorption.

f) Policy and economic options: carbon, energy and environmental taxes, emissions market, etc.

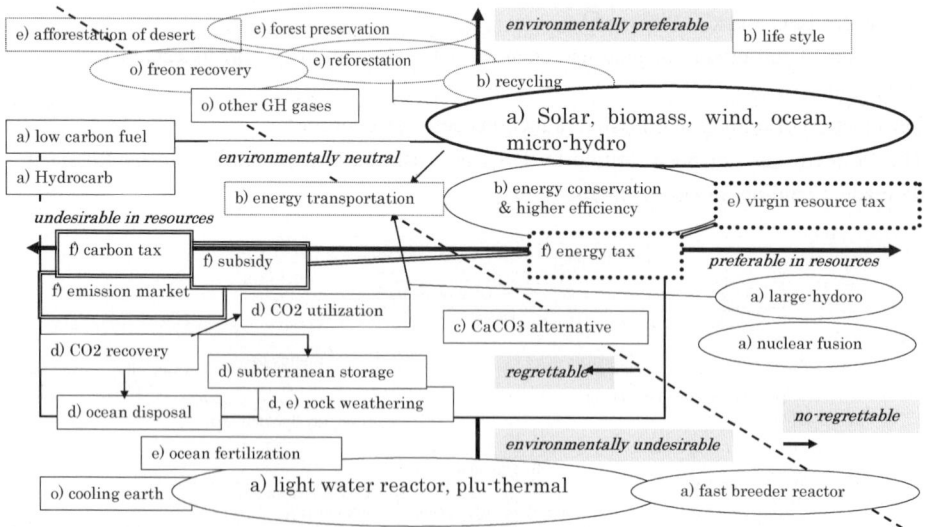

Fig. 2. Map of CO_2 measures from view point of environment and resources [4].

First of all, it should be stressed that various organic chemical and biological processes so far proposed should be evaluated from the view points of total energy systems. As shown in Fig. 1, in most of the proposed processes, non CO_2 emission energy source is finally converted through "black box" into artificial energy. It was clarified that these systems with energy use (usually, the energy is solar, which can be used as other primary energy source instead) should be evaluated from the energy point of view as primary and secondary energy systems; i.e. they should be classified into the categories of a) and b) if it were feasible. On the other hand, the measures in category d) are to be evaluated as those of carbon fixation processes, because they essentially require no artificial energy.

2.1. Regrettable and No-regrettable Measures

Secondly, as shown in Fig. 2 [4], the various measures as above were classified into two categories; regrettable and no-regrettable. We should take resources problems and other environmental problems into account. The measures which give positive effects on them are classified into non-regrettable measures: those that will not be regretted if global warming does not in the end take place, or if its effect is minimal. Other evaluation factors are their scale and stability. No-regrettable measures with larger capacity and reliability should be undertaken, as early as possible, however, most of these measures have their own difficulties, from technological, economical, and/or international points of view.

2.2. Energy Resources with Non or Less Carbon Contents

To convert primary energy source from higher carbon content resources, e.g., coal to non or less carbon content resources, e.g., LWR atomic or natural gas would be the simplest way to meet the agreement. But from the viewpoints of reserves, these resources would be exhausted within 21st century, if all energy were converted into them, as shown in Table 1.

For example, the measure of change from coal to natural gas or nuclear fuel is classified into regrettable one from the viewpoints of resources, because the amount of natural gas or uranium resources is much less than that of coal.

Table 1 CO_2 Release from Fuels and Resources

	carbon	coal	petro.	N.G	Hydrogen/Uran.	Total
H.V. [kcal/kg]	7800	7000	10000	13000	$34000/10^8$	
(est. from H/C ratio)		9800	11200	14200		
H/C ratio [atm/atm]		0.9	1.8	3.9		
H/C ratio [kg/kg]		0.075	0.15	0.325		
CO_2 release[g-C/kcal]	0.128	0.103	0.0781	0.0564	0/0	
prov. res. [Ttoe]: R		0.329	0.142	0.131	0/0.045	0.65
est.res.[Ttoe]: E		3.42	0.311	0.295		4.03
E. (incl. low grad coal)		5.30	(0.824)	0.295?		6.42
anu. prod. [Gtoe] : P		2.29	3.15	1.91	0/0.60	T=7.95
Years (R/P ratio)		144	40	69	-/75	81.4
(E/P ratio)		2310	99(262)	154(?)	-/300?	743(808)
(E/Tratio)		667	39(104)	37(?)	-/23?	{831}

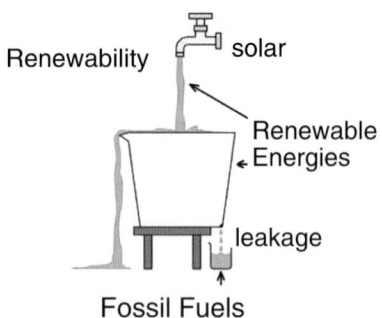

Fig. 3. Concept of renewable energy

On the other hand, the development of high efficient technology in fossil fuel conversion or energy use is classified into no-regrettable options, thus we recognized the importance of investigation on coal conversion technologies [5-8]. When coal is used, though the emission of CO_2 per unit energy production is large, the value of 1% improvement of efficiency is more than the case of NG or atomic from the view point of CO_2.

3. RELIABLE AND PREFERABLE MEASURES

Most important, reliable and no-regrettable measures are two: move to renewable energies and energy saving/conservation. The concept of renewable energy is shown in Fig. 2. The trials of developments of new route to solar energies, for example production of polycrystalline silicon is important [9, 10]. The conversion of waste oil to fuel has also been investigated [11]. The study on coal conversion is also developed to the biomass conversion study.

Improvement of conventional processes for saving energy has great impact on CO_2 emission reduction. In our case, researches on reduction of energy use and conversion from batch to continuous of the powder production were conducted [12, 13].

On the other hand, the measure of the afforestation is also classified into no-regrettable one from the viewpoint of environment. Furthermore, most of desert area is said to be formed by partly somewhat human activities. The activities of afforestation of arid land without crop production are also a no-regrettable measure [14-17].

4. COMPREHENSIVE MEASURES THEIR EVALUATION AND INCENTIVES

All of measures do not stand alone. They should be comprehensively incorporated into the energy systems. Thus the all of the possible measures should be evaluated in the present or possibly future energy systems.

The first point is combination between renewable energies and fossil or other conventional energies [18].   The combination between renewable energy is also important [19].   It is also promising to use renewable energies for afforestation, especially in developed countries by combination with production of food or other agricultural/forestry products [20, 21].

The critical and relative evaluation of these various options including hybrid measures is necessary [22,23].   One of our results in Fig. 4. [24] clearly shows the relative roll of solar, afforestation and biomass.   We must put the right man in the right place.

Various policy and economic options have also been proposed as shown in Fig. 2. However, we should consider the effects of these options on the energy utilization system. For example, when a kind of carbon tax is introduced, the regrettable measure of the change from coal to natural gas or nuclear fuel will proceed.   The energy tax will not cause the problem, however, the introduction of renewable energy will not be enhanced by the tax. Some additional options such as subsidy should be introduced together with the energy tax. Taking these conditions into account, a tax to all virgin resources is thought to be most suitable, though critical evaluation of amount of resources is essential.

5. CONCLUSION

The fossil fuel energy resources are limited to at most several hundred years, which may be shortened by the increase in the energy use, especially in the developing countries.   While the term is clearly short comparing the history of the human beings, we have not taken any serious action against it.   On the other hand, climate change issue is one of the hottest environmental subjects in our society.   But if we consider our several hundred years' future, the society with substantial CO_2 emission is unrealistic because of the fossil fuel depletion. The conclusion on the strategy we should take now is simple: development of renewable energies (or some alternative energy with much more reserves) or reduction of energy use (e.g., energy conservation with higher energy use efficiencies), and environmentally

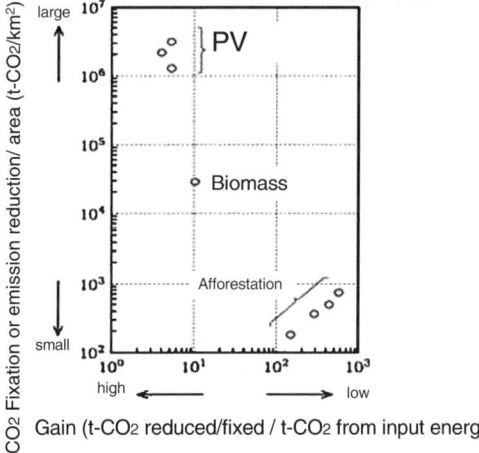

Fig. 4. Relative importance among PV, biomass and afforestation from view points of CO_2.

acceptable technologies, though various kinds of CO_2 abatement technologies including fixation technologies are proposed, some of which are classified into regrettable measures against CO_2 problem.

REFERENCES

[1]   T. Kojima, The Carbon Dioxide Problem, Gordon and Breach Science Publishers, London, 1998.

[2]   N. Saikia, A. Usami, S. Kato and T. Kojima, Resources Processing, 51(2004)35.

[3]   M. Matsumura and T. Kojima, J. Hazardous Materials, B97(2003)99.

[4]   T. Kojima, Proceedings of the 5th ASME/JSME Joint Thermal Engineering Conference, San Diego, USA, 1999, in CDROM, AJTE99-6411.

[5]   H. Liu and T. Kojima, Energy and Fuels, 18(2004)908.

[6]   H. Liu and T. Kojima, Energy and Fuels, 18(2004)913.

[7]   H. Liu, M. Kaneko, C. Luo, S. Kato and T. Kojima, Fuel, 83(2004)1055.

[8]   H. Liu, C. Luo, M. Kaneko, S. Kato and T. Kojima, Energy and Fuels, 17(2003)961.

[9]   T. Kojima, Trends in Chemical Engineering, 2(1994)159.

[10] T. Kojima, T. Uchiyama, D. Murata, S. Kato, Y. Watanabe and H. Shibuya, Kagakukogaku Ronbunshu, 30(2004)306.

[11] S. Kato, K. Kunisawa, T. Kojima and S. Murakami, J. Chem. Eng. Jpn., 37(2004)863.

[12] T. Kojim, S. Komiya and S. Uemiya, Kagakukogaku Ronbunshu, 31(2005)51.

[13] T. Hanabusa, S. Uemiya and T. Kojima, Chem. Eng. Sci., 54(1999)3335.

[14] T. Kojima, N. Asaka, J. Ishida, H. Hamano and K. Yamada, J. Arid Land Studies, 14S(2004)223.

[15] T. Kojima, Y. Tanaka, S. Kato, K. Tahara, N. Takahashi, K. Yamada, Science in China, Series D, 45S(2003)142.

[16] K. Yamada, T. Kojima, Y. Abe, M. Saito, Y. Egashira, N. Takahashi, K. Tahara and J. Low, J. Chem. Eng. Jpn., 36(2003)328.

[17] K. Yamada, T. Kojima, Y. Abe, A. Williams and J. Law, J. Arid Land Studies 9(1999)143.

[18] T. Kojima and K. Tahara, Energy Conversion and Management, 42(2001)1839.

[19] Y. Nishigami, H. Sano and T. Kojima, Applied Energy 67(2000)383.

[20] S. Sinha, S. Kumar, H.Hamano, K.Tahara and T.Kojima, World Res. Review, 12(2000)509.

[21] J. Shrestha and T. Kojima: J. Arid Land Studies, 7S(1998)253.

[22] K. Tahara T. Kojiman and A. Inaba, J. Arid Land Studies, 7S(1998)117.

[23] K.Tahara, S.Sinha, R.Sakamoto, T.Kojima, K.Taneda, A.Funasaki, T. Ohtaki, and A.Inaba, World Resources Review, 13(2001)52.

[24] T. Kojima, K. Tahara and A. Inaba, Relative importance of solar PV, bio-mas energy and afforestation as measures for CO2, Problem Proc. Intern. Conf. on Role of Renewable Enery Technology for Rural Development, Oct. 12-14, 1998, Kathmandu, Nepal, pp.81-85, 1999.1.

BIOLOGICAL AND BIOCHEMICAL
REACTION ENGINEERING

Studies in Surface Science and Catalysis, volume 159
Hyun-Ku Rhee, In-Sik Nam and Jong Moon Park (Editors)

Enzymatic hydrolysis of waste paper: process optimization using response surface methodology

K.H. Chu[a], J.P. Chambenoit[b] and E.Y. Kim[c]

[a]Department of Chemical and Process Engineering, University of Canterbury, Private Bag 4800, Christchurch, New Zealand. Email: khim.chu@canterbury.ac.nz

[b]ENSIL, Université de Limoges, Parc ESTER, 16 rue Atlantis, F-87068 Limoges Cedex, France

[c]Department of Chemical Engineering, The University of Seoul, Seoul 130-743, Korea

A Box-Behnken design was employed to investigate statistically the main and interactive effects of four process variables (reaction time, enzyme to substrate ratio, surfactant addition, and substrate pretreatment) on enzymatic conversion of waste office paper to sugars. A response surface model relating sugar yield to the four variables was developed on the basis of the experimental results. The model could be successfully used to identify the most efficient combination of the four variables for maximizing the extent of sugar production.

1. INTRODUCTION

Replacing part of the fossil-based transportation fuel with biomass-derived products can be a significant contribution to environmental sustainability. For example, blending biomass-derived ethanol into gasoline will result in significant reductions in greenhouse gas emissions. Besides the obvious environmental benefits, biomass-derived ethanol would reduce significantly the world's dependence on increasingly scarce supplies of oil. Over the last few decades research efforts have focused on converting agricultural byproducts and other inexpensive and plentiful cellulosic materials such as wood biomass to ethanol [1-4].

The cellulose fraction of biomass is first hydrolyzed to simple sugars which can then be fermented into ethanol. Cellulose may be converted to fermentable sugars by acid or enzymatic hydrolysis. The purpose of this research was to explore hydrolysis of waste office paper using the enzyme cellulase. The potential of waste paper as a cellulosic substrate for sugar production has been recognized in a number of earlier investigations [5-7]. Enzymatic saccharification however suffers from drawbacks such as low sugar yields and enzyme inactivation. These shortcomings tend to compromise the economic viability of the enzymatic approach. Various methods have been tested to enhance sugar production and protect enzyme activity. Examples of these measures include pretreatment of substrate to make its cellulose more susceptible to enzyme attack and addition of surfactants to enhance enzyme stability. A recent review on pretreatment methods is available elsewhere [8].

Enzymatic hydrolysis is affected by numerous process factors including pH, temperature, reaction time, and enzyme loading. These factors often interact with one another. In the current study the main and interactive effects of four factors (reaction time, enzyme to substrate ratio, surfactant addition, and pretreatment of substrate by phosphoric acid) on the extent of sugar production were evaluated using a Box-Behnken experimental design. A response surface model, developed on the basis of the experimental results, was successfully used to locate the most efficient combination of the four factors for maximizing the extent of sugar production. This integrated modeling and optimization approach, known as response surface methodology, is a powerful tool for analyzing the effect of multiple variables or factors on a given process rapidly and efficiently with a minimal number of experiments while keeping a high degree of statistical significance in the results.

2. MATERIALS AND METHODS

Cellulase and all chemicals used in this work were obtained from Sigma. Hydrolysis experiments were conducted by adding a fixed amount of 2 x 2 mm office paper to flasks containing cellulase in 0.05 M acetate buffer (pH = 4.8). The flasks were placed in an incubator-shaker maintained at 50 °C and 100 rpm. A Box-Behnken design was used to assess the influence of four factors on the extent of sugar production. The four factors examined were (i) reaction time (h), (ii) enzyme to paper mass ratio (%), (iii) amount of surfactant added (Tween 80, g/L), and (iv) paper pretreatment condition (phosphoric acid concentration, g/L), as shown in Table 1. Each factor is coded according to the equation

$$x_i = \frac{(X_i - X_0)}{\Delta X} \tag{1}$$

where x_i is the coded value of the ith factor, X_i is the natural value of the ith factor, X_0 is the factor's natural value at the center point level, and ΔX is the step change value. The Box-Behnken design required a total of 29 trials, including five trials at the center point level, which allowed analysis of the error of the measurement. The design matrix of the Box-Behnken design is not shown here for the sake of brevity. A statistical analysis software package (Design Expert) was used for data analysis. Sugar concentrations were measured using the dinitrosalicylic acid method.

Table 1
Coded and actual levels of the four factors

Coded level	Actual levels			
	Reaction time (h)	Enzyme to paper ratio (%)	Surfactant (g/L)	H_3PO_4 pretreatment (g/L)
-1	4	2	0	0
0	26	10	4	4
1	48	18	8	8

3. RESULTS AND DISCUSSION

Cellulase enzyme complexes consist of three major types of proteins that synergistically catalyze the breakdown of a cellulosic substrate. Because the enzymes are strictly substrate-specific in their action, any change in the structure or accessibility of the substrate can have a considerable influence on the course of the hydrolysis reaction. A pretreatment method based on exposing cellulosic substrate to phosphoric acid solution [9] and addition of the nonionic surfactant Tween 80 [10, 11] were used in this work to enhance the rate and extent of enzymatic hydrolysis of office paper. The phosphoric acid pretreatment was intended to disrupt the structure of office paper to increase exposure of its cellulose to cellulase while the addition of Tween 80 was intended to minimize irreversible sorption of cellulase to non-cellulosic components. The influence of these two factors and two other important process factors, reaction time and enzyme to paper ratio, on the extent of sugar production was characterized using a Box-Behnken design.

The resulting data of the Box-Behnken design were used to formulate a statistically significant empirical model capable of relating the extent of sugar yield to the four factors. A commonly used empirical model for response surface analysis is a quadratic polynomial of the type

$$y = b_0 + \sum_i b_i x_i + \sum_i b_{ii} x_i^2 + \sum_i \sum_j b_{ij} x_i x_j \qquad (2)$$

where y is the predicted response, x_i, x_i^2, and $x_i x_j$ are the first-order, second-order, and interaction terms, respectively, and b_0, b_i, b_{ii}, and b_{ij} ($i < j$) are the coefficients obtained by multiple regression of experimental data. Fitting Eq. (2) to the results of the Box-Behnken design yields the following model:

$$y = 41.31 + 15.81x_1 + 7.25x_2 - 1.10x_3 + 3.89x_4 - 6.18x_1^2 - 4.08x_2^2 - 1.07x_3^2 - 3.98x_4^2 - $$
$$2.55x_1x_2 - 0.72x_1x_3 - 0.01x_1x_4 - 0.09x_2x_3 - 1.19x_2x_4 - 0.14x_3x_4 \qquad (3)$$

where y is the sugar yield, x_1 is the reaction time, x_2 is the enzyme to paper ratio, x_3 is the surfactant concentration, and x_4 is the H_3PO_4 concentration. Here, the factors are specified in their coded units. The correlation coefficient, R^2, which is a measure of how well the model can be made to fit the raw data, was 0.923, indicating an adequate model fit. Fig. 1 shows sugar yields calculated by the model plotted against the corresponding experimental data. It is evident that most of the model calculations lie very close to the line of perfect prediction.

A t test was conducted to assess the statistical significance of the model terms in Eq. (3) and the results are summarized in Table 2. The statistical analysis showed that all cross terms with the exception of x_1x_2 were found to be statistically insignificant at $p \leq 0.10$. In addition, all terms related to x_3, the Tween 80 surfactant concentration, had no significant effect on the model. Eriksson et al. [11] found that various surfactants (Tween 80 included) were able to reduce nonspecific binding of cellulase to the lignin part of lignocellulosic biomass, resulting in enhanced sugar production. The insignificant effect of Tween 80 observed here suggests that office paper contains very little lignin. Elimination of the apparently insignificant terms from Eq. (3) did not yield a better fitting model. As a result, Eq. (3) was used for optimization studies without removing the extraneous terms.

A genetic algorithm was used to identify the most efficient combination of X_1-X_4 within the tested boundaries of Eq. (3) to convert office paper to sugars. Genetic algorithms are a stochastic optimization technique inspired by the biological concept of natural selection and evolution [12]. According to Eq. (3), a maximum sugar yield of 53.7 mg/L could be attained with $X_1 = 48$ h, $X_2 = 13.8$ %, $X_3 = 2.5$ g/L, and $X_4 = 5.7$ g/L. Experimental validation of the optimum X_1-X_4 combination gave a sugar yield of 47.2 mg/L. The good agreement between the measured result and the predicted value corroborates the effectiveness of the response surface approach as a practically useful tool for bioprocess modeling and optimization.

Table 2
Statistical significance of the model terms in Eq. (3)

Model term	Coefficient	p
x_1	15.81	< 0.0001
x_2	7.25	< 0.0001
x_3	-1.10	0.177
x_4	3.89	0.0002
x_1^2	-6.18	< 0.0001
x_2^2	-4.08	0.002
x_3^2	-1.07	0.327
x_4^2	-3.98	0.002
$x_1 x_2$	-2.55	0.078
$x_1 x_3$	-0.72	0.602
$x_1 x_4$	-0.01	0.993
$x_2 x_3$	-0.09	0.946
$x_2 x_4$	-1.19	0.398
$x_3 x_4$	-0.14	0.917

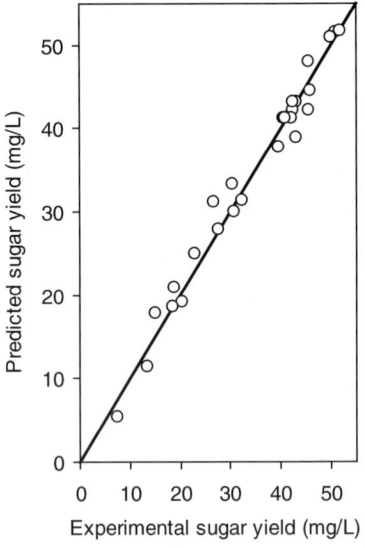

Fig. 1. Predicted vs measured sugar yields.

ACKNOWLEDGEMENT

One of the authors (J.P.C.) was a visiting student at the University of Canterbury during completion of this work. The hospitality of the Department of Chemical and Process Engineering is gratefully acknowledged.

REFERENCES

[1] J. Sheehan and M. Himmel, Biotechnol. Prog., 15 (1999) 817.
[2] R. Wooley, M. Ruth, D. Glassner and J. Sheehan, Biotechnol. Prog., 15 (1999) 794.
[3] Y. Sun and J. Cheng, Bioresource Technol., 83 (2002) 1.
[4] A. Demirbaş, Int. J. Green Energy, 1 (2004) 79.
[5] J.P.H. van Wyk and P.B. Leogale, Biotechnol. Lett., 23 (2001) 1849.
[6] J.P.H. van Wyk, Energy Fuels, 16 (2002) 1277.
[7] J.P.H. van Wyk and M. Mohulatsi, Bioresource Technol., 86 (2003) 21.
[8] N. Mosier, C. Wyman, B. Dale, R. Elander, Y.Y. Lee, M. Holtzapple and M. Ladisch, Bioresource Technol., 96 (2005) 673.
[9] T. Nikolov, N. Bakalova, S. Petrova, R. Benadova, S. Spasov and D. Kole, Bioresource Technol., 71 (2000) 1.
[10] J. Wu and L.-K. Ju, Biotechnol. Prog., 14 (1998) 649.
[11] T. Eriksson, J. Börjesson and F. Tjerneld, Enzyme Microb. Technol., 31 (2002) 353.
[12] D.E. Goldberg, Genetic Algorithms in Search, Optimization and Machine Learning, Addison-Wesley, New York, 1989.

Studies in Surface Science and Catalysis, volume 159
Hyun-Ku Rhee, In-Sik Nam and Jong Moon Park (Editors)

Characterization of Substrate-specificity for Cutinase Produced by Novel *Pseudozyma aphidis* Isolated from Pea Leaves in South Korea

Yang-Hoon Kim[a,b] and Jeewon Lee[a]*

[a]Department of Chemical and Biological Engineering, Korea University, Anam-Dong 5-1, Sungbuk-Ku, Seoul 136-713, South Korea

[b]R&D Center, Daesang Corporation, 125-8 Pyokyo, Majang, Ichon-Si, Gyeonggi-Do 467-813, South Korea

1. INTRODUCTION

Cutinase is a hydrolytic enzyme that degrades cutin, the cuticular polymer of higher plants (i.e. a polyester composed of hydroxy and epoxy fatty acids, usually n-C16, n-C18). Unlike the other lipolytic enzymes, such as lipases and esterases, cutinase is able to have enzymatic activity without interfacial activation. Some microorganisms such as *Fusarium oxysporum* f. sp. *pisi* can utilize cutin as a sole carbon source by producing extracellular cutinolytic enzymes. Several bacterial cutinases have been isolated and characterized from the phyllospheric fluorescent *Pseudomonas putida, Pseudomonas mendocina*, and *Corynebacterium* sp. In recent years, the esterification and transesterification activities of cutinases and esterases have been largely exploited and can be applied advantageously in chemical synthesis [1]. These enzymes have also been applied as lipolytic enzymes in laundry or dishwashing detergent compositions to efficiently remove immobilized fats. Potential uses include its application in the oleochemistry industry and in the degradation of plastics. Previously, we reported that cutinase have degraded successfully the environmental toxic pollutants (i.e. endocrine disrupting chemicals and organophosphate insecticide) without cellular toxic effects compared with yeast esterase [2,3]. Recently, wide studies (i.e. degrading microbes, degradation enzymes, and their genes) on the biodegradation of plastics have been carried out in order to overcome the environmental problems associated with synthetic plastic waste [4,5]. Phytopathogenic *Fusarium* secreted polycaprolactone(PCL) depolymerase, which was identified as cutinase, that hydrolyzes the insoluble polyester to water-soluble products for carbon and energy [6]. Our previous studies have shown that the rapid access to the substrate and chain length-specific hydrolysis is intrinsic properties of cutinase only, which were never found in the hydrolysis by lipases and esterases [7]. These distinct properties of cutinase would be of great usefulness in rapid and easy isolation of various natural cutinases with different microbial origins, each of which might provide a novel industrial application with a specific enzymatic function. In the present study, a novel cutinase-producing microorganism, *Pseudozyma aphidis* was isolated from pea leaves in mid-west area of South Korea. We also observed that *Pseudozyma aphidis* was efficiently capable of degrading polycaprolactone contained in extracellular culture medium.

2. MATERIALS AND METHODS

For the isolation of cutinase-producing microorganisms, the samples were taken from pea leaves in the Chung-Cheong-Nam-Do of South Korea. Leaves above 1 m above ground were removed from branches of the trees, collected in plastic bags and stored in a cold room (4°C) for 3-4 h before used for microbe isolation. They were sliced into pieces, put loosely into flasks, and extracted for 30 min with 10 mL of 0.15 M sterile NaCl. Without further treatment, the supernatant was plated on NBY (0.2% yeast extract; 0.8% nutrient broth (Difco)) agar and the plates incubated for 2 days at 30°C. Plates were maintained on NBY medium and kept at 4°C until used. A single colony of each strain, grown on NBY agar plate, was inoculated into a test tube (with 5-ml medium volume) and cultured at 30°C. After 24 h, 1.5 mL of culture broth was inoculated into flasks containing 100 ml of NBY, NBYG (NBY + glucose 10 g l^{-1}), NBYC (NBY + cutin 10 g l^{-1}), and NBYP (NBY + PCL (Aldrich Chemical Company, Inc.) 10 g l^{-1}) media, and the flask culture continued for 12 days in the shaking incubator at 500 rpm. The cells were harvested every 24 h by centrifugation for 20 min at 4°C, and the extracellular fractions were used to estimate cutinolytic activity. In order to investigate the degradation of cutin and PCL, the water-soluble total organic carbon (TOC) concentration in cell-free culture broth was determined at 30°C using TOC analyzer (Rose Mount DC-18, Dohrmann, Germany). In addition, both cutinase activities were determined by a spectrophotometric assay with p-nitrophenyl butyrate (PNB) and p-nitrophenyl palmitate (PNP) (Sigma, St. Louis, MO, USA) as substrates as previously described [7]. Initial maximum velocity ($V_{0,}$ initial maximum ΔOD_{405} per second) was measured at 405 nm using Bio-Rad 96-well microplate reader (Bio-Rad, Hercules, CA).

3. RESULTS

Our previous studies have shown that the rapid access to the substrate and chain length-specific hydrolysis are intrinsic properties of cutinase only, which were never found in the hydrolysis by lipases and esterases. These distinct properties of cutinase would be of great usefulness in rapid and easy isolation of various natural cutinases with different microbial origins, each of which might provide a novel industrial application with a specific enzymatic function. In the present study, 730 strains were tested for their ability to hydrolyze the insoluble plant polyester (cutin) using PNB-PNP hydrolysis assay. Among these microorganisms, a yeast strain BMEL-2 showed best growth on NBYC medium (NBY medium + 1% cutin) (Fig. 1) but relatively poor growth on NBY medium.

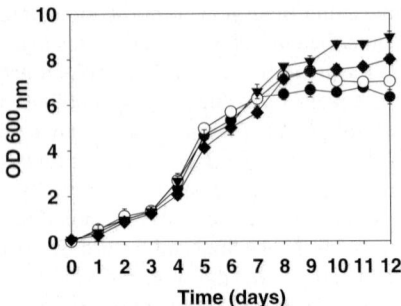

Fig.1. Time-course variation in culture turbidity (OD600) in NBY medium (●), NBY + 1% glucose

medium (○), NBY + 1% cutin (▼), NBY + 1% PCL (◆).

BMEL-2 was found to produce a highly inducible cutinase when grown in medium containing purified cutin (1%) from apple cv. Golden delicious but did not produce any cutinase under NBY and NBY + glucose (1%) medium (Fig. 2a). This corresponds to the previous reports by Kolattukudy *et al.* [8] that the cutinase activity was significantly repressed by the presence of glucose in the medium. Therefore, the results in Figs. 1 and 2a strongly suggest that BMEL-2 effectively hydrolyzed cutin using effectively induced cutinase in extracellular culture broth. During the BMEL-2 cultivation, the cutinolytic enzyme activity on PNB rapidly increased after 4 days and was maintained at high level until the end of cultivation (Fig. 2a). Fig. 2b showed that PNP was also hydrolyzed by cell-free broth from cutin-containing culture. On the basis of previous results [8] that *Fusarium solani* produces several other types of hydrolases like esterases and lipases, depending on the inducers present in the medium [8], it seems that esterase or lipase was co-induced with cutinase from BMEL-2. From Fig. 2c, it is also noticeable that TOC level in NBYC medium containing cutin was increased more than 5-fold compared to cutin-free NBY medium, which implies that cutin was significantly degraded and converted into water soluble products by cutinase.

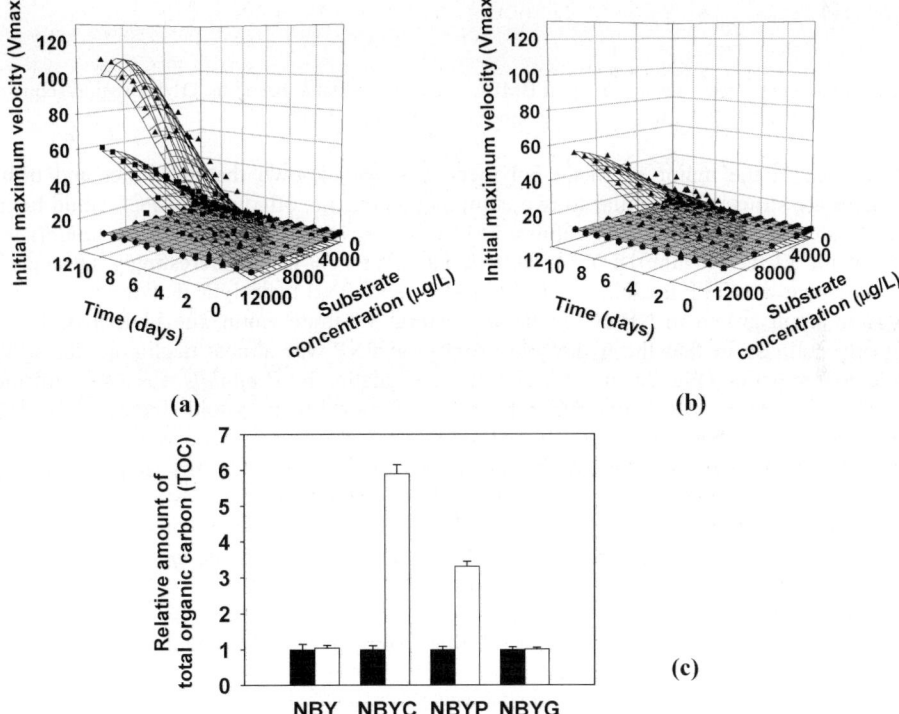

Fig. 2. Initial maximum rate of (a) PNB and (b) PNP hydrolysis by extracellular culture broth of *Pseudozyma aphidis* that was cultivated in NBY (◆), NBYC (▲), NBYP (■), and NBYG (●) medium. (c) TOC concentrations measured in extracellular culture broth of *Pseudozyma aphidis* grown in NBY, NBYC, NBYP, and NBYG medium. (Initial TOC concentration of fresh medium=1. Black and white columns represent TOC concentrations in fresh medium and 12-day culture broth, respectively.)

The isolated strain, BMEL-2, was classified and identified through 25S rDNA full sequencing at the Korea Collection for Type Cultures (KCTC) center in the Korea Research Institute of Bioscience and Biotechnology (KRIBB). As a result, BMEL-2 was identified as *Pseudozyma aphidis* as shown in Fig. 3.

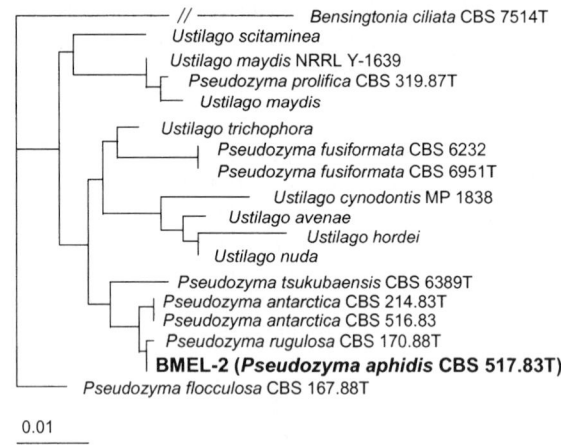

Fig. 3. Phylogenetic relationship of strain BMEL-2 and related taxa, based on D1/D2 region sequence of the large-subunit rDNA. Bar, 0.01 nucleotide substitution per position.

PCL, one of the major synthetic polymers, has been known that its dimer and trimer structures are structurally similar to two cutin monomers, i.e. oligomers of PCL could be the natural inducers and substrate of cutinase [6]. In the present study, *Pseudozyma aphidis* was cultured for 12 days in the NBYP medium containing 1 % PCL. Interestingly, although *P. aphidis* grown in NBYP medium seemed to show lower PNB hydrolytic activity compared to the same strain grown in NBYC containing natural substrate, cutin, the PNB hydrolase is evidently cutinase in that the hydrolytic activity on PNP was almost negligible during the whole culture period (Fig. 2a and 2b). The PCL degradation by *P. aphidis* was also confirmed through measuring TOC concentration, that is, more than 3-fold increase in TOC concentration was observed in NBYP medium while TOC concentration was never changed in NBY medium (Fig. 2c). For the first time, we report in this paper the isolation of novel cutinolytic *Pseudozyma* strain that is capable of degrading synthetic plastic PCL as well as natural substrate cutin. Feasibility of industrial application would be explored through further detailed characterization of purified cutinase from the novel *Pseudozyma aphidis*.

REFERENCES
[1] Y.H. Kim, J. Lee, J.Y. Ahn, M.B. Gu and S.H. Moon, Appl. Environ. Microbiol. 68 (2002) 4684.
[2] Y.H. Kim, J. Lee and S.H. Moon, Appl. Microbiol. Biotechnol. 63 (2003) 75.
[3] Y.H. Kim, J.Y. Ahn, S.H. Moon and J. Lee, Chemosphere (2005) In press.
[4] C.A. Murphy, J.A. Cameron, S.J. Huang, R.T. Vinopal, Appl. Microbiol. Biotechnol. 50 (1998) 692.
[5] M. Shimao, Curr. Opin. Biotechnol. 12 (2001) 242.
[6] C.A. Murphy, J.A. Cameron, S.J. Huang, R.T. Vinopal, Appl. Environ. Microbiol. 62 (1996) 456.
[7] Y.H. Kim, J. Lee and S.H. Moon, J. Microbiol. Biotechnol. 13 (2003) 57.
[8] T.S. Lin and P.E. Kolattukudy, *J. Bacterol.* 113 (1978) 2.

Studies in Surface Science and Catalysis, volume 159
Hyun-Ku Rhee, In-Sik Nam and Jong Moon Park (Editors)

Modeling and simulation of anaerobic filter process: two-dimensional distribution of acidogens and methanogens

M.W. Lee[a], H.W. Lee[b], J.Y. Joung[b] and J.M. Park[a,b],*

[a]Advanced Environmental Biotechnology Research Center, School of Environmental Science and Engineering, [b]Department of Chemical Engineering, Pohang University of Science and Technology, Pohang 790-784, Korea

1. INTRODUCTION

Recently, anaerobic filter process has been widely used for the treatment of various wastewaters because of several advantages over aerobic process such as lower nutrient requirement, less surplus sludge production, and energy recovery via methane production [1]. Although many types of anaerobic filter process have already been successfully commercialized, the details of their complicated process dynamics are still challenging issues to be clarified for their stable operations.

In general, anaerobic conversion of organic pollutants into methane gas can be described by two consecutive microbial reactions, i.e., acidogenesis and methanogenesis. In a single-stage anaerobic filter process, these reactions take place simultaneously in biofilm, where two responsible microbial groups, acidogens and methanogens, are distributed spatially. Hence, if the information on this microbial distribution in the biofilm is provided, the complex dynamics of the anaerobic filter process can be elucidated more reasonably. In this study, a simplified mathematical model to describe an anaerobic filter process was proposed, and the two-dimensional distributions of acidogens and methanogens in the biofilm were investigated by numerical simulations.

2. MODEL DEVELOPMENT AND NUMERICAL SIMULATIONS

To describe a single-stage anaerobic filter process, a dynamic model was developed with following assumptions: the biofilm of anaerobic filter process can be portrayed with flat plate geometry; the biofilm consists of acidogens, methanogens, and inert particulates; feed wastewater contains only sucrose and acetate, each of which is the rate-limiting substrate of acidogens and methanogens, respectively; within biofilm, mass transport of substrates are subjected to Fick's law in two dimension (biofilm depth and reactor height); the growth of microorganisms and the substrate utilization follow Monod relationship; the loss of microbial density due to cell maintenance and the consequential increase of inert particulate fraction are considered; the thickness and the overall density of biofilm are constant at a fixed operating condition, which implies that the presented model can describe the transient dynamics only for a matured biofilm system. Table 1 summarizes the details of the model including model parameters and variables, introduced dimensionless variables, and the finally deduced dimensionless forms of the model equations.

---

* Corresponding author. E-mail: jmpark@postech.ac.kr

Table 1
Details of the mathematical model to describe anaerobic filter process

Model parameters and variables		
Symbol	Meanings	Values
D_{b1}, D_{b2}	Dispersion coefficients in bulk liquid (m^2/day)	$3.30\times10^{-5}, 1.08\times10^{-4}$
D_{f1}, D_{f2}	Molecular diffusivities in biofilm (m^2/day)	$2.64\times10^{-5}, 8.64\times10^{-5}$
k_{L1}, k_{L2}	Mass transfer coefficients in external liquid film (m/day)	0.934, 8.468
μ_A, μ_M	Maximum specific growth rates (day^{-1})	11.4, 0.825
Y_A, Y_M	Microbial yield coefficients (g-VSS/g-substrate)	0.20, 0.2
m_A, m_M	Cell maintenance coefficients (day^{-1})	3.42, 0.2475
K_{S1}, K_{S2}	Saturation coefficients (g/m^3)	150, 480
C_f	Conversion factor of sucrose to acetate (g-acetate/g-sucrose)	1.05
L_f, X_{MAX}	Thickness and maximum density of biofilm (m, g/m^3)	0.002, 20000
L, V_R	Length and volume of reactor (m, m^3)	0.05, 5×10^{-6}
a	Surface area of biofilm per unit reactor volume (m^2/m^3)	240
ε, u	Bed void fraction and superficial liquid velocity (−, m/day)	0.36, 9.26×10^{-2}

Dimensionless variables			
$\eta = y/L_f$	$\zeta = z/L$	$S_{fi}^* = S_{fi}/K_i$	$S_{bi}^* = S_{bi}/K_i$
$\tau = ut/L\varepsilon$	$\alpha_i = k_{Li}aL/u$	$\beta_i = D_{fi}L\varepsilon/L_f^2 u$	$Pe_i = Lu/D_{bi}$
$Bi_i = k_{Li}L_f/D_{fi}$	$\phi_A^2 = L_f^2 \dfrac{\mu_{mA}X_A}{Y_A K_1 D_{f1}}$	$\phi_M^2 = L_f^2 \dfrac{\mu_{mM}X_M}{Y_M K_2 D_{f2}}$	$sf = L_f^2/L^2$

Dimensionless forms of the model equations		
	Governing equations	Boundary conditions
Solid phase	$\dfrac{\partial S_{fi}^*}{\partial \tau} = \beta_i\left(\dfrac{\partial^2 S_{fi}^*}{\partial \eta^2} + sf\dfrac{\partial^2 S_{fi}^*}{\partial \zeta^2}\right) - R_i^*$	$\eta = 0, \dfrac{\partial S_{fi}^*}{\partial \eta}\Big|_{\eta=0} = -Bi_i\left(S_{bi}^* - S_{fi}^*\big|_{\eta=0}\right)$
	$R_1^* = \beta_1\phi_A^2 \dfrac{S_{f1}^*}{1+S_{f1}^*}$	$\eta = 1, \dfrac{\partial S_{fi}^*}{\partial \eta}\Big|_{\eta=1} = 0$
	$R_2^* = \beta_2\phi_M^2\left(\dfrac{S_{f2}^*}{1+S_{f2}^*} - C_f\dfrac{Y_M}{Y_A}\dfrac{\mu_{mA}X_A}{\mu_{mM}X_M}\dfrac{S_{f1}^*}{1+S_{f1}^*}\right)$	
Bulk liquid phase	$\dfrac{\partial S_{bi}^*}{\partial \tau} = \dfrac{1}{Pe_i}\dfrac{\partial^2 S_{bi}^*}{\partial \zeta^2} - \dfrac{\partial S_{bi}^*}{\partial \zeta} - \alpha_i\left(S_{bi}^* - S_{fi}^*\big|_{\eta=0}\right)$	$\zeta = 0, \dfrac{\partial S_{bi}^*}{\partial \zeta}\Big|_{\zeta=0+} = -Pe_i\left(S_{bi}^*\big|_{\zeta=0-} - S_{bi}^*\big|_{\zeta=0+}\right)$
		$\zeta = 1, \dfrac{\partial S_{bi}^*}{\partial \zeta}\Big|_{\zeta=1} = 0$

The values of most model parameters were adopted from literature [2].
Subscripts: 1 → sucrose; 2 → acetate; A → acidogens; M → methanogens.

By using the presented model, three different numerical simulations were conducted to investigate the effects of feed wastewater compositions on the performance of anaerobic filter process, especially focused on the microbial distribution within the biofilm. The tested feed compositions of sucrose and acetate were 9000 and 0 g/m^3 for simulation 1, 6000 and 3000 g/m^3 for simulation 2, 3000 and 6000 g/m^3 for simulation 3, respectively. To obtain steady-state solutions, the model equations were solved by using finite difference method with separate time steps for soluble substrates and microorganisms. Soluble substrate profiles were firstly determined with the assumption of quasi-steady-state of microorganisms, and then the microbial profiles were determined with a relatively large time step. A homogeneous initial microbial distribution of $X_{A0} = X_{M0} = 500$ g/m^3 was assumed for each numerical simulation. All calculations were performed using the MATLAB software.

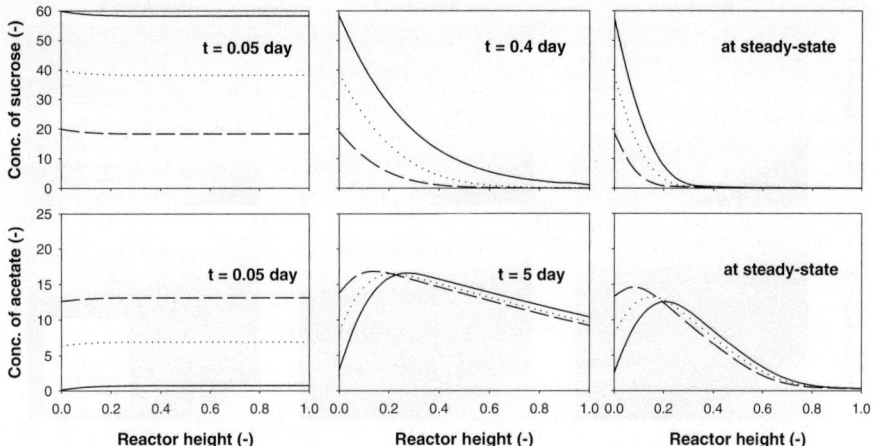

Fig. 1. Time courses of the concentration profiles of sucrose and acetate in bulk liquid phase (solid line: simulation 1, short-dotted line: simulation 2, long-dotted line: simulation 3).

3. RESULTS AND DISCUSSION

Fig. 1 represents the changes of the concentration profiles of sucrose and acetate in the bulk liquid phase according to the simulation-time proceeding. In general, the sucrose profile reached a steady-state more rapidly than the acetate profile, which is naturally expected since the growth rate of acidogens was assumed relatively higher than that of methanogens in this study. A similar result was also obtained for the substrate concentration profiles within the biofilm (data not shown). The anaerobic filter process showed almost complete removal efficiencies regardless of the applied feed compositions because a sufficient HRT (1.5 day) was assumed in this study. However, the reactor height where the acidogenesis reaction was completed and the time when a steady-state was achieved were clearly different depending on the feed compositions. As the sucrose content in the feed wastewater increased, the location of the complete acidogenesis was shifted toward the end of the reactor and it took much longer time to reach a steady-state.

Fig. 2 shows the two-dimensional steady-state microbial distributions of acidogens and methanogens in the biofilm for the different feed compositions. With regard to the direction of the biofilm depth, it was identified that acidogens were dominant near the surface of the biofilm while methanogens were prevailing near the substratum of supporting media for all cases. This result may support that the conventional approaches separating the biofilm into two homogeneous stratified layers (acidogenic and methanogenic layers) are somehow applicable to the modeling of anaerobic filter process [3,4]. However, the simulated two-dimensional microbial distributions clearly revealed that both acidogens and methanogens could coexist in some narrow regions within the biofilm and the microbial distribution along the reactor height was even more significant. In order to describe the anaerobic filter process more reasonably, this heterogeneity of the biofilm should be carefully considered, especially when the process is operated close to a plug-flow mode like the case of this study.

Although the presented model could describe the anaerobic filter process more reasonably, it has still some limitations such as the fixed thickness and the fixed overall density of the

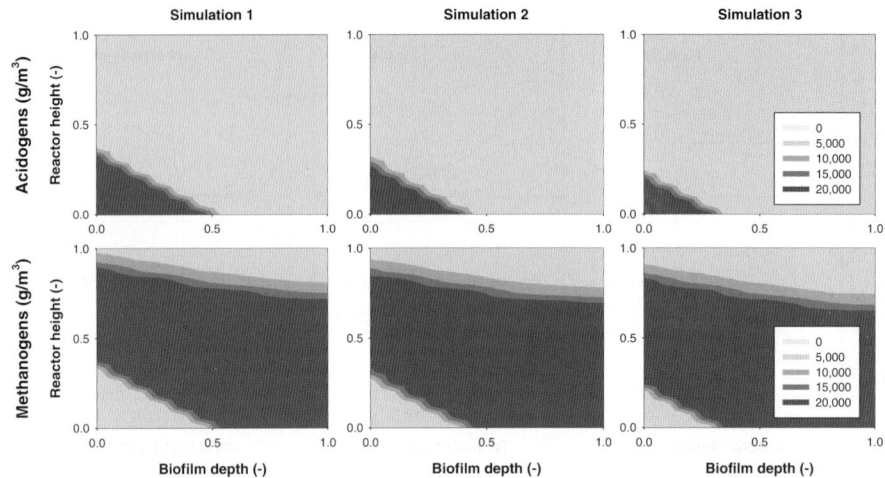

Fig. 2. Steady-state microbial distributions within the biofilm for different feed compositions.

biofilm which cannot but be presumed. Recently, several advanced modeling approaches considering the development and the detachment of biofilm have been reported, though confined to a microscopic view point [5-7]. If it is possible to incorporate a scheme considering the dynamic changes of the biofilm thickness into the presented model, the whole dynamics of the anaerobic filter process could be described more clearly with greatly enhanced model flexibility.

4. CONCLUSION

To elucidate the complex dynamics of anaerobic filter process, a simplified dynamic model capable of predicting the two-dimensional distribution of acidogens and methanogens in biofilm was proposed. The simulation results clearly revealed that the biofilm should be considered as a heterogeneous system in which acidogens and methanogens are spatially distributed strictly depending on operating conditions such as feed wastewater composition.

ACKNOWLEDGMENT

This work was financially supported by the Korea Science and Engineering Foundation through the Advanced Environmental Biotechnology Research Center at POSTECH.

REFERENCES

[1] M.W. Lee, J.Y. Joung, D.S. Lee, J.M. Park and S.H. Woo, Ind. Eng. Chem. Res., 44 (2005) 3973.
[2] C.H. Wu, J.S. Huang, J.L. Yan and C.G. Jih, Biotechnol. Bioeng., 57 (1998) 367.
[3] M. Canovas-Diaz and J.A. Howell, Biotechnol. Bioeng., 32 (1988) 348.
[4] N. Gupta, S.K. Gupta and K.B. Ramachandran, Chem. Eng. J., 65 (1996) 37.
[5] C. Picioreanu, M.C.M. Van Loosdrecht and J.J. Heijnen, Biotechnol. Bioeng., 58 (1988) 101.
[6] D.R. Noguera, G. Pizarro, D.A. Stahl and B.E. Rittmann, Wat. Sci. Tech., 39 (1999) 123.
[7] C. Picioreanu, M.C.M. Van Loosdrecht and J.J. Heijnen, Biotechnol. Bioeng., 72 (2001) 205.

Studies in Surface Science and Catalysis, volume 159
Hyun-Ku Rhee, In-Sik Nam and Jong Moon Park (Editors)

Effect of Pretreatment on Lactic Acid Fermentation of Bean Curd Refuse with Simultaneous Saccharification

Katsuhiko Muroyama, Ryohei Atsumi and Atsushi Andoh

Dept. of Chemical Engineering, Kansai University, 3-3-35 Yamate-cho, Suita, Osaka, 565-8680, Japan

1. INTRODUCTION

Ca. 800,000 tons of bean curd refuse are annually discharged as by-products in the tofu production in Japan. The expense for disposing of the bean curd refuse may be accounted for an amount of 16 billion yen per year. In the past, the bean curd refuse was effectively used as livestock feeds. Presently, however, major part of it is burned uselessly. In this research, therefore, we aim to produce lactic acid from the bean curd refuse through a lactic acid fermentation under the presence of a cellulase (termed as simultaneous saccharification and fermentation, SSF), establishing a high lactic acid productivity. The SSF process offers various advantages such as the use of a single-reaction vessel for hydrolysis, short processing time, reduced product inhibition and, increased productivity [1-4]. In the fermentation processes using various biomass substrates, many pretreatment methods have been examined [5-10]. In the present study we examined the effects of pretreatments for the bean curd refuse substrate using HCl aqueous solution with heating in an auto clave on the lactic acid production efficiency through SSF. It is shown that an 87% conversion of lactic acid from the pretreated bean curd refuse substrate was attained at a finally attained lactic acid concentration of about 46 g/l.

2. MATERIALS AND METHODS

2.1 Substrate and enzyme

Dry bean curd refuse was used as the substrate in the lactic acid fermentation with simultaneous saccharification (SSF). The dry bean curd refuse was preliminarily sieved under a mesh size of 250 μ m. It contained 12.3% water, 4.0% ash, 0.8% lipid, 29.3% protein, 53.6% carbohydrate, respectively, in weight basis. The cellulase derived from *Aspergillus niger* with an enzymatic activity of 25,000 units/g (Tokyo Kasei Industry Inc.) was employed as the saccharification enzyme.

2.2 Microorganism and media

A lactic acid bacterium, *Lactobacillus paracasei* (LA1), which was supplied by Prof. K. Nakasaki of Shizuoka University, was used throughout this work. A pre-agar-culture medium contained the following components in one liter: 10 g glucose, 10 g yeast extract, 5 g bactopeptone, 2 g beef extract, 2 g $CH_3COONa \cdot 12H_2O$, 0.2 g $MgSO_4 \cdot 7H_2O$, 0.01 g $MnSO_4 \cdot 4H_2O$, 0.01 g $FeSO_4 \cdot 7H_2O$, 0.01 g NaCl, 0.5 g Tween80, 5 g $CaCO_3$, 12.5 g agar. The medium for the main culture contained the following components per 1 L fermentation broth: 2.5 g yeast extract, 1.25 g bactopeptone, 6.25 g $CaCO_3$. At first the inoculum for fermentation was prepared with the pre-agar-culture in Petri dishes at 37°C for 48 h and it was further cultured in a 500 ml flask at 37°C for 24 h.

2.3 Analysis method

The concentrations of lactic acid and other organic acid species were determined by HPLC CTO10A

(Shimadzu, Kyoto, Japan) with Shim-pack SCR-102H column. The cell concentration in the broth was determined from the measurement of an optical density at a wavelength of 660nm. Enzymatically or chemically solubilized sugars were assayed using the Somogyi-Nelson method for evaluating reducing end groups and the phenol-sulfuric acid method for evaluating total sugars [11-12]. Total organic carbons (TOC) solubilized from bean curd refuse was measured with a TOC analyzer (TOC-5000, Shimadzu).

2.4 Pretreatment method

The dilute acid of HCl was used as a pretreatment agent. In the first method, bean curd refuse was pretreated without heating. An amount of 0.5 g bean curd refuse was dipped in 10 ml of 1 mol/l HCl aqueous solution for 1, 3, 6 and 24 hours. In the next, 0.5 g bean curd refuse was dipped in 10 ml distilled water or HCl solution in the concentration range of 0.01-1 mol/l and held in a steam-heated auto clave at 121℃ for 1, 15 or 30 min. After neutralization with 5 mol/l NaOH solution, the concentrations of reducing sugar, total sugar and TOC were measured for the pretreated solution.

2.5 Lactic acid fermentation of HCl-pretreated bean curd refuse substrate with ESS

Table 1 shows the dry weight of substrate, and amounts of HCl aqueous solution for pretreatment, cellulase and suspension broth for the lactic acid fermentation with ESS. The initially supplied amount of bean curd refuse in dry weight basis was changed from 10 to 150 g to examine the influence of substrate loading. The amount of cellulase was increased against initial substrate loading. And also, the amount of 0.1 mol/l HCl was increased against an increased amount of bean curd substrate as shown in Table 1. Batch fermentation operations were conducted by using a thermo-controlled fermentor of 2 L broth volume, equipped with a pH stat device (EYLA MBF-300ME, Tokyo Rikakikai Co.Ltd.).

Table 1 Operating condition in the lactic acid fermentation with SSF using HCl-pretreated substrate

Substrate weight [g]	Volume of 0.1 mol/l HCl for pretreatment [ml]	Cellulose weight [g]	Volume of suspension broth [L]
10	200	1	1
50	1000	2.5	2
150	1000	7.5	2

3. RESULTS AND DISCUSSION

3.1 Results for pretreatment of bean curd refuse using HCl solutions with/without heating

Table 2 shows the variations in the concentrations of reducing sugar, total sugar, and TOC for the pretreated substrate solution using 1 mol/l HCl aqueous solution without heating. The concentrations of reducing sugar, total sugar and TOC respectively increased with increasing processing time. However, their values still remained at lower values even after 24 hours operation. Table 3 shows the variations in the concentrations of reducing sugar, total sugar and TOC, for the pretreated substrate with heating using water or 0.01-1 mol/l HCl aqueous solutions. The concentrations of reducing sugar and total sugar and TOC significantly increased with increasing HCl concentration in their range of 0 to 0.1 mol/l, while they respectively remained at similar values at HCl concentrations beyond 0.1mol/l. Furthermore, the solution after pretreatment with 1 mol/l HCl aqueous solution at 121℃ was colored brown due to the formation of furfural and HMF, etc. Moreover, considering the salt inhibition by NaCl generated after neutralization on the lactic acid fermentation, we concluded that pretreatment using 0.1 mol/l HCl with heating at 121℃ for 30 min was suitable.

3.2 Results of lactic acid fermentation with ESS

Figure 1 shows the time course of lactic acid production through SSF in a batch reactor of 1L volume for an initial load of 10 g/l bean curd refuse with or without pretreatment using 0.1 mol/l HCl aqueous solution with heating at 121℃ for 30 min. The finally attained lactic acid yield on a carbon basis and its corresponding concentration are respectively 20.1% and 1.46 g/l for the non-pretreated bean curd refuse, while they are respectively 87.4% and 6.14 g/l for the pretreated one. The pretreatment of the bean curd refuse using 0.1mol/l HCl at 121℃ could achieve the dissolution of cellulosic components

Table 2 The dissolved component concentrations for the pretreated substrates using 1 mol/l HCl without heating.

Time [hour]	TOC[ppm]	Reducing sugar [g/l]	Total sugar [g/l]
1	1843	0.08	2.60
3	2347	0.25	3.32
6	2482	0.28	3.51
24	2841	0.35	3.82

Table 3 Variations in the dissolved component concentrations for the pretreated substrate with heating using distilled water or HCl aqueous solution of 0.01-1 mol/l.

Pretreatment condition		TOC[ppm]	Reducing sugar [g/l]	Total sugar [g/l]
121℃, 1min	H₂O	1048	0.34	0.69
	0.01 mol/l HCl	1434	0.30	0.44
	0.1 mol/l HCl	5755	2.95	6.77
	1 mol/l HCl	6320	6.36	6.75
121℃, 15min	H₂O	1237	0.30	1.11
	0.01 mol/l HCl	2143	0.34	2.21
	0.1 mol/l HCl	5955	3.49	7.11
	1 mol/l HCl	6265	6.31	6.78
121℃, 30min	H₂O	1820	0.25	1.83
	0.01 mol/l HCl	3158	0.31	3.82
	0.1 mol/l HCl	5795	4.19	7.03
	1 mol/l HCl	6585	6.52	6.78

into lower molecular weight substances or soluble fragments, significantly increasing the enzymatic saccharification activity. Figure 2 shows the time course of lactic acid production through SSF in a batch reactor of 2L volume for a 25 g/l of initial bean curd refuse substrate, pretreated using 0.1 mol/l HCl with heating at 121℃ for 30 min. The finally attained lactic acid yield and its corresponding concentration are respectively 96.9% and 16.86 g/l. Finally, the concentration of the initially loaded bean curd refuse was increased up to 75 g/l and the lactic acid fermentation was conducted. In this case, after the pretreatment, the viscosity of the treated solution became very high and the enhancement effect of pretreatment on the lactic acid yield was decreased as shown by their fermentation results in Fig. 3. Then, 0.2 mol/l HCl aqueous solution was examined to enhance the dissolution effect of pretreatment and to decrease the viscosity of the treated solution. Figure 4 shows the time course of lactic acid production through SSF in a batch reactor of 2L volume for the 75 g/l initial load of bean curd refuse substrate, pretreated by the 0.2 mol/l HCl aqueous solution with heating

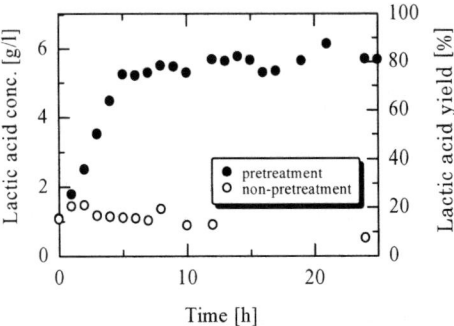

Fig. 1 Time course of lactic acid yield and its concentration in SSF with or without pretreatment using 0.1 mol/l HCl with heating at 121℃ for 30 min. Fermentation conditions: 37℃, pH=5.0, initial load of bean curd refuse(BCR) 10 g, cellulase amount=1 g in 1L suspension.

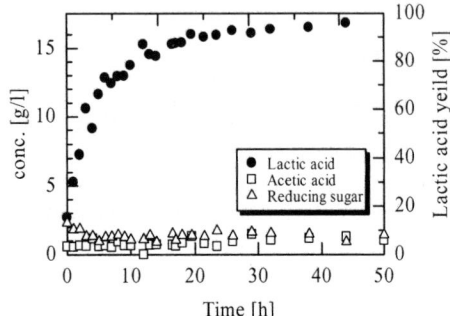

Fig. 2 Time course of lactic acid yield and its concentration in SSF with pretreatment using 0.1 mol/l HCl at 121℃ for 30 min. Fermentation conditions: temperature=37℃, pH=5.0, initial load of BCR=50 g, cellulase amount=2.5 g in 2L suspension.

at 121℃ for 30 min. The finally attained lactic acid yield and the corresponding concentration are as high as 87.0% and 45.8 g/l, respectively. As shown in Fig. 4, the concentration of reducing sugar concentration was high compared with that in Fig. 3 where the same 150g initial and a higher concentration of lactic acid was attained due to the higher reducing sugar concentration maintained by enhanced enzymatic saccharification on the extensively pretreated bean cured refuse using a 0.2 mol/l HCl aqueous solution. It should be noted, however, that increased amount of HCl in the pretreatment would result in a higher salt concentration in the substrate solution after neutralization, and cause a salt inhibition in the lactic acid fermentation [6].

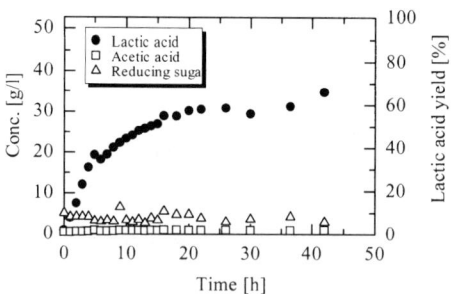

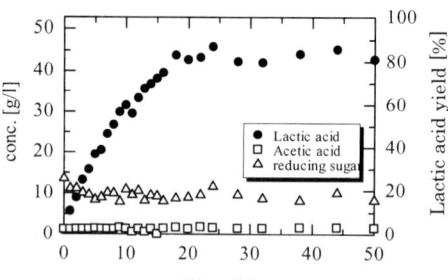

Fig. 3 Time course of lactic acid yield and its concentration in SSF with pretreatment using 0.1 mol/l HCl aqueous solution at 121℃ for 30 min. Fermentation conditions: temperature=37℃, pH=5.0, initial load of BCR=150 g, cellulase amount=7.5 g in 2L suspension.

Fig. 4 Time course of lactic acid yield and its concentration in SSF with pretreatment using 0.2 mol/l HCl at 121℃ for 30 min. Fermentation conditions : 37℃, pH=5.0, initial load of bean curd refuse 150 g, cellulase amount=7.5 g in 2L suspension.

4. CONCLUSION

For lactic acid fermentation of bean curd refuse with simultaneous saccharification, it is found that the pretreatment of the substrate using 0.1 or 0.2 mol/l HCl aqueous solution with heating at 121℃ for 30 min efficiently solubilized the raw material and significantly enhanced the enzymatic saccharification followed by the lactic acid fermentation. The amount of initial load of bean curd refuse in dried state could be increased up to 75 g/l in a batch fermentation, and the finally attained lactic acid yield and its concentration were as high as 87.0% and 45.8 g/l, respectively.

5. ACKNOWLEDGEMENTS

This research was financially supported in part by the Kansai University Grant-in-Aid for Promotion of Advanced Research in Graduate Course, 2004.

6. REFERENCES

[1] Nakasaki, K., and T. Adachi, Environ. Sci. Technol., **13(5)**, 570-578, 2000
[2] Sreenath H K, A.B. Moldes, R.J. Straub, and Koegel R G, J. Biosci. Bioeng. **92**, 518-523, 2001
[3] Moritz, J.W. and S.J.B. Duff, Biotechnol. Bioeng. **49**, 504-511, 1996
[4] Iyer, V.P. and Y.Y. Lee, Biotechnol. Lett. **21**, 371-373, 1999
[5] Hideaki, Y., Technological Trends in Energy Generation from Biomass (in Japanese), CMC Press, 2001
[6] Ishikawa, K., Master's thesis in Kochi University of Technology, 2002
[7] Inoi, T., T. Akabane, A. Saito, Y. Kurokawa, and S. Matsuoka, Ferment. Eng. **63**, 227-230, 1985
[8] Yu, Z., and H. Zhang, Biomass Bioenergy, **24**, 257-262, 2003
[9] Varga, E., Z. Szengyel, and K. Reczey, Appl. Biochem. Biotechnol. Part a, **98/100**, 73-87, 2002
[10] Nakamura, Y., M. et al., Chemical engineering thesis collection, **17**, 504-510, 1991
[11] Somogyi, M, J. Biol. Chem. **195**, 19, 1952
[12] The Society for Biotechnology Japan, Experiments of Biotechnology, Baihukan, 1992

Studies in Surface Science and Catalysis, volume 159
Hyun-Ku Rhee, In-Sik Nam and Jong Moon Park (Editors)

Enzymatic Degradation and Detoxification of Diethyl Phthalate by *Fusarium oxysporum* f. sp. *pisi* Cutinase

Yang-Hoon Kim[a,b], Seung Wook Kim[a,c], and Jeewon Lee[a]*

[a]Department of Chemical and Biological Engineering, Korea University, Anam-Dong 5-1, Sungbuk-Ku, Seoul 136-713, South Korea

[b]R&D Center, Daesang Corporation, 125-8 Pyokyo, Majang, Ichon-Si, Gyeonggi-Do 467-813, South Korea

[c]Applied Rheology Center, Korea University, Anam-Dong 5-1, Sungbuk-Ku, Seoul 136-713, South Korea

1. INTRODUCTION

Phthalate esters are used in the production of polyvinyl chloride to make it flexible and workable and, to a lesser degree, in paints, lacquers, and cosmetics. Phthalates have been found in sediment, water, and air and have also been detected in foods, as they can migrate out of food-packaging materials [1]. Diethyl phthalate (DEP) is a phthalic ester that is present mainly in packaging materials (papers, paperboards, etc.) for aqueous, fatty, and dry foods [2]. DEP is known to cause decrease in the number of live offspring born to female animals that were exposed to DEP. DEP can be mildly irritating when applied to the skin of animals. It can also be slightly irritating when put directly into the eyes of animals [3].

Cutinase is a hydrolytic enzyme that degrades cutin, the cuticular polymer of higher plants [4]. Unlike the other lipolytic enzymes, such as lipases and esterases, cutinase does not require interfacial activation for substrate binding and activity. Cutinases have been largely exploited for esterification and transesterification in chemical synthesis [5] and have also been applied in laundry or dishwashing detergent [6].

In the present study, we investigated the efficacy of two lypolytic enzymes, fungal cutinase and yeast esterase, in the enzymatic degradation of DEP. During enzymatic DEP degradation for an extended period (3 days), degradation products that are not currently classified as endocrine-disrupting chemicals (EDCs) in animals, in humans or in vitro were detected, and their time-course compositional changes were also monitored. To evaluate the other types of potential toxicity of the DEP-derived products, the cellular toxicity of degradation products was measured in detail, using various recombinant bioluminescent bacteria.

2. MATERIALS AND METHODS

The purified fungal cutinase from *F. oxysporum f.* sp. *pisi* was kindly provided by Prof. C.M.J. Sagt in Utrecht University. The commercial product of esterase from *Candida cylindracea* was purchased from Boehringer Mannheim (Germany). The enzymatic degradation of DEP

(99%, Acros) was begun by adding 50 µl of the concentrated DEP solution (500 g l^{-1} in pure methanol) to 50 ml of the enzyme solution (10 or 100 mg l^{-1} in Tris-HCl buffer (10 mM, pH 8.0)). All chemical compounds in solvents were analyzed using GC/MS (HP6890 series GC-MSD; Hewlett Packard). Bioluminescent recombinant *Escherichia coli* strains were used to estimate the toxic effect of degradation products, if any. Strain GC2 (*lac::luxCDABE*) has the luxCDABE gene from *Xenorhabdus luminescens* under the control of the lac promoter, giving constitutive expression. Several other bioluminescent bacteria were used for evaluating possible modes of toxicity: DPD2794 (*recA::luxCDABE*) for detecting genotoxicity, TV1061 (*grpE::luxCDABE*) which is sensitive to protein damage, DPD2511 (*katG::luxCDABE*), sensitive to oxidative damage, and DPD2540 (*fabA:: luxCDABE*), sensitive to membrane damage [7]. The recombinant culture and test sample mixtures were prepared in triplicate in a highly sensitive 96-well microplate luminometer (Microlite^{TM}, CA) at 30°C, and the emitted bioluminescence (BL) was measured with a regular time interval for 6 h. The relative BL, i.e. [maximum BL measured with the supplement of final degradation products] · [maximum BL measured with the supplement of only buffer solution neither containing DEP nor enzyme]$^{-1}$, was used as a parameter indicating toxicity of each test sample.

3. RESULTS

The DEP-degradation rate of fungal cutinase was surprisingly high, i.e. almost 73% of the initial DEP (500 mg l^{-1}) was decomposed within 4.5 h and nearly 41% of the degraded DEP disappeared within the initial 15 min. With the yeast esterase, despite the same concentration (10 mg l^{-1}), more than 73% of the DEP remained even after 3 days of treatment (Fig. 1a). As shown Fig. 1b, the degradation kinetics of DEP was estimated for initial 15 min. Since the initial degradation results were well fitted by linear regression with high correlation coefficient in the semi-logarithmic plot of Fig. 1b, it seems reasonable to assume that the initial DEP breakdown follows a first-order kinetics, i.e. dN/dt = − kN where N and k represent residual DEP amount and degradation constant, respectively.

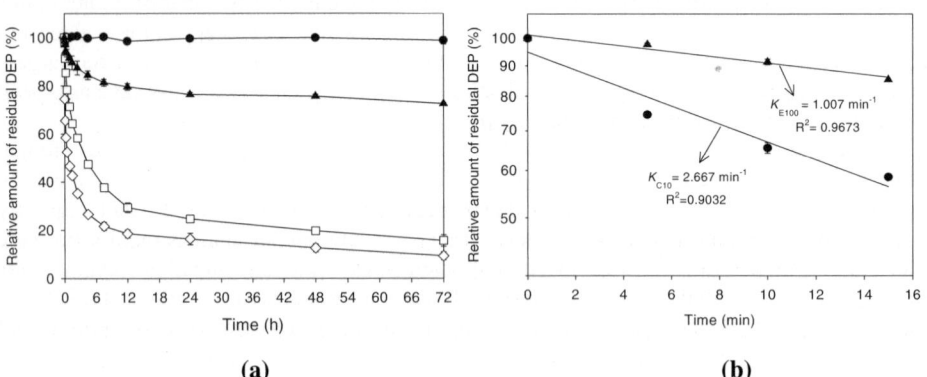

(a)   (b)

Fig. 1. (a) Time-course variation in residual amount of DEP during the enzymatic degradation of DEP, explained in terms of relative amount (i.e. initial DEP amount: 100%): ●, DEP (500 mg l^{-1}) in Tris-HCl buffer (10 mM, pH 8.0) without enzyme; ▲, DEP (500 mg l^{-1}) degraded by esterase (10 mg l^{-1}); □, DEP (500 mg l^{-1}) degraded by esterase (100 mg l^{-1}); ◇, DEP (500 mg l^{-1}) degraded by cutinase

(10 mg l⁻¹). (b) Semi-logarithmic plot of time-course variation in relative residual DEP amount for initial 15 min after the enzymatic degradation by cutinase (10 mg l⁻¹, ●) and esterase (100 mg l⁻¹, ▲) started.

The estimation of the initial degradation constants, k_{C10} and k_{E100}, for cutinase (10 mg l⁻¹) and esterase (100 mg l⁻¹), respectively, demonstrated that the initial DEP degradation by cutinase seemed to proceed almost 2.6 times faster.

During the enzymatic degradation of DEP, several DEP-derived compounds were detected, and time-course changes in composition were also monitored. In the enzymatic degradation of DEP, three organic chemicals were produced from DEP: 1,3-isobenzofurandione (IBF), ethyl methyl phthalate (EMP), and dimetyl phthalate (DMP). Apparently, IBF is a product by ester hydrolysis, followed by spontaneous oxo-bridge formation. EMP and DMP were produced probably via transesterification reaction in 0.1% methanol. The final chemical composition after 3 days was significantly dependent on the enzyme used. By fungal cutinase, DEP was converted mostly into 1,3-isobenzofurandione (IBF) that was a major degradation product. However, in the DEP degradation by yeast esterase, EMP was produced in abundance in addition to IBF and DMP (Fig. 2).

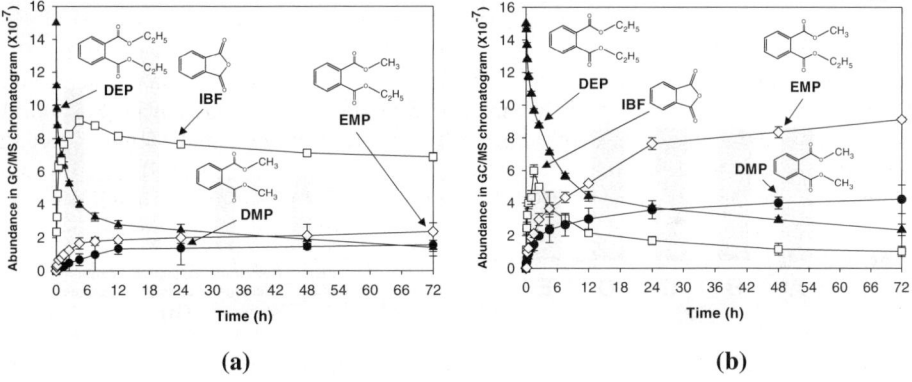

(a)   (b)

Fig. 2. Time-course variation in the amount of chemical compounds produced during DEP degradation by fungal cutinase (10 mg l⁻¹) (a) and yeast esterase (100 mg l⁻¹) (b), analyzed through GC/MS chromatography.

Although toxicity detection using recombinant bioluminescent *E. coli* is not directly related to a toxic effect in animals and/or humans, it is now generally accepted as a potential method for environmental monitoring of various industrial wastes and pollutants, including genotoxicants and stress inducers [7,8]. First, general toxicity due to nonspecific cellular stresses was analyzed by using strain GC2. Under increased cellular stress that represses cell growth, the BL emitted by the GC2 strain is subject to being decreased. As shown in Fig. 3a, the final products of degradation by *C. cylindracea* esterase significantly decreased the relative BL, and therefore, the presence of a toxic component(s) seems evident. Supplementing the bioluminescent culture with DEP (500 mg l⁻¹) did not decrease the BL (Fig. 3); hence, DEP did not seem to be toxic to bacterial cells, although phthalate esters like DEP have been known to potentially interfere with human endogenous hormones [7]. As is evident in Fig. 3a, no harmful effect was detected with the enzyme itself, cutinase (100 mg l⁻¹), or esterase (100 mg l^{-1}). The toxic effects of the final degradation products were also investigated, using various recombinant bioluminescent bacteria. As shown in Fig. 3b, DEP (500 mg l^{-1}) and the products of its degradation by cutinase never caused any cellular damage, which is consistent with the results shown in Fig. 3a. Fig. 3b shows that the degradation products from yeast esterase exerted oxidative damage as well as protein damage on the bacteria. Cellular stress by oxidative hazards like active oxygen species may be a great threat to the structures and functions of proteins, nucleic acids, lipids, and membranes, whether they are added externally or produced in intracellular region. Also, the increase of BL due to the induction of the *grpE* promoter indicated that defective protein synthesis took place in the presence of the degradation products from yeast esterase. For in situ DEP degradation with cutinase, the addition of the purified cutinase or fungus itself to DEP-contaminated sites does not seem to be practical. A plausible approach to the practical application of cutinase for DEP degradation may be to develop a microbial gene expression system for producing recombinant cutinase. The recombinant yeast system may have a great advantage in that extracellular production of recombinant enzyme is possible, and therefore, the cell-free medium containing a large amount of recombinant cutinase can be applied directly to the in situ degradation of DEP without costly purification.

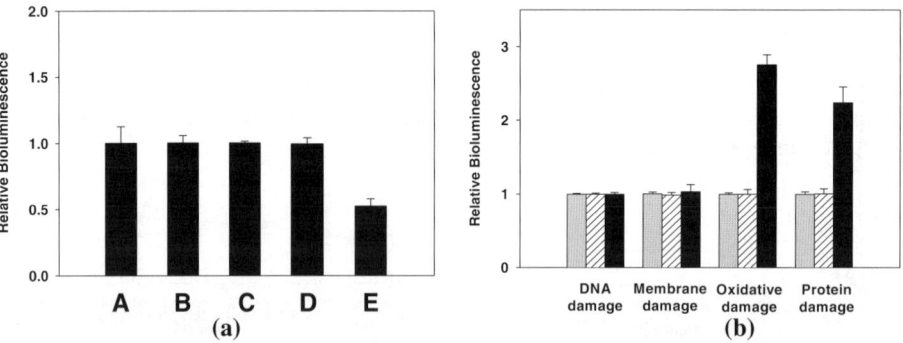

Fig. 3. (a) Results of estimation of relative bioluminescence (BL) by using strain GC2, showing the general toxicity of the test samples: A, Tris-HCl buffer containing DEP (500 mg l^{-1}); B, Tris-HCl buffer containing cutinase (100 mg l^{-1}); C, Tris-HCl buffer containing esterase (100 mg l^{-1}); D, final products of DEP degradation by cutinase; and E, final products of DEP degradation by esterase. (b) Results of the estimation of relative BL, using strains DPD2794, TV1061, DPD2511, and DPD2540. (gray columns: Tris-HCl buffer containing DEP (500 mg l^{-1}); striped and black columns: final products of DEP degradation by cutinase and esterase, respectively.

REFERENCES

[1] H.H. Sung, W.Y. Kao and Y.J. Su, Aquat. Toxicol. 64 (2003) 25.
[2] C.D. Cartwright, S.A. Owen, I. Thompson and R.G. Burns, FEMS Microbiol. Lett. 186 (2004) 27.
[3] Toxicological Profile for Diethyl Phthalate, U. S. Department of Health & Human Services, Public Health Service, Agency for Toxic Substances and Disease Registry, (1993).
[4] P. Kolattukudy, R. Purdy and I. Maiti, Method Enzymol. 71 (1981) 652.
[5] C. Carvalho, M. Aires-Barros and J. Cabral, Biotechnol. Bioeng. 66 (1999) 17.
[6] J. Flipsen, A. Appel, H. Van der Hijden and C. Verrips, Enzyme Microb. Technol. 23 (1998) 274.
[7] M.B. Gu, J. Min and E.J Kim, Chemosphere, 46 (2002) 289.
[8] J. Min, E. Kim, R. LaRossa and M. Gu, Mutat. Res. 442 (1999) 61.

Studies in Surface Science and Catalysis, volume 159
Hyun-Ku Rhee, In-Sik Nam and Jong Moon Park (Editors)

141

Biosorption of heavy metals and cyanide complexes on biomass

Seung Jai Kim[a, b], Jae Hoon Chung[a], Tae Young Kim[a], and Sung Yong Cho[a]

[a]Department of Environmental Engineering, [b]Environmental Research Institute,
Chonnam National University, Gwangju, 500-757, Republic of Korea

1. INTRODUCTION

Biosorption is a process that utilizes biological materials as adsorbents [Volesky, 1994], and this method has been studied by several researchers as an alternative technique to conventional methods for heavy metal removal from wastewater.

In this work, the waste brewery yeast and *Aspergillus niger* were used for the adsorption of lead, copper and cadmium, and their cyanide complexes. Biosorption equilibrium was studied in a batch reactor with respect to pH, initial concentration of heavy metal and metal-cyanide complex. Biosorption equilibrium over the temperature range of 288K – 308K was investigated and the biosorption heat was evaluated.

2. EXPERIMENTAL

Aspergillus niger was obtained from KCTC (Korean Collection for Type Cultures) and grown for five days at 23℃ in conical flasks which were kept in a rotary shaker agitated at 125rpm. The harvested biomass was pretreated and washed with generous amounts of deionized water. Waste yeast biomass obtained from a beer brewery was washed several times with deionized water and then dried in a vacuum drying oven at 80℃ for 48 h. The dried biomass was ground, sieved and stored in a sealed bottle with a silica gel to prevent resorption of moisture. Biosorption equilibria were studied for various pH and temperature. Heavy metal solutions were prepared by dissolving metal nitrates in deionized water. To study the effect of solution pH on heavy metal adsorption, the pH of the solution was adjusted between 2.5 and 6.0 with 1N HNO_3 or NaOH solution. The experiments were not conducted above pH 6.0 (5.0 for Pb) to avoid possible hydroxide precipitation. The effect of temperature on the adsorption equilibrium was studied at 288K, 298K, 303K and 308K. For heavy metal-cyanide complex experiments the molar concentration ratio of the cyanide to heavy metal in the solution was 4, and the initial pH of the solution was adjusted to 12. Samples were withdrawn at predetermined time intervals, filtered, and the heavy metal ion concentration was measured using ICP (Shimadzu ICPS7500).

3. RESULTS AND DISCUSSION

3.1. Adsorption equilibrium

Langmuir and Freundlich equations were applied to represent equilibrium adsorption data of heavy metals and metal-cyanide complexes. To find parameters for each adsorption isotherm equation, a linear least-squares method and pattern search algorithm were used. Langmuir and Freundlich isotherm parameters are obtained and listed in Table 1. The equilibrium isotherms were favorable type and the Langmuir equation represents our experimental data very well. The biosorption capacity of heavy metal as metal-cyanide anion complexes for two biosorbents was decreased significantly compared to that of metal ion only. The biosorption capacity of *Aspergillus niger* for each heavy metal was much greater than that of brewery yeast as can be seen in Table 1.

Table 1.
Isotherm parameters and correlation coefficients.

Biosorbent		Langmuir			Freundlich		
		q_m	b	R^2	K	n	R^2
Brewery yeast	Pb (pH5)	87.44	0.054	0.99	10.42	2.759	0.92
	Cu (pH5)	44.93	0.060	0.98	5.622	2.522	0.89
	Cd (pH5)	12.03	0.029	0.98	1.383	2.714	0.90
	Pb-CN (pH12)	21.76	0.012	0.98	3.030	1.339	0.84
	Cu-CN (pH12)	14.36	0.021	0.98	0.776	2.219	0.79
	Cd-CN (pH12)	9.442	0.022	0.99	0.613	1.650	0.82
Aspergillus niger	Pb (pH5)	303.5	0.070	0.99	57.27	3.222	0.91
	Cu (pH5)	134.2	0.107	0.95	37.03	4.250	0.80
	Cd (pH5)	81.40	0.064	0.98	24.97	3.997	0.88
	Pb-CN (pH12)	91.58	0.009	0.96	3.464	1.867	0.91
	Cu-CN (pH12)	33.43	0.040	0.97	1.512	1.433	0.85
	Cd-CN (pH12)	26.75	0.020	0.98	0.542	1.694	0.81

Langmuir eq. : $q(mg/g) = \dfrac{q_m bCe}{1+bCe}$ (1)   Freundlich eq. : $q(mg/g) = KC^{1/n}$ (2)

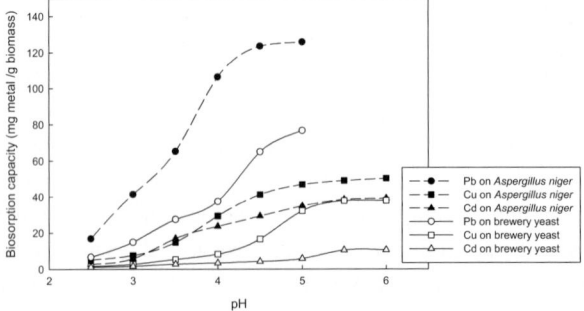

Figure 1. Effect of initial pH on the biosorption capacity (C_0: 200mg/L, biomass conc.: 1g/L, 25 ℃)

3.2. Effects of pH

The effect of pH on metal biosorption was studied at 25 ℃ by varying the solution pH from 2.5 to 6.0 (for Pb 5.0). The plot of metal adsorption capacity (mg/g) versus pH was shown in Figure 1. From the figure it is observed that the adsorption on biomass was highly pH dependent. The biosorption capacity increased with increasing pH and the effect of pH was in the order of Pb > Cu > Cd. The effect of pH on the biosorption capacity can be interpreted by the competition of the hydronium ions and metal ions for binding sites. At low pH values, the ligands on the biomass are closely associated with the hydronium ions, but when the pH is increased, the hydronium ions are gradually dissociated and the positively charged metal ions are associated with the free binding sites on the biomass.

3.3. Effects of initial metal concentration

The equilibrium time required for adsorption of metals on biomass was studied for various initial metal concentrations and the results were shown in Figure 2. The adsorption increases rapidly with time in the initial period of adsorption and approaches an equilibrium at about 120min for all the concentrations studied (10-100 mg/L). The slow but gradual increase of metal biosorption after 120min indicates that the adsorption occurs through a continuous formation of adsorption layer in the final period of adsorption.

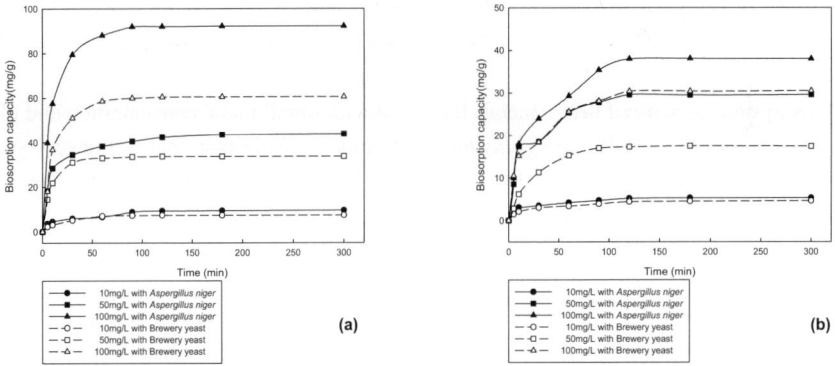

Figure 2. Biosorption capacities for different initial concentrations (pH 5.0, biomass conc.: 1g/L, 25 ℃), (a) Pb, and (b) Pb-CN.

3.4. Heat of adsorption

The heat of adsorption can be evaluated from adsorption equilibrium data and Eq. (3). If b values are known for different temperatures, the biosorption heats can be calculated from the plot of ln b versus $1/T$ [Ozer, 2003].

$$b = b_0 \exp\left[-\frac{\Delta H}{RT}\right] \qquad (3)$$

Where b_0 is a constant, ΔH (kcal·mol^{-1}) is the heat of adsorption, R is a universal gas constant $(1.987 \text{ cal·mol}^{-1}\text{·K}^{-1})$ and T is the absolute temperature (K). Heats of adsorption obtained are listed in Table 2. The values of biosorption heats for heavy metals show that the reaction is endothermic. The heat of physical adsorption is less than 1kcal mol^{-1}, and that of chemical adsorption is 20-50 kcal·mol^{-1} [Smith, 1981]. Since the heat of adsorption for heavy metals in this study are 3.30-5.35 kcal·mol^{-1}, we believe that both physical and chemical adsorptions are involved in the biosorption.

Table 2.
Adsorption heat of heavy metal on the biosorbent.

Biosorbent	Heavy metal	Biosorption heat (kcal·mol^{-1})
	Pb	4.06
Brewery yeast	Cu	3.30
	Cd	5.35
	Pb	4.44
Aspergillus niger	Cu	3.68
	Cd	4.61

4. CONCLUSION

In this study, biosorption of heavy metals and their cyanide complexes on *Aspergillus niger* and brewery yeast was investigated and the following conclusions are obtained.

- The biosorption capacity of heavy metals increased with initial metal concentration and pH.
- The biosorption capacity of *Aspergillus niger* was much greater than that of brewery yeast, and the biosorption capacity of metal-cyanide anion complexes was significantly lower than that of metal ion only.
- The equilibrium isotherms were favorable type and the Langmuir equation represents our experimental data very well.
- The biosorption reactions of heavy metals and metal-cyanide complexes were endothermic, and the heats of adsorption were in the range of 3.3-5.3 kcal·mol^{-1}, which imply that both physical and chemical adsorptions are involved.

ACKNOWLEDGEMENTS
This study was financially supported by research fund of Chonnam National University in 2003.

REFERENCES
[1] A. Özer and D. Özer, Comparative study of the biosorption Pb(II), Ni(II) and Cr(VI) ions onto S. cerevisiae: determination of biosorption heats, Journal of Hazardous Materials **B100**, 219-229, (2003).
[2] B. Volesky, Advances in biosorption of metals: Selection of biomass type. FEMS Microbiology reviews, **14**, 291-301(1994).
[3] Smith, J.M., Chemical Engineering Kinetic, 3rd Edition, McGraw Hill, New York (1981).

Studies in Surface Science and Catalysis, volume 159
Hyun-Ku Rhee, In-Sik Nam and Jong Moon Park (Editors)   145

Design of an aerated wetland for the treatment of municipal wastewater

Sung-Chul Kim and Dong-Keun Lee

Dept. of Chemical and Biological Engineering, Environmental Biotechnology National Core Research Center, Environmental and Regional Development Institute, Gyeongsang National University, Kajwa-dong 900, Jinju, Gyeongnam 660-701, Korea

1. INTRODUCTION

Constructed wetlands are an eco-friendly alternative municipal and industrial wastewater treatment [1-5]. The environment within a constructed wetland is mostly either anoxic or anaerobic, because there is no direct contact between the water column and the atmosphere. Some excess oxygen is supplied to the wastewater by the roots of the emergent plants, but this oxygen is likely to be used up in the biofilm growing directly on the roots and rhizomes, and is unlikely to penetrate very far into the water column itself. Therefore typical constructed wetland systems are not good for the treatment of raw wastewaters, because the pollutant loadings are too high to be treated successfully by the biological elements of the wetland. Experience shows that ammonia removal in a wastewater wetland is likely to be the limiting design factor, because nitrification is always limited by oxygen availability and natural aeration is a reactively show process in wastewater wetlands. If ammonia can be removed from the water column in the wetland to the desired level, then other pollutants will generally be removed to acceptable levels as well. By making the wetland aerobic, faster and more efficient biological nitrification is expected to occur within the wetland. Accordingly the required wetland size will be reduced greatly, and thereby the application of the constructed wetland becomes more practical for the treatment of raw municipal wastewaters.

In this study, a constructed wetland was designed to remove BOD_5 together with total nitrogen (T-N) from municipal wastewater. The designed wetland was composed of the aerobic tank and anaerobic/anoxic one which was connected in series immediately after the aerobic one, and could treat 100 m^3 raw municipal wastewater every day.

2. EXPERIMENTAL

The constructed wetland was composed of the two tanks connected in series; the one is the aerobic tank and the other is the anaerobic/anoxic one (Figure 1). The former tank could remain aerobic owing to the continuous supply of air through the natural draft system (Figure 2) whose driving force for airflow was the temperature difference between the

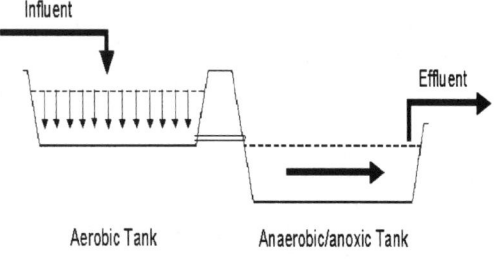

Figure 1. The schematic diagram of the constructed wetland.

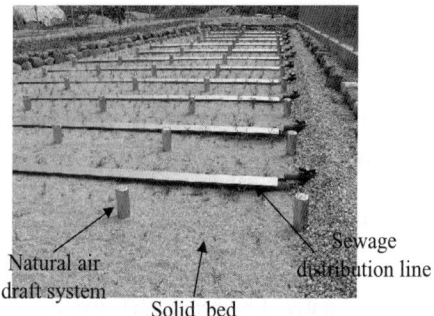

Figure 2. The natural air draft system installed inside the aerobic tank of the constructed wetland.

Figure 3. The detailed photograph of the wastewater distribution lines onto the surface of the aerated constructed wetland.

ambient air and the air inside the tank [6]. The aerobic tank had 290 m^2 surface area and 1.2m depth, and accordingly the theoretical hydraulic residence time (HRT) in the tank was 3.48days. From the bottom of the tank spherical gravels (25mm diameter) were packed upto the depth of 0.9m above which tiny sands (2.5mm diameter) were packed again with 0.3m depth. Reeds were planted on the sand level.

At immediate after the aerobic tank was installed the anaerobic/anoxic one whose surface area and depth were 580m^2 and 1.5m, respectively. The theoretical HRT of the tank was 8.7days. Municipal wastewater was distributed onto the surface of the aerobic tank by using the wastewater distribution lines (Figure 3). The distributed wastewater flows vertically under gravity force.

3. RESULTS AND DISCUSSION

The influent and effluent concentrations of BOD$_5$ and T-N in the constructed wetland are shown in Figure 4. The average influent concentrations of BOD$_5$ and T-N were 80.0 mg/L and 37.9 mg/L, respectively. After being treated at the aerobic tank, less than 80 % of BOD$_5$ was removed and the average effluent BOD$_5$ concentration from the aerobic tank was 10.8 mg/L. At the aerobic tank about 30.5 % of T-N was removed, and more than 96 % of the organic

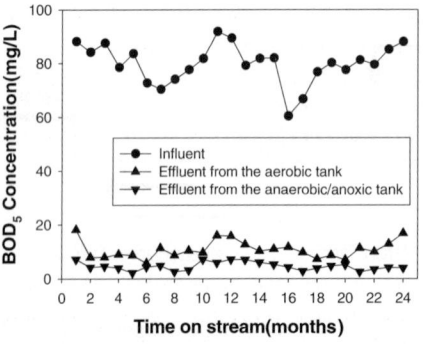
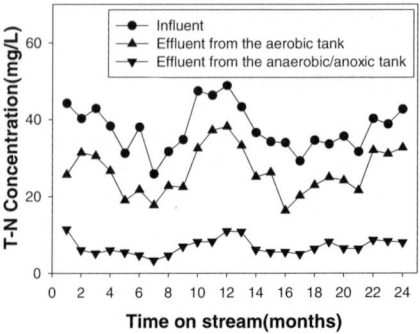

Figure 4. The influent and effluent concentrations of BOD$_5$ (left) and T-N (right) in the constructed wetland.

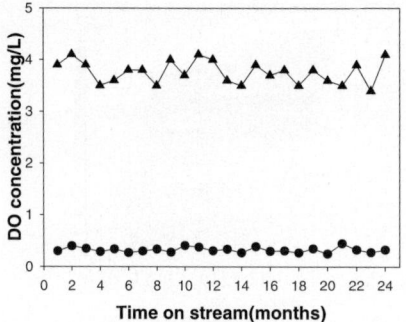

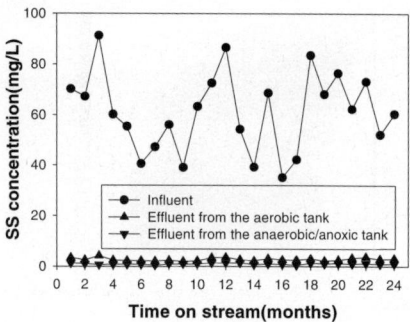

Figure 5. DO concentration in the influent (●) and the effluent (▲) from the aerobic tank of the constructed wetland.

Figure 6. The influent and effluent concentrations of SS in the constructed wetland.

nitrogen and ammonia nitrogen were nitrified successfully into nitrate nitrogen, which is indicative of the successful biological nitrification.

The changes in DO concentration can be clearly seen from the data in Figure 5. The DO concentrations in the influent wastewater were about 0.3mg/L. DO concentrations in the effluents were, however, more than ten times higher than the corresponding values in the influent. This result indicates that the natural air draft system supplies in the aerobic tank with sufficient oxygen for the biochemical oxidation.

Additional removal of BOD_5 and T-N could be obtained at the anaerobic/anoxic tank of the constructed wetland. About 57 % of the remaining BOD_5 was removed at the anaerobic/anoxic tank. More than 74 % of the remaining T-N was denitrified, and the average concentration of T-N at the final effluent was 6.9 mg/L. The maximum available capacity for nutrient uptake in plants was far less (of the order of 5 %) than the loading rate of nutrients to the constructed wetland.

Figure 6 shows the efficiencies of SS removal in the aerobic and the anaerobic/anoxic tank. More than 95% of the initial SS was removed in the aerobic tank. Additional 3.4% removal of SS could proceed in the anaerobic/anoxic tank again.

Biofilm concentrations attached onto the surface of the sands and gravels in the aerobic tank were measured at 120 different location of the constructed wetland at different times (6 months, 12 months, 18 months and 24 months). The concentrations of the biofilm increased continuously upto 6 month operation, but thereafter remained almost unchanged upto 24 months. Figure 7 and 8 show the concentrations and distributions of biofilms attached onto the the surface of the sands and gravels of the aerobic tank, respectively. The attached biofilm was well distributed over the aerobic tank, and the total volume of the attached biofilm occupied just about 0.5% of the initial void volume of the aerobic tank.

4. CONCLUSION

The constructed wetland with the aerobic tank and anaerobic/anoxic tank connected in series was employed for the treatment of raw municipal wastewater. More than 94% of the initial BOD and 80% of the initial T-N could be removed. Successful biological nitrification was

148

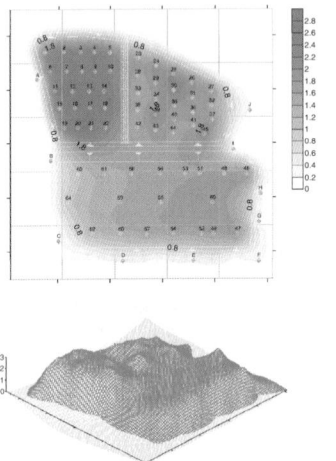 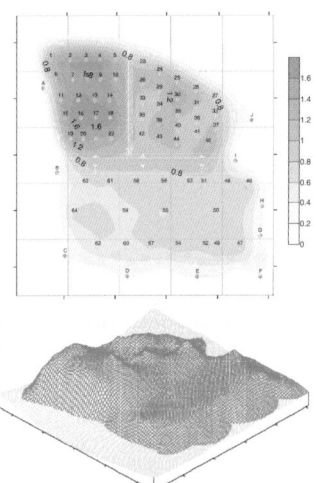

Figure 7. The amounts and distribution of biofilm attached onto the surface of the sands after 2 year operation.

Figure 8. The amounts and distribution of biofilm attached onto the surface of the gravels after 2 year operation.

accomplished at the aerobic tank. The remaining BOD and organic SS were further consumed during the subsequent biological denitrification at the anaerobic/anoxic tank.

ACKNOWLEDGEMENT

This work was supported by a grant from the KOSEF/MOST to the Environmental Biotechnology National Core Research Center (grant #: R15-2003-012-02002-0).

REFERENCES

[1] Department of Land and Water Conservation (DLWC). *The Constructed Wetlands Manual*, New South Wales, Australia, 1998.
[2] R. Crites and G. Tchbanoglous, *Small and Decentralized Management Systems*, McGraw-Hill, New York, 1998.
[3] K.M. Leonard, and G.W. Swanson, *Water Science and Technology*, 43 (2001) 301.
[4] S.C. Reed, R.W. Crites and E.J. Middlebrooks. *Natural Systems for Waste Management and Treatment*, McGraw-Hill, New York, 1988.
[5] U.S. Environmental Protection Agency, *Subsurface Flow Constructed Wetlands for Wastewater Treatment; A Technology Assessment*, Report EPA/ 832-R/93/001.Washington, D.C, 1994.
[6] E.D. Schroeder and G. Tchbanoglous, *Journal of Water Pollution Control Fedration*, 48 (1976) 771.

Studies in Surface Science and Catalysis, volume 159
Hyun-Ku Rhee, In-Sik Nam and Jong Moon Park (Editors)

Fermentative hydrogen production from food waste

Y. Kim, J.H. Jo, D.S. Lee and J.M. Park

Advanced Environmental Biotechnology Research Center, School of Environmental Science and Engineering, POSTECH, San 31, Hyoja-dong, Nam-gu, Pohang, Gyeongbuk 790-784, Republic of Korea

1. INTRODUCTION

Hydrogen is considered as a clean and efficient energy source of the future. It has a higher specific energy content per unit mass (122 kJ g^{-1}) than methane and it produces only harmless water when it burns [1]. Most hydrogen is currently produced by physicochemical processes [2]; but these processes are uneconomic and undesirable since they require external energy sources. In recent years, biological hydrogen production processes have received great interests since they are not only environmental friendly, but also able to utilize renewable energy sources which are inexhaustible [2]. Biological hydrogen production can be classified into two categories: photo fermentation and dark fermentation. Especially, dark fermentation generating hydrogen from organic compounds is a more feasible process for the sustainable hydrogen production. It can produce hydrogen continuously without sunlight and establish high hydrogen production rate due to high cell growth rate [3]. In the mean time, food waste has been a great concern in Korea. It amounts to 11,397 tons per day accounting for 22.5% of municipal solid waste [4]. However major recycling methods such as composting and feed stuffing are not appropriate as food waste has high salinity and landfill is no longer allowed. Since food waste has a high organic content, it can be used for hydrogen production in dark fermentation. Therefore, dark fermentation can facilitate waste recycling, which can be a feasible alternative to the food waste recycling methods.

2. EXPERIMENTAL

A fed-batch type reactor with 5 liter working volume has been operated for the production of hydrogen. Fig. 1 shows a schematic diagram of the experimental set-up. Food waste collected from a dining hall of POSTECH (Pohang, Korea) was utilized as an organic substrate and Table 1 shows the characteristics of food waste. The seed sludge was obtained from an anaerobic digester of Pohang wastewater treatment plant in Korea. The collected sludge was heat-treated for 20 minutes at 95℃ to inactivate methanogens and harvest spore-forming anaerobic bacteria. The reactor was inoculated with 1.5 liter of the heat-treated sludge and filled with the feed to the working volume. The reactor was operated at 35℃ with 150 rpm agitation. During the start-up period, the reactor was operated in a batch mode for the germination of endospores. After the acclimation of sludge it was converted into a fed-batch operation mode with hydraulic retention time (HRT) of 12.5 d which was step-wisely shortened. The pH was maintained between 5.2-5.5 with 5 N NaOH to inhibit methanogenesis. The gas and liquid samples were taken daily during the operation period. Biogas composition

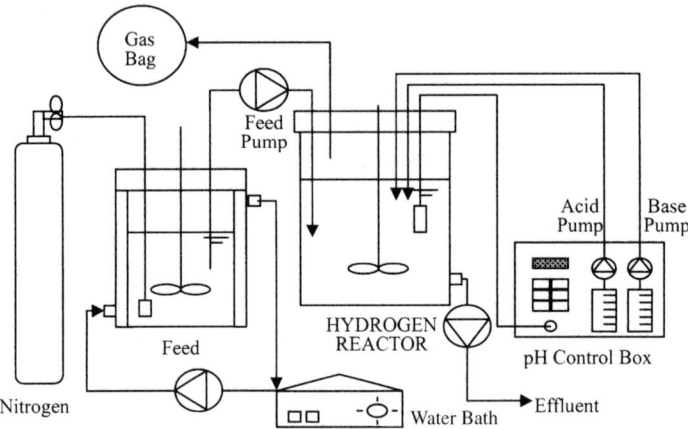

Fig. 1. Schematic diagram of the experimental set-up

was analyzed by a gas chromatography (GC, Hewlett Packard) equipped with a thermal conductivity detector (TCD) and a caboxen-1010 capillary column. The oven, injector and detector were kept at 120℃, 150℃ and 200℃, respectively. Volatile fatty acids (VFAs) and alcohols were also analyzed by a GC equipped with a flame ionization detector (FID) and a HP-INNOWax capillary column. Both the injector and detector were kept at 250℃, while oven was held at 60℃ for 1 min, heated to 220℃ at 10℃ min^{-1} and maintained at 220℃ for 1 min. Chemical oxygen demand (COD), total solids (TS), volatile solids (VS) and alkalinity of the samples were determined according to the standard methods [5].

Table 1
Characteristics of food waste

Item	Unit	Value
Physical characteristics		
Density	g l^{-1}	1008.1
Moisture Content	%	76.0
TS	%	24.0
VS	%	22.6
VS/TS		0.94
Elementary analysis		
Carbon, C	% TS	47.9
Hydrogen, H	% TS	6.8
Oxygen, O	% TS	40.4
Nitrogen, N	% TS	3.7
Sulfur, S	% TS	0.3
C/N		13.3

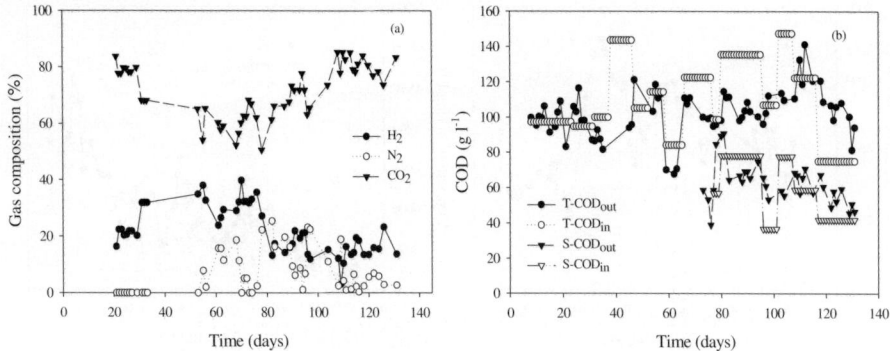

Fig. 2. Biogas composition (a) and COD concentrations in the influent and the effluent (b)

3. RESULTS AND DISCUSSION

The reactor has been operated for six months. The variations in the volume and composition of the biogas are shown in Fig. 2 (a). Hydrogen and carbon dioxide were the main biogas produced. After a steady state hydrogen production was observed at HRT of 12.5 d, HRT increased to 6.25 d at day 75. The hydrogen content which reached up to 39.7% at HRT of 12.5 d was abruptly reduced to 21.9%. Since the hydrogen production continued to decrease, HRT was increased to 12.5 d at day 104. However, the microbial activity was not restored. After the dilution ratio of food waste was increased by 3 times at day 126 on the assumption that the decline was caused by a substrate inhibition, the highest hydrogen content was slightly increased to 23.4%. The maximum hydrogen rate and the maximum hydrogen yield are 0.5 l-H_2 l^{-1} d^{-1} at day 76 and 1040.6 l-H_2 kg-$COD^{-1}_{consumed}$ at day 56, respectively. Methane gas was not observed showing methanogens were well inhibited by a heat-treatment and low pH. A detected nitrogen gas was might be originated from the leak of an air bag and/or air intrusion from the feeding system.

Fig. 2 (b) shows the changes of COD concentrations in the influent and the effluent. The food waste was collected at intervals of 1 to 2 weeks resulting in the high fluctuation of COD content in the influent. Therefore, it was so difficult to investigate the changes in COD caused by the HRT variations that the amount of food waste for a month was prepared from day 117. From the day 117, COD concentration started to decrease drastically since the dilution ratio of food waste to water was changed from 1:1 to 1:3. The highest COD removal efficiency was 35% at day 45 at HRT of 12.5 d. The residual COD was due to VFAs and alcohols in the effluent. COD in the effluent ranged from 141.2 to 70.1 g l^{-1}. Since the dilution ratio was changed to 1:3, it has not yet reached steady state.

The formation of hydrogen is accompanied with VFAs or solvent production during an anaerobic digestion process. Therefore, the distribution of VFA concentrations and their fractions is a useful indicator for monitoring hydrogen production. Fig. 3 shows the variations in alcohol and VFAs. Most of the VFAs were analyzed as acetate and butyrate, and most of the alcohols were analyzed as ethanol. The propionate concentration was below the analytical limit. It indicates that the anaerobic pathway in the reactor is not propionic-type fermentation but butyrate-type fermentation. *Clostridium butyricum* is considered to be the dominant

152

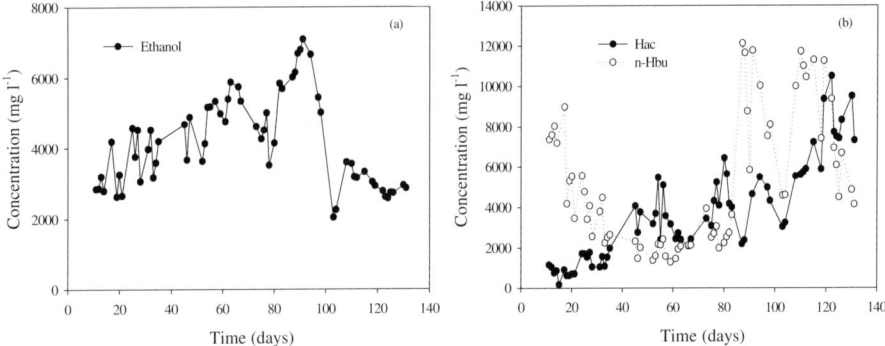

Fig. 3. The variations in concentration of alcohol (a) and VFAs (b)

organism for producing hydrogen because this organism is reported to be responsible for butyrate type fermentation. However, the pathway may be changed to butanol fermentation when pH reaches about 4.0, due to accumulation of VFAs. Therefore, pH is one of the key parameters as the hydrogen production process has a high potential to fail once the pathway is changed to butanol [6-7].

4. CONCLUSION

In this study, we have tested the possibility and the feasibility of hydrogen production from food waste by fermentation. The experimental results showed that hydrogen production could be possible from food waste. But we obtained a relatively low hydrogen production rate. In addition, COD concentrations are very high due to slowly biodegradable matters. Therefore, a further study is needed to maximize hydrogen production by optimizing operational parameters such as HRT, temperature, pH and agitation speed. It can be achieved in a statistical way based on Response Surface Methodology (RSM). A two-stage process could be developed to treat high organic solid in the effluent further in the near future.

ACKNOWLEDGEMENTS
This work was financially supported by the Korea Science and Engineering Foundation through the Advanced Environmental Biotechnology Research Center at Pohang University of Science and Technology.

REFERENCES
[1] J-J. Lay, YJ Lee and T. Noike, Water Res., 33 (1999) 2579
[2] D. Das, T. N. Veziroglu, Int. J. Hydrogen Energy, 26 (2001) 13
[3] T. Noike, O. Mizuno, Hydrogen Water Sci. Technol., 42 (2000) 155
[4] Ministry of Environment. Internet Homepage of Ministry of Environment http://www.me.go.kr, 2004
[5] APHA, Standard methods for the examination of water and wastewater, 16th edition, American Public Health Association, Washington, DC, 1985
[6] R. Nandi and S. Sengupta, Crit. Rev. Microbiol., 24 (1998) 61
[7] Y. Ueno, T. Kawai, S. Sato, S. Otsuka and M. Morimoto, J. Fement. Bioeng., 79 (1995) 395

Studies in Surface Science and Catalysis, volume 159
Hyun-Ku Rhee, In-Sik Nam and Jong Moon Park (Editors)

153

Biodiesel from Transesterification of Cottonseed Oil by Heterogeneous catalysis

H. Chen and J.-F. Wang[*]

Department of Chemical Engineering, Tsinghua University, Beijing 100084, China

1. INTRODUCTION

During the past 20 years, diesel fuel consumption in developing countries has increased rapidly, and it is important to find substitute energy sources because of the limited crude oil resource and serious pollution in the world. Fatty Acid Methyl Esters (FAMEs, known as biodiesel; $RCOOCH_3$), obtained by transesterification of natural triglycerides, have assumed importance as a potential diesel fuel extender for their several advantages: biodegradability, non-toxicity, and renewability of the source. Furthermore, biodiesel has similar fuel characteristics to diesel, and either blends or 100% alternative fuel can be used without major modifications to diesel engines. Among the plant oils, cottonseed oil is a suitable raw material. The production of cotton seed oil is nearly 40 million ton/yr in the world and over 10 million ton/yr in China. Thus, it is of significance to look for a useful outlet for cottonseed oil.

The conversion of the triglyceride involves three consecutive reactions with intermediates:

Triglycerides + CH_3OH $\rightleftharpoons$ Diglycerides + R_1COOCH_3 (1a)
Diglycerides + CH_3OH $\rightleftharpoons$ Monoglycerides + R_2COOCH_3 (1b)
Monoglycerides + CH_3OH $\rightleftharpoons$ Glycerol + R_3COOCH_3 (1c)

Generally, the above transesterification reactions are catalyzed by strong acids or alkalis [1, 2]. In the homogeneous catalytic process by acids or alkalis, neutralization is required of the product. This post-treatment produces waste water, and increases equipment investment and production cost. Recently, more attention has been paid to the heterogeneous catalysis process [3] for an easier production process and to reduce pollution of the environment.

Of all the solid base catalysts, oxides of alkaline earth (MgO, CaO et.) are the most used. They have high basicity and stability. Magnesium aluminum mixed oxides derived from hydrotalcites exhibit strong basicity capable of catalyzing various reactions. As compared to MgO or CaO, metal oxides as solid acid usually need some chemical transformation to enhance the acidity. We note that solid super acids such as ZrO_2 -SO_4^{2-} or TiO_2-SO_4^{2-} also have strong acidity and high stability. In this work, the activity and stability of the above solid catalysts for the transesterification reaction were studied systematically.

---

[*] Corresponding author. E-mail: wangjf@flotu.org

2. EXPERIMENTAL

2.1. Preparation of catalysts

MgO samples were prepared by heating basic magnesium carbonate for 3 h. CaO samples were made from calcium carbonate pretreated in N_2 at 1173 K for 2 h. MgO-Al$_2$O$_3$ samples were prepared by heating hydrotalcite at 823 K for 4 h.

The solid super acids (ZrO$_2$-SO$_4$$^{2-}$, TiO$_2$-SO$_4$$^{2-}$) were prepared by mounting H$_2$SO$_4$ on TiO$_2$·nH$_2$O, Zr (OH)$_4$, respectively, followed by calcining at 823 K.

2.2. Reaction procedure and analysis

The transesterification reactions were conducted in a sealed 250 ml autoclave equipped with a stirrer. The molar ratio of methanol to oil was 12:1, reaction temperature was 200°C-230°C, and the ratio of catalyst to oil was about 2 wt%. Samples were taken out from the reaction mixture and biodiesel portions were separated by centrifuge.

The concentration of biodiesel (fatty acid methyl esters) and glycerides were analyzed by liquid chromatography (Shimadzu-10A HPLC). An ODS-2 column (250×4.6mm) was used for the separation. The flow rate of the mobile phase (acetone: acetonitrile=1:1) was set to 1 ml/min. Peaks were identified by comparison with reference standards. Standards of methyl esters, monoglycerides, diglycerides and triglycerides were bought from Fluka.

3. RESULTS AND DISCUSSION

The activities of the heterogeneous catalysts toward transesterification were measured.

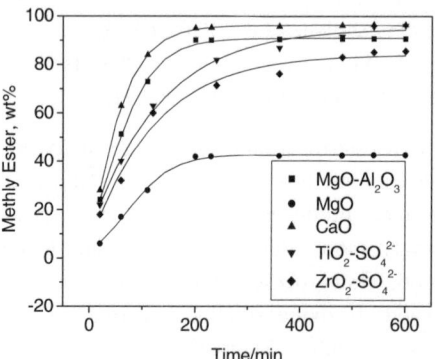

Fig. 1. Methyl ester concentration versus reaction time for solid bases and acids catalysts. Temperature: 230°C, catalyst:2wt%, CH$_3$OH/oil = 12:1.

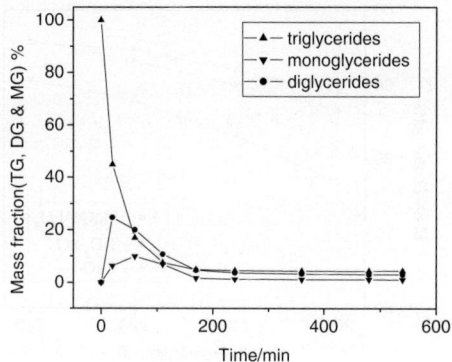

Fig. 2. Reactant and intermediate product profiles for MgO-Al$_2$O$_3$ catalyst. Temperature: 230°C, catalyst: 2wt%, CH$_3$OH/oil = 12:1.

Figure 1 shows the methyl ester yield as a function of the reaction time for the transesterification of cottonseed oil catalyzed by the solid bases and acids. Of the solid bases, CaO and MgO-Al$_2$O$_3$ were more effective for this reaction. Over 90% methyl ester was obtained in less than 3 hours. The reactions catalyzed by the solid acids were slower than that by the solid bases, which need about 9 hours to reach high concentration of ME. The different plateau values of the solid bases may be caused by the deactivited of the catalysts.

Figures 2 and 3 illustrate the reactant and intermediate product profiles at different reaction times for MgO-Al$_2$O$_3$ and TiO$_2$-SO$_4^{2-}$ catalysts, respectively. The results in Fig. 2 show that over 95% triglyceride (TG) is converted to methyl ester and intermediate products within 3 h and the intermediate products of monoglyceride (MG) and diglyceride (DG) remain at a very low concentration level of 1.5% and 4%, respectively.

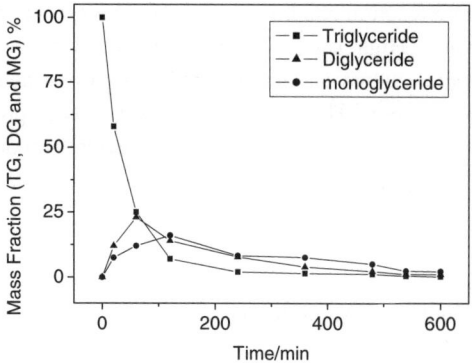

Fig. 3. Reactant and intermediate product profiles for TiO$_2$-SO$_4^{2-}$ catalyst. Temperature: 230°C, catalyst: 2wt%, CH$_3$OH/oil = 12:1.

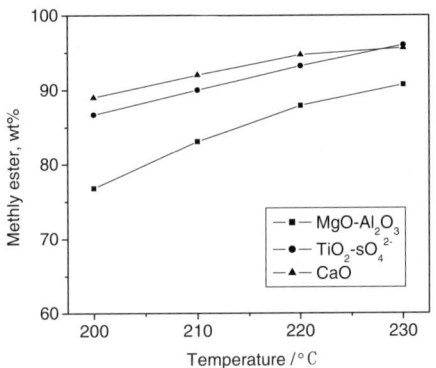

Fig. 4. Methyl ester concentration at different temperatures for catalysts $MgO-Al_2O_3$ and $TiO_2-SO_4^{2-}$. Catalyst: 2wt%, $CH_3OH/oil = 12:1$.

Figure 3 shows the results using $TiO_2-SO_4^{2-}$ catalyst. It can be seen that a high TG conversion and low MG and DG concentrations were obtained only under the condition of a long reaction time of about 9 h.

The reaction temperature can have a large effect on the activities of the catalysts. Figure 4 shows the concentration of methyl ester with the solid base and acid catalysts at different temperatures after 10 hours' reaction. The results show that the concentration of methyl ester with $MgO-Al_2O_3$ catalyst decreased more quickly than that with $TiO_2-SO_4^{2-}$ and CaO catalysts. This means that the activation energy of the reverse reactions with $MgO-Al_2O_3$ catalysts is higher.

4. CONCLUSION

Experiments showed that high methyl ester yields can be achieved with solid bases and super acids under moderate reaction conditions. The solid bases were more effective catalysts than the solid super acids. High stability can be achieved by an ordinary inexpensive preparation process, and the catalyst can be separated easily from the reaction products in the heterogeneous catalysis process. The costly catalyst removal process can be avoided compared with the homogeneous process. Therefore, the heterogeneous process using a solid catalyst should be more economical for biodiesel production.

REFERENCE
[1] H. Noureddini, D. Harkey, V. Medikonduru, J. Am. Oil Chem. Soc., 12 (1998) 1775.
[2] E. Crabbe, C.N. Hipolito, G. Kobayashi, K. Sonomoto and A. Ishizaki, Process Biochemistry, 37 (2001) 65.
[3] G. J. Suppes, M. A. Dasari, E. J. Doskocil, P. J. Mankidy and M. J. Goff, Applied Catalysis A: General, 257 (2004) 213.

Studies in Surface Science and Catalysis, volume 159
Hyun-Ku Rhee, In-Sik Nam and Jong Moon Park (Editors)

Oxygen evolution rate of photosynthetic microalga *Haematococcus pluvialis* depending on light intensity and quality

Y. C. Jeon, C. W. Cho and Y.-S. Yun [*]

Div. of Environmental and Chemical Engineering and Dept. of Bioprocess Engineering, Chonbuk National University, 664-14 Duckjin-Dong, Jeonju, Jeonbuk 561-756, Republic of Korea
* ysyun@chonbuk.ac.kr

ABSTRACT

The green microalga *Haematococcus pluvialis* has a potential for production of an antioxidant astaxanthin (3,3'-dihydroxy-β,β'-carotene-4,4'-dione) from carbon dioxide as a carbon source. Therefore, this alga has been studied as a model microorganism for fixation of carbon dioxide and for production of value-added chemicals. In this study, a system for the measurement of photosynthetic activity of microalgae was developed and used for evaluating the effects of light quality and spectra on the oxygen evolution rate of *H. pluvialis*. The photosynthetic activity was measured by the system using various types of sun light, green, and red light filters. As increasing the light intensity, the photosynthetic activity increased and eventually reached a maximum value, e.g., 59.7 ± 2.3 mg g^{-1} h^{-1} under red light condition at 0.123 g L^{-1} of cell concentration. However, each photosynthesis-irradiation curves were different at the same light intensity because these microalgae absorb selectively the light that could be used easily for photosynthesis.

1. INTRODUCTION

Photoautotrophic microalgal mass culture has been extensively studied with various purposes such as production of biomass as a source of fine chemicals or foods. Among these algal products, astaxanthin is of great commercial interest [1, 3] because of its high price (approximately US\$ 2,500 kg^{-1}). Although some kinds of plants, bacteria, and yeast are known to synthesize astaxanthin, the green alga *Haematococcus pluvialis* is capable of accumulating a superior amount of astaxanthin to other sources [4]. Therefore, *H. pluvialis* is mainly used for the commercial production of natural astaxanthin [4, 5].

Light, the essential energy source, is considered as one of the most important factors affecting the microalgal photosynthetic activity [6]. Since light cannot be stored in the culture system, it must be continuously supplied to support microalgal growth. In addition, light cannot deeply penetrate into the dense microalgal suspension due to absorption and scattering of light, which makes it difficult to maintain high growth and CO_2 fixation rates [7]. In this study, the experiment was focused on photosynthetic activity of *H. pluvialis* because photosynthetic activity was primarily affected by light. To study the effects of light intensity and quality on the photosynthetic activity of *H. pluvialis*, the specially designed equipment was used for the measurement of microalgal photosynthetic oxygen evolution rate.

158

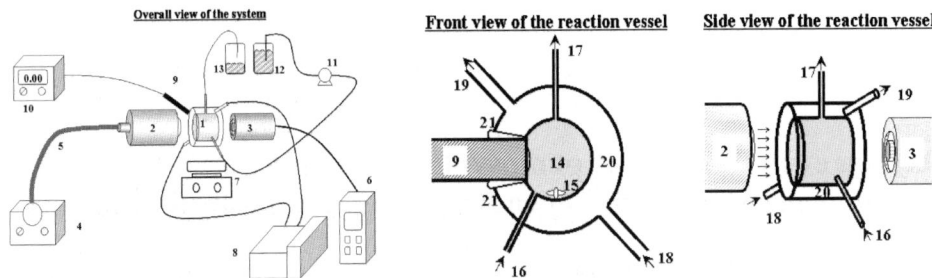

Fig. 1.   Schematic diagram of the photosynthetic activity measurement system and the reaction vessel: 1, reaction cell; 2, convex lens; 3, quantum sensor; 4. quartz halogen illuminator; 5, goose-neck-type optical fiber; 6, data logger; 7, magnetic stirrer; 8, water bath; 9, dissolved oxygen electrode; 10, dissolved oxygen meter; 11, peristaltic pump; 12, sample reservoir; 13, waste reservoir; 14, microalgal suspension; 15, magnetic bar; 16, inlet of sample; 17, outlet of sample; 18, inlet of cooling water; 19, outlet of cooling water; 20, cooling water jacket; and 21, septum.

2. MATERIALS AND METHODS

2.1. Microalgal strain and cultivation

The green microalga *Haematococcus pluvialis* UTEX 16 obtained from the Culture Collection of Algae at the University of Texas, Austin was used in this study. The alga was cultivated in 0.25 L flask with 0.1 L of sterilized OHM medium [8]. The culture medium was sterilized at 121 ℃ for 5 min before use. The culture flask was agitated on a shaker at 170 rpm and 25 ℃ with sterilized air bubbling. Light was supplied continuously at 20 μE m^{-2} s^{-1} on average with 20 W warm-white fluorescent tubes (GE-Korea, Seoul, Korea).

2.2. Preparation of algal suspensions

The *H. pluvialis* having green color under exponential phase was used to be centrifuged at 3000g for 5 min at a room temperature and washing with the fresh medium. After three cycles of centrifugation and washing, the suspension was diluted consecutively to make different concentrations of algal suspensions using fresh medium. The measurement of dry cell weight was carried out by drying 5ml of the suspension at 60 ℃ in a drying oven for 24 h after being filtered through a pre-dried and pre-weighed 0.45 μm nitrocellulose membrane filter (Millipore, USA). The algal suspensions with different cell concentrations were shaken for 2 h at 170 rpm and 25 ℃ in a dark condition in order to remove any residual effects of previously exposed light. When needed, nitrogen gas was bubbled to decrease the concentration of dissolved oxygen. The resulting suspensions were used to measure the oxygen production rate.

2.3. Measurement of photosynthetic activity

The oxygen evolution rate was measured by using the photosynthetic activity measurement system (Fig. 1). When light was illuminated to the reaction vessel, algal cells began to evolve oxygen and the linearity between dissolved oxygen and time was observed just after a few

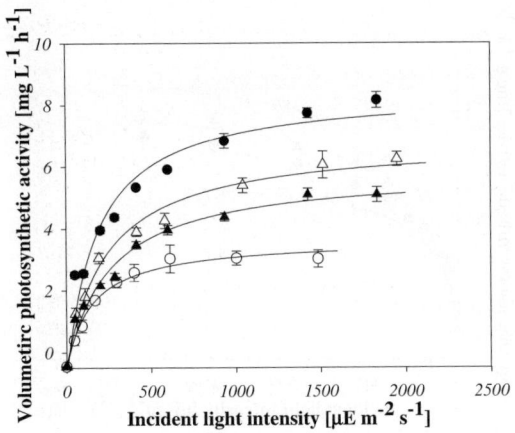

Fig. 2. Effects of light intensity and quality, and cell concentration on the volumetric oxygen evolution rate by algal photosynthesis. Light types and cell concentration used: (●): simulated daylight and 0.215 g L^{-1}; (▲): simulated daylight and 0.123 g L^{-1}; (△): red light and 0.123 g L^{-1}; and (○): green light and 0.123 g L^{-1}.

minutes. The volumetric oxygen production rate was measured from the slope. The measurement system reproduced consistent results. Detailed information on the measurement system and measuring procedure are available in the previous report [9].

3. RESULTS AND DISCUSSION

Algal photosynthetic activity was measured at various incident light intensities and cell concentrations (Fig. 2). When the incident light intensity was less than 500 µE m^{-2} s^{-1}, the volumetric activity rose linearly according to the incident light intensity. The volumetric activity was saturated at a higher level of incident light. No light inhibition (similar to substrate inhibition in heterotrophic microbial culture) was observed in this experiment. The volumetric activity was significantly affected by the algal cell concentration. As increasing the algal concentration, the maximum volumetric activity increased. Additionally, a denser density suspension required a higher light intensity for its maximal activity.

As can be seen in Fig. 3, the specific activity reached the maximal value at the high light intensity similar to the volumetric activity. The specific activity decreased as increasing the cell concentration, which was opposite to the dependence of cell concentration on the volumetric activity. The volumetric and specific activities are likely to closely relate to the volumetric and specific growth rate of algal culture, respectively. High-density culture usually gives a higher volumetric productivity of biomass but does not have higher specific growth rates than low-density culture because the light is significantly attenuated at high cell concentrations [10, 11].

To investigate the effect of light spectra on the photosynthetic activity, three types of light (red, green, and simulated daylight) were provided for algal photosynthesis. There was no difference in activity order between volumetric and specific activities at the same algal concentration of 0.123 g L^{-1}. Meanwhile, the effect of light source was obviously observed. As shown in Fig. 2 and 3, red light was more effectively utilized for photosynthetic activity than green light and even simulated daylight which covered whole range of PAR. This was

160

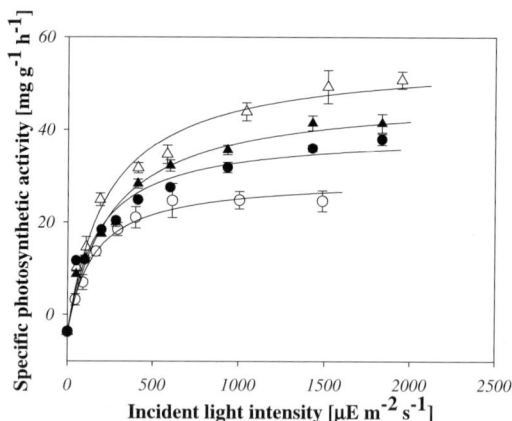

Fig. 3. Effects of light intensity and quality, and cell concentration on the specific oxygen evolution rate by algal photosynthesis The light types and cell concentrations were (●): simulated daylight and 0.215 g L^{-1}; (▲): simulated daylight and 0.123 g L^{-1}; (△): red light and 0.123 g L^{-1}; and (○): green light and 0.123 g L^{-1}

probably because of the wavelength-specific pigment of green alga for light harvesting. Green algae like *H. pluvialis* used in this study have Chl *a* as a major pigment which absorbs preferentially red (around 650 nm) and blue (around 420 nm) lights but green light (around 550 nm) cannot be efficiently utilized by green algae.

The results of this study may be helpful to understand the light-dependent photosynthetic activity of *H. pluvialis*. At the same time, the information presented here can be used for enhancing the volumetric biomass productivity of the astaxanthin-accumulating alga.

References

[1] K. W. Glombitza and M. Koch, Secondary Metabolites of Pharmaceutical Potential. In: R. C. Cresswll, T. A. V. Rees and N. Shah (eds.), Algal and Cyanobacterial Biotechnology, Longman Scientific and Technical, UK, 1989.
[2] E. W. Becker, Microalgae for Human and Animal Consumption, In: M. A. Borowitzka and L. J. Borowitzka (eds.), Microalgal Biotechnology, Cambridge University press, Cambridge, 1998.
[3] E. W. Becker, Microalgal Biotechnology and Microbiology, Cambridge University Press, Cambride, 1994.
[4] R. T. Lorenz and G. R. Cysewski, TIBTECH, 18 (2000) 160.
[5] Y. E Choi, Y.-S. Yun and J. M. Park, Biotechnol. Prog., 18 (2002) 1170.
[6] D.O. Hall and K.K. Rao , Photosynthesis, Cambridge University Press, Cambridge, 1987.
[7] R. J. Geider and B. A. Osborne, Algal Photosynthesis, Chapman and Hall, New York, 1992.
[8] J. Fábregas and A. Domínguez, Appl. Microbiol. Biotechnol., 53 (2000) 530.
[9] Y.-S. Yun and J. M. Park, Biotechnol. Bioeng., 83 (2003) 303.
[10] L. V. Liere and L.R. Mur, J. Gen. Microbial., 115 (1979) 153.
[11] Y.-S. Yun, J. M. Park and Volesky, Korean J. Biotechnol. Bioeng., 14 (1999) 328.

Studies in Surface Science and Catalysis, volume 159
Hyun-Ku Rhee, In-Sik Nam and Jong Moon Park (Editors)

Broadening of the optimal pH range for reactive dye biosorption by chemical modification of surface functional groups of *Corynebacterium glutamicum* biomass

Min Hee Han, Sung Wook Won and Yeoung-Sang Yun

Div. of Environmental and Chemical Engineering and Dept. of Bioprocess Engineering, Chonbuk National University, 664-14 Duckjin-Dong, Jeonju, Jeonbuk 561-756, Republic of Korea
ysyun@chonbuk.ac.kr

ABSTRACT

Biosorption has been demonstrated to be a useful alternative to conventional treatment systems for the removal of dyes from dilute aqueous solution. This study deals with a renewable, low cost biosorbent derived from waste biomass of *Corynebacterium glutamicum*. The biosorbent has been proved to have higher (or comparable) capacity of dye uptake than conventional sorbents like activated carbons and ion-exchange resins. This study focuses on the underlying mechanisms on dye binding to the biosorbent. The binding sites were identified to be primary amine groups present in the biomass. Chemical modification of the biomass, FTIR and potentiometric titration studies revealed that carboxyl and phosphate groups play a role in repulsion of dye molecules, which inhibits the dye binding. Based on the biosorption mechanism, the performance of biosorbent could be enhanced by the removal of inhibitory carboxyl groups from the biomass for practical application of the biosorbents.

1. INTRODUCTION

Textile industries consume large volumes of water and chemicals for the wet processing of textiles. The presence of very low concentrations of dyes in effluent discharged from these industries is highly visible and undesirable [1]. Due to their chemical structure, dyes are resistant to fading when exposed to light, water and many chemicals [1-2]. According to the survey of the Ecological and Toxicological Association of the Dyestuffs Manufacturing Industry (ETAD), over 90% of dyes have LD50 values greater than 2000 mg kg^{-1} [2].

Various physical, chemical, and biological methods have been used for the treatment of dye-containing wastewater. However, these conventional technologies have disadvantages like poor removal efficiency and high running cost. Therefore, low-cost sorbents which can bind dye molecules and be easily regenerated have been extensively searched and tested [3-7].

In this study, the waste biomass of *C. glutamicum* was used and evaluated as a biosorbent for the treatment of an anionic reactive dye Reactive Orange 16 (RO16). Especially, the surface functional groups of this biomass were modified so that the optimal pH range was shifted from very acidic (< pH 3) to mild acidic (pH 4-5) condition. By using this method, cost of strong acids supply would be expected to be reduced. The *C. glutamicum* biomass is generated in a great quantity from the full-scale fermentation process for amino acid production. Amino acid fermentation industries have been troubled with a huge amount of biological solid waste, which is mainly composed of the biomass of *C. glutamicum*. Although this fermentation byproduct is potentially recyclable, until now most of it has been dumped at sea. Therefore, the feasibility for reuse of the solid waste as a value-added biosorbent deserves to be assessed.

2. MATERIALS AND METHODS

The fermentation wastes (*Corynebacterium glutamicum* biomass) were obtained in a dried powder form from a lysine fermentation industry (BASF-Korea, Kunsan, Korea). The protonated biomass was prepared by treating the raw biomass with a 1 N HNO_3 solution for 24 h, thereby replacing the natural mix of ionic species with protons. The resulting *C. glutamicum* biomass was dried and stored in a desiccator and used as a biosorbent for the sorption experiments.

The potentiometric titration was carried out in order to determine the functional groups present in the biomass surface. During the titration experiments, the CO_2-free condition was always maintained to avoid the influence of inorganic carbon on the solution pH. Detailed potentiometric titration procedure and estimation method of functional groups are available in the previous reports [4,6].

In order to remove carboxyl groups which inhibit the binding of reactive dye, 4 g of the biomass was suspended in 400 mL of anhydrous methanol and 5 mL of concentrated hydrochloric acid for 6 h at 150 rpm. Using this method, carboxyl groups can esterifies as follows [8]: $B\text{-}COOH + CH_3OH \rightarrow B\text{-}COOCH_3 + H_2O$

To evaluate the sorption capacity of the biomass, biosorption isotherms of RO16 were obtained at different solution pHs. The initial concentration was varied from 0 to 5000 mg L^{-1}, which resulted in different final dye concentrations after the sorption equilibrium was achieved.

To study the equilibrium relationship between the dye uptake and the final pH, the pH edge experiments were carried out according to previously reported method [5].

3. RESULTS AND DISCUSSION

3.1. Analysis of functional groups

The biomass titration curve shows its distinct characteristics depending upon types and amounts of functional groups in the biomass. The proton isotherm was obtained from the titration results according to the method previously reported [5] for the protonated *C. glutamicum* biomass. By applying the proton-binding model [4], it was found that the biomass contained two types of negatively charged groups and one type of positively charged group. The estimated pK_H values and contents of the functional groups are summarized in Table 1.

Table 1. Estimated parameters of proton-binding model

functional groups	1st group	2nd group	3rd group
charge	–	–	+
pK_H (-)	3.57 (0.08)	6.90 (0.06)	9.14 (0.07)
b (mmol g^{-1})	0.32 (0.01)	0.56 (0.02)	0.68 (0.02)

The first group is likely to be carboxyl site considering that carboxyl groups occurring in the biomaterials in the pH range 3 to 4. The third groups able to have positive charges were believed to be primary amine sites. Since the sulfonate groups in RO16 is negatively ionized in normal pH range, the binding sites for RO16 ions should be positively charged amine groups. The three kinds of functional groups were also confirmed by FTIR study of the protonated biomass as shown in the previous report [7].

3.2. Binding mechanisms

The effect of pH on biosorption of RO16 is shown in Fig. 1. As decreasing the solution pH, the uptake of dye increased and, at the pH less than 3 the uptake kept constant at a maximum value. The pH effect did not coincide with the fact that the binding sites (amine groups) are fully charged at pH < 8 (Fig. 2). Carboxyl and phosphate groups have negative charges around neutral pH. Therefore, it was noted that these groups could inhibit the binding of anionic RO16 to amine sites.

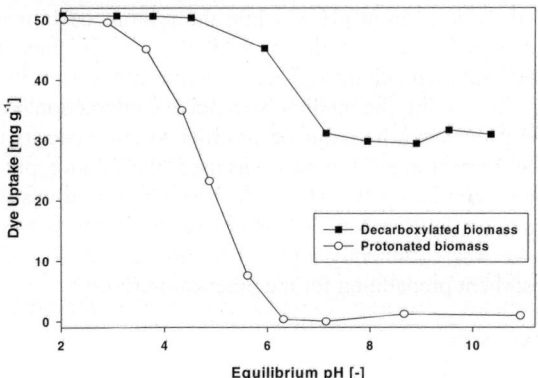

Fig. 1. Effect of pH on the uptake of RO16 by protonated biomass and the biomass from which carboxyl groups were removed.

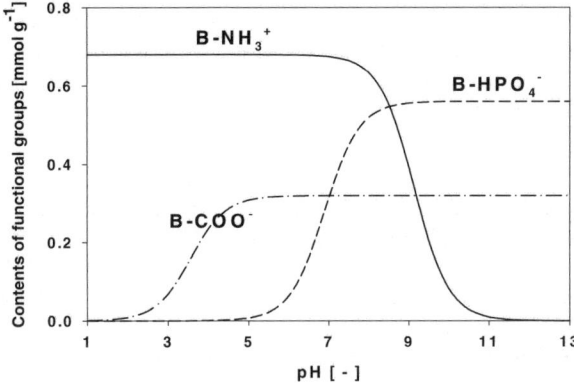

Fig. 2. Speciation of functional groups present in the biomass depending on the solution pH.

164

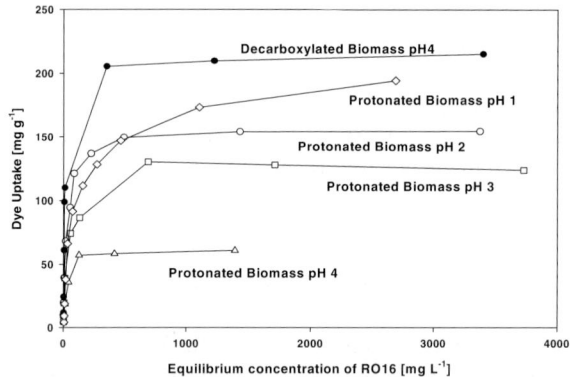

Fig. 3. Isotherms of RO 16 biosorption by the protonated biomass and the decarboxylated biomass.

Then, if the carboxyl groups are removed from the biomass, it is expected that the repulsive interaction around pH 3 can be eliminated. As expected, the decarboxlyated biomass effectively adsorbed the dye even at pH 5, while the protonated biomass did not (Fig. 2). Therefore, it can be considered that the dye binding to the biomass takes place via electrostatic interaction between anionic dye and positive amine sites in the biomass and the negatively charged groups inhibit the binding by repulsive interaction. For the practical point of view, the carboxyl groups can be removed in order to eliminate its inhibitory effect. As shown in Fig. 3, when the protonated biomass was used, the solution pH should be decreased down to pH 1 for the maximum uptake. However, it is not easy or costly to decrease the pH of wastewater to such a low level. When the decarboxylated biomass was applied, even at pH 4 the maximum uptake was sufficiently high. Therefore, chemical modification could be a useful tool in the biosorbent preparation for the practical purpose.

ACKNOWLEDGEMENTS

This work was financially supported by the KOSEF through the AEBRC at POSTECH and partially by grant No. R08-2003-000-10987-0 from the Basic Research Program of the KOSEF.

4. REFERENCES

[1] K. R. Ramakrishna and T. Viraraghavan, Water Sci. Technol., 36 (1997) 189.
[2] P. Nigam, G. Armour, I. M. Banat, D. Singh and R. Marchant , Bioresource Technol., 72 (2000) 219.
[3] T. Robinson, G. McMullan, R. Marchant and P. Nigam, Bioresource Technol., 77 (2000) 247.
[4] Y.-S. Yun, D. Park, J. M. Park and B. Volesky, Environ. Sci. Technol., 35 (2001) 4353.
[5] Y.-S. Yun and B. Volesky, Environ. Sci. Technol., 37 (2003) 3601.
[6] D. Park, Y.-S. Yun and J. M. Park, Environ. Sci. Technol., 38 (2004) 4860.
[7] S. W. Won, S. B. Choi, B. W. Chung, D. Park, J. M. Park and Y.-S. Yun, Ind. Eng. Chem. Res., 43 (2004) 7865.
[8] A. Kapoor and T. Viraraghavan, Bioresource Technol., 61 (1997) 221.

Studies in Surface Science and Catalysis, volume 159
Hyun-Ku Rhee, In-Sik Nam and Jong Moon Park (Editors)

Model-based optimization of a sequencing batch reactor for advanced biological wastewater treatment

D.S. Lee[a], G. Sin[b], G. Insel[c], I. Nopens[b] and P. A. Vanrolleghem[b]

[a]Advanced Environmental Biotechnology Research Center, School of Environmental Science and Engineering, POSTECH, San 31, Hyoja-dong, Nam-gu, Pohang, Gyeongbuk 790-784, Republic of Korea
[b]BIOMATH, Ghent University, Coupure Links 653, B-9000 Gent, Belgium
[c]Istanbul Technical University, Environmental Engineering Department, 34469, Maslak, Istanbul, Turkey

1. INTRODUCTION

Sequencing batch reactor (SBR) technology for nutrient removal has received great attention from the wastewater treatment community [1-2]. A SBR process has a unique cyclic batch operation for biological wastewater treatment. Most of the advantages of SBR processes may be attributed to their single-tank designs and the flexibility allowing them to meet many different treatment objectives [3-5]. Due to the ever-stricter demands on effluent discharge quality, an existing SBR plant may require optimization in terms of nutrient removal, but this often needs a better understanding and quantification of the biological processes occurring in each phase of the SBR operation. A calibrated activated sludge model is a practical tool to try numerous operating scenarios within a short evaluation time when an upgrade of the SBR is considered.

In this study, Activated Sludge Model No. 2d (ASM2d) is employed to model a lab-scale SBR [6]. Then, based on a survey of the relevant literature and a preliminary model-based analysis of the system [1,7], the following degrees of freedoms were identified and used for the SBR optimization: oxygen set-point in the aerobic phase (S_O) and the lengths of anaerobic (T_{AN}), aerobic (T_A), and feeding (T_F) phases. A grid of scenarios is formulated as full-factorial experimental design to simulate the effect of the key degrees of freedom in the SBR system. Effluent quality in combination with a robustness index for each of the scenarios is used to select the best operational strategy for the SBR system.

2. SEQUENCING BATCH REACTOR

The data used in this study were collected from a lab-scale SBR system as shown in Fig. 1. A fill-and-draw SSBR system with a 4-liter working volume was operated in an 8 h cycle mode and each cycle consists of 2 h anaerobic (initially anoxic), 4 h 30 min aerobic, 1 h 30 min settling and fill/draw stages. Temperature was controlled at the reactor that was jacked with water for temperature control at $20\pm1\,^{\circ}\mathrm{C}$ using a water circulation system. Clarified supernatant of 2 liters was withdrawn from the reactor at the end of settling stage and fresh (synthetic) wastewater of 2 liters was pumped into the reactor during the filling stage. Solid

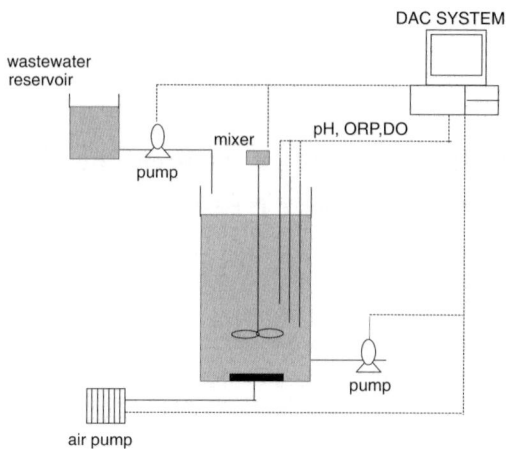

Fig. 1. Schematic diagram of a lab-scale sequencing batch reactor

retention time (SRT) was maintained at about 12 days by wasting mixed liquor suspended solids (MLSS) at the end of the aerobic phase. Loading amounts of COD (as CH_3COOH), NH_4^+-N, and PO_4^{3-}-P per cycle in a standard condition were 600, 40, and 15 mg l^{-1} respectively.

The controls of duration/sequence of stages and on/off of peristaltic pumps, mixer, air supply were automatically achieved by an in-house developed data acquisition and control (DAC).

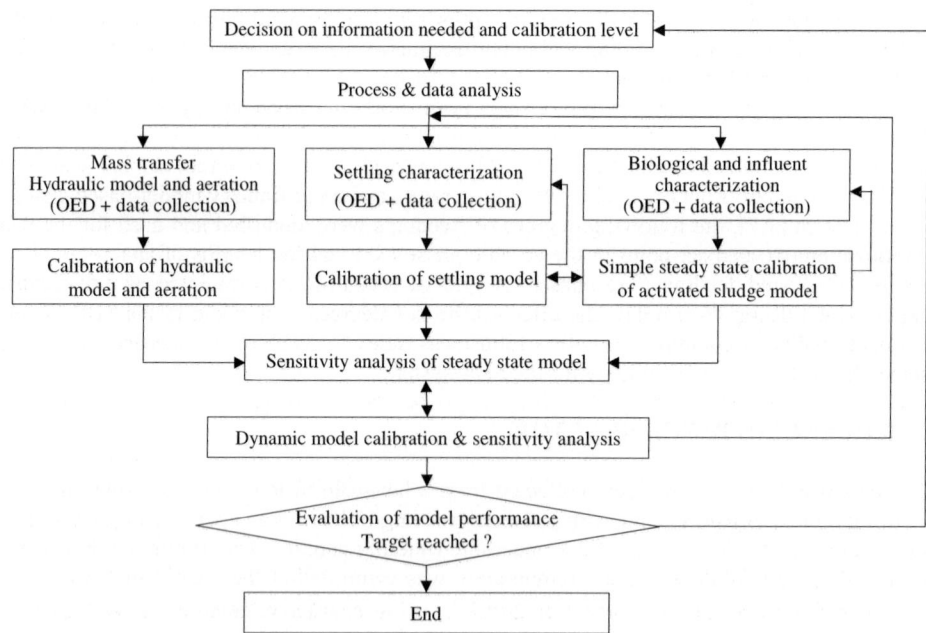

Fig. 2. A systematic methodology for the calibration of activated sludge models

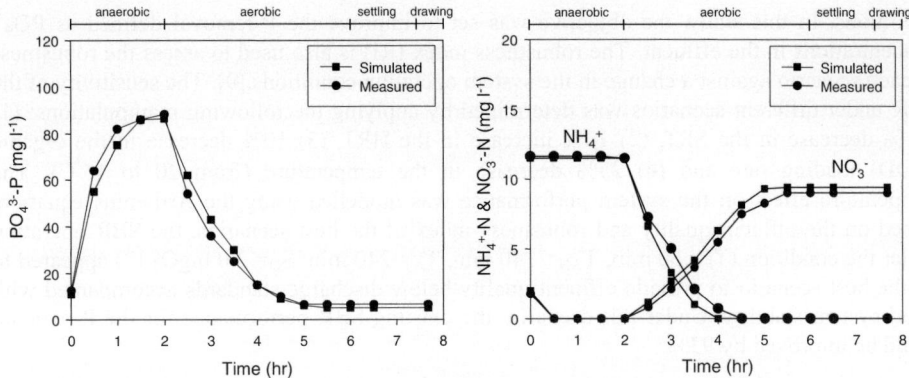

Fig. 3. Measurements and simulation results for nutrient concentrations in a SBR cycle

The DAC system consisted of computer, interface cards, meters, transmitters, and solid state relays (SSR). Electrodes of pH (Ingold), Oxidation-Reduction Potential (Cole-Parmer), and Dissolved Oxygen (Ingold) were installed and connected to individual meter. The status of reactor and the value of electrode signal were displayed in a computer monitor, and stored in data file.

3. RESULTS AND DISCUSSION

In the model simulations, the settling and decanting phase were characterized by a reactive point-settler model. The simulations were carried out using matlab 6.5 simulation platform. A systematic model calibration methodology as described in Fig. 2 was applied to the SBR. Fig. 3. shows the simulation results from the calibrated model. The model predicted the dynamics of the SBR with good accuracy.

A grid of scenarios considering the degrees of freedom and the constraints of the system mentioned above was formulated as a full-factorial experimental design [8]. The combination of these degrees of freedom under the SBR operation constraints results in 108 scenarios, which is expected to be sufficient to provide significant insight into the optimal operational scheme for the SBR system. In this way, the optimal scenario of the SBR operation can be searched using the predefined criteria. The grid of scenarios presented in Table 1 is simulated for 36 days, equal to 3 times the system SRT. The scenario analysis results (SCA) indicate that the best system performance for P-removal is obtained under different operating

Table 1
The grid of scenarios to simulate the effect of key degrees of freedom on the SBR system

Degrees of freedom			
S_O (mg O_2 l^{-1})	T_F (min)	T_{AN} (min)	T_A(min)
[0.5, 1.0, 1.5, 2.0]	[10,20,30]	[100, 120,140]	[210,240,270]
[0.5, 1.0, 1.5, 2.0]	[10,20,30]	[100, 120,140]	[130,140,150]
[0.5, 1.0, 1.5, 2.0]	[10,20,30]	[100, 120,140]	[130,140,150]
[0.5, 1.0, 1.5, 2.0]	[10,20,30]	[100, 120,140]	[130,140,150]

conditions. In this study, the objective was set to improve the P removal defined as PO_4^{3-} concentrations in the effluent. The robustness index (RI) is also used to assess the robustness of each scenario against a change in the system operation conditions [9]. The sensitivity of the SBR under different scenarios was determined by applying the following manipulations: (1) 10 % decrease in the SRT, (2) 10% increase in the HRT, (3) 10% decrease in the organic (COD) loading rate and (4) 25% decrease in the temperature (from 20 to $15^{\circ}C$). The temperature effect on the system performance was modelled using the Arrhenius equation. Based on the effluent quality and robustness index of the best scenarios, the SBR operation under the condition (T_F= 60 min, T_{AN}= 140 min, T_A= 240 min, S_O= 2.0 mgO$_2$ l^{-1}) appeared to be the best scenario to provide effluent quality below discharge standards accompanied with good system stability. Under this scenario, the existing SBR performance for the P-removal could be improved by 93%.

4. CONCLUSION

A systematic approach to determine the optimal operation strategy for nitrogen and phosphorus removal of sequencing batch reactors has been developed and applied to successfully to a lab-scale SBR. In this optimisation study, the dissolved oxygen concentrations in the aerobic phase and the variable length of the filling, anaerobic and aerobic sequences are selected as key manipulating variables. Based on ASM2d model, each operation scenario is evaluated to improve the effluent quality and the robustness of the process operation. The selected best scenario has been implemented to the lab-scale SBR reactor.

ACKNOWLEDGEMENTS
This work was financially supported by the Korea Science and Engineering Foundation through the Advanced Environmental Biotechnology Research Center at Pohang University of Science and Technology.

REFERENCES
[1] P. Wilderer, R.L. Irvine and M.C. Goronszy, Sequencing Batch Reactor Technology, IWA Scientific and Technical Report No: 10, IWA Publishing, London, 2001.
[2] D.S. Lee and J.M. Park, J. Biotechnol., 75 (1998) 229.
[3] C. Demuynck, P.A. Vanrolleghem, C. Mingneau, J. Leissens and W. Verstraete, Water Res., 30 (1994) 169.
[4] D.S. Lee, C.O. Joen and J.M. Park, Water Res., 35 (2001) 3968.
[5] S. Mace and J. Mata-Alvarez, Ind. Eng. Chem. Res., 41 (2002) 5539.
[6] M. Henze, W. Gujer, T. Mino, T. Matsuo, M.C. Wentzel, G.v.R. Marais and M.C.M. van Loosdrecht, Activated Sludge Model No. 2d, ASM2d, Water Sci. Tech., 39(1), 165.
[7] N. Artan, P. Wilderer, D. Orhon, R. Tasli and E. Morgenroth, Water SA, 28 (2002) 423.
[8] G. Sin, G. Insel, D.S. Lee and P.A. Vanrolleghem, Water Sci. Tech., 50(10), 97.
[9] P.A. Vanrolleghem and S. Gillot, Water Sci. Tech, 45 (2002) 117.

Studies in Surface Science and Catalysis, volume 159
Hyun-Ku Rhee, In-Sik Nam and Jong Moon Park (Editors)

Enhancement of the catalytic activity of the wheat germ, cell-free, protein synthesis system using a fortified translation extract

Young Seoub Park[a], Hun Su Chu[b], Cha Yong Choi[b] and Gyoo Yeol Jung[a, c*]

[a]Division of Molecular and Life Sciences, Pohang University of Science and Technology, San 31, Hyoja-dong, Nam-gu, Pohang, 790-784, Republic of Korea

[b]School of Chemical and Biological Engineering, Seoul National University, Shillim-dong, Gwanak-gu, Seoul, 151-742, Republic of Korea

[c]Department of Chemical Engineering, Pohang University of Science and Technology, San 31, Hyoja-dong, Nam-gu, Pohang, 790-784, Republic of Korea

1. ABSTRACT

A wheat germ, cell-free, translation extract was fractionated into three concentrated parts using ammonium sulfate: the 0 - 40 % saturated fraction, the 40 - 60 % saturated fraction, and the ribosome fraction. These fractions were tested for their ability to enhance the translational activity of the wheat germ, cell-free extract for dihydrofolate reductase. The fortified cell-free system supplemented with the 0 – 40 % ammonium sulfate fraction enhanced the efficiency of protein synthesis by 50 %.

2. INTRODUCTION

Cell-free translation system, used for the identification of cloned genes and gene expression, has been investigated extensively as a preparative production system of commercially interesting proteins after the development of continuous-flow cell-free translation system. Many efforts have been devoted to improve the productivity of cell-free system [1], but the relatively low productivity of cell-free translation system still limits its potential as an alternative to the protein production using recombinant cells. One approach to enhance the translational efficiency is to use a condensed cell-free translation extract. However, simple addition of a condensed extract to a continuous-flow cell-free system equipped with an ultrafiltration membrane can cause fouling. Therefore, it needs to be developed a selective condensation of cell-free extract for the improvement of translational efficiency without fouling problem.

Previously, it has been reported that the amounts of eukaryotic initiation factors in wheat germ extract prepared by a common method were deficient for the translation of some kinds of mRNAs including α-amylase mRNA and β-globin mRNA [2]. Therefore, it can be expected that the activity of wheat germ extract prepared by a common method can be enhanced by the simple addition of extract containing deficient initiation factors. In this study, a wheat germ extract was further purified partially by ammonium sulfate fractionation

---

*Corresponding author
(Tel.: +82-54-279-2391; FAX.: +82-54-279-5528; e-mail: gyjung@postech.ac.kr)

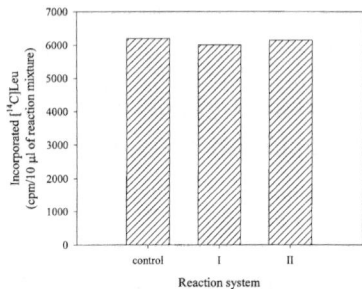

Fig. 1. Reconstruction of the cell-free protein synthesizing system with the partially purified wheat germ extracts. Control; normal wheat germ cell-free system, (I); 0 - 40 % ammonium sulfate fraction 3 μl, 40 - 60 % ammonium sulfate fraction 4 μl, and ribosome 3 μl were added to 25 μl reaction mixture, (II); 0 - 40 % ammonium sulfate fraction 4 μl, 40 - 60 % ammonium sulfate fraction 4 μl, and ribosome 1.5 μl were added to 25 μl reaction mixture.

followed by concentration, and then each fraction expected to have concentrated initiation factors was added to a normal wheat germ cell-free protein synthesis system. The addition effect of fractionated extract on the translational efficiency and the enhancement of the catalytic activity of this fortified cell-free system were investigated.

3. MATERIALS AND METHODS

3.1. Plasmid

The plasmid vector pTEV-DHFR was used in this study. Internal ribosome entry sequence originated from tobacco etch virus (TEV) was located at upstream of dihydrofolate reductase (DHFR) gene.

3.2. Preparation of S170 supernatant and washed ribosomes

Wheat germ extract was prepared from wheat germ gifted from Daehan Flour Mills Co. Ltd. (Inchon, Korea), according to the procedures of Anderson *et al.* [3], and then divided into two fractions as follows. A 2 ml wheat germ extract was centrifuged at 170,000 g for 2 h at 4 °C. The upper three-fourths of the supernatant (designated as S170) was withdrawn and stored in liquid N_2. The pellet was resuspended in wheat germ extract buffer as described in reference [3], and the resulting suspension was recentrifuged at 170,000 g for 2 h at 4 °C. The pellet was added to 500 μl wheat germ extract buffer and the resulting suspension was clarified by centrifugation at 6,000 g for 5 min at 4 °C, and then stored in liquid N_2 (washed ribosomes). The ribosomal concentration of the solution was 14.2 mg/ml (based on A_{260} of 1480 = 100 mg/ml).

3.3. Preparation of 0 - 40 % and 40 - 60 % ammonium sulfate fractions from S170 supernatant

The S170 supernatant (220 ml) was made to 40 % saturation with solid ammonium sulfate, stirred for 20 min, and then the precipitate was collected by centrifugation at 15,000 g for 15 min. The precipitate was suspended in small volume of buffer B-50 at pH 7.6 containing 20 mM HEPES/KOH, 0.1 mM EDTA, 1 mM dithiothreitol, 10 % (v/v) glycerol, and 50 mM potassium acetate. The 60 % saturated ammonium sulfate solution was prepared similarly. Protein concentrations for 0 – 40 % and 40 - 60 % ammonium sulfate fractions were 4.2 mg/ml and 4.7 mg/ml, respectively.

3.4. Translation reaction

Translation reaction mixture contained 20 % (v/v) wheat germ extract and the followings (final concentrations): 60 mM HEPES/KOH (pH 7.6), 1 mM ATP, 100 μM GTP, 300 μM spermidine, 2 mM magnesium acetate, 40 mM potassium acetate, 160 μM each amino acid (except leucine), 50 U creatine kinase/ml, 20 mM creatine phosphate, 12.5 μg wheat germ tRNA/ml and 20 μg mRNA/ml. Translation reactions were performed with 4 μCi [^{14}C]leucine/ml. After 2 h incubation, aliquots (10 μl) were removed and the amount of leucine incorporated in DHFR was measured by the analysis of trichloroacetic acid-precipitable radioactivity. Sizes of the cell-free synthesized proteins were analyzed by discontinuous SDS-PAGE as described by Laemmli [4].

4. RESULTS AND DISCUSSION

For the preparation of fractionated extracts possessing the various initiation factors, wheat germ extract was separated into ribosome fraction, 0 - 40 % ammonium sulfate fraction, and 40 - 60 % ammonium sulfate fraction as described in Materials and Methods. In order to test the activities of the fractions, wheat germ cell-free systems were reconstructed by a combinatorial approach varying the amount of each fraction and compared with the normal wheat germ cell-free extract. Fig. 1 shows that two sets of combinations displaying similar activities with normal wheat germ system. The translational activities of the reconstructed system containing 3 - 4 μl of 0 - 40 % ammonium sulfate fraction, 4 μl of 40 - 60 % ammonium sulfate fraction, and 1.5– 3 μl of ribosome fraction were almost the same to that of the normal wheat germ system, which indicates that each fraction plays its corresponding role.

The catalytic activities of the fortified wheat germ cell-free systems supplemented with each fraction were investigated (Fig. 2). As shown in Fig. 2, only 0 – 40 % ammonium sulfate fraction showed an enhancement in DHFR protein synthesis. This enhancement of protein experimental results and the fact that the various eukaryotic initiation factors are contained in synthesis was also confirmed by SDS-PAGE and autoradiography (Fig. 3). From the above 0– 40 % ammonium sulfate fraction [5, 6], it can be concluded that the amount of initiation factors in a conventionally prepared wheat germ cell-free extract is deficient for the translation of DHFR with internal ribosome entry site. Therefore, it needs to supplement a wheat germ cell-free extract with the fraction containing the limited initiation factors for the efficient protein translation, and this fortified cell-free system can be easily made by simple

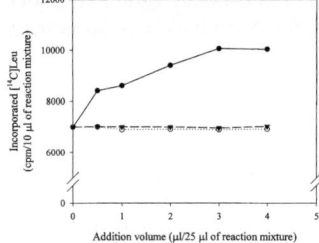

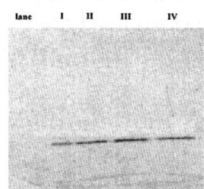

Fig. 2. The addition effect of partially purified wheat germ extract. Symbols represent 0 - 40 % ammonium sulfate fraction (-●-), 40 - 60 % ammonium sulfate fraction (-○-) and ribosome fraction (-▼-), respectively.

Fig. 3. Autoradiograph of SDS-PAGE of *in vitro* translated dihydrofolate reductase (DHFR) in the wheat germ cell-free protein synthesis systems with (II) 4 μl of ribosome fraction, (III) 4 μl of 0 - 40 % ammonium sulfate fraction, or (IV) 4 μl of 40 - 60% ammonium sulfate fraction, respectively. Lane I is control: dihydrofolate reductase produced in the normal wheat germ cell-free protein synthesis system.

addition of small amount of the corresponding fraction.

The 0 – 40 % ammonium sulfate fraction has been known to contain eIF-1A, eIF-1B, eIF-2A, eIF-2B, eIF-4A, eIF-4B, eIF-4F and eIF-3. On the other hand, eIF-1C, eIF-1D, eIF-1E and eIF-2D are contained in 40 - 60 % ammonium sulfate fraction [6]. Among these initiation factors, three eukaryotic initiation factors, eIF-4A, eIF-4B and eIF-4F, play a key role in ATP-dependent binding of mRNA to 40S ribosomal subunits [5]. The eIF-4A is an ATP-dependent single stranded RNA binding protein whose ATPase activity is stimulated by the presence of single stranded RNA. The eIF-4F is an ATP-independent cap binding protein and possesses ATP-dependent mRNA unwinding activity, similar to eIF-4A. The eIF-4B is required for 40S ribosomal subunit binding to mRNA and it stimulates significantly eIF-4A, eIF-4F RNA-dependent ATPase and ATP-dependent RNA helicase activity. Therefore, it can be suggested that the concentrations of the initiation factors, particularly factors for the binding of mRNA to ribosomes, may play a key role in the translation of DHFR mRNAs. It can be also inferred that these initiation factors as eIF-4F and eIF iso-4F in a conventionally prepared extract are deficient for the optimal translation.

Our previous approach for the enhancement of the protein productivity of a cell-free translation system was to concentrate wheat germ extract as a whole, which resulted in a poor improvement in the translational efficiency [7]. Total protein concentration of the cell-free system using condensed wheat germ extract, especially, was 10 mg/ml while that of the fortified system in this study was 6 mg/ml. This should be greatly advantageous in the application to the continuous flow system. As shown in this study, the simple fortification of cell-free extract with the purified fraction showed better enhancement of the translational efficiency of a wheat germ cell-free translation system. Therefore, the catalytic activity of the wheat germ cell-free protein synthesis system could be enhanced using a fortified translation extract supplemented with the fractionated extract expected to possess the deficient initiation factors. Furthermore, the fortified extract shown in the present study may not cause the fouling trouble in the continuous-flow cell-free system.

5. ACKNOWLEDGMENTS

This work was supported by POSTECH BSRI Research Fund-2005 (grant number 1RB0511201) and the KOSEF through the AEBRC at POSTECH. Y.S. Park was supported by Brain Korea 21 program.

REFERENCES

[1]   H. Nakano, T. Tanaka, Y. Kawarasaki, and T. Yamane, Biosci. Biotechnol. Biochem., 58 (1994) 631.
[2]   R.T. Timmer, L.A. Benkowski, J.M. Ravel, and K.S. Browning, Biochem. Biophys. Res. Commun., 210 (1995) 370.
[3]   C.W. Anderson, J.W. Straus, and B.S. Dudock, Methods Enzymol., 101 (1983) 635.
[4]   U.K. Laemmli, Nature, 227 (1970) 680.
[5]   S.R. Lax, S.J. Lauer, K.S. Browning, and J.M. Ravel, Methods Enzymol., 118 (1986) 109.
[6]   B.J. Walthall, L.L. Spremulli, S.R. Lax, and J.M. Ravel, Methods Enzymol., 60 (1979) 193.
[7]   G.Y. Jung, E.Y. Lee, Y.-E. Kim, B.W. Jung, S.-H. Kang, and C.Y. Choi, J Biosci Bioeng, 89 (2000) 192.

Studies in Surface Science and Catalysis, volume 159
Hyun-Ku Rhee, In-Sik Nam and Jong Moon Park (Editors)

Comparison of whole cell biocatalytic reaction kinetics for recombinant *Escherichia coli* with periplasmic-secreting or cytoplasmic-expressing organophosphorus hydrolase

Dong Gyun Kang[a], Sang Hwan Seo[b], Suk Soon Choi[b] and Hyung Joon Cha[a] *

[a]Department of Chemical Engineering, Pohang University of Science and Technology, Pohang 790-784, Korea

[b]Department of Environmental Engineering, Semyung University, Jecheon 390-711, Korea

1. INTRODUCTION

Organophosphate compounds are widely used in many pesticides (Paraoxon, Parathion, Coumaphos, and Diazinon) and chemical nerve agents (Sarin and Soman) [1]. Organophosphorus hydrolase (OPH) from *Pseudomonas diminuta* or *Flavobacterium* sp. is a homodimeric organophosphotriesterase that requires metal ion as a cofactor and can degrade a broad spectrum of toxic organophosphates [2]. This enzyme can hydrolyse various phosphorus-ester bonds including P-O, P-F, P-CN, and P-S bonds. The application of OPH for bioremediation is of great interest due to its high turnover rate.

Recombinant *Escherichia coli* expressing OPH can degrade a variety of organophosphates [3]. The ability of *E. coli* to grow to much higher densities than *P. diminuta* or *Flavobacterium* enables the development of large-scale detoxification processes [4]. However, recombinant *E. coli* cells produce low yields of OPH due to the low solubility of this protein [2]. In addition, the *E. coli* cell membrane can be a substrate diffusion barrier affecting whole cell biocatalytic efficiency [5]. Several strategies attempted to enhance OPH production yield or whole cell biocatalytic efficiency include fusion with a soluble partner to increase solubility [6], co-expression with *Vitreoscilla* hemoglobin [7] and display on the cell surface [8].

Periplasmic secretion of target proteins via translocation across the cytoplasmic membrane in Gram-negative bacteria such as *E. coli* can be a potential strategy for reducing the substrate diffusion barrier in whole cell biocatalyst systems [9]. Previously, we have successfully shown that functional secretion of OPH molecules into the periplasmic space was achieved using the twin-arginine translocation (Tat) pathway [10]. In particular, we have used the twin-arginine signal sequence of the *E. coli* enzyme trimethylamine *N*-oxide (TMAO) reductase (TorA) because OPH molecules require metal ion cofactors. Using this system, whole cell OPH activity was approximately 2.8-fold higher due to successful translocation of OPH into the periplasmic space. This study has shown that Tat-driven periplasmic secretion of OPH is a potential strategy to overcome traditional substrate diffusion limitations in whole cell biocatalyst systems.

In the present work, for detail kinetic studies, we compared biocatalytic reaction kinetics for four types of whole cell biocatalyst systems; whole cells with periplasmic-secreting OPH under *trc* or T7 promoters and whole cells with cytoplasmic-expressing OPH under *trc* or T7 promoters.

2. MATERIALS AND METHODS

2.1. Bacterial strains

E. coli strain BL21(DE3) (F⁻ *ompT hsdS_B(r_B⁻ m_B⁻) gal dcm*) (Novagen, Madison, WI, USA) was used as a host for recombinant OPH expression. Recombinant plasmids pTOH and pEOH that contain OPH gene fused with hexa-histidine affinity tag under *trc* and T7 promoter, respectively, as control vectors and pTTOH and pETOH that contain OPH gene fused with Tat signal sequence and hexa-histidine affinity tag under *trc* and T7 promoter, respectively, were used (Fig. 1).

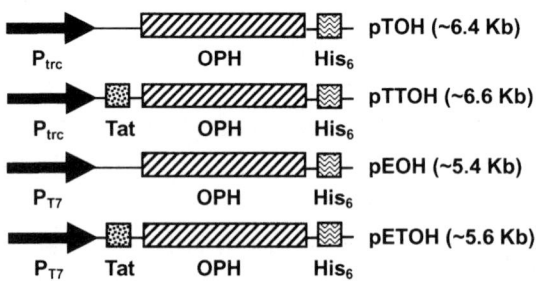

Fig.1. Gene maps of recombinant plasmids pTOH, pTTOH, pEOH, and pETOH. Abbreviations: P_{trc}, *trc* promoter; P_{T7}, T7 and *lac* hybrid promoter; Tat, twin-arginine TorA signal sequence of TMAO reductase; OPH, organophosphorus hydrolase gene; His₆, hexa-histidine affinity tag.

2.2. Culture condition

Plasmids bearing strains were cultured in Luria broth (LB) or M9 minimal medium (12.8 g l⁻¹ Na₂HPO₄.7H₂O, 3 g l⁻¹ KH₂PO₄, 0.5 g l⁻¹ NaCl, 1 g l⁻¹ NH₄Cl, 3 mg l⁻¹ CaCl₂, 1 mM MgSO₄) containing 0.5% (wt/vol) glucose, 50 μg ml⁻¹ of ampicillin, and 0.5 mM CoCl₂ at the final concentration. Sub-cultures were grown overnight in 6 ml LB and used to inoculate 50 ml M9 to a starting OD₆₀₀ of 0.2. Cells were cultured in 250 ml Erlenmeyer flasks at 250 rpm and 37°C. When cultures reached a cell density (OD₆₀₀) of 1.2, 0.5 mM IPTG was added to induce recombinant protein expression.

2.3. Analytical assays

Whole cell OPH activity was measured by following the increase in absorbancy of *p*-nitrophenol from the hydrolysis of substrate (0.1 mM Paraoxon) at 400 nm (ε_{400} = 17,000 M⁻¹ cm⁻¹). Samples of culture (1 ml) were centrifuged at 10,000 g and 4°C for 5 min. The cells were washed, resuspended with distilled water, and 100 μl was added to an assay mixture containing 400 μl 250 mM CHES [2-(N-cyclohexylamino)ethane-sulfonic acid] buffer, pH 9.0, 100 μl 1 mM Paraoxon, and 400 μl distilled water. One unit of OPH activity was defined as μmoles Paraoxon hydrolyzed per min. Each value and error bar represents the mean of two independent experiments and its standard deviation.

2.4. Whole cell bioconversion kinetics

The values for kinetic parameters, V_{max} and K_m, were determined by analyzing Lineweaver-Burk plots [11] over the range 0.1 to 1 mM Paraoxon in 100 mM CHES buffer, pH 9.0. Linear regression analysis for Lineweaver-Bulk plot was performed using SigmaPlot software (Systat Software, USA).

3. RESULTS AND DISCUSSION

We performed whole cell biocatalytic reaction with various substrate (Paraoxon) concentrations in order to compare reaction kinetics for four types of whole cell *E. coli* systems (cytoplasmic-expressing cells under *trc* promoter regulation, cytoplasmic-expressing cells under T7 promoter regulation, periplasmic-secreting cells under *trc* promoter regulation, and periplasmic-secreting cells under T7 promoter regulation). Whole cell reaction rates (*v*) were plotted against substrate concentrations (S) as depicted in Fig. 2. All data shown here were based on unit whole cell concentration (1 mg-dry cell weight ml^{-1}). Both reactions showed Michael-Menten kinetic patterns and the periplasmic-secreting strain exhibited higher OPH biocatalytic reaction rates than the cytoplasmic-expressing strain in all ranges of Paraoxon concentrations. Also, the strains under T7 promoter regulation showed higher biocatalytic reaction rates than the strains under Trc promoter regulation regardless of cellular location of OPH. Interestingly, the biocatalytic rate in the periplasmic-secreting strain was much increased with substrate concentrations compared to that in the cytoplasmic-expressing strain and showed a 2-fold higher conversion rate with 1 mM Paraoxon. These results demonstrated that the developed whole cell biocatalyst system with periplasmic-secretion could more efficiently degrade organophosphates, especially high concentration of environmental toxic organophophate compounds.

From the linear Lineweaver-Burk plots (Fig. 3), kinetic parameters V_{max} and K_m were determined for four whole cell systems and shown in Table 1. In the case of T7 promoter system, the periplasmic-secreting whole cells under T7 promoter showed 2.7-fold higher maximum bioconversion rate (V_{max}, 0.373 mM min^{-1} per unit cell concentration (1 mg-dry cell weight ml^{-1})) than that (0.138 mM min^{-1}) of the cytoplasmic-secreting whole cells under T7 promoter. However, the periplasmic-secreting whole cells under T7 promoter had higher Michaelis-Menten constant; implying that substrate affinity of periplasmic-secreting whole cell biocatalyst was lower. When we calculated overall reaction efficiency, V_{max}/K_m, the periplasmic-secreting whole cells under T7 promoter exhibited highly enhanced bioconversion efficiency by 1.8-fold. In the *trc* promoter system, we obtained similar comparative results between the periplasmic-secreting and cytoplasmic-expressing whole cell systems. Also, T7 promoter system showed higher V_{max} than the trc system. However, overall reaction efficiency, V_{max}/K_m, was similar regardless of promoter types.

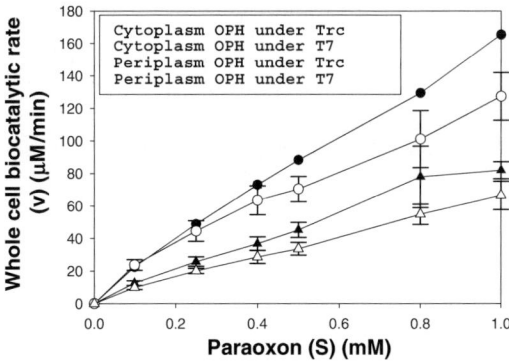

Fig. 2. Whole cell biocatalytic reactions for four types of recombinant whole cell systems. Bioconversion reactions were performed in 'resting cell' condition. All data were based on unit cell concentration (1 mg-dry cell weight ml^{-1}). Each value and error bar represents the mean of two independent experiments and its standard deviation.

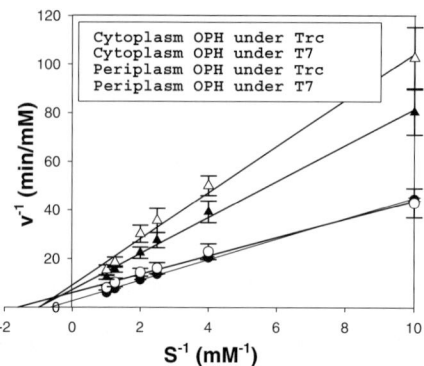

Fig. 3. Lineweaver-Burk plot analyses for four types of recombinant whole cell biocatalytic reactions.

Previously, we have shown that functional secretion of OPH molecules into the periplasmic space induced about 2.8-fold higher specific whole cell OPH activity [10]. From the detail reaction kinetic studies in this work, we showed that this periplasmic space-secretion strategy provided much improved bioconversion capability and efficiency (~1.8-fold) for Paraoxon as a model organophosphate compound. From these results, we confirmed that Tat-driven periplasmic secretion of OPH can be successfully employed to develop a whole cell biocatalysis system with notable enhanced bioconversion efficiency and capability for environmental toxic organophosphates.

Table 1. Kinetic parameters for Paraoxon-biocatalytic reaction using the OPH-expressing whole cells. All data were based on unit cell concentration (1 mg-dry cell weight ml^{-1}).

	Cytoplasmic-expressing Cells		Periplasmic-secreting Cells	
	Trc	T7	Trc	T7
V_{max} (mM min^{-1})	$0.109 \pm 11.3\%$	$0.138 \pm 4.9\%$	$0.166 \pm 15.1\%$	$0.373 \pm 6.4\%$
K_m (mM)	$1.040 \pm 2.1\%$	$1.027 \pm 6.9\%$	$0.629 \pm 1.1\%$	$1.578 \pm 8.2\%$
V_{max}/K_m (min^{-1})	0.105	0.135	0.264	0.237

REFERENCES

[1] W.J. Donarski, D.P. Dumas, D.P. Heitmeyer, V.E. Lewis and F.M. Raushel, Biochem., 28 (1989) 4650.
[2] W.W. Mulbry and J.S. Karns, J. Bacteriol., 171 (1989) 6740.
[3] C.M. Serdar and D.T. Gibson, Bio/Technology, 3 (1985) 567.
[4] W. Chen and A. Mulchandani, Trends Biotechnol., 16 (1998) 71.
[5] E. Rainina, E Efremenco, S. Varfolomeyev, A.L. Simonian and J.R. Wild, Biosens. Bioelectron., 11 (1996) 991.
[6] H.J. Cha, C.F. Wu, J.J Valdes, G. Rao and W.E. Bentley, Biotechnol. Bioeng., 67 (2000) 565.
[7] D.G. Kang, J.Y.H. Kim and H.J. Cha, Biotechnol. Lett., 24 (2002) 879.
[8] L. Li, D.G. Kang and H.J. Cha, Biotechnol. Bioeng., 85 (2004) 214.
[9] M.A. Kaderbhai, C.C. Ugochukwu, S.L. Kelly and D.C. Lamb, Appl. Environ. Microbiol., 67 (2001) 2136.
[10] D.G. Kang, G.B. Lim and H.J. Cha, J. Biotechnol., 118 (2005) 379.
[11] M.L. Shuler and F. Kargi, Bioprocess engineering: Basic concepts, Prentice Hall, New Jersey, 1992.

CATALYSIS AND CATALYTIC
REACTION ENGINEERING

Studies in Surface Science and Catalysis, volume 159
Hyun-Ku Rhee, In-Sik Nam and Jong Moon Park (Editors)

Catalytic hydrogen supply from a decalin-based chemical hydride under superheated liquid-film conditions

Shinya Hodoshima[a], **Atsushi Shono**[a], **Kazumi Satoh**[a] and **Yasukazu Saito**[a]

[a]Department of Industrial Chemistry, Faculty of Engineering, Tokyo University of Science
1-3 Kagurazaka, Shinjuku-ku, Tokyo 162-8601, Japan

Efficient hydrogen supply from decalin was only accomplished by the "superheated liquid-film-type catalysis" under reactive distillation conditions at moderate heating temperatures of 210-240°C. Carbon-supported nano-size platinum-based catalysts in the superheated liquid-film states accelerated product desorption from the catalyst surface due to its temperature gradient under boiling conditions, so that both high reaction rates and conversions were obtained simultaneously.

1. INTRODUCTION

Any storage medium of hydrogen for fuel cells requires the following characteristics: (1) High storage capacity of hydrogen; (2) Safe & economical; (3) Facile reversibility. A catalysis pair of decalin dehydrogenation / naphthalene hydrogenation (Eq. (1)), proposed by our group as an organic chemical hydride [1-5], satisfies the DOE target value on storage (6.5 wt%, 62.0 kg-H_2 / m^3) with the hydrogen content of decalin (7.3 wt%, 64.8 kg-H_2 / m^3). Both decalin and naphthalene have been accepted socially as safe commodity chemicals, e.g., organic solvent and insect killer at home. As the catalytic process for naphthalene hydrogenation has been commercialized since the 1940s [6], this decalin / naphthalene pair would be suitable for operating fuel cells, provided that catalytic hydrogen supply from decalin be attained quite efficiently under moderate reaction conditions.

Since endothermic decalin dehydrogenation is restricted thermodynamically at around 250°C [7], a key technology to generate hydrogen from decalin efficiently at this temperature range is how to use catalysts. In the present paper, a novel concept of "superheated liquid-film-type catalysis" [1-5] was attempted to decalin dehydrogenation with carbon-supported platinum-based nano-particles under reactive distillation conditions [8], which made it possible to attain high reaction rates and conversions at 210-240°C.

$$\text{(decalin)} \rightleftharpoons \text{(naphthalene)} + 5\,H_2 \qquad (1)$$

2. EXPERIMENTAL

2.1. Catalyst preparation

Prior to the catalyst preparation, a base-pretreatment toward activated carbon (KOH-activation, BET specific surface area: 3100 m^2 / g, average pore size: 2.0 nm, Kansai Netsukagaku Co. Ltd.) by immersing an aqueous solution of NaOH (pH 12, 14) for 24 h was carried out in order to promote the anion exchange between the ligand chloride of impregnated metal precursers (K_2PtCl_4) and the aqueous hydroxide ion inside the micropores of activated carbon [1-5]. A carbon-supported platinum catalyst (Pt / C, 5 wt-metal%) was prepared by an impregnation method [1-5]. Carbon-supported platinum-tungsten (Pt-W / C, 5 wt-Pt%, mixed molar ratio of Pt / W: 5) [1-3] and platinum-rhenium (Pt-Re / C, 5 wt-Pt%, mixed molar ratio of Pt / Re: 4) [3,5] composite catalysts were prepared by a dry-migration method [5].

2.2. Evaluation of catalytic dehydrogenation activities for decalin

Under boiling and refluxing conditions by heating at 210-240°C and cooling at 25°C, catalytic decalin dehydrogenation was performed for 2.5 h at atmospheric pressure in a batch-wise reactor (Eggplant-type flask, 50 ml) [1-5], containing a constant amount of the carbon-supported platinum-based catalysts (0.75 g or 0.30 g) and various charged amounts of decalin. Reaction rates were obtained from time courses of evolved hydrogen, whereas conversions after 2.5 h were checked by both total amounts of evolved hydrogen and liquid compositions, consisting of naphthalene and unconverted decalin, at the final stage of reaction measured with gas-chromatographic analysis. In this way, a relationship of the dehydrogenation activities with the charged amount of decalin was investigated to optimize reaction conditions.

3. RESULTS AND DISCUSSION

3.1. Catalytic decalin dehydrogenation over Pt / C catalyst under reactive distillation conditions

In the batch-wise reactor, hydrogen evolution from decalin with a constant amount (0.30 g or 0.75 g) of carbon-supported platinum catalyst was performed at various charged amounts of decalin under reactive distillation conditions by heating at 210°C and cooling at 5°C. **Fig. 1(A)(B)** shows relationships of catalytic dehydrogenation activities with the charged amount of decalin. A sensitive correlation between the catalyst amounts (0.75 g and 0.30 g) and the catalyst / decalin amount ratios was elucidated with respect to 2.5 h conversions of decalin into hydrogen and naphthalene (**Fig. 1(A)**). A determined catalyst amount definitely required a certain adequate amount of decalin, where just-immersed catalyst states (liquid-film) were realized, being different from either sand-bath or suspended states. In the case of decalin scarcity, the catalyst surface became dry soon due to evaporation into the open space inside the reactor, which brought us a large initial dehydrogenation rate only at the beginning. On the other hand, plenty amounts of decalin could disperse the catalyst homogeneously in the solution, giving rather low conversions and reaction rates. At the case of adequate amount (1.0 ml or 3.0 ml decalin), the catalysts appeared just wet but not suspended, which was designated previously as "superheated liquid-film state" [1-5]. Even at a thermodynamically-unfavorable temperature of 210°C, the carbon-supported catalysts gave high 2.5 h conversions (41.0% at 0.30 g / 1.0 ml, 39.1% at 0.75 g / 2.0 ml), since a thin liquid-film of substrate covered the catalyst under boiling conditions (**Fig. 1(A)**). In contrast to the suspended state, where the catalyst-layer temperature was equal to the boiling point, the liquid-film state resulted in a temperature gradient due to "superheating" [9]. Since the active-site temperature was higher than the boiling point, the dehydrogenation rates were accelerated and the adsorption equilibrium of naphthalene was shifted toward desorption. In addition, it is to be noted that a much larger initial reaction rate per one gram of platinum and one milliliter of charged decalin was obtained from the liquid-film state at the 0.30 g / 1.0 ml ratio than the

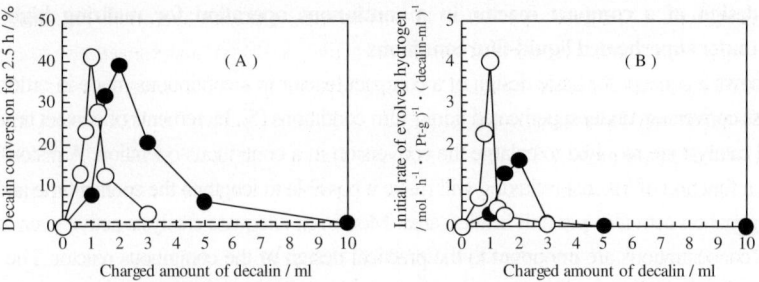

------: Equilibrium conversion (1.2%) for decalin dehydrogenation at 210°C and 1 atm
Catalyst: Carbon-supported platinum nano-particles (Pt / C, 5 wt-metal%) 0.30 g (○) and 0.75 g (●)
Reaction conditions: Boiling and refluxing by heating at 210°C and cooling at 5°C.

Fig. 1   Relationship of catalytic dehydrogenation activities with charged amounts of decalin

0.75 g / 3.0 ml case (**Fig. 1(B)**). Therefore, the superheated liquid-film state is controllable by the extent of superheating. High performance is guaranteed with sharp temperature gradients at the catalyst-solution interface, leading to efficient desorption of reaction products and enhanced dehydrogenation processes. A compact reactor is therefore possible to design with improved catalytic performances for dehydrogenation.

3.2. Improvement of dehydrogenation activities for decalin by carbon-supported composite catalysts under superheated liquid-film conditions

Fig. 2(A)(B) shows time courses of amounts of evolved hydrogen and decalin conversions with carbon-supported platinum-based catalysts under superheated liquid-film conditions. Enhancement of dehydrogenation activities for decalin was realized by using these composite catalysts. The Pt-W / C composite catalyst exhibited the highest reaction rate at the initial stage, whereas the Pt-Re / C composite catalyst showed the second highest reaction rate in addition to low in sensitivity to retardation due to naphthalene adsorbed on catalytic active sites [1-5], as indicated in **Fig. 2(A)(B)**.

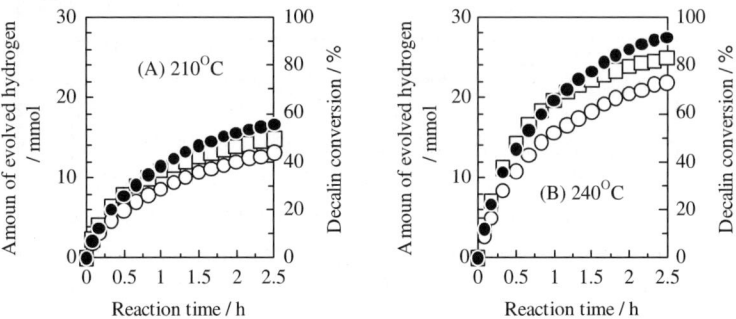

Catalyst: Pt / C (5 wt%, ○) 0.30 g, Pt-W / C (5 wt-Pt%, Pt / W = 5, □) 0.30 g and Pt-Re / C (5 wt-Pt%, Pt / Re = 4, ●) 0.30 g
Charged amount of decalin solution: 1.0 ml (superheated liquid-film state)
Reaction conditions: Boiling and refluxing by heating at 210 and 240°C and cooling at 5°C

Fig. 2   Time courses of hydrogen evolved from decalin and decalin conversion
with carbon-supported platinum-based catalysts under superheated liquid-film conditions

180

3.3. Basic design of a compact reactor in a continuous operation for realizing high one-pass conversion under superheated liquid-film conditions

Fig. 3 shows a concept for basic design of a compact reactor in a continuous mode in order to realize high one-pass conversion under superheated liquid-film conditions [5]. Increments of contact time between reactant and catalyst are required to enlarge the conversion in a continuous operation. A piston-flow type reactor with a function of internal refluxing will make it possible to lengthen the contact time and improve the dehydrogenation activities per unit heating area. Moreover, adequate catalysts active even at the high naphthalene concentrations are important to the practical design of the continuous reactor. The Pt-W / C composite catalyst exhibited the highest rate constant, whereas the Pt-Re composite catalyst was the most suppressive toward retardation due to naphthalene adsorption. The Pt-W and the Pt-Re / C catalysts should be thus set close to (catalyst bed (1) in **Fig. 3**) and distant from the inlet (catalyst bed (2)), respectively.

3. CONCLUSIONS

Efficient hydrogen supply from decalin at moderate temperatures of below $250^{\circ}C$ was accomplished by utilizing the superheated liquid-film-type catalysis under reactive distillation conditions in the present study. The composite catalysts in the liquid-film states improved dehydrogenation activities for decalin.

REFERENCES
[1] C. Liu, M. Sakaguchi and Y. Saito, J. Hydrogen Energy Systems Soc. Jpn., 22(1) (1997) 27.
[2] S. Hodoshima, H. Arai, Y. Saito, Int. J. Hydrogen Energy, 28(2) (2003) 197.
[3] S. Hodoshima, H. Arai, S. Takaiwa, Y. Saito, Int. J. Hydrogen Energy, 28(11) (2003) 1255.
[4] S. Hodoshima and Y. Saito, J. Chem. Eng. Jpn., 37(3) (2004) 391.
[5] S. Hodoshima, S. Takaiwa, A. Shono, K. Satoh and Y. Saito, Appl. Catal. A, 283(1-2) (2005) 235.
[6] A.W. Weitkamp, Adv. Catal., 18 (1968) 1.
[7] O.R. Stull, E.F. Westrum Jr., and G.C. Sinke, The Chemical Thermodynamics of Organic Compounds, John Wiley, New York, 1969. p. 15-20.
[8] V.H. Agreda, L.R. Partin and W.H. Heise, Chem. Eng. Progr., 86 (1990) 40.
[9] A. Heidrich and M. Jakob, Heat Transfer, John Wiley, New York, 1957, Vol. 1, p. 618.

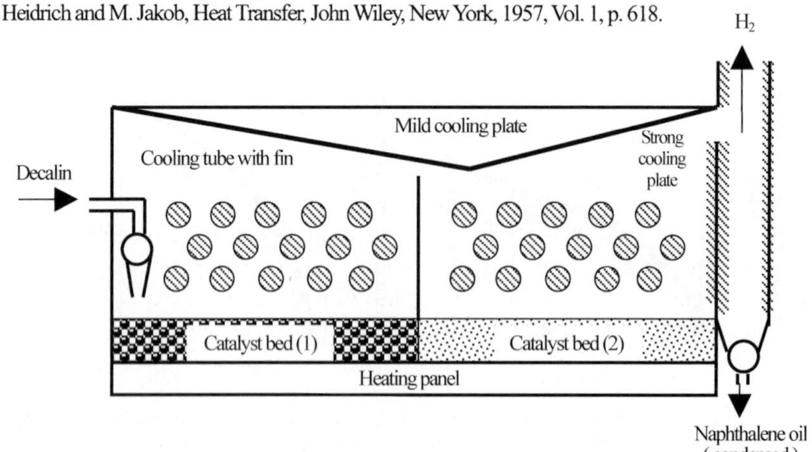

Fig. 3   A continuous reactor for decalin dehydrogenation of piston-flow type under superheated liquid-film conditions

Studies in Surface Science and Catalysis, volume 159
Hyun-Ku Rhee, In-Sik Nam and Jong Moon Park (Editors)

Solid-Liquid Reaction Catalyzed by a Liquid Containing Rich Phase-Transfer Catalyst ---- Synthesis of Hexyl Acetate

Hsu-Chin Hsiao[*], Chung-Wei Hsu[+] and Hung-Shan Weng[+]

[*]Department of Chemical Engineering, Tung Fung Institute of Technology, Kaohsiung 829, Taiwan, R. O. C.

[+]Department of Chemical Engineering, National Cheng Kung University, Tainan 701, Taiwan, R. O. C.

Abstract

This work is for the purpose of evaluating the feasibility of synthesizing hexyl acetate (ROAc) from n-hexyl bromide (RBr) and sodium acetate (NaOAc) by a novel phase-transfer catalysis (PTC) technique. In this new technique, the solid-liquid reaction was catalyzed by a catalyst-rich liquid phase. Experimental results reveal that the use of this technique for synthesizing ROAc gives a far higher reaction rate than the solid-liquid-PTC does and a slightly faster rate than tri-liquid PTC. The amount of water added greatly influences the type of the reaction system and the reaction rate. The kind of catalyst affects the conversion of RBr and the fractional yield of ROAc significantly. The catalyst with a longer chain length gives a better performance. However, it will more easily dissolve in the organic phase and hence is more difficult to be recovered and reused after reaction. Tetra-n-butylammonium bromide is the best choice. As the experimental results of reusing catalyst phase show that the conversion decreases due to part of catalyst dissolving into the organic phase, an effort for improving this drawback should be made in the future.

1. INTRODUCTION

Although the use of phase-transfer catalysis (PTC) for manufacturing esters has the merits of a mild reaction condition and a relatively low cost [1], PTC has its limitations, such as the low reactivity of carboxylic ion by liquid-liquid PTC [2], a slow reaction rate by solid-liquid PTC, and the difficulty of reusing the catalyst by both techniques.

This work was initiated for the purpose of evaluating the feasibility of synthesizing hexyl acetate (ROAc) from n-hexyl bromide (RBr) and sodium acetate (NaOAc) by a novel PTC technique. In this new technique, the solid-liquid reaction was catalyzed by a catalyst-rich liquid phase in a batch reactor. Because there a solid phase and two liquid phases coexist, it is called as a SLL-PTC system [3]. Actually, this liquid phase is the third liquid phase in the tri-liquid PTC system. It might be formed when the phase-transfer catalyst is insoluble or slightly soluble in both aqueous and organic phases. Both aqueous and organic reactants can easily transfer to this phase where the intrinsic reaction occurs [4, 5].

In this study, tetra-n-alkylammonium bromides ((n-Alkyl)$_4$NBr, QBr) were used as the phase-transfer catalysts. The reaction will proceed in the catalyst-rich liquid phase in a way similar to that in the tri-liquid-phase phase transfer catalysis [4, 5]:

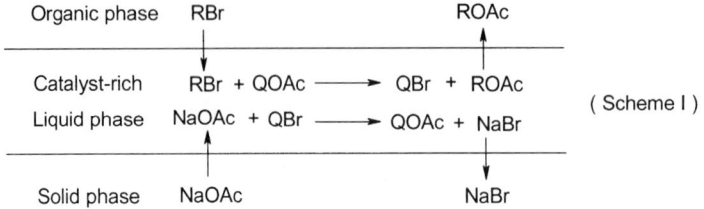

(Scheme I)

2. EXPERIMENTAL METHODS

The catalyst-rich liquid phase was prepared by forming a third liquid phase in a way similar to that has been described previously [5, 6]. A 125-mL three-neck round-bottom flask was employed as the reactor. The procedure is similar to the previous work [7].

3. RESULTS AND DISCUSSION

3.1. Formation of the Catalyst-Rich Liquid Phase

The quantity of water added to the phase-transfer catalytic reaction system is a main factor for forming a catalyst-rich liquid phase (the third liquid phase). In the aqueous-organic two phase system, in which the aqueous phase contains phase-transfer catalyst and sodium acetate, when the amount of water is decreased, the volume of aqueous phase will decrease accordingly, even a third liquid phase will appear. The third liquid phase consists of water, sodium acetate, organic solvent and phase-transfer catalyst. If the concentration of sodium acetate is high, most of the phase-transfer catalyst will be salted out to the third liquid phase, hence a catalyst-rich liquid phase is formed. After the appearance of the third liquid phase, the volume of aqueous phase will decline significantly with decreasing the amount of water added, while that of the third liquid phase only shrinks slightly. This fact implies that the catalyst in the third liquid phase attracts a fixed amount of water and will not be affected by the water content in the aqueous phase. Figure 1 reveals the above-mentioned facts. If the amount of water added is continuously decreased, a solid phase (sodium acetate) will appear in the aqueous phase.

3.2. Effect of the Amount of Water Added on the Reaction

Figure 2a reveals that the conversion of RBr increases with increasing the amount of water up to the amount of water being 3 mL, then it declines. Note that the sharp increase of the conversion curves is due to the fast diffusion of RBr from the organic phase to the catalyst-rich liquid phase where only part of RBr reacts with QOAc.

Except that RBr can be hydrolyzed to ROH, ROAc (the main product) can be hydrolyzed to form ROH and HOAc (the byproducts) too, so the addition of water will reduce the fractional yield of ROAc. On the contrary, another byproduct, NaBr, can adsorb water to form the hydrated ions. These two contradictory facts result in a peculiar phenomenon (Figure 2b).

3.3. Effect of the Kind of Quaternary Ammonium Salt

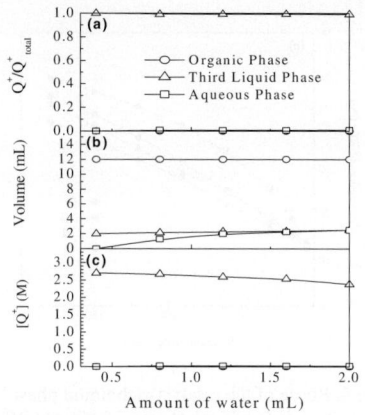

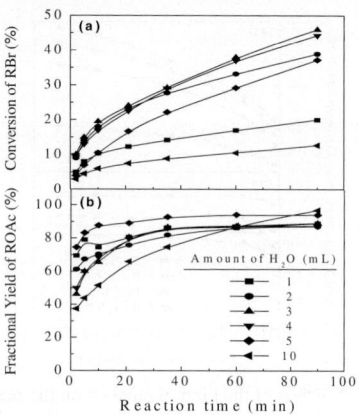

Figure 1. Effect of the added amount of water on the volume and Q^+ concentration of each phase. System: 10 mL Octane + 0.014mole NaOAc + 0.006mol Bu$_4$NBr @ 80^{0}C, 1000 rpm

Figure 2. Effect of the added amount of water on the conversion of RBr and the yield of ROAc. System: 25 mL Octane + 0.025 mol RBr + 0.035 mol NaOAc + 0.015 mol Bu$_4$NBr @ 80^{0}C, 1000 rpm

Figure 3a shows that the use of quaternary ammonium salt with a longer carbon chain gives a higher reaction rate. This is because that a longer carbon chain makes the catalyst more lipophilic. Consequently, the catalyst is easier to react with RBr and more catalyst will dissolve into the organic phase where the catalyst has more chance to react with RBr. However, Figure 3b shows the fractional yield of ROAc decreases obviously with the length of carbon chain in the catalyst. This fact might be attributed to the fact that ROAc produced will stay longer at the catalyst-rich liquid phase which is more lipophilic hence has more chance to be hydrolyzed to ROH because this phase contains water. It should be pointed out that although the catalyst with a longer carbon chain has a higher activity and gives a higher conversion of RBr, it is more difficult to be recovered for reuse due to a larger solubility in the organic phase in addition to a lower fractional yield of ROAc.

3.4. Reuse of Catalyst

The feasibility of reusing the phase transfer catalyst via the reuse of the third liquid phase in a tri-liquid catalytic system has been investigated in our laboratory [6,7]. In the present study, the catalyst-rich liquid phase was repeatedly used for four times and the changes in the conversion of RBr and the fractional yield of ROAc were observed.

As shown in figure 4, the conversion of RBr decreased significantly when the catalyst phase was reused first time (2nd run), however, no obvious change in the subsequent runs (3rd and 4th runs). The decrease in the conversion might be due to the following reasons: (a) NaBr formed during the reaction would accumulate in the catalyst-rich liquid phase hence the solubility of NaOAc decreased; (b) ROAc produced made the polarity of the organic phase increased hence more catalyst would dissolve into the organic phase; and (c) Part of NaBr produced would precipitate and formed solid particles which would hinder the mass transfer of RBr to the catalyst-rich phase. This drawback should be overcome before the technique of solid-liquid-liquid catalysis to be commercialized.

184

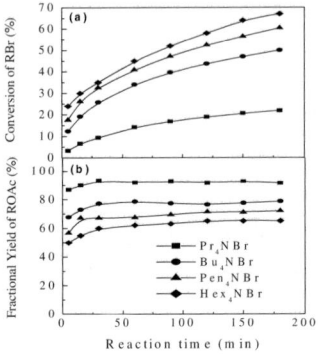

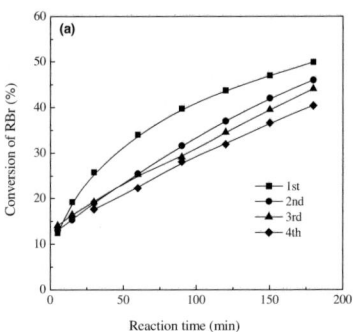

Figure 3. Effect of the kind of catalyst on the reaction System: 2 mL H_2O + 25 mL Octane + 0.025 mol RBr + 0.060 mol NaOAc + 0.025 mole catalyst @ 80^{0}C, 600 rpm

Figure 4. Reuse of the catalyst-rich liquid phase. System: 2 mL H_2O + 25 mL Octane + 0.025 mol RBr + 0.060 mol NaOAc + 0.025 mol Bu_4NBr @ 80^{0}C, 600 rpm

4. CONCLUSIONS

Based on the above discussions, we can draw the following conclusions:

(1) The use of the SLL-PTC technique for synthesizing ROAc gives a far higher reaction rate than the SL-PTC does and a slightly faster rate than tri-liquid PTC.

(2) For forming a SLL reaction system, the amount of water added should be small.

(3) For reusing the catalyst, the organic solvent should be nonpolar.

(4) Although the catalyst with a longer chain length benefits the reaction rate, it is not easy to be recovered. Tetra-n-butylammonium bromide is the best choice.

(5) The amount of catalyst added should not be too high, otherwise the fractional yield of ROAc would decrease.

(6) The amount of NaOAc added should not be too high, otherwise the reaction rate of RBr would decrease.

(7) A lower concentration of RBr benefits the fractional yield of ROAc.

(8) The conversion of RBr will decline when the catalyst-rich liquid phase is reused, mainly due to the accumulation of NaBr and lose of catalyst. This drawback should be overcome before the technique of solid-liquid-liquid catalysis to be commercialized.

References

1. C. M. Stark, C. Liotta and M. Halpen, Phase Transfer Catalysis, Fundamental, Applications and Industrial Perspectives, Chapman & Hall, New York, 1994.

2. D. Albanese, D. Landini, A. Maia and M. Penso, Ind. Eng. Chem. Res., 40(2001); 2396- 2401.

3. G. Jin, T. Ido and S. Goto, Catalysis Today, 79(2003), 471-478.

4. H.-S. Weng and W.-C. Huang, J. Chin. Inst. Chem. Eng., 18 (1987), 109-115.

5. D.-H. Wang and H.-S. Weng, Chem. Eng. Sci., 43 (1988), 2019-2022.

6. H.-S. Weng, C.-M. Wang and D.-H. Wang, Ind. Eng. Chem. Res., 36 (1997), 3613-3618.

7. H.-C. Hsiao and H.-S. Weng, Chem. Eng. Comm., 191 (2004), 694-704.

Studies in Surface Science and Catalysis, volume 159
Hyun-Ku Rhee, In-Sik Nam and Jong Moon Park (Editors)
185

Preparation of Mono-Dispersed MFI-type Zeolite Nanocrystals via Hydrothermal Synthesis in a Water/Surfactant/Oil Solution

Teruoki Tago[*], **Kazuyuki Iwakai, Mieko Nishi and Takao Masuda**

Chemical Engineering, Division of Chemical Process Engineering, Graduate School of Engineering, Hokkaido University, N13W8, Kita-Ku, Sapporo, 060-8628, Japan
[*]tago@eng.hokudai.ac.jp

1. Introduction

Zeolites, crystalline alumino-silicates, have been widely used in industry as heterogeneous catalysts and adsorbents. Moreover, zeolite crystals contain micropores with diameters almost equal to those of lighter hydrocarbons. The uniform size and shape of the micropores allows zeolites to be used as shape-selective catalysts. However, as compared with the size of micropores exhibiting a molecular-sieving effect, the crystal size of zeolites is very large, approximately 1~3 μm. In catalytic reactions using a zeolite, reactant molecules diffuse into the micropores of the zeolite, and adsorb on the active sites, where chemical reactions proceed. The diffusion rates of reactants, such as hydrocarbon molecules, within the zeolite crystals are relatively low in comparison with the reaction rates. This resistance to mass transfer causes the limitation of the reaction rates, low selectivity of intermediates and coke deposition, leading to the short lifetime of the catalysts. In order to overcome these problems, it will be necessary to achieve lower resistance to mass transfer of the reactants in the crystals; to this end, nanometer-sized zeolite crystals would be a promising solution.

The zeolite nanocrystals have attracted the considerable attention of many researchers [1–5]. The syntheses of several types of zeolites with different nanometer sizes, such as silicalite-1, ZSM-5, A-type and Y-type, have been reported. Recently, micellar solutions or surfactant-containing solutions have been used for the preparation of zeolite nanocrystals [4,5]. We have also successfully prepared silicalite nanocrystals via hydrothermal synthesis using surfactants. In this study, we demonstrate a method for preparing mono-dispersed silicalite nanocrystals in a solution consisting of surfactants, organic solvents and water.

2. Experimental

Two kinds of solution were prepared in advance. Solution A was a water solution containing an Si source, which was obtained by hydrolyzing metal alkoxide (tetraethylorthosilicate, TEOS) with a dilute tetrapropylammoniumhydroxide (TPA-OH)/water solution at room temperature. The molar ratio of Si to the template was 3. In preparation of ZSM-5 zeolite nanocrystals, aluminium isopropoxide as an Al source and sodium chloride were added into solution A. Solution B was an organic solution containing surfactant. Nonionic surfactants, polyoxyethylene (15) cetylether (C-15), polyoxyethylene (15) nonylphenylether (NP-15), and polyoxyethylene (15) oleylether (O-15), and ionic surfactants, sodium bis(2-ethylhexyl) sulfosuccinate (AOT) and cetyltrimethyl ammonium bromide (CTAB), were employed in this study. Cyclohexane (C_6H_{12}) was used as an organic solvent except in the case of CTAB, because CTAB/cyclohexane microemulsion could not be formed. 1-hexanol was used as an organic solvent in the case of CTAB. Solution A was added to solution B, and the mixture was magnetically stirred at 50 degree C for 1 h. The mixture was then placed in a Teflon-sealed stainless bottle, heated to 100-120 degree C, and held at the desired temperature for 50 h with stirring to produce MFI-type zeolite nanocrystals. The precipitate thus obtained was centrifuged, washed thoroughly with propanol, dried at 100 degree C overnight, and calcined under an air flow at 500 degree C to remove the surfactant and the template. The morphology and crystallinity of the silicalite nanocrystals were investigated as functions of the types of surfactants and the concentration of the Si source using a field-emission scanning electron microscope (FE-SEM; JEOL JSM-6500F) and an X-ray diffractometer (XRD; JEOL JDX-8020). The pore structure of the nanocrystals was evaluated by a N_2 adsorption and desorption method (BEL Japan Belsorp mini). An NH_3-TPD experiment [6] was carried out to determine the acidity of the ZSM-5 zeolite nanocrystals.

3. Results and Discussion

3.1. Effect of the types of surfactants on the morphology of silicalite samples

The morphology of samples prepared using various surfactants was investigated. Figures 1 and 2 show FE-SEM photographs and X-ray diffraction patterns of the obtained

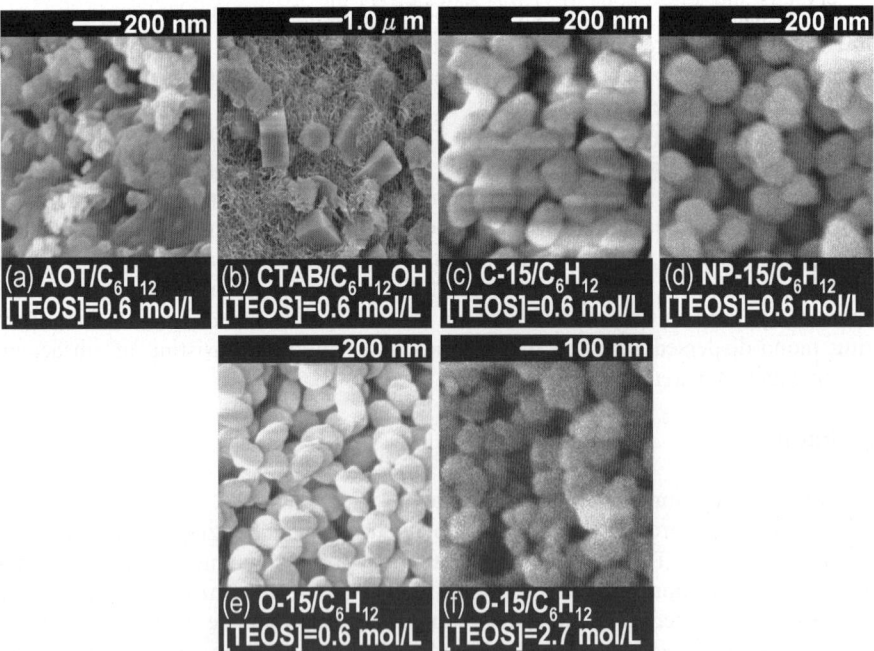

Fig.1. Effect of surfactant type on the morphology of samples prepared in surfactant/organic-solvent solution. The concentration of the Si source, [TEOS], was 0.63 mol/L in (a)–(e) and 2.73 mol/L in (f).

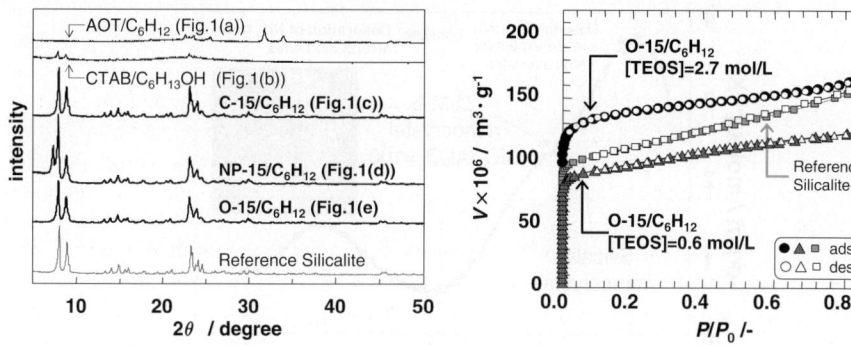

Fig. 2. X-ray diffraction patterns of nanocrystals. Fig. 3. N_2 adsorption and desorption isotherm of nanocrystals.

samples, respectively. The sample prepared in 1-hexanol solution containing AOT (AOT/cyclohexane) was irregular in shape. In the X-ray diffraction pattern of the sample prepared in AOT/cyclohexane solution, some peaks appeared, and these corresponded to sodium sulfate rather than silicalite. It was considered that the TPA-OH molecules could not act as a template in the synthetic solution. This result was ascribed to the fact that the surfactant of AOT and the template of TPA-OH have opposite ionic charges. When using CTAB/1-hexanol, the silicalite crystals, which were approximately 1.0 μm in size, were embedded in the amorphous SiO_2 based on the SEM observation. The coexistence of silicalite crystals and amorphous SiO_2 was revealed by the X-ray diffraction analysis (Fig. 2). Since the pH of the synthetic solution was alkaline, the surface of the SiO_2 produced by hydrolysis of TEOS possessed a negative charge. Accordingly, it was considered that the CTAB and TPA-OH were independently adsorbed on the surface of SiO_2 because of their cationic ionicity. Therefore, SiO_2 species adsorbing TPA-OH and CTAB would change into silicalite crystals and amorphous SiO_2, respectively.

On the other hand, in the case of the nonionic surfactants C-15, NP-15 and O-15 (the nonionic surfactant/cyclohexane system), mono-dispersed silicalite nanocrystals were obtained as shown in Fig. 1(c), 1(d) and 1(e), respectively. The X-ray diffraction patterns of the samples showed peaks corresponding to pentasile-type zeolite. The average size of the silicalite nanocrystals was approximately 120 nm. These results indicated that the ionicity of the hydrophilic groups in the surfactant molecules played an important role in the formation and crystallization processes of the silicalite nanocrystals.

3.2. Effect of concentration of the Si source on the size of silicalite nanocrystals

To control the size of the silicalite nanocrystals, the effect of the concentration of the Si source in the solution was investigated. As shown in Fig. 1(e) and 1(f), mono-dispersed silicalite nanocrystals with different size could be obtained. The average sizes of the silicalites were 120 nm and 80 nm at the TEOS concentrations of 0.6 mol/l and 2.7 mol/l, respectively. The finding that the size of silicalite was dependent on the TEOS concentration was ascribed to the number of silicalite nuclei produced at the initial stage of hydrothermal synthesis. The increase in the TEOS concentration led to an increase in the number of silicalite nuclei, on which the nonionic surfactants were adsorbed. Since the aggregation of the silicalite nuclei was inhibited by the adsorbed surfactants on the surface during hydrothermal treatment, mono-dispersed nanocrystals could be prepared.

Figure 3 shows the N_2 adsorption isotherms of the silicalite nanocrystals. The steep

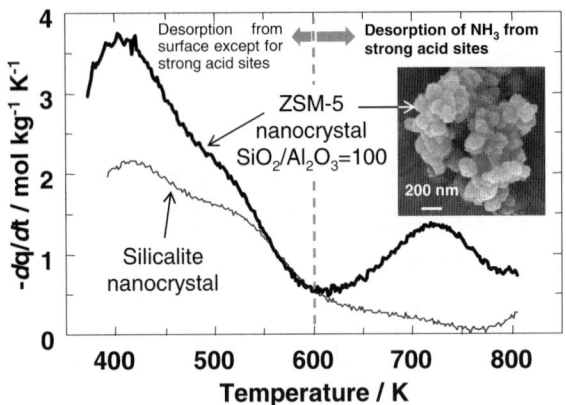

Fig. 4. NH₃-TPD spectrum and FE-SEM photograph of ZSM-5 zeolite nanocrystals.

increase in the amount of adsorbed N_2 in the region of low relative pressure (P/P_0), followed by the flat curve, corresponded to the filling of micropores. This result indicated that the nanocrystals contained considerable volume of micropores volume in their framework and were well crystallized.

3.3. Synthesis and characterization of ZSM-5 zeolite nanocrystals

In order to prepare ZSM-5 zeolite nanocrystals, an Al source of aluminium isopropoxide was added into solution A, and hydrothermal synthesis of the solution A containing Si and Al sources was carried out in an O-15/cyclohexane solution at 120 degree C for 50 h. Figures 4 show ac-NH₃-TPD spectra and a SEM photograph of the ZSM-5 zeolite nanocrystals. Nanocrystals with a diameter of approximately 150 nm were observed, and the NH₃-TPD spectrum showed desorption of NH₃ above 600 K, indicating that the nanocrystals possessed strong acid sites.

4. Conclusion

Mono–dispersed silicalite and ZSM-5 type zeolite nanocrystals with a diameter of 80–120 nm were successfully prepared in a surfactant-oil-water solution. The ionicity of the surfactants used in the preparation affected the crystallinity and structure of the silicalite crystals, and silicalite nanocrystals could be obtained when using a nonionic surfactant. By adding an Al source into the synthetic solution, ZSM-5 type zeolite nanocrystals with strong acid sites could be obtained.

References

[1] M. Tsapatsis, et al., Microporous Materials, 5 (1996) 381.
[2] R. Singh and P. K. Dutta, Langmuir, 16 (2000) 4148.
[3] S. Mintova, et al., Science, 283 (1999) 958.
[4] H. Hosokawa and K. Oki, Chem. Lett., 32 (2003) 586.
[5] T. Tago, et al., Chem.Lett., 33 (2004) 1040.
[6] T. Masuda, et al., Appl.Catal., 165 (1997) 57.

Studies in Surface Science and Catalysis, volume 159
Hyun-Ku Rhee, In-Sik Nam and Jong Moon Park (Editors)

Methane reforming with carbon dioxide to synthesis gas over Mg-promoted Ni/HY catalyst

H. Jeong[a,b], K. Kim[b], D. Kim[b] and I.K. Song[a]*

[a]School of Chemical and Biological Engineering, Institute of Chemical Processes,
Seoul National University, Shinlim-dong, Kwanak-ku, Seoul 151-744, South Korea

[b]Korea Institute of Energy Research, Jang-dong, Yuseong-ku, Daejeon 305-343, South Korea

*Corresponding author (Tel:+82-2-880-9227, E-mail: inksong@snu.ac.kr)

1. INTRODUCTION

Methane reforming with carbon dioxide to synthesis gas has attracted much attention recently [1]. Numerous supported catalysts based on nickel and novel metals have been investigated for this reaction. One of the serious problems in this reaction is the carbon deposition formed via Boudouard reaction ($2CO \rightarrow C+CO_2$) or methane decomposition ($CH_4 \rightarrow C+2H_2$), which eventually leads to severe catalyst deactivation [2,3]. High price of noble metals renders their application quite questionable. Therefore, it is more practical from the industrial point of view to develop an improved non-noble metal-based catalyst. That is, it is necessary to develop supported catalysts comprising non-noble metal and promoter showing considerable catalytic activity without suffering the catalyst deactivation by carbon deposition. In this study, we prepared the Ni/HY, Ni-Mn/HY, and Ni-Mg/HY catalysts and investigated their catalytic activity and stability in the methane reforming with carbon dioxide, with an aim of minimizing carbon deposition on the catalyst surface and improving stability and performance of the catalysts.

2. EXPERIMENTAL

Metal-promoted Ni/HY catalysts were prepared by co-impregnation method with a solution containing known concentration of metal nitrate salt. Mn and Mg were used as a promoter, and their content was fixed at 5 wt.%. All the prepared catalysts were reduced with a mixed stream of hydrogen (10 ml/min) and nitrogen (30 ml/min) at 500°C for 3 h prior to the reaction. The supported catalysts were characterized by XRD, BET, FT-IR and TGA analyses. The methane reforming with carbon dioxide was conducted in a continuous fixed-bed reactor (stainless steel tube with inner diameter of 8 mm) at temperatures ranging from 400 to 800°C and at an atmospheric pressure. The powder catalyst (1 g) was packed in the tubular reactor. The reactor (with length of 300 mm) was placed in a vertical tube furnace and connected to a feed stream ($N_2/CH_4/CO_2 = 2/1/1$). Total flow rate of reactant gas stream was 200 ml/min (GHSV=3500 h^{-1}).

190

3. RESULTS AND DISCUSSION

3.1. Characterization

Physicochemical properties of the prepared catalysts are summarized in Table 1. It was observed that pore diameter of Ni/HY catalyst was drastically decreased upon addition of promoter. The Ni/HY catalyst containing Mg promoter showed higher surface area and larger pore volume than any other catalysts. This result implies that the density of meso-pores in the Ni-Mg/HY catalyst was higher than that in the other catalysts. Moreover, nickel particle size determined from X-ray line broadening of Ni peak ($2\theta=44.3°$) was the smallest in the Ni-Mg/HY catalyst, indicating that nickel particles were highly dispersed on the surface of reduced Ni-Mg/HY catalyst. The above result demonstrates that the addition of Mg promoter reduced the Ni particle size, and consequently, yielded highly dispersed Ni particles in the Ni-Mg/HY catalyst.

Table 1
Physicochemical properties of the prepared catalysts

Catalyst	S_{BET} (m^2/g)	D_p (nm)	D_{micro} (nm)	V_p (cm^3/g)	D_{Ni} (nm)
Ni/HY	280.7	19.8	0.61	0.299	23.7
Ni-Mg/HY	293.3	4.3	0.43	0.428	14.9
Ni-Mn/HY	239.5	4.4	0.47	0.345	16.2

The catalyst deactivation in this reaction is basically related to the deposition of inactive carbon species on the active metal sites. The amounts of coke formation on the Ni-Mg/HY and Ni/HY catalysts after 72 h reaction were determined by TGA measurements, which were carried out in an oxygen-containing atmosphere (Fig. 1). TGA results showed that the weight losses of Ni/HY and Ni-Mg/HY catalysts were ca. 32% and 18%, respectively, indicating the formation of relatively large amounts of carbon on the Ni/HY catalyst. The catalytic reaction was also conducted over the Ni-Mg/HY catalyst with a feed stream of $N_2/CH_4/CO_2 = 5/3/1$ in order to accelerate the catalyst deactivation by methane decomposition. Under such an unfavorable condition, however, the deactivation rate was only 0.52% per hour, although the amounts of carbon deposition were increased (the weight loss determined by TGA was 21%).

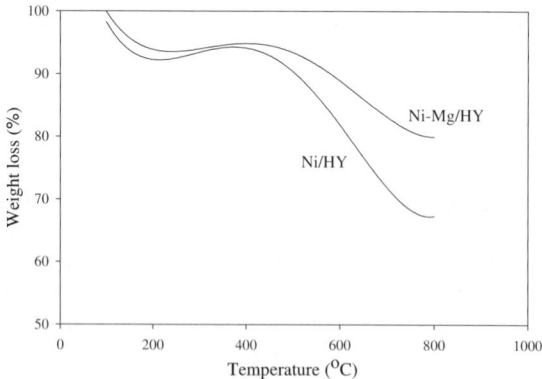

Fig. 1. TGA profiles of Ni/Mg/HY and Ni/HY catalysts after reaction at 700°C for 72 h.

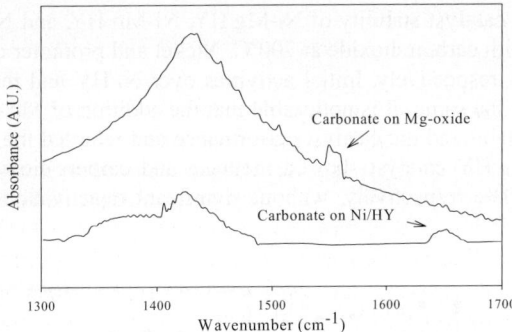

Fig. 2. FT-IR spectra of Ni-Mg/HY and Ni/HY catalysts after reaction at 700°C for 72 h.

Fig. 2 showed the IR spectra of Ni-Mg/HY and Ni/HY catalysts obtained after the reaction at 700°C for 72 h. It was observed that carbonate species appearing at ca. 1650 cm^{-1} were formed on the Ni/HY catalyst [4]. On the other hand, the Ni-Mg/HY catalyst showed a new and strong band at ca. 1550 cm^{-1}; the IR band at 1550 cm^{-1} can be assigned to carbonate species on magnesium oxide [5]. It should be noted that the IR band for carbonate species consistently grew and shifted to lower wavenumber in the Ni-Mg/HY catalyst. The nature of carbonate species between the two catalysts was totally different. Compared to carbonate species on the Ni/HY, carbonate species on the magnesium oxide was more reactive. Therefore, it is believed that Mg promoter in the Ni-Mg/HY catalyst was highly efficient in stabilizing the supported catalyst in the methane reforming with carbon dioxide.

It was previously reported that magnesium oxide with a moderate basicity formed reactive surface carbonate species, which reacted with carbon deposited on the support by the methane decomposition [6]. Upon addition of Mg to the Ni/HY catalyst, reactive carbonate was formed on magnesium oxide and carbon dioxide could be activated more easily on the Mg-promoted Ni/HY catalyst. Reactive carbonate species played an important role in inhibiting the carbon deposition on the catalyst surface.

3.2. Catalytic activity and stability in the methane reforming with carbon dioxide

Effects of nickel loading on the catalytic performance of Ni/HY catalyst are summarized in Table 2. At the initial stage of reaction, conversions and product yields were increased with increasing nickel loading up to 13 wt.%. At higher nickel loading more than 13 wt.%, the catalytic performance of Ni/HY catalyst was slowly decreased. The 13 wt.% Ni/HY catalyst exhibited 93% methane conversion and 82% hydrogen yield after 1 h reaction.

Table 2
Effects of Ni loading on the performance of Ni/HY catalyst at 700°C after 1 h reaction

Ni loading	Conversion (%)		Yield (%)	
(wt.%)	CH_4	CO_2	H_2	CO
5	35	39	32	45
10	84	74	66	69
13	93	86	82	87
20	78	72	76	77

192

Fig. 3 showed the catalyst stability of Ni-Mg/HY, Ni-Mn/HY, and Ni/HY catalysts in the methane reforming with carbon dioxide at 700°C. Nickel and promoter contents were fixed at 13 wt.% and 5 wt.%, respectively. Initial activities over Ni/HY and metal-promoted Ni/HY catalysts were almost the same. It is noticeable that the addition of Mn and Mg to the Ni/HY catalyst remarkably stabilized the catalyst performance and retarded the catalyst deactivation. Especially, the Ni-Mg/HY catalyst showed methane and carbon dioxide conversions more than ca. 85% and 80%, respectively, without significant deactivation even after the 72 h catalytic reaction.

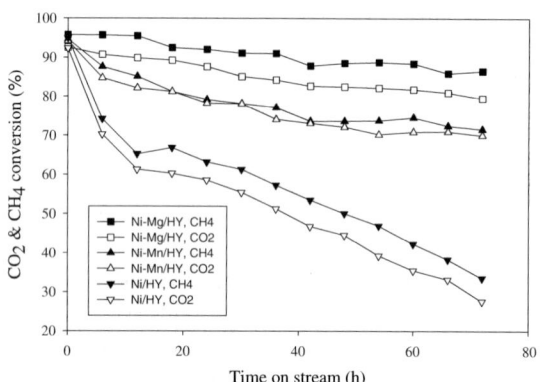

Fig. 3. Catalyst stability with time on stream in the methane reforming with carbon dioxide (T=700°C, GHSV=3500 h^{-1}).

4. CONCLUSIONS

It is concluded that the addition of magnesium to the Ni/HY catalyst in the methane reforming with carbon dioxide stabilized the catalytic performance by suppressing the carbon formation. Among the catalysts tested in this work, the Ni-Mg/HY catalyst showed the highest coke resistance and the most stable catalytic activity. The Ni-Mg/HY catalyst exhibited methane and carbon dioxide conversions more than ca. 85% and 80%, respectively, even after the 72 h reaction. TGA analyses revealed that the Ni-Mg/HY catalyst showed the slow deactivation rate in the methane reforming with carbon dioxide, even when methane decomposition was increased. FT-IR results demonstrated that the reactive carbonate species formed on the magnesium oxide played an important role in suppressing carbon deposition on the catalyst surface.

REFERENCES

[1] A.T. Ashcroft, A.K. Cheetham and M.L.H. Green, Nature, 352 (1991) 225.
[2] S.H. Seok, S.H. Han and J. S. Lee, Appl. Catal. A, 215 (2001) 31.
[3] S. Srihiranpullop and P. Praserthdam, Korean J. Chem. Eng., 20 (2003) 1017.
[4] Z.L. Zhang, X.E. Verykios, S.M. McDonald and S. Affrossman, J. Phys. Chem., 100 (1996) 744.
[5] M. Baldi, E. Finocchio, C. Pistarino and G. Busca, Appl. Catal. A, 173 (1998) 61.
[6] O.V. Krylov, A.K. Mamedov and S.R. Mirzabekova, Catal. Today, 42 (1998) 211.

Studies in Surface Science and Catalysis, volume 159
Hyun-Ku Rhee, In-Sik Nam and Jong Moon Park (Editors)

Regeneration of high silica zeolite used for Beckmann rearrangement

Takeshige Takahashi, Mayumi Nakanishi and Takami Kai

Dept. of Appl. Chem. and Chem. Eng., Kagoshima Univ., Kagoshima, 890-0065 JAPAN
takahashi@cen.kagoshima-u.ac.jp

Abstract

Beckmann rearrangement of cyclohexanone over the regenerated TS-1 and SSZ-31 was carried out to clarify the effect of regeneration number and regeneration atmosphere on e-caprolactam selectivity and catalyst deactivation rate. The rearrangement rate constant for the both catalysts decreased with the number of regeneration, but the e-caprolactam selectivity and deactivation factor observed from TS-1 was almost constant. On the other hand, e-caprolactam from SSZ-31 regenerated in Ar atmosphere decreased with the number of regeneration. These behaviors indicated that the coke precursor deposited on the SSZ-21 surface changed to hard coke in the pore, whereas the coke precursor produced on the pore mouth for TS-1 was easily removed in Ar atmosphere.

1. Introduction

Recently, vapor phase Beckmann rearrangement of cyclohexanone oxime was carried out over many kinds of solid catalysts, such as high silica MFI zeolite [1], β type zeolite [2] or meso-porous materials [3]. Sato and Ichihashi [1,4] reported that high silica MFI zeolite without acidity was the most suitable for the Beckmann rearrangement, but the catalyst deactivation rate was too large to use the catalyst for a fixed bed reactor. Takahashi et al.[5,6] found that the catalyst deactivation of the Beckmann rearrangement was the adsorption of coke precursor, such as oligomer of the ε-caprolactam (CL).

In the present study, the regeneration was carried out in argon or air atmosphere. The rearrangement was performed over the regenerated high silica TS-1 zeolite and SSZ-31 zeolites to elucidate the effect of the regeneration number of on CL selectivity and catalyst deactivation rate. The characterization of the used catalyst was also carried out to clarify the deactivation mechanism for Beckmann rearrangement.

2. Experimental

High silica SSZ-31 zeolite (Si/Al=962) and TS-1 (Si/Ti= 45~300) was prepared by authentic methods [7]. The pore size of SSZ-31 (0.57*0.86 nm) is larger than that of TS-1 (0.54*0.56 nm). The zeolites changed to proton type immersing in ammonium nitrate solution and followed by calcination at 773K for 5h. The physical properties and acidity of the zeolites is listed in Table 1

The zeolite was compressed to small pellet and sieved to 32 to 48 mesh. The rearrangement was carried out in a fixed bed reactor operated at atmospheric pressure.

194

Table 1 Properties of zeolites

Catalyst	Si/Ti ratio [-]	Acidity [mmol/g]	Surf.area[m²/g]	Pore size [nm]
TS-1 (45)	48	0.055		
TS-1 (90)	96	0.016		
TS-1 (200)	221	0.0092	250 ~ 300	0.54*0.56
TS-1 (300)	305	0.0040		
SSZ-31	Si/Al = 962	0.026	330	0.57*0.86

The zeolite fixed into the reactor was dried by carbon dioxide at 623K for 1h. The CHO dissolved into methanol (50 mass%) was supplied by a micro feeder at constant flow rate to evaporator. The vapor diluted by carbon dioxide (20mol%) was fed to the reactor. The reactor effluent was collected at prescribed time intervals up to 2h and analyzed by a gas chromatograph equipped glass with column (40m).

The used zeolite was regenerated by air atmosphere or argon atmosphere at 723K for 15h. The rearrangement was repeated over the regenerated zeolite. When the zeolite was regenerated in argon atmosphere, the volatile coke precursor was removed from the catalyst surface. On the other hand, when the regeneration was carried out in air atmosphere, the coke precursor and non-volatile hard coke were simultaneously removed. The acidity and surface area of the used zeolite were measured to obtain the information for the catalyst deactivation. Furthermore, coke content of the used zeolite was measured by a thermal gravimetric balance. The preciously weighed used zeolite was placed into the cell of the gravimetric balance. After the cell was dried at 773K for 2h in argon atmosphere, air was fed into the balance at the same temperature. The coke content was calculated from the weight loss and the remained zeolite.

3. Results and discussion

Figure 1 shows the relationship between CHO conversion, CL selectivity and process time (time on stream) over TS-1s with different Si/Ti ratio and SSZ-41. The result over ZSM-5 (Si/Al ratio=90) is also represented in Figure 1. The CHO conversion decreases with process time, whereas the CL selectivity is almost constant during the process time. The deactivation of SSZ-31 is largest among the zeolites. The CL selectivity over SSZ-31 is lowest among the zeolites. The catalyst deactivation of TS-1(45) is larger than that of TS-1(200). These results suggest that the acidity and micro pore size of the zeolite simultaneously affected the catalyst deactivation.

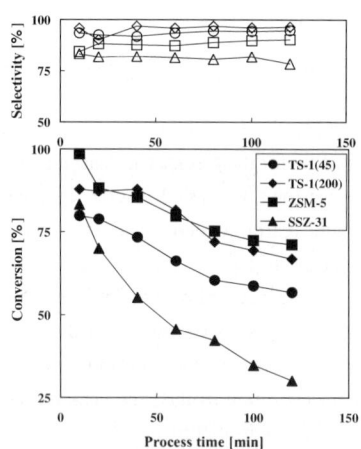

Figure 1 Relationship between CHO conversion, CL selectivity and process time (673K)

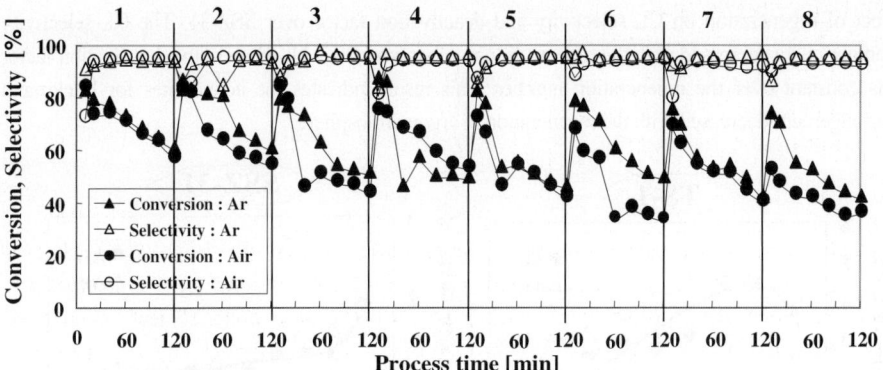

Figure 2 Relationship between CHO conversion, CL selectivity and regeneration

Figure 2 shows the effect of regeneration on CHO conversion and CL selectivity over TS-1 zeolite (200) at 623K. The used catalyst was regenerated with air atmosphere or argon atmosphere at 673K for 17h. The regeneration was continued 8 or 10 times for the rearrangement. The thermal gravimetric analysis revealed that the most of coke or coke precursor deposited on the zeolite was removed under the conditions as described above. Although the conversion decreases with process time, the selectivity is almost constant for the both atmospheres. The initial conversion for each regeneration decreases with the repetition number. This result indicates that the concentration of active sites decreases by the regeneration. The catalyst deactivation for the regeneration is known to be represented by Equation (1) [5].

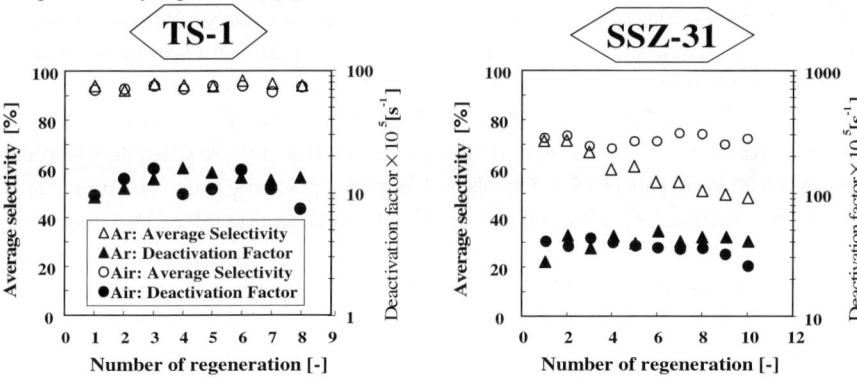

Figure 3 Relationship between average selectivity, deactivation factor and number of regeneration (TS-1).

Figure 4 Relationship between average selectivity, deactivation factor and number of regeneration (SSZ-31).

$$k = k_0 \cdot \exp(-b \cdot t) \quad (1)$$

k and k_0 are the rate constant at any process time and initial state, respectively, b is deactivation constant and t is process time.

Figure 3 shows the effect of regeneration number on average CL selectivity and deactivation constant over TS-1 zeolite regenerated with air and argon. The average selectivity and catalyst deactivation constant did not change with the regeneration atmosphere. Figure 4 demonstrates the effect of regeneration on CL selectivity and deactivation factor over SSZ-31. The CL selectivity monotonously decreased with the regeneration number over SSZ-31, whereas the deactivation factor was constant over the regeneration number. This result indicates the active sites for Beckmann rearrangement decreased with the regeneration in argon atmosphere.

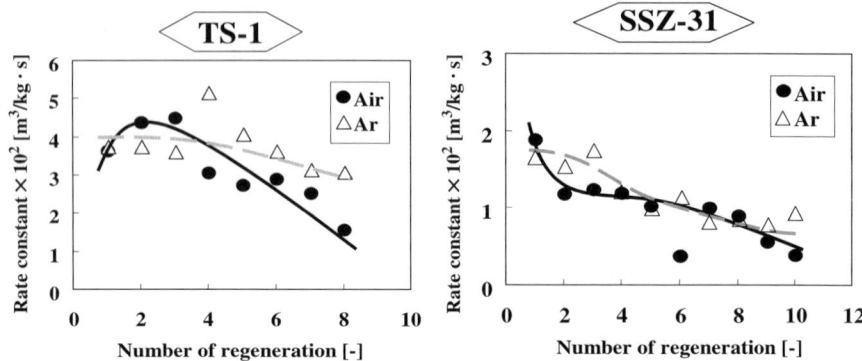

Figure 5 Effects of regeneration gases on the relationship between rate constant and number of regeneration (TS-1)

Figure 6 Effects of regeneration gases on the relationship between rate constant and number of regeneration (SSZ-31)

Figures 5 and 6 show the relationship between rate constant and regeneration number over TS-1 and SSZ-31, respectively. The rate constant regenerated by argon was almost constant for all regeneration number, whereas the constant for TS-1 regenerated by air gradually decreased. On the other hand, the constant for SSZ-31constantly decreased with the regeneration number. When the coke was removed from the catalyst surface, the activity almost recovered to the initial activity. However, the catalytic activity gradually decreased with the regeneration by the destruction of active sites called silanol nest. Especially, the catalyst was significantly deactivated by the regeneration of air.

Since the micro pore of TS-1 and SSZ-31 was smaller than that of CL or CHO, the CHO could not penetrate in to micro pore of TS-1. Then the rearrangement proceeded at the outer surface of the zeolite. On the other hand, since the pore size of SSZ-31 was larger than that of TS-1, the reactant penetrated into the inside of pore and converted to CL, but as the molecular size of product was larger than that of reactant, the product remained into the pore. As a result, the product changed to hard coke. When the zeolite had strong acid sites, the coking rate should increase. On the other hand, the rearrangement occurred on the weak acid sites located at the outer surface over TS-1. The produced coke was easily removed in argon atmosphere. These results suggest that the zeolite with medium pore size and weak acidity is suitable for Beckmann rearrangement.

Literature cited

[1] H. Sato, K.Hirose and M.Kitamura, *Nippon Kagaku Kaishi*, **1989**, 548(1989)

[2] T.Tatsumi and L.X.Dai, *Proceedings of 12th Inter.Zeolite Conference*, **Vol.II**, 1455(1998)

[3] D.Shouro, Y.Moriaya, T.Nakajima and S.Mishima, *Appl.Catal., A General*, **198**, 275(2000)

[4] H.Ichihashi and M.Kitamura, *Catal.Today*, **73**, 23(2002)

[5] T.Takahashi, M.N.A.Nasution and T.Kai, *Appl.Catal.,A, General*, **210**, 339(2001)

[6] Takahashi,T. and T.Kai, *J.Jpn.Petrol.Inst.*, **47**, 190 (2004)

Studies in Surface Science and Catalysis, volume 159
Hyun-Ku Rhee, In-Sik Nam and Jong Moon Park (Editors)

197

Y zeolite from kaolin taken in Yen Bai-Vietnam: synthesis, characterization and catalytic activity for the cracking of n-heptane

Ta Ngoc Don[a], Vu Dao Thang[a,c], Pham Thanh Huyen[*b,c], Pham Minh Hao[a], Nguyen Khanh Dieu Hong[b]

[a] Department of organic chemistry, tandon-ocd@mail.hut.edu.vn;

[b] Department of organic and petrochemical technology, pthuyen@mail.hut.edu.vn;

[c] Laboratory of petrochemical and catalysis materials,
Faculty of chemical technology, Hanoi university of technology, 1 Dai Co Viet str., Hanoi, Vietnam

ABSTRACT

This paper is concerned with the synthesis of Y zeolite with SiO_2/Al_2O_3 ratio of 4.5 from kaolin taken in Yen Bai-Vietnam and their catalytic activity for the cracking of n-heptane. The synthesized sample (NaY1) showed the Y zeolite crystallinity of 53% and P1 zeolite crystallinity of 32%, and exhibited good thermal stability up to 880°C. The activity and the stability of HY1 turned out to be lower than those of standard sample (HYs), but the toluene selectivity was higher. The conversion of n-heptane to toluene might be due to the metal oxide impurities, which was present in the raw materials and this indicates the potential application of this zeolite for the conversion of n-paraffin to aromatics.

1. INTRODUCTION

Y zeolites synthesized from pure chemicals have now been used as the main composition of FCC catalysts [1-4]. However, the application of Y zeolites synthesized from kaolin in the catalytic processes is still limited. The refinery and petrochemical industry is being built in Vietnam, so the synthesis of Y zeolites from domestic materials and minerals is necessary [4]. In this paper, the initial results in the synthesis of Y zeolites with SiO_2/Al_2O_3 ratio of 4.5 from kaolin taken in Yen Bai-Vietnam and their catalytic activity for the cracking of n-heptane are reported .

2. EXPERIMENTAL

NaY zeolite was synthesized with the gel composition of $3.5Na_2O.Al_2O_3.7SiO_2.150H_2O$ from Vietnamese kaolin, liquid glass and organic template, NaOH, NaCl and distilled water. The crystallization was carried out at 95°C for 24 hours. Synthesized sample (noted as NaY1) was converted to acid form (noted as HY1) and subsequently calcined at 650°C in nitrogen. Standard NaY zeolite (noted as NaYs) synthesized from pure chemicals was treated simultaneously and the obtained acid form was noted as HYs. NaY1, NaYs, HY1 and HYs were characterized by XRD (Siemens D5005), IR (Shimadzu FTIR 8100), SEM (JSM 5410 LV), BET (Coulter SA3100), cation exchanging capacity (CEC) and benzene adsorption (AC_6H_6). The catalytic tests for the cracking of n-heptane on HY1 and HYs were carried out by flow method at 450-600°C and reaction time of 5-30 minutes.

3. RESULTS AND DISCUSSION

3.1. Synthesis and characterization of Y zeolite.

Only single crystalline phase was observed in the XRD pattern of NaYs (Fig.1), and this indicates that the crystallinity of NaYs is 100%. NaY1 was found to contain mainly the crystalline phase of NaY. The presence of the NaP_1 crystalline phase and trace of quartz crystalline phase are also observed. Since there is no kaolinite crystalline phase, it is confirmed that the kaolin in the raw materials was completely converted. NaY zeolite in NaY1 has the PDF 43-0168 with the formula of $Na_2Al_2Si_{4.5}O_{13}.xH_2O$ similar to that of NaYs. It is crystallized in cubic form with $a_o=24.676$ Å and SiO_2/Al_2O_3 ratio of 4.5 [3]. NaP_1 zeolite in NaY1 has the PDF 39-0219 with the formula of $Na_6Al_6Si_{10}O_{32}.12H_2O$, and it is crystallized in tetragonal form with $a_o=10.043$ Å and SiO_2/Al_2O_3 ratio of 3.33 [4].

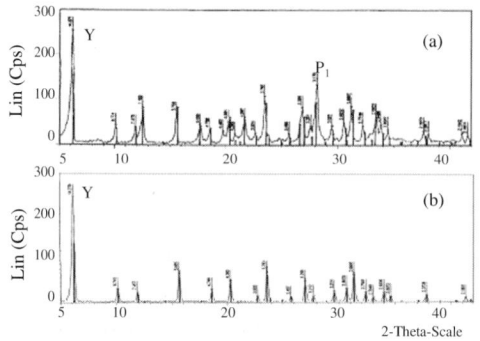
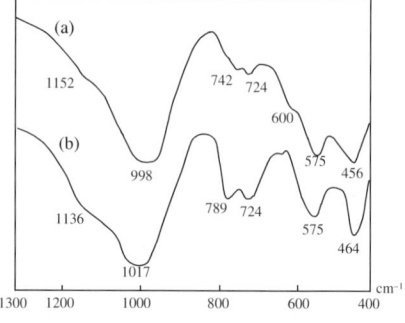

Fig.1. XRD patterns of NaY1 (a) and NaYs (b)   Fig. 2. IR spectra of NaY1 (a) and NaYs (b)

The IR spectra of NaY1 and NaYs (Fig.2) are very similar to each other. In the IR spectrum of NaY1, beside the bands at 575 and 724 cm^{-1} assigned for the Y zeolite, there are additional bands at 600 and 742 cm^{-1} representing P_1 zeolite [5]. The other bands are shifted to lower wave lengths compared to that in IR spectrum of NaYs and thus the SiO_2/Al_2O_3 ratio in NaY1 is lower than that in NaYs. This is in accordance with the result obtained from XRD analysis. The band at 575 cm^{-1} in NaY1 spectrum has a much stronger intensity than the band at 600 cm^{-1}. This again indicates that NaY1 sample contains mainly Y zeolite.

Fig. 3. SEM micrographs of NaY1 (left-hand side) and NaYs (right-hand side).

Fig. 3 shows that NaY1 contains two types of crystals: Y zeolite in cubic form and P_1 zeolite in spherical form with diameter of about 5 μm and 3 μm, respectively. On the other hand, NaYs contains only crystals of Y zeolite in spherical form with diameter of about 0.5 μm. It is clearly demonstrated that the diameter of zeolite synthesized from kaolin is much larger than that of zeolite synthesized from pure chemicals.

The data of CEC and AC_6H_6, thermal stability and SiO_2/Al_2O_3 ratio... are presented in Table 1 and these results show that NaY_1 is a porous material which contains a large number of negative charges in the network. NaY1 seems to meet the requirement as a good catalyst for the conversion of hydrocarbon, and thus we carried out ion exchange in NH_4Cl solution and then calcination at 650°C to obtain the acidic form HY1 and HYs. The Na^+ exchange levels of HY1 and HYs are 87.15 and 95.84%, respectively. Calcination at 650°C not only increases the Na^+ exchange levels but also brings about the collapse of the P_1 structure and keeps the stable structure of Y zeolite in HY1 [6]. The BET surface area of HY1 is 284 m^2/g with a crystallinity of 40%. The conversion of P_1 zeolite to the amorphous phase which supports Y zeolite makes HY1 suitable for the cracking of hydrocarbon [1,6].

Table 1.
Characteristics of NaY1 and NaYs zeolites.

Sample	CEC, meq $Ba^{2+}/100g$	AC_6H_6 %	Y zeolite crystallinity	P1 zeolite crystallinity	Thermal stability, °C	SiO_2/Al_2O_3 ratio	Fe_2O_3 wt.%	FeO wt. %	TiO_2 wt.%
NaY1	216	18,0	53	32	880	4.38	0.40	0.05	0.08
NaYs	235	20,2	100	0	900	4.52	-	-	-

3.2. Catalytic activity in the cracking of n-heptane

Conversion (C) of n-heptane, composition of products and selectivities of toluene and gas products at different temperatures are presented in Table 2 and Fig. 4. Clearly, the conversion of n-heptane and the selectivity of toluene increase with temperature, whereas the selectivity of gas products decreases. At the same temperature the conversion and selectivity of gas products on HY1 are slightly lower than that on HYs, but the selectivity of toluene is higher.

Table 2.
Conversion (C) of n-heptane and composition of products at different temperatures with the reaction time of 20 minutes

Sample	Reaction temperature °C	n-heptane conversion (C), %	Product composition, % wt.				Selectivity (S), %	
			n-heptane	Toluene	Gas products	Other liquid product	Toluene	Gas products
HY1	450	18.1	81.9	5.5	10.8	1.8	30.39	59.67
	500	32.6	67.4	10.6	19.3	2.7	32.52	59.2
	550	45.9	54.1	22.7	20.3	2.9	49.46	44.23
	600	51.7	48.3	26.3	21.2	4.2	50.87	41.01
HYs	450	21.4	78.6	3.7	14.9	2.8	17.29	69.63
	500	35.8	64.2	8.6	23.9	3.3	24.02	66.76
	550	51.8	48.2	18.1	29.6	4.1	34.94	57.14
	600	58.7	41.3	23.4	29.8	5.5	39.86	50.77

As shown in Table 2, during the cracking process, two main types of product are obtained: gas products, including light paraffins and olefins, and toluene. These products are formed by the breaking of C-C bonds to form light paraffins and olefins, the oligomerization of light olefins and then the cyclization of obtained oligomers, the hydrogen transfer between cyclized oligomers and olefins to form aromatics. Moreover, because of the large amount of toluene in the products, dehydro-cyclization of n-heptane directly to toluene might occur. The reforming process is carried out on bifunctional catalyst [2] and the metal sites promote the dehydrogenation and cyclization. Hence, the impurities in HY1, such as Fe_2O_3, FeO, TiO_2 with the content of 0.4; 0.05 and 0.08 wt%, respectively (table 1), can promote the formation of toluene and thus the selectivity of toluene on HY1 catalyst is higher than that on HYs.

2-methylhexane and 3-methylhexane are also found in the liquid product. So another reaction, isomerization of n-heptane, might have taken place. However, the activation energy of isomerization is higher than that of dehydro-cyclization [2], the content of isomers in the product is much lower than that of other products.

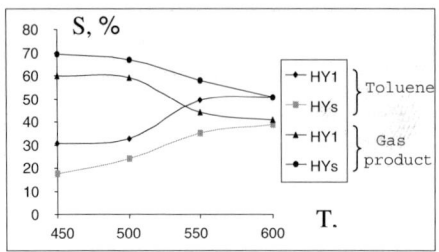

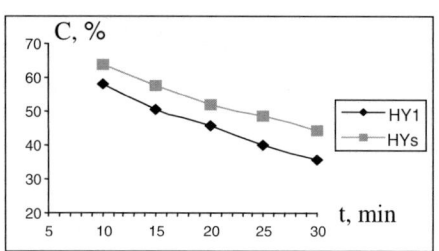

Fig.4. Selectivities to toluene and gas products at different temperatures (with reaction time of 20 minutes)

Fig.5. Influence of reaction time (t) on the conversion of n-heptane (C) at 550°C.

The stability of catalyst is one of the most important criteria to evaluate its quality. The influence of time on stream on the conversion of n-heptane at 550°C is shown in Fig. 5. The conversion of n-heptane decreases faster on HY1 than on HYs with time, so the question is "Could the formation of coke on the catalyst inhibit diffusion of reactant into the caves and pores of zeolite and decrease the conversion?" According to Hollander [8], coke was mainly formed at the beginning of the reaction, and the reaction time did not affect the yield of coke. Hence, this decrease might be caused by some impurities introduced during the catalyst synthesis. These impurities could be sintered and cover active sites to make the conversion of n-heptane on HY1 decrease faster.

4. CONCLUSION

NaY zeolite with SiO_2/Al_2O_3 ratio of 4.5 has been synthesized successfully from Vietnamese kaolin under hydrothermal condition, at 95°C and atmospheric pressure, synthesis time of 24 hours, under the presence of organic template. Y zeolite synthesized from kaolin can be used for the cracking of hydrocarbons to obtain lighter ones and for the conversion of n-paraffin to aromatics. The metal oxide impurities introduced during the synthesis of catalyst cannot be avoided. However, these impurities can contribute to the increase of catalytic activity to some extent for the formation of toluene in the cracking of n-heptane. These results indicate that Y zeolite can be synthesized from plentiful raw materials and minerals in Vietnam to apply as catalysts for the petrochemical and refining industry.

ACKNOWLEDGEMENTS
Financial support of this work by Project VLIR/HUT IUC/PJ1 from the VLIR/HUT research fund in the co-operation between Hanoi University of Technology, Vietnam and Flemish Universities, Belgium is gratefully acknowledged.

REFERENCES
[1]. Scherzer J., *Catal. Rev. - Sci. Eng.*, 31(3), **1989**, pp. 215-354.
[2]. Gates B.C., Kazer J.R. and G.C.A Schuit, *Chemistry of catalytic processes*, McGraw-Hill, New York, **1979**.
[3]. Bergeret G. et al., *Unit cell data*, J. Phys. Chem., 87, **1983**, p. 1160 (ICDD Grant-in-Aid, 1991).
[4]. Barlocher Ch., Meier W. M., *Unit cell data*, J. Kristallchem., 135, **1972**, p. 339.
[5]. K. Kurita, K. Tomita, T. Tada, S. Ishii, F. S. Nishimura, K. Shimoda. J. Polym. Sci., Part A – Polym. Chem. Vol. 31, **1993**, pp. 485-491.
[6]. Breck D.W., *Zeolite Molecular Sieves*, A Wiley. publication, New York, **1974**.
[7]. Ta Ngoc Don, *Ph. D Thesis*, **2002**, Hanoi, Vietnam.
[8]. Hollander Ir. M.A., *PhD-projects of the OSPT* , **1996**.

Studies in Surface Science and Catalysis, volume 159
Hyun-Ku Rhee, In-Sik Nam and Jong Moon Park (Editors)

Anoxic Hydrogen Production over CdS-based Composite Photocatalysts under Visible Light Irradiation (λ≥420nm)

Jum Suk Jang*, Sang Min Ji*, Jae Sung Lee*,
Wei Li[§], and Se Hyuk Oh[§]

*Department of Chemical Engineering and School of Environmental Engineering,
Pohang University of Science and Technology (POSTECH) San 31 Hyoja-dong, Pohang
790-784, Republic of Korea

[§]General Motors R&D Center, Warren, Michigan 48090

A nano-bulk composite (NBC) photocatalyst based on bulk CdS and nanosize TiO_2 nanoparticles was fabricated by precipitation method and sol-gel synthesis. Its photocatalytic hydrogen production rate from aqueous solution containing hole scavengers such as sulfide and sulfite under visible light irradiation (λ≥420nm) was higher than that of single-phase CdS photocatalyst by a factor of ca. 4.5. The formation of nanoscale heterojunctions between two phases seemed to cause an efficient electron-hole separation and contribute to the unprecedented high rate of hydrogen production under visible light.

1. INTRODUCTION

Hydrogen production by use of semiconductor photocatalysts has recently received much attention in view of solar energy utilization [1-2]. Thus visible light-driven photocatalysts that could produce hydrogen from water splitting under solar light have been actively sought. CdS has an ideal band gap energy (2.4 eV) and band positions that can drive both oxidation and reduction of water under visible light irradiation [3]. Yet, pure CdS is usually not very active in hydrogen production, and furthermore, decomposes to Cd^{2+} and S^0 under light irradiation, because holes photogenerated in the valence band tend to react with CdS itself [4-5]. In attempt to improve the photoactivity and photostability of CdS, CdS has been combined with other materials such as ZnO, TiO_2 and $LaMnO_3$ or intercalated into layered compounds [6-7]. As a new method of promoting CdS photocatalysis, here we fabricated a nano-bulk composite (NBC) photocatalyst consisting of bulky CdS with a high crystallinity interfacing with nanosized TiO_2 particles. This configuration of the composite photocatalyst exhibited an unprecedented high rate of hydrogen production under visible light (λ≥420nm) from water containing sulfide and sulfite as hole scavengers.

2. EXPERIMENTAL

To prepare CdS photocatalyst, a stoichiometric amount of Na_2S aqueous solution was added drop-by-drop to $Cd(NO_3)_2$ dissolved in isopropyl alcohol. Precipitated powder was filtered and dried, then was calcined at $800°C$ for 1 h under He flow to increase the crystallinity of CdS. To fabricate CdS-TiO_2 NBC photocatalyst, the bulky CdS catalyst was stirred in isopropyl alcohol and tetra-titanium isopropoxide (in a mole ratio of Ti to CdS of 4) and H_2O was added drop-by-drop. The prepared composite powder was filtered and dried again, then was calcined at $400°C$ for 1-3 h under air flow to increase the crystallinity of TiO_2 in CdS-TiO_2 NBC photocatalysts. The crystalline phases of the products were determined by X-ray diffractometer (Mac Science Co., M18XHF). The optical properties were analyzed by UV-Visible diffuse reflectance spectrometer (Shimadzu, UV 2401). Morphologies of photocatalysts were investigated by transmission electron microscope (JEOL JEM 2010F, Field Emission Electron Microscope). The photocatalytic reactions were carried out at room temperature under normal pressure in a closed circulation system using a Hg-arc lamp (500 W) equipped with UV cut off filter ($\lambda \geq$ 420 nm). Before reaction, 1wt% of Pt was deposited on photocatalysts by photodeposition method under visible light ($\lambda \geq 420nm$). The reaction was performed with 0.1g of photocatalyst in 100mL aqueous solution containing 0.1M Na_2S and 0.02M Na_2SO_3. The evolved amounts of H_2 were analyzed by gas chromatography (TCD, molecular sieve 5-Å column and Ar carrier).

3. RESULTS AND DISCUSSION

In preliminary experiments, it was established that the configuration of nano-TiO_2/bulk CdS composite yielded a photocatalyst of superior activity in hydrogen evolution. Thus, it was much more active than nano-CdS-bulk TiO_2 or nano-TiO_2/nano-CdS coprecipitates. The latter two catalysts were even less active than the single phase CdS of a high crystallinity. Figure 1 shows XRD patterns of (a) TiO_2 calcined at $400°C$, (b) CdS calcined at $800°C$ for 1hr under He flow and (c) CdS-TiO_2 with NBC configuration. The crystallinity of CdS in CdS-TiO_2 NBC photocatalyst was as good as the single phase CdS, considering its concentration (30wt%) in the composite. This sample showed well-developed hexagonal CdS phase. The TiO_2 phase was anatase both in single and NBC photocatalysts. Figure 2 shows the UV-diffuse reflectance (DR) spectra for these catalysts. From UV-DR spectrum of CdS-TiO_2 composite photocatalyst, it can be confirmed that CdS-TiO_2 NBC maintains each crystal phase of CdS and TiO_2 just like a physical mixture of two compounds. The morphology of CdS-TiO_2 NBC photocatalyst was irregular with CdS particles of ca. 1~2 µm size decorated with TiO_2 nanoparticles of ca. 10~20 nm as shown in figure 3(a)~(c). Energy dispersive X-ray spectroscopy (EDAX) spectrum showed that

NBC photocatalyst was composed of CdS and TiO$_2$ (figure 3(d)).

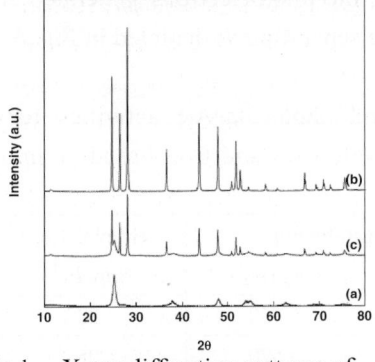

Fig,1. X-ray diffraction patterns of (a) TiO$_2$-400°C, (b) CdS-800°C, (c) CdS-TiO$_2$.

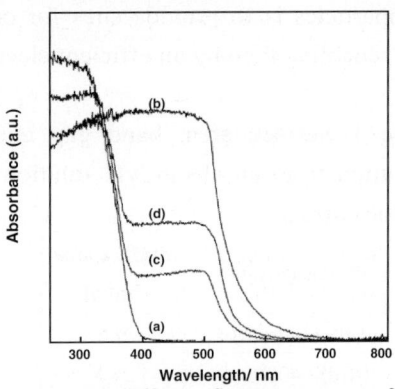

Fig. 2. UV-VIS diffuse reflectance spectra of (a) TiO$_2$-400°C, (b) CdS-IPA-800°C, (c) CdS-TiO$_2$, (d) CdS-TiO$_2$ physically mixed (PM).

From this result, it can be confirmed that nano-bulk heterojunctions have been formed between two components. Shown in figure 4 are the results of photocatalytic splitting of aqueous solution containing 0.1 M Na$_2$S and 0.02M Na$_2$SO$_3$ as sacrificial reagents under visible light irradiation (using a cutoff filter of $\lambda \geq 420$ nm for all catalysts). The CdS-TiO$_2$ NBC photocatalyst shows a higher photocatalytic activity than that of single phase CdS photocatalyst having similar crystallinity. As previously reported [8], there was an apparent induction period of 1-2 h before the steady state hydrogen evolution rate of bulky CdS photocatalyst was established. But, CdS-TiO$_2$ NBC photocatalyst did not show such an induction period.

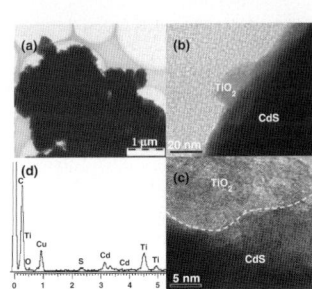

Fig. 3. TEM images and EDAX data of CdS-TiO$_2$ NBC photocatalyst.

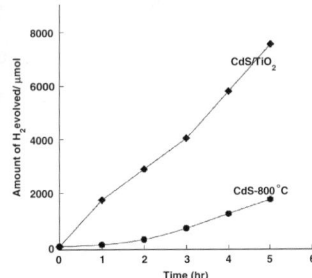

Fig. 4. The amount o f H$_2$ evolution vs reaction time.

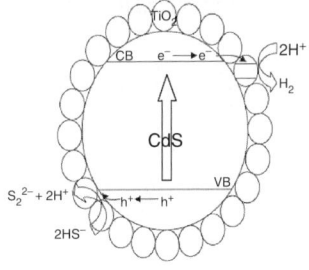

Fig.5. Proposed Mechanism of NBC photocatalyst.

The average rates of hydrogen evolution during 5hr are compared in Table 1 together with other physical properties. From the rate data for three CdS photocatalysts with different crystallintiy, it can be concluded that the high crystallinity of CdS with hexagonal phase is essential for high photocatalytic activity. The formation of CdS-TiO$_2$ NBC photocatalyst is an effective strategy to further increase the photocatalytic activity of bulk CdS. The possible role of TiO_2 nanoparticles is to provide sites for collecting the photoelectrons generated from CdS, enabling thereby an efficient electron-hole separation as depicted in Fig. 5.

Table 1 Surface area, band gap energies and photocatalytic activities for H_2 evolution from an electrolyte solution over single CdS and CdS-based composite photocatalysts.

Photocatalysts	Surface area [m²/g]	Bandgap Energy		H_2 evolution [μmolh^{-1}]
		$E_g(eV)$	$\lambda_{ab}(nm)$	
CdS (Aldrich)	9.6	2.25	550	32
CdS- 400°C	28.8	2.18	570	25
CdS- 800°C	< 1	2.18	570	347
CdS-TiO₂	97.0	2.25	550	1562

4. CONCLUSION

The CdS photocatalyst of a high crystallinity shows a high photocatalytic activity for hydrogen production from water under visible light irradiation. The fabrication of CdS-TiO₂ NBC photocatalyst is a successful strategy to develop even more active CdS-based photocatalysts. The NBC structure formed with bulky CdS decorated with nanosized TiO₂ particles results in an efficient charge separation, caused by fast diffusion of photoelectrons generated from bulky CdS toward surrounding TiO₂ nanoparticles, leading to high photocatalytic activity of hydrogen production.

ACKNOWLEDGEMENT

This work was supported by General Motors R& D Center, the Hydrogen Energy R&D Center, National Research Laboratory, the Brain Korea 21 Project, National R & D Project for Nano Science and Technology.

REFERENCES

[1] Kim, H.G., Hwang, D.W., and Lee, J.S., J. Am. Chem. Soc., **126**(29) (2004) 8912.

[2] Kato, H., Asakura, K., and Kudo, A., J. Am. Chem. Soc., **125**(10) (2003) 3082.

[3] Buhler, N., Meier, K., and Reber, J., J. Phys. Chem., **88** (1984) 3261.

[4] Frank, A. J., and Honda, K., J. Phys. Chem., **86** (1982) 1933

[5] Meissner, D., Memming R., and Kastening, B., J. Phys. Chem., **92** (1988) 3476

[6] Spanhel, L., Weller, H., and Henglein, A., J. Am. Chem. Soc., **109** (1987) 6632

[7] Shangguan, W., and Yoshida, A., J. Phys. Chem., **106** (2002) 12227

[8] Serpone, N., and Borgarello, E., Inorg. Chim. Acta, **90** (1984) 191

Studies in Surface Science and Catalysis, volume 159
Hyun-Ku Rhee, In-Sik Nam and Jong Moon Park (Editors)

Asymmetric Ring Opening of Some Terminal Epoxides Catalyzed by Dimeric Type Novel Chiral Co(Salen) Complexes

Santosh Singh Thakur[a], Wenji Li[a], Seong-Jin Kim[b] and Geon-Joong Kim[a]*

[a]Department of Chemical Engineering, Inha University, 253 Yonghyun-Dong, Nam-Gu, Incheon 402-751, Korea Email : *kimgj@inha.ac.kr

[b]RStech Corp. #305 Venture Town, 1688-5, Sinil-dong, Daedeok-gu, Daejeon, Korea, 306-230

The asymmetric ring opening (ARO) of racemic terminal epoxides with H_2O via hydrolytic kinetic resolution provides an efficient synthetic route to prepare optically pure terminal epoxides. The dimeric type chiral Co(salen)AlX$_3$ complex has great potential to catalyze HKR of terminal epoxides in a highly reactive and enantioselective manner in comparison to their monomeric analogy.

1. INTRODUCTION

Enantiopure compounds in the area of pharmaceuticals, plant-protecting agent, and fragrances are rapidly expanding and holding world market [1]. From an economic and ecological viewpoint the most efficient way to prepare such compounds (or building blocks thereof) is enantioselective catalysis [2]. Terminal epoxides are versatile starting materials for the preparation of bioactive molecules and optically pure terminal epoxides is very useful in the chiral drug industry [3]. Hydrolytic kinetic resolution (HKR) technology is the prominent way to synthesize optically pure or enriched terminal epoxides among various available methods [4]. Dimeric and oligomeric Co(salen) complexes exhibit improved reactive and highly enantioselective property to the asymmetric ring opening of terminal epoxides with H_2O than their monomeric analogy [5-8]. Herein we report the synthesis of Co(salen)AlX$_3$ dimer complex and their catalytic activity for the asymmetric ring opening of terminal epoxides with H_2O.

2. EXPERIMENTAL

2.1. Synthesis and Characterization of Co(Salen) Monomer and Dimer Complex

To a solution of hexa hydrated aluminum chloride (1.99g, 8.28 mmol, 1.0 equiv.) in tetrahydrofuran (25 mL), precatalyst (R,R)-salenCo (5.0 g, 8.28 mmol,1.0 equiv.) was added and stirred in at open atmosphere at room temperature. As soon as the chiral Co (salen) was added color of the solution changes from brick red to dark olive green. The mixture was stirred for 1 h .The resulting solution was concentrated under reduced pressure. The crude solid was worked up with H_2O and CH_2Cl_2 . Yield = 98-99 % as a dark green solid powder. The complex have been analyzed by ^{27}Al NMR with reference to $[Al(D_2O)_6]^{3+}$ and

UV visible spectrophotometer. Al $(NO_3)_3.9H_2O$ monomer was prepared in same manner .In case of synthesis of dimer complex, 2 equivalent of the Co(salen) was taken with respect to $AlX_3.nH_2O$.

2.2. General Procedure for Hydrolytic Kinetic Resolution.

An oven dried 25 mL flask equipped with a stir bar was charged with (R, R) catalysts (0.2 mmol, 0.2-0.5 mol %) and (±)-terminal epoxides (40 mmol, 1.0 equiv.), and the mixtures were stirred at room temperature. H_2O (22 mmol. 0.55 equiv.) was added drop wise. The reaction was mildly exothermic.

2.3. Characterization.

The HKR reaction mixture was stirred up to the occurrence of optically pure terminal epoxides and checked periodically by Chiral GC (Hewlett-Packard 6890 Series II instruments equipped with FID detector) using a chiral column (CHIRALDEX A-TA and G-TA, 20m x 0.25 mm i.d. (Astec)) and by chiral HPLC (Regis (S,S)Whelk-O1 column at 254 nm or Chiralcel® OD column ,24 cm x 0.46 cm i.d.; Chiral Technologies, Inc.). All ^1H NMR,^{13}C NMR and ^{27}Al NMR (I=5/2) were recorded using 400MHz FT NMR spectrophotometer (VARIANUNITYNOVA400) at ambient temperature. Optical rotation measurements were conducted using a Jasco DIP 370 digital polarimeter. UV spectra were recorded on UV-Vis spectrophotometer (Optizen 2120 UV) interfaced with PC using Optizen view 3.1 software for data analysis. Vibrational Circular Dichroism (VCD) and IR were measured in ChiralirTM ABB Bomem Inc using Bomem GRAMS-32 software. Terminal epoxides obtained from Aldrich, Fluka or TCI.

3. RESULTS AND DISCUSSION

The formation of monomer and dimer of (salen)Co AlX_3 complex can be confirmed by ^{27}Al NMR. Monomer complex **1a** show ^{27}Al NMR chemical shift on δ=43.1 ppm; line width =30.2 Hz and dimer complex **1b** δ=37.7 ppm; line width =12.7 Hz. Further instrumental evidence may be viewed by UV-Vis spectrophotometer. The new synthesized complex showed absorption band at 370 nm. The characteristic absorption band of the precatalyst Co(salen) at 420 nm disappeared (Figure 1). It has long been known that oxygen atoms of the metal complexes of the Schiff bases are able to coordinate to the transition and group 13 metals to form bi- and trinuclear complex [9]. On these proofs the possible structure is shown in Scheme 1.

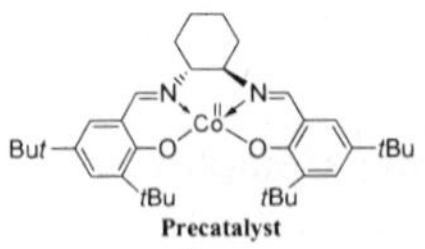

Precatalyst
1
(R,R)-(-)-N,N´-bis(3,5-di-tert-butyl salicyl-idene)-1,2-cyclohexane diamino cobalt(II)

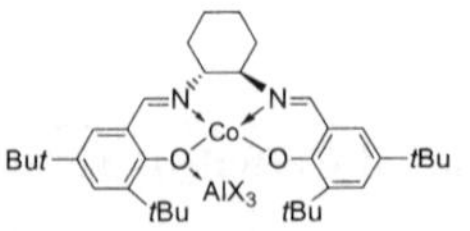

X; a, b = NO_3
a´, b´ = Cl

1a
1a´

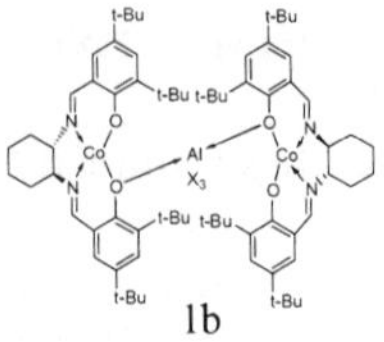

1b
1b´

Scheme 1. Schematic drawing of the monomer and dimer chiral (salen)Co-AlX_3 complexes.

The catalytic activities of the dimer catalyst **1b** for HKR of the diverse and valuable racemic terminal epoxides are shown in Table 1. Due to high selectivity factor (k_{rel} values) every epoxide underwent resolution in excellent yield and high ee % employing 0.2-0.5 mol% of catalyst in solvent free condition in most of the cases. In a similar condition, catalyst **1a** loading per [Co] basis gives < 25% ee and < 10% isolated products.

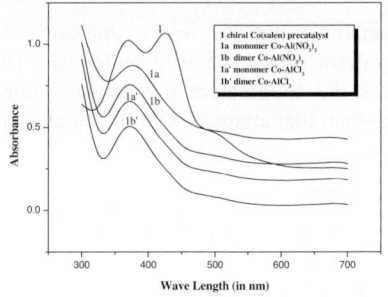

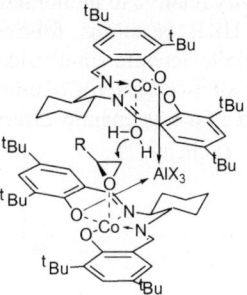

Fig. 1. UV-Vis absorption spectra of the precatalyst chiral Co(salen) and monomer and dimer complex.

Scheme 2. Possible working model for the HKR of terminal epoxides catalyzed by Co-AlX₃ dimer complex.

Table 1
The hydrolytic kinetic resolution of terminal epoxides catalyzed by the monomer 1a and dimer 1b

Entry	Recovered[a] Epoxide	Catalyst	Catalyst Loading[b]	Time (h)	% Yield(ee)[c]
1		1a	0.4	12	23(9.8)
		1b	0.2	2	42(99.6)
2		1a	0.4	12	25(8.5)
		1b	0.2	3	41(99.3)
3		1a	0.4	12	21(7.3)
		1b	0.2	4	43(99.5)
4		1a	0.4	12	19(8.0)
		1b	0.2	5	40(99.1)
5		1b	0.4	4	42(99.6)
6		1b	0.5	7	43(99.3)

[a]Isolated yield is based on racemic epoxides (theoretical maximum=50%). [b]in mol% loading on a per [Co] basis w.r.t. racemic epoxides. [c]ee% was determined by chiral GC or chiral HPLC.

The HKR of epichlorohydrin (ECH) was studied as a representative reaction for kinetic studies. For dimer catalyst 1b and 1b′ and corresponding monomers 1a and 1a′, show the two-term rate equation involving both intra and intermolecular components[10].

Plots of rate/[catalyst] vs. [catalyst] should be linear with slopes equal to k_{inter} and y-intercepts corresponding to k_{intra}. Analysis of such plots with rate data obtained with dimer catalyst 1b and 1b′ revealed linear correlations with positive slopes and nonzero y-intercepts, consistent with participation of both inter- and intramolecular pathways in the HKR.   Similar analysis of rate data obtained with monomeric catalyst 1a and 1a′revealed y-intercepts of zero, reflecting the absence of any first-order pathway (Figure 2). Thus, the dimer catalyst provides appropriate relative proximity and orientation, which eventually reinforces the reactivity and selectivity relative to monomeric analogy.

The HKR reactions follow the cooperative bimetallic catalysis where epoxide and nucleophile activate simultaneously by two different (salen)Co-AlX$_3$ catalyst molecules. The linking of two (salen)Co unit through the Al induces the cooperative mechanism, albeit through a far less enantio-discriminating transition state than that attained with the catalyst 1a and 1a′ (Scheme 2).

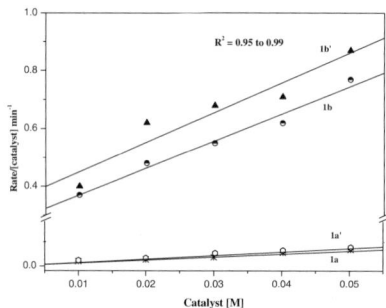

Fig. 2. Initial rate kinetics for the enantioselective ring opening of the ECH with H$_2$O catalyzed by the monomer and dimer chiral (salen)CoAlX$_3$ catalysts.

4. CONCLUSIONS

In the present study the dimer (salen)CoAlX$_3$ showed enhanced activity and enantioselectivity. The catalyst can be synthesized easily by readily commercially available precatalyst Co(salen) in both enantiomeric forms. Potentially, the catalyst may be used on an industrial scale and could be recycled. Currently we are looking for the applicability of the catalyst to asymmetric reaction of terminal and meso epoxides with other nucleophiles and related electrophile-nucleophile reactions.

REFERENCES

[1] R.A.Sheldon, Chirotechnology, Industrial Synthesis of Optically Active Compounds. New York: Dekker; 1993. 416 p.
[2] Asymmetric Catalysis on Industrial Scale (Eds.: H.U. Blaser, E. Schmidt), Wiley-VCH Verlag, Germany , 2004, 454 p.
[3] H. C. Kolb, M. S.Van Nieuwenhze, K. B. Sharpless, Chem. Rev., 94 (1994) 2483.
[4] a) M. Tokunaga, J. F. Larrow, F. Kakiuchi, E. N. Jacobsen, Science, 277 (1997) 936; b) M. E. Furrow, S. E. Schaus, E. N. Jacobsen, J. Org. Chem., 63 (1998) 6776.
[5] J. M. Ready, E. N. Jacobsen, J. Am. Chem. Soc. 123 (2001) 2687.
[6] J. M. Ready, E. N. Jacobsen, Angew. Chem. Int. Ed. 41 (2002) 1374.
[7] D. E. White, E. N. Jacobsen, Tetrahedron: Asymmetry 14 (2003) 3633.
[8] S. S. Thakur, W. Li, S. J. Kim, G.-J. Kim, Tetrahedron Lett. 46 (2005) 2263.
[9] H. Aoi, M. Ishimori, T. Tsuruta, Bull. Chem. Soc. Jpn., 48 (1975) 1897.
[10] R. G. Konsler, J. Karl, E. N. Jacobsen, J. Am. Chem. Soc. 120 (1998) 10780.

Studies in Surface Science and Catalysis, volume 159
Hyun-Ku Rhee, In-Sik Nam and Jong Moon Park (Editors)

EXAFS characterization of supported PtRu/MgO prepared from a molecular precursor and organometallic mixture

S. Chotisuwan[1, 2, 3], J. Wittayakun[1*] and B. C. Gates[3]

[1]School of Chemistry, Suranaree University of Technology, Thailand
[2]Faculty of Science and Technology, Prince of Songkla University, Pattani Campus, Thailand
[3]Dept. of Chemical Engineering and Materials Science, University of California, Davis, USA
[*] jatuporn@sut.ac.th

ABSTRACT

Supported bimetallic PtRu/MgO catalysts were prepared by adsorption of MgO with $Pt_3Ru_6(CO)_{21}(\mu_3\text{-}H)(\mu\text{-}H)_3$ in CH_2Cl_2 cluster or a mixture of $Pt(acac)_2$ and $Ru(acac)_3$ in toluene (acac = acetylacetonate). High metal dispersion was obtained in the catalysts after treated in He or H_2 at 300°C to remove ligands. Changes of the cluster before and after treatment, monitored by infrared (IR) spectroscopy indicated that it was not intact during adsorption on MgO. However, it was still metallic carbonyl species and was not extractable from support by CH_2Cl_2 solvent. After ligand removal, PtRu/MgO catalysts from both precursors were characterized by extended X-ray absorption fine structure (EXAFS) spectroscopy. Only the catalyst prepared from the cluster precursor contained Pt-Ru bonds with lower Pt-Ru contributions than the original precursor due to strong metal-support interactions. Although activity for ethylene hydrogenation at low temperatures was observed on the both catalysts, it was slightly lower in the catalyst prepared from the cluster precursor.

KEYWORDS

Bimetallic catalyst, Pt-Ru, magnesia, ethylene hydrogenation, carbonyl cluster, acetylacetonate

1. INTRODUCTION

Highly dispersed supported bimetallic catalysts with bimetallic contributions have been prepared from molecular cluster precursors containing preformed bimetallic bond [1-2]. For examples, extremely high dispersion Pt-Ru/γ-Al_2O_3 could be prepared successfully by adsorption of $Pt_2Ru_4(CO)_{18}$ on alumina [2]. By similar method, Pt-Ru cluster with carbonyl and hydride ligands, $Pt_3Ru_6(CO)_{21}(\mu_3\text{-}H)(\mu\text{-}H)_3$ (**A**) was used in this work to adsorb on MgO support. The ligands were expectedly removable from the metal framework at mild conditions without breaking the cluster metal core.

The goal of this work was to prepare and characterize PtRu/MgO catalysts from cluster **A** which contained Pt-Ru bonds and compare with that prepared from a mixed solution of $Pt(acac)_2$ and $Ru(acac)_3$. The characterization methods included IR and EXAFS spectroscopy. Ethylene hydrogenation was used to test the catalytic activity of both PtRu/MgO catalysts.

2. EXPERIMENTAL

2.1 Preparation of PtRu/MgO catalysts

The PtRu/MgO catalysts containing 1.0 wt% Pt and 1.0 wt% Ru were prepared by slurrying MgO power with a CH_2Cl_2 solution of cluster **A** which was synthesized by a procedure described elsewhere [3] or a mixed solution of $Pt(acac)_2$ and $Ru(acac)_3$ in toluene. After solvent removal by evacuation, the ligands of adsorbed cluster precursor were eliminated by heating in He flow at 300°C for 2 h while acac ligands of the second precursor were removed in H_2 flow at the same conditions.

2.2 Catalyst characterization

2.2.1 IR spectroscopy

IR spectra of adsorbed precursors on alumina were recorded before and after solvent removal with a Bruker IFS-66v spectrometer with a resolution of 4 cm^{-1}.

2.2.2 EXAFS spectroscopy and data analysis

EXAFS experiments were performed at the National Synchrotron Light Source (beamline X18B), Brookhaven National Laboratory, Upton, New York, USA. Details of EXAFS measurement and condition were as reported elsewhere [4]. The EXAFS data processing was carried out with the software ATHENA [5] and analyzed with the software EXAFSPAK [6]. EXAFS data were analyzed with phase shift and backscattering amplitudes calculated by FEFF7.0 [7]. The fittings were done both in r space (r is interatomic distance from the absorber atom) and k space (k is the wave vector) with application of k^0, k^1, and k^3 weightings. The EXAFS data of PtRu/MgO prepared from cluster precursor treated in He scanned at Pt L_{III} edge were Fourier transformed over the ranges $3.00 < k < 13.40$ and $0.0 < r < 4.0$ Å. The EXAFS data scanned at Ru K edge were Fourier transformed over the ranges $4.25 < k < 14.35$ and $0.0 < r < 4.0$ Å. The EXAFS data of PtRu/MgO prepared from mixtures of acac complexes treated in H_2 scanned at Pt L_{III} edge were Fourier transformed over the ranges $3.55 < k < 14.50$ and $0.0 < r < 5.0$ Å. The EXAFS data scanned at Ru K edge were Fourier transformed over the ranges $3.20 < k < 14.50$ and $0.0 < r < 5.0$ Å.

2.3 Catalysis Test

Ethylene hydrogenation was carried out in a once-through flow reactor. The effluent gas mixture was analyzed with an online gas chromatograph (Hewlett-Packard HP 6890) equipped with an Al_2O_3 capillary column and a flame ionization detector. Testing conditions included $P_{hydrogen}$ = 200 Torr, $P_{ethylene}$ = 40 Torr, catalyst mass of 10 to 20 mg and temperature varied from -50 to -25°C.

3. RESULTS AND DISCUSSION

3.1 IR Evidence for strong adsorption between cluster precursor and MgO surface

IR intensity characterizing isolated OH groups at 3,766 cm^{-1} [8] decreased after adsorption indicating that they involved in interactions between cluster and surface support. The interactions could be between oxygen of CO ligands and support surface functional groups such as hydroxyl and O^{2-} sites forming carbonates, carboxylates adsorbed species presumably the peak at 1,595 cm^{-1}. It was reported that these species on MgO surface could occur in the vibrational range of 1,700-1,200 cm^{-1} [9, 10].

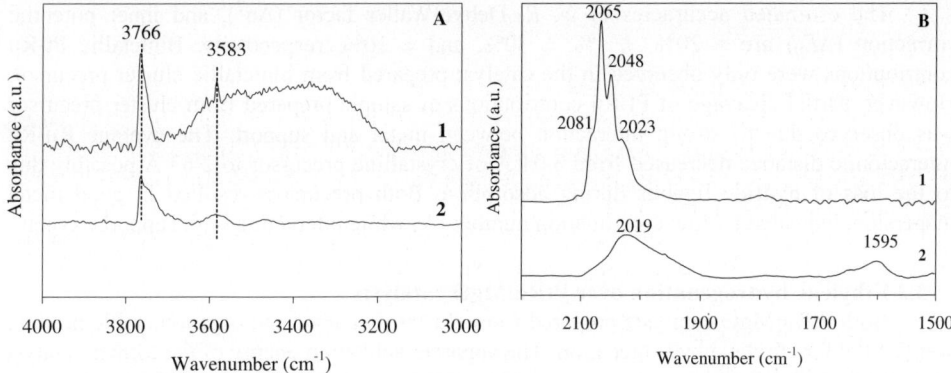

Fig.1. IR spectra in (A) v_{OH} region; (1) bare MgO; and (2) adsorbed precursor on MgO; (B) v_{C-O} region; (1) cluster A in CH_2Cl_2; and (2) adsorbed precursor on MgO.

In addition, hydrogen bonding was still observed at 3,583 cm^{-1} (Fig.1 A). IR peaks in v_{C-O} region (Fig.1 B) shifted to lower frequency after adsorption due to precursor-support interaction and differed from that of cluster solution implying that precursor was not intact after adsorption on support but still in the form of metal carbonyl species.

3.2 EXAFS evidence for bimetallic structure on PtRu/MgO after decarbonylation
EXAFS results and fitting parameters compared between two precursors after ligand removal are displayed in Table 1.

Table 1　Coordination number (N) and bond distance (R) of cluster core of crystalline A determined by XRD [3] and EXAFS data of PtRu/MgO prepared from cluster A and acac precursors after ligand removal

Edge	Shell	Cluster A		PtRu/MgO prepared from cluster A		PtRu/MgO prepared from acac	
		N	R (Å)	N	R (Å)	N	R (Å)
Pt L$_{III}$	Pt-Pt	2.0	2.64	1.3	2.69	1.0	3.07
	Pt-Ru	4.0	2.80	0.9	2.69	-	-
	Pt-O$_{support}$						
	Pt-O$_s$	-	-	2.3	2.03	0.4	1.98
	Pt-O$_l$	-	-	0.4	3.09	2.6	2.51
Ru K	Ru-Ru	2.0	3.04	2.6	2.63	3.5	2.69
	Ru-Pt	2.0	2.80	1.1	2.69	-	-
	Ru-O$_{support}$						
	Ru-O$_s$	-	-	1.4	2.09	2.5	2.01
	Ru-O$_l$	-	-	0.2	2.87	0.9	2.53

Notation: Subscript s and l refer to short and long, respectively.

The estimated accuracies of N, R, Debye-Waller factor ($\Delta\sigma^2$), and inner potential correction (ΔE_0) are $\pm$ 20%, $\pm$ 1%, $\pm$ 30%, and $\pm$ 10%, respectively. Bimetallic Pt-Ru contributions were only observed in the catalyst prepared from bimetallic cluster precursor. However, partial cleavage of Pt-Ru contributions of sample prepared from cluster precursor was observed due to strong interaction between metal and support. The average Ru-Ru interactomic distance decreased from 3.04 Å of crystalline precursor to 2.63 Å possibly due to the loss of hydride ligands during adsorption. Both precursors resulted in good metal dispersion, indicated by low coordination number (N) while interacting with support oxygen.

3.3 Ethylene hydrogenation over PtRu/MgO catalysts

Both PtRu/MgO catalysts prepared from cluster precursor and organometallic mixture were active for ethylene hydrogenation. The apparent activation energy of the former catalyst obtained from the Arrhenius plot during -40 to -25°C was 5.2 kcal/mol and that of the latter catalyst obtained during -50 to -30°C was 6.0 kcal/mol. The catalytic activity in terms of turn over frequency (TOF) was calculated on the assumption that all metal particles were accessible for reactant gas. Lower TOF of catalyst prepared from cluster A at -40°C, 57.3 x 10^{-4} s^{-1} was observed probably due to Pt-Ru contribution compared to that prepared from acac precursors, 83.9 x 10^{-4} s^{-1}.

4. CONCLUSIONS

Highly dispersed Pt-Ru/MgO catalyst contain Pt-Ru contribution could be prepared successfully by adsorption with $Pt_3Ru_6(CO)_{21}(\mu_3\text{-}H)(\mu\text{-}H)_3$. Metals could interact with surface oxygen forming M-O$_{support}$ bondings and caused structural distortion of the cluster core. Only the bimetallic cluster precursor gave Pt-Ru catalyst with Pt-Ru bonds. However, strong M-O$_{support}$ interactions on MgO led to partial cleavage of Pt-Ru contributions. High dispersion of (Pt+Ru)/MgO prepared from mixture of acac compounds was obtained after ligand removal but none of Pt-Ru bonds were observed. Both PtRu/MgO catalysts were active for ethylene hydrogenation.

REFERENCES

[1] O. S. Alexeev, G. W. Graham, D-W. Kim, M. Shelef and B. C. Gates, Phys. Chem. Chem. Phys., 1 (1999) 5725.

[2] O. S. Alexeev, G. W. Graham, M. Shelef and B. C. Gates, J. Phys. Chem. B, 106 (2002) 4697.

[3] R. D. Adams, T. S. Barnard, Z. Li, W. Wu and J. Yamamoto, Organometallics, 13 (1994) 2357.

[4] O. S. Alexeev, G. Panjabi, B. C. Gates, J. Catal., 173 (1998) 196-209.

[5] B. Raval, http://feff.phys.washington.edu/~ravel/software/exafs/aboutathena.html (download 2003).

[6] G. N. George, J. S. George and I. J. Pickering, http://www-ssrl.slac.stanford.edu/exafspak.html (download 2004).

[7] J. J. Rehr, J. Mustre de Leon, S. I. Zabinsky and R. C. Albers, J. Am. Chem. Soc., 113 (1991) 5135.

[8] A. A. Davydov, Infrared spectroscopy of adsorbed species on the surfaces of transition metal oxides, Wiley, Chichester, 1990.

[9] G. Busca, and V. Lorenzelli, Mater. Chem., 7 (1982) 89-126.

[10] D. G. Rethwisch, and J. A. Dumesic, Langmuir 2 (1986) 73-79.

Studies in Surface Science and Catalysis, volume 159
Hyun-Ku Rhee, In-Sik Nam and Jong Moon Park (Editors)

Selective Conversion of Methane to C_2 Hydrocarbons using Carbon Dioxide as an Oxidant over CaO-MnO/CeO$_2$ Catalyst

Nor Aishah Saidina[1] Amin and Istadi[2]

[1]Chemical Reaction Engineering Group (CREG), Faculty of Chemical and Natural Resources Engineering, Universiti Teknologi Malaysia, 81310 UTM Skudai, Malaysia

[2]Chemical Reaction Engineering & Catalysis Groups, Dept. of Chemical Engineering, Diponegoro University, Semarang, Indonesia

ABSTRACT

Carbon dioxide rather than oxygen seemed to be an alternative oxidant for the catalytic reaction of methane to produce C_2 hydrocarbons via oxidative coupling of methane (CO_2 OCM). The proper amount of medium and strong basic sites and the reducibility of the catalyst enhanced the CH4 conversion and C_2 hydrocarbon yield, which may be due to the synergistic effect among CeO_2, CaO and MnO in the catalyst. The C_2 hydrocarbons selectivity and yield of 75.6% and 3.9%, respectively were achieved over the 12.8CaO-6.4MnO/CeO$_2$ catalyst. The catalyst showed a good stability for 20 h time on stream in the CO_2 OCM process.

1. INTRODUCTION

Direct conversion of methane to ethane and ethylene (C_2 hydrocarbons) has a large implication towards the utilization of natural gas in the gas-based petrochemical and liquid fuels industries [1]. CO_2 OCM process provides an alternative route to produce useful chemicals and materials where the process utilizes CO_2 as the feedstock in an environmentally-benefiting chemical process. Carbon dioxide rather than oxygen seems to be an alternative oxidant as methyl radicals are induced in the presence of oxygen. Basicity, reducibility, and ability of catalyst to form oxygen vacancies are some of the physico-chemical criteria that are essential in designing a suitable catalyst for the CO_2 OCM process [2]. The synergism between catalyst reducibility and basicity was reported to play an important role in the activation of the carbon dioxide and methane reaction [2].

In this paper, the selective conversion of methane to C_2 hydrocarbons over ternary CaO-MnO/CeO$_2$ catalysts in the CO_2 OCM process are presented. The synergistic effect between catalyst reducibility and distribution of basic sites are highlighted. The most promising catalyst was then tested towards its stability.

2. EXPERIMENTAL

The first ternary metal oxide catalyst of CaO-MnO/CeO$_2$ was prepared by simultaneous impregnation method, while the second ternary metal oxide of CaO/MnO-CeO$_2$ catalyst was prepared by combination of co-precipitation and impregnation method. The catalysts composition used in this paper were based on multi-responses optimization result [3]. H$_2$-TPR was carried out using Micromeritics 2900 TPD/TPR equipped by TCD. A catalyst amount of about 0.05 g was purged with Ar (25 cm^3 min^{-1}) at 773 K for 1 h and was cooled down to room temperature. The flow of 6% H$_2$ in Ar (25 cm^3 min^{-1}) was then switched into the system, and the sample was heated up to 1223 K from room temperature at a rate of 5 K min^{-1}. CO$_2$-TPD was carried out using Micromeritics 2900 TPD/TPR equipped by TCD. The catalyst samples of about 0.05 g each were initially calcined at 1073 K in a flow of argon (25 cm^3 min^{-1}) for 1 h. The chemisorption of CO$_2$ was carried out at 373 K by flowing CO$_2$ (25 cm^3 min^{-1}) for 1 h. The excess of CO$_2$ was purged in a flow of Ar (25 cm^3 min^{-1}) for 1 h. The sample was then heated to 1223 K at a linear heating rate of 5 K min^{-1} in a flow of Ar (25 cm^3 min^{-1}). The amount of H$_2$ uptake and the amount of CO$_2$ desorbed in both characterizations were detected using TCD. The performances of the catalysts were tested using a fixed-bed quartz reactor at the following conditions: reactor temperature = 1123 K; CH$_4$/CO$_2$ = 1/2, feed flow rate = 100 ml min^{-1}; catalyst loading = 2 g. Before reaction, the catalyst was recalcined at 1123 K in air flow for 1 h and was flushed with high purity nitrogen at 1123 K for another 1 h. The products and the unreacted gases were analyzed by an online GC equipped with a thermal conductivity detector and PORAPAK N packed-column.

3. RESULTS AND DISCUSSION

3.1. Catalyst Characterization

The H$_2$-TPR peaks of CeO$_2$-based catalysts are presented in Fig. 1. The reduction of CeO$_2$ appears at about 300 °C and two broad peaks are observed at about 540 (weaker intensity) and 770 °C [4]. The TPR peak at 540 °C is assigned to the reduction of surface-capping oxygen of ceria [4], while the peak at 770 °C can be ascribed to the reduction of bulk oxygen of CeO$_2$. This reduction is associated to an increased reducibility of the bulk mixed oxide, which is evident in Figs. 1(c) and 1(d) as the H$_2$-TPR peaks shift from 770 °C to 760 °C and 740 °C for the CaO/CeO$_2$ and CaO-MnO/CeO$_2$ catalysts, respectively. The introduction of manganese strongly modifies the reduction behavior of CeO$_2$ by shifting the main peak of H$_2$ consumption to a lower temperature [5]. The TPR peak shifts show that the reducibility of both catalysts is different. The promotion of Ce^{4+} reduction is related to higher mobility of oxygen from chemisorbed CO$_2$ due to the introduction of manganese. The reduction of surface capping oxygen of CaO-MnO/CeO$_2$ catalyst was shifted to higher temperature (from 540 to 580 °C) due to strong interaction effect of MnO$_x$ and CeO$_2$ support. The interaction also generates a new strong H$_2$-TPR peak at about 510 °C which is attributed to reduction of Mn$_2$O$_3$ or MnO$_2$ to Mn$_3$O$_4$. Meanwhile, the new small peak is also appeared at about 396 °C due to the interaction effect. The H$_2$-TPR spectra as depicted in Fig.1 reveal that the CaO-MnO/CeO$_2$ catalyst has moderate reducibility with high medium and strong basic sites (CO$_2$ TPD). It is shown that the CaO/MnO-CeO$_2$ catalyst has higher MnO$_x$ content than the CaO-MnO/CeO$_2$ catalyst as indicated by high intensity of TPR peak at 390 °C. The very small peak within the temperature range of 380-400 °C indicates that the catalyst synthesis method brings about a little proportion of MnO$_2$ form in the surface of CaO-MnO/CeO$_2$ catalyst which may be due to formation of solid solution of Ca$_{1-x}$Mn$_x$O. The strong H$_2$-TPR peak at about 510 °C is assigned to the reduction of Mn$_2$O$_3$ or MnO$_2$ to Mn$_3$O$_4$, while the peak at about 580 °C is attributed to final reduction from Mn$_3$O$_4$ to MnO. The color change of the catalyst from black to brown and/or grey during the reaction is in agreement with the results of H$_2$-TPR peaks [6].

The CO$_2$-TPD curves demonstrating the base strength distribution of different CeO$_2$-based catalysts are presented in Fig. 2. From the figure, the difference in the distribution of basic sites for each catalyst indicates that the basicity and base strength distributions are significantly influenced by CaO and its interaction with MnO$_x$ in the CeO$_2$-based catalysts. The

CO_2 TPD spectra obviously exhibit that the CaO/CeO_2 catalyst gives the largest number of very strong basic sites, followed by $CaO/MnO-CeO_2$ and $CaO-MnO/CeO_2$ catalysts. According to Figs. 2(b-d), the doping CaO on the CeO_2-based catalysts results in the creation of a large number of medium, strong and very strong basic sites at the expense of the weak basic sites as compared to the pure CeO_2 catalyst [7]. The CO_2-TPD spectra in Figs. 2(c) and 2(d) show different peaks owing to the total number of medium and strong basic sites. Impregnation of calcium and manganese nitrate solutions to the CeO_2 catalyst leads to higher distribution of medium and strong basic sites rather than when the $MnO-CeO_2$ solid solution is used as the support. The CO_2-TPD peak of medium basic sites becomes more intense significantly for CaO/CeO_2 and $CaO-MnO/CeO_2$ catalysts, but not for $CaO/MnO-CeO_2$ catalyst.

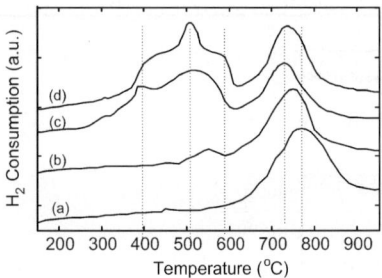

Fig. 1. H_2-TPD spectra of different catalysts. (a) CeO_2; (b) $12.8CaO/CeO_2$; (c) $12.8CaO/6.4MnO-CeO_2$; (d) $12.8CaO-6.4MnO/CeO_2$

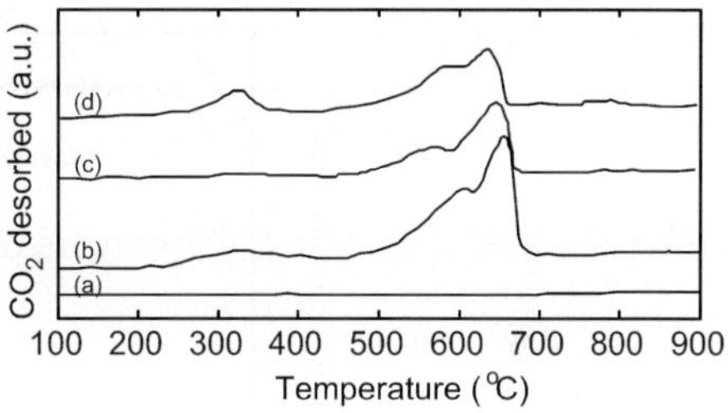

Fig. 2. CO_2-TPD spectra of different catalysts. (a) CeO_2; (b) $12.8CaO/CeO_2$; (c) $12.8CaO/6.4MnO-CeO_2$; (d) $12.8CaO-6.4MnO/CeO_2$

3.2. Catalyst Activity and Correlation with Catalyst Basicity and Reducibility

The screening results of CeO_2-based catalysts over binary and ternary metal oxides in Table 1 indicate that the $12.8CaO-6.4MnO/CeO_2$ catalyst is the most potential [3,8] with CH_4 conversion, C_2 selectivity and yield being 5.1%, 75.6%, and 3.9%, respectively. Addition of CaO to the pure CeO_2 catalyst results in a significant increase in the C_2 hydrocarbon selectivity. The enhancement in the C_2 yield is possibly due to the synergistic effect of MnO and CeO_2 where the reducibility of the catalyst are increased as revealed in the H_2-TPR results (Fig. 1). Our present study indicates that there exists a correlation between basic sites distribution, catalyst reducibility and catalytic activity toward C_2 hydrocarbons production as exposed in CO_2-TPD result (Fig. 2). The CaO species is suggested to play an important role in CO_2 chemisorption on the catalyst surface due to the role in distribution of medium and strong basic sites of the catalyst [9,10]. Proper amount of catalyst basicity, particularly medium and strong basic sites, greatly enhances the selectivity to C_2 hydrocarbons [2,9]. The MnO species evidently increases the reducibility of CeO_2 due to increasing oxygen mobility of the CeO_2 catalyst which enhances its reducibility and produces more oxygen vacancies [2]. The catalyst also shows high stability during 20 h time on stream for CO_2 OCM as revealed in Fig. 3.

Previously, we performed single- and multi-response optimization works in order to address optimal catalyst composition (%CaO and %MnO) and optimal operating conditions (temperature and CO_2/CH_4 feed ratio [3]. The maximum C_2 selectivity and yield of 76.6% and 3.7%, respectively were achieved in multi-responses optimization over the 12.8% $CaO-6.4\%$ MnO/CeO_2 catalyst corresponding to the optimum reactor temperature being 1127 K and CO_2/CH_4 ratio being 2 [3]. The recent contribution on the catalyst technology of CO_2 OCM was developed by He *et al.* [11] using nano-CeO$_2$/ZnO catalyst achieved C$_2$ yield 4.8% and selectivity 83.6% at 1098 K. Due to still low yield in CO$_2$ OCM, further improvements are required including the exploitation of some non-conventional technologies.

Table 1. Catalysts performance results of CeO$_2$-based catalysts [3,8]

Catalysts	CH$_4$ conversion (%)	C$_2$ selectivity (%)	C$_2$ yield (%)
CeO$_2$	13.2	9.1	1.2
12.8CaO/CeO$_2$	2.7	75.0	2.0
6.4MnO/CeO$_2$	8.8	3.1	0.3
12.8CaO/6.4MnO-CeO$_2$	5.3	62.2	3.3
12.8CaO-6.4MnO/CeO$_2$	5.1	75.6	3.9

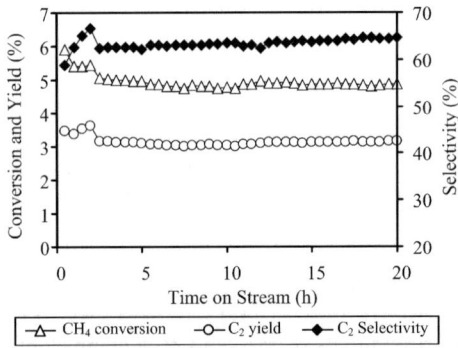

Fig.3. Stability test of 12.8CaO-6.4MnO/CeO$_2$ catalyst in CO$_2$ OCM reaction

4. CONCLUSIONS

Addition of CaO and MnO to the CeO$_2$ catalyst increased the CH$_4$ conversion and C$_2$ hydrocarbons yield in CO$_2$ OCM. The enhancement may be due to the synergistic effects between highly number of medium and strong basic sites and high reducibility of catalyst. The 12.8CaO-6.4MnO/CeO$_2$ catalyst exhibited better performance and high stability for the CO$_2$ OCM process.

REFERENCES
[1]   T. Suhartanto, A.P.E. York, A. Hanif, H. Al-Megren, M.L.H. Green, Catal. Lett., 71 (2001) 49.
[2]   Y. Wang, Y. Ohtsuka, J. Catal., 192 (2000) 252.
[3]   Istadi, N.A.S. Amin, Chem. Eng. J., 106 (2005) 213.
[4]   M. Boaro, M. Vicario, C.D. Leitenburg, G. Dolcetti, A. Trovarelli, Catal. Today, 77 (2003) 407.
[5]   J. Kaspar, P. Fornasiero, M. Graziani, Catal. Today, 50 (1999) 285.
[6]   L.E. Cadus, O. Ferretti, Appl. Catal. A, 233 (2002) 239.
[7]   Y. Wang, Y. Ohtsuka, Appl. Catal. A, 219 (2001) 183.
[8]   Istadi, N.A.S. Amin, J. Nat. Gas Chem., 13 (2004) 23.
[9]   V.R. Choudhary, S.A.R. Mulla, B.S. Uphade, Fuel, 78 (1999) 427.
[10] Y. Wang, Y. Takahashi, Y. Ohtsuka, J. Catal., 186 (1999) 160.
[11] Y. He, B. Yang, G. Cheng, Catal. Today 98 (2004) 595.

Studies in Surface Science and Catalysis, volume 159
Hyun-Ku Rhee, In-Sik Nam and Jong Moon Park (Editors)

Reduced elementary reaction model of the propane pyrolysis

Motoaki Kawase, Hiroshika Goshima, and Kouichi Miura

Department of Chemical Engineering, Kyoto University
Kyotodaigaku-Katsura, Nishikyo-ku, Kyoto 615-8510, Japan

1. INTRODUCTION

Chemical vapor deposition (CVD) of carbon from propane is the main reaction in the fabrication of the C/C composites [1,2] and the C-SiC functionally graded material [3,4,5]. The carbon deposition rate from propane is high compared with those from other aliphatic hydrocarbons [4]. Propane is rapidly decomposed in the gas phase and various hydrocarbons are formed independently of the film growth in the CVD reactor. The propane concentration distribution is determined by the gas-phase kinetics. The gas-phase reaction model, in addition to the film growth reaction model, is required for the numerical simulation of the CVD reactor for designing and controlling purposes. Therefore, a compact gas-phase reaction model is preferred. The authors proposed the procedure to reduce an elementary reaction model consisting of hundreds of reactions to a compact model objectively [6]. In this study, the procedure is applied to propane pyrolysis for carbon CVD and a compact gas-phase reaction model is built by the proposed procedure and the kinetic parameters are determined from the experimental results.

2. EXPERIMENTAL

Experiments of propane pyrolysis were carried out using a thin tubular CVD reactor as shown in Fig. 1 [4]. The inner diameter and heating length of the tube were 4.8 mm and 30 cm, respectively. Temperature was around 1000°C. Propane pressure was 0.1–6.7 kPa. Total pressure was 6.7 kPa. Helium was used as carrier gas. The product gas was analyzed by gas chromatography and the carbon deposition rate was calculated from the film thickness measured by electron microscopy. The effects of the residence time and the temperature

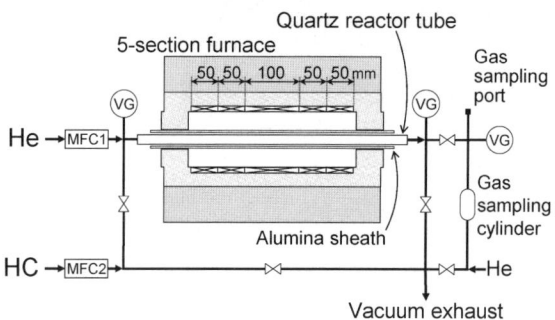

Fig. 1. A thin tubular CVD reactor.

were measured in the ranges of 5–20 ms and 800–1050°C.

3. NUMERICAL SIMULATION

Numerical simulation of the gas-phase reaction was carried out by using the Plug package of the CHEMKIN Collection [7]. This package was designed for simulating a plug flow reactor and able to calculate three-body reactions. 651 elementary reactions including 145 species from H to C_8H_8 involved in propane pyrolysis were taken from the NIST Chemical Kinetics Database [8].

4. RESULTS AND DISCUSSION

4.1. Reduction of elementary reaction model
An elementary reaction model can be reduced by the following 7 steps: (1) limitation of the range of species considered, (2) screening reactions having high reaction rates, (3) grouping isomers at partial equilibrium or having similar behavior, (4) excluding minor components in reversible reactions at partial equilibrium, (5) excluding intermediates of consecutive reactions, (6) excluding elementary reactions insignificant for material balance, and (7) ignoring the elementary reactions that do not affect the deposition rate [6]. Although the first 6 steps are all-purpose, the final step is only applicable in case of CVD, the rate of which has been formulated. The proposed procedure was applied to propane pyrolysis. 453 elementary reactions of 651 reactions were chosen by step 1. By carrying out the numerical simulation of the CVD experiments by the 453-reaction model, the other reduction steps were made. By extracting the reactions having high reaction rates based on the simulation results, the 26-reaction model was developed.

4.2. Effects of reaction temperature
Fig. 2 shows examples of the results obtained by the 453-reaction model. The effects of the reaction temperature were investigated by experiments and numerical simulation. It is

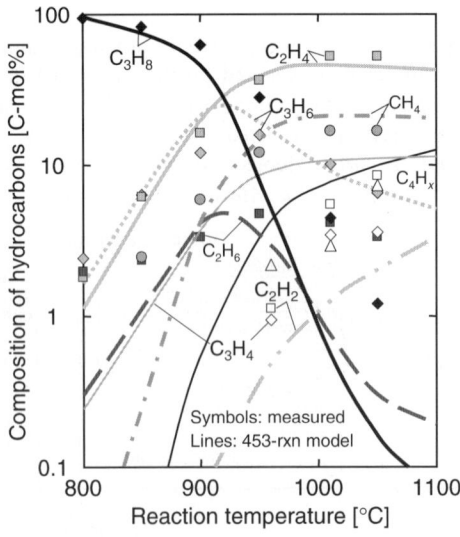

Fig. 2. Effects of temperature on the product gas composition. ($P_t = 6.7$ kPa, $\tau = 20$ ms)

obvious that various hydrocarbons are formed from propane. In general, the experimental results are well reproduced by the numerical simulation. The calculated ethylene and propylene fractions are in good agreement with the experimental results at all temperatures. At high temperatures, however, the numerical simulation underestimates the fractions of ethane and propane. Particularly, the error in C_3H_4 and C_4H_x components was great, since it is possible that the database includes some uncertain reaction rate constants and those measured under conditions different from the present conditions.

4.3. Refinement of the reduced model

In order to obtain a good agreement between the numerical simulation and experiments, the kinetic parameters in the reaction model have to be refined. As it is difficult to refine the huge model, the 26-reaction model as shown in Fig. 3 was used for fitting with the experimental data. Fig. 4 shows the measured and calculated hydrocarbon compositions at the exit of the reactor. The results calculated by the 26-reaction model are similar to the results by the 453-reaction model. However, the calculated results are not in very good agreement with the measured gas compositions. By referring to all the original articles, only the rate constants of reactions r2, r6, r9, r14, r16, r17, r19, and r21 shown in Fig. 3 were found to be measured or estimated under conditions different from this study. These reaction rate constants were refined by fitting the product gas composition with the experimental results. By repeating the correction of the rate constant of the most influential reaction to the material balance of the species of the most mismatched concentration, the reaction rate constants were refined. The determined values are listed in Table 1. Fig. 4 shows the gas composition calculated by the models before and after the refinement. The agreement between the measured and calculated composition was greatly improved by the refinement of the rate constants.

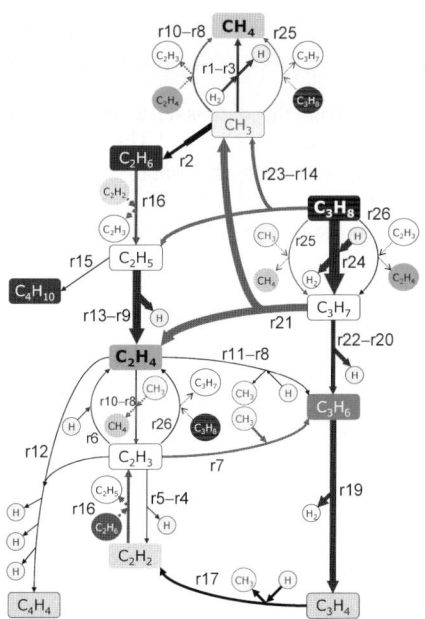

Fig. 3. Reduced elementary reaction model of the propane pyrolysis. (1010°C, $\tau = 20$ ms)

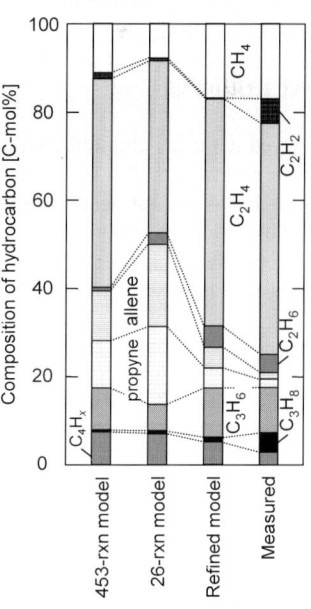

Fig. 4. Calculated and measured hydrocarbon compositions at the exit of the reactor. (1010°C, $P_t = 6.7$ kPa, $p_A = 1.5$ kPa, $\tau = 20$ ms)

Table 1. Refined reaction rate constants.

Reaction	Rate constant @1010°C
r2	$2.9 \times 10^6 \text{ m}^6/(\text{mol}^2\text{s})$
r6	$8.8 \times 10^{10} \text{ m}^3/(\text{mol}\text{s})$
r9	$1.9 \times 10^8 \text{ m}^3/(\text{mol}\text{s})$
r14	$1.9 \times 10^7 \text{ m}^3/(\text{mol}\text{s})$
r16	$9.5 \times 10^3 \text{ m}^3/(\text{mol}\text{s})$
r17	$2.7 \times 10^8 \text{ m}^3/(\text{mol}\text{s})$
r19	$8.7 \times 10^7 \text{ m}^3/(\text{mol}\text{s})$
r21	$2.9 \times 10^8 \text{ s}^{-1}$

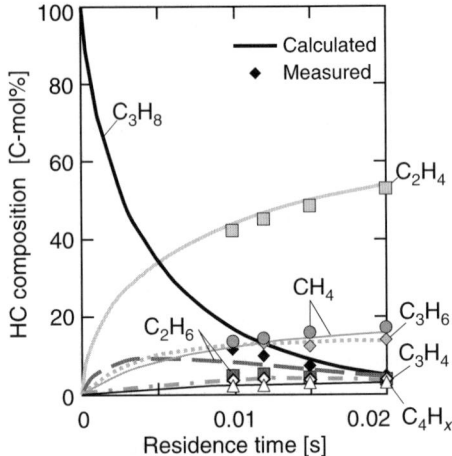

Fig. 5.   Effects of residence time on the gas composition from propane pyrolysis. ($T = 1010°C$, $P_t = 6.7$ kPa, $p_{A0} = 1.5$ kPa)

4.4. Effects of the residence time

Fig. 5 shows the effects of the gas residence time on the hydrocarbon composition.   The numerical simulation was carried out by the refined model.   The calculated results are in good agreement with the experimental results.   It should be noted that the measured gas composition only at a residence time of 20 ms was used for the parameter fitting. Nevertheless, the calculated hydrocarbon composition agrees with the experimental results at residence times shorter than 20 ms.

5. CONCLUSIONS

By reducing an elementary reaction model taken from the database, a comprehensive gas-phase reaction model of propane pyrolysis was derived objectively.   The reaction rate constants that were not accurate under the conditions of interest were found and refined by fitting with the experimental results.   The obtained reaction model well represented the effects of the gas residence time and temperature on the product gas composition observed in experiments under pyrocarbon CVD conditions.

REFERENCES

[1] H. O. Pierson and J. F. Smatana, Proc. 2nd Int. Conf. Chem. Vapor Deposition., 2 (1970) 487.
[2] A. Becker and K. J. Huettinger, Carbon, 46(4) (1998) 225.
[3] M. Kawase, T. Tago, M. Kurosawa, H. Utsumi, and K. Hashimoto, Chem. Eng. Sci., 54 (1999) 4427.
[4] M. Kawase, T. Nakai, H. Goshima, and K. Miura, Abst. 18th Int. Symp. Chem. React. Eng., Chicago, Jun, 2004, MS#154.
[5] Y. J. Kim and J. Y. Lee, Carbon, 41(7) (1994) 1041.
[6] M. Kawase, H. Goshima, T. Nakai. and K. Miura, Proc. 10th Asia Pacific Confed. Chem. Eng. Congr., Kitakyushu, Sep, 2004, #3P-08-048.
[7] R. J. Kee, et al., Chemkin Collection, Release 3.7.1, Reaction Design, Inc., San Diego, CA (2003).
[8] National Institute of Standards and Technology; Chemical Kinetics Database on the Web, Standard Reference Database 17, Version 7.0, Release 1.1 (2000). http://kinetics.nist.gov/.

Studies in Surface Science and Catalysis, volume 159
Hyun-Ku Rhee, In-Sik Nam and Jong Moon Park (Editors)

MASS TRANSFER AND CHEMICAL REACTION ASPECTS CONCERNING ACETALDEHYDE OXIDATION IN AGITATED REACTOR

Suprapto[a), Dyah Suci Perwitasari[b)] and Ali Altway[a)]

[a)]Chemical Engineering Department, Institut Teknologi Sepuluh Nopember (ITS) Surabaya 60111, Indonesia, E-mail: reaktorkimia@yahoo.com

[b)] Chemical Engineering Department, UPN Veteran Surabaya-Indonesia

Abstract

Experimental investigation has been carried out concerning acetaldehyde oxidation with oxygen/air using homogenous catalyst mangan acetate. Two steps of work have been carried out. Firstly, hydrodynamic of reactor has been evaluated by measuring gas-liquid mass transfer coefficient using oxygenation dynamic method. This work gave empirical correlation of the coefficient in term of rotation speed and gas flow rate. Secondly, the evaluation of acetaldehyde oxidation reaction was carried out in mechanically agitated reactor at atmospheric pressure, where impeller rotation speed, air flow-rate, and temperature were varied. This research showed that increasing impeller rotation speed could not increase acetaldehyde conversion. On the other hand, the selectivity of acetic acid decreased with increasing impeller rotation speed. The observation also showed that the conversion of acetaldehyde and selectivity of acetic acid increased with the increasing of gas flow rate and temperature. The maximum conversion was 32.5 % and the highest selectivity was 70.5 %.

1. INTRODUCTION

Liquid phase oxidation of acetaldehyde using air and homogeneous catalyst mangan acetate is one of processes in industrial scale production of Acetic Acid wide spread in the world. To utilize molasses, by product of sugar industry, abundantly produced in Indonesia, in 1989 Acetic Acid Plant have been built using the above process. In United States, production of acetic acid from acetaldehyde was applied since 1960 [6]. We observe that even the process of acetaldehyde oxidation to produce Acetic Acid has been applied in industry long time ago, however detailed phenomena occurring in the reactor has not been studied by previous researchers. The performance of process was affected by kinetic and hydrodynamic factors. Hydrodynamic and mass transfer characteristic in stirred tank gas-liquid reactor were determined by the interaction between impeller rotation speed and gas flow rate factors as has been studied by several workers [4, 5, 7, 10]. Venugopal et al. [12] studied the kinetic of acetaldehyde oxidation and reported that the reaction follows second order kinetic.

The present research was focused on the study of acetaldehyde oxidation using air with aqueous mangan acetate catalyst in mechanically stirred tank reactor.

2. METHODOLOGY

The research has been carried out in two steps:
1. Evaluation of reactor hydrodynamics by measuring gas-liquid mass transfer coefficient

2. Evaluation of acetaldehyde oxidation reactor especially by determining reaction conversion and selectivity.

2.1. Equipment and Method

This research used mechanically agitated tank reactor system shown in Fig. 1. The reactor, 102 mm in diameter and 165 mm in height, was made of transparant pyrex glass and was equipped with four baffles, 120 mm in length and 8 mm in width, and six blades disc turbine impeller 45 mm in diameter and 12 mm in width. The impeller was rotated by electric motor with digital impeller rotation speed indicator. Waterbath thermostatic, equipped with temperature controller was used to stabilize reactor temperature. Gas-liquid mass transfer coefficient k_la was determined using dynamic oxygenation method as has been used by Suprapto et al. [11].

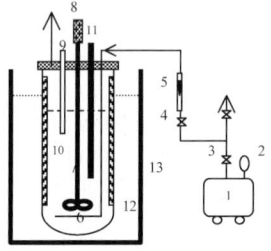

1. Air compressor
2. Pressure indicator and controller
3. Main control valve
4. Air flow Needle Valve
5. Flowmeter
6. Air Distributor
7. Agitator

8. Motor
9. Liquid sampling tap
10. Baffle
11. Thermometer
12. Reactor
13. Thermostatic Water-
    bath

Fig. 1. Equipment Apparatus

The evaluation of acetaldehyde oxidation process was carried out by aeration of acetaldehyde solution and analyzing the concentration of acetic acid using gas chromatography HP 5890 with detector FID equipped with PEG Column in 15 minutes time interval. The gas flow rate (Qg), impeller rotation speed (N) and temperature (T) were varied.

2.2. Mathematical Model Development

Acetaldehyde oxidation reaction comprise of a main reaction and several side reaction as follow:

Main reaction : $\quad CH_3CHO + \frac{1}{2} O_2 \xrightarrow{k1} CH_3COOH$

Side reaction : $\quad CH_3CHO + O_2 \xrightarrow{k2} CH_3OH + CO_2$

$\quad\quad\quad\quad\quad CH_3OH + CH_3COOH \xrightarrow{k3} CH_3COOCH_3 + H_2O$

These reactions were carried out in semi batch stirred tank reactor. The governing equations for this process were :

$$da'/d\tau = E\,\xi\,(a^* - a') - (\tfrac{1}{2} + \alpha)\,a'b' \qquad (1)$$

$$db'/d\tau = -(1 + \alpha + \gamma)\,a'b' \qquad (2)$$

$$dc'/d\tau = a'b' - \beta c'd' \qquad (3)$$

$$dd'/d\tau = \alpha\,a'b' - \beta c'd' \qquad (4)$$

$$de'/d\tau = \beta c'd' \qquad (5)$$

$$a_k\,y_{AG}^2 - b_k\,y_{AG} + c_k = 0 \qquad (6)$$

where $a^* = C_A^*/C_{Bi} = (1 - \delta)\,a_p$, $a' = C_A/C_{Bi}$; $b' = C_B/C_{Bi}$; $c' = C_C/C_{Bi}$; $d' = C_D/C_{Bi}$; $e' = C_F/C_{Bi}$; $\alpha = k_2/k_1$; $\beta = k_3/k_1$; $\xi = k_La/(k_1 C_{Bi})$, δ = fraction of gas film resistance, $a_p = P_{AG}/(H_A$

C_{Bi}), H_A= Henry constant, k_La= volumetric gas-liquid mass transfer coefficient., C_{Bi} = initial concentration of B in bulk of liquid, and P_{AG}= partial pressure of A in bulk of gas, $a_k = E$ $\varphi (1 - \delta) (P/(H_A C_{Bi})$, $b_k = E \varphi (1 - \delta)[P/(H_A C_{Bi})] + y_{Ai}/(1-y_{Ai}) + 1$, $c_k = y_{Ai}/(1-y_{Ai})$, $\varphi = k_La V C_{Bi}/G$, P = operating pressure, G = molar gas flow rate, A= O_2, B = CH_3CHO, C = CH_3COOH , D = CH_3OH , F = CH_3COOCH_3. Equation (6) can be written as $y_{AG} = c_k/(b_k - a_k y_{AG})$, where y_{AG} mole fraction of A in bulk gas phase can be determined iteratively, y_{Ai} = mole fraction of A in gas inlet. Equations (1) to (6) were solved using fourth order Runge-Kutta method [1, 8]. The value of enhancement factor, E, was predicted using equation of Van Krevelen and Hoftijzer [2].

The computation result yield acetaldehyde concentration as function of time. The value of kinetics parameters, k_1, k_2, k_3 were adjusted to minimize the sum of square of error between the predicted and measured concentration using Hooke Jeeve method [3].

3. RESULTS AND DISCUSSION

The measurement of liquid side gas – liquid mass transfer coefficient k_La, showed that the value of k_La increase with increasing rotation speed (N) and gas flow rate (Qg). In the present research, the effect of impeller rotation on mass transfer coefficient was more significant than the effect of gas flow rate. The following correlation was obtained $k_La =1.7 x 10^{-3} N^{0.98} Qg^{0.22}$. Pedersen et al. [9] who carried out the measurement of k_La using tracer ^{85}Kr, gave almost the same prediction as the present work.

The value of k_La predicted above and kinetic data obtained by Venugopal et al. [12] were used for simulation of acetaldehyde oxidation reaction. The present study obtained the expression of kinetic konstants as follows: $k_1 = 6.64.10^{10} \exp(-12709/RT)$, $k_2 = 244.17 \exp(-1.8/RT)$ and $k_3 = 3.11.10^7 \exp(-13639/RT)$ m^3.kmol^{-1}.s^{-1}. The value of k_1, obtained in this research are almost the same as that obtained by Venugopal. Venugopal neglected the side reactions. The value of E and Hatta Number $\sqrt{M}$ were greater than 3, so that the reaction system can be considered as pseudo-first order reaction with respect to oxygen and the process was controlled by mass transfer aspect.

Fig. 2 and 3 showed that increasing gas flow rate (Qg) and temperature (T) at 500 rpm rotation speed (N), will increase the acetaldehyde conversion. Fig. 2 also showed that the reaction is in kinetic regime at low temperature, while at higher temperature (Fig.3), the reaction approach the equilibrium condition or in thermodynamic region.

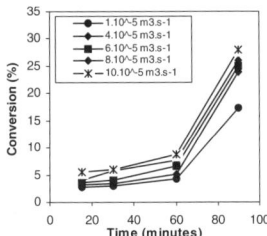
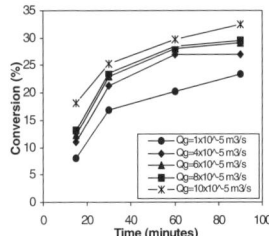
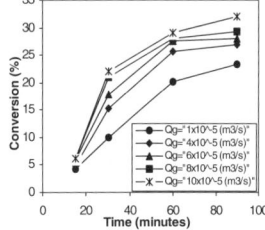

Fig. 2. The effect of Qg on conversion (N= 500 rpm, 35°C)

Fig. 3. The effect of Qg on conversion (N= 500 rpm, 55°C)

Fig. 4. The effect of Qg on conversion (N= 900 rpm,55°C)

224

In the range of 500 to 900 rpm, Fig. 3 and 4 showed that impeller rotation speed does not affect significantly on reaction conversion for relatively long reaction time (above 60 minutes). Apparently, the hydrodynamic condition of liquid in the reactor was sufficiently turbulent by aeration.

Increasing of temperature and gas flow rate can enhance the selectivity, however the selectivity decrease with increasing impeller rotation speed (Fig. 5 and 6). In general, increasing selectivity was followed by decreasing of conversion, as shown in Fig. 7.

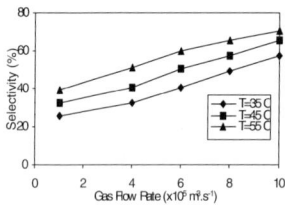 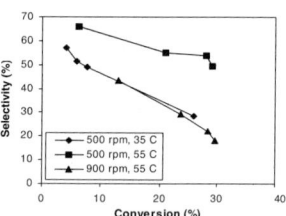

Fig. 5. The Effect of T and Qg on Selectivity (for N= 500 rpm)

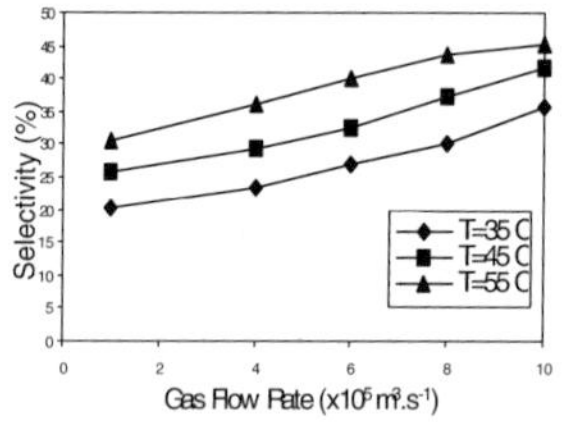

Fig. 6. The Effect of T and Qg on Selectivity (for N= 900 rpm)

Fig. 7. Relation between conversion and selectivity

4. CONCLUSION

Liquid phase oxidation reaction of acetaldehyde with Mn acetate catalyst can be considered as pseudo first order irreversible reaction with respect to oxygen, and the reaction occurred in liquid film. The value of kinetic constant as follow : $k_1 = 6.64.10^{10}$ exp(-12709/RT), $k_2 = 244.17$ exp(-1.8/RT) and $k_3 = 3.11.10^7$ exp(-13639/RT) m^3.kmol^{-1}.s^{-1}. The conversion can be increased by increasing gas flow rate and temperature, however the effect of impeller rotation on the conversion is not significant. The highest conversion 32.5% was obtained at the rotation speed of 900 rpm, temperature 55 °C, and gas flow rate 10^{-4} m^3.s^{-1}. The selectivity of acetic acid was affected by impeller rotation speed, gas flow rate and temperature. The highest selectivity of acetic acid was 70.5% at 500 rpm rotation speed, temperature of 55 °C and 10^{-4} m^3.s^{-1} gas flow rate.

REFERENCES

[1] Conte. S.D. and Carl de Boor. Elementary Numerical Analysis. An Algorithmic Approach. 3rd ed., Mc.Graw Hill Book Co., Singapore, 1981.
[2] Danckwerts.P.V., Gas-Liquid Reactions, Mc. Graw Hill Book Co., New York, 1970.
[3] Edgar.T.F. and Himmelblau.D.M, Optimization of Chemical Process, Mc.Graw Hill Book Co, Singapore, 1989.
[4] Gezork. K.M., Bujalski.W., Cooke.M, and Nienow.A.W, Trans IchemE., Vol.79, Part A (2001), 965-972
[5] Hiraoka.S., Kato.Y., Tada.Y., Kai.S.C. and Inoul.N., Proc. ISMIP3, (1999), 203-209
[6] Kent. J.A. Riegel's Handbook of Industrial Chemistry, 8thed, Van Nostrand Reihold Co., New York, 1983.
[7] Lu.W.M., Wu.H.Z., Chou.C.Y., and Chang.C.Y., Proc.ISMIP3, (1999), 403-409.
[8] Mathews. J.H.. Numerical Methods for Mathematics. Sciences and Engineering, 2nd ed., Prentice Hall, New York, 1992
[9] Pedersen.A.G., Andersen.H., Nielsen.J., and Villadsen.J.,Chem.Eng.Sci. , 49(6) (1994), 803-810.
[10] Ranade, V.V., Rerrand.M., Xuereb.C., Sauze.N.L., and Bertrand.J. ,TransIChemE., 79, Part A, 957-964.
[11] Suprapto, Wehrer.A, Ronze.D., Zoulalian.A., Odours & VOC'S J., 1 (5) (1996), 374-378.
[12] Venugopal. B., Kumar.R. and Kuloor.N.R., Ind.Eng.Chem.Process Dev., 6(1)(1967), 139-146.

Studies in Surface Science and Catalysis, volume 159
Hyun-Ku Rhee, In-Sik Nam and Jong Moon Park (Editors)
© 2006 Elsevier B.V. All rights reserved

225

Selective oxidation of hydrogen sulfide to elemental surfur and ammonium thiosulfate using VO_x /TiO_2 catalysts

Moon-Il Kim, Wol-Don Ju, Kyung-Hoon Kim, Dae-Won Park* and Seong-Soo Hong[a]

Division of Applied Chemical Engineering Pusan National University, Busan 609-735 Korea
[a]Division of Applied Chemical Engineering, Pukyong National University, Busan, Korea
E-mail : dwpark@pusan.ac.kr (D. W. Park)

VO_x/TiO_2 catalysts were prepared by precipitation-deposition and impregnation method using TiO_2 support obtained by sol-gel method from titanium isopropoxide. The performance of the VO_x/TiO_2 catalysts was investigated for the selective oxidation of hydrogen sulfide in the stream containing both ammonia and excess water. All the catalysts showed good dispersion of vanadium on TiO_2 surface and they had high H_2S conversion without production of sulfur dioxide. The catalysts prepared by the precipitation-deposition method had better catalytic activity compared to those by the impregnation method.

1. INTRODUCTION

In general, SO_x emission problem has been caused by H_2S released from crude oil and natural gas refineries. Hydrogen sulfide from stationary source is usually recovered as elemental sulfur by the Claus process [1, 2]. H_2S contained in the coke oven gas of the steel smelting process is scrubbed and concentrated using aqueous ammonia solution. The concentrated H_2S separated from ammonia solution is generally transferred to the Claus plant, and remaining aqueous ammonia solution is incinerated without further treatment. Since the separation of H_2S is not perfect, the remaining aqueous ammonia contains about 2 % of H_2S which in turn can cause the SO_x emission problem during incineration. Hence, new technologies are being examined to remove H_2S in excess water and ammonia stream. One approach is the selective catalytic oxidation of H_2S to elemental sulfur and ammonium thiosulfate (ATS: $(NH_4)_2S_2O_3$) as reported in our previous work [3-5]. In a prvious work [6] , we prepared highly active TiO_2 by sol-gel method. This method has an advantage in preparing high purity metal oxides at low calcinations temperatures. In this study, we examined the performance of VO_x/TiO_2 catalyst, prepared by precipitation-deposition method and impregnation method, for the selective oxidation of H_2S in the stream containing both of ammonia and water.

2. EXPERIMENTAL

The preparation method of titania support was described in the previous paper [6]. Titanium tetraisopropoxide (TTIP 97%, Aldrich) was used as a precursor of titania. Supported VO_x/TiO_2 catalysts were prepared by two different methods. The precipitation-deposition catalysts (P-VO_x/TiO_2) were prepared following the method described by Van Dillen et al. [7], in which the thermal decomposition of urea was used to raise homogeneously the pH of a suspension formed by the support in NH$_4$VO$_3$ solution. The pH of the solution was adjusted to 4 by addition of oxalic acid, and then continuously stirred at 90 °C for 10 h for 1 wt.% V, 20 h for the catalyst up to 10 wt.% V. Finally, all catalysts were dried and calcined at 500 °C for 10 h. The impregnation procedure used was the incipient wetness technique, in which the aqueous solution of NH$_4$VO$_3$ was slowly added to the support to obtain the desired metal content in the final catalyst.

The reaction test was carried out at atmospheric pressure using a vertical continuous flow fixed bed reactor. The content of effluent gas was analyzed by a gas chromatograph (HP 5890).

3. RESULTS AND DISCUSSION

A series of TiO$_2$-supported vanadium oxide catalysts were prepared by impregnation and precipitation-deposition method with different vanadate loadings. Fig. 1 shows X-ray diffraction patterns of impregnated catalysts (I-VO$_x$/TiO$_2$) calcined at 500 °C. P-VO$_x$/TiO$_2$ catalysts also showed nearly the same XRD patterns as I-VO$_x$/TiO$_2$ catalysts. From the XRD results, all the catalysts show XRD peaks due to anatase titania with an intense peak at 2Θ = 25.3 ° corresponding to the [101] plane of titania. No reflection is observed at 2Θ = 20.3 ° corresponding to the most intense reflection of V$_2$O$_5$. This result indicates that vanadium oxide is present in a highly dispersed amorphous state on titania.

Fig. 2 and 3 shows temperature programmed reduction (TPR) profiles for I-VO$_x$/TiO$_2$ and P-VO$_x$/TiO$_2$ catalyst, respectively. In I-VO$_x$/TiO$_2$ catalyst, the maximum peak temperature increased from 450 to 490 °C when vanadium loading increased from 1 wt.% up to 7 wt.%. TPR results for P-VO$_x$/TiO$_2$ catalysts (Fig. 3) show that the maximum peak temperature increased only about 20 °C (from 450 to 470 °C) with the vanadium loading from 1 to 7 wt.%. Therefore, the P-VO$_x$/TiO$_2$ can be reduced more easily than I-VO$_x$/TiO$_2$. It means that the vanadates in P-VO$_x$/TiO$_2$ are more highly dispersed than those in I-VO$_x$/TiO$_2$. However, I-VO$_x$(10%)/TiO$_2$ and P-VO$_x$(10%)/TiO$_2$ show the maximum peak temperature at about 540 °C, much higher than other low loading catalysts. It is reported that vanadate can be covered on the surface of TiO$_2$ with a monolayer when it is loaded less than 7 wt.% [8]. From the TPR results, I-VO$_x$(10%)/TiO$_2$ and P-VO$_x$(10%)/TiO$_2$ seem to have microcrystalline V$_2$O$_5$ particles [9] since they have higher than monolayer coverages of vanadate.

VO$_x$/TiO$_2$ catalysts were tested in the selective oxidation of H$_2$S to elemental sulfur and ammonium thiosulfate. Table 1 and 2 show the H$_2$S conversion and selectivity to the products for the two differently prepared VO$_x$/TiO$_2$ catalysts at 260 °C with H$_2$S/O$_2$/NH$_3$/H$_2$O/He = 5/2.5/5/60/27.5 and GHSV of 12,000 h^{-1}. For all the catalysts, surface area decreased as the vanadia loading increased. P-VO$_x$/TiO$_2$ catalysts prepared by the precipitation-deposition method showed a little higher surface areas than those prepared by the impregnation. In P-VO$_x$/TiO$_2$ catalysts (Table 1), the conversion of H$_2$S for 1 to 5 wt.% of vanadia loading were higher than 92 % without emission of sulfur dioxide. Although there was no appreciable change in the H$_2$S conversion for the three catalysts, the decrease of the H$_2$S conversion was notable for P-VO$_x$/TiO$_2$ catalysts of 10 wt.% of vanadia loading. Higher than monolayer surface coverage of vanadate formation of bulk V$_2$O$_5$ phase might be a cause for this big decrease of H$_2$S conversion. Low value of the H$_2$S conversion for P-VO$_x$(10%)/TiO$_2$ might also be related to its low surface area. Time variant H$_2$S conversion data also showed that VO$_x$/TiO$_2$ with higher vanadium loading had more severe activity decrease with process time. I-VO$_x$/TiO$_2$ catalysts in Table 2 also show that the H$_2$S conversion decreased with the increase of vanadia loading. P-VO$_x$(10%)/TiO$_2$ and I-VO$_x$(10%)/TiO$_2$ showed lower selectivity to sulfur (S-S) than other smaller vanadium loading catalysts. I-VO$_x$/TiO$_2$ catalysts have a little lower H$_2$S conversions than P-VO$_x$/TiO$_2$ catalysts. Higher surface area and higher vanadium dispersion are considered to be main reasons of better catalytic performance of P-VO$_x$/TiO$_2$ catalysts.

4. CONCLUSION

VO$_x$ supported on TiO$_2$ showed good catalytic activity in the selective oxidation of H$_2$S to ammonium thiosulfate and elemental sulfur. VO$_x$/TiO$_2$ catalysts prepared by the precipitation-deposition method can achieve higher vanadium dispersions, and higher H$_2$S conversions compared to those prepared by the impregnation method.

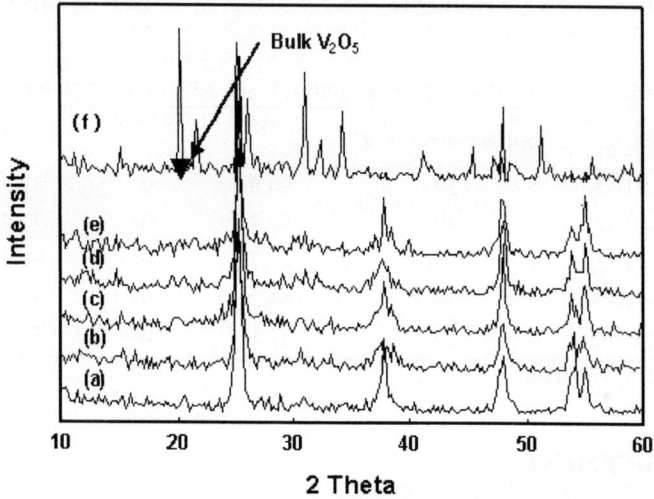

(a) 1 wt% (b) 3 wt% (c) 5 wt% (d) 7 wt% (e)10 wt% (f) V$_2$O$_5$

Fig. 1. XRD patterns of VO$_x$/TiO$_2$ catalysts with different vanadium loading

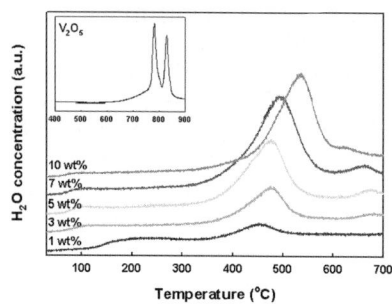 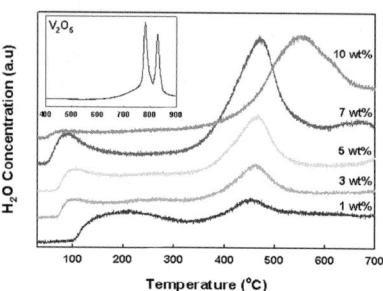

Fig. 2. TPR profiles of I-VO$_x$/TiO$_2$ catalysts   Fig. 3. TPR profiles of P-VO$_x$/TiO$_2$ catalysts

228

Table 1. Conversion of H_2S and selectivities to SO_2, S and ATS for P-VO_x/TiO_2 catalysts.

Catalyst	Surface area(m^2/g)	X-H_2S (%)	S-SO_2 (%)	S-S (%)	S-ATS (%)
P-VOx(1%)/TiO_2	94	93.0	0	79.1	20.9
P-VOx(3%)/TiO_2	69	92.8	0	73.6	26.4
P-VOx(5%)/TiO_2	55	92.5	0	78.2	21.8
P-VOx(7%)/TiO_2	51	86.3	0	72.5	27.5
P-VOx(10%)/TiO_2	9	78.4	0	45.9	54.1

P : Precipitation-deposition method
Reaction condition : $H_2S/O_2/NH_3/H_2O/He$ = 5/2.5/5/60/27.5, T= 260 ℃, GHSV= 12,000 h^{-1}, time= 6 h

Table 2. Conversion of H_2S and selectivities to SO_2, S and ATS for I-VO_x/TiO_2 catalysts.

Catalyst	Surface area(m^2/g)	X-H_2S (%)	S-SO_2 (%)	S-S (%)	S-ATS (%)
I-VOx(1%)/TiO_2	59	90.8	0	87.6	12.4
I-VOx(3%)/TiO_2	49	89.2	0	77.4	22.6
I-VOx(5%)/TiO_2	43	87.7	0	69.3	30.7
I-VOx(7%)/TiO_2	34	84.2	0	71.5	28.5
I-VOx(10%)/TiO_2	29	82.1	0	59.6	40.4

I : Impregnation method

ACKNOWLEDGEMENT

This work was supported by Pusan National University Research Grant.

REFERENCES

[1] J.A. Lagas, J. Borsboom, and P.H. Berben, Oil & Gas J., Oct. 10, 68 (1988).
[2] J. Wieckowska, Catal. Today, 24 (1995) 105.
[3] D. W. Park, B. K. Park, D. K. Park, H. C. Woo, Appl. Catal. A : Gen., 223 (2002) 215.
[4] B. K. Park, D. W. Park, I. Kim, H. C. Woo, Catal. Today, 87 (2003) 11.
[5] D. W. Park, B. G. Kim, M. I. Kim, I. Kim, H. C. Woo, Catal. Today, 93 (2004) 235
[6] G.H. Lee, M. S. Lee, G. D. Lee, Y. H. Kim, S. S. Hong, J. Ind. Eng. Chem., 6 (2002) 572.
[7] J.A. Van Dillen, J.W. Geus, L.A.M. Hermans, J. Van Der Meijden, Proc. 6th Int. Congr.on Catal., London (1976).
[8] S. Krishnamoorthy, J, P, Baker, M. D. Amiridis, Catal. Today, 40 (1998) 39.
[9] G. Deo, I. E. Wachs, J. Catal., 146 (1994) 335.

Studies in Surface Science and Catalysis, volume 159
Hyun-Ku Rhee, In-Sik Nam and Jong Moon Park (Editors)

Oxidative Catalytic Absorption of NO in Aqueous Ammonia Solution with Hexamminecobalt Complex

Zhi-Ling Xin, Xiang-Li Long, Wen-De Xiao*

UNILAB Research Center of Chemical Reaction Engineering,
East China University of Science and Technology, Shanghai, 200237, P. R. China
Tel&Fax: 86-21-64252814, E-mail: wdxiao@ecust.edu.cn

ABSTRACT

The hexamine cobalt (II) complex is used as a coordinative catalyst, which can coordinate NO to form a nitrosyl ammine cobalt complex, and O_2 to form a μ -peroxo binuclear bridge complex with an oxidability equal to hydrogen peroxide, thus catalyze oxidation of NO by O_2 in ammoniac aqueous solution. Experimental results under typical coal combusted flue gas treatment conditions on a laboratory packed absorber- regenerator setup show a NO removal of more than 85% can be maitained constant.

Keywords: Flue gas denitration, NO oxidation, ammine cobalt complex, activated carbon

INTRODUCTION

NO accounts for more than 90% of NOx, a major pollutant from the boiler flue gas, causing acid rain, greenhouse effect and ozone layer destruction. As NO is insoluble a water, several oxidative additives have been proposed to increase the absorption of NO in aqueous solution as used in the wet scrubbing flue gas desulfurization processes [1], such as OCl_2^-[2], H_2O_2[3], and yellow phosphorous[4], which are worthless for industrial application. The chelated Fe (II), such as $Fe(EDTA)^{2+}$ [5], has been investigated in detail in 1970s and 1980s as a coordinative absorbent of NO, but NO is turned into some unusable N-S components by reduction of the simultaneously absorbed SO_2, and Fe (II) is easily oxidized by the dissolved O_2 to Fe (III), inactive to NO coordination, which prevents it from commercialization.

Recent years, the authors have innovatively proposed a method by using the aqueous ammonia liquor containing hexamine cobalt (II) complex to scrub the NO-containing flue gases[6-9], since several merits of this complex have been exploited such as: (1) activation of atmospheric O_2 to a peroxide to accelerate the O_2 solubility, (2) coordination of NO, as NO is a stronger ligand than NH_3 and H_2O of Co(II) complexes to enhance the NO absorption and (3), catalysis of NO oxidation to further improve the absorption both of O_2 and NO. Thus, a valuable product of ammonium nitrate can be obtained.

This paper presents the experimental results, with a focus on studies of the regeneration of the hexamine cobalt complex additive by using the activated carbon in a laboratory packed-bed absorber.

THEORETICAL

NO may coordinate with $Co(NH_3)_6^{2+}$ by substituting NH_3 to form a nitrosyl pentammine complex[10],

$$[Co(NH_3)_6]^{2+}(I) \xrightarrow{+NO} [Co(NH_3)_5NO]^{2+}(II) + NH_3 \tag{1}$$

O_2 will react with $Co(NH_3)_6^{2+}$ to form a μ -peroxo binuclear bridge complex [11],

$$2[Co(NH_3)_6]^{2+} \xrightarrow{+O_2} [(NH_3)_5Co-O-O-Co(NH_3)_5]^{4+}(III) + 2NH_3 \tag{2}$$

whose O-O bond is anaogous to that of hydrogen peroxide, and which can function as a highly strong oxidant as H_2O_2. Therefore, in presence of O_2, NO may react with complex III to form a μ -peroxo binuclear nitrosyl complex(IV),

$$(III) \xrightarrow{+2NO} \begin{bmatrix} (NH_3)_4Co-O-O-Co(NH_3)_4 \\ \quad\quad| \quad\quad\quad\quad | \\ \quad\quad NO \quad\quad\quad ON \end{bmatrix}^{4+}(IV) + 2NH_3 \tag{3}$$

The O-O bond cleavage of complex (IV) will produce two mole of nitroxyl complex,

$$\begin{bmatrix} (NH_3)_4Co-O \vdots O-Co(NH_3)_4 \\ \quad\quad | \quad\quad\quad\quad\quad | \\ \quad\quad NO \quad\quad\quad\quad ON \end{bmatrix}^{4+} \xrightarrow{+2NH_3} 2[Co(NH_3)_5NO_2]^{2+}(V) \tag{4}$$

which will regenerate the initial complex,(I) in the excess of ammonia, to produce ammonium nitrite and nitrate, and thus make repeat of the above mentioned cascade reactions possible.

$$2(V) \xrightarrow{+4NH_3+H_2O} 2(I) + NH_4NO_2 + NH_4NO_3 \tag{5}$$

This mechanism implies that the presence of O_2 will reinforce the NO absorption performance by the aqueous $Co(NH_3)_6^{2+}$ complex, and the total reaction for NO control is given by,

$$2NO + O_2 + 2NH_3 + H_2O = NH_4NO_2 + NH_4NO_3 \tag{6}$$

Certainly, this is a green chemical reaction with a 100% atom utility, which recovers the waste NO and O_2 to produce a valuable fertilizer with the help of ammonia.

The above proposed process can be expected to easily put into practice as ammonia is abuandant as the main feedstock for fertilizer. Nevertheless, there is also a problem that $Co(NH_3)_6^{2+}$ is apt to be oxidized to $Co(NH_3)_6^{3+}$ which is unable to form the peroxo binuclear complex and ineffective to O_2 solubility enhancement, thus reaction (4) is inhibited. But Co^{2+} will be relatively stable, and Co^{3+} may be reduced to Co^{2+} by H_2O [12]. As a result, a regenration method has also been proposed by using the activated carbon as the catalyst[7], in which $Co(NH_3)_6^{3+}$ dissociation into Co^{3+} and NH_3 occurs on the activated carbon surface followed by reduction of Co^{3+} with H_2O into Co^{2+}, O_2 and H^+.

EXPERIMENTAL

Experiments were performed in a packed column with 1000 mm in length and 18 mm in diameter. The absorber was operated with a continuous influent gas feed rate of 0.2L/min from the bottom and a continuous scrubbing solution fed at a superficial flow rate of 13 m^3/m^2.hr at the top, as described in [8]

Co^{2+} concentration was determined by spectrophotometer (Varian Cary 500) at 692 nm wave length, with the sample diluted with a 9 mol/L concentrated HCl solution. NO content in gas phase was obtained by an on-line Fourier transform infrared spectrometer (Nicolet E.S.P. 460 FT-IR) equipped with a gas cell and a quantitative package, Quant Pad.

RESULTS AND DISCUSSION

In order to confirm the proposed mechanism described above, in which O_2 may have a positive effect on NO absorption, the comparative experiments have been carried out. The results are shown in Fig. 1, from which one can see that the presence of O_2 will greatly improve the NO removal performance. In the absence of O_2, NO coordination occurs according to Eq. (2), a reversible reaction limited by equilibrium, the NO removal decreases from the initial 100% to about 60% in one hour. In the presence of O_2 however, contribution of Eq. (2) is little, the most coordination of NO is certainly attributed to the cascade reactions from Eq.(3) to Eq.(6), and the final reaction of Eq. (7), which will not be constrained by the reaction equilibrium, and thus the NO removal can be maintained 100% in 2-3 hours.

On the other hand, it can be inferred from Fig. 1 that the NO removal efficiency will decrease as the reaction proceeds. The NO removal declines from 100% at the very beginning to 93% after 4 hours with 5.2% O_2 present in the gas phase. This phenomenon may be due to the oxidation of $Co(NH_3)_6^{2+}$ into $Co(NH_3)_6^{2+}$, meaning deactivation of the catalyst.

Fig.2 shows the experimental results of $Co(NH_3)_6^{3+}$ reduction, which is done in a stirred reactor at 80 °C and pH 4.1. It can be seen that activated carbon can promote the $[Co(NH_3)_6]^{3+}$ reduction significantly. The $[Co(NH_3)_6]^{3+}$ conversion reaches 81.2% with6.7g/l activated carbon in the aqueous solution while only 8.18% of $[Co(NH_3)_6]^{3+}$ is reduced when there is no activated carbon added. Thus, a regeneration column packed with the activated carbon should be equipped with the absorber.

Fig.3 illustrates the NO removal performance in the absorber-regenrator set-up. It was operated without regeneration of the absorbent solution in the early 11 hours, and then the regeneration was started up. The absorption temperature is 50°C, O_2 composition is 5.2% (vol), NO content is 748 ppmv, and the $[Co(NH_3)_6]^{2+}$ concentration is 0.02mol/l.The flow rate of scrubbing solution fed into the regeneration column was 25ml/min, and the initial regeneration temperature was 50°C. The system had an initial NO removal efficiency of100%, which declines to about 44.7% after 16h. When regeneration was switched on at 50°C, NO removal rises to 54.6%, and continuously to 86.3% when the regeneration temperature is increased to 70-80°C. This test shows that the regeneration method works very well.

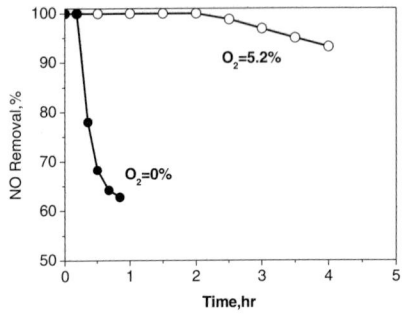

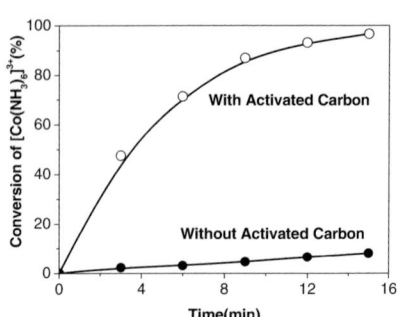

Figure 1. Effect of O_2 on NO removal
(0.01M $Co(NH_3)_6^{2+}$, 50℃, NO=640ppm)

Figure 2. Effect of activated carbon on reduction
of $Co(NH_3)_6^{3+}$, (pH=4.1, t=80℃)

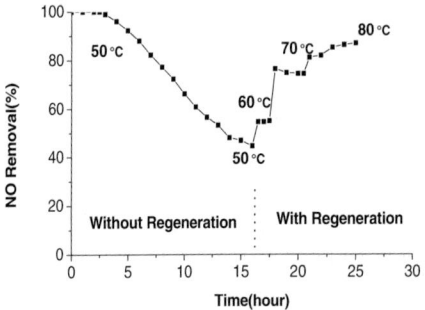

Figure 3 NO removal of the laboratory
absorber-regenrator

Conclusions

$Co(NH_3)_6^{2+}$ is effective for NO absorption, and O_2 can enhance NO absorption as the result of oxidation reaction. Activated carbon is also active to maintain stability of NO absorption by turning the side reaction of Co^{2+} complex oxidized into Co^{3+} complex backward, and the suitable regeneration conditions are obtained. A laboratory packed absorber- regenerator setup show a NO removal of more than 85% can be maitained constant.

REFERENCES

[1] W. D., Xiao, CNP 02136906(2002)

[2] C.L. Yang, H. Shaw, Environ. Progress, 17(1998) 80.

[3] D. Thomad, J. Vanderschuren, Chem. Eng. Sci., 51(1996)2649.

[4] E.K. Pham, S.G. Chang, Nature, 369(1994)139.

[5] E. Sada, H. Kumazawa, H. Hikosada, Ind. Eng. Chem. Fundam., 25(1986)238

[6] X. L. Long, Ph. D. Dissertation, East China Univ. of Sci. and Technol.,2002.

[7] Z. L. Xin, Master Thesis, East China Univ. of Sci. and Techn., 2003

[8] X. L. Long, W. D. Xiao, W.K. Yuan, Ind. Eng. Chem. Res., 43(2004)4048

[9] X. L. Long, W. D. Xiao, W.K. Yuan, Chemosphere, 59(2005)811

[10] S. Xiang, X. Yan, L. Cao, Inorganic Chemistry Series(4), Sci. Techn. Press, Beijing, 1998.

[11] M. Mori, J. A. Weil, M., Ishiguro, J. Ameri. Chem. Soc., 90(1969)615.

[12] A. Dai, Inorganic Chemistry Series(12) Sci. Techn. Press, Beijing, 1987.

Studies in Surface Science and Catalysis, volume 159
Hyun-Ku Rhee, In-Sik Nam and Jong Moon Park (Editors)

Production of TFE by Catalytic Pyrolysis of Chlorodifluoro-methane (CHClF$_2$)

Dae Jin Sung[a], Dong Ju Moon[b*], Joonho Kim[a], Sangjin Moon[a] and Suk-In Hong[a*]

[a]Department of Chemical & Biological Engineering, Korea University, 5-ka Anam-dong, Sungbuk-ku, Seoul 136-701, Korea (*sihong@korea.ac.kr)

[b]Hydrogen Energy Research Center, Korea Institute of Science and Technology, 39-1 Hawolgok-dong, Sungbuk-ku, Seoul 136-791, Korea (*djmoon@kist.re.kr)

Abstract

The catalytic pyrolysis of R22 over metal fluoride catalysts was studied at 923K. The catalytic activities over the prepared catalysts were compared with those of a non-catalytic reaction and the changes of product distribution with time-on-stream (TOS) were investigated. The physical mixture catalysts showed the highest selectivity and yield for TFE. It was found that the specific patterns of selectivity with TOS are probably due to the modification of catalyst surface. Product profiles suggest that the secondary reaction of intermediate CF$_2$ with HF leads to the formation of R23.

1. INTRODUCTION

Chlorodifluoromethane (CHClF$_2$, R22) has been used as a raw material for the manufacturing of fluorinated compounds such as tetrafluoroethylene (C$_2$F$_4$, TFE) and hexafluoropyropylene (C$_3$F$_6$, HFP). TFE has been used as a monomer for manufacturing fluorinated compounds such as polytetrafluoroethylene (PTFE) and Teflon. The most widely used process for the commercial production of TFE is the direct pyrolysis of R22 [1]. To improve the relatively poor yield of TFE, especially at high reaction temperature, researchers introduced catalysts [2, 3]. However, the significant increase of TFE yield has not yet been reported, and little work has been done on the catalytic pyrolysis of R22 using catalysts.

In our previous work [4, 5], results on the catalytic pyrolysis of R22 over Cu-promoted catalysts were reported. In this work, various metal fluoride catalysts were introduced to improve the relatively poor yield of TFE.

2. EXPERIMENTAL

Aluminum fluoride (AlF$_3$), calcium fluoride (CaF$_2$) and their physical mixture (denoted as "Mixed" hereafter for abbreviation) were prepared. The amount of copper was adjusted to be 10 wt.%. The reaction was carried out in the fixed-bed reaction system under the reaction temperature of 923 K, space velocity of 15,000 h^{-1}, 10% R22 in N$_2$-balacne and atmospheric pressure. Acid scrubber contained 750 cc of 0.5 M NaOH solution was used for the analysis of halogen halides such as HCl and HF formed during the pyrolysis of R22.

3. RESULTS AND DISCUSSION

3.1. Comparison of catalytic pyrolysis with non-catalytic pyrolysis

Table 1 shows the product distributions for the catalytic pyrolysis of R22 over various metal fluoride catalysts after the reaction at 923 K for 200 min. TFE was a major product and byproducts such as trifluoromethane (CHF_3, R23) and trifluoroethylene (CF_2CHF, TrFE) were observed during the reaction. The conversion of R22 over metal fluoride catalysts such as AlF_3, CaF_2 and Mixed was much higher than that in the non-catalytic pyrolysis under the same reaction conditions, while these catalysts showed low selectivity for TFE and even lower than that in the non-catalytic case. However, the TFE yield in the catalytic pyrolysis was much higher than that in the non-catalytic case. It was found that Mixed catalyst showed the highest TFE yield of 28.7% among the metal fluoride catalysts after reaction for 200 min under the tested conditions.

Table 1. Product distribution for the pyrolysis of R22 over prepared catalysts

Catalyst	R22 conversion (%)	Selectivity (%)			Y_{TFE} (%)	F (wt.%)
		TFE	R23	TrFE		
Non-catalytic	25.6	87.6	4.4	-	22.4	
AlF_3	36.4	75.2	9.9	8.5	27.4	
CaF_2	28.8	83.6	2.6	6.6	24.1	
Mixed	33.9	84.5	2.3	2.6	28.7	
$Cu-AlF_3$	29.9	86.5	4.2	4.6	25.8	0.94
$Cu-CaF_2$	28.6	93.1	1.7	-	26.6	0.77
Cu-Mixed	28.4	96.8	1.6	-	27.5	-1.36

The conversion of R22 over Cu-promoted catalysts was higher than that in the non-catalytic pyrolysis. The Cu-promoted catalyst was also more selective for TFE than Cu-unpromoted catalysts. This result indicates that the addition of Cu increases the selectivity for TFE. There were no XRD peaks corresponding to carbon, chlorine and Cu after the reaction. The Cu content in the $Cu-AlF_3$ and $Cu-CaF_2$ catalysts before and after the reaction decreased from 7.6 and 8.2 wt.% to 0.1 and 0.2 wt.%, respectively. The decrease of Cu content after the reaction can be explained by the metal sintering at relatively high temperature and/or the formation of coke on the catalyst surface. If high selectivity for TFE over Cu-promoted catalysts were due to the catalytic effect of Cu loading, TFE selectivity would be expected to decrease as the surface is progressively covered by carbon formed during the pyrolysis. However, the selectivity for TFE over the Cu-promoted catalysts increased with TOS and was higher than that over the Cu-unpromoted catalysts after the reaction for 200 min. Therefore, it was considered that the enhanced TFE yield due to adding the Cu can be explained by the enhanced heat transfer from the heat source or tube outside to gaseous reactant via the inside catalyst particles, resulting in increasing the overall rate of pyrolysis reaction.

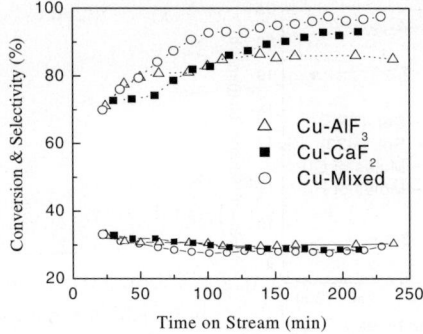

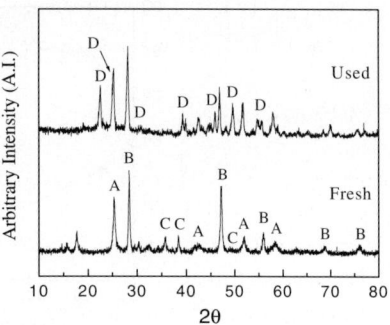

Fig. 1. Change of R22 conversion and selectivity for TFE with TOS (solid line: R22 conversion; dot line: selectivity for TFE).

Fig. 2. XRD of Cu-Mixed catalyst before and after reaction (A: AlF$_3$; B: CaF$_2$; C: CuO; D: CaAlF$_5$).

3.2. Effect of time-on-stream on the product distribution

The change of R22 conversion and selectivity for TFE with TOS over the Cu-promoted catalysts are shown in Fig. 1. The conversion of R22 slightly decreased at initial TOS and reached at steady state conversion of about 29%. However, the selectivity for TFE increased with increasing TOS. Among the Cu-promoted catalysts, Cu-Mixed catalyst exhibited the highest selectivity for TFE. These changes in conversion and selectivity with TOS might be attributed to the modification of solid surface during the pyrolysis of R22. As shown in Fig. 2, it was found that new peaks corresponding to CaAlF$_5$ over the Cu-Mixed catalyst are detected after the reaction. The average amount of F atoms adsorbed during the pyrolysis can be estimated from the EPMA analysis at various points after reaction (Table 1). XRD shows that the used catalysts were completely transformed into the metal fluoride catalysts. Therefore, the amount of F atoms adsorbed during the pyrolysis could be estimated from the difference between F amount calculated by the stoichiometric ratio of AlF$_3$ or CaF$_2$ and F amount measured by EPMA analysis. It was found that the bimetallic Cu-Mixed catalyst exhibited the lowest and negative average amount of adsorbed F atoms, which indicates the formation of substoichio-metric metal fluorides such as AlF$_x$ (x<3), CaF$_x$ (x<2) and CaAlF$_x$ (x<5). It was reported that the more F atoms were adsorbed on the catalyst surface, the higher the probability of interaction of intermediate difluorocarbene (:CF$_2$) with adsorbed F and H atoms, resulting in the higher selectivity for R23 [6]. It is concluded that the increase of the selectivity and yield for TFE over Cu-Mixed catalyst may be attributed to the surface modifications such as the formation of the bimetallic fluoride and the variation of F content on the surface of catalyst by the attack of HF produced during the pyrolysis of R22.

The effect of TOS on the product distribution during the pyrolysis of R22 over Cu-AlF$_3$ catalyst is shown in Fig. 3. The amount of halogen ion trapped in NaOH solution was determined by IC. The concentration of Cl formed during the pyrolysis of R22 was higher than the concentration of F at all TOS. This result is a consequence of the facile cleavage of the C-Cl bond in comparison to the C-F bond. Bond dissociation energy for the C-element of R22 is followed by the order: C-Cl<C-H<C-F.

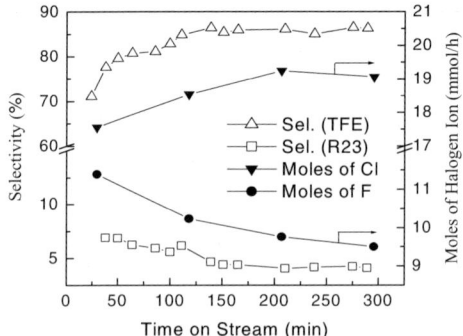

Fig. 3. Product distribution with TOS during the pyrolysis of R22 over Cu-AlF$_3$ catalyst.

It is considered that the increase of the amount of Cl with TOS is closely related with the increase of the TFE selectivity during the reaction, and that the decrease of amount of F with TOS is related with the decrease of R23 selectivity. At initial TOS, the low concentration of Cl is due to the low rate of formation of difluorocarbene(:CF$_2$), resulting in the low selectivity of TFE, while the high concentration of F may be relevant to the high probability of reaction between :CF$_2$ and HF produced by C-F bond dissociation, resulting in the high selectivity of R23. These results suggest that :CF$_2$ formed during the pyrolysis plays an important role in the formation of R23 as well as TFE. Thus, it is concluded that the coupling or dimerization of :CF$_2$ with another :CF$_2$ on the catalyst surface and/or in homogeneous gas-phase leads to the formation of TFE, and that the secondary reaction of :CF$_2$ with HF produced by the decomposition of R22 leads to the formation of R23. The possible reaction pathway for the production of TFE and R23 during the pyrolysis of R22 can be written as follows:

$$CHClF_2 \rightleftharpoons :CF_2 + HCl \tag{1}$$
$$:CF_2 + :CF_2 \rightleftharpoons C_2F_4 \tag{2}$$
$$:CF_2 + HF \rightleftharpoons CHF_3 \tag{3}$$

4. CONCLUSIONS

It was concluded that the enhanced selectivity and yield for TFE over Cu-Mixed catalyst may be attributed to the surface modifications by the attack of HF produced during the pyrolysis of R22. The results suggest that R23 is formed by the secondary reaction between intermediate CF$_2$ and HF.

REFERENCES

[1] E. Broyer, A. Y. Bekker and A. B. Ritter, Ind. Eng. Chem. Res., 27 (1988) 208.
[2] K.S. Revell, Preparation of Tetrafluoroethylene, Brit. Patent No. 983 222 (1963).
[3] Daikin Kogyo, A Process for Manufacturing Tetrafluoroethylene, Brit. Patent No. 1061 377 (1964).
[4] D.J. Sung, D.J. Moon, S. Moon, J. Kim and S.I. Hong, Appl. Catal. A 292 (2005) 130.
[5] D.J. Sung, D.J. Moon, Y.J. Lee and S.I. Hong, Int. J. Chem. React. Eng., 2 (2004) A6.
[6] P.P. Kulkarni, S.S. Deshmukh, V.I. Kovalchuk, J.L. d'Itri, Catal. Lett., 61 (1999) 161.

Studies in Surface Science and Catalysis, volume 159
Hyun-Ku Rhee, In-Sik Nam and Jong Moon Park (Editors)

237

Hydrothermal synthesis of titanium dioxides using acidic and basic peptizing agents and their photocatalytic activity on the decomposition of orange II

Jun Ho Kim[a], Gun-Dae Lee[a], Seong Soo Park[a] and Seong-Soo Hong[a]*

[a]Division of Applied Chemical Engineering, Pukyong National University,,100 Yongdang-dong, Nam-ku, Busan, 608-739, Korea

TiO_2 nanoparticles were prepared using the hydrolysis of titanium tetraisopropoxide (TTIP) using HNO_3 and TENOH as peptizing agents in the hydrothermal method. The photocatalytic degradation of orange II has been studied using a batch reactor in the presence of UV light. When the molar ratio of HNO_3/TTIP is 1.0, the rutile phase appeared on the titania and the photocatalytic activity decreased with an increase of HNO_3 concentration. However, no rutile phase was shown even though the molar ratio of TENOH/TTIP increased up to 1.0. The titania particle prepared at TENOH/TTIP molar ratio=0.1 shows the highest activity on the photocatalytic decomposition of orange II.and the photocatalytic activity decreases according to an increase in TENOH/TTIP molar ratio.

1. INTRODUCTION

Nanocrystalline powders of titania continues to attract much interest because of its wide variety of applications, such as, optical devices, pigment, photocatalyst, etc.[1]. The synthesis of nanocrystalline particles with controlled size and composition is of technological importance because they have more active sites for achieving enhanced performance. Hydrothermal synthesis, in which chemical reactions can take place in aqueous or organo-aqueous media under simultaneous generation of pressure upon heating, has been used to prepare nano-crystalline titania at low temperatures[2]. Hydrothermal synthesis, in which chemical reactions can take place in aqueous or organo-aqueous media under simultaneous generation of pressure upon heating, has been used to prepare nano-crystalline titania at low temperatures. High crystalline anatase or rutile nano-particles were synthesized by hydrothermal treating TENOH or HNO_3 peptized titania sols, respectively[3]. The peptizing agents have an influence on the physical properties of titania, such as, particle size, shape, and the ratio of anatase to rutile phase.

In this paper, we prepared nanosized TiO_2 particles by hydrolysis of TTIP (titanium tetraisopropoxide) using HNO_3 and tetraethyl ammonium hydroxide(TENOH) as peptizers in the hydrothermal method. The physical properties of prepared nanosized TiO_2 particles were investigated. We also investigated the effect of peptizing agents on the physical properties of nanosized TiO_2 particles, and examined the activity of TiO_2 particles as a photocatalyst for the decomposition of orange II.

---

* To whom correspondence should be addressed

E-mail: sshong@pknu.ac.kr

2. EXPERIMENTAL

TiO$_2$ precipitates were obtained by adding 0.5 M isopropanol solution of the TTIP dropwise into deionized water and rigorously stirred for 30 min. The precipitates were washed with deionized water using a centrifuge. Portions of the white precipitates were peptized by adding diluted solutions of HNO$_3$(HNO$_3$/ TTIP=0.2-2.0) or tetraethyl ammonium hydroxide(TENOH/TTIP=0.1-1.0) at room temperature.

This solution of 100 ml was transferred to a 250 ml Teflon container held in a stainless-steel vessel. After the vessel was tightly sealed, it was heated at $120 \sim 200\,^{\circ}\text{C}$ for 5 h. After hydrothermal treatment, the TiO$_2$ particles were separated in a centrifuge at 10,000 rpm for 10 min and were then washed in distilled water. The particles were dried at $105\,^{\circ}\text{C}$ for 12 h and were then calcined at $300 \sim 700\,^{\circ}\text{C}$ for 3 h.

The major phase of the obtained particles was analyzed by X-ray diffraction (Rigaku D/MAXIIC) using Cu-Kα radiation.

A biannular quartz glass reactor with a lamp immersed in the inner part of the reactor was used for all the photocatalytic experiments. The batch reactor was filled with 500 ml of an aqueous dispersion in which the concentration of titania and of orange II were 1.0 g/L and 100 mg/L, respectively and magnetically stirred to maintain uniformity both concentration and temperature. A 500W high-pressure mercury lamp (Kumkang Co.) was used. The circulation of water in the quartz glass tube between the reactor and the lamp allowed the lamp to stay cool and to warm the reactor to the desired temperature. Nitrogen was used as a carrier gas and pure oxygen was used as an oxidant. The samples were immediately centrifuged and the quantitative determination of orange II was performed by a UV-vis spectro- photometer (Shimazu UV-240).

3. RESULTS AND DISCUSSION

Fig. 1 shows the XRD patterns of the TiO$_2$ particles prepared using different concentration of nitric acid(a) and TENOH(b). These particles are only dried at $105\,^{\circ}\text{C}$ without any calcination.

The evolution(Fig. 1(a)) of the intensities for the (101) reflection of anatase and the (110) reflection of rutile, is function of HNO$_3$ concentration for different solutions. It can be observed that a rutile phase appears from HNO$_3$/TTIP=1.0. According to Gopal et al. [4], the formation of anatase and rutile TiO$_2$ is determined by the nucleation and the growth of TiO$_2$ clusters. If the condensation starts before completion of hydrolysis, either amorphous or metastable anatase TiO$_2$ will form. Acid environment promotes hydrolysis rate and at the same time decreases the condensation rate. In the strong acid solution, the condensation rate is very slow enough and the formation of rutile phase is kinetically favored, so rutile TiO$_2$ can be obtained at low temperature.

However, no rutile phase is shown even though the molar ratio of TENOH/TTIP increased up to 1.0(Fig. 1(b)). In addition, the anatase particles with a high degree of crystallinity are formed compared to the HNO$_3$-peptized particles. It is thought that the existence of OH- groups in the peptized sols can enhance the growth of anatase particles.

The crystallite size of the particles prepared at different HNO$_3$ and TENOH concentration can be determined by the Scherrers equation [5] and is listed in Table 1. The titania particles prepared using

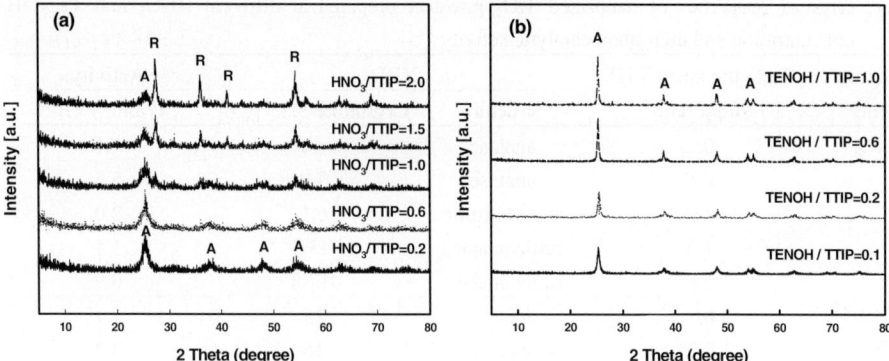

Fig. 1. XRD pattern of nanosized TiO$_2$ powders prepared at different HNO$_3$ concentration(a) and TENOH concentration(b); synthesized at 180℃(a) and 160℃(b), dried at 105℃.

HNO$_3$ as a peptizing agent shows the smaller size compared to those prepared using TENOH as a peptizing agent. The crystallite size of the anatase phase shows similar value from HNO$_3$/TTIP=0.2 to 1.0 but that of rutile phase increases with an increase of HNO$_3$ concentration. One can see that the crystallite size of the anatase phase is increased from 15 to 30 nm as the molar ratio of TENOH/TTIP increases from 0.1 to 1.0.

It is well known that photocatalytic oxidation of organic pollutants follows Langmuir-Hinshelwood kinetics[6]. Therefore, this kind of reaction can be represented as follows.

$$-dc/dt = kC \qquad (1)$$

In addition, it can be integrated as follows.

$$C = C_o \exp(-kt) \qquad (2)$$

Where C_o is initial concentration of the orange II and k is a rate constant related to the reaction properties of the solute which depends on the reaction conditions, such as reaction temperature, pH of solution. The photocatalytic activity increases with an increase in this value.

When blank test in the absence of TiO$_2$ photocatalyst was carried out, orange II was decomposed to about 18% after 3 h reaction by photolysis reaction. However, the presence of TiO$_2$ prepared at HNO$_3$/TTIP molar ratio=0.2 decomposed completely orange II within 1 h. The photocatalytic activity for the decomposition of orange II on the titania particles prepared at different HNO$_3$/TTIP and TENOH/ TTIP molar ratio was examined and the result is shown in Table 1. The titania particle prepared at HNO$_3$/TTIP molar ratio=0.2 shows the highest activity on the photocatalytic decomposition of orange II. However, the photocatalytic activity steeply decreases on the titania particles prepared at HNO$_3$/TTIP molar ratio=1.0. This result suggests that the photocatalytic activity of the decomposition of orange II depends on crystal structure. Titanium dioxide can take on any of the following three crystal structures: rutile, anatase, or brookite. An anatase-type titanium dioxide generally exhibits a higher photocatalytic activity than the other types of titanium dioxide in regard to the decomposition of

Table 1. Physical properties of nanosized TiO_2 powders prepared at different HNO_3 and TENOH concentration and their photocatalytic activity

Peptizing agents	Peptizing agent/TTIP Molar ratio[a]	XRD		Activity
		Structure	Crystallite Size[b] [nm]	$^{c}k'$(min^{-1}) x10^{-2}
HNO_3	0.2	anatase	6.5	7.1
	0.6	anatase	7.0	5.5
	1.0	anatase/rutile	6.1	2.0
	1.5	rutile/anatase	-/13.3^{d}	1.1
	2.0	rutile/anatase	-/17.2^{d}	0.9
TENOH	0.1	anatase	15	7.1
	0.2	anatase	16	4.7
	0.6	anatase	24	2.9
	1.0	anatase	30	2.2

[a]: R ratio(H_2O/TTIP)=150, synthesis temperature=180 ℃ (HNO_3) and 160 ℃ (TENOH), dried at 105 ℃.

[b]: obtained by Scherrer equation.

[c]: apparent first-order constants(k') of orange II.

[d]: rutile structure

organic pollutants by suppressing the electron-hole recombination[2]. It is thought that the photocatalytic activity decreases on the titania
particles prepared at HNO_3/TTIP molar ratio>1.0 owing to the formation of rutile phase of titania.

The titania particle prepared at TENOH/TTIP molar ratio=0.1 shows the highest activity on the photocatalytic decomposition of orange II. However, the photocatalytic activity gradually decreases with an increasing TENOH/TTIP molar ratio. This result suggests that the photocatalytic activity of the decomposition of orange II depends on the particle size. That is, the activity increases with a decreasing particle size. It is consistent with the result that the photocatalytic reaction has a small particle-size effect, wherein the photocatalytic activity increases with a decrease of particle size. It can be also confirmed that a small particle has a large illuminated surface area by the reason that the particles have a constant density at the same structure.

ACKNOWLEDGEMENTS
This work was supported by a grant of the Brain Busan 21 Project in 2003.

REFERENCES
[1] A. Wold, *Chem. Mater.*, 5, 280 (1993).

[2] Y. B. Ryu, M. S. Lee, S. S. Park, G.-D. Lee and S. S. Hong, *React. Kin. & Catal. Lett.*, 84, 101 (2005).

[3] J.Yang, S. Mei, J.M.F. Ferreira, *Mater. Sci. Eng., C,* 15, 183 (2001)

[4] M. Gopal, W.J. Mobey Chan, L.C. Jonghe, *J. Mater. Sci.*, 32, 6001 (1997).

[5] B.D. Cullity, *"Elements of X-Ray Diffraction"*, 2nd edn., Addison-Wesley, Reading, MA 1978, 102.

[6] C. S. Turchi, D. F. Ollis, *J. Catal.*, 122, 178 (1990).

Studies in Surface Science and Catalysis, volume 159
Hyun-Ku Rhee, In-Sik Nam and Jong Moon Park (Editors)

241

Photocatalytic decomposition of ethylene in a wire-net reactor

Yasushi Sugawara[a], Katsuyasu Sugawara[b] and Takuo Sugawara[b]

[a]Industrial Material Group, Akita Prefectural Industrial Technology Center,
4-11 Sanuki, Arayamachi, Akita, Japan
[b]Dept. of Materials-process Engineering and Applied Chemistry for Environments,
Akita University, 1-1 Gakuenmachi, Tegata, Akita, Japan

1. Introduction

We have developed a compact photocatalytic reactor [1], which enables efficient decomposition of organic carbons in a gas or a liquid phase, incorporating a flexible and light-dispersive wire-net coated with titanium dioxide. Ethylene was selected as a model compound which would rot plants in sealed space when emitted. Effects of the titanium dioxide loading, the ethylene concentration, and the humidity were examined in batches. Kinetic analysis elucidated that the surface reaction of adsorbed ethylene could be regarded as a controlling step under the experimental conditions studied, assuming the competitive adsorption of ethylene and water molecules on the same active site.

2. Experimental

A flexible wire-net (mesh-size of 20 or 350 mesh) photocatalyst, of which length was 450mm and width 390mm, was prepared by using commercial wire-net and commercial titanium dioxide coating suspension (TAYCA Corporation, TKC-304; The crystal form is anatase and the particle size is around 6nm.). First, the wire-net was washed by ethanol, and dipped in titanium dioxide coating suspension for 20seconds. Then it was dried at 110℃ for 20minutes. Titanium dioxide loading on a wire-net photocatalyst was controlled by the number of cycles of dipping and drying.The photodegradation experiments were performed batch-wise.The schematic diagram of experimental apparatus is shown in Fig.1. The whole configuration consists of a light source (Toshiba, FL20S BLB), a spiral-type wire-net photocatalyst, and two glass tubes for supply of ethylene and two silicone corks for gas-sealing. Total volume of batch reactor is 2.2L. Ethylene was first supplied in the reactor, and 20minutes later, a black light lamp (main spectrum around 352nm) turned on. Ethylene concentration was measured with a GC (Shimazu Model GC-15A) equipped with a flame ionization detector, by using N_2 gas as a carrier gas. Experimental conditions are shown in Table1.The effect of humidity for decomposition of ethylene was also studied.The moisture content was varied from 10,200 to 28,300 ppm under three concentration levels of ethylene (30, 50, and 100 ppm).

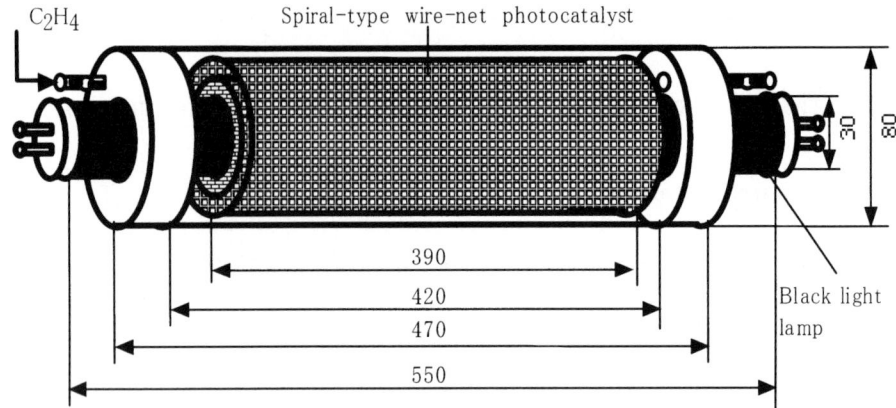

Fig.1. Schematic diagram of experimental apparatus

3. Results and discussion

Fig.2 shows an effect of initial irradiation cycle on decomposition performance of ethylene over a 20mesh wire-net photocatalyst with titanium dioxide loading of 1.88g. In the first cycle, ethylene of 50ppm was completely decomposed after irradiation with a black light lamp for 70minutes. As ethylene of 50ppm was prepared from a standard ethylene cylinder (99.5%) with adding indoor air (oxygen 21%), it was clear that ethylene was decomposed completely by the wire-net photocatalyst in the presence of oxygen. Initial activity change was examined with a newly prepared wire-net photocatalyst, and it was found that time for complete conversion decreased with a repeated irradiation cycle, becoming constant after a few cycles.This result suggests that some photo-inhibitive compounds in the coating suspension adsorbed on a wire-net photocatalyst were removed after several irradiation cycles with a black light lamp; as a result, time for decomposition of ethylene decreased.

Fig.3 shows an effect of titanium dioxide loading on the decomposition of ethylene over a 20mesh wire-net photocatalyst. The decomposition rate increased in proportion to the amount of titanium dioxide loaded: The specific initial rate ($r_0=(-dc/dt)_{t=0}$) was $(1.8\pm0.2)\times10^{-2}$ ppm s^{-1}g-catalyst^{-1}.

Table 1   Experimental conditions

Mesh size of wire net	Titanium dioxide loading(g) (Specific weight (g/m^2))		Initial concentration of ethylene(ppm)
20	0.92	(3.5)	52
20	1.88	(7.2)	35, 50, 107
20	2.70	(10.4)	50
350	1.88	(5.1)	35, 52, 105

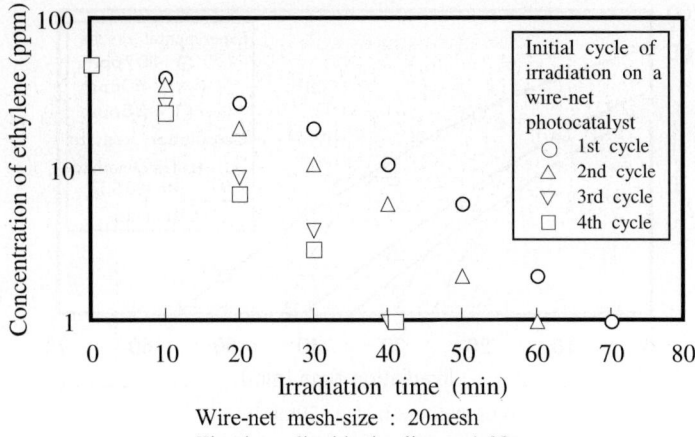

Fig.2. Effect of initial irradiation cycle on decomposition performance of ethylene

on a wire-net photocatalyst

The photocatalytic activity of 20mesh wire-net photocatalyst was observed to be nearly equal to that of 350mesh one under the same amount of titanium dioxide loading (1.88 g).

Fig.4 shows an effect of initial concentration of ethylene on decomposition rate over a 20mesh wire-net photocatalyst with titanium dioxide loading of 1.88g. When ethylene of 107ppm was decomposed, decomposition occurred rapidly at an initial stage and then became slow gradually as the case of 50ppm loading. When concentration was 35ppm, rapid decomposition occurred for 20 minutes, and then became slow. Calculated results based on

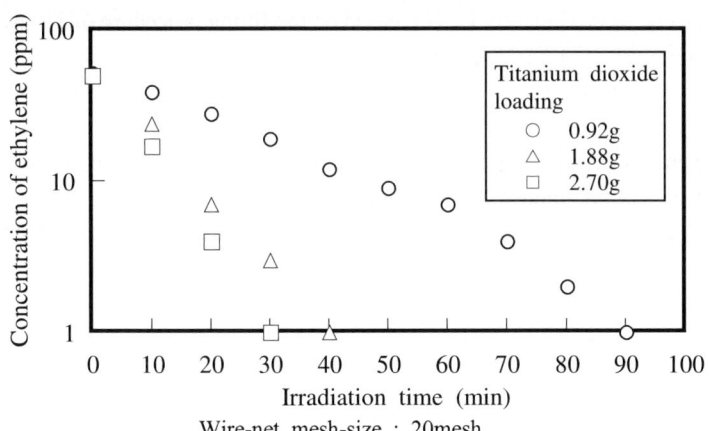

Fig.3. Effect of titanium dioxide loading on decomposition of ethylene

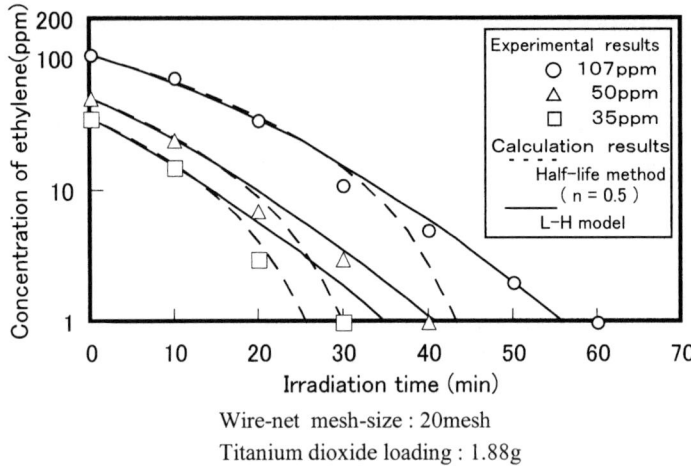

Wire-net mesh-size : 20mesh
Titanium dioxide loading : 1.88g
Humidity : 17000ppm ~ 19500ppm

Fig.4. Effect of initial concentration of ethylene on decomposition rate

the half-life method by which the order of reaction was determined to be the 0.5th to the ethylene concentration are drawn in the figure as broken lines.

Here, we found that the lower the humidity, the shorter the decomposition time under the various degrees of humidity from 10,200 to 28,300ppm.

Kinetic analysis based on the Langmuir-Hinshelwood model was performed on the assumption that ethylene and water vapor molecules were adsorbed on the same active site competitively [2]. We assumed then that overall photocatalytic decomposition rate was controlled by the surface reaction of adsorbed ethylene. Under the water vapor concentration from 10,200 to 28,300ppm, and the ethylene concentration from 30 to 100 ppm, the reaction rate equation can be represented by Eq.(1), based on the fitting procedure of $1/r$ vs. $1/C_{C2H4}$:

$$r = 3.70 \times \frac{3.18 \times 10^{-2} C_{C2H4}}{1 + 3.18 \times 10^{-2} C_{C2H4} + 4.95 \times 10^{-5} C_{H2O}} \qquad (1)$$

where C_{C2H4} (ppm) is concentration of ethylene, C_{H2O} (ppm) is concentration of water vapor, and r (ppm · min^{-1}) is decomposition rate of ethylene.

Calculated results are shown as solid lines in Fig.4 under the respective experimental conditions, and the assumption that the rate was controlled by the surface reaction of adsorbed ethylene was reasonable under the experimental conditions studied.

References

[1] Y. Sugawara, K. Asari, K. Sugawara and T. Sugawara, Kagaku Kogaku Ronbunshu, 29 (2003) 572.
[2] S. Yamazaki, S. Tanaka and H.Tsukamoto, J. Photochem. Photobiol. A Chem., 121 (1999) 55.

Studies in Surface Science and Catalysis, volume 159
Hyun-Ku Rhee, In-Sik Nam and Jong Moon Park (Editors)

Systematic assessment of alumina-supported cobalt-molybdenum nitride catalyst: Relationship between nitriding conditions, innate properties and CO hydrogenation activity

Y.J. Lee, L.J. Wong and A.A. Adesina*

Reactor Engineering and Technology Group, School of Chemical Engineering and Industrial Chemistry, University of New South Wales, Sydney, NSW Australia 2052
*Corresponding author: a.adesina@unsw.edu.au; fx: 61-2-9385-5966

1. INTRODUCTION

Recent work [1] has shown that nitrides of transition metals (Co, Mo, Fe and Ni) characterized by excellent mechanical strength and high thermal stability are also endowed with platinum-like electronic attributes as well as admirable anti-coking and poison-resistance properties. In this research, preparation of high surface area alumina supported Co-Mo nitride system has been investigated via temperature-programmed reaction [2] in order to find the relationship between nitriding conditions of oxide precursor and innate properties of the bimetallic nitrides and to secure optimal recipe for a highly active catalyst for the Fischer-Tropsch reaction. The general nitridation reaction may be written:

$$MO_y + aNH_3 \longrightarrow MN_{1-z} + yH_2O + bH_2 \qquad (1)$$

where M = Co or Mo, $0<z<1$, $a>0$, $y>0$ and b may be positive or negative depending on whether H_2 is produced or required (in the NH_3 feed stream) respectively. Thus, the $H_2:NH_3$ ratio in the nitriding gas is another preparation variable. Stoichiometrically, the associated $H_2:NH_3$ ratio is given by:

$$\left| \frac{b}{a} \right| = \left| \frac{\frac{3}{2}(1-z) - y}{(1-z)} \right| \qquad (2)$$

which for MoO_3 yields 4.5 for a fully nitrided ($z=0$) catalyst. Thus, the lower and upper limits of the $H_2:NH_3$ ratio in the nitriding gas were assigned 1:1 and 5:1 accordingly. Similarly, the limiting values for the nitriding temperature and time were informed by thermodynamic and morphological considerations as 773-973 K and 1-4 hr respectively.

2. EXPERIMENTAL PROCEDURE

Bimetallic Co-Mo oxide specimens were prepared via co-impregnation of calculated amounts of cobalt nitrate and ammonia heptamolybdate on γ-alumina to achieve a total metal loading of 20wt% with an equimolar Co:Mo ratio. Nitridation of catalysts was carried in a fixed bed reactor with a total mixture of 50 ml min^{-1} passed downwards into the reactor. A heating rate of 5 K min^{-1} was employed to bring the reactor to the desired nitridation temperature (773-973 K). Catalyst properties such as BET surface area, particle dispersion, acid site strength and concentration, surface elemental composition, bulk phase analysis were determined from N_2 adsorption at 77K, H_2 chemisorption, NH_3-TPD, H_2-TPR and X-Ray Diffraction (XRD) data. The progress of nitridation reaction was monitored thermogravimetrically (TGA). CO hydrogenation activity was measured in a fixed bed reactor using a feed containing H_2:CO ratio= 5 at 553 K and atmospheric pressure.

3. RESULTS AND DISCUSSIONS

Table 1 summarises the surface features for all eight catalysts. BET areas varied between 173-237 m^{2}g^{-1}. H_2-TPR carried out using 5 K min^{-1} indicated major H_2 consumption between 573-1123 K on all catalysts as shown in Fig. 1. The peaks in the lower temperature region located between 600-800 K correspond to the reduction of surface NH_X species or the oxynitride, while the higher temperature may be ascribed to the nitride conversion to the metallic phase [3]. Fig. 2 shows that the NH_3-TPD spectra are characterized by peak temperatures below 670 K suggesting the presence of weak Lewis acid sites on the catalyst surface. Although the catalyst profiles shown belong to samples 5 to 8, catalyst samples 1 to 4 also possess peak temperatures (< 670 K). However, there is a variation in heat of adsorption and acid site density with respect to nitridation conditions; an indication of surface heterogeneity. As may be seen in Fig. 3, catalysts nitrided at 973 K generally exhibit higher heat of adsorption for NH_3, typically between 60-100 kJ mol^{-1} while those prepared at 773 K have lower acid site strength ($-\Delta H_{des}$ < 40 kJ mol^{-1}). The corresponding surface acidity ranged from 170-190 μmol g^{-1} for high temperature catalysts but samples nitrided at 773 K possessed relatively lower acid site concentration. In addition, feed gas composition also influenced the surface acidity. Nitridation with H_2:NH_3=1 gave higher acid site concentration than those synthesized with H_2:NH_3=5 at the same temperature. XRD analysis in Fig 4 points to the presence of Co_3Mo_3N and γ-Mo_2N as the primary phases [4], although a strong signal for $Co_{5.47}N$ was evident for samples nitrided at 973 K suggesting further reduction at this temperature.

Table 1
BET and H_2-chemisorption parameters

Sample	Nitrid-ation Temp (K)	H_2:NH_3 ratio		Time (hr)	BET surface area (m^2 g^{-1})	Particle dispersion (%)	Active particle diameter (nm)
1	773	1	1	1	201	1.4	83
2	973	1	1	1	184	1.7	65
3	773	5	1	1	237	1.2	95
4	973	5	1	1	183	1.3	87
5	773	1	1	4	233	0.9	122
6	973	1	1	4	212	0.9	119
7	773	5	1	4	173	3.0	38
8	973	5	1	4	209	1.1	98

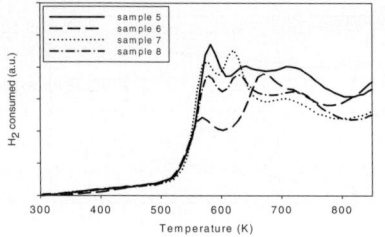

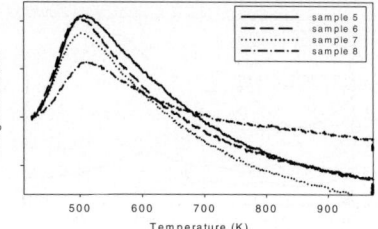

Figure 1. H$_2$-TPR profiles of catalysts

Figure 2. NH$_3$-TPD spectra of catalysts

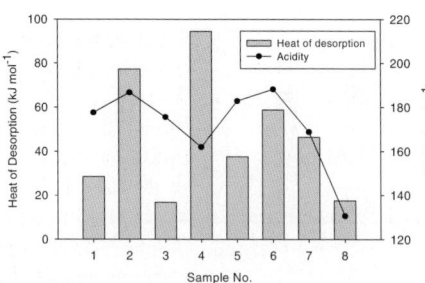

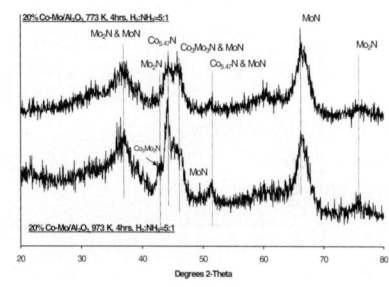

Figure 3. Parameters from NH$_3$-TPD

Figure 4. XRD analysis

Thermogravimetric data obtained from the ThermoCahn TGA (model 2121) were analysed to understand the solid-state nitridation kinetics. Fig. 5 which displays the thermal profiles for different heating rates indicates the presence of 2 discernible peaks under each condition with the exception of the thermogram for 2 K min^{-1}. The appearance of the peak shifted to the right (on the temperature-axis) with increased heating rate. The location of the peaks seemed to be linearly correlated with heating rate as seen in Fig. 6. The parallelism between the low and high temperature peaks suggests that similar solid-state mechanism was involved, most likely the substitutionary incorporation of N into the oxide structure. We believe that the nitridation is therefore a 2-step process involving the conversion of the oxide phase to an oxynitride phase which may then be subsequently transformed to the metal nitride. The existence of different phase compositions is therefore a reflection of the extent of oxynitride conversion under the particular nitridation conditions.

Consequently, additional runs were carried out using different H$_2$:NH$_3$ ratios isothermally at 773 K. The results shown in Fig. 7 clearly demonstrate the dependency of rate on NH$_3$ partial pressure as evidenced from the decrease in peak magnitude (and time of appearance) with increased H$_2$:NH$_3$ ratio. In fact, application of the power law expression;

$$r = kP_{NH_3}^a \tag{3}$$

to the data confirms the square-root dependency of rate on NH$_3$ partial pressure (a=0.45) with an associated pseudo-rate constant, k=0.88 g min^{-1} and correlation coefficient of 0.991. This implies that NH$_3$ was dissociated in its interaction with the metal oxide.

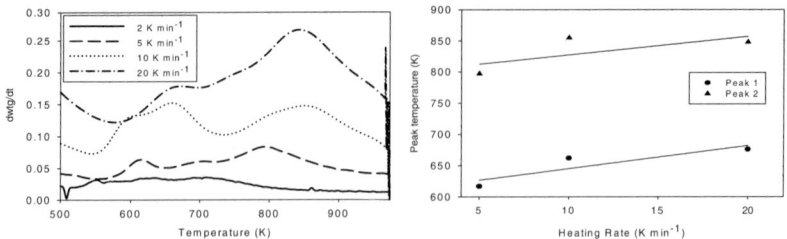

Figure 5. Nitridation at different heating rates   Figure 6. Peak temperature vs heating rate

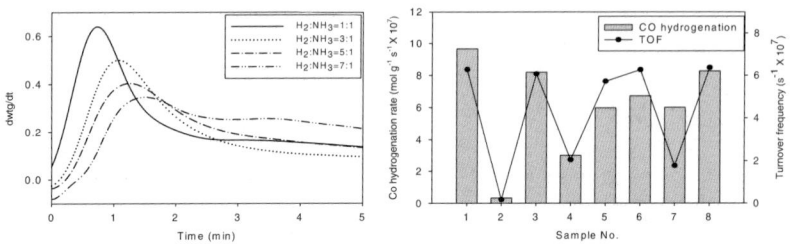

Figure 7. Nitridation at different $H_2:NH_3$ Figure 8. CO hydrogenation reaction rate

The activity of all catalysts were evaluated for the CO hydrogenation reaction. The histogram shown in Fig. 8 reveals that the bimetallic Co-Mo nitride system has appreciable hydrogenation activity with exception of samples 2 and 4. This apparent anomaly was probably due to the relatively high heat of adsorption for these two catalysts, which offered strong CO chemisorption but with unfavourable product release.

4. CONCLUSIONS

Physicochemical attributes of catalysts were mostly controlled by nitridation temperature although there was a little influence on catalyst reducibility and acidity, better nitride species were formed at 973 K and TGA results revealed that complete nitridation occurs between 750-973 K and the feed gas stream containing $H_2:NH_3=1:1$ is preferably better mixture for the nitridation of Co-Mo bimetallic catalysts.

ACKNOWLEDGEMENTS
The authors are grateful to the Australian Research Council for financial support of this investigation

REFERENCES
[1] H.K. Park, J. K. Lee, J.K. Yoo, E.S. Ko, D. S. Kim, K. L. Kim, Appl. Catal. A:Gen., 150 (1997) 21
[2] Y.J. Lee, T.H. Nguyen, A. Khodakov, A.A. Adesina, J. Mol. Catal. A: Chem., 211 (2004) 191
[3] M. Nagai, Y. Goto, H. Ishii, S. Omi, App. Catal. A: General., 192 (2000) 189
[4] K. Hada, J. Tanabe, S. Omi, M. Nagai, J. Catal. 207 (2002) 10

Studies in Surface Science and Catalysis, volume 159
Hyun-Ku Rhee, In-Sik Nam and Jong Moon Park (Editors)

A study on the SO_2 reduction by CO over SnO_2-ZrO_2 catalysts

G. B. Han, J. H. Jun, N. K. Park, J. D. Lee, C. H. Park, S. O. Ryu and T. J. Lee*

School of Chemical Engineering and Technology, Yeungnam University, 214-1 Dae-Dong, Gyeongsan-si, Gyeongsangbuk-do 712-749, Korea

FAX: +82-53-810-4631. E-mail: tjlee@yu.ac.kr

SO_2 which is an air pollutant causing acid rain and smog can be converted into elemental sulfur in direct sulfur recovery process (DSRP). SO_2 reduction was performed over catalyst. In this study, SnO_2-ZrO_2 catalysts with Sn/Zr mole ratio were prepared by a co-precipitation method and CO was used as reduction agent. The reactivity profile of SO_2 reduction was investigated at the various reaction conditions. SnO_2-ZrO_2 (Sn/Zr=2/1) catalyst showed the best performance for SO_2 reduction. As a result, SO_2 conversion and sulfur yield were about 100% and 97%, respectively, under the optimized conditions such as 325 ℃ and 10000 $cm^3/g_{-cat.}\cdot h$.

1. INTRODUCTION

The integrated gasification combined cycle (IGCC), in which a combustible gaseous fuel is produced by a gasification of almost any type of coal, is considered as an attractive technology to generate electricity from coal because of its higher thermal efficiency, lower pollution, and better economics than a conventional pulverized coal-fired power plant.

Since a coal gas exiting from a gasifier contains relatively large amount of sulfur in the form of hydrogen sulfide (H_2S), it is removed by a regenerable metal oxide sorbent. Under the reducing condition, the metal oxide sorbents are converted into metal sulfides during a sulfidation process. The metal sulfides are restored to their initial state, metal oxides, by reaction with oxygen and then the removed sulfur is subsequently converted into SO_2 during a regeneration process. Sulfur dioxide can be converted to element sulfur by using the reductants such as CO, H_2, CH_4 and carbon over the catalyst in the direct sulfur recovery process (DSRP).

It was reported that various catalysts and reductants have been utilized in DSRP for the SO_2 reduction [1-5]. In this study, appropriate catalysts were developed for DSRP. The SnO_2-ZrO_2 catalyst was prepared by co-precipitation method and used for the catalysts in DSRP. The characteristics of SO_2 reduction were investigated with CO as a reducing gas.

2. EXPERIMENTAL

2-1. Preparation of catalysts

SnO_2 and ZrO_2 catalysts were prepared by a precipitation method, while SnO_2-ZrO_2 catalysts with Sn/Zr molar ratios corresponding to 2/8, 3/5, 5/5 and 2/1 were prepared by a co-preparation method.

Tin chloride pentahydrate ($SnCl_4\cdot5H_2O$) and zirconyl nitrate hydrate ($ZrO(NO_3)_2\cdot6H_2O$) were used as precursors of catalysts. These precursors were dissolved in the distilled water with stirring. Ammonium hydroxide solution was dropped into an aqueous solution of the precursor until its pH reached 9-10 and then the white precipitate was formed. This precipitated slurry was dried at 110 ℃ overnight and was calcined at 600 ℃ for 4 hours in the electric furnace. The prepared catalysts were analyzed by X-ray diffraction (XRD; Rigaku, D/MAX-2500).

2-2. Reactivity test

The prepared SnO_2-ZrO_2 catalyst was ground and the particles of 75-150 μm diameter were collected through sieving. The selected catalyst was used for the reaction test. The test was performed in a fixed-bed quartz reactor with a 1/2 inch inner diameter in the temperature range of 300-550 ℃. About 0.5 gram of catalyst was packed into the reactor. Space velocity was 10,000 cm³/g-cat.·h and the mole ratio of CO to SO_2 was fixed to 2:1. The outlet gases from the reactor were automatically analyzed by a gas chromatograph equipped with a thermal conductivity detector (TCD). Porapac T and Hayasep Q were used as column materials to detect CS_2, COS, H_2S, CO_2, and SO_2.

3. RESULTS AND DISCUSSION
3-1. Structure of catalysts

Structures of SnO_2-ZrO_2 catalysts at several different mole ratios were analyzed by XRD and XRD patterns of catalysts were compared before and after the reaction as shown in Fig. 1. XRD patterns corresponding to SnO_2 and ZrO_2 were confirmed in the database of JCPDS card. After the reaction occurred, the changes in the peak intensity for the fresh catalysts were observed. Since the extent of intensity change for SnO_2 was larger than that for ZrO_2, it was indicated that SnO_2 would have a greater influence on the reactivity of catalyst. Reactivity study on SnO_2-ZrO_2 catalyst for the SO_2 reduction in DSRP was performed with CO on the basis of XRD results.

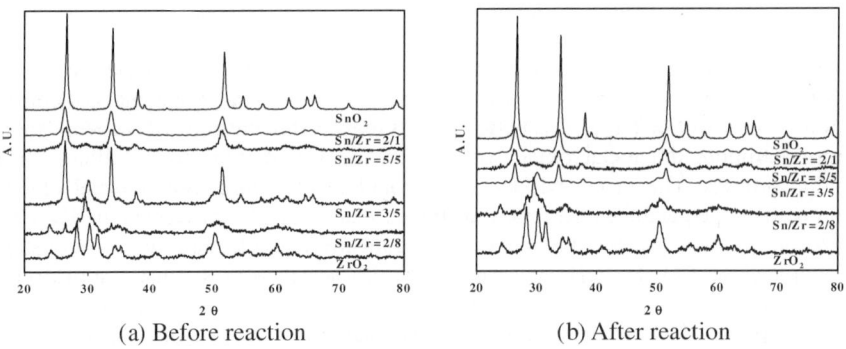

(a) Before reaction   (b) After reaction

Fig. 1. XRD patterns of SnO_2-ZrO_2 catalysts with Sn/Zr molar ratio.

3-2. SO_2 reduction over catalysts

Since SnO_2 and ZrO_2 are main active components of SnO_2-ZrO_2 catalyst, they were used for the catalysts in order to investigate their reactivity on the SO_2 reduction. The SO_2 reduction by CO over SnO_2 was conducted to evaluate the performance of SnO_2 catalyst. Space velocity was maintained to 10000 cm³/g-cat.·h and the molar ratio of [CO]/[SO_2] was fixed to 2.0 and temperature was varied in the range of 350-550 ℃ during the SO_2 reduction. Fig. 2 shows the effect of the reaction temperature on the SO_2 conversion and sulfur yield. The light-off temperature was 375 ℃ and the SO_2 conversion and sulfur yield tended to be on the increase as the reaction temperature was raised. Relatively low SO_2 conversion and sulfur yield were obtained in this study. About 45% of SO$_2$ conversion and 32% of sulfur yield were achieved at 550 ℃.

The performance of ZrO$_2$ catalyst for the SO$_2$ reduction by CO was also evaluated in this study. Experiment was carried out by varying temperature from 450℃ to 550℃ at space velocity of 10000 cm^3/g$_{-cat.}$·h with [CO]/[SO$_2$] molar ratio of 2.0. Fig. 3 shows the effect of the reaction temperature on the SO$_2$ conversion and the sulfur yield for ZrO$_2$ catalyst. The light-off temperature was 490 ℃ and the optimal reaction temperature was in the range of 550-650 ℃. In this study, 100% of SO$_2$ conversion and 92% of sulfur yield were achieved in the temperature range of 550-650 ℃. Our experimental results showed that ZrO$_2$ catalyst gave higher SO$_2$ conversion and sulfur yield than those of SnO$_2$ even though ZrO$_2$ catalyst reacted at higher temperature than SnO$_2$.

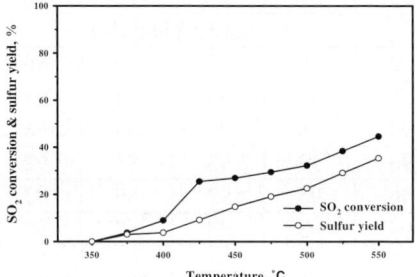

Fig. 2. The effect of reaction temperature on SO$_2$ reduction by CO over SnO$_2$ catalyst.

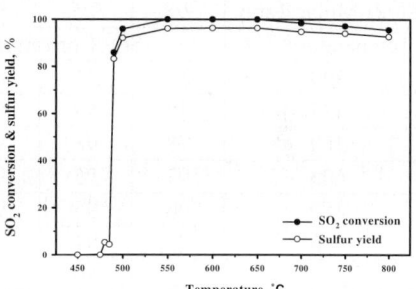

Fig. 3. The effect of reaction temperature on SO$_2$ reduction by CO over ZrO$_2$ catalyst.

From the results, the desirable nature of the two catalysts was complemented to obtain synergy effect for better performance. SnO$_2$+ZrO$_2$ catalyst was prepared by physically mixing SnO$_2$ and ZrO$_2$ with Sn/Zr molar ratio of 2/1. The SO$_2$ reduction was carried out by varying temperature from 350℃ to 600℃ at space velocity of 10000 cm^3/g$_{-cat.}$·h with [CO]/[SO$_2$] molar ratio of 2.0. Fig. 4 shows the effect of temperature on the SO$_2$ conversion and sulfur yield with SnO$_2$+ZrO$_2$ catalyst. Its light-off temperature was 375 ℃ and its SO$_2$ conversion and sulfur yield was about 71% and 64% at 600 ℃, respectively. The SO$_2$ conversion and sulfur yield over SnO$_2$+ZrO$_2$ catalyst are higher than those over pure SnO$_2$. The reactivity of SO$_2$ reduction by CO over SnO$_2$+ZrO$_2$ catalyst was higher than that of pure SnO$_2$ catalyst and, however, it was lower than that of pure ZrO$_2$ catalyst.

On the basis of the experimental results, SnO$_2$-ZrO$_2$ catalysts were prepared by the co-precipitation method. That is, SnO$_2$ and ZrO$_2$ were mixed chemically rather than physically. The effects of the Sn/Zr molar ratio and reaction temperature on the SO$_2$ conversion and sulfur yield for SnO$_2$-ZrO$_2$ catalysts were

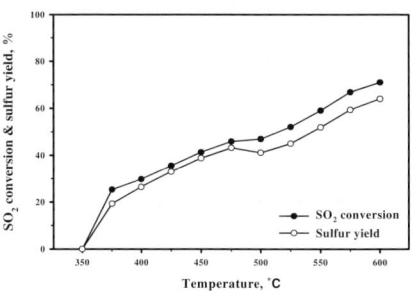

Fig. 4. The effect of reaction temperature on SO$_2$ reduction by CO over SnO$_2$+ZrO$_2$(Sn/Zr=2/1) catalyst.

shown in Table 1. The Sn/Zr molar ratios were varied over the temperature range of 300-550 °C at space velocity of 10000 cm³/g$_{cat.}$·h with [CO]/[SO$_2$] molar ratio of 2.0. In addition, Sn/Zr molar ratios of SnO$_2$-ZrO$_2$ catalysts were 1/4, 3/5, 1/1, and 2/1. The SnO$_2$-ZrO$_2$ catalyst in the Sn/Zr molar ratio of 1/2 showed the best performance at 325 °C. At 325 °C, 100% of SO$_2$ conversion and 97% of sulfur yield were achieved with it. It was observed that the SO$_2$ conversion and sulfur yield were proportional to Sn content in the catalyst but inversely proportional to reaction temperature.

Table 1
The effect of reaction temperature on SO$_2$ reduction over SnO$_2$-ZrO$_2$ catalysts with Sn/Zr molar ratio

Sn/Zr Molar Ratio	2/8	3/5	5/5	2/1	2/8	3/5	5/5	2/1
Temperature[°C]	SO$_2$ Conversion (%)				Sulfur Yield (%)			
300	17	71	80	92	7	53	69	86
325	48	89	87	100	39	86	82	98
350	78	98	97	99	75	96	93	96
400	97	95	97	96	93	93	93	92
450	96	95	94	96	95	92	89	93
500	95	91	92	94	93	88	87	90
550		89	90	92		85	86	89

4. CONCLUSION

SnO$_2$-ZrO$_2$ catalysts were developed for removal of SO$_2$ by DSRP and the effects of the various reaction conditions on the SO$_2$ reduction by CO over these catalysts were also investigated. It was confirmed that SnO$_2$-ZrO$_2$ catalysts were the composite metal oxides in which SnO$_2$ and ZrO$_2$ were mixed chemically rather than physically. For the SO$_2$ reduction by CO over SnO$_2$ catalyst, light-off temperature was 375 °C and the SO$_2$ conversion and sulfur yield was about 45% and 32%, respectively, at 550 °C. In case of the SO$_2$ reduction by CO over ZrO$_2$ catalyst, light-off temperature was 490 °C and SO$_2$ conversion and sulfur yield was about 100% and 92%, respectively, at 600 °C.

SnO$_2$-ZrO$_2$ catalyst in the Sn/Zr molar ratio of 1/2 showed the best performance. Light-off temperature was started from below 300 °C and the 100% of SO$_2$ conversion and 97% of sulfur yield were achieved at 325 °C. In our result, light-off temperature was lower than that of pure SnO$_2$ and pure ZrO$_2$, however, the SO$_2$ removal efficiency was higher than those of pure SnO$_2$ and pure ZrO$_2$. It was believed that SnO$_2$-ZrO$_2$ catalyst achieved the synergy effect from the desirable nature of the two catalysts.

REFERENCES

[1] S. C. Paik, H. Kim and J. S. Chung, Catalysis Today, 38 (1997) 193.
[2] H. Kim, D. W. Park, H. C. Woo and J. S. Chung, Applied Catalysis B: Environmental, 19 (1998) 233.
[3] Z. Zhaoliang, M. Jun, Y. Xiyao, Journal of Molecular Catalysis A: Chemical, 195 (2003) 189.
[4] B. S. Kim, J. D. Lee, N. K. Park, S. O. Ryu, T. J. Lee and J. C. Kim, HWAHAK KONGHAK, 41 (2003) 572.
[5] W. Liu, C. Wadia, M. F. Stephanopoulos, Catalysis Today, 28 (1996) 391.

Studies in Surface Science and Catalysis, volume 159
Hyun-Ku Rhee, In-Sik Nam and Jong Moon Park (Editors)

Role of H$_2$O$_2$ in photocatalytic reaction over TiO$_2$-loaded Cr, Ti-substituted MCM-41 in visible light

Sung Gab Kim, Hee Hoon Jeong, Jin Hwan Park, Seong Soo Park and Gun Dae Lee[*]

Division of Chemical Engineering, Pukyong National University, 100 Yongdang-dong, Nam-gu, Busan 608-739, Republic of Korea

1. INTRODUCTION

Photocatalytic decomposition of organic pollutants is a promising process for air and water decontamination. Recently, the development of new photocatalysts that have high activity under visible light has received great attention. Mesoporous silica MCM-41 with well ordered pores in the size range of 2–10 nm, is used for a range of applications in the field of heterogeneous catalysis. The chromium substituted MCM-41 was found to serve as the best support for titania to achieve the highest degradation rates of formic acid and 2,4,6-trichlorophenol in visible light [1]. In this study, transition metal (Cr, Ti)-substituted MCM-41 samples were prepared, and then TiO$_2$ is loaded on these materials with the aim of extending the photocatalytic activity toward the visible region. The 4-nitrophenol (4-NP), which is one of the most refractory pollutants present in industrial wastewater, was used as a probe organic compound for photodecomposition in the visible light. The objective of the present investigation was to demonstrate the role of H$_2$O$_2$ in the photocatalytic reaction.

2. EXPERIMENTAL

Cr-Ti-substituted MCM-41 (Si/Cr = 80, Si/Ti = 40) was prepared as previously reported [1]. And then the required amount of TiO$_2$ was loaded on the support using titanium isopropoxide. P-25 TiO$_2$ (Degussa) was used as a reference catalyst. X-ray diffraction (XRD) analyses of the catalyst were carried out on a Philips X'pert diffractometer using Cu Kα radiation. UV-vis diffuse reflectance spectroscopy (DRS) was performed on Varian Cary 100 with PTFE (polytetrafluoroethylene) as standard. To demonstrate the effect of H$_2$O$_2$ on the adsorption of 4-NP on catalyst, the adsorption experiments were carried out in the absence and presence of H$_2$O$_2$ (5mmol/L). The photocatalytic activity measurements were performed in a batch-type annular Pyrex reactor using a 200-W halogen lamp as the light source.

3. RESULTS AND DISCUSSION

3.1. Characterization of catalysts

XRD results showed that the MCM-41 samples after modifications and TiO$_2$ loading exhibit patterns similar to that of siliceous MCM-41. However, the modification and TiO$_2$ loading lead to some decrease in the higher order Bragg reflections and the broadening of the peaks, indicating decrease in the structural integrity. The UV-vis diffuse reflectance spectra of the samples in the range of 200-500 nm are shown in Fig. 1. As shown in Fig. 1, the behavior of visible light absorption of transition-metal-substituted TiO$_2$-loaded MCM-41 samples was drastically different from that of neat TiO$_2$ (P-25). Cr-Ti-MCM-41 material exhibits some absorption in visible light. Significant absorption occurs in visible light in TiO$_2$-loaded Cr-Ti-MCM-41 and the absorption both in UV and visible light increased with increasing TiO$_2$ loading.

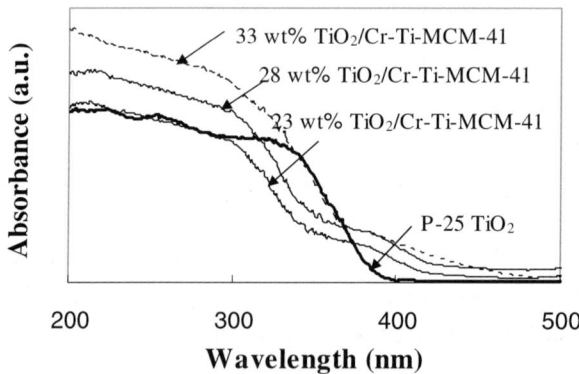

Fig. 1. UV-vis diffuse reflectance spectra of neat TiO$_2$, chemically-modified and TiO$_2$-loaded MCM-41 samples.

3.2. Effect of H$_2$O$_2$ on the photocatalytic decomposition of 4-NP on TiO$_2$/Cr-Ti-MCM-41 in visible light

Fig. 2 shows the effect of H$_2$O$_2$ on the photocatalytic decomposition of 4-NP over TiO$_2$/Cr-Ti-MCM-41 in visible light. The addition of H$_2$O$_2$ did not lead to the complete decomposition of 4-NP, but resulted in a marked increase in photocatalytic activity of TiO$_2$/Cr-Ti-MCM-41. Photocatalytic reactions have been reported to exhibit appreciable rate enhancement by the addition of H$_2$O$_2$ in reaction system [2]. Such enhancements can be explained by several facts ; H$_2$O$_2$ is a better electron acceptor than molecular oxygen (reaction 1) and H$_2$O$_2$ may be split photolytically to produce $\cdot$OH radical directly (reaction 2). In this work, however, the contribution of reaction (2) seems to be negligible since visible light was used.

$$e^- + H_2O_2 \rightarrow \cdot OH + OH^- \qquad\qquad (1)$$

$$H_2O_2 + h\nu \rightarrow 2 \cdot OH \qquad\qquad (2)$$

The photocatalytic decomposition of 4-NP can be written in terms of Langmuir-Hishelwood (L-H) kinetics [3].

$$r = \frac{kKC}{1+KC} \quad or \quad \frac{1}{r} = \frac{1}{kKC} + \frac{1}{K} \qquad\qquad (3)$$

From plotting of 1/r versus 1/C, the reaction rate constant, k and adsorption constant, K can be obtained. Fig. 3 indicates that photocatalytic decomposition of 4-NP is in good agreement with L-H model. In the present work, the values of k and K in the presence of H_2O_2 were found to be higher than those in the absence of H_2O_2, as shown in Table 1.

In addition, the results of adsorption experiment in Fig. 4 revealed that H_2O_2 promotes the adsorption of 4-NP on the Cr-Ti-MCM-41 surface. From considering above results, it can be said that H_2O_2 increases the reaction rate by the promotion of adsorption of reactant and the removing of surface-trapped electrons.

4. CONCLUSIONS

In the presence of H_2O_2, TiO_2/Cr-Ti-MCM-41 showed a marked increase in photocatalytic activity for decomposition of 4-NP over TiO_2/Cr-Ti-MCM-41 in visible light. H_2O_2 increases the reaction rate by the promotion of adsorption of reactant and the removing of surface-trapped electrons.

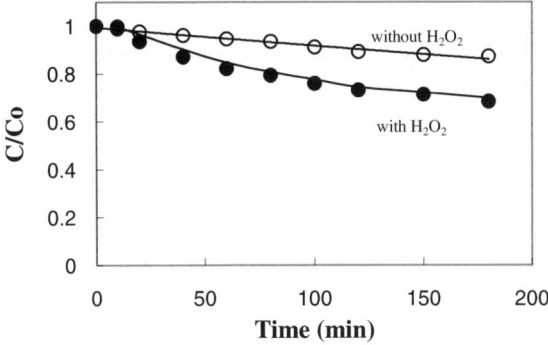

Fig. 2. Effect of H_2O_2 (5mmol/L) on the photocatalytic decomposition of 4-NP (C_o=25ppm) over TiO_2/Cr-Ti-MCM-41 in visible light.

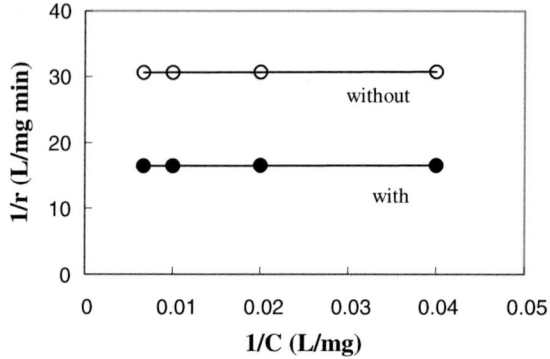

Fig. 3. Linear fitting of the 1/r vs. 1/C.

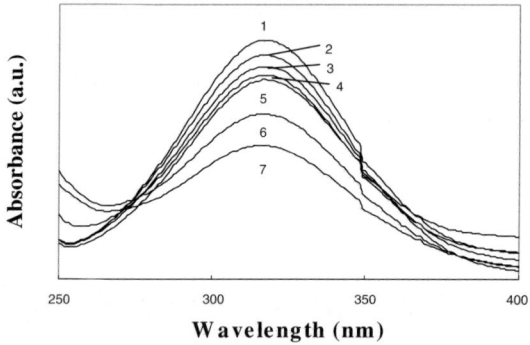

Wavelength (nm)

Fig. 4. Change of UV spectrum during the adsorption of 4-NP on catalysts under various conditions (1; 4-NP, 2; 4-NP on TiO_2/MCM-41, 3; 4-NP on Cr-Ti-MCM-41, 4; 4-NP on MCM-41, 5; 4-NP+H_2O_2 on MCM-41, 6; 4-NP+H_2O_2 on TiO_2/MCM-41, 7; 4-NP+H_2O_2 on Cr-Ti-MCM-41).

Table 1. Rate constant (k) and adsorption constant (K) for the photocatalytic decomposition of 4-NP on TiO_2/Cr-Ti-MCM-41 in visible light

	In the absence of H_2O_2	In the presence of H_2O_2 (5mmol/L)
k (mg/L min)	0.033	0.061
K (L/mg)	5.155	10.103

REFERENCES

[1] L. Davydov, E.P. Reddy, P. France and P.G. Smirniotis, J. Catal., 203 (2001) 157.

[2] D. F. Ollis, E. Pelizzetti and N. Serpone, Environ. Sci. Technol., 25 (1991) 1523.

[3] V. Augugliaro, L. Palmisano, M Schiavello and A. Sclafani, Appl. Catal., 69 (1991) 323.

Studies in Surface Science and Catalysis, volume 159
Hyun-Ku Rhee, In-Sik Nam and Jong Moon Park (Editors)

Contributions of three types of Ga sites in propane aromatization over Ga_2O_3/Ga-MOR catalysts

A. Satsuma[a]*, A. Gon-no[a], K. Nishi[a†], S. Komai[a], and T. Hattori[b]

[a] Dept. of Applied Chemistry, Nagoya University, Chikusa, Nagoya 464-8603, Japan
[b] Dept. of Applied Chemistry, Aichi Institute of Technology, Toyota 470-0392, Japan
[†] Present address; Dept. of Chemistry, National Defense Academy, Yokosuka 239-8686, Japan.

Contributions of three types of Ga sites to propane conversion into aromatics were examined by using model catalysts, i.e., gallosilicate of MOR structure with deposited Ga_2O_3 particles. The rates of propane conversion and aromatics formation were correlated with the densities of three types of Ga sites determined by NH_3-TPD, and it was shown that the propane conversion and the aromatics formation were limited by Ga sites on Ga_2O_3 surface.

INTRODUCTION

It is sometimes quite difficult to identify the site which primarily contributes to catalytic reaction. This is because the catalytic reaction consists of several reaction steps, each of which may require different sites. One of such examples is the transformation of light alkanes into higher hydrocarbons such as aromatics and branched-chain aliphatics over Ga-containing zeolites. The reaction consists of several steps such as the dehydrogenation of light alkanes, the oligomerization of alkenes, the dehydrogenation of cycloalkanes, and so on. And it is widely accepted that the Bronsted acid sites of zeolite provide the catalytic sites for oligomerization and Ga species are responsible for dehydrogenation activity. It should be noted, however, there can be three types of Ga species: Ga ions incorporated into silicate framework, dispersed extraframework Ga species, and Ga_2O_3 particles. Present study aims at identifying the catalytic sites limiting the conversion of propane and the formation of aromatics over gallosilicate of MOR structure (Ga-MOR) with deposited Ga_2O_3 particles.

EXPERIMENTAL

Ga-MOR was hydrothermally synthesized by using tetraethylammonium bromide as a template. Ga content was controlled by removing Ga through HCl treatment in the same way as the dealumination of zeolite [1]. The numeral at the end of catalyst name stands for SiO_2/Ga_2O_3 ratio in Ga-MOR thus prepared. Ga_2O_3/Ga-MOR catalyst was prepared by impregnating Ga-MOR into Ga(NO$_3$)$_3$ aqueous solution, followed by evaporation-to-dryness. Ga content was determined by ICP, and MOR structure was confirmed by X-ray diffraction. Acidic properties were characterized by the temperature programmed desorption (TPD) and the differential heat of adsorption (Tokyo-riko HTC-450) of NH$_3$, and Ga$_2$O$_3$ particles size was measured by transmission electron microscopy (TEM, HITACHI H-800). The catalytic test was carried out using a continuous flow reaction apparatus at atmospheric pressure [2]. The products were analyzed by on-line FID-GC equipped with Protocol DH fused silica capillary column and VZ10 packed column.

RESULTS AND DISCUSSION

TPD profile of NH$_3$ over parent Ga-MOR showed three desorption peaks, L-peak at 450-500K, H1-peak at 600-700K, and H2-peak at 900-1000K, as shown by profile a in Fig. 1. Decrease of Ga content by HCl-treatment preferentially reduced L- and H2-peaks (profile d), and the deposition of Ga$_2$O$_3$ recovered L- and H2-peaks (profiles b and c). H1-peak was ascribed to the desorption from zeolitic acid sites generated by framework Ga in the same manner as Ga-MFI [2,3] and Ga-MCM-41 [4], and L- and H2-peaks from large Ga$_2$O$_3$ particles of ca. 90 nm and finely dispersed Ga species of ca. 5 nm. TEM photographs indicated that large Ga$_2$O$_3$ particles and fine Ga$_2$O$_3$ particles were more abundant on the catalysts with larger L- and H2-peaks, respectively. The generation of such strong acid sites as H2-peak was also reported on Ga-MFI [3]. The deconvolution analysis of TPD profiles was carried out to determine the amounts of L-, H1-, and H2-sites on the assumption that the profile is expressed by the sum of a desorption profile from Ga$_2$O$_3$/SiO$_2$ (c in Fig. 2) and two Gaussian functions (d and e), as shown in Fig. 2.

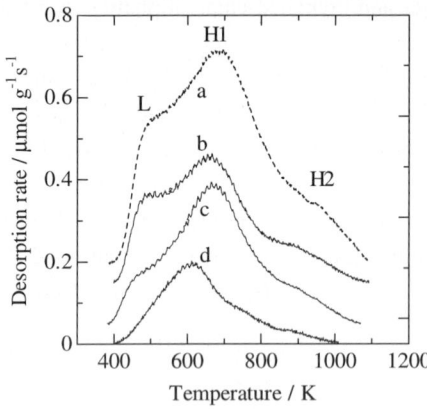
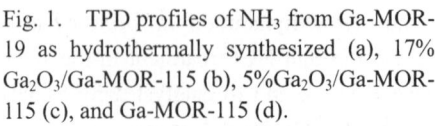
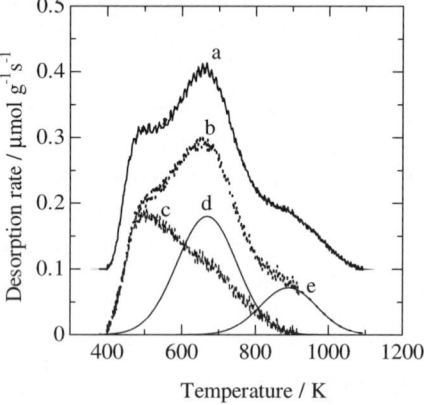

Fig. 1. TPD profiles of NH$_3$ from Ga-MOR-19 as hydrothermally synthesized (a), 17% Ga$_2$O$_3$/Ga-MOR-115 (b), 5%Ga$_2$O$_3$/Ga-MOR-115 (c), and Ga-MOR-115 (d).

Fig. 2. Deconvolution of TPD profile from 17%Ga$_2$O$_3$/Ga-MOR-115 (a) by using a profile from Ga$_2$O$_3$/SiO$_2$ (c) and two Gaussian functions (d and e). Profile b is the sum of c, d and e.

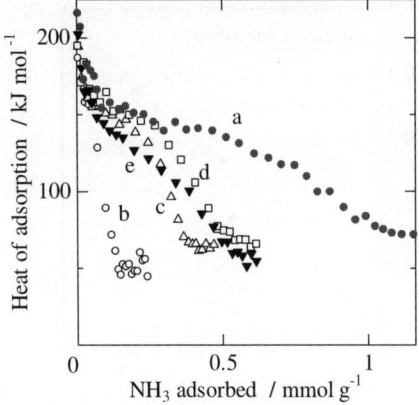

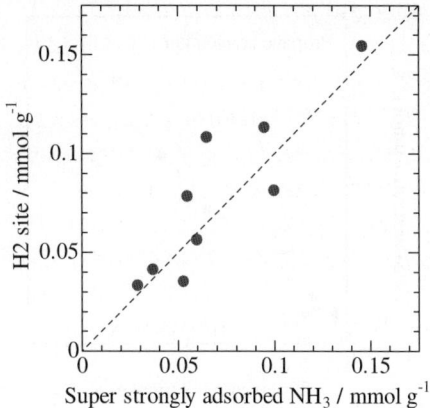

Fig. 3. Differential heat of adsorption of NH3 at 473 K over Ga-MOR-19 (a), Ga-MOR-115 (b), 2% (c), 5% (d) and 17% (e) Ga$_2$O$_3$/Ga-MOR-115.

Fig. 4. Comparison of the amount of H2 sites with that of super strongly adsorbed NH$_3$ with the heat of adsorption above 155 kJ mol^{-1}.

The adsorption of NH$_3$ was also measured with a microcalorimeter, and some of the results are shown in Fig. 3. Figure 4 compares the amount of H2 site determined by the deconvolution analysis with that of super strongly adsorbed NH$_3$ with the heat of adsorption above 155 kJ mol^{-1}. A good correlation shown in the figure indicates the validity of the amounts of Ga sites determined by the deconvolution analysis.

The propane aromatization was conducted under the differential condition by using Ga$_2$O$_3$/Ga-MOR catalysts thus characterized. The contributions of L, H1, and H2 sites to the propane conversion and the aromatics formation were estimated by assuming that the observed reaction rates are the sum of the reaction rate on each site which is equal to the product of the turnover frequency (TF$_{i,j}$) and the amount of active sites per weight of catalyst (A$_j$):

$$R_{C3} = TF_{C3,L} A_L + TF_{C3,H1} A_{H1} + TF_{C3,H2} A_{H2} \qquad (1)$$
$$R_{Ar} = TF_{Ar,L} A_L + TF_{Ar,H1} A_{H1} + TF_{Ar,H2} A_{H2} \qquad (2)$$

where R$_i$ is the observed reaction rate of propane conversion (sub C3) and aromatics formation (sub Ar) per weight of catalyst. The turnover frequencies were determined through a least square fitting by using the observed rates of propane conversion and aromatics formation and the amounts of three types of Ga sites determined above for nine catalysts. Figure 5 compares the rates calculated by substituting thus obtained TF's for Eqs. (1) and (2) with the observed rates. The good agreements for both propane conversion and aromatics formation indicate the validity of this analysis.

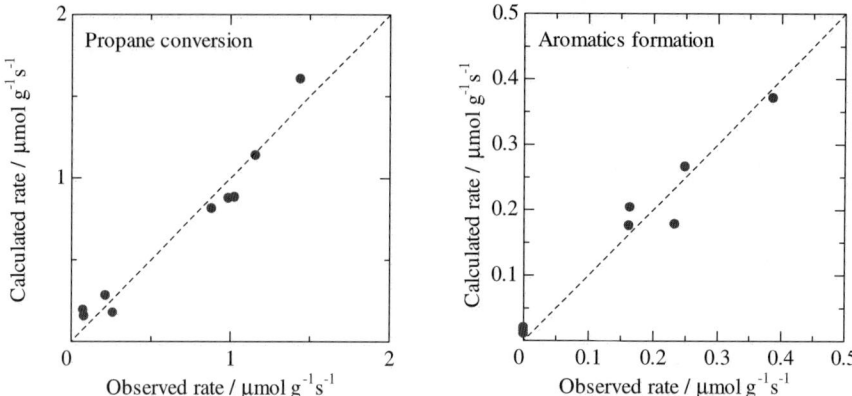

Fig. 5. Comparison between the calculated rates and the observed rates.

Table 1 summarizes the turnover frequencies of propane conversion and aromatics formation on each site calculated by the least square fitting (the correlation coefficients R^2 were 0.955 and 0.958, respectively). As shown, the rate of propane conversion is much affected by both L and H2 sites, and that of aromatics formation only by H2 sites. The contribution of H1 site, i.e., zeolitic acid sites generated by framework Ga, is not large for both rates. However, this result does not necessarily mean that zeolitic acid sites do not participate in the propane conversion into aromatics. It means that, as the most important finding, the rate is limited by the amounts of L and H2 sites in propane conversion and by that of H2 site in aromatics formation, or, in other words, the rate determining step proceeds on these sites. Thus, one can expect to enhance the rates by increasing these sites. It is also interesting to note that the contributions of the sites to the aromatics formation are different from those to the propane conversion. This difference may suggest the possibility of controlling product distribution in the transformation of light alkanes over Ga containing zeolites and silicates by changing the structure of extraframework Ga species.

Table 1   Turnover frequencies of propane conversion and aromatics formation over L, H1 and H2 sites of Ga_2O_3/Ga-MOR catalysts.

	Turnover frequency / ks^{-1}		
	L site	H1 site	H2 site
Propane conversion	2.6	0.0	3.0
Aromatics formation	0.3	-0.4	2.4

REFERENCES

[1] M. Sawa, M. Niwa, and Y. Murakami, Appl. Catal., 53, 169 (1989).

[2] K. Nishi, S. Komai, K. Inagaki, A. Satsuma, and T. Hattori, Appl. Catal. A, 223, 187 (2002).

[3] T. Miyamoto, N. Katada, J-.H. Kim, and M. Niwa, J. Phys. Chem., 102, 6738 (1998).

[4] A. Satsuma, Y. Segawa, H. Yoshida, and T. Hattori, Appl. Catal. A, 264, 229 (2004).

Studies in Surface Science and Catalysis, volume 159
Hyun-Ku Rhee, In-Sik Nam and Jong Moon Park (Editors)
© 2006 Elsevier B.V. All rights reserved

Catalytic removal of diesel soot particulates over LaMnO$_3$ perovskite-type oxides

Seong-Soo Hong*and Gun-Dae Lee

Division of Applied Chemical Engineering, Pukyong National University,100 Yongdang-dong, Nam-ku, Pusan, 608-739, Korea

Catalytic combustion of diesel soot particulates over LaMnO$_3$ perovskite-type oxides prepared by malic acid method has been studied. In the LaMnO$_3$ catalyst, the partial substitution of alkali metal ions into A site enhanced the catalytic activity in the combustion of diesel soot particulates and the activity was shown in following order;Cs>K>Na. In the La$_{1-x}$Cs$_x$MnO$_3$ catalyst, the catalytic activity increased with an increase of x value and showed constant activity at the substitution of x>0.3

1. INTRODUCTION

One of principal problems in larger urban centers is the presence of particulate material in the atmosphere due to the emission of diesel engine[1]. One of the most dangerous components of diesel exhaust is particulate, which consists of agglomerates of small carbon particles with a number of different hydrocarbons and sulfates adsorbed on their surface. A potential way to face the related environmental problem is that of filtering the particulate and burning it out in catalyzed traps before any emission of diesel exhausts in the environment.

The combustion temperature of soot particulates can be lowered by the addition of an oxidation catalyst in the form of fuel additives[2], by spraying metal salt solution on an accumulated soot or by the impregnation of filter walls with an oxidation catalyst. For the last option, oxides of supported metals are considered to be the most promising candidates.

Several researchers have focused their attention on the application of oxide materials to lower the oxidation temperature of soot particulates. It was reported that active soot oxidation catalysts are PbO, Co$_3$O$_4$, V$_2$O$_5$, MoO$_3$, CuO, and perovskite type oxides[3].

In this paper, we prepared LaMnO$_3$ perovskite-type oxides using the malic acid method and investigated their physical properties. It has been also investigated the effect of partial substitution of metal ions into La and Mn sites and the reaction conditions on the activity for the combustion of soot particulates.

2. EXPERIMENTAL

The preparation method of perovskite-type oxides was taken from the previous paper[4]. Malic acid was added into mixed aqueous solution of metal nitrates in a desired proportion so as for the molar ratio of malic

---
*To whom correspondence should be addressed

E-mail: sshong@pknu.ac.kr

acid to the total metal cations to be unity. The solution was then evaporated to dryness with stirring, and further dried at 150℃. The precursor was ground and then calcined in air at 200℃ for 30 min, 350 ℃ for 30 min and 600℃ for 12h. Experiments were performed with a model carbon(Printex-U) which was obtained from Degussa AG.. The properties(primary particle size, BET surface area, oxidation rate) of this model soot were similar to those of a real diesel soot particulate. The catalyst and the soot(5wt.%) were well mixed in an agate mortar with a pestle for more than 20 min. Although the soot/catalyst contact was known to affect significantly experimental results, this mixing procedure gave reproducible results under present experimental conditions. The catalyst/soot mixture(0.2g) was placed in a quartz-tube reactor, preheat-treated at 300℃ for 2 h, and then cooled down to 200℃ in a He stream. After that, the temperature programmed reaction(TPR) was started with the linear heating rate of 1℃/min in a gaseous mixture of O_2(4%) and He(balance)(flow rate; 100cm^3/ min). The outlet gases were analyzed by gas chromatograph(HP 5890) at intervals of about 20min. The soot was almost oxidized into CO_2. Very small amount of CO was detected but it was neglected. In addition, the carbon mass balances were generally better than 97%.

3. RESULTS AND DISCUSSION

The carbon removal reaction supposedly takes place at two-phase boundary of a solid catalyst, a solid reactant(carbon particulate) and gaseous reactants(O_2, NO). Because of the experimental difficulty to supply a solid carbon continuously to reaction system, the reaction have been exclusively investigated by the temperature programmed reaction(TPR) technique in which the mixture of a catalyst and a soot is heated in gaseous reactants.

The thermogravimetric analysis of catalyst/carbon mixtures was carried out to elucidate the combustion properties of carbon particulates with a catalyst and the result is shown in Fig. 1. While noncatalytic oxidation of carbon particulates occurs above 500°C, ignition temperature remarkably decreases in the presence of a catalyst. The combustion is initiated at 520°C in the absence of a catalyst but the ignition temperature decreases to 330°C in the presence of $La_{0.8}Cs_{0.2}MnO_3$ catalyst. With increasing temperature, the charged carbon particulate is progressively consumed and finally exhausted.

The effect of substitution of metal ion into A and B sites of $LaMnO_3$ on the oxidation of soot particulate was examined and the result is shown in Table 1. The substitution of alkali metal ions decreases the ignition temperature of soot particulates and the catalytic activity is shown in the order of Cs>K>Na. From simple geometric considerations, unit cells may expand upon the substitution of larger alkali metal ions for smaller magnesium ions. Therefore, the substitution of larger alkali metal ions brings about the formation of a large number of ion vacancies in their lattice. It suggests that the partial substitution of Cs gives rise to easy reduction of oxides and forms oxide ion vacancies on the surface, and then increases the adsorption rate of active oxygen on the catalyst surface. This result can be verified by the TPR(temperature programmed reduction) result(Fig. 2). As shown in Fig. 2, they show one reduction peak and the reduction peak appears at lowest temperature in the $La_{0.7}Cs_{0.3}MnO_3$ catalyst. This result suggests that the substitution of Cs into A site gives rise to easy reduction of oxides and forms oxide ion vacancies on the surface and then increases the adsorption rate of active oxygen on the catalyst surface.

Table 1. Perovskite-type oxides prepared by malic acid method and their catalytic performances

Catalyst	$T_{ig}(°C)^a$	$T_m(°C)^b$
$LaMnO_3$	445	-
$La_{0.7}Cs_{0.3}MnO_3$	266	340
$La_{0.7}K_{0.3}MnO_3$	323	379
$La_{0.7}Na_{0.3}MnO_3$	345	403
$La_{0.9}Cs_{0.1}MnO_3$	295	367
$La_{0.8}Cs_{0.2}MnO_3$	293	360
$La_{0.6}Cs_{0.4}MnO_3$	280	345
$La_{0.5}Cs_{0.5}MnO_3$	278	346

[a] Ignition temperature estimated by extrapolating the steeply ascending portion of the CO_2 formation curve to zero CO_2 concentration

[b] Temperature which shows the maximum CO_2 concentration

In the $La_{1-x}Cs_xMnO_3$ catalyst, the T_m decreases with an increase of x value and shows an almost constant value upon substitution of x>0.3. It is thought that the oxygen vacancy sites of perovskite oxide increase with an increase of amount of Cs and the oxidation activity also increases. This result is also verified by the TPR result of these catalysts(Fig. 3). As shown in Fig. 3, the reduction peak appears at low temperature with an increase of x value and no change is shown at more than x=0.3. It can thus be concluded that the catalytic performance of these oxides increases as the amount of Cs in the crystal lattice increases. However, the substitution of Cs to more than x=0.3 leads to excess Cs, which is present on the surface of mixed oxides might have no effect on the catalytic activity

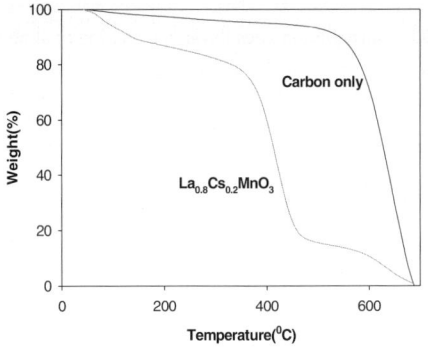

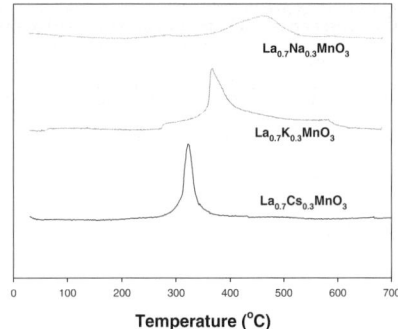

Fig. 1. TG spectra of carbon particulates with catalyst; heating rate=1 K/min.

Fig. 2. TPR profiles measured for various $La_{0.8}Cs_{0.2}MnO_3$ perovskite type oxides; heating rate=10 K/min, gas mixture= 5% H_2/He.

264

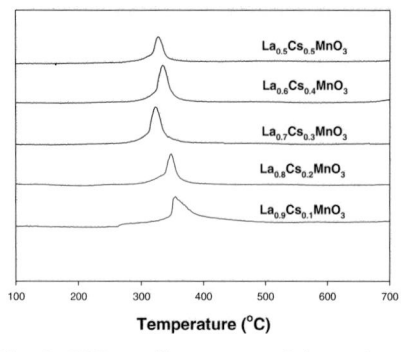

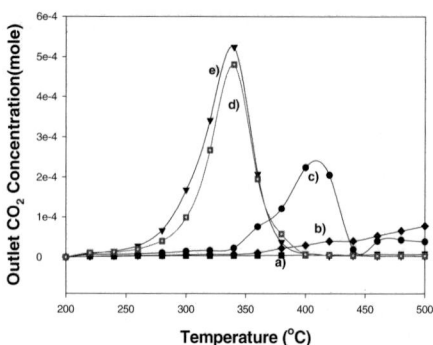

Fig. 3. TPR profiles measured for various
perovskite type oxides; heating rate=10
K/min, gas mixture= 5% H_2/He.

Fig. 4. Temperature programmed reaction on
combustion of carbon particulates over
$La_{0.7}Cs_{0.3}MnO_3$ catalyst: heating rate=
1 K/min, NO= 500 ppm, O_2=4%, a)carbon,
b)carbon+oxygen, c)carbon+catalyst,
d)carbon+NO+oxygen+catalyst,
e)carbon+oxygen+catalyst.

Fig. 4 shows outlet CO_2 concentration over $La_{0.7}Cs_{0.35}MnO_3$ catalyst at the various reactant compositions. In the presence of carbon without the catalyst and oxygen(Fig. 4(a)), carbon dioxide cannot be produced at even high temperature. In the absence of oxygen(Fig. 4(c)), carbon dioxide can be produced relatively high temperature, which is thought to be due to the carbon particulate oxidation by lattice oxygen. The similar tendency was shown in the previous report with the presence of perovskite-type catalyst and carbon particulate[5]. In the absence of the catalyst(Fig. 4(b)), a small amount of carbon dioxide can be produced at very high temperature. This result indicates that the catalyst play an important role on the combustion of carbon particulates.

In the presence of NO and oxygen(Fig. 4(d)), outlet CO_2 concentration goes through a maximum at about 350℃ and similar result is obtained in the absence of NO(Fig. 4(e)). It is demonstrated that NO has little effect on the catalytic oxidation of carbon particulate.

In this study, catalytic combustion of diesel soot particulates over $LaMnO_3$ perovskite-type oxides prepared by malic acid method has been carried out. In the $LaMnO_3$ catalyst, the partial substitution of alkali metal ions into A site enhanced the catalytic activity in the combustion of diesel soot particulates and the activity was shown in following order;Cs>K>Na.

REFERENCES
[1] G. Mul, F. Kapteijn and J.A. Muulijn, Appl. Catal., B, 12 (1997) 33.
[2] N. Miyamoto, Z. Hou and H. Ogawa, SAE paper 881224(1988).
[3] S.S. Hong, J.S. Yang and G.D. Lee, Reac. Kinet. & Catal. Lett., 66 (1999) 305.
[4] J.S. Yang, G.D. Lee, B.H, Ahn and S.S. Hong, J. Ind. & Eng. Chem., 4 (1998) 263.
[5] Y. Teraoka, K. Nakano, W.F. Shangguan and S. Kagawa, Catal. Today, 27 (1996) 107.

Studies in Surface Science and Catalysis, volume 159
Hyun-Ku Rhee, In-Sik Nam and Jong Moon Park (Editors)

Preparation, characterization, and catalytic activity of $H_3PW_{12}O_{40}$ heteropolyacid catalyst supported on mesoporous γ-Al_2O_3

Pil Kim, Heesoo Kim, Jongheop Yi, In Kyu Song*

School of Chemical and Biological Engineering, Institute of Chemical Processes,
Seoul National University, Shinlim-dong, Kwanak-ku, Seoul 151-744, South Korea

*Corresponding author (Tel:+82-2-880-9227, E-mail: inksong@snu.ac.kr)

1. INTRODUCTION

Heteropolyacids (HPAs), also known as polyoxometalates, are early transition metal oxygen anion clusters that exhibit a wide range of molecular sizes, compositions, and architectures [1] They are inorganic acids and at the same time can act as strong oxidizing agents [2]. Although HPA catalysts have tremendous merits such as pseudo-liquid phase behavior and tunable catalytic property, one of the disadvantages of bulk HPA catalysts is that their surface area is very low (< 10 m^2/g). To overcome the low surface area of HPA catalysts, many attempts have been made on the preparation of supported HPA catalysts [3]. Reported in this work are preparation, characterization, and catalytic activity of $H_3PW_{12}O_{40}$ catalyst supported on mesoporous γ-Al_2O_3.

2. EXPERIMENTAL

Mesoporous Al_2O_3 was prepared by a post hydrolysis method [4] at atmospheric pressure and at room temperature. 1.89 g of $Al(sec$-$BuO)_3$ (aluminum source) and 8.21 g of stearic acid (surfactant) were separately dissolved in 7.65 ml of sec-butyl alcohol, and the two solutions were then mixed. Some amounts of water were dropped into the mixture at a rate of 1 ml/min, until a white precipitate was formed. The crude product was calcined at 450°C for 3 h with an air stream (50 ml/min) to yield the final support. $H_3PW_{12}O_{40}$ was purchased from Aldrich Chem. Co. $H_3PW_{12}O_{40}$ catalyst supported on mesoporous γ-Al_2O_3 was prepared by an impregnation method. ICP analysis revealed that the composition of the supported catalyst was 6.5 wt.% $H_3PW_{12}O_{40}/\gamma$-Al_2O_3. Adsorption isotherms of nitrogen were obtained with an ASAP-2010 (Micromeritics) apparatus. Pore size distribution was determined by the BJH method applied to the desorption branch of the nitrogen isotherm. The interaction between $H_3PW_{12}O_{40}$ species and Al_2O_3 support was examined by FT-IR measurements (Jasco, FT-IR 460 Plus). Vapor-phase 2-propanol conversion reaction was carried out in a continuous flow fixed-bed reactor at atmospheric pressure. Supported catalyst with a size of 150-200 μm was charged into a tubular quartz reactor, and activated with a stream of air at 330°C for 1 h. 2-Propanol was sufficiently vaporized by passing a pre-heating zone and fed into the reactor together with air carrier (20 ml/min). The catalytic reaction was also conducted over the bulk $H_3PW_{12}O_{40}$ catalyst under consistent conditions for reference and comparison purpose. Contact time was maintained at 2.30 g-$H_3PW_{12}O_{40}$-h/2-propanol-mole. The product stream was periodically sampled and analyzed with an on-line GC (HP 5890II).

3. RESULTS AND DISCUSSION

3.1. Characterization

Phase transformation of the synthesized Al_2O_3 was examined by XRD measurements with a variation calcination temperature. It was observed that as-synthesized Al_2O_3 exhibited a mixed phase of bayerite and pseudo-boehmite. γ-Al_2O_3 phase was formed at the thermal treatment of 270°C, and this phase was still maintained at 450°C. This result indicates that γ-Al_2O_3 obtained by the thermal treatment at 450°C can serve as a support for an active catalyst. Thermal decomposition temperature of $H_3PW_{12}O_{40}$ catalyst is known to be ca. 550°C [5]. Therefore, it is inferred that $H_3PW_{12}O_{40}$ catalyst supported on γ-Al_2O_3 will be stable as long as the supported catalyst is operated below 450°C. Nitrogen adsorption-desorption results showed that γ-Al_2O_3 exhibited type IV isotherm with type H2 hysteresis loop. Compared to γ-Al_2O_3, $H_3PW_{12}O_{40}$/γ-Al_2O_3 showed the similar patterns of isotherm and hysteresis loop with no significant change. It was also observed that both γ-Al_2O_3 and $H_3PW_{12}O_{40}$/γ-Al_2O_3 showed very narrow pore size distribution center at around 4 nm, indicating that pore structure of γ-Al_2O_3 was maintained even after the supporting of $H_3PW_{12}O_{40}$ catalyst. In the SAXS patterns of γ-Al_2O_3 and $H_3PW_{12}O_{40}$/γ-Al_2O_3, both samples exhibited a continuous decay without any peaks indicating the existence of size-controlled porosity. This demonstrates that γ-Al_2O_3 and $H_3PW_{12}O_{40}$/γ-Al_2O_3 samples retained randomly connected pore structure [4]. BET surface areas of $H_3PW_{12}O_{40}$, γ-Al_2O_3, and $H_3PW_{12}O_{40}$/γ-Al_2O_3 samples were 5.8, 333, and 308 m^2/g, respectively. Pore volumes of γ-Al_2O_3 and $H_3PW_{12}O_{40}$/γ-Al_2O_3 were found to be 0.52 and 0.305 cm^3/g, respectively. These results imply that $H_3PW_{12}O_{40}$ was successfully supported on γ-Al_2O_3.

Fig. 1 shows the TEM images of γ-Al_2O_3 and $H_3PW_{12}O_{40}$/γ-Al_2O_3. Both samples showed worm hole-like pore structure with uniform pore size, in good agreement with the results of nitrogen sorption and SAXS measurements. In particular, no visible evidence representing agglomerates of $H_3PW_{12}O_{40}$ species was found in the TEM image of $H_3PW_{12}O_{40}$/γ-Al_2O_3, indicating that $H_3PW_{12}O_{40}$ species were finely dispersed on the γ-Al_2O_3 support.

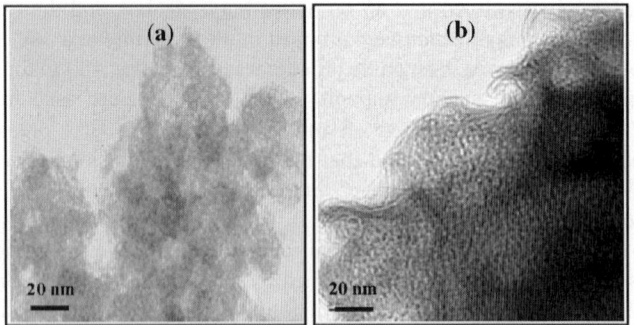

Fig. 1. TEM images of (a) γ-Al_2O_3 and (b) $H_3PW_{12}O_{40}$/γ-Al_2O_3.

Fig. 2 shows the IR spectra of $H_3PW_{12}O_{40}$, γ-Al_2O_3, and $H_3PW_{12}O_{40}$/γ-Al_2O_3 samples. The primary structure of $H_3PW_{12}O_{40}$ can be identified by four characteristic IR bands appearing at 700-1200 cm^{-1}. As shown in Fig. 2(a), IR bands appearing at 1080, 982, 889, 810 cm^{-1} can be attributed to P-O symmetric stretch, W=O asymmetric stretch, W-O-W inter- and intra-octahedral stretches, respectively. The γ-Al$_2$O$_3$ support showed no characteristic IR bands within the range from 700 to 1200 cm^{-1}. Although the exact peak positions representing W=O and W-O-W bands were not clearly resolved in the H$_3$PW$_{12}$O$_{40}$/γ-Al$_2$O$_3$ catalyst, W=O and W-O-W bands appeared as shoulder peaks. A weak IR band appearing at 1065 cm^{-1} in the H$_3$PW$_{12}$O$_{40}$/γ-Al$_2$O$_3$ catalyst can be assigned to the P-O band. An important point is that P-O band in the H$_3$PW$_{12}$O$_{40}$/γ-Al$_2$O$_3$ catalyst appeared at smaller wavenumber than that in the bulk H$_3$PW$_{12}$O$_{40}$ catalyst. A shift of P-O band observed in the supported catalyst implies that rather a strong interaction was formed between H$_3$PW$_{12}$O$_{40}$ catalyst and γ-Al$_2$O$_3$ support. The above result also indicates that H$_3$PW$_{12}$O$_{40}$/γ-Al$_2$O$_3$ catalyst maintained the primary structure of H$_3$PW$_{12}$O$_{40}$ catalyst [6].

Fig. 3 shows the XRD patterns of bulk H$_3$PW$_{12}$O$_{40}$ and supported H$_3$PW$_{12}$O$_{40}$/γ-Al$_2$O$_3$ catalysts. Unsupported H$_3$PW$_{12}$O$_{40}$ catalyst exhibited characteristic XRD pattern of the HPA. However, H$_3$PW$_{12}$O$_{40}$/γ-Al$_2$O$_3$ catalyst only showed characteristic XRD pattern of γ-Al$_2$O$_3$ without any peaks indicating crystalline phase of H$_3$PW$_{12}$O$_{40}$. This result demonstrates that H$_3$PW$_{12}$O$_{40}$ catalyst was finely dispersed on the γ-Al$_2$O$_3$ support.

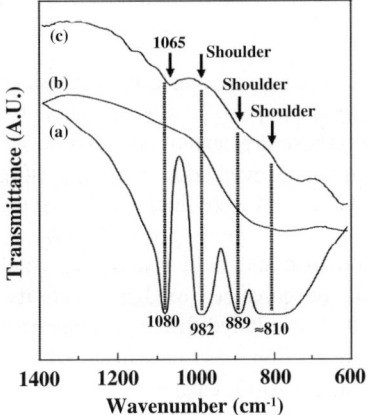

Fig. 2. IR spectra of (a) H$_3$PW$_{12}$O$_{40}$, (b) γ-Al$_2$O$_3$, and (c) H$_3$PW$_{12}$O$_{40}$/γ-Al$_2$O$_3$.

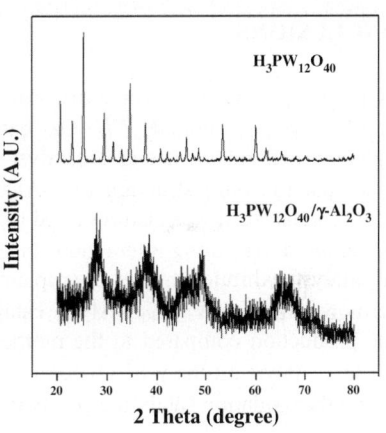

Fig. 3. XRD patterns of H$_3$PW$_{12}$O$_{40}$ and H$_3$PW$_{12}$O$_{40}$/γ-Al$_2$O$_3$.

3.2. Catalytic performance in the 2-propanol conversion

Fig. 4 shows the catalytic performance of H$_3$PW$_{12}$O$_{40}$ and H$_3$PW$_{12}$O$_{40}$/γ-Al$_2$O$_3$ catalysts in the vapor-phase 2-propanol conversion reaction. The supported H$_3$PW$_{12}$O$_{40}$/γ-Al$_2$O$_3$ catalyst exhibited higher 2-propanol conversion than the unsupported catalyst, resulting from high dispersion of H$_3$PW$_{12}$O$_{40}$ on the γ-Al$_2$O$_3$ support. It should be noted that acetone yields over the H$_3$PW$_{12}$O$_{40}$/γ-Al$_2$O$_3$ were much higher than those over the mother catalyst. It is known that acetone is formed by the redox function of HPA, while propylene is produced by the acid catalytic function of HPA catalyst [7]. The enhanced oxidation activity of the supported catalyst may be understood in a similar manner as described in a literature [7]. It is concluded that high dispersion of H$_3$PW$_{12}$O$_{40}$ catalyst on the γ-Al$_2$O$_3$ support via strong interaction was responsible for the enhanced oxidation catalytic activity of the H$_3$PW$_{12}$O$_{40}$/γ-Al$_2$O$_3$ catalyst.

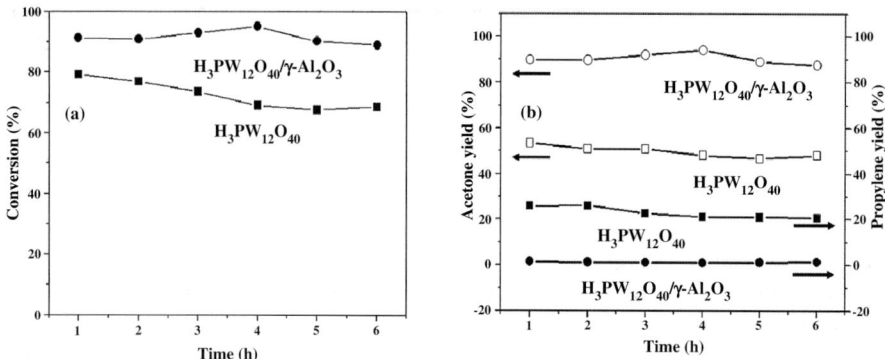

Fig. 4. Catalytic performance of $H_3PW_{12}O_{40}$ and $H_3PW_{12}O_{40}/\gamma$-Al_2O_3 catalysts at 330°C: (a) 2-propanol conversions and (b) product yields.

4. CONCLUSIONS

A mesoporous γ-Al_2O_3 with randomly ordered pore structure was successfully synthesized by a post hydrolysis method. $H_3PW_{12}O_{40}$ catalyst supported on γ-Al_2O_3 was prepared by an impregnation method for use as a catalyst in the vapor-phase 2-propanol conversion. It was observed that the pore structure of γ-Al_2O_3 was maintained even after the supporting of $H_3PW_{12}O_{40}$ catalyst. It was also revealed that $H_3PW_{12}O_{40}$ species were finely dispersed on the γ-Al_2O_3 support via strong interaction. In the catalytic reaction, the supported $H_3PW_{12}O_{40}/\gamma$-Al_2O_3 catalyst exhibited higher 2-propanol conversion than the bulk $H_3PW_{12}O_{40}$ catalyst. Furthermore, the $H_3PW_{12}O_{40}/\gamma$-Al_2O_3 catalyst showed the enhanced oxidation activity for acetone production compared to the mother catalyst. It is concluded that high dispersion of $H_3PW_{12}O_{40}$ catalyst on the γ-Al_2O_3 support via strong interaction was much more efficient in enhancing the oxidation catalytic activity of the $H_3PW_{12}O_{40}/\gamma$-Al_2O_3 catalyst.

ACKNOWLEDGEMENTS

The authors acknowledge the support from Seoul National University (400-20040199).

REFERENCES

[1] M. Misono, Catal. Rev. -Sci. Eng., 29 (1987) 199.
[2] I.K. Song, H.S. Kim and M.-S. Chun, Korean J. Chem. Eng., 20 (2003) 844.
[3] K. Nowinska, R. Formaniak, W. Kaleta and A. Waclaw, Appl. Catal. A, 256 (2003) 115.
[4] Y. Kim, B. Lee and J. Yi, Korean J. Chem., Eng. 19 (2002) 908.
[5] T. Okuhara, N. Mizuno and M. Misono, Adv. Catal., 41 (1996) 113.
[6] W. Kuang, A. Rives, M. Fournier and R. Hubaut, Appl. Catal. A, 250 (2003) 221.
[7] J.K. Lee, I.K Song, W.Y. Lee and J.-J. Kim, J. Mol. Catal. A, 104 (1996) 311.

Studies in Surface Science and Catalysis, volume 159
Hyun-Ku Rhee, In-Sik Nam and Jong Moon Park (Editors)

NiO-TiO$_2$ Catalyst Modified with WO$_3$ for Ethylene Dimerization

Jong Rack Sohn[*] **and Jong Soo Han**
Dept. of Applied Chemistry, Kyungpook National University, Taegu, 702-701, Korea
[*] jrsohn@knu.ac.kr

1. Introduction

Heterogeneous catalysts for the dimerization and oligomerization of olefins have been known for many years. A considerable number of papers have dealt with the problem of nickel-containing catalysts for ethylene dimerization [1-3].

Recently, Hino and Arata reported zirconia-supported tungsten oxide as an alternative material in reaction of requiring strong acid sites [4]. Several advantages of tungstate, over sulfate, as a dopant include that it does not suffer from dopant loss during thermal treatment and it undergoes significantly less deactivation during catalytic reaction. As an extension of our study on ethylene dimerization, we also prepared other catalyst systems by combining nickel hydroxide to give low valent nickel after decomposition with TiO$_2$ modified with WO$_3$ which is known to be an acid [5]. In this paper, characterization of NiO-TiO$_2$/WO$_3$ and catalytic activity for ethylene dimerization are reported.

2. Experimental

The catalysts containing various tungsten oxide contents were prepared by adding an aqueous solution of ammonium metatungstate[(NH$_4$)$_6$(H$_2$W$_{12}$O$_{40}$)-nH$_2$O] to the Ni(OH)$_2$-Ti(OH)$_4$ powder followed by drying and calcining at high temperatures for 1.5 h in air. This series of catalysts are denoted by their weight percentage of NiO and WO$_3$. The Raman spectra were recorded on a Spex Ramalog spectrometer with holographic gratings. The 5145-Å line from a Spectra-Physics Model 165 argon-ion laser was used as the exciting source. Catalysts were checked in order to determine the structure of the support as well as that of tungsten oxide by means of a Jeol Model JDX-8030 diffractometer, employing Cu Kα (Ni-filtered) radiation. Chemisorption of ammonia was also employed as a measure of the acidity of catalysts [6]. The catalytic activity for ethylene dimerization was determined at 20 °C using a conventional static system following a pressure change from an initial pressure of 290 Torr. A fresh catalyst sample of 0.2 g was used for every run and the catalytic activity was calculated as the number of moles of ethylene consumed in the initial 10 min.

3. Results and Discussion
3.1 Raman Spectra

The Raman spectra of WO$_3$, 25-NiO-TiO$_2$/30-WO$_3$, 25-NiO-TiO$_2$/15-WO$_3$, 25- NiO-TiO$_2$/5-WO$_3$, and TiO$_2$ under ambient conditions are presented in Fig. 1. The WO$_3$ structure is made up distorted WO$_3$ octahedra. The major vibrational modes of WO$_3$ are located at 808, 714, and 276 cm^{-1}, and have been assigned to the W=O stretching mode, the W=O bending mode, and the W-O-W deformation mode, respectively [7]. The Raman spectrum of the 25-

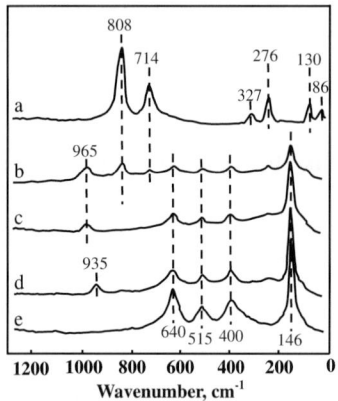

Fig. 1. Raman spectra of (a) WO$_3$, (b) 25-NiO-TiO$_2$/30-WO$_3$, (c) 25-NiO-TiO$_2$/15-WO$_3$, (d) 25-NiO-TiO$_2$/5-WO$_3$, and (e) TiO$_2$.

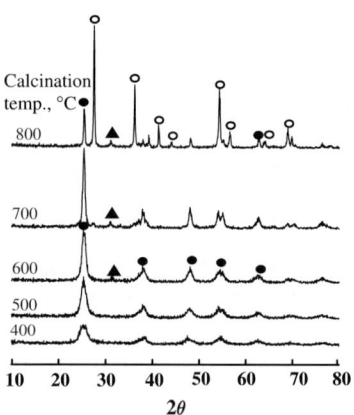

Fig. 2. X-ray diffraction patterns of 5-NiO-TiO$_2$/15-WO$_3$ calcined at different temperature for 1.5 h : •, anatase TiO$_2$; ○, rutile TiO$_2$; ▲, NiTiO$_3$.

NiO-TiO$_2$/5-WO$_3$ sample shows a weak and broad band at ~ 935 cm^{-1} which is characteristic of tetrahedrally coordinated surface tungsten oxide species [8]. In addition to this 935 cm^{-1} band, the feature of titania at 640, 515, 400, and 146 cm^{-1} is also present. These Raman bands have been assigned to the anatase modification [9]. As the loading is increased, the W=O stretching modes shifts upward to 965 cm^{-1}. The 965 cm^{-1} band in the Raman spectrum of the 25-NiO-TiO$_2$/15-WO$_3$ sample is assigned to the octahedrally coordinated polytungstate species [8]. The shift from 935 cm^{-1} for 25-NiO-TiO$_2$/5-WO$_3$ to 965 cm^{-1} for 25-NiO-TiO$_2$/15-WO$_3$ is in agreement with reported Raman spectra [8], suggesting the presence of different two-dimensional tungsten oxide species.

3.2 Crystalline Structures of Catalysts

For NiO-TiO$_2$/WO$_3$ catalysts the crystalline structures of samples were different from that of support, TiO$_2$. For the 5-NiO-TiO$_2$/15-WO$_3$, as shown in Fig. 2, TiO$_2$ is anatase phase up to 700 ℃. However, from 600 ℃ nickel titanium oxide (NiTiO$_3$) was observed due to the reaction between TiO$_2$ and NiO and its amount increased with increasing calcination temperature. X-ray diffraction data indicated a two-phase mixture of the anatase and rutile TiO$_2$ forms at 800 ℃. It is assumed that the interaction between tungsten oxide and TiO$_2$ hinders the transition of TiO$_2$ from anatase to rutile phase [5]. The presence of tungsten oxide strongly influences the development of textural properties with temperature in comparison with pure TiO$_2$. No phase of tungsten oxide was observed up to 25 wt % at any calcination temperature, indicating a good dispersion of tungsten oxide on the surface of TiO$_2$ support due to the interaction between them. Moreover, for the samples of 5-NiO-TiO$_2$/15-WO$_3$ and 25-NiO-TiO$_2$/15-WO$_3$ the transition temperature from the anatase to rutile phase was higher by 200 ℃, than that of pure TiO$_2$. As shown in Fig. 3, for 25-NiO-TiO$_2$/15-WO$_3$ TiO$_2$ was amorphous to X-ray diffraction up to 400 ℃, with a anatase phase of TiO$_2$ at 500-650 ℃, a two-phase mixture of the anatase and rutile TiO$_2$ forms at 700 ℃, and rutile TiO$_2$ forms at 800 ℃.

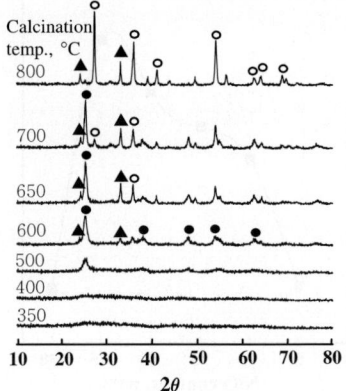

Fig. 3. X-ray diffraction patterns of 25-NiO-TiO₂/15-WO₃ calcined at different temperature for 1.5 h : ●, anatase TiO₂; ○, rutile TiO₂; ▲, NiTiO₃.

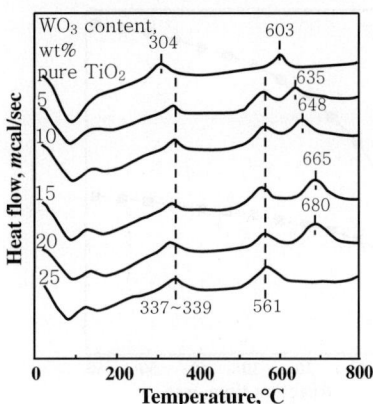

Fig. 4. DSC curves of precursors for TiO₂ and 25-NiO-TiO₂/WO₃ having different WO₃ contents.

3.3 Thermal Analysis

As shown in Fig. 4, for pure TiO₂ the DSC curve showed a broad endothermic peak around 100 ℃ due to water elimination, and two exothermic peaks at 304 and 603 ℃ due to the phase transition of TiO₂ from amorphous to anatase, and from anatase to rutile, respectively [5]. In the case of NiO-TiO₂/WO₃ a broad endothermic peak appeared around 170 ℃ is due to the evolution of NH_3 and H_2O decomposed from ammonium metatungstate, and an exothermic peak at 561 ℃ is due to the NiTiO₃ formation from NiO and TiO₂. As shown in Fig. 4, the exothermic peak due to the phase transition of TiO₂ appeared at 304 ℃ for pure TiO₂, while for NiO-TiO₂/WO₃ it was shifted to higher temperatures, 337-339 ℃. The exothermic peak due to the phase transition from anatase to rutile appeared at 603 ℃ for pure TiO₂, while for 25-NiO-TiO₂/WO₃ the shift to higher temperature increased with increasing WO₃ content. It is considered that the interaction between WO₃ and TiO₂ delays the transition of TiO₂ from amorphous to anatase phase and from anatase to rutile phase [5,10].

3.4 Catalytic Activities for Ethylene Dimerization

NiO-TiO₂/15-WO₃ catalysts were tested for their effectiveness in ethylene dimerization. It was found that over 25-NiO-TiO₂/15-WO₃ and 5-NiO-TiO₂/15-WO₃, ethylene was continuously consumed, as shown by the results presented in Fig. 5. It is known that for ethylene dimerization the variations in catalytic activities are closely correlated to the acidity of catalysts[6]. The acidity of some samples after evacuation at 400 ℃ is listed in Table 1 together with their surface area. The catalytic activity of NiO-TiO₂/15-WO₃ containing different NiO contents were examined; the results are shown as a function of NiO content in Fig. 6. The catalytic activity increased with increasing the NiO content, reaching a maximum at 25 wt%. In view of Table 1 and Fig. 6, the catalytic activities substantially run parallel to the acidity. It is remarkable that the samples which were not modified with WO₃ were inactive as catalysts for ethylene dimerization. Therefore, it is concluded that the high catalytic activity of NiO-TiO₂/WO₃ is closely correlated to the increase of acidity and acid strength by the addition of WO₃.

272

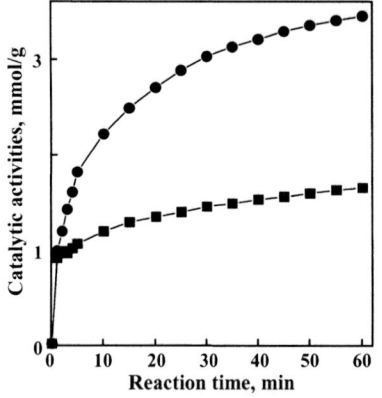

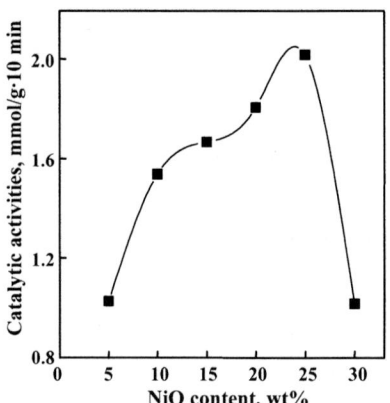

Fig. 5. Time-course of ethylene dimerization over catalysts evacuated at 400 ℃ for 1 h: (■) 5-NiO-TiO₂/15-WO₃; (●) 25-NiO-TiO₂/15-WO₃.

Fig. 6. Catalytic activities of NiO-TiO₂/15-WO₃ for ethylene dimerization as a function of NiO content.

Table 1. Specific surface area and acidity of NiO-TiO₂/15-WO₃ catalysts containing different NiO contents and calcined at 400 ℃ for 1.5 h

NiO content (wt %)	Surface area (m^2/g)	Acid amount (μmol/g)
Pure TiO₂	72	78
5	267	199
10	316	222
15	318	226
20	320	233
25	323	234
30	317	213
40	312	202

Acknowledgements

This work was supported by the Brain Korea 21 Project in 2003. We wish to thank Korea Basic Science Institute (Daegu Branch) for the use of X-ray diffractometer.

References

[1] F. Bernardi, A. Bottoni and I. Rossi, J. Am. Chem. Soc. 120 (1998) 7770.
[2] J. R. Sohn and W. C. Park, Appl. Catal. A:General 239 (2003) 269.
[3] J. S. Oh, B. Y. Lee and T. H. Park, Korean J. Chem. Eng. 21(1) (2004) 110.
[4] K. Arata, Adv. Catal. 37 (1990) 165.
[5] J. R. Sohn and J. H. Bae, Korean J. Chem. Eng. 17 (2000) 86.
[6] J. R. Sohn and S. H. Lee, Appl. Catal. A:General 266 (2004) 89.
[7] A. Anderson, Spectrosc. Lett. 9 (1976) 809.
[8] M. A. Vuurman, I. E. Wachs and A. M. Hirt, J. Phys. Chem. 95 (1991) 9928.
[9] J. Engweiler, J. Harf and A. Baiker, J. Catal. 159 (1996) 259.
[10] J. R. Sohn, S. G. Cho, Y. I. Pae and S. Hayashi, J. Catal. 159 (1996) 170.

Studies in Surface Science and Catalysis, volume 159
Hyun-Ku Rhee, In-Sik Nam and Jong Moon Park (Editors)
273

MoVW-mixed oxide as a partial oxidation catalyst for methanol to formaldehyde

B.Ramachandra[a*], Jung Sik Choi[b], Keun-Soo Kim[b], Ko-Yeon Choo, Jae-Suk Sung and Tae-Hwan Kim*

[a]Department of Chemistry, National Institute of Technology Karnataka, Surathkal-575025, India, Fax: +91-824-2476090, email: ram@nitk.ac.in

[b]Department of Chemical Engineering Chungnam National University, Yuseong-gu, Daejeon, Korea

Korea Institute of Energy Research, P.O.Box No.103, Yuseong-gu, Daejeon, Korea, fax: Fax: +82-42-860-3102, email: thkim@kier.re.kr

ABSTRACT

Mixed oxide with composition $Mo_{0.65}V_{0.25}W_{0.10}O_x$ was synthesized and its catalytic activity for the selective oxidation of methanol to formaldehyde was investigated. The characterization by scanning electron microscopy, X-ray diffraction, energy dispersive X-ray and Fourier transform infra-red spectroscopy reveals that the prepared catalyst is inhomogeneous nanocrystalline Mo_5O_{14}-type oxide with minor amount of MoO_3-type and MoO_2-type material. Thermal activation treatment of the catalyst at 813K resulted in better crystalline sample. The overall structural analysis suggests that the catalytic performance of the MoVW-mixed oxide catalyst in partial oxidation of methanol is related to the formation of the Mo_5O_{14}-type material.
Key words: Formaldehyde, Methanol, Mixed oxide, Partial oxidation.

1. INTRODUCTION

Transition metal oxides show a broad structural variety due to their ability to form phases of varying metal to oxygen ratios reflecting multiple stable oxidation states of the metal ions

[1-3]. Metal oxides exhibiting strong crystallographic anisotropy may show differing catalytic properties for different exposed crystal faces. One possible reason responsible for surface structure sensitivity may be the differently strong M=O bonds at the different surface planes. The stronger the M=O bond the more basic is its function with respect to hydrocarbon activation. This compound sensitivity may result from its special geometric, electronic or lattice diffusion properties providing selective active sites, an optimum match of catalyst and substrate electronic states and fast red ox kinetics. Structure and compound sensitivity for oxidation reactions serve as for the development and fundamental understanding of catalysts and their catalytic properties [4]. Methanol is very reactive and because the possible reaction products are linked to different reaction channels, the conversion of methanol to formaldehyde was used as a test reaction to investigate the catalytic properties of the mixed oxide. Pure MoO_3 catalysts exhibit high selectivity to formaldehyde. The real metal–oxygen stoichiometry and defect structure of molybdenum oxides thus may play an important role in selective partial oxidation reactions.

In this paper, the preparation, characterization and the catalytic performance of the $Mo_{0.65}V_{0.25}W_{0.10}O_x$–mixed oxide as a partial oxidation catalyst for the methanol to formaldehyde reaction was studied.

2. EXPERIMENTAL

Aqueous solutions of ammonium heptamolybdate (AHM), ammonium metatungstate (AMT), and ammonium metavanadate (AMV) having the respective transition metal concentrations were mixed in order to obtain the catalyst with a composition of Mo, W and V of 65, 10 and 25 wt%, respectively. This solution was dried by evaporation and decomposed under nitrogen at 700 K. The greenish black compound was obtained. The thermal activation treatment was done at 813 K in nitrogen for 2 h. The characterization of the catalyst was done by BET (Micrometrics ASAP-2010), SEM and EDX (Hitachi S-4700 FE SEM with EDX facilities.), XRD (Rigaku (D/Max2000-Ultima plus; X-ray radiation, CuKα) and FT-IR (Nicolet Magna IR 550) analysis. The partial oxidation of the methanol was carried out in a fixed bed reactor at atmospheric pressure. The activity of the catalyst was examined by taking 10 g of the 80 wt% of the catalyst prepared with SiC. The methanol to oxygen ration was kept at 1.47 and the nitrogen gas was kept at 60 sccm. The temperature of the reactor was varied in the range of 573 K to 673 K. The feed mixtures were prepared by injecting liquid methanol in to nitrogen flow. Analyses of the reaction products were done using online gas chromatograph.

3. RESULTS AND DISCUSSION

The BET surface area of the prepared catalyst is 5.9 g/m^2. The SEM, Fig. 1. picture shows agglomerates of platelet-like crystallites of a few hundreds of nanometers in size. However, there remained parts of the sample, which still showed irregular particle shapes. The EDX analysis, Fig. 2. gave the elemental distribution of Mo, V and W as 64.91, 24.42 and 10.67 wt% respectively. The mixed oxide characterized by an inhomogeneous elemental distribution on the length scale of few microns. It can be seen that the material is not well crystalline. This may be due to the different solubilities of the ammonium precursors led to elemental inhomogenities during the drying process.

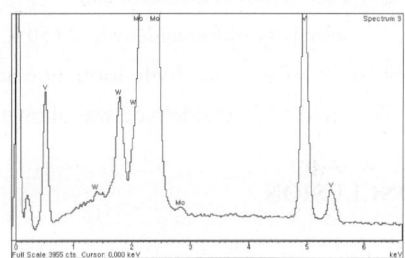

Fig. 1. SEM image of the catalyst activated at 813 K.

Fig. 2. EDX analysis of the thermally activated MoVW-mixed oxide catalyst.

The XRD pattern of the catalyst, Fig. 3. can be understood as thermal treatment lead to the crystallization of the catalyst and mixture of a majority of nanocrystalline Mo$_5$O$_{14}$-type oxide with minor amounts of nanocrystalline MoO$_3$ and MoO$_2$-type material [5]. The crystallization of the catalyst takes place only in a small temperature range and above which decomposes. The FTIR pattern, Fig. 4. shows the peak at 711 cm^{-1} suggests that there exits a multi phase component like Mo (or V or W)-O- Mo bond [6].

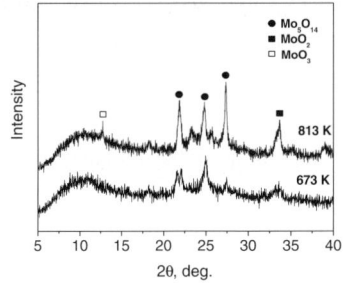

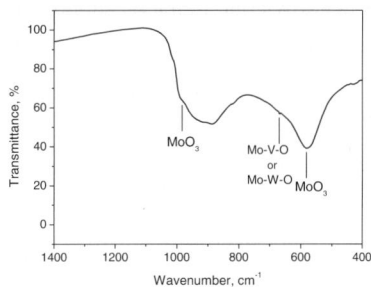

Fig. 3. XRD of the catalyst before and after the thermal treatment.

Fig. 4. FT-IR spectra of the MoVW catalyst.

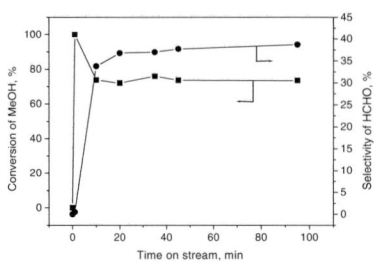

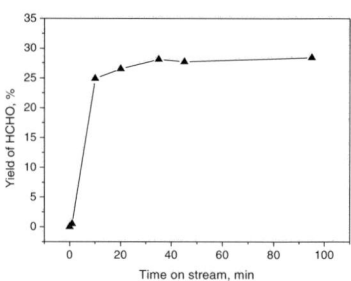

Fig. 4. Conversion of methanol and   Fig. 5. Yield of formaldehyde.
selectivity of formaldehyde at 650 K.

The selectivity of formaldehyde formation and conversion of methanol at 650 K is shown in Fig.5. The yield of formaldehyde was shown in Fig.6.

4. CONCLUSION

The present work demonstrates that the mixed oxide catalyst with inhomogeneous nanocrystalline Mo_5O_{14}-type oxide with minor amount of MoO_3- and MoO_2-type material. Thermal treatment of the catalyst shows a better performance in the formation of the crystals and the catalytic activity. The structural analysis suggests that the catalytic performance of the MoVW- mixed oxide catalyst in the partial oxidation of methanol is related to the formation of the Mo_5O_{14} type mixed oxide.

ACKNOWLEDGEMENTS

We gratefully acknowledge financial support from the CDRS Center, MOST, Korea. B.R is grateful to NITK, India and to KOSEF, South Korea for the Brain Pool fellowship.

REFERENCES

[1] J.B. Goodenough, in H.F. Barry, P.C.H. Mitchell (Eds.), in: Proceedings of the 4th International Conference on the Chemistry and Uses of, Climax Comp., Ann Arbor, MI, 1982, p. 1.
[2] H. Gruber, E. Krautz, Phys. Stat. Sol. (A) (1980) 615.
[3] E. Canadell, M.-H. Wangbo, Chem. Rev. (1991) 965.
[4] J. Haber, in: G. Ertl, H. Knözinger, J. Weitkamp (Eds.), Handbook of Heterogeneous Catalysis, Vol. 5 (Wiley VCH, Weinheim, 1997) p. 2253ff.
[5] H.Werner, O. Timpe, D. Herein, Y. Uchida, N. Pfaender, U. Wild, R. Schlögl and H. Hibst, Catal. Lett., 44, 153 (1997).
[6] K. Oshihara, Y. Nakamura, M. Sakuma and W. Ueda, Catal. Today., 71, 153 (2001).

Studies in Surface Science and Catalysis, volume 159
Hyun-Ku Rhee, In-Sik Nam and Jong Moon Park (Editors)

Benzene hydroxylation to Phenol with Iron impregnated Activated carbon Catalysts

Jung-Sik Choi*, Tae-Hwan Kim[a], Ko-Yeon Choo[a], Jae-Suk Sung[a], M.B. Saidutta[b]*,
B. Ramachandra[c] and Young-Woo Rhee

Department of Chemical Engineering, Chungnam National University, Yuseong-gu, Daejeon, Korea, E-mail : jschoi@kier.re.kr

[a]Korea Institute of Energy Research (KIER), P.O.Box No.103, Yuseong-gu, Daejeon, Korea

[b]Department of Chemical Engineering, National Institute of Technology Karnataka, Surathkal-575025, India, E-mail : mbskrec36@yahoo.co.in

[c]Department of Chemistry, National Institute of Technology Karnataka, Surathkal-575025, India

ABSTRACT

Iron impregnated on activated carbon was used as catalyst for the direct synthesis of phenol from benzene. The effect of Sn addition to the catalyst was studied. The prepared catalysts were characterized by BET, SEM and XRD analysis. The catalyst 5.0Fe/AC showed good activity in the conversion of benzene and addition of Sn seemed to improve the selectivity of phenol in the reaction.

Key words: Activated carbon, Benzene hydroxylation, Phenol, Transition metal

1. INTRODUCTION

Phenol is the starting material for numerous intermediates and finished products. About 90% of the worldwide production of phenol is by Hock process (cumene oxidation process) and the rest by toluene oxidation process. Both the commercial processes for phenol production are multi step processes and thereby inherently unclean [1]. Therefore, there is need for a cleaner production method for phenol, which is economically and environmentally viable. There is great interest amongst researchers to develop a new method for the synthesis of phenol in a one step process [2]. Activated carbon materials, which have large surface areas, have been used as adsorbents, catalysts and catalyst supports [3,4]. Activated carbons also have favorable hydrophobicity/ hydrophilicity, which make them suitable for the benzene hydroxylation. Transition metals have been widely used as catalytically active materials for the oxidation/hydroxylation of various aromatic compounds.

In this work, catalysts containing iron supported on activated carbon were prepared and investigated for their catalytic performance in the direct production of phenol from benzene with hydrogen peroxide and the effect of Sn addition to iron loaded on activated carbon catalyst were also studied.

2. EXPERIMENTAL

The catalysts used in benzene hydroxylation were prepared by impregnation method with each precursor. For impregnating iron on activated carbon (Aldrich Co. Norit RO 0.8), iron nitrate nonhydrate salt was used as metal precursor. The impregnation of Sn was carried out with tin chloride. The salts were dissolved in appropriate solvents. The salt solution and activated carbon were mixed together using a rotary equipment. After contacting the metal salt solution and activated carbon for half an hour, vacuum evaporation of the excess solvent was done. The catalyst samples were then dried at 355 K in an air oven overnight and calcined in a rotary kiln in nitrogen atmosphere at 825 K for 5 hours. The catalysts were characterized by BET, SEM and XRD. Benzene hydroxylation reactions were carried out in a jacketed stainless steel reactor at 338 K for 5 hours with 30% H_2O_2 as oxidant. The molar ratio of reactants benzene : H_2O_2 : solvent was 1:3:4.25 in each reaction. In all the experimental runs 0.1 g of catalyst was used. Analysis of the reaction products were done using HPLC.

3. RESULTS AND DISCUSSION

Nitrogen adsorption experiments showed a typical type I isotherm for activated carbon catalysts. For iron impregnated catalysts the specific surface area decreased from 1088 m^2/g (0.5 wt% Fe) to 1020 m^2/g (5.0 wt% Fe). No agglomerization of metal tin or tin oxide was observed from the SEM image of 5Fe-0.5Sn/AC catalyst (Fig. 1). In Fig. 2 iron oxides on the catalyst surface can be seen from the X-Ray diffractions. The peaks of tin or tin oxide cannot be investigated because the quantity of loaded tin is very small and the dispersion of tin particle is high on the support surface.

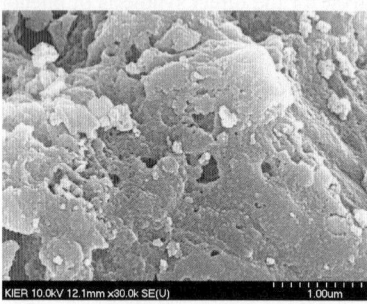

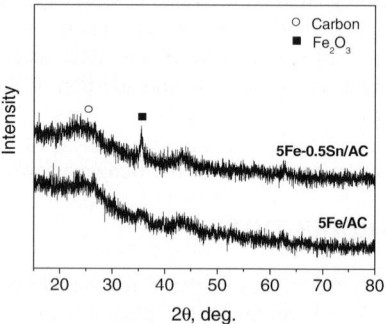

Fig. 1. SEM image of 5Fe-0.5Sn/AC catalyst. Fig. 2. XRD patterns of used catalysts.

In the present study, catalysts containing 0.5 wt% transition metals were prepared and reaction was carried out. The results of these studies are presented in Fig. 3. In screen test of various transition metals loaded AC catalysts acetone was used as the solvent. It was also found that Fe containing catalyst gave the highest yield of phenol. Hence, for further study catalysts containing Fe loaded on activated carbon were chosen.

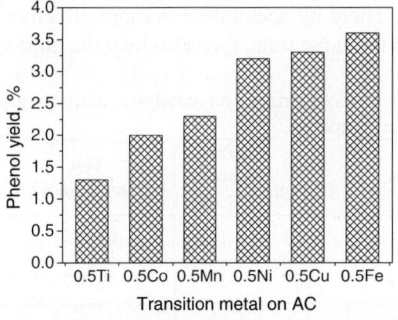

Fig. 3. Phenol yield on various transition metals loaded AC catalysts.

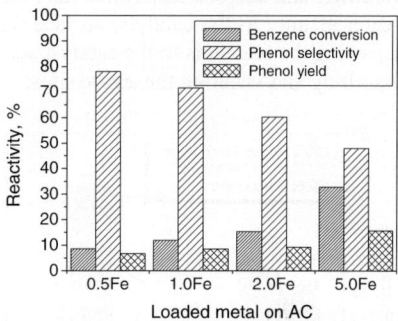

Fig. 4. Reactivity of Fe/AC

Catalysts were prepared with 0.5, 1.0, 2.0 and 5.0 wt% of iron loaded on activated carbon. Benzene hydroxylation with hydrogen peroxide as oxidant was carried out. The conversion of benzene, selectivity and yield of phenol for these catalysts are shown in Fig. 4. As the weight of loaded metal increased the benzene conversion increased by about 33% but the selectivity to phenol decreased. The yield of phenol that was obtained with 5.0Fe/AC was about 16%.

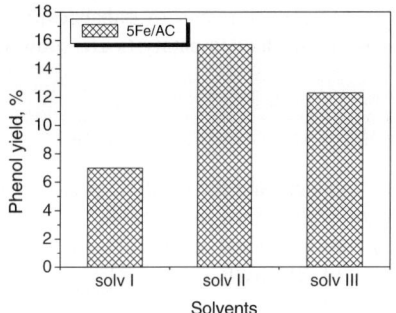

Fig. 5. The effect of different solvents
※ solv I , acetone(20.8g); solv II, acetonitrile (14.7g), solv III, acetonitrile(20.8g)

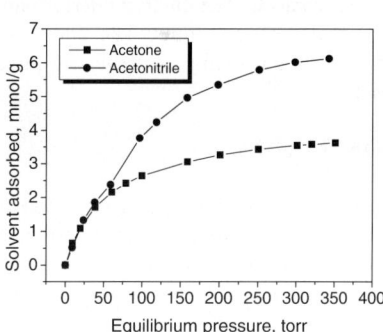

Fig. 6. The adsorption behaviors of 5.0Fe/AC with acetone and acetonitrile.

To study the effect of solvents, reactions carried out with acetonitrile (benzene : acetonitrile = 1:4.65, 1:6.58 mole ratio), and acetone (benzene : acetone = 1:6.58 mole ratio) as solvents on 5.0 wt% iron impregnated activated carbon catalyst and the results were compared. The results are presented in Fig. 5. It can be seen that acetonitrile as a solvent leads to more yield of phenol on these catalysts.

The performance of various solvents can be explained with the help of the role of these solvents in the reaction. These solvents help in keeping both benzene and hydrogen peroxide in one phase. This helps in the easy transport of both the reactants to the active sites of the catalyst. The acetonitrile, and acetone adsorption data on these catalysts (Fig. 6), suggests that acetonitrile has a greater affinity to the catalytic surface than acetone. There by acetonitrile is more effective in transporting the reactants to the catalyst active sites. At the same time, they also help the products in desorbing and vacating the active sites.

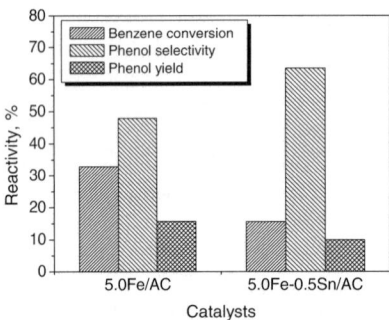

Fig. 7. The effect of Sn addition to 5.0Fe/AC.

Table 1. Comparison on catalytic activity and physical property.

Cat.	BET (cm^2/g)	Pore size* (Å)	TOF (mmol/mmol·h)
0.5Fe	1088	2.8	62.7631
1.0Fe	1167	3.1	40.0423
2.0Fe	1079	2.9	21.6787
5.0Fe	1020	2.8	14.7476
5.0Fe-0.5Sn	1009	2.7	9.2972

* Median pore radius by H-K equation

To improve selectivity towards phenol 0.5 wt% of Sn was added as a promoter while preparing 5.0Fe/AC catalyst. The catalytic performance of 5.0Fe-0.5Sn/AC catalyst was investigated under similar reaction conditions. The addition of Sn to Fe/AC catalyst seems to enhance phenol selectivity by 33% (Fig. 7). TOF and physical properties of iron loaded catalysts are shown in Table 1.

4. CONCLUSION

The preparation of iron impregnated activated carbon as catalysts and the catalytic performance of these catalysts were studied in benzene hydroxylation with hydrogen peroxide as oxidant. 5.0Fe/AC catalyst containing 5.0 wt% iron on activated carbon yielded about 16% phenol. The addition of Sn on 5.0Fe/AC catalyst led to the enhancement of selectivity towards phenol.

REFERENCES

[1] T. Miyake, M. Hamada, Y. Sasaki and M. Oguri, App. Catal. A: Gen., 133, 33 (1995).
[2] M. Stockmann, F. Konietzni, J.V. Notheis, J. Voss, W. Keune and W.F. Maier, App. Catal. A: Gen., 208, 343 (2001).
[3] F. Rodrguez-Reinoso, Carbon, 36, 159 (1998).
[4] J.S. Choi, T.H. Kim, M.B. Saidutta, J.S. Sung, K.I. Kim, R.V. Jasra, S.D. Song and Y.W. Rhee, J. Ind. Eng. Chem., 10, 445 (2004).

Studies in Surface Science and Catalysis, volume 159
Hyun-Ku Rhee, In-Sik Nam and Jong Moon Park (Editors)

Liquid film state under reactive distillation conditions for the dehydrogenation of decalin on platinum supported on active carbon and boehmite

S. Sugiyama,[a] C. Shinohara,[a] D. Makino,[a] S. Kawakami[a] and H. Hayashi[a]

[a]Department of Chemical Science and Technology, Faculty of Engineering,
The University of Tokushima, Minamijosanjima, Tokushima 770-8506, Japan

ABSTRACT

The dehydrogenation of decalin to naphthalene has been investigated on Pt/C, Pt/Al(OH)O and Pt/Al$_2$O$_3$ catalysts. The maximum conversion of decalin on 3.9% Pt/C, which did not repel decalin, was observed at 483 K under the conditions of 0.3 g of the catalyst and 1ml of decalin, which was corresponded to the liquid film state under reactive distillation conditions. However such a maximum was not observed on Pt/Al(OH)O and Pt/Al$_2$O$_3$, which repelled decalin. Furthermore it was found that the reaction temperature, at which the maximum hydrogen evolution was observed on Pt/C, was shifted from the boiling point of decalin to that of naphthalene with increasing the amount of naphthalene in the reaction solution.

1. INTRODUCTION

As a key technology for hydrogen supply to fuel cell vehicles, the combination of catalytic dehydrogenation of decalin with hydrogenation of naphthalene has been proposed. Although the catalytic hydrogenation of naphthalene to decalin was established in 1940s, the active catalysts for the dehydrogenation of decalin have been recently developed. It has been suggested that catalytic dehydrogenation of decalin could be accomplished efficiently with Pt/C in the liquid film state under reactive distillation conditions [1]. In the present system, the employment of the liquid film state under reactive distillation conditions was suggested to obtain an efficient dehydrogenation of decalin [1,2]. However the nature of the liquid film state is not yet clear although the liquid film state has been generally defined as the specific ratio of the weight of catalyst and the volume of decalin at a reaction temperature greater than boiling point of decalin [1]. In the present study, the effects of the support and the reaction temperature are investigated to obtain further information on the liquid film state.

2. EXPERIMENTAL

The preparation procedure of 3.9 wt. % Pt/C was shown in our previous paper [2]. The commercially available 5 wt. % Pt/C (Aldrich) and 1 wt. % Pt/Al$_2$O$_3$ (Aldrich) were also employed. Boehmite (Al(OH)O) was prepared from the corresponding sol, which was obtained from hydrolysis of aluminum isopropoxide [3]. Platinum was supported on boehmite (1.5 wt. % Pt/Al(OH)O) with the essentially identical procedure for Pt/C but the support was not pretreated with NaOH. Batch-wise catalytic dehydrogenation of decalin was performed in a two-necked Kjeldahl flask with a short neck equipped with a reflux condenser. Details of the reaction procedure have been described elsewhere [2]. In the dehydrogenation of decalin, the volume of Pt/Al(OH)O and Pt/Al$_2$O$_3$ was adjusted to be the same that of Pt/C. Thus 0.3 g of Pt/C and 1.0g of Pt/Al(OH)O and Pt/Al$_2$O$_3$ were employed since the density of active carbon was approximately third that of those aluminum catalysts. The surface areas of 5 wt.% Pt/C, 1 wt.% Pt/Al$_2$O$_3$ and 1.46 wt.% Pt/Al(OH)O were 1445, 286 and 170 m^2/g, respectively. X-ray absorption fine structure (XAFS) near Pt L$_3$-edge was measured (2.5 GeV) with a storage ring current of 340 mA at the High Energy Research Organization (Proposal No. 2003G065). Data from XAFS were analysed as previously described [2].

3. RESULTS AND DISCUSSION

Figure 1 shows the effects of the volume of decalin on the conversion of decalin on 3.9 wt. % Pt/C (0.3g) (a), 1 wt. % Pt/Al$_2$O$_3$ (1.0 g) (b) and 1.46 wt. % Pt/Al(OH)O (1.0 g) (C) at 483 K. Under those conditions, 0.0117, 0.010 and 0.0146 g of Pt were contained in the systems with Pt/C, Pt/Al$_2$O$_3$ and Pt/Al(OH)O, respectively. The conversion of decalin on Pt/C showed to be a maximum at 1 ml of decalin (Fig.1 (a)). This point is generally accepted as the liquid film state under reactive distillation conditions, at which the catalyst was just wet but not suspended at all through the dehydrogenation and covered with a thin film of liquid substrate. If such reactive distillation conditions are attained, the dehydrogenation proceeds more efficiently than liquid- and gas-phases [1].

However the employment of Al$_2$O$_3$ and Al(OH)O as a support afforded the dissimilar results as shown in Figs. 1 (b) and (c), respectively. Active carbon does not repel decalin while boehmite and alumina repel extensively decalin, indicating that the affinity between the liquid substrate and the support should be considered in the preparation of the catalysts.

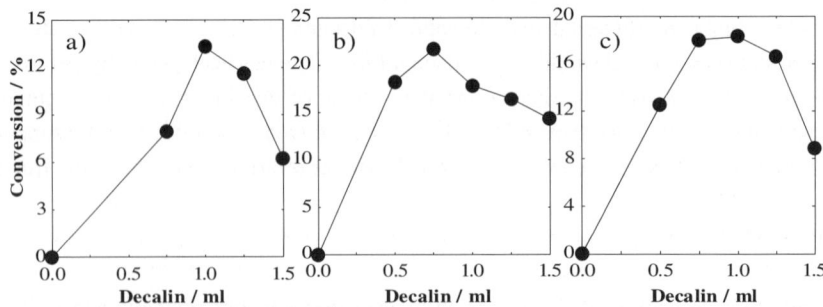

Fig. 1   Relationship between the conversion of decalin and the catalyst weight of 3.9 wt. % Pt/C (0.3g), 1 wt. % Pt/Al$_2$O$_3$ (1.0 g) and 1.46 wt. % Pt/Al(OH)O (1.0 g) at 483 K.

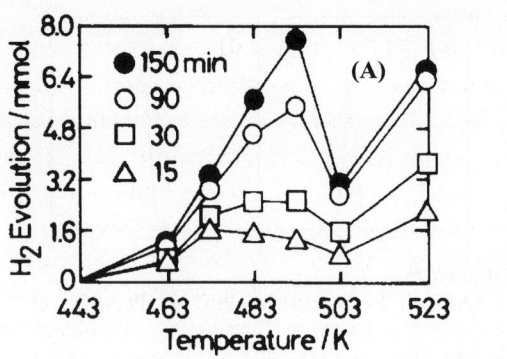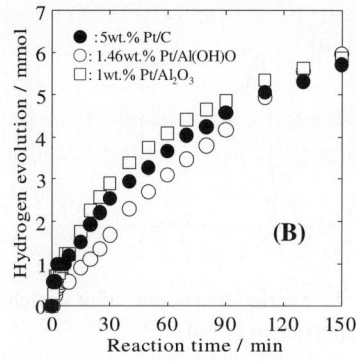

Fig. 2 Hydrogen evolution at various reaction times on 5 wt. % Pt/C (A) and on various catalysts at 483 K (B).

It should be noted that the ratio of decalin against naphthalene is changed during the dehydrogenation, indicating that the suitable reaction temperature for the liquid film state may be also changed. Therefore various reaction temperatures have been employed for the dehydrogenation of decalin on 5 % Pt/C. As shown in Fig. 2 (A), the maximum hydrogen evolution at 15 min was observed at approximately 473 K. Under the present conditions, an excess volume of decalin over that of naphthalene is present in the liquid substrate while the reaction temperature, at which the maximum hydrogen evolution is observed, is essentially corresponded to the boiling point of decalin (466 and 458 K for cis- and trans-isomers, respectively). With increasing the reaction time, the temperature, at which the maximum evolution was observed, was shifted from 473 to 493 K. The increase of the reaction time results in the enhancement of naphthalene ratio in the liquid substrate while the reaction temperature, at which the maximum hydrogen evolution is observed, is essentially corresponded to the boiling point of naphthalene (491 K). Thus the control of the reaction temperature based on the boiling point of those substrates with increasing the reaction time is important for the achievement of the liquid film state under reactive distillation conditions.

In our previous paper [2], it has been revealed that catalytic active sites on Pt/C, which was prepared with the active carbon support treated with NaOH and showed great activities for the dehydrogenation of decalin to naphthalene, are non-metallic Pt species. In order to examine the effects of the supports on the catalytic activities and active sites over those catalysts the dehydrogenation of decalin to naphthalene has been investigated on 5 wt. % Pt/C, 1 wt.% Pt/Al_2O_3 and 1.46 wt. % Pt/Al(OH)O. The dehydrogenation behaviors on the commercially available Pt/C, Pt/Al_2O_3 and Pt/Al(OH)O were described in Fig. 2 (B). Although the rather similar contents of Pt species in those systems and the dissimilar surface areas of those three catalysts are evident, it is of interest to note that hydrogen evolution from decalin after 150 min are rather identical. Therefore the nature of Pt species is strongly influenced by the supports. From XAFS analyses, the nearest-neighbor distances of Pt-Pt in Pt/Al_2O_3 (Fig. 3 (b)) and Pt/Al(OH)O (Fig. 3 (c)) were 0.283 and 0.279 nm, which were

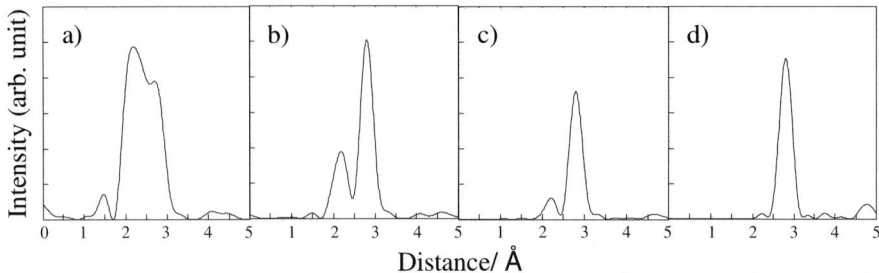

Fig. 3 Fourier transforms of k³-weighted EAXFS of Pt L₃-edge: a) Pt/C, b) Pt/Al₂O₃, c) Pt/Al(OH)O, d) Pt foil.

corresponding to that (0.280 nm) in Pt foil (Fig. 3 (d)), indicating that the nature of Pt species in those catalysts is essentially identical to that of metallic Pt. However the rather shorter Pt-Pt distance (0.259 nm) was obtained from the commercially available Pt/C (Fig. 3 (a)). As shown in our previous study [2], the nearest-neighbor distance of Pt-Pt in active Pt/C prepared with active carbon treated with NaOH solution was 0.265 nm while that in non-active Pt/C prepared with the support not-treated with NaOH was 0.280 nm. Those results from the active Pt/C indicate that the high dispersion of Pt affords a more covalent species, resulting in the great activities for the dehydrogenation of decalin. Based on those previous results, present XAFS analyses of the commercially available Pt/C reveal that Pt species with covalent nature are also present on the commercial Pt/C catalysts. It should be noted that metallic Pt species in Pt/Al₂O₃ and Pt/Al(OH)O afforded the corresponding activities to those on Pt/C catalysts, indicating that the interaction of Pt species and the support together with the affinity between the support and the substrate contributes to the activities together with the nature of the catalytic active species. Such an interaction in Pt/Al₂O₃ may affords the greater conversion on Pt/Al₂O₃ than that on Pt/C shown in Fig. 1.

In conclusion, the affinity between the substrate and the support together with the adjustment of the reaction temperature should be considered in order to attain the liquid film state for the dehydrogenation of decalin although the liquid film state has been generally defined as the specific ratio of the weight of the catalyst and the volume of decalin at a reaction temperature greater than boiling point of decalin. Furthermore it is also suggested that interaction between metallic Pt species and the support in Pt/Al₂O₃ and Pt/Al(OH)O together with covalent nature of Pt species in Pt/C catalysts may contribute to the dehydrogenation.

REFERENCES

[1] S. Hodoshima, H. Arai, S. Takikawa and Y. Saito, Int. J. Hydrogen Energy, 28 (2003) 1255.

[2] C. Shinohara, S. Kawakami, T. Moriga, H. Hayashi, S. Hodoshima, Y. Saito and S. Sugiyama, Appl. Catal. A: Gen., 226 (2003) 251.

[3] S. Sugiyama, S. Kawakami, S. Tanimoto, M. Fujii, H. Hayashi, F. Shibao and K. Kusakabe, J. Chem. Eng. Jpn., 36 (2003) 109

Studies in Surface Science and Catalysis, volume 159
Hyun-Ku Rhee, In-Sik Nam and Jong Moon Park (Editors)

Elucidation of reduction behaviors for Co/TiO$_2$ catalysts with various rutile/anatase ratios

Bunjerd Jongsomjit[*], Tipnapa Wongsalee, and Piyasan Praserthdam

Center of Excellence on Catalysis and Catalytic Reaction Engineering
Department of Chemical Engineering, Chulalongkorn University
Bangkok 10330 Thailand, [*]corresponding author: bunjerd.j@chula.ac.th

The present study revealed effects of various rutile/anatase ratios in titania on the reduction behaviors of titania-supported cobalt catalysts. It was found that the presence of rutile phase in titania could facilitate the reduction process of the cobalt catalyst. As a matter of fact, the number of reduced cobalt metal surface atoms, which is related to the overall activity during CO hydrogenation increased.

1. INTRODUCTION

Supported cobalt (Co) catalysts are preferred for Fischer-Tropsch synthesis (FTS) based on natural gas [1] due to their high activities for FTS, high selectivity for long chain hydrocarbons and low activities for the competitive water-gas shift (WGS) reaction. Many inorganic supports such as silica, alumina, titania and Zeolites have been extensively studied for supported Co catalysts for years. It is known that in general, the catalytic properties depend on reaction conditions, catalyst compositions, types of inorganic supports and the degrees of metal dispersion as well. It is reported that during the past decades, titania-supported Co catalysts have been investigated widely by many authors, especially for the application of FTS in a continuously stirred tank reactor (CSTR) [2-4]. However, it should be noted that titania itself has different crystalline phases such as anatase, brookite and rutile phase. Thus, the differences in compositions of crystalline phases could result in changes on physical and chemical properties of titania, then consequently for the dispersion of cobalt. In order to give a better understanding of those, the focus of this present study was to investigate the cobalt dispersion on titania consisting various ratios of rutile/anatase. The Co/TiO$_2$ was prepared and then characterized using different characterization techniques.

2. EXPERIMENTAL

2.1 Material preparation

Preparation of titania support

The various ratios of rutile:anatase in titania support were obtained by calcination of pure anatase titania (obtained from Ishihara Sangyo, Japan) in air at temperatures between 800-1000°C for 4 h. The high space velocity of air flow (16,000 h^{-1}) insured the gradual phase transformation to avoid rapid sintering of samples. The ratios of rutile:anatase were determined by XRD according to the method described by Jung et al. [5] as follows:

$$\% \,Rutile = \frac{1}{[(A/R)0.884 + 1]} \times 100$$

Where, A and R are the peak area for major anatase ($2\theta = 25^\circ$) and rutile phase ($2\theta = 28^\circ$), respectively.

Preparation of catalyst samples
A 20 wt% of Co/TiO_2 was prepared by the incipient wetness impregnation. A designed amount of cobalt nitrate [$Co(NO_3)\bullet 6H_2O$] was dissolved in deionized water and then impregnated onto TiO_2 containing various ratios of rutile:anatase obtained from above. The catalyst precursor was dried at $110^\circ C$ for 12 h and calcined in air at $500^\circ C$ for 4 h.

2.2   Catalyst nomenclature
The nomenclature used for the catalyst samples in this study is following:
Rn: titania support containing **n%** of rutile phase (R)
Co/Rn: titania support containing **n%** of rutile phase (R)-supported cobalt

2.3   Catalyst characterization
X-ray diffraction: XRD was performed to determine the bulk crystalline phases of catalyst. It was conducted using a SIEMENS D-5000 X-ray diffractometer with CuK_α ($\lambda = 1.54439$ Å). The spectra were scanned at a rate of 2.4 degree/min in the range $2\theta = 20\text{-}80$ degrees.
Scanning electron microscopy and energy dispersive X-ray spectroscopy: SEM and EDX were used to determine the catalyst morphologies and elemental distribution throughout the catalyst granules, respectively. The SEM of JEOL mode JSM-5800LV was applied. EDX was performed using Link Isis series 300 program.
Transmission electron microscopy (TEM): The dispersion of cobalt oxide species on the titania supports were determined using a JEOL-TEM 200CX transmission electron spectroscopy operated at 100 kV with 100k magnification.
Hydrogen chemisorption: Static H_2 chemisorption at $100^\circ C$ on the reduced cobalt catalysts was used to determine the number of reduced surface cobalt metal atoms. This is related to the overall activity of the catalysts during CO hydrogenation. Gas volumetric chemisorption at $100^\circ C$ was performed using the method described by Reuel and Bartholomew [6]. The experiment was performed in a Micromeritics ASAP 2010 using ASAP 2010C V3.00 software.
Temperature-programmed reduction: TPR was used to determine the reduction behaviors of the catalyst samples. It was carried out using 50 mg of a sample and a temperature ramp from 35 to $800^\circ C$ at $5^\circ C$/min. The carrier gas was 5% H_2 in Ar. A cold trap was placed before the detector to remove water produced during the reaction.

3. RESULTS AND DISCUSSION

In this present study, we basically showed dependence of the number of reduced cobalt metal surface atoms on dispersion of cobalt oxides along with the presence of rutile phase in titania. Both XRD and SEM/EDX results (not shown) revealed good distribution of cobalt oxides over the titania support. However, it can not differentiate all samples containing various ratios of rutile/anatase phase. Thus, in order to determine the dispersion of cobalt oxide species on titania, a more powerful technique such as TEM was applied with all samples. The TEM micrographs for all samples are shown in Figure 1. The dark spots represented cobalt oxides species present after calcination of samples dispersing on titania consisting various ratios of rutile:anatase. It can be observed that cobalt oxide species were highly dispersed on the titania supports for Co/R0, Co/R3, and Co/R19 samples resulting in an appearance of smaller cobalt oxide patches present. However, the degree of dispersion for cobalt oxide species essentially decreased with increasing the rutile phase in titania from 40 to 99% as seen for Co/R40, Co/R96, and Co/R99 samples resulting in the observation of larger cobalt oxide patches. It was suggested that the presence of rutile phase in titania from 0 (pure anatase phase) to 19% exhibited the highly dispersed forms of cobalt oxide species for the calcined samples. It is known that the active form of supported cobalt catalysts is cobalt metal (Co^0). Thus, reduction of cobalt oxide species is essentially performed in order to transform cobalt oxide species obtained after calcination process into the active cobalt metal atoms for catalyzing the reaction. Therefore, the static H_2 chemisorption on the reduced cobalt samples was used to determine the number of reduced Co metal surface atoms. This is usually related to the overall activity of the catalyst during carbon monoxide (CO) hydrogenation [7].

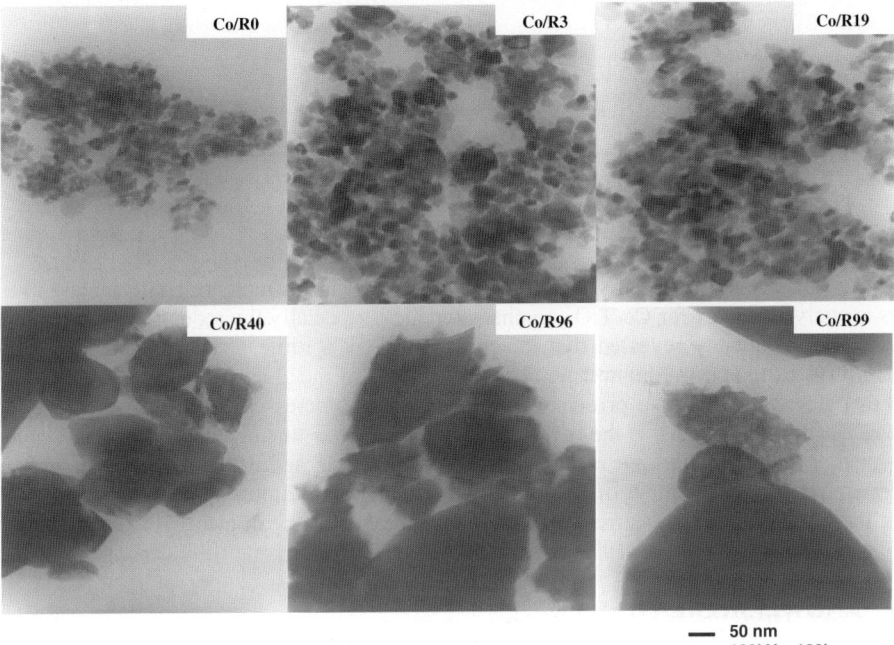

Figure 1 TEM micrographs of samples on various rutile/anatase ratios in titania

The resulted H_2 chemisorption for all samples revealed that the number of the reduced cobalt metal surface atoms increased with the presence of rutile phase in titania up to a maximum at 19% of rutile phase (Co/R19) before decreasing with the greater amounts of rultile phase as shown in **Table 1**. Considering the number of cobalt metal atoms for Co/R0 (pure anatase titania), the number was apparently low even though highly dispersed cobalt oxides species. This was suggested that highly dispersed forms of cobalt oxide species be not only the factor that insures larger number of reduced cobalt metal surface atoms in Co/TiO_2 [8]. On the other hand, it can be observed that the number of reduced cobalt metal surface atoms for Co/R40 and Co/R96 (with the low degree of dispersion of cobalt oxide species as seen by TEM) was larger than that for Co/R0. This was due to the presence of rutile phase in Co/R40 and Co/R96. It should be mentioned that the largest number of reduced cobalt metal surface atoms for the Co/R19 sample was attributed to both highly dispersed cobalt oxide species and the presence of rutile phase in titania. In addition, the resulted TPR as also shown in Table 1 confirmed that the presence of rultie phase could facilitate the reduction of cobalt oxide species by lowering the reduction temperatures. As a result, the number of the reduced cobalt metal surface atoms increased.

Table 1 Resulted H_2 chemisorptiion and reduction temperatures for various Co/TiO$_2$ samples

Samples	Total H$_2$ Chemisorption (μmol H$_2$/g cat.)	Reduction Temperature ($^\circ$C)
Co/R0	0.93	370
Co/R3	1.55	270
Co/R19	2.44	320
Co/R40	1.66	285
Co/R96	1.71	275
Co/R99	0.69	275

4. SUMMARY

The present research showed a dependence of various ratios of rutile:anatase in titania as a catalyst support for Co/TiO$_2$ on characteristics, especially the reduction behaviors of this catalyst. The study revealed that the presence of 19% rutile phase in titania for Co/TiO$_2$ (Co/R19) exhibited the highest number of reduced Co metal surface atoms which is related the number of active sites present. It appeared that the increase in the number of active sites was due to two reasons; i) the presence of rutile phase in titania can facilitrate the reduction process of cobalt oxide species into reduced cobalt metal, and ii) the presence of rutile phase resulted in a larger number of reduced cobalt metal surface atoms. No phase transformation of the supports further occurred during calcination of catalyst samples. However, if the ratios of rutile:anatase were over 19%, the number of active sites dramatically decreased.

ACKNOWLEDGMENT

The financial support from the Thailand Research Fund (TRF) is greatly appreciated.

REFERENCES

[1] H.P. Wither, Jr., K.F. Eliezer, and J.W. Mechell, Ind. Eng. Chem. Res., 29 (1990)1807.
[2] J.L. Li, G. Jacobs, T. Das, and B.H. Davis, Appl. Catal. A., 233 (2002) 255.
[3] G. Jacobs, T. Das, Y.Q. Zhang, J.L. Li, G. Racoillet, and B.H. Davis., Appl. Catal. A., 233 (2002) 263.
[4] J.L. Li,, L.G. Xu, R. Keogh, and B.H. Davis, Catal. Lett., 70 (2000) 127.
[5] K.Y. Jung, and S.B. Park, J. Photochem. Photobio. A: Chem., 127 (1999) 117.
[6] R.C. Reuel, and C.H. Bartholomew, J. Catal., 85(1984) 63.
[7] B. Jongsomjit, C. Sakdamnuson, J.G. Goodwin, Jr., and P. Praserthdam, Catal. Lett., 94 (2004) 209.
[8] B. Jongsomjit, T. Wongsalee, and P. Praserthdam, Mater. Chem. Phys., 92 (2005) 572.

Studies in Surface Science and Catalysis, volume 159
Hyun-Ku Rhee, In-Sik Nam and Jong Moon Park (Editors)
289

Low Temperature Ammonia Oxidation Catalyst Studies

P. Y. Lui[a] and K. L. Yeung[a,*]

[a]Department of Chemical Engineering, Hong Kong University of Science and Technology,
Clear Water Bay, Kowloon, Hong Kong, People Republic of China

1. INTRODUCTION

Many malodorous compounds are not only nuisance, but also a health threat under prolonged exposure [1]. Ammonia (NH_3) is emitted from landfill and sewage treatment plant and associated with many agricultural activities (e.g. poultry and piggery). Ammonia is also a problem in public toilets, hospitals and nursing homes. Selective catalytic oxidation (SCO) can convert NH_3 to N_2 at mild temperature (i.e. 473-673 K) as shown in equation 1, however nitrous oxides (N_2O) and nitrogen oxides (NO_x) are often produced (*cf.* Eqn. 2 & 3).

$$4NH_3 + 5O_2 \rightarrow 4NO + 6H_2O \qquad \text{Eqn. 1}$$
$$4NH_3 + 4O_2 \rightarrow 2N_2O + 6H_2O \qquad \text{Eqn. 2}$$
$$4NH_3 + 3O_2 \rightarrow 2N_2 + 6H_2O \qquad \text{Eqn. 3}$$

It has been reported that titanium supported vanadium catalyst is active for ammonia oxidation at temperatures above 523 K [2,3]. Also, supported vanadium oxides are known to be efficient catalyst for the catalytic reduction of nitrogen oxides (NO_x) in the presence of ammonia [4]. This work investigates the nanostructured vanadia/TiO_2 for low temperature catalytic remediation of ammonia in air.

2. EXPERIMENTAL

The vanadia/TiO_2 catalysts with different vanadium loadings were prepared by impregnating nanostructured anatase TiO_2 (Hombikat UV 100; BET surface area 300 m^2/g) with NH_4VO_3 solutions. The catalysts were dried in air at 388 K for 24 h before pretreatment with O_2 or 100 ppm O_3/O_2 gas at 473 K for 4 h. The catalyst activity was measured in a flow reactor. Twenty-five milligrams of catalyst and 200 sccm of dry air containing 2185 ppm of NH_3 were used for the reaction. The ammonia conversion to nitrogen was conducted at 348 K and monitored using a gas chromatograph (SRI 8610C).

---

* Author to whom correspondence should be addressed

Tel: 852-2358-7123; Fax: 852-2358-0054; e-mail: kekyeung@ust.hk

3. RESULTS AND DISCUSSION

The vanadium content, the average particle size (i.e., TiO_2) and the BET surface area of the vanadia/TiO_2 catalysts are summarized in Table 1. The average particle size of TiO_2 was determined from the X-ray diffraction peak broadening of (101) anatase TiO_2 peak, while the BET surface area was measured by nitrogen physisorption. The TiO_2 support has an average crystal size of about 9 nm. The original TiO_2 has a surface area of 300 m^2/g, which decrease with increasing vanadium loading. The VO_x surface density in moles of vanadium atoms per m^2, was calculated from the vanadia content and the BET surface area.

Table 1
Physicochemical properties of nanostructured vanadia/TiO_2

Sample name	V content (wt.%)	Particle size (nm)	BET surface area (m^2g^{-1})	VO_x surface density (μmol $V_2O_5 m^{-2}$)
TiO_2 support	---	9.59	300.0	---
V1Ti-c*	1	9.31	255.4	0.384
V3Ti-c	3	8.75	248.3	1.186
V10Ti-c	10	9.15	222.3	4.416
V15Ti-c	15	9.36	185.4	7.941
V1Ti-o#	1	8.75	255.3	0.384
V3Ti-o	3	8.75	254.9	1.155
V10Ti-o	10	8.94	216.5	4.534
V15Vi-o	15	9.05	185.3	7.945

*c: pretreated by O_2 calcination at 473 K; #o: pretreated by ozonation at 473 K

A monolayer of vanadium with 7.9 μmol V_2O_5 per m^2 [5] was obtained for 15 wt.% vanadia/TiO_2. This was independently verified by both micro-Raman and XPS experiments. No V_2O_5 crystals were detected by XRD up to monolayer coverage indicating that the deposited vanadia is well dispersed on the TiO_2. The Raman band at 920 cm^{-1} indicated that bridging V-O-V bonds are formed at higher vanadium loading (Fig. 1.).

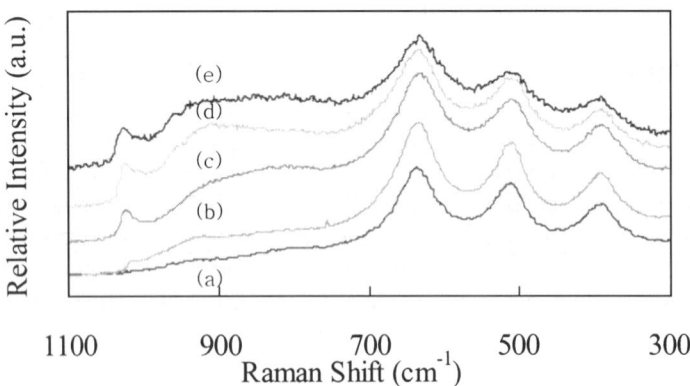

Fig. 1. Raman spectra of (a) 1, (b) 3, (c) 5, (d) 10, (e) 15 wt.% vanadia supported on TiO_2 activated by calcination under dehydrated conditions.

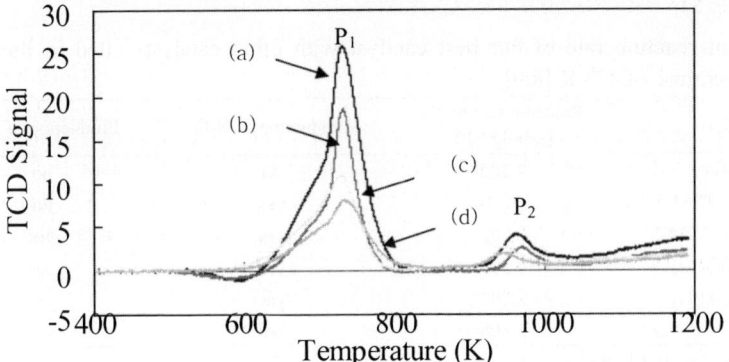

Fig. 2. TPR profile of (a) 3, (b) 5, (c) 10, (d) 15 wt.% vanadia supported on TiO_2 activated by pretreatment in O_2 at 473 K.

Temperature programmed reduction (TPR) was also conducted and the results are shown in Fig. 2. The weight of the catalyst was adjusted to give a constant amount of vanadia and thus eliminate the peak shift associated with water production [6]. However, the TPR profiles are subject to change with experimental parameters such as ramping rate, reducing gas concentration and gas flowrate [7]. In our work, two peaks corresponding to monomeric and polymer surface vanadia species (P_1) and crystalline vanadia (P_2) were identified. P_1 can be further deconvoluted into two peaks with peak P_{1a} located at 733 K attributed to monomeric surface vanadia. Together with the micro-Raman experiments, it can be concluded that surface polymeric species are more easily reduced than the monomeric surface vanadia.

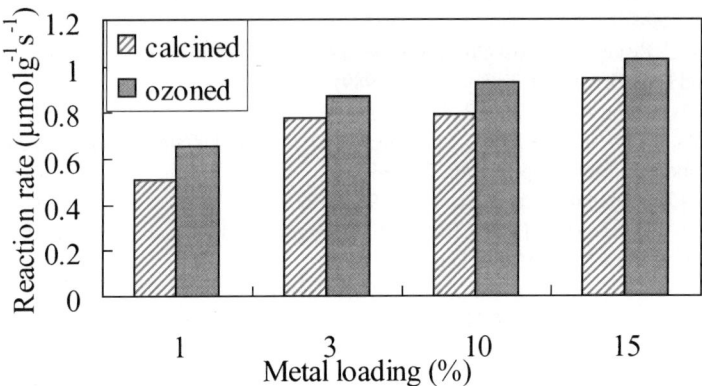

Fig. 3. Comparison of average NH_3 oxidation rate of 1wt. %, 3wt. %, 10wt. % and 15wt. % VO_x supported on TiO_2 at 348 K over 3 h.

Table 2

Comparison of reaction rate of our best catalyst with other catalysts cited in literature at reaction temperature of 473 K [8,9].

Catalyst	Reaction rate to N_2 ($\mu molg^{-1}s^{-1}$)	Production of N_2O	Production of NO_x
V10Ti-o	7.500*	no	no
0.6%Pt-Fe-ZSM-5	1.518	yes	no
0.6%Pd-Fe-ZSM-5	0.202	yes	yes
Reduced PtO/Al$_2$O$_3$	0.506	yes	yes
10% Ag/Al$_2$O$_3$	2.795	yes	yes
10%Cu-10% Ag/Al$_2$O$_3$	1.808	yes	yes

* reaction temperature of 348K

Fig. 3. compares the ammonia conversion for nanostructured vanadia/TiO$_2$ catalysts pretreated with O$_2$ and 100 ppm O$_3$/O$_2$ gases. The reactions were conducted at 348 K for 3 h. No N$_2$O and NO$_x$ byproducts were detected in the reactor outlet. It is clear from the figure that higher vanadium content is beneficial to the reaction and ozone pretreatment yields a more active catalyst. Unlike the current catalysts, which require a reaction temperature of at least 473 K, the new catalyst is able to perform at much lower temperature. Also, unlike these catalysts, complete conversion to nitrogen was achieved with the new catalysts. Table 2 shows that the reaction rate of the new catalysts compared favorably with the established catalysts.

ACKNOWLEDGEMENT

The authors would like to acknowledge the financial support from the Hong Kong Innovation and Technology Fund (ITS/176/01C).

REFERENCES

[1] G. Busca and C. Pistarino, J. Loss Prevent. Proc. Ind., 16 (2003) 157
[2] F. Cavani and F. Trifiro, Catal. Today, 4 253 (1989).
[3] Y.J Li and J. N Armor, Appl. Catal. B, 13 (1997) 131.
[4] M.D. Amiidis, I. E. Wachs, G. Deo, J-M. Jehng and D. S. Kim, J. Catal. ,161 (1996) 247.
[5] G.C. Bond and S.F. Tahir, Appl. Catal., 71 (1991) 1.
[6] J.W. Ha and J.R. Regalbuto, Catal. Lett., 29 (1994) 189.
[7] A. Jones and B.D. McNicol (eds.), Temperature Programmed Reduction for Solid Materials and Characterization, New York: M. Dekker, 1986
[8] R.Q. Long and R. T.Yang, Catal. Lett., 1-4 (2002) 353.
[9] M. Yang, C. Wu, C. Zhang and H. He, Catal. Today, 90 (2004) 263.

Studies in Surface Science and Catalysis, volume 159
Hyun-Ku Rhee, In-Sik Nam and Jong Moon Park (Editors)
293

Terephthalic Acid Hydropurification over Pd/C Catalyst

J.-H. Zhou*, G.-Z. Shen, J. Zhu, W.-K. Yuan

UNILAB，State Key Laboratory of Chemical Engineering, ECUST, Meilong Rd. 130, Shanghai, 200237, P. R. China   ***jhzhou@ecust.edu.cn**

1. INTRODUCTION

Terephthalic acid (p-TA or TA), a raw material for polyethylene terephthalate (PET) production, is one of the most important chemicals in petrochemical industry. Crude terephthalic acid (CTA), commonly produced by homogeneous liquid phase p-xylene oxidation, contains impurities such as 4-carboxybenzaldehyde (4-CBA, 2000-5000 ppm) and several colored polyaromatics that should be removed to obtain purified terephthalic acid (PTA). PTA is manufactured by hydropurification of CTA over carbon supported palladium catalyst (Pd/C) in current industry [1].

Although the process is of significance, it has not well studied. Since the initial development of the CTA hydropurification process in 1960s', only a few papers have been published, mainly regarding catalyst deactivation [2]. Recently, Samsung Corporation, in collaboration with Russian scientists, developed a novel carbon material-CCM supported palladium-ruthenium catalyst and its application to this process [3]. However, pathways and kinetics of CTA hydrogenation, which are crucial to industrialization, are not reported hitherto.

In this paper, the pathways, side reactions, as well as kinetics of CTA hydrogenation process over Pd/C will be presented.

2. EXPERIMENTAL

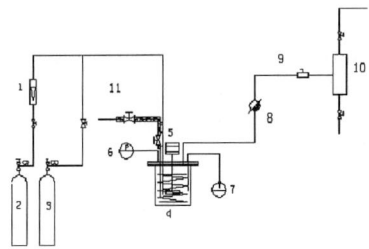

1 flow meter; 2 N$_2$ cylinder; 3 H$_2$ cylinder; 4 autoclave; 5 mixer; 6 pressure sensor; 7 temperature sensor; 8 check valve; 9 back pressure valve; 10 buffer tank; 11 special sampling device

Fig.1. Schematic of the experimental setup

Experiments were carried out in a batch autoclave (Type 4571, Parr Corp., USA) with varying hydrogen partial pressures and temperatures. An industrial 0.5%Pd/C (CBA-300)

catalyst, manufactured by Engelhard Corporation, was used for CTA purification. Considering barriers encountered in smooth sampling, only was the 4-CBA added into the autoclave as reactant. Actually, our preliminary experiments indicate that the behavior of reaction was almost not affected by TA, even though the latter had a small effect on the reaction rate. Fig.1 shows the schematic diagram of the experiment. Samples were analyzed by HPLC [4].

3. RESULTS AND DISCUSSION

3.1. Mechanism of the hydropurification of CTA over Pd/C

It has been commonly understood that during the hydropurification process, CTA is first dissolved in water at about 280 ℃; then the CTA stream, mixed with dissolved hydrogen, flows through the Pd/C bed, wherein 4-CBA is hydrogenated to p-toluic acid (PT) and is washed out in the subsequent process. The reaction scheme can be illustrated in Fig.2.

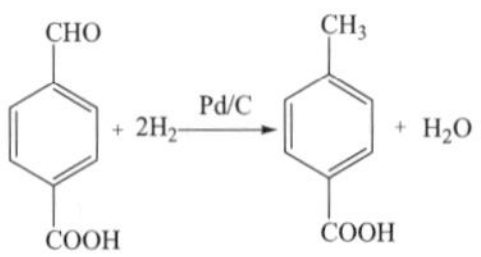

Fig.2. Reaction occurs during the purification. 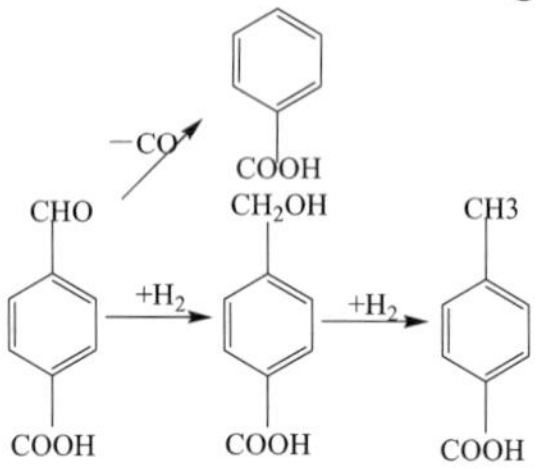   Fig.3. Mechanism of the hydropurification of CTA over Pd/C.

However, our experiments, combined with our thermodynamic analysis, have revealed that the previously mentioned view is not the case, but more complex than expected. Rather, the purification processes undergo two reactions in parallel: hydrogenation and decarbonylation. CTA hydrogenation is actually a complex reaction system as illustrated in Fig.3. Hydrogenation is a tandem process, in which 4-CBA is hydrogenated first to 4-hydroxymethylbenzoic acid (4-HMBA) and subsequently to PT. Simultaneously, 4-CBA is decarbonylated to benzoic acid (BA) over the Pd/C. Fortunately, it is much easier for the intermediate hydrogenation product (4-HMBA with a solubility > 1 g/100 g H_2O) and decarbonylation product (BA with a solubility of 0.29 g/ 100 g H_2O) to be dissolved in water. Thus it is also much easier to wash them out than PT whose solubility is 0.035g/100gH_2O at room temperature. Therefore, paradoxically, this concomitant side reaction does not harm the purification itself. Furthermore, analysis of samples from an industrial reactor of Yangtze petrochemical Corporation, SINOPEC, further confirmed the above path and pathways for the CTA purification system.

3.2. Kinetic feature

As has been elucidated, 4-CBA undergoes complex reactions in the course of purification, during which the hydrogenation inevitably competes with the decarbonylation. At the same time, although the BA is easy to be washed out, the concomitant CO during decarbonylation is a poison to Pd/C catalyst.

Experiments were carefully designed to study the effect of the oxygen concentration in the system on the competition between hydrogenation and decarbonylation, as being shown in

Fig.4. First, the reaction system was filled with about 2 MPa nitrogen at room temperature then heated to 220 ℃ (Curve A in Fig.4). Understandably, there must be some dissolved oxygen in solution under such a condition. Secondly, the system was flushed with nitrogen three times under the room temperature, filled with about 2 MPa nitrogen and 0.2 MPa hydrogen and afterwards heated to 220 ℃ (Curve B in Fig.4). Most of the dissolved oxygen in the solution was displaced by nitrogen during the flush. Thus, the oxygen concentration under such a treatment was much lower than that under condition A. Thirdly, the system was flushed with hydrogen three times at room temperature, filled with 2MPa hydrogen and then heated to 220 ℃ (Curve C in Fig.4). There was almost no oxygen in the reaction system under condition C. It is indicated from Fig.4 that the decarbonylation reaction was significantly influenced by the oxygen concentration in the solution. The dissolved trace amount of oxygen facilitated decarbonylation whereas the hydrogen in the system inhibited it, which may attributes to the consumption of CO by the dissolved oxygen. The more oxygen was in the reaction system, the more rapid the decarbonylation occurred.

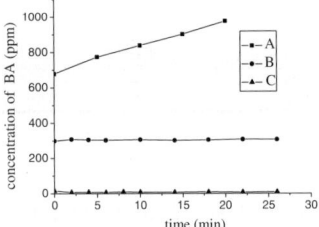

Fig.4. Decarbonylation of 4-CBA under different oxygen concentrations (220℃, 2 g 20-60 mesh catalyst)

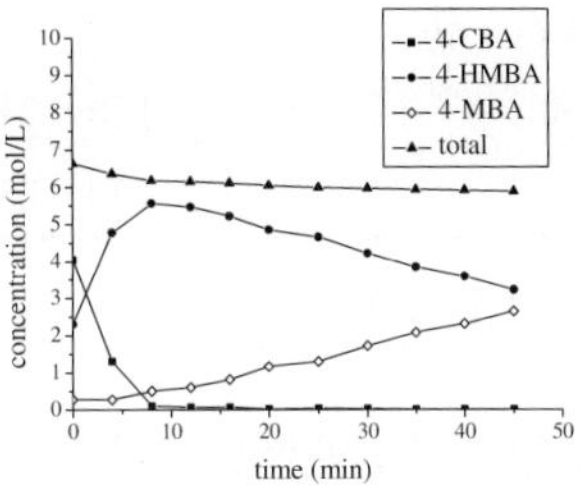

Fig.5. Evolution of the reactant and product with time (60-100 mesh catalyst, 120 ℃, 0.7MPa hydrogen, 1000rpm)

The hydrogen process feature was also analyzed. Fig.5 presents a typical plot of the change of reactant and product distribution with time during the hydrogenation. It is obvious that the hydrogenation of 4-CBA was very rapid, even though the reaction temperature was only 120 ℃ which was much lower than the industrial one (about 280℃), the 4-CBA conversion reached 99% in 7 minutes. The intermediate 4-HMBA's concentration first increased to a maximum, and followed with a gradual lowering. This fact is likely to verify the proposed mechanism of the CTA hydrogenation process. Hydrogenation of 4-CBA to 4-HMBA was much more rapid than the subsequent hydrogenation of 4-HMBA to PT. Our thermodynamic analysis indicated that reaction free energy for 4-CBA to 4-HMBA and that from 4-HMBA to PT were about -11.5 and -93.26 kJ/mol, respectively. This means that the equilibrium constant of the former reaction is much lower than that of the latter one, which was unexpected.

3.3. Apparent kinetics of 4-CBA hydrogenation

The 4-CBA's concentration that is left in the system is of most interest in industrialization. For this reason, the dependence of the rate of 4-CBA disappearance on reaction conditions was carefully investigated. Reaction conditions included hydrogen atmosphere, temperature and catalyst particle size. It is believed that the hydrogen atmosphere has little effect on hydrogenation reaction within the industrially acknowledged boundary. However, the hydrogenation rate is sensitive to agitation speed, temperature and catalyst particle size, implying obvious diffusion resistance existing in the hydropurification process. As a result, apparent kinetic equations for the 4-CBA's hydrogenation under different conditions can be obtained from the experimental data. The activation energy and reaction order were estimated based on the power-law kinetics under various conditions, as have been given in Table 1. When the catalyst particle size was decreased from 4-8 mesh to 140-180 mesh, the apparent reaction order did not change much (from 1.23 to 1.17). On the other hand, the apparent reaction activation was increased from 28.0 to 45.0 kJ/mol, which is significant. The greater the particle size is, the smaller the activation is, which attribute to the fact that the diffusion activation is normally smaller than the reaction activation.

Table 1 Reaction order and activation energy of 4-CBA hydrogenation

Catalyst particle (mesh)	Temperature range (℃)	E_a (kJ/mol)	Reaction order n
140－180	110－150	45.0	1.17
60－100	120－150	31.1	1.15
20－60	170－215	29.5	1.20
4－8	250－280	28.0	1.23

4. CONCLUSION

1）Experimental studies, combined with thermodynamic analysis, indicate that the CTA hydropurification process is a complex reaction system including both parallel and tandem reactions wherein 4-CBA hydrogenation is exothermic and its paralleled decarbonylation is endothermic.

2) The 4-CBA decarbonylation does not weaken much the function for purification. However, it undoubtedly competes with its hydrogenation. The competition is greatly influenced by the amount of the dissolved oxygen in the system. The more is the dissolved oxygen, the easier the decarbonylation takes place. However, the hydrogen has an opposite effect.

3) Unlike what has long been accepted, hydrogenation of 4-CBA is composed of a series of reactions rather than a single one. 4-CBA is hydrogenated first to 4-HMBA in a very rapid manner and then to 4-PT in a much slower manner. However, the equilibrium constant of the first reaction was much smaller than that of the second one, which is not expected.

4) Although hydrogenation of 4-CBA over Pd/C is very fast, there is strong diffusion resistance. Furthermore, apparent kinetic equations on different catalyst particle sizes have been obtained from experimental data.

Acknowledgement: This work is financially supported by NSFC(No. 20490200), Shanghai Science & Technology Committee (No.036505010) and SINOPEC.

References:
[1] I. P. Wheaton, S. A. Cerefice, US Patent No. 4 476 242 (1984).
[2] L. Shen, W. Mao, Shiyou Huagong, 20(1991), 234.
[3] S. H. Jhung, A. V. Romanenko, K. H. Lee, et al., Applied Catalysis A: General, 225(2002), 131.
[4] L. G. Chen, J. H. Zhou, Journal of East China University of Sci. and Tech., 30 (2004), 719.

Studies in Surface Science and Catalysis, volume 159
Hyun-Ku Rhee, In-Sik Nam and Jong Moon Park (Editors)
© 2006 Elsevier B.V. All rights reserved

Preparation and catalytic activity of $H_3PMo_{12}O_{40}$ catalyst molecularly immobilized on polystyrene support

H. Kim[a], P. Kim[a], J. Yi[a], K.-Y. Lee[b], S.H. Yeom[c] and I.K. Song[a]*

[a]School of Chemical and Biological Engineering, Institute of Chemical Processes, Seoul National University, Shinlim-dong, Kwanak-ku, Seoul 151-744, South Korea

[b]Department of Chemical and Biological Engineering, Korea University, Annam-dong, Sungbuk-ku, Seoul 136-701, South Korea

[c]Department of Environmental and Applied Chemical Engineering, Kangnung National University, Kangwondo 210-702, South Korea

*Corresponding author (Tel:+82-2-880-9227, E-mail: inksong@snu.ac.kr)

1. INTRODUCTION

Heteropolyacids (HPAs) have been widely employed as catalysts in various homogeneous and heterogeneous reactions [1-3]. One of the great advantages of HPA catalysts is that their catalytic properties can be tuned by changing the identity of charge-compensating counter-cations, heteroatoms, and framework polyatoms. Another advantage is their great thermal stability, which makes HPAs well suitable for catalytic reactions that may require harsh environments. To overcome low surface area of HPAs ($<10m^2/g$), HPA catalysts have been supported on carbon [4], inorganic materials [5], and polymer matrix [6]. Polystyrene bead has found successful applications in the field of biochemistry to form inorganic-organic hydride materials. It is expected that polystyrene can also serve as a catalyst support due to its feasible nature of surface modification. In this work, $H_3PMo_{12}O_{40}$ (PMo_{12}) HPA catalyst was immobilized on the aminated polystyrene (PS) bead, by taking advantage of the overall negative charge of heteropolyanion. The supported catalyst was applied to the vapor-phase 2-propanol conversion reaction as an oxidation catalyst.

2. EXPERIMENTAL

$H_3PMo_{12}O_{40}$ (PMo_{12}) catalyst and aminated PS bead (2.0 m-mole NH_2/g) were purchased from Aldrich Chemical Co. PMo_{12} catalyst immobilized on the aminated PS bead was prepared according to the method reported in a literature [7]. PMo_{12} species were immobilized on the PS bead as charge-matching components in an acetonitrile (ACN) medium, as shown in Fig. 1. The solid product was repeatedly washed several times with water until the washing solvent became colorless, and then it was dried overnight at $80^{\circ}C$ to yield the final form (PMo_{12}-PS). The loading of PMo_{12} was 23.1 wt.%. Infrared spectra of supports and supported catalysts were obtained with a FT-IR spectrometer (Nicolet, Impact 410). Supports and supported catalysts were further characterized by SEM-EDX (Jeol, JSM-6700F) and XRD (Mac Science, M18XHF) measurements. Vapor-phase 2-propanol conversion reaction was

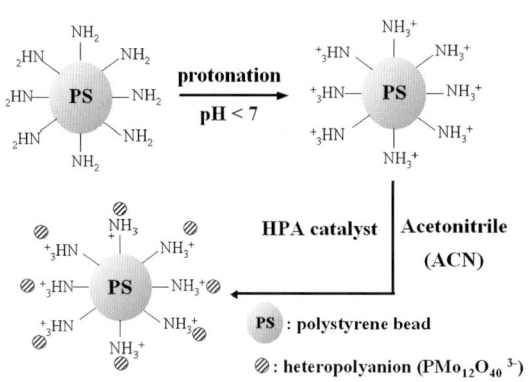

Fig. 1. Scheme for preparation of PMo$_{12}$ catalyst immobilized on PS bead.

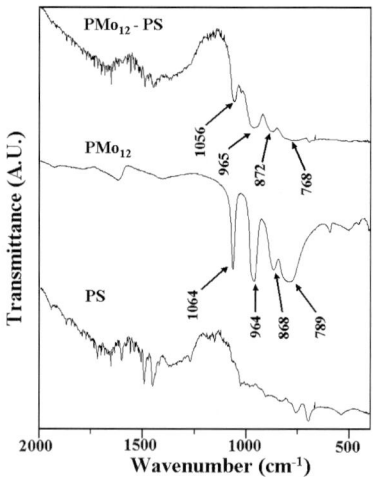

Fig. 2. FT-IR spectra of support and catalysts.

carried out in a continuous flow fixed-bed reactor at an atmospheric pressure. Known amounts of each catalyst were charged into a tubular quartz reactor, and pretreated at 523K with an air stream (20 ml/min) for 1 h. The catalytic reaction was carried out at 493K. 2-Propanol (6.55 x 10^{-3} mole/h) was sufficiently vaporized and fed into the reactor continuously together with air carrier (20 ml/min). In each run, contact time was maintained at 7.10 g-PMo$_{12}$-h/2-propanol-mole. The product stream was periodically sampled and analyzed with an on-line GC (HP 5890II).

3. RESULTS AND DISCUSSION

3.1. Characterization

Immobilization of PMo$_{12}$ on the PS bead was confirmed by FT-IR analyses, as shown in Fig. 2. The primary structure [8] of PMo$_{12}$ could be identified by the four characteristic IR bands appearing within the range 700-1200 cm^{-1}. Characteristic IR bands of the unsupported PMo$_{12}$ catalyst appeared at 1064(P-O band), 964(Mo=O band), 868, and 789 cm^{-1} (Mo-O-Mo bands). The four characteristic IR bands of PMo$_{12}$ species in the PMo$_{12}$-PS catalyst were also observed at the slightly shifted positions compared to those of the unsupported PMo$_{12}$ catalyst. This result implies that PMo$_{12}$ catalyst was successfully immobilized on the PS bead.

Fig. 3 shows the SEM images of PS support and PMo$_{12}$-PS catalyst. The surface of PMo$_{12}$-PS catalyst was very clean, and no visible evidence representing PMo$_{12}$ agglomerates was found in the PMo$_{12}$-PS catalyst. This result indicates that PMo$_{12}$ species were finely dispersed on the surface of PS bead. Fine dispersion of PMo$_{12}$ catalyst on the PS support was also confirmed by EDX analysis, as shown in Fig. 3(c). N$_2$ adsorption-desorption analysis revealed that PS bead had no pore-like feature.

Fig. 4 shows the XRD patterns of PS bead, unsupported PMo$_{12}$, and PMo$_{12}$-PS samples. PS bead showed no characteristic XRD patterns due to its amorphous nature. Unsupported PMo$_{12}$ catalyst showed the characteristic XRD patterns of the HPA catalyst. On the other hand, however, PMo$_{12}$-PS catalyst showed no characteristic XRD patterns of PMo$_{12}$, as observed for

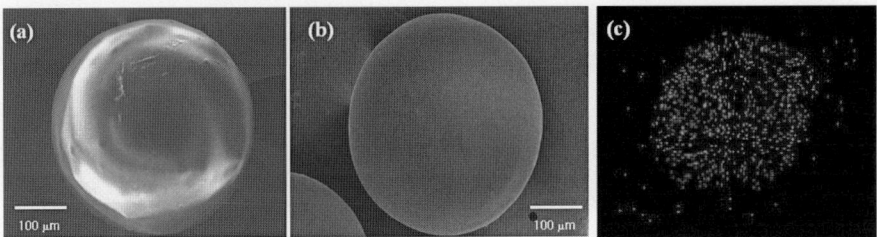

Fig. 3. (a) SEM image of PS bead, (b) SEM image of PMo$_{12}$-PS catalyst, and (c) EDX image of PMo$_{12}$-PS catalyst obtained by mapping on molybdenum.

PMo$_{12}$-polymer composite film catalyst [9]. This demonstrates that PMo$_{12}$ catalyst was not in a crystal state but in an amorphous-like state, indicating that PMo$_{12}$ catalyst was molecularly dispersed on the PS support via chemical interaction. As attempted in this work, it is believed that heteropolyanions (PMo$_{12}$O$_{40}$$^{3-}$) were strongly immobilized on the cationic sites of the PS bead as charge-compensating components.

3.2. Catalytic activity in the 2-propanol conversion reaction

Fig. 5(a) shows the typical catalytic performance of unsupported PMo$_{12}$ and supported PMo$_{12}$-PS catalysts in the vapor phase 2-propanol conversion reaction. The PMo$_{12}$-PS catalyst exhibited higher 2-propanol conversion than the unsupported PMo$_{12}$ catalyst. The enhanced catalytic performance of PMo-PS$_{12}$ catalyst was due to the fine dispersion of PMo$_{12}$ species on the PS bead. Fig. 5(b) shows the propylene and acetone yields over unsupported PMo$_{12}$ and supported PMo$_{12}$-PS catalysts. It is known that propylene is formed by the acid function of HPA catalyst, while acetone is formed by the redox function of HPA [10]. The PMo$_{12}$-PS catalyst showed the remarkably enhanced acetone yield and the suppressed propylene yield compared to the unsupported PMo$_{12}$ catalyst. In other words, the PMo$_{12}$-PS catalyst showed an excellent oxidation activity compared the mother catalyst. It should be noted that PMo$_{12}$ species were chemically and molecularly immobilized on the aminated PS bead as charge-matching components. Therefore, the PMo$_{12}$-PS catalyst showed the enhanced oxidation activity by suppressing acid catalytic activity compared to the unsupported PMo$_{12}$ catalyst.

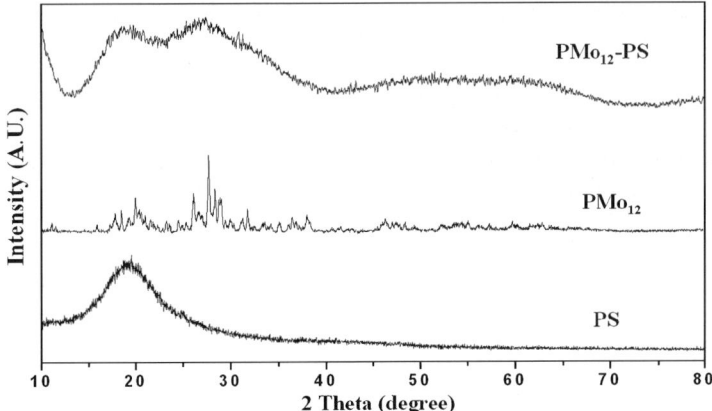

Fig. 4. XRD patterns of support and catalysts.

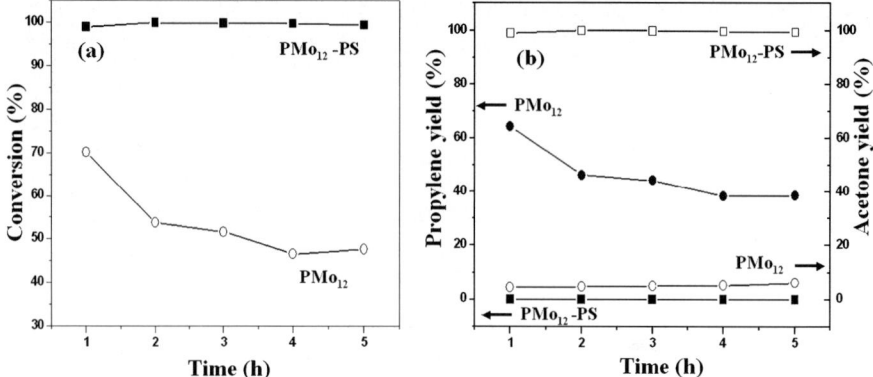

Fig. 5. (a) 2-Propanol conversions and (b) product yields over unsupported PMo_{12} and supported PMo_{12}-PS catalysts.

4. CONCLUSIONS

PMo_{12} species were successfully immobilized on the surface of aminated PS bead as charge-compensating components, by taking advantage of the overall negative charge of $PMo_{12}O_{40}^{3-}$. It was found that PMo_{12} species in the PMo_{12}-PS catalyst were chemically and molecularly immobilized on the PS support. The supported PMo_{12}-PS catalyst exhibited higher 2-propanol conversion than the unsupported PMo_{12} catalyst. Furthermore, the supported catalyst showed the enhanced oxidation activity for acetone formation and the suppressed acid catalytic activity for propylene formation compared to the mother catalyst. Thus, the PMo_{12}-PS catalyst served as an efficient oxidation catalyst.

ACKNOWLEDGEMENTS

The authors acknowledge the support from Korea Science and Engineering Foundation (KOSEF R01-2004-000-10502-0).

REFERENCES

[1] M. Misono, Catal. Rev. -Sci. Eng., 29 (1987) 199.
[2] I.V. Kozhevnikov, Catal. Rev. -Sci. Eng., 37 (1995) 311.
[3] I.K. Song, H.S. Kim and M.-S. Chun, Korean J. Chem. Eng., 20 (2003) 844.
[4] S.R. Mukai, T. Sugiyama and H. Tamon, Appl. Catal. A, 256 (2003) 99.
[5] K. Nowinska, R. Formaniak, W. Kaleta and A. Waclaw, Appl. Catal. A, 256 (2003) 115.
[6] M. Hasik, W. Turek, E. Stochmal, M. Lapowski and A. Pron, J. Catal., 147 (1994) 544.
[7] S.S. Lim, G.I. Park, J.S. Choi, I.K. Song and W.Y. Lee, Catal. Today, 74 (2003) 299.
[8] J.F. Keggin, Nature, 131 (1933) 908.
[9] S.S. Lim, G.I. Park, I.K. Song and W.Y. Lee, J. Mol. Catal. A, 182-183 (2002) 175.
[10] J.K. Lee, I.K Song, W.Y. Lee and J.-J. Kim, J. Mol. Catal. A, 104 (1996) 311.

Studies in Surface Science and Catalysis, volume 159
Hyun-Ku Rhee, In-Sik Nam and Jong Moon Park (Editors)

Synthesis of Perfluoroalkyl Iodides over Metal Catalysts in Gas Phase

Young Ju Ko[a], Nam Cook Park[a], Jae Soon Shin[a], Honggon Kim[b], Byoung Sung Ahn[b], Dong Ju Moon[b*] and Young Chul Kim[a*]

[a]Faculty of Applied Chemical Engineering and Nano Technology Research Center, Chonnam National University, Gwang-ju 500-757, Korea ([*]youngck@jnu.ac.kr)

[b]Hydrogen Energy Research Center, Korea Institute of Science and Technology, 39-1 Hawolgok-dong, Sungbuk-ku, Seoul 136-791, Korea ([*]djmoon@kist.re.kr)

Abstract

This study relates to a continuous process for the preparation of perfluoroalkyl iodides over nanosized metal catalysts in gas phase. The water-alcohol method provided more dispersed catalysts than the impregnation method. The Cu particles of about 20 nm showed enhanced stability and higher activity than the particles larger than 40 nm. This was correlated with the distribution of copper particle sizes shown by XRD and TEM. Compared with silver and zinc, copper is better active and stable metal.

1. INTRODUCTION

Perfluoroalkyl iodides having about 6 to 12 carbon atoms are useful as intermediates for numerous applications relating in general to the field of fluorinated surface-active substances and more particularly the bases for extinguisher formulations, hydrophobic and oleophobic finishes for the treatment of textiles, and more recently for applications of a medical sector [1-3]. Perfluoroalkyl iodides are prepared industrially by telomerization of tetrafluoroethylene (C_2F_4, taxogen) and pentafluoroethyl iodide (C_2F_5I, telogen) in accordance with the Eq. (1).

$$C_2F_5I + n(CF_2=CF_2) \rightarrow C_2F_5(C_2F_4)_n\text{-}I \qquad (1)$$

Industrially, the perfluoroalkyl iodides by telomerization are mostly made by a batch system using peroxide initiators. However, the difficulty of mass production, and the production of hydrogen-containing byproducts in the process are disadvantageous [4]. In this study, a continuous process for the preparation of perfluoroalkyl iodides over nanosized metal catalysts in gas phase and the effects of the particle size on the catalytic activities of different the preparation methods and active metals were considered.

2. EXPERIMENTAL

The various silica-alumina supported metal catalysts were prepared by impregnation, and water-alcohol methods. The used support in this paper was silica-alumina (Norton Corp., SA3232) whose surface area was $30m^2/g$. The reactants, pentafluoroethyl iodide (C_2F_5I, 97%) and tetra-fluoro ethylene (C_2F_4, 97%), were obtained from Fluorochem, England. The XRD technique (MAC Science Co., Model M18XHF) was used to determine the particle sizes of the supported metal clusters. TEM (JEN-2000FX II) was also used to measure the particle sizes of the catalysts. The reaction was carried out in a fixed bed flow reactor and the products were analyzed by a gas chromatography (GC-8A, Schimadzu Co.). The reaction temperature was 320~440℃, the molar ratio of reactants was C_2F_5I : C_2F_4 = 2 : 1, which was shown the highest C_4F_9I yield compared with 2.5:1, 3:1 and above molar ration, and SV was 900 h^{-1}.

3. RESULTS AND DISCUSSION

As can be seen in table 1, with different preparation methods and active metals, the average size of the copper particle for the catalysts A and D were 20.3 nm and 50.0 nm. While those of the catalysts B and C were 51.3 nm and 45.4 nm, respectively. CuO, non-supported metal oxide, made by impregnation is sintered and cluster whose particle size was 30 μm. The water-alcohol method provided more dispersed catalysts than the impregnation method.

Table 1. Characteristics of the catalysts used in this work

Catalyst	Active Metal	Preparation Method	Loading amount of metal (wt%)	Particle Size (nm)
A	Cu/Al_2O_3	Water-alcohol	10	20.3
B	Ag/Al_2O_3	Water-alcohol	10	51.3
C	Zn/Al_2O_3	Water-alcohol	10	45.4
D	Cu/Al_2O_3	Impregnation	10	50.0
E	CuO	Impregnation	10	30(μm)

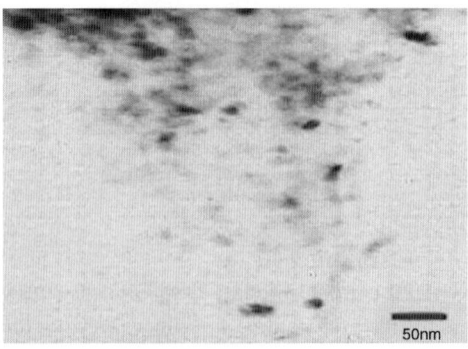

Fig. 1. TEM image of Cu/Al_2O_3 obtained by the water-alcohol method, average particle size = 20.3 nm.

Figure 1 is a TEM photograph of the Cu (10wt%)/Al$_2$O$_3$ catalyst prepared by water-alcohol method, showing the dispersed state of copper and was confirmed the particle sizes from XRD data. Figure 2 is X-ray diffraction patterns of above-mention catalysts, was used to obtain information about phases and the particle size of prepared catalysts. Metal oxide is the active species in this reaction. Particle sizes were determined from the width of the XRD peaks by the Debye-Scherrer equation.

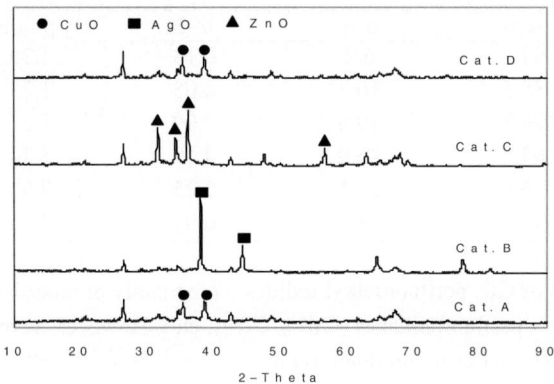

Fig. 2. X-ray diffraction patterns of different catalysts.

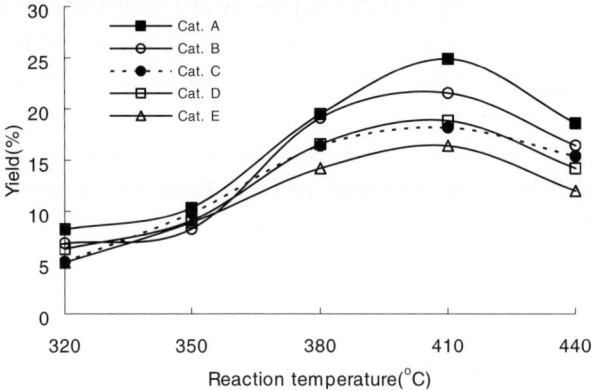

Fig. 3. The C$_4$F$_9$I yields of different catalysts.

Fig. 3 shows the C$_4$F$_9$I yield with the above-mention catalysts in the reaction temperature range of 320~440℃. Totally, the most stable and active catalyst was Cu/Al$_2$O$_3$ (Cat. A) whose particle size was 20 nm and the catalyst was made by water-alcohol method.

In the effect of active metal, copper was more stable and active than silver and zinc. In the effect of particle size, the smaller particle is the more stable and active. It seems that the catalysts with smaller particle size have more active sites compared with those of larger particle size in same space. Through out the Fig. 3 and Table 2, in the effect of reaction temperature, telomerization reactions with the catalysts seem to show totally similar trends. By 410℃, as reaction temperature increased, the C_4F_9I yield also increased up to 27% and after that, the C_4F_9I yield got lower.

Table 2. The product distributions and the conversion at the different reaction temperature using Cat A

Reaction Temperature	$C_2F_5(CF_2-CF_2)_nI$ (% by weight)				Conversion of C_2F_4
	n=0	n=1	n=2	By products	
320℃	92.2	6.2	0.13	1.39	42.0
350℃	88.3	10.3	0.05	1.25	47.2
380℃	79.1	19.6	0.09	1.25	55.9
410℃	63.5	26.9	4.19	5.34	82.7
440℃	66.0	21.1	5.25	7.50	77.6
470℃	44.6	18.6	6.97	29.6	73.5

From the results of GC, perfluoroalkyl iodides were mainly produced with a little amount of by-products such as perfluoroalkanes (C_4F_{10}, C_6F_{14}, etc). However, as reaction temperature is above 410℃, the amount of by-products get more.

As can be seen in Table 2, the conversion of C_2F_4 was getting increase by 410℃ and then getting decrease. The results of the product distributions also show that the best reaction temperatures are between 380℃ and 410℃. The yield of $C_2F_5(CF_2-CF_2)_2I$ increased with increasing reaction temperature up to 470℃. However much unexpected by-products were produced. To get the desirable perfluoroalkyl iodides having about 6 to 12 carbon atoms, the yield of $C_2F_5(CF_2-CF_2)_nI$ is mainly considered in lower by-products for more economical process. In addition, the value of n is well controlled and necessarily the recycling of the product is also needed by using developed process. Therefore the more study will be continued in the future.

4. CONCLUSIONS

The water-alcohol method provided more dispersed catalysts than the impregnation method. The case of copper, the Cu particles of about 20 nm showed enhanced stability and higher activity than the particles larger than 40 nm. In the effect of active metal, copper was better stable and active than silver and zinc. In case of telomerization reaction, the most suitable reaction temperature was 380~ 410℃.

REFERENCES
[1] R. Bertocchio and G. Lacote, US patent No.5268516 (1993).
[2] Funnakoshi, Y. and Miki, Jun, EP patent No.1380557 (2002).
[3] W. A. Blanchard, US patent No.3226449 (1965).
[4] Konrad Von Werner, US patent No.5639923 (1997).

Studies in Surface Science and Catalysis, volume 159
Hyun-Ku Rhee, In-Sik Nam and Jong Moon Park (Editors)

Determination of surface chemical states of CoO_x/TiO_2 catalysts for continuous wet TCE oxidation

Won-Ho Yang[a] and Moon Hyeon Kim

Department of Environmental Engineering, Daegu University,
15 Naeri, Jillyang, Gyeongsan 712-714, Korea

[a]Department of Occupational Health, Catholic University of Daegu
330 Keumnak, Hayang, Gyeongsan 712-702, Korea

ABSTRACT

A 5 wt.% CoO_x/TiO_2 catalyst gave the most promising activity for continuous catalytic wet oxidation of trichloroethylene at 310 K with a unsteady-state behavior up to 1 h. The catalyst after the oxidation possessed a Co $2p_{3/2}$ main peak at 779.8 eV, while the peak was obtained at 781.3 eV for a fresh sample. Only reflections for Co_3O_4 were indicated for these samples upon XRD measurements. The simplest model for nanosized Co_3O_4 particles existing with the fresh catalyst could reasonably explain the transient activity behavior.

1. INTRODUCTION

Wastewaters containing chlorinated hydrocarbons (CHCs) are very toxic for aquatic system even at concentrations of ppm levels [1]; thus, appropriate treatment technologies are required for processing them to non-toxic or more biologically amenable intermediates. Catalytic wet oxidation can offer an alternative approach to remove a variety of such toxic organic materials in wet streams. Numerous supported catalysts have been applied for the removal of aqueous organic wastes *via* heterogeneous wet catalysis [1,2].

Although catalytic wet oxidation of acetic acid, phenol, and p-coumaric acid has been reported for Co-Bi composites and CoO_x-based mixed metal oxides [3-5], we could find no studies of the wet oxidation of CHCs over supported CoO_x catalysts. Therefore, this study was conducted to see if such catalysts are available for wet oxidation of trichloroethylene (TCE) as a model CHC in a continuous flow fixed-bed reactor that requires no subsequent separation process. The supported CoO_x catalysts were characterized to explain unsteady-state behavior in activity for a certain hour on stream.

2. EXPERIMENTAL

2.1. Catalyst preparation

A 5 wt.% CoO_x/TiO_2 catalyst was prepared *via* an incipient wetness technique in which an aqueous solution of $Co(NO_3)_2 \cdot 6H_2O$ (Aldrich, 99.999%) was impregnated onto a shaped TiO_2 (Millennium Chemicals, commercially designated as DT51D, 30/40 mesh), as described in detail elsewhere [6]. Other supported metal oxide catalysts, such as FeO_x, CuO_x, and NiO_x, were obtained in a fashion similar to that used for preparing the CoO_x catalyst.

2.2. Continuous catalytic wet oxidation

A continuous flow fixed-bed reactor system was used for the wet oxidation of TCE at 310 K over the catalysts, as described earlier [7]. The quantity of pure TCE (Aldrich, 99.5%) corresponding to 30 ppm was directly injected into a Teflon-coated stainless steel reservoir containing 0.45 m^3 of water through an injection port, and this was rigorously mixed by a large volume stirrer (BelArt, Model 37028) for 1 h. The 30-ppm solution with constant stirring was fed into a catalyst bed using a Cole-Parmer high performance precision pump system, and the pumping rate of the solution was set to be 50 cm^3 min^{-1}, corresponding to a weight hourly space velocity of 7,500 h^{-1}. All the catalysts prepared were calcined *in situ* in the reaction system at 843 K for 1 h in flowing air (Prexair, 99.999%) at 100 cm^3 min^{-1} using a Brooks Model 5850E mass flow controller, prior to being used for the wet oxidation reaction at 310 K. Samples taken from a sampling port were analyzed using an Agilent 6890N gas chromatograph equipped with a flame ionization detector.

2.3. Spectroscopic characterization

Co 2p XPS spectra of a sample of 5 wt.% CoO_x/TiO_2 after either calcination at 843 K in flowing air for 1 h or subsequent wet TCE oxidation at 310 K for on-stream hours near 6 following sufficient drying at ambient conditions were collected using a VG Scientific ESCALAB 220iXL X-ray photoelectron spectrometer and were compared with those for reference Co compounds such as CoO, Co_3O_4 and $Co(OH)_2$ to identify surface chemical structures of the CoO_x catalysts. All XPS spectra were corrected using the C 1s peak at 284.8 eV. X-ray diffractograms of all the samples used for XPS measurements were obtained *ex situ* using a Rigaku D/MAX2500 PC diffractometer.

3. RESULTS

All the supported catalysts used gave TCE conversions less than 20% for the wet oxidation at 310 K, except for the 5 wt.% CoO_x/TiO_2, which had a steady-state conversion of 45% *via* a transient behavior in activity up to 1 h on stream (Fig. 1). Subsequently, there was negligible TCE conversion for the bare TiO_2 during continuous operating hours near 6.

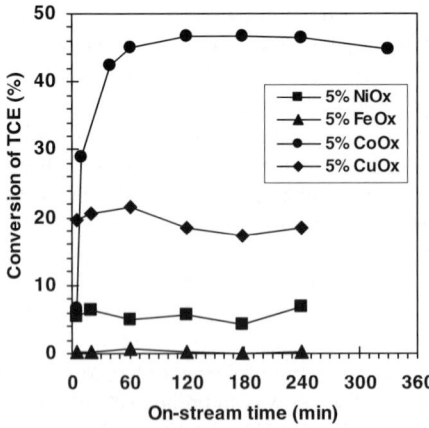
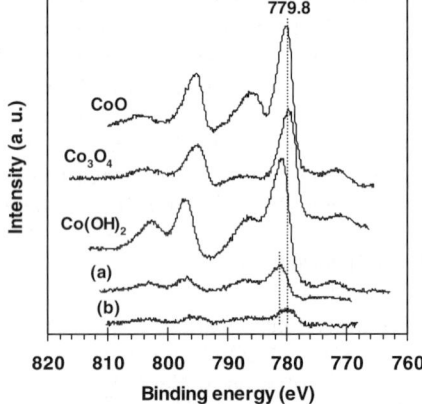

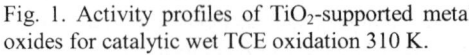

Fig. 1. Activity profiles of TiO_2-supported metal oxides for catalytic wet TCE oxidation 310 K.

Fig. 2. Co 2p XPS spectra of 5 wt.% CoO_x/TiO_2 and reference Co chemicals. (a) fresh; (b) spent.

Co $2p_{3/2}$ binding energy for Co species in the fresh catalyst (Spectrum a) appeared at 781.3 eV that was much higher, by 1.0 ~ 1.4 eV, than those obtained with the reference CoO and Co_3O_4, as shown in Fig. 2. The spent catalyst represented a 779.8-eV main peak for Co $2p_{3/2}$ and this binding energy was very similar to that of Co_3O_4 among the reference Co compounds employed here. The reference $Co(OH)_2$ possessed a binding energy at 781.2 eV for Co $2p_{3/2}$. All the XPS results are listed in Table 1 with an earlier study of Brik *et al.* [8] for cobalt titanates. XRD measurements exhibited the phase for Co_3O_4 even for the fresh catalyst. An average crystallite size of Co_3O_4 for both fresh and spent catalysts was shown to be 29 nm, based on the XRD peak at $2\theta = 31.26°$ by the crystallographic (220) plane. Finally, ICP measurements for the spent catalyst yielded the loss of *ca.* 0.2 wt.% CoO_x.

Table 1
Co $2p_{3/2}$ binding energies for CoO_x/TiO_2 catalysts and reference Co compounds

Catalyst	Binding energy (eV)	Ref.
Fresh 5% CoO_x/TiO_2	781.3	This study
Spent 5% CoO_x/TiO_2	779.8	This study
CoO	780.3	This study
Co_3O_4	779.9	This study
$Co(OH)_2$	781.2	This study
$CoTiO_3$	781.2	8
Co_2TiO_4	781.0	8

4. DISCUSSION

The activity profiles of the catalytic TCE oxidation at 310 K depend significantly on the kinds of metal oxides used here and the range of indicated TCE conversions is almost zero to 45% based on the measurements after the continuous feed hours greater than 1 h. The 5 wt.% CoO_x/TiO_2 catalyst maintains highest steady-state activity for the wet oxidation, although this catalyst showed a transient activity behavior at the initial reaction period. The bare TiO_2 plays no catalytic role in removing TCE in the aqueous solution, which is in good agreement with our previous study [7] for TiO_2-assisted photocatalytic decomposition of aqueous TCE. Consequently, this supported CoO_x catalyst may be the most promising for the continuous wet oxidation of TCE at very low temperatures, such as 310 K.

Characterization of fresh and spent CoO_x/TiO_2 samples by XPS measurements was attempted to distinguish between surface chemical states of the CoO_x species active for the wet TCE oxidation at 310 K. The Co_3O_4 employed here has very weak satellite structures for Co $2p_{3/2}$ because both magnetic Co^{2+} ions and diamagnetic low-spin Co^{3+} ions are present, compared to the CoO which gives strong shake-up peaks for Co $2p_{3/2}$ by Co^{2+} ions occupying the octahedral sites. Thus, it is very easy to differentiate the Co $2p_{3/2}$ spectral difference between these two cobalt chemicals. The spinel oxides consisting of Co and Ti possess significantly higher binding energy for Co 2p than CoO and Co_3O_4, as has been indicated for $CoTiO_3$ and Co_2TiO_4 [8]. A multitude of XPS, NIR and XRD studies of Co species such as CoO, Co_3O_4, Co_2TiO_4, and $CoTiO_3$ have been reported [8-10], and based on those earlier results and Co 2p XPS line positions of the reference Co compounds used in this work, the Co $2p_{3/2}$ main peaks (*i.e.*, 779.8, and 781.3 eV) observed for 5 wt.% CoO_x/TiO_2 after and before reaction can be assigned. The 779.8 eV peak in the XPS spectrum of the spent sample may indicate the presence of CoO_x very similar to the surface chemical states of Co_3O_4. The 781.3 eV main peak with the fresh catalyst calcined at 843 K is associated probably with $CoTiO_x$ species such as Co_2TiO_4 and $CoTiO_3$, which is in good agreement with an earlier XPS result [10] that the calcination of CoO_x/TiO_2 catalysts at 673 K in an oxygen-rich flow could lead to the formation of surface $CoTiO_3$ in addition to Co_3O_4 particles thereby giving significantly higher Co $2p_{3/2}$ binding energies near 781.2 ± 0.2 eV.

Based on these XPS and XRD results, coupled with the activity profile of the 5 wt.% CoO_x/TiO_2 catalyst for the wet oxidation of TCE at 310 K, and the previous discussion, to a very good approximation two types of model CoO_x particles can be proposed to exist with the fresh catalyst: Type A, Co_3O_4 particles completely encapsulated by $CoTiO_x$, exhibiting inactivity in the wet oxidation but becoming highly active with removal of the $CoTiO_x$ in the wet stream containing TCE; and Type B, Co_3O_4 particles partially covered by $CoTiO_x$ although XPS and XRD measurements had no direct evidence for this cobalt species, perhaps because of amounts too small to be appreciably detected by those techniques. The Co 2p XPS spectrum of the fresh catalyst consisting predominantly of the Type A Co_3O_4 particles with a large crystallite size of 29 nm would be very similar to that for $CoTiO_x$ due to such complete encapsulation but XRD measurements may give discernable peaks for the phase structure of Co_3O_4 crystallites because of very thin $CoTiO_x$ overlayer on the outer surface of the Co_3O_4 particles and their crystallite size sufficient to be visible by XRD. Such cobalt titanate may leach out into the wet reaction solution, thereby causing exposure of the Co_3O_4 to the surface along with a very small amount of CoO_x loss in weight as stated previously and creating rapid increase in the catalytic activity at the initial period during the course of reaction.

5. CONCLUSIONS

A 5% CoO_x/TiO_2 catalyst is quite active for the wet TCE oxidation at very low reaction temperatures, such as 310 K, and our proposed model of different forms of CoO_x species existing with the fresh catalyst can reasonably explain the unsteady-state catalytic behavior at the initial period during the wet catalysis.

REFERENCES

[1] A. Pintar, Catal. Today, 77 (2003) 451.

[2] S. Haumodi, F. Larachi and A. Sayari, J. Catal., 177 (1998) 247.

[3] S. Imamura, M. Nakamura, N. Kwabata, J.I. Yoshida and S. Ishida, Ind. Eng. Chem. Prod. Res. Dev., 25 (1986) 34.

[4] D. Mantzavinos, R. Hellenbrand, A.G. Livingston and I.S. Metcalfe, Appl. Catal. B, 7 (1996) 379.

[5] A. Fortuny, C. Bengona, J. Font and A. Fabregat, J. Hazard. Mater. B, 64 (1999) 181.

[6] M.H. Kim, J.R. Ebner, R.M. Friedman and M.A. Vannice, J. Catal., 208 (2002) 381.

[7] K.W. Park, K.H. Choo and M.H. Kim, Proceedings of Asian Waterqual'03, Bangkok, Thailand, 2003, Session 3Q3G14, p. 1.

[8] Y. Brik, M. Kacimi, M. Ziyad and F. Bozon-Verduraz, J. Catal., 202 (2001) 118.

[9] M. Voβ, D. Borgmann and G. Wedler, J. Catal., 12 (2002) 10.

[10] S.W. Ho, J.M. Cruz, M. Houalla and D.M. Hercules, J. Catal., 135 (1992) 173.

Studies in Surface Science and Catalysis, volume 159
Hyun-Ku Rhee, In-Sik Nam and Jong Moon Park (Editors)

Ammonia oxidation over Cu-based metal oxides under microwave irradiation

Takashi Aida [*] **and Yasubumi Kikuchi**

Department of Chemical Engineering, Tokyo Institute of Technology,
2-12-1 O-okayama, Meguro-ku, Tokyo 152-8552, Japan
[*] taida@chemeng.titech.ac.jp

Ammonia oxidation over copper based oxides (CuO, La_2CuO_4, $CuTa_2O_6$ and Cu-MOR) under microwave irradiation was investigated. Effects of the catalysts, amount and particle size of CuO, diameter of the reactor, microwave power and cycle number were examined. CuO was found to be most effective among the catalysts tested due to its high efficiency for microwave absorption. The highest temperature reached for the catalyst was 792 K. And the absorption efficiency became higher when the amount and particle size of the catalyst, the diameter of the reactor and the microwave power were greater, and no significant change was observed when the microwave irradiation was repeated. The results of the NH_3 oxidation under microwave irradiation were almost the same as those obtained by conventional heating by an electric furnace except the negligible NO formation at high temperatures.

1. INTRODUCTION

Removal of low concentration volatile organic compounds (VOCs) is of world wide interest. Odorous compounds including organic amines are a type of the in-house VOCs. There are several ways to remove such an odorous compound, for example, adsorption by activated charcoal and degradation by photocatalysis. Our group has reported a variation of trapping reactor using adsorbent/catalyst bed with rapid heating by microwave irradiation for low concentration VOC removal. Na^+ ion exchanged mordenite adsorbent and supported platinum catalyst were employed for the removal of several hundred ppm level ethylene from the air in the presence of water vapor [1, 2]. Cobalt oxide was also examined as a heating medium for microwave absorption as well as the oxidation catalyst [3].

In this study, ammonia oxidation to nitrogen over copper based oxides was investigated with microwave irradiation. Ammonia was chosen because it can be a representative of organic amines for the first step. CuO is known to absorb microwave efficiently [4], and is also well known for its strong affinity to ammonia and amines. Perovskite type oxides are also known to absorb microwave under multi-mode [5]. So, in this paper, two perovskite type oxides containing copper as well as CuO were tested. One is La_2CuO_4 with Cu at B cite and the other is $CuTa_2O_6$ with Cu at A cite. Ion exchanged mordenite with cupper was also tested because mordenite shows microwave absorbance as described in our previous papers [1, 2]. Effects of the various catalysts, amount and particle size of CuO, diameter of the reactor, microwave power and cycle number were examined.

2. EXPERIMENTAL

Commercial copper oxide (Kanto Chemical Co. Ltd., research grade) was employed as a catalyst/adsorbent as received. La_2CuO_4 was prepared by calcination of a physical mixture of CuO and La_2O_3 with the stoichiometric ratio at 1273 K for 24 h in air. $CuTa_2O_6$ was prepared also by calcination of the physical mixture of CuO and Ta_2O_5 with the stoichiometric ratio at 1273 K for 24 h in air. Cu ion exchanged mordenite (Cu-MOR) was prepared by ion-exchange method with aqueous solution of copper nitrate. Experimental set-up is the same as in the previous reports [3, 6], except the employment of NO_x analyzer to determine NO concentration in the product gas and a single mode microwave heater (Green Motif I: IMCR-25003, Tokyo-Denshi, Inc., 2.45 GHz, 0-300 W). Temperature of the catalyst bed was measured by K-type thermocouple inserted in the catalyst bed, and its metal jacket was grounded to avoid the electric charge. Pylex glass reactors with different diameters of 8 and 16 mmϕ-i.d. were used. Powder CuO was the sample as received. CuO and other catalysts were pelletized into 60/80 mesh particles. Concentrations of N_2, N_2O and unreacted NH_3 were determined by gas chromatography with TCD. Experiments of temperature measurement under microwave irradiation were conducted with 2 g of catalyst packed in the reactor without a gas feed. Standard condition for ammonia oxidation was 2 g of catalyst/adsorbent, 100 mL-NTP·min^{-1} for the total gas flow rate with 0.22 vol.% NH_3 and 2.0 vol.% O_2 diluted by He to atmospheric pressure.

3. RESULTS AND DISCUSSION

Closed symbols in Fig. 1 show the effect of reaction temperature on ammonia oxidation over CuO by heating with a conventional electric furnace. The reaction started at about 400 K and the conversion of NH_3 became 1 at temperatures higher than 500 K. Fig. 1 also indicates that selectivity to N_2 was high at low temperatures but it decreased as the temperature increased. Both N_2O and NO increased instead of N_2, except at 623 K, at which N_2O decreases. NO was detected above 583 K, and it monotonously increased by the temperature. High reaction temperature seems to tend deeper oxidation to NO_x. Considering that oxidation of N_2 to N_2O and NO is difficult in the tested temperature range,

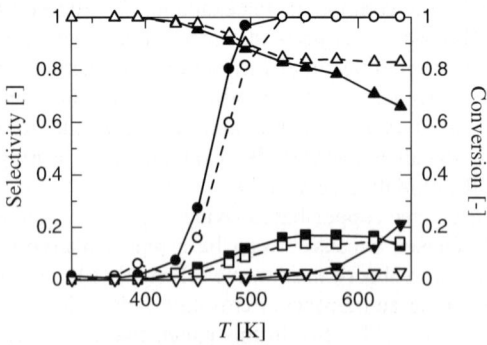

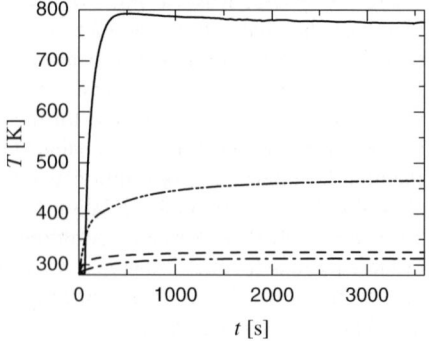

Fig. 1 Effect of temperature on NH_3 oxidation. Closed symbols: electric furnace, open symbols: microwave heating, ●,○: NH_3 conversion, ▲,△: N_2 selectivity, ▼,▽: NO selectivity, ■,□: N_2O selectivity.

Fig. 2 Microwave heating performance of the catalysts used. CuO(———), La_2CuO_4(– ·· – ··), $CuTa_2O_6$(------), Cu-MOR(– ·· –).

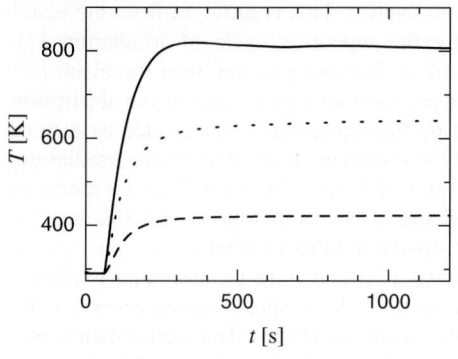

Fig. 3   Effect of power of the microwave irradiation.   100 W(·········), 200 W (----), 300 W (——).

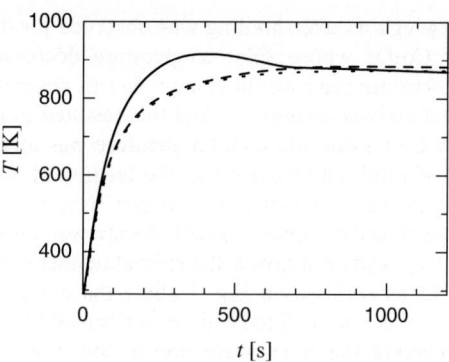

Fig. 4   Effect of cycle number in the repetitive microwave irradiation. 1st cycle (——), 2nd cycle (----), 3rd cycle (·········).

successive oxidation of N_2 is doubtful.   Hence, the result indicates the parallel paths from NH_3 to N_2O and/or NO along that to N_2.   From separately conducted experiments on the effect of contact time at 448 K, we can clearly say N_2 is not the intermediate to N_2O and NO. Other materials were also tested and the activity for NH_3 removal was in the order of CuO > $CuTa_2O_6$ > La_2CuO_6.   Note that NO_x formation was negligible for $CuTa_2O_6$.

Fig. 2 shows the temperature as a function of irradiation time of Cu based material under microwave irradiation.   CuO reached 792 K, whereas La_2CuO_4, $CuTa_2O_6$ and Cu-MOR gave only 325, 299 and 312 K, respectively.   The performances of the perovskite type oxides were not very significant compared to the expectation from the paper reported by Will et al. [5].   This is probably because we used a single mode microwave oven whereas Will et al. employed multi-mode one.   The multi-mode microwave oven is sometimes not very sensitive to sample's physical properties, such as electronic conductivity, crystal sizes. From the results by electric furnace heating in Fig. 1, at least 400 K is necessary for NH_3 removal.   So, CuO was employed in the further experiments although other materials still reserve the possibility as active catalysts when we employ a multi-mode microwave oven.

The effect of the amount of CuO in the same size reactor tube on the final temperature was examined and it was found that the more the catalyst, the higher the final temperature.   The effect of particle size was also investigated, and the smaller the diameter of the particle, the higher the final temperature when the same amount of catalyst was used.   The effect of the reactor tube diameter was also studied, and the larger the diameter, the higher the final temperature.   Of course, the heat escaping from the catalyst layer per catalyst weight changes by the above conditions.   The heat would decreases with the increase of the amount of catalyst and the reactor diameter, but the difference should be not very significant.   The result on the particle size effect was opposite to the one predicted only by the heat escape. There should be another reason for such a sensitive temperature variation tendency.   Fig. 3 shows the effect of microwave power.   The temperature changes were 127, 347 and 508 K for 100, 200 and 300 W, and $\Delta T/Q$ were 1.27, 1.74 and 1.69 K/W, respectively, and this indicates that the temperature changes are not linear to the power.   Considering the nonlinearity of $\Delta T/Q$, the efficiency of the microwave absorption is a function of temperature and this could be the reason of the sensitive temperature variation tendency.   Fig. 4 shows the effect of repetitive irradiation of microwave.   Results of 1st, 2nd and 3rd cycles were illustrated in the figure.   The final temperatures for the three runs were almost the same although an overshooting was observed for the first cycle.   This is different from the result of Co_3O_4 whose final temperature decreased by the repetitive cycle of irradiation [3]. Electronic state would change due to desorption of surface oxygen, and then the electronic conductivity changes.   And this resulted in the significant change in microwave absorption for Co_3O_4 and the final temperature was affected by the repetition.   The surface oxygen of Co_3O_4 did not recovered by the feeding of only 2 % of oxygen under microwave irradiation. In the case of CuO, the surface oxygen desorption did not affect the final temperature significantly.   Only a small change was disappearance of overshooting for the 2nd and 3rd runs.   CuO can give a stable final temperature compared to other Co_3O_4.

Open symbols in Fig. 1 show the effect of reaction temperature on ammonia oxidation over CuO in the fixed bed reactor heated by microwave.   The temperature was controlled by changing the microwave power and other factors were constant.   The performance was almost the same as that obtained by the electric furnace heating below 550 K.   An interesting finding was that NO formation was not significant at temperatures higher than 583 K, and this tendency was different from that of the electric furnace heating.   Yang et al. [5] reported that direct decomposition of NO occurs on supported iron oxide on zeolites by microwave irradiation and the activity is higher than the conventional heating.   Electronic and magnetic state of the catalyst surface may change by the irradiation of microwave because it induces screw current in the semi-conducting catalysts.   And this could be the reason why NO formation was negligible at high temperatures on the microwave-irradiated surface. La_2CuO_4, $CuTa_2O_6$ and Cu-MOR were also tested under the single mode microwave irradiation but didn't show significant activities for NH_3 removal because the catalyst bed could not reach a sufficiently high temperature.   In order to improve the activities of the catalysts under microwave irradiation, addition of a promoting material such as CuO is promising.   CuO can be heated to a high temperature, and once the catalyst bed temperature is raised up, the absorption efficiency of microwave of the catalysts could be improved.   And then, high performance is expected by the synergetic effect.

We have also tried the trapping reactor system, in which ammonia is trapped on the catalyst/adsorbent and microwave is irradiated intermittently.   However, due to the small specific surface area and the small ammonia adsorption capacity on the employed CuO, the trapping system was not effective compared to the continuous irradiation.   Further study should be made to develop a material having high ammonia adsorption capacity and high efficiency for microwave absorption.   Supported CuO on high surface area material or preparation of high surface area CuO can be effective.

REFERENCES

[1] S.-I. Kim, A. Tsuda, T. Aida and H. Niiyama, *Proceedings of Regional Symposium on Chemical Engineering 2004 (Bangkok, Thailand, December, 2004)*, #AS-321 (2004).
[2] S.-I Kim, T. Aida and H. Niiyama, *Sep. Purif. Technol.*, in press (2005).
[3] S.-I. Kim, Y. Watabe, T. Aida and H. Niiyama, *Proceedings of APCChE 2004*, Kita-Kyushu, Japan, 2004, P03-08-049, and accepted for publication in special issue for APCChE 2004 in *J. Chem. Eng. Jpn.*
[4] S.G. Deng, Y.S. Lin, *Chem. Eng. Sci.*, **52**, 1563 (1997).
[5] H. Will, P. Scholz, B. Ondruschka, *Topics in Catal.*, **29**, 175 (2004).
[6] S.-I. Kim, T. Aida and H. Niiyama, *Kagaku Kogaku Ronbunshu*, **29**, 345 (2003).
[7] Z. Yang, J. Zhang, X. Cao, Q. Liu, Z. Xu and Z. Zou, *Appl. Catal. B: Environ.*, **34**, 129 (2001).

Studies in Surface Science and Catalysis, volume 159
Hyun-Ku Rhee, In-Sik Nam and Jong Moon Park (Editors)
© 2006 Elsevier B.V. All rights reserved

New Synthesis of Optically Active (L)-Alaninol over Palladium Supported Catalysts

Chang-Kyo Shin[a], Soo-Hyun Kim[a], Kyung-Hye Chang[a], Hyeong-Chul Lee[b] and Geon-Joong Kim[a*]

[a]Department of Chemical Engineering, Inha University, 253 Yonghyun-Dong, Nam-Gu, Incheon 402-751, Korea E-mail : kimgj@inha.ac.kr
[b]Department of Food & Nutrition, Song Won College, 199-1 Kwangchun-Dong, Su-Gu, Gwangju 502-210, Korea

Pd metals immobilized on SBA-15 and NaY were applied as catalysts in the synthesis of amino alcohol. These catalysts afford a high level of enantioselectivity in the asymmetric hydrogenation of α-keto alcohol to corresponding amino alcohol. The large palladium metal exhibited higher catalytic activity and enantioselectivity than well dispersed one over porous supports in the hydrogenation.

1. INTRODUCTION

Almost enantioselective catalysts are known as soluble metal complexes containing some type of chiral ligands. It appears that such catalysts are effective because of the chiral environment created around the active metal center by the chiral ligands. The heterogeneous catalysts are usually produced by attaching ligands to an insoluble matrix and then using these insoluble ligands to complex with the active metal species. The other type of chiral heterogeneous catalyst can be either a supported metal which has been treated with a chiral modifier or an active metal on a chiral support.   Asymmetric syntheses of α-amino acids from their corresponding α-keto acids have been reported [1].   Hiskey and Northrop[2] have demonstrated the synthesis of optically pure α -amino acids by catalytic hydrogenation and subsequent hydrogenolysis of the Schiff bases of α-keto acids with chiral α-methyl benzylamine. Harada [3] reported the syntheses of optically active amino acids in a way principally similar to those done by Hiskey but by the use of α-phenylglycine in alkaline aqueous solution (optical purity 40-65%). These reactions are interesting because they are essentially a kind of asymmetric transamination reaction performed by catalytic hydrogenation and hydrogenolysis.   In this study, Pd metals immobilized on SBA-15 and NaY were applied as efficient catalysts in the synthesis of amino alcohol. These catalysts afford a high level of enantioselectivity in the asymmetric hydrogenation of α-keto alcohol to corresponding amino alcohol.   Indeed, this reaction has been investigated and reported only using Pd metals immobilized on active carbons.[4]

2. EXPERIMENTAL

The (S)-α-Methylbenzylamine(2.42g, 0.02 mole) in methanol(30 mℓ) was added to

314

acetol(0.01 mole) in cold methanol (40 ml). The mixture was allowed to stand for 30 min at room temperature. To the solution was added palladium catalysts, and then it was hydrogenated for 12 hrs at room temperature under different hydrogen pressure, respectively. The catalyst was removed by filtration and washed with hot water. The combined solution was evaporated to 20 ml. To the concentrated solution was added 30 % aqueous ethanol (50 ml) and palladium hydroxide supported on charcoal. The hydrogenolysis was carried out at room temperature for 12 h. The filtrate was concentrated to 5 ml in *vacuo* to obtain L-Alaninol. The conversion and %ee were determined by GC analysis using chiral column (GTA-CD column; Astec, FID).

3. RESULTS AND DISCUSSION

In this work, Hiskey-type reaction was carried out in order to screen the effect of inorganic support. Initially comparative investigations were carried out under the given reaction conditions to establish the suitability of the prepared Pd-containing catalyst for hydrogenation. The optical purities of the resulting amino acids were dependant on the kinds of supports and the enantiomeric excess values vary according to the composition of zeolitic materials. The Pd catalyst supported on zeolite support exhibited higher selectivity and conversion than carbon support. Relatively high %ee of 80% was obtained on the 10%Pd/SBA-15, and the highest optical purity of 86% was obtained over Pd/NaY. 10%Pd/Active Carbon gave a relatively low enantioselectivity of around 65 %ee for the synthesis of alaninol from acetol. The unsupported Pd black itself also gave a low enantioselectivity, less than 52 %ee. When the acidic zeolite was used as a support such as HY, a slight decrease in the optical yield was investigated as compared to Na type. The reaction pathway is also represented in Scheme 1.

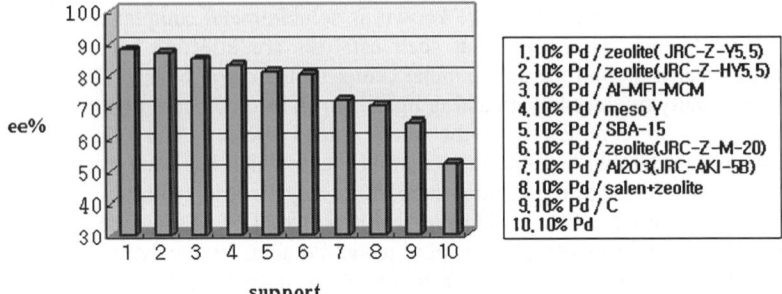

Fig. 1. The effect of support materials on the enantiomeric excess(ee)% of product.

Fig. 2 shows the effect of the loading amount on the conversion of reactant and %ee of produced alcohol. In this case, the reaction was conducted with a hydrogen pressure of 3.5 atm. Optical yields mainly depended on hydrogen pressure and palladium size. The Pd metals were observed to be apparently aggregated with the increased amount of Pd on the support, and the mean size of metal particle became larger with the elevation in H_2-treatment temperature. The %ee increased with the increase in the loading amount of palladium on the support up to 10 wt%. Nitta et al. have reported that the catalyst with the larger crystallite size gave the higher optical yield in the enantio-differentiating hydrogenation of methyl acetoacetate. The catalyst with a larger crystallite size had regularly-arranged metal atoms on the catalyst surface providing sites for a strong and regular adsorption of the modifier, propitious to obtain a high optical yield as reported by them [5]. When this fact was taken into account, the results in Fig. 1 indicate that larger metal would provide the appropriate surface for the enantio-differentiating hydrogenation of acetol to (L)-Alaninol.

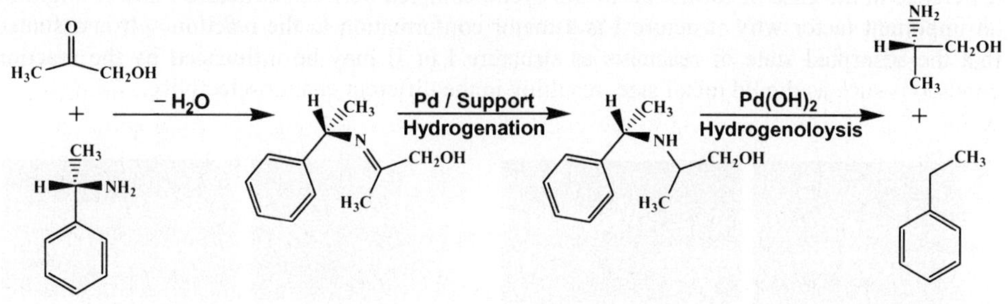

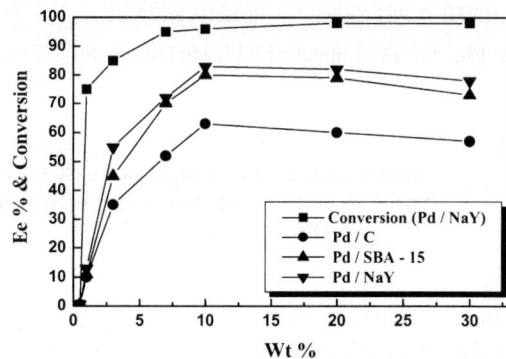

Scheme 1. Reaction pathway to synthesize chiral amino alcohol

Fig. 2. The effects of loading amount of pd metal on the enantioselectivity for L-Alaninol synthesis

Fig. 3 shows the TEM images of Pd metal supported on the mesoporous materials. The Pd metals were observed to be apparently aggregated, and the mean size of metal particle became larger with the increased amount of Pd on the supports. The result obtained in our work indicates that larger crystallite size of Pd would provide the suitable surfaces for the effective enantio-differentiation in the hydrogenation of acetol to synthesize a chiral alaninol. The effect of hydrogen pressure on the product %ee was also investigated. Optical yields mainly depended on hydrogen pressure. The maximum %ee of the product was obtained at the hydrogen pressure of 3.5 atm.

The enantioselective mechanism proposed in the literature stated that the structure I might be the most predominant structure and structure II might be a minor structure. Structure I resulted in (S)-amino alcohol when (S)-amine additive was used. On the other hand, structure II resulted in (R)-amino alcohol when (S)-amine additive was used. When the alkyl group of keto alcohol is methyl, conformation of reactant might be composed mainly of structure I, therefore resulting in highly optically active alaninol as indicated in Scheme 2. However, according to the experimental results, structure I can be a major conformation in this reaction.

316

The structure I might form a five-membered cyclic structure on Pd metal and then the structure would be adsorbed at the less bulky side of the molecule. On the other hand, structure II might not form such a cyclic structure because of the steric hindrance. The difference in the ease of formation of the cyclic complex between structure I and II might be an important factor why structure I is a major conformation in the reaction. It is assumed that the adsorbed state of reactants as structure I or II may be influenced by the reaction conditions such as the Pd metal size, resulting in the different enantioselectivity.

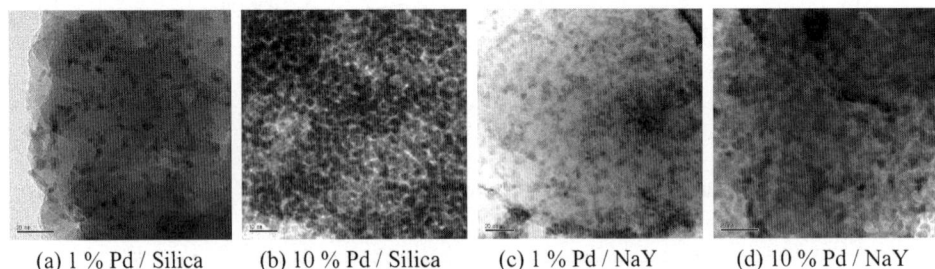

| (a) 1 % Pd / Silica | (b) 10 % Pd / Silica | (c) 1 % Pd / NaY | (d) 10 % Pd / NaY |

Fig. 3. TEM image of Pd-loaded Supports

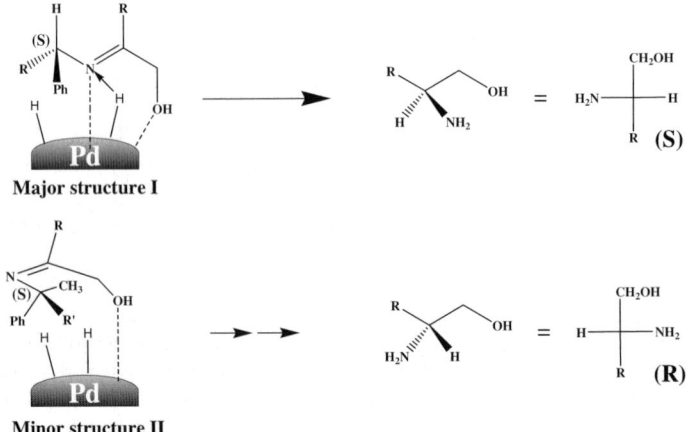

Scheme 2. Major structure I, Minor structure II.

REFERENCES

[1] K. Harada and K. Matsumoto, J. Org. Chem. 32 (1967) 1794.
[2] R.G.Hiskey and R.C.Northrop, J. Am. Chem. Soc. 83 (1961) 4798.
[3] K. Harada, Nature, 212 (1966) 1571.
[4] Robert L.Augustine, Heterogeneous catalysis for the synthetic chemist, MARCEL DEKKER, INC., New York, 1966, p.332.
[5] Y.Nitta, F.Sekine, T.Imanaka, and S. Teranishi, Bull.Chem.Soc.Jpn. 54 (1981) 980.

Studies in Surface Science and Catalysis, volume 159
Hyun-Ku Rhee, In-Sik Nam and Jong Moon Park (Editors)

Application of Ferrierite Catalyst to Polyolefin Degradation

Jong-Ki Jeon[1], Hyun Ju Park[2], Joo-Sik Kim[2], Ji Man Kim[3], Jinho Jung[4],
Jin-Heong Yim[5], Kwang-Eun Jeong[6], Son-Ki Ihm[6], Young-Kwon Park[2*]

[1]Dept. of Chemical Engineering, Kongju National University, Gongju 314-701, Korea

[2]Faculty of Environmental Engineering, University of Seoul, Seoul 130-743, Korea

[3]Dept. of Chemistry, Sungkyunkwan University, Suwon 440-746, Korea

[4]Div. of Environmental Science & Ecological Eng., Korea University, Seoul 136-713, Korea

[5]Div. of Advanced Materials Eng., Kongju National University, Gongju 314-701, Korea

[6]Dept. of Chemical and Biomolecular Eng., Korea Advanced Institute of Science and
Technology, Daejeon 305-701, Korea
*Corresponding author: catalica@uos.ac.kr

The catalytic degradation of polypropylene was carried out over ferrierite catalyst using a thermogravimetric analyzer as well as a fixed bed batch reactor. The activation of reaction was lowered by adding ferrierite catalyst, which was similar with that from ZSM-5. Ferrierite produced less gaseous products than HZSM-5, where the yields of i-butene and olefin over ferrierite were higher than that over HZSM-5. In the case of liquid product, main product over ferrierite is C_5 hydrocarbon, while products were distributed over mainly C_7-C_9 over HZSM-5. Ferrierite showed excellent catalytic stability for polypropylene degradation.

1. INTRODUCTION

Catalytic pyrolysis of waste plastics is one of the most promising way to convert polymers into valuable products of low molecular weights. Among the waste plastics, catalytic pyrolysis of waste polyolefins is of great interest because the products have a high potential to be used as fuels or chemical resource [1]. Catalysts mainly used in the pyrolysis were Y and ZSM-5 molecular sieves, SAPO, silica-alumina, sulfated zirconium oxide and used FCC catalyst [2]. Recently, zeolite ferrierite (FER) has emerged as a promising catalyst in many industrially important reactions like isomerization of n-butenes to isobutenes and the catalytic reduction of NO with methane [3]. Ferrierite is a molecular sieve with small pores and an orthorhombic framework, containing one-dimensional channels of 10-membered rings (5.4 x 4.2 Å) and of 8-membered rings (3.5 x 4.8 Å) which are perpendicularly intersected among them. However ferrierite has not been applied for polyolefin degradation. In this study, ferrierite was first attempted to decompose polyolefin.

2. EXPERIMENTAL

Polypropylene (PP) in powder forms was supplied from Samsung Total Co., Ltd. in Korea, and ferrierite was supplied by Zeolyst Company. The catalyst was mixed with an equal weight of PP. Catalytic degradation of PP was investigated under nitrogen environment using a thermogravimetric analyzer (CAHN, TG-2121) at linear heating rates of 5, 10, and 20 °C/min. Before each run, the sample (6 - 7 mg) was purged with nitrogen at a flow rate of 50 ml/min at least for 1 hr. The catalytic degradation of PP was also carried out in a fixed bed batch reactor (60 ml volume). A well mixed sample with different ratio of PP to catalyst was charged into the reactor. The air of the reactor was removed by purging with a nitrogen flow of 50 ml/min before the reaction. The catalytic pyrolysis of polyolefin was carried out at 400 °C for 1 hr. The degradation products were analyzed by a gas chromatography (Young Lin-M600D) which was equipped with a FID and TCD. Ferrierite was characterized using various analytical methods such as XRD, BET surface area, and ammonia temperature programmed desorption (NH_3-TPD).

3. RESULTS AND DISCUSSION

Table 1 shows the result of activation energy of polypropylene degradation over catalysts. The activation energy was very lowered by addition of zeolite catalysts. The activation energy over ferrierite was similar with that over a good catalyst for polymer degradation, HZSM-5, indicating that ferrierite could be a good catalyst for polypropylene degradation. Table 2 shows the product yield of polypropylene degradation. It may be expected that ferrierite gives higher yield of gaseous products than HZSM-5 from the NH_3-TPD results in Fig. 1, because ferrierite exhibits higher amount of acid sites than HZSM-5. However, ferrierite produced less gaseous products than HZSM-5, as shown in Table 2. As the pore size of ferrierite is smaller than that of HZSM-5, it may be difficult for reaction intermediates to diffuse into the pore and crack into gas. Table 3 shows the gas product distribution over catalysts. Higher yields of olefin were obtained over ferrierite. i-Butene yield was also higher than that of HZSM-5. These results suggest that ferrierite is more useful for producing valuable olefin products than HZSM-5. The yield of gaseous products can be increased up to 78 % by increasing the amount of ferrierite in the reaction mixture (see Table 4), which is comparable to that of HZSM-5.

Table 1
Activation energy of PP degradation

Catalyst	Without catalyst	HZSM-5	Ferrierite
Activation Energy(kJ/mol)	129.6	79.1	83.9

Table 2
Product yield of PP degradation

Catalyst	HZSM-5	Ferrierite
Yield(wt%)		
Gas	80.0	51.6
Oil	20.0	48.0

Catalyst/PP = 1/5 (wt/wt)

Table 3
Product distribution of gas product

| Catalyst | Product distribution (mol %) | | | | | | | |
	Methane	Ethene	Ethane	Propene	Propane	Butene	Butane	i-Butene
Ferrierite	0.8	4.8	0.7	30.8	11.6	28.2	2.8	20.3
HZSM-5	0.6	3.3	1.1	8.2	30.4	12.5	40.5	3.4

Catalyst/PP = 1/5 (wt/wt)

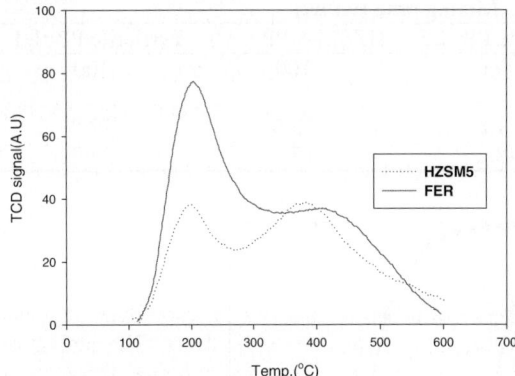

Fig. 1. NH$_3$ TPD of HZSM-5 and ferrierite

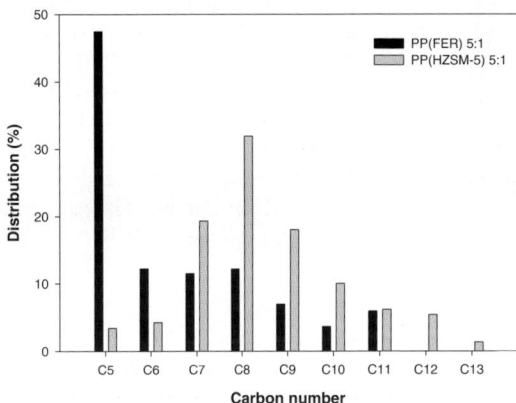

Fig. 2. Product distribution of liquid product (mol%)

Fig. 2 shows the liquid product distributions over catalysts. Main product over ferrierite is C$_5$ hydrocarbon, while products were distributed over mainly C$_7$-C$_9$ over HZSM-5. Table 4 shows the effect of mixing ratio on product distribution. While HZSM-5/PP ratio does not affect product distribution, higher amount of gas is obtained with increasing ferrierite/PP ratio. This is ascribed to the increased possibility of polypropylene diffusion into pore as the amount of ferrierite is increased.

320

As shown in Fig. 3, TGA experiments were carried out to evaluate the stability of ferrierite. Ferrierite was used 5 times without regeneration. Until the 3rd run, ferrierite showed almost similar degradation curves. The 4th and the 5th runs resulted in a little shift to higher temperatures. This implies that ferrierite may have high catalytic stability for the PP degradation.

Table 4
Effect of mixing ratio on polypropylene degradation

	Mixing ratio (wt/wt)			
	HZSM-5/PP = 1/5	Ferrierite/PP=1/5	HZSM-5/PP = 1/1	Ferrierite/PP=1/1
Conv.(%)	99.8	99.6	100	100
Sel.(wt%)				
Gas	80.0	51.8	83.0	77.8
Oil	20.0	48.2	17.0	22.2

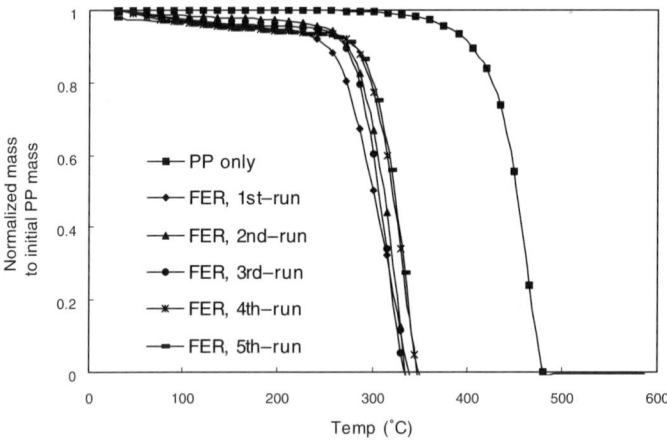

Fig 3. TGA curves of PP in the presence of ferrierite

4. CONCLUSIONS

The activation energy of polypropylene degradation was lowered with ferrierite catalyst. The yield of i-butene as well as the yield of olefin over ferrierite was higher than that over HZSM-5. In the case of liquid product, main product over ferrierite is C_5 hydrocarbon, while products were distributed over mainly C_7-C_9 over HZSM-5. The amount of gaseous product increased with increasing ferrierite/PP ratio. Ferrierite showed high catalytic stability for the polypropylene degradation.

REFERENCES
[1] W. Kaminsky, J.S. Kim, J. Anal. Appl. Pyrolysis, 51 (1999) 127.
[2] Y.K. Park, J.S. Kim, J. Choi, J.K. Jeon, S.D. Kim, S.S. Kim, J.M. Kim, K.S Yoo, J. Korea Soc. Waste Management, 20 (2003) 56.
[3] A.A. Belhekar, R.K. Ahedi, S. Kuriyava, S.S Shevade, B.S. Rao, R. Anand, Z. Tvaruzkova, Catal. Comm., 4 (2003) 295.

Studies in Surface Science and Catalysis, volume 159
Hyun-Ku Rhee, In-Sik Nam and Jong Moon Park (Editors)

CO oxidation over Au/CeO$_2$ prepared by a co-precipitation method

E.D. Park*

Department of Chemical Engineering, Division of Chemical Engineering and Materials Engineering, Ajou University, Wonchun-Dong Yeongtong-Gu Suwon , 443-749, Republic of Korea. (edpark@ajou.ac.kr)

The catalytic activity for CO oxidation in the absence and presence of water vapor was examined over co-precipitated 0.95% Au/CeO$_2$ pretreated at different conditions. Oxidizing environment generally appeared to be more favorable than reducing one. Calcinations at low temperatures such as 373 K and 473 K showed the highest activity in dry condition. However, in wet condition, the maximum activity was observed when the catalyst was calcined at 773 K. The presence of metallic gold was observed only after drying process at 373 K. Nano-sized metallic gold and strong interaction between gold and ceria appeared to be important for high catalytic activities in CO oxidation.

INTRODUCTION

Before Haruta *et al.* [1] reported that gold particles smaller than 10nm could be formed on Co$_3$O$_4$, α-Fe$_2$O$_3$, NiO, and Be(OH)$_2$ through co-precipitation, and that these Au catalysts were active in the oxidation of CO at sub-ambient temperatures, gold has been regarded as far less active as a catalyst than platinum-group metals. A number of nano-sized gold catalysts have been prepared by different preparation methods and have been reported to have extraordinary activities for various reactions such as oxidations and hydrogenations. Among them, Au/CeO$_2$ catalyst has attracted interests due to its potential applications for some reactions. Flytzani-Stephanopoulos's group have studied the water-gas shift reaction over gold-cerium oxide catalysts and found some advantages over commercial Cu-ZnO [2]. Carrettin et al [3] reported that nanocrystalline CeO$_2$ could increase the activity of Au for CO oxidation by two orders of magnitude compared with conventional ceria. Bera and Hegde [4] prepared ceria-supported gold catalyst by the solution combustion method and conducted various reactions such as CO and hydrocarbon oxidation and NO reduction. They found different activities between as-prepared and the heat-treated 1%Au/CeO$_2$. Recently, some interesting works for the application of ceria-supported gold catalysts on preferential oxidation of carbon monoxide in hydrogen-rich gas mixture for the polymer electrolyte fuel cell have been reported [5-7]. In this study, Au/CeO$_2$ was prepared with a co-precipitation method and was applied to CO oxidation. To enhance oxidation activity in the dry and wet conditions, a number of pretreatment conditions were tested and different tendencies were observed.

EXPERIMENTAL

Au/CeO$_2$ catalysts were prepared by a co-precipitation method. Two aqueous solutions, one containing AuCl$_3$ (Aldrich) and the other containing Ce(NO$_3$)$_3$ (Aldrich) were mixed under continuous stirring. The pH of this solution was raised to 10 by adding an aqueous solution of 1M NaOH drop by drop. After 1 h stirring at 343 K, precipitate was filtered, washed several times to remove residual chloride ion, dried at 353 K, and stored as fresh samples. The gold contents were determined by an ICP. Experiments were carried out in a small fixed bed reactor with catalysts that had been retained between 45 and 80 mesh sieves. A standard gas of 1.0 vol% CO and 1.0 vol% O$_2$ balanced with helium was passed through the catalyst bed at atmospheric pressure. For activity tests under the wet condition, the reaction gases were directed through a water vapor saturator immersed in a bath at 283 K and fed to a reactor through a glass line warmed by a heating tape. The conversion of CO was determined through gas chromatographic analysis (HP5890A, molecular sieve 5A column) of the effluent from the reactor. The transmission electron microscope (TEM) and X-ray diffraction (XRD) was conducted to measure the gold particle size and bulk crystalline structure of catalysts, respectively. The surface area was measured with ASAP 2010.

RESULTS AND DISCUSSION

CO conversions over Au/CeO$_2$ catalyst were measured in the dry and wet condition as shown in Fig. 1. Similar to other supported gold catalysts, Au/CeO$_2$ catalyst showed higher CO conversions in the presence of water vapor than in the absence of it at the same temperature. Catalytic activities for CO oxidation over Au/CeO$_2$ catalysts prepared at different calcinations temperature were compared in the dry and wet condition as shown in Fig. 2. Au/CeO$_2$ catalyst calcined at 473 K showed the highest initial CO conversion in the absence of water vapor. However, the CO conversion decreased steadily and reached a steady-state value over this catalyst.

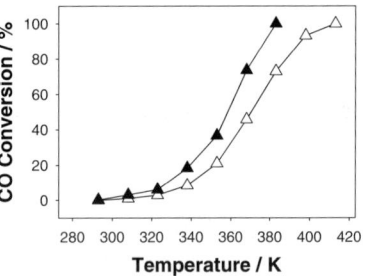

Fig. 1. CO conversions at different reaction temperatures in the dry (open points) and wet condition (filled points) over 0.95wt% Au/CeO$_2$ catalyst calcined at 573 K. F/W = 1,000 ml/min/g$_{cat.}$.

The steady-state CO conversion in the dry condition decreased with calcinations temperatures in the order; 473 K, 373 K > 573 K > 773 K, 673 K > 873 K. In the presence of water vapor, Au/CeO$_2$ catalyst calcined at 773 K showed the highest CO conversion. The steady-state CO conversion in the wet condition decreased with calcinations temperatures in the order; 773 K >> 873 K, 673 K > 373 K > 473 K, 573 K. It is interesting that Au/CeO$_2$ catalyst just dried at 373 K had comparable catalytic activities with other calcined ones at higher temperatures. The promotional effect of water vapor on the catalytic activity was manifest on Au/CeO$_2$ catalyst calcined at 773 K. The gold particle size of Au/CeO$_2$ calcined at 373 K and 773 K was determined to be 4.1±1.7 and 4.7±1.6 nm, respectively. The BET surface area of Au/CeO$_2$ catalyst calcined at 373 K and 773 K was determined to be 113 and 93 m^2/g, respectively. Only slight increase of the average gold particle size with increasing calcinations temperatures can be due to strong interactions between gold and ceria.

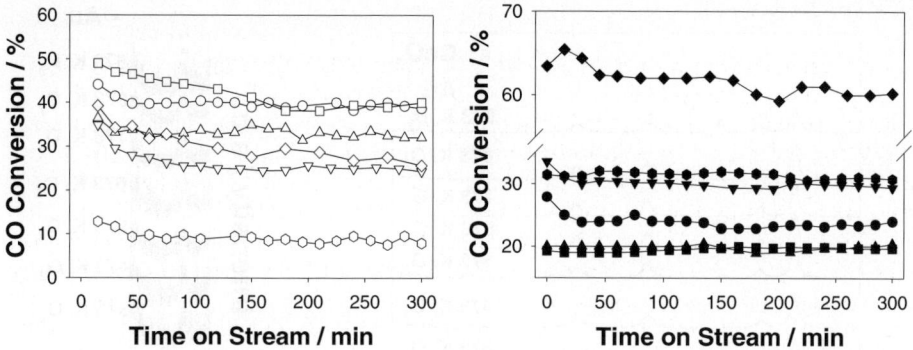

Fig. 2. CO conversions at 363 K in the dry condition (open points) and 353 K in the wet condition (filled points) over 100mg and 50 mg of Au/CeO$_2$ catalyst containing 0.95 wt% Au prepared at different calcination temperatures (373 K (circle) , 473 K(square), 573 K(triangle up), 673 K (triangle down), 773 K (diamond), 873 K (hexagon)). The reactants of 100 ml/min, 1 vol% CO and 1 vol% O$_2$ in He, were fed to the catalyst.

Some gold catalysts have been reported to have enhanced catalytic activity for CO oxidation when it reduced at low temperatures. The effect of a reductive pretreatment with hydrogen on CO conversion in dry and wet condition was investigated. Figure 3 shows that no noticeable difference in CO conversion in the dry condition was found between Au/CeO$_2$ catalysts calcined and reduced at 573 K. However, the much lower CO conversion was obtained in the wet condition over reduced Au/CeO$_2$ catalyst compared with calcined one. The BET surface area of Au/CeO$_2$ catalyst reduced at 573 K was determined to be 105 m^2/g. XRD patterns were obtained to determine if there was any change in bulk crystalline structures. For all catalysts prepared, peaks representing crystalline CeO$_2$ and metallic gold were obtained. Figure 4 shows that crystalline CeO$_2$ and metallic gold can be formed even through a drying process at 373 K. As the calcination temperature increased, the peak intensity corresponding to crystalline CeO$_2$ also increased.

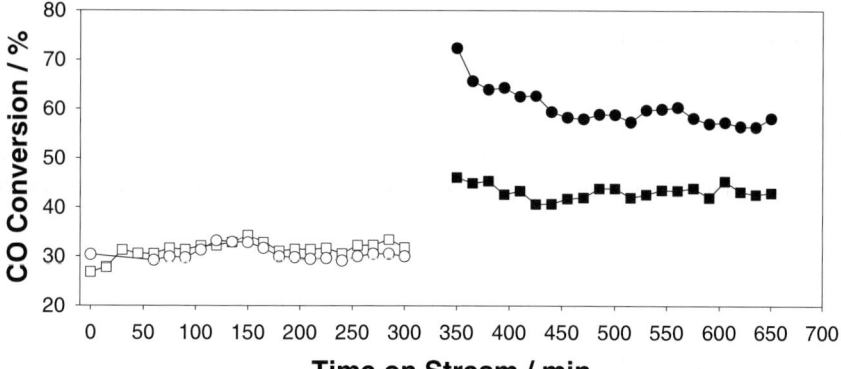

Fig. 3. CO conversions at 363 K in the dry condition (open points) and in the wet condition (filled points) over 0.95wt% Au/CeO$_2$ catalyst calcined (circle) or reduced (square) at 573 K. F/W = 1,000 ml/min/g$_{cat.}$.

324

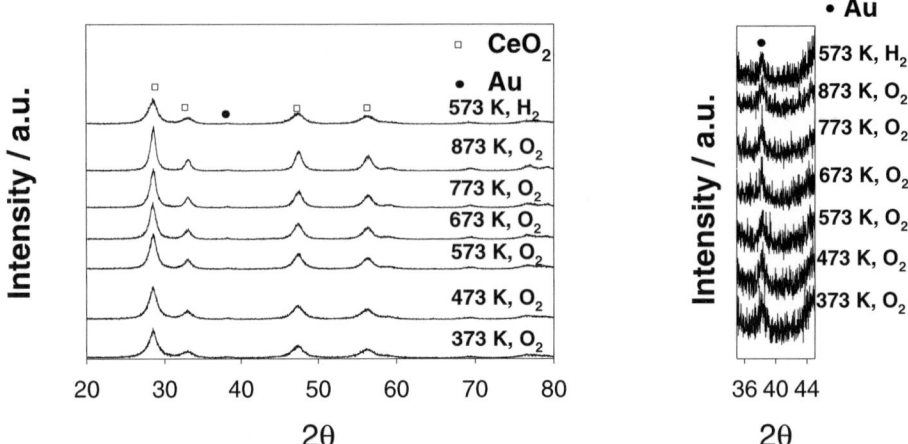

Fig. 4. XRD patterns of Au/CeO$_2$ catalyst containing 0.95 wt% Au pretreated at different conditions.

These results lead us to the conclusion that nano-sized metallic gold and strong interaction between gold and ceria contribute enhanced catalytic activity for CO oxidation in the absence and presence of water vapor.

CONCLUSION

In the dry condition, Au/CeO$_2$ catalyst, co-precipitated and then calcined at low temperatures such as 373 K or 473 K, showed the highest CO conversion. However, Au/CeO$_2$ catalyst calcined at 773 K showed the highest CO conversion in the presence of water vapor. Oxidizing environment appeared to be more favorable than reducing condition. The metallic gold can be formed after drying process at 373 K for co-precipitated Au/CeO$_2$ catalysts. As the calcinations temperature increased, the crystalline size of CeO$_2$ also increased. Nano-sized metallic gold and strong interaction between gold and ceria appeared to be important for high catalytic activities in CO oxidation.

ACKNOWLEDGEMENT
This work is supported by the Research Initiation Program at Ajou University (20041340).

REFERENCES
[1] M. Haruta, Catal. Today, 36 (1997) 153.
[2] Q. Fu, W. Deng, H. Saltsburg, M. Flytzani-Stephanopoulos, Appl. Catal. B: Environ. 56 (2005) 57.
[3] S. Carrettin, P. Concepcion, A. Corma, J.M. Lopez, and V. Puntes, Angew. Chem. Int. Ed., 43, (2004) 2538.
[4] P. Bera, and M.S. Hegde, Catal. Lett., 79(1~4) (2002) 75.
[5] W.-S. Shin, C.-R. Jung, J. Han, S.-W. Nam, T.-H. Lim, S.-A. Hong, and H.-I. Lee, J. Ind. Eng. Chem., 10(2) (2004) 302.
[6] A. Luengnaruemitchai, S. Osuwan, and E. Gulari, Int. J. Hydrogen Energy 29 (2004) 429.
[7] G. Panzera, V. Modafferi, S. Candamano, A. Donato, F. Frusteri, and P.L. Antonucci, J. Power Source, 135 (2004) 177.

Studies in Surface Science and Catalysis, volume 159
Hyun-Ku Rhee, In-Sik Nam and Jong Moon Park (Editors)

Preparation of highly active CuCl catalyst for the direct process of methylchlorosilane production

W.-X. Luo, G.-R. Wang and J.-F. Wang[*]

Department of Chemical Engineering, Tsinghua University, Beijing 100084, China

1. INTRODUCTION

Organosilanes, especially dimethyldichlorosilane (M_2), are important chemicals used in the silicone industries. The direct reaction of silicon with an organic halide to produce the corresponding organosilanes as a gas-solid-solid catalytic reaction was first disclosed by Rochow [1]. In the reaction, a copper-containing precursor first reacts with silicon particles to form the catalytically active component, which is a copper-silicon alloy, the exact state of which is still under discussion. As the reaction proceeds, Si in the alloy is consumed, which is followed by the release of copper. This copper diffuses into the Si lattice to form new reaction centers until deactivation occurs. The main reaction of the direct process is:

$$Si_{(s)} + 2CH_3Cl_{(g)} \xrightarrow{Cu} (CH_3)_2SiCl_{2(g)} \qquad \Delta_r H = -291.6\,\text{kJ/mol} \qquad (1)$$

Byproducts, such as methyltrichlorosilane (M_1), trimethylchlorosilane (M_3), methyldichlorosilane (MH), and some residuals (R) having a boiling point above $70^\circ C$ are also produced. One of the most efficient ways to prepare the active copper-silicon alloy is by the uniform mixing of CuCl and silicon particles. The following reaction takes place when the mixture is heated to the reaction temperature:

$$CuCl + Si \longrightarrow SiCl_4 + Cu^* \qquad \text{and} \qquad Cu^* + Si \longrightarrow Cu_3Si \qquad (2)$$

It is evident that a highly dispersed CuCl catalyst will combine better with silicon and the Cu will diffuse faster into the Si lattice in the subsequent reaction steps. However, most works on this aspect have focused on the post treatment of a ready-made catalyst, which is a complicated process using a large amount of energy.

CuCl, especially in a single crystal form, is extensively used as an optical material for its special optical properties. Orel et al. [2] first proposed a new method to obtain CuCl particles by the reduction of Cu^{2+} with ascorbic acid. Several dispersants were used in the reduction and monodispersed CuCl particles can be obtained by selecting the proper dispersant and reduction conditions. In this work, the above method was used to modify the traditional process of CuCl preparation, namely, by reducing the Cu^{2+} with sodium sulfite to obtain the highly active CuCl catalyst to be used in the direct process of methylchlorosilane synthesis.

---

[*] Corresponding author. E-mail: wangjf@flotu.org

2. EXPERIMENTAL

2.1 Catalyst preparation

The CuCl precipitate was obtained by mixing an aqueous solution of cupric sulfate and sodium sulfite to produce the reactions:

$$2CuSO_4 \cdot 5H_2O + 4NaCl \rightarrow 2CuCl_2 + 2Na_2SO_4 \qquad (3)$$

$$2CuCl_2 + Na_2SO_3 + H_2O \rightarrow 2CuCl + Na_2SO_4 + H_2SO_4 \qquad (4)$$

XRD analysis of the solid product showed three main peaks at 28.5°, 47.4° and 56.3°, which indicated that pure crystalline CuCl was formed [3]. Several well-known dispersants: polyvinyl pyrrolidone (PVP), sodium hexameta phosphate (SHP), the sodium salt of EDTA (EDTA-Na), sodium dodecyl sulfonate (SDS), and sodium dodecyl benzene sulfonate (SDBS), were introduced to obtain a highly dispersed catalyst. The X-ray patterns obtained with these were basically the same as the patterns obtained with the solids prepared in the other experiments described here.

2.2 Catalyst evaluation

Catalytic measurements were carried out in a stirred-bed reactor of stainless steel. The details of the reactor and experimental procedures have been described elsewhere [4].

2. 3 SEM analysis

The morphology of the CuCl precipitates were observed by scanning electron microscopy (SEM). The SEM microscopy was carried out with a beam energy of 15kV.

3. RESULTS AND DISCUSSION

3.1 Effect of additives on the morphology and size of the CuCl particles

The influence of the additives on the morphology and size of the CuCl particles are shown in Fig. 1. It can be seen that the various additives all have obvious effects on the reducing and dispersing of the CuCl precipitates. However, different dispersing agents promote dissimilar morphologies of the CuCl crystal. CuCl obtained without adding any dispersant were obtained as massive pyramid-like crystals. The addition of PVP produces CuCl agglomerates of small spherical particles. The addition of SDBS produces CuCl precipitates with a small trigonal-flake morphology. The mechanism on how a dispersant influences the morphology of a CuCl particle is complex and needs additional research.

3.2 Effect of CuCl morphology and size on the activity in the Rochow reaction

The activities of the different CuCl catalysts are illustrated in Fig. 2. It can be seen that all CuCl precipitates prepared with the addition of a dispersant show high activity. The activity of the CuCl prepared without any dispersant is here used as the standard activity. It is of interest that the highest activity, which is nearly twice the standard activity, was obtained

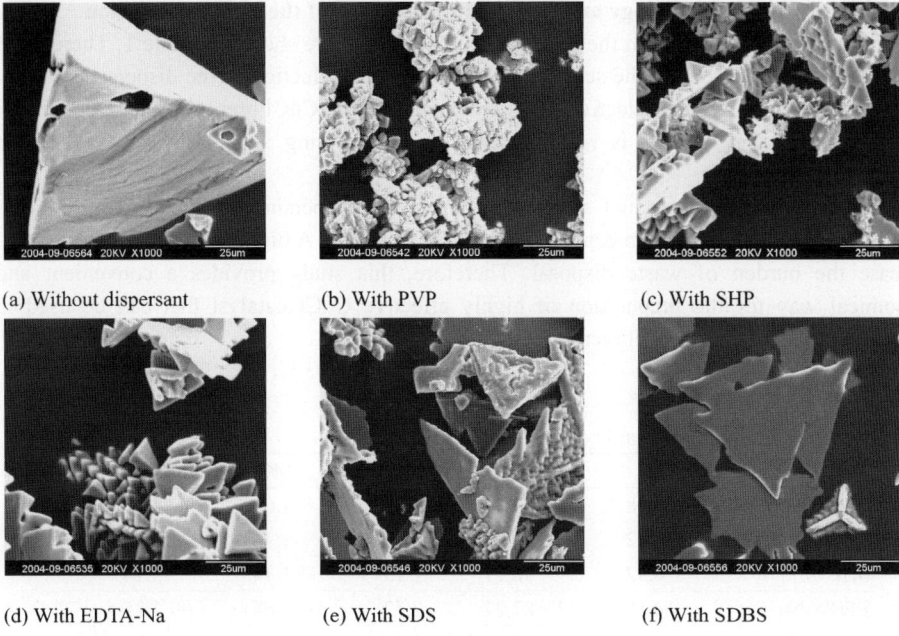

(a) Without dispersant   (b) With PVP   (c) With SHP

(d) With EDTA-Na   (e) With SDS   (f) With SDBS

Fig. 1 Effect of various dispersants on CuCl morphology

when SDBS used. In general, a better dispersed catalyst gave a higher activity as expected.

Reaction (2) was also studied using the different catalysts. Before exposure to CH_3Cl, the contact masses were subject to XRD analysis. In the XRD patterns of the catalysts with higher activities for the Rochow direct process, the XRD peaks of the Cu-containing species were weaker and broader when normalized to the silicon peaks (silicon was used in excess). This suggests that some undetectable species were formed and the catalytic species were well-dispersed. This agrees well with the view of Lieske and co-workers [5].

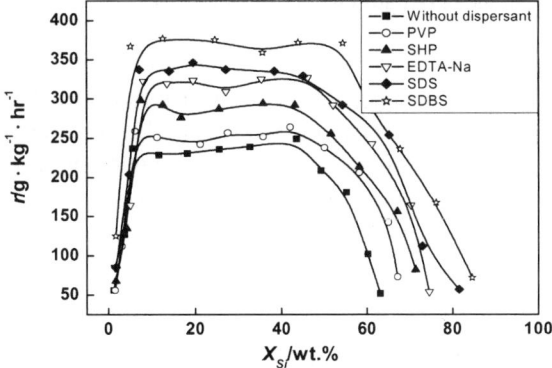

Fig. 2 Effect of adding dispersants on reactivity

328

3.3 Effect of CuCl morphology and size on the selectivity of the Rochow reaction

Product selectivities from the different CuCl catalysts are shown in Table 1. There is no visible detrimental effect on the selectivities due to the introduction of the dispersants during CuCl preparation. The best selectivity was obtained with the CuCl prepared with the addition of SDBS. Therefore, SDBS is recommended as a dispersing agent that can be used in practical production.

The reactivity and product selectivity increase as dispersing agents were introduced. Simultaneously, a higher silicon conversion was also obtained. A higher silicon conversion will decrease the burden of waste disposal. Therefore, this study provides a convenient and economical way for the preparation of highly effective CuCl catalyst that can be used in practical production using the direct process.

Table 1
Effect of adding dispersants on the product selectivity

Dispersant added	M_1	M_2	M_3	MH	R
Without dispersant	5.08	86.31	2.72	1.61	4.28
PVP	5.02	85.17	2.96	1.72	5.13
SHP	4.92	86.17	2.15	1.74	5.14
EDTA-Na	4.18	87.21	2.72	1.64	4.25
SDS	4.56	87.28	2.14	1.09	4.93
SDBS	3.25	91.23	1.56	1.23	2.73

4. CONCLUSION

Highly active CuCl catalysts for the direct process of methylchlorosilane synthesis were prepared by reducing Cu^{2+} with a sodium sulfite solution in the presence of dispersing agents. Several well-known dispersants, e.g. SDBS, were used in this study. When SDBS was used, a catalyst in the form of small flakes was obtained that gave the best performance in reactivity, product selectivity and silicon conversion. This provides a convenient way to prepare the CuCl catalyst for use in industrial production.

REFERENCES
[1] E.G. Rochow, Preparation of organosilicon halides, US Patent No. 2 380 995 (1945).
[2] Z. C. Orel, E. Matijevi and D V Goia, Colloid Polym. Sci., 281 (2003) 754.
[3] Joint Committee on Powder Diffraction Standards (JCPDS), Standard Powder Diffraction File, International Center for Diffraction Data, Swarthmore, PA.
[4] W. X. Luo, G. L. Zhang, G. R. Wang and J. F. Wang, J. Tsinghua. Univ (Sci. & Tech.), to be published.
[5] M. Stoyanova, G. Stein and H. Lieske, In: H. A. Øye, H. Rong, L. Nygaard, G. Schüssler and J. K. Tuset (Eds.), Silicon for the chemical industry VI, Leon, Norway, (2002) 299.

Studies in Surface Science and Catalysis, volume 159
Hyun-Ku Rhee, In-Sik Nam and Jong Moon Park (Editors)

329

Synthesis of dimethyl carbonate by transesterification of ethylene carbonate and methanol using quaternary ammonium salt catalysts

D.W. Park, E.S. Jeong, K.H. Kim[*], K.V. Bineesh, J.W. Lee[1], and S.W. Park

Division of Chemical Engineering, Pusan National University, Busan 609-735, Korea
[1]Department of Chemical Engineering, Sogang University, Seoul 121-742, Korea
E-mail : khkim@pusan.ac.kr (K.-H. Kim)

1. INTRODUCTION

Dimethyl carbonate (DMC) is achieving increasing importance in the chemical industry. DMC, an environmentally benign and biodegradable chemical, has attracted substantial research efforts in recent years. DMC can be synthesized from cyclic carbonate compounds such as ethylene carbonate or propylene carbonate. In this reaction, the cyclic carbonate is transesterified with methanol to DMC and a corresponding glycol. There are some reports on the synthesis of DMC from ethylene carbonate and methanol[1-7]. In these studies, alkali metals[1], free organic phosphines supported on partially cross-linked polystyrene[2], and heterogeneous catalysts such as zeolite[3,6], basic metal oxides[4,5], and hydrotalcite[7] are used as catalysts for the reaction. However, with some of these catalysts, activity or selectivity was not so high.

In our previous works[8,9] on the synthesis of various 5-membered cyclic carbonates, quaternary ammonium salts such as tetrabutylammonium halides showed excellent catalytic activities in relatively mild reaction conditions, under atmospheric pressure and below 140 ℃. In this work, several kinds of quaternary ammonium salts have been used for the transesterification reactions of the ethylene carbonate with methanol to DMC and ethylene glycol.

2. EXPERIMENTAL

Quaternary ammonium salt catalysts based on different alkyl cations such as tetrapropylammonium (TPA^+), tetrabutylammonium (TBA^+), tetrahexylammonium (THA^+), tetraoctylammonium (TOA^+), tetradodecylammonium ($TDodA^+$), and those with different anions such as Cl^-, Br^-, and I^- were used.

The transesterification reaction was carried out in a 50 mL stainless steel autoclave equipped with a magnetic stirrer. For each typical reaction, quaternary ammonium salt (2 mmol), propylene carbonate (25 mmol) and excess methanol (200 mmol) were charged into the reactor, and the CO_2 was introduced at room temperature to a preset pressure. The reaction was started by stirring when the desired temperature and pressure were attained. The reaction was performed in a batch operation mode. The analysis of the products and reactants was performed by using a gas chromatograph (HP 6890N) equipped with a FID and a capillary column (HP-5, 5 % phenyl methyl siloxane). Selectivities to DMC and EG are calculated on the basis of EC as a limited reactant.

3. RESULTS AND DISCUSSION

The synthesis of DMC from EC and methanol was carried out in a batch reactor using various quaternary ammonium salt catalysts under carbon dioxide pressure.

$$(CH_2O)_2CO + 2\ CH_3OH \rightarrow (CH_3O)_2CO + (CH_2OH)_2$$

DMC and EG were main products of the transesterification reaction. No by-product such as dimethyl ether and glycol monoethyl ether was observed in the resulting products. Only small peaks of ethylene oxide from the decomposition of EC could be detected at longer reaction time and at high temperature.

Fig. 1 shows time variant conversion of EC, selectivity and yield to DMC at 140 ℃ with tetrabutyl ammonium chloride (TBAC) catalyst under initial CO_2 pressure of 300 psig. As the reaction time proceeded, the conversion of EC increased up to 2 h, and then it remained nearly constant. It was reported that the reaction time needed to arrive at equilibrium depended highly on the type of catalysts[10]. LiOH, KOH and K_2CO_3 catalysts reached equilibrium in less than 1 h, however, KBr and KI arrived at equilibrium after more than 6 h[10]. The selectivities of DMC and EG did not change very significantly. However, DMC selectivity was higher than EG selectivity. According to a mechanism proposed by Fang and Xiao[7], DMC was produced via several reaction steps involving the formation of CH_3O^- and its reaction with EC. Ethylene glycol (EG) was suggested to form by the reaction of methanol and the by-product intermediate produced in the step of DMC synthesis. Therefore, the lower selectivity of EG can be explained by a lower reaction rate of this consecutive reaction.

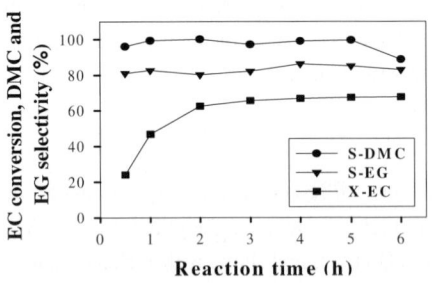

Fig. 1. Time variant conversion of EC and selectivities to DMC and EG at 140 ℃.

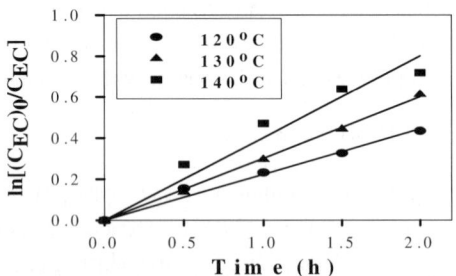

Fig. 2. Linear plots of $\ln[(C_{EC})_0/C_{EC}]$.

In order to understand the effects of the cation structure in the transesterification between methanol and EC, quaternary ammonium chloride catalysts of different alkyl cations such as TPAC, TBAC, THAC, TOAC, and TDodAC were used at 140 ℃. Table 1 shows EC conversions after 1 h and 6 h, and selectivites to DMC and EG after 6 h. The EC conversion at 1 h and DMC selectivity increased as the size of alkyl cation increased from TPAC to TDodAC. However, the EC conversion at 6 h did not vary significantly because the reaction was close to equilibrium. It has been generally reported that quaternary ammonium salt having bulkier alkyl chain length exhibited higher catalytic activity in many reactions[11]. In the case of the effect of anions of the quaternary salts, the conversion of EC increased in the order of I^-< Br^-< Cl^-, which is consistent with the order of nucleophilicity of the anions.

Table 1. Effect of catalyst structure on EC conversion and selectivities to DMC and EG

Catalyst	EC conversion (%)		Selectivity (%) (t = 6 h)	
	t = 1 h	t = 6 h	DMC	EG
TPAC	40.6	68.1	81.7	76.8
TBAC	45.2	67.5	88.7	82.9
THAC	48.3	69.6	96.0	86.6
TOAC	49.7	70.5	97.1	92.1
TDodAC	51.0	70.9	88.5	81.7
TBAB	42.1	66.9	92.5	92.1
TBAI	39.8	66.6	90.4	91.0

Reaction condition : EC = 25 mmol, MeOH = 200 mmol, Catalyst = 2 mmol, T = 140 ℃.

Table 2. Conversion of EC and selectivities to DMC and EG at different reaction temperatures and pressures with TBAC catalyst

Temp. (℃)	Pressure (psig)	Cat. amount (mmol)	Conversion (%)	Selectivity (%)	
				DMC	EG
120	300	2	58.1	91.4	92.0
130	300	2	65.4	84.2	81.3
140	300	2	67.5	88.7	82.9
140	250	2	65.2	81.6	82.1
140	350	2	78.4	90.8	88.7
140	400	2	76.7	82.5	79.2
140	300	1	65.5	80.9	81.3
140	300	3	73.9	79.3	78.5
140	300	4	74.2	90.4	75.6

Reaction condition : EC = 25 mmol, MeOH = 200 mmol, Reaction time = 6 h.

The effects of reaction temperature, pressure and catalyst amount on the catalytic activity were also studied with TBAC. The results are summarized in Table 2. The conversion of EC increased with the increase of reaction temperature and the amount of catalyst. The conversion of EC and the selectivity of DMC increased as the pressure increased from 250 psig to 350 psig. But, at the pressure over 350 psig, the EC conversion decreased. Although CO_2 is not required for this reaction, its presence alters the reaction profile. It is reported that high pressure of CO_2 can inhibit the decomposition of EC to ethylene oxide and CO_2[12].

The transesterification reaction between EC and MeOH is expressed as EC + 2 M → D + G : where M, D, and G represent MeOH, DMC, and EG, respectively. If the reaction is assumed elementary, the reaction rate can be expressed as follows :

$$-r_{EC} = -[dC_{EC}/dt] = kC_{EC}C_M^2$$

Since MeOH was used in excess amount compared to EC, C_M can be assumed constant during the reaction. Therefore, the reaction rate equation can be written as a pseudo first order with respect to the concentration of EC.

$$-r_{EC} = k'C_{EC}, \text{ where } k' = kC_M^2$$

Fig. 2 shows the plot of $\ln[(C_{EC})_0/C_{EC}]$ vs. time during first 2 h. Quite good straight lines were obtained, and the pseudo first-order reaction rate constants for 120, 130 and 140 $^{\circ}$C were 0.002421, 0.002481 and 0.002545 h^{-1}, respectively. From the Arrhenius plot of the first order reaction rate constants, one can estimate the activation energy as 41.5 kJ/mol.

4. CONCLUSIONS

In the synthesis of DMC from the transesterification of EC and methanol, quaternary ammonium salt catalysts showed good catalytic activity. The main byproduct was ethylene glycol. The quaternary salt with the cation of bulkier alkyl chain length and with more nucleophilic anion showed better reactivity. High temperature and large amount of catalyst increased the conversion of EC. The EC conversion and DMC selectivity increased as the pressure of CO_2 increased from 250 to 350 psig.

ACKNOLEDGEMENT

This work was supported by Applied Rheology Center, and Brain Korea 21, and by the Korea Science and Engineering Foundation (R-01-2005-000-10005-0).

REFERENCES

[1] J. F. Knifton, R. G. Duranleau, J. Mol. Catal., 67 (1991) 389.
[2] T. Kondoh, Y. Okada, F. Tanaka, S. Asaoka, S. Yamanoto, US Patent 5, 436, 362 (1995).
[3] Y. Urano, M. Kirishiki, Y. Onda, H. Tsuneki, US Patent 5, 430, 170 (1995).
[4] T. Welton, Chem. Rev., 99 (1999) 2071.
[5] T. Tatsumi, Y. Watanabe, K. A. Koyano, Chem. Commun., (1996) 2281.
[6] Y. Watanabe, T. Tatsumi, Micropor. Mesopor. Mater., 22 (1998) 399.
[7] Y. Fang, W. Xiao, Sepn. Puri. Technol., 34 (2004) 255.
[8] D. W. Park, B. S. Yu, E. S. Jeong, I. Kim, M. I. Kim, K. J. Oh, S. W. Park, Catal. Today, 98, (2004) 499.
[9] D. H. Shin, J. J. Kim, B. S. Yu, M. H. Lee, D. W. Park, Korean J. Chem. Eng., 20 (2003) 71.
[10] B. S. Ahn, B. G. Lee, H. S. Kim, M. S. Han, proc. 10th APCChE, Paper No. 645, 17-21 Oct. (2004), Kitakyushu, Japan.
[11] C. M. Starks, C. L. Liotta, M. Halpern, "Phase Transfer Catalysis", Chapmann & Hall, New York, 1994.
[12] K. Nishihira, K. Mizutare, S. Tanaka, EP Patent 425 (1991) 197.

Studies in Surface Science and Catalysis, volume 159
Hyun-Ku Rhee, In-Sik Nam and Jong Moon Park (Editors)

Vapor phase propylene epoxidation kinetics[*]

G. Qian, Y.H. Yuan, W. Wu and X.G. Zhou*

State Key Laboratory of Chemical Engineering,
East China University of Science and Technology, Shanghai 200237, China

1. INTRODUCTION

Since the discovery of the catalyst of Au over TiO_2 support for vapor phase C_3H_6 epoxidation [1], great efforts have been made to understand the reaction mechanism in order to improve the catalyst performance [2,3]. Currently the Au catalyst suffers from low activity and fast deactivation, and is thus far from commercialization. Perhaps it is why at present no publication on the reaction kinetics can be found in the literature.

Vapor phase C_3H_6 epoxidation involves at least five molecules, with C_3H_6, O_2 and H_2 as reactants, and PO and H_2O as products. Moreover, both Au and TiO_2 are involved in the reaction, and the perimeter of the surface contact between Au and TiO_2 plays decisive roles.

C_3H_6 adsorbs weakly on TiO_2 via γ-hydrogen bonding with surface hydroxyl groups of TiO_2 support and is completely reversible at temperatures between 300 and 400 K [3]. It was shown that C_3H_6 adsorbs very weakly on clean Au surfaces [4], while Haruta reported that C_3H_6 can preferentially adsorb on the surface of Au particles under reaction conditions [1].

On smooth Au surfaces, the adsorbed oxygen, which is only possible in molecular form [5], is not selective to form PO [6]. Therefore partial oxidation is contributed by oxygen adsorbed on TiO_2. However, there are no direct experimental evidence whether O_2 is adsorbed dissociatively or non-dissociatively. It is generally accepted that O_2 adsorbs on TiO_2 in a molecular form [7] and is activated at the Au/TiO_2 interface [1].

H_2 adsorption is weak on the anatase surfaces [8]. No dissociative adsorption of H_2 takes place over the smooth surfaces of Au at temperatures below 473 K [9,10]. On small Au particles, adsorption is possible at low temperature. Dissociative adsorption of H_2 can be accelerated by the negatively charged molecular oxygen species at steps, edges, corners of Au particles [5].

H_2O adsorbs strongly on both Au and TiO_2. However, the influence of water on the catalyst or reaction is negligible [2]. PO is known to strongly interact with TiO_2, and adsorption, desorption, and decomposition of PO are independent on the presence of gold [3].

The exact epoxidation mechanism is still not quite clear. However, in all possible mechanisms, the interaction between Au and TiO_2 is essential. According to a mechanism suggested by Hayashi et al [1], molecular oxygen adsorbed on TiO_2 is activated, probably to a

---

[*] Supported by NSFC (20276018)

negatively charged molecular oxygen species, which forms hydropeoxo- or peroxo-like species directly through reaction with H_2. The oxygen species adsorbed on Ti site then reacts with C_3H_6 mainly adsorbed on the surface of Au particles to produce PO. Moulijn et al [3] suggested that deactivation of Au/TiO_2 catalysts is due to the formation of propoxy species over (Brønsted) acid sites, which blocks the selective epoxidation sites on the TiO_2 surface.

In this article, a dynamic reaction kinetics for propylene epoxidation on Au/TiO_2 is presented. Au/TiO_2 catalyst is prepared and kinetics experiments are carried out in a tube reactor. Kinetic parameters are determined by fitting the experiments under different temperatures, and the reliability of the proposed kinetics is verified by experiments with different catalyst loading.

2. EXPERIMENTS

Au/TiO_2 catalysts is prepared by D-P method according to the procedure detailed in [11], which has a gold loading of 2.4wt% and a BET surface area of $61.1 m^2/g$. The average size of the Au particle is about 3.5nm and the shape of the Au particle is almost semispherical, with a large perimeter between Au and TiO_2.

In a stainless steel tube with 4mm i.d., Au/TiO_2 catalysts (80-120 mesh) are loaded on glass fibers and quartz sands. A thermal couple is installed at the outlet of the catalyst bed to determine the reaction temperature. C_3H_6 and PO concentrations are measured by an on-line GC (Agilent 4890D, Porapak Q column, FID detector, N_2 carrier gas), whereas H_2 concentration is not determined. Since the catalyst powders are quite small, 80-120 mesh, the flow rate can not be very high owing to the resistance of the catalyst bed and the inlet pressure limited by the pressure regulator of the propylene column. During the experiments, the inlet gas composition is fixed at $H_2/O_2/C_3H_6/N_2 = 10/10/10/70$ and the gas flow rate fixed at 2L/hr.

3. REACTION MECHANISM AND REACTOR MODELING

Adsorption of C_3H_6, H_2, O_2 and PO and reaction between the admoleculars or adatoms may involve different sites. However, by assuming that the controlling step of the overall reaction is the formation of hydropeoxo- or peroxo-like species at the interfacial edges of the Au particles and the TiO_2 support, it does not matter to which site C_3H_6, H_2, or O_2 are adsorbed, because the adsorptions are all in equilibrium. As a result the Eley-Rideal mechanism is applicable for C_3H_6, H_2, and O_2. Consequently, without considering H_2 combustion, the following reaction mechanism is assumed [1,3],

$$\left\{ \begin{array}{l} H_2(g) + O_2(g) + * \xrightarrow{k_1} H_2O_2 * \\ \\ H_2O_2 * + C_3H_6(g) \xrightarrow{k_2} C_3H_6O * + H_2O \end{array} \right. \quad \left| \begin{array}{l} C_3H_6O * \underset{k_{-3}}{\overset{k_3}{\rightleftarrows}} C_3H_6O + * \\ \\ C_3H_6O * \xrightarrow{k_4} D * \end{array} \right. \tag{1}$$

where * denotes the active Au or Ti site at the interfacial edge and D is the propoxy species that blocks the active Ti site and causes deactivation. H_2 combustion is assumed independent of the above reactions:

$$H_2(g) + 0.5O_2(g) \xrightarrow{k_5} H_2O \tag{2}$$

A one-dimensional isothermal plug-flow model is used because the inner diameter of the reactor is 4 mm. Although the apparent gas flow rate is small, axial dispersion can be neglected because the catalyst is closely compacted and the concentration profile is placid. With the assumption of Langmuir adsorption, the reactor model can be formulated as,

$$
\left\{
\begin{aligned}
&\frac{\partial y_{PO}}{\partial t} + u\frac{\partial y_{PO}}{\partial l} = k_3\theta_{PO}/m - k_{-3}y_{PO}\theta_0 \\[4pt]
&\frac{\partial y_{H_2}}{\partial t} + u\frac{\partial y_{H_2}}{\partial l} = (k_1 y_{H_2}y_{O_2}\theta_0 - k_5 y_{H_2}y_{O_2})m \\[4pt]
&\frac{\partial y_{O_2}}{\partial t} + u\frac{\partial y_{O_2}}{\partial l} = (-k_1 y_{H_2}y_{O_2}\theta_0 - \tfrac{1}{2}k_5 y_{H_2}y_{O_2})m \\[4pt]
&\frac{\partial y_{PE}}{\partial t} + u\frac{\partial y_{PE}}{\partial l} = -k_2\theta_{H_2O_2}y_{PE}
\end{aligned}
\right.
\qquad
\begin{aligned}
&\frac{\partial \theta_0}{\partial t} = -k_1 m y_{H_2}y_{O_2}\theta_0 + k_3\theta_{PO}/m - k_{-3}y_{PO}\theta_0 \\[4pt]
&\frac{\partial \theta_{PO}}{\partial t} = k_2 y_{PE}\theta_{H_2O_2} + k_{-3}y_{PO}\theta_0 \\
&\qquad\quad -k_3\theta_{PO}/m - k_4\theta_{PO}/m \\[4pt]
&\frac{\partial \theta_D}{\partial t} = k_4\theta_{PO}/m \\[4pt]
&\theta_{H_2O_2} = 1 - \theta_0 - \theta_{PO} - \theta_D
\end{aligned}
\tag{3}
$$

with initial and boundary conditions:

$$t = 0, \quad y_{PO} = \theta_{PO} = \theta_D = 0, \quad \theta_0 = 1, \quad y_{PE} = y_{H_2} = y_{O_2} = 0. \tag{4}$$

$$l = 0, \quad y_{PO} = \theta_{PO} = \theta_D = 0, \quad \theta_0 = 1, \quad y_{PE} = y_{H_2} = y_{O_2} = 0.1 \tag{5}$$

where $m = P/RT$. P is the total pressure, T the reaction temperature, R the gas constant, y molar fraction, θ_i the active site coverage by component i, and l the packed length of the catalyst. The PDEs are transformed to ODEs (initial value problem) by finite differentiation in l direction before numerical integration.

4. RESULTS AND DISCUSSIONS

The rate constants are determined by fitting the PO concentrations that change with time, as shown in Fig. 1. With the rate constants at different reaction temperatures, the activation energies and the pre-exponential factors are determined by plotting $\ln k$ against $1/T$.

The kinetic parameters are listed in Table 1. The linearity of $\ln k \sim 1/T$ plot is revealed by the correlation coefficient R^2. For all reactions but the deactivation, the rate constants follow the Arrhenius' law satisfactorily, implying catalyst deactivation may involve more than one elementary steps.

Fig. 2 compares the experiments (at 50°C) with the calculations by using the plug-flow model without adjusting the kinetic parameters. The predictions are quite satisfactory except for large catalyst loading. This is an indication that in this reaction more than one elementary steps may be controlling, and more elaborate kinetic models are to be constructed.

Table 1. Activation energies

	k_1	k_2	k_3	k_{-3}	k_4	k_5
E, kJ/mol	50.16	61.50	85.37	148.74	39.44	89.01
R^2	0.9049	0.9695	0.9662	0.9646	0.7814	0.9708

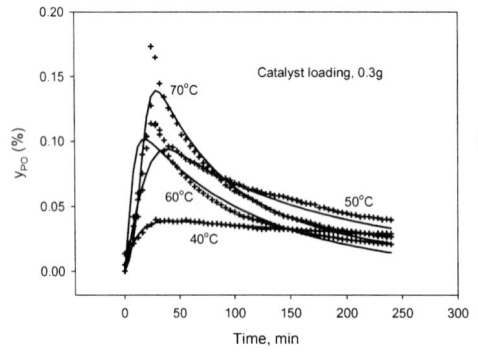 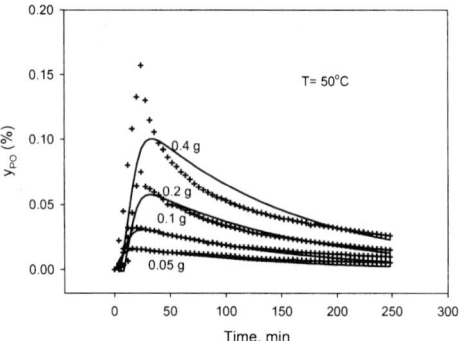

Fig. 1 Fitting of the outlet molar PO fraction different reaction temperatures

Fig.2 Prediction of the exit molar PO fraction with different catalyst loadings

Film diffusion may influence the overall reaction because of the low gas flow rate. As the bulk concentrations change little with time along the length of the reactor, an assumption of constant difference between bulk and catalyst surface concentrations is used in this study and the rate constants will change with gas flow rates. Nevertheless, the activation energies will remain constant, and the proposed reaction kinetics still provides useful hint for understanding the reaction mechanism and optimizing the reactor and operation conditions.

REFERENCE

[1]   T. Hayashi, K. Tanaka, and M. Haruta, J. Catal., 178 (1998) 566.
[2]   T.A. Nijhuis, B.J. Huizinga, and J.A. Moulijn, Ind. Eng. Chem. Res., 38 (1999) 884.
[3]   G. Mul, A. Zwijnenburg, B. Linden, and J. A. Moulijn, J. Catal. 201 (2001) 128.
[4]   K.A. Davis and D.W. Goodman, J. Phys. Chem. B 104 (2000) 8557.
[5]   A.G. Sault, R.J. Madix and C.T. Campbell, Surf. Sci. 169 (1986) 347.
[6]   M. Haruta and M. Daté, Appl, Catal, A: Gen., 222 (2001) 427.
[7]   M. Haruta, and M. Daté, Appl. Catal. A: Gen., 222 (2001) 427.
[8]   M. Calatayud and C. Minot, Surf. Sci., 552 (2004) 169.
[9]   M. Haruta, The Chemical Record, 3 (2003) 75.
[10] L. Stobinski, and R. Dus, Surf. Sci., 298 (1993) 101.
[11] Y.H. Yuan, X.G. Zhou, W. Wu, Y.R. Zhang, W.K. Yuan, Catal. Today, 105 (2005) 544.

Studies in Surface Science and Catalysis, volume 159
Hyun-Ku Rhee, In-Sik Nam and Jong Moon Park (Editors)

Optimization of catalytic pellet formation for CO_2 hydrogenation

Sung-Chul Lee[b], Ja-Kyung Cho[a], Dong-Jin Kim[a] and Suk-Jin Choung[a]

[a] College of Environment and Applied Chemistry,
Kyung Hee University, Kiheung, Yongin 449-701, Republic of Korea

[b] Energy Lab., CRD, Samsung SDI, Suwon 443-731, Republic of Korea

1. INTRODUCTION

The Fischer-Tropsch synthesis (FTS) has been extensively investigated since the discovery of methane production over nickel. In the past three decades, the hydrogenation of carbon monoxide (CO) was actively studied all over the world because it is an important step in utilizing coal and natural gas as carbon sources [1, 2]. On the other hand, hydrogenation of carbon dioxide (CO_2) received much less attention partly because of unfavorable thermodynamic considerations. In recent years, however, various kinds of chemical processes to turn CO_2 into valuable chemical compounds have been tried, attracting much attention as a promising solution in the future. In recent years, iron based catalysts showed remarkably high activity values and selectivity of long-chain hydrocarbons and light olefins in CO_2 hydrogenation [3-5]. Though iron based catalysts have been reported to show excellent capability of CO_2 hydrogenation, there has been little research on the preparation and formation of catalytic pellet. Especially, the formation of catalyst is one of the major problems encountered in the industrial application of iron catalyst. In order to improve the strength of catalysts as they are formed, there is no doubt about the need for adding binder, which is not generally chemically inert to the catalyst, particularly at high temperature. This study is focused on developing a better understanding of the pellet formation and effect of pellet size in CO_2 hydrogenation. For the optimization of catalytic pellet size is determined by theoretical approaches. The other is the binder effect for pellet preparation obtained by ^{27}Al-NMR and temperature programmed decarburization (TPDC).

2. EXPERIMENTAL

Fe-K/γ-Al_2O_3 catalysts are prepared by the impregnation of γ-Al_2O_3 with aqueous solutions of $Fe(NO_3)_3 \cdot 9H_2O$ and K_2CO_3 [4]. Alumina, silica, and PVA binders of 5 wt.% of the total catalyst weight are added during catalyst preparation. The nominal catalyst compositions were $1.00Fe/0.35K/ 5.00Al_2O_3$. The impregnated catalysts were homogeneously mixed, and a small amount of water and binders are added. Catalytic pellets are extruded through a 1.5 mm-diameter die. These pellets are dried at 393K for 12 h and calcined at 773K for 24 h in air.

CO_2 hydrogenation is carried out in a bench scale fixed bed reactor (1.6 cm-ID×60 cm-High). The reaction and internal standard gases (CO_2, H_2, N_2, He) were taken from cylinders and their flow rates were controlled by MFC (mass flow controller, Brooks Co.). The reaction conditions of Fe-K/γ-Al_2O_3 catalyst is selected by literatures [4, 6, 9]. Reaction temperature is controlled at 573K by an electric heater and reaction pressure is maintained at 10atm by BPR (back pressure regulator, Tescom Co.). 21.0 g of catalysts are filled up and the flow rate of the mixed gas is 2,000 ml/g-*cat*.h at STP. The composition of the gas is $H_2/CO_2 = 3/1$. The liquid products were separated from gas products in the gas-liquid separator and condenser. The exit gas flow rate was measured by a digital bubble flow meter to evaluate the reaction conversion.

3. RESULTS AND DISCUSSION

CO_2 hydrogenation reaction was induced using the one-dimensional reactor at a constant inlet temperature (573K). Temperature profiles were recorded at the indicated axial positions in the reactor bed. Fig. 1 shows typical temperature profiles in the axial direction. It was quite different from the temperature profiles of every stage. The initial reaction temperature in 1st-stage bed reached 600K although the inlet temperature was 573K. It decreased as the reaction time became longer. However, the reaction temperature in 2nd-stage bed kept in the temperature range of 573-583K, and then it started to increase rapidly. Then it had a steady state after 80 hours. As the reaction continued, the maximum temperature of the reactor shifted to the lower position of the reactor.

The binder is not generally chemical inert to catalysts, particularly at high calcined temperature. Hence, binder-catalyst interactions can have a strong influence on the activity and selectivity. The influence is mainly due to changes in catalytic acidity, metal-support interaction and the structure. The catalytic activities for Fe-K/γ-Al_2O_3 with and without various binders are compared in Fig. 2. Initially, activities slowly increase until the value reaches the maximum, and then they have steady state with increasing time-on-stream. When catalyst with PVA binder is used under reaction conditions, the conversion of CO_2 is ca. 35%. In case of catalyst with alumina binder, the conversion of CO_2 is 41%. However, when catalyst with silica is used, it shows a very low activity compared with those obtained over other catalysts. The acidity of catalysts influences the catalytic activity and selectivity. A large amount of strong acid sites results in high selectivity of C_5+ hydrocarbons. CO_2 conversion is also affected by the amount of medium acid sites. The coordination environment of aluminum for all samples is examined by ^{27}Al MAS NMR spectroscopy. NMR spectra show resonances corresponding to sixfold coordinate non-framework aluminum (δ of ca. -50 ppm), four coordinate non-framework aluminum (δ of ca. 0 ppm) and four coordinate framework aluminum (δ of ca. $+45$ ppm) [7, 8].

The spectrum of Fe-K/γ-Al_2O_3 itself (Fig. 3(a)) displays a double ^{27}Al resonance with $\delta \approx$ -50, 0, $+45$ ppm. In the case of Fe-K/γ-Al_2O_3 with silica, the amount of sixfold coordinate non-framework aluminum increases. However, in Fig 3(c), the ^{27}Al MAS NMR spectra of Fe-K/γ-Al_2O_3 with alumina binder clearly reveal a narrow peak at $+45$ ppm. It has been suggested that a new framework of $(-Al-O-Fe-)_n$ is formed so that alumina binder reacts with $Fe(OH)_3 \cdot xH_2O$. The incorporation of Fe^{3+} into the framework of $Fe-K/\gamma-Al_2O_3$ would lead to strong acid sites. From this reason, it can be seen that higher hydrocarbon selectivity of $Fe-K/\gamma-Al_2O_3$ with alumina binder is increased as the amount of strong acid site increases.

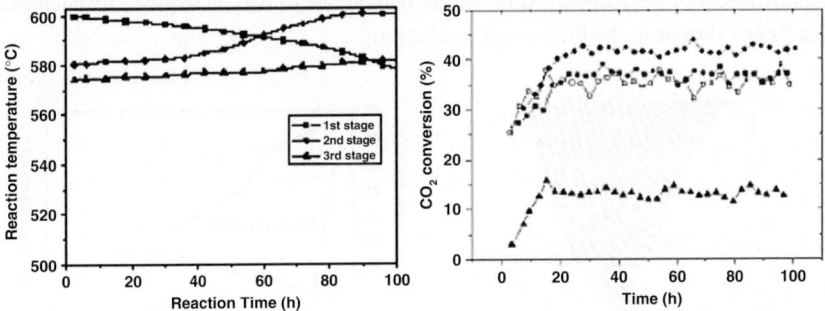

Fig. 1. Temperature profiles of the CO_2 hydrogenation reactor

Fig. 2. Catalytic activity of CO_2 hydrogenation.

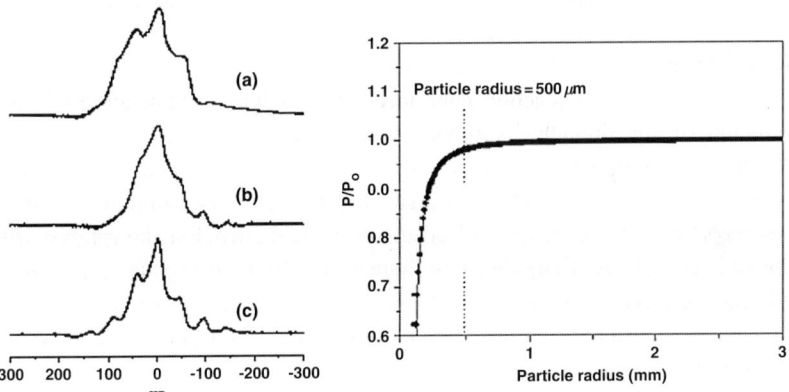

Fig. 3. ^{27}Al MAS NMR of $Fe-K/\gamma-Al_2O_3$ with various binders: (a) PVA binder, (b) silica binder, (c) alumina binder.

Fig. 4. Effect of pressure drop with catalytic particle size.

Although the chemical properties of catalyst arc analogue, the pellet size affects on catalytic activity and selectivity in the packed reactor. In gas phase reactions, the concentration of the reacting species is proportional to total pressure. The equation used most to calculate pressure drop in a packed porous bed is the Ergun equation. In calculating the pressure drop using the Ergun equation, the minimum particle size in packed reactor is determined. The maximum particle size is calculated from internal diffusion equation. The fig. 3 and 4 shows the relationship between pressure drop or reactant concentration and catalytic particle size. With the larger particle sizes, diffusion resistances may reduce the rate of reaction in the center of the particles and hence decrease the activity of the catalyst. Therefore, the optimum particle size is calculated. Samples of cylindrical catalyst with diameter of 0.5 – 7 mm were prepared with Fe-K/γ-Al$_2$O$_3$, to determine the influence of the pellet size. The catalytic activities and selectivities of CO$_2$ hydrogenation with pellet size are show in Fig. 6. Significant differences were observed in weight percentage of metals retained by each catalyst. These results are similar to the theoretical predictions.

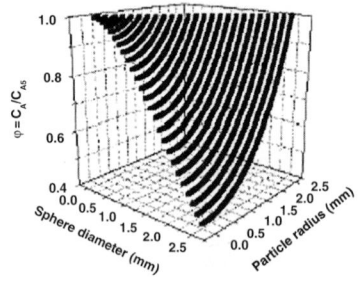

Fig. 5. Reactant concentration with catalytic particle size.

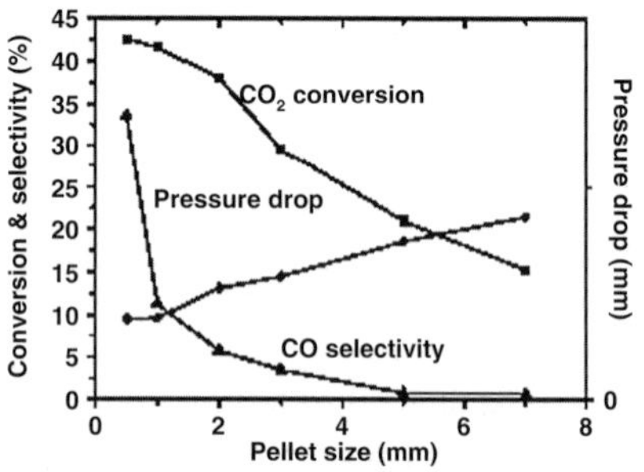

Fig. 6. The activity and selectivity with catalytic pellet size.

4. CONCLUSIONS

The structure of iron aluminate is formed in large pores of γ-Al$_2$O$_3$ due to added binder in Fe-K/γ-Al$_2$O$_3$ with alumina binder. Because of the formation of iron aluminate, the structure of Fe-K/γ-Al$_2$O$_3$ with alumina binder is dramatically changed. The incorporation of Fe^{3+} into the framework of Fe-K/γ-Al$_2$O$_3$ would lead to strong acid sites. For the catalyst with silica binder, the activity and selectivity decrease dramatically by reducing the acidity. It has been postulated that the influence of inorganic binders on the catalytic activity and selectivity is consistent with their influence on the acidity of catalysts. The significance of pellet size is compatible with theoretical predictions.

REFERENCES

[1] H. Kusama, K. Okabe, H. Arakawa, Appl. Catal. A, 207 (2001) 85.
[2] M. Wieser, N. Fujii, T. Yoshida, T. Nagasawa, Eur. J. Biochem., 257 (1998) 495.
[3] S. Y. Yokoyama, Energy Conversion & Management , 38 Suppl., (1997) S569.
[4] J. Hwang, K. Jun, K. Lee, Appl. Catal., 208 (2001) 217.
[5] M. Beyer, C. Berg, G. Albert, U. Achatz, S. Joos, G. Niednerschatteburg, V. Bondybey, J. Am. Chem. Soc., 119 (1997) 1466.
[6] T. Riedel, G. Schaub, K. Jun, K. Lee, Ind. Eng. Chem. Res., 40 (2001) 1355.
[7] H. Zhao, K. Hiragushi, Y. Mizota, J. Eur. Ceram. Soc., 22 (2002) 1483.
[8] K. Okumura, M. Hashimoto, T. Mimura, M. Niwa, J. Catal., 206 (2002) 23.
[9] J. Kim, H. Kim, S. Lee, M. Choi, K. Lee and Y. Kang, Korean J. of Chem. Eng., 18 (2001) 463.

Studies in Surface Science and Catalysis, volume 159
Hyun-Ku Rhee, In-Sik Nam and Jong Moon Park (Editors)

Reaction of Butadiynediyl dimetal Complexes with Fe$_2$(CO)$_9$: Formation of Various Complexes with the C$_4$ bridge

M.-C. Chung[a*], H.-G. Ahn[a], and M. Akita[b]

[a]Dept. of Chemical Engineering, Sunchon National University, 315 Maegok-dong, Suncheon, Jeonnam 540-742, Korea. E-mail: mchung@sunchon.ac.kr
[b]Chemical Resources Laboratory, Tokyo Institute of Technology, 4259 Nagatsuta, Midoriku, Yokohama 226-8503, Japan.

The interaction of butadiynediyl dimetal complexes [Fp*-C≡C-C≡C-M, Fp*=FeCp*(CO)$_2$, M= Fp*, Rp, SiMe$_3$, Rp= RuCp(CO)$_2$] with diiron nonacarbonyl, Fe$_2$(CO)$_9$, results in the formation of a mixture of products, as is also observed in the case of their interaction with organic acetylenes. Interesting polymetallic complexes, μ_3-η^3-propargylidene-ketene compounds, zwitterionic cluster compounds, and μ_3-η^3-propargylidene-cyclobutene compounds were isolated from the reaction mixtures and successfully characterized. The product distributions were found to be dependent on the metal fragment (M) at the other end of the C$_4$ rod. The results of the reaction are described

1. INTRODUCTION

The organometallic chemistry of iron carbonyl has been extensively studied since the first synthesis of Fe(CO)$_5$ was independently carried out by Mond and Berthelot [1]. The reactions of iron carbonyl with alkynes have also been studied in detail, but it has been found that the reaction affords a variety of products, depending on the alkyne substituents and reaction conditions [2]. Polyynediyl complexes with the structure, M-(C≡C)$_n$-M, were found to be versatile precursors for cluster compounds containing a one-dimensional linear carbon chain [3]. We carried out a synthetic study of polynuclear complexes containing carbon rich metal fragments [4]. Polynuclear complexes with a longer carbon chain are interesting from the viewpoint of material science. For example, polyynediyl complexes are regarded as a molecular wire, because the two terminal metal fragments can communicate with each other through the π-conjugated systems arising from the interaction of the p-orbitals of the carbon atoms and d orbitals of the terminal metals [5]. Herein, we describe the results of our investigation into the interaction of butadiynediyl dimetallic complexes with diiron nonacarbonyl.

2. EXPERIMENTAL

All reactions and manipulations were carried out under an inert atmosphere (N$_2$ or Ar gas) using the Schlenk technique. Solvents were freshly distilled under an Ar atmosphere using the standard procedures (Na/K/benzophenone or CaH$_2$). Chromatography was performed on alumina (aluminum oxide, activity II-IV(Merck art 1097). The ^{1}H- and ^{13}C-NMR spectra were recorded on a Bruker AC-200 spectrometer (^{1}H, 200 MHz) and Nippon

Denshi EX-400 spectrometer (^{1}H, 400 MHz), respectively. The IR and FD-MS spectra were obtained on a JASCO FT-IR 5300 spectrometer and a Hitachi M80 mass spectrometer, respectively. Single crystal data were obtained from a Nigaku Denshi AFC-7R, 4 axis X-ray diffractometer. We previously reported the synthetic procedures for compounds 1 ~6 [4, 6]. The synthesis of compound 7 was performed as follows. A THF solution (35 ml) of Fp*-C≡C-C≡C-SiMe$_3$ (310 mg, 0.843 mmol) containing excess Fe$_2$(CO)$_9$ was stirred for 3 hrs at room temperature. After removal of the volatile materials, the residues was extracted with CH$_2$Cl$_2$, and passed through an alumina pad. The filtrate was evaporated in a vacuum, and then product 7 was collected by alumina chromatography (eluted with Hexane-CH$_2$Cl$_2$: 5:1~1:3). Compound 7 (with a yield of 29.4% on the basis of Cp*) was obtained as yellow crystals, which crystallized from THF-Hexane solution at -30 ℃.

3. RESULT AND DISCUSSION

Butadynyl and butadynediyl complexes bearing the FeCp*(CO)$_2$ (Fp*) terminus were prepared by conventional methods(scheme 1). Butadynyl complexes, Fp*-C≡C-C≡C-H, were readily obtained by the alkynylation of Fp*-I with Li-C≡C-C≡C-SiMe$_3$, followed by desilylation. The Cu-catalyzed metalation of complex 2 with M-X (M =Fp*, Rp) gave the butadiyne complexes 3.

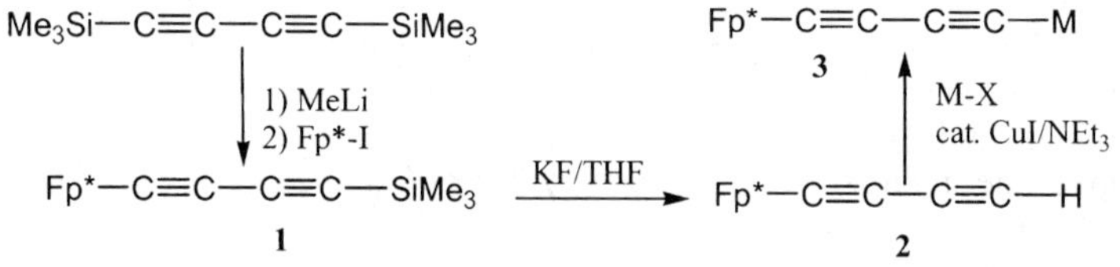

Scheme 1

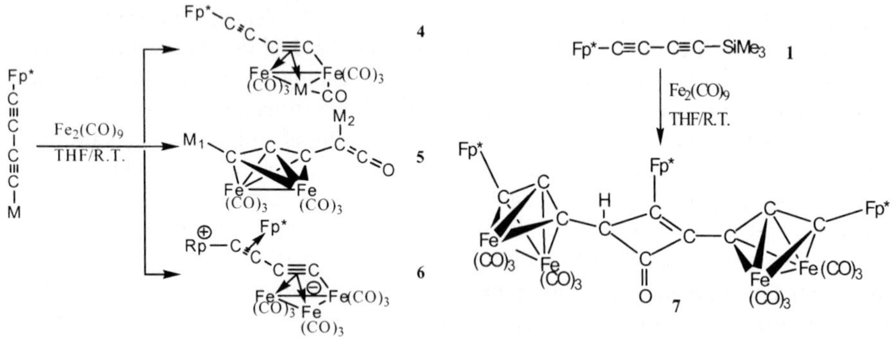

Scheme 2

The reactions of the butadiynediyldimetal(Fe, Ru) complexes with Fe$_2$(CO)$_9$ at room temperature afforded mixtures of products, from which three types of products, viz. the μ$_3$-acetylide cluster compound 4, the μ$_3$-η^3-propargylidene-ketene compound 5 and zwitterionic cluster compound 6, were isolated. While the reaction with an excess amount of Co$_2$(CO)$_8$ results in addition to the sterically congested Fp*-C≡C part [6]. The distributions of the products were dependent on the metal fragments situated at the other end of the conjugated carbon rod. The cluster compounds so obtained were characterized by spectroscopic and crystallographic methods. Compound **1** was assigned the trinuclear μ_3-η^1: η^2:η^2-acetylide cluster structure based on its ^{13}C-NMR and IR(1580 cm^{-1}, bridging CO) spectra, and this assignment was also supported by the molecular ion peaks(FD-MS). Compound **5** contains the ketene functional group in addition to the diiron μ_3-η^3:η^3-propargylidene structure. The ^{13}C-NMR signals of the ketene functional group [$\delta c(C_\alpha=C=O)$: -4.2 ~ -6.8]; [$\delta c(C=C_\beta=O)$: 156 ~ 160] appear in slightly higher fields than those of the corresponding signals of organic ketene compounds. The structure of compound **6** consists of two parts, the triiron μ-η^1: η^2:η^2-acetylide cluster part and the dinuclear μ-η^2:η^2-acetylide complex part. When the 18 electron rule is taken into account, it can be concluded that the former part is anionic and the latter part is cationic. Therefore, it was inferred that complex **6** is a zwitterionic species. However, the reaction of butadiynediyl compound 1 with Fe$_2$(CO)$_9$ gave the μ_3-η^3-propargylidene-ketene-cyclobutene compound **7**. The ^{1}H-NMR spectrum of compound **7** contains three Cp* signals indicating the formation of a higher nuclearity cluster compound, and it is notable that the SiMe$_3$ signal disappeared. The molecular structure of **7** was determined by X-ray crystallography (Figure 1). The structure can be described as cyclobutenone with two propagylidene diiron structures at the 2- and 4-positions and an Fp* group at the 3-position.

The cyclobutenone structure is characterized by (i) the C5=C6 distance (1.42(2) Å) being shorter than the other C-C distances (1.45 ~ 1.47 Å) in the four-membered ring, (ii) the sp^2-hybrized C5, C6 and C10 atoms indicating the planarity of the structures around the carbon atoms, and (iii) the sp^3-hybridization of the C4 atom. Ward et al. reported in organic synthesis

that the dimethylketene dimerized rapidly in THF at -30 ℃ to give the hydroxy-cyclobutene [7]. The formation mechanism of compound **7** can be considered to occur in the following steps; (i) the addition of the Fe$_2$(CO)$_m$ species to the C≡C bond adjacent to Fp*, which contributes to the steric and electronic influence of the SiMe$_3$ caps, results in the formation of adducted triiron complexes as an intermediate and subsequently produces the

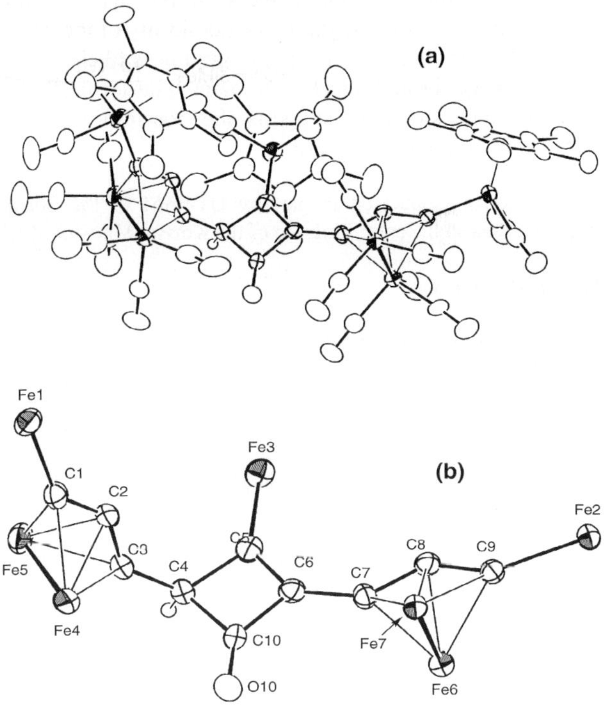

Fig. 1. ORTEP drawing for **7** with non-hydrogen atoms propargylidene-ketene compound (a) Overview Structure (b) Core Structure

from the interaction of the unbounded C≡C bond and the CO species in the adducted triiron complexes, the propargylidene-ketene compound has the same structure as compound **5** , (ii) desilylation by the substitution of the silyl terminal species with the H$^+$ ion originating from H$_2$O or the solvent gives the hydrogen-ketene form,

(iii) the [2+2] cycloaddition of the C=C in the ketene part gives the hydroxyl-cyclobutenone compound by enolization, (iv) the subsequent reaction of Fp*- with the hydroxy group followed by the elimination of the Fp*-OH complexes to give compound **7**.

Table 1. Selected structure parameters for compound **7**

Bond Lengths (Å)				Angles (^{o})	
C3-C4	1.47(2)	C6-C10	1.4591)	$\angle$ C5-C4-C10	83.6(2)
C4-C5	1.54(2)	C4-C10	1.56(1)	$\angle$ C6-C5-C4	93.0(8)
C5-C6	1.40(2)	C10-O10	1.18(1)	$\angle$ C10-C6-C5	93.1(8)
C6-C7	1.45(1)	C5-Fe3	1.92(1)	$\angle$ C4-C10-C6	90.3(8)

4. CONCLUSION

The interaction of the butadiynediyl complexes with $Fe_2(CO)_9$ results in the formation of a mixture of products as also observed in the case of their interaction with organic acetylenes. From the reaction mixtures, the interesting polymetallic complexes **5**, **6** and **7** were isolated, in addition to the conventional acetylide cluster compound **4**. Complex **5** contains the ketene functional group as well as the μ_3-propargylidene diiron structure, which arises from the interaction of the non-coordinated C≡C part with the diiron moiety. The formation of the zwitterionic complex **6** suggests the possibility of the migration of a metal fragment along the C_4-rod. The driving force for the migration is likely to be charge delocalization over the M-C_4-M' linkage through electron transfer. The formation of compound **7**, which contains the cyclobutenone structure, suggests the possibility of [2+2] ketene cycloaddition and the elimination of the Fp*-OH group

This work was supported in part by GT-FAM of the Sunchon National University through the Regional Research Center Program of the Korean Ministry of Commerce, Industry and Energy.

REFERENCES

[1] G. Wilkinson, E.W. Abel, F.G.A. Stone, Comprehensive organometallics chemistry, Pergamon, Oxford, 1982, Vol 4, Chapter 4.
[2] G. Wilkinson, E.W. Abel, F.G.A. Stone, Comprehensive organometallics chemistry, Pergamon, Oxford, 1995, Vol 7, Chapter 4.
[3] P.J. Kim, M. Masai, K. Sonogashira, N. Hagihara, Inorg. Nucl. Chem. Lett. 6 (1970), 181.
[4] (a) M. Akita, M.-C. Chung, A. Sakurai, S. Sugimoto, M. Terada, M. Tanaka, Y. Moro-oka, Organometallics, 16 (1997), 4882
(b) M. Akita, A. Sakurai, M.-C. Chung, Y. Moro-oka, J. Organomet. Chem., 670 (2003)), 2.
[5] (a) F. Paul, C. Lapinte, Coord. Chem. Rev. 178-180 (1998), 431
(b) B.C.Shekar, J. Lee, S.W. Rhee, Korean J. of Chem. Eng., 21, No.1, ((2004), 267.
[6] M.-C. Chung, A. Sakurai, M. Akita, Y. Moro-oka, Organometallics, 18(1999), 4684.
[7] C.C McCarrney, R.S. Ward, J. Chem. Soc., Perkin I, (1975), 1601.

Studies in Surface Science and Catalysis, volume 159
Hyun-Ku Rhee, In-Sik Nam and Jong Moon Park (Editors)

345

Effect of solvent on reaction rate constant of reaction between carbon dioxide and glycidyl methacrylate using Aliquat 336 as a catalyst

Sang-Wook Park[a]* and Jae-Wook Lee[b]

[a] Division of Chemical Engineering, Pusan National University, Busan 609-735, Korea
*E-mail: swpark@pusan.ac.kr

[b] Deapartment of Chemical Engineering, Sogang University, Seoul 121-742, Korea

Absorption rates of carbon dioxide were measured in organic solutions of glycidyl methacrylate at 101.3 kPa to obtain the reaction kinetics between carbon dioxide and glycidyl methacrylate using tricaprylylmethylammonium chloride(Aliquat 336) as catalysts. The reaction rate constants were estimated by the mass transfer mechanism accompanied by the pseudo-first-order fast reaction. An empirical correlation between the reaction rate constants and the solubility parameters of solvents, such as toluene, N-methyl-2-pyrrolidinone, and dimethyl sulfoxide was presented.

1. INTRODUCTION

Recently, the chemistry of carbon dioxide has received much attention[1], and its reaction with oxiranes leading to five-membered cyclic carbonate is well-known among many examples[2]. These carbonates can be used as aprotic polar solvent and sources for polymer synthesis. In the oxirane-CO_2 reaction, high pressure (5-50 atm) of CO_2 has been thought to be necessary[2]. The oxirane-CO_2 reactions under atmospheric pressure have been reported[3] only recently. Many organic and inorganic compounds including amines, phosphines, quaternary ammonium salts, and alkali metal salts are known to catalyze the reaction of CO_2 with oxirane[2]. Most purpose of these papers have been to show the reaction mechanism, the pseudo-first-order reaction rate constant with respect to the concentration of oxirane, and the catalyst dependence of its conversion. In the mass transfer accompanied by a chemical reaction, the diffusion may have an effect on the reaction kinetics. It is considered worthwhile to investigate the effect of diffusion on the reaction kinetics of the gas-liquid heterogeneous reaction such as the oxirane-CO_2 reaction.

In this study, the absorption rates of carbon dioxide into the solution of GMA and Aliquat 336 in such organic solvents as toluene, N-methyl-2-pirrolidinone(NMP), and dimethyl sulfoxide(DMSO) was measured to determine the pseudo-first-order reaction constant, which was used to obtain the elementary reaction rate constants.

2. THEORY

The overall reaction between CO_2 and GMA to form five-membered cyclic carbonate is presented as follows [2]:

$$\text{(1)}$$

The overall reaction (1) in this study is assumed to consist of two steps as follows:
(i) A reversible reaction between GMA(B) and Aliquat 336(QX) to form an intermediate complex (C_1), (ii) An irreversible reaction between C_1 and carbon dioxide(A) to form QX and five-membered cyclic carbonate(C).

$$B + QX \overset{k_1}{\underset{k_2}{\longleftrightarrow}} C_1 \tag{2}$$

$$A + C_1 \overset{k_3}{\rightarrow} C + QX \tag{3}$$

The reaction rate of CO_2 under the condition of steady-state approximation to formation of C_1 is presented as follows:

$$r_A = \frac{k_1 k_3 C_A C_B Q_o}{k_2 + k_3 C_A + k_1 C_B} = \frac{C_B Q_o}{\dfrac{1}{k_1} + \dfrac{k_2}{k_1 k_3 C_A} + \dfrac{C_B}{k_3 C_A}} \tag{4}$$

where Q_o is the total concentration of catalyst and sum of the concentration of species QX and C_1 exited in the reaction.

If the value of k_1 is very large and then $1/k_1$ approaches to 0, Eq. (4) is arranged to

$$r_A = \frac{C_A C_B Q_o}{\dfrac{k_2}{k_1 k_3} + \dfrac{C_B}{k_3}} \tag{5}$$

The mass balances accompanied by reaction (5) of species A and B, and the boundary conditions based on the film theory are given as follows:

$$D_A \frac{d^2 C_A}{dz^2} = \frac{C_A C_B Q_o}{\dfrac{k_2}{k_1 k_3} + \dfrac{C_B}{k_3}} \tag{6}$$

$$D_B \frac{d^2 C_B}{dz^2} = \frac{C_A C_B Q_o}{\dfrac{k_2}{k_1 k_3} + \dfrac{C_B}{k_3}} \tag{7}$$

$$z = 0; \quad C_A = C_{Ai}; \quad \frac{dC_B}{dz} = 0 \tag{8}$$

$$z = z_L; \quad C_A = 0; \quad C_B = C_{Bo} \tag{9}$$

where C_{Ai} and C_{Bo} are the solubility of CO_2 in solvent and the fed concentration of GMA, respectively.

The enhancement factor of CO_2 defined as ratio of the flux of CO_2 with chemical reaction to that without chemical reaction is shown as follows:

$$\beta = -\frac{da}{dx}\bigg|_{x=0} \tag{10}$$

where $a = C_A/C_{Ai}$ and $x = z/z_L$

β can not be obtained because the reaction rate constants such as k_1, k_2, and k_3 are unknown. Therefore, these constants are obtained using the pseudo-first-order reaction methods as following procedures.

If C_B is very larger C_{Ai} to be constant as C_{Bo}, r_A in Eq. (5) is shown as

$$r_A = k_o C_A \tag{11}$$

where $k_o = \dfrac{C_{Bo} Q_o}{\dfrac{k_2}{k_1 k_3} + \dfrac{C_{Bo}}{k_3}}$ (12)

Eq. (12) is rearranged as follows:

$$\frac{Q_o C_{Bo}}{k_o} = \frac{k_2}{k_1 k_3} + \frac{C_{Bo}}{k_3} \tag{13}$$

The mass balance of species A with a pseudo-first-order reaction is

$$D_A \frac{d^2 C_A}{dz^2} = k_o C_A \tag{14}$$

β with the pseudo-first-order reaction is expressed from the analytic solution of Eq. (14) as follows:

$$\beta = \frac{m}{\tanh m} \tag{15}$$

where $m = \sqrt{k_o D_A} / k_L$.

The k_o can be obtained from the measured value of β and Eq.(15).

3. EXPERIMENTAL

The absorption rates of CO_2 were measured in such non-aqueous solvents as toluene, NMP, and DMSO with GMA concentration ranging from 0.5 to 3 kmol/m^3 in a semi-batch flat-stirred agitated vessel constructed of pyrex glass of 0.075 m inside diameter and of 0.13 m in height. The apparatus and the experimental procedure are the same as those described by Park et al[4].

4. RESULTS AND DISCUSSION

The k_o was obtained using the measured β and Eq. (15) with the given values of D_A and k_L. Fig. 1 shows plots of $Q_o C_{Bo}/k_o$ against C_{Bo}, and the plots satisfied straight lines, and k_3 and k_2/k_1 were obtained from the slope and intercept of the straight line of Eq.(13).

The rate constants in organic reaction in a solvent generally reflect the solvent effect. Various empirical measures of the solvent effect have been proposed and correlated with the reaction rate constant [5]. Of these, some measures have a linear relation to the solubility parameter of the solvent. The logarithms of k_3 and k_2/k_1 were plotted against the solubility parameter of toluene, NMP and DMSO[6] in Fig. 2. As shown in Fig.2, the plots satisfied the linear relationship. The solvent polarity is increased by the increase of solubility parameter of the solvent. It may be assumed that increase of unstability and solvation of C_1 due to the increase of solvent polarity make the dissociation reaction of C_1 and the reaction between C_1 and CO_2 such as SN_1 by solvation[7] easier, respectively, and then, k_2/k_1 and k_3 increases as increasing the solubility parameter as shown in Fig. 2.

348

5. CONCLUSIONS

The overall reaction between CO_2 and GMA was assumed to consist of two elementary reactions such as a reversible reaction of GMA and catalyst to form an intermediate and an irreversible reaction of this intermediate and carbon dioxide to form five-membered cyclic carbonate. Absorption data for CO_2 in the solution at 101.3 N/m^2 were interpreted to obtain pseudo-first-order reaction rate constant, which was used to obtain the elementary reaction rate constants. The effects of the solubility parameter of solvent on k_2/k_1 and k_3 were explained using the solvent polarity.

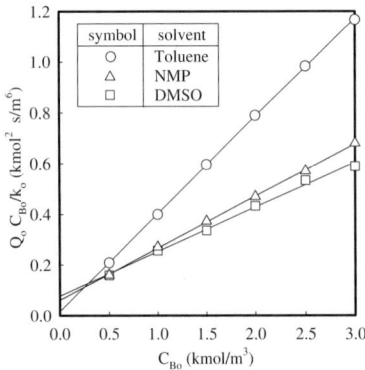

Fig.1. Q_oC_{Bo}/k_o vs. C_{Bo} for various solvents in the reaction of CO_2 with GMA using Aliquat 336 at 85°C.

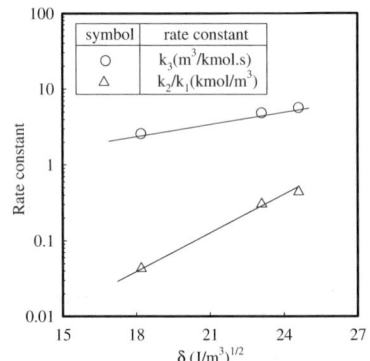

Fig.2. Relationship between reaction rate constant and solubility parameter of solvent in the reaction of CO_2 with GMA using Aliquat 336 at 85°C.

ACKNOWLEDGEMENTS

This research was supported by the Basic Research Program of the Korea Science and Engineering Foundation(KOSEF) through ARC and Brain Korea 21 Project in 2004.

REFERENCES

[1] S. Inoue, In organic and bioorganic chemistry of carbon dioxide; S. Inoue and N. Yamazaki, Eds. Kodansha Ltd., Tokyo, 1982.

[2] N.Kihara, N. Hara, T. Endo, J. Org Chem., 58(1993) 6198.

[3] G. Rokicki, Makromol.Chem., 186(1985) 331.

[4] S. W. Park, I.J. Sohn, D.W. Park, K.J. Oh, Sep. Sci. Technol., 38(2003) 1361.

[5] H.F. Herbrandson, and F.B. Neufeld, J. Org. Chem., 31 (1966) 1140.

[6] J. Brandrup and E.H. Immergut, Polymer Handbook, Second Ed., John Wiley & Sons, New York, 1975.

[7] R. T. Morrison and R. N. Boyd, Organic Chemistry, Fourth Ed., Allyn and Bacon, Inc., Toronto, 1983.

Studies in Surface Science and Catalysis, volume 159
Hyun-Ku Rhee, In-Sik Nam and Jong Moon Park (Editors)
© 2006 Elsevier B.V. All rights reserved

Reaction Characteristics of Immobilized Ru-BINAP Catalysts in Asymmetric Hydrogenation of Dimethyl itaconate

Sung Hyun Ahn[a], Yeung Ho Park[a,*] and Pierre A. Jacobs[b]

[a]Division of Materials Science and Chemical Engineering, Hanyang University, 1271 Sa-1-dong, Sangnok-gu, Ansan-si, Gyeonggi-do 426-791, Korea, e-mail : *parkyh@hanyang.ac.kr
[b]Faculty of Agricultural and Applied Biological Sciences, Centre for Surface Chemistry and Catalysis, Katholieke Universiteit Leuven, Kasteelpark Arenberg 23, 3001 Leuven, Belgium

Chiral Ru-BINAP catalyst was immobilized on the alumina support using a phosphotungstic acid as the anchoring agent. The activity and enantioselectivity of the immobilized catalyst were slightly lower than those of the homogeneous catalyst but the immobilized catalyst showed a stable reaction rate even at high S/C ratios. The catalytic performance of the catalyst was tested at various reaction conditions and it was found that the activity and enantioselectivity of the immobilized catalyst could be enhanced by changing the reaction temperature and solvent and adding triethylamine.

1. INTRODUCTION

Chiral catalysts have received much attention due to the importance of pure enantiomers in the production for the fine chemicals and pharmaceuticals. Most of research work has been mainly performed with homogeneous catalysts since they can produce chiral compounds with good reactivity and selectivity. But homogeneous catalyst needs to be immobilized on inorganic or polymer support before their application to the existing industrial processes due to the difficulties in their recovery and reuse. A number of approaches have been attempted to develop supported homogeneous catalysts, but they suffer from numerous problems, including laborious syntheses for the ligand modification, reduced catalytic activity and selectivity relative to the homogneous catalyst, and excessive leaching of the catalytic species from the support [1]. Recently, Augustine [2-3] developed a dependable immobilization method which an immobilized metal species on modified support and can avoid those problems. But the understanding on this technique is not sufficient to be used for various industrial applications since only few reports were published regarding the reaction characteristics of the catalysts supported by Augustine's method.

In this work, various Ru-BINAP catalysts immobilized on the phosphotungstic acid(PTA) modified alumina were prepared and the effects of the reaction variables (temperature, H_2 pressure, solvent and content of triethylamine) on the catalytic performance of the prepared catalysts were investigated in the asymmetric hydrogenation of dimethyl itaconate (DMIT).

2. EXPERIMENTAL

All reactions were carried out using standard schlenk techniques under inert atmosphere. The phosphotungstic acid(Strem), gamma-alumina(Strem), dimethyl itaconate(Aldrich), triethylamine(Aldrich) and [RuCl₂((R)-BINAP)](Strem) were used without purification. The [RuCl((R)-BINAP)(p-cymene)]Cl and [RuCl((R)-BINAP)(Benzene)]Cl were synthesized by procedures in the literature [4]. All solvent were degassed with nitrogen before use. An immobilized catalyst was prepared using the recipe developed by Augustine [2].

Catalytic hydrogenation reactions were carried out in a 50 mL autoclave. Reaction temperature applied were between 30 and 70°C. Hydrogen pressures between 100 and 300 psig were applied. Under an inert atmosphere, the immobilized catalyst, dimethyl itaconate and degassed solvent were introduced into the reactor equipped with a stirring bar. The reactor was connected to the hydrogenation apparatus and purged with hydrogen (50 psig) five times before stirring was started. The reaction was initiated by introducing a hydrogen at a pressure of 100-300 psig. Samples were collected to determine the reaction rate and enantioselectivity. The reaction mixture was analyzed by GC on a Chiraldex G-TA column. The Ruthenium content remaining in a reaction mixture after hydrogenation and subsequent filteration of immobilized catalyst was also analyzed by ICP-AES.

3. RESULTS AND DISCUSSION

Fig. 1. shows the ^{31}P MAS NMR chemical shifts for the immobilized and homogeneous catalyst. The chemical shifts at the -15.2 and -13.7 ppm correspond to PTA while the chemical shifts in the range from 20 and 40 ppm correspond to phosphine oxide. The chemical shifts at the 66 and 118 ppm seems to be those of BINAP ligand, which is confirmed by the spectrum of Ru-BINAP catalyst. This spectrum shows that PTA exist in large amount on the surface of immobilized catalyst and that BINAP ligand is intact after immobilization.

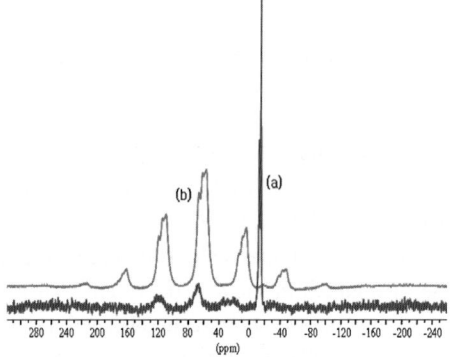

Fig. 1. ^{31}P MAS NMR spectrum of (a)Ru-BINAP/PTA/γ-Al₂O₃, and (b)Ru-BINAP cmplex

In order to find the characteristics of the immobilized catalyst, asymmetric hydrogenation of the prochiral C=C bond was performed as a model reaction. Firstly, three different homogeneous Ru-BINAP complexes including [RuCl₂((R)-BINAP)], [RuCl((R)-BINAP)(p-cymene)]Cl and [RuCl((R)-BINAP)(Benzene)]Cl were immobilized on the PTA-modified alumina. Reaction test of immobilized catalysts showed that [RuCl₂((R)-BINAP)] was the most active and selective so all the experiment were done using this catalyst afterwards.

Table 1
Catalytic performance of immobilized Ru-BINAP catalyst[a]

Catalyst	S / C[b]	DMIT (ml)	DMIT (M)	TOF (h^{-1})[c]	% ee
Homogeneous Ru-BINAP	250	0.5	0.11	333	95
Immobilized Ru-BINAP	250	0.5	0.11	167	88
	500	1	0.23	200	88
	1,000	2	0.44	250	87.5
	5,000	10	1.75	227	87.5

[a]Reaction conditions : 0.014 mmol Ru, H$_2$ 100psig, temp. 50°C, 30 ml EtOH, [b]Molar ratio of substrate to catalyst. [c]Turnover frequency.

Reaction experiments were performed at the substrate to catalyst ratios between 250 and 5000 (Table 1). The immobilized catalyst showed a rather constant values of TOF and enantioselectivity in spite of the increase in the S/C ratio, even though these values were slightly lower than those of the homogeneous Ru-BINAP catalyst. After the reaction, the Ru content in the reaction mixture was measured by ICP-AES and was found to be under 2 ppm, the detecting limit of the instrument, indicating the at Ru metal didn't leach significantly during the reaction. These results show that the immobilized Ru-BINAP catalyst had stable activity and enantioselectivity and that the Ru metal complex formed a stable species on the alumina support.

Table 2
Effect of reaction conditions on the asymmetric hydrogenation of dimethyl itaconate over immobilized Ru-BINAP catalyst

Entry	Solvent	S/C[b]	N/C[c]	H$_2$ (psig)	Temp. (°C)	Time[d] (h)	% Conv.	TOF[e] (h^{-1})	% ee[f]
1	EtOH	500	0	100	30	2.5	10	20	-
2	EtOH	500	0	100	50	2.5	100	200	88
3	EtOH	500	0	100	70	0.5	100	1000	90
4	EtOH	250	0	100	50	1.5	100	167	88
5	EtOH	250	0	300	50	0.5	100	500	82
6	MeOH	250	0	100	50	0.6	100	417	91
7	IPA	250	0	100	50	4	92	58	80
8	THF	250	0	100	50	4	100	63	85
9	Toluene	250	0	100	50	4	3.5	2	-
10	EtOH	250	5	100	50	1	100	250	95
11	EtOH	250	10	100	50	1.5	100	167	93
12	EtOH	250	50	100	50	1.5	100	167	91

[a]Reaction conditions : 0.014 mmol Ru, 3.5-7 mmol DMIT, 30 ml solvent. [b]Molar ratio of substrate to catalyst. [c]Molar ratio of triethylamine to catalyst. [d]Time allowed for reaction to proceed. [e]Turnover frequency. [f]The (S) enantimer was preferentially formed.

In order to find the factors that affect the activity and enantioselectivity of the immobilized catalyst in asymmetric hydrogenation, the effects of reaction temperature, hydrogen pressure, solvents and triethylamine (NEt_3) additive on the reaction rate and % ee of products were investigated and the results obtained are shown in Table 2. As the reaction temperature increased the reaction rate increased drastically, while ee value of products increased a little. Although the increase in the hydrogen pressure gave a higher reaction rate, it lowered the enantioselectivity. Reduced enantioselectivity with increased hydrogen pressure was also reported in the hydrogenation over homogeneous Ru-BINAP catalyst [5]. In case of the solvent effect, the reaction rate and % ee of products with protic solvents was higher than those obtained with aprotic and aromatic solvent. These results could be explained by the differences in the capability of solvents in assisting the release of the product from the catalyst through proton transfer. The addition of NEt_3 in the reaction mixture also increased the reaction rate and % ee of products in the given reaction over immobilized Ru-BINAP catalyst. But the reaction rate and enantioselectivity decreased with the increase of N/C ratio. Similar results were reported on asymmetric hydrogenation of carboxylic acids over homogeneous Ru-BINAP catalysts [6-8], where the reasons for the higher rate and enantioselectivity were ascribed to enhancement of the rate of carboxylate exchange on metal-substrate complex or prevention of protonolysis of Ru-H bond. Understanding the exact role of NEt_3 in the asymmetric hydrogenation of ester of this work needs further investigation.

4. CONCLUSIONS

Ru-BINAP catalyst immobilized on PTA-modified alumina was prepared and tested in the asymmetric hydrogenation of DMIT. It showed a stable activity at the high S/C ratios and no metal leaching, even though its activity and enantioselectivity were slightly lower than that of a homogeneous catalyst. The reaction temperature, hydrogen pressure, the polarity of solvents significantly affected the reaction rate and enantioselectivity. Adding small amount of NEt_3 into the reaction mixture enhanced the catalytic activity and enantioselectivity. From these results, it was found that optimizing the reaction temperature, changing solvent and adding NEt_3 could enhance the activity and selectivity of the immobilized catalyst further.

REFERENCES
[1] G. Ertl, H. Knozinger and J. Weitkamp (eds.), Handbook of Heterogeneous Catalysis, Wiley-VCH, Weinheim, 1997, Vol. 1, pp. 231-240.
[2] R. Augustine, S. Tanielyan, S. Anderson and H. Yang, Chem. Commun. (Cambrige), (1999) 1257.
[3] S.K. Tanielyan and R.L. Augustine, US Patent No. 6 005 148 (1999).
[4] K. Mashima, K. Kusano, N. Sato, Y. Matsumura, K. Nozaki, H. Kumobayashi, N. Sayo, Y. Hori, T. Ishizaki, S. Akutagawa and H. Takaya, J. Org. Chem., 59 (1994) 3064.
[5] T. Ohta, H. Takaya, M. Kitamura, K. Nagai and R. Noyori, J. Org. Chem., 52 (1987) 3174.
[6] C. J. A. Daley, J. A. Wiles and S. H. Bergens, Can. J. Chem., 76 (1998) 1447.
[7] M. Saburi, H. Takeuchi, M. Ogasawara, T. Tsukahara, Y. Ishii, T. Ikariya, T. Takahashi and Y. Uchida, J. Organometal. Chem., 428 (1992) 155.
[8] A.S.C. Chan, US Patent No. 5 202 474 (1993).

Studies in Surface Science and Catalysis, volume 159
Hyun-Ku Rhee, In-Sik Nam and Jong Moon Park (Editors)

$H_3PW_{12}O_{40}$ catalyzed liquid phase nitration of aromatics - a green process without using H_2SO_4

Lianhai Lu[a, b], Junna Xin[a], Chang-Soo Woo[b], Tianxi Cai[a,*] and Ho-In Lee[b,*]

[a] State Key Laboratory of Fine Chemicals and School of Chemical Engineering, Dalian University of Technology, Dalian 116012, China

[b] School of Chemical and Biological Engineering & Research Center for Energy Conversion and Storage, Seoul National University, Seoul 151-744, Korea

$H_3PW_{12}O_{40}$ with Keggin structure showed excellent activity for liquid phase nitration of various aromatic compounds, suggesting the possibility of a green nitration process by simply replacing H_2SO_4 in current nitration unit. The $H_3PW_{12}O_{40}$ could be recycled to use with constant activity. Effects of various reaction parameters were investigated extensively.

1. INTRODUCTION

Nitration is an important chemical reaction widely used in commercial manufacturing of various nitro-aromatics. Concentrated H_2SO_4 has been currently used as the most effective and efficient catalyst in a large-scale liquid phase nitration process. In the traditional nitration process, concentrated H_2SO_4 functions as both an acidic catalyst and a dehydration media. Although the reaction system is highly efficient, there are two serious problems which need to be overcome urgently. At first, H_2SO_4 in the reaction system and recycling unit results in inevitable equipment corrosion. And a large amount of waste H_2SO_4 discharged from the reaction system makes serious environmental pollution even though great efforts have been made to recover it. Therefore, replacing H_2SO_4 by environmentally friendly solid acid catalyst has been attracting worldwide attentions with an aim of solving the above problems [1-4]. Montmorillonite and mixed metal oxide [5, 6], mordenite [7], supported heteropoly acid (HPA) [8], zeolite [9-11], lanthanide (III) complex of aromatic sulfonic acid [12] and modified clay [13] have been studied as nitration catalyst to replace H_2SO_4. Various reaction systems, including liquid phase, gas phase, phase transfer [14] and ionic liquid system [3] have been involved in these studies. Supported HPAs have been reported to be effective for gas phase nitration reaction. However, the gas phase nitration method has a few limitations in practice: Firstly, the reaction temperature should be carefully set up to be about 150 °C above which nitric acid will seriously decompose. Secondly, it can not be used for aromatics with high boiling point. Finally, lifetime of the catalyst in a fixed-bed reaction system is still far behind satisfactory level. Therefore, it is very important to explore new route for application of HPA in nitration reaction. It has been well known that HPA is soluble in aqueous solution and in many polar organic solvent, and its Keggin structure is stable in acidic media even under severe reaction conditions. We thus tried to directly apply HPA in the liquid phase nitration of several typical aromatic compounds to avoid using traditional H_2SO_4.

2. EXPERIMENTAL

HPA catalyzed liquid phase nitration was carried out in a Teflon-lined stainless autoclave of 200 mL equipped with a magnetic stirrer. Reactants and HPA were quantitatively added to the autoclave, which was sealed and heated in an oil-bath. Products were analyzed by GC with OV-101 30 m capillary column and FID detector by using calibrated area normalization and internal standard method. All products were confirmed by GC-MASS analysis.

3. RESULTS AND DISCUSSION

Benzene, toluene and chlorobenzene were selected as model compounds using either 65% nitric acid or fumed nitric acid as nitration reagent. Reaction parameters such as temperature, time, catalyst concentration and molar ratio of reactants have been investigated. Fig. 1(a) shows the effect of reaction temperature on the catalytic activity of HPA in liquid phase nitration of benzene when 65% nitric acid was used. The highest activity was obtained at about 100 °C with almost 100% selectivity for mononitrobenzene. Fig. 1 (b) shows the results of toluene and chlorobenzene by using fumed nitric acid. Rather high conversion was achieved for both reactions. Keggin structure of HPA is stable during the reaction which was confirmed by IR spectra and also by activity test. HPA recovered from the reaction mixture showed the same activity as the fresh one.

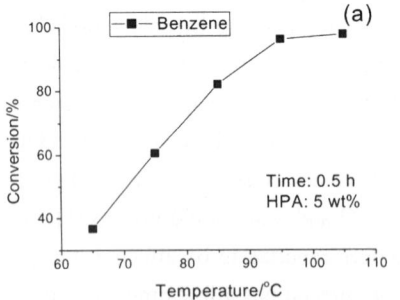 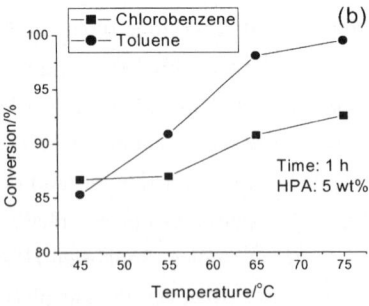

Fig. 1. Temperature effect for liquid phase nitration of (a) benzene, (b) toluene and chlorobenzene.

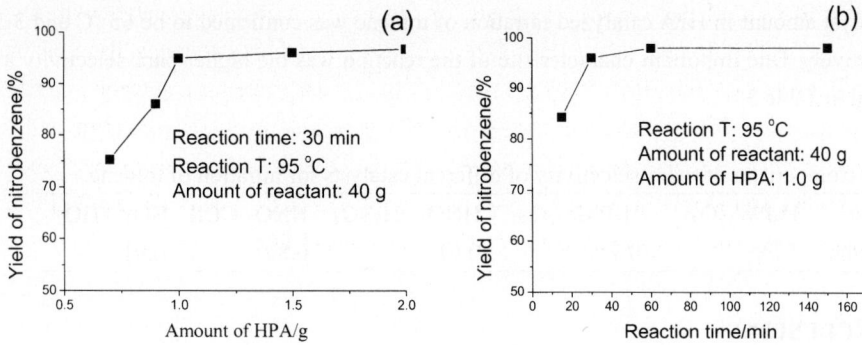

Fig.2. Effect of (a) HPA amount and (b) reaction time on nitration of benzene.

The effects of catalyst amount and reaction time were investigated as shown in Fig 2. While other conditions were kept constant, 2.5 wt% HPA (1 g in 40 g reaction mixture) showed fairly good activity. Further increase of the catalyst amount does not have serious effect on the activity. One hour was enough for the reaction to complete as illustrated in Fig. 2 (b).

Table 1. Effect of reaction temperature on nitration of toluene

T/℃	Regioselectivity			para/ortho	Total yield of mononitrotoluene/%
	ortho/%	meta/%	para/%		
45	51.4	4.5	44.1	0.86	85.3
55	48.0	4.7	47.3	0.99	90.9
65	45.8	4.8	49.4	1.08	98.1
75	45.8	4.6	49.6	1.08	99.5

Fumed nitric acid (>95%), nitric acid/toluene (n/n)= 1.4, catalyst amount: 3.3 g, time: 3 h.

Table 2. Effect of HPA amount on nitration of toluene

HPA amount/g	Regioselectivity			para/ortho	Total yield of mononitrotoluene/%
	ortho/%	meta/%	para/%		
0	52.2	4.4	43.4	0.83	84.1
1.7	48.7	4.8	46.5	0.95	91.0
3.3	45.8	4.8	49.4	1.08	98.1
4.6	47.8	5.2	47.0	0.98	96.6

Reactant amount 40 g, fumed nitric acid (>95%), nitric acid/toluene (n/n)=1.4, reaction temperature: 65 ℃, reaction time: 3 h.

HPA catalyzed nitration of toluene was summarized in Table 1 and 2. The optimal temperature and catalyst amount in HPA catalyzed nitration of toluene was confirmed to be 65 °C and 3.3 g respectively. One important characteristic of the reaction was the higher para-selectivity as indicated in Table 3.

Table 3. Comparison of regio-selectivity of different catalysts for nitration of toluene

Catalyst	$H_3PW_{12}O_{40}$	$H_3PMo_{12}O_{40}$	$HNO_3 -H_2SO_4$	HNO_3-CCl_4	SO_4^{2-}/TiO_2
para/ortho	1.08	0.75	0.63	0.80	0.91

4. CONCLUSIONS

Keggin type $H_3PW_{12}O_{40}$ is a stable, recyclable and effective catalyst for H_2SO_4-free liquid phase nitration of bezene, chlorobenzene and toluene with nitric acid as a nitration agent. Higher para-selectivity of nitrotoluene was obtained, and the result implies that HPA can effectively catalyze the liquid phase nitration of various aromatics as an environmentally friendly nitration process.

ACKNOWLEDGEMENTS

This work was financially supported by Natural Science Foundation of Liaoning, China and by BK21 Program of the Ministry of Education, Korea.

REFERENCES

[1] K. Qiao and C. Yokoyama, Chem. Lett., 33(7) (2004) 808.
[2] B. M. Choudary, M. Sateesh and M. L. Kantam, Chem. Commun., (1) (2000) 25.
[3] K. K. Laali and V. J. Gettwert, J. Org. Chem., 66 (2001) 35.
[4] L.-T. Chen, H-M. Xiao and J-J. Xia, J. Phys. Chem. A, 107(51) (2003) 11440.
[5] H. Sato, K. Hirose and K. Nagai, Appl. Catal. A: General, 175 (1998) 201.
[6] H. Sato, K. Nagai and H. Yoshioka, Appl. Catal. A: General, 180 (1999) 359.
[7] L. Bertea, H.W. Kouwenhoven and R. Prins, Appl. Catal. A: General, 129 (1995) 229.
[8] J. Chen, W. Cheng, H. Liu Q. Lin and L. Lu, Chinese Chem. Lett., 13(4), 311(2002)
[9] S. Bernasconi, G. D. Pirngruber and R. Prins, J. Catal., 224 (2004) 297.
[10] X.-H. Peng, H. Suzuki and C.-X. Lu, Tetrahedron Lett., 42 (2001) 4357.
[11] X.-H Peng and H. Suzuki, Org. Lett., 3(22) (2001) 3431.
[12] T.N. Parac-Vogt, K. Deleersnyder and K. Binnemans, J. Alloys and Comp., 374 (2004) 46.
[13] C. B. Manoranjan, K. M. Lakshmi and S. Mutyala. EP 949,240 A1, 1999-10-13.
[14] A. Zaraiskii, P. Kachurin, O. I., Velichko and L.I. Russ. J. Org. Chem., 39(11) (2003) 1576.

Studies in Surface Science and Catalysis, volume 159
Hyun-Ku Rhee, In-Sik Nam and Jong Moon Park (Editors)

Comparison of Structural Properties of SiO$_2$, Al$_2$O$_3$, and C/Al$_2$O$_3$ Supported Ni$_2$P catalysts

Yong-Kul Lee and S. Ted Oyama

Environmental Catalysis and Nanomaterials Laboratory, Department of Chemical Engineering, Virginia Tech, Blacksburg, Virginia, 24061, USA

This paper describes the catalytic activity of nickel phosphide supported on silica, alumina, and carbon-coated alumina in the hydrodesulfurization of 4,6-dimethyldibenzothiophene. The catalysts are made by the reduction of phosphate precursors. On the silica support the phosphate is reduced easily to form nickel phosphide with high catalytic activity, but on the alumina support interactions between the phosphate and the alumina hinder the reduction. The addition of a carbon overlayer on alumina decreases the interactions and leads to the formation of an active phosphide phase.

1. INTRODUCTION

Metal phosphides are a novel class of catalysts for deep hydrotreating which have received much attention due to their high activity in hydrodesulfurization (HDS) and hydrodenitrogenation (HDN) of petroleum feedstocks [1-6]. Previous studies of supported Ni$_2$P catalysts were carried out mostly with SiO$_2$ supports [1,2,4,5]. It is the objective of this work to investigate γ-Al$_2$O$_3$ as a support and to compare the structural properties of the Ni$_2$P with SiO$_2$-supported samples. Particular attention will be placed on understanding the effect of carbon coatings on the γ-Al$_2$O$_3$ support and the catalytic and structural behavior, as the γ-Al$_2$O$_3$ support is reported to inhibit the formation of Ni$_2$P due to the high interaction with phosphorus [7,8]. The present study also includes the use of X-ray absorption fine structure (XAFS) spectroscopy to study the structure of the dispersed phosphide phases.

2. EXPERIMENTAL

The supports used in this study were SiO$_2$ (Cabot, Cab-O-Sil) of high surface area (EH5, 350 m^2 g^{-1}) and γ-Al$_2$O$_3$ (AkzoNobel, 230 m^2 g^{-1}) and were used as received. A carbon-coated γ-Al$_2$O$_3$ support (C-Al$_2$O$_3$) was prepared by pyrolysis of ethylene at high temperature. Ethylene was decomposed onto 10 g of the γ-Al$_2$O$_3$ sample at 973 K at a flowrate of 200 μmol s^{-1} The supported Ni$_2$P catalysts were prepared on these supports with excess phosphorus (Ni/P=1/2) and a loading of 1.16 mmol Ni/g support (12.2 wt% Ni$_2$P/ C-Al$_2$O$_3$). Previous studies [4,5] had shown that this composition and loading level gave high activity and stability in hydroprocessing reactions. A sample prepared with the carbon-coated γ-Al$_2$O$_3$ was denoted as Ni$_2$P/C-Al$_2$O$_3$. The synthesis of the catalyst involved two steps [4,5]. In the first step, a supported nickel phosphate precursor was prepared by incipient wetness impregnation of a solution of nickel nitrate and ammonium phosphate, followed by calcination at 673 K. In the second step, the supported metal phosphate was reduced to a phosphide by temperature-programmed reduction (TPR). In catalyst preparation, larger batches using up to 5.50 g of supported nickel phosphate were prepared in a similar manner by reduction to 853 K, 1230 K, and 1873 K for Ni$_2$P/SiO$_2$, Ni$_2$P/Al$_2$O$_3$, and Ni$_2$P/C-Al$_2$O$_3$, respectively. Sulfided Ni-Mo/Al$_2$O$_3$ (CR 424) was used as reference.

The prepared catalysts were characterized by x-ray diffraction (XRD), N$_2$ adsorption and CO chemisorption. Also, X-ray absorption spectroscopy (XAS) at the Ni K edge (8.333 keV) of reference and catalyst samples was carried out in the energy range 8.233 to 9.283 keV at beamline X18B of the

National Synchrotron Light Source (NSLS) at Brookhaven National Laboratory (BNL).

Hydrotreating was carried out at 3.1 MPa (450 psig) and 613 K (340 °C) in a three-phase upflow fixed-bed reactor (Figure 1). The feed liquid was prepared by combining different quantities of 1.0 wt% tetralin (Aldrich, 99%), 0.02 wt% N as quinoline (Aldrich, 99%), 0.05 wt% S as 4,6-dimethyldibenzothiophene (4,6-DMDBT, Fisher, 95%), 0.3 wt% S as dimethyldisulfide (DMDS, Aldrich, 99%), n-octane (Aldrich, 99%), and balance n-tridecane (Alfa Aesar, 99%). Liquid product compositions were determined with a Hewlett-Packard 5890A chromatograph equipped with a 50 m dimethylsiloxane column (Chrompack, CPSil 5B) of 0.32 mm i.d. Reaction products were identified by matching retention times with commercially available standards.

In this paper, the HDS conversion is defined as

$$HDS\ Conversion\,(\%) = 100 \times (1 - \frac{MCHT + DMBCH + 3,3DMBP}{4,6DMDBT_{in}})$$

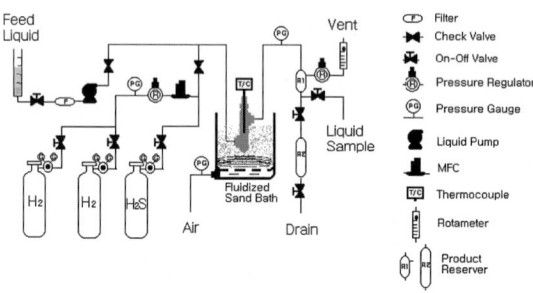

Figure 1. Experimental set-up for hydrotreating.

3. RESULT AND DISCUSSION

Figure 2 shows the TPR profiles of the calcined phosphate precursors of the Ni_2P/SiO_2, Ni_2P/Al_2O_3, and $Ni_2P/C-Al_2O_3$ catalysts.

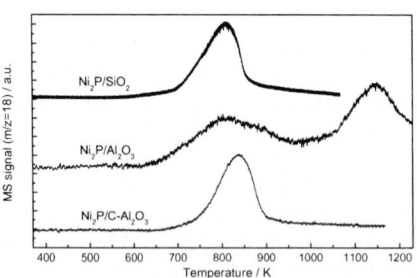

Fig. 2. Temperature-programmed reduction profiles for Ni_2P/SiO_2, Ni_2P/Al_2O_3, and $Ni_2P\ /C-Al_2O_3$.

The Ni_2P/SiO_2 and $Ni_2P/C-Al_2O_3$ catalysts have distinct reduction peaks at around 850 K and 873 K respectively, while the Ni_2P/Al_2O_3 has two broad reduction peaks at around of 810 and 1150 K. This indicates that the nature of the oxidic precursor of the Ni_2P/Al_2O_3 is different from that of the Ni_2P/SiO_2 or $Ni_2P/C-Al_2O_3$. As will be shown in the EXAFS analysis below, the reduction of the oxidic precursor for the Ni_2P/Al_2O_3 goes through Ni at a lower temperature of 853 K, followed by Ni_2P

at a higher temperature 1230 K.

Figure 3 shows the Fourier transforms of the Ni K-edge EXAFS spectra for the freshly prepared samples before (A) and after reduction (B) and bulk reference samples (C,D). For the oxidic precursor (A) of the Ni_2P/SiO_2 catalyst the Fourier transform gives three main peaks centered at 0.09, 0.16, and 0.27 nm, respectively. These peaks are located at almost the same positions as those of the bulk nickel phosphate reference (C). Unlike the case of the Ni_2P/SiO_2, the oxidic precursor (A) of the Ni_2P/Al_2O_3 gives two main peaks, a smaller peak at 0.16 nm and a larger peak at 0.25 nm, being similar in appearance to those of the bulk nickel oxide reference (C). This indicates the formation of the NiO phase during the impregnation and calcination step. This is related to the high interaction between the Al_2O_3 support and phosphorus, which results in the formation of an $AlPO_4$ phase during the calcination [7,8]. For the oxidic precursor of the $Ni_2P/C-Al_2O_3$ the Fourier transform gives three main peaks of which locations are almost identical to those of the bulk nickel phosphate reference (C). This suggests that the carbon coating readily reduce the interaction between the support and phosphorus.

For the bulk Ni_2P the Fourier transform (D) gives two main peaks, a smaller peak at 0.171 nm, and a larger peak at 0.228 nm, corresponding to the Ni-P and Ni-Ni bonding, respectively. For the Ni_2P/SiO_2 (B) there are also two main peaks corresponding to the Ni_2P phase. For the Ni_2P/Al_2O_3 a lower reduction temperature of 853 K led to the formation of Ni metal, while a higher temperature of 1230 K gave rise to a Ni_2P phase. For the $Ni_2P/C-Al_2O_3$ (B) there are two main peaks, again, corresponding to a Ni_2P phase.

It is thus likely that the carbon coating on the alumina support lessened the interaction between Al_2O_3 and P and inhibited the formation of nickel phosphate after calcination and also lowered the reduction temperature. Structural models of the supported Ni_2P catalysts are shown in Figure 4.

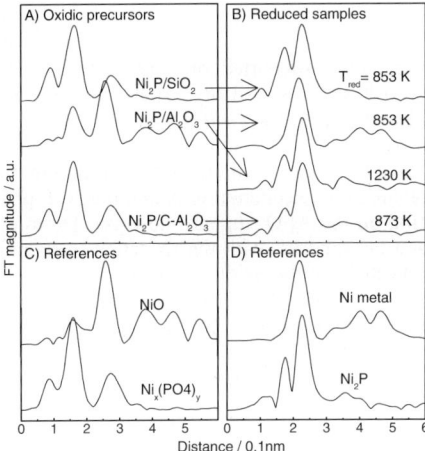

Fig. 3. EXAFS analysis results for A) oxidic precursors before reduction, B) fresh samples after reduction, and C), D) bulk reference samples.

Table 1 compares the physical properties of the catalyst samples and the catalytic activities in the HDS of 4,6-DMDBT at 613 K and 3.1 MPa. The catalytic activity for the catalyst samples in the HDS of 4,6-DMDBT followed the order, Ni_2P/Al_2O_3 < $Ni_2P/C-Al_2O_3$ (~ Ni-MoS/Al_2O_3) < Ni_2P/SiO_2 under 0.35 % S, 0.02 % N, and 1 % tetralin. The order correlated well with the amount of CO uptake. These results thus suggest that the HDS activity of the Ni_2P catalysts highly depend on the dispersion of the Ni_2P phase.

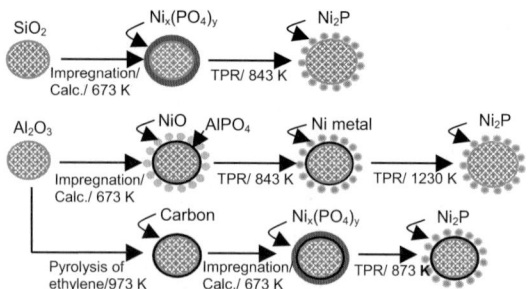

Fig. 4. Proposed structure model for Ni$_2$P/SiO$_2$, Ni$_2$P/Al$_2$O$_3$, and Ni$_2$P/C-Al$_2$O$_3$.

Table 1. Physical properties and catalytic activities of the catalyst samples

Samples	CO uptake / μmol g^{-1}	BET area / m^2 g^{-1}	4,6-DMDBT HDS conversion [b] / %
Ni$_2$P/SiO$_2$	102	240	99
Ni$_2$P/Al$_2$O$_3$	59	81	52
Ni$_2$P/C-Al$_2$O$_3$	69	92	71
Ni-Mo-S/Al$_2$O$_3$	287 [a]	155	75

[a] Atomic oxygen uptake, [b] Based on equal site loadings of 230 μmol in the reactor

4. CONCLUSION

Nickel phosphide (Ni$_2$P) catalysts supported on SiO$_2$, Al$_2$O$_3$, and carbon-coated Al$_2$O$_3$, were successfully prepared. The Ni$_2$P/SiO$_2$ showed high HDS activity for 4,6-DMDBT compared to a commercial Ni-Mo-S/Al$_2$O$_3$ catalyst at 613 K (340 °C) and 3.1 MPa based on equal sites (230 μmol) loaded in the reactor. Unlike the case of SiO$_2$ support, the Al$_2$O$_3$ support gave rise to strong interactions with phosphorus and formed AlPO$_4$ and NiO during the initial synthesis steps of impregnation followed by calcination. Reduction to obtain the Ni$_2$P phase required high temperature (1230 K). The carbon coating on the Al$_2$O$_3$ support, however, led to a considerable decrease in the interaction between Al$_2$O$_3$ and phosphorus, with the reduction temperature being lowered to 873 K. The EXAFS analysis also indicates that the oxidic precursor of nickel phosphate is readily formed on the carbon-coated Al$_2$O$_3$ support. The Ni$_2$P/C-Al$_2$O$_3$ catalyst also showed comparable HDS activity for 4,6-DMDBT with the Ni-Mo-S/Al$_2$O$_3$ catalyst. The HDS activity of the supported Ni$_2$P catalysts was closely related to the dispersion of the Ni$_2$P on the supports. Therefore, it will be of great interest to increase the dispersion of Ni$_2$P phase on the carbon-coated Al$_2$O$_3$ support.

ACKNOWLEDGMENT
Support for this work came from the U.S.Department of Energy, Office of Basic Energy Sciences, through Grant DE-FG02-963414669 and Brookhaven National Laboratory under grant 3972.

REFERENCES
[1] W. R. A. M. Robinson, J. N. M. van Gastel, T. I. Korányi, S. Eijsbouts, J. A. R. van Veen, and V. H. J. de Beer, J. Catal. 161 (1996) 539.
[2] W. Li, B. Dhandapani, and S. T. Oyama, Chem. Lett. (1998) 207.
[3] C. Stinner, R. Prins, and Th. Weber, J. Catal. 191 (2000) 438.
[4] S. T. Oyama, X. Wang, Y.-K. Lee, and F. G. Requejo, J. Catal. 210 (2002) 207.
[5] S. T. Oyama, X. Wang, Y.-K. Lee, and W.-J. Chun, J. Catal. 221 (2004) 263.
[6] S. T. Oyama, J. Catal. 216 (2003) 343.
[7] J. M. Campelo, M. Jaraba, D. Luna, R. Luque, J. M. Marinas, and A. A. Romero, Chem. Mater. 13 (2003) 3352.
[8] J. Quartararo, J. Amoureux, and J. Grimblot, J. Mol. Catal. A 162 (2000) 353.

CHEMICAL REACTION ENGINEERING IN MICROELECTRONICS

Studies in Surface Science and Catalysis, volume 159
Hyun-Ku Rhee, In-Sik Nam and Jong Moon Park (Editors)

Synthesis of nanodot arrays using self-assembled niobium oxide nanopillar mask by reactive ion etching

Ik Hyun Park, Jang Woo Lee and Chee Won Chung

Department of Chemical Engineering, Inha University, 253 Yonghyun-Dong, Nam-Ku, Incheon 402-751, Republic of Korea

1. INTRODUCTION

The formation of nanostructures such as nanodot arrays has drawn a great attention due to the feasible applications in a variety of functional structures and nanodevices containing optoelectronic device, information storage, and sensing media [1-3]. The various methods such as self-assembled nanodots from solution onto substrate, strain-induced growth, and template-based methods have been proposed for the fabrication of nanodot arrays on a large area, [4-6]. However, most of these works can be applied to the small scale systems due to the limited material systems.

We reported a novel method for the fabrication of nanodot arrays by reactive ion etching (RIE) using self-assembled tantalum oxides which were formed at the bottom of anodic aluminum oxide (AAO) [7,8]. However, tantalum oxides as an etching mask had the slanted sidewall slope so that the nanodots fabricated by etching increased in diameter. In general, a good etching mask shows high etch selectivity to films to be etched and has high aspect ratio with the vertical sidewall. We propose the use of self-organized niobium oxide nanopillars as an etching mask for the fabrication of nanostructure. The niobium oxide nanopillar masks, formed by the electrochemical anodization of Al/Nb films, have very high aspect ratio with vertical sidewall and show good etch selectivity for the formation of nanodots.

In this report, the fabrication of Si nanodot arrays using niobium oxide nanopillars as an etching mask was performed by an inductively coupled plasma reactive ion etching (ICPRIE).

2. EXPERIMENTAL

The overall procedure for the formation of Si nanodot arrays on a wafer was schematically presented at the previous report [8]. The specimens consisted of the Al(20nm)/Nb(8nm)/Si (30nm) layers on a thermally oxidized Si wafer. Thin Si film was deposited on the SiO_2/Si wafer by low pressure chemical vapor deposition and the successive deposition of Nb and Al films by DC sputtering was followed. The full anodization of Al and Nb films leads to the formation of niobium oxide pillar arrays at the bottom of porous alumina. The selective removal of porous alumina layer by chemical etching was performed to reveal niobium oxide pillar arrays. After anodizing the Al/Nb/Si/SiO_2/Si substrate and removing alumina, the specimen was transformed to the structure of NbO_x(40nm)/Si(30nm)/SiO_2/Si. Experimental details of the anodization process of Al/Nb layer were reported by Vorobyova and Mozalev [9]. Finally, direct pattern-transfer to the underlying Si film was carried out by reactive ion etching for the formation of Si nanodot arrays on a wafer. The etch equipment used in this study was a high density inductively coupled plasma reactive ion etch (ICPRIE) system. The detailed information on the equipment can be found elsewhere [8].

The etch rates were measured by a surface profiler and field emission scanning electron microscopy (FESEM), and the etch profiles were observed by FESEM. In this study, a Cl_2/Ar gas chemistry was chosen to obtain high etch selectivity of Si film to niobium oxide mask since Cl_2 gas was known to be a good etch gas for Si films. The etch rate, etch selectivity and etch profile of niobium oxide nanopillars and Si films were explored by varying the Cl_2 concentration, coil RF power and dc bias voltage to substrate.

3. RESULTS AND DISCUSSION

The formation of Si nanodot arrays on a substrate was performed by ICPRIE of Si films using self-assembled niobium oxide pillars as an etching mask. The etch rates of niobium oxide pillars and Si films, and the etch selectivity of Si films to niobium oxide were investigated by varying etch parameters in a Cl_2/Ar gas. The main etch parameters used in this study were the concentration of Cl_2 gas, coil rf power, and dc-bias to substrate.

Figure 1(a) shows the etch rates of niobium oxide pillar and Si film, and the etch selectivity of Si to niobium oxide as a function of Cl_2 concentration. The etch condition was fixed at coil rf power of 500 W, dc-bias to substrate to 300 V and gas pressure of 5 mTorr. As the Cl_2 concentration increased, the etch rate of niobium oxide pillar gradually decreased while Si etch rate increased. It indicates that the etch mechanism of niobium oxide in Cl_2/Ar gas is mainly physical sputtering. As a result, the etch selectivity of Si film to niobium oxide monotonously increased. The effect of coil rf power on the etch rate and etch selectivity was examined as shown in Fig. 1(b). As the coil rf power increased, the etch rates of niobium oxide and Si increased but the etch rate of niobium oxide showed greater increase than that of Si. It is attributed to the increase of ion density with increasing coil rf power. Figure 1(c)

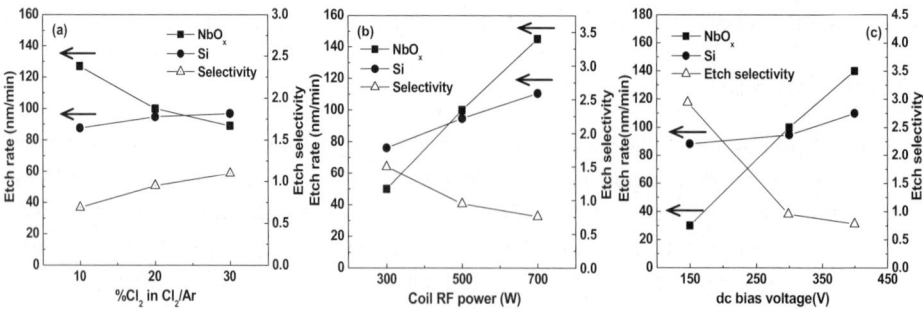

Fig. 1. Etch rates of niobium oxide pillar and Si film, and etch selectivity of Si to niobium oxide pillar for the variation of (a) Cl_2 concentration, (b) coil rf power, and (c) dc-bias voltage to susceptor

presents the change of the etch rate and etch selectivity by varying dc-bias to substrate. The dc-bias applied to substrate controls the ion energy bombarding on to the substrate. With increasing the dc-bias to substrate, the etch rate of niobium oxide showed a great increase while Si etch rate slightly increased. Therefore, the etch selectivity of Si to niobium oxide decreased with increasing dc-bias to substrate. From the etch results obtained by the variation of three etch parameters, it is thought that high etch selectivity of Si to niobium oxide mask can be attained by adjusting etch parameters.

FESEM micrographs of the niobium oxide pillar arrays after removal of AAO templates were shown in Fig. 2(a). The dense and pillar structure of niobium oxides is observed, and the size and height of nanopillar arrays are almost constant over the entire surface. The diameter of pillar arrays is in the range of 30~40 nm, depending on the thickness of anodized films and anodizing conditions. The detailed experimental conditions can be found elsewhere [9]. Figures 2(b) and 2(c) were the FESEM images of nanodot arrays etched by varying an etch time. The etch conditions were 20% Cl_2 in a Cl_2/Ar gas mix, coil rf power of 500 W, gas pressure of 5 mTorr, and dc-bias voltage of 300 V. It is observed that the average diameters of nanodots etched by this etching condition decreased by increasing the etch time.

In order to confirm the formation of nanodots, the fabricated nanodot arrays on a substrate were examined using energy dispersive spectroscopy (EDS). The EDS analysis of niobium oxide arrays on Si film before etching (Fig. 2(a)) was shown in Fig. 3(a). The Si peak as well as Nb and O peaks was observed because niobium oxide on Si film was so thin.

Figures 3(b) and 3(c) were the EDS results of the etched nanodot arrays shown in Figs. 2(b) and 2(c). The EDS of Fig. 3(b) was almost identical to that of Fig. 3(a). It means that niobium oxide masks were still on Si film although the Si dots were formed. Fig. 3(c) was the EDS result of the nanodot arrays etched for longer etch time than Fig. 3(b). The Nb peak disappeared due to the increased etching and it was confirmed that the nanodots consisted of only Si. The diameters of Si nanodots were approximately 20~30 nm. It was demonstrated that the optimal etching condition could form close-packed and highly ordered Si nanodot arrays without niobium oxide mask. It is expected that this novel technique of forming

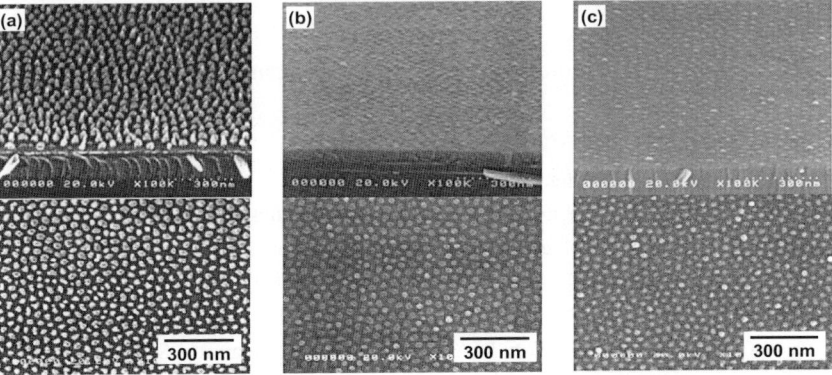

Fig. 2. FESEM images of (a) niobium oxide arrays before etching, and of Si nanodot arrays etched for (b) 20 s and (c) 30 s at 20% Cl_2, 500 W coil rf power, 300 V dc-bias voltage and 5 mTorr gas pressure

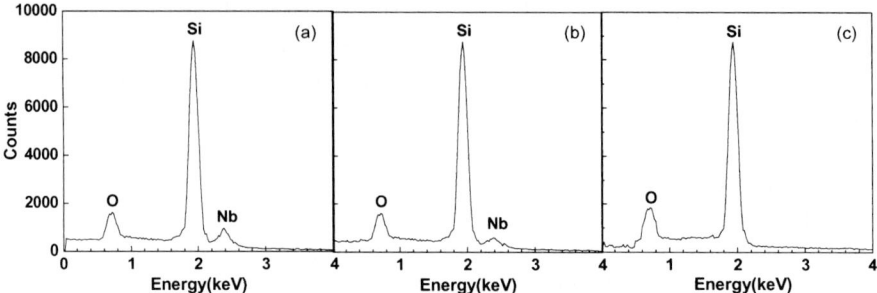

Fig. 3. EDS analysis of (a) niobium oxide arrays before etching, and of Si nanodot arrays etched for (b) 20 s and (c) 30 s at 20% Cl_2, 500 W coil rf power, 300 V dc-bias voltage and 5 mTorr gas pressure

nanodot arrays is not limited to any specific material systems. In addition, Si nanodots smaller than 20~30 nm can be formed from small niobium oxide pillars (e.g., 20~30 nm) obtained by the optimized AAO process.

4. CONCLUSIONS

The formation of Si nanodot arrays has been achieved by ICPRIE using self-organized niobium oxide pillar masks, which are formed at the interface of Al/Nb layer during AAO process. The proper anodizing conditions were chosen for highly ordered niobium oxide nanopillar arrays. The etching conditions were explored as a function of the concentration of Cl_2 gas, coil rf power and dc-bias to susceptor for high etch selectivity of Si film to niobium oxide pillar mask and for good etch profile of Si dots with nanometer-size. Once the niobium oxide pillar arrays are formed on a Si film, the shape and diameter of the resultant Si nanodots are dependent of the etching conditions. In this study, highly ordered Si nanodot arrays of 20~30 nm were formed over a large area by the optimized etching conditions. It is anticipated that the proposed novel technique can be applied to form nanodot arrays composed of a variety of materials on a wafer scale.

ACKNOWLEDGMENTS

This work was supported by Korea Research Foundation Grant.

REFERENCES

[1] A. M. Morales and C. M. Lieber, Science, 279 (1998) 208.
[2] S. Pan, R. Villeneuve, J. D. Joannopoulos and H. A. Hauss, Phys. Rev. Lett., 80 (1998) 960.
[3] K. Matsumoto, Y. Gotoh, T. Maeda, J. A. Dagata and J. S. Harris, Appl. Phys. Lett., 76 (2000) 239.
[4] R. Notzel, J. Temmyo and T. Tamamura, Nature, 369 (1994) 131.
[5] H. Masuda , K. Yasui and K. Nishio, Adv. Mater., 12 (2000) 1031.
[6] M. Park , C. Harrison, P. M. Chaikin, R. A. Register and D.H. Adamson, Science, 276 (1997) 1401.
[7] A. I. Vorobyova and E. A. Outkina, Thin Solid Films, 324 (1998) 1.
[8] S. H. Jeong, Y. K. Cha, I. K. Yoo, Y. S. Song and C. W. Chung, Chem. Mater., 16 (2004) 1612.
[9] A. Mozalev, M. Sakairi, I. Saeki and H. Takahashi, Electrochim. Acta, 48 (2003) 3155.

Studies in Surface Science and Catalysis, volume 159
Hyun-Ku Rhee, In-Sik Nam and Jong Moon Park (Editors)

Chemical analysis of etching residues in metal gate stack for CMOS process

Wan Sik Hwang, Hui Hui Ngu, and Won Jong Yoo
Silicon Nano Device Laboratory, Department of Electrical and Computer Engineering,
National University of Singapore, Singapore, 119260 email: eleyoowj@nus.edu.sg

Vladimir Bliznetsov
Institute of Microelectronics, 11 Science Park Road, Singapore Science Park II, Singapore
117685

1. INTRODUCTION

The rapid progress in complementary metal oxide semiconductor (CMOS) processing is being made mainly by scaling down of active areas such as channel length and gate dielectric thickness. However, as device continues to shrink, metal / high-k gate stacks have been extensively studied to replace conventional poly-Si/SiO_2 gate stacks [1-3]. There are considerable challenges to overcome for the integration of metal / high-k gate stacks although several promising candidate materials have been identified: e.g. TaN, TiN, and HfN as metal gates and Hf-based compounds as gate dielectrics [1-3]. One of the main challenges is the etching of these new materials [4-5], since a significant amount of nonvolatile byproducts is generated and remains on the etched surface; resulting in the degradation of surface properties of active areas. Furthermore, during the subsequent cleaning process to remove the residues, these active areas might be damaged or eroded. In this work, possible issues of metal gate stack etching were identified and their mechanism had been studied. Angle resolved x-ray photoelectron spectroscopy (XPS) was used to analyze etching residues on both the etch front and sidewall of gate stacks. As considerable challenges in wet processing arise due to implementation of these new metal gate materials, wet cleaning process for these new metals was discussed here as well.

2. EXPERIMENTS

The etching experiments were performed using either inductively coupled plasma (ICP) with Cl_2/O_2 or decoupled plasma source (DPS) with Cl_2/Ar / HBr. However, the experiments were primarily performed in ICP unless mentioned otherwise. Metal nitride films (TaN, TiN, and HfN) were deposited by reactive sputtering. After etching, the profile was analyzed using a cross-sectional scanning electron microscope (SEM) and residues remaining on the gate stacks were analyzed using angle resolved XPS. For surface roughness measurement, atomic force microscopy (AFM) was used within the tapping mode and data were acquired on 1um × 1um frames having 256×256 data points. Dilute hydrofluoric acid (DHF) was used for wet cleaning process.

3. RESULTS AND DISCUSSION

Etched surface was investigated as a function of etching time in fig. 1. Agglomerated residues were observed, in addition to the increase of residues amount on the surface with HfN etching time. The cross sectional images of AFM in the inset show that surface properties also degraded with etching time. It is believed that the nonvolatile residues generated during etching play a role as a local mask, resulting in increase of surface roughness.

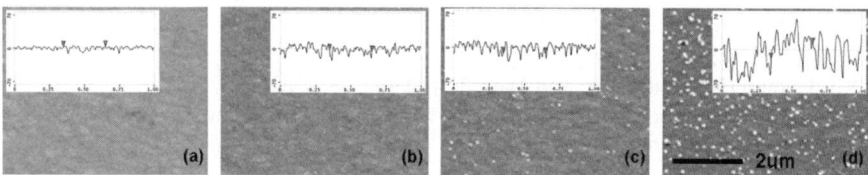

Fig. 1. SEM images of etched HfN surface in Cl_2 with etching time; (a) 10s, (b) 15s, (c) 20s, and (d) 25s. The inset shows an evolution of surface topography (height) using AFM with various etching time; X: 0.25 um/div, Z: 70nm/div. The experiments were performed at a pressure of 10mTorr, source power of 400W, and bias voltage of -200V.

It was reported that high selectivity can be achieved with addition of small amount of O_2, which increases the etch rate of metal electrode [4] while suppresses the etch rate of dielectrics [5]. Figure 2 shows the comparison of residues formation after etching between in (a) Cl_2 and (b) Cl_2/O_2. Agglomerated residues were observed after etching in Cl_2/O_2, indicating that addition of O_2 (1%) enhances residues formation on the etched surface.

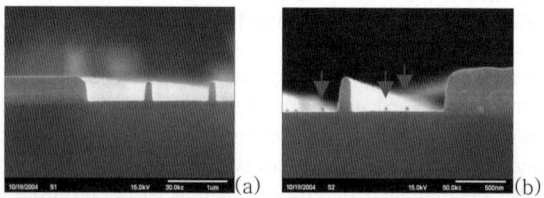

Fig. 2. SEM images of TaN gate stack with photoresist masks after etching (a) in pure Cl_2 and (b) Cl_2/O_2. The experiments were performed at a pressure of 10mTorr, source power of 400W, and bias voltage of -200V.

More Cl and O were observed by adding 1% O_2 in Cl_2 and the amounts of residues were removed by wet cleaning process. It is expected that the residues generated during etching on the gate stack can be removed by wet cleaning process such as dilute HF.

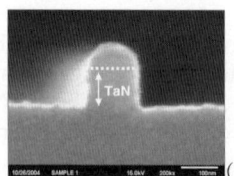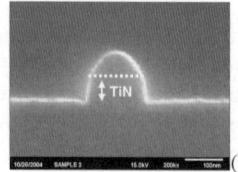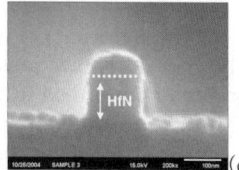

Fig. 3. SEM images of (a) TaN, (b) TiN and (c) HfN gate stack with SiO_2 mask after etching in Cl_2. The experiments were performed in the same condition as in fig. 1.

The residues formation (in fig. 3) can be correlated to the boiling temperature of the byproducts (239 °C for $TaCl_5$, 136 °C for $TiCl_4$, and 317 °C for $HfCl_4$) which are most favorable, according to Gibb's free energy. As expected, the significant amount of residues were formed on the surface during HfN etching due to high boiling temperature of $HfCl_4$ while no degradation of surface properties was observed during TiN etching due to low boiling temperature of $TiCl_4$.

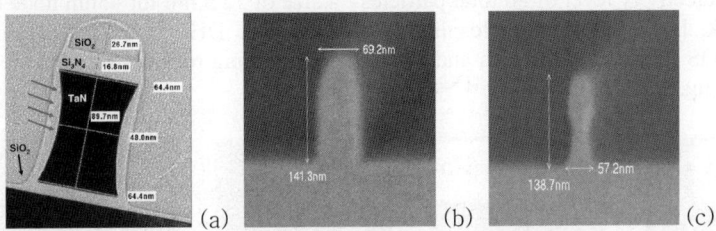

(a)   (b)   (c)

Fig. 4. (a) TEM image of TaN metal electrode gate stack after Cl_2 etching, revealing thick residues formation on the sidewall (etching was done in DPS); SEM image of TaN metal electrode (b) before and (c) after DHF cleaning.

The cross sectional TEM image of TaN metal electrode gate stack after etching is shown in fig. 4 (a). Thick residues were observed on the sidewall of the gate stack after etching. To analyze the residues on the sidewall, angle resolved XPS was done with substrate tilting [4] and analysis was performed *ex-situ*. The samples for XPS analysis were kept in vacuum to minimize the possibility of the change in chemical compositions of the residues induced by the exposure to air. Gaussian-Lorentzian curve fitting method is used for XPS data analysis. Figure 5 shows the results of XPS analysis with top view (electron analyzer faces substrate horizontally so only electrons coming from etched surface and top of mask were detected) and side view (substrate was tilt for the electron analyzer to face sidewall of gate so only electrons coming from sidewall of gate stack were detected).

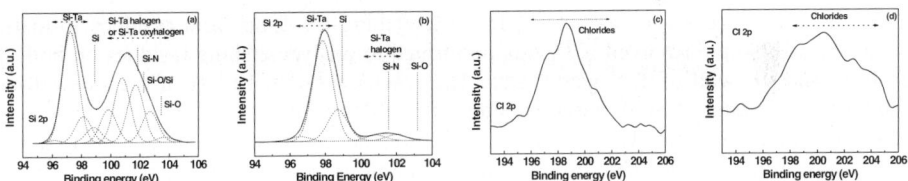

Fig. 5. XPS spectra from the residues on the gate stack after Cl_2 etching: (a) and (b) Si 2p, (c) and (d) Cl 2p: (a) and (c) top view: (b) and (d) side view.

One strong peak detected at the binding energy lower than the designated Si peak in figs. 5 (a) and (b) is believed to be originated from Si-Ta bond. Peak position of Si will be shifted to lower binding energy when Si reacts with materials showing lower electronegavity (such as Ta at 1.5) than Si (at 1.9). On the other hand, peak position of Si will be shifted to higher binding energy when Si reacts with materials showing higher electronegavity value (such as O at 3.44, N at 3.04, and Cl at 3.16) than Si (at 1.9) [7]. Figures 5 (c) and (d) show that more oxychlorides residues were observed on the sidewall of the gate than etched surface between gate stack. Since XPS results show that most of the sidewall and etched surface consist of Si-Ta based residues, it is expected that these residues can be removed by DHF. In addition, Si-Ta based residues are incorporated into passivation films during plasma etching, and this helps to achieve anisotropic profile.

Currently wet cleaning process is an essential step to remove particles and residues generated during plasma etching and megasonic energy helps to remove particles effectively. However, the wet cleaning processing has risen to considerable challenges in the front end of the line (FEOL) of metal gate due to the International Technology Roadmap for Semiconductors (ITRS) guidelines for maximum material loss of 0.04nm/cleaning cycle and maximum defectivity level of 86 total particles / a size of 22.5 nm for 45nm node technology. In this work, DHF is used for cleaning process because DHF is widely employed in the current CMOS fabrication to clean and/or remove remaining residues on the gate stack after plasma etching.

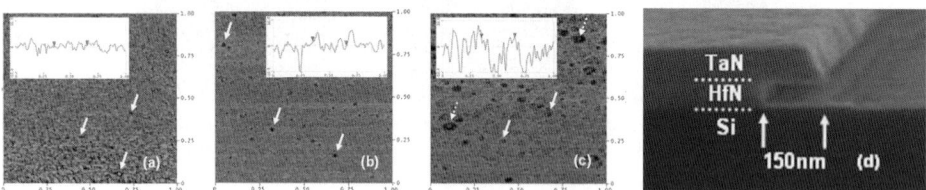

Fig. 6. AFM images of etched surface of HfN films after 1% DHF dipping with the time; (a) 5 s, (b) 15 s, (c) 40 s; (d) SEM image of metal gate stack after etching 5min in 1% DHF, showing HfN film is laterally etched.

Figure 4 (c) shows that thick residues consisting of Si-Ta have been successfully removed by DHF cleaning; indicating that DHF works effectively also for new metal gate stacks. However, figure 6 shows that pitting was observed only on the etched surface of HfN. In addition, the pit size increases with cleaning time. It was also found that HfN film can be even laterally etched as shown in fig. 6 (d), indicating that HfN film is highly soluble in DHF.

4. CONCLUSIONS

In this work, the issues for metal gate stack etching resulting from residues formation were addressed. Angle resolved XPS was performed to analyze etching residues on both the etch front and sidewall of TaN gate stacks. The results show that most of the sidewall and etched surface consist of Si-Ta based residues after TaN metal gate etching with SiO_2 mask. These residues are incorporated into passivation films during plasma etching so as to help to achieve anisotropic profile. The results of wet cleaning process for new metal gate materials show that DHF is effective for TaN gate, whereas it results in the pit formation on HfN gate. This was confirmed by the AFM and SEM analysis.

REFERENCES

[1] G. D. Wilk, J. Appl. Phys. 89 (2001) 5243.
[2] C. Ren, H. Y. Yu, J. F. Kang, Y. T. Hou, M. -F. Li, W. D. Wang, D. S. H. Chan and D.-L. Kwong, IEEE Electron Dev. Lett. 25 (2004) 123.
[3] K. Pelhos, V. M. Donnelly, A. Kornblit, M. L. Green, R. B. Van Dover, L. Manchanda, Y. Hu, M. Morris, and E. Bower, J. Vac. Sci. Technol. A 19 (2001) 1361.
[4] W. S. Hwang, J. H. Chen, W. J. Yoo, and V. Bliznetsov, J. Vac. Sci. Technol. A 23 (2005) 964.
[5] J. H. Chen, W. J. Yoo, Zerlinda Y. L. Tan, Y. Q. Wang, and Daniel S. H. Chan, J. Vac. Sci. Technol. A 22 (2004) 1552.
[6] N. Fukushima, H. Katai, T. Wada, and Y. Horiike, Jpn. J. Appl. Phys. 35 (1996) 2512.
[7] D. R. Lide, *Handbook of Chemistry and Physics*; CRC Press, (2003-2004) 84th Ed.

Studies in Surface Science and Catalysis, volume 159
Hyun-Ku Rhee, In-Sik Nam and Jong Moon Park (Editors)

Structural and optical properties of InGaN/GaN triangular-shaped quantum wells grown by metalorganic chemical vapor depostion

Rak Jun Choi and Yoon-Bong Hahn

School of Chemical Engineering and Technology, Nanomaterials Processing Research Center, Chonbuk National University, Chonbuk National University, Jeonju 561-576, Korea

1. INTRODUCTION

III-nitrides and their alloys have received much attention due to their tremendous potential for fabricating light-emitting diodes (LEDs) and laser diodes (LDs) that operate in the red to ultraviolet (UV) energy ranges [1-3]. InGaN alloy is very important for applications of the III-nitride materials in LEDs and LDs because the alloy constitutes an active region in the form of quantum well (QW) and emits light by recombination of electrons and holes injected into the InGaN active layer. The optical and structural properties of $In_xGa_{1-x}N$/GaN QWs are quite sensitive to the growth conditions of the InGaN layer. However, the quality of InGaN films is dependent on growth variables such as growth temperature, flow rate of Ga and/or In composition, well and barrier widths in the QW regions. Therefore, in order to effectively optimize the growth conditions and tune emission wavelengths, it is necessary to study and understand the effects of growth variables and QW structures on the optical and structural properties of $In_xGa_{1-x}N$/GaN QW structures. Furthermore, understanding the emission mechanism of the $In_xGa_{1-x}N$/GaN QWs is essential for further improving the performance of optical devices.

The crystal quality of the InGaN QWs becomes poor mainly due to the lattice-constant mismatch and the difference of the thermal expansion coefficient between InN and GaN with increasing the In composition [4,5]. Therefore, in order to improve the external quantum efficiency (η_{ext}) of the InGaN-based LEDs and LDs, it is important to elucidate and optimize the effects of the various growth conditions for the InGaN active layer on the structural and optical properties. Recently, we reported a fabrication of efficient blue LEDs with InGaN/GaN triangular shaped QWs and obtained a substantial improvement of electrical and optical properties of the devices [6,7].

In this paper, we report the structural and optical properties of $In_xGa_{1-x}N$/GaN triangular shaped MQWs obtained under various conditions of growth variables. In addition, the emission mechanism of the InGaN QWs is intensively discussed.

2. EXPERIMENTAL

A 5-period $In_xGa_{1-x}N$/GaN triangular shaped QW structure was grown on the n-GaN layer by metalorganic chemical vapor deposition (MOCVD) under various conditions of growth temperature, flow rate of In, and well and barrier widths. Trimethylgallium (TMGa), trimethylindium (TMIn), ammonia (NH_3), and silane (SiH_4) were used as the precursors of Ga, In, N, and Si, respectively. Before growing the nitride films, the substrates loaded into the reactor were thermally cleaned in hydrogen atmosphere at 1200 ℃ for 10 min. A GaN nucleation layer of 25 nm thickness was grown on the cleaned substrate at 560 ℃, and a 1-μm-thick GaN:Si with 2×10^{18} cm^{-3} of carrier concentration was grown above the buffer layer at 1130 ℃. A 5-period $In_xGa_{1-x}N/GaN$ triangular shaped QW structures was grown on the n-GaN layer at various growth conditions such as growth temperature, flow rate of In, well width, and barrier width. The triangular band structure in the QWs was obtained by gradating the flow rate of In linearly with time in the course of growing the well layer. Details of the growth method are available elsewhere [7,8]. A 100 nm-thick undoped GaN capping layer was deposited finally at 1050 ℃ on the MQWs. Photoluminescence (PL) measurements were carried out using a He-Cd laser operating at 325 nm. Structural properties were analyzed by high resolution X-ray diffraction (HRXRD) and high resolution transmission electron microscopy (HRTEM).

3. RESULTS AND DISCUSSION

Figure 1 shows the PL spectra of $In_xGa_{1-x}N/GaN$ 5MQW structures grown at 760, 775, and 795 ℃, which were named TA, TB, and TC, respectively. The flow rate of TMIn was increased linearly with time up to 32 μmol/min for the first 30 s, maintained constant for 6 s, and then decreased linearly down to zero for the last 30 s. The peak energy linearly decreased to show a red shift with decreasing growth temperature (T_g). The PL intensity also increased with T_g, but full widths at half maximum (FWHM) decreased with T_g. Figure 2 shows the HRXRD patterns for the (0002) reflection from the $In_xGa_{1-x}N/GaN$ MQW structures of samples IA, IB, and IC, respectively. From HRXRD and HRTEM, the estimated average In compositions of the $In_xGa_{1-x}N$ well region are 24, 28, and 32 % for samples IA, IB, and IC,

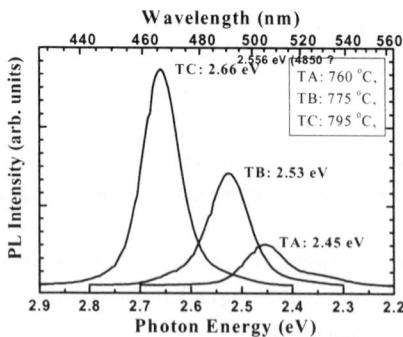

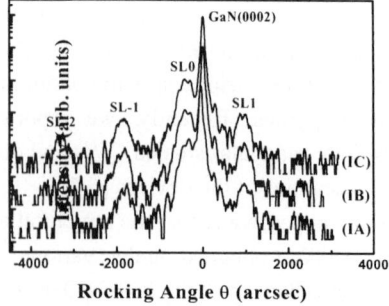

Fig. 1. PL spectra of InGaN/GaN MQW grown at various temperatures.

Fig. 2. High resolution XRD patterns of InGaN/GaN MQW as a function of flow rate of TMIn.

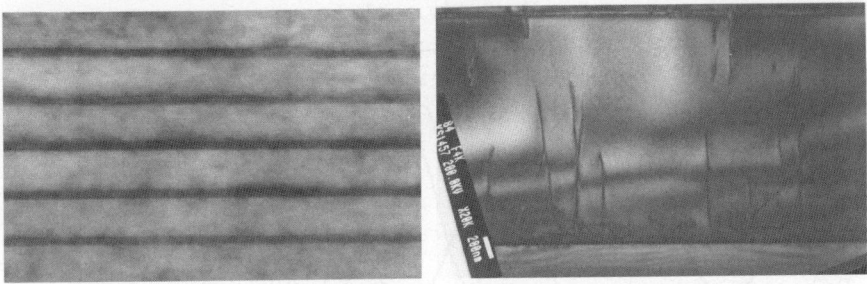

Fig. 3. Cross section of high resolution TEM (top) and cross-sectional bright-field TEM images of InGaN/GaN triangular-shaped multiple quantum well structure (sample TA).

respectively. Also, the thicknesses of the well and barrier are close to 25 and 94 Å for all the samples. The FWHMs of SL-1 are 250, 258, and 278 arcsec for samples IA, IB, and IC, respectively. That is, the In composition and the FWHM of SL-1 increased with decreasing growth temperature. The amount of In incorporation into the InGaN alloy decreases with increasing T_g due to high volatility of In. Therefore, the decrease in the PL intensity and increase of FWHM in the HRXRD are presumably due to crystalline imperfection caused by indium segregation.

Figure 3 shows the typical images of cross-sectional high-resolution TEM (left) and bright-field TEM (right) images from sample TA. The HRTEM image shows a contrast between the well and barrier, exhibiting five periods of InGaN/GaN QWs. The bright-field TEM image also shows a relatively low threading dislocation (TD) density developed along the c-axis from the underlying layer into the $In_xGa_{1-x}N$/GaN active layer. The TD density estimated from the TEM image is about 1.5×10^8. Although not illustrated, as the flow rate of In increased, the peak energy showed a red-shift. This result is well described with In band gap engineering, which means the potential depth of In composition in the well region. However, as the In composition increased, the PL intensity decreased due to the crystalline imperfection caused by In segregation. The peak energy was almost independent of the barrier width in the triangular shaped MQWs, but it showed a red-shift with barrier width in the rectangular shaped ones. This is probably due to the relaxation of the piezoelectric fields in the triangular shaped QWs.

Figure 4 shows the PL spectra of $In_xGa_{1-x}N$/GaN 5MQW structures grown with varying the emission peak energy (i.e., wavelength) from 2.88 to 2.47 eV. The PL intensity showed a highest peak energy of 2.66 eV and the optical property showed a parabolic shape centred at 2.66 eV. It is believed that the optical property in QWs with suitable In composition is improved due to formation of uniform QD-like In composition fluctuation. However, optical property becomes deteriorated due to the crystalline imperfections such as defects and/or impurities caused by increased misfit strain in the QW regions with a relatively higher In composition. Therefore, it is necessary to develop a technique that is able to effectively incorporate much In-content into the well regions.

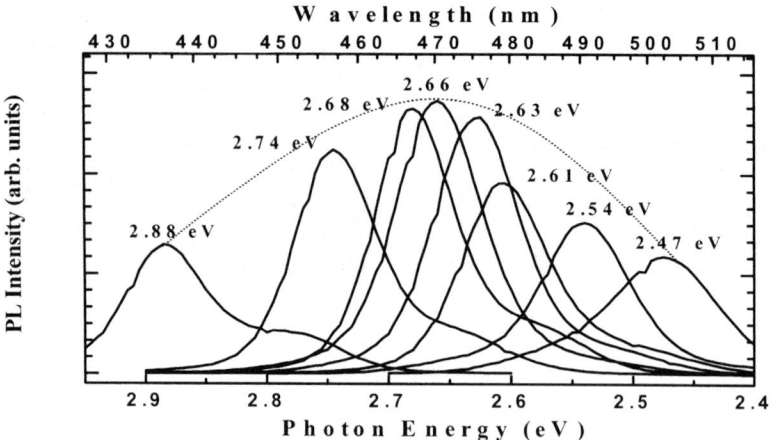

Fig. 4. PL spectra as a function of emission peak energy (i.e., wavelength) measured at
room temperature.

4. CONCLUSIONS

The structural and optical properties of InGaN/GaN triangular-shaped MQW structures were
investigated as functions of growth variables such as growth temperature, flow rate of In
and/or In composition, well width, and barrier width. The peak energy showed a red shift
with decreasing T_g. The peak energy also showed a red shift as the flow rate of In increased.
The peak energy was almost independent of the barrier width in the triangular-shaped MQWs,
but it showed a red shift with an increase in the barrier width in the rectangular-shaped ones.
Also, the PL spectra of $In_xGa_{1-x}N$/GaN 5MQW structures grown by varying the emission
peak energy (wavelength) from 2.88 to 2.47 eV showed strong peak intensities with a
parabolic shape centred at 2.66 eV.

ACKNOWLEDGMENTS

This work was supported by the Brain Korea 21 Project in 2005.

REFERENCES
[1] S. Strite and H. Morkoc, J. Vac. Sci. Technol., **B10** (1992)1237.

[2] S. Nakamura, Science, **281** (1998) 956.

[3] S. J. Pearton, J. C. Zolper, R. J. Shul, and F. Ren, J. Appl. Phys., **86** (1999) 1.

[4] I.-H. Ho and G. B. Stringfellow, Appl. Phys. Lett., **69** (1996) 2701.

[5] L. T. Romano, M. D. McCluskey, C. G. Van de Walle, J. E. Northrup, D. P. Bour, M. Kneissl, T.
Suski, and J. Jun, Appl. Phys. Lett., **75** (1999) 3950.

[6] R. J. Choi, H. W. Shim, M. S. Han, E. K. Suh, H. J. Lee, and Y. B. Hahn, Appl. Phys. Lett., **82**
(2003) 2764.

[7] R. J. Choi, H. J. Lee, Y. B. Hahn, H. K. Cho, Korean J. Chem. Eng., 21, (2004) 292-295.

[8] R. J. Choi, E.-K. Suh, H. J. Lee, and Y. B. Hahn, Korean J. Chem. Eng., **22** (2005) 298-302.

Studies in Surface Science and Catalysis, volume 159
Hyun-Ku Rhee, In-Sik Nam and Jong Moon Park (Editors)

Atomic layer deposition of hafnium silicate thin films using HfCl$_2$[N(SiMe$_3$)$_2$]$_2$

Won-Hee Nam and Shi-Woo Rhee

Laboratory for Advanced Molecular Processing(LAMP), Department of Chemical Engineering, Pohang University of Science and Technology(POSTECH), San 31, Hyoja-dong, Nam-gu, Pohang, Kyung-buk, 790-784, Republic of Korea

1. INTRODUCTION

Silicon dioxide has been used as a primary gate dielectric material in metal-oxide-semiconductor field effect transistors (MOSFETs) for more than 30 years. However, as the dimension of MOSFET devices are scaled down to sub-0.1μm, higher dielectric constant materials are needed to allow the use of physically thicker gate dielectric with electrically equivalent oxide thickness (EOT). Also amorphous structure material with highly stable interface on silicon is required. Among many high-k materials, hafnium silicate material is considered to be the most promising due to its thermodynamic stability in direct contact with silicon up to high temperatures [1,2]. As a deposition process, Atomic layer deposition (ALD) is favored to control the deposition process in an angstrom range. Since ALD is based on self-limiting reactions on the substrate surface, the growth rate depends only on the number of deposition cycles.

Until now, several methods have been reported for ALD of hafnium silicate thin films. However, all these methods require the combination of a hafnium precursor with another silicon precursor and a single precursor for hafnium silicate has been tried recently [3,4].

In our previous work, we reported that a new ALD precursor HfCl$_2$[N(SiMe$_3$)$_2$]$_2$, which contains Si in ligands, can be used to deposit hafnium silicate films using H$_2$O as an oxidant but the Si content in the film was relatively low [4]. In the present work, two different approaches were performed to increase Si content and to improve the film properties. One is to use hydrogen peroxide as a stronger oxidant than water, and the other is to use tetra-n-butyl orthosilicate (TBOS, Si(O^{n}Bu)$_4$) as an additional Si source.

2. EXPERIMENTAL

HfCl$_2$[N(SiMe$_3$)$_2$]$_2$ was synthesized with the reaction of anhydrous HfCl$_4$ and Na[N(SiMe$_3$)] in toluene [5]. The films were grown in a cold-wall flow-type ALD reactor on (100) oriented p-Si substrates in the temperature range of 150-400℃. Prior to deposition, Si substrate was etched in dilute HF solution to remove the native oxide and then rinsed in deionized water. The pressure in the reactor was fixed at about 0.5 torr. Argon (99.99995%) was used as a carrier and purging gas. The valve on/off time was varied to control source injection and purge time. $HfCl_2[N(SiMe_3)_2]_2$ was evaporated at 100°C. Feed lines and shower head are all heated to prevent condensation. Water and hydrogen peroxide (50wt.%) was kept at room temperature and was supplied by its own vapor pressure without any bubbling system.

The film thickness and refractive index were calculated using spectroscopic ellipsometry. X-ray photoelectron spectroscopy (XPS) was used for composition analysis. Auger electron spectroscopy (AES) and secondary ion mass spectroscopy (SIMS) was used to investigate the depth profiles of the film.

3. RESULTS AND DISCUSSION

It was shown that hafnium silicate thin films could be deposited by ALD using $HfCl_2[N(SiMe_3)_2]_2$ with H_2O in the temperature range of 150-400℃. As the deposition temperature increased, the Si content in the film increased but growth rate linearly decreased. However at low temperature, the Si content was as low as Si/(Hf+Si)=0.05 at 150℃, which was close to hafnium dioxide [4]. The surface mechanism of this ALD process was analyzed by in-situ FT-IR [6].

Figure 1 shows the growth rate and silicon content of hafnium silicate films obtained by ALD using H_2O_2 (50wt.%). To show the effect of oxidant, the result of ALD using H_2O was plotted together. Similar growth rate means that the film grows by the exchange reaction of hydroxyl groups with ligands of $HfCl_2[N(SiMe_3)_2]_2$ at low temperature. However, due to strong oxidation effect of hydrogen peroxide, silicon content in the film increased at low temperature and the composition ratio reached to Si/(Hf+Si)=0.2 at 150℃. Since the growth rate decreases mainly due to the decrease of the density of hydroxyl groups on the surface as the deposition temperature increases, the composition of the film shows the same result at high temperature.

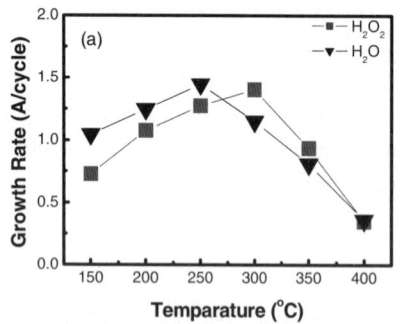 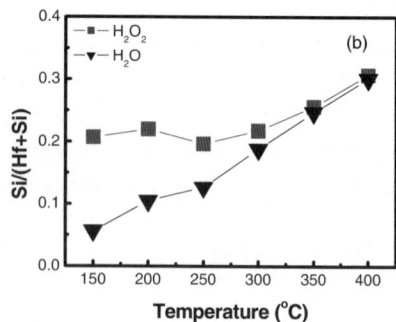

Fig. 1. (a) Growth rate using $HfCl_2[N(SiMe_3)_2]_2$ and H_2O_2 (50wt.%) and (b) composition ratio Si/(Hf+Si) in the film as a function of deposition temperature

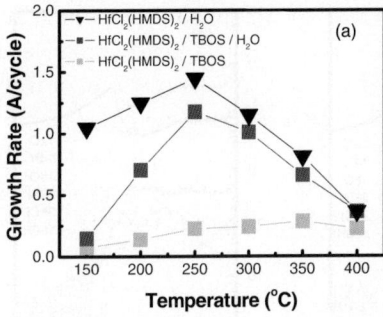 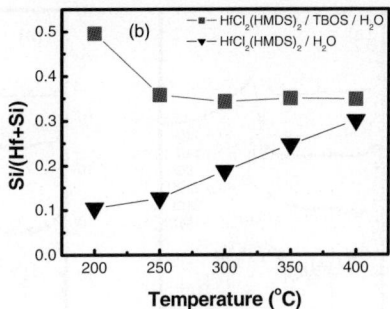

Fig. 2. (a) Growth rate using HfCl$_2$[N(SiMe$_3$)$_2$]$_2$, TBOS and H$_2$O, and (b) composition ratio Si/(Hf+Si) in the film as a function of the deposition temperature

In other studies, the combination of alkoxides and chlorides, or alkoxides and amides without oxidant have been used to deposit hafnium silicate thin films by ALD [3,7].

Since HfCl$_2$[N(SiMe$_3$)$_2$]$_2$ has both chlorines and methylsilazanes as ligands, it may react with another Si alkoxides to form hafnium silicate without oxidant. Tetra-n-butyl orthosilicate (TBOS) was chosen for an alkoxide precursor. Figure 2 shows the growth rate and silicon content of hafnium silicate films obtained by ALD using TBOS as an additional Si precursor with and without H$_2$O. The growth rate was measured after 300 cycles. For the ALD using HfCl$_2$[N(SiMe$_3$)$_2$]$_2$ and TBOS without H$_2$O, the film thickness was the same after 300 cycles and 500 cycles, and the film growth stopped after about 100 Å. This may possibly due to steric hindrance of absorbed molecules. By introduction of H$_2$O as oxidant, hafnium silicate films could be obtained. The low growth rate at low temperature may be due to the thermal and chemical stability of TBOS. TBOS is not thermally decomposed below 650 ℃ [7] and ALD growth with H$_2$O did not take place up to 400 ℃. Silicon composition ratio was Si/(Hf+Si)=0.5 at 200 ℃ and the increased silicon may originate from TBOS.

To investigate composition profile in the film for various processes, SIMS analysis was performed. From the uniform distribution of silicon in figure 3, it is evident Si was not outdiffused from Si substrates, but was incorporated into the films from the precursor. The use of H$_2$O$_2$ instead of H$_2$O decreased the level of impurities such as carbon and chlorine which deteriorate the electrical properties of dielectric films. In figure 3(c), the content of silicon, oxygen and carbon increased, whereas the content of chlorine was decreased. This indicates that TBOS reacts with absorbed HfCl$_2$[N(SiMe$_3$)$_2$]$_2$ molecules during ALD process and TBOS is a source of Si, O and carbon. From AES analysis, not shown in the figure, oxidation state of hafnium silicate film was O/(Hf+Si)=2.0 with TBOS, but O/(Hf+Si)=1.6 without TBOS. The oxygen vacancies of the dielectric film act as fixed oxide charges [9]. The decrease of oxygen vacancies will improve the electrical properties of dielectric films

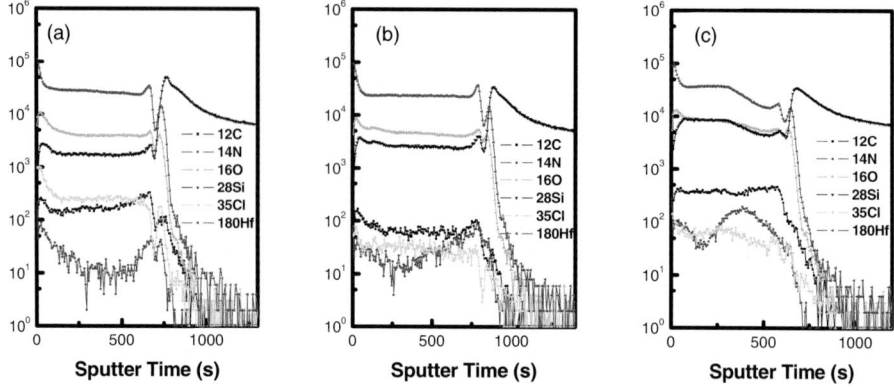

Fig. 3. SIMS depth profiles of hafnium silicate films deposited at 200℃ (a) using $HfCl_2[N(SiMe_3)_2]_2$ and H_2O, (b) using $HfCl_2[N(SiMe_3)_2]_2$ and H_2O_2, and (c) $HfCl_2[N(SiMe_3)_2]_2$, TBOS and H_2O

4. CONCLUSIONS

The low temperature ALD process for hafnium silicate films using $HfCl_2[N(SiMe_3)_2]_2$ and H_2O was modified to improve the film properties by two different methods. With hydrogen peroxide, the silicon content in the film increased to Si/(Hf+Si)=0.2 at 200℃ and the impurity levels decreased due to its strong oxidation effect. By introducing TBOS as an additional Si precursor, the silicon content in the film increased to Si/(Hf+Si)=0.5 at 200℃ and the hafnium silicate films became fully oxidized with O/(Hf+Si)=2.0.

ACKNOWLEDGMENTS

This research was supported by the Korea Science and Engineering Foundation (KOSEF) through the National Research Laboratory Project, and the system IC2010 program of the Korea government.

REFERENCES

[1]. G. D. Wilk, R. M. Wallace and J. M. Anthony, Appl. Phy. Rev., 89 (2001) 5243.

[2]. G. D. Wilk and R. M. Wallace, Appl. Phy. Lett., 74 (1999) 2854.

[3]. W. Kim, S. Rhee, N., J. Lee and H. Kang, J. Vac. Sci. Technol. A, 22(4) (2004) 1285.

[4]. W. Nam and S. Rhee, Electrochem. Solid-state Lett., 7(4) (2004) C55.

[5]. R. A. Andersen, Inorg. Chem., 18 (1979) 2928.

[6]. S. Kang and S. Rhee, J. Vac. Sci. Technol., A22(6) (2004) 2392.

[7]. J. Kim and K. Yong, J. Electrochem. Soc., 152(4) (2005) F45.

[8]. E. J. Kim and W. N. Gill, J. Electrochem. Soc., 142 (1995) 676.

[9]. V. K. Bhat and A. Subrahmanyam, Semicond. Sci. Technol., 15 (2000) 883.

Studies in Surface Science and Catalysis, volume 159
Hyun-Ku Rhee, In-Sik Nam and Jong Moon Park (Editors)

Inductively coupled plasma reactive ion etching of Co$_2$MnSi magnetic films for magnetic random access memory

Byul Shin, Ik Hyun Park and Chee Won Chung

Department of Chemical Engineering, Inha University, 253 Yonghyun-Dong, Nam-Ku, Incheon 402-751, Republic of Korea

1. INTRODUCTION

Magnetic random access memory (MRAM), based on magnetic tunnel junction (MTJ) and CMOS, is a prominent candidate among prospective semiconductor memories because it can provide nonvolatility, fast access time, unlimited read/write endurance, low operating voltage, and high storage density [1-3]. The etching of MTJ stack as well as the deposition of MTJ stack with good magnetic properties is one of the key processes for the realization of high density MRAM. Since the magnetic materials and metals in MTJ stack rarely react with chemically active species in a plasma, it is known that dry etching of magnetic materials and metal films is hard to achieve. Initially, ion milling which utilizes the physical sputtering of Ar ion with high energy was applied to etch the magnetic materials. Since ion milling is limited by factors such as its slow etch rate, sidewall redeposition and etching damages, new etching technique needs to be developed. [4-6].

In this study, the reactive ion etching of Co$_2$MnSi magnetic films with TiN hard mask was investigated in an inductively coupled plasma (ICP) of a Cl$_2$/O$_2$/Ar gas mix. The effects of gas concentration on etch characteristics of Co$_2$MnSi films were explored.

2. EXPERIMENTAL

For the samples used to investigate the etch rates of the Co$_2$MnSi film and the TiN hard mask, the dc magnetron sputtering was employed to deposit Co$_2$MnSi thin films on thermally oxidized Si wafers. The TiN film was deposited on the Si wafers by reactive sputtering using a Ti target and N$_2$ gas. They were patterned by photolithography using the 800 nm-thick conventional photoresist. For the etch profiles study of Co$_2$MnSi films, the samples were deposited on thermally oxidized Si wafers, and then thin TiN films were deposited on top of them. These specimens were then patterned by photolithography and the TiN films were etched by reactive ion etching in a C$_2$F$_6$/Cl$_2$/Ar gas mix. Finally, the photoresist mask was removed by O$_2$ plasma ashing, leaving a patterned TiN mask atop the blanket Co$_2$MnSi films.

The etch experiments were performed using a commercial ICP etcher equipped with an ICP source and a load lock. The substrate susceptor was cooled by He through a chilled fluid. The coil, which was connected to a 13.56 MHz RF power supply, was located on the lid of a ceramic chamber to generate a high density plasma. A bias voltage induced by RF power at

13.56 MHz was capacitively coupled to the substrate susceptor to control the ions' energy in the plasma.

A Dektak surface profilometer was used to measure the etch rates. The profiles of the etched films were observed by field emission scanning electron microscopy (FESEM). In addition, x-ray photoelectron spectroscopy (XPS) was utilized to examine the existence of possible etch products or redeposited materials, and to elucidate the etch mechanism of Co_2MnSi magnetic films in a $Cl_2/O_2/Ar$ plasma.

3. RESULTS AND DISCUSSION

In this study, a Cl_2/Ar gas mix was used as an etching gas since Cl_2 gas was known to be one of the effective etch gases in the etching of magnetic films [5-7]. When the magnetic films were etched using a photoresist mask in a Cl_2/Ar plasma, troublesome results such as sidewall redeposition and shallow etch slope occurred [7]. In an effort to overcome these problems, the high density plasma reactive ion etching of Co_2MnSi films was performed using a hard mask in an $O_2/Cl_2/Ar$ gas mix.

In order to find the proper concentration of Cl_2 gas in an $O_2/Cl_2/Ar$ gas mix, the etch rates of Co_2MnSi films were examined in a Cl_2/Ar. Figure 1 shows the change in etch rate as a function of the concentration of Cl_2 gas in a Cl_2/Ar. The other etch conditions were; coil rf power of 1100 W, dc-bias voltage of 300 V, and gas pressure of 5 mTorr. As the concentration of Cl_2 gas increased, the etch rate rapidly dropped below 60 nm/min. The decrease in etch rate with increasing Cl_2 concentration was attributed to the reduction of ion bombardment onto the specimen due to the decrease of energetic argon ions and/or to the hindrance of the increased chlorine radicals and ions. It is thought that the etching of Co_2MnSi films does not obey the mechanism of reactive ion etching. Figure 2 shows the etch profiles of Co_2MnSi films etched at different Cl_2 concentrations. As the Cl_2 concentration increased, the redeposition on the sidewall of the patterns decreased but the etch slope became slanted.

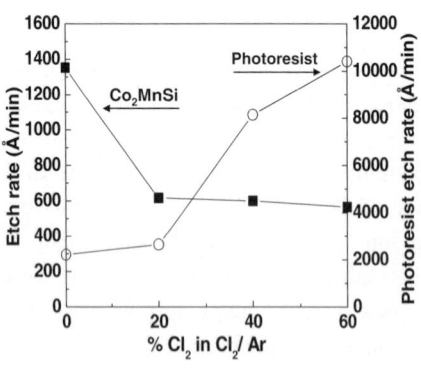

Fig. 1. Etch rates of Co_2MnSi and photoresist as a function of Cl_2 concentration in Cl_2/Ar

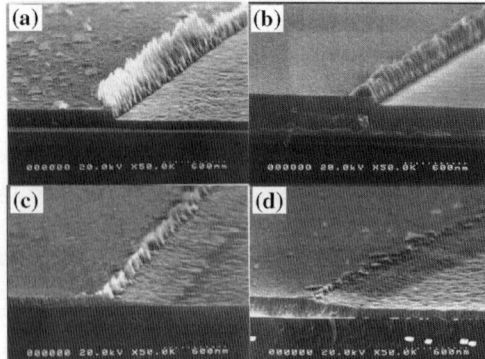

Fig. 2. FESEM micrographs of etched Co_2MnSi using photoresist mask as a function of Cl_2 concentration in Cl_2/Ar; (a) Ar only, (b) 20% Cl_2/Ar, (c) 40% Cl_2/Ar and (d) 60% Cl_2/Ar

The etching was carried out at the constant concentration of 10% and 20% Cl$_2$ in order to obtain the proper etch rates of magnetic films. Figure 3 shows the etch rates of the Co$_2$MnSi film and TiN hard mask on varying O$_2$ concentration in O$_2$/10% Cl$_2$/Ar gas mix. The other etch parameters were constant. The coil power was 1100 W, the dc-bias was 300 V and the gas pressure was 5 mTorr. Although only 10% O$_2$ concentration was added in the 10% Cl$_2$/Ar gas mix, the etch rates of the Co$_2$MnSi films decreased below 20 nm/min. As the O$_2$ gas added up to 20%, the etch selectivity of the Co$_2$MnSi films increased from 0.7 to 5. The increase of etch selectivity with increasing O$_2$ concentration is caused by the fact that the etch rate of the TiN film was more suppressed by O$_2$ addition than that of the Co$_2$MnSi film. Figure 4 shows the XPS spectra of Co$_2$MnSi films etched in 20% O$_2$/10% Cl$_2$/Ar gas mix. The spectrum of as-deposited Co$_2$MnSi film before etching showed Co, Mn and Si peaks and O peak due to the oxidation of films in atmosphere. The spectra of etched surfaces after etching for 0.5 and 1 mins were very similar to that before etching. The Cl peak was not detected even after 1 min etching. It means that neither etch products nor etch residues at this etch condition remained on the sidewall as well as the surface of the films. The possible etch products are CoCl$_X$, MnCl$_X$, SiCl$_4$ and some of them has very high boiling points. The high density plasma helps these nonvolatile products to be desorbed on the film surface and enables one to use the thin hard mask which is effective for clean etch profile. From etch rates results and XPS analysis, it is considered that the etching of Co$_2$MnSi films mainly proceeds by the physical sputtering.

Figure 5 shows the FESEM images of etched Co$_2$MnSi films using a TiN hard mask. The etch condition was; coil power of 1100 W, dc-bias of 300 V and gas pressure of 5 mTorr. The Co$_2$MnSi films etched at 20% Cl$_2$/Ar produced the redeposited materials and the sidewall angle of the etched film was slanted. However, the amount of the redeposition was greatly reduced, comparing to the etching using a photoresist. The clean etch profile was obtained at 60% Cl$_2$/Ar. As the O$_2$ gas was added to a 10% Cl$_2$/Ar, the etch profile became clean and more vertical sidewall was obtained. The etch profile etched in 10% O$_2$/10% Cl$_2$/Ar gas mix

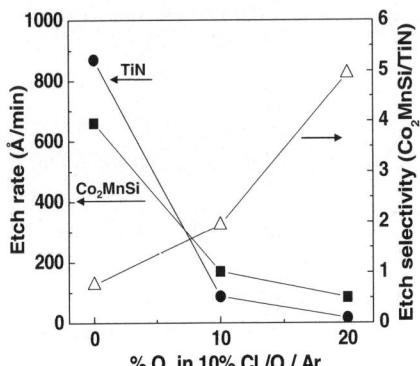

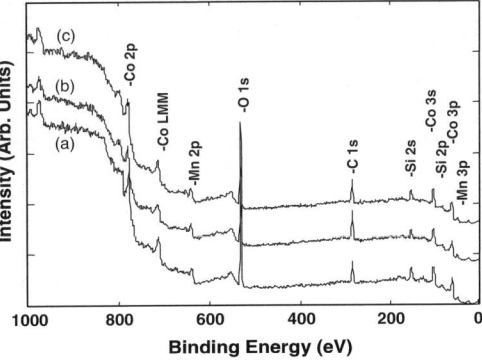

Fig. 3. Etch rate and etch selectivity of Co$_2$MnSi and TiN as a function of O$_2$ concentration

Fig. 4. XPS spectrum of (a) as-deposited Co$_2$MnSi films and XPS spectra of Co$_2$MnSi films etched for (b) 0.5 min, (c) 1 min

showed vertical slope. These results were attributed to the high selectivity of the Co$_2$MnSi film to the TiN hard mask owing to O$_2$ addition. The FESEM image of etched magnetic tunnel junction (MTJ) stack containing the Co$_2$MnSi free layer was seen in Fig. 5(d). The 0.5 x 0.5 μm^2 square arrays of the MTJ stack with Co$_2$MnSi films were successfully defined in 20% O$_2$/10% Cl$_2$/Ar gas mix.

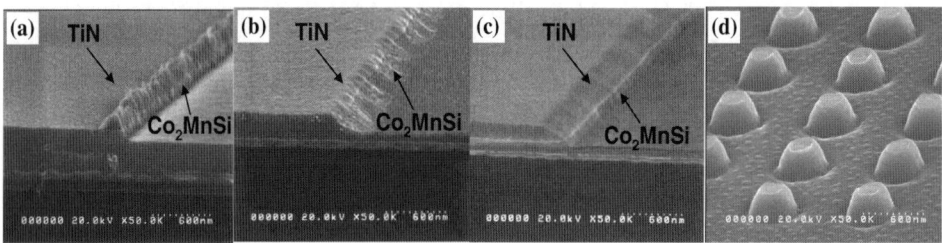

Fig. 5. FESEM micrographs of Co$_2$MnSi films etched in (a) 20% Cl$_2$/Ar, (b) 60% Cl$_2$/Ar, (c) 10% O$_2$/10% Cl$_2$/Ar and (d) FESEM micrograph of etched MTJ stack arrays using TiN mask at 20% O$_2$/10%Cl$_2$/Ar. TiN hard mask were used

4. CONCLUSIONS

Inductively coupled plasma reactive ion etching of Co$_2$MnSi magnetic thin films was carried out using the photoresist and TiN hard mask in a Cl$_2$/O$_2$/Ar gas mix. As the Cl$_2$ gas increased, the etch rate monotonously decreased and etch residues decreased but the etch slope was slanted. The etch rate of the Co$_2$MnSi magnetic films decreased with increasing Cl$_2$ or O$_2$ concentrations. The addition of O$_2$ gas resulted in the enhancement in etch selectivity of Co$_2$MnSi films to TiN hard mask. A SEM observation of Co$_2$MnSi films etched with increasing O$_2$ gas confirmed the improvement of the etch profiles. The highly anisotropic etching of Co$_2$MnSi thin films was accomplished in the etch chemistry of 20% O$_2$/10% Cl$_2$/Ar and optimized etch conditions. It can be concluded that the etching of Co$_2$MnSi films mainly proceeded by physical sputtering in this study. The clean and vertical etch profiles of MTJ stack with Co$_2$MnSi free layer were achieved using thin TiN hard mask in a Cl$_2$/O$_2$/Ar plasma.

REFERENCES

[1] K. Nordquist, S. Pendharkar, M. Durlam, D. Resnick, S. Tehrani, D. Mancini, T. Zhu and J. Shi, J. Vac. Sci. Technol., B 15 (1997) 2274.
[2] S. Tehrani, J.M. Slaughter, E. Chen, M. Durlam and J. Shi, M. DeHerrera, IEEE Trans. Magn., 35 (1999) 2814.
[3] R. C. Sousa and P. P. Freitas, IEEE Trans. Magn., 37 (2001) 1973.
[4] K. Nagahara, T. Mukai, N. Ishiwata, H. Hada and S. Tahara, Jpn. J. Appl. Phys., 42 (2003) L499.
[5] N. Matsui, K. Mashimo, A Egami, A. Konishi, O. Okada and T. Tsukada, Vacuum, 66 (2002) 479.
[6] B. Shin, Y. S. Song, S. J. Park, T. W. Kim and C. W. Chung, Phys. Stat. Sol.(a), 201 (2004) 1644.
[7] K. B. Jung, H. Cho, Y. B. Hahn, E. S. Lambers, S. Onishi, D. Johnson, A. T. Hurst, J. R. Childress, Y. D. Park and S. J. Pearton, J. Appl. Phys, 85 (1999) 4788.

Studies in Surface Science and Catalysis, volume 159
Hyun-Ku Rhee, In-Sik Nam and Jong Moon Park (Editors)

Effect of plasma-induced damage on electrical properties of InGaN/GaN multiple quantum well light-emitting diodes

Chi Won Ok and Yoon-Bong Hahn

School of Chemical Engineering and Technology, Nanomaterials Processing Research Center, Chonbuk National University, Jeonju 561-576, Korea

1. INTRODUCTION

In recent years, there has been remarkable progress for light-emitting diodes (LEDs) and laser diodes (LDs) by utilizing wide band gap III-nitride semiconductors which have a direct band gap, suitable for blue light-emitting devices. Especially, InGaN/GaN multiple quantum well (MQW) LEDs have attracted more attention because InGaN constitutes the active region in the form of quantum well (QW) and emits light by the recombination of electrons and holes injected into the InGaN [1-3]. All of the LEDs and a majority of the LDs have a ridge waveguide structure in which the mesas are formed by a dry etching technique. Plasma etching techniques have been predominantly used in the patterning of the III-nitrides. However, plasma-induced damage often occurs under conditions of high ion flux and energetic ion bombardment [4,5]. Plasma-induced damage can include lattice defects (or dislocations) and formation of dangling bonds on the surface mainly due to energetic ion bombardment, sidewall damage, hydrogen passivation, polymer deposition and etch products deposition, and unequal removal rate of group III and group V elements. Several research groups reported the dry etch damage results for InN, InGaN and InAlN [6-8], GaN Schottky diodes [9,10], and GaN MESFET [11]. However, very little work has been reported for the InGaN/GaN MQW LED structures. The plasma-induced damage can affect the electrical properties of the LEDs and LDs, such as forward turn-on and reverse breakdown voltages. In this paper, we report the ICP-induced damage of the InGaN/GaN MQW LEDs grown by metal-organic chemical vapor deposition in terms of etch rates, surface morphology, and forward and reverse voltages.

2. EXPERIMENTAL

The QWs and LED structures were grown on c-plane sapphire substrates by a low-pressure metal-organic chemical vapor deposition (MOCVD) system. Trimethylgallium (TMGa), trimethylindium (TMIn), ammonia (NH_3), and silane (SiH_4) were used as the precursors of Ga, In, N, and Si, respectively. A GaN nucleation layer of 25 nm thickness was grown on the cleaned substrate at 560 °C, and a 3-μm-thick GaN:Si was then grown at 1130 °C. For fabrication of the InGaN/GaN MQW LED chips, the processing procedures were summarized as: 1) SiO_2 film was deposited by PECVD onto the epiwafer as the etch mask before ICP mesa etching, 2) ICP etching was carried out to form mesa structure with Cl_2/Ar discharges, 3) Au(5 nm)/Ni(7 nm) bi-layer for p-ohmic metal was deposited by e-beam evaporation and lift-off, then alloyed at 500 °C, and 4) Ti(30 nm)/Al(70 nm) bi-layer for n-type contact was deposited and patterned by lift-off. Etching was performed in a planar-type inductively coupled plasma (ICP) system (Vacuum Science ICP etcher, VSICP-1250A), in which the ICP source operated at 13.56 MHz. The ion energy was controlled by the radio frequency (rf) chuck power at 13.56 MHz, applied to the backside helium-cooled chuck. The Cl_2/Ar mixture with total gas flow rate of 40 sccm was injected into the etcher through electronic mass flow controllers. Surface morphologies and etch depth profiles of the fabricated MQW LEDs were analyzed with atomic force microscopy (AFM) operating in tapping mode with Si tip and scanning electron microscopy (SEM), respectively. The forward and reverse I-V characteristics of the InGaN/GaN MQW LED were measured with a semiconductor parameter analyzer (HP 4145B).

3. RESULTS AND DISCUSSION

Figure 1 shows the effect of dc bias (or ion energy) on the forward turn-on (V_F) and reverse breakdown (V_B) voltages (top), along with the etch rate and the surface roughness (bottom). V_F and V_B are the voltages assigned at 20 mA injection current and at 10   A leakage current, respectively, and they are obtained from the I-V measurements. The root mean square (rms) roughness was obtained from the AFM analysis for the etched surface of n-GaN in the mesa structure. The etch rate and the surface roughness increased with the ion energy, typical in ICP etching systems. Little effect on the forward voltage was observed below dc biases of - 180 V. However, the dc bias was increased from - 180 to - 245 V, V_F and V_B somewhat increased and decreased, respectively. This is attributed to a substantial increase in surface roughness (Fig. 1, bottom) and a damage accumulation on the sidewall due to energetic ion bombardment.

Figure 2 shows the SEM micrographs of etch profiles of the LED structures at various dc biases. The depth profile became more anisotropic as the dc bias increased, due to the perpendicular nature of the incident ion energies. However, it is seen that etch products are deposited on the sidewall and covered the MQW active layer. The sidewall contamination became worse at higher dc biases probably because of redeposition of etch products on the sidewall. Hence, the physical degradation of the sidewall along with the rougher surface morphology may help explain the deterioration of the forward and reverse voltages above - 180 V dc bias.

The effects of the ICP source power and etch gas concentration on the the etch rate and electrical properties also have been studied. V_F decreased slightly up to 800 W with the ICP power, showing a minimal effect on the turn-on voltage, but increased at 1000 W because of the plasma damage and the increase in surface roughness. By contrast, the breakdown voltage showed severe degradation at lower and higher ICP powers. Higher ICP powers resulted in faster etch rates and more anisotropic etch profiles, but more degradation and rougher surface morphology were observed due to increased ion scattering and more interact-

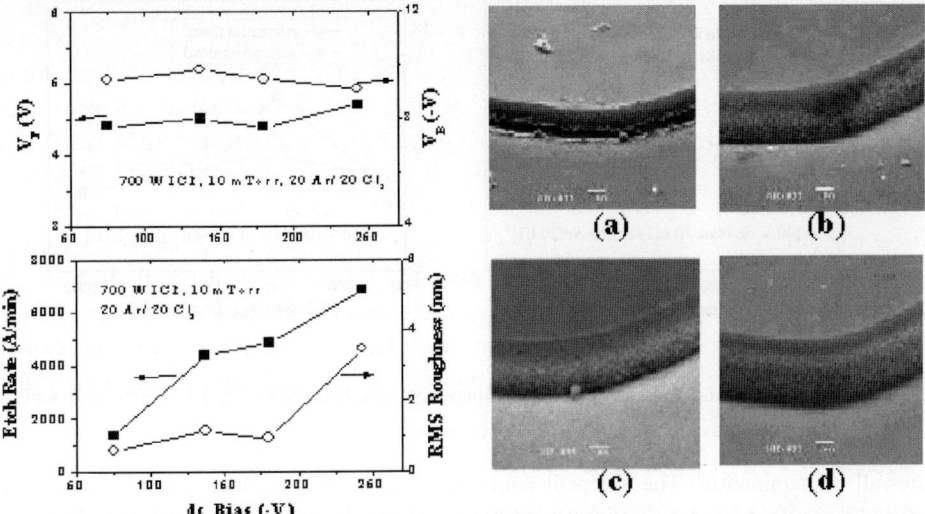

Fig. 1. Effect of rf chuck power on forward reverse voltages (top) and etch rate and surface roughness (bottom).

Fig. 2. SEM micrographs of etch profiles of and InGaN/GaN MQW LED structures, etched at (a) -75, (b) -140, (c) -180, (d) -240 V dc bias.

ions of reactive neutrals with the sidewall of the mesa. The etch rate increased with the Cl_2 concentration because of increased chlorine neutrals, which are dominant reactive species for etching. The etched surface of n-GaN showed the worst roughness under lower chlorine percentage conditions (< 25 % Cl_2), but relatively insensitive to the plasma chemistry up to 75 %, and slightly increased at > 75 %. The forward voltage showed a similar trend to the surface roughness, but the reverse voltage showed the worst degradation at 75 % Cl_2 mainly because of a sidewall contamination.

In order to improve the electrical properties of the MQW LEDs defected by the plasma-induced damage, annealing was carried out under N_2 at 930 °C for 30 s after the ICP mesa etching. The turn-on and breakdown voltages were measured after contact metallization. Figure 3 shows the comparison of the values of V_F (top) and V_B (bottom) for the cases of with and without annealing after etching at 100 W rf chuck power, 10 mTorr and 50 % Cl_2 at various ICP powers. We can see substantial improvements, especially the forward voltage, with the annealing after the mesa etching.

4. CONCLUSIONS

The plasma-induced damage in InGaN/GaN MQW LEDs has been studied in terms of etch behavior and electrical properties. Physical degradation of the sidewall along with the rougher surface morphology deteriorated the forward and reverse voltages at a higher dc bias voltage. The experimental results led to a conclusion that the forward turn-on voltage is more sensitive to the surface roughness and the reverse breakdown voltage is strongly affected by

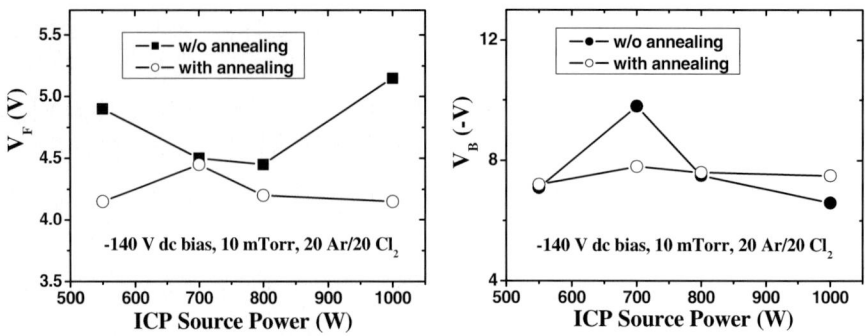

Fig. 3. Forward (top) and reverse (bottom) voltages of InGaN/GaN MQW LED with and without annealing after etching as a function of ICP source power.

sidewall contamination. The sidewall contamination by etch products on the active layer induced a severe degradation of the breakdown voltage. However, a thermal annealing after the mesa etching showed a substantial improvement of the electrical property, especially the forward voltage.

ACKNOWLEDGMENTS

This work was supported by the Korea Research Foundation (Grant KRF-2003-005-C00016)

REFERENCES

[1] S. Strite and H. Morkoc, J. Vac. Sci. Technol., **B10** (1992)1237.

[2] S. Nakamura, Science, **281** (1998) 956.

[3] S. J. Pearton, J. C. Zolper, R. J. Shul, and F. Ren, J. Appl. Phys., **86** (1999) 1.

[4] R. J. Choi, H. J. Lee, Y. B. Hahn, H. K. Cho, Korean J. of Chem. Eng., **21**, (2004) 292-295.

[5] H. J. Park, H.-W. Ra, K. S. Song, Y. B. Hahn, Korean J. Chem. Eng., **21**, (2004) 1235-1239.

[6] S. J. Pearton, Appl. Surf. Sci., 117/118, 597 (1997).

[7] F. Ren, J. Lothian, Y. K. Chem, J. D. MacKenzie, S. M. Donovan, C. B. Vartuli, C. R. Abernathy, J. W. Lee, and S. J. Pearton, J. Electrochem. Soc., 143, 1217 (1996).

[8] X. A. Cao, A. P. Zhang, G. T. Dang, F. Ren, S. J. Pearton, R. J. Shul, and L. Zhang, J. Vac. Sci. Technol. A, 18, 1144 (2000).

[9] R. J. Shul, L. Zhang, A. G. Baca, C. G. Willison, J. Han, S. J. Pearton, and F. Ren, J. Vac. Sci. Technol. A, 18, 1139 (2000).

[10] D. G. Kent, K. P. Lee, A. P. Zhang, B. Luo, M. E. Overberg, C. R. Abernathy, F. Ren, K. D. MacKenzie, S. J. Pearton, and Y. Nakagawa, Solid-State Electron., 45, 1837 (2001).

[11] R. J. Shul, L. Zhang, A. G. Baca, C. G. Willison, J. Han, S. J. Pearton, K. P. Lee, and F. Ren, Solid-State Electron., 45, 13 (2001).

Studies in Surface Science and Catalysis, volume 159
Hyun-Ku Rhee, In-Sik Nam and Jong Moon Park (Editors)

Characteristics of Tin Oxide Thin Films on a Poly Ethylene Terephthalate Substrate Prepared by Electron Cyclotron Resonance-Metal Organic Chemical Vapor Deposition

Y. H Kook[a,b], D. J Byun[b], B. J Jeon[a] and J. K Lee[a]

[a]Eco Nano Research Center, Korea Institute of Science and Technology, P.O.Box 131, Cheongryang Seoul 130-650, Republic of Korea
[b]Department of Material Science & Engineering, Korea University[*]

1. INTRODUCTION

Room temperature deposition of transparent conducting oxide (TCO) is considered to be an important subject for potential applications for the solar cells, flexible electronic devices, and flat panel display [1]. The TCO in the form of thin films on a plastic substrate replaced by glass provides many advantages due to lighter weight, smaller volume, lower cost and flexibility. The ECR-MOCVD (Electron Cyclotron Resonance - Metal Organic Chemical vapor Deposition) method has some unique plasma characteristics such as high ion density and electron temperature. This technique uses the microwave power to ionize the source gases in the ECR zone with a magnetic field of 875 Gauss for resonance. The ions are directed to the substrate by the divergent magnetic field [2]. In this work, the relationship between characteristics of tin oxide film and ECR-MOCVD process parameters such as microwave power, magnet current, and deposition time is investigated.

2. EXPERIMENTS

Then microwaves at frequency of 2.45 GHz (magnetic flux density, 875 Gauss) were introduced through a rectangular guide into the plasma chamber to generate plasma. Substrate of PET (polyethylene terephthalate) film with dimension of 10×10 cm is rotated during the tin oxide film deposition. The precursor TMT (tetra-methyl tin) with purity 99.999% was used as the organometallic source. The carrier gas Ar flowed through the bypass line of the TMT bubbler until the reactor was stabilized; then it flowed through the TMT bubbler which was maintained at -14°C in order to carry TMT vapor into the reactor. At the same time, O_2 gas was introduced directly into the reaction chamber. Characterization analyses of the transparent conductive oxide thin films are carried out by using SEM (scanning electron microscopy)- morphology, TEM (transmission electron microscopy), 4PPM (4 points probe method) - resistance and resistivity, UV-visible spectrophotometer (UV) - transmittance and reflectance. The chemical structure of the films was observed with Fourier transform infrared (FT-IR).

3. RESULTS & DISCUSSION

Fig.1 (a) shows scanning electron micrograph of SnO_2 films on PET film substrate deposited by ECR-MOCVD. The optimized processing conditions were 1600 W of microwave power, 160 A of magnetic current power, 10 sccm of H_2, 20 sccm of O_2, 0.25 sccm of TMT, 3 sccm of Ar, 8 cm of the distance from magnet to TMT feeding point, 10 cm of the distance from TMT feeding point to substrate, 20 mTorr of working pressure, and 5 min of deposition time. Through optimization of the process variables, the tin oxide film with very fine grains of 30~50 nm with film thickness of 140 nm were obtained at the optimum condition. Fig.1 (b) shows electron diffraction pattern (Mag. 6100) of SnO_2 films on PET film substrate deposited by ECR-MOCVD. The diffraction pattern of SnO_2 film reveals the characteristic 110, 101, 211 and 112 rings of polycrystalline structure of tin oxide.

Fig.2 shows the infrared absorption spectrum of the tin oxide film. In order to analyze the molecular structure of the deposited film, we deposited the tin oxide film on a KBr disc with thickness of 1 mm and diameter of 13 mm. Various peaks formed by surface reaction are observed including O-H stretching mode at 3400 cm^{-1}, C=C stretching mode at 1648 cm^{-1}, and SnO_2 vibration mode at 530 cm^{-1}. The formation of sp^2 structure with graphite-like is due to ion bombardment with hydrogen ions at the surface and plasma polymerization of methyl group with sp^3-CH_2.

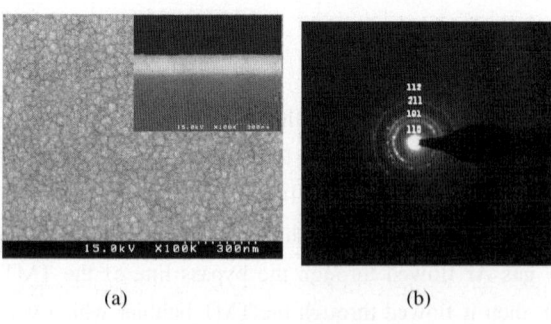

(a)   (b)

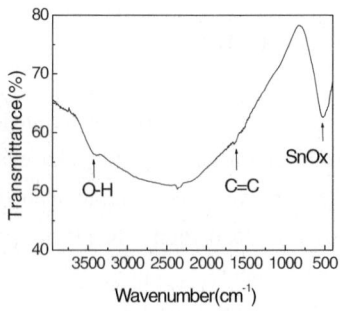

Wavenumber(cm^{-1})

Fig.1. Scanning electron micrography (a) and electron diffraction pattern (b) of SnO_2 film on PET substrate prepared by ECR-MOCVD.

Fig.2. FT-IR spectrum of SnO_2 film on a KBr disc.

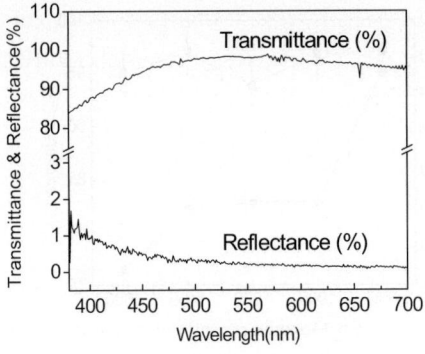
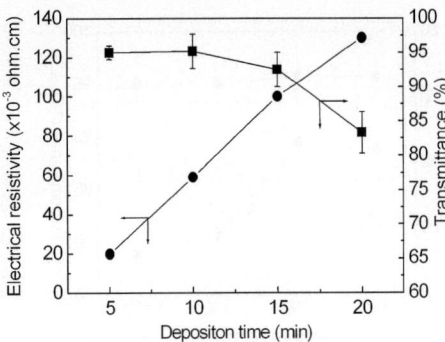

Fig. 3. Transmittance and reflectance of SnO₂ film prepared by ECR-MOCVD.

Fig. 4. Effect of deposition time on transmittance and electrical resistivity of SnO₂ film.

Fig. 3 shows the transmittance and reflectance of the deposited film on the PET substrate in the visible range of 380-780 nm. The transmittance and reflectance of the films prepared at the optimized processing condition exhibited the values of 83-98 % and 0.1-0.5 % respectively. Fig. 4 shows the effect of deposition time on electrical resistivity and transmittance of tin oxide film. The electrical resistivity of the film for 5 min deposition time showed 2×10^{-2} Ω·cm. The transmittance of the films exhibited their average values of 84~96 % at range of 380-780 nm. After 10 min of deposition time, some cracks on the surface of SnO₂ film were observed. The electrical resistivity increases with the deposition time. The corresponding films thickness is 140nm at 5 min and 500 nm at 20 min of deposition time.

Fig. 5 shows effect of magnet current on electrical resistivity and transmittance of tin oxide film. The electrical resistivity decreases from 4.4×10^{-2} to 1.2×10^{-2} Ω·cm with an increase of magnet current, while variation of transmittance with the magnet current was insignificant in our experimental range. The increment of magnet current brings on an increase of ionization because magnetic intensity increases the electron path and raises the possibility of inelastic collisions. These ions, arriving at the surface with a high energy, can sufficiently import momentum to the surface atoms of the growth film. Thus, low electrical resistivity of deposited film result in the formation of films with tightly packed columns. Fig. 6 shows the effect of microwave power on electrical resistivity and transmittance of SnO₂ film. As shown in the figure, the electrical resistivity sharply decreases and then it keeps constant values at the range of 1000 to 1400 W. The lowest electrical resistivity 2×10^{-2} Ω·cm was obtained at 1600 W. So the electrical resistivity showed a decrease of ten times as compared with the value at 800 W.

388

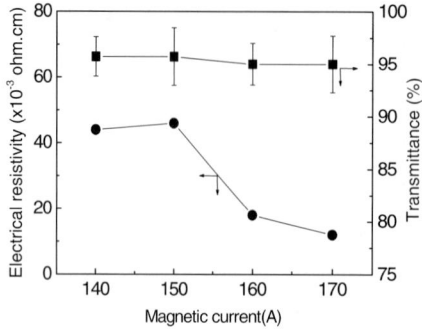 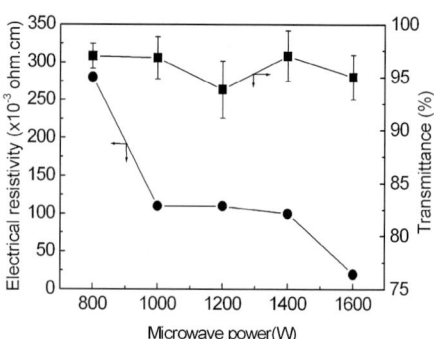

Fig. 5. Effect of magnet current on electrical resistivity and transmittance of SnO_2 film. Microwave power of 1600 W.

Fig. 6. Effect of microwave power on electrical resistivity and transmittance of SnO_2 film. Magnet current of 160 A.

An increase in microwave power drives an increase of ionization density, delivers the higher ion energy on the PET substrate, and thus enhances the surface reaction. The electrons gain more energy for ionization of molecules; the electron energy is almost transferred to the molecules during the inelastic collision, leading to reduction in the electron energy [3].

4. CONCLUSION

Tin oxide films on PET film were prepared by ECR-MOCVD at room temperature and effects of process parameters on the characteristics of the films were investigated. Electrical resistivity of the films showed 2×10^{-2} $\Omega \cdot$cm. The transmittance and reflectance of the films exhibited their average values of 83-98 % and 0.1-0.5 % at wave length range of 380-780 nm, respectively. In our experimental range, increase in microwave power and magnet current brought on formation of SnO_2 film with low electrical resistivity.

REFERENCE

[1] Y.-S Kim, Surface and Coatings Technology 173 (2003) 299.

[2] J. K. Lee, H. D. Ko, J. Hyun, D. J. Byun, B. W. Cho and D. K. Park, in: E. Manias, G.G. Malliaras (Eds.), Polymer/Metal Interfaces and Defect Mediated Phenomena in Ordered Polymers, Boston, U.S.A., December 2-6, 2002, Materials Research Society Symposium Proceeding 734 (2002), B9.49.1

[3] M.A. Lieberman and A.J. Lichtenberg, "Principles of plasma discharges and materials processing", John Wiley & Sons, Inc., 1994.

ENVIRONMENTAL
REACTION ENGINEERING

Studies in Surface Science and Catalysis, volume 159
Hyun-Ku Rhee, In-Sik Nam and Jong Moon Park (Editors)

Mineralization of Indigo Carmine at Neutral pH Using a Nanocomposite as a Heterogeneous Photo–Fenton Catalyst

Jiyun Feng, Xijun Hu, and Po Lock Yue[*]
Department of Chemical Engineering,
Hong Kong University of Science and Technology, Clear Water Bay, Kowloon, Hong Kong
[*] keplyue@ust.hk

Mineralization of 0.2 mM Indigo Carmine using bentonite clay-based Fe nanocomposite (Fe-B) as a heterogeneous photo-Fenton catalyst at different initial solution pH in the presence of 10 mM H_2O_2 and 8W UVC was studied in detail. It was found that the Fe-B catalyst exhibited a reasonably good photo catalytic activity at a neutral solution pH of 7.0 (TOC removal can reach 60%), suggesting that it becomes feasible for the heterogeneous photo Fenton process to treat the original wastewater without the pre-adjustment of the initial solution pH. The rapid decrease in solution pH in the first 60 minutes is caused by the formation of acidic intermediates.

1. Introduction

It is well known that one of the most important advanced oxidation processes in wastewater treatment is the photo Fenton reaction. It is also well accepted that initial solution pH can significantly influence the efficiency of the photo Fenton reaction, and the optimal solution pH for the homogeneous as well as heterogeneous photo Fenton reactions has been determined to be 2.8-3.2 [1-3]. However, it should be pointed out that pre-adjustment of the initial solution pH of wastewater is a costly process, which uses a lot of chemicals and human resources. Therefore, it has academic significance and practical industrial application to explore whether a heterogeneous photo Fenton catalyst works at a neutral initial solution pH or not. We report here the mineralization of Indigo Carmine at neutral pH using a bentonite clay based Fe nanocomposite (Fe-B) as a heterogeneous photo Fenton catalyst.

2. Experimental

The Fe-B nanocomposite was synthesized by the so-called pillaring technique using layered bentonite clay as the starting material. The detailed procedures were described in our previous study [4]. X-ray diffraction (XRD) analysis revealed that the Fe-B nanocomposite mainly consists of Fe_2O_3 (hematite) and SiO_2 (quartz). The bulk Fe concentration of the Fe-B nanocomposite measured by a JOEL X-ray Reflective Fluorescence spectrometer (Model: JSX 3201Z) is 31.8%. The Fe surface atomic concentration of Fe-B nanocomposite determined by an X-ray photoelectron spectrometer (Model: PHI5600) is 12.25 (at%). The BET specific surface area is 280 m^2/g. The particle size determined by a transmission electron microscope (JOEL 2010) is from 20 to 200 nm.

The mineralization of 0.2 mM Indigo Carmine was conducted in a batch photo reactor (0.5 L) in the presence of 8W UVC light and 10 mM H_2O_2 at different initial solution pHs and 30 °C, as described in our previous study [4]. The total organic carbon (TOC) of the reaction solution was measured by a Shimadzu 5000A TOC Analyzer equipped with an auto-sampler.   The solution pH as a function of time was measured by a pH meter (Model: Thermal Orion 420).   To monitor Fe leaching from Fe-B catalyst, the Fe ion concentration in reaction solution as a function of time was measured by ICP.

3. Results and Discussion

Before the mineralization experiment, dark adsorption of 0.2 mM Indigo Carmine on the 1.0 g/L Fe-B nanocomposite at different initial solution pHs was investigated at 30 °C, and the result is shown in Fig. 1.   It is seen that the initial solution pH can significantly influence the adsorption of the Indigo Carmine on the surface of the Fe-B nanocomposite.   As the initial solution pH increases, the amount of Indigo Carmine adsorbed on the surface of the Fe-B nanocomposite markedly decreases.   For example, at the initial solution pH of 3.0, the TOC removal after 120 minutes attains around 20 % while it is only about 10% when the initial solution pH is 7.0.   The results indicate that the TOC removal due to adsorption cannot be ignored.   In addition, the difference in TOC removal due to the different initial solution pHs helps us gain a deeper understanding of the mineralization of Indigo Carmine.

Fig. 2. displays the TOC removal of 0.2 mM Indigo Carmine as a function of time under different initial solution pHs in the presence of 10 mM H_2O_2, 1.0 g Fe-B/L, and 8W UVC. As the initial solution pH increases from 3.0 to 7.0, the mineralization kinetics becomes slower, indicating that the Fe-B nanocompsoite showed a decreased photo catalytic activity. For example, the difference between the TOC removal at initial solution pH of 3.0 and 7.0 is about 25%, which agrees well with previous studies [1-3].

However, it should be noticed that even we started the reaction at a neutral solution pH of 7.0, the TOC removal of 0.2 mM Indigo Carmine can also be achieved over 60% after 120 minutes reaction, implying that the Fe-B nanocomposite exhibited a reasonable good activity when the initial solution pH is neutral.   The result also reveals that pre-adjustment of initial solution pH may not be necessary, and makes it feasible for the Fe-B nanocomposite to be applied to real industrial application for wastewater treatment.

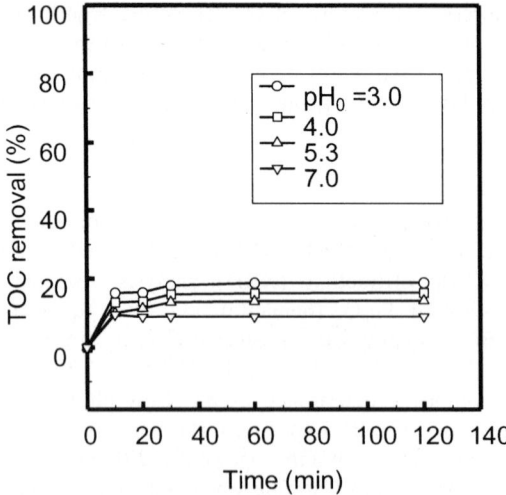

Fig. 1. TOC removal of 0.2 mM Indigo Carmine due to dark adsorption on the Fe-B nanocomposite.

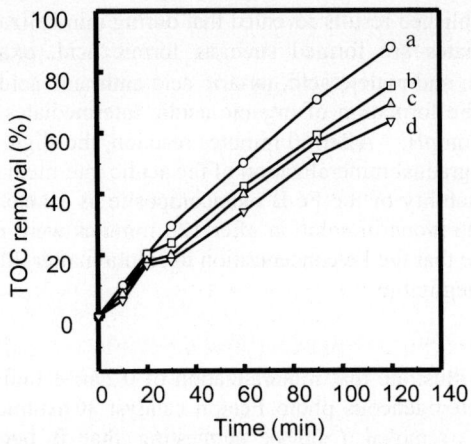

Fig. 2. TOC removal of 0.2 mM Indigo Carmine in the presence of 1.0 g Fe-B/L, 10 mM H_2O_2, 8W UVC, and at 30 °C.   Initial solution pH: (a) 3.0, (b) 4.0, (c) 5.3, and (d) 7.0.

In order to explain the results shown in Fig. 2., the solution pH as a function of time was measured by a pH meter, and the results are displayed in Fig. 3.   As can be seen from the data in the figure, when the initial solution pH is over 3.0, as reaction time increases, the solution pH decreases very fast to around 3.5 within 10 minutes, then reaches a lowest value of around 3.2 at 60 minutes, followed by a slight increase.   What causes this interesting phenomenon?   We believed that the rapid decrease in solution pH is due to the fact that acidic intermediates are formed during the mineralization of 0.2 mM Indigo Carmine. Because of its complicated molecular structure, theoretically, it cannot be oxidized completely into CO_2 and H_2O in one step.

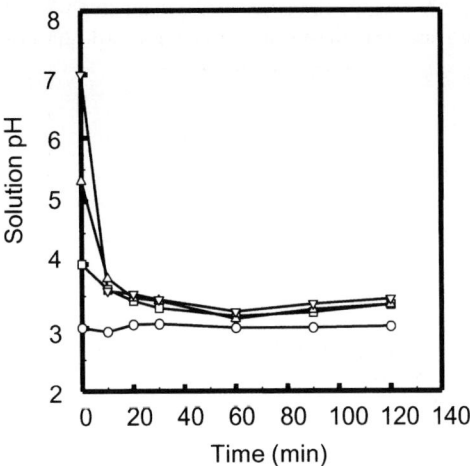

Fig. 3. Solution pH as a function of time during mineralization of 0.2 mM Indigo Carmine Carmine in the presence of 10 mM H_2O_2, 1.0 g Fe-B/L, 8W UVC, and at 30 °C.

In fact, previous published results revealed that during mineralization of Indigo Carmine, many acidic intermediates are formed such as formic acid, oxalic acid, glycolic acid containing two carbons; and maleic acid, tartaric acid anthranic acid containing Cn ($n \geq 4$), etc (5). Apparently, the formation of organic acidic intermediates definitely resulted in a rapid decrease in solution pH. After 60 minutes reaction, the slight increase in solution pH could be caused by the gradual mineralization of the acidic intermediates into CO_2 and H_2O.

To determine the stability of the Fe-B nanocomposite as a heterogeneous photo Fenton catalyst, the Fe concentrations in solution after 120 minutes were measured by ICP. The results obtained indicate that the Fe concentration in solution after 20 minutes reaction is less than 1 mg/L, which is negligible.

4. Conclusions

Our results clearly illustrate that mineralization of 0.2 mM Indigo Carmine using Fe-B nanocomposite as a heterogeneous photo Fenton catalyst at neutral solution pH can reach reasonable good TOC removal (>60%), suggesting that it becomes feasible for the heterogeneous photo Fenton process to treat the original wastewater without the pre-adjusting the solution pH. The rapid decrease in solution pH in the first 60 minutes is caused by the formation of acidic intermediates.

Acknowledgements
The financial support of a grant (ITS176/01C) from Hong Kong Innovation Technology Funding is gratefully acknowledged.

References
[1] J. J. Pignatello, Environ. Sci. Technol., 26 (1992) 944-951.
[2] J. Fernandez, J. Bandara, A. Lopez, Ph.Buffar and J. Kiwi, Langmuir., 15 (1) (1999) 185-192.
[3] JY. Feng, XJ. Hu, P. L. Yue, HY. Zhu and GQ. Lu, Ind. Eng. Chem. Res., 42 (2003) 2058-2066.
[4] JY. Feng, XJ. Hu and P. L. Yue, Environ. Sci. & Technol., 38 (2004) 265-275.
[5] M. Vautier, C. Guillard and J-M. Herrmann, J. Catalysis., 201 (2001) 46-59.

Studies in Surface Science and Catalysis, volume 159
Hyun-Ku Rhee, In-Sik Nam and Jong Moon Park (Editors)
© 2006 Elsevier B.V. All rights reserved

Catalytic wet peroxide oxidation of dyehouse effluents with Cu/Al$_2$O$_3$ and copper plate

Sung-Chul Kim and Dong-Keun Lee

Dept. of Chemical Engineering, Environmental Biotechnology National Core Research Center, Environmental and Regional Development Institute, Gyeongsang National University, Kajwa-dong 900, Jinju, Gyeongnam 660-701, Korea

1. INTRODUCTION

The effluents from the textile dyeing industries impose serious environmental problems because of their color and their high chemical oxygen demand (COD). The discharge of highly colored waste is not only aesthetically displeasing, but it also interferes with the transmission of light and upsets the biological processes which may then cause the direct destruction of aquatic communities present in the receiving stream. The removal of color and COD from dyehouse wastewater to meet the discharge standards is currently a major problem in the textile industry.

As a potential alternative to incineration and biological treatments catalytic wet oxidation has been the subject of numerous investigations to reduce the amount of organic effluents in wastewaters [1-5]. The reaction is carried out under different conditions, depending on the type of oxidant (O$_2$, O$_3$, H$_2$O$_2$). Catalytic wet oxidation with H$_2$O$_2$ (CWPO) is a more efficient process due to the strong oxidizing properties of hydrogen peroxide, and therefore the reaction is performed in mild conditions.

In the present work CWPO of dyehouse effluent was carried out in a batch reactor with 1L capacity and in a pilot plant scale continuous flow reactor with 5m^3/day treatment capacity. Cu/Al$_2$O$_3$ and Copper plate were used as the catalysts.

2. EXPERIMENTAL

A real effluent, produced from the washing process of a certain dyeing industry, was employed for the catalytic wet oxidation. In order for dyeing textile substrates the industry had used the aqueous solution of the mixture of reactive black 5, reactive blue 19 and reactive red 198. In addition small amounts of some penetrating agents together with NaOH were contained in the effluent. The dark black reddish effluents had TOC values of 6,900 mg/L~8,600 mg/L and their color units were in the range between 4,900~5,500. Cu/Al$_2$O$_3$ catalyst was prepared by incipient wetness method. Cu plate was a commercial product and had 1 cm length, 1 cm width and 0.1 cm thickness. The catalytic wet peroxide oxidation of

Figure 1. Pilot plant scale continuous flow reactor for wet oxidation of dyehouse effluent.

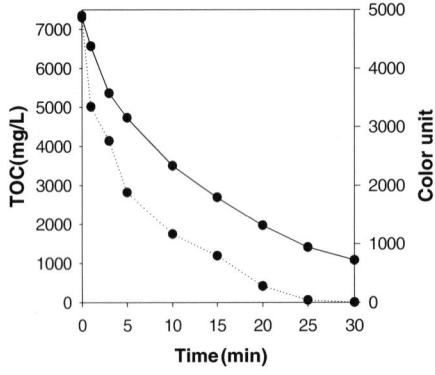

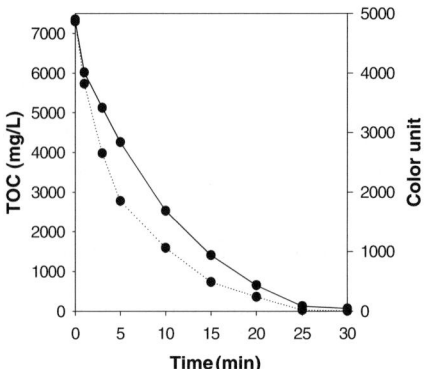

Figure 2. Removal of TOC(—) and color(---) during the catalytic wet oxidation of the real dyehouse effluent with 10g 10wt% Cu/Al$_2$O$_3$

Figure 3. Removal of TOC(—) and color(---) during the catalytic wet oxidation of the real dyehouse effluent with 10g copper plate.

dyehouse effluent was performed in a batch reactor with 1L capacity and in a continuous flow reactor with 5m^3/day treatment capacity (Figure 1). The reactions were conducted at atmospheric pressure and 80℃. Liquid sample were immediately filtered and analyzed for total organic carbon (TOC), hydroxyl radical (HO·), color unit and residual materials in water.

3. RESULTS AND DISCUSSION

Figures 2 and 3 show the time dependences of the removal of TOC and color during the catalytic wet oxidation with 20ml 0.5N H$_2$O$_2$ in the presence of 10wt% Cu/Al$_2$O$_3$ and copper plate catalyst in the batch reactor, respectively. While more than 80% TOC and most color were removed after 30min reaction with the 10wt% Cu/Al$_2$O$_3$ catalyst, almost complete removal of both the TOC and color could be obtained with the copper plate within 30min reaction. The fast reaction rates of the catalytic wet oxidation with H$_2$O$_2$ as opposed to the uncatalyzed thermal oxidation and catalytic wet oxidation with air seem to be due to the decomposition of H$_2$O$_2$ to give two hydroxyl radicals which react with the dyes in water.

In Figure 4 is shown the removal of TOC together with the concentration of H$_2$O$_2$ consumed and HO· produced during the reaction in the presence of the copper plate. The removal of TOC was shown to be strongly related to the consumption of H$_2$O$_2$ which will be decomposed into HO·. Similar results were also obtained from 10wt% Cu/Al$_2$O$_3$ catalyst.

Separate experiments of H$_2$O$_2$ decomposition in the absence of any reactive dye were carried out at the same reaction condition. The concentration of H$_2$O$_2$ was the same as that in Figure 4. The measured changes in the concentration of H$_2$O$_2$ and HO· are plotted in Figure 5. As seen, in accordance with the consumption of H$_2$O$_2$ the formation of HO· occurs during the reaction. An interesting feature in Figure 5 is that the rates of both the H$_2$O$_2$ consumption and HO· production were increased significantly by the action of the catalysts, especially copper plate, which must have played an important role on the activation of H$_2$O$_2$ decomposition and the subsequent HO· formation. The subtracted amounts of HO·, corresponding to the differences between HO· formed in Figure 5 and HO· remained in Figure 4 must have participated in the oxidation of organic compounds in the effluents.

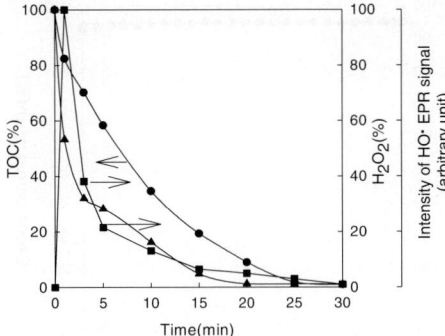

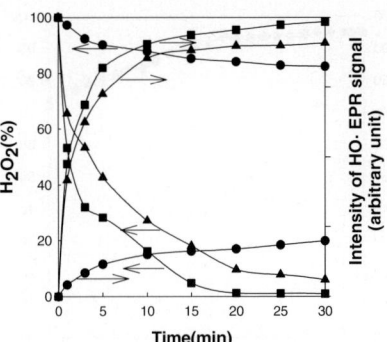

Figure 4. Correlation between TOC removal($\bullet$), H_2O_2 consumption($\blacktriangle$) and HO· formation($\blacksquare$) during the catalytic wet peroxide oxidation of dyehouse effluent with 10g copper plate.

Figure 5. Time dependence of H_2O_2 conversion and HO· formation during H_2O_2 decomposition in the absence of the catalyst($\bullet$), in the presence of 10g 10wt% Cu/ Al_2O_3($\blacktriangle$) and 10g copper plate($\blacksquare$). (The reactive dye was excluded in these experiments)

On the basis of the results obtained in a batch reactor analysis, optimum operation conditions of a pilot plant scale continuous flow reactor were determined. The flow rates of the dyehouse effluent and 0.5N H_2O_2 into the reactor were kept to be 3.5L/min and 10mL/min, respectively. 5 kilograms of the catalysts were loaded inside the reactor. Figure 6 shows the performance of the 10wt% Cu/Al_2O_3 catalysts toward wet oxidation of the dyehouse effluent having 8,400 mg/L TOC and 5,300 color unit. More than 95% TOC and color could be removed up to 70 h continuous operation. After that time, however, the removal efficiency of the both TOC and color decreased continuously. The main reason for the decrease of the catalytic efficiency of Cu/Al_2O_3 was proved to be due to the considerable copper loss during the reaction which might have occurred from the leaching out of the copper component through the formation of copper hydroxide precipitates. After 470 h reaction, the copper concentration in the used catalyst was measured again with ICP spectrometer. The percentage of copper loss was estimated to be 76.8%. When using the copper plate instead of the 10wt% Cu/ Al_2O_3 as catalyst, more than 95% conversion of initial TOC and color could be successfully and stably maintained for the whole reaction time (Figure 7). Less than 1% copper loss from the copper plate was measured, which indicates that the copper plate is very practical catalyst for the wet oxidation of dyehouse effluents.

4. CONCLUSION

Catalytic wet oxidation with H_2O_2 using Cu/Al_2O_3 and copper plate was employed for the treatment of the dark black reddish dyehouse effluents. The removal efficiency of both TOC and color was strongly related to the consumption of H_2O_2 and formation of hydroxyl radical. During the continuous operation of the wet oxidation the Cu/ Al_2O_3 catalysts lost their activity significantly. The activity loss was proved to be due to the leaching out of copper component

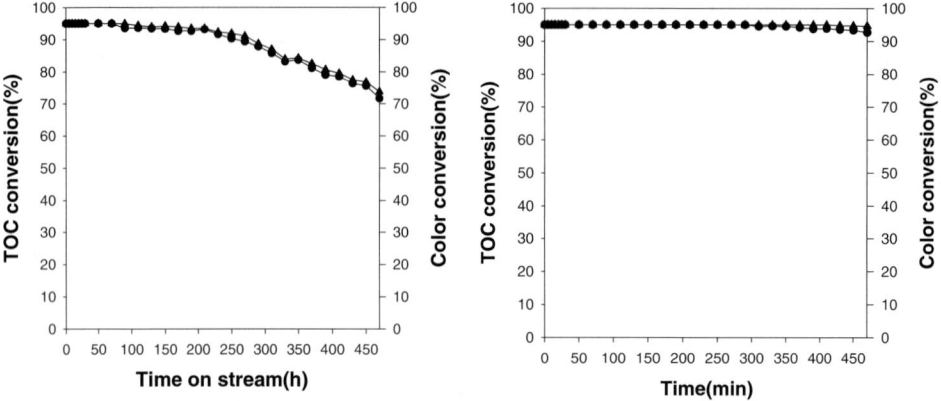

Figure 6. Removal of TOC(●) and color(▲) during catalytic wet oxidation of a real dyehouse effluent with 10wt% Cu/Al₂O₃ in the continuous flow system.

Figure 7. Removal of TOC(●) and color(▲) during catalytic wet oxidation of a real dyehouse effluent with copper plate in the continuous flow system.

from the catalysts. The copper plate, however, could oxidize more than 95% conversion of TOC and color of dyehouse effluent, and was stable against copper leaching.

ACKNOWLEDGEMENTS

This work was supported by Korea Research Foundation Grant (KRF-2004-041-D00185).

REFERENCES

[1] F. Luck, Catal. Today, 27 (1996) 195.
[2] J. Levec, A. Pintar, Catal. Today, 24 (1995) 51.
[3] P. Gallezot, N. Laurain, P. Isnard, Appl. Catal. B, 9 (1996) L11.
[4] D. Duprez, F. Delanoe, J. Barbier Jr, P. Isnard, G. Blanchard, Catal. Today, 29 (1996) 317.
[5] D.-K. Lee, D.-S. Kim, Catal. Today, 63 (2000) 249.

Studies in Surface Science and Catalysis, volume 159
Hyun-Ku Rhee, In-Sik Nam and Jong Moon Park (Editors)
397

A novel process to produce chlorine-free fuel gas and char from waste PVC and waste glass

H.-J. Sung, T. Hirotani, A. Honya, R. Noda and M. Horio

Department of chemical engineering, Graduate School of Bio-Applications and Systems Engineering, Tokyo University of Agriculture and Technology, 24-16-2 Naka-machi, Koganei-shi, Tokyo 184-8588, Japan

1. INTRODUCTION

In Japan the amount of waste poly(vinyl chloride) (PVC) has been sharply increasing since disposal of building materials such as pipe and roof liners with long service life has been started recently. It was estimated to be approximately one million tons in FY 2000, and it is predicted to increase further to approximately 1.8 million tons in FY 2020. Since the landfill sites run short and the incineration to reduce its volume produces dioxins and corrodes boiler tube by hydrogen chloride, development of a safe and inexpensive technology for the recycling of waste PVC is one of the most urgent issues.

Authors have proposed a novel process not to dispose to landfill sites both waste PVC and waste glass but to utilize them to produce fuel and neutralize each other at the same moment. It has been successfully demonstrated that hydrogen chloride produced during flash pyrolysis of PVC was completely neutralized by the fixed glass bed and thus chlorine-free fuel was produced [1-2]. To carry forward our proposed process we need to know the kinetics of the neutralization process. Also we have to solve the problem of formation of metal chlorides in the product char during pyrolysis of PVC, which is a critical issue for its thermal utilization. Consequently, in the present study the evaluations of neutralization kinetics of glass cullets and the decomposition of $CaCl_2$ in char by steam were conducted.

2. NEUTRALIZATION KINETICS OF GLASS CULLETS

2.1 Experimental procedure for evaluation of neutralization kinetics

A single quartz tube of 30 mm i.d. and 710 mm long (heating area, 470 mm) with a porous quartz plate in its middle was used as a differential reactor. Experimental conditions are listed in Table 1. The waste glass cullets tested were soda-lime glasses (13.5 Na_2O– 12.1 CaO– 72.1 SiO_2– 0.91 K_2O–1.38 Al_2O_3) containing 0.26 wt% chromium oxide. Three HCl concentrations (0.07, 0.14 and 0.28 vol%) were used to evaluate the effect of HCl concentration on the neutralization kinetics. After quenching the reactor, the reacted glass samples were washed with 50 ml of pure water in a shaker for 1 h. Chlorine and inorganic

Table 1

Experimental conditions for neutralization kinetics

	Unit	Series A		
Reaction temperature	[K]	823	823	823
Reaction time	[s]	60 to 3600	60 to 3600	60 to 3600
HCl concentration	[vol%]	0.07	0.14	0.28
Total flow rate	[m·s^{-1}]	5.1×10^{-2}	5.1×10^{-2}	5.1×10^{-2}
Glass cullets size	[μm]	355-600	355-600	355-600
Glass cullets quantity	[kg]	About 9.5×10^{-4}	About 9.5×10^{-4}	About 9.5×10^{-4}

Table 2

Experimental conditions for decomposition of $CaCl_2$ by steam

	Unit	Series B	Series C
Sample		Model PVC pellets	Wallpaper
PVC content	[wt%]	71±3	26±2
$CaCO_3$ content	[wt%]	29±3	34±6
Feed sample	[kg]	About 0.8×10^{-4}	About 1.0×10^{-4}
Cl /Ca	[mol/mol]	3.92	1.22
Time	[min]	10 to 60	10 to 180
Temperature	[K]	1073	1073
Feed water	[kg/s]	6.7×10^{-7}	6.7×10^{-7}

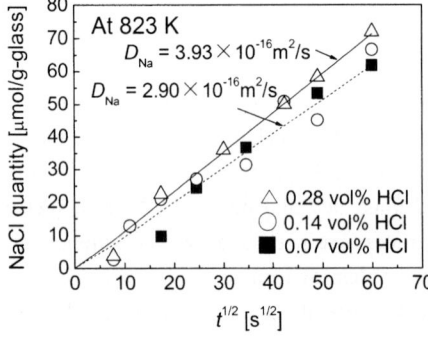

Fig. 1 NaCl Formation on glass cullets and Na diffusion coefficient at 823K

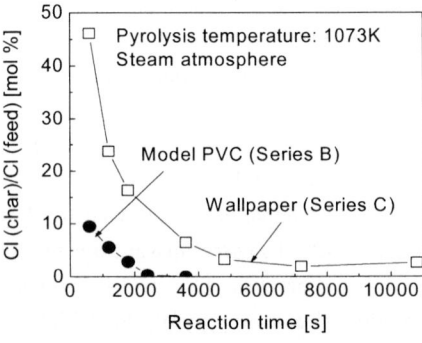

Fig. 2 Influence of reaction time on decomposition of $CaCl_2$ by steam

metals of the leachates were analyzed by ion chromatography (CDD-10AVP, Shimadzu Corp.) and ICP-MS (7500ce, Agilent Technologies Inc.), respectively.

2.2 Analysis of neutralization kinetics of glass cullets

Hydrogen chloride quantity captured by sodium of glass cullets at 823K as a function of square root of time is shown in Fig. 1. The amount of hydrogen chloride captured as sodium chloride was proportional to square root of time for most of the region, and thus the neutralization rates were controlled by diffusion. On the other hand partial pressure of hydrogen chloride did not affect the formation of sodium chloride even though its partial pressure increased by four times.

To determine diffusion coefficient D of sodium in glass cullets, the flux of sodium on glass surface was calculated. The diffusion coefficient is defined by the equation

$$F = -D \partial c / \partial x \qquad (1)$$

Where F is the flux of diffusing species and $\partial c / \partial x$ is its gradient of concentration c in the x direction. From the equation of continuity,

$$\frac{\partial c}{\partial t} = \frac{\partial(-F)}{\partial x} = D\frac{\partial^2 c}{\partial x^2} \qquad (2)$$

where t is the time and D is independent of concentration.

Assuming that the initial sodium concentration of glass particle with a radius R is C_i and its surface sodium concentration C_s is zero at $t > 0$, dimensionless terms can be written as follows:

$$\phi_c = \frac{C - C_s}{C_i - C_s} \qquad (3) \qquad\qquad \xi = \frac{r}{R} \qquad (4) \qquad\qquad \theta = \frac{D \cdot t}{R^2} \qquad (5)$$

where C is concentration of sodium within glass [mol·g^{-1}], r is distance from center [m], and D is diffusion coefficient [m^2·s^{-1}] and t is time [s].

Introducing a new variable of Θ equal to $\phi_c \cdot \xi$, the diffusion equation (Eq. (2)) can be transformed into

$$\frac{\partial \Theta}{\partial \theta} = \frac{\partial^2 \Theta}{\partial \xi^2} \qquad (6)$$

Solving Eq. (6) subject to boundary condition (7): at $\theta > 0$ and $\xi = 1$, $\Theta = 0$ and at $\theta > 0$ and $\xi = 0$, $\Theta = 0$ and initial condition (8): at $\theta \leq 0$ and $0 \leq \xi \leq 1$, $\xi = \Theta$, we obtain the following solutions:

$$C(r,t) = -C_i \left(\frac{2R}{\pi r}\right) \sum_{n=1}^{\infty} \frac{(-1)^n}{n} e^{-\frac{(n\pi)^2 Dt}{R^2}} \sin\left(\frac{n\pi r}{R}\right) \qquad (9)$$

$$F(t) = -4\pi r^2 D \frac{dC}{dt}\Big|_{r=R} = 8\pi D C_i R \sum_{n=1}^{\infty} (-1)^n \cos(n\pi) e^{-\frac{(n\pi)^2 D}{R^2} t} \qquad (10)$$

where $C(r, t)$ is the distribution of sodium in glass, $F(t)$ is the flux of sodium on glass surface.

Sodium diffusion coefficients in glass cullets calculated from fitting Eq. (10) with the profiles of sodium were $2.9 - 3.9 \times 10^{-16}$ m^2/s at 823K (see Fig. 1). Comparison of experimental results with predicted results for Na neutralized with HCl at 823K was fairly consistent and indicated that D_{Na} calculated would well represent the neutralization kinetics.

3. DECOMPOSITION OF CALCIUM CHLORIDE BY STEAM

400

3.1 Experimental conditions for steam decomposition of calcium chloride

Experimental conditions are listed in Table 2. After pyrolysis of samples at 1073K, the product char was washed with pure water and chlorine of the leachates was analyzed by ion chromatography. Refer to the reference [3] for a detailed experimental procedure.

3.2 Steam decomposition of calcium chloride in the product char

In our previous work [3], it was found that the residual chlorine in the product char after pyrolysis of wallpaper under a N_2 atmosphere increased sharply with increasing pyrolysis temperature. Accordingly, we conducted the steam decomposition at high pyrolysis temperature of 1073K. As shown in Fig. 2, in the case of model PVC pellets (series B) the residual chlorine was completely removed within 60 minutes from char which was remained in the state of powder after pyrolysis. On the other hand, in the case of wallpaper (series C) residual chlorine was detected in char even after 180 minutes. Maintaining of the original shape after pyrolysis of wallpaper is supposed to interrupt decomposition of $CaCl_2$ by steam. However, when the char was ground and pyrolyzed again with steam, chlorine in it was completely removed.

To design a pyrolyzer, steam flow required for decomposition of calcium chloride has to be evaluated. The steam flow depends on the equilibrium constant of decomposition reactions as follows:

$CaCl_2$ (s) + $2H_2O$ (g) → $Ca(OH)_2$ (s) + 2HCl (g) at $\leq 853K$ (11)
$CaCl_2$ (s) + H_2O (g) → CaO (s) + 2HCl (g) at > 853K   (12)

Equilibrium constant K for the reactions in the temperature range of 573K to 1373K increased with increasing reaction temperature indicating pyrolysis at higher temperature enhanced the decomposition of larger amount of calcium chloride, and thus the steam flow required for decomposition would be reduced at higher reaction temperature. However, since residual chlorine in the product char is reduced at lower pyrolysis temperature, two-stage pyrolysis is proposed to reduce steam consumption. At first stage PVC is dechlorinated at 573K reducing calcium chloride formation and at second stage formed calcium chloride is decomposed at 1073K.

4. CONCLUSIONS

The amount of hydrogen chloride captured as sodium chloride was proportional to square root of time and sodium diffusion coefficients in glass cullets calculated were $2.9 - 3.9 \times 10^{-16}$ m^2/s at 823K. Also, chlorine-free char can be produced by steam decomposition, even though particle size issue remains.

References
[1] M.Horio, R. Noda and H.-J. Sung, Method and Equipment to Produce Fuel Gas from Halogen Content Combustible Materials and Alkali Content Substances, Japan Patent No. 2004-155872 (2004)
[2] H.-J. Sung, R. Noda and M. Horio, JCEJ, 38 (2005) 220.
[3] T. Hirotani, H-J. Sung, R. Noda and M. Horio, Proceedings of 24th International Conference on Incineration and Thermal Treatment Technologies, U.S.A., 2005

Studies in Surface Science and Catalysis, volume 159
Hyun-Ku Rhee, In-Sik Nam and Jong Moon Park (Editors)
401

Modified ASM2d model including pH effect on enhanced biological phosphorus removal

Sang Kyu Park, Min Woo Lee, Dae Sung Lee and Jong Moon Park[*]

Dept. of Chemical Engineering, School of Environmental Science and Engineering, Pohang University of Science and Technology, San 31, Hyoja-dong, Nam-gu, Pohang, Gyeongbuk 790-784, Republic of Korea

1. INTRODUCTION

In most biological processes, pH is very important factor. Microorganisms show the best activity at their own optimum pH values and are inhibited if pH significantly deviates from their optimum value. Therefore, a biological model should include the effect of pH to describe relevant biological processes properly. The standard for modeling biological wastewater treatment plants is the so-called ASMs [1]. The ASM No.1 (ASM1) describes biological removal of biological oxygen demand (BOD) and nitrogen. The ASM No.2 (ASM2) and the ASM No. 2d (ASM2d) include biological phosphorus removal processes. The ASM No.3 is modified version of ASM1 with a new concept on storage of organic substrates and corrects some defects of ASM1. These activated sludge models were developed by IWA task group and has been widely used to simulate biological wastewater treatment plants.

Phosphorus is a major cause for eutrophication. The most widely used method for phosphorus removal in wastewater is biological methods. Biological phosphorus removal is achieved by the microorganisms called polyphosphate accumulating organisms (PAOs). There are several factors that influence the performance of enhanced biological phosphorus removal (EBPR) process and pH is known as one of the most important one. The ratio of phosphate release to acetate uptake is directly influenced by pH and it has a significant effect on the performance of EBPR process [2]. Filipe et al. (2001) showed that pH is a key factor for competition between PAOs and glycogen accumulation organisms (GAOs) [3,4]. GAOs have similar carbon conversion processes as PAOs but do not perform phosphorus uptake and release; therefore, when only GAOs are enriched in the culture, EBPR process does not occur.

2. MODEL MODIFICATION

The ASM2d contains 19 components and 21 biological processes that represent hydrolysis, nitrification, denitrification and phosphate removal. There are three groups of microorganisms in the ASM2d – heterotrophs, PAOs and Autotrophs. Each group of microorganisms has a characteristic role for biological wastewater treatment. For example, heterotrophs perform denitrification under an anaerobic condition and autotrophs perform nitrification under an aerobic condition. However, the ASM2d does not contain the necessary processes of pH or major components such as GAOs and glycogen for EBPR process. The processes of GAOs are based on from the processes of PAOs with the exception of the processes related to polyphosphate. Glycogen also plays a important role in EBPR process.

Therefore, to improve model performance, several processes related to GAOs and glycogen were added to the ASM2d.

As mentioned above, pH directly influences substrate uptake rates of PAOs and GAOs. Filipe et al. (2001a) suggested that the ratio of phosphate release to acetate uptake by PAOs is directly influenced by pH with following the relationship [3].

$$P\!\!\Big/\!\!_{Hac} = 0.25 + \alpha_{PAO} = 0.16\,pH_{OUT} - 0.55 \qquad (1)$$

Filipe et al. (2001b) also suggested a parameter that influences the glycogen metabolism of GAOs under anaerobic condition. The parameter is described as below [4].

$$\alpha_{GAO} = 0.057\,pH_{OUT} - 0.34 \qquad (2)$$

In this study, these two parameters were employed to ASM2d with proper assumption and conversion. The parameter described in equation (1) was applied to anaerobic storage of PHA by PAOs in ASM2d. Then the parameter in equation (2) was applied to the anaerobic storage of PHA process by GAOs

3. MATERIALS AND METHODS

Two sequencing batch reactors (SBRs) with 8 hours cycle employed. The sludge was inoculated from the aerobic sludge of a wastewater treatment plant at Pohang in Korea. Each cycle consists of 2 hours anaerobic phase, 4 hours aerobic phase, 1.5 hour settling and 30 min drawing phase. Influent was supplied at the first 10 min of the anaerobic phase. Excess sludge was removed at the end of the aerobic phase to control sludge retention time (SRT) as 10 days. The hydraulic retention time (HRT) was set to 16 hours. The temperature was maintained at $20\,^{\circ}\mathrm{C}$ with a water bath. The pH of the first SBR was controlled to 7.0 with 1 N H_2SO_4 and 1 N NaOH and the other SBR was not pH-controlled. A synthetic wastewater was used as the influent. The synthetic wastewater contains 1.525 g NaAc, 306.5 mg NH_4Cl, 184 mg $MgSO_4 \cdot 7H_2O$, 132.5 mg KH_2PO_4, 60.5 mg Na_2CO_3, 57 mg NaCl, 21 mg $CaCl_2$, 2 mg yeast extract and 0.56 ml trace element solution per 1 L of tap water. The trace solution contains 900 mg $FeCl_3$, 150 mg HBO_3, 30 mg $CuSO_4 \cdot 5H_2O$, 180mg KI, 60 mg $MnCl_2 \cdot 4H_2O$, 120 mg $ZnSO_4 \cdot 7H_2O$, 150 mg $CoCl_2 \cdot 6H_2O$, 36 mg MoO_3 and 10 g EDTA per 1 L of distilled water. Ammonium and COD were measured by the standard method [5]. Nitrite, nitrate and phosphate were measured by ion chromatography (DIONEX DX-120). TOC was measured by TOC analyzer (Shimadzu, TOC-500). MLSS was measured after drying the filtered biomass at $105\,^{\circ}\mathrm{C}$ oven. The simulation program was coded with MATLAB 6.5 and simulating condition was the same with experimental conditions.

4. RESULTS AND DISCUSSION

To achieve a steady state, simulation time was set to 60 days which is six times of the SRT. The settling was assumed to ideal settling. A stepwise calibration nethdology was applied to the SBR process and some key parameters in ASM2d were optimized (results not shown).

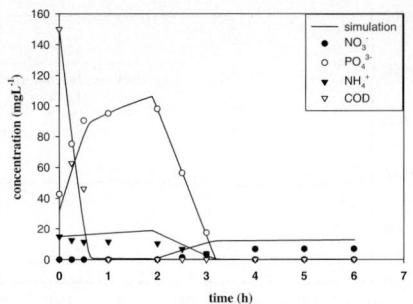

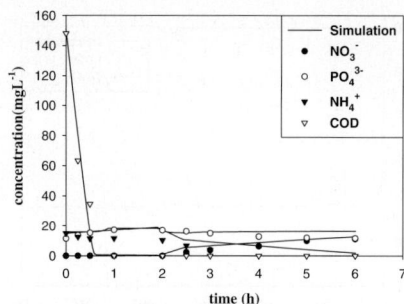

Fig.1. Simulation result for pH-uncontrolled SBR   Fig.2. Simulation result for pH controlled-SBR

The simulation data of pH 7.0 and uncontrolled pH was compared to the experimental data. (Fig. 1 and 2) The settling was assumed to ideal settling. The pH of pH-uncontrolled SBR was about 8.5. The phosphate profile showed relatively the same result with a general EBPR process. Under pH-uncontrolled condition, phosphate was released to the bulk solution in the anaerobic phase and removed in the subsequent aerobic phase both in simulation and in the experiment. This is the typical result of a good EBPR process. Under pH-controlled condition, phosphate was neither released nor removed both in the simulation and in the experiment.

This result shows that the EBPR process was completely broken. However the profile of ammonium in anaerobic phase a little differs from experimental data in both cases. This result was occurred by the absence of ammonium-consuming process in anaerobic phase in ASM2d.

The simulation of the SBRs for a long period was also performed. Simulation was performed under four different conditions. Simulation time was 300 days for all conditions and each data collected at the end of aerobic phase of each cycle. Fig. 3 and Fig. 4 show the simulation result of variation of PAOs and GAOs concentrations under pH-uncontrolled condition and pH-controlled condition. Under pH-uncontrolled condition, PAOs grow dominantly and GAOs wash out. In contrast, the PAOs concentration decreases and GAOs grow to dominant microorganism under pH-controlled condition pH in simulation result. In the simulation result, phosphorus removal rate was significantly decreased when PAOs concentration decreased below around 1000 mgL^{-1}. The effect of the change of pH control strategy was also simulated and their results were shown in Fig. 5. Fig. 5(a) shows the simulation data that assumes the reactor started under uncontrolled pH condition and pH control started after 100 days. In this result, PAOs were inhibited with pH control and GAOs start to grow.

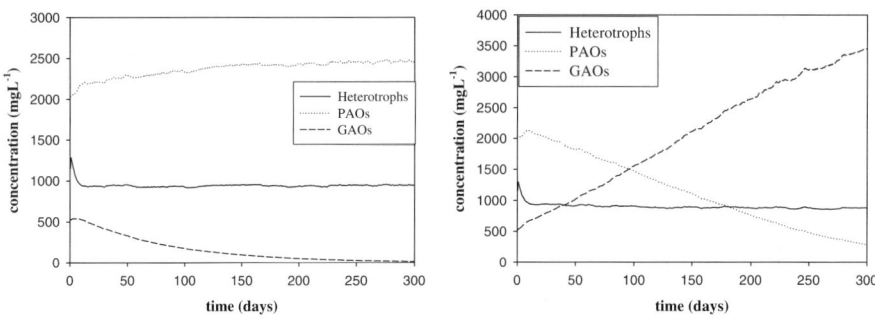

Fig. 3. Long period simulation at uncontrolled pH   Fig. 4. Long period simulation at controlled pH

404

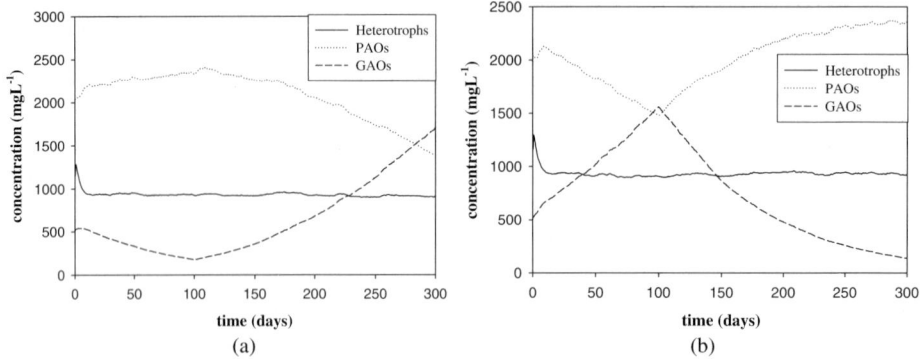

Fig. 5. Long period simulation with change of pH control scheme

In Fig. 5(b), the reactor started with pH control and stopped pH control after 100 days. Then PAOs concentration increased significantly after stopping pH control and GAOs concentration decreased.

These simulation data should be proved by experiment. However, it is still unknown that what microorganisms are PAOs or GAOs. Therefore the precision of model for EBPR process should be determined with indirect methods. Then with the concentration decrease of PAOs, the GAOs concentration was increased rapidly. However, this simulation result should be supported by the experimental data with the identification of PAOs and GAOs.

5. CONCLUSIONS

The original ASM2d model was modified with the parameter that expresses pH effect and the processes related to glycogen and GAOs. The performance of the modified model was tested with experimental result of lab scale SBRs. The simulation result could describe the experimental result properly. The concentration variation of PAOs and GAOs was predicted by the modified model under different pH conditions.

ACKNOWLEDGEMENT

This work was financially supported by the Korea Science and Engineering Foundation through the Advanced Environmental Biotechnology Research Center at POSTECH.

REFERENCES

[1] M. Henze, W. Gujer, T. Mino, M.C.M. van Loosdrecht, *Activited sludge models ASM1, ASM2, ASM2d and ASM3*, IWA Publishing, 2000
[2] J.G.F. Smolders, J. van der Meij, M.C.M. van Loosdrecht, J.J. Heijnen, *Biotechnol. Bioeng.*, **43**, (1994).
[3] C.D.M. Filipe, G.T. Daigger, C.P.L. Grady Jr., *Biotechnol. Bioeng.*, **76(1)**, (2001a).
[4] C.D.M. Filipe, G.T. Daigger, C.P.L. Grady Jr., *Biotechnol. Bioeng.*, **76(1)**, (2001b).
[5] APHA, *Standard methods for the examination of water and wastewater*, 16th edition, American Public Health Association, Washington, DC, 1985

Studies in Surface Science and Catalysis, volume 159
Hyun-Ku Rhee, In-Sik Nam and Jong Moon Park (Editors)

Effect of Na-alginate and bead diameter on lactic acid production from pineapple waste using immobilized *Lactobacillus delbrueckii* ATCC 9646

Ani idris* and Suzana Wahidin

Department of Bioprocess Engineering, Faculty of Chemical and Natural Resoources Engineering, Universiti Teknologi Malaysia, 81310 UTM, Skudai, Johor Bahru.
(*Corresponding author: E-mail: ani@fkkksa.utm.my)

1. INTRODUCTION

Application of immobilized living cells is a new and rapidly growing area in biotechnology. Cell immobilization can improve production rates of lactic acid while reducing medium requirements and inhibitions. Entrapment in calcium alginate bead is the most widely used procedure for immobilization [1]. Prasad and Mishra (1995) immobilized *Saccharomyces cerevisiae* in calcium alginate matrix, Mohamed et al. (2000) entrapped cells of *Bacillus amyloliquefaciens* in calcium alginate and Gough et al. (1998) immobilized *Kluyveromyces marxianus* in a calcium alginate matrix. The mild condition for immobilization and its simplicity are some of the reasons calcium alginate was chosen as the immobilization matrix [2-4]. In this study, *Lactobacillus delbrueckii* strain has been entrapped in the bead matrix through which substrates and products diffuse in and out easily. Stability of the beads is important to maintain high conversion of substrate to product. The concentration of sodium alginate and bead diameter were found to have a pronounced effect on the stability of the beads, which will affect lactic acid production [5].

Several authors have studied the effect of Na-alginate concentration on lactic acid production by immobilized organisms. Bead diameter is another factor that affects the lactic acid fermentation using immobilized *Lactobacillus delbrueckii*. Goksungur and Guvenc (1999) in their early work with beet molasses used various Ca-alginate bead sizes ranking from 1.3 to 3.2 mm diameter. It was reported that the highest lactic acid production was obtained with cell entrapped in the 1.3 to 1.7mm Ca-alginate beads [6]. Abdel-Naby et al. (1992) reported maximum lactic acid was produced with cells entrapped in 2.0mm Ca-alginate beads. When the bead diameter is increased beyond 3.0mm, the production of lactic acid is increased.

Goksungur and Guvenc (1999) reported that maximum lactic acid production, 5.93% was obtained with beads prepared at 2.0% w/v of Na-alginate concentration. Abdel-Naby et al. (1992) investigated lactic acid production using Ca-alginate immobilized beads and found that maximum lactic acid produced with beads containing 3.0% Ca-alginate concentration and obtained lower yields with beads made of 4.0 and 6.0% Ca-alginate due to diffusion problems [7]. Although there has been much work on lactic acid, none of them have used pineapple waste as the substrate for fermentation using the immobilization technique.

Since, the pineapple canning industry is one of the many food industries producing large quantities of solid and liquid wastes and due to the stringent environmental regulations regarding to waste disposal a special interest has developed in using the pineapple waste, which is rich in nutrients such as glucose. Thus the ability to utilize this liquid effluent into useful by products such as lactic acid will help reduce or eliminate sources of pollution [1].

2. MATERIAL AND METHODS

2.1 Immobilization cell

L. delbrueckii cells grown in 25ml MRS broth were mixed with an equal volume (1:1 v/v) of Na-alginate solution and were stirred for 5 minutes. The composition for 1L MRS medium are as follows: 5g yeast extract; 5g meat extract; 10g peptone; 2g K_2HPO_4; 5g diammonium citrate; 20g glucose; 2g sodium acetate; 0.58g $MgSO_4.7H_2O$; 0.25g $MnSO_4.4H_2O$ and 1ml Tween-80.

The mixed solution obtained was then placed in a syringe and allowed to drop into a sterile 0.2M $CaCl_2$ solution that was stirred continuously. Alginate drops solidified upon contact with $CaCl_2$, forming beads and thus entrapping bacteria cells. The beads were allowed to harden for 30 minutes at $37^\circ C$ and then washed with sterile saline solution to remove excess calcium ions and untrapped cells.

2.2 Fermentation conditions

2.2.1 Effect of Na-alginate concentration and bead diameter

The submerged fermentations were carried out in 250 mL Erlenmeyer flasks containing 100 mL of pineapple waste with 31.3 gL^{-1} of glucose concentration. Flushing the flasks with nitrogen and sealing them with tight fitting rubber stoppers maintained anaerobic conditions. The fermentation flasks were placed in a incubator shaker with an agitation rate of 150 rpm.

i) The effect of Na-alginate concentration on fermentation was conducted at various concentrations ranging from 1.0%, 2.0%, 4.0%, 6.0% and 8.0% w/v for 72 hours. Initial pH of the fermentation medium was 6.5; with 5 g bead of 1.0 mm bead diameter at $37^\circ C$.

ii) The effect of bead diameter was studied for various bead diameter 1.0, 3.0 and 5.0mm. The diameter of the bead was measured using imaging software connected to a microscope. These flasks were incubated at $37^\circ C$; with 2% w/v of sodium alginate and 5 g bead.

2.2.2 Analytical methods

Lactic acid and glucose concentrations were determined by HPLC. A 250 mm x 4.6 mm ID Spherisob Octyl Column manufactured by Waters was used with a UV detector (wavelength 210 nm) to measure lactic acid concentration. The adsorbed substances were eluted with 0.2 M H_3PO_4 at flow rate of 0.5 mL/min at room temperature. For glucose, a 4 mm diameter, 300 mm long IDμ Bondapak/ Carbohydrate column manufactured by Waters with RI detector were used. The carrier solution used was acetonitrile:water (80:20) at a flow rate 1.0 mL/min at room temperature. The concentration of living cells entrapped in Ca-alginate beads was determined by dissolving three beads in 10 mL of 0.3 M sodium citrate solution (adjusted to pH 5.0 with 1 M citric acid) for 20 minutes with continuous stirring at room temperature. For determining the cell number entrapped in Ca-alginate beads and leaked cells from the gel beads, bacterial counts were done by plating on MRS agar and incubating them at 37 $^\circ C$ for 48 hours.

3. RESULTS AND DISCUSSION

Lactic acid bacteria were immobilized in Ca-alginate beads prepared from different concentration of Na-alginate (1.0%, 2.0%, 4.0%, 6.0% and 8.0% w/v) and their fermentation efficiencies were investigated in liquid pineapple waste containing 31.3 gL^{-1} of glucose initially. Fig. 1 shows the growth pattern for the five concentrations of sodium alginate. The lag phase of bacterial growth for 1, 4, 6 and 8% Na-alginate concentration are longer; 24 hr compared to the 2% Na-alginate concentration, which is only 8 hr. Increasing the Na-alginate concentration above 2% only prolong the lag phase and the bacteria does not exhibit improved growth.

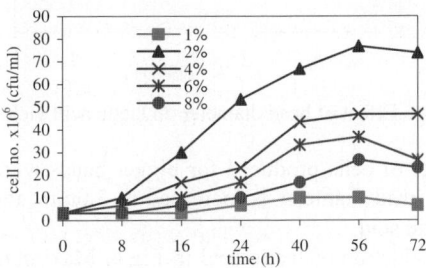

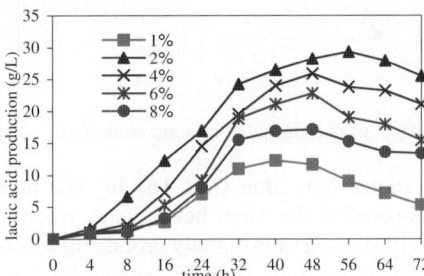

Fig.1. Effect of sodium alginate concentration on cell concentration

Fig. 2. Effect of sodium alginate concentration on lactic acid production

This longer lag phase could be due to the bacteria requiring to adapt with their environment. The exponential growth can be seen in all the flasks accept for the flask containing 1.0% of Na-alginate concentration. The 2.0% Na-alginate conc. produces more cell number compared to other samples. The exponential phase begins after 8 hours and the cells grow gradually until 56 hr where the death phase begins. Thus, the presence of only 2.0% Na-alginate concentration in the calcium alginate beads creates the optimum condition for *L. delbrueckii*.

The effect of Na-alginate concentration on the lactic acid production is depicted in Fig.2. The highest lactic acid production is obtained for the 2.0% of Na-alginate concentration with a yield of 29.39 gL^{-1}. Increasing the Na-alginate concentration above 2.0%, decreases the lactic acid production due to the lower diffusion efficiency of the beads. However when only 1.0% of Na-alginate concentration is used, the beads were disrupted in the medium at the end of fermentation. Fig.3 shows the growth pattern for three different sizes of bead diameter. The 1.0mm bead produced more cell number (73.3 x 10^6 cfu mL^{-1}) compared to the 3.0 mm (50.0 x 10^6 cfu mL^{-1}) and 5 0mm (26.7 x 10^6 cfu mL^{-1}) beads. The lag phase of bacterial growth for 3.0mm and 5.0mm are longer than 1mm bead diameter. The 1mm bead diameter went into exponential phase growth at the 8th hr until 24th hr before the stationary phase started. The high cell growth promotes lactic acid production, which started at about the same time. Different patterns were observed for the 3.0mm and 5.0mm beads, where the exponential

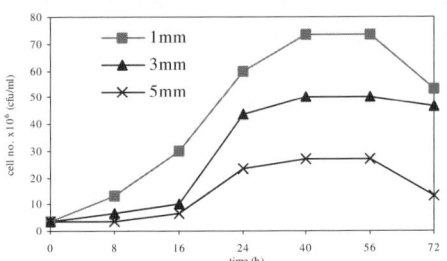

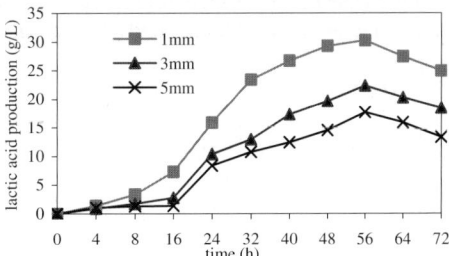

Fig.3. Effect of bead diameter on cell concentration

Fig.4. Effect of bead diameter on lactic acid production

408

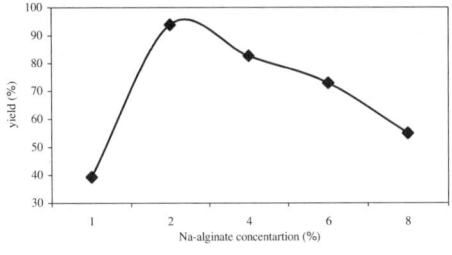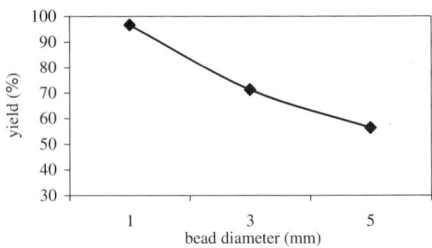

Fig.5.Effect of Na-alginate on lactic acid yield   Fig.6.Effect of bead diameter on lactic acid yield

growth started only after from 16th hr. The number of cells produced for bigger beads were less compared to the 1mm bead. Thus, when the bead diameter is increased to 3.0mm, the bacteria grew even more slowly producing less lactic acid.

A similar trend is also observed for the production of lactic acid in Fig.4. Maximum lactic acid concentration is attained for the 1mm bead diameter with a yield of 30.27 gL^{-1}. A further increase in the bead diameter to 5mm results in a decrease of lactic acid production to 17.65 gL^{-1}. Abdel-Naby et al. (1992) had studied the effect of bead diameter for lactic acid production and found the optimum lactic acid yield was obtained using a 2mm bead diameter. Lactic acid production increased as bead diameter continues to decrease.

Fig. 5 shows the pattern of lactic acid yield during the fermentation process at various Na-alginate concentrations. The results show the highest yield of lactic acid was obtained when 2.0% of Na-alginate concentration was used in lactic acid fermentation process. Increasing Na-alginate concentration beyond this value do not result in any increase of lactic acid yield. These results seem to be in agreement with those obtained by Goksungur and Guvenc [6] where optimum Na-alginate concentration is 2.0%. Too low Na-alginate concentration results in very soft beads whilst increased Na-alginate to above 2.0% hardens the beads, thus causing diffusion problems to occur. At high Na-alginate concentration, the bacteria do not get enough nutrients (food) as the substrate has difficulty in diffusing through the beads. However when only 1.0% Na-alginate concentration is used, the beads which are too soft as mentioned earlier are easily broken because their mechanical strength are lower and the bacteria leaks out from the beads [8].

Effect of bead diameter on lactic acid yield is clearly revealed in Fig.6. The optimum bead diameter for the fermentation of lactic acid for cell entrapped in Ca-alginate is 1.0mm with a yield of 30.27 gL^{-1} and 96.7%. Increasing bead diameter beyond this value did not improve lactic acid production. Smaller bead diameter yields more lactic acid production, due to an increase in the surface volume ratio [9]. A further increase in bead diameter to 5.0mm results in a decrease of lactic acid production to 17.65 gL^{-1} or 50.7%.

REFERENCES
[1] I. Ani, W. Suzana and H.B. Mat, *Water and Environmental Management Series*, (2003).
[2] B. Prasad and I.M. Mishra, *J. of Bioresource Technology*, **53**, (1995).
[3] S.Gough, N. Barron, V.I. Lozinsky and A.L. Zulbov, *J. of Bioprocess Engineering*, **19**, *(1998)*.
[4] A. Mohamed, M. Reyad and F. abdel-Fattah, J. of Biochemical Engineering, 5, (2000).
[5] J. Rajagopalan., S.T. Pillutla, V. Sonal, *J. of Fermentation and Bioengineering*, **73**,4 (1992).
[6] Y. Goksungur and U. Guvenc, *J. of Chem. Technol. and Biotechnol.*, **74**, (1999).
[7] M. Abdel-Naby, K. Mok, and C. Lee, *UNIDO Proceedings*, (1992).
[8] P. Boyaval and J. Goulet, *J. Enzyme Microbiol Technol.*, **10**, (1988).
[9] D. Guoqiang, R. Kaul and B. Mattiasson, *Appl. Microbiol. Biotechnol.*, **36**, (1991).

Studies in Surface Science and Catalysis, volume 159
Hyun-Ku Rhee, In-Sik Nam and Jong Moon Park (Editors)

409

CO_2 separation by membrane/absorption hybrid method

Kazuhiro Okabe[a]**, Miho Nakamura**[a]**, Hiroshi Mano**[a]**, Masaaki Teramoto**[b]**,
and Koichi Yamada**[a]

[a] Research Institute of Innovative Technology for the Earth,
9-2 Kizugawadai, Kizu-cho, Kyoto 619-0292 Japan

[b] Kyoto Institute of Technology, Matsugasaki, Sakyo-ku, Kyoto 606-8585, Japan

1. INTRODUCTION

We have been studying the novel process for CO_2 separation named membrane/absorption hybrid method. The advantages of this process are that high gas permeance and selectivity were obtained. The concept of this process is shown in Fig. 1. Both feed gas and absorbent solution are supplied to the inside of hollow fibers. While the liquid flows upward inside the hollow fibers, absorbent solution absorbs CO_2 selectively and it becomes a rich solution. Most of rich solution permeates the membrane to the permeate side maintained at reduced pressure, where it liberated CO_2 to become a lean solution. Compared to a conventional gas absorption

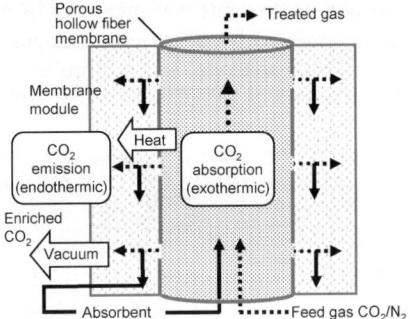

Fig. 1. Concept of membrane/absorption hybrid method

process, this process needs only one unit in which both absorption and stripping of CO_2 occur. Another advantage may be that a part of heat liberated in gas absorption is transferred to the permeate side through the membrane and effectively used for stripping. This advantage makes this system very efficient in energy utilization.

In the present study, we fabricated hollow fiber membrane modules and performed experiments at several conditions. The energy consumption of this process is compared to those of conventional gas absorption processes and membrane gas separation processes.

2. EXPERIMENTS

We used diethanolamine (DEA), 2,3-diaminopropionic acid monohydrochloride (DAPA), β-alanine and taurine, as absorbents of CO_2 without further purification. An equimolar amount of CsOH was added to amino acids aqueous solutions to deprotonate the protonated amino group. Feed gases and carrier solutions were supplied at 303 K and membrane module was heated at 323 K. The concentration of absorbent was 2 mol/l in every solution. The molecular formula of CO_2 absorbents were shown in Table 1.

Table 1 CO_2 absorbents

Amine or amino acids	Formula
Diethanolamine (DEA)	$HN(CH_2CH_2OH)_2$
2,3-Diamino-propionic acid (DAPA)	$H_2NCH_2CH(NH_2)COOH$
β-alanine	$H_2NCH_2CH_2COOH$
Taurine	$H_2NCH_2CH_2SO_3H$

The polyethersulfone capillary ultrafiltration membranes (Daicel Chemical Industries, Ltd., inner diameter: 0.8 mm, outer diameter: 1.3 mm, length: 40 cm, molecular weight cut-off: 150 000, water permeability: 3×10^{-7} m^3 m^{-2} s^{-1} kPa^{-1} at 298 K) were used. The length and area of membrane consisting of seven hollow fibers are 40 cm and 70 cm^2.

The schematic diagram of the experimental setup is shown in Fig. 2 and the experimental conditions are shown in Table 2. Each gas was controlled its flow rate by a mass flow controller and supplied to the module at a pressure slightly higher than the atmospheric pressure. Absorbent solution was supplied to the module by a circulation pump. A small amount of absorbent solution, which did not permeate the membrane, overflowed and then it was introduced to the upper part of the permeate side. Permeation and returning liquid fell down to the reservoir and it was recycled to the feed side. The dry gas through condenser was discharged from the vacuum pump, and its flow rate was measured by a digital soap-film flow meter. The gas composition was determined by a gas chromatograph (Yanaco, GC-2800, column: Porapak Q for CO_2 and (N_2+O_2) analysis, and molecular sieve 5A for N_2 and O_2 analysis). The performance of the module was calculated by the same procedure reported in our previous paper [1].

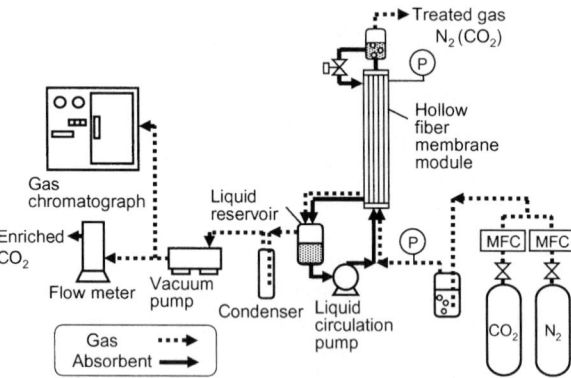

Fig. 2. Schematic diagram of experimental apparatus

3. RESULTS AND DISCUSSION

The experimental results are summarized in Table 2. CO_2 permeance (R_{CO2}), selectivity ($\alpha_{CO2/N2}$) and CO_2 recovery (Y) increased with decreasing CO_2 mole fraction in feed gas. CO_2 in the feed gas was successfully concentrated to 97-99 % by the single-stage operation. CO_2

permeance was 1.88-9.42 × 10^{-4} mol m^{-2} s^{-1} kPa^{-1}, these permeances were 5 to 10 times higher than conventional polymeric membranes. Selectivity was over 1000 and also CO_2 recovery was larger than 65 % when the CO_2 mole fraction in feed gas was 0.05. Since higher selectivities ($\alpha_{CO2/N2}$) were obtained when amino acid solutions were used as absorbents. Amino acids were considered to be promising as absorbents of membrane/absorption hybrid method.

Table 2 Experimental conditions and results

Absorbent Solution	$y_{CO2,f}$ [-]	v_g [ml min^{-1}]	v_L [ml min^{-1}]	P_f [kPa]	P_p [kPa]	$y_{CO2,p}$ [-]	Y [%]	R_{CO2} [mol m^{-2} s^{-1} kPa^{-1}]	$\alpha_{CO2\,N2}$ [-]
DEA	0.05	700	23	110	4.9	0.971	76.4	9.42 × 10^{-4}	1020
DEA	0.15	700	12	107	5.4	0.988	33.2	3.11 × 10^{-4}	560
DAPA	0.05	700	23	112	4.6	0.998	75.8	9.11 × 10^{-4}	34700
DAPA	0.15	700	23	117	5.2	0.999	66.4	7.10 × 10^{-4}	9700
ß-Alanine	0.05	700	23	125	3.8	0.998	65.1	6.58 × 10^{-4}	31200
ß-Alanine	0.15	700	23	126	4.8	0.998	23.6	1.88 × 10^{-4}	7400
Taurine	0.05	700	23	116	5.1	0.999	67.8	7.46 × 10^{-4}	32500
Taurine	0.15	500	12	115	5.9	0.999	48.4	4.65 × 10^{-4}	8600

$y_{CO2,f}$: CO_2 mole fraction in feed gas $y_{CO2,p}$: CO_2 mole fraction in permeate side
v_g : Volumetric flow rate of feed gas   Y : CO_2 Recovery
v_L : Volumetric flow rate of carrier solution R_{CO2} : CO_2 permeance
P_f: Total pressure in feed side $\alpha_{CO2/N2}$: Selectivity of CO_2 over N_2
P_p : Total pressure in permeate side

The total energy consumptions were estimated for each experiment and the results are shown in Table 3.

Table 3 Energy consumption

Absorbent	$y_{CO2,f}$ [-]	Energy for CO_2 recovery [kWh kg-CO_2^{-1}]				
		Liquid pump	Blower	Vacuum pump	Separation Total	Separation and Liquefaction
DEA	0.05	0.015	0.040	0.120	0.176	0.32
DEA	0.15	0.006	0.020	0.113	0.139	0.27
DAPA	0.05	0.020	0.050	0.125	0.195	0.32
DAPA	0.15	0.008	0.027	0.116	0.151	0.28
ß-Alanine	0.05	0.019	0.123	0.141	0.283	0.41
ß-Alanine	0.15	0.018	0.118	0.122	0.257	0.39
Taurine	0.05	0.022	0.068	0.117	0.207	0.34
Taurine	0.15	0.008	0.030	0.106	0.144	0.27

The energy of liquid circulation pump was calculated by Eq. (1), and the energies of blower and vacuum pump were calculated by Eq. (2).

$$E = \rho\, g\, q\, h\, \eta^{-1}$$ (1)

Here, ρ is density of liquid; g is gravitational acceleration; q is flow rate of liquid; h is height which solution pumped up to, it was assumed 10 m high in this calculation; and η is efficiency of a pump.

$$E = n R T \eta^{-1} ln(P_h/P_l) \qquad (2)$$

Here, n is mole of evacuated in unit time; R is gas constant; T is temperature; η is efficiency of a pump; and P_h/P_l is rate of pressure. The energy for liquefaction of CO_2 was assumed 0.13 kWh kg-CO_2^{-1} when CO_2 mole fraction was higher than 0.99. The energy consumption for each device and total energies were summarized in Table 3. The energy of vacuum pump was relatively high compared to other devices. The total energy for separation was equal to or two times higher than liquefaction energy. The total energies of separation by membrane/absorption hybrid method and liquefaction were calculated at 0.27 to 0.41.

The total energies of this work and conventional methods were shown in Fig. 3. In the chemical absorption method which was assumed to be attached to the present power plant, its total energy was calculated at about 0.51 kWh kg-CO_2^{-1}. This value was derived from the data of pilot plant test carried out by The Kansai Electric Power Company [2]. With regard to a polymer membrane method, Mano et al. reported that the CO_2 separation and liquefaction energy was 0.41 kWh kg-CO_2^{-1} when a cardo type polymer membrane with a CO_2/N_2 selectivity of 35 was used [3]. Our preliminary estimation of total energy consumption is 0.27 kWh kg-CO_2^{-1}, which is about 53% to the chemical absorption method and 66 % to the polymer membrane method. Therefore, it may be concluded that the hybrid process is very energy-efficient compared to the conventional gas absorption process and is promising for the CO_2 recovery with low energy consumption.

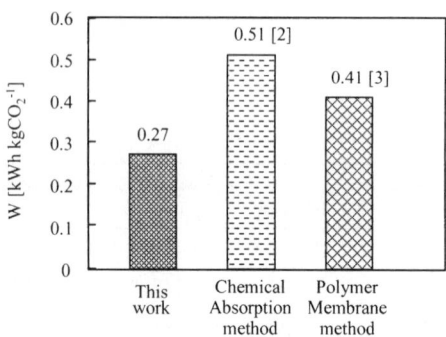

Fig. 3. Total energies of this work and conventional methods

ACKNOWLEDGEMENT
A part of this work was supported by New Energy and Industrial Technology Development Organization (NEDO).

REFERENCES
[1] M. Teramoto, S. Kitada, N. Ohnishi, H. Matsuyama, N. Matsumiya, J. Mem. Sci., 234 (2004) 83.
[2] Home page of The Kansai Electric Power Company, Inc., *http://www.kepco.co.jp/rd/topics/t 3.htm* (2003).
[3] H. Mano, S. Kazama, K. Haraya, Greenhouse Gas Control Technologies, J. Gale and Y. Kaya, eds., Oxford: Pergamon, (2003) 1551.

Studies in Surface Science and Catalysis, volume 159
Hyun-Ku Rhee, In-Sik Nam and Jong Moon Park (Editors)
© 2006 Elsevier B.V. All rights reserved

Effect of ozone pretreatment on low temperature CO oxidation catalysts

K. Y. Ho[a] and K. L. Yeung[b,*]

[a]Environmental Engineering Program and Department of Chemical Engineering, the Hong Kong University of Science & Technology, Clear Water Bay, Kowloon, Hong Kong

[b]Department of Chemical Engineering, the Hong Kong University of Science & Technology, Clear Water Bay, Kowloon, Hong Kong

1. INTRODUCTION

Titanium dioxide supported gold catalysts exhibit excellent activity for CO oxidation even at temperatures as low as 90 K [1]. The key is the high dispersion of the nanostructured gold particles over the semiconducting TiO_2 support. The potential applications of ambient temperature CO oxidation catalysts include air purifier, gas sensor and fuel cell [2]. This work investigates the effects of ozone pretreatment on the performance of Au/TiO_2 for CO oxidation.

2. EXPERIMENTAL

Nanostructured TiO_2 was prepared by a modified sol-gel method [3]. Five milliliters of titanium isopropoxide (TIP, 98 %, Acros) was rapidly mixed with an alcohol solution containing 1.5 mL water in 30 mL isopropanol (IPA, 99.7 %, BDH) at room temperature. The uniform-sized titania gel-spheres formed by rapid hydrolysis of TIP was collected by vacuum filtration. The dried powder was recrystallized by thermal treatment [4] to produce nanometer-sized, anatase TiO_2 of controlled crystal and aggregate sizes, crystallinity and surface chemical property [5]. Well-dispersed gold catalysts on TiO_2 were prepared from an aqueous solution of hydrogen tetrachloroaurate (III) trihydrate ($HAuCl_4$, ACS, Aldrich) at a neutral pH. Two gold catalysts were pretreated at 473 K for 5 h in air and in 100 ppm O_3/O_2 mixture, respectively. Table 1 lists the gold loading, average TiO_2 crystal size, BET surface area of the catalyst, pretreatment condition and the average gold particle size for three catalyst samples prepared for this study.

The catalysts were tested for CO oxidation in a flow reactor using a 2.5 % CO in dry air mixture at a fixed flow rate of 200 sccm. Thirty milligrams of the catalyst were used for each experimental run. The reaction was conducted at 298, 323, 373 and 473 K with 75 minutes duration at each temperature. The carbon monoxide conversion to carbon dioxide was monitored by an online gas chromatograph equipped with a CTR-1 column and a thermal conductivity

---
* Author to whom correspondence should be addressed
  Tel: 852-2358-7123; Fax: 852-2358-0054; e-mail: kekyeung@ust.hk

detector. After completing the temperature program, the gas inlet and outlet to the reactor were shut off to isolate the catalyst. The catalyst was allowed to cool down to room temperature and left overnight for 16 h. The reaction was resumed next day by flowing in the reaction mixture at room temperature.

Table 1. Properties of gold catalysts

Sample	Au loading[#] (wt.%)	TiO_2 crystal size[+] (nm)	BET area ($m^2 g^{-1}$)	Pre-treatment	Au particle size before (after) reaction[Δ] (nm)
Au/Nano-TiO_2	0.59	12 (anatase)	68	O_2 at 473 K for 5 h	2.5 (5.7)
Au/Nano-TiO_2	0.58	13 (anatase)	70	100 ppm O_3/O_2 at 473 K for 5 h	2.7 (2.4)
Au/P25[*]	0.78	22 (anatase), 37 (rutile)	52	100 ppm O_3/O_2 at 473 K for 5 h	2.5 (2.8)

[*] Degussa P25 titanium dioxide
[#] Determined by inductively coupled plasma-mass spectrometry of acid digested catalyst samples
[+] Calculated from X-ray diffraction peak broadening at (101) for anatase and (110) for rutile TiO_2
[Δ] Mean particle diameter measured from transmission electron microscopy pictures of gold catalysts

3. RESULTS AND DISCUSSION

Fig. 1 plots the results of the reaction experiment conducted on the three gold catalysts. The figure shows that within the range of 298 to 473 K, the reaction conversion is a weak function of temperature. Catalyst deactivation is observed except for the catalyst pretreated by ozone/oxygen mixture. Besides the nanostructured TiO_2 prepared by the sol-gel process, commercial P25 TiO_2 was also used as the support for the gold catalyst. Both the Au/Nano-TiO_2 samples treated in oxygen and ozone/oxygen mixtures are more active than the Au/P25 catalyst.

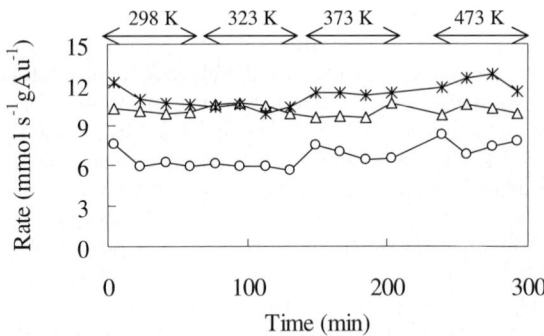

Fig. 1. Rate of CO oxidation of ($*$) O_2-treated Au/Nano-TiO_2, ($\triangle$) O_3/O_2-treated Au/Nano-TiO_2 and ($\bigcirc$) O_3/O_2-treated Au/P25 from 298 to 473 K.

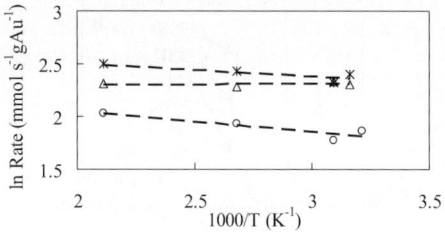

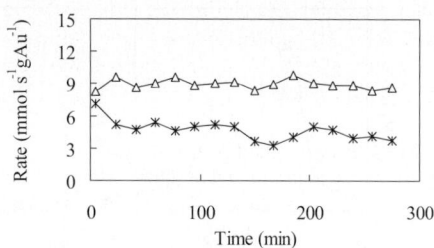

Fig. 2. Arrhenius plots for the rate of CO oxidation over (∗) O_2-treated Au/Nano-TiO$_2$, (△) O_3/O_2-treated Au/Nano-TiO$_2$ and (○) O_3/O_2-treated Au/P25.

Fig. 3. Rate of CO oxidation of (∗) O_2-treated Au/Nano-TiO$_2$, (△) O_3/O_2-treated Au/Nano-TiO$_2$ at 298 K.

The Arrhenius plots of the CO conversion rate in Fig. 2 indicate that the activation energy for the Au/Nano-TiO$_2$ catalysts is nearly zero. Haruta et al. [6] also reported similar observations. They suggest that this occurs when the CO adsorbed on gold particles reacts with adsorbed O_2 on the support at the interfacial junction between the two surfaces.

Fig. 3 plots the reaction rates of the catalysts versus time at room temperature. The reaction was conducted 16 h from the first set of experimental run. The O_2-treated Au/Nano-TiO$_2$ lost half of its activity after 45 minutes of reaction, while O_3/O_2-treated Au/Nano-TiO$_2$ displays a constant activity. It is clear from these results that Au/Nano-TiO$_2$ catalyst pretreated with ozone exhibits better and more stable catalyst performance. The transmission electron microscope pictures of the O_2-treated Au/Nano-TiO$_2$ shown in Fig. 4 display an increase in the gold particle size to as large as 17 nm after the reaction, while the O_3/O_2-treated catalyst did not show any changes in the gold particle size. Coarsening of gold particles [7,8] is one of the reasons for catalyst deactivation and the ozone pretreatment is seen to inhibit this process. XPS analysis of the fresh catalysts shows that the O_3/O_2-treated gold catalyst has broader Au 4f peaks indicating the presence of nonmetallic gold that may be responsible for the enhanced stability of this catalyst (Fig. 5).

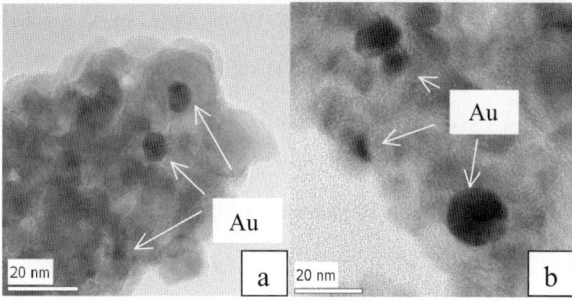

Fig. 4. High resolution TEM of O_2-treated Au/Nano-TiO$_2$ (a) before and (b) after CO oxidation.

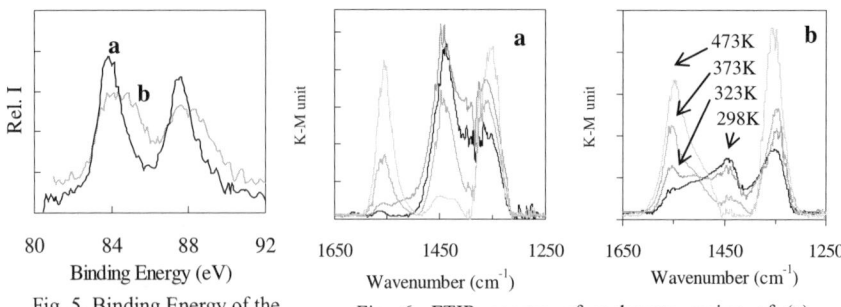

Fig. 5. Binding Energy of the Au 4f region for (a) O_2-treated Au/Nano-TiO$_2$, (b) O_3/O_2-treated Au/Nano-TiO$_2$.

Fig. 6. FTIR spectra of carbonate region of (a) O_2-treated Au/Nano-TiO$_2$ and (b) O_3/O_2-treated Au/Nano-TiO$_2$.

The accumulation of carbonates is another reason for gold catalyst deactivation [9]. The in-situ FTIR experiments in Fig. 6 show that the carbonate build-up is slower in the O_3/O_2-treated gold catalyst (Fig. 6a) compared to the air-treated sample (Fig. 6b). Also, the air-treated catalyst displays a strong band at 1435 cm^{-1} corresponding to the non-coordinated carbonate. Although our understanding of the process is incomplete, it is clear from the results that O_3 pretreatment inhibits the deactivation of gold catalyst.

ACKNOWLEDGEMENT

The authors would like to acknowledge the financial support from the Hong Kong Innovation and Technology Fund (ITS/176/01C).

REFERENCES

[1] F. Boccuzzi, A.M. Manzoli, P. Lu, T. Akita, S. Ichikawa, M. Haruta, J. Catal., 202 (2001) 256.
[2] C.W. Corti, R.J. Holliday, D.T. Thompson, Gold Bull., 35 (2002) 111.
[3] A.J. Maira, K.L. Yeung, C.Y. Lee, P.L. Yue, C.K. Chan, J. Catal., 192 (2000) 185.
[4] A.J. Maira, K.L. Yeung, J. Soria, J.M. Coronado, C. Belver, C.Y. Lee, V. Augugliaro, Appl. Catal. B: Environ., 29 (2001) 327.
[5] A.J. Maira, J.M. Coronado, V. Augugliaro, K.L. Yeung, J.C. Conesa, J. Soria, J. Catal., 202 (2001) 413.
[6] M. Haruta, M. Daté, Appl. Catal. A, 222 (2001) 427.
[7] G.C. Bond, D.T. Thompson, Cat. Rev. – Sci. Eng., 41 (1999) 319.
[8] M. Haruta, Catal. Today, 36 (1997) 153.
[9] P. Konova, A. Naydenov, Cv. Venkov, D. Mehandjiev, D. Andreeva , T. Tabakova, J. Mol. Catal. A 213 (2004) 235.

Studies in Surface Science and Catalysis, volume 159
Hyun-Ku Rhee, In-Sik Nam and Jong Moon Park (Editors)

Mechanism of Scaling on the Oxidation Reactor Wall in TiO$_2$ Synthesis by Chloride Process

E Zhou^{a,b}, Zhangfu Yuan^{*,a} , Zhi Wang^a, Xiaoqiang Wang^a and Jiajun Ke^a

^{a} Institute of Process Engineering, Chinese Academy of Sciences P O Box 353, Hai Dian District, Zhong Guan Cun, Beijing 100080, P.R. China
^{b}Graduate School, Chinese Academy of Sciences, Beijing 100080, P.R. China

Abstract: The formation of wall scale was mostly due to being deposited and sintered of TiO$_2$ particles formed in the gas phase reaction of TiCl$_4$ with O$_2$. The gas phase oxidation of TiCl$_4$ was in a high temperature tubular flow reactor with quartz and ceramic rods put in center respectively. Scale layers were formed on reactor wall and two rods. Morphology and phase composition of them were characterized by scan electron microscopy and X-ray diffraction. The state of reactor wall has little effect on scaling formation. With uneven temperature distribution along axial of reactor, the higher the reaction temperature is, the thicker the scale layer and the more compact the scale structure is.
PACS: 61.46.+w; 81.15.Gh; 81.20.-n
Key words: Scale, reactor wall, TiO$_2$, oxidation

1. INTRODUCTION

Titania particles are synthesized by the gas-phase oxidation of titanium tetrachloride in a high temperature tubular flow reactor on a large industrial scale. In the industrial operation, a serious problem is that a hard TiO$_2$ coating forms on the inner surface of oxidation reactor. The deposition of the TiO$_2$ on the hot reactor wall will cause the generation of hard scale layer and alteration of the reactor efficient dimension and heat exchange efficiency. The worst of it is that the tube is blocked and the production must be shut down. Although some techniques have been adopted to remove the scales, such as mechanical scraping, sand blasting and gas-film surrounding [1-3], the fundamentals of the process are not well understood and the mechanisms of the scale formation have not been further investigated.

In this paper, TiCl$_4$ was oxidized in the flow reactor at various temperature and gas flow rate. The wall scales were characterized by scan electron microscopy and X-ray diffraction. The effects of reactor wall surface state, radial growth of scale layer and reactor axial temperature distribution on scaling formation were discussed. At the same time, the mechanism of scaling on the reactor wall was explored furthermore.

2. EXPERIMENTAL

The used raw materials were TiCl$_4$, oxygen, Ar and NaOH. The reaction apparatus consisted of gas purifiers, reactant preheaters, reactor, water cooler, separator and an off-gas treatment unit. The reactor was a 27 mm in I.D. (32 mm in O.D.), 1430 mm in length quartz tube that was heated by a horizontal electrical furnace. A quartz rod and a ceramic rod, 6 mm

---

* Corresponding author.   Tel: +8610 62527440; Fax: +8610 62527440.
E-mail address: yuanzhf@home.ipe.ac.cn

in diameter, 1000 mm in length were arranged in the reactor center respectively. Prepurified dry argon gas was used as a carrier gas and bubbled into a flask containing liquid $TiCl_4$ in water bath (maintained at about $363\pm1K$). This mixture of Ar and $TiCl_4$ vapor was combined with another Ar stream for dilution and preheated O_2 stream (at about 573K) flowed into the reactor tube. The gas flow rate ranged from 1.5 to 3.5 L/min. Exiting from the furnace and reactor tube, the gas stream was cooled by dry N_2 gas and was neutralized by a 1 M NaOH solution. All of the gas flows were monitored by rotameters. Lines transporting $TiCl_4$ were maintained at 423K to prevent condensation of $TiCl_4$ on the lines.

The weight fractions of the anatase and rutile phases in the samples were calculated from the relative intensities of the strongest peaks in XRD patterns corresponding to anatase and rutile as described by Spurr and Myers [4].

3. RESULTS AND DISCUSSION

3.1. Reactor wall surface state

With being deposited and sintered of the TiO2 particles on the reactor wall, the scale layer, which was affected by surface state of wall, such as smoothness or ruggedness, would be formed on its surface.

At the end of oxidation reaction lasting for 50 min, the ceramic rod had been covered with a hard scale layer composed by whisker and bud. At the site 200mm away from gas entrance, the whisker was longest, which indicated that the reaction was most severest there. On the one hand, in this region, supersaturation degree of gas phase TiO_2 was greatest, so, high concentrated minimal particles were produced in homogeneous nucleation. On the other hand, the rough surface of ceramic rod was liable to be the nucleation centers in heterogeneous nucleation. Under the same reaction conditions, such as reacting temperature, reactants concentration and residence time, particles would still be sintered on the glossy surface of quartz rod and had the same morphology as that on the rough ceramic rod surface.

As a result, scale layer was always formed on the smooth surface of reactor wall as well as irregular. Microstructure was blanketed as soon as the surface was coated with a TiO_2 scale film. It could be deduced that once scales were formed, the effect of surface geometrical shape would be more important than its microstructure.

3.2. Radial growth of scale layer

After gas-phase oxidation reaction finished, the reactor wall surface was coated with a thick rough scale layer. The thickness of scale layer along axial direction was varied. The scale layer at front reactor was much thicker than that at rear. The SEM pictures were shown in Fig.1were scale layers stripped from the reactor wall surface. Fig. 1(a) was a cross sectional profile of scale layer collected from major scaling zone. Seen from right side of scale layer, particles-packed was loose and this side was attached to the wall surface. Its positive face was shown in Fig. 1(b). Seen from left side of scale layer, compact particles-sintered was tight and this side was faced to the reacting gases. Its local amplified top face was shown in Fig. 1(c). The XRD patterns were shown in Fig. 2(a) were the two sides of scale layer. Almost entire particles on sintered layer were characterized to be rutile phase. While, the particle packed layer was anatase phase.

For the accumulation of heat, scale surface temperature increased with the increase of scale layer thickness. The thicker the layer was, the temperature was closer to gases temperature; the thinner the layer was, the temperature was closer to wall surface temperature. Obviously, if the scale layer was thin enough, the scale temperature was low enough and neoformative particles were incompact and could not be fused or sintered. If they were cleared instantaneously, the radial growth of scale could be controlled. Otherwise, the ultrafine particles were ease to be coagulated and aggregated at high temperature and a particle-packed layer was formed on reactor wall surface. Once this layer reached a given thickness, scale surface temperature was adequate high, the deposited particles would be sintered each other and the scale layer would rapidly build-up and would be difficult to remove.

(a) Scale layer profile; (b) Particles packed layer; (c) Sinteried layer
Fig.1.   SEM pictures of scale layer

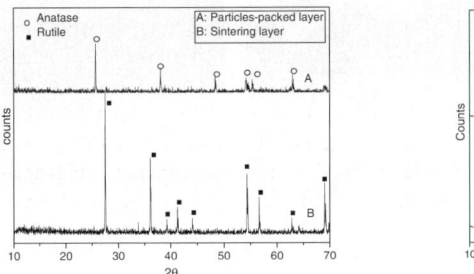

(a) Scales from reactor wall surface   (b) Scales from rods
Fig.2.   XRD patterns of scale layers

Besides, without addictive $AlCl_3$ as a crystal conversion agent, phase composition of most neogenic TiO_2 particles was anatase in our experiment. Conversions active energy from anatase to rutile was 460 kJ/mol [5], with temperature arose, crystal conversion rate as well as mass fraction of rutile would increase [6,7]. Hence, after a lot of heat accumulated, phase composition of particle-sintered layer was rutile.

3.3. Axial temperature distribution in reactor

Along the length of the tube, there was an about 600 mm isothermal region. After experiment finished, the most severe scaled region was at site 200-300 mm away from $TiCl_4$ entrance where temperature was just lower than isothermal region. Then the scale became smooth step-by-step from front to rear of reactor.

The SEM pictures were shown in Fig. 3 were scale layers stripped from different sections of quartz rod. Fig. 3(a) was a cross sectional profile of scale from the severe scaled point. Level (1) was quartz rod substrate; scale tightly appressed to level (1) was named level (2). It could be seen from Fig. 3(a) that there was no obvious interface between the scale and quartz substrate. The positive face of this scale block was shown in Fig. 3(b) and the scale stripped from the smooth scaled point was shown in Fig. 3(c). Compared with Fig. 3(b), there were less agglomerations and shorter whisker columns in Fig. 3(c). The XRD patterns were shown in Fig. 2(b) were scales on different sections of quartz rod. It was obvious that the scale from front rod was characterized to be rutile phase and scale from rear rod was partial anatase phase.

Due to existence of an isothermal region, temperature of both entrance and outlet was rather lower than that of intermediate section. Where temperature was high, the reaction was sever and fast. So, at site 200-300 mm away from entrance, the temperature was highest, the scale layer was thickest and the whisker column was longest there. The reaction route in this zone could be described as *phase reaction → homogeneous nucleation → coagulation*. When the concentration of TiO_2 monomers and clusters were improved with the increase of temperature, the probability of homogeneous nucleation was enhanced in the same way. Scale surface shown in Fig. 3(b) was just produced by this route. In contrast, the dominating reaction route should be *gas phase reaction → heterogeneous growth* in the lower temperature region [8]. TiO_2 monomers and clusters were deposited on the reactor wall and a thin continuous film was formed. Especially because of heterogeneous growth of TiO_2 monomers, the film grew thick and compact, which could be seen in Fig. 3(c). In above two processes, the coagulation and aggregation of TiO_2 particles resulted in a hard scale being formed.

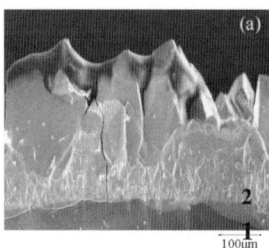  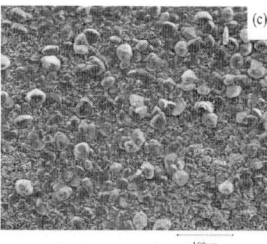

(a) Scale profile 1.subtrate, 2. Scale layer; (b) Scale stripped from front rod; (c) Scale stripped from rear rod.

Fig.3.　SEM pictures of scale layers

4. CONCLUSION

The scales on the reactor wall were generated by the gas-phase oxidation of $TiCl_4$. The state of reactor wall had little effect on scaling formation. the higher the reaction temperature was, the thicker the scale and the compacter the scale structure were. It was concluded that the dominated processes were gas phase reaction, homogeneous nucleation and fine particles/clusters coagulation at high temperature region. However, at the low temperature region, the dominated processes were gas phase reaction and heterogeneous growth.

REFERENCES

[1]　M.K. Akhtar, S. Vemury, S.E. Pratsinis, AIChE J., No. 40 (1994) 1183.
[2]　S. Morooka, T. Okubo, K. Kusakabe, Powder Tech., No. 63 (1990) 105.
[3]　H. Jang, C. Li, Z. Lu, D. Cong, Y. Zhu, J. ECUST., No. 27 (2001) 152.
[4]　R.A. Supurr, H. Myers, Anal. Chem., No. 29 (1957) 760.
[5]　A. Kobata, K. Kuaskabe, S. Morooka, AIChE J., No. 37 (1991) 347.
[6]　W.S. Coblenz, J.M. Dynys, R.M. Cannon, Mater. Sci. Res., No. 13 (1980) 141.
[7]　K. Machenzie, Trans. J. Br. Ceram. Soc., No. 74(1975) 29.
[8]　Y. Zhu, A. Chen, C. Li, J. ECUST., No. 25 (1999) 382.

Studies in Surface Science and Catalysis, volume 159
Hyun-Ku Rhee, In-Sik Nam and Jong Moon Park (Editors)
© 2006 Elsevier B.V. All rights reserved

Conversion of Methane to Hydrogen and Carbon black by D.C. Plasma Jet

Sun-Hee Park and Dong-Wha Park

Department of Chemical Engineering and Regional Research Center for Environmental Technology of Thermal Plasma, Inha University, 253 Yonghyun-dong, Nam-gu, Incheon 402-751, Korea

1. Introduction

The steam reforming of natural gas process is the most economic near-term process among the conventional processes. On the other hand, the steam reforming natural gas process consists of reacting methane with steam to produce CO and H_2. The CO is further reacted or shifted with steam to form additional hydrogen and CO_2. The CO_2 is then removed from the gas mixture to produce a clean stream of hydrogen. Normally the CO_2 is vented into the atmosphere. For decarbonization, the CO_2 must be sequestered[1,2]. The alternative method for hydrogen production with sequestration of carbon is the thermal decomposition of methane.

The purposes of the study were hydrogen production at high efficiency and carbon black production of high quality. Thermal plasma conditions (high temperatures and a high degree of ionization) can be used to accelerate thermodynamically favorable chemical reactions or to provide the energy required for endothermic reforming processes [3]. When methane is heated to high temperatures, the methane is decomposed or cracked into carbon and hydrogen [4]:

$$nCH_4 \rightarrow nC + 2nH_2 \qquad (1)$$

The main gaseous product is the hydrogen. Carbon can be either sequestered or used as a material commodity under less severe CO_2 restraints. It can also be used as a reducing agent in metallurgical processes. From the point of view of carbon sequestration, it is easier to separate, handle, transport, and store solid carbon than CO_2 [5].

2. Experimental

A thermal plasma system has been developed for the decomposition of methane. A schematic diagram of the experimental apparatus is shown in Fig. 1. The system consists primarily of D.C. plasma torch, plasma reactor and filter assembly. Plasma was discharged between a tungsten cathode and a copper anode using N_2 gas. All the experiments were carried out at atmospheric pressure at 6 kW input electric power and N_2 flow rate of 10 to 12 l/min. The feed gas (CH_4) flow rates were varied from 3 to 15 l/min depending on the operating conditions, shown in Table.1.

The major gaseous components were analyzed by a gas chromatograph equipped with a TCD and a molecular sieve 13X column. The specific surface areas of carbon produced were measured by the BET method(ASAP 2010, Micromeritics). The morphology and particle size of the formed carbon were investigated by the scanning electron microscopy(S-4200, Hitachi

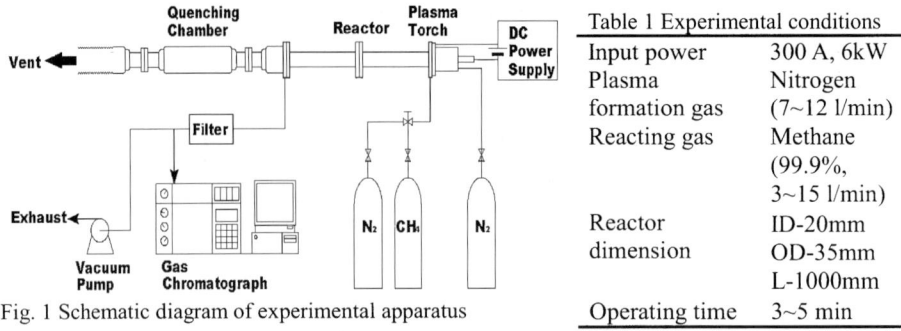

Fig. 1 Schematic diagram of experimental apparatus

Table 1 Experimental conditions	
Input power	300 A, 6kW
Plasma formation gas	Nitrogen (7~12 l/min)
Reacting gas	Methane (99.9%, 3~15 l/min)
Reactor dimension	ID-20mm OD-35mm L-1000mm
Operating time	3~5 min

Co.) and carbon particle size analyzer(LS230, COULTER Co.).

3. Results and discussion

Before the experiment, we carried out the calculation of thermodynamic equilibrium composition during the decomposition of methane by D.C. plasma jet. Chemical equilibrium compositions were calculated by a software program based on Gibbs' free energy minimization[6]. The calculations were performed from 0 to 2000℃. In the case of the steam reforming, as expected, major products were H_2, CO and CO_2 at above 800℃. In the thermal decomposition, the major products were H_2 and solidified carbon at the similar temperature range, which can be easily separated from each other. These calculations imply that high purity of hydrogen can be easily achieved by thermal decomposition without separation process. Fig. 2 shows thermodynamic equilibrium composition in steam reforming process and in thermal decomposition of methane as a function of temperature at atmospheric pressure.

The feed gas (CH_4) and all gaseous products were analyzed by gas chromatography (GC) and the methane conversion and hydrogen yield were calculated from the area of each peak. The results with respect to methane flow rate are shown in Fig 3. The methane conversion was defined as:

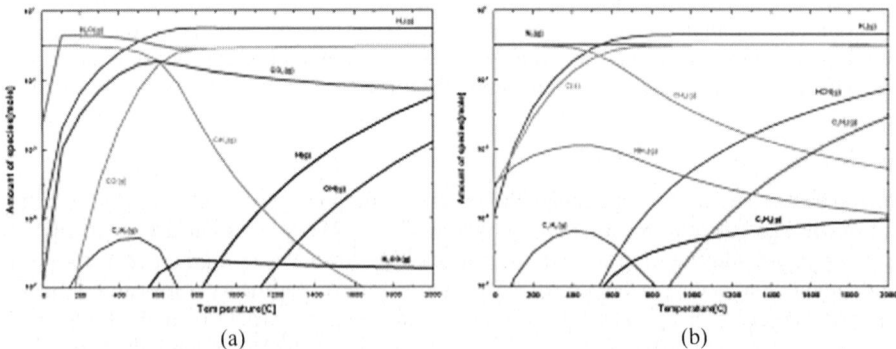

(a)   (b)

Fig. 2 Equilibrium chemical composition
(a) in steam reforming process   (b) in thermal decomposition of methane

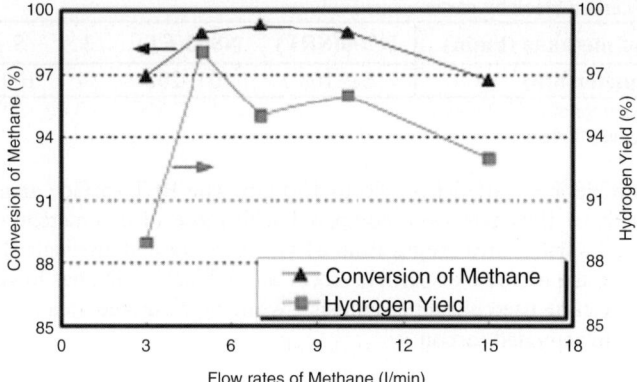

Fig. 3 Methane conversion and hydrogen yield as a function of methane flow rates

$$\text{Conversion of CH}_4 \ (\%) = \frac{\text{In CH}_4 - \text{Out CH}_4}{\text{In CH}_4} \times 100$$

where In CH_4 : amount of the methane gas before the plasma reaction from the area of GC peak, Out CH_4 : amount of the methane gas after the plasma reaction from the area of GC peak.

As shown in Fig. 3, methane conversion was very high (above 90%) over the entire range investigated. At very high methane flow rate (when the flow rate was 15 l/min), the conversion slightly decreased, probably due to the decrease of residence time. Because the hydrogen selectivity is very high, shown in Fig. 3, the hydrogen yield was very close to the methane conversion except at 3 l/min CH_4 flow rate. No other gaseous product was detected even by FT-IR analysis, which means all the products from methane decomposition by thermal plasma are hydrogen and solidified carbon.

The morphology of the carbon was characterized by SEM and the particle size was determined by particle size analyzer. Figure 4 shows the SEM images of the carbon black. It was observed that the carbon black particles exhibit a sphere particle with nano-sized diameter.

For accurate determination of the particle size, the particle size analyzer was used and particle sizes of carbons synthesized are compared with commercial carbon black (N700 and N800, KCB) in Table 2. It was confirmed from particle size analysis that particle size of carbon increased with increase of methane flow rate, as also appeared from SEM analysis.

Fig. 4 SEM images of the carbon black (sampled in reactor)
(a) CH_4: 3 l/min (b) CH_4: 5 l/min (c) CH_4: 7 l/min (d) CH_4: 10 l/min

424

Table 2 Size of carbon black by particle size analyzer

Flow rates of methane (l/min)	N700(SRF)	N800(FT)	3	5	7	10
Particle diameter(nm)	61~100	101~200	84	109	113	116

N700(SRF) : Semi Reinforcing Furnace
N800(FT) : Medium Thermal

The carbon particle size varied from 80 to 120 nm. The BET surface areas of carbon as a function of methane flow rate were compared with those of commercial carbon blacks in Table 3. The BET surface area ranges from 81 to 193 m^2/g with methane flow rates and this decrease is due to the increase of particle size. Carbon black which has lower surface area of 30 to 100 m^2/g can be used in rubber industry, while high surface area (> 700 m^2/g) carbon black is applied to activated carbon.

Table 3 BET Surface Area of carbon blacks by CH_4 flow rate

	IRB # 7	N-234	Ensaco 20	Exp.1	Exp.2	Exp.3	Exp.4
Method	Furnace		MMM	Plasma			
Feed gas(l/min)	Coal tar, PFO		Pyrolysis	3(CH_4)	5(CH_4)	7(CH_4)	10(CH_4)
BET Surface Area (m^2/g)	80	125	65	193	191	100	81

IRB#7, N-234, Ensaco250 : Used to rubber industry   PFO : Pyrolysis fuel oil

4. Conclusion

Direct thermal decomposition of methane was carried out, using a thermal plasma system which is an environmentally favorable process. For comparison, thermodynamic equilibrium compositions were calculated by software program for the steam reforming and thermal decomposition. In case of thermal decomposition, high purity of the hydrogen and solidified carbon can be achieved without any contaminant.

The methane conversion and hydrogen yield were investigated as a function of with respect to methane flow rate and both of the two were very high more than 90%. Particle size and surface area of synthesized carbon were strongly dependent on methane flow rate. Hydrogen produced from thermal plasma can be applied to fuel cell due to its high purity and carbon black can be applied for the synthesis of rubber industry.

It could be concluded that thermal plasma process for methane decomposition is very effective for the production of high purity of the hydrogen as well as synthesis of the carbon black.

5. Acknowledgements

This work was supported by INHA UNIVERSITY Research Grant.

References

[1] M. Steinberg and H. C. Cheng, Int. J. Hydrogen Energy, 14, 797 (1989).
[2] M. Steinberg, Int. J. Hydrogen Energy, 24, 771 (1999).
[3] L. Brombeerg and R. Ramprasad, Int. J. Hydrogen Energy, 25, 1157 (2000).
[4] I. Rusu and J. M. Cormier, Int. J. Hydrogen Energy, 28, 1039 (2003).
[5] N. Z. Muradov, Int. J. Hydrogen Energy, 18, 211 (1993).
[6] Factsage, software program, Version 5.3.1, GTT-Technologies, Germany.

Studies in Surface Science and Catalysis, volume 159
Hyun-Ku Rhee, In-Sik Nam and Jong Moon Park (Editors)

Influence of reducing power on selective oxidation of H_2S over V_2O_5 catalyst in IGCC system

Jong Dae Lee, No-Kuk Park, Ki Bo Han, Si Ok Ryu and Tae Jin Lee[*]

National Research Laboratory, School of Chemical Engineering and Technology, Yeungnam University, Gyeongsan 712-749, Korea.

FAX: +82-53-810-4631. E-mail: tjlee@yu.ac.kr

ABSTRACT:

Among the transition metal oxides tested for the selective oxidation of H_2S in this study, V_2O_5 showed the highest conversion of H_2S and the selectivity of elemental sulfur. Conversion of H_2S on V_2O_5 in the presence of water vapor was 75%, which was 17% lower than that under the water-free condition, at 200 °C due to the degraded reduction of V_2O_5, caused by the decrease of reducing power. The degraded reduction of V_2O_5 could be lessened by physically mixing with zeolite-NaX. Conversion of H_2S on a physically mixed V_2O_5/zeolite-NaX was about 83%, which was 7% higher than that on an unmixed V_2O_5 under the same condition. On the other hand, conversion of H_2S on a chemically mixed V_2O_5/zeolite-NaX was about 20%. It could be concluded that the selective oxidation of H_2S could be improved by the alleviated reduction of V_2O_5 as it was physically mixed with zeolite-NaX. However, the conversion of H_2S on V_2O_5 in presence of coal derived synthesis gas was 30%. It could be enhanced by the increase of oxygen or water content. From these results, it could be concluded that the reducing power of reactants was the most important factor in the selective oxidation of H_2S and the selective oxidation of H_2S can be applied to the hot gas cleanup in IGCC system.

1. INTRODUCTION

The integrated coal gasification combined cycle (IGCC) system is considered as one of the most environmentally sustainable technologies for power generation from coal. The IGCC system consists of coal gasifier, gas cleanup system, and power generation facilities [1]. In the coal gasification process the sulfur contained in coal is converted to hydrogen sulfide (H_2S). Since hydrogen sulfide is a very corrosive material, the high concentration of H_2S in hot coal gas can cause not only air pollution but serious damages to gas turbines. Therefore, it is necessary to reduce the H_2S content of gasified fuel gas to several hundred ppmv [2-4]. We decided to apply a selective oxidation of H_2S for reducing H_2S content of gasified fuel gas.

Our study was focused on the influence of reducing power on the selective oxidation of H_2S over the various transition metal oxides, which would be proceeded by the redox mechanism [5,6]. The redox mechanism and the reducing power [7] in selective oxidation of H_2S can be defined as follows:

Redox mechanism;
$Mat\text{-}O + xH_2S \rightarrow Mat\text{-}O_{1-x} + H_2O + xS$ (Reduction of the matal oxide catalyst)
$Mat\text{-}O_{1-x} + x/2O_2 \rightarrow Mat\text{-}O$ (Oxidation of the reduced catalyst by oxygen)

Reducing power;

$$\text{Reducing Power} = \frac{[H_2S] + [CO] + [H_2]}{[O_2] + [H_2O] + [CO_2]}$$

2. EXPERIMENTAL

The selective oxidation of H_2S was studied on MoO_3, Fe_2O_3, MnO_2, WO_3, V_2O_5, and V_2O_5 physically mixed with zeolite-NaX under the water-free condition, in the presence of water vapor, and in the environment of coal-derived synthesis gas. Zeolite-NaX, MoO_3, Fe_2O_3, MnO_2, WO_3 and V_2O_5 were obtained from Aldrich Chemical Company Inc. Catalytic activity for the selective oxidation of H_2S was tested by a continuous flow reaction in a fixed-bed quartz tube reactor with 0.5 inch inside diameter. Gaseous H_2S, O_2, H_2, CO, CO_2 and N_2 were used without further purification. Water vapor (H_2O) was introduced by passing N_2 through a saturator.

Catalytic activity for the selective oxidation of H_2S was tested by a continuous flow reaction in a fixed-bed quartz tube reactor with 0.5 inch inside diameter. Gaseous H_2S, O_2, H_2, CO, CO_2 and N_2 were used without further purification. Water vapor (H_2O) was introduced by passing N_2 through a saturator. Reaction test was conducted at a pressure of 101 kPa and in the temperature range of 150 to 300 °C on a 0.6 gram catalyst sample. Gas flow rates were controlled by a mass flow controller (Brooks, 5850 TR) and the gas compositions were analyzed by an on-line gas chromotograph equipped with a chromosil 310 column and a thermal conductivity detector.

The compositions of reactant flow are listed in Table 1. The composition A is for the basic reaction test under water-free reaction condition, the composition B is for the reaction test under the influence of water, and the composition C is for the reaction test with the coal-derived synthesis gas.

Table 1. Reaction Conditions

	A	B	C
H_2S	1	1	1
O_2	0.5	0.5	0.5
H_2O	-	10	10
H_2	-	-	11.7
CO_2	-	-	6.8
CO	-	-	19
N_2	Balance	Balance	Balance

3. RESULTS AND DISCUSSION

As described in the Experimental section, the reactant flow B contained 10 vol.% water vapor. Figure 1 exhibits the conversion of H_2S and selectivity of elemental sulfur on MoO_3, Fe_2O_3, MnO_2, WO_3 and V_2O_5, in which V_2O_5 displays a higher activity than others. The conversion of H_2S and selectivity of elemental sulfur on V_2O_5 at 275 °C were respectively 69 and 80.6%.

Meanwhile, as shown in Figure 2, the conversion of H_2S on V_2O_5 for the reactant composition B at 225 °C was 17% lower than that for the reactant composition A. As we thought that the deterioration of activity of V_2O_5 under the influence of water vapor was not entirely due to the thermodynamic equilibrium, we conducted TPR/TPO experiments by varying the water contents of reactant flow.

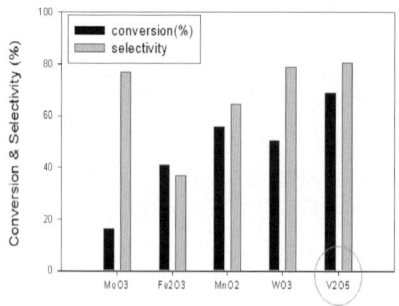
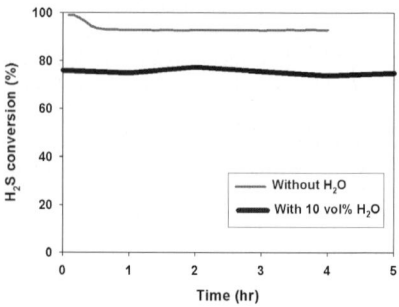

Figure 1. Conversion of H₂S and selectivity of
elemental sulfur on various metal oxides at 275 ℃.

Figure 2. The reactivity of V₂O₅ with and without
H₂O at 225 ℃.

Figure 3 shows that the reduction of V_2O_5 was degraded by the increase of water content. The V_2O_5 sample was reduced to V_2O_3 by TPR experiment, as analyzed by X-ray diffraction. Figure 4 shows that the reoxidation of the V_2O_3 was enhanced by the increase of water content.

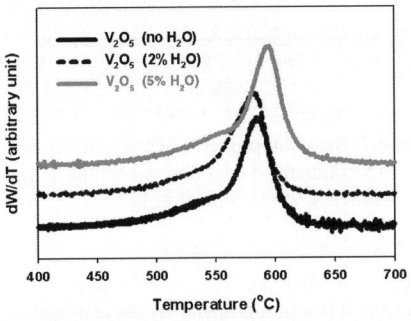

Figure 3. TPR experiments of V_2O_5 by varying
the water contents of reactant flow.

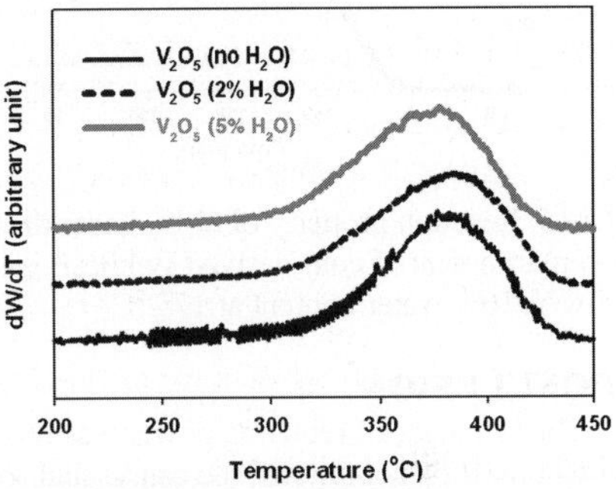

Figure 4. TPO experiments of V_2O_5 by varying
the water contents of reactant flow.

It can be suggested that the lower activity of V_2O_5 under the influence of water vapor was caused not only by the shift of thermodynamic equilibrium but also by the reduction of V_2O_5. It was believed that the decrease of the V_2O_5 reduction property, caused by the decrease of the reducing power, leaded to the deterioration of the activity of V_2O_5 catalyst in the selective oxidation of H_2S under the influence of water vapor.

In reaction condition C, Figure 5 shows that the optimal molar ratio of O_2/H_2S is 0.6, different from the stoichiometric ratio obtained under water-free condition. While COS was detected for the ratios less than 0.6, SO_2 was detected for ratios greater than 0.6.

Figure 6 and Figure 7 show the reaction properties for the 10% and 18% water contents over V_2O_5 under the environment of coal-derived synthesis gas. The reaction property of V_2O_5 was enhanced by the decrease of reactant reducing power, controlled by the increase of water content.

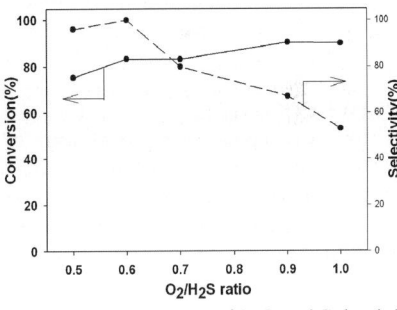

Figure 5. Conversion of H_2S and Selectivity of
elemental sulfur for the various O_2/H_2S.

V_2O_5 was excellent for the selective oxidation of H_2S for the reactant composition simulating coal-derived synthesis gas with 18% water content, supporting its potential use in the hot gas cleanup in IGCC system. The yield of elemental sulfur was 98% under the environment of coal-derived synthesis gas with 18% water content at 175 °C.

From these results, it was believed that the decrease of the V_2O_5 reoxidation property, caused by the increase of the reducing power, leaded to the deterioration of the activity of V_2O_5 catalyst in the selective oxidation of H_2S under the environment of coal-derived synthesis gas. And the decreased reoxidation property of the V_2O_5 was enhanced by increasing oxygen and/or water content.

428

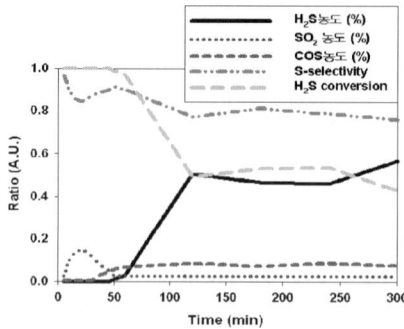
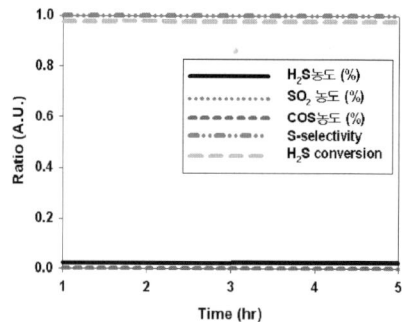

Figure 6. Reaction property of V_2O_5 under the environment of coal-derived synthesis gas with 10% water content at 175 °C.

Figure 7. Reaction property of V_2O_5 under the environment of coal-derived synthesis gas with 18% water content at 175 °C.

4. CONCLUSION

The influence of reducing power was observed in selective oxidation of H_2S on V_2O_5 in three different reaction conditions; the causes and solutions of lessened reaction property by the changed reducing power could be summarized as follows:

1. The V_2O_5 catalyst was suitable for the hydrated reactant flows.
2. The presence of water vapor leaded to the decrease of reactivity caused by the deterioration of the V_2O_5 reduction capacity.
3. The decrease of the V_2O_5 reoxidation property, caused by the increase of the reducing power, leaded to the deterioration of the activity of V_2O_5 catalyst in the selective oxidation of H_2S under the environment of coal-derived synthesis gas. And the decreased reoxidation property of the V_2O_5 was enhanced by increasing oxygen and/or water content.
4. The control of reducing power could be a suitable solution for the deterioration of the activity of V_2O_5 in the selective oxidation of H_2S.

ACKNAWLEDGMENT

This work was supported by the Ministry of Science and Technology (MOST) through the Korea Institute of Science and Technology Evaluation and Planning (KISTEP) as a Nation Research Laboratory (NRL) project.

REFERENCES

1. O. Shinada, A. Yamada and Y. Koyama, Energy Conversion and Management, 43 (2002) 1221.
2. S.O. Ryu, N.K. Park, C.H. Chang, J.C. Kim and T.J. Lee, Ind. Eng. Chem. Res. 43 (2004) 1466.
3. J.H. Pi, J.D. Lee, N.K. Park, S.O. Ryu and T.J. Lee, J. Korean Ind. Eng. Chem., 14 (2003) 813.
4. T.J. Lee, W.T. Kwon, W.C. Chang and J.C. Kim, Korean J. Chem. Eng., 14 (1997) 513.
5. M.Y. Shin, C.M. Nam, D.W. Park and J.S. Chung, Appl. Catal. A: General, 211 (2001) 213.
6. S.W. Chun, J.Y. Jang, D.W. Park, H.C. Woo and J.S. Chung, Appl. Catal. B: Environmental, 16 (1998) 235.
7. R.P. Gupta and S.K. Gangwal, Topical Report to DOE/METC, 15, (1992).

Studies in Surface Science and Catalysis, volume 159
Hyun-Ku Rhee, In-Sik Nam and Jong Moon Park (Editors)

Pyrolysis characteristics of polyethylene using waste catalysts

Ki Min Park[a], Tae Young Kim[a], Hwan Beom Kim[a], Seung Jai Kim[a, b]and Sung Yong Cho[a]

[a]Department of Environmental Engineering, [b]Environmental Research Institute, Chonnam National University, 300 Yongbong-dong, Buk-gu, Gwangju, 500-757, Republic of Korea

1. INTRODUCTION

Since the World War II, plastics have been used extensively because of special qualities such as strength, durability, manufacturing, chemical stability, heat stability and cheap cost of production. Most of the waste plastics are landfilled or incinerated; however, these methods are facing great social resistance because of environmental problems such as air pollution and soil contamination, as well as economical burden due to the increase of space and disposal costs [1,2]. Accordingly, recycling of the waste plastics has become an important issue worldwide. This method can be classified as energy recovery, material recycling and chemical recycling. One prevalent alternative method is the production of converted fuel and chemicals by means of the thermal or catalytic degradation of polymers [3]. These products can be used as valuable raw materials. Furthermore, the catalytic hydrocarbon production from polymer has a great advantage compared to the thermal degradation process, due to the high-quality hydrocarbon production with low energy requirement [4].

In this work, LDPE and HDPE were used as the waste plastics and ZSM-5 and RFCC were used as the waste catalyst. The effects reaction temperature and catalyst concentration on the production of liquid products were investigated in a semi-batch reactor.

2. EXPERIMENTAL

The pyrolysis of the plastics was carried out in a semi-batch reactor which was made of cylindrical stainless steel tube with 80mm in internal diameter and 135mm in height. A schematic diagram of the experimental apparatus is shown in Fig. 1, which includes the main reactor, temperature controller, agitator, condenser and analyzers.

The plastic samples used in this study were palletized to a form of 2.8~3.2㎜ in diameter. The molecular weights of LDPE and HDPE were 196,000 and 416,000, respectively. The waste catalysts used as a fine powder form. The ZSM-5 was used a petroleum refinement process and the RFCC was used in a naphtha cracking process. The BET surface area of ZSM-5 was 239.6 m^2/g, whose micropore and mesopore areas were 226.2 m^2/g and 13.4 m^2/g, respectively. For the RFCC, the BET surface area was 124.5 m^2/g, and micropore and mesopore areas were 85.6 m^2/g and 38.89 m^2/g, respectively. The experimental conditions applied are as follows: the amount of reactant and catalyst are 125 g and 1.25-6.25 g, respectively. The flow rate of nitrogen stream is 40 cc/min, and the reaction temperature and heating rate are 300-500 ^oC and 5 ^oC/ min, respectively. Gas products were vented after cooling by condenser to -5 ^oC. Liquid products were collected in a reservoir over a period of reaction time and were measured by weight. Solid yield was determined from the weights of the initial solid and the residual solid after reaction. Accordingly, gas yield was obtained from the mass balance between the initial reactant amount and liquid and solid products after reaction. Produced oil was analyzed by Simdis-GC (HP-6890 series) using mixed n-paraffin standard (Standard carbon C5-C44, Sigma-Aldrich Co.).

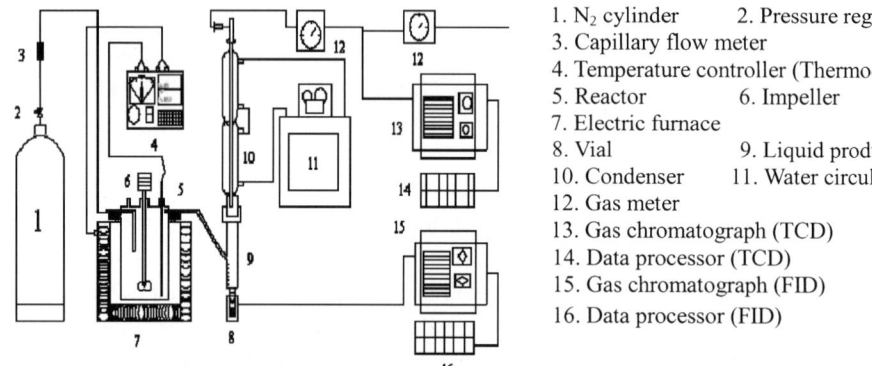

1. N$_2$ cylinder   2. Pressure regulator
3. Capillary flow meter
4. Temperature controller (Thermocouple)
5. Reactor   6. Impeller
7. Electric furnace
8. Vial   9. Liquid product trap
10. Condenser   11. Water circulator
12. Gas meter
13. Gas chromatograph (TCD)
14. Data processor (TCD)
15. Gas chromatograph (FID)
16. Data processor (FID)

Fig. 1. Schematic diagram of the experimental apparatus

3. RESULTS AND DISCUSSION

Effect of temperature was studied without catalyst. Isothermal pyrolysis was observed in the range of 400-500 °C. Fig. 2 shows the initial temperature of degradation and the cumulative product yield for LDPE and HDPE. The initial temperature of degradation for LDPE and HDPE is about 420 °C. And the cumulative liquid product yield for LDPE is greater than that for HDPE. To investigate the catalytic degradation of the plastics using ZSM-5 and RFCC, the catalytic degradations were performed under the same experimental conditions as that of thermal-catalytic degradation. Fig. 3 shows the yield distributions of degradation products as a function of lapsed time. As can be seen in this figure, the rate of degradation of LDPE using ZSM-5 is faster than that of RFCC, due to the greater specific surface area of the ZSM-5 compared to that of RFCC. Fig. 4 shows the rate of degradation for HDPE. The rate of degradation for HDPE using the catalysts is faster than that for thermal degradation. Comparing the data in Fig. 3 and Fig. 4, the rate of degradation for LDPE is faster than that for HDPE. As can be seen in Figs. 3 and 4, the degradation product yield for ZSM-5 catalyst increased sharply in the early stage of reaction, whereas they increased smoothly for thermal or RFCC degradation. If a catalyst is deactivated rapidly by the catalytic degradation, the pattern of product yield-lapse time curve for the catalytic degradation will be similar to that for the thermal degradation. The curve patterns of thermal and RFCC degradations are similar, however, the differences in lapsed between the thermal degradation and ZSM-5 catalytic degradation are gradually increased with the increase of product yield. This can be explained by the fact that the ZSM-5 deactivation rate is decreased with progress of degradation reaction.

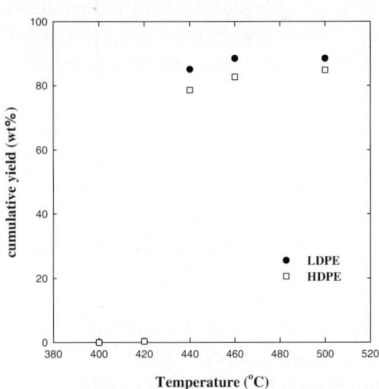

Table 1. Liquid products at different catalysts.

Catalyst		Carbon number (wt%)		
		C_8~C_{12}	C_{13}~C_{24}	C_{25}<
Pure	LDPE	44.5	54.0	1.5
	HDPE	44.0	55.5	0.5
ZSM-5	LDPE	97.4	2.2	0.4
	HDPE	98.5	1.4	0.1
RFCC	LDPE	49.7	46.9	3.4
	HDPE	47.1	44.8	8.1

Fig. 2. Cumulative liquid product yield from thermal degradation of LDPE and HDPE
at different temperature

Figs. 5 and 6 show the initial temperature of degradation for LDPE and HDPE with or without catalyst. The initial temperature of degradation for LDPE using ZSM-5 is 360 ℃, in the case of HDPE is 380 ℃. Thermal degradation temperature for LDPE and HDPE is about 420 ℃ as shown in Fig. 2. The catalyst degradation temperature is lower than that of thermal degradation and the difference is 40~60 ℃. And as can be seen in Figs 5 and 6, the catalytic degradation process shows shorter lapse time for a given product yield than the thermal degradation process, due to the activity of the catalysts in the catalytic degradation. The liquid product yield for different catalytic conditions is shown in Table 1. As can be seen in this table, the amount of gasoline region products (C_8-C_{12}) by the ZSM-5 degradation is much greater than that of the others.

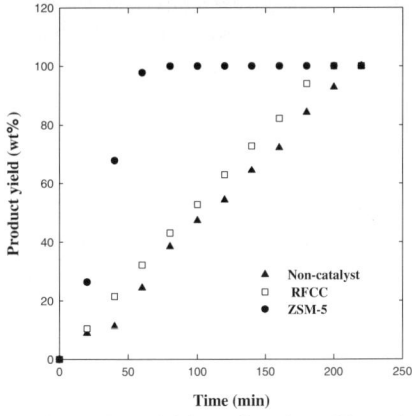

Fig. 3. Product yield as a function of lapsed
time for thermal and catalytic degradation of
LDPE

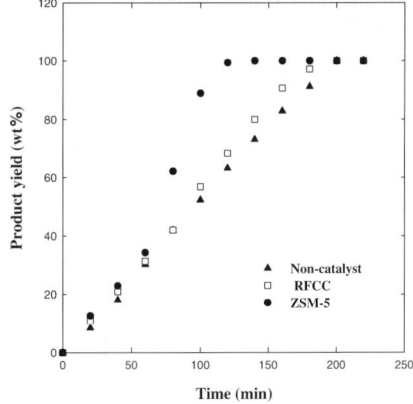

Fig. 4 Product yield as a function of lapsed
time for thermal and catalytic degradation of
HDPE

432

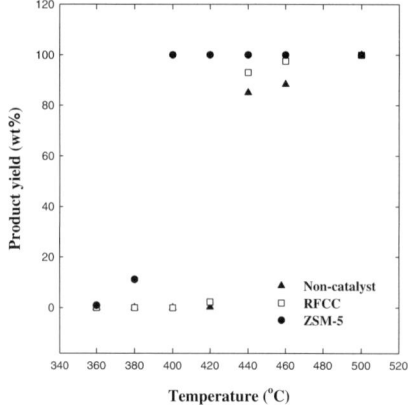 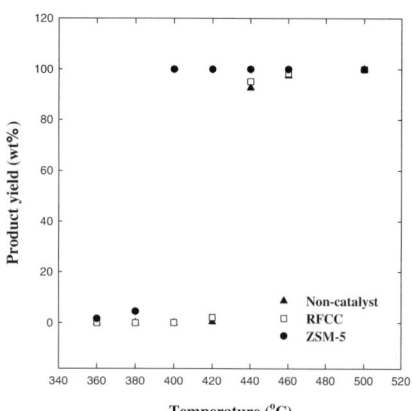

Fig. 5. Cumulative liquid product yield from thermal and catalytic degradation of LDPE at different temperature

Fig. 6. Cumulative liquid product yield from thermal and catalytic degradation of HDPE at different temperature

4. CONCLUSIONS

1. The rate of degradation of LDPE and HDPE was increased linearly with lapsed time and the rate of degradation of LDPE was faster than that of HDPE.

2. The product yield of LDPE and HDPE for the same degradation time is in the order of ZSM-5 catalyst > RFCC catalyst > thermal, and the cumulative yield of liquid product from LDPE was greater than that from HDPE.

3. The ZSM-5 deactivation rate is decreased with progress of degradation reaction. And also the catalytic degradation of the plastics occurred at lower reaction temperature than the thermal degradation by about 40-60 °C.

ACKNOWLEDGEMENTS

This research was financially supported by the KRF (Y00-316).

REFERENCES

[1] Smith, R.: Overview of Feedstock Recycling of Commingled Waste Plastics, Presentation at the consortium for fossil fuel Liquefaction Science, 9th Annual Meeting, Pipestem. WV, August, (1995).

[2] D. C. Kim, W. R. Yoon and K. B. Choi, "A Study on the Development of Energy Recovery from Specified Waste (II)", KIER REPORT, 982218, 13-15 (1998).

[3] J. M. Arandes, M. Olazar and J. Bilobao, "Transformation of several Plastic Waste into Fuels by Catalytic crackintg", Ind. Eng. Chem. Res., 36, 4528 (1997)

[4] D.S Scott, S.R. Czernik, J. Piskorz, D.St, A.G. Radlein, "Fast pyrolysis of plastic wastes", Energy fuels 4, pp. 407-411 (1990).

Studies in Surface Science and Catalysis, volume 159
Hyun-Ku Rhee, In-Sik Nam and Jong Moon Park (Editors)

Degradation of polystyrene using natural clay catalysts

Kyo-Hyun Cho, Byung-Sik Jang, Kyung-Hoon Kim and Dae-Won Park*

Division of Chemical Engineering, Pusan National University, Busan 609-735, Korea
E-mail : dwpark@pusan.ac.kr

1. INTRODUCTION

Plastic waste disposal has been recognized as worldwide environmental problem. Therefore, in recent years, increased attention has been paid to the recycling of synthetic polymer waste [1]. This can contribute to solving pollution problem and the reuse of cheap and abundant waste products. The waste plastics are thermally or catalytically degraded into gases and oils, which can be used as resources in fuels or chemicals. The oils produced by catalytic degradation are known to contain a relatively narrow distribution of hydrocarbons [2,3]. An excellent summary on the catalytic recycle of polymers was reported by Uemichi [3]. Natural clays are known to possess a high ability to catalyze reactions in either polar or non-polar media. Acid activation of clay materials is one of the most effective methods to produce catalysts with high acidity, surface area, porosity and thermal stability [4-6]. Halloysite, montmorillonite and pyrophyllite are a good example of clay materials. In our previous works, we studied the degradation of polystyrene using several catalysts such as natural clinoptillorite zeolite [7] and ultra-stable Y zeolite [8]. The object of this study is to compare the performance of the three natural clay materials in the catalytic degradation of polystyrene. They have advantages of easy availability, large pore size, and low cost.

2. EXPERIMENTAL

2.1. Material and catalysts

PS, in powder form, was supplied by LG Chemical Co. (Grade 50IS, Mn=98000-99000, melt index=7.5 g/10 min, density=1.03 g/cm^3). Several types of solid acid catalysts such as natural halloysite (NH; $Al_2O_3 \cdot 2SiO_2 \cdot 4H_2O$) and pyrophyllite (NP; $Al_2Si_4O_{10}(OH)_2$), occurring in south-east area of Korea, and montmorillonite (K-30; reagent of Fluka) were evaluated through the degradation experiments of PS. NH and NP were ion-exchanged three consecutive times with 1M NH_4Cl solution at 70~80℃ for 20 h. The catalysts exchanged with NH_4^+ were dried at 120℃ overnight, and then calcined in air at 500℃ to obtain the proton(H^+)-exchanged catalysts like HH and HP.

2.2. Apparatus and procedure

The catalytic degradation of PS was carried out in a semi-batch reactor where nitrogen is continuously passed with a flow rate of 30 mL/min. A mixture of 3.0 g of PS and 0.3 g of the catalyst was loaded inside a Pyrex vessel of 30 mL and heated at a rate of 30 °C/min up to the desired temperature. The distillate from the reactor was collected in a cold trap(-10 °C) over a period of 2 h. The degradation of the plastic gave off gases, liquids and residues. The residue means the carbonaceous compounds remaining in the reactor and deposited on the wall of the reactor. The condensed liquid samples were analyzed by a GC (HP6890) with a capillary column (HP-1MS).

3. RESULTS AND DISCUSSION

Table 1 shows chemical compositions of clay catalysts measured by XRF analysis. SiO_2 and Al_2O_3 are main components of the three clay catalysts with minor amount of Na_2O, Fe_2O_3 and others. The Si/Al ratio increased from HH <HP <K-30.

Table 1

Chemical compositions of clay catalysts

Catalyst	Composition (wt.%)							Si/Al ratio
	SiO_2	Al_2O_3	Na_2O	Fe_2O_3	Others	LOI	Total	
HH	53.3	33.7	1.4	0.9	4.4	6.0	99.7	2.7
HP	74.3	21.1	0.1	0.3	1.8	2.0	99.6	6.0
K-30	78.0	10.5	0.2	0.2	2.4	5.9	99.1	12.7

Table 2 shows pore volumes (micropores, meso-+macropores), pore sizes, and specific surface areas of the catalysts. K-30 has the biggest pore volume and surface area among the three clay catalysts. HP has slightly higher micropores than HH, however, the latter has much higher meso- and macropores than the former. Table 3 lists the gaseous, liquid products, and residues obtained in the catalytic degradation of PS at 400 °C for 2 h. The amount of gaseous products was calculated by subtracting the weight of liquid products, residues and catalyst from the total weight of PS sample and fresh catalyst initially loaded to the reactor. The amount of residue was measured by dissolving it with n-hexane. In all cases, the liquid oils were main products. The amount of residue decreased as HP>HH>K-30, which is consistent with the order of pore volume and surface area. It means that K-30 can increase the approach of the intermediately degraded fragments of polystyrene for their cracking to lower carbon number aromatics.

Table 2
Pore volume distribution and surface area

Catalyst	Micro-pore volume[a] (cm^3/g)	Meso-+macro- pore volume (cm^3/g)	Total pore volume[b] (cm^3/g)	BET surf. area (cm^2/g)	Pore size (Å)
HH	0.00037	0.143	0.143	27	10-12
HP	0.00096	0.077	0.078	20	15-20
K-30	0.01980	0.575	0.595	330	14-40

a : t-plot method,   b : P/Po~0.99

Table 3
Yield(wt.%) of product in the catalytic degradation of PS at 400 ℃ for 2 h

Catalyst	Gas (wt.%)	Liquid (wt.%)	Residue (wt.%)
HH	9.09	78.27	12.14
HP	5.14	78.5	16.36
K-30	10.2	79.0	10.8

Composition of some major liquids products, formed in the degradation of PS at 400 °C for 2 h is presented in Table 4. The catalytic degradation over HP exhibited the highest selectivity of styrene(58.80%) and the lowest selectivity of ethylbenzene(12.37%) and benzene(0.15%) compared to the degraded aromatic products with K-30(styrene 43.52%, ethylbenzene 21.10%), and HH(styrene 56.79%, ethylbenzene 13.27%). The increase of pore volume and specific surface area resulted in an increase of ethylbenzene and a decrease of styrene. The large pore may be considered to facilitate the hydrogenation of styrene fragment to produce ethylbenzene [9]. In a separate experiment, thermal degradation showed the highest selectivity to styrene(70.10%) and the lowest selectivity to ethylbenzene(8.80%). One can also see that the amount of styrene dimer and trimer were also observed for all the three clay catalysts. When the degradation temperature increased from 400 to 450℃, the production of styrene increased, but that of ethylbenzene decreased.

4. CONCLUSION

The performance of clay materials (Halloysite, Pyrophyllite, Montmorillonite K-30) in the degradation of polystyrene (PS) was investigated in this study. The catalysts showed good catalytic activity for the degradation of PS with high selectivity to aromatics liquids. Styrene is the major product, and ethylbenzene is the second most abundant one in the liquid product.

436

The increase of surface area and pore volume enhanced the production of ethylbenzene by promoting further hydrogenation of the degraded styrene fragments. High degradation temperature favored the selectivity to styrene monomer.

Table 4

Selectivity of some major products formed in the degradation of PS at 400 ℃ for 2 h

Aromatics	HH	HP	K-30
Benzene	0.21	0.15	2.99
Toluene	8.35	8.12	8.22
Ethylbenzene	13.27	12.37	21.10
Styrene	56.79	58.80	43.52
iso-Propylbenzene	2.65	2.64	5.02
α-Methylstyrene	9.60	9.69	8.59
n-Propylbenzene	0.14	0.14	0.17
Others[a]	0.65	0.63	1.34
C_{11}-C_{15}	1.08	0.76	2.81
C_{16}-C_{21}	6.53	6.41	6.12
C_{22}-C_{30}	0.73	0.29	0.12

[a] Other aromatic compounds having C_5-C_{10}

ACKNOWLEDGEMENTS

This work is financially supported by the Ministry of Education and Human Resources Development(MOE) and the Ministry of Commerce, Industry and Energy(MOCIE) through the fostering project of the Industrial-Academic Cooperation Centered University.

REFERENCES

[1] R.C. Mordi, R. Fields, J. Dwyer, J. Anal. Appl. Pyrol., 29 (1994) 45.

[2] H. Ohkita, R. Nishiyama, Y. Tochihara, T.Mizushima, Ind. Eng. Chem. Res., 32 (1993) 3112.

[3] Y. Uemichi, Catalysis (Japan)., 37 (1995) 286.

[4] F. Hojabri, J. Appl. Chem. Biotechnol., 21 (1971) 87.

[5] D. Njopwouo, G. Roques, R.A. Wandji, Clay Miner., 21 (1987) 145.

[6] R. Fahn, K. Fenderl, Clay Miner., 18 (1983) 447.

[7] S.Y. Lee, J.H. Yoon, J.R. Kim, D.W. Park, J. Anal. Appl. Pyrol., 64 (2002) 71.

[8] J.W. Tae, B.S. Jang, K.H. Kim, D.W. Park, React. Kinet. Catal. Lett., 84(1) (2005) 167.

[9] U. Flessner, D.J. Jones, J. Roziere, J. Zajac, L. Storaro, M. Lenarda, M. Pavan, A. Jimenez, E. Rodriguez-Castellon, M. Trombetta, G. Busca, J. Mol. Catal. A., 168 (2001) 247.

Studies in Surface Science and Catalysis, volume 159
Hyun-Ku Rhee, In-Sik Nam and Jong Moon Park (Editors)

Catalytic degradation of high density polyethylene over post grafted MCM-41 catalyst: kinetic study

Seungdo Kim[1], Eun Suk Jang[1], Joo-Sik Kim[2], Jong-Ki Jeon[3],
Jin-Heong Yim[4], Ji Man Kim[5], Jinho Jung[6], Young-Kwon Park[2*]

[1]Dept. of Environmental System Engineering, Hallym University, Chuncheon, 200-702, Korea
[2]Faculty of Environmental Engineering, University of Seoul, 90 Jeonnong-dong, Dongdaemun-gu, Seoul 130-743, Korea
[3]Dept. of Chemical Engineering, Kongju National University, Gongju 314-701, Korea
[4]Div. of Advanced Materials Eng., Kongju National University, Gongju 314-701, Korea
[5]Dept. of Chemistry and Sungkyunkwan Advanced Institute of Nano Technology, Sungkyunkwan University, Suwon 440-746, Korea
[6]Div. of Environmental Science & Ecological Eng., Korea University, Seoul, 136-713, Korea
*Corresponding author: catalica@uos.ac.kr

The catalytic activities of Al-MCM-41-P, synthesized through the post-synthetic grafting method, and Al-MCM-41-D, synthesized through the direct sol-gel method, in the pyrolysis of high-density polyethylene (HDPE) were compared in the present work. Thermogravimetric analyses under static conditions were also carried out to estimate the activation energies of the catalytic decomposition reactions of HDPE. Higher catalytic activities and lower activation energies of the catalytic decomposition reactions of HDPE were observed when Al-MCM-41-P was applied compared to when Al-MCM-41-D was applied, which may be due to the fact that Al-MCM-41-P has more acid sites compared to Al-MCM-41-D.

1. INTRODUCTION

The common ways of disposing of waste plastics—land-filling and incineration—have posed serious environmental problems. This explains why a new technology for the disposal of waste plastics, pyrolysis, has drawn much attention: it has been regarded as more environmentally sound than the conventional treatment methods. Furthermore, pyrolysis can convert waste plastics into valuable gas, oil, and solid products [1]. As such, the pyrolysis of waste plastics has contributed greatly to addressing the problem of how to dispose of the troublesome wastes. Pyrolysis, however, has a number of disadvantages, among them low selectivity and high energy input. To address these disadvantages, or to improve the product selectivity of the process and to reduce its energy input, catalytic pyrolysis was introduced. Various kinds of zeolites have been used in researches on this process, namely: ZSM-5, Y-type, zeolite beta, mordenite, etc. [2]. Among these, it is ZSM-5 that has been intensively investigated owing to its strong acidity. The use of ZSM-5 in the pyrolysis of waste plastics, however, has revealed technical problems, such as the formation of a relatively high quantity of gas products and coke near the pore entrance. The formation of these products could be minimized by using mesoporous materials, such as MCM-41, instead of ZSM-5. MCM-41 has a high potential to be an effective pyrolysis catalyst of waste plastics due to its high surface area, tunable uniform mesopores (from 20 to 100 Å), and moderate acid strength [3]. To synthesize Al-MCM-41, various routes have been developed to incorporate aluminum into the framework of MCM-41. One of these, direct sol-gel synthesis, has been applied extensively for this purpose. Al-MCM-41 from direct sol-gel synthesis (Al-MCM-41-D) has been used in many research areas [4]. It exhibited not only unfavorable hydrothermal deterioration but also low concentration and strength of Brønsted acid sites even with high aluminum content. Post-synthesis modification has been developed to promote structural stability and to incorporate various metals easily into the siliceous MCM-41 support [5].

In this study, post-synthetic metal grafting was used to incorporate Al into MCM-41 (Al-MCM-41-P), which was probably the first attempt at the catalytic degradation of waste plastics. The objective of this study was to investigate the kinetic aspect of high-density polyethylene (HDPE) over Al-MCM-41-P.

2. EXPERIMENTAL

The powdered HDPE that was used in this study was supplied by Samsung Total Co., Ltd. in Korea. Mesoporous materials, particularly MCM-41, were synthesized following the procedure described elsewhere [4], using sodium silicate as the silica source and cetyltrimethylammonium bromide as the structure-directing agents. Al-MCM-41 (Si/Al = 15, 30, 60) catalysts were prepared using the post-synthetic metal grafting method [5]. A state-of-the-art thermobalance was introduced to evaluate the catalytic pyrolysis kinetics of HDPE. The device was designed to constitute a reliable isothermal environment in the reactor, to minimize the thermal decomposition during the sample insertion, and to continuously monitor the weight loss with respect to time. A high stainless-steel tube (5.5 cm i.d. × 1.0 m) was used as a pyrolysis reactor. The reactor was heated to a reaction temperature under a stream of nitrogen with a linear velocity of 8.3 cm/sec, yielding a laminar flow (Re < 100 at most reaction temperatures). The samples were suspended in a 100-mesh stainless-steel wire basket. The sample basket was connected to an electric balance (Presisa Model 205-A) by a 0.3-mm-diameter nichrome wire. The balance was quickly lowered with the use of a winch to insert the samples in a reaction zone after the reaction temperature was stabilized. The weight loss of a sample with respect to time was recorded continuously over time by an online personal computer. About 10 mg of the HDPE sample was thoroughly mixed with an equal quantity of Al-MCM-41. The kinetic parameters were estimated from four isothermal temperatures, namely: 420, 460, 480, and 520 ℃. For the kinetic analysis, the following two assumptions were introduced: (1) the reaction would be of the first order; and (2) catalytic decomposition would be dominant. The kinetic equation, hence, was formulated as follows:

$$\frac{dM_p}{dt} = k(T)M_p,\tag{1}$$

where M_p is the mass of HDPE at a time (mg), $k(T)$ is the reaction constant (sec^{-1}) at the reaction temperature T, which will be expressed by the Arrhenius relationship. At an isothermal temperature, the reaction constant k was derived from the one best fitted to the experimental kinetic result. The activation energy and frequency factor were then evaluated from the slope and intercept of the Arrhenius plot, respectively, as shown below:

$$\ln k = \ln A - \left(\frac{E}{R}\right)\frac{1}{T},$$ (2)

where A, E, and R were the frequency factor (sec^{-1}), activation energy (J/mol), and gas constant (8.314 J/mol·K), respectively.

3. RESULTS AND DISCUSSION

Fig. 1 shows the thermal decomposition curves of HDPE mixed with Al-MCM-41, with respect to time, at isothermal operating temperatures. Lag periods were formed at the initial stage of decomposition, possibly due to the heat transfer effect, which could delay the decomposition of a sample until the latter reaches the operating temperatures. As the reaction temperature increased, the reaction time became noticeably shorter. The shortening of the reaction time was clearly observed when the reaction occurred at the reaction temperatures between 420 and 460 ℃. The HDPE on Al-MCM-41-P decomposed faster than that on blank and that on Al-MCM-41-D, as shown in Fig. 1(b).

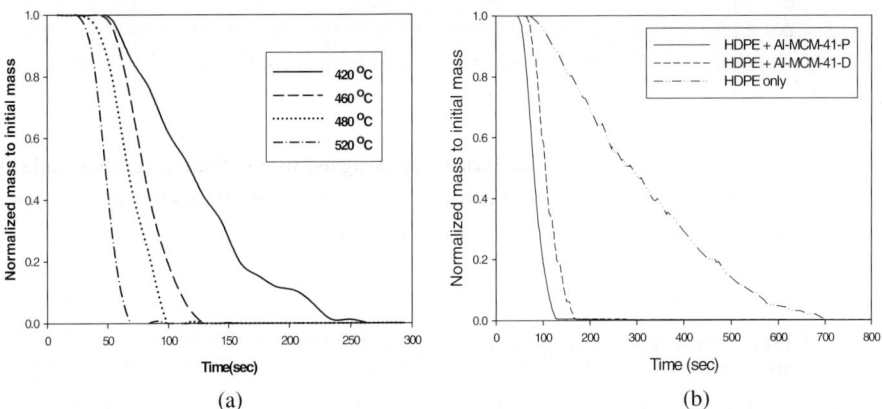

(a)   (b)

Fig. 1. Pyrolysis kinetic results of HDPE mixed with Al-MCM-41
  (a) Al-MCM41-P at isothermal operating temperatures of 420, 460, 480, and 520 ℃
  (b) Al-MCM41-P, -D and blank (HDPE only) at isothermal operating temperature of 460 ℃

Table 1 shows a summary of the apparent activation energies for various catalytic conditions. The apparent activation energy of HDPE mixed with pure MCM-41 is significantly lower than that of HDPE only, indicating that pure MCM-41 is likely to demonstrate catalytic activity. As the Al content increased, the apparent activation energy significantly decreased. Al-MCM-41-P demonstrated activation energies lower than those demonstrated by Al-MCM-41-D at the same Si/Al.

As shown in Fig. 2, the NH$_3$ TPD technique provides information on acid sites over catalysts. While Al-MCM-41-P and Al-MCM-41-D have almost the same acid strengths due to their similar temperature peak of around 250°C, Al-MCM-41-P has more acid sites compared to Al-MCM-41-D. It can be gleaned from this result that the catalytic activity of Al-MCM-P is better than that of Al-MCM-D not because of Al-MCM-P's acid strength but because it has more acid sites. The oil products over

Al-MCM-41-P were distributed mainly over C5-C20 hydrocarbons and were quite similar to those over Al-MCM-41-D in spite of the differences in the number of their acid sites.

Table 1. Kinetic parameters for the HDPE degradation over MCM-41

Sample	Activation Energy (kJ/mol)	ln A (A:sec^{-1})
HDPE only (No catalyst)	206.841	28.592
Si-MCM-41	169.759	22.834
Al-MCM-41-(15)-D	122.601	16.699
Al-MCM-41-(30)-D	139.845	19.298
Al-MCM-41-(60)-D	144.907	20.415
Al-MCM-41-(15)-P	100.548	12.886
Al-MCM-41-(30)-P	119.394	15.953
Al-MCM-41-(60)-P	131.118	18.539

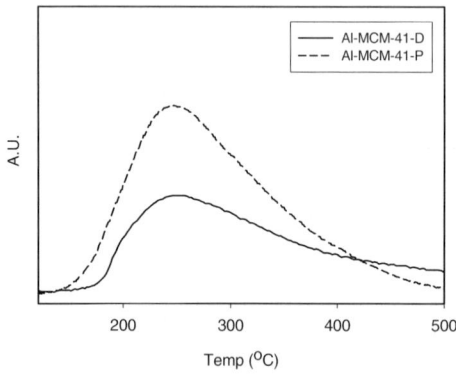

Fig. 2. NH$_3$ TPD of Al-MCM-41-D and Al-MCM-41-P

4. CONCLUSIONS

The two methods of incorporating Al into MCM-41 provided HDPE pyrolysis products of the similar quality. Al-MCM-41-P, however, demonstrated higher catalytic activity compared to Al-MCM-41-D, as indicated by its lower activation energies. The higher catalytic activity of Al-MCM-41-P could be attributed to the fact that it has more acid sites compared to Al-MCM-41-D. Thus, the use of Al-MCM-41-P is expected to be more effective than the use of Al-MCM-41-D in the pyrolysis of HDPE.

References

[1] W. Kaminsky, J.S. Kim, *J. Anal. Appl. Pyrolysis* **51**, 127 (1999).
[2] Y.K. Park, J.S. Kim, J. Choi, J.K. Jeon, S.D. Kim, S.S. Kim, J.M. Kim, K.S. Yoo, *J. Korea Soc. Waste Management*, **20**, 56 (2003).
[3] J. Aguado, J.L. Sotelo, D.P. Serrano, J.A. Calles, J.M. Escola, *Energy Fuels* **11**, 1225 (1997).
[4] J.M. Kim, J.H. Kwak, S. Jun and R. Ryoo, J. Phys. Chem., **99**, 16742 (1995).
[5] R. Ryoo, S. Jun, J.M. Kim, M.J. Kim, Chem. Commun. 2225 (1997).

Studies in Surface Science and Catalysis, volume 159
Hyun-Ku Rhee, In-Sik Nam and Jong Moon Park (Editors)

Synergistic Roles of NO and NO_2 in Selective Catalytic Reduction of NO_x by NH_3

Joon Hyun Baik[a], Jeong Hyun Roh[a], Sung Dae Yim[b], In-Sik Nam[*,a], Jong-Hwan Lee[c], Byong K. Cho[c] and Se H. Oh[c]

[a]Department of Chemical Engineering/School of Environmental Science and Engineering, Pohang University of Science and Technology (POSTECH), San 31, Hyoja-dong, Pohang 790-784, Korea

[b]Fuel Cell Research Center, Korea Institute of Energy Research (KIER), 71-2 Jang-dong, Daejoen 305-343, Korea

[c]General Motors R&D Center, Warren, Michigan 48090-9055, USA

1. Introduction

It has been reported that NOx reduction activity of SCR catalysts can be improved by pre-oxidation of NO to NO_2 before the SCR reaction in the catalytic reactor [1], which can be done by installing an oxidation catalyst or a nonthermal plasma device upstream of the SCR catalysts [2]. However, the role of NO_2 during the catalytic reduction of NOx has not been clearly understood yet. According to the literature, the rate of reaction between NO_2 and NH_3 is much faster than that between NO and NH_3 over Fe-exchanged TiO_2-pillared clay [3] and HZSM5 catalysts [4]. Kiovsky et al. [5] also observed that NO_2 was more reactive than NO toward NH_3 over HM, and an increase of the NO_2/NO ratio from 2 to 12 in the feed gas stream significantly enhanced NOx reduction activity. However, Koebel et al. [6] reported that the reaction with equal amounts of NO and NO_2 by NH_3 was much faster than that with NO (or NO_2) only over V_2O_5-WO_3/TiO_2 catalyst. Long and Yang [7] reported that the reactivity of FeZSM5 with NO and NO_2 increases in the following order: NO+NO_2 > NO_2 ≫ NO. These literature reports suggest that there may indeed be a synergistic role of NO_2 with NO in the feed gas stream for catalytic reduction of NOx by NH_3, even though there is no comprehensive quantitative analysis on the role of NO_2 for the SCR reaction. In this study, the synergistic effect of NO and NO_2 in the feed gas stream on the catalytic NOx reduction activity has been systematically examined to better understand the roles of NO and NO_2 in the SCR process.

2. Experimental

Three representative SCR catalysts – V_2O_5/TiO_2, FeZSM5 and CuZSM5 – were used in this study. The V_2O_5/TiO_2 catalyst was prepared by the wet impregnation method with an aqueous solution of NH_4VO_3 (Aldrich) on TiO_2 support (Hombikat UV-100). The CuZSM5 and FeZSM5 catalysts were prepared by the wet ion exchange method using $(CH_3COO)_2Cu \cdot H_2O$ (0.01 M, Aldrich) and $FeCl_2$ (0.05 M, Aldrich) solutions with 15 g of ZSM5 (Tosoh, HSZ-830NHA) at room temperature. The contents of V, Cu and Fe are 2.83, 3.06 and 2.67 wt.%, respectively.

The synergistic effects of NO and NO_2 on the NOx reduction activity of the SCR catalysts were examined using a packed-bed flow reactor containing 20/30 mesh size of the catalyst by varying the NO/NO_2 feed ratio. In initial tests of CuZSM5 catalysts, the effect of NO_2 was not detectable at the reactor space velocity of 100,000 h^{-1}. When the reactor space velocity was increased to 500,000 h^{-1}, the synergistic effects of NO and NO_2 became clearly detectable over the entire temperature range from 200 to 450 $^{\circ}$C. The concentrations of NO, NH_3 and NO_2 were determined by on-line chemiluminescence NO-NOx analyzer (Thermo Environmental Instrument, Model 42H), NH_3 analyzer with NDIR (Rosemount Analytical, Model 880A) and NO_2 analyzer with electrochemical cell (Testo, Model 350M), respectively. The details of the experimental system and procedures have been already described elsewhere [8].

3. Results and Discussion

Fig. 1 shows the conversions of NO, NO_2 and NOx as a function of the NO_2/NOx feed ratio to the SCR reactor. For all three catalysts (Fig. 1b, 1d and 1f), the synergistic effects of NO and NO_2 on NOx conversion are most pronounced at low temperatures, gradually disappearing with the rising catalyst temperatures. Interestingly, this synergistic effect completely disappears above 400 $^{\circ}$C for both V_2O_5/TiO_2 and FeZSM5 catalysts, while vanishing above 300 $^{\circ}$C for the CuZSM5 catalyst. It is of practical importance to note for the CuZSM5 catalyst that the effect of NO_2 is significant only at low temperatures below 300 $^{\circ}$C, above which the kinetic behaviors of NO and NO_2 are essentially identical. The optimum NO_2/NOx feed ratio for the best synergistic effect is 0.75 for both FeZSM5 and CuZSM5 catalysts, while it varies from 0.5 to 0.75 for the V_2O_5/TiO_2 catalyst.

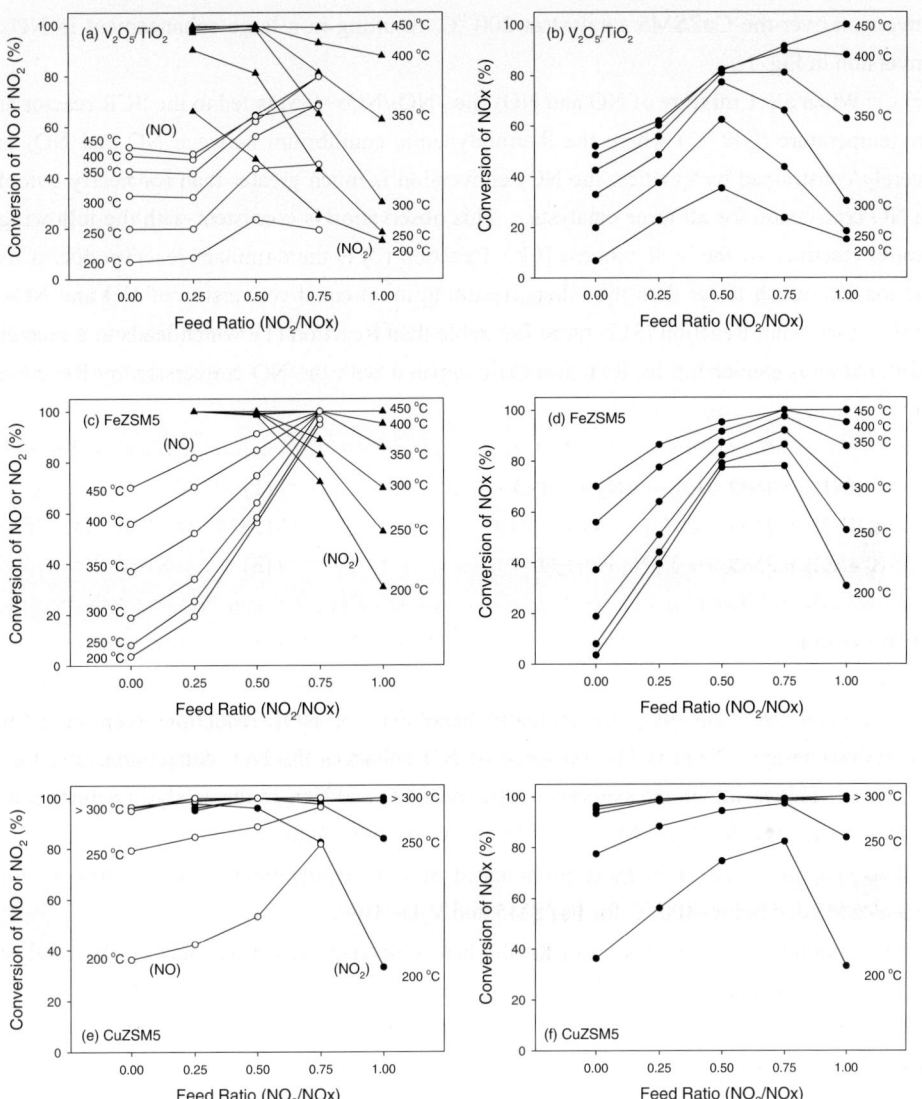

Fig. 1. Effect of NO_2 on the reduction of NOx by NH_3.   Feed gas composition: 500 ppm NOx (NO + NO_2), 500 ppm NH_3, 5 % O_2, 10 % H_2O and N_2 balance; Reactor SV: 500,000 h^{-1}.

The NO and NO_2 conversion data (Fig. 1a, 1c and 1e) indicate that the presence of NO improves the NO_2 conversion while NO_2 improves the NO conversion, with some minor exceptions in the case of V_2O_5/TiO_2 catalyst.   It is particularly noteworthy in Fig. 1e that a small amount of NO (i.e., $NO_2/NOx=0.75$) can make a big improvement in the NO_2

conversion over the CuZSM5 catalyst at 200 °C, resulting in a large enhancement in NOx conversion in Fig. 1f.

When a 1:1 mixture of NO and NO_2 (i.e., $NO_2/NOx=0.5$) is fed to the SCR reactor at low temperature (200 °C) where the thermodynamic equilibrium between NO and NO_2 is severely constrained by kinetics, the NO_2 conversion is much greater than (or nearly twice) the NO conversion for all three catalysts. This observation is consistent with the following parallel reactions of the SCR process [6]: Reaction (2) is the dominant reaction due to its reaction rate much faster than the others, resulting in an equal conversion of NO and NO_2. On the other hand, Reaction (3) is more favorable than Reaction (1), which leads to a greater additional NO_2 conversion by Reaction (3) compared with the NO conversion by Reaction (1).

$$4NH_3 + 4NO + O_2 \rightarrow 4N_2 + 6H_2O \qquad (1)$$
$$4NH_3 + 2NO + 2NO_2 \rightarrow 4N_2 + 6H_2O \qquad (2)$$
$$4NH_3 + 3NO_2 \rightarrow 3.5N_2 + 6H_2O \qquad (3)$$

4. Conclusion

- In general, NO and NO_2 are mutually beneficial for NOx reduction over the SCR catalysts tested. That is, the presence of NO enhances the NO_2 conversion, and vice versa. This results in the synergistic effects of NO and NO_2 in the catalytic reduction of NOx with NH_3 over CuZSM5, FeZSM5 and V_2O_5/TiO_2 catalysts.
- The synergistic effect is most pronounced at low temperatures – below 300 °C for CuZSM5 and below 400 °C for FeZSM5 and V_2O_5/TiO_2.
- The optimum NO/NO_2 feed ratio for the best synergistic effect depends on the catalyst and its temperature.

References

[1] S. Bröer and T. Hammer, Appl. Catal. B: Environ. 28 (2000) 101.
[2] S.J. Schmieg, B.K. Cho and S.H. Oh, Appl. Catal. B: Environ. 49 (2004) 113.
[3] R.Q. Long and R.T. Yang, J. Catal. 190 (2000) 22.
[4] S.A. Stevenson and J.C. Vartuli, J. Catal. 208 (2002) 100.
[5] J.R. Kiovsky, P.B. Koradia and C.T. Lim, Ind. Eng. Chem. Prod. Res. Dev. 19 (1980) 218.
[6] M. Koebel, G. Mania and M. Elsener, Catal. Today 73 (2002) 239.
[7] R.Q. Long and R.T. yang, J. Catal. 207 (2002) 224.
[8] J.H. Baik, S.D. Yim, I.-S. Nam, J.-H. Lee, B.K. Cho and S.H. Oh, Top. Catal. 30-31 (2004) 37.

Studies in Surface Science and Catalysis, volume 159
Hyun-Ku Rhee, In-Sik Nam and Jong Moon Park (Editors)

A mathematical model for the design of extruded honeycomb reactor for selective catalytic reduction of NOx

Jeong-Yeop Koh[a], Jin Woo Choung[a], Joon Hyun Baik[a], In-Sik Nam[a*], Sung-Won Ham[b] and Jeong-Bin Lee[c]

[a]Dept. of Chem. Eng./School of Environ. Sci. & Eng., Pohang University of Science & Technology (POSTECH), San 31, Hyoja-Dong, Pohang 790-784, Korea

[b]Dept. of Chem. Eng., Kyungil University, 33 Buhori, Hayang, Kyungsan 712-701, Korea

[c]Korea Electric Power Research Institute, 103-16 Moonji-Dong, Taejon 305-380, Korea

1. INTRODUCTION

The selective catalytic reduction (SCR) of NO by NH_3 is a well-developed and commercially available technology for the control of nitrogen oxides (NOx) emission from stationary sources including utility boilers [1-2]. One of the most important features for the technology is the development of low-pressure drop reactors such as honeycomb and PPR (Parallel Passage Reactor) along with the high performance of SCR catalysts for the direct installation of the reactor to the actual operating conditions of a utility boiler. The honeycomb-type reactor contains unique advantages over a packed-bed reactor, such as low-pressure drop and high surface areas per unit volume. Hence, many studies on the design and modeling of the honeycomb reactor for the SCR process have been conducted [3-6].

In the present study, honeycomb reactors containing a variety of cell dimensions of 10, 20, 50 and 100 CPSI (cells per in^2) have been extruded with V_2O_5/TiO_2 catalyst for the reduction of NO_x by NH_3. A reactor model considering pressure drop as well as interphase and intraphase diffusions for the extruded honeycomb reactor has been developed for the optimal reactor design of the commercial SCR process. The modeling has been carried out through establishing intrinsic reaction kinetics based upon the Eley-Rideal mechanism [6].

2. EXPERIMENTAL

The honeycomb reactor has been fabricated with KEPOSE-TiO_2 [7] according to the following steps: (1) dry mixing of the solid raw materials, (2) wet mixing and plasticizing with water and organic and inorganic additives, (3) kneading and extrusion of the monolith honeycomb by 1"×1" and 3" ×3" vacuum extruder, (4) drying in the controlled constant humidity and temperature, and (5) calcining at 500°C for 5h. The prepared honeycomb reactor was then dipped into the aqueous solution of NH_4VO_3 diluted in oxalic acid to load 2 wt.% V_2O_5, and the reactor was dried and calcined again at the identical conditions.   The NO removal activity over the honeycomb reactor with 25mmx25mmx100mm dimension has been examined within a range of reactor space velocity from 10,000 to 40,000 h^{-1}, which is defined as the ratio of the total flow rate to the reactor volume occupied by the honeycomb under simulated flue gas condition.   The details of the experimental procedures and conditions have been described elsewhere [2].

3. HONEYCOMB REACTOR MODEL

A mathematical model to predict the performance of extruded honeycomb reactors has been derived on the basis of the model developed by Chae et al. [6].   A material balance for bulk flow within a channel becomes,

$$-u\frac{dC_{NO}^b}{dx} = k_{m,NO}A_e(C_{NO}^b - C_{NO}^s) \tag{1}$$

$$-u\frac{dC_{NH3}^b}{dx} = k_{m,NH3}A_e(C_{NH3}^b - C_{NH3}^s) \tag{2}$$

with the initial conditions: $C_{NO}^b = C_{NO}^{b,0}$ & $C_{NH3}^b = C_{NH3}^{b,0}$ at $x=0$.

At any axial position x of the honeycomb reactor, a material balance over the catalyst layer of thickness dy yields the following equations:

$$D_{e,NO}\frac{d^2C_{NO}}{dy^2} = -r_{NO} \tag{3}$$

$$D_{e,NH3}\frac{d^2C_{NH3}}{dy^2} = -r_{NH3} \tag{4}$$

with the following boundary conditions:

$$\frac{dC_{NO}}{dy}=0 \ \& \ \frac{dC_{NH3}}{dy}=0 \quad at \ y=0, \qquad C_{NO}=C_{NO}^s \ \& \ C_{NH3}=C_{NH3}^s \quad at \ y=R$$

A material balance at axial position x of the honeycomb reactor over the external gas film yields,

$$k_{m,NO}(C_{NO}^b - C_{NO}^s) = D_{e,NO}\left[\frac{dC_{NO}}{dy}\right]_s \tag{5}$$

$$k_{m,NH3}(C_{NH3}^b - C_{NH3}^s) = D_{e,NH3}\left[\frac{dC_{NH3}}{dy}\right]_s \tag{6}$$

The consideration of the pressure drop over the monoliths containing a variety of CPSI (cells per in^2) for the modeling of honeycomb reactor may be required, since Δp of the reactor strongly depends on CPSI of monolith. Eqn. (7) for the pressure drop of the honeycomb was employed to develop the reactor model describing the performance of the honeycomb fabricated in the present work [8]. K_c and K_e indicate contraction and expansion loss coefficient at the honeycomb inlet and outlet, respectively and σ is the ratio of free flow area to frontal area.

$$\Delta P = \frac{28.4Q(d+\theta)^2 \mu}{h^2 d^4}L + \frac{\rho Q^2(d+\theta)^4 \mu}{2h^4 d^4}(K_c + K_e) \tag{7}$$

where $K_c = -0.42\sigma^2 + 0.02\sigma + 1.18$ and $K_e = 0.94\sigma^2 - 2.71\sigma + 1.0$.

4. MODELING OF EXTRUDED HONEYCOMB REACTOR

In Fig. 1, a comparison can be observed for the prediction by the honeycomb reactor model developed with the parameters directly obtained from the kinetic study over the packed-bed flow reactor [6] and from the extruded honeycomb reactor for the 10 and 100 CPSI honeycomb reactors. The model with both parameters well describes the performance of both reactors although the parameters estimated from the honeycomb reactor more closely predict the experiment data than the parameters estimated from the kinetic study over the packed-bed reactor. The model with the parameters from the packed-bed reactor predicts slightly higher conversion of NO and lower emission of NH$_3$ as the reaction temperature decreases. The discrepancy also varies with respect to the reactor space velocity.

The kinetic parameters estimated by the experimental data obtained from the honeycomb reactor along with the packed bed flow reactor as listed in Table 1 reveal that all the kinetic parameters estimated from both reactors are similar to each other. This indicates that the honeycomb reactor model developed in the present study can directly employ intrinsic kinetic parameters estimated from the kinetic study over the packed-bed flow reactor. It will significantly reduce the effort for predicting the performance of monolith and estimating the parameters for the design of the commercial SCR reactor along with the reaction kinetics.

Table 1. Kinetic parameters obtained from packed-bed and honeycomb Reactors

Kinetic parameters	Packed-bed reactor	Honeycomb reactor
$Ea_{,NO}$ (kJ/mol)	40.2	53.6
$Ea_{,NH3}$ (kJ/mol)	215.9	212.5
$\Delta H_{,NH3}$ (kJ/mol)	54.4	50.6
$k_{0,NO}$	4.46×10^5	1.46×10^7
$k_{0,NH3}$	4.68×10^8	4.74×10^8
$K_{0,NH3}$	2260	2280

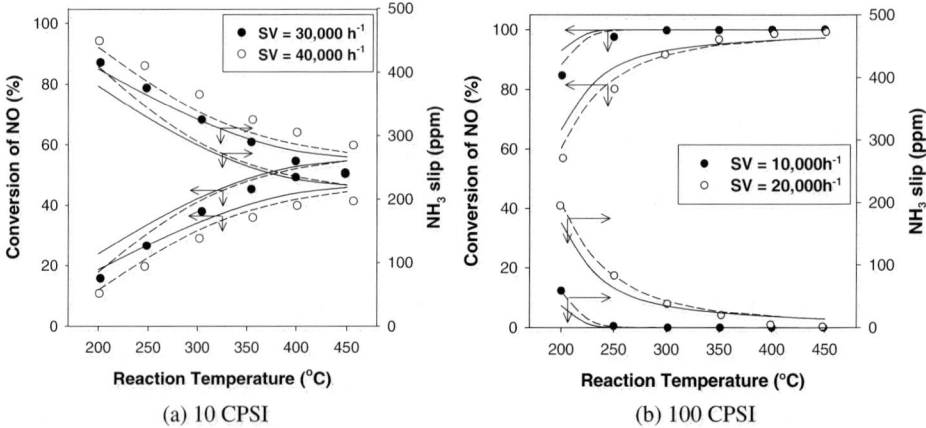

(a) 10 CPSI

(b) 100 CPSI

Fig. 1. Prediction of the model for 10 and 100 CPSI honeycomb reactors extruded with the V_2O_5/sulfated TiO_2 catalyst. (—, prediction with the parameters estimated from the experimental data over a packed-bed flow reactor; ---, prediction with the parameters estimated from the experimental data over a honeycomb reactor).

REFERENCES

[1] H. Bosch and F. Janssen, Catal. Today, 2 (1988) 369.

[2] S. T. Choo, I.-S. Nam, S. W. Ham and J. B. Lee, Korean J. Chem. Eng., 20 (2003) 273.

[3] L. C. Young and B. Finlayson, AIChE J., 22 (1976) 331.

[4] M. A. Buzanowski and R. T. Yang, Ind. Eng. Chem. Res., 29 (1990) 2074.

[5] E. Tronconi, A. Cavanna and P. Forzatti, Ind. Eng. Chem. Res., 37 (1998) 2341.

[6] H. J. Chae, S. T. Choo, H. Choi, I.-S. Nam, H. S. Yang and S. L. Song, Ind. Eng. Chem. Res., 39 (2000) 1159.

[7] S. T. Choo, S. W. Ham, I. S. Nam, I. Y. Lee, D. W. Kim, J. B. Lee, M. H. Uhm and P. S. Gi, V_2O_5 Based Catalyst for Removing NOx from Flue gas and Preparing Method therefore, US Patent No. 6 380 128 B1 (2002).

[8] H. Choi, S. W. Ham, I.-S. Nam and Y. G. Kim, Ind. Eng. Chem. Res., 35 (1996) 106.

Studies in Surface Science and Catalysis, volume 159
Hyun-Ku Rhee, In-Sik Nam and Jong Moon Park (Editors)

Reactivity of absorbent prepared from oil palm ash for flue gas desulfurization: Effect of SO_2 concentration and reaction temperature

A. R. Mohamed, N. F. Zainudin, K. T. Lee and A. H. Kamaruddin

School of Chemical Engineering, Engineering Campus, Universiti Sains Malaysia, Seri Ampangan, 14300 Nibong Tebal, Seberang Perai Selatan, Pulau Pinang, Malaysia

Oil palm ash was utilized as an absorbent for dry-type flue gas desulfurization. The absorbents were prepared using water hydration method with the addition of other chemicals such as CaO and $CaSO_4$. The absorbents were then subjected to synthetic flue gas under various SO_2 feed concentration (500 to 2000 ppm) and reaction temperature (65°C to 400°C). It was found that higher feed SO_2 concentration reduces the time the absorbent could maintain 100% removal of SO_2. On the other hand, higher reaction temperature was found to increase the reactivity of the absorbent. However, reaction temperature above 300°C was found to have negative effect on the reactivity of the absorbent.

1. INTRODUCTION

Environmental issues have received considerable attention in the recent years as various industrial activities throughout the world continue to pollute the environment and threaten the natural eco-system. One such issue is the release of sulfur dioxide (SO_2) from power generating plants burning solid and liquid fuels such as crude oil and coal. SO_2 is recognized world wide as one of the main pollutants due to its acidic and toxic characteristic. Apart from being the primary cause of acid rain, which damages buildings, vegetation and water ground cycle, SO_2 also cause the formation of secondary particles in the atmosphere that impairs visibility. SO_2 is also considered to be toxic to humans by inhalation. Process currently under active development for the removal of SO_2 is using absorbent synthesized from siliceous material such as coal fly ash [1-4]. This study, on the other hand, presents finding on absorbent prepared from oil palm ash, an abundant agricultural solid waste in tropical countries in Malaysia and Thailand [5]. The absorbents synthesized were then tested for its reactivity in absorbing SO_2 by exposing it to synthetic flue gas under various SO_2 feed concentration (500 to 2000 ppm) and reaction temperature (65°C to 400°C).

2. MATERIALS AND METHODS

2.1. Absorbent preparation

The raw materials used to prepare the absorbents were oil palm ash, CaO, and $CaSO_4$. The oil palm ash was provided by United Palm Oil Mill, Nibong Tebal, Penang, Malaysia. The chemical composition of the oil palm ash is 40.0% SiO_2, 12.1% K_2O, 10.0% CaO, 8.2% P_2O_5, 6.4% MgO, 6.1% Al_2O_3, 5.4% C, 2.5% Fe_2O_3, 2.0% others and 7.3 % ignition loss. The CaO and $CaSO_4$ were obtained from BDH, England. The procedure used to prepare the absorbents comprised the following steps. A fixed amount of CaO (5 g) was added to 100 ml of water at 70°C. Upon stirring, the temperature of the slurry increased to about 90°C. 15 g of oil palm ash and 1 g of $CaSO_4$ were then added in the slurry simultaneously. The slurry was then maintained at 95°C for a period of 24 h for the hydration process to occur. The resulting slurry was filtered and dried at 200°C for 2 h. The absorbents in powder form were then palletized, crushed, and sieved to produce the required particle size range of 250-300 μm. These preparation parameters were obtained from an optimization study reported elsewhere [6]. The BET surface area of the resulting absorbent was analyzed using Autosorb Quantochrome 1C and was found to be 88.4 m^2/g as compared to 8.6 m^2/g for oil palm ash, 5.6 m^2/g for CaO and 4.9 m^2/g for $CaSO_4$.

2.2. Desulfurization experiments

Desulfurization experiments were performed using a fixed bed test rig. The reaction zone was contained in a 0.008 m inner diameter stainless steel tube fitted in a furnace for isothermal operation. The absorbent (0.7 g) was packed in the center of the reactor supported by glass wool. A thermocouple was used to monitor the temperature of the absorbent bed continuously. A stream of feed gaseous mixture containing 500 to 2000 ppm of SO_2, 500 ppm of NO, 12% of CO_2, 5% of O_2 and balance N_2 at a reaction temperature of 65 to 400°C was passed through the absorbent. Prior to that, the N_2 gas stream was humidified using a humidification system, which consists of two 250 ml conical flask immersed in a water bath at constant temperature. The total flow rate of the gas stream was controlled at 150 ml/min using mass flow controller. The schematic diagram of the experimental test rig is shown elsewhere [7]. The concentration of SO_2 in the flue gas was measured using a Portable Flue Gas Analyzer IMR2800P before and after the absorption process. The concentration of SO_2 was recorded continuously until 90 min. Two replicate measurements were made for each activity test and the relative standard deviation was found to be less than 3.5%. For clarity, only the averages are presented in this paper. The desulfurization activity in this study is reported as the breakthrough curves of the desulfurization reaction (C/C_o vs. t) where C is the outlet concentration of SO_2 (ppm) from the reactor, C_o is the initial feed concentration of SO_2 (ppm) and t is the reaction time (min).

3. RESULTS AND DISCUSSION

The absorbent and the various raw materials used (oil palm ash, CaO and $CaSO_4$) were subjected to synthetic flue gas consisting of 500 ppm SO_2, 500 ppm NO, 5% O_2, 12% CO_2 and balance N_2 at 100°C. The breakthrough curves of the desulfurization activity of the various materials are shown in Fig. 1. From Fig. 1, it was found that the reactivity of the absorbent in removing SO_2 outperformed the reactivity of the raw materials. The absorbent was found to be capable of removing 100% of the SO_2 from the feed gas for the first 22 min of reaction time. Beyond 22 min, the reactivity of the absorbent started to gradually decrease. On the other hand, the raw materials used to synthesize the absorbent was found to have negligible desulfurization activity. This result is most likely due to the fact that the surface area of the absorbent is much higher than those of the raw materials.

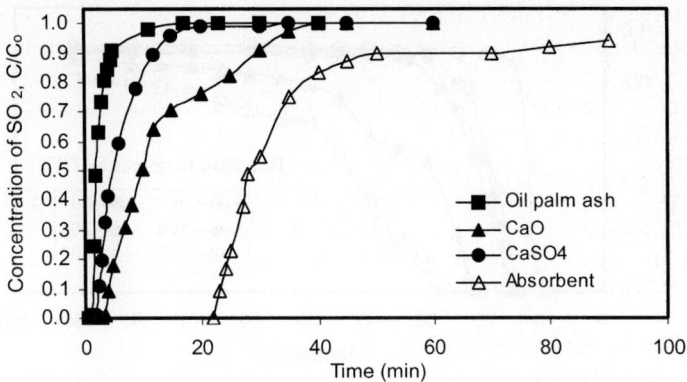

Fig. 1. Desulfurization breakthrough curves for the absorbent and raw material

The duration the absorbent could maintain 100% removal of SO_2 was then studied under different feed concentration of SO_2 and reaction temperature. Fig. 2 shows the desulfurization breakthrough curves of the absorbent when the feed concentration of SO_2 was increased from 500 ppm to 2000 ppm. Reaction temperature was fixed at 100°C. It was found that the time the absorbent could maintain 100% removal of SO_2 reduces from 22 min to 7 min. This phenomena is most probably due to exposing fixed amount of absorbents to more molecules of SO_2 at a higher feed concentration of SO_2.

Fig. 3 shows the desulfurization activity of the absorbent at a reaction temperature ranging from 65°C to 400°C. The feed concentration of SO_2 was fixed at 1000 ppm. When the reaction temperature was increased from 65°C to 80°C and then to 100°C, the changes in the reactivity of the absorbent could not really be observed probably due to the small increment in the reaction temperature. However, when the reaction temperature was further increased to 200°C, there is a significant increase in the reactivity of the absorbent. Similarly, when the reaction temperature was increased from 200°C to 300°C, the reactivity of the absorbent also increased. The increase in the reactivity of the absorbent at higher reaction temperature is due to the increase in the reaction rate constant at higher reaction temperature.

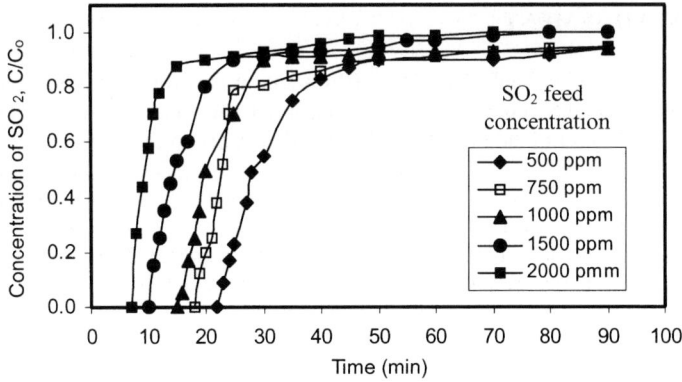

Fig. 2. Desulfurization breakthrough curves for the absorbent at various feed concentration of SO_2

452

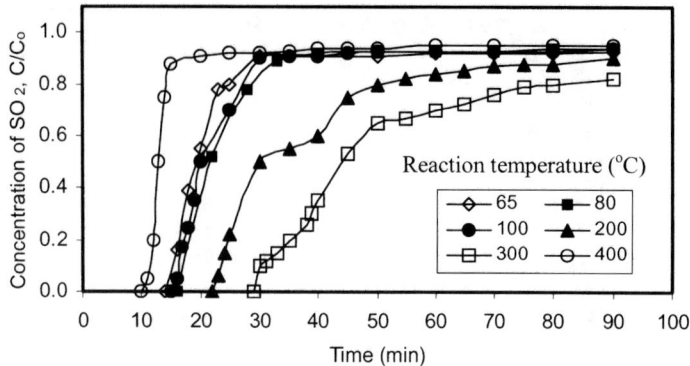

Fig. 3. Desulfurization breakthrough curves for the absorbent at various reaction temperature

On the other hand, when the reaction temperature was increased further to 400°C, the reactivity of the absorbent significantly dropped. It was previously reported that for absorbent prepared from coal fly ash, when the absorbent was dried at temperature above 400°C, the reactivity of the absorbent dropped due to the decomposition of the active materials in the absorbent [8]. Since the effect of drying the absorbent above 400°C is similar to exposing the absorbent to reaction temperature above 400°C, therefore it can be concluded that the active materials in absorbent prepared from oil palm ash also decompose at reaction temperature above 400°C resulting in lower reactivity. Apart from that, another possible explanation for the drop in the reactivity of the absorbent at 400°C could be due to the sintering of the absorbent that decreases the surface area of the absorbent.

4. CONCLUSION

This study has demonstrated that absorbent prepared from oil palm ash does have a high reactivity towards SO_2 absorption. The reactivity of the absorbent was found to increase with higher reaction temperature up to 300°C and lower feed concentration of SO_2.

5. ACKNOWLEDGEMENT

The authors would like to thank JSPS-VCC (Program on Environmental Science, Engineering and Ethics), Ministry of Science, Technology and Innovation Malaysia (Project No. 08-02-05-2040EA001) and Yayasan Felda for the funding and support on this project.

REFERENCES
[1] J. Fernandez and M. J. Renedo, Energ. Fuel., 17 (2003) 1330.
[2] R. B. Lin, S. M. Shih and C. F. Liu, Ind. Eng. Chem. Res., 42 (2003) 1350.
[3] M.J. Renedo and J. Fernandez, Ind. Eng. Chem. Res., 41 (2002) 2412.
[4] T. Ishizuka, T. Yamamoto, T. Murayama, T. Tanaka and H. Hattori, Energ. Fuel., 15 (2001) 438.
[5] J. Gou and A. C. Lua, Mater. Lett., 55 (2002) 334.
[6] N.F. Zainudin, M.Sc. Thesis, Universiti Sains Malaysia, 2005.
[7] K.T. Lee, S. Bhatia and A.R. Mohamed, Chem. Eng. Sci., 60 (2005) 3419.
[8] H. Tsuchiai, T. Ishizuka, T. Ueno, H. Hattori and H. Kita, Ind. Eng. Chem. Res., 34 (1995) 1404.

Studies in Surface Science and Catalysis, volume 159
Hyun-Ku Rhee, In-Sik Nam and Jong Moon Park (Editors)

Investigation of ozonolysis of phenol using γ-alumina based catalysts

A. Godde[a], S. Heng[b], K. L. Yeung[a,*] and J. C Schrotter[c]

[a]Department of Chemical Engineering, the Hong Kong University of Science & Technology, Clear Water Bay, Kowloon, Hong Kong

[b]Environmental Engineering Program and Department of Chemical Engineering, the Hong Kong University of Science & Technology, Clear Water Bay, Kowloon, Hong Kong

[c]Anjou Recherche, Veolia Water Research Centre, Chemin de la Digue, BP 76, 78603 Maisons-Lafitte Cedex, France

1. INTRODUCTION

Ozone finds application for color and taste removal from water and for treatment of refractory organic pollutants in wastewater. The use of catalysts is expected to result in more efficient ozone utilization and better mineralization of organic pollutants. The $\gamma\text{-}Al_2O_3$ supported metal and metal-oxide catalysts are the most common catalysts used in industrial ozone treatments of water and wastewater. However, there is yet to be an agreement as to the exact role of the $\gamma\text{-}Al_2O_3$ during the ozone reaction. The $\gamma\text{-}Al_2O_3$ serves as support for the active catalyst (e.g. CuO_x and TiO_x) and is believed to promote the adsorption and dissolution of ozone. The strong affinity of $\gamma\text{-}Al_2O_3$ for carboxylic compounds means that some organic pollutants and most intermediate ozone byproducts are strongly adsorbed on the support. Adsorption traps these compounds in the reaction zone for a longer residence time leading to better conversion and mineralization. This work investigates the role of $\gamma\text{-}Al_2O_3$ supported catalyst for ozone treatment of phenol in water.

2. EXPERIMENT

The commercial $\gamma\text{-}Al_2O_3$ pellets provided by Sasol were crushed and sieved to obtain 425 μm $\gamma\text{-}Al_2O_3$ powder. Copper catalyst was prepared by impregnating the $\gamma\text{-}Al_2O_3$ powder with copper nitrate (99 %, Aldrich) [1], while $TiO_2/\gamma\text{-}Al_2O_3$ was obtained by mixing the $\gamma\text{-}Al_2O_3$ powder with a measured quantity of nano-TiO_2 particles. The catalyst was dried at 393 K and calcined in air at 773 K for 8 h. The final catalysts contained 10 wt. % loading. Table 1 resumes the properties of each catalyst from XRD and BET characterizations whereas the isoelectric point pH was determined by a saturation technique [2].

---

* Author to whom correspondence should be addressed
Tel: 852-2358-7123; Fax: 852-2358-0054; email: kekyeung@ust.hk

454

Table 1 Physicochemical properties of the catalysts

Catalyst	Structure	Particle Size	BET Surface Area	Isoelectric Point pH
γ-Al$_2$O$_3$	-	1.8 nm	210 m^2/g	8.2
CuO γ-Al$_2$O$_3$	Cu (II) oxide	3.4 nm	190 m^2/g	8.4
TiO$_2$ γ-Al$_2$O$_3$	Anatase TiO$_2$	2.6 nm	190 m^2/g	4.9

Phenol (99.9%, Aldrich) was dissolved in distilled deionized water to a concentration of 250 mg/L (i.e. 190 mg/L of organic carbon). The catalytic ozonolysis of phenol was conducted for γ-Al$_2$O$_3$ and γ-Al$_2$O$_3$ supported copper oxide and titanium oxide catalysts. The reaction was carried out in a semi-batch, stirred-tank reactor using 1 g/L of catalyst with the ozone bubbled through a porous glass diffuser. The ozone feed was kept constant at 15 mg/min. The phenol concentration was analyzed by UV-Vis spectrometer and the organic content of the solution was measured by a total organic carbon (TOC) analyzer (Shimadzu 5000 A). The amount of adsorbed species on the catalysts was carried out using thermogravimetric analysis (TGA). These measurements provide an exact accounting of the carbon in both liquid and solid phases

3. RESULTS AND DISCUSSION

Figure 1 shows that phenol conversion has a negligible improvement with the addition of γ-Al$_2$O$_3$ powder, but the TOC level shows a significant drop when γ-Al$_2$O$_3$ is added as shown in Fig. 2. Experiments showed that phenol adsorption on γ-Al$_2$O$_3$ is negligible, but carboxylic compound formed during ozonolysis of organics are strongly adsorbed. This is confirmed by a simple experiment where γ-Al$_2$O$_3$ was added after 1, 2 and 3 h of ozone treatments as shown in Figs. 1 and 2. It is clear that the lower TOC level is simply due to the adsorption of ozone reaction byproducts on γ-Al$_2$O$_3$.

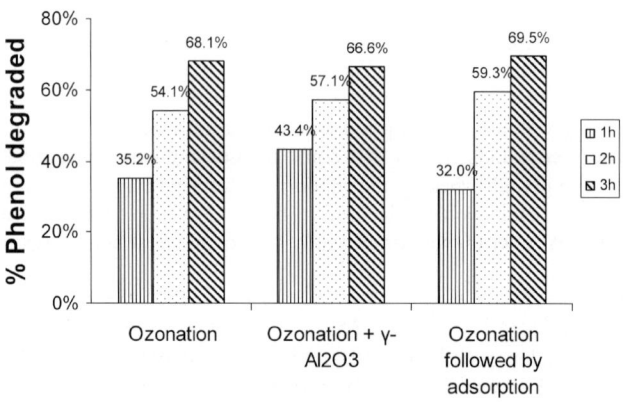

Fig 1: Phenol conversion during ozonolysis with and without γ-Al$_2$O$_3$.

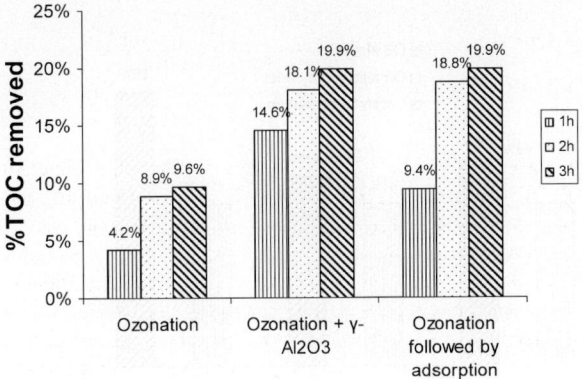

Fig.2: TOC reduction during ozonolysis with and without γ-Al₂O₃.

Keeping the overall ozone feed at 15 mg/min, experiments were conducted at a higher ozone concentration but lower flowrate. Fig. 3 plots the TOC reduction as a function of reaction time. At these conditions, higher TOC reduction was obtained in shorter time. The presence of γ-Al₂O₃ and γ-Al₂O₃ supported catalysts yields better results.

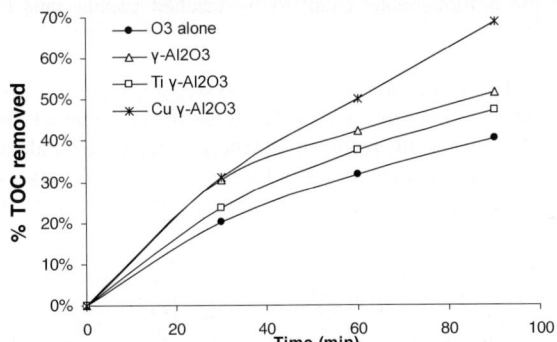

Fig. 3: TOC reduction as a function of time.

The acidic byproducts generated during ozonolysis of phenol are well known [3,4]: maleic acid (pKa = 3.15), pyruvic acid (pKa = 2.65) and oxalic acid (pKa$_1$ = 1.25 and pKa$_2$ = 4.21) being the most common. As the pH during ozonolysis decreases from 6 to 3.4, these acids were mostly in their conjugated base form RCOO⁻ and were more easily attracted to the positively charged catalyst surface. The TiO₂/γ-Al₂O₃ catalyst has a lower isoelectric point pH than the γ-Al₂O₃ alone. This means that the TiO₂/γ-Al₂O₃ catalyst is less positively charged than the γ-Al₂O₃. So at pH = 3.4, which is close to isoelectric point pH of the TiO₂/γ-Al₂O₃ catalyst, it has a lower adsorption capacity resulting in a lower TOC removal.

456

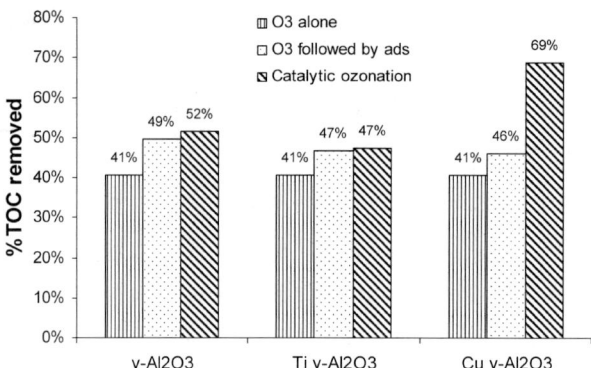

Fig. 4: TOC removal with γ-Al₂O₃ and γ-Al₂O₃ supported catalysts during catalytic ozonolysis and ozonolysis followed by adsorption.

It is clear from Fig. 4 that the lower TOC content is again mainly due to the adsorption of reaction byproducts on γ-Al₂O₃ and TiO₂-γ-Al₂O₃, while CuO-γ-Al₂O₃ exhibits catalytic activity. However, further experiments suggested that the enhanced TOC reduction displayed by CuO/γ-Al₂O₃ catalyst was due to the homogeneous catalysis by leached copper ions (i.e. 18 mg/L) present in the solution [5].

These results strongly suggest that for phenol ozonolysis, the γ-Al₂O₃ support plays an important role as a contactor. The adsorption of byproducts on γ-Al₂O₃ has the immediate effect of reducing the TOC level that could be easily mistaken for an enhanced catalytic degradation of organic pollutants. Besides removing unwanted byproducts, adsorption on γ-Al₂O₃ also served to trap these organic species within the reaction zone for a longer time leading to their eventual mineralization.

Acknowledgement

The authors would like to thank Veolia Water Hong Kong, Anjou Recherche and the Consulate General of France in Hong Kong for funding this work.

References

[1] F. J. Beltran, F. J. Rivas, Montero de Espinosa, *Ind. Eng Chem Res.*, 42 (2003), 3218.
[2] J. A. Schwartz, C.I. Contescu, *Surface of Nano-Particles and Porous Materials*, Marcel Dekker, 1999.
[3] N. Al Hayek, B. Legube, M. Dore, *Environ. Technol. Lett.*, 10 (1989), 415.
[4] Zhu et al. *J Zhejiang Univ SCI*, 5(12) (2004), 1543.
[5] Y. Pi, M. Ernst, J.C. Schrotter, *Ozone Sci. Eng.*, in press.

Studies in Surface Science and Catalysis, volume 159
Hyun-Ku Rhee, In-Sik Nam and Jong Moon Park (Editors)
457

Adsorption-desorption characteristics of modified activated carbons for volatile organic compounds

Ki-Joong Kim[a], Chan-Soon Kang[a], Young-Jae You[a], Min-Chul Chung[a], Seung Won Jeong[b], Woon-Jo Jeong[c], Myung-Wu Woo[a], Ho-Geun Ahn[a†]

[a]Dept. of Chemical Engineering and [b]Jeonnam Techno Park, Sunchon National University, 315 Maegok-dong, Suncheon-si, Jeonnam, 540-742 Korea.
[c]Dept. of Inform. & Comm., Chosun College Sci. & Techn., Seosuk-dong, Gwangju, Korea.

Modification techniques for activated carbon were used to increase the removal capacity by surface adsorption and to improve the selectivity to volatile organic compounds (VOCs). Modified activated carbons (MACs) were prepared by modifying the purified activated carbon with various acids or bases. The effects of adsorption capacity and modified contents on the textural properties of the MACs were investigated. Furthermore, VOC adsorption and desorption experiments were carried out to determine the relationship between the adsorption capacity and the chemical properties of the adsorbents. High adsorption capacity for the selected VOCs was obtained over 1wt%-H_3PO_4/AC (1wt%-PA/AC). As a result, MAC was found to be very effective for VOC removal by adsorption with the potential for repeated use through desorption by simple heat treatment.

1. INTRODUCTION

Several techniques for VOC removal have been investigated such as thermal incineration, catalytic oxidation, condensation, absorption, bio-filtration, adsorption, and membrane separation. VOCs are present in many types of waste gases and are often removed by adsorption [1]. Activated carbon (AC) is commonly used as an adsorbent of gases and vapors because of its developed surface area and large pore volumes [2]. Modification techniques for AC have been used to increase surface adsorption and hence removal capacity, as well as to improve selectivity to organic compounds [3].

In this study, the surface properties of modified AC (MAC) and its adsorption capacity for VOCs were investigated. The effect of modified contents on adsorption performance was studied. Furthermore, adsorption and desorption of VOCs was carried out to evaluate the adsorption capacity and desorption characteristics after saturated adsorption.

2. EXPERIMENTAL

This subject is supported by the Ministry of Environment as "The Eco-technopia 21 project".
[†] corresponding author Tel: +82-61-750-3583. E-mail: hgahn@sunchon.ac.kr

2.1. Preparation of MAC

The AC used in this study was a granular type (30~35 mesh) prepared from coconut shell. The purified AC (PAC) was prepared by boiling the AC for 5 hr in a water bath. The acidic and alkaline solutions for preparation of MACs were made with HNO_3 (NA), H_2SO_4 (SA), HCl (HA), H_3PO_4 (PA), CH_3COOH (AA), KOH (PH), and NaOH (SH). The AC was modified into each solution according to the conventional wet process. The specific surface area of the adsorbents was measured by BET method (ASAP 2020, Micrometrics, USA).

2.2. Measurement of VOC adsorption

The variation of concentration in the course of adsorption was continuously obtained as a thermal conductivity detector (TCD) signal. Model gases were BTX (benzene, toluene and o-, m-, p-xylene), alcohols (methanol, ethanol, iso-propanol), and methylethylketone (MEK). The concentration of VOCs for adsorption was controlled by vaporizing VOCs in the saturator with helium, which was maintained in the range of 10,000ppm ~ 15,000ppm. The temperature in the saturator was maintained with a constant temperature vessel. Total VOC flow rate was 40ml/min as it passed through a U-type adsorbent column. Before adsorption experiment, the MAC was pretreated for 1 hr in an adsorbent column of 250℃, and 0.2g of the adsorbent was used for each experiment. The VOC concentrations were monitored with TCD of gas chromatograph. In addition, the MAC desorption characteristics were investigated by temperature programmed desorption (TPD) technique.

3. RESULTS AND DISCUSSION

3.1. Modified effect of activated carbon

The prepared MAC adsorbents were tested for benzene, toluene, o-, m-, p-xylene, methanol, ethanol, iso-propanol, and MEK. The modified content of all MACs was 5wt% with respect to AC. The specific surface areas and amounts of VOC adsorbed of MACs prepared in this study are shown in Table 1. The amounts of VOC adsorbed on 5wt%-MAC with acids and alkali show a similar tendency. However, the amount of VOC adsorbed on 5wt%-PA/AC was relatively large in spite of the decrease of specific surface area excepting in case of o-xylene, m-xylene, and MEK. This suggests that the adsorption of relatively large molecules such as o-xylene, m-xylene, and MEK was suppressed, while that of small molecules was enhanced. It can be therefore speculated that the phosphoric acid narrowed the micropores but changed the chemical nature of surface to adsorb the organic materials strongly.

The variation of amount of VOC adsorbed and the variation of BET surface area with modified contents were shown in Fig. 1. The optimum modified content was 1wt% for benzene, toluene, p-xylene, methanol, ethanol and iso-propanol, but the amount of o-xylene, m-xylene, and MEK adsorbed were decreased with increasing modified contents. Interestingly, the amount of benzene, p-xylene, and ethanol adsorbed on 1wt%-PA/AC was 1.5 to 2 times that on purified AC. The BET surface area of 1wt%-PA/AC ($1109m^2/g$) took the maximum value.

Table 1. The prepared MACs and their specific surface area and VOC adsorption on PAC and various 5wt%-MACs

Sample	BET Surface Area (m^2/g)	Aromatics					Alcohols			MEK
		Ben- zene	Tol- uene	o- xylene	m- xylene	p- xylene	MeOH	EtOH	iso- propanol	
PAC	892	5.1	3.6	4.7	6.3	4.2	5.3	1.9	1.6	4.4
NA/AC	894	4.7	3.1	5.2	3.9	3.9	4.4	1.9	1.6	3.1
AA/AC	867	5.7	3.6	4.5	6.2	4.2	5.4	1.6	1.9	3.5
HA/AC	718	4.0	2.3	3.9	3.1	3.5	3.7	1.5	1.3	3.0
SA/AC	840	5.1	3.2	4.4	3.9	4.4	5.1	1.9	1.7	3.5
PA/AC	719	6.4	4.1	4.9	4.1	5.2	5.7	2.2	1.9	4.0
PH/AC	668	4.5	2.9	4.1	3.6	4.4	4.3	1.8	1.4	3.2
SH/AC	636	3.8	2.7	4.0	4.5	3.9	4.7	2.0	1.5	3.1

The reason for enhancement of adsorption performance of PA/AC was considered to be due to combination effect of increase of BET surface area and chemical modification by the treatment with PA. Consequently, 1wt%-PA/AC was determined to be a best candidate as an adsorbent for removing benzene, toluene, p-xylene, methanol, ethanol, and iso-propanol. Therefore, 1wt%-PA/AC was used as the adsorbent to investigate the adsorption isotherm, adsorption and desorption performance.

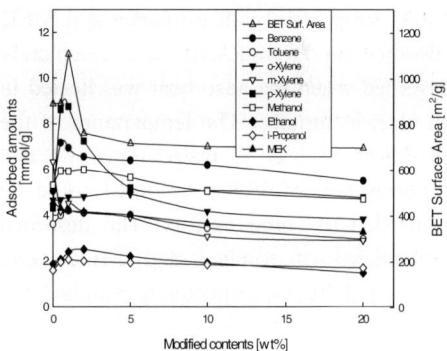

Fig. 1. Effect of modified contents on PA/AC.

3.2. Adsorption isotherms

Langmuir and Freundlich adsorption isotherms for toluene and MEK are shown in Figs. 2

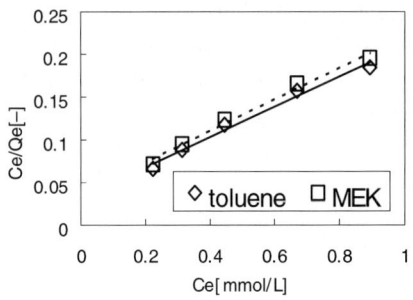

Fig. 2. Langmuir adsorption isotherm of toluene and MEK on 1wt%-PA/AC.

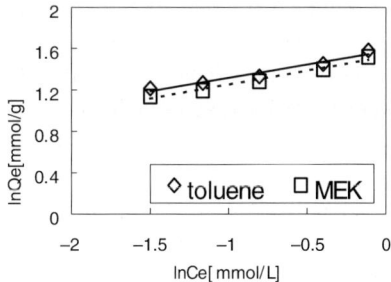

Fig. 3. Freundlich adsorption isotherm of toluene and MEK on 1wt%-PA/AC.

460

and 3, respectively. Langmuir and Freundlich isotherms expressed relatively well the adsorption of toluene and MEK, indicating the dependence on both physical and chemical adsorption. This analysis indicated that the adsorption capacity of MEK on 1wt%-PA/AC was larger than that of toluene.

3.3. Desorption performance by TPD

Fig. 4 shows the TPD experimental results for 1wt%-PA/AC with toluene and MEK. The desorption characteristics of 1wt%-PA/AC with toluene and MEK were determined by raising the room temperature by 5℃ /min to 300℃. Maximal desorption concentrations and temperatures of both toluene and MEK were 1.13, 1.25 and 150℃, 100℃, respectively. The toluene and MEK adsorbed on 1wt%-PA/AC was completely desorbed when the adsorbent was heated in an electric furnace. The temperature of the adsorbent was programmed, and concentration of VOCs desorbed could be controlled at some extents. The desorbed efficiencies of toluene and MEK were 98.1% and 99.1%, respectively. Similar TPD patterns were obtained even though the adsorption-desorption operation was repeated. It was therefore considered that the adsorbents will be effective for repetition of adsorption-desorption processes.

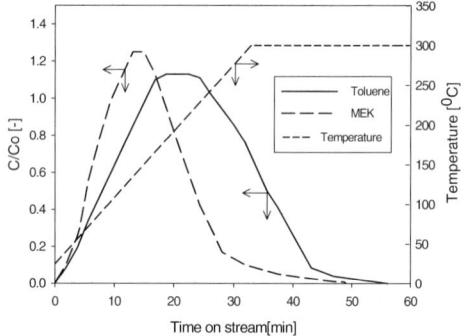

Fig. 4. TPD spectra from the desorption of toluene and MEK on 1wt%-PA/AC.

4. CONCLUSIONS

Adsorption capacity and desorption characteristics of modified activated carbon (MAC) prepared with various acids and bases were investigated. Among the prepared MACs, PA/AC showed the greatest adsorption capacity for benzene, toluene, p-xylene, methanol, ethanol and iso-propanol due to chemical modification of its surface despite the decreased specific surface area. Also, the maximum BET surface area with modified content of PA/AC showed 1wt%-PA/AC. The amount of VOC adsorbed on 1wt%-PA/AC was larger than that on purified AC excepting that of o-xylene, m-xylene, and MEK. The adsorbed toluene and MEK were easily desorbed by heat treatment to 300℃, suggested the possibility for repeated use. The findings confirmed the potential of 1wt%-PA/AC as a promising adsorbent for controlling VOC emissions with low concentrations.

REFERENCES

[1] C.L. Chuang, P.C. Chiang, *Chemosphere*, 53, 17 (2003).
[2] D.J. Kim, W.G. Shim and H. Moon, *Korean J. Chem. Eng.*, 18, 518 (2001).
[3] L. Guo, A.C. Lua, *J. Porous Materials*, 7, 491 (2000).

Studies in Surface Science and Catalysis, volume 159
Hyun-Ku Rhee, In-Sik Nam and Jong Moon Park (Editors)

Selective adsorption of phenanthrene on activated carbons for surfactant reuse in soil washing process

C. K. Ahn[a], Y. M. Kim[a], J. M. Park[a] and S. H. Woo[b]

[a]School of Environmental Science and Engineering, Pohang University of Science and Technology, San 31, Hyoja-dong, Nam-gu, Pohang, Kyungbuk 790-784, Republic of Korea

[b]Dept. of Chemical Engineering, Hanbat National University, San 16-1, Dukmyung-dong, Yuseong-gu, Daejeon 305-179, Republic of Korea

1. INTRODUCTION

Polycyclic aromatic hydrocarbons (PAHs) are widespread soil contaminants and major environmental concerns due to their carcinogenic properties [1, 2]. PAHs have extremely low water solubility and are strongly sorbed to soil. A potential technology for remediation of PAH-contaminated soils is a soil washing with surfactant solutions. While the use of surfactants significantly enhances the performance of soil remediation, operation costs are increased up to 50% as surfactant dosages are increased [3]. Current approaches to reduce surfactant usage include ultrafiltration, precipitation, foam fractionation, and photochemical treatment [3]. However, these methods are limited since the fluxes are decreased by gel-layer formation or ineffectiveness above critical micelle concentration [3]. In this study, selective adsorption of contaminants by activated carbons is proposed to reuse the surfactants in the soil-washing process for the first time. The adsorption isotherms of pure chemicals (Triton X-100 and phenanthrene) onto three granular activated carbons (GACs) were investigated. The adsorption selectivity of phenanthrene in mixed solution was examined at various concentrations of phenanthrene and Triton X-100. The selectivity results were discussed with pore size distribution of activated carbons and molecular sizes of phenanthrene and the surfactant monomer.

2. EXPERIMENTAL

Charcoal-based GACs (Darco 20~40, 12~20, and 4~12 mesh) were used as adsorbents for phenanthrene and Triton X-100. The pore size distribution and specific surface area of GACs were obtained from N_2 gas adsorption at 77.3 K by using surface area analyzer (ASAP 2010, Micromeritics). Surface functional groups of GACs were determined by using X-ray photoelectron spectroscopy (EscaLab 220-IXL). In isotherm experiments, 0.1 g of each GAC was added to a 250-ml Erlenmeyer flask and then filled with 100 ml at various concentrations of compounds at 25 °C. In selective adsorption tests, the flasks containing surfactant solutions with phenanthrene (0~100 mg l^{-1}) were shaken at 100 rpm at 25 °C for 48 h to reach equilibrium state. Phenanthrene and Triton X-100 were analyzed by using HPLC (Dionex) with an UV detector at 250 nm and 230 nm for phenanthrene and Triton X-100, respectively. The analytical column was a reversed-phase Supelcosil LC-PAH column (150 mm × 4.6 mm).

3. RESULTS AND DISCUSSION

Effectiveness of selective adsorption of phenanthrene in Triton X-100 solution depends on surface area, pore size distribution, and surface chemical properties of adsorbents. Since the micellar structure is not rigid, the monomer enters the pores and is adsorbed on the internal surfaces. The size of a monomer of Triton X-100 (27 Å) is larger than phenanthrene (11.8 Å) [4]. Therefore, only phenanthrene enters micropores with width between 11.8 Å and 27 Å. Table 1 shows that the area only for phenanthrene adsorption is the highest for 20~40 mesh. From XPS results, the carbon content on the surfaces was increased with decreasing particle size. Thus, 20~40 mesh activated carbon is more beneficial for selective adsorption of phenanthrene compared to Triton X-100.

Table 1
The surface properties of activated carbons used in this study

Activated carbon	A_{tot} $(m^2 g^{-1})$	A_{ext} $(m^2 g^{-1})$	A_{phe} $(m^2 g^{-1})$	A_{TX} $(m^2 g^{-1})$	$A_{phe} - A_{TX}$ $(m^2 g^{-1})$	C (%)	O (%)
L	274.0	22.5	62.7	31.8	30.9	84.57	15.43
M	529.7	277.0	494.5	363.6	130.9	88.43	11.57
S	621.4	342.2	617.0	448.9	168.1	90.32	9.68

L: 4~12 mesh, M: 12~20 mesh, S: 20~40 mesh, A_{tot}: total surface area, A_{ext}: external surface area, A_{phe}: surface area for phenanthrene adsorption, A_{TX}: surface area for Triton X-100 adsorption

The adsorption isotherms for each chemical, Triton X-100 or phenanthrene, on the activated carbons were shown in Figs. 1 and 2. The adsorption isotherms are expressed as q_e [g g^{-1}], the amount of compounds adsorbed per unit mass of adsorbent, as a function of C_e [g l^{-1}], the concentration in solution at equilibrium [5, 6]. The best-fit parameters for Freundlich isotherms ($q_e = K_F C_e^{1/n}$) or linear isotherms ($q_e = K_L C_e$) were summarized in Table 2. The partition coefficient of phenanthrene was much higher than Triton X-100, indicating that selective adsorption of phenanthrene on GACs is possible.

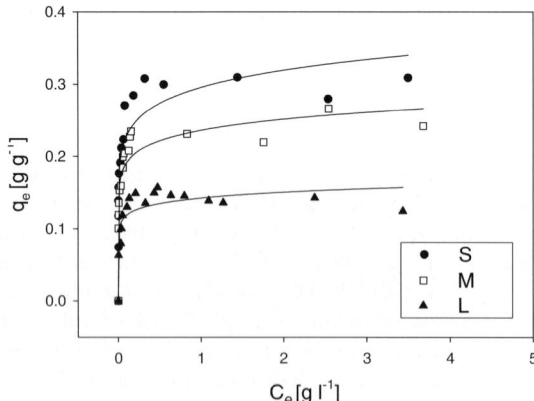

Fig. 1. Adsorption isotherm of Triton X-100 on the activated carbons.

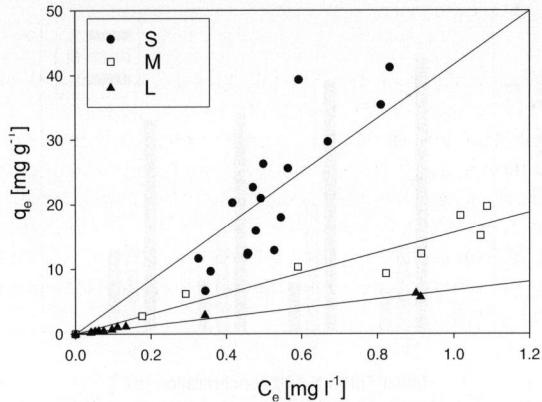

Fig. 2. Adsorption isotherm of phenanthrene on the activated carbons.

Table 2
Summary of isotherm equation parameters

| Activated carbon | Triton X-100 | | | Phenanthrene | | | | |
| | Freundlich | | | Freundlich | | | Linear | |
	K_F	$1/n$	r^2	K_F	$1/n$	r^2	K_L	r^2
L	0.143	0.079	0.810	6.53	0.888	0.990	6.71	0.985
M	0.236	0.092	0.910	15.6	0.996	0.901	15.9	0.853
S	0.296	0.112	0.875	53.7	1.47	0.770	41.6	0.638

Fig. 3 shows the selectivity obtained from batch adsorption tests at various concentrations of phenanthrene. The selectivity is defined as the ratio of partition coefficient of phenanthrene to that of Triton X-100. In all cases, the selectivity for phenanthrene to Triton X-100 was larger than 1, suggesting that adsorption by using these GACs is an effective method to reuse surfactants. The highest selectivity was obtained with the 20~40 mesh. The selectivity was increased with decreasing particle size of GACs and decreasing initial phenanthrene concentration.

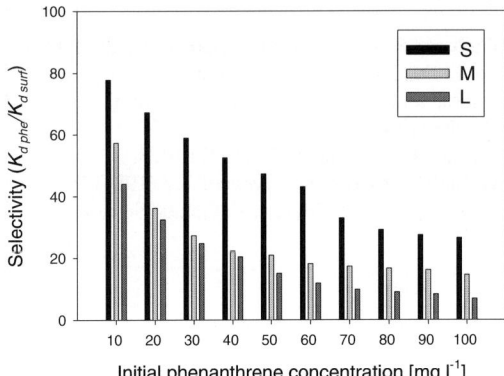

Fig. 3. Effect of initial phenanthrene concentration on selectivity at 5g Triton X-100 l^{-1}.

464

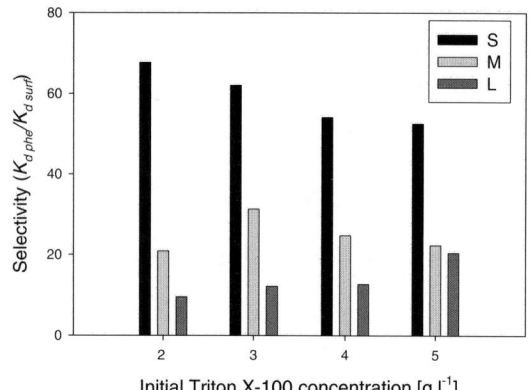

Fig. 4. Effect of initial Triton X-100 concentration on selectivity at 40 mg phenanthrene l^{-1}.

The effect of initial surfactant concentration was different according to the particle size of GACs (Fig. 4). In large particle size, selectivity was increased with increasing surfactant concentration unlike small particle size. This is because excess surfactant did not adsorb onto activated carbon due to small A_{TX}.

4. CONCLUSIONS

Phenanthrene dissolved in Triton X-100 solution was separated by sorption with three GACs with different particle size (4~12, 12~20, and 20~40 mesh). The highest adsorption selectivity was obtained with the 20~40 mesh over a wide concentration range of phenanthrene and Triton X-100. The results demonstrate that the selective adsorption is potentially effective to reuse surfactants in a soil-washing process for the remediation of contaminated soils.

ACKNOWLEDGMENT
This work was supported by grants from the Korea Science and Engineering Foundation (KOSEF) through the Advanced Environmental Biotechnology Research Center (AEBRC).

REFERENCES
[1] S. K. Samanta, O. V. Singh and R. K. Jain, Trends Biotechnol., 20 (2002) 243.
[2] S. H. Woo, C. O. Jeon and J. M. Park, Korean J. Chem. Eng., 21 (2004) 412.
[3] D. F. Lowe, C. L. Oubre, and C. H. Ward, Reuse of surfactants and cosolvents for NAPL remediation, Lewis Publishers, 1999.
[4] R.J. Robson and E.A. Dennis. J. Phy. Chem., 81 (1977) 1075.
[5] R. W. Walters and R. G. Luthy, Environ. Sci. Technol., 18 (1984) 395.
[6] C. M. Gonzalez-Garcia, M. L. Gonzalez-Martin, V. Gomez-Serrano, J. M. Bruque, and L. Labajos-Broncano, Carbon, 39 (2001) 849.

Studies in Surface Science and Catalysis, volume 159
Hyun-Ku Rhee, In-Sik Nam and Jong Moon Park (Editors)
© 2006 Elsevier B.V. All rights reserved

465

Synthesis of Monolithic Titania-Silica Aerogel for PCO Reactions

S. Cao[a], K. L. Yeung[b,*] and P. L. Yue[b]

[a]Environmental Engineering Program and Department of Chemical Engineering, the Hong Kong University of Science & Technology, Clear Water Bay, Kowloon, Hong Kong

[b]Department of Chemical Engineering, the Hong Kong University of Science & Technology, Clear Water Bay, Kowloon, Hong Kong

1. Introduction

Titanium dioxide has been widely used as a catalyst for the photocatalytic oxidation (PCO) of organic pollutants in air streams [1]. TiO_2 powder is coated onto plates, ceramic honeycombs and reticulated glass supports to prevent the loss of catalyst and to increase the contact and irradiated areas. However, the presence of a support can significantly reduce light penetration and adds to the total cost and weight of the reactor. This work investigates the preparation, characterization and performance of a lightweight freestanding titania-silica aerogel for the remediation of airborne pollutants.

2. Experimental

The titania-silica aerogel was prepared by a modified sol-gel method. Tetramethyl orthosilicate (TMOS, 98%, Aldrich) and titanium isopropoxide (TIP, 98+%, ACROS) were used as the precursors. Ethanol (99.9%, Merck) was chosen as the solvent. TMOS in ethanol was prehydrolyzed by HNO_3 solution. Organic ligand was added into TIP/ethanol solution in order to inhibit its hydrolysis rate. The prehydrolyzed TMOS and modified TIP were mixed together in 1 Ti:1 Si atomic ratio and the resulting solution was further hydrolyzed with ammonia solution. The clear solution was transferred into a plastic mold and kept at room temperature. A solid alcogel is formed after several days, further ageing was performed before the solvent was removed through supercritical drying. Two supercritical drying methods were employed: (1) a high-temperature ethanol supercritical drying at 553 K and (2) a low-temperature CO_2 extraction/ supercritical drying at 298 K/323 K. The resulting aerogel sample (raw areogel) was then calcined in air at 723 K for 1 h.

Photocatalytic oxidation tests were performed in a photoreactor [2]. The flat, rectangular stainless steel reactor has dimensions of 578 mm × 113 mm and inlet/outlet ports at the two ends. The aerogel catalyst placed in a recess located at the center of the reactor was uniformly irradiated by

* Author to whom correspondence should be addressed
  Tel: 852-2358-7123; Fax: 852-2358-0054; e-mail: kekyeung@ust.hk

five fluorescent black lamps (365 nm, 6 W) through a 6.25-mm-thick Pyrex® glass cover. Measured amounts of volatile organic compounds, including trichloroethylene (TCE, 99.5%, Aldrich) and isopropanol (IPA, 99.7%, BDH), were injected into flowing dry synthetic air (O_2 22%, N_2 78%) and fed to the reactor at 200 ~ 400 sccm. The reactor outlet was monitored by a gas chromatograph (6890, Hewlett Packard) equipped with a GS-GASPRO capillary column (0.32 mm × 30 m) and a flame ionization detector.

3. Results and Discussions

Figure 1 shows that crack-free monolithic titania-silica aerogels can be obtained by both high-temperature ethanol supercritical drying and low-temperature CO_2 supercritical drying. The samples are lightweight (ca. 0.5 g cm^{-3}) and freestanding. Table 1 lists the BET surface area, pore volume and average pore diameter calculated from N_2 physisorption data. The titania-silica aerogels possess larger surface area and pore volume compared to traditional TiO_2 powders (i.e., Degussa P25). Also, the pore structure is predominantly mesoporous, containing small quantities of micropores. Ethanol supercritical drying results in roughly half the pore size and double the surface area compared to aerogel prepared by CO_2 supercritical drying.

Table 1
BET surface area, pore volume and average pore diameter calculated from N_2 adsorption-desorption isotherms

Sample	S_{BET} (m^2/g)	$S_{micropore}$ (m^2/g)	$V_{P (D < 100\,nm)}$ (cm^3/g)	$V_{P (D < 2\,nm)}$ (cm^3/g)	Average pore diameter (nm)
Ti-Si aerogel (ethanol supercritical drying)	550	38	1.13	0.010	8.2
Ti-Si aerogel (CO_2 supercritical drying)	306	27	1.38	0.009	18.1

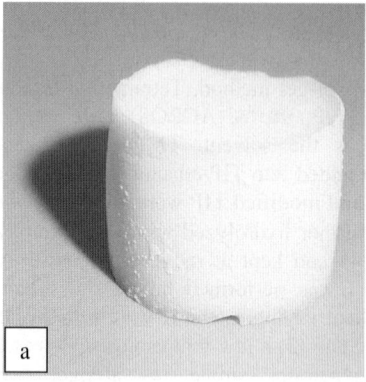 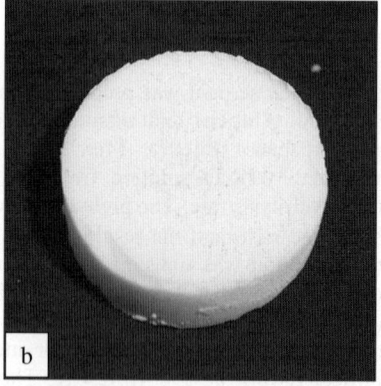

Fig. 1. Crack-free monolithic titania-silica aerogel photos, (a) aerogel prepared by high-temperature ethanol supercritical drying. (b) aerogel prepared by low-temperature CO_2 supercritical drying.

The supercritical drying process also affects the crystallization of the titania phases in the aerogel materials. X-ray diffraction (Fig. 2a) shows that 8.9 nm anatase TiO_2 crystals are formed during ethanol supercritical drying of titania-silica alcogels. Further heat treatment at 723 K led to a slight coarsening of the TiO_2 crystals (10.5 nm). Titania-silica aerogel prepared by CO_2 supercritical drying is X-ray amorphous and amorphous TiO_2 is difficult to crystallize (Fig. 2b). Only after ten hours of heat treatment at 1023 K was 1.45 nm anatase TiO_2 produced.

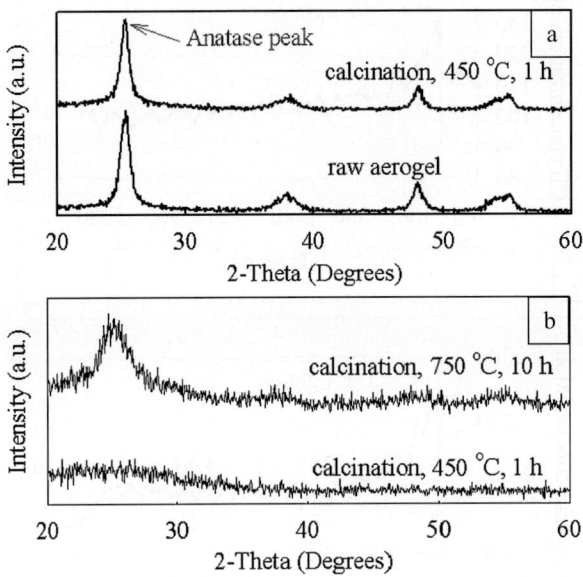

Fig. 2. XRD patterns of titania-silica aerogels, (a) aerogel prepared by high-temperature ethanol supercritical drying. (b) aerogel prepared by low-temperature CO_2 supercritical drying.

Photocatalytic oxidations of trichloroethylene and isopropanol were carried out for the monolithic titania-silica aerogels. The samples weighed 0.137 g, had a diameter of 12 mm, a thickness of 3 mm and a Ti/Si ratio of 1. The aerogel prepared by low-temperature CO_2 supercritical drying did not contain anatase TiO_2 phase and was not active for PCO reactions. However, the aerogel prepared by high-temperature ethanol supercritical drying was active. Figure 3a plots the TCE concentration in the reactor outlet as a function of time. An initial increase in the outlet TCE concentration upon UV irradiation can be explained by the desorption of adsorbed TCE molecules. The TCE concentration decreases rapidly and reaches a steady-state value of 100 ppm equivalent to 10 percent conversion. Isopropanol weakly adsorbed on the aerogel catalyst and readily mineralized to carbon dioxide and water (Fig. 3b). A steady-state conversion of 27 % was calculated. The calculated reaction rates for TCE and IPA were 4.3×10^{-5} mmol·s^{-1}·g^{-1} TiO_2 and 1.4×10^{-4} mmol·s^{-1}·g^{-1} TiO_2, respectively. Comparison was made using

84 mg Degussa P25 TiO_2 under the same conditions. The reaction rates were about an order of magnitude smaller and had values of 5.5×10^{-6} mmol·s^{-1}·g^{-1}TiO$_2$ and 1.1×10^{-5} mmol·s^{-1}·g^{-1}TiO$_2$ for TCE and IPA, respectively. These results showed that the freestanding, titania-silica aerogel is active for photocatalytic oxidation of airborne organic pollutants.

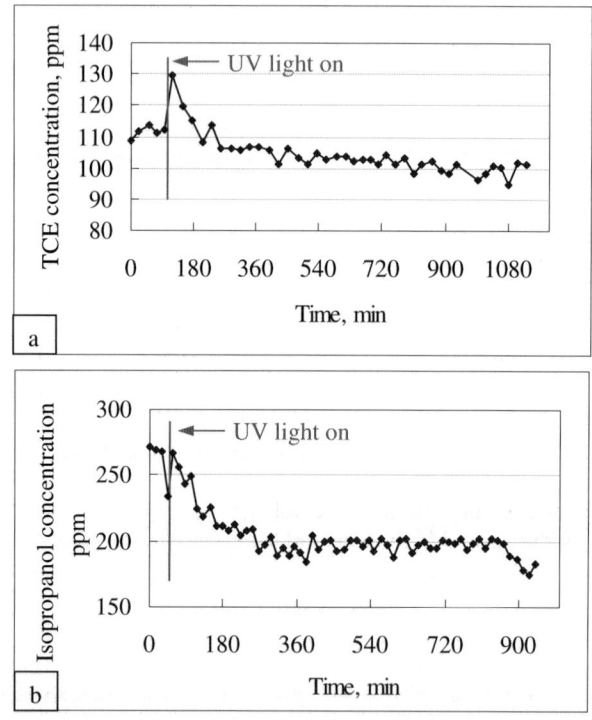

Fig. 3. Photocatalytic oxidation test of titania-silica aerogels, (a) TCE (air: 400 sccm, TCE: 113 ppm), (b) isopropanol (air: 200 sccm, isopropanol: 268 ppm).

References

[1] J. Peral, X. Domenech, D. F. Ollis, J. Chem. Technol. Biotechnol., 70 (1997) 117.
[2] A. J. Maira, K. L. Yeung, C. Y. Lee, P. L. Yue, C. K. Chan, J. Catal., 192 (2000) 185.

Studies in Surface Science and Catalysis, volume 159
Hyun-Ku Rhee, In-Sik Nam and Jong Moon Park (Editors)
469

The Photocatalytic Effects of TiO₂ based catalysts modified by transition metals for removal of pollutants in liquid phase

ᵃChul W. Lee, ᵃSuk B. Hong, ᵇChae H. Shin and ᵃWon M. Lee

ᵃDepartment of Chemical Engineering, Hanbat National University, Dukmyung-dong, Yuseong-gu, Daejeon 305-719, Korea.
ᵇDepartment of Chemical Engineering, Chungbuk National University, Cheongju, Chungbuk.

1. Backgrounds and experiments

Recently, it is reported that TiO_2 particles with metal deposition on the surface is more active than pure TiO_2 for photocatalytic reactions in aqueous solution because the deposited metal provides reduction sites which in turn increase the efficiency of the transport of photogenerated electrons (e⁻) in the conduction band to the external system, and decrease the recombination with positive hole (h⁺) in the balance band of TiO_2, i.e., less defects acting as the recombination center[1,2,3]. The catalytic converter contains precious metals, mainly platinum less than 1 wt%, partially, Pd, Re, Rh, etc. on cordierite supporter. Thus, in this study, solutions leached out from wasted catalytic converter of automobile were used for precious metallization source of the catalyst. The TiO_2 were prepared with two different methods i.e., hydrothermal method and a sol-gel method. The prepared titanium oxide and commercial P-25 catalyst (Deagussa) were metallized with leached solution from wasted catalytic converter or pure H_2PtCl_6 solution for modification of photocatalysts. They were characterized by UV-DRS, BET surface area analyzer, and XRD[4].

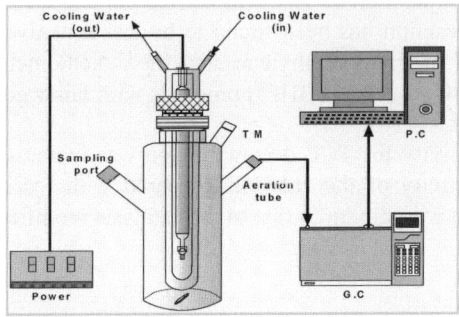

Fig. 1. Photocatalytic reaction system for liquid phase.

Band gap energies of the catalysts were calculated from adsorption edge wavelength equation [Eg(eV)= 1240/λ(nm)].

The prepared photocatalysts were tested to know the reactivity and quantum efficiency in the aqueous solution with trichloroethylene(TCE) as a reactant in photocatalytic batch reactor. Also these results were compared the reactivity to the case of P25 catalyst. The liquid phase photocatalytic reaction system was shown in Fig. 1.

2. Results and discussion

Modified hotocatalysts were prepared using commercial P25 and synthesized TiO_2. The modification was carried out in two different methods, i.e. platinization with H_2PtCl_6 solution and metallization with leached solution from wasted catalytic converter.

The basic structure of TiO_2 wasn't changed by platinization and metallization under this preparation conditions. Particle size of modified TiO_2 catalysts were about 30nm bigger than P25 based catalyst. Also the surface areas of P25 based catalysts were larger than those of TiO_2 based catalysts prepared by sol-gel method.

The band gap energy changes of the catalysts are shown in Table 1. The band gap energies of P25 and TiO_2 without precious metals were about 3.0 eV, which corresponds to 400 nm radiation energy. It is interesting that band gap energy is decreased with the increase of platinum content. For P25-600R and TiO2-600R which were doped with 600 ppm of platinum using leached solution from wasted automobile catalytic converter on the commercial P25 and synthesized TiO_2 respectively, it was revealed that the lowest band gap energy of 1.8 eV corresponds to 700nm of absorption wavelength of photon. Lower band gap energy of P25-500R and TIO2-500R is due to increased transition metal contents as the catalysts which prepared using leached solution contains 0.3 wt% Al, 0.7 wt% Fe, 0.2 wt% Mg, 0.03 wt% Ce, respectively and trace amount of Rh, Si, Zr and La. It is reasonable that UV diffuse reflectance spectra of modified photocatalysts were varied and their band gap energy decrease on increase of loading amounts of transition metals.

The basic structure of the catalysts was not changed on the conditions of modification from the XRD patterns shown in Fig. 2. From the ICP analysis, it was observed that impregnated concentration of transition metals on the surface of TiO_2 were consistent with leached solution concentration.

The photocatalytic decomposition of TCE was carried out and the results were shown in Fig. 3. Photocatalytic reaction has been found to be less sensitive to the conditions such as the concentration of trichloroethylene, and the stoichiometric decomposition ($Cl_2C=CHCl + 3/2O_2 + H_2O \rightarrow 2CO_2 + 3HCl$) proceeds with fairly good reproducibility by prepared photocatalysts.

The photocatalytic reactivity for TCE decomposition was increased by platinization and the photocatalytic activity of the catalysts prepared with leached solution from wasted automobile catalyst was similar to that of the catalysts modified with H_2PtCl_6.

Table1. The Band gap energy changes of modified photocatalysts

Catalyst	Band gap energy (eV)	Wavelength (nm)
P25	3.05	407
P25/500	2.60	478
*P25-500R	1.76	705
P25-1000	1.91	648
TiO2	3.07	403
TiO2-500	2.99	414
*TiO2-500R	1.80	690
TiO2-1000	2.51	495

*R: prepared from wasted catalyst

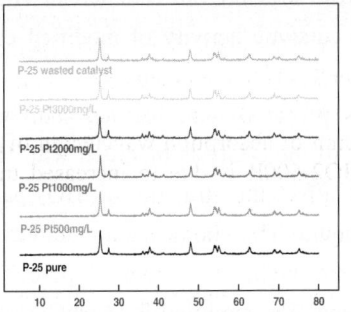 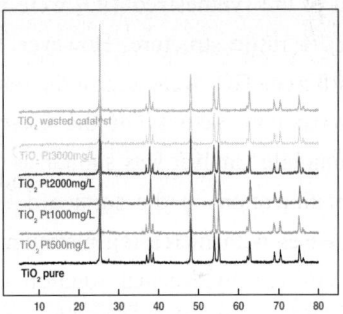

Fig. 2. XRD patterns of modified photocatalysts

3. Conclusion

Modified hotocatalysts were prepared using commercial and synthesized TiO_2. The modification was carried out in two different methods, i.e. platinization with H_2PtCl_6 solution and metallization with leached solution from wasted catalytic converter. They were characterized by UV-DRS, BET, and XRD and tested their catalytic performance for decomposition and oxidation of TCE in liquid phase.

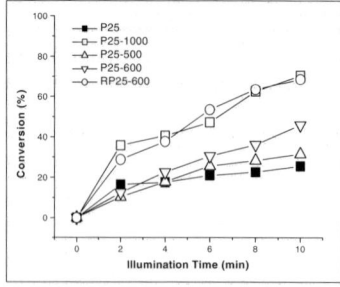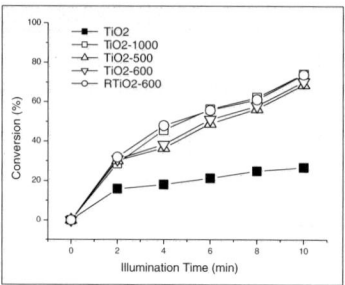

Fig. 3. Catalytic performance of modified P-25 catalysts and prepared TiO₂ catalysts for photocatalytic decomposition of TCE (0.1g/L)

The band gap energy of modified catalysts decreased down to 1.6 eV, and the basic structure and physical properties of the catalysts were not changed during modification process. All of the synthesized TiO_2 were anatase structure but commercial TiO_2 were contained 30% rutile structure. However, the catalytic activity of modified catalysts using two different TiO_2 were almost the same in this reaction conditions.

The photocatalytic activity of the catalysts prepared with leached solution from wasted automobile catalyst was similar to that of the catalysts modified with H_2PtCl_6. Although platinization on the surface doesn't affect the structure of TiO_2, band gap energy decreases with increasing platinum amount. The photocatalytic activity of the catalysts platinized by leached solution is 50% higher than that of pure TiO_2 for photocatalytic decomposition of trichloroethylene. The modified P-25 catalyst (RP-25) showed the highest activity for the photocatalytic decomposition of TCE.

References

[1]. D. Bahnemann, D. Bockelmann and R. Goslish, Solar Energy Materials, 24 (1991) 564.
[2]. M. R. Prairie, L. R. Evans, B. M. Stange and S. L. Martinez, Environ. Sci. Technol., 27 (1993) 1776.
[3]. R.Mathews et. al., J. Chem. Soc., 80 (1984) 457.
[4] Byung-Yong Lee, Sang-Hyuk Park, Sung-Chil Lee, Misook Kang, Chang-Ho Park, Suk-Jin Choung, Korean J. Chem. Eng., 20 (2003) 812

Contact address : wmlee@hanbat.ac.kr

Studies in Surface Science and Catalysis, volume 159
Hyun-Ku Rhee, In-Sik Nam and Jong Moon Park (Editors)

Effect of surface modification by low temperature plasma on the photocatalytic activity of TiO$_2$ thin film

D. L. Cho[a*], J. H. Lee[a], B. H. Kim[a], J.-H. Kim[a], and G. S. Cha[b]

[a]Faculty of Applied Chemical Engineering, Research Center for Photonic Materials and Devices, Chonnam National University, 500-757 Gwangju, Korea

[b]Environmental Engineering Department, Gwangju University, 503-703, Korea
*E-mail: dlcho@chonnam.ac.kr

TiO$_2$ films prepared by sol-gel method were surface-modified by treating the surface with low temperature plasmas. Effects of the modifications on the photocatalytic activity of the films were investigated. H$_2$+Ar and N$_2$+Ar plasma treated films showed photocatalytic activities up to 1.5 times higher under UV-A and up to 3.0-3.2 times higher under fluorescent light than an untreated film based on MB degradation capability. The improvements seem to attribute to the formation of oxygen vacancies and/or Ti^{3+} species through reduction or partial nitration during the treatments.

1. INTRODUCTION

TiO$_2$ is a leading material among various photocatalytic materials. Because of its high band-gap energy (3.2 eV), however, it shows photocatalytic activity only in the ultraviolet region and can use only a small fraction (less than 5%) of solar light. Therefore, efforts are given to make it respond to the light of longer wavelengths.

It is reported that doping TiO$_2$ lattice with various transition metals can improve the photocatalytic activity of TiO$_2$ and extend its absorption edge into the visible-light region [1]. However, the doped TiO$_2$ suffers from thermal instability, increase of carrier-recombination centers, or requirement of an expensive ion-implantation facility. Doping with non-metallic substances such as S and N is also known to be effective and may be better than doping with transition metals [2,3]. And, it is well known that absorption spectral broadening towards the visible range appears in n-type TiO$_{2-x}$ semi-conductors prepared by reduction of single crystals of rutile TiO$_2$ at temperatures of 1000 °C or higher.

In this work, TiO$_2$ films prepared by sol-gel method were surface-modified by treating the surface with low temperature plasmas. And, effects of the modifications on the photocatalytic activity of the films under UV-A and fluorescence light were investigated.

2. EXPERIMENTAL

TiO$_2$ thin films for surface modification were prepared by sol-gel method and dip-coating process on slide glasses. The films were surface-modified by treating the surface with low temperature plasmas, H_2+Ar, N_2+Ar, H_2+N_2, H_2+Ar/N_2+Ar and O_2 (Azusanso, 99.999%) plasmas, in a home-made tubular R.F. plasma reactor [4]. R.F. discharge power was varied from 100 to 300W. Gas pressure was varied from 100 to 400 mTorr. Treatment time was aried from 10 to 120 minutes. After the modifications, chemical compositions and structures were analyzed with ESCA (ESCALAB 250, VG Scientifics), SEM (S-4700, Hitachi) and XRD (X'pert PRO, Philips).

photocatalytic activity was evaluated based on the degradation capability of methylene blue (MB, Daejung, 97%) under irradiation of black-light bulb UV-A (365nm, 15W, 8.5 W/m^2, F15T8BLB, Sankyo Denki) and fluorescent light (180×10^{-3} W/m^2 in UV-A detector, HD 9021, Delta Ohm). A TiO_2 thin film was immersed into a 5.0 ppm MB solution and concentration was measured with a UV-VIS spectrophotometer (UV-2450, Shimadzu) every one hour.

3. RESULTS AND DISCUSSION

After the modifications, light absorbance of a TiO_2 film at wavelengths longer than 350 nm was increased up to almost 4.0 times (see Fig. 1). As a consequence, photocatalytic activity under both UV-A and fluorescent light was highly improved. Extent of the improvement was dependent on treatment conditions. The optimum treatment conditions are listed in Table 1.

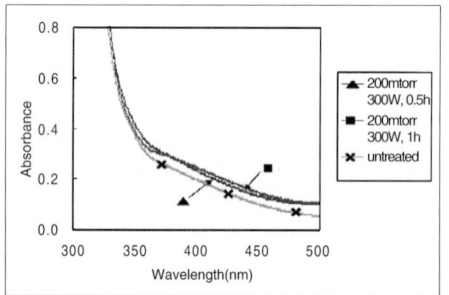 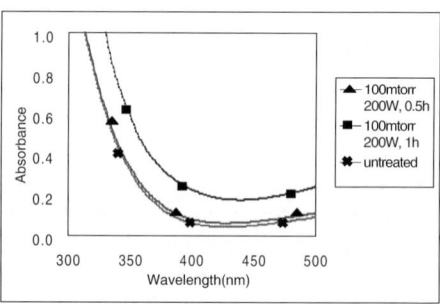

Fig. 1. UV-Vis absorption spectra of H_2+N_2 plasma and O_2 plasma treated TiO_2 films

Table 1. The optimum treatment conditions of various plasmas

Type of plasma	Discharge power (W)	Gas pressure (mtorr)	Treatment time (min)
H_2+Ar	200	100	30
N_2+Ar	300	200	50
H_2+N_2	300	200	50
H_2+Ar/ N_2+Ar	300	200	20
O_2	200	100	70

Fig. 2 and 3 show degradation of MB by untreated, H_2+Ar plasma treated and N_2+Ar TiO_2 films under the irradiation of UV-A and fluorescent light. H_2+Ar plasma treated films show photocatalytic activities up to 1.5 times higher under UV-A and up to 3.2 times higher under fluorescent light than an untreated film. N_2+Ar plasma treated TiO_2 films showed almost the same trend; 1.5 times higher under UV-A and up to 3.0 times higher under fluorescent light.

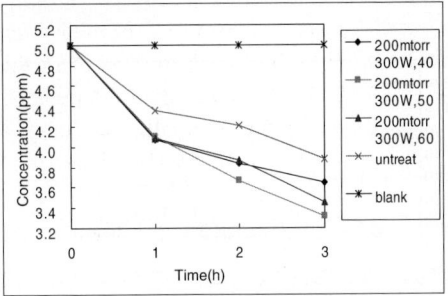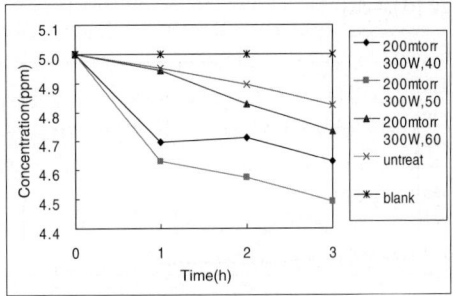

Fig. 2. Concentration changes of MB under UV-A (left) and fluorescent light (right) by untreated and H_2+Ar plasma treated TiO_2 films (blank refers to the reference line for the calibration of concentration changes due to the evaporation of water by lights during the measurements)

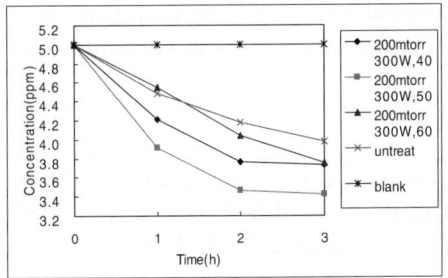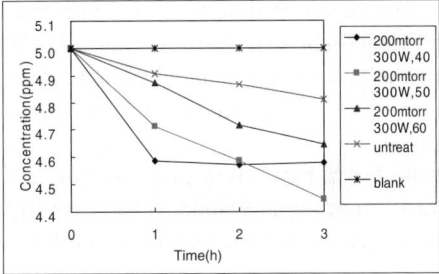

Fig. 3. Concentration changes of MB under UV-A (left) and fluorescent light (right) by untreated and N_2+Ar plasma treated TiO_2 films

The films were analyzed with SEM, XRD and ESCA to investigate reason for the improvements. There was no noticeable change in the surface morphology and crystalline structure after the modifications. However, chemical state of the atoms in the surface layer was changed. In ESCA analysis, untreated film showed Ti_{2p} signals at binding energies of 463.5 eV ($Ti_{2p1/2}$) and 457.8 eV ($Ti_{2p3/2}$) (Fig. 4), signals of Ti^{4+}, and O_{1s} signals at 531.4 eV and 529.2 eV, signals of O atoms in the form of TiOH and TiO_2. After H_2+Ar plasma treatment, however, Ti_{2p} signals appeared at 465.6 eV, 463.7 eV ($Ti_{2p1/2}$), 458.4 eV, 458.1 eV and 457.3 eV ($Ti_{2p3/2}$) (Fig. 4) and O_{1s} signals appeared at 531.9 eV, 529.5 eV and 530.7 eV. This implies that Ti^{3+} ions were created by the modification and some O atoms are

476

bonding to newly formed Ti^{3+} species. An N_2+Ar plasma treated film showed a similar pattern of Ti_{2p} and O_{1s} signals. The difference was that N_{1s} signal newly appeared at 400 eV and O_{1s} signals in the form of TiOH did not increase while it increased after H_2+Ar plasma treatment. Therefore, the improvements of photocatalytic activity seem to attribute to the formation of oxygen vacancies and/or Ti^{3+} species through reduction or partial nitration during the treatments, which is known to be effective for the red-shift of the absorption edge [3].

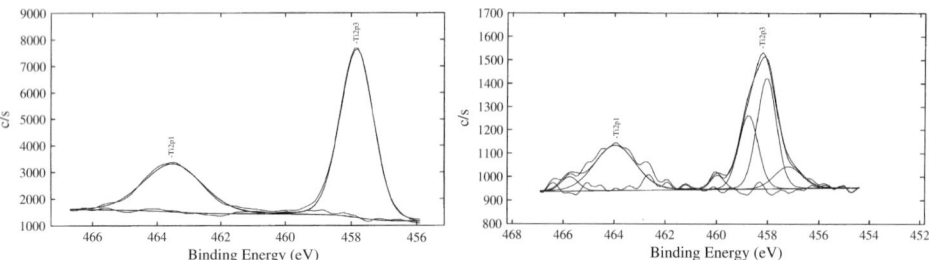

Fig. 4. Ti_{2p} ESCA spectra of untreated (left) and H_2+N_2 plasma treated TiO_2 films

To enhance the formation of Ti^{3+} species by reduction and nitration together, treatment with H_2+N_2 plasma and H_2+Ar/N_2+Ar (H_2+Ar treatment followed by N_2+Ar) were tried. However, there was no further improvement. H_2+Ar/N_2+Ar plasma treatment reduced the optimum treatment time to 20 minutes but improvement of photocatalytic activity under fluorescent light was just 2.0 times. H_2+Ar pretreatment seems to make substitutional doping of N easier through reduction reaction.

O_2 plasma treatment was also effective for the improvement of photocatalytic activity. Although the improvement was not as high as H_2+Ar or N_2+Ar plasma treatments, O_2 plasma treated films show photocatalytic activities up to 1.3 times higher under UV-A and up to 1.7 times higher under fluorescent light than an untreated film. According to ESCA analysis, the improvements seem to attribute to the cleaning effect of O_2 plasma.

ACKNOWLEDGEMENT

This work was supported by Regional Research Center for Photonic Materials and Devices under grant R12-2002-054.

REFERENCES

[1] E. Borgarello, J. Kiwi, M. Grätzel, E. Pelizzetti, and M. Visca, J. Am. Chem. Soc., 104 (1982) 2996

[2] R. Asahi, T. Morikawa, T. Ohwaki, K. Aoki, Y. Taga, Science, 293 (2001) 269.

[3] L. Miao, S. Tanemura, H. Watanabe, Y. Mori, K. Kaneko, S. Toh, Crystal Growth, 260 (2004) 118.

[4] D. L. Cho, S.-H. Kim, Y. I. Huh, and D. Kim, Macromolecular Research, 6 (2004) 553.

Studies in Surface Science and Catalysis, volume 159
Hyun-Ku Rhee, In-Sik Nam and Jong Moon Park (Editors)

477

Multivariate process monitoring and diagnosis of a full-scale industrial wastewater treatment plant

D.S. Lee[*], D.-H. Park, Y.-M. Kim, H.-S. Cho and S.H. Hong

Advanced Environmental Biotechnology Research Center, School of Environmental Science and Engineering, Department of Chemical Engineering, POSTECH, San 31, Hyoja-dong, Nam-gu, Pohang, Gyeongbuk 790-784, Republic of Korea

1. INTRODUCTION

With increasingly stringent regulations of effluent quality, the on-line monitoring of wastewater treatment processes becomes very important to enhance their performance by detecting disturbances leading to abnormal operations in an early stage. Traditionally, wastewater treatment plants have been monitored by using time series charts where operators can view the different variables as historical trends and judge deviation from the norm. However, as the number of variables increases from modern industrial wastewater treatment plants with well-equipped computerized measurement devices, it becomes difficult or impossible to interpret all measurement data simultaneously. Therefore, a more systematic way to handle and analyze data is needed to effectively extract relevant information for monitoring and supervision. In recent years, multivariate statistical process control such as principal component analysis (PCA) has become increasingly popular in many chemical and biological processes [1]. Those techniques can be used to extract the state of the system from the huge volume of the stored data via applications of statistical methods. They have also been applied to biological WWTPs operation [2-3]. In this work, PCA methodology is applied for a full-scale industrial wastewater treatment plant to monitor abnormal process behaviors and to identify the major sources of process disturbances in real time.

2. FULL-SCALE INDUSTRIAL WASTEWATER TREATMENT PLANT

The cokes wastewater treatment plant in Korea, is a conventional activated sludge unit as is shown in Fig. 1. It is designed for the removal of toxic organic pollutants and nitrogenous compounds from cokes-making plants. Most of the chemical oxygen demand (COD) in the cokes wastewater originates from phenol, which is a toxic inhibitory substrate but is also a useful carbon source for acclimatized microorganisms. In addition, cyanides and toxic aromatic hydrocarbons such as cresol, indole, and toluene contribute to the wastewater COD. Since wastewater with this composition is inhibitory to biodegradation, pretreatment steps, such as ammonia stripping are employed to render the wastewater more amendable to biodegradation. To alleviate the impact of high concentrations of deleterious substances on the biological treatment, an equalization tank is installed after the preliminary treatment stage and before the anoxic tanks of the activated sludge process. The hydraulic retention time of

---

[*]Corresponding author: dslee@postech.ac.kr

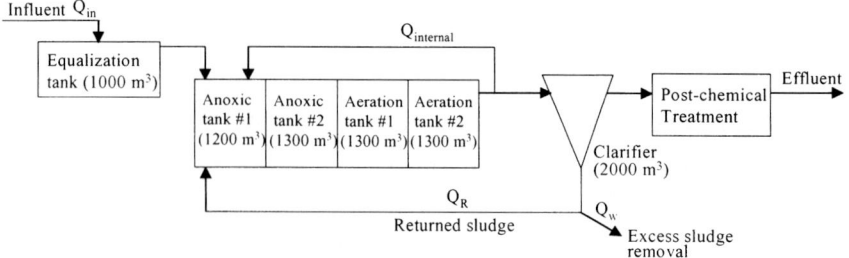

Fig. 1. Schematic diagram of a full-scale wastewater treatment plant

the plant is approximately 3.7 days. The effluent from the secondary settler is passed through chemical treatment units to remove particular ions and to reduce the level of suspended solids and organic matter. Fifteen-months of operational data were collected to develop a reliable monitoring and diagnosis system. All the samples were analyzed for suspended solids (SS), volatile suspended solids (VSS), COD, sludge volume index (SVI), cyanide (CN), thiocyanate, phenol, ammonium-N, nitrate, nitrite and total nitrogen. The dissolved oxygen concentration, pH, and temperature of the influent and the reactors, the influent flow rate, the recycle flow rate, and mixed-liquor volatile suspended solids were also measured at each sampling time. This measuring campaign resulted in 455 operational data sets with 24 measured variables in total.

3. ADAPTIVE PRINCIPAL COMPONENT ANALYSIS

Most real wastewater treatment plants are time-varying due to changes of influent characteristics, temperature and microorganism activity. When a time-invariant PCA model is used to monitor processes with time-varying behaviors, false alarms often result, which significantly compromise the reliability of the monitoring system [4]. To overcome the problem of changing process conditions, an adaptive PCA model based on a moving window can be developed. A window is a fixed-length data set of data matrix $\mathbf{X}$. When new process data set are available, another window is created by omitting the first data set in the window and by adding the new data set to the window. In this approach, a new covariance structure is identified for each new data set and all data sets inside the window frame will have a constant influence on the model until it leaves the window. The model at time k is based on the following covariance matrix:

$$\mathbf{X}^\mathrm{T}\mathbf{X}(k) = \sum_{n=0}^{w} \mathbf{x}(k\text{-}i)^\mathrm{T}\mathbf{x}\,(k\text{-}i) \tag{1}$$

where $\mathbf{X}^\mathrm{T}\mathbf{X}(k)$ is the covariance matrix at time k and w is the length of the window.

4. RESULTS AND DISCUSSION

Initially, the whole data set was analyzed by the linear PCA. By examining the behaviors of the process data in the projection spaces defined by small number of principal components, it

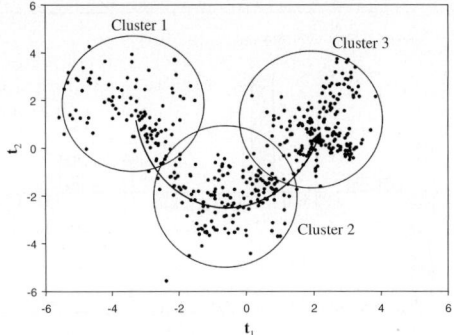

Fig. 2. Score plot for all operation data set

is often possible to extract very useful information. Fig. 2 shows a score plot of the data in the first two principal components. Three distinctive clusters are obtained and these clusters with cluster centers are also shown in Fig. 2. The transitions between the clusters were due to gradual variations of cyanide and phenol compositions in the influent. A monitoring model with adaptive covariance structure was subsequently developed to reduce the problem of changing process conditions. In this application, different window lengths ranging from 50 to 250 samples were tested. The time span of the moving window was optimally set to 200 samples to allow detection of slower disturbances as well as fast ones. The criterion for the selection of the window size was how fast and correctly the model could detect known disturbances in the validation data sets. When a new block of operation data becomes available, the covariance matrix is updated over the selected window. Since the number of significant principal components can change over time, it is necessary to determine the number of principal components recursively. The number of significant principal components is calculated using the cumulative percent variance (CPV) method [5]. The CPV is a measure of the percent variance captured by the first R principal components:

$$CPV(R) = \frac{\sum_{l=1}^{R} \lambda_l}{trace(\mathbf{V})} \times 100\% \qquad (2)$$

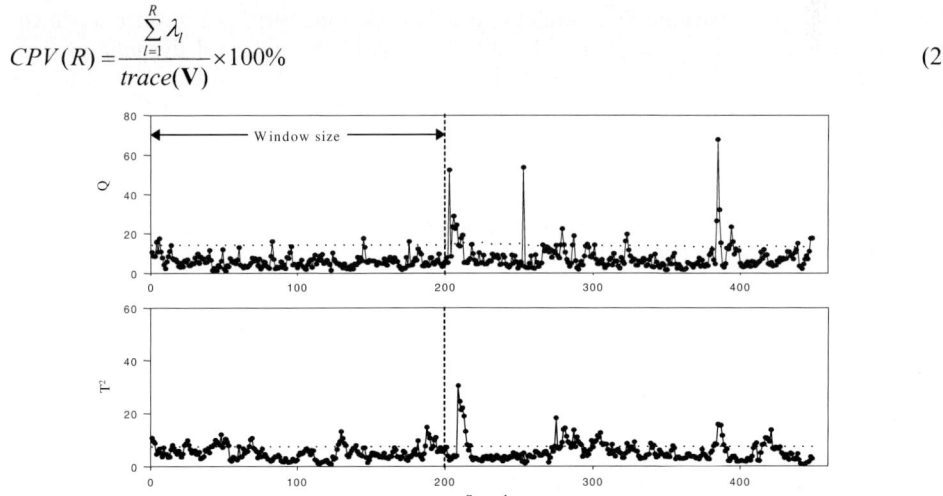

Fig. 3 shows Q and T^2 monitoring charts (dotted line: 99% confidence limit)

480

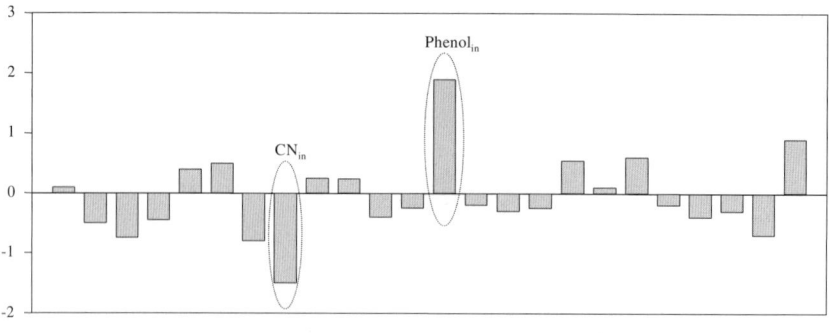

Fig. 4. Contribution plot for sample 212

where the λ_l are the eigenvalues of $\mathbf{X}$ and $\mathbf{V}=\mathbf{EE}^T/(\mathbf{I}\text{-}1)$. The number of principal components was chosen when CPV reached a predetermined limit (80%). A potential adaptation problem was that the model could adapt not only to normal process evolution, but also to disturbances and failures. To prevent this, the model updating was skipped when Q and T^2 values of new operation data were exceeding certain limits. Fig. 3 shows Q and T^2 values calculated from the adaptive PCA model as well as their 99% confidence. Both Q and T^2 values are mostly well inside the confidence limits; this implies that as the covariance structure adapts to new process conditions, the updating model effectively captures the variability of the process and significantly reduce false alarms. For sample 212, a disturbance was clearly alarmed in the T^2 values. To identify the disturbance for the sample 212, a contribution plot of the residuals is shown in Fig. 4. It is clear from Fig. 4 that cyanide and phenol concentrations in the influent mainly contributed to the disturbance.

5. CONCLUSION

The application of a dynamic PCA strategy to a full-scale industrial wastewater treatment plant has demonstrated the feasibility and effectiveness of the proposed dynamic process monitoring approach. The methodology is relatively simple, only based on the historical operation data sets, and can easily be applied to most biological wastewater treatment processes.

ACKNOWLEDGEMENTS
This work was financially supported by the Korea Science and Engineering Foundation through the Advanced Environmental Biotechnology Research Center at Pohang University of Science and Technology.

REFERENCES
[1] J.F. MacGregor, T. Kourti, Control Eng. Pract., 3 (1995) 403.
[2] D.S. Lee, J.M. Park and P.A. Vanrolleghem, J. Biotechnol., 116 (2005) 195.
[3] C. Rosen, J.A. Lennox, Water Res., 35 (2001) 3402.
[4] D.S. Lee, P.A. Vanrolleghem, Biotechnol. Bioeng., 82 (2003) 489.
[5] W. Li, H. Yue, S. Valle-Cervantes and S.J. Qin, J. Process Cont., 10 (2000) 471.

FLUIDIZED BED AND MULTIPHASE REACTORS

Studies in Surface Science and Catalysis, volume 159
Hyun-Ku Rhee, In-Sik Nam and Jong Moon Park (Editors)

Effect of superficial gas velocity on growth of the green microalga *Haematococcus pluvialis* in airlift photobioreactor

Sorawit Powtongsook [1,2] **Kamonpan Kaewpintong**[3] **Artiwan Shotipruk**[3,*] **and Prasert Pavasant**[3]

[1]National Center for Genetic Engineering and Biotechnology, Thailand Science Park, Pathum Thani 12120, THAILAND

[2]Center of Excellence for Marine Biotechnology, Faculty of Science, Chulalongkorn University, Bangkok 10330, THAILAND

[3]Department of Chemical Engineering, Faculty of Engineering, Chulalongkorn University, Bangkok 10330, THAILAND

*e-mail:artiwan.s@chula.ac.th

ABSTRACT

The green microalga *Haematococcus pluvialis* NIES144 was cultured in a 3 L airlift photobioreactor at $27\pm1°C$ and light intensity of 1,000 lux. The effect of different superficial gas velocities on growth of *H. pluvialis* was evaluated within the range between 0.4 and 3 $cm·s^{-1}$. It was found that growth of *H. pluvialis* was strongly affected by the superficial gas velocity. The maximum cell density and specific growth rate were obtained in the culture with the superficial gas velocity of 0.4 $cm·s^{-1}$. These values are 77×10^4 cells mL^{-1} (2.79 g L^{-1} dry weight) and 0.45 day^{-1}, respectively. Higher levels of gas velocity did not show any benefits, and in fact they were found to drastically slow down the growth, due to the increased shear stress. Further more an increase in superficial gas velocity could significantly change the cell morphology from motile vegetative cells to non-motile green cells or cysts.

1. INTRODUCTION

Astaxanthin (or 3'3-dihydroxy-β,β'-carotene-4,4'-dione) is an important pigmentation source in aquaculture that has been shown to possess higher antioxidant activity than any other carotenoids [1]. The freshwater alga *Haematococcus pluvialis* is considered the most capable microorganism that produces the largest amount of astaxanthin compared to other astaxanthin producing microorganisms such as yeast and bacteria [2]. For this reason, the production of astaxanthin from *H. pluvialis* has gained a great deal of attention.

Typical microalgal cultivation can usually be divided into open and close systems. A large-scale algal culture is usually an open tank or pond which is easy to operate and consumes less resource than a close system. Nevertheless, the difficulty in cultivation of *H. pluvialis* is its slow growth. The requirement for low growth temperature and the susceptibility to contamination also make it difficult to obtain optimum growth when cultivated in an open system [3]. As a result, cultivation in a closed system such as in a photobioreactor is required to achieve high cell densities. A variety of bioreactors including both mechanically and nonmechanically agitated systems have been used. However, mechanically stirred bioreactors are not suitable for the cultivation of *H. pluvialis* because the cells are sensitive to shear stress. On the other hand, mixing in nonmechanically agitated bioreactors such as bubble and airlift bioreactors is accomplished by aeration with

482

compressed air or using an air pump, thus the shear force is greatly reduced. Airlift bioreactor has particularly gained attention for cultivation of shear sensitive microorganisms as it offer several advantages such as well defined fluid flow pattern, low energy requirement, and simple construction [4].

The aims of the present work were to culture *H. pluvialis* in the airlift bioreactor in order to examine the effect of superficial gas velocities on growth of *H. pluvialis*.

2. MATERIALS AND METHODS
2.1 Source of the microalga and culture medium

Haematococcus pluvialis NIES-144 was obtained from the National Institute of Environmental Studies, Tsukuba, Japan. The F1 medium [5] consisted of (per litre) 0.41 g KNO_3, 0.03 g Na_2HPO_4, 9.78 mg $CaCl_2 \cdot 2H_2O$, 2.46 mg $MgSO_4 \cdot 7H_2O$, 0.008 mg $CuSO_4 \cdot 5H_2O$, 0.08 mg $Na_2MoO_4 \cdot 2H_2O$, 0.66 mg $MnCl_2 \cdot 4H_2O$, 0.05 mg Cr_2O_3, 0.0078 mg $CoCl_2 \cdot 6H_2O$, 2.21 mg $FeSO_4 \cdot 5H_2O_2$, 0.03 mg SeO_2, vitamin B1 16 mg, vitamin B6 1.2 mg and vitamin B12 12 µg. The medium was autoclaved at 121°C for 20 min prior to use.

2.2 Bioreactor design and culture condition

The airlift photobioreactor used was simply a cylindrical acrylic bubble column with a draft tube that was centrally installed within the outer column having a working volume of 3 L. Both the draft tube and the outer column were made. The dimension of the outer tube was 60 cm in height and 10 cm in diameter, and that of the draft tube was 40 cm in height. The ratio between the downcomer and riser cross section areas (A_d/A_r) was 3.2. Prior to each experiment, the bioreactors were sterilized by sparging ozone through the 0.45 µm Gelman autoclavable filter and a flow meter into the water at the base of the bioreactor for 1 h. Ozone was subsequently substituted by compressed air for another 3-4 hours to remove ozone residual. The bioreactor was filled with 3000 ml of culture with approximately 250 mL of starter innoculum. This accounted for an initial density of *H. pluvialis* of 2×10^4 cells mL^{-1}. Continuous illumination was provided by cool white fluorescent lamp (18W) installed vertically along the length of the column. The distance between the lamps and the column was set at 3 cm. The average illumination intensity incident to the bioreactor outer surface was approximately 1,000 lux (50 µmole photon m^{-2} s^{-1} as measured with a digital lux meter (DIGICON LX-50). The temperature was controlled at 27±1°C. The bioreactor was aerated by rising air bubbles introduced into the base of the bioreactor through the sparger assembly. Air was supplemented with 1% CO_2, sterilized using 0.2 µm air-filter. The effect of different superficial velocities (U_g=0.4, 2, 2.5, 3 cm s^{-1}) were determined.

2.3 Growth measurement

Cell density was measured by daily counting of triplicate algal samples using a haematocytometer under the microscope. The recorded cell number data were separated into vegetative cells and resting cells (aplanospore or cyst). Specific growth rate (µ) was calculated from the increase in the number of cells during logarithmic phase of the growth.

3. RESULTS AND DISCUSSION

Figure 1 shows that high superficial gas velocity in the airlift bioreactor obviously inhibited growth of *H. pluvialis*. In this study, the best superficial gas velocity for growth of *H. pluvialis* was found at the lower limit of the experiment (0.4 cm s^{-1}), which providing the maximum cell density and maximum specific growth rate of 77×10^4 cells mL^{-1} (2.79 g L^{-1} of cell dry weight) and 0.45 day^{-1}, respectively. Due to the equipment constraints, the air flow could not be accurately adjusted below 0.4 cm s^{-1}. It was also difficult to adjust the air flow rate in the range between 0.4 and 2 cm s^{-1}, therefore it could not be concluded at this point that this 0.4 cm s^{-1} of superficial gas velocity was the optimal level. It should however be noticed that the maximum cell density obtained in this study was significantly higher than the cell densities reported in literatures as shown in Table 1. Further increase of superficial gas velocity to 0.4 cm s^{-1} gave no beneficial effect for growth. However, increase in superficial gas velocity in this study seemed not support growth.

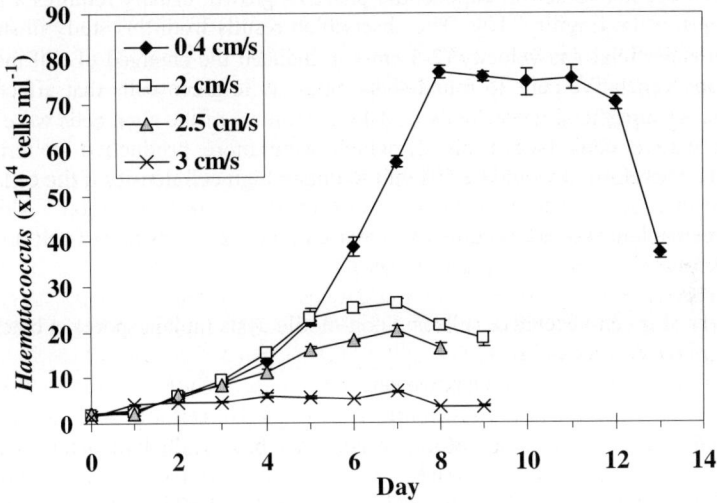

Figure 1 Growth curve of *H. pluvialis* at different superficial gas velocities (U_g=0.4, 2, 2.5, 3 cm s^{-1}) in 3L airlift bioreactor (A_d/A_r=3.2).

Table 1 Comparison of the result from this study with published literatures.

Reference	Reactor type	Reactor volume (ml)	Medium	Condition	Illumination (x1000 lux)	Aeration (L/h)	Maximum density (x10^4 cell/ml)
[2]	Tube	70	OHM	Mixotrophic	2	15	37
[6]	Flask	200	Basal	Mixotrophic	8.6		51
[3]	Stirred tank	3710		Heterotrophic		100	14.5
[1]	Airlift	30000	Bold's	Autotrophic	2.5	80-90	25
This work	Airlift	3000	Control	Autotrophic	1	27	77

In general, increasing aeration rate induces mixing, liquid circulation and mass transfer between gas and liquid phases [7]. A higher mass transfer also facilitates the removal of gases such as oxygen, preventing the accumulation of these gases which might adversely affect the growth rate [8]. High superficial gas velocity was therefore expected to strongly influence mixing of the liquid medium, increasing the absorption of CO_2 and improving light availability of the algal culture which probably stimulating photosynthesis and growth. However, the cell of *H. pluvialis* in the airlift system was negatively affected by an increase in superficial gas velocity. This was possibly due to the shear stress caused by the high aeration rate. Hence, the cell of *H. pluvialis* was suppose to be highly shear sensitive and even with the shear caused by aeration could badly deteriorate the growth.

As mentioned by Gudin and Chaumont [9], the key problem of microalgal cultivation in photobioreactors is cell damage due to shear stress. However, microscopic observation suggested that the inhibition of growth in this study was due to the transformation of vegetative cell into cyst, rather than cell death. To explain in more detail, *H. pluvialis* cells responded to high superficial gas velocity by releasing the flagella, altering metabolism, discontinuing cell division and finally transforming into resting stage aplanospore. The culture of green vegetative cell in exponential phase of growth usually requires a low liquid velocity because of the fragility [10]. The observation results from this study illustrated that an increase in superficial gas velocity (2-3 cm·s^{-1}), induced the changed of cell morphology from oval-shape vegetative cells to round-shape non-motile green cells that affected to cell concentration. At superficial gas velocity of 0.4 cm·s^{-1}, most of the algal cells were remain in the green vegetative cells (see Table 2) which were more productive in term of cell multiplication. Therefore, it would be difficult to obtain high cell density if the cell could not be maintained in vegetative form particularly in high shear stress condition. Although the non-motile green aplanospores can grow by increase in the size, their rate of cell division was very slow, this made the cell density almost constant.

Table 2 Density of green vegetative cells and non-motile cysts (aplanospore) in batch cultures of *H. pluvialis* on day 7.

Superficial gas velocity (cm s^{-1})	Vegetative cell ($\times 10^4$ cell mL^{-1})	Cyst ($\times 10^4$ cell mL^{-1})	% of cyst
0.4	56.17	1.17	2.04
2	6.67	19.67	74.68
2.5	3.4	16.77	83.14
3	0.63	6.21	90.79

CONCLUSIONS

This work demonstrated that an airlift system was suitable for the cultivation of *Haematococcus pluvialis*, one of the most effective microorganisms that could produce high potential antioxidant carotenoid, astaxanthin. Aeration was shown to be crucial for a proper growth of the alga in the airlift bioreactor, but it must be maintained at low level, and the most appropriate superficial velocity was found to be at the lower limit of the pump, *i.e.* 0.4 cm s^{-1}.

REFERENCES

[1] M. Harker, A. J. Tsavalos and A. J. Yong, J. Fermentation and Biotechnology., 82 (1996) 113-118
[2] J. Fabregas. A. Otero and A. Dominguez, J. Biotech., 89 (2001) 65-67.
[3] F. Chen, H. Chen and X. Gong, J. Biores. Tech., 62 (1997) 19-24.
[4] J. A. Asenjo and J. C. Merchuk (eds), Bioreactor system design, New York: Dekker, 1995.
[5] J. Fabregas, A. Dominquez, D.G. Alvarez, T. Lamela and A. Otero, Biotech. Lett. 20 (1998) 623-626.
[6] A. Tjahjono, T. Kakizono, Y. Hayama, N. Nishio and S. Nagai, J. Fermentation and Bioengineering, 77 (1994) 352-357.
[7] J. C. Merchuk and Y. Stein, J. AIChE., 27 (1981) 377-388.
[8] H. L. Tung, C.C. Tu, Y. Y. Chang and W. T. Wu, Bioproc. Eng., 18(1998) 323-328.
[9] C. Gudin and D. Chaumont, J. Biores. Tech., 38 (1991) 145-151.
[10] N. Hata, J. C. Ogbonna, Y. Hasegawa, H. Taroda and H. Tanaka, J. Appl. Phycol., 13 (2001) 395-402.

Studies in Surface Science and Catalysis, volume 159
Hyun-Ku Rhee, In-Sik Nam and Jong Moon Park (Editors)

485

Wet granulation of nano-particles in a rotating fluidized bed

S. Watano, T. Tokuda and H. Nakamura

Department of Chemical Engineering, Osaka Prefecture University
1-1 Gakuen-cho, Sakai, Osaka 599-8531, Japan (watano@chemeng.osakafu-u.ac.jp)

1. ABSTRACT

In this study, a novel rotating fluidized bed (RFB) has been developed for the uniform fluidization of nano-particles. The fluidization behaviors of nano-sized titanium dioxide (TiO_2) particle having its primary size of 21nm were analyzed by measuring the pressure-drop against the fluidization air velocity and by visualization with a high speed video camera. Wet granulation of nano TiO_2 particle was also attempted and properties of granules prepared under various operating parameters were evaluated. Performance of the RFB as a reactor of nano-particle catalyst with the fluidization technique and also as a processor to tailor nano-particles was also discussed.

2. INTRODUCTION

Fine particles including nano-particles have become of major interest lately. Many industrial sectors such as pharmaceutical, agriculture, foods, chemicals, ceramics and electronics are expected to find applications and take advantage of the new functionalities and many desirable properties attributed to the ultra small size.

Fluidization is one of the most promising techniques for handling fine particles because of its advantages of high heat and mass-transfer rates, temperature homogeneity and high flowability of particulate materials. However, as pointed out by Geldart in his classification map [1], fine particles in Group C (small particle size and low particle density) fluidize poorly due to their strong inter-particle forces, exhibiting channeling, lifting as a plug and forming "rat holes" when aerated. Therefore, development of reliable techniques to improve the fluidization of cohesive fine particles is required. Although several external devices have been suggested to improve the flowability of cohesive fine particles, handling of fine particles is still extremely difficult and wet processings, such as coating and granulation of fine particles, have been regarded as nearly impossible.

In this study a novel rotating fluidized bed (RFB) has been developed for the uniform fluidization of nano-particles. Titanium dioxide (TiO_2) having its primary size of 21nm was used as a model nano-particle and its fluidization behaviors under various centrifugal accelerations were analyzed by measuring the pressure-drop against the fluidization air velocity. The fluidization of nano TiO_2 was also visualized by using a high speed video camera. Wet granulation of nano-particle (TiO_2) was also attempted and properties of granules prepared under various operating parameters were evaluated.

3. EXPERIMANTAL

A schematic diagram of the experimental apparatus is shown in Fig. 1. A rotating fluidized bed composes of a plenum chamber and a porous cylindrical air distributor (ID400×D100mm) made of stainless sintered mesh with 20μm openings [2-3]. The horizontal cylinder (air distributor) rotates around its axis of symmetry inside the plenum chamber. There is a stationary cylindrical filter (ID140×D100mm, 20μm openings) inside the air distributor to retain elutriated fine particle. A binary spray nozzle mounted on the metal filter sprays binder mist into the particle bed. A pulse air-jet nozzle is also placed inside the filter, which cleans up the filter surface in order to prevent clogging.

Figure 2 illustrates the particle flow mechanism in fluidized bed. In a conventional fluidized bed, an air distributor is mounted horizontally and particles are introduced onto the distributor and then lifted up by a vertical airflow (drag force and buoyancy against the gravity force). In a rotating fluidized bed, particles are introduced inside the vessel and are forced to the wall by a centrifugal force due to the rotation of the vessel. Air flows radially inward through the air distributor, and the forces on the particle are balanced by the airflow (drag force and buoyancy) and the centrifugal force. Unlike the conventional fluidized bed, a rotating fluidized bed can impart a high centrifugal force, which enables fine particles to behave as Geldart Group A particle.

For a model particle, titanium dioxide (TiO_2, Nippon Aerosil, P-25) having its primary size of 21nm was used, which was normally impossible to be uniformly fluidized.

In wet granulation experiments, dried TiO_2 particles were fed into the cylindrical air distributor (vessel), then the air distributor rotated and fluidization air was supplied. After a predetermined amount of binder liquid was sprayed, drying of products was conducted.

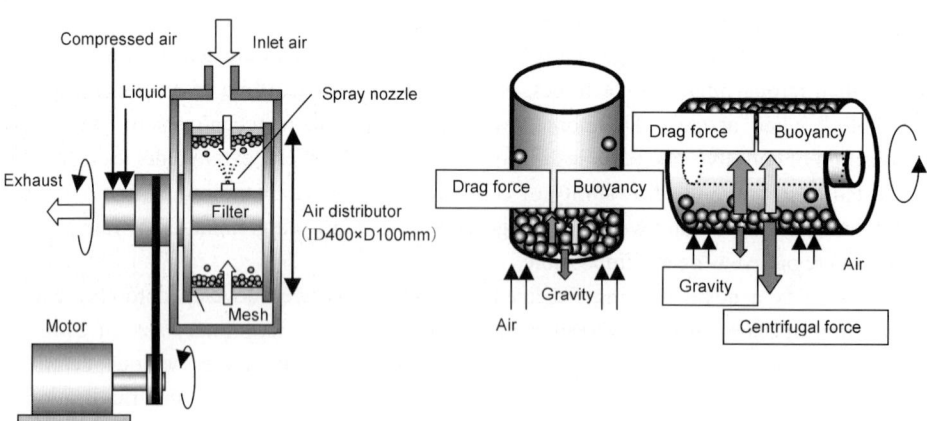

Fig.1. Schematic diagram of RFB   Fig.2. Particle flow mechanism in RFB

4. RESULTS AND DISCUSSION

4.1 Fluidization behavior of nano-particle

Figure 3 shows relationship between pressure drop across particle bed and air flow velocity at different centrifugal accelerations. The pressure drop shows a rise to peak at minimum fluidization velocity, followed by a constant value regardless of the increase in the gas velocity. Due to the larger radial acceleration, pressure drop increases with an increase in centrifugal acceleration. Seen from this figure, fluidization behavior of nano-particles is almost similar to that of A- particles of Geldart's classification and minimum fluidization velocity increases almost linearly with the centrifugal acceleration.

Figure 4 indicates a digital picture of particle fluidization observed by a high speed video camera (120,000frames/s). During the fluidization, stationary layer, channeling and rat-holes were never observed. Circulation speed was pretty high and mixing between particles seemed well conducted. It was thus confirmed that nano-particles could be uniformly fluidized in a rotating fluidized without causing any troubles. This implies the possibility of using RFB as a reactor utilizing nano-particle catalysts.

4.2 Wet granulation of nano-particle

In order to tailor the physical properties of nano-particles, wet granulation was conducted. Hydroxypropylecellulose (HPC-L) 5% aqueous solution was sprayed onto nano-sized TiO_2 particles through a binary nozzle. Fluidization air velocity was set at approximately 1.5 times as large as the minimum fluidization air velocity.

Figure 5 indicates temporal change in mass median diameter of granules under various binder feed speeds. The mass median diameter increased gradually with operation time, and its increase speed was larger when the binder feed speed was high. This implied the adhesion between nano- particles was favorably taken place by liquid bridges generated by sprayed binder mist. Tapped density of nano-particles also increased with operation time, indicating

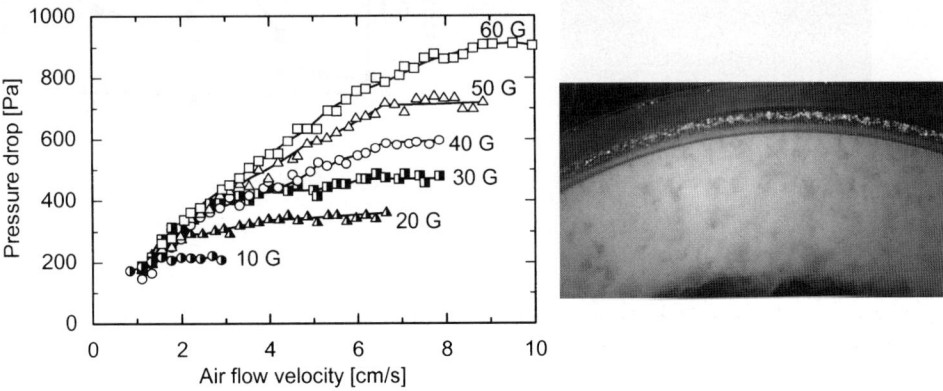

Fig.3. Fluidization behaviors of nano particle
(TiO_2: 21nm)

Fig.4. Visualization of nano-particle fluidization
(Top of the vessel, observation from front side)

488

nano-particles were well densified by agglomeration (Fig.6). Also, according to a SEM (Scanning Electron Microscopy) observation, shape of granules was nearly spherical (Fig.7).

Figure 8 investigates flowability change by measuring the particle discharge speed from a stainless funnel. The discharge speed dramatically increased with operation time, showing the flowability was improved greatly. It is noteworthy that 1 g/s of the discharge speed is almost the same value that glass beads having 100-micron diameter discharge through the same funnel. From this result, it is easily understood that the flowability of cohesive nano-particle is greatly improved by the wet granulation. The wet granulation technique of nano-particles is expected to improve physical properties of nano-particles, leading to increase the efficiency of transportation, storage, etc.

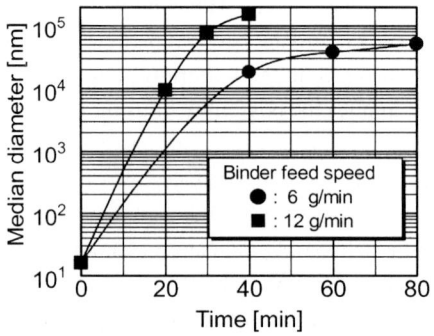

Fig.5. Temporal change in mass median diameter

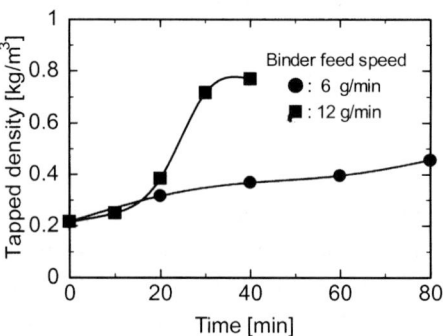

Fig.6. Tapped density v.s. operation time

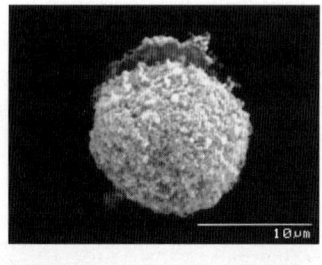

Fig.7. SEM photograph of granule

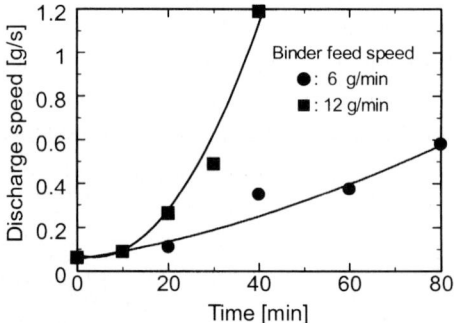

Fig.8. Discharge speed v.s. operation time

REFERENCES

[1] D.Geldart, Powder Technol., 7 (1973) 285.
[2] S. Watano, Y. Imada, K. Hamada, Y. Wakamatsu, Y. Tanabe, R.N. Dave, and R. Pfeffer, Powder Technol., 131 (2003) 250.
[3] S. Watano, H. Nakamura, K. Hamada, Y. Wakamatsu, Y.Tanabe, R.N. Dave and R.Pfeffer, Powder Technol., 141 (2004)172.

Studies in Surface Science and Catalysis, volume 159
Hyun-Ku Rhee, In-Sik Nam and Jong Moon Park (Editors)
© 2006 Elsevier B.V. All rights reserved

Direct mass production technique of dimethyl ether from synthesis gas in a circulating slurry bed reactor

F. Ren, J.-F. Wang[*] and H.-S. Li

Beijing Key Lab of Green Reaction Engineering and Technology, Tsinghua University, Beijing 100084, China (e-mail:wangjf@flotu.org)

1. INTRODUCTION

There is a need to seek an environmentally benign, technically feasible and economical alternative fuel because of the limited crude oil reserves and serious pollution all over the world. Recently, dimethyl ether (DME) is proved to be used as an alternative clean fuel in transportation, power generation and household use for its excellent behavior in compression ignition for combustion, cetane number of over 55 and zero sulfur content, and is praised as a super-clean fuel in the 21th century. It has a promising foreground of application. Therefore, the efficient synthesis of DME from syngas derived from natural gas, coal or biomass has drawn much attention.

Compared with the traditional two-step process, an inherent advantage of the one-step DME synthesis from syngas is the reaction-reaction synergy effect that can break through the equilibrium limitation for the methanol synthesis reaction [1]. Slurry bed reactors can be used for the process of direct synthesis of DME. Compared with fixed bed reactors, slurry bed reactors have many advantages due to the presence of liquid medium, such as easier to remove reaction heat, almost isothermal reaction condition, high conversion of syngas and simple reactor structure. Direct synthesis of DME from syngas in slurry reactors has been considered as the most promising technology for DME synthesis.

2. DEVELOPMENT OF KEY TECHNIQUES

2.1 Bifunctional catalyst

The process of direct synthesis of DME includes reactions of methanol synthesis and methanol dehydration, which are catalyzed by two different catalysts. Although the technology for the production of methanol is generally considered mature, most of them are gas phase process, and the performances of these catalysts are restricted remarkably in liquid phase process. Development of high performance bifunctional catalyst system is very

---

[*] Corresponding author. E-mail: wangjf@flotu.org

important for liquid-phase direct synthesis of DME process. A new copper-based methanol synthesis catalyst, coded LP201, with high catalytic activity and thermal stability was developed by the optimization of the preparation method [2]. The comparison for methanol synthesis between LP201 and the commercial catalysts (C301, C306) in a slurry reactor is shown in Fig. 1. It can be seen that the activity of LP201 is much higher than that of the commercial catalysts. LP201 is not only suitable for methanol synthesis, but also suitable for direct synthesis of DME in a slurry reactor if it is combined with an Al_2O_3 dehydration catalyst developed by ourselves which coded TH16. The bifunctional catalyst comprising LP201 and TH16 has a high catalytic activity and good stability, and is a suitable catalytic system for the direct synthesis of DME in slurry reactors.

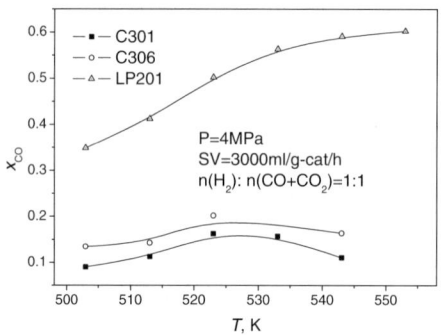

Fig. 1. Comparison between LP201 and the commercial catalysts in a slurry reactor

2.2 Circulating slurry bed reactor

Bubble columns and mechanically stirred reactors are the most common reactor types for slurry systems in laboratories, but they have many disadvantages from an industrialization perspective. Mechanically stirred reactors usually used for laboratorial studies are difficult to scale-up. In order to achieve good mixing and mass transfer between the gas and slurry phases, bubble column must be operated at a high space velocity, which leads to a relative low one-through conversion of the syngas.

For the sake of developing commercial reactors with high performance for direct synthesis of DME process, a novel circulating slurry bed reactor was developed. The reactor consists of a riser, down-comer, gas-liquid separator, gas distributor and specially designed internals for mass transfer and heat removal intensification [3]. Due to density difference between the riser and down-comer, the slurry phase is circulated in the reactor. A fairly good flow structure can be obtained and the heat and mass transfer can be intensified even at a relatively low superficial gas velocity.

2.3 Application in a pilot plant

We started the fundamental research on the direct synthesis of DME from synthesis gas, including catalyst preparation and reactor and process development, since 1998. In 2002, we began to develop mass production technique for direct synthesis of DME in a circulating slurry bed reactor, cooperated with Chongqing Yingli Fuels & Chemicals Co. Ltd. A pilot plant with a capacity of 3000 t/a was built up in Chongqing. In this process, the syngas with the molar ratio of CO to H_2 about 1, which was obtained using methane reforming with CO_2 and stream followed by removal of CO_2 and H_2, was compressed and fed into the DME slurry reactor filled with catalysts and inert liquid paraffin. After reaction, the effluent from the reactor is DME, by-product CO_2, small amount of methanol and un-reacted syngas. DME and other by-products were absorbed with solvent and purified, while the un-reacted reactants in the vent gas were recycled.

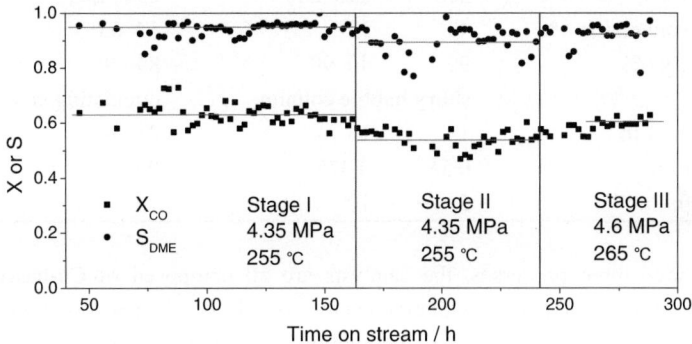

Fig. 2. Conversion of CO and selectivity to DME in the pilot plant

The industrial experiments were carried out in three stages. The only difference between stages I and II is that the syngas passes the slurry bed reactor without recycling the tail gas in stage I, while in stage II the tail gas was recycled. Compared with stage II, the temperature and pressure in stage III were promoted to increase the conversion of CO. Some of the conditions and results in the pilot plant are shown in Fig. 2 and Table 1. It can be seen that in stage I the average one-through conversion of CO and selectivity to DME in the organic product can reached 63% and 95%, respectively. Because the tail gas was recycled in stage II, the high fraction of methane leads to a decrease in the average conversion of CO and selectivity to DME. In stage III, a little elevation of the reaction temperature and pressure can increase the average onversion of and selectivity to DME to 61% and 92%, respectively. These results are almost the same as those obtained in laboratory, where mechanic stirred autoclave reactor was used. It also proves that the circulating slurry bed reactor provides excellent mass and heat transfer in the synthesis of DME. Moreover, during the operation time, the catalysts show no noticeable deactivation.

3. COMPARISON OF DIFFERENT INDUSTRIAL EXPERIMENTS

Up to now, different processes for mass production of DME in a slurry bed reactor have been developed by APCI, USA [4], JFE, Japan [5], and Tsinghua University and Chongqing Yingli

Fuels & Chemicals Co. Ltd., China. Table 1 shows the comparisons of these three different processes for DME production by direct synthesis from synthesis gas.

Table 1
Comparisons of processes for mass production of DME in a slurry bed reactor

Parameters	NKK	APCI	Tsinghua University
$n(H_2)/n(CO)$	1	0.7	1
Catalyst	–	Cu-Zn-Al+Al2O3	LP201+ TH16
Pressure / MPa	5	5–10	4.35– 4.6
Temperature / ℃	260	250–280	255– 265
One-pass CO conversion / %	40	22	54–63
DME Selectivity / %	90	40–90	89– 95
Reactor type	slurry bubble column		circulating slurry reactor
Height of reactor / m	15	15.24	21.56
Inner diameter / m	0.55	0.475	0.6
Designed output /t.d^{-1}	5	10	10

In the above three processes, the catalysts are all composed of Cu-based methanol synthesis catalyst and methanol dehydration catalyst of Al_2O_3. The reactors used by JFE and APCI are slurry bubble column, while a circulating slurry bed reactor was used in the pilot plant in Chongqing. It can be found from Table 1 that conversion of CO obtained in the circulating slurry bed reactor developed by Tsinghua University is obvious higher and the operation conditions are milder than the others.

4. CONCLUSIONS

The bifunctional catalyst comprising LP201 and TH16 has a high catalytic activity and a good thermal stability, and is a suitable catalytic system for the direct synthesis of DME in a slurry bed reactor. The novel circulating slurry bed reactor is efficient and easy to scale up, and the one-through conversion of syngas gas is higher than that in a bubble column. The results obtained in the pilot plant prove that the new technology of direct synthesis of DME using the circulating slurry bed reactor developed by Tsinghua University is a significant breakthrough in the industrialization of the process of direct synthesis of DME.

REFERENCES
[1]. Wang, Z L, Wang J F, Diao J, et al. Chemical Engineering & Technology, 5 (2001) 507
[2]. Wang Z Q, Pan W X, Li J L, et al. Chinese Journal of Catalysis, 7 (2003) 485
[3]. Wang T F, Wang J F, Zhao B, et al. Chemical Engineering Communications, 8 (2004) 1024
[4]. Air Products and Chemicals, Inc. Liquid Phase Dimethyl Ether Demonstration in the LaPorte Alternative Fuels Development Unit., January 2001.
[5].Ogawa T, Inoue N, Shikada T, et al. Journal of Natural Gas Chemistry, 12 (2003) 219

Studies in Surface Science and Catalysis, volume 159
Hyun-Ku Rhee, In-Sik Nam and Jong Moon Park (Editors)
© 2006 Elsevier B.V. All rights reserved

Production TiCl$_4$ Using Combined Fluidized Bed by Titanium Slag Containing High-Level CaO and MgO

Cong Xu, Zhangfu Yuan*, Xiaoqiang Wang, Jianfeng Fan, Jing Li and Zhi Wang

Institute of Process Engineering, Chinese Academy of Sciences,
P. O. Box 353, Beijing 100080, P.R China (yuanzhf@home.ipe.ac.cn)

1. INTRODUCTION

Titanium tetrachloride has been widely utilized as intermediate materials for producing titanium white and titanium sponge, two major products in titania industry. For present commercial production, TiCl$_4$ has mainly been obtained by chloridizing high-grade titania feed-stock (HGTF) such as rutile and high titanium slag in bubble bed. The total content of CaO and MgO in HGTF is required to be lower than 0.5~1.0 wt% in order to reduce the particle agglomeration[1-3]. The storage capacity of worldwide rutile suitable to bubble bed, however, composes only 7.0wt% of gross reserves of titanium resource. In China, at least 90.5wt% of the titanium resources are located in PANZHIHUA, which all belong to magnetite with high-level of CaO and MgO[4]. Some researchers have studied on the problem [5-9], but these studies have mainly focused on the improvement of the reaction conditions in bubble bed and little research about a new reactor has been done. In this paper, our work is devoted to develop a new reactor which can be used in the chloridization of materials with high-level CaO and MgO and examine its anti-agglomeration effect and reaction yield.

2. EXPERIMENTAL PROCEDURE

A new reactor, combined fluidized bed, which can chloridize materials with high-level CaO and MgO, is shown in Fig. 1-a. The reactor may be composed of one or more parts, each of which consists of a riser tube and a semi-circulating fluidized bed (SCFB), and a structure such as riser-SCFB-riser. The method of anti-agglomeration for this reactor is to break up the liquid bridge between the particles by the shear force generated from the turbulence before the bridge is strengthened.

In the riser tube, the gas velocity of chlorine, is greater than both of the terminal velocities of the slag particle and the petrocoke particle, makes the particles to be at a pneumatic transport state. No agglomeration occurs in the riser tube. At the top of the riser tube, a

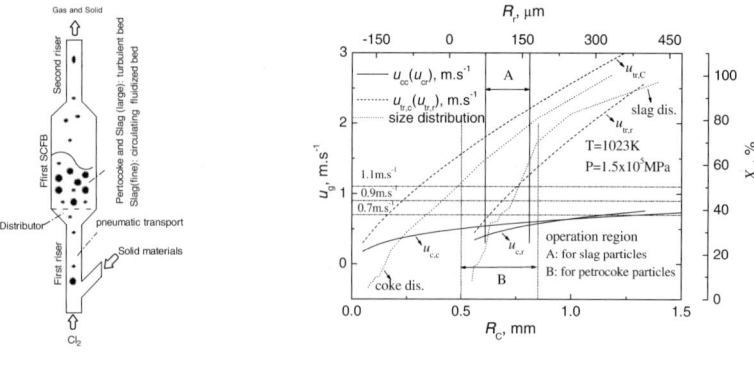

a) combined fluidized bed   b) operation range

Fig.1 Diagram of the combined fluidized bed and operation range

distributor is set up to distribute the gases and particles. In the semi-circulating fluidized bed, the gas velocity is controlled to be lower than the transport velocity ($u_{tr,c}$) of petrocoke particle and higher than the transition velocity (u_{cc}) as shown in Fig. 1-b. As for the high titanium slag particles, the case is different from the petrocoke particles in that only a part of the particles (large particles) are at turbulent fluidization ($u_{cr}<u_g<u_{tr,r}$), and others (fine particles) are at circulating fluidization ($u_g>u_{tr,r}$). That is, all of the petrocoke particles and a part of the slag particles (large) form a turbulent bed, at meanwhile, a circulating fluidized bed formed by the other part of the slag particles (fine) is superposed on the turbulent bed. Such combination may be called semi-circulating fluidized bed, in which the turbulent fluidization brings on higher shear force than the conventional bubble fluidization. On the other hand, because the residence time of the slag particles in the semi-circulating fluidized bed is shorter than in the bubble bed, the source of the liquid $CaCl_2$ and $MgCl_2$ can quickly leave the reaction area and get isolated to reduce the danger of the agglomeration. Moreover, the mass transfer rate as well as heat transfer rate is superior to semi-circulating bed than in the bubble bed.

A quartz reactor is employed in the experiment, in which the first riser tube is 710mm (height)×22mm (i. d.), the semi-circulating fluidized bed is 636mm (height)×59mm (i. d.), and the second riser tube is 320mm (height)×22mm (i. d.).The distributor has a hole of 10mm diameter in the centre and the opening fraction is 0.03. Two kinds of high titanium slag, in which the total content of CaO and MgO is 2.03 wt% and 9.09wt%, are utilized in this work, and the corresponding chemical constitution is shown in Table 1. The material 2 is obtained from PANZHIHUA. In the experiment, 75~165μm of high titanium slag particles and 0.5~0.85mm of petrocoke particles are used, as shown in Figure 1-b).

Table 1

Chemical constitution of high titanium slag(%)

No	TiO$_2$	SiO$_2$	Al$_2$O$_3$	ΣFe	V	CaO	MgO	Cr	MnO	Others
1	91.55	1.20	1.23	2.53	0.18	0.29	1.74	0.032	1.12	0.128
2	76.76	4.70	2.15	5.02	0.083	1.72	7.37	-	0.832	1.365

3. EXPERIMENTAL RESULTS AND DISCUSSIONS

3.1. Examinations on anti-agglomeration

No agglomeration occurs in the combined fluidized bed, much of the $CaCl_2$ and $MgCl_2$ can be found in the solids collected by the cyclone. An experiment is designed to obtain mass balance data of Mg before and after reaction. As suggested in Fig. 2, although the content of MgO in the slag is as high as 7.37wt%, the mass amount discharged from reactor approximately equals to the that fed into the reactor even after 1.5h. This means Mg is not accumulated in the reactor.

3.2. Chloridizing efficiency

From Table 2 it can be seen that at 923K and 973K the conversion (X_{TiO2}) of TiO_2 lies between 20%~45%, the conversion (X_{Cl2}) of chlorine is less than 36% and the production capacity(R_{TiCl4}) of $TiCl_4$ achieve about 26.0 t-$TiCl_4 \cdot m^{-2} \cdot d^{-1}$ at most. In conventional bubble bed, 95% of conversion for TiO_2, 1% (by volume) of the residual chlorine in off-gas and the production capacity of 25~40 t-$TiCl_4 \cdot m^{-2} \cdot d^{-1}$ are generally required[7-10]. In contrast to the bubble bed, it is no doubt that the conversion of TiO_2 and chlorization is comparatively low. The value of R_{TiCl4} under some cases, e.g. L-3 and L-6~8, can achieve the lower-limit of production capacity of $TiCl_4$ for the bubble bed. This suggests that the low-temperature chloridization can be adopted to obtain $TiCl_4$ using materials with high-level of CaO and MgO, meanwhile avoid the agglomeration.

The conversion of TiO_2 and chlorine at higher temperature (at 1023K and 1073K) increased, but the conversions of 50~85% for TiO_2 and less than 80% for chlorine are still not comparable to that in the bubble bed. It can be seen that the R_{TiCl4} at the higher temperature is 40~75 $t \cdot m^{-2} \cdot d^{-1}$, which is 1~3 times the R_{TiCl4} in the bubble bed. That is, the production capacity of the combined fluidized bed is superior to that of the bubble bed.

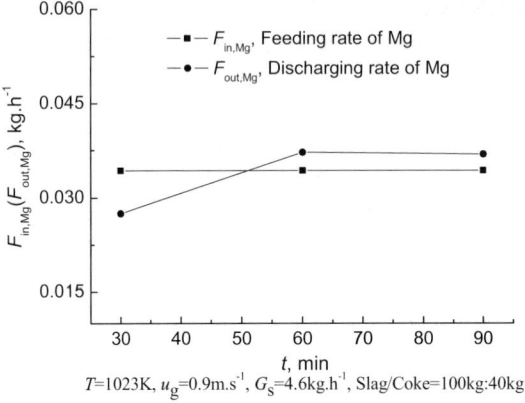

Fig.2 Mass balance of Mg before and after chlorination

496

Table 2

Effect of temperature on reaction efficiency.

No.	T K	u_g m/s	G_s kg $\cdot$ h^{-1}	Slag/Petrocoke kg/kg	X_{Cl2} %	X_{TiO2} %	$X_{Petrocoke}$ %	R_{TiCl4} t $\cdot$ m^{-2} $\cdot$ d^{-1}
L-1	923	0.9	4.6	100:30	20.62	23.54	12.96	13.99
L-2	973	0.7	4.6	100:30	23.51	21.31	11.06	12.67
L-3	973	0.9	4.6	100:30	35.45	45.04	24.23	26.77
L-4	973	1.1	4.6	100:30	25.40	30.87	15.87	18.35
L-5	973	0.7	5.8	100:30	32.91	23.80	12.49	18.09
L-6	973	0.9	5.8	100:30	32.17	30.55	15.92	23.21
L-7	973	1.2	5.8	100:30	29.88	34.90	18.36	26.51
L-8	973	0.9	7.0	100:30	33.75	25.65	13.76	23.72
H-1	1023	0.7	4.6	100:30	71.02	71.09	39.78	52.47
H-2	1023	0.9	4.6	100:30	63.53	82.86	44.02	61.16
H-3	1023	1.1	4.6	100:30	33.61	54.04	28.98	39.89
H-4	1023	0.9	5.8	100:30	58.04	60.04	32.85	55.88
H-5	1073	0.9	5.8	100:30	78.36	81.06	43.93	75.44

4. CONCLUSIONS

In this paper, a method for producing $TiCl_4$ from high titanium slag with high-level of CaO and MgO has been proposed using a new reactor, the combined fluidized bed. The agglomeration caused by the molten $MgCl_2$ and $CaCl_2$ can be effectively prevented by the enhancement of the shear forces in the new reactor. The low-temperature chlorination (less than 973K) in the combined fluidized bed can be utilized in the production of $TiCl_4$ in comparison to the low limit of production capacity of $TiCl_4$ in the bubble bed. The production capacity of $TiCl_4$ for the new reactor in the high-temperature chloridization (more than 973K) is 1~3 times the production capacity for the conventional bubble bed.

ACKNOWLEDGEMENTS
This work was supported by National Natural Science Foundation (No. 20306030) of China.

REFERENCES
[1]   B. Li, Z. Yuan, W. Li and C. Xu. China Particuology, No.2 (2004) 84.
[2]   Z. Yuan, C. Xu, S. Zheng and J. Zhou. Modern Chemical Industry, No. 23 (2003) 1.
[3]   J. Wang. Complex Utilization of Mineral, No.1 (1997) 23
[4]   G. Den. Vanadium and Titanium, No.1 (1994) 36.
[5]   F.L. Yang and V. Hlavacek. AIChE J., No.46 (2000) 355.
[6]   G. Tardos, D Mazzone and R Pfeffer. Can. J. Chem. Eng, No.63 (1985) 377.
[7]   A.J. Morris and R.F. Jensen. Metall. Trans. B, No.7 (1975) 1976.
[8]   C. Lin and T. Lee. J. Chin. I. Ch. E., No.17 (1986) 2.
[9]   I. Barin and W. Schuler. Metall. Trans. B, No. 11 (1980) 199.
[10] W. Wen. Journal of Guangdong Nonferrous Metals, No.9 (1999) 18.

Studies in Surface Science and Catalysis, volume 159
Hyun-Ku Rhee, In-Sik Nam and Jong Moon Park (Editors)

Fluidization quality of fluidized catalyst beds involving a decrease in gas volume

T. KAI[a], K. TORIYAMA[a], T. TAKAHASHI[a] and M. NAKAJIMA[b]

[a]Department of Applied Chemistry and Chemical Engineering, Kagoshima University,
1-21-40 Korimoto, Kagoshima 890-0065, Japan (t.kai@cen.kagoshima-u.ac.jp)

[b]Ishikawajima Plant Engineering & Construction Co., Ltd.
3-3-3 Harumi, Chuo-ku 104-0053, Tokyo

ABSTRACT

The fluidization quality significantly decreased when the reaction involving a decrease in the gas volume was carried out in a fluidized catalyst bed. In the present study, we carried out the hydrogenation of CO_2 and used relatively large particles as the catalysts. Since the emulsion phase of the fluidized bed with these particles does not expand, we expected that the bed was not affected by the gas-volume decrease. However, we found that the fluidization quality decreased and the defluidization occurred. We studied the effects of the reduction rate of the gas volume and the maximum gas contraction ratio on the fluidization behavior.

1. INTRODUCTION

It is reported [1] that the fluidization quality was drastically decreased when the hydrogenation of CO_2 was carried out in a fluidized catalyst bed (FCB). Recently, the phenomena occurring in the bed were directly observed [2] and it was found that the upper part of the emulsion phase was defluidized and this packed particles was lifted up through the column like a moving piston.

Bavarian and Fan [3, 4] reported a similar phenomenon occurring in a three-phase fluidized bed. In their case, the hydraulic transport of a packed bed occurred at the start-up of a gas-liquid-solid fluidized bed. Although the cause was different from the case reported in the present study, similar phenomena were observed in both cases.

In the case of a FCB, the gas volume decreases when the reaction involving a decrease in the volume is carried out at constant temperature and under constant pressure. If the gas in the emulsion phase cannot be compensated by the gas supply from bubbles, the emulsion phase is condensed and bubbles cannot rise through the emulsion phase. Finally, defluidization in the bed occurs. This part of the packed bed will be lifted up like a moving piston.

In the case of a FCB with small particles, the emulsion phase expands [5, 6, 7] when the bed is fluidized. This would make the bed sensitive to the decrease in the gas volume in the emulsion phase. If this assumption is true, we can postulate that the fluidization quality is hardly affected by the gas-volume reduction when the particles, which induce a small emulsion phase expansion, are used. The emulsion phase expansion decreases with increasing particle size and density [6]. In the present study, therefore, the particles used were larger and heavier than that generally used in the FCB. We carried out the hydrogenation of CO_2 in a

FCB using these catalyst particles. We investigated the effects of the gas-volume reduction rate and the maximum contraction ratio on the fluidization behavior during the reaction.

2. EXPERIMENTAL

In the present study, we carried out the hydrogenation of CO_2. We did not use any inert components in the feed. We changed the value of η by changing the molar ratio of H_2 to CO_2 in the feed gases, α. The parameter η is defined as the volume ratio of the product gas to the reactant gas when the reaction completely proceeds under a constant pressure. The extent of the gas-volume reduction is affected by the stoichiometric relation of the reaction and the content of the inert components in the feed. As given in Eq.(1), η is the function of only the parameter α and the expression of η is affected by α. When $\alpha > 4$, the limiting reactant is H_2, while it is CO_2 in the case of $\alpha < 4$.

$$\eta = (\alpha - 1)/(\alpha + 1) \quad (\alpha \geq 4), \qquad \eta = (\alpha/2 + 1)/(\alpha + 1) \quad (\alpha < 4) \tag{1}$$

In the case of $\alpha = 4$, α is a stoichiometric value and η is 0.6. This is the lowest value for this reaction system.

The column made of glass with a 50-mm inner diameter was used as the reactor in order to observe the inside of the column. A transparent electrical resistant material was coated on the outer surface of the glass tube and it worked as an electrical heater.

The catalyst was prepared by impregnating porous alumina particles with a solution of nickel and lanthanum nitrates. The metal loading was 20 wt% for nickel and 10 wt% for lanthanum oxide. The catalyst particles were "A group" particles [8], whereas they were not classified as the "AA group" [9]. The average particle diameter was 120 μm, and the bed density was 1.09 kg m^{-3}. The minimum fluidization velocity was 9.6 mm s^{-1}. The settled bed height was around 400 mm. The superficial gas velocity was 40-60 mm s^{-1}. The reaction rate was controlled by changing the reaction temperature.

The gas composition at the outlet of the reactor was determined using gas chromatography. The selectivity of methane was almost 100%. We directly observed the fluidization behavior and photographed it using a video camera with recording onto a videotape. We also measured the expansion of the emulsion by the bed collapse method [10] during the reaction.

3. RESULTS AND DISCUSSION

3.1. Effect of temperature on fluidization quality

The fluidization quality was affected by the reaction conditions. Although we used the catalyst particles of which expansion was very small, the defluidization was observed. Fig.1 shows the conversion of CO_2 when the temperature was raised from 453 to 523 K. The conversion was higher for higher value of α, while the conversion increased with increasing temperature for all α. In the case of $\alpha = 4.2$, the fluidization quality began to decrease around 473 K. In addition, when the temperature reached 493 K, the motion of the emulsion phase began to be slow-moving and channeling was sometimes observed. The intermittent generation of a defluidized part was observed at above 503 K. Above this temperature, the bed was defluidized and this part was lifted up through the column by the fluidizing gas. This packed particles collapsed from its bottom at 503 K while it rose in the column. On the other

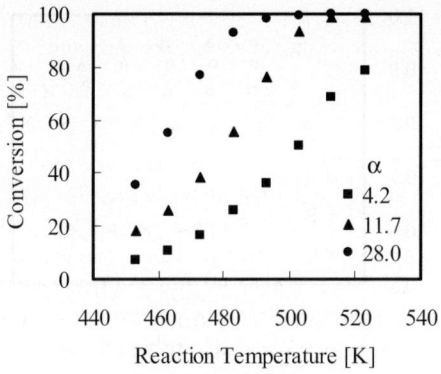

Fig. 1. Relationship between reaction temperature and conversion of CO_2.

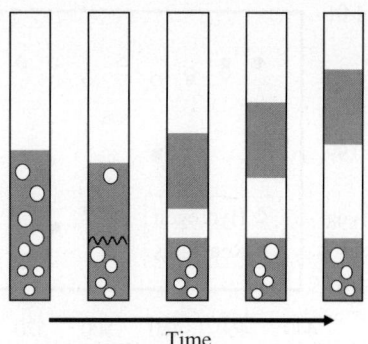

Time

Fig. 2. Schematic time-series pictures of the formation and movement of a defluidized part.

hand, it did not break when the temperature was 523 K. It was difficult to predict when the defluidization occurred, while it invariably occurred within several minutes at this temperature level. Fig.2 is a time-series schematic picture of the defluidization based on the recorded video pictures. The defluidized part of the bed formed a packed bed and it was lifted up through the column. The rising velocity of this part was almost equal to the superficial gas velocity of the feed. The length of this part was 150 - 200 mm. Since the bed height was 400 mm, the defluidization first occurred at the height of around 200 mm.

When $\alpha=8.8$ at 523 K, similar phenomena were observed. On the other hand, when the value of α was 37, the good fluidization state was maintained even at 523 K. There was no direct relationship between the conversion and fluidization quality. This is because that the effect of the gas-volume reduction was low at high α values.

3.2. Contraction rate of the bed

The apparent reaction rate constant for the first order reaction, k, was calculated from the conversion of CO_2. Since the gas-volume reduction rate increased with k, a poor fluidization was induced by high reaction rate. We investigated the effect of the rate of the gas-volume change on the fluidization quality. The rate of the gas-volume change can be defined as $r_c=\varepsilon_A(dx_A/dt)$, where ε_A is the increase in the number of moles when the reactants completely react per the initial number of moles. This parameter is given by $\eta - 1$. When the parameter, ε_A, is negative, the gas volume decreases as the reaction proceeds.

By considering the small region in the emulsion phase where the distribution of the gas concentration can be regarded as flat, the equation for a batch reactor can be adapted and the following relationship is finally obtained.

$$-r_c = -k\varepsilon_A \qquad (2)$$

Fig.3 shows the emulsion phased expansion measured by the bed collapse method [10] under the reaction conditions. In this case, the value of α was 3.9. The expansion ratio when the bed was fluidized by only H_2 shows that the emulsion phase slightly expanded, and that the ratio was not influenced by the temperature. On the other hand, when H_2 and CO_2 were supplied as fluidizing gases, the expansion ratio decreased with the reaction temperature when

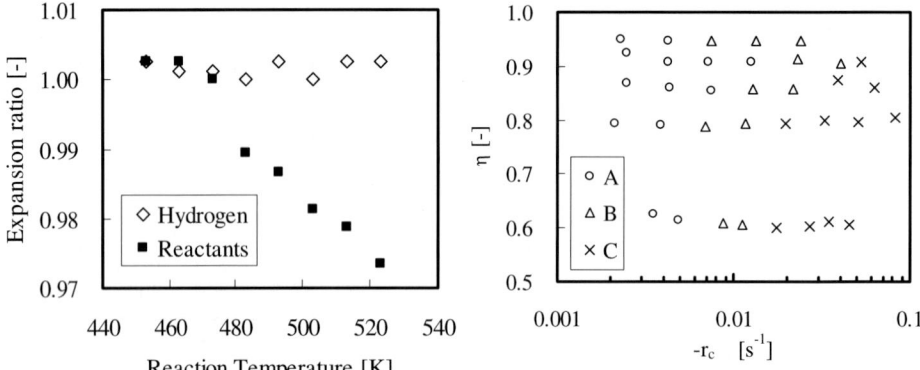

Fig. 3. Effect of reaction temperature on the emulsion-phase expansion.

Fig. 4. Evaluation of fluidization qualities by $-r_c$ and η.

the temperature was greater than 473 K and it became smaller than that of a settled bed. This result means that the gas-volume reduction in the emulsion phase due to the reaction was not compensated by the gas supply from the bubble phase.

3.3. Evaluation of the fluidization quality

Based on direct observations, we classified the fluidization quality into three regions: A (good fluidization), B (poor fluidization) and C (defluidization). In region A, a good fluidization state occurred, which is usually observed in cold model experiments. In region B, we sometimes observed channeling. In region C, defluidization was observed and channeling intermittently occurred and a defluidized particles were lifted up through the column. In order to stably operate the FCB, the conditions should be chosen to avoid regions B and C.

Fig.4 shows the fluidization quality when $-r_c$ and η are used as indices. When $-r_c$ was low, the good fluidization was maintained regardless of the value of η. The fluidization quality gradually decreased with increasing $-r_c$. When $-r_c > 0.02$ s^{-1}, the fluidization quality sharply decreased and the possibility of defluidization increased. Under these conditions, the effect of η was clearly observed. When η was near unity, defluidization was not observed even when $-r_c$ was high. This criterion is useful for determining the operating conditions to establish a good fluidization when the reaction involving a decrease in the gas volume was carried out in a FCB.

REFERENCES

[1] Kai, T. and S. Furusaki., Chem. Eng. Sci., 42 (1987) 335.
[2] Kai, T., K. Nishie, T. Takahashi and M. Nakashima, Kagaku Kogaku Ronbunshu, 30 (2004) 256.
[3] Bavarian, F. and L.-S. Fan, Ind. Eng. Chem. Res., 30 (1991) 408.
[4] Bavarian, F. and L.-S. Fan, Chem. Eng. Sci., 46 (1991) 3081.
[5] Formisani, B., R. Girimonte and G. Pataro, Powder Technol. , 125 (2002) 28.
[6] Kai, T., A. Iwakiri and T. Takahashi, J. Chem. Eng. Japan, 20 (1987) 282.
[7] Abrahamsen, A. R. and D. Geldart, Powder Technol., 26 (1980) 47.
[8] Geldart, D., Powder Technol., 7 (1973) 285.
[9] Kai, T., T. Tsutsui and S. Furusaki, Ind. Eng. Chem. Res., 43 (2004) 5474.
[10] Rietema, K., Proc. Int. Symp. on Fluidization, pp. 154-163 (1967).

Studies in Surface Science and Catalysis, volume 159
Hyun-Ku Rhee, In-Sik Nam and Jong Moon Park (Editors)

CO_2 capture characteristics of dry sorbents in a fast fluidized reactor

Chang-Keun Yi[a], S.H. Jo[a], Y. Seo[a], S.D. Park[a], K.H. Moon[b], J.S. Yoo[b], J.B. Lee[c], C.K. Ryu[c]

[a]Korea Institute of Energy Research, Daejeon, 305-343, Korea(e-mail:ckyi@kier.re.kr)
[b]Doosan Heavy Industries & Construction, Inc., Changwon, 641-792, Korea
[c]Korea Electric Power Research Institute, Daejeon, 305-380, Korea

1. ABSTRACT

One of the advanced concepts for capturing CO_2 is an absorption process that utilizes dry regenerable sorbents. Pure sodium bicarbonate from Dongyang Chemical Company and spray-dried sorbents were used to examine the characteristics of CO_2 reaction in a flue gas environment. The chemical characteristics were investigated in a fast fluidized reactor of 0.025 m i.d., and the effects of several variables on sorbent activity, including gas velocity (1.5 to 3.5 m/s), temperature (40 to 70 °C), and solid concentration (15 to 25 kg/m^2/s)], were examined in a fast fluidized-bed. Spray-dried Sorb NX30 showed fast kinetics in the fluidized reactor.

2. INTRODUCTION

Combusting fossil fuels to generate electricity increases CO_2 concentration in the earth's atmosphere. Capturing CO_2 in flue-gas streams is essential to control CO_2 emissions and is best performed at large point sources of emissions, such as power stations, oil refineries, steel works and paper mills, as well as petrochemical, fertilizer, and gas processing plants. Reducing the cost of this process is critical step in the management of greenhouse gases. Therefore, much attention has been recently directed towards developing cost-effective and energy-efficient CO_2 separation techniques to capture the CO_2 that is released from fossil fuel–fired power plants and boilers [1]. One such technique is to capture CO_2 using an absorption process that utilize dry regenerable sorbents. The process for CO_2 capture with dry regenerable sorbents consists of carbonation reactor and regeneration reactor. The following reaction occurs in each reactor:

Carbonation: $M_2CO_3 + CO_2 + H_2O \rightarrow 2\ MHCO_3 + Heat$ (1)
Regeneration: $2MHCO_3 \rightarrow M_2CO_3 + CO_2 + H_2O$ (2)
where M is an alkali metal.

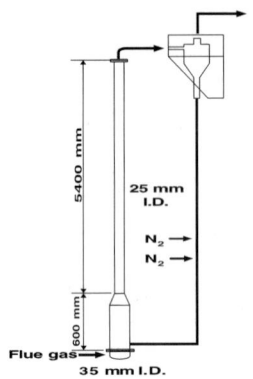

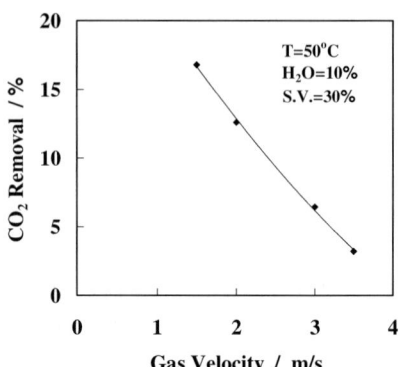

Fig. 1. Schematic of a fast fluidized bed reactor

Fig. 2. Effect of gas velocity on CO_2 removal in a fast fluidized-bed reactor

The fluidized reactor can be a bubbling fluidized type or a fast fluidized type, depending on the gas velocity and the reactivity of the sorbents. We adopted a fast fluidized bed type reactor for carbonation and regeneration reactions in order to identify the chemical characteristics of sorbents in a fast fluidized reactor of 0.025 m i.d.

3. EXPERIMENTS

Figure 1 shows a schematic of the fast fluidized-bed reactor. The general features of the system include a fast fluidized reactor with a mixing zone in the lower part, a cyclone, and a standpipe. The fast fluidized-bed is a 5.4m tall pipe of 0.025 m i.d. The lower mixing zone is a 0.6 m tall section of 0.035m i.d. After the water is condensed and removed, the CO_2 concentration in effluent gas is measured by a continuous gas analyzer. For the fluidized-bed CO_2 capture process, the sorbent should have high chemical reactivity and high attrition resistance. Also, it should be regenerable over multicycle use or have a continuous solid circulation mode between carbonation and regeneration. A commercial grade $NaHCO_3$ from Dongyang Chemical Company and Sorb sorbents were used to measure chemical reactivities in a fast fluidized bed. The Sorb sorbents were made by Korea Electric Power Research Institute by being processed through a comminution of raw materials, a preparation of colloidal slurry, a spray drying, and calcination process [2]. The physical properties of the sorbents used in this study were presented in Table 1. The reaction temperature was maintained within a range of 40 to 70 °C, the gas velocity was changed from 1.5 to 3.5 m/s, and the solid circulation rate was changed from 15 to 25 $kg/m^2/s$. The initial charge in the reactor was 7 kg throughout the entire experimental procedure. The simulated gas composition was 10 % CO_2, 10 % H_2O and balanced N_2.

4. RESULTS AND DISCUSSIONS

4.1. Reactivity of $NaHCO_3$ in a fast fluidized bed
Figure 2 shows the effects of gas velocity on CO_2 removal under the following conditions:

carbonation temperature at 50 °C, steam at 10%, and slide valve opening at 30%. The CO_2 removal was 16.5 % at a gas velocity of 1.5 m/s and was 4 % at a gas velocity of 3 m/s. The CO_2 removal decreases as the gas velocity is increased because of the reduced contact time between the sorbent and the gas. Figure 3 shows the effects of reactor temperature on CO_2 removal when gas velocity is 2 m/s, steam is 10 %, and the slide valve opening is 30 %. As the carbonation temperature is increased, the CO_2 removal decreases, as is expected in thermodynamics. Figure 4 shows the effects of solid circulation rate on CO_2 removal. The rate of solid circulation is dependent on the slide valve opening. As it is increased, the CO_2 removal also increases, probably because solid holdup increases in the riser reactor and the contact frequency between the active sorbent and CO_2 gas increases as the solid circulation rate is increased. The figure shows that the CO_2 removal is increased to 25% at a solid circulation rate of 25 kg/m^2/s.

4.2. Reactivity of spray dried Sorb sorbents in a fast fluidized bed

The reactivities of spray-dried sorbents were examined in a fast fluidized bed. The reactor was operated at a carbonation temperature of 50 °C, and a gas velocity of 2 m/s with an initial sorbent inventory of 7 kg to compare CO_2 concentration profiles in effluent gas for spray-dried Sorb NH series and NX30 sorbent. Figure 5 shows the comparison of CO_2 concentration profiles in effluent gas of Sorb NHR, NHR5, and NX30 in a fast fluidized-bed reactor. The CO_2 removals of Sorb NHR and NHR5 were initially maintained at a level of 100 % for a short period of time and quickly dropped to a 10 to 20 % removal level.

Table 1 Physical Properties of Sorbents

Properties	NaHCO$_3$	Sorb NHR	Sorb NHR5	Sorb NX30
Bulk density, g/cm^3	0.67	0.41	0.44	0.75
Mean particle diameter, μm	62	164	120	104.6
Attrition index, AI	56	52	65	1
BET surface area, m^2/g	1.43	-	26	90.5

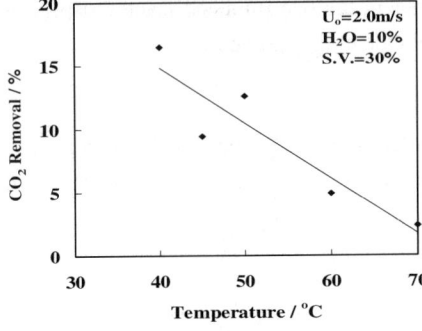 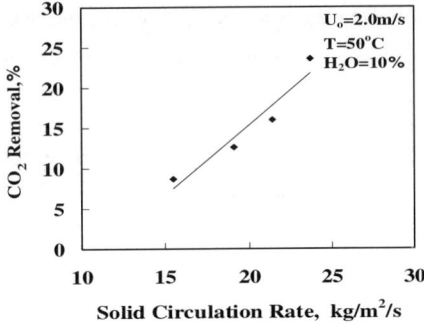

Fig. 3. Effect of carbonation temperature on CO_2 removal in a fast fluidized-bed reactor.

Fig. 4. Effect of solid circulation rate on CO_2 removal in a fast fluidized-bed reactor

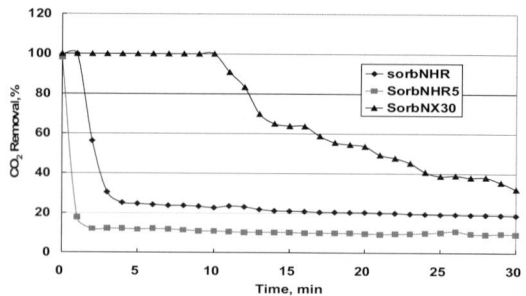

Fig. 5. Comparison of sorbent kinetics and sorption capacities
of spray dried sorbents, Sorb NHR, NHR5, and NX30

However, the CO_2 removal of the Sorb NX30 sorbent was maintained at 100 % for 12 minutes. This is the first time that a dry sorbent was able to capture all of the 10 % CO_2 in the flue gas within 3 seconds of residence time in the fast fluidized reactor. The sorption capacity of Sorb NX30 is also much greater than that of the two other sorbents and of $NaHCO_3$. Sorb NX30 showed fast kinetics in a fast fluidized reactor, capturing all of the 10 % CO_2 within 3 seconds in the fast fluidized reactor. The most important factor affecting CO_2 removal is closely related to water sorption capability of the sorbents during carbonation reaction. As can be seen in Table 1, Sorb NX30, that is most porous and is able to sorb water effectively, showed a best CO_2 removal among all the sorbents.

5. CONCLUSIONS

The reactivities of pure $NaHCO_3$ solid, Sorb NHR, NHR5, and NX30 sorbents were examined in a fast fluidized bed reactor. The CO_2 removal of the pure $NaHCO_3$ solid increased from 3 % to 25 % when the variables were altered. Removal increased as gas velocity was decreased, as the carbonation temperature was decreased, or as the solid circulation rate was increased. The CO_2 removal of Sorb NHR and NHR5 was initially maintained at 100 % for a short period of time but quickly dropped to a 10 to 20 % removal. However, the Sorb NX30 sorbent showed fast kinetics in the fast fluidized reactor, capturing all of the 10 % of the CO_2 in the flue gas within 3 seconds in the fast fluidized reactor.

ACKNOWLEDGEMENT
This research was supported by the Carbon Dioxide Reduction & Sequestration Center, one of of the 21st Century Frontier R&D Programs funded by the Ministry of Science and Technology of Korea.

REFERENCES
[1] Y. Liang, D.P. Harrison, R.P. Gupta, D.A. Green and W.J. McMichael, Energy & Fuels 18 (2004) 569.
[2] C.K. Ryu, J.B. Lee, T.H. Eom, J.M. Oh, and C.K. Yi. Proceeding in 21st Annual International Pittsburgh Coal Conference, Pittsburgh, PA, CD-Rom (2004) 47-4.

Studies in Surface Science and Catalysis, volume 159
Hyun-Ku Rhee, In-Sik Nam and Jong Moon Park (Editors)

Numerical simulation of particle fluidization behaviors in a rotating fluidized bed

Hideya Nakamura and Satoru Watano

Department of Chemical Engineering, Osaka Prefecture University
1-1 Gakuen-cho, Sakai, Osaka 599-8531, Japan (watano@chemeng.osakafu-u.ac.jp)

1. INTRODUCTION

Fluidized bed has many advantages as gas-solid reactors or powder handling processors because of its advantages of high heat and mass transfer, temperature homogeneity, and mixing property. However, conventional fluidized bed has some limitations; it is very difficult to operate at high gas velocity, etc. Operation at high gas velocity leads to formation of large bubbles or slugs, and passage through powder bed. In this case, the gas-solid contact becomes rather poor and reaction efficiency decreases. Another limitation is that it is very difficult to fluidize fine particles, according to a Geldart classification [1], since it exhibits channeling and lifting as a plug and forming rat holes when aerated.

Recently, a rotating fluidized bed (RFB) has attracted special interest, since it can overcome the limitations mentioned above. In a RFB, particles are forced toward the rotating cylindrical air distributor due to the large centrifugal force and formed annular bed near at the air distributor. Air flows inward through the air distributor, and then particles are balanced by drag and centrifugal forces, leading to achieve uniform fluidization. Since the minimum fluidization velocity increases with an increase in the vessel rotational speed, a RFB can prevent from forming large bubbles at any higher gas velocities by controlling the rotational speed. Therefore, it is expected that the gas-solid contact and reaction efficiency are greatly improved in a RFB. In addition, a RFB can fluidize very fine particles, such as Geldart group-C particles, and even nano-particles [2], since it can impart high centrifugal force and drag force to particles, which are larger than the cohesive forces between particles.

So far, some researchers have analyzed particle fluidization behaviors in a RFB, however, they have not well studied yet, since particle fluidization behaviors are very complicated. In this study, fundamental particle fluidization behaviors of Geldart's group B particle in a RFB were numerically analyzed by using a Discrete Element Method (DEM)- Computational Fluid Dynamics (CFD) coupling model [3]. First of all, visualization of particle fluidization behaviors in a RFB was conducted. Relationship between bed pressure drop and gas velocity was also investigated by the numerical simulation. In addition, fluctuations of bed pressure drop and particle mixing behaviors of radial direction were numerically analyzed.

2. NUMERICAL MODEL AND EXPERIMENTAL

In this study, particle fluidization behaviors in a RFB were numerically analyzed by using a DEM-CFD coupling model [3]. The particle motion was calculated by DEM, which calculates the motion of each particle by integrating the Newton's equations for individual particle step by step, allowing for the external forces acting on a particle. Equations of transitional and rotational motions for individual particles are as follows;

$$m\frac{d^2 X}{dt^2} = F_c + F_d + F_{cen} + F_{cori} + mg \tag{1}$$

$$\frac{d\omega_p}{dt} = \frac{T}{I} \tag{2}$$

where X, m, t are the position vector, mass of a particle, time, and F_c, F_d, F_{cen}, F_{cori} indicate the contact, drag, centrifugal, coriolis forces, respectively. Also, g, ω_p, T and I are the gravitational acceleration, angular velocity, angular moment by contact and inertia moment, respectively. The contact forces were calculated by a soft sphere model shown in Fig. 1 [3]. The drag forces were calculated by using Ergun's and Wen-Yu's equations [3]. The fluid motion was calculated by locally averaged equations of continuity and motion. The basic equations were given as follows;

$$\frac{\partial \varepsilon}{\partial t} + \nabla \cdot (\varepsilon u) = 0 \tag{3}$$

$$\frac{\partial}{\partial t}(\varepsilon u) + \nabla \cdot (\varepsilon u u) = -\frac{\varepsilon}{\rho_f}\nabla P + f_i \tag{4}$$

where ε, u, P, ρ_f and f_i indicate voidage, gas velocity, pressure, gas density and interaction term between particle and fluid. In this study, the calculations were conducted in a pseudo three-dimensional fluidized bed having a thickness equivalent to one particle diameter, and particle and fluid motions were calculated in a two-dimensional cylindrical coordinate. Spherical polyethylene particles were used as model particles, which had diameter of 0.5 mm. In this study, model particle used was Geldart's group B particle, not as A or C, in order to reduce computational load. The number of particles and the vessel diameter were set to 16,000 and 120 mm. A schematic diagram of a rotating fluidized bed is shown in Fig. 2. A RFB is basically consists of a plenum chamber and a cylindrical thin air distributor (ID 120 mm × D 5 mm) that rotates around its axis of symmetry inside the chamber.

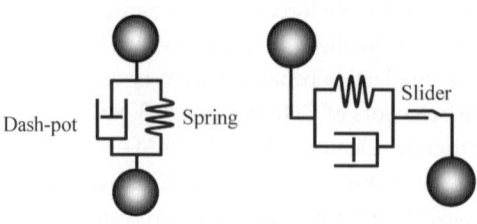

(a) Normal direction   (b) Tangential direction

Fig. 1. Models of contact force
(Soft sphere model)

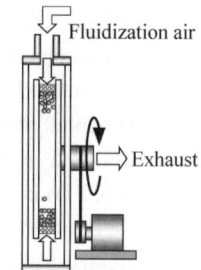

Fig. 2. Schematic diagram of
a rotating fluidized bed (RFB)

3. RESULTS AND DISCUSSIONS

Fig. 3 shows the calculated and experimental results of particle fluidization behaviors in a RFB. A high-speed video camera (FASTCAM MAX, Photoron CO., Ltd.) was used for visualization of actual particle fluidization behavior. The bubbling fluidization behaviors, such as the bubble formation, eruption and particle circulation with rotational motion, could be well simulated, and these behaviors were also observed in the experimental results.

Fig. 4 indicates the calculated bed pressure drops as a function of superficial gas velocity (U_0) at various centrifugal accelerations. The bed pressure drop was calculated by numerical integration of Ergun's equation for packed bed pressure drop, based on the simulated bed voidage, gas and particle velocity. With an increase in the gas velocity, the bed pressure drop showed a rise to peak at minimum fluidization velocity (U_{mf}), followed by a slight constant value with fluctuation at each centrifugal accelerations. Fig. 5 also shows minimum fluidization velocity as a function of centrifugal acceleration estimated by the present model and the analytical equation by Kao et. al. [4]. It was found that the calculated U_{mf} by the present model showed good agreement with those by the analytical model by Kao et. al [4]. From these results, validity of the present model was quantitatively confirmed.

Rotational direction

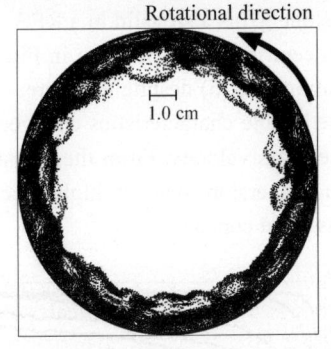

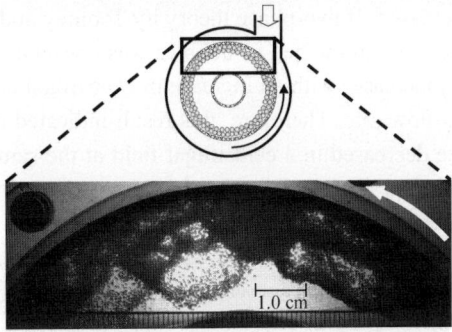

(a) Calculated particle fluidization behavior
(10 G, U_0=1.15 m/s)

(b) Actual fluidization behavior
(10 G, U_0=0.90 m/s)

Fig. 3. Calculated and experimental results of particle fluidization behaviors in a RFB

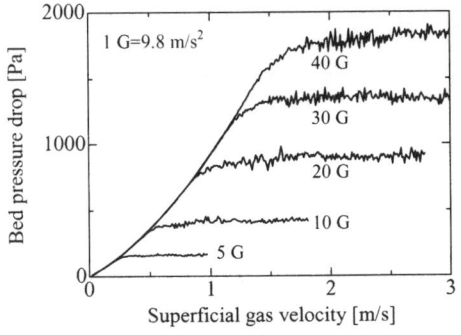

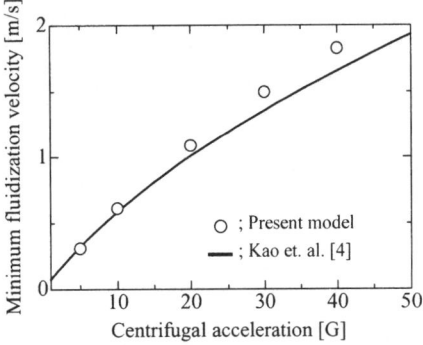

Fig. 4. Calculated bed pressure drop vs. superficial gas velocity

Fig. 5. Estimated minimum fluidization velocity vs. centrifugal acceleration

508

Fig. 6 shows the FFT spectrum for calculated bed pressure drop fluctuations at various centrifugal accelerations. The excess gas velocity, defined by (U_0-U_{mf}), was set at 0.5 m/s. Here, "1 G" means numerical result of particle fluidization behavior in a conventional fluidized bed. In Fig. 6, the power spectrum density function has typical peak in each centrifugal acceleration. However, as centrifugal acceleration increased, typical peak shifted to high frequency region. Therefore, it is considered that periods of bubble generation and eruption are shorter, and bubble velocity is faster at higher centrifugal acceleration.

Fig. 7 shows temporal changes in the degree of particle mixing, M, at various centrifugal accelerations. The degree of particle mixing, M, is defined as the following equation;

$$M = 1.0 - \sigma/\sigma_0 \tag{5}$$

where, σ and σ_0 are the standard deviation based on the number fraction of tracer particles at each time and initial condition, respectively. Here, the degree of particle mixing can be expressed; the "zero" corresponds to completely segregated condition, and the "one" corresponds to uniformly mixed condition. The outside concentric layer of particle bed was defined as tracer particles, and particle mixing behavior of radial direction was numerically analyzed. In Fig. 7, as centrifugal acceleration increased, mixing speed of particles was decreased. If two-phase theory by Toomey and Johnstone [5] is also valid in a RFB, gas flow rate through the bed as bubbles was constant at each centrifugal acceleration in Fig. 7, since U_{mf} increases with an increase in centrifugal acceleration (Fig. 5) despite the increase in total gas flow rate. Therefore, this result indicated that the bubble characteristics changed, bubble size decreased in a centrifugal field at the same excess gas velocity. From these results, it is indicated that the smaller bubbles exist as higher dispersion state at higher centrifugal acceleration in a RFB, which leads to improve the gas-solid contact.

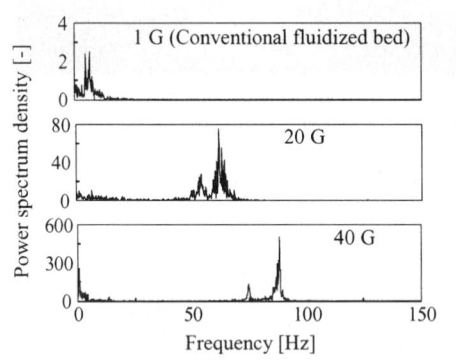

Fig. 6. FFT spectrums for fluctuation of bed pressure drop (U_0-U_{mf} =0.5 m/s)

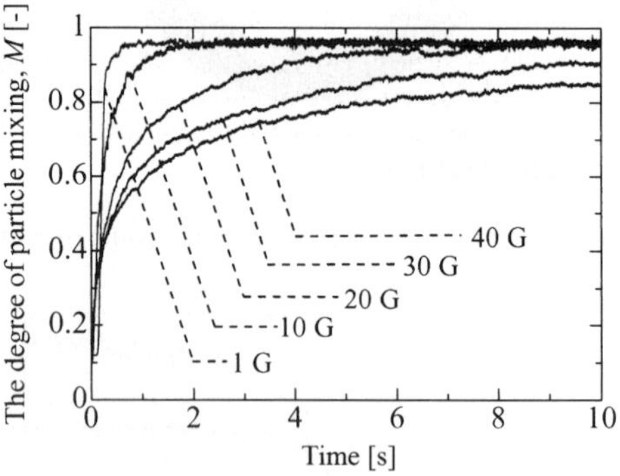

Fig. 7. Temporal changes in the degree of particle mixing (U_0-U_{mf} =0.5 m/s)

REFERENCES
[1] D. Geldart, Powder Technol., 7 (1973) 285.
[2] H. Nakamura, S. Watano and K. Hamada, Proc. APCChE 2004, Kitakyusyu, Japan, 2004.
[3] Y. Tsuji, T. Kawaguchi and T. Tanaka, Powder Technol., 77 (1993) 79.
[4] J. Kao, R. Pfeffer and G.I. Tardos, AIChE J., 33 (1987) 858.
[5] R.D. Toomey and H.F. Johnstone, Chem. Eng. Prog., 48 (1952) 220.

Studies in Surface Science and Catalysis, volume 159
Hyun-Ku Rhee, In-Sik Nam and Jong Moon Park (Editors)
509

Characteristics of particle mixing and detection of poor fluidization in a fluidized bed ash cooler

Jong-Min Lee[a], Dong-Won Kim[a], Eun-Mo Lee[b], Jae-Sung Kim[a] and Jong-Jin Kim[a]

[a]Power Generation Laboratory, Korea Electric Power Research Institute, 103-16 Munji-dong, Yusung-gu, Daejeon 305-380, Republic of Korea (jmlee@kepri.re.kr)

[b]Mechanical Engineering, Chungnam National University, 220 Gung-dong, Yusung-gu, Daejeon 305-764, Republic of Korea

1. INTRODUCTION

A CFB boiler has a fluidized bed ash cooler (FBAC), of which type is a bubbling fluidized bed, for discharging bed material such as sand or coal ash particles [1,2]. If the coal used as a fuel contains a large quantity of ash, the discharging process is necessary to removal of accumulated ash in the combustor of the CFB boiler. However, interruption of good fluidization in the FBAC is frequently happened because of agglomeration of the particles in the bed. Consequently, this unstable operation may, in the worst case, result in an unscheduled boiler shut down [3]. A method capable of observing small changes in the behavior of fluidized bed is not only useful for early detection of unwanted changes, but also for the control of intentionally enforced changes [4,5]. However, a method of observing changes of fluidization may have to be applied together a method of clearing up poor fluidization problems. It can be accomplished by the analysis of a practical system and operation conditions. In this study, we have studied and introduced the simple detection and solution techniques with analyzing the mixing property and the occurrence of poor fluidization in a simulated fluidized bed ash cooler system.

2. EXPERIMENT

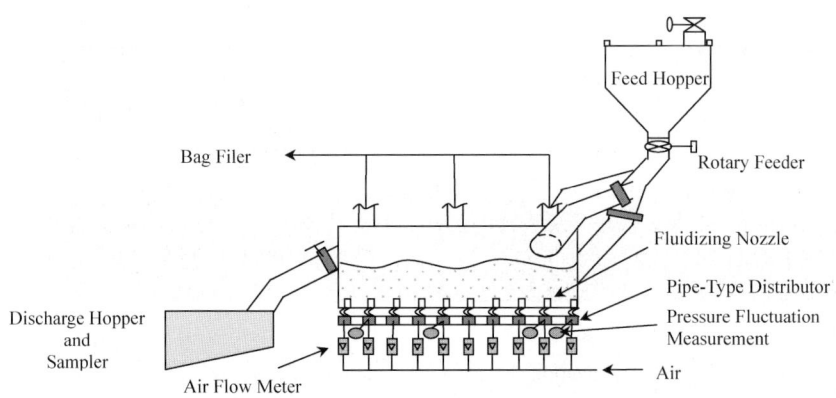

Fig 1. Schematic diagram of a simulated fluidized bed ash cooler

The simulated FBAC consists of an acrylic main reactor (0.5m-H x 0.5m-W x 1.0m-L), an air distributor system, particles feeding system including a feed hopper, a discharging sampler, a bag filter for capture of the elutriated fine particles and, pressure and flow rate measurement systems (Fig. 1). The air distributor system has ten air headers. An individual air header is connected with 5 air nozzles and can regulate the airflow rate. The opening ratio of the distributor is 2.1% and each nozzle has four holes for uniform air supply. To measure the pressure fluctuation at an individual air header, high frequency pressure transmitters were mounted at the approach and the exit headers of the FBAC.

In this study, we observed the mixing characteristics and the pressure fluctuation when the introducing particles were larger (860, 1460μm) than bed materials (320μm, u_{mf} = 0.034m/s) with the variation of the feeding rate and the fluidizing velocity. We also developed the simple method for detection of poor fluidization using analysis of the pressure fluctuation and solved the poor fluidization by regulating local fluidization velocity at the individual air headers.

3. RESULTS AND DISCUSSION

3.1 Poor fluidization phenomenon
Fig. 2 shows the poor fluidization phenomenon at the entrance of the FBAC when coarse particles (860μm) are fed into the bed (320μm). Fig. 2 (a) and (b) show the status before fluidization and good fluidization of the FBAC, whereas Fig, 2 (c) and (d) show the onset of poor fluidization and its growth respectively. The bridge of the large particles at the bed surface can be observed, and causes to form the poor fluidization area at the entrance of the FBAC. This poor fluidization is somewhat different from jetsam and flotsam separation. The large particles, fed into the FBAC, stayed at the bed surface, whereas the fine particles were fluidized under the bridge of the large particles. However, the poor fluidization could not be observed when the same particles of the bed were fed. This may be from the reason that the discharging rate of the large particles is larger than the time needed for absorption into the bed. Additionally, the large particles made a local bed height around the entrance of the FBAC increase, which means that the pressure drop of the area increased. Consequently bubbles coming from lower part of the bed bypassed gradually to other areas having lower pressure drop. Also bubbles were broken underneath the bridge of the large particle and went through the lump of the large particle. This causes the defluidization to intensify at the entrance of the FBAC.

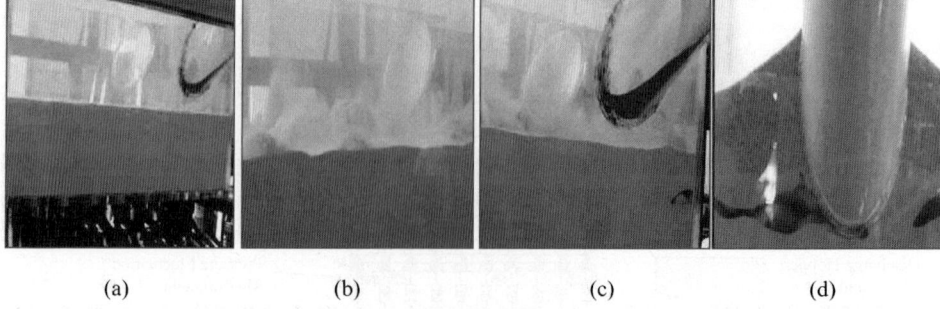

(a)   (b)   (c)   (d)

Fig. 2. Photos for a) before fluidization, b) good fluidization, c) onset of defluidization at the entrance to the introducing particles and d) linkage of large particles of the bed surface

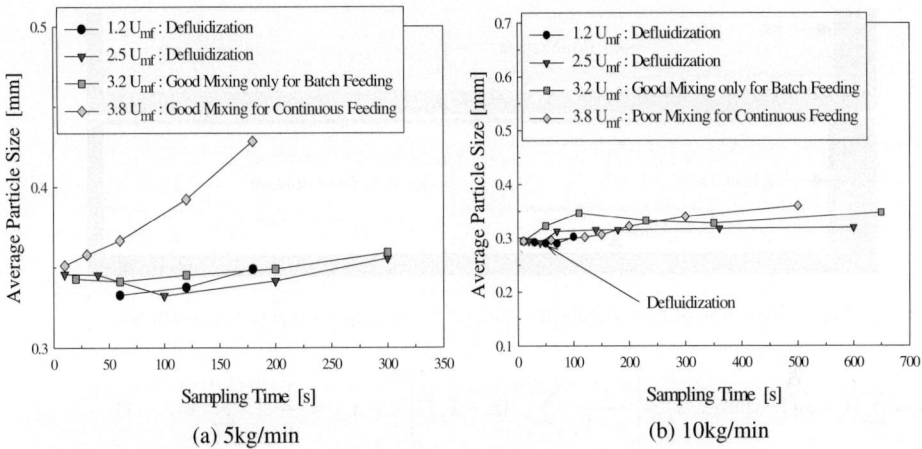

(a) 5kg/min

(b) 10kg/min

Fig. 3. Mixing characteristics with variation of fluidizing velocities and of discharging rate

3.2 Particle mixing and detection of poor fluidization

The extent of the mixing can be noticed by increase of the fraction of the large particles in the samples after injection of large particles into the bed. It can be also noticed by increase of the average particle size of the sample particles as shown in Fig. 3. The average size of the sampled particles at the feeding rate of 5kg/min shows that the large particles were not mixed with the bed materials under 2.5 u_{mf}, and the good mixing could be observed above 3.8 u_{mf} in a continuous feeding of the particles. However, the defluidization could be observed at the feeding rate of 10kg/min though the airflow is large enough to mix the large particles with the bed materials. This means that the defluidization is affected not only by airflow rates but also by the particles discharging rates.

In this study, the accumulated standard deviation error (ss_t) for a given period of time was introduced to simply detect the local poor fluidization. The accumulated standard deviation error, ss_t, is defined as Eq. (1). In Eq. (1), n means the number of sampling data for a given period of time and j means the number of sampling data set during the measurement of the pressure fluctuation.

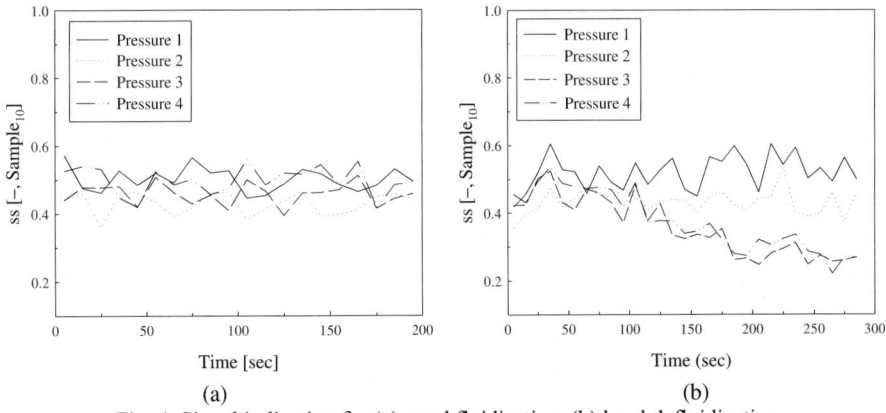

(a)

(b)

Fig. 4. Signal indication for (a) good fluidization, (b) local defluidization

512

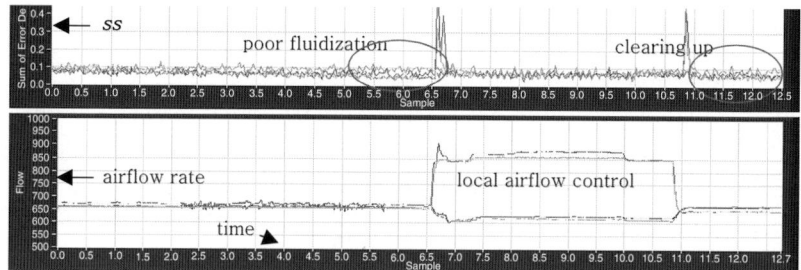

Fig. 5. Detection of poor fluidization and its clearing up for real time experiment

$$ss_t = \sum_{j=1}^{t} \left(s_j / \sqrt{n} \right), \text{ where, } s_j = \left[\frac{1}{n-1} \sum_{i=n\times(j-1)+1}^{n\times(j-1)+n} (x_i - \overline{x}_j)^2 \right]^{1/2}, \; \overline{x}_j = \frac{1}{n} \sum_{i=n\times(j-1)+1}^{n\times(j-1)+n} x_i \quad (1)$$

Fig. 4 shows the accumulated standard deviation error, ss, with the measuring time of the pressure fluctuation for a good fluidization case (a) and a locally poor fluidization case (b). Pressure 1 and 2 were measured at the exit and the center air-headers of the FBAC respectively. Pressure 3 and 4 were measured at the entrance air-headers of the FBAC. As shown in Fig. 4 (a) and (b), the accumulated standard deviation error, ss, stayed in a limited range if the bed is in good fluidization state, but ss for the entrance of the FBAC decreased steadily if the bed is in local poor fluidization state. This may be from the decrease of the bubble explosion force and frequency, which have influence on the standard deviation error of the pressure fluctuation, at the bed surface due to the bubble break and bypass around the poor fluidization area. Therefore, we can easily detect the local poor fluidization through this simple method. Additionally, as detecting the local poor fluidization, we could also regulate the overall or local airflow rate to clear up the local poor fluidization, as shown in Fig. 5. The accumulated standard deviation error, deviated from a limited range due to poor fluidization, shows to return into a limited value after regulations of local airflow rates.

4. CONCLUSIONS

The detecting and the clearing poor fluidization in a simulated FBAC were studied with analyzing the mixing property and the pressure fluctuation. The bridge of the large particles at the bed surface could be observed, and this caused to form the defluidization area at the entrance of the FBAC. The poor fluidization was affected not only by airflow rates but also by the particles discharging rates as well as particle size distribution in the FBAC. The local poor fluidization could be detected by analysis of the accumulated standard deviation error. Additionally, the regulation of the overall or local airflow rate made clearing up the local poor fluidization possible.

REFERENCES

[1] J.M. Lee, J.S. Kim, J.J. Kim and P.S. Ji, 13th Korea-US Jonit Workshop on Energy & Environment, Sep., Nevada, USA (1999) 41.
[2] J.M. Lee and J.S. Kim, KJChE, 16(5) (1999) 640.
[3] J.R. Ommen, Ph. D. Thesis, Delft University Press, the Netherlands, (2001).
[4] J.C. Schouten, R.C. Schouten and C.M. Van den Bleek, Chem. Eng. Sci. 54, (1999) 2103.
[5] J. Wether, Powder Technol. 102 (1999) 15.

Studies in Surface Science and Catalysis, volume 159
Hyun-Ku Rhee, In-Sik Nam and Jong Moon Park (Editors)

Separation characteristics of Phenoxyacetic Acids in Fixed and Semi-Fluidized Beds

Tae Young Kim[a], Seung Jai Kim[a,b] and Sung Yong Cho[a]

[a]Department of Environmental Engineering, Chonnam National University, 300 YongBong-Dong, Buk-Gu, Gwangju 500-757, Korea (e-mail:tykim001@chonnam.ac.kr)

[b]Environmental Research Institute, Chonnam National University, 300 YongBong-Dong, Buk-Gu, Gwangju 500-757, Korea

ABSTRACT

Adsorption equilibrium of CPA and 2,4-D onto GAC could be represented by Sips equation. Adsorption equilibrium capacity increased with decreasing pH of the solution. The internal diffusion coefficients were determined by comparing the experimental concentration curves with those predicted from the surface diffusion model (SDM) and pore diffusion model (PDM). The breakthrough curve for packed bed is steeper than that for the fluidized bed and the breakthrough curves obtained from semi-fluidized beds lie between those obtained from the packed and fluidized beds. Desorption rate of 2,4-D was about 90 % using distilled water.

1. INTRODUCTION

Phenoxyacetic acids are very important chemicals because of their wide distribution and extensive use as plant growth regulators. 2,4-diphenoxyacetic acid (2,4-D) is used as agricultural herbicides against broad leaf weeds in cereal crop farm as well as on pastures and lawns, in parks, and on golf courses. Chlorophenoxyacetic acid (CPA) is used worldwide as a plant growth regulator for agricultural and non-agricultural purposes. Typically, it is employed on a large scale for weed control on cereal crops and lawns [1,2]. Adsorption of phenoxyacetic acids onto solid adsorbents is very important, since it can effectively remove pollutants from aqueous stream. In order to design effective activated carbon adsorption units and develop mathematical models which can accurately describe their operation characteristics, sufficient information on both the adsorption and desorption of individual pollutant under different operating conditions is required. The main purpose of this work is to study the adsorption and desorption characteristics of the phenoxyacetic acids experimentally as well as theoretically to remove the acids from aqueous solution in fixed and fluidized beds.

2. EXPERIMENTAL

The concentration of the phenoxyacetic acids (CPA, 2,4-D) was measured using a spectrophotometer (Shimadzu 1601). The wavelengths, corresponding to a maximum absorbance of CPA and 2,4-D, were found to be 273 nm and 284 nm, respectively. The adsorbent used in this study was an activated carbon, Filtrasorb-400, manufactured by Calgon Co. (USA). Single species equilibrium adsorption data were obtained by measuring the adsorbate concentration in an aqueous solution of the phenoxyacetic acids. Batch adsorption experiments were conducted in a Carberry-type batch adsorber. A single-species adsorption experiment was carried out in a fixed bed, semi-fluidized bed and fluidized bed. A desorption experiment of the phenoxyacetic acids from the adsorbents was performed using distilled water.

3. RESULTS AND DISCUSSION

Figs. 1 and 2 show the adsorption capacity of CPA and 2,4-D for different pH of the solution. As can be seen in those figs, the adsorption amounts decreased with increasing pH of the solution. Single-species isotherm data were correlated by well-known the Langmuir, Freundlich and Sips equations. To find the parameters for each adsorption isotherm, the linear least square method and the pattern search algorithm (NMEAD) were used. The value of the mean percentage error has been used as a test criterion for the fit of the correlation. The model parameters and the average percent differences between measured and calculated values are given in Table 1. As shown in the table, the Sips equation gives the best fit of our data among the three. From this result, we believe that the Sips equation is suitable for representing single-component equilibrium adsorption data of CPA and 2,4-D onto activated carbon.

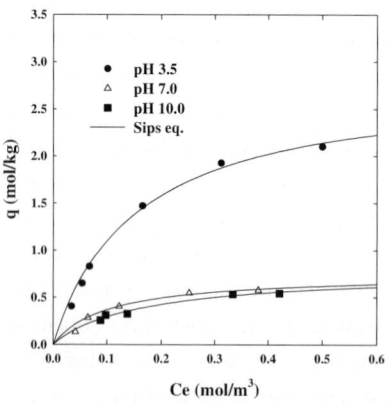

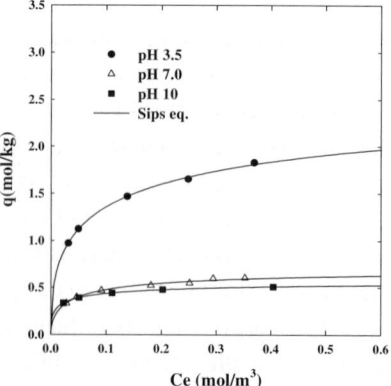

Fig. 1. Adsorption isotherms of the CPA for different pH (298 K).

Fig. 2. Adsorption isotherms of the 2,4-D for different pH (298 K).

Table 1. Adsorption isotherm parameters of CPA and 2,4-D with different pH (298 K).

Isotherm	Parameters	pH 3.5		pH 7.0		pH 10.0	
		CPA	2,4-D	CPA	2,4-D	CPA	2,4-D
Langmuir	q_m	2.77	1.86	0.38	0.62	0.77	0.50
	b	6.59	33.9	9.90	40.7	5.86	83.8
	error (%)	2.66	3.11	1.98	2.94	4.99	2.71
Freundlich	k	3.18	2.37	0.90	0.77	0.85	0.59
	n	2.09	3.97	2.44	4.45	2.14	6.82
	error (%)	7.32	1.35	5.26	2.34	5.38	2.01
Sips	q_m	2.27	3.03	0.80	0.75	0.78	0.63
	b	6.86	2.36	10.3	7.32	5.88	6.58
	n	0.98	2.16	0.78	1.62	1.01	2.16
	error (%)	2.31	1.02	1.71	1.70	5.06	0.27

There are several correlations for estimating the film mass transfer coefficient, k_f, in a batch system. In this work, we estimated k_f from the initial concentration decay curve when the diffusion resistance does not prevail [3]. The value of k_f obtained from the initial concentration decay curve is given in Table 2. In this study, the pore diffusion coefficient, D_p, and surface diffusion coefficient, D_s, are estimated by pore diffusion model (PDM) and surface diffusion model (SDM) [4]. The estimated values of k_f, D_p, and D_s for the phenoxyacetic acids are listed in Table 2.

Table 2. Kinetic parameters of the phenoxyacetic acids in a batch reactor (298 K, pH 3.5).

Adsorbates	$k_f \times 10^{-5}$ m/sec	$D_s \times 10^{-13}$ m^2/sec	$D_p \times 10^{-9}$ m^2/sec
CPA	7.67	6.80	2.90
2,4-D	5.00	1.90	1.42

Fig. 3 shows the breakthrough curves for the packed, semi-fluidized, and fluidized beds. It is seen that the breakthrough curve obtained from the semi-fluidized bed lies between those obtained from the packed and fluidized beds, since the semi-fluidized bed possesses the features of both the fluidized and packed beds. This figure also shows that the shape of the breakthrough curve for the packed bed is steeper than that for the fluidized bed.

For the successful application of an adsorption system, an efficient regeneration of the used adsorbent is very important from the economic point of view. In general, there are many regeneration techniques such as thermal, steam, acid or base and solvent regenerations. The choice of a certain regeneration method depends upon the physical and chemical characteristics of both the adsorbate and adsorbent. In this study, distilled water was used as desorbate for GAC. As can be seen in Fig. 4, the desorption rate of 2,4-D was 90% using distilled water only. The effluent pH increased in the initial stage of adsorption, and

516

decreased to the pH of the initial solution as adsorption proceeded, and then increased as desorption proceeded, as shown in Fig. 4.

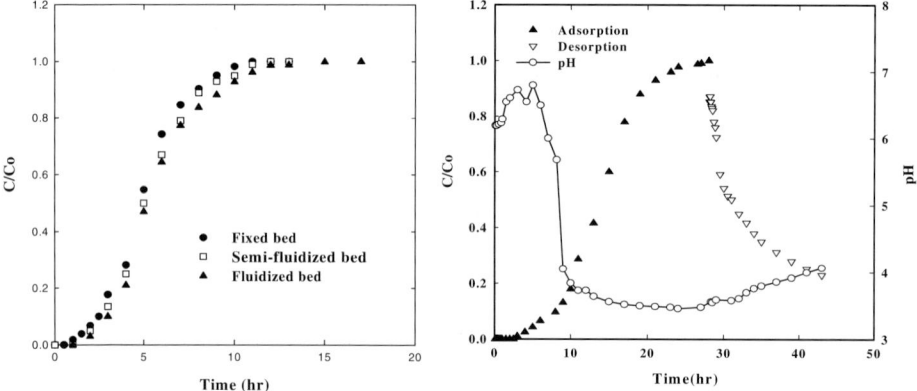

Fig. 3. Comparison of breakthrough curves of CPA for different bed types.

Fig. 4. pH variations during adsorption and desorption processes for 2,4-D.

4. CONCLUSIONS

1. The Sips isotherm was suitable for CPA and 2,4-D among the various isotherms, and the equilibrium adsorption capacity for CPA and 2,4-D onto GAC decreased with increasing pH of the solution.

2. The differences in the rate of adsorption are primarily attributable to the differences in the equilibrium capacity at the various pHs, and the pore diffusion model simulated our data satisfactorily.

3. The shape of the breakthrough curve for a packed bed is steeper than that for the fluidized bed and the breakthrough curves obtained from semi-fluidized bed lies between those obtained from the packed and fluidized beds.

4. Desorption of P-CPA was about 90% using distilled water only.

ACKNOWLEDGEMENT

This research was financially supported by the KRF(Y00-316).

REFERENCES

[1] B. Enric, B. Birame, I. Sirés, J. A. Garrido, R. M. Rodríguez, C. Arias, P. L. Cabot and C. Comninellis, *Electrochimica Acta*, **49**, 4487 (2004).

[2] J. P. Chen. and X. Wang, " *Sep. & Purifi. Tech.*, **19**, 157 (2000.)

[3] T. Y. Kim, S. J. Kim and S. Y. Cho, *Korean J. of Chem. Eng.*, **18**(5) 775 (2001)

[4] H. Moon, and C. Tien, *Ind. Eng.Chem. Res.*, **26**, 2024 (1987)

Studies in Surface Science and Catalysis, volume 159
Hyun-Ku Rhee, In-Sik Nam and Jong Moon Park (Editors)
© 2006 Elsevier B.V. All rights reserved

Surface Analyses of Cobalt Catalysts for the Steam Reforming of Tar derived from Biomass Gasification

Kazuhiko Tasaka[a], Takeshi Furusawa[b], Kiyoshi Ujimine[a], and Atsushi Tsutsumi[a]

[a] Department of Chemical System Engineering, The University of Tokyo,
7-3-1 Hongo, Bunkyo-ku, Tokyo 113-8656, Japan (e-mail:tsutsumi@chemsys.t.u-tokyo.ac.kr)
[b] Department of Applied Chemistry, Utsunomiya University,
7-1-2 Yoto, Utsunomiya 321-8585, Japan

Abstract

The surface analyses of the Co/MgO catalyst for the steam reforming of naphthalene as a model compound of biomass tar were performed by TEM-EDS and XPS measurements. From TEM-EDS analysis, it was found that Co was supported on MgO not as particles but covering its surface in the case of 12 wt.% Co/MgO calcined at 873 K followed by reduction. XPS analysis results showed the existence of cobalt oxide on reduced catalyst, indicating that the reduction of Co/MgO by H_2 was incomplete. In the steam reforming of naphthalene, film-like carbon and pyrolytic carbon were found to be deposited on the surface of catalyst by means of TPO and TEM-EDS analyses.

Keywords: cobalt catalyst, naphthalene, steam reforming, biomass, gasification, tar

1. Introduction

Biomass gasification offers the potential for producing a fuel gas that can be used for power generation system or synthesis gas applications. The volatile matter contains a considerable amount of tar which is a complex mixture of aromatics. Despite extensive research efforts tar formation which causes the pipe plugging and the reduction of conversion efficiency is still a major problem in biomass gasification systems [1-6].

In previous research, we prepared several kinds of Co/MgO catalysts and investigated the catalytic performance for the steam reforming of naphthalene as a model compound of biomass tar [7, 8]. Having larger exposed Co metal surface area, catalysts calcined at 873 K showed higher activity than the catalysts calcined at higher temperature for the same Co loading. It was found that 12 wt.% Co/MgO catalyst calcined at 873 K showed the best performance. However, little is known about the factor in the activity of the catalyst and the reaction mechanism of the catalytic steam reforming of tar derived from biomass gasification.

In this study, we focused on the surface analyses of 12 wt.% Co/MgO by TEM-EDS and XPS techniques, and discussed the presented results.

2. Experimental

2.1. Preparation of the catalyst
Cobalt-supported-on-MgO catalysts were prepared by impregnating MgO (JRC-MGO-4 1000A, 14-16 m^2 g^{-1}) with aqueous solutions of $Co(NO_3)_2 \cdot 6H_2O$ (Kanto Co.), followed by drying. The material was then calcined in air for 8 h at 873 K or 1173 K. Before catalytic tests, catalysts were reduced at 1173 K in H_2/Ar (50/50 vol.%).

2.2 Catalytic tests
The steam reforming of naphthalene was conducted on the fixed bed of catalyst (bed temperature: 1173 K, GHSV: $3000h^{-1}$). The detail of the experiment was described in previous papers [7, 8].

2.3. Surface analyses of the catalysts
TEM observation and elemental analysis of the catalysts were performed by means of a transmission electron microscope (JEOL, JEM-2010F) with energy dispersion spectrometer (EDS). The surface property of catalysts was analyzed by an X-ray photoelectron spectrometer (JEOL, JPS-90SX) using an Al $K\alpha$ radiation (1486.6 eV, 120 W). Carbon $1s$ peak at binding energy of 284.6 eV due to adventitious carbon was used as an internal reference. Temperature programmed oxidation (TPO) with 5 vol.% O_2/He was also performed on the catalyst after reaction, and the consumption of O_2 was detected by thermal conductivity detector. The temperature was ramped at 10 K min^{-1} to 1273 K.

3. Results and discussion

3.1. Surface analyses of the catalysts by TEM-EDS and XPS
The TEM images of 12 wt.% Co/MgO calcined at 873 K (Catalyst I) before and after reduction are shown in Fig.1(a) and (b), respectively. Although Co metal phase was detected in reduced Co/MgO by X-ray diffraction measurements (XRD) [7, 8], no Co metal particle was observed on both catalysts. EDS elemental analysis showed that primary particles contain both Mg and Co elements, whose concentrations were about the same as loaded amounts. Figure 2 shows TEM image of 12 wt.% Co/MgO calcined at 1173 K (Catalyst II).

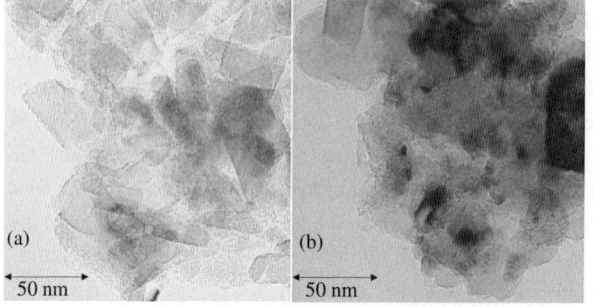
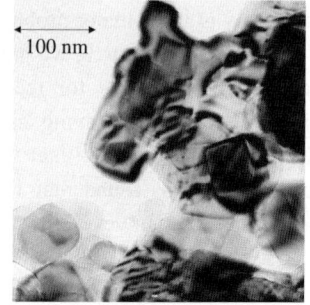

Fig.1 TEM images of (a) Catalyst I, (b) Catalyst I after reduction   Fig. 2 TEM image of Catalyst II

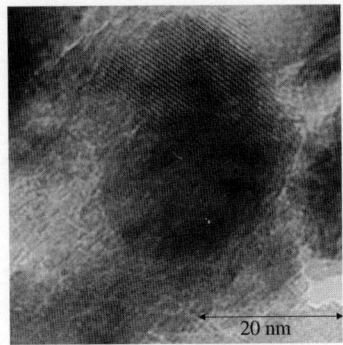

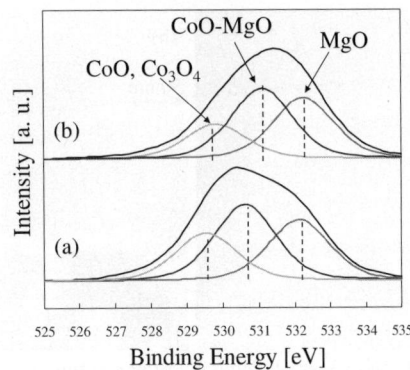

Fig. 3 Magnified TEM image of Catalyst I after reduction

Fig. 4 Oxygen 1*s* XPS spectra including curve-fitted components for (a) Catalyst I, (b) Catalyst I after reduction

In Fig. 2, a marble-like pattern was observed, which is attributable to solid solution phase of CoO and MgO, because XRD measurement on Catalyst II showed the existence of CoO-MgO solid solution phase [7, 8]. On the other hand, for Catalyst I, no solid solution phase of CoO-MgO was observed. In addition, XRD pattern of Catalyst I indicated the existence of CoO or Co_3O_4. These results suggest that in the case of Catalyst I, Co is loaded on the surface of MgO as CoO or Co_3O_4 phase. Magnified TEM image of Catalyst I after reduction is shown in Fig. 3. In this figure, crystalline lattice image was observed. It is likely that the observed lattice corresponds to the metal phase of Co, because XRD measurement on Catalyst I after reduction showed the existence of Co metal phase [7, 8].

The oxygen 1*s* XPS spectra for Catalyst I before and after reduction are shown in Fig. 4(a) and (b), respectively. In this figure, the intensities are normalized to the oxygen 1*s* associated with MgO. Curve-fitting technique was applied to the XPS spectrum. According to the literature [9, 10], three curve-fitted components at 529.5, 531, and 532.1 eV can be assigned to CoO and Co_3O_4, CoO-MgO solid solution, and MgO, respectively. Even in the catalyst after reduction, CoO and Co_3O_4 phases were still observed, although oxygen associated with CoO and Co_3O_4 decreased after reduction. On the other hand, binding energy of oxygen associated with CoO-MgO solid solution increased. This result indicates that CoO-MgO solid solution was stabilized in reduction at 1173 K.

3.2. Carbon deposition on the catalysts after reaction

The TEM images of deposits observed on Catalyst I used for the steam reforming of naphthalene are shown in Fig. 5. Two types of deposits were observed and they were proved to be composed of mainly carbon by EDS elemental analysis. One of them is film-like deposit over catalysts as shown in Fig. 5(a). This type of coke seems to consist of a polymer of C_nH_m radicals. The other is pyrolytic carbon, which gives image of graphite-like layer as shown in Fig. 5(b). Pyrolytic carbon seems to be produced in dehydrogenation of naphthalene. TPO profile is shown in Fig. 6. The peaks around 600 K and 1000 K are attributable to the oxidation of film-like carbon and pyrolytic carbon, respectively [11-13]. These results coincide with TEM observations.

520

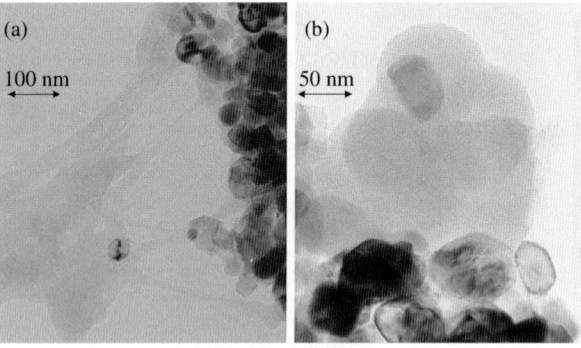

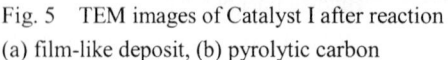

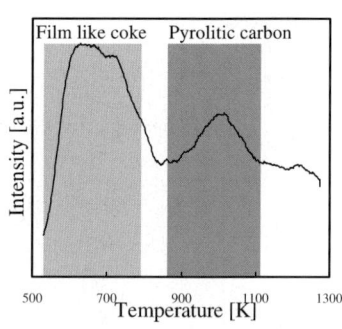

Fig. 5   TEM images of Catalyst I after reaction   Fig. 6 TPO profile of Catalyst I
(a) film-like deposit, (b) pyrolytic carbon   after reaction

According to the results of TPO analyses for all tested catalysts (not shown here), most of the coke existed as film-like carbon over the catalysts whose lifetime were short.   On the other hand, pyrolytic carbon existed over all catalysts.   These results show that the deactivation of Co/MgO is caused by film-like carbon deposition.

4. Conclusions

TEM-EDS and XPS analyses were conducted on Co/MgO catalysts.   The results of surface analyses showed that Co metal is not supported on the MgO as particles, but covers MgO surface in the case of 12 wt.% Co/MgO calcined at 873 K followed by reduction. After the reduction of catalyst at 1173 K, both cobalt oxide and CoO-MgO solid solution are observed on the surface of catalyst.   In the steam reforming of naphthalene, two types of coke deposited on the surface of catalyst are observed.   These are assigned to film-like and graphite type carbon by TPO analysis.

REFERENCES

[1] I. Narváez, A. Orío, M. P. Aznar, J. Corella, Ind. Eng. Chem. Res. 35 (1996) 2110.

[2] J. Gil, M. P. Aznar, M. A. Caballero, E. Francés, J. Corella, Energy and Fuels 11 (1997) 1109.

[3] J. Gil, J. Corella, M. P. Aznar, M. A. Caballero, Biomass and Bioenergy 17 (1999) 389.

[4] N. Padban, W. Wang, Z. Ye, I. Bjerle, I. Odenbrand, Energy and Fuels 14 (2000) 603.

[5] D. Sutton, B. Kelleher, J. R. H. Ross, Fuel Processing Technology 73 (2001) 155.

[6] L. Devi, K. J. Ptasinski, F. J. J. G. Janssen, Biomass and Bioenergy 24 (2003) 125.

[7] T. Furusawa, A. Tsutsumi, Appl. Catal. A, 278 (2005) 195.

[8] T. Furusawa, A. Tsutsumi, Appl. Catal. A, 278 (2005) 207.

[9] B. A. Sexton, A. E. Hughes, T. W. Turney, J. Catalysis, 97 (1986) 390.

[10] C. D. Wagner, D. A. Zatko, R. H. Raymond, Anal. Chem., 53 (1980) 52.

[11] J. R. Rostrup-Nielsen, J-H. B. Hansen, J. Catal.,144 (1993) 38.

[12] D. L. Trimm, Catal. Today, 49 (1999) 3.

[13] C. H. Bartholomew, Appl. Catal. A, 212 (2001) 17.

Studies in Surface Science and Catalysis, volume 159
Hyun-Ku Rhee, In-Sik Nam and Jong Moon Park (Editors)

Spatial Profiles of Gas Holdup in a Novel Internal-loop Airlift Reactor

T.-W. Zhang, J.-F. Wang*, Z. Luo and Y. Jin

Department of Chemical Engineering, Tsinghua University, Beijing 100084, China

1. INTRODUCTION

Internal-loop airlift reactors (ALRs) are widely used for their self-induced circulation, improved mixing, and excellent heat transfer [1]. This work reports on the design of an ALR with a novel gas-liquid separator and novel gas distributor. In this ALR, the gas was sparged into the annulus. The special designed gas-liquid separator, at the head of the reactor, can almost completely separate the gas and liquid even at high gas velocities.

2. EXPERIMENTAL

The experimental apparatus is of the concentric-tube type, with an enlarged degassing zone, as shown in Fig. 1. The reactor is 5.6 m in height and 376 mm in diameter. A draft tube of $\Phi220\times7$ mm and 5 m in height was inserted into the column, and the clearance between the column base and the bottom of the draft tube was 7 cm. The tubes installed on the slope of the separator are 35 mm in diameter and the average height is 0.3 m. The un-aerated water level was controlled at 100 mm above the top of these separator tubes. The air was feed into the annulus and the flux was controlled by a calibrated rotameter. The gas sparger consists of a base plate with 20 holes onside, 20 stainless steel tubes and 20 sintered steel tubes. The 20 holes are distributed on two concentric circles at the base plate, 12 at the outer circle and 8 at the inner circle. Each sintered steel tube is connected with one stainless tube and the other end of the stainless tube is connected with the hole on the base plate. The experimental systems were a two-phase-system (air, water) and a three-phase-system (air, water, glass beads). The density and average diameter of the glass beads were 2700 kg/m^3 and 0.1 mm, respectively. Different solid loadings were used to investigate the influence of the solid phase on the flow behavior.

A conductivity probe was used to measure the local gas holdup. A typical signal is shown in Fig. 2. The lower figure corresponds to the raw data, while the upper figure is the result after filtering to eliminate high-frequency white noise. The peaks in the signal correspond to when the tip is in gas bubbles. The probe was calibrated and set to a threshold value so that signals larger than the threshold value can be used to indicate the time when the probe tip is in

---

* Corresponding author. E-mail: wangjf@flotu.org

522

the gas bubble. The gas holdup is the fraction of time when the signal is above the threshold value. A cross-sectional average gas holdup can be obtained from the following equation.

$$\bar{\varepsilon}_g = \frac{1}{A_r} \int_{r_1}^{r_2} \varepsilon_g(r) 2\pi r dr \qquad (1)$$

The local liquid velocity in the riser was measured by a backward scattering LDA system (system 9100-8, model TSI). Details have been given by Lin et al. [2].

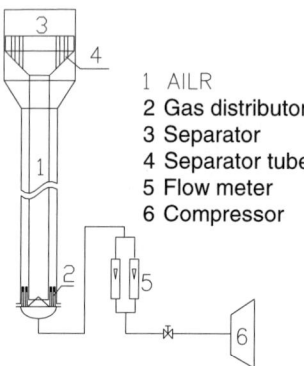

1 AILR
2 Gas distributor
3 Separator
4 Separator tube
5 Flow meter
6 Compressor

Fig. 1. Experimental set-up   Fig. 2. Conductivity probe signal

3. RESULTS AND DISCUSSION

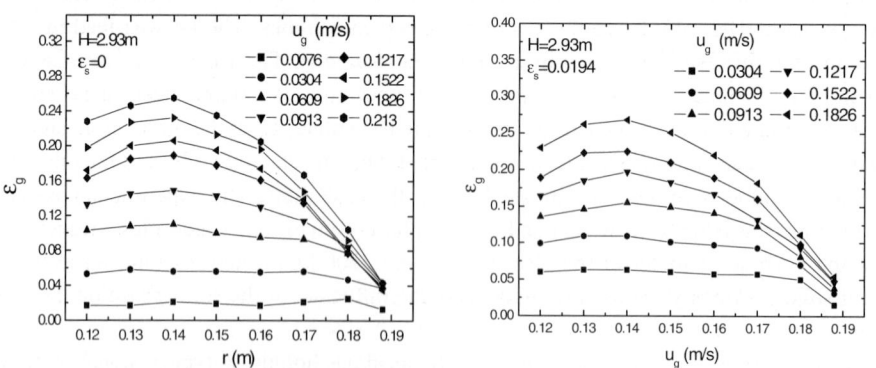

Fig. 3. Radial profile of the gas holdup at different solid holdups

Figure 3 shows the radial profile of the gas holdup in the riser with increasing superficial gas velocity under different solid holdups. The gas holdup increases with increasing superficial gas velocity at the different solid holdups. At a low superficial gas velocity, the liquid velocity is low and the effect of velocity gradient on bubbles is not remarkable, therefore the bubbles tend to uniformly distribute in the radial direction. Thus, the radial profile of the gas holdup at a low gas velocity is flatter.

Figure 4 shows the effect of the solid holdup on the cross-sectional average gas holdup in the riser at different superficial gas velocities. At all superficial gas velocities, the gas holdup generally increases with increasing solid holdup, and there is a decrease when the solid holdup increases from 0.19 vol% to 0.39 vol%. The solid particles have three different effects on the flow behavior: increasing the bubble break rate, increasing the bubble coalescence rate and increasing the flow resistance [3, 4]. Different effects will be dominant at different solid holdups. At low solid holdups, the effect on the bubble break rate is dominant, which leads to a decrease in the bubble size decreases and an increase in the gas holdup. With increasing solid holdup, the effect on the bubble coalescence rate becomes dominant, which leads to an increase in the bubble size and a decrease in the gas holdup. With a further increase in the solid loading, the effect of increasing the flow resistance becomes dominant. This causes a decrease in the liquid velocity and bubble rise velocity, which leads to an increase in the gas holdup.

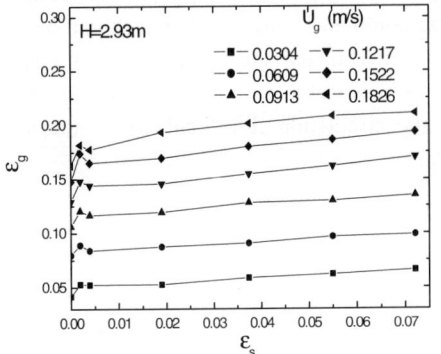

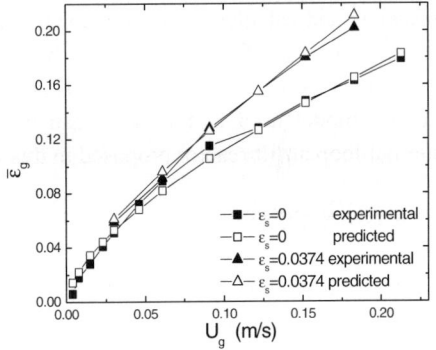

Fig. 4. Effect of solid holdup on gas holdup at different superficial gas velocities

Fig. 5. Comparison of calculated and measured gas holdup

4. MODEL VERIFICATION

A model based on energy balance was developed by Garcia-calvo et al. [5]. Energy input during gas expansion is dissipated in the flow field and in the phase interfaces, therefore:

$$E = F + S \qquad (2)$$

The inputted energy due to gas expansion in the riser can be determined by:

$$E = P_b U_g \ln(1 + \frac{\rho_h g H_r}{P_0}) \qquad (4)$$

where P_b and P_0 are the pressures at the bottom and top of the riser and U_g is the superficial gas velocity at the bottom of the riser.

The dissipated energy due to flow resistance and turbulence F consists of four parts: the energy consumption in the riser, f_r, that in the separator tubes, f_t, that in the down-comer, f_d, and that due to the flow direction changes of the U-bending channel at the top and bottom of the reactor, f_u. Therefore, F can be written as:

$$F = \sum_i f_i + f_u, \quad i = r, t, d \qquad f_i = 0.5\alpha\lambda_i\rho_h\frac{H_i}{d_i}u_{l,i}^3 A_i, \quad i = r, t, d \qquad f_u = \beta f_r \qquad (7)$$

In the above equations, α is a coefficient with the value of 1.0 for single phase flow and 2.0 for multi-phase flow [6], and β is an adjustable coefficient and has a value of 2.1 by fitting the experimental results for the two phase flow. The flow resistant coefficient is determined by the Blasius equation.

The dissipated energy on the gas-liquid interface S is calculated as:

$$S = \varepsilon_g u_{slip}\rho_h gH_r \qquad (9)$$

where u_{slip} is the bubble slip velocity.

In this model, energy balances are set up for the reactor and the separator tube separately, and two equations are obtained. The gas holdup can then be obtained from combining these two equations. Details can be found in Zhang et al. [7]. The comparison between the measured and calculated cross-sectional mean gas holdups is shown in Fig. 5. It can be seen that there is a satisfactory agreement between the experimental and calculated gas holdup in the different operating conditions. Therefore, it is reasonable to conclude that the energy balance model used in this work can describe the circulation flow behavior in the novle internal-loop airlift reactor proposed in this work.

5. CONCLUSION

A specially built conductivity probe was used to investigate the gas holdup in a novel internal-loop airlift reactor. The gas holdup generally increases with increasing solid holdup due to increased flow resistance. A model based on energy balance was developed that can be used to predict the average gas holdup in this novel interal-loop airlift reactor.

REFERENCES
[1] K. Koide, M. Kimura, H. Nitta and H. Kawabata, J. Chem. Eng. Japan, 21(1988) 393.
[2] J. Lin, M.H. Han, T.F. Wang and J.F. Wang, Ind. Eng. Chem. Res., 43 (2004) 5432.
[3] A.G. Livingston and S.F. Zhang, Chemical Engineering Science, 48 (1993) 1641.
[4] J.W.A. de Swart, R.E. Van Vliet, R. Krishna, Chem. Eng. Sci., 51(1996) 4619.
[5] E. Garcia-Calvo, A. Rodriguez, A. Prados and J. Klein, Chem. Eng. Sci., 54(1999) 2359.
[6] M.A. Young, R.G. Carbonell, and D. F. Ollis, AIChE J., 1991,37(3): 403.
[7] T.W. Zhang, J.F. Wang, Z. Luo and Y. Jin, Chem. Eng. J., 2005,109 (1-3): 115-122.

Studies in Surface Science and Catalysis, volume 159
Hyun-Ku Rhee, In-Sik Nam and Jong Moon Park (Editors)

CFD simulation of hydrodynamics of gas-liquid flow in an oxidation airlift reactor

Rong-Chun Shen, Zhong-Ming Shu *, Fa-Rui Huang, Ying-Chun Dai
State Key Laboratory of Chemical Engineering, East China University of Science and Technology, Shanghai 200237, P.R.China (e-mail:zmshu@ecust.edu.cn)

1. INTRODUCTION

Airlift loop reactor (ALR), basically a specially structured bubble column, has been widely used in chemical industry, biotechnology and environmental protection, due to its high efficiency in mixing, mass transfer, heat transfer etc [1]. In these processes, multiple reactions are commonly involved, in addition to their complicated aspects of mixing, mass transfer, and heat transfer. The interaction of all these obviously affects selectivity of the desired products [2]. It is, therefore, essential to develop efficient computational flow models to reveal more about such a complicated process and to facilitate design and scale up tasks of the reactor. However, in the past decades, most involved studies were usually carried out in air-water system and the assumed reactor constructions were oversimplified which kept itself far away from the real industrial conditions [3][4].

CFD modeling was conducted as a real industrial ALR taken as a background –a cyclohexane oxidation airlift loop reactor. The CFD software FLUENT6.0 was used to study two-phase flow in the reactor.

2. CONSTRUCTION OF ALR AND COMPUTATION METHOD

A schematic diagram of cyclohexane oxidation airlift loop reactor is illustrated in Fig.1. This reactor consists of outer vessel (riser), concentric draft-tube(downcomer) and gas

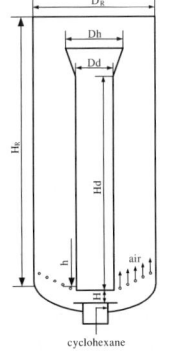
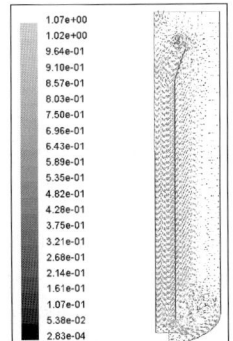
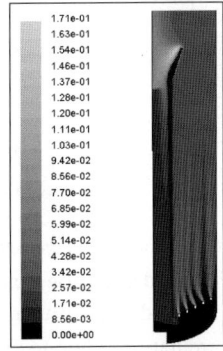

Fig.1 Schematic diagram of ALR   Fig.2 Liquid velocity profile   Fig.3 Gas hold-up profile

sparger, etc. Air is injected into the reactor through the ring type sparger, and rises in the annulus of the reactor. The driving force, induced by density difference between the riser and the downcomer, generates the loop liquid circulation. The feeding gas flowrate and liquid flowrate for the reactor is $0.097m^3/s$ and $0.134m^3/s$ separately under room temperature.

The Eulerian multiphase model is used to predict the dispersed gas-liquid flow in the airlift loop reactor. It involves a set of momentum and continuity equations for each phase. Model equation coupling is achieved through the pressure and interphase exchange coefficients [5].

3. RESULTS AND DISCUSSION

3.1 Distribution of liquid velocity and gas hold-up

Fig.2 and Fig.3 show the typical liquid velocity and gas hold up distribution in the ALR. From the figures, one notices that the cyclohexane circulates in the ALR under the density difference between the riser and the downcomer. An apparent large vortex appears near the air sparger when the circulating liquid flows from the downcomer to the riser at the bottom. In the riser, liquid velocity near the draft-tube is much larger than that near the reactor wall, the latter moved somewhat downward. The gas holdup is nonuniform in the reactor, most gas exists in the riser while only a little appears in the dowmcoer.

3.2 Effect of draft-tube horn-mouth diameter on liquid velocity and gas hold-up

In the Fig.4, it can be seen that the gas hold-up in both riser and downcomer decreases with increasing the draft-tube horn-mouth diameter and approaches the maximum when the draft-tube horn-mouth diameter is 1.05m. However, due to the gas hold-up decreases more in the downcomer, the gas hold-up difference between the downcomer and the riser increases. Therefore, the apparent density difference between the riser and the downcomer enhances, causing higher liquid superficial velocity in the downcomer and in the riser with increasing the horn-mouth diameter. Fig.5 also shows that the existence of horn-mouth promotes the ability to separate gas from liquid and decreases the amount of gas entrained into the downcomer.

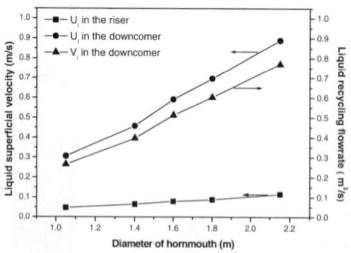 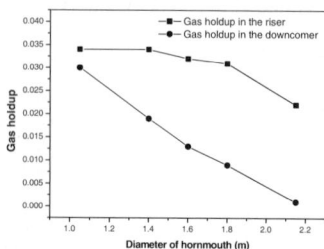

Fig.4 Effect of horn-mouth diameter on liquid velocity   Fig.5 Effect of horn-mouth diameter on gas hold-up

3.3 Effect of draft-tube diameter on liquid velocity and gas hold-up

Fig.6 and Fig.7 illustrate the effect of draft-tube diameter on liquid superficial velocity, liquid circulating flowrate and gas hold-up. Results show that the liquid superficial velocity in the riser increases with increasing the draft-tube diameter while the liquid velocity in the downcomer decreases. As the draft-tube diameter increases, the liquid flowrate in the downcomer does not decrease, but increases. This variation trend can be explained by the larger flow area of the downcomer when draft-tube diameter increases. Meanwhile, gas hold-up in the riser and in the downcomer decreases with increasing the draft-tube diameter. With a larger draft-tube diameter, the annulus flow space of the riser understandably decreases, so the liquid superficial velocity rises accordingly. A higher liquid velocity causes a shorter residence time of the bubbles in the riser, and thereby lowering the gas hold-up in the riser. As the liquid velocity in the downcomer decreases with increasing the draft-tube diameter, less gas is carried in form of bubbles into the downcomer, so the gas hold-up in the downcomer also drops.

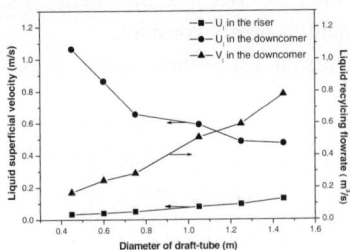
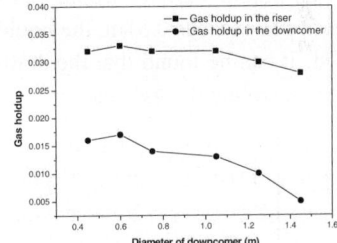

Fig.6 Effect of downcomer diameter on   Fig.7 Effect of downcomer diameter on
liquid velocity and flowrate   gas holdup

Fig.8 illustrates the liquid velocity distribution at the bottom section of the reactor when the draft-tube diameter is 0.45m, 1.05m and 1.45m respectively. Results show that the liquid velocity at the outlet of the draft-tube lowers when the draft-tube diameter is raised, to subsequently influence the shape and size of the vortex at the bottom of the gas sparger.

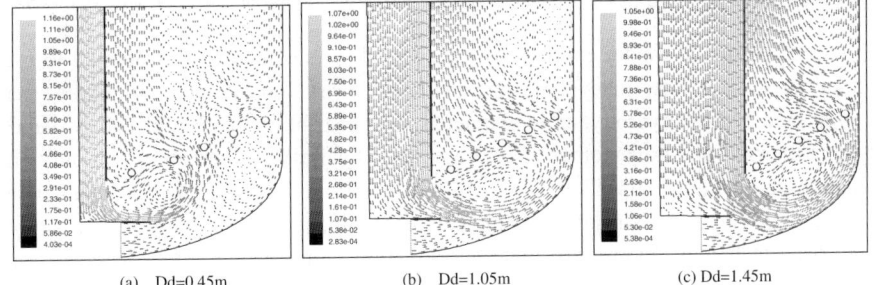

(a) Dd=0.45m   (b) Dd=1.05m   (c) Dd=1.45m

Fig.8 Distribution of liquid velocity

3.4 Effect of draft-tube height position on liquid velocity and gas hold-up

The height of the draft-tube also influences the flow characteristics in the ALR. Fig.9 and Fig.10 show that the liquid superficial velocity increases with increasing the height (H) of the draft-tube, while the liquid superficial velocity remains approximately unchanged when H exceeds 0.51m. With a higher position of draft-tube, the flow area at the outlet of the draft-tube becomes larger, so the liquid velocity at the outlet decreases.

Since the local turbulence intensity and the energy dissipation in the reactor are lower, under the condition that the gas hold-up keeps approximately constant in the riser and in the

528

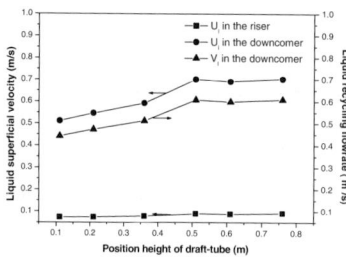

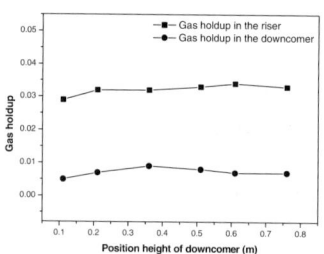

Fig.9 Effect of downcomer height on liquid velocity and flowrate

Fig.10 Effect of downcomer height on gas holdup

downcomer, the liquid velocity would certainly increase. Because the energy dissipation no more changes as H exceeds 0.51m, the liquid velocity is nearly constant.

In Fig.11, it can be found that the draft-tube height also influences the shape and size of the liquid vortex around the gas sparger.

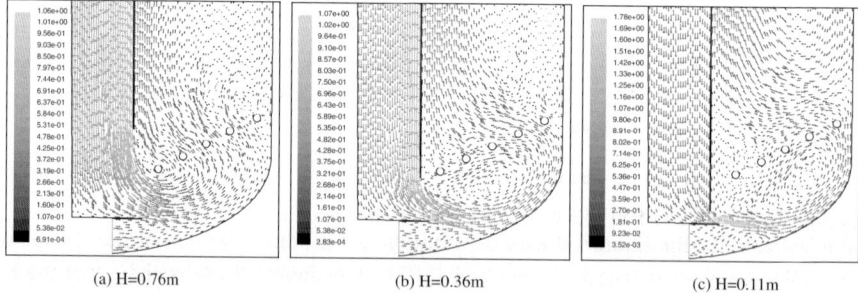

(a) H=0.76m   (b) H=0.36m   (c) H=0.11m

Fig.11 Distribution of liquid velocity

4. CONCLUSIONS

Eulerian two-fluid model coupled with dispersed $k - \varepsilon$ equations was applied to predict gas-liquid two-phase flow in cyclohexane oxidation airlift loop reactor. Simulation results have presented typical hydrodynamic characteristics, distribution of liquid velocity and gas hold-up in the riser and downcomer were presented. The draft-tube geometry not only affects the magnitude of liquid superficial velocity and gas hold-up, but also the detailed liquid velocity and gas hold-up distribution in the reactor, the final construction of the reactor lies on the industrial technical requirement. The investigation indicates that CFD of airlift reactors can be used to model, design and scale up airlift loop reactors efficiently.

REFERENCES

[1] M.Y. Chisti. Airlift Bioreactors, Elsevier, London, 1989.

[2] Ch. Vial,   S. Poncin, G.. Wild, N. Midoux, Chem. Eng. Sci., 57(2003), 4745-4762.

[3] D. Pfleger, S. Gomes, N. Gilbert, H.-G. Wagner, Chem.Eng.Sci.,54(1999), 5091-5099.

[4] V. Buwa and V. Ranade. The Can. J. of   Chem. Eng., 81(4) (2003): 402-411.

[5] J. Sanyal, S. Vasquez, S. Roy, and M.P. Dudukovic. Chem. Eng. Sci., 54(1999): 5071-5083.

Studies in Surface Science and Catalysis, volume 159
Hyun-Ku Rhee, In-Sik Nam and Jong Moon Park (Editors)

Pyrolysis for the Recycling of Polystyrene Plastic (PSP) Wastes in a Swirling Fluidized-Bed Reactor

Suk Hwan Kang[a], Sung Mo Son[a], Pyung Seob Song[a], Yong Kang[a*] and Myoung Jae Choi[b]

[a]School of Chemical Engineering, Chungnam National University,
Daejeon 305-764, Korea(South) ([*]kangyong@cnu.ac.kr)

[b]Advanced Chemical Technology Division, KRICT, Daejeon 305-600, Korea(South)

ABSTRACT

Pyrolysis of waste polystyrene plastics (PSP) was investigated for the effective recycling. To obtain the kinetic information, a non-isothermal process (TGA/DTA) was used. The thermal decomposition was conducted in a swirling fluidized-bed reactor (0.762 m ID x 2.5 m in height). The apparent activation energy for the thermal decomposition of waste PSP was in the range of 50.4–72.1 kJ/mol according to the conversion level. The mean reaction order was 0.36. In the swirling fluidized-bed reactor, the optimum temperature, gas velocity and swirling ratio were 500°C, 0.4 m/s and 0.3, respectively, for the maximum yields of oil and styrene monomer.

1. INTRODUCTION

Recycling to monomers, fuel oils or other valuable chemicals from the waste polymers has been attractive and sometimes the system has been commercially operated [1-4]. It has been understood that, in the thermal decomposition of polymers, the residence time distribution (RTD) of the vapor phase in the reactor has been one of the major factors in determining the products distribution and yield, since the products are usually generated as a vapor phase at a high temperature. The RTD of the vapor phase becomes more important in fluidized bed reactors where the residence time of the vapor phase is usually very short. The residence time of the vapor or gas phase has been controlled by generating a swirling flow motion in the reactor [5-8].

In the present study, the pyrolysis of a waste polystyrene plastic (PSP) has been investigated in a swirling fluidized-bed reactor to develop an effective reactor. Effects of the reaction time, temperature, ratio of the swirling gas and the gas velocity on the yields of an oil and a styrene monomer have been discussed.

2. EXPERIMENTAL

Experiments were carried out in a stainless steel fluidized-bed reactor whose diameter was 0.762 m and 2.5 m in height, respectively, as can be seen in Fig. 1. The nitrogen was injected into the reactor through the perforated-type distributor installed between the main column section and the nitrogen box through which the main stream of nitrogen (primary) was fed to the reactor. The distributor contained 147 holes with triangular pitches and it was covered with a 400 mesh screen to prevent the bed material from weeping. Silica sand particle whose density and mean diameter were 2500 kg/m^3 and 0.24 mm, respectively, were used as the bed materials. To generate the swirling flow pattern of the gas-solid mixture interior of the reactor, a swirling gas (secondary) was fed tangentially into the reactor at the wall of the reactor. The height of the swirling gas (nitrogen) injection port was 0.2 m from the distributor to avoid the end effect.

The waste polystyrene plastic (PSP) was melted at 250°C prior to being injected into the reactor, to create particles with a mean diameter of 2 mm. About fifty grams of feed material (waste PSP) was fed into the reactor at a given operating condition. The oil product or styrene monomer was obtained by means of a condenser. The yields of the oil and the styrene monomer were determined by the following equations [5].

$$\text{Yield of oil ; } Y_{oil} = \frac{W_o}{W_f} \times 100 \text{ . Yield of styrene monomer ; } Y_{SM} = \frac{W_{SM}}{W_f} \times 100 = \frac{W_{SM} \times Y_{oil}}{W_o} \times 100 \quad (1)$$

The outlet gas, which was obtained at the gas detection port, was analyzed by means of an on-line gas analyzer (GC-MS, HP-5890 plus, column; DB-1HT; GC-TDC; GC-FID). The yield of the gas product was determined by Eq (2).

$$\text{Yield of gas ; } Y_{Gas} = \frac{W_{Gas}}{W_f} \times 100 = \frac{C_{Gas} \times Q_{N_2} \times time}{W_f} \times 100 \quad (2)$$

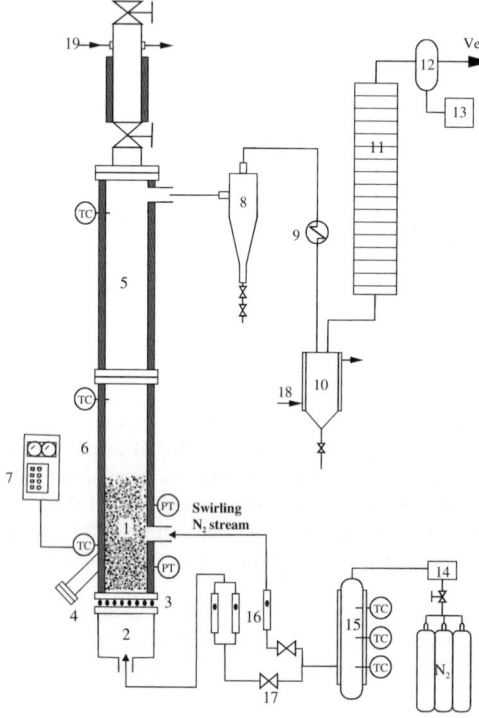

Fig. 1. Experimental apparatus :

1. Fluidized bed reactor
2. Distributor box
3. Distributor
4. Drain
5. Freeboard
6. Electric heater
7. T-controller
8. Cyclone
9. Heat exchanger
10. Condenser
11. Mist filter
12. Gas sample bag
13. GC
14. Regulator
15. Preheater
16. Flowmeter
17. Control valve
18. Water jacket
19. Purge gas

3. RESULTS AND DISCUSSION

Typical variations of the temperature and the conversion rate in terms of the weight loss of waste PSP during the non-isothermal pyrolysis process (TGA/DTG) can be seen in Fig. 2. In Fig. 2, the waste PSP would be decomposed in the range from 370℃ to 485℃ at a heating rate of 10℃/min. The heating rate was either 10, 20 or 30℃/min. One maximum peak in the conversion rate could be obtained at a reaction temperature of 430℃. The activation energy was obtained from the plot of ln(dX/dt) versus (1/T) by using Friedman's method [9]. The calculated activation energy at different conversion levels of the waste PSP can be seen in Fig. 3. In this Figure, the value of the activation energy tends to increase gradually with an increasing conversion level. The activation energy was distributed in the range from 50.4 to 72.1 kJ/mol according to the change of the conversion level [5]. The mean value of the reaction order could be obtained as 0.36. The melting of the PSP, initially started to be decomposed by cutting the relatively short carbon-hydrogen bond and the benzene radical which is attached to the main chain. With an increasing reaction time, however, the carbon-carbon bond of the main chain could be decomposed gradually. Thus, the activation energy would be distributed.

Effects of the reaction time on the yield of the oil in a swirling fluidized-bed reactor can be seen in Fig. 4. In this figure, 82 wt.% of Y_{oil} has been obtained in the reactor at 400℃ during a reaction of 34 min, while it can be increased up to 91 wt.% for a reaction time of 20 min at 550℃.

Effects of the reaction temperature on the yields of the oil and styrene monomer can be seen in Fig. 5. In this figure, the increase of the reaction temperature can increase the values of voil yield (Y_{oil}) as well as the SM yield (Y_{SM}). However, the increase of the reaction temperature from 500 to 550℃ leads to a slight decrease of the oil yield. This can be due to the fact that the increase of the reaction temperature leads to an increase of the vapor phase products instead of the liquid phase. It could be stated that the optimum reaction temperature would be 500℃ for the recovery of the oil from the waste PSP within these experimental conditions. Also, the values of the yields increase gradually by increasing the ratio of the swirling gas (R_S). With an increasing R_S, the size and rising velocity of the bubbles decrease while the number of bubbles increases due to an increase of the turbulence intensity and a more effective periodic gas flow for the breaking up of the bubbles. In other words, the bubble holdup in the bed tends to increase with the increasing R_S. However, the merits due to an increase of R_S become marginal with a further increase from 0.3, especially at a high reaction temperature.

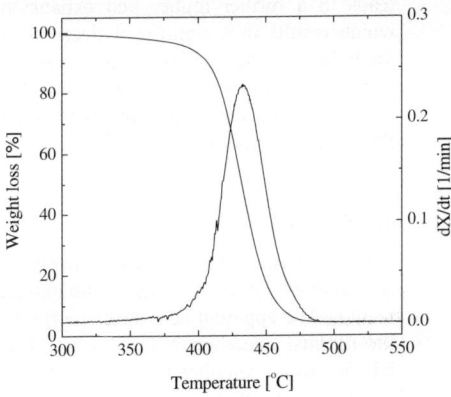

Fig. 2. Typical TGA curve and rate of conversion for the pyrolysis of waste PSP (heating rate = 10℃/min).

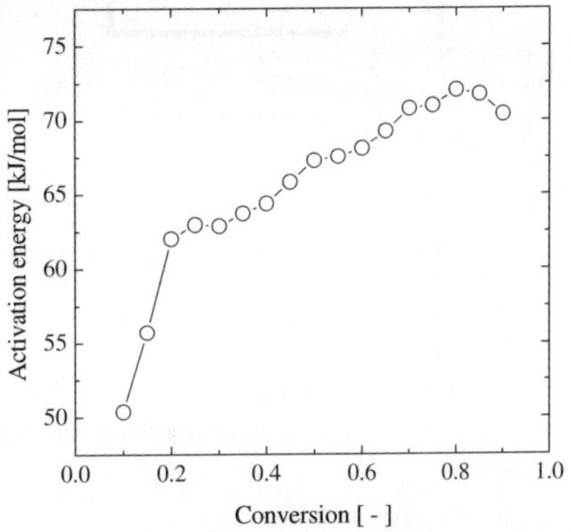

Fig. 3. Calculated activation energies at different conversion levels of waste PSP.

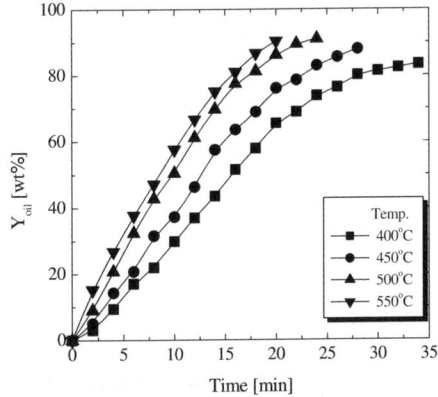

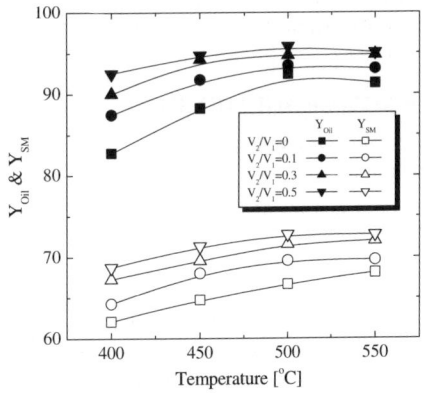

Fig. 4. Typical example of the yield of oil in a fluidized-bed reactor(V_2/V_1=0, U_G=0.3m/s).

Fig. 5. Effects of temperature on the yields of oil and styrene monomer in a swirling fluidized-bed reactor(U_G=0.3m/s).

Effects of the gas velocity (U_G) on the yields of the oil and styrene monomer can be seen in Fig. 6. Note in this figure that the values of the oil yield increase with an increasing U_G at a relatively lower

532

gas velocity, but decrease gradually with a further increase in the gas velocity from 0.4 to 0.6 m/s. This could be due to the fact that the time period of the thermal decomposition of the waste PSP decreases with an increasing gas velocity in a swirling fluidized-bed reactor. Since the minimum fluidization velocity of the bed materials is 0.09 m/s, they cannot be fully fluidized at a relatively lower gas velocity. Thus, in these conditions, an increase of the gas velocity could lead to a more effective fluidization condition for a higher value of the oil yield. The optimum gas velocity for the maximum oil yield would be 0.4 m/s within these experimental conditions. It has been understood that, in a relatively lower range of U_G, the fluidization motion of the bed materials becomes more vigorous with an increasing U_G, but in a higher range of U_G, an increase of U_G leads to a further higher bed expansion, which results in a significant decrease of the holdup of the bed materials. These can directly decrease the heat transfer coefficient and the mixing intensity in the fluidized-bed reactor.

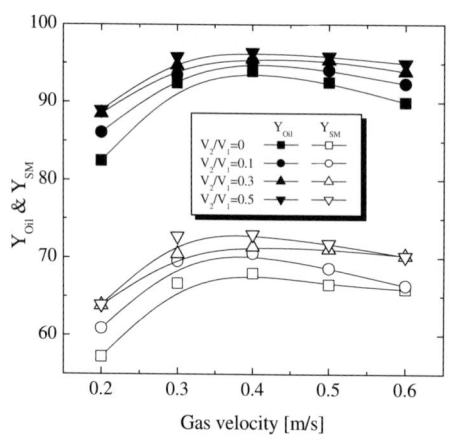

Fig. 6. Effects of gas velocity on the yields of oil and styrene monomer in a swirling fluidized-bed reactor ($T=500°C$).

4. CONCLUSION

The waste PSP was effectively decomposed in a swirling fluidized-bed reactor. The apparent activation energy for the thermal decomposition of waste PSP, which was calculated by Friedman's method, was distributed in the range from 50.4 to 72.1 kJ/mol according to a change of the conversion level. The mean value of the reaction order was 0.36. The optimum temperature for the maximum yields of the oil and styrene monomer in the swirling fluidized-bed reactor was found to be 500°C. The optimum gas velocity was 0.4 m/s within these experimental conditions. With an increasing swirling ratio of the gas phase (R_S) up to 0.3, the yields of the oil and styrene monomer increase by increasing the turbulence intensity due to the swirling motion of the gas-solid mixture in the reactor.

ACKNOWLEDGEMENTS

This work was supported by the 21C Frontier R&D Program, Industrial Waste Recycling R&D Center (Project 2A-B-1-1).

REFERENCE

[1] T. Hirose, Y. Takai, N. Azuma and A. Ueno, J. Mater, Res., 13 (1998) 77.
[2] F. Sasse and G. Emig, Chem. Eng. Technol., 21 (1998) 777.
[3] J. Mertin, A. Kirsten, M. Predel and W. Kaminsky, J. Anal. Appl. Pyrolysis, 49 (1999) 87.
[4] Y. Liu, J. Qian and J. Wang, Fuel Processing Technology, 63 (2000) 45.
[5] C. G. Lee, Y. J. Cho, P. S. Song, Y. Kang, J. S. Kim and M. J. Choi, Catalysis Today, 79 (2003) 453.
[6] B. Sreenivasan and V. R. Raghavan, Chem. Eng. Process, 41 (2002) 99.
[7] Y. Kang, P. S. Song, J. S. Yun, Y. Y. Jeong and S. D. Kim, Chem. Eng. Commun., 177 (2000) 31.
[8] S. H. Kang, D. K. Park, H. C. Jung, P. S. Song, Y. Kang and M. J. Choi, J. Korean Ind. Eng. Chem., 15 (2004) 855.
[9] H. L. Friedman, J. Poly. Sci., 6 part C (1963) 183.

Studies in Surface Science and Catalysis, volume 159
Hyun-Ku Rhee, In-Sik Nam and Jong Moon Park (Editors)

Features of liquid-continuous impinging streams and their influences on kinetics

Yuan Wu, Yuxin Zhou and Chuanping Bao

Wuhan Institute of Chemical Technology, Wuhan 430073, PR China

1. INTRODUCTION

Impinging streams (IS) was presented first for gas-solid systems [1, 2], and then extended many sides, including taking liquid as the continuous phase. Discovery of liquid-continuous impinging streams (LIS) promoting micromixing effectively in 1990s is the most important progress in IS field. It broadens the area IS-applicable, which included systems involving chemical reactions, for micromixing is very important for processes occurring at molecular scale. Since then the emphasis of investigation on IS has diverted into LIS [3].

2. PERFOMANCE OF IS RELATED TO ITS CONTINUOUS PHASE

Because of various gather status, liquids are quite different in properties from gases. For IS the following are important: (1) In density, liquid is greater than gas by about 10^3, (2) In viscosity, liquid is greater than gas by about 10^2, and (3) Molecules of gas have relatively large free-path, while those in stationary liquid can only vibrate and/or rotate round their balanced positions with extremely small displacement. These affect strongly the performances of IS.

2.1. Transfer coefficient

Gas-continuous impinging streams (GIS) enhances transfer by the following [1-5]: (1) very large relative velocity round the impingement plane, (2) penetration of particles to and fro between opposed streams, and (3) strong turbulence in impingement zone. In LIS, the effects above become very weak or even no exist because of small difference in densities of the two phases and large viscosity of liquid. As the results, no significant enhancement of transfer can be expected. In fact, experimental data about salts solving have shown that LIS does not enhance transfer obviously [3].

2.2. Interaction between streams

The operating impinging velocity, u_0, in LIS equipment is usually smaller although, the interaction between the opposed streams, including collision, pressing, shearing etc., is much stronger than in GIS, because of high density of liquid and thus strong momentum transfer. This leads to the features of LIS to be described below.

3. FEATURES OF LIS

3.1. Micromixing

Mahajan et al [6] studied micromixing in a two impinging jets (TIJ) mixer under the conditions of both free and submerged impingement. The parallel-competitive reaction system proposed by Bourne [7] was used to bound the micromixing time, t_M. The reported results are: $t_M < 200$ ms in the range of $Re > 700$ and free IS has smaller t_M. However, the use of Reynolds number to interpret data is questionable. Energy for micromixing is provided by impingement of opposed streams and is proportional to u_0^2. If one wants to keep Re the same in scaling-up, very small u_0 must be used for large system, resulting in poor micromixing. This shows that the method based on the π-law for scaling-up is invalid for not only most of systems involving chemical reaction(s) but also for some physical processes, like IS.

Wu et al [8] studied the same topic in the submerged circulative impinging stream reactor (SCISR) shown in Fig. 2. The major results are: (1) $t_M = 192 - 87$ ms in the range of $u_0 = 0.184 - 0.326$ m·s^{-1}, (2) t_M and u_0 meet about the relationship of $t_M \propto u_0^{-1.5}$, and (3) In the range of conditions tested, the values for t_M calculated are systematically greater than those measured by 2 to 3 times, as shown in Table 1, suggesting that the existing theory needs improvement. Values for t_M in STR are $500 - 1000$ ms. So, LIS promoting micromixing very efficiently.

3.2. Pressure fluctuation

Table 1
Characteristic time constant for micromixing calculated and measured

No	T, K	u_0, m·s^{-1}	$t_{M,cal}$, ms	$t_{M,ex}$, ms	$t_{M,cal} / t_{M,ex}$
1	298	0.184	594	192	3.09
2	303	0.245	285	136	2.10
3	308	0.255	205	95	2.16
4	308	0.326	142	87	1.63

The measured results [9] show that there in SCISR under the conditions of $u_0 = 0.25 - 0.3$ m·s^{-1} exists strong pressure fluctuation. The frequency is concentrated round 1000 Hz and the maximum amplitude of up to 1.58 kPa, implying conversion of energy from.

3.3. A qualitative analysis for influences of strong micromixing and fluctuation

A completely uniform assumption was usually used in kinetic studies. This is not easy for

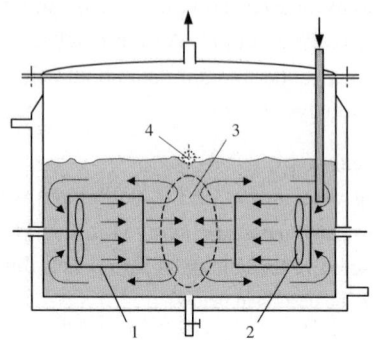

Fig. 2. A brief view of SCISR　1－Drawing tube　2－Propeller　3－Impingement zone　4－Outlet

liquids. Simply, macro- is mixing between fluid elements, while micro- is between molecules. Poor micromixing is characterized by existing of segregation scale greater than Kolmogoroff length. Since elements tend to follow streamlines, molecules in segregated different elements are difficult to contact each other. This leads to deviation of the uniform assumption. Also, according to kinetic theory, only collisions between molecules with energy high enough can induce reaction. In LIS, the fluctuation resulted from interaction between the opposed streams may lead to more violent molecule collision and change molecule energy distribution so that more molecules achieve higher level of activation and thus react with others by collision. A reasonable inference is that strong micromixing and fluctuation favor process kinetics.

4   EXPERIMENTAL RESULTS FOR LIS PROMOTING KINETICS

4.1 Crystal-growth rate of disodium phosphate

A comparative study [10] is made for crystal-growth kinetics of Na_2HPO_4 in SCISR and a fluidized bed crystallizer (FBC). The details of the latter can be found in [11]. Experiments are carried out at rigorously controlled super-saturations without nucleation. The overall growth rate coefficient, K, are determined from the measured values for the initial mean diameter, d_{p0}, masses of seed crystals before and after growth. The results show that the values for K measured in ISC are systematically greater than those in FBC by 15 to 20%, as can be seen in Table 2. On the other hand, the values for the overall active energy measured in ISC and FBC are essentially the same.

4.2 Saponification kinetics

Table 2
Comparison of crystal growth rate coefficients measured in ISC and FBC

$d_{p0} \times 10^4$, m	T, ℃	Averaged $K \times 10^6$, m·s^{-1}		K_{IS}/K_{FB}
		K_{IS}	K_{FB}	
	32.7	6.61	5.16	1.281
2.51	35.9	8.77	7.54	1.163
	38.6	11.43	9.53	1.199
	33.9	7.42	6.30	1.178
3.48	36.5	9.10	7.86	1.158
	38.7	13.83	11.58	1.194
	34.2	8.67	7.30	1.188
5.22	37.2	13.50	10.50	1.286
	39.6	17.38	14.70	1.182
	33.6	10.92	8.25	1.169
7.36	37.0	16.81	14.35	1.171
	38.8	21.37	19.15	1.116
	33.4	12.66	9.69	1.307
8.95	36.7	17.02	14.54	1.171
	39.0	25.52	22.23	1.148

Kinetics of the reaction below is studied in both SCISR and a stirred tank reactor (STR).

$CH_3COOC_2H_5$ (A) + NaON (B) = CH_3OONa (R) + C_2H_5OH (S)

Both the reactors are operated in batch, and the concentrations of components involved are measured online by electro-conductivity. Data interpretation is made by the kinetic equation of second order. The results obtained in the range of 25-45°C are given in Table 3. Again, the values for the rate constant measured in SCISR, k_{IS}, are systematically higher than those in STR, k_{ST}, by about 20%, and no significant difference between the values for the active energy measured in SCISR and STR has been found.

Table 3

Comparative data for the rate constant of reaction (4)

T, °C	k, $m^3 \cdot kmol^{-1} \cdot s^{-1}$		k_{IS} / k_{ST}
	k_{IS}	k_{ST}	
25	0.175	0.137	1.28
35	0.239	0.196	1.22
45	0.758	0.651	1.16

5 CONCLUSION REMARKS

The studies yield that the overall crystal-growth rate coefficient of Na_2HPO_4 and the rate constant of saponification of alcohol ester measured in LIS device are systematically higher than those in fluidized-bed crystallizer or stirred tank reactor. These results support the inference that enhanced micromixing in LIS increases probability of collision between molecules, and relatively strong fluctuation leads to a part of molecules getting more energy to achieve higher level of activation, and both the two effects promote kinetic processes.

REFERENCES

[1] I. T. Elperin, J. Eng. Physics (in Russ), (1961) No.6, 62.

[2] Y. Wu, Chem. Indus. Eng. Prog. (in Chinese), 20 (2001) No.11, 8.

[3] Y. Wu Yuan, Indus. Eng. Prog. (in Chinese), 22 (2003) 1066.

[4] A. Tamir, Impinging Stream Reactors—Fundamentals and application, Elsevier, Amsterdam, 1994.

[5] Y. W, Chem. Eng. (in Chinese), 26 (1998) No.4, 14.

[6] A. Mahajan, D. J. Kirwan, AIChE J, 42 (1996): 1801.

[7] J. R. Bourne, Chem. Eng. Commun., 16 (1982) 79.

[8] Y. Wu, Y. Xiao and Y. Zhou, Chinese J. of Chem. Eng., 11 (2003) 420.

[9] H. Sun, Y. Wu, J. Zhang and C. Xu, Pressure fluctuation in the submerged circulative impinging stream reactor, a paper To be published.

[10] Y. Wu, Y. Zhou, X. He and C. Bao, Crystal-growth rate of di-sodium phosphate in impinging stream crystallizer, a paper to be published.

[11] Y. Wu, Chem. Eng. (in Chinese), 13 (1985) No. 4, 41.

[12] C. Bao, Y. Zhou and Y. Wu, Kinetics of alcohol ester saponification in impinging stream reactor, a paper to be published.

Studies in Surface Science and Catalysis, volume 159
Hyun-Ku Rhee, In-Sik Nam and Jong Moon Park (Editors)

Recovery of Copper Powder from Wastewater in Three-Phase Inverse Fluidized-Bed Reactors

Pyung Seob Song [a,b], Suk Hwan Kang [b] ,Wang Kyu Choi[a], Chong Hun Jung[a], Won Zin Oh[a] and Yong Kang[b,*]

[a]Division of R&D on Decontamination & Decommissioning Technology, KAERI,
Daejeon 305-600, Korea.
[b]School of Chemical Engineering, Chungnam National University,
Daejeon 305-764, Korea(Corresponding author: e-mail: kangyong@cnu.ac.kr)

ABSTRACT

The recovery of copper powder from wastewater of electronic industries was investigated in three-phase inverse fluidized-bed electrode reactors(0.102m ID x 1.0m). Effects of gas and liquid velocities, current density, distance between the two electrodes and amount of fluidized particles on the recovery of copper powder were examined. The addition of a small amount of gas or fluidized particles into the reactor resulted in the decrease in the powder size of copper recovered as well as increase in the copper recovery. The value of copper recovery exhibited a maximum with increasing gas or liquid velocity, amount of fluidized particles or distance between the two electrodes but increased with increasing current density.

1. INTRODUCTION

Numerous etching and plating processes occurring in most electronic industries have generated wastewater including heavy metal ions such as copper, nickel, silver, zinc ions etc., which have to be treated for the protection of environment[1-4]. In addition, the heavy metals are very expensive and useful resources, thus, they should be recovered. The common methods of eliminating heavy metal ions are anode oxidation, electrolysis, electro-dialysis, electro-flotation, electro-coagulation and cathode reduction [5-6]. Among them, the electrolysis method employing a fluidized bed reactor could be the most economic for recovering heavy metal powder, since the recovery in the fluidized bed reactor can be higher than that of any other reactor. In the present study, effects of the operating variables such as the gas(U_G) and liquid velocities(U_L), current density(I), distance(L_{AC}) between the two electrodes and amount of fluidized solid particles(W) on the recovery of copper powder from wastewater were investigated in three-phase inverse fluidized-bed reactors.

2. EXPERIMENTAL

Experiments were carried out in an acrylic column (0.102m ID x 1.0m), as can be seen in Fig. 1. The wastewater was introduced to the top of the reactor through a pipe from the wastewater reservoir. A perforated plate, which was used as a distributor, was situated between the main column section and a 0.2m high stainless steel L/S separator. Oil-free compressed air was fed to the column through a pressure regulator, filter and a calibrated rotameter. It was admitted to the column through four 3.0mmID perforated pipes drilled horizontally in the grid. The pipes were evenly spaced across the grid having 12 holes with a diameter of 1mm. The particles, whose diameter was 0.5mm, were made of polystyrene and divinylbenzene to adjust the density of 1000kg/m^3. The temperature of wastewater was maintained at 19~20℃ by means of a temperature control system and the pH was 2~2.5. To measure the concentration of copper ion in the wastewater, 4 sampling taps with solenoid valves were installed at the wall of the column. The anode plate (0.06m x 0.6m) was made of lead alloy (Pb:Sb=90:10) and a stainless steel plate with the same dimension was used as a cathode. The two electrode plates, which were severed from the distributor at the bottom of the column, were connected to the AV-meter to control the electric current density between the two electrodes. The composition of pre-treated wastewater from PC industry is listed in Table 1. The recovery of copper was analyzed by means of AAS(Atomic Absorption Spectrometry) and quantitative analysis method. And the recovered copper powder was analyzed by means of XRD(X-Ray Deflection).

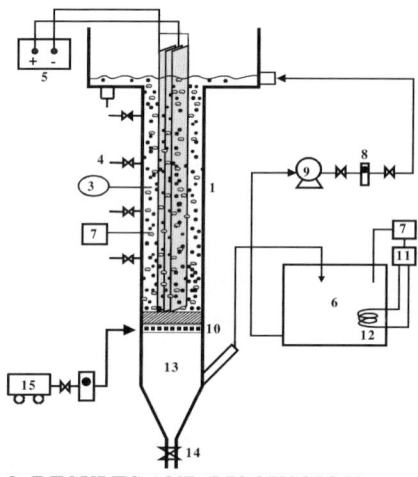

Fig. 1. Experimental apparatus of a fluidized-bed electrode reactor :

1. Reactor	2. Electrode
3. pH meter	4. Sampling tap
5. A-V meter	6. Wastewater reservoir
7. Thermocouple	8. Rotameter
9. Liquid metering pump	10. Distributor
11. T-controller	12. Cooler
13. L/S separation section	14. Drain valve
15. Compressor	

Table 1. Composition of wastewater from electronic industry after pre-treatment

Element	Cu	Fe	Mn	Ni	Zn
Content	6.31wt.%	0.29ppm	ND	0.43ppm	13.11ppm

3. RESULTS AND DISCUSSION

Typical examples of XRD analysis of copper powder recovered in the inverse fluidized bed electrode reactors can be seen in Fig. 2, where the recovered copper powder was almost pure. Effects of fluidized particles on the size distribution of copper powder recovered in the reactors can be seen in Fig. 3. Note that the addition of a small amount of fluidized particles could decrease the size of recovered copper powder, but a further increase of particle amount could increase the size of copper powder, compared with that without fluidized particles. This can be due to that the added particles(up to 1.0wt.%) can contact with the cathode plate frequently, which could be resulted in the effective cut of the copper powder growing perpendicular to the surface of the cathode plate.

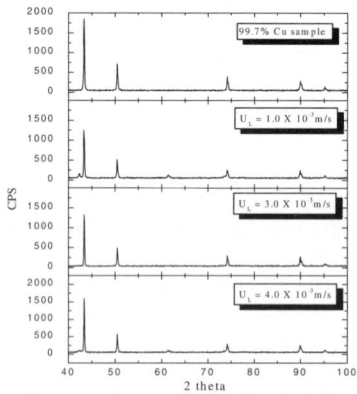

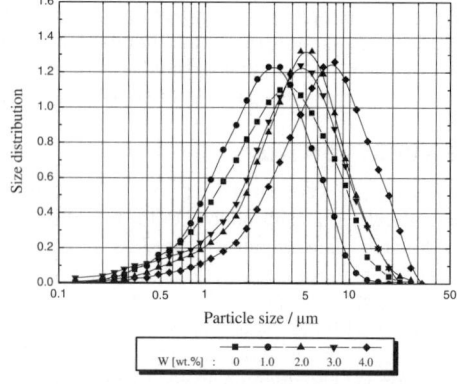

Fig. 2. Typical example of XRD analysis of copper powder recovered in inverse fluidized bed reactors(L_{AC}=0.015m, I=3.0 A/dm^2, W=1.0 wt.%).

Fig. 3. Effects of amount of fluidized particles on the size distribution of copper powder recovered in inverse fluidized bed reactors(L_{AC}=0.015m, I=3.0 A/dm^2, U_L=0.001m/s).

However, the more amount of added particles can cause the copper powder to grow parallel to the cathode plate because of relatively higher shear force acting on the cathode plate, owing to the collision of fluidized particles, thus, the growing copper powder could not be cut easily from the cathode plate. Therefore, the size of copper powder recovered tended to increase.

Effects of the addition of gas and gas velocity on the recovery of copper in the reactors can be seen in Fig. 4. In this figure, the value of copper recovery(R) increased almost linearly with increasing reaction time(t) during the first 120min, but after that time, the increase trend of R with t decreased considerably. Note that the addition of a small amount of gas(air) into the inverse fluidized bed reactor could enhance the copper recovery considerably. Actually, when a small amount of air(U_G=0.001m/s) was injected into the reactor, the value of R increased up to 27%, and the value of R increased up to 15% when U_G=0.002m/s. This can be due to that the injection of a small amount of air could generate the turbulence and help to discharge very small gas bubbles generated during the reaction. It has been understood that the very small gas bubbles generated from the electrode reaction are attached at the surface of the cathode. This could be a important resistance for the mass transfer of copper ion to the cathode plate. The turbulence generated by the injection of a small amount of air could act to separate the very small gas bubbles from the cathode plate. In addition, the separated small gas bubbles can be coalesced easily with injected air bubbles rising and going out from the reactor. However, a further increase of air amount (higher than U_G is 0.002m/s) could decrease the copper recovery due to the relative decrease of liquid holdup between the two electrodes. Because, the decrease in the liquid holdup can lead to the decrease in the mass transport of copper ions by penetration in the reactor.

Effects of current density(I) on the recovery of copper in the reactor can be seen in Fig. 5. As can be seen, the value of R increased gradually with increasing current density, since the mass transfer rate of copper ion is proportional to the current density. Effects of amount of fluidized particles on the recovery of copper can be seen in Fig. 6. Note that the addition of a small amount of fluidized particles (W=1.0wt.%) to the reactor could increase the copper recovery up to 10~25%. It has been understood that the contacting of fluidized solid particles with the cathode plate could clean the surface as well as decrease the diffusion layer of copper ion, which results in the increases of reaction rate and current efficiency, thus, the recovery of copper could be increased.

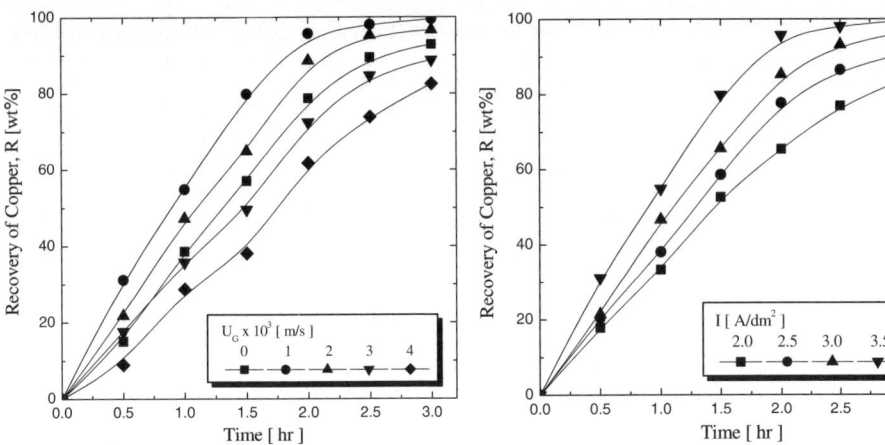

Fig. 4. Effects of U_G on the recovery of copper in inverse fluidized bed electrode reactors(W= 1.0wt.%, I=3.5A/dm^2, U_L=0.001m/s, L_{AC}= 0.015m).

Fig. 5. Effects of I on the recovery of copper in inverse fluidized bed reactors(W=1.0wt.%, L_{AC}= 0.015m, U_L=0.001cm/s, U_G=0.001m/s).

However, when the amount of added particles increased(W=2.0 or 3.0wt.%), the effective surface area of cathode plate decreased due to the considerable increase of solid holdup between the two electrodes, thus, the amount of copper recovery decreased. In this experimental conditions, the distance between the two electrodes(L_{AC}) also influenced the recovery of copper, as can be seen in Fig. 7. In this figure, the value of R was maximum when the distance(L_{AC}) was 1.5cm, in all the cases studied.

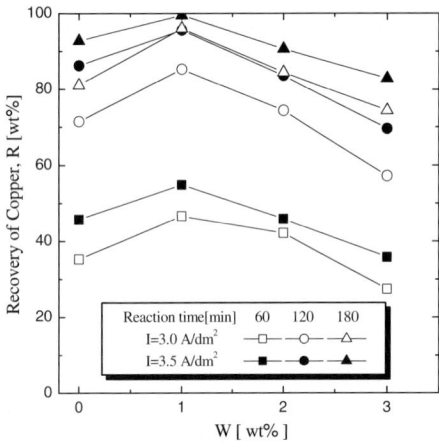

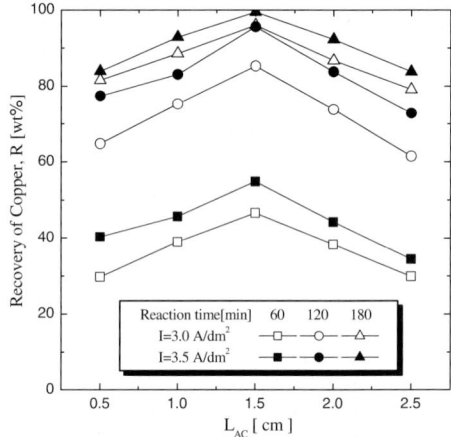

Fig. 6. Effects of amount of fluidized solid particles on the recovery of copper in an inverse fluidized bed electrode reactor(L_{AC}=0.015m, U_G= 0.001m/s, U_L= 0.001m/s).

Fig. 7. Effects of distance between the two electrodes on the recovery of copper in an inverse fluidized bed electrode reactor(W=1.0wt.%, U_L= 0.001m/s, U_G= 0.001m/s).

4. CONCLUSION

The copper was efficiently recovered more than 95-98% from the wastewater of electronic industries within 2-3 hrs as a powder by employing the three-phase inverse fluidized-bed reactors. The addition of a small amount of gas(U_G= 0.001) or fluidized particles(W=1.0wt.%) into the inverse fluidized bed reactors resulted in the increase of the copper recovery and decrease in the size of copper powder recovered. The value of copper recovery exhibited a maximum value with increasing gas or liquid velocity, amount of fluidized particles or distance between the two electrodes, but it increased gradually with increasing current density up to 3.5A/dm^2 between the two electrodes. The optimum conditions for the maximum recovery of copper powder were U_G=0.001m/s, U_L= 0.001m/s, W=1.0wt.%, I=3.5A/dm^2 and L_{AC}=0.015m within this experimental conditions.

REFERENCE

[1] K. Juttner, U. Galla and H. Schmieder, Electrochemica Acta, 45 (2000) 2575.
[2] M. A. Rabah, Hydrometallurgy, 56 (2000) 75.
[3] C. W. Won, Y. Kang and H. Y. Sohn, Metallurgical Transaction B, 24B (1993) 192.
[4] C. W. Won and Y. Kang, J. Materials, 49 (1997) 41.
[5] R. D. Weijden, J. Mahabier, A. Abbadi and M. A. Reuter, Hydrometallurgy, 64 (2002) 131.
[6] K. Horikawa and I. Hirasawa, Korean J. Chem. Eng., 17 (2000) 629.

Studies in Surface Science and Catalysis, volume 159
Hyun-Ku Rhee, In-Sik Nam and Jong Moon Park (Editors)

A mechanistic model for propane steam reforming on a bimetallic Co-Ni catalyst in fluidized bed reactor

Kelfin Martino Hardiman, Cheng-Han Hsu and Adesoji Adediran Adesina

Reactor Engineering and Technology Group, School of Chemical Engineering and Industrial Chemistry, University of New South Wales, Sydney, New South Wales 2052, Australia

1. INTRODUCTION

Steam reforming of hydrocarbon is an important route for the production of industrial hydrogen and synthesis gas ($CO+H_2$) used as a vital feedstock in petroleum and petrochemical plants [1]. The reaction is, however, accompanied by severe carbon deposition. In a previous publication [2], we demonstrated that a bimetallic Co-Ni catalyst exhibited better coking resilience and favourable product selectivities during steam reforming of propane compared to conventional monometallic Ni-catalysts. Furthermore, the properties of coke deposits were probed in order to explore relevant dependency on process variables. This paper reports a kinetic investigation of the propane reforming reaction in fluidized bed reactor over the bimetallic catalyst under conditions of both low and high steam-to-carbon (S:C) ratios via Langmuir-Hinshelwood (LH) and Eley-Rideal (ER) kinetic models with single and dual-site adsorption-desorption mechanisms. These models were subsequently discriminated on the basis of statistical and thermodynamic Boudart-Mears-Vannice (BMV) criteria.

2. EXPERIMENTAL DETAILS

The steam reforming runs were performed in a fluidized bed system described in Hardiman et al. [2] with reaction temperature in the range of 773-873 K at atmospheric pressure. The reactor was made of a quartz tube (ID=20 mm, H=490 mm) fitted with a 3-mm thick sintered quartz plate (50 μm holes). The reactor was loaded with 1 g of catalyst with a total fluidizing flow rate of 300 ml min^{-1}. Product gases were sent to a condenser and then analyzed by a gas chromatograph. Prior to each run, the catalyst was reduced using hydrogen flow at 873 K for 2 h.

3. STEAM REFORMING RESULTS

Figs. 1 and 2 show the effect of steam partial pressure on reforming rate, $-r_{SR}$, at low and high S:C ratio respectively at 3 different temperatures. The increase in reaction rate with steam partial pressure is evident under both S:C ratios. However, rates at low S:C ratio appeared to be higher than under similar conditions at high S:C ratio due to higher hydrocarbon partial pressure employed. This would suggest that reaction rate has positive order dependencies on both propane and steam partial pressures. Since carbon deposition is a prerequisite for steam, reforming and also causes catalyst decay, it would appear that steam and propane may be chemisorbed on different sites on the catalyst.

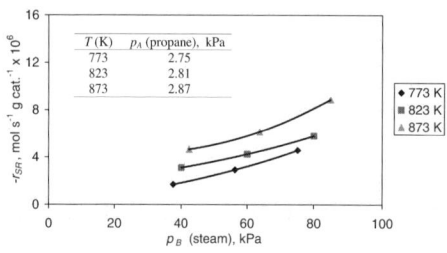 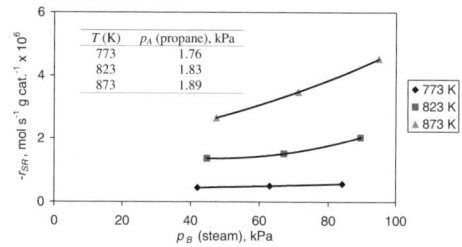

Fig. 1. Reforming rates for low S:C ratio (0.8-1.6)   Fig. 2. Reforming rates for high S:C ratio (2.4-4.8)

4. MECHANISTIC CONJECTURE

The steam reforming of hydrocarbons may be given by:

$$C_x H_y + 2S_1 \leftrightarrow C_{x-m_1} H_{y-m_2} - S_1 + C_{m_1} H_{m_2} - S_1$$

$$H_2 O + 2 S_2 \leftrightarrow OH - S_2 + H - S_2$$

$$C_{m_1} H_{m2} - S_1 + OH - S_2 \leftrightarrow C_{m_1} H_{m_2} O - S_1 + H - S_2$$

$$C_{m_1} H_{m2} O - S_1 + S_1 \leftrightarrow CHO - S_1 + C_{m_1 - 1} H_{m_2 - 1} - S_1$$

$$CHO - S_1 + S_2 \leftrightarrow CO - S_1 + H - S_2$$

$$CO - S_1 + OH - S_2 \leftrightarrow COO - S_1 + H - S_2 \quad \text{(in excess H}_2\text{O)}$$

$$CO - S_1 \leftrightarrow CO \uparrow + S_1$$

$$COO - S_1 \leftrightarrow CO_2 \uparrow + S_1$$

$$H - S_2 + H - S_2 \leftrightarrow H_2 \uparrow + 2 S_2$$

where m_1, m_2, x and y are integers and S_1 & S_2 are 2 types of sites. Formal treatment using quasi-steady-state approximation and concept of most abundant reactive intermediate yield the expressions in Table 1.

Table 1
Langmuir-Hinshelwood (LH) and Eley-Rideal (ER) kinetic models

No.	Model (M)	Remarks
1	$\dfrac{k_{rxn} p_A^{1/2} p_B^{1/2}}{\left(1 + \sqrt{K_A p_A}\right)\left(1 + \sqrt{K_B p_B}\right)}$	Dual-site adsorption of propane (A) and steam (B)
2	$\dfrac{k_{rxn} p_A^{1/2} p_B}{\left(1 + \sqrt{K_A p_A}\right) + \left(1 + K_B p_B\right)}$	Dual-site adsorption with non-dissociative water adsorption
3	$\dfrac{k_{rxn} p_A^{1/2} p_B^{1/2}}{\left(1 + \sqrt{K_A p_A} + \sqrt{K_B p_B}\right)^2}$	Adsorption of both propane (A) and steam (B) on the same site with bimolecular rate determining step
4	$\dfrac{k_{rxn} p_A^{1/2} p_B}{1 + \sqrt{K_A p_A}}$	Eley-Rideal model

5. MODEL DISCRIMINATION

Nonlinear regression of the data provides the parameter estimates (shown in Table 2) associated with the models listed in Table 1.

Table 2
Estimates of kinetic and reaction constants

M	T (K)	k_{rxn} $(\times 10^8)$	K_A $(\times 10^3)$	K_B $(\times 10^3)$	R^2
	773	5.47±0.030	30.08±0.015	192.16±0.096	0.999
1	823	6.64±0.144	0.83±0.018	2.16±0.047	0.957
	873	11.25±0.238	0.42±0.089	0.18±0.038	0.958
	773	2.39±0.034	174.91±2.470	0.15±0.002	0.972
2	823	8.22±0.237	72.32±2.091	1.48±0.043	0.943
	873	17.03±0.528	12.90±0.400	6.33±0.196	0.939
	773	10.60±0.262	0.16±0.040	1.16±0.029	0.951
3	823	19.70±0.887	0.55±0.025	1.44±0.065	0.912
	873	40.50±1.887	1.50±0.069	2.66±0.124	0.909
	773	0.18±0.008	46.50±1.990		0.916
4	823	0.65±0.025	24.76±0.947		0.925
	873	1.61±0.033	7.15±0.148		0.959

The positive values obtained in practically all cases indicate that all these models may be plausible representations of the data and indeed, the correlation coefficients, R^2, are greater than 0.9. Thus, statistical compliance is not a sufficient basis for model discrimination. Specifically, the thermodynamic consistency of the estimates, as proposed by Boudart et al. [3], is appropriate further scrutinizing criterion during kinetic modelling and has been gainfully employed in other reactions [4-6].

The BMV criterion [3] requires

$$10 \le -\Delta S \le 12.2 - 0.0012\Delta H \tag{1}$$

where ΔH and ΔS may be obtained from

$$\ln K = -\frac{\Delta H}{R}\frac{1}{T} + \frac{\Delta S}{R} \tag{2}$$

for each of A and B.

Tables 3 and 4 host the estimated thermodynamic constants and Arrhenius parameters for all models respectively. As shown in Table 3, Models 2 & 3 did not meet the BMV standard. Model 4 showed marginal values. In view of the unambiguous satisfaction of the BMV criterion for A and B in Model 1, it was selected as the most meaningful description of the present data. The corresponding activation energy, E_A, values are also listed in Table 4, from where the estimate for Model 4 is given as 40 kJ mol^{-1}. In general, activation energy values during steam reforming decrease with increased carbon number as reported by Praharso et al. [6]. For propane reforming, depending on Ni-catalyst, values between 60-70 kJ mol^{-1} have been reported [6]. The relatively low E_A value obtained here may be due to synergistic effect of the new bimetallic catalyst. In fact, we had reported a power-law model-based E_A value of 38 kJ mol^{-1} for the same Co-Ni catalyst [2]. Additionally, excellent mixing obtained under the hydrodynamic conditions used (Fr≈1) enhanced better conversion while reducing the propensity for deactivation since the deactivation kinetics have a negative order with respect to steam partial pressure.

Table 3
Thermodynamic coefficients and BMV assessment

Parameter	M	Estimate	BMV criterion	Parameter	M	Estimate	BMV criterion
ΔH_A (kJ mol^{-1})	1	-243.23		ΔH_B (kJ mol^{-1})	1	-394.03	
	2	-145.15			2	212.56	
	3	125.02			3	46.00	
	4	-104.21			4	-99.28	
ΔS_A (J mol^{-1} K^{-1})	1	347.24	Yes	ΔS_B (J mol^{-1} K^{-1})	1	525.49	Yes
	2	200.97	Yes		2	-202.34	No
	3	-89.26	No		3	-2.73	No
	4	159.38	Borderline		4	152.54	Borderline

Table 4
Arrhenius parameters

T (K)	-ln k				E_A (kJ mol^{-1})			
	M1	M2	M3	M4	M1	M2	M3	M4
773	16.7	17.6	16.1	20.1				
823	16.5	16.3	15.4	18.8	40.0	110.7	75.0	123.4
873	16.0	15.6	14.7	17.9				

6. CONCLUSIONS
We have shown that the steam reforming of propane may be adequately described by LH mechanism involving different adsorption sites for steam and hydrocarbon. The associated model satisfied both statistical compliance and the BMV thermodynamic criterion.

ACKNOWLEDGEMENTS
The authors appreciate the Australian Research Council (ARC) for supporting this project. KMH is a recipient of University Postgraduate Award (UPA) scholarship at the University of New South Wales.

REFERENCES
[1] J.R. Rostrup-Nielsen, Catal. Today, 63 (2000) 159.
[2] K.M Hardiman, T.T Ying, A.A. Adesina, E.M. Kennedy and B.Z. Dlugogorski, Chem. Eng. J., 102, (2004) 119.
[3] M. Boudart, D.E. Mears and M.A. Vannice, Ind. Chim. Belg., 32 (1967) 281.
[4] A.A. Adesina, J. Nig. Soc. Chem. Eng., 7 (1988) 258.
[5] J. Xu and G.F. Froment, AIChE J., 35 (1989) 88.
[6] Praharso, A.A. Adesina, D.L. Trimm and N.W. Cant, Chem. Eng. J., 99, (2004) 131.

Studies in Surface Science and Catalysis, volume 159
Hyun-Ku Rhee, In-Sik Nam and Jong Moon Park (Editors)

Photocatalytic oxidation of TCE in a TiO$_2$-coated activated carbon fluidized bed reactor

Sung-Chul Kim and Dong-Keun Lee

Dept. of Chemical and Biological Engineering, Environmental Biotechnology National Core Research Center, Environmental and Regional Development Institute, Gyeongsang National University, Kajwa-dong 900, Jinju, Gyeongnam 660-701, Korea(e-mail:d-klee@gsnu.ac.kr)

1. INTRODUCTION

Although halogenated organic compounds were widely used in industry, they are considered to be very dangerous environmental pollutants [1]. Trichloroethylene (TCE) have been widely used in dry cleaning, metal degreasing and as chemical intermediates, therefore they are easily found in a various water supplies. Due to its high toxicity and volatility, TCE in water may be removed by activated carbon and air stripping, which do not degrade them but relocate it in another environment. In recent years, photocatalytic oxidation with TiO$_2$ powder to destroy organic pollutants from contaminated water has received considerable attention and extensively been studied [2,3]. TiO$_2$ powder has, however, some detrimental shortcomings for practical application. TiO$_2$ powder is not only difficult to be separated from water after being used, but also reduces photocatalytic efficiency due to light scattering.

In this study TiO$_2$ was coated mainly at the exterior surface of granular activated carbon, and this TiO$_2$-coated activated carbon was employed for the adsorption followed by photocatalytic decomposition of TCE.

2. EXPERIMENTAL

In order to locate TiO$_2$ mainly at the exterior surface in the vicinity of macropores of the activated carbon, a modified sol-gel preparation method was employed.

The basic experiments for the adsorption and photocatalysis of TCE were performed with a water-jacketed borosilicate glass vessel (1L capacity) incorporating a quartz window for ultraviolet (UV) light illumination. Dosage of granular activated carbon without TiO$_2$ (GAC) and with TiO$_2$ (GAC-Ti) was 1.0 mg/L if not mentioned otherwise. The light source was a 4-Watt black light lamp (370 nm, F4T5-BLB). Analysis of TCE was performed by GC (HP 6890) with FID and purge & trap (Tekmar 300).

Adsorption and photocatalytic oxidation of TCE with TiO$_2$-coated activated carbon was also carried out in a cylindrical continuous flow fluidized bed reactor with 65cm height and 68cm inside diameter (Figure 1).

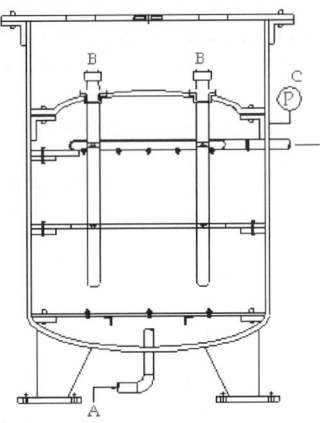

Figure 1. Schematics of the fluidized bed reactor (A: inlet, B: UV-lamp, C: Pressure gauge, D: outlet).

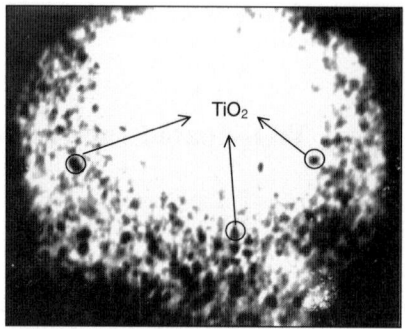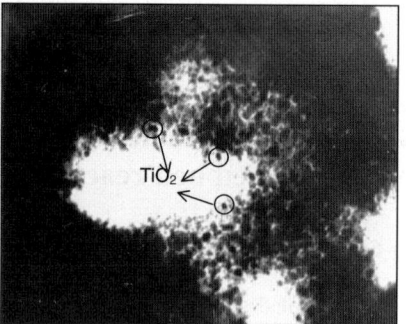

Figure 2. TEM micrographs of TiO₂-coated activated carbon (left: ×40,000 magnification, right: ×80,000 magnification)

Four low pressure mercury lamps (15 W, 254 nm) were installed inside the reactor. 4.5kg of TiO₂-free activated carbon (GAC) or TiO₂-coated activated carbon (GAC-Ti) was loaded inside the reactor and aqueous solution of TCE (0.05mg/L concentration) was fed into the bottom of the reactor. The exit stream of the reactor was analysed for TCE concentration. Successful fluidization without carryover of the activated carbon particles could be obtained at the flow rates of upward-flowing water between 150~200cm^3/sec.

3. RESULTS AND DISCUSSION

Figure 2 shows the TEM micrographs of the TiO₂-coated granular activated carbon with 0.6 wt% TiO₂ loading used in this study. Macropores are well developed and their average diameter was estimated to be 1.0 μm, and fine particles of TiO₂ are known to be dispersed at the exterior surface in the vicinity of the macropores. XRD analysis showed that TiO₂ particles are in anatase crystalline form.

Figure 3 demonstrates the effects of GAC-Ti (0.6 wt%) on the removal of TCE in the presence and absence of light. The initial concentration of TCE was 0.05 mg/L. TCE disappears very rapidly on exposure to GAC-Ti and light. This rapid disappearance, however, is not observed in the absence of light. Almost complete removal of TCE under illumination can be achieved in approximately 20 min. This extremely rapid removal rate of TCE on GAC-Ti (0.6 wt%) might be due to the photocatalytic oxidation of TCE on the surface of TiO₂ particles. As could be seen in Figure 2, fine TiO₂ particles were located mainly in the vicinity of the pore entrances of the activated carbon. TCE might adsorb onto the surface of TiO₂ particles and/or on the surface of the activated carbon.

After being exposed to TCE aqueous solution in the absence and presence of light for different times, GAC-Ti was subjected to solvent extraction with 1L methanol for 6 h to determine if residual TCE was remained, and the results are shown in Figure 4. The extracted concentrations of TCE from GAC-Ti which was exposed to TCE aqueous solution in the presence of light were much lower than the corresponding concentrations obtained from GAC-Ti (0.6 wt%) which was exposed to TCE aqueous solution in the absence of light. The extracted concentration of TCE from the illuminated GAC-Ti (0.6 wt%) for 30 min was 0.7 μg/L. This concentration is more than twenty times lower than the extracted one (16.3 μg/L) from GAC-Ti (0.6 wt%) without light. In addition, the extracted concentration from GAC-Ti without light increases gradually with increasing exposure time, while the extracted concentration from the illuminated GAC-Ti (0.6 wt%) goes through maximum and then

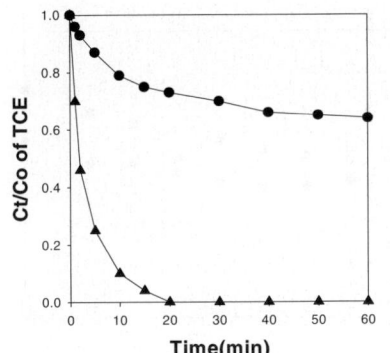

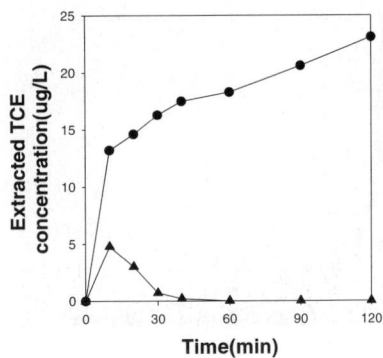

Figure 3. Effects of GAC-Ti (0.6 wt%) on the removal of TCE in the absence(●) and presence(▲) of light .

Figure 4. Concentration of TCE extracted from GAC-Ti (0.6 wt%) after being exposed to TCE aqueous solution for different times in the absence (●) and presence (▲) of light.

decreases rapidly. After being exposed for longer than 60 min, no residual TCE was detected from GAC-Ti (0.6 wt%). In the GAC-Ti (0.6 wt%) sample, part of TCE in the aqueous solution will directly adsorb onto the surface of TiO_2 particles, and will then photocatalytically be oxidized quickly. The other part of TCE seems to adsorb on the surface of the activated carbon in the vicinity of TiO_2 particles, and the adsorbed TCE migrates onto neighbouring TiO_2 particles for its subsequent photocatalytic degradation. That is, the surfaces of the activated carbon in the vicinity of the TiO_2 particles provided sites for continuing adsorption of TCE, which results in the synergistic effect of the GAC-Ti (0.6 wt%) samples on the efficient removal of TCE.

Figure 5 shows the effects of TiO_2 loading on the TCE removal efficiencies on the GAC-Ti samples. With increasing TiO_2 loading, the TCE concentrations drop more rapidly up to the first 2min. After 2min reaction, however, the TCE concentrations in the solution are higher during the reaction with the GAC-Ti sample having higher TiO_2 loadings. In the case of the GAC-Ti (0.6 wt%) compete removal of TCE from the solution is achieved within 15min, while it takes 40min with the GAC-Ti (1.5 wt%) sample. Ti concentrations present in the activated carbon particle were analyzed with a SEM/EDX (X-ray energy dispersive analysis). Most of TiO_2 were mainly located at the exterior surface of the activated carbon particle irrespective of the weight loadings of TiO_2. Too much quantities of TiO_2 may result in the partial blocking of the macropores and micropores of the activated carbon. The pore volume and surface area of the GAC-Ti (1.5 wt%) were 0.36 cm^3/g and 615 m^2/g, respectively which were much lower than those of the GAC (0.48 cm^3/g pore volume and 996 m^2/g surface area) and the GAC-Ti (0.6 wt%)(0.43 cm^3/g pore volume and 862 m^2/g surface area). In the case of the GAC-Ti (1.5 wt%) sample part of the pores of the activated carbon must have been blocked during the coating processes of TiO_2.

TCE solution (0.05mg/L concentration) was fed into the reactor at the fluidizing flow rates, and their concentrations at the exit stream of the reactor were measured. Figure 6 illustrates TCE removal efficiencies obtained with the 4.5 kg GAC and with 4.5 kg GAC-Ti under the illumination of UV-light at different fluidizing velocities. The concentrations of TCE at the exit of the reactor after 30 min operation with GAC were 0.019 and 0.022 mg/L at the flow rates of 155 cm^3/sec and 199 cm^3/sec, respectively. About 60 % of TCE in the feed solution can be said to be removed. Then exit concentrations of TCE increase gradually, and

548

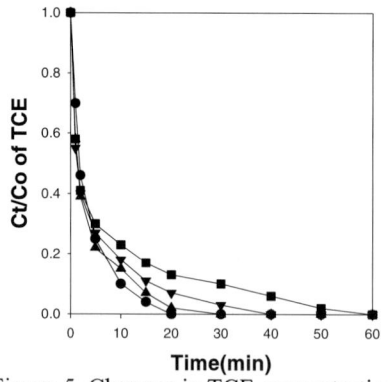

Time(min)

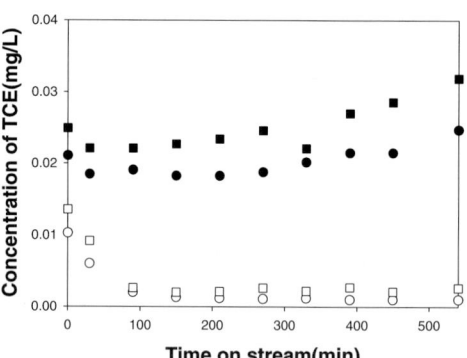

Time on stream(min)

Figure 5. Changes in TCE concentration during the reaction with the GAC-Ti (0.6 wt%)(●), GAC-Ti (1.1 wt%)(▲), GAC-Ti (1.5 wt%)(▼), GAC-Ti (1.9 wt%)(■) in the presence of light.

Figure 6. Concentration of TCE at the exit of the fluidized bed reactor with GAC and GAC-Ti at different volumetric flow rates (●: 155cm³/sec, GAC; ■: 199 cm³/sec, GAC; ○: 155cm³/sec, GAC-Ti; □: 199 cm³/sec, GAC-Ti).

reach at around 0.025 and 0.032 mg/L after 540 min operation, respectively. These results indicate that the surface of the activated carbon becomes saturated due to the continuing uptakes of TCE. On the contrary, more than 90% of TCE in the feed solution is steadily removed up to 540 min operation with GAC-Ti. This stable and high removal efficiency is believed to be due to the adsorption of TCE on the surface of the activated carbon followed by continuous migration onto the surface of TiO_2 particles where subsequent fast oxidation of TCE proceeds.

4. CONCLUSION

TCE in water could successfully be degraded by using TiO_2-coated granular activated carbon. Continuous migration and subsequent photocatalytic oxidation on the surface of TiO_2 accelerated TCE removal efficiency greatly, and made the application of the TiO_2-coated granular activated carbon more practical. Photocatalytic oxidation of TCE with the TiO_2-coated activated carbon in the continuous flow fluidized bed reactor yielded more than 90% removal efficiency stably up to 540min operation.

ACKNOWLEDGEMENTS
This research was supported by the grant from the MOST/KOSEF to the Environmental Biotechnology National Core Research Center (grant #: R15-2003-012-02002-0), Korea.

REFERENCES
[1] H. Takashima, L. Ren and Y. Kanno, Catal. Communications., 5 (2004) 317.
[2] D. -K. Lee and I. -C. Cho, Stud. Surf. Sci. Catal. 133 (2001) 303.
[3] D.F. Ollis, E. Pellizzetti and N. Serpone, Photocatalysis: Fundamentals and Applications, Wiley, New York, 1989

Studies in Surface Science and Catalysis, volume 159
Hyun-Ku Rhee, In-Sik Nam and Jong Moon Park (Editors)

Attrition and CO_2 adsorption of dry sorbents in fluidized bed

Ki Chul Cho [a]**, Eun Yong Lee** [a]**, Jeong Suk Yoo** [b]**, Young Hean Choung** [a]**,Sang Woo Park**[c]
and Kwang Joong Oh [a*]

[a] Department of Environmental Engineering, Pusan National University, San 30 jangjeon-
dong, Guemjeong-gu, Busan 609-735, Republic of Korea
[b] Doosan Heavy Industries & Construction Co. Ltd., R&D Center, 555 Gwigok-dong,
Changwon, Gyeongnam 641-792 , Republic of Korea
[c] Department of Chemical Engineering, Pusan National University, 30 jangjeon-dong,
Guemjeong-gu, Busan 609-735, Korea (e-mail: kjoh@pusan.ac.kr)

Abstract

Characteristics of attrition and adsorption were investigated to remove CO_2 in fluidized bed
using activated carbon, activated alumina, molecular sieve 5A and molecular sieve 13X. For
every dry sorbent, attrition mainly still occurs in the early stage of fluidization and attrition
indexs(AI) of molecular sieve 5A and molecular sieve 13X were higher than those of
activated carbon and activated alumina. Percentage loss of adsorption capacity of molecular
sieve 5A and molecular 13X were 14.5% and 13.5%, but that of activated carbon and
activated alumina were 8.3% and 8.1%, respectively. Overall attrition rate constant (Ka) of
activated alumina and activated carbon were lower than other sorbents.

Introduction

In 1997, more than 160 nations negotiated binding limitations on green house gases for
developed nations in Kyoto, Japan. In Kyoto protocol, the developed nations agreed to limit
their green house gas emissions relative to the levels emitted in 1990. A number of
techniques have been used for separation of carbon dioxide from flue gas streams. Among
these techniques, chemical absorption has been commercially operated. The chemical
absorption process can control carbon dioxide with high removal efficiency, but it needs
intensive energy for carbon dioxide removal and has a problem with corrosion. Therefore,
fluidized bed is known as a proper process to control high volumes of flue gases, and dry
sorbents can be used to cut down the operating cost, so the use of dry sorbents in a fluidized
bed may remove carbon dioxide economically. However, this technique is required for
developments of the fluidized bed process with dry sorbents which have low attrition and
high adsorption capacity for carbon dioxide.
Therefore, in this study, activated carbon, activated alumina, molecular sieve 5A, and
molecular sieve 13X were used as dry sorbents to control carbon dioxide in a fluidized bed.
In addition, the attrition and percentage loss of adsorption capacity of the dry sorbents were
investigated.

Experimental
Apparatus and procedure for attrition experiment.
Three-hole air jets used by Gwyn on the basis of a research of Forsythe and Hertwig were used to investigate attrition of dry sorbent with fluidization.

Air was supplied from a compressor, moisture and particles in the air are removed passing a trap, and air flow rate was controlled by mass flow controller (5850E, Brooks Co.). The dry sorbents after the attrition were collected, then the particle sizes of them were measured by sieves of 60, 80, and 140 mesh.

Apparatus and procedure for adsorption experiment.
An experimental fluidized bed reactor has a 2.5 cm in diameter and 230 cm in height, and the distributor has 32 holes and each hole was 2 mm in diameter. 200 mesh net was put on the distributor to prevent particles from falling down. The cyclone was made by standard proportion to collect fine particles. Air flow rate was controlled by a flow meter, CO_2 (99.9%) flow rate was controlled by mass flow controller and then 10% CO_2 inlet concentration was maintained by mixing in a mixing chamber. CO_2 outlet concentration was also measured by CO_2 analyzer (CD 95, Geotechnical instruments, England).

Results and Discussion
Fig. 1 shows that minimum fluidization velocities of activated carbon, activated alumina, molecular sieve 5A and molecular sieve 13X are 8.0 cm/s, 8.5 cm/s, 6.2 cm/s and 6.5 cm/s, respectively. Also, theoretical calculation values of minimum fluidization velocity and terminal velocity of each dry sorbent were summarized in Table 1.

The remaining weight of dry sorbents with time is shown in Fig. 2. For every dry sorbent, attrition mainly still occurs in the early stage of fluidization and AI test on the basis of the weight after 5 hours shows that AIs of molecular sieve 5A and molecular sieve 13X presented 2.1~4.0 times higher than those of activated carbon and activated alumina. Therefore, the use of molecular sieve 5A and 13X in a fluidized bed can cause high maintenance cost and problems in the operating the process.

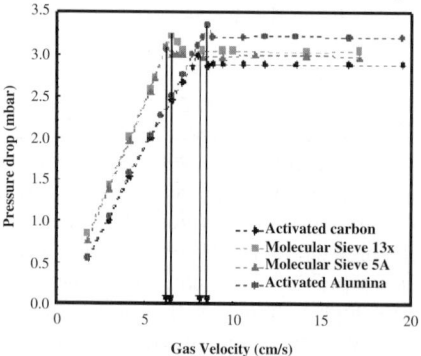

Fig. 1. Minimum fluidiztion velocity for dry sorbents with gas velocity.

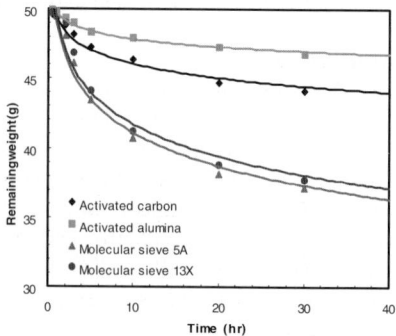

Fig. 2. The remaining weight with time for attrition of dry sorbents.

Table 1. minimum fluidization velocity and terminal velocity of dry sorbents

Dry sorbent	Mean	Density	$U_{mf,exp.}$	$U_{mf,cal.}$	$U_{t,cal}$

	Size(μm)	(g/cm³)	(cm/s)	Wen&Yu	Richardson	Chitester	(cm/s)
Activated carbon	326.0	2.79	8.0	9.4	10.8	13.1	191.7
Molecular sieve 5A	326.0	2.47	6.2	8.3	9.6	11.7	177.9
Molecular sieve 13X	326.0	2.51	6.5	8.5	9.8	11.8	179.7
Activated Alumina	326.0	2.88	8.5	9.7	11.2	13.5	195.4

CO_2 adsorption capacities with dry sorbents before and after attrition were shown in Fig.3. We found variation of CO_2 adsorption capacity during operation by examining effect of attrition on adsorption capacity. So, adsorption experiments for each sorbent fluidized for 30hours were carried out. As a result, percentage losses of adsorption capacity of molecular sieve 5A and molecular 13X were 14.5% and 13.5%, but those of activated carbon and activated alumina were 8.3% and 8.1% respectively. This is because retention time of molecular sieve 5A and molecular 13X decreased due to elutriation of particle generated from attrition.

Gas velocity is an important operating condition in the fluidized bed process and it can highly affect the attrition of dry sorbents. Therefore, the weight remaining in the bed with fluidization time for gas velocity of 20.59 cm/s, 25.74 cm/s, and 30.89 cm/s was measured to estimate the attrition of dry sorbent with gas velocity. As shown in Fig. 4, attrition mainly occurred in the early stage of fluidization. The attrition rate with time decreased and the regression equations fit natural log functions. In addition, Fig. 4 shows that the attrition of dry sorbents is highly affected by gas velocity in the fluidized bed process.

Table 2 summaries overall attrition rate constants (K_a) and physical properties for each dry sorbent. As shown in Table 2, K_a of activated alumina was the lower than any other sorbent, but was similar to activated carbon. However, we used activated carbon as dry sorbent to control CO_2 because it is the most cost-effective among others.

Table 3 summaries K_a for activated carbon. The value of K_a presented 4.032×10^{-5}, 9.184×10^{-5}, 1.980×10^{-4} with each gas velocity and the value of Ka increased with increasing gas velocity.

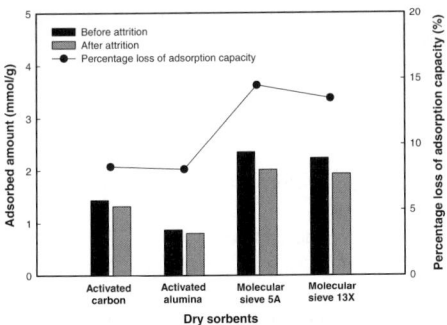

Fig. 3. Percentage loss of adsorption capacity by attrition with dry sorbents.

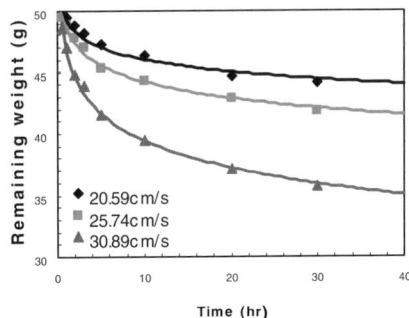

Fig. 4. The remaining weight with time for attrition of 40/60 mesh activated carbon at different gas velocities.

Table 2. Summary of attrition rate constants for dry sorbents at gas velocity of 20.59 cm/s

Sorbent	BET (m²/g)	Pore Volume (cm³/g)	Pore Diameter (Å)	Mean Size (um)	Density (g/cm³)	K_a (cm²/s³)
Activated Carbon	1142	0.58	20.3	326	2.79	4.032×10^{-5}
Activated Alumina	275	0.34	48.8	326	2.88	3.824×10^{-5}
Molecular Sieve 5A	394	0.31	31.6	326	2.47	7.169×10^{-5}
Molecular Sieve 13X	480	0.34	28.6	326	2.51	7.863×10^{-5}

Table 3. Summary of attrition rate constants for activated carbon at different gas velocities.

Gas velocity (cm/s)	K_a (cm²/s³)
20.59	4.032×10^{-5}
25.74	9.184×10^{-5}
30.89	1.980×10^{-4}

Conclusion

In the fluidized bed process, attrition caused dry sorbent to be carryover. This mainly occurred in the early stage of fluidization and was highly affected by gas velocity. The amount of attrition of molecular sieve 5A and molecular sieve 13X were larger than those of activated carbon and activated alumina. In addition, percentage losses of adsorption capacities of molecular sieve 5A and molecular 13X were 14.5% and 13.5%, whereas those of activated carbon and activated alumina were 8.3% and 8.1%, respectively. This is because retention time of molecular sieve 5A and molecular 13X decreased due to elutriation of particle generated from attrition. Also, K_a of activated alumina and activated carbon were the lower than those of Molecular sieve 13X and 5A. Consequently, molecular sieve 5A and molecular 13X could cause high maintenance cost for dry sorbent and problems in the operation of fluidized bed process.

Acknowledgement

This research was supported by a grant (DA2-201) from Carbon Dioxide Reduction & Sequestration Research Center, one of the 21st Century Frontier Programs funded by the Ministry of Science and Technology of Korean government.

References

[1] Hoffman, J. S., Pennline, H. W.; J. Energ. & Environ. Res., U.S.A, Pittsburgh, 1(1), (2001), pp. 90-100.
[2] Forsythe, W. L. and Hertwig, W. R., "Attrition characteristics of fluid cracking catalyst," Ind. Eng. Chem.,(1949), 41, pp. 1200~1206.
[3] Daizo Kunii, Octave Levenspiel, Fluidization Engineering second edition, (1991).
[4] Lin cy, Wey My, Korean Journal of Chemical Engineering, Republic of Korea, Seoul, 22(1), (2005), pp. 154-160.
[5] S. J. Son, J. S. Choi and K. Y. Choo, Korean Journal of Chemical Engineering, Republic of Korea, Seoul, 22(2), (2005), pp. 291-297.
[6] S.S. Lee, J.S. Yoo, G.H. Moon, S.W. Park, D.W. Park, K.J. Oh, Am. Chem. Soc., Div. Fuel Chem. U.S.A., Washington, 48(2), (2003).

Studies in Surface Science and Catalysis, volume 159
Hyun-Ku Rhee, In-Sik Nam and Jong Moon Park (Editors)
© 2006 Elsevier B.V. All rights reserved

Bio-oil upgrading over Ga modified zeolites in a bubbling fluidized bed reactor

Hyun Ju Park[1], Young-Kwon Park[1*], Joo-Sik Kim[1], Jong-Ki Jeon[2], Kyung-Seun Yoo[3], Jin-Heong Yim[4], Jinho Jung[5], Jung Min Sohn[6]

[1]Faculty of Environmental Eng., University of Seoul, Seoul 130-743, Korea (catalica@uso.ac.kr)
[2]Dept. of Chemical Engineering, Kongju National University, Gongju 314-701, Korea
[3]Dept. of Environ.Eng., Kwangwoon University, Seoul, Korea
[4]Div. of Advanced Materials Eng., Kongju National University, Gongju 314-701, Korea
[5]Div. of Environmental Science & Ecological Eng., Korea University, Seoul, 136-713, Korea
[6]Dept. of Mineral Resource & Energy Eng., Chonbuk Nat'l Univ., Jeonju, 561-756, Korea

Abstract

Catalytic upgrading of bio-oil was carried out over Ga modified ZSM-5 for the pyrolysis of sawdust in a bubbling fluidized bed reactor. Effect of gas velocity (U_o/U_{mf}) on the yield of pyrolysis products was investigated. The maximum yield of oil products was found to be about 60% at the U_o/U_{mf} of 4.0. The yield of gas was increased as catalyst added. HZSM-5 shows the larger gas yield than Ga/HZSM-5. When bio-oil was upgraded with HZSM-5 or Ga/HZSM-5, the amount of aromatics in product increased. Product yields over Ga/HZSM-5 shows higher amount of aromatic components such as benzene, toluene, xylene (BTX) than HZSM-5.

Introduction

Worldwide concern about the diminishing trend of primary energy sources and environmental problems have prompted many observers to call for a decreased reliance on fossil fuels. Renewable sources of energy are constantly examined as alternatives for fossil fuels. Renewable energy is of growing importance in satisfying environmental concerns of fossil fuel usage and its contribution to the greenhouse effect. Biomass forms are some of the main renewable energy resources available [1]. Biomass has received considerable attention both as a source of energy and as an organic chemical feedstock [2,3]. The energy potential of biomass has increasingly become recognized as a means to help meet world energy demand. The utilization of biomass and other alternative fuel sources, rather than existing fossil fuels, would offer more environmentally acceptable processes for energy production and will aid in conserving the limited supplies of fossil fuels [4-6]. Among the processes, pyrolysis using fluidized bed reactor has known to be a promising process to convert biomass materilas into useful liquid products that can be used as alternative fules or valuable chemicals [7]. However, pyrolytic oils are not always completely volatile and contain high levels of oxygen, this being the major factor responsible for the high viscosity and corrosiveness. The upgrading of pyrolitic oils is a necessary process and involves the removal of oxygen by catalyst such as ZSM-5, Y zeolite. However, there are few reports about the metal supported or ion exchanged HZSM-5 for bio-oil upgrading. In this study, catalytic pyrolysis of biomass over Ga modified zeolite was carried out using bubbling fluidized bed reacotr. Also, to obtain optimum reaction condition, effect of temperature, L/D ratio, U_o/U_{mf} and oxygen concentration on the yield of sawdust pyrolysis has been determined in a bubbling fluidized bed reactor.

Experimental

The sawdust used in this work was a radiata pine. HZSM-5 and HY pellet was purchased from Zeobuilder. Ga was impregnated over zeolites by excess water evaporation. The catalytic upgrading was carried out in a bubbling fluidized bed as shown in Fig. 1. The reactor had a diameter of 0.0762 m and a height of 0.8 m. The catalyst bed was located after fluidized bed. The pyrolysis vapor formed in the fluidized bed was then upgraded by passing them over a fixed bed of catalyst. For the capturing the bio-oil mist, an electrostatic precipitator was equipped. The degradation products were analyzed by a gas chromatography (Young Lin-M600D) equipped with an FID and a TCD. The upgraded oil was also analyzed by GC-MS, FT-IR, and elemental analysis.

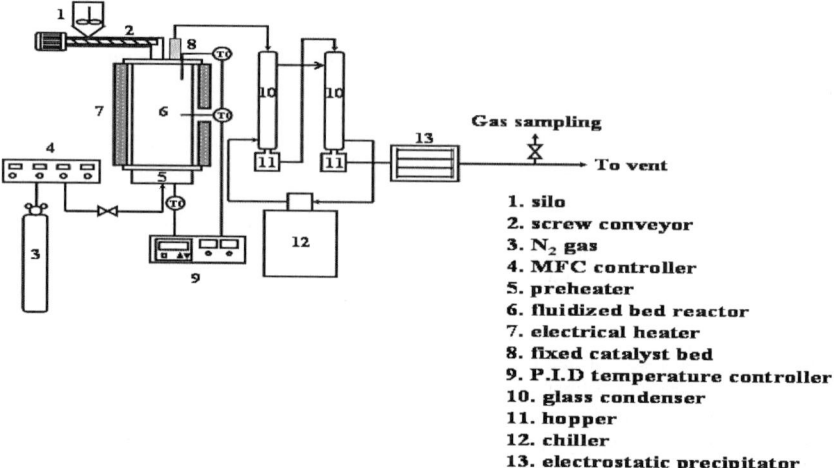

1. silo
2. screw conveyor
3. N$_2$ gas
4. MFC controller
5. preheater
6. fluidized bed reactor
7. electrical heater
8. fixed catalyst bed
9. P.I.D temperature controller
10. glass condenser
11. hopper
12. chiller
13. electrostatic precipitator

Fig. 1. Schematic diagram of catalytic fluidized bed reactor

Results and Discussion
Effect of gas velocity (U_o/U_{mf}) on the yield of pyrolysis products was shown in Fig. 2. It was observed that the maximum yield of oil products was found to be about 60% at the U_o/U_{mf} of 4.0. These results implicate that heat transfer rate of bed particles to sawdust is enhanced with the increase of physical mixing between sawdust and bed material. In addition, the increase of gas velocity also reduces the residence time of pyrolytic products, which can be decomposed to smaller chemicals by secondary reactions in freeboard. As can be seen in Fig. 2, however, increase of U_o/U_{mf} above 4.0 reduces the oil yield, even though the heat transfer rate and solid mixing are enhanced. This might be the increase of entrainment of sawdust due to the increase of bubble size in the fluidized bed reactor. It should be also noted that segregation between sawdust and bed particles might be happened due to the large difference of solid density. This leads to the slow pyrolysis and the yield of oil decreased.

As shown in Table 1, the yield of gas was increased as catalyst added. This is due to the oxygen removal over catalyst transforming into H_2O, and CO_2(Table 2 and Table 3). Especially when catalyst was added, the yield of CO_2 and H_2O were remarkably increased. HZSM-5 shows the better gas yield than Ga/HZSM-5 because HZSM-5 has higher cracking ability than Ga/HZSM-5 due to its higher acidity from NH_3 TPD results(not shown). Fig. 3 shows that the peaks of bio-oil with catalyst are different with that of bio-oil without catalyst. The peak of 1718 cm^{-1} due to C=O stretching vibration means the presence of carboxylic acid and its derivatives or ketone and aldehyde. Also peak of 1640 cm^{-1} means the presence of aromatics[9]. Without catalyst, the aromatic peak was not obtained. It is inappropriate for bio-oil to use fuel because carboxylic acid, ketone and aldehyde are too reactive component. So, it is desirable to convert them into stable aromatic compounds. When bio-oil was upgraded with HZSM-5 or Ga/HZSM-5, the peak intensity of aromatics increased. Table 1 also shows the aromatic distribution in bio-oil. No aromatics were observed without catalyst. Product yields over Ga/HZSM-5 shows higher amount of aromatic components such as benzene, toluene, xylene (BTX) than HZSM-5. Ga seems to play a role in converting intermediate oils to aromatics.

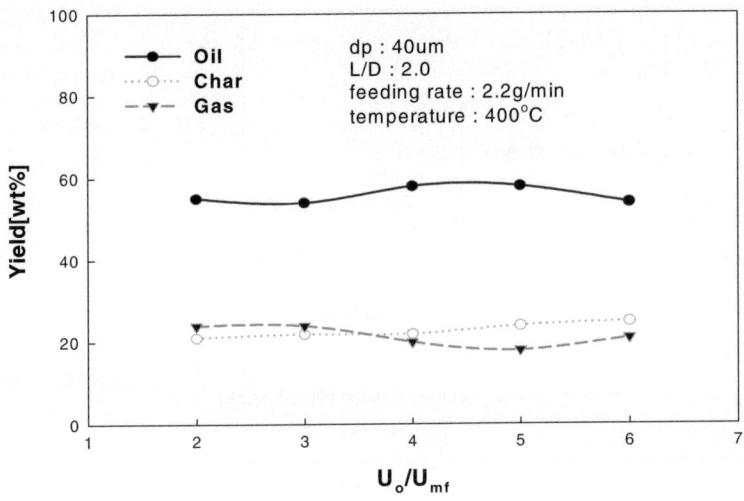

Fig. 2. Effect of U_o/U_{mf} on the products yield

Table 1. Yield of products through catalytic pyrolysis (wt%)

Products	Without catalyst	HZSM-5		Ga/HZSM-5	
Oil	58.0	43.7		51.3	
Aromatics in oil	-		12.9		23.4
BTX in oil	-		5.3		12.0
Gas	19.6	35.2		21.9	
Char	22.4	21.1		20.8	

Table 2. Water content and higher heating value of bio-oil

	Without catalyst	HZSM-5	Ga/HZSM-5
Water content (wt%)	29.1	67.6	60.7
Higher Heating Value (MJ/kg)	15.4	22.5	23.0

Table 3. Product distribution of gas (mol %)

	CO	CO_2	C_1-C_4
Without Catalyst	66.4	14.5	19.1
HZSM-5	53.9	27.2	18.9
Ga/HZSM-5	55.2	29.4	15.4

556

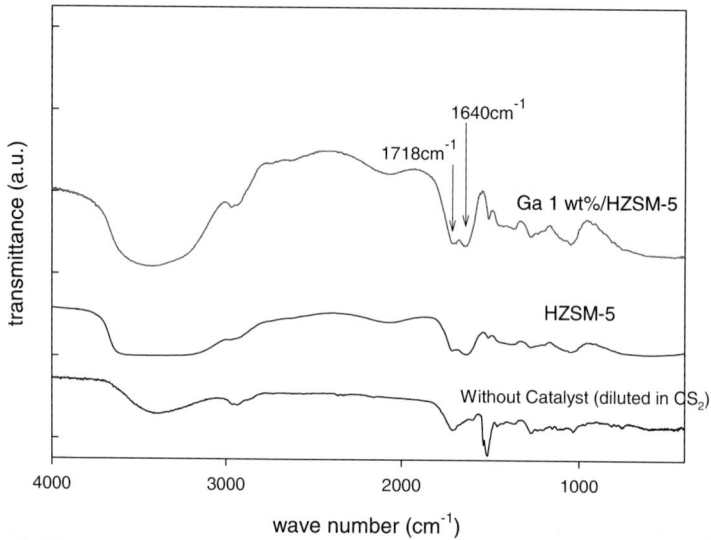

Fig. 3. FT-IR spectra of bio-oil

Conclusion

The maximum yield of oil products was found to be about 60% at the U_o/U_{mf} of 4.0. The yield of gas was increased as catalyst added. HZSM-5 shows the larger gas yield than Ga/HZSM-5. When bio-oil was upgraded with HZSM-5 or Ga/HZSM-5, the amount of aromatics in product increased. Product yields over Ga/HZSM-5 shows higher amount of aromatic components such as benzene, toluene, xylene (BTX) than HZSM-5. Ga seems to play a role in converting intermediate oils to aromatics.

Acknowledgement

This work was supported by the grant number (R01-2002-000-00374-0) from the basic research program of the Korea Science and Engineering Foundation.

References

[1] AV. Bridgwater, D. Meier, D. Radlein, *Org. Geochem.*, **30**, 1479 (1999).
[2] P. Mckendry, *Bioresour Technol.* **83**, 47 (2002).
[3] S. Arvelakis, E.G. Koukios, Biomass Bioenergy, **22**, 331 (2002).
[4] J.M. Encinar, J.F. Gonzalez, J. Gonzalez, Fuel Process Technol., **68**, 209 (2000).
[5] A.P. Horne, P.T. Williams, Fuel, **75**, 1051 (1996).
[6] E. Biagini, F. Lippi, L. Petarca, L. Tognotti, *Fuel*, **81**, 1041 (2002).
[7] Z. Luo, S. Wang, K. Cen, Renewable Energy, **30**, 377 (2005).
[8] U. Arena and M.L. Mastellone, Powder Technology 120, 130 (2001)
[9] S. Vitolo, B. Bresci, M. Seggiani, M.G. Gallo, Fuel, **80**, 17 (2001).

Studies in Surface Science and Catalysis, volume 159
Hyun-Ku Rhee, In-Sik Nam and Jong Moon Park (Editors)

557

Effect of Uniformity of Gas Distribution on Fluidization Characteristics in Conical Gas Fluidized Beds

Seong Yong Son, Dong Hyun Lee[+] and Sang Done Kim[*]

Department of Chemical Engineering, Sungkyunkwan University,
300 Chunchun, Jangan, Suwon 440-746, Korea (e-mail: dhlee@skku.edu)
[*]Department of Chemical and Biomolecular Engineering and Energy & Environment
Research Center, Korea Advanced Institute of Science and Technology,
Daejeon 305-701, Korea

1. Introduction

Fluidized bed reactors are widely used in food and chemical industries. Fluidized beds such as gas-solid, liquid-solid and gas-liquid-solid three phase reactors are more effective for increasing mixing and heat and mass transfers than the other reactor types [1]. One of the problems in stable operation of the fluidized beds is particle segregation that leads poor fluidization by accumulation of relatively large or high density particles on the distributor plate during operation of the beds [2]. Even though conical fluidized beds are widely used but its hydrodynamic properties are not fully understood. The reported studies on conical fluidized beds are mainly focused on the fluidization characteristic of coarse and fine particles, formulating correlation of gas dispersion coefficient and pressure drop characteristics in conical fluidized beds [2-4]. Gas distributors are the most important component of a fluidized bed reactor to provide uniform gas distribution to particles that provides stable fluidization. If there is a slight change of gas distribution in the grid zone, it immediately affects the fluidization quality above the grid zone. When pressure drop of a gas distributor is too small, partial defluidization may occur. However, if the pressure drop is too high, power consumption would rise. Therefore, in this study, the fluidization characteristics of the bed pressure drop ($-\triangle P_{bed}$), minimum bubbling velocity (U_{mb}), bubble frequency (N_B), the maximum velocity of full defluidization (U_{mfd}), the minimum velocity of full fluidization (U_{mff}) and the minimum velocity of partial fluidization (U_{mpf}) with the uniform gas distributors having different opening fractions were determined. The difference of fluidization characteristics with different opening fraction is analyzed and compared with the data of Son et al. [3] with the nonuniform gas distributors.

2. Experimental

Experiments were carried out in a conical shape gas fluidized bed (0.1 m-i.d. × 0.6 m-high) that made of a transparent acryl column with an apex angle of 20°. The details of the conical fluidized beds can be found elsewhere [3]. Air velocity ($U_g = 0 - 1.4$ m/s) were measured by a flowmeter. The particle used in this study was 1.0 mm glass beads with a density of 2,500

kg/m^3. The static bed height in the bed was 0.2 m. As shown in Table 1, the opening fraction of the gas distributors was ranged from 0.92 to 3.87 % with the same number of holes on the distributor. Four uniform distributors were designed to have similar opening fractions as those used in previous experiments with the nonuniform distributors as shown in Fig. 1. The circles in the center of Figs. 1-(a) and (b) indicate the nozzles, which were used to conduct granulation tests in the later experiments. A differential pressure transducer was installed in the conical fluidized bed to measure pressure in the bed. Three optical probes were installed axially at 0.05, 0.10 and 0.15 m above the distributor to measure bubble properties. The optical probe where a gap between the two tips was 5 mm and installed at 0.1 m above the distributor to measure the bubble size and bubble rise velocity.

Table 1. Opening fraction and hole dimension and of the uniform distributor

Distributor Type	Hole dimension	Opening %	Remarks
No. 1	Ø 2.0mm × 23EA	0.92	Son et al. [3]
No. 2	Ø 2.7mm × 23EA	1.68	
No. 3	Ø 3.5mm × 23EA	2.82	
No. 4	Ø 4.1mm × 23EA	3.87	

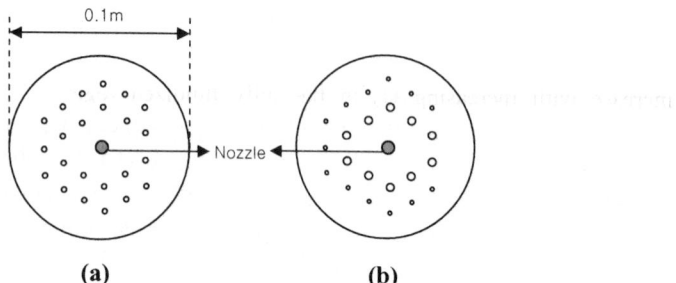

(a)   (b)

Fig. 1. Schematic diagrams of (a) uniform distributor and (b) nonuniform distributor.

3. Results and Discussion

The effect of gas velocity (U_g) on the pressure drops of the distributor ($-\triangle P_{dist}$) with the uniform and the nonuniform distributors by Son et al. [3] is shown in Fig. 2. As can be seen, $-\triangle P_{dist}$ increases with increasing U_g and decreasing opening fraction of the distributor [5].

Bubble passage with time along the bed height in the conical beds having a uniform distributor (opening fraction, F_{open} = 0.92 %) with increasing and decreasing U_g is shown in Fig. 3. As can be seen, the initial minimum bubble velocity is higher with increasing than decreasing U_g as reported in the conical beds by Toyohara and Kawamura [2].

A comparison of the minimum bubble velocities (U_{mb}) with decreasing and increasing U_g with the uniform and nonuniform distributors is shown in Fig. 4. As can be seen, U_{mb} are identical with increasing U_g through the bed with the uniform and nonuniform distributors, but it is somewhat different with decreasing U_g. Initially, the bed is charged with particles at dense state in a fixed bed. When this bed is fluidized with increasing U_g, aggregation force is needed to loose the interlocked particles from the dense state. Therefore, the effect of the opening fraction on U_{mb} is more pronounced with decreasing gas velocity than increasing gas velocity with the different uniformity. As can be seen, the minimum bubble velocity increases initially and then reaches a maximum value with increasing opening fraction of the uniform and nonuniform distributors. However, with decreasing U_g, the maximum bubble disappearing velocity is slightly different with the uniform and nonuniform distributors having the similar opening fractions.

The effect of gas velocity on the bed pressure drop ($-\triangle P_{bed}$) with a uniform distributor (F_{open} = 1.68 %) in the beds with decreasing and increasing U_g is shown in Fig. 5. As can be seen, - $\triangle P_{bed}$ maintains almost a constant value until the minimum velocity of full fluidization (U_{mff}) and then it decreases with decreasing U_g. As shown, U_{mfd} is the maximum velocity of full defluidization, U_{mpf} is the minimum velocity of partial fluidization, and U_{mff} is the minimum velocity of full fluidization [6].

Variation of the individual velocities (U_{mfd}, U_{mpf}, U_{mff} and U_{mb}) with opening fraction of the distributor in the conical fluidized beds with the uniform distributor is shown in Fig. 6. With increasing U_g, U_{mb} determined by the optical probe method is well agrees with U_{mff}. However, with decreasing U_g, there is some discrepancy between U_{mb} and U_{mfd} since the optical probe in the lowest part is located 0.05 m above the distributor.

Variation of bubble size, bubble frequency, and the standard deviation of $-\triangle P_{bed}$ with variation of U_g in the conical fluidized beds with a uniform gas distributor (F_{open} = 3.87 %) is shown in Fig. 7. As can be seen, the standard deviation of - $\triangle P_{bed}$ and the bubble size increase with increasing U_g in the fully fluidized region. However, bubble frequency remains unchanged with variation of U_g that may imply the bubble size will increase as much as the volumetric gas flow increases. As shown, the bubble size dramatically increases with increasing U_g. Also, it is confirmed that the increase of standard deviation of $-\triangle P_{bed}$ is closely related to bubble size.

4. Conclusion

Minimum bubbling velocity (U_{mb}) in the bed of both the uniform and nonuniform distributors with increasing U_g is identical but it is somewhat different with decreasing U_g. With increasing U_g, U_{mb} determined by the optical probe method is well accord with U_{mff} but, there is some disagreement between U_{mb} and U_{mfd} with decreasing U_g. The standard deviation of - $\triangle P_{bed}$ and the bubble size increase with increasing U_g in the fully fluidized region. However, the bubble frequency does not change with gas velocity.

REFERENCES
[1] M. G. Suh, J. H. Suh and J. S. Kang, Korean J. Food Sci. Technol., **25**, 210-213 (1993).
[2] H. Toyohara and Y. Kawamura, Kakagu Kogaku Ronbunshu, **15**, 773-780 (1989).
[3] S. Y. Son, D. H. Lee and S. D. Kim, Korean J. Chem. Eng., **22**, 315-320 (2005).
[4] Y. Nishi, Kagaku Kogaku Ronbunshu, **5**, 202-204 (1979).
[5] S. H. Lee, D. H. Lee and S. D. Kim, Korean J. Chem. Eng., **18**, 387-391 (2001).
[6] Y. Peng and L.T. Fan, Chem. Eng. Sci., **52**, 2277-2290 (1997).

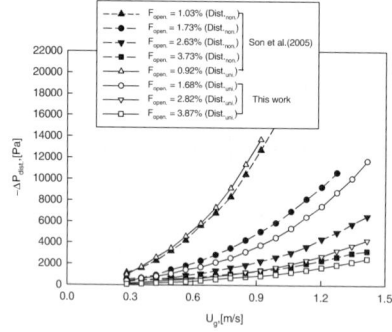

Fig. 2. Variation of -$\triangle P_{dist}$ with gas velocity.

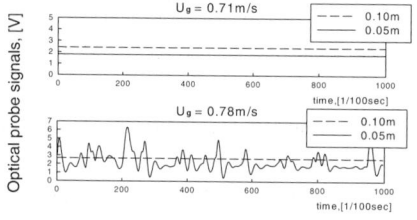

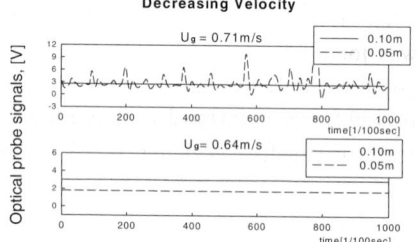

Fig.3. Optical probe signals of bubble passages.

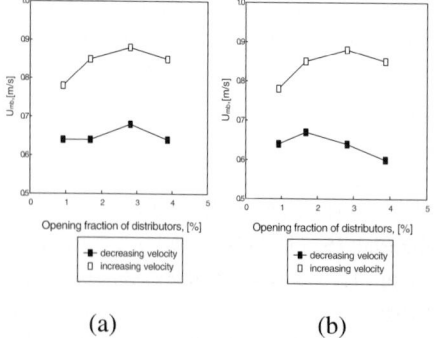

(a)   (b)

Fig.4. Comparison of the bubble disappearing and the initial bubble formation velocities with (a) uniform and (b) nonuniform distributors.

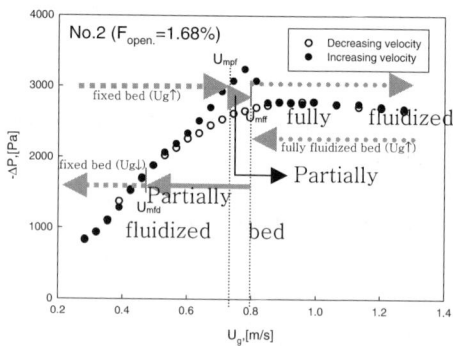

Fig. 5. Variation of bed pressure drop as a function of gas velocity.

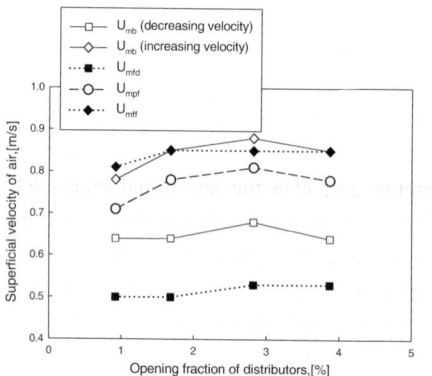

Fig. 6. Variation of the velocities (U_{mfd}, U_{mpf}, U_{mff} and U_{mb}) with the uniform distributor.

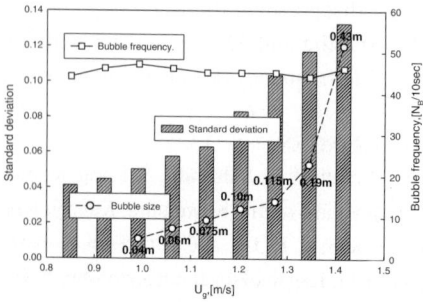

Fig. 7. Variation of bubble size, bubble frequency and the standard deviation of -$\triangle P_{bed}$.

Studies in Surface Science and Catalysis, volume 159
Hyun-Ku Rhee, In-Sik Nam and Jong Moon Park (Editors)
561

Application of fluidized reactor in photocatalytic decomposition of gaseous acetic acid and ammonia

Yeon Hee Son, Min Kyu Jeon, Ji Young Ban, Sung Chul Lee, Misook Kang*, Suk Jin Choung+

*Industrial Liaison Research Institute, Kyung Hee University

+Department of Chemical Engineering, School of Environmental Applied Chemistry, , Yongin, Gyeonggi, 449-701, South Korea

Abstract: To enhance the performance for the decomposition of acetic acid and ammonia, a fluidized photo-catalytic system was designed and prepared for this study. TiO_2 and $Al\text{-}TiO_2$ photocatalysts were used which was prepared by sol-gel method, and these were characterized by XRD, SEM, XPS, and TPD analyses to elucidate the kinetics of acetic acid and ammonia decomposition. The catalytic activities for acetic acid and ammonia decomposition were enhanced in a fluidized photocatalytic reactor compared to that in a common steady liquid reactor, and also the additional enhanced activity was found when $Al\text{-}TiO_2$ was used as a photocatalyst when compared with that of conventional TiO_2 photocatalyst; the both of conversions to N_2 in ammonia decomposition and CO_2 in acetic acid decomposition reached above 90 % until 600 minutes at air bubbling of 1L/min condition with 0.5 g/L $Al\text{-}TiO_2$ catalysts in a fluidized photo-system. On the other hand, undesirable NO_2^- and NO_3^- were detected as by-products of ammonia photo-degradation at about 1-2 ppm, which was detected by FT-IR spectra and ion chromatogram. As a conclusion, the newly devised fluidized photo-catalytic reactor system was able to decompose the ammonia and acetic acid up to 90% conversion inspite of the draw-lack of little bit   production of undesirable NO_2- and NO_3-ions.

1. INTRODUCTION

Ammonia and acetic acid in waste water give rise serious pollution problems which bring about eutrophication of rivers, lakes, etc [1, 2]. These have been treated by the conventional method of biological techniques, adsorption, and thermal incineration. A band of researchers have suggested that the ammonia molecules could be transferred to N_2 using a photocatalytic redox mechanism as shown follows: $4NH_3 + 3O_2 \rightarrow 2N_2 + 6H_2O$. However, it has been reported that when metal-incorporated TiO_2 photocatalysts were used to ammonia decomposition, with the catalytic activity enhancement, considerable amounts of N_2O, NO_2, and NO_3 could be formed. Conventionally, the photocatalytic application has limitation to use in industrial application because the quantitative performances were very poor in spite of its

562

high conversion. In order to increase the capacity of photocatalytic decomposition, a three-phase fluidized photoreactor had been designed in this study. At these days, three-phase reactors, involving the contact of gases, liquid, and solid phases in the system, have been emphasized with an increasing importance in a wide range of industrial applications[3, 4]. Therefore, the main objective of present study had been set to enhance the photocatalytic activities of acetic acid and ammonia photocatalytic decompositions by using a three-phase fluidized reactor which can be operated at atmospheric condition. We have prepared the catalysts of TiO_2 and Aluminum inserted TiO_2 by using a sol-gel method and these are applied for the decomposition of the acetic acid and ammonia solution.

2. EXPERIMENTAL

The conventional sol-gel method was employed for the preparation of TiO_2 and Al (5.0- and 10.0-mol-%)-TiO_2 catalysts. And these catalysts were characterized by XRD[model PW 1830 from Philips], SEM[model JEOL-JSM35CF], XPS[ESCA 2000], and TPD analyses for the elucidation of surface phenomena.

The decompositions of acetic acid and ammonia were carried out using a continuous three-phase fluidized photoreactor designed in the laboratory. The reactor column has an outer diameter of 7.5 cm and inner diameter of 6.5 cm with a height of 118 cm and was made of pyrex materials. In the experiments for acetic acid and ammonia decompositions, TiO_2 and Al-TiO_2 powders of 0.5g/L were added into a pyrex cylinder reactor with 3 L volume of water. The flowed concentration of acetic acid and ammonia in the feed stream were fixed 300 ppm and 80 ppm, respectively, and the injected rate was 1~2 L/min. As a light source, the UV-lamp ($39 W/cm^2$, 90 cm length x 2.0 cm diameter, YongWha lamp, Korea) of 254 nm was used.

Analyses of the concentration of acetic acid and ammonia before and after the reactor were performed by using a gas chromatograph (GC17A) with FID/TCD detectors (HP-1 capillary column). The removal in percentage value was based on the disappearance of pollutants during the decomposition process. All experiments were performed at room temperature and atmosphere pressure. A Fourier Transform Infrared (FT-IR) spectrometer (Shimadzu, FTIR-8400) was also used to conduct the in-situ analysis of gaseous products during the ammonia decomposition up to 60 mins.

3. RESULTS AND DISCUSSION

It has been well known that the main structure of TiO_2 for photocatalytic reaction is the anatase type. However it could be changed easily to rutile type by heating. In this XRD analysis result, it was found out that the newly developed rutile structure could be found before 700 ℃ in the case of pure TiO_2. However, in the case of the Al-TiO_2 the anatase structure could maintain stable until 700 ℃. Therefore, the increase in the aluminum amount resulted in a corresponding increase in the thermal stability of anatase structure. In addition, the 10.0-mol-% Al-TiO_2 catalyst exhibited slightly small particle size compared with those of the other catalysts in SEM images. Quantitative XPS analysis for O1s was performed on the TiO_2 and 10.0 mol-% Al-TiO_2 particles as shown in **Fig. 1**. In particular, the secondary peak values which could be assigned as the bulk oxide (O^{2-}) and hydroxyl (OH) species on the TiO_2 surface, which related to the hydrophilic property of catalyst[4]. This XPS analysis results explains two facts; the first is that the oxygen species in Al-TiO_2 was changed to higher oxidation states compared to that in pure TiO_2 from the observation of peak shift. The second is that the Al-TiO_2 was more hydrophilic than pure conventional TiO_2 from the observation of larger O1s secondary peak.

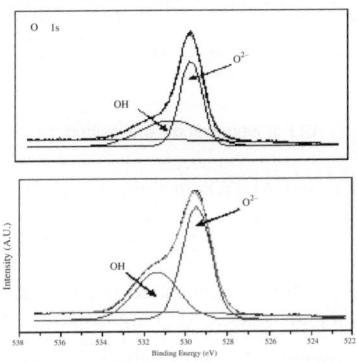

Fig.1. XPS spectroscopy for O1s
a) TiO_2 and b) 10 mol-% Al-TiO_2

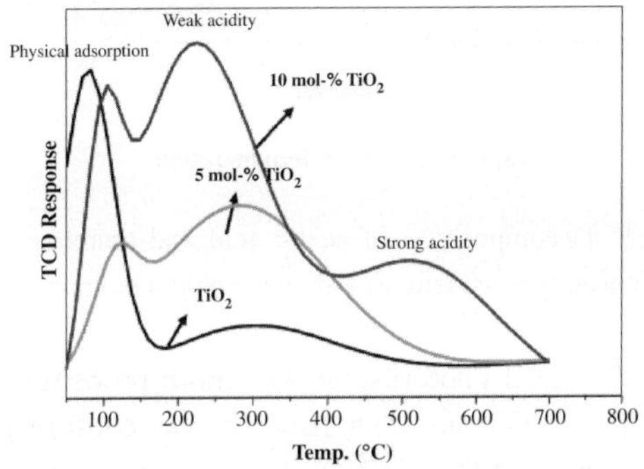

Fig.2. NH_3-TPD profiles of TiO_2 and Al-TiO_2

To confirm the effect of the addition of aluminum into the TiO_2 framework, the NH_3-TPD test was performed. **Fig. 2** shows the resulting profiles. Generally, these profiles consist of two peaks. The low and high temperature peaks correspond to the weak and strong acid sites, respectively. In the case of pure TiO_2, only one peak assigned to physical adsorption was found at around 100℃, on the other hand, three peaks at around 150, 250, and 550℃ could be found in the case of10.0-mol-% Al-TiO_2 sample. This indicates that the newly developed acid sites, more seriously at strong acid sites, on the TiO_2 photo- catalyst framework are generated by the addition of aluminum. In addition, we have compared the amount of acetic acid desorbed in photocatalysts, which was calculated based on the TG curve. As the result, with an increase of Al amount, the adsorbed acetic acid amount was significantly increased. About 14% increase of adsorption capacity could be found at the sample of 10-mol% Al-TiO_2 catalyst.

Fig. 3 gives the conversions for acetic acid and ammonia decomposition over TiO_2 and Al-TiO_2 in a three-phase fluidized photoreactor. In the case of acetic acid decomposition (Inlet condition of 300 ppm), the conversion increased with aluminum addition. In particular, the conversion to CO_2 reached about 90% and then it was kept until 600 mins on Al-TiO_2 catalyst. On the other hand, in b), the ammonia removal (Inlet condition of 80ppm) also enhanced on Al-TiO_2 compared to that conventional TiO_2 catalyst; the conversion to N_2 reached above 95% in Al-TiO_2. We have also observed that the ammonia conversion in a conventional batch type steady photoreactor could be obtained up to 70%. From this result, we could confirmed that

564

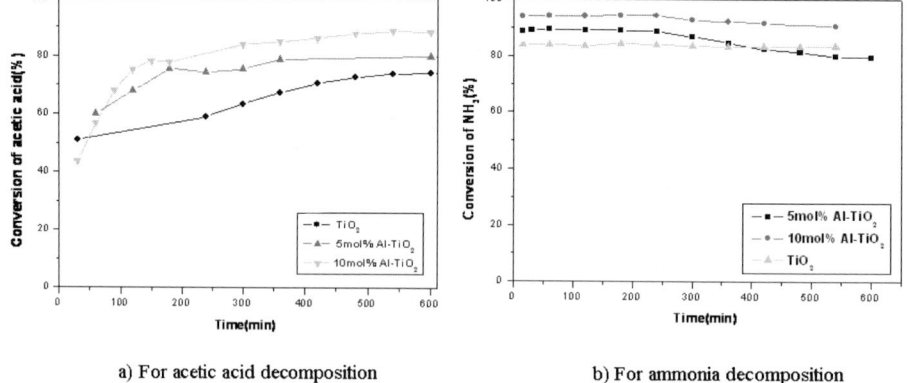

a) For acetic acid decomposition   b) For ammonia decomposition

Fig.3. Decomposition of acetic acid and ammonia over TiO_2 and Al-TiO_2 in three–phase fluidized photocatalytic system. a) For acetic acid decomposition and b) For ammonia decomposition

the fluidized photoreactor was more powerful to remove high concentrated acetic acid and ammonia. On the other hand, it was confirmed from FT-IR spectra that undesirable small amounts of NO_2 and NO_3 products were identified. However, the production of these by-products were much depressed when on Al-TiO_2 compared to that when on TiO_2. In addition, it was confirmed from ion chromatograph result that the total amounts of NO_2 and NO_3 were just about 1-2 ppm.

4. CONCLUSION

These results have confirmed that the VOC removal is more useful in a fluidized photoreactor than that in a steady photoreactor, and in particular when Al-TiO_2 photocatalyst is used.

ACKNOWLEDGEMENT
This work was supported by Korea Institute Environmental Science and Technology (2005-01003-0034-1). The authors are grateful for financial support.

REFERENCE
[1] P. I. Riggan, R. N. Lockwood, and E. N. Lopez, Environ. Sci. Technol., 19, 971 (1985).
[2] Y. Li and J. N. Armor, Appl. Catal. B., 13, 131 (1997).
[3] J. S. Dalton, P. A. Janes, and N. G. Jones, Environ. Pollution, 120, 415 (2002).
[4] B-Y Lee, S-H. Park, M. Kang, S-C. Lee, S-J. Choung, Appl. Catal. A, 253, 371 (2003)
[5] M. Chakchouk, G. Deiber, J. N. Foussard, and H. Debellefontaine, Environ. Technol., 16, 645 (1995).
[6] B-Y. Lee, S-H. Park, M. Kang, S-C. Lee, C-H. Park, S-J. Choung, Korean J. Chem. Eng., 20, 812 (2003)

Studies in Surface Science and Catalysis, volume 159
Hyun-Ku Rhee, In-Sik Nam and Jong Moon Park (Editors)

Gasification of tire scrap and sewage sludge in a circulating fluidized bed with a draft tube

B. H. Song[a] and S. D. Kim[b]

[a]Department of Chemical Engineering, Kunsan National University
Gunsan, Jeonbuk 573-701, Korea (e-mail:bhsong@kunsan.ac.kr)

[b]Department of Chemical and Biomolecular Engineering and Energy & Environment
Research Center, Korea Advanced Institute of Science and Technology,
Daejeon 305-701, Korea

1. ABSTRACT

Sewage sludge and waste tire scrap were co-gasified at 650 - 850°C in an internal circulating fluidized bed with a draft tube. The effect of bed temperature on the composition of two different product gases from draft tube zone and annulus reaction zone, and on the carbon conversion was investigated. Caloric value of the product gas from gasification of waste tire scrap is 15 MJ/m^3 through the annulus zone at 800°C. Caloric value of the product gas decreases below 5 MJ/m^3 when wet sludge is co-gasified. The entrainment of fine particles is large especially in the annulus zone for the waste tire gasification.

2. INTRODUCTION

Recently an internal circulating fluidized bed system with a draft tube has been studied to sustain two reaction zones in one fluidized bed [1, 2]. A feed gas to one reaction zone may bypass to the other zone through the passage of solids at lower section of bed. Ahn et al. [3] showed that the gas bypassing could be largely reduced by using orifices rather than the gap height as a solids passage. Whereas, gas bypassing from annulus to draft tube is relatively high so that a part of steam should go into draft tube and also some gasification occurs there. A physical gas separator, a baffle over the top of the draft tube can be adapted to reduce the mixing of each gas produced from the two zones. A simple baffle is enough to reduce the separation cost for product stream in the internal circulation system. Nowadays both sewage sludge and waste tire are produced largely in Korea. Fortunately they are normally not mixed with another wastes, it seems to be a promising way to gasify them to produce fuel gas containing hydrogen. To get a basic design data of a waste gasification process, the effect of reaction temperature on the product gas composition and carbon conversion for the gasification of sewage sludge and tire scrap have been determined in a circulating fluidized bed with a draft tube having a baffle gas separator.

3. EXPERIMENTAL

The fluidized bed (0.15 m-i.d. × 2.0 m-high) with a draft tube (0.05 m-i.d × 0.48 m-high) is shown in Fig. 1. Four orifices (0.01 m-i.d.) were evenly spaced along the circumference of the draft tube. The air box comprised two plenums to supply two independent gases to the draft tube and annulus. A baffle was installed right above the top of draft tube to separate two emanating gas streams from the two zones. An electric heater of 3 kW was installed at the wall of the main column to heat the reactor up to ignition temperature of the solid fuel. The mean diameter of the bed particles (silica sand) was 0.4 mm and that of tire scrap was in the range of 0.6 - 1.2 mm. The gas velocity to the draft tube was varied from at 5 to 15 times of minimum fluidization gas velocity (U_{mf}). The sewage sludge from the Iksan waste water treatment plant has moisture of 85% as shown in Table 1. The wet sludge was fed to the top of draft tube as a form of droplet with help of a cavity pump. The feed rate of wet sludge was 1.46 kg h^{-1} and that of tire scrap was controlled as 0.7 – 1.5 kg h^{-1}. Steam feed rate is up to 1.85 kg h^{-1}. A part of the product gas was continuously transported to gas chromatography for its composition analysis.

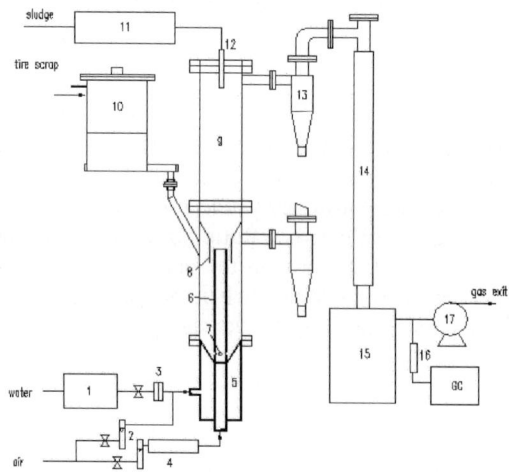

Fig 1. Schematic diagram of the ICFB gasifier. 1. steam generator, 2. flowmeter, 3. orifice flowmeter, 4. air preheater, 5. gas plenum, 6. draft tube, 7. orifice, 8. gas separator, 9. freeboard, 10. screw feeder, 11. mono-pump, 12. sludge feed nozzle, 13. cyclone, 14. condenser, 15. collector, 16. filter, 17. ID fan.

Table 1. Analyses of waste tire and sewage sludge (as received).

	Waste tire	Sewage sludge
Proximate analysis, wt%		
Volatile	61.7	6.4
Fixed carbon	33.1	2.3
Moisture	0.7	84.5
Ash	4.5	6.8
HHV, kcal/kg	7057	3115*

*: calculated by Dulong's formula

4. RESULTS AND DISCUSSION

Initially, gasification of waste tire scrap was carried out in the reactor. The obtained product gas has higher heat value of 15 MJ/m^3 from the annulus zone and 4 MJ/m^3 from the draft tube zone. The product gas from the annulus at 850°C contains up to 15% hydrogen (wet basis). However, the heat value of the product gas decreases below 5 MJ/ m^3 and content of hydrogen reduced below 5% when the sludge was introduced simultaneously with the waste tire scrap, as can be seen in Fig. 2. This may be due to the fact that the wet sludge contained so large amount of moisture that a portion of heat needed for gasification reaction is spent to vaporize the water. The calculated carbon conversion is found to be quite low below 35% especially for the co-feeding of tire and sludge. The carbon conversion over 42% is obtained in the case of sole feed of tire scrap. Such a low carbon conversion is mainly due to the large entrainment of unburned fine particles of waste tire scrap. The entrainment of fines in the annulus zone should be reduced in the present system. Caloric values (CV, higher heat value) of the product gas from gasification of the waste tire are compared with other studies of coal gasification in Fig. 3. A quite high CV gas can be obtained from the annulus region at 850°C. Whereas, CV of the product gas from the draft tube varies little around 3 MJ/m^3 that is similar to the value from the conventional fluidized bed coal gasification [4, 5]. Some gasification occurs even in the draft tube combustion zone because the portion of steam fed to the annulus bypasses to draft tube. Fig. 3 also shows the effect of configuration of fluidized bed gasifiers on the product gas. The orifice type provides higher CV gas compared to gap height type as a solid passage between the two zones. This is because the draft tube with orifices can reduce the gas bypass from draft tube to annulus more efficiently [3]. Jang [6] used the same system as the present study for coal gasification.

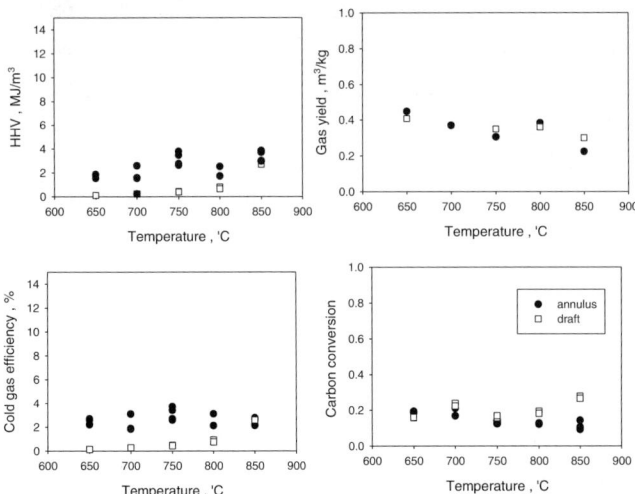

Fig. 2. Performance of the gasification of sludge/tire with variation of temperature. (tire feed = 0.77 - 1.1, sludge = 1.46 kg/h, H$_2$O/C = 1.7 - 4.2, O$_2$/C = 1.35 - 3.19)

568

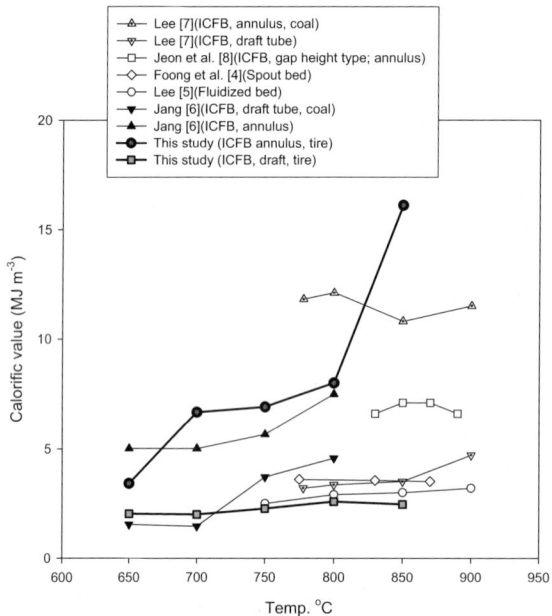

Fig. 3. Caloric value of the product gas from different fluidized bed gasification systems.

ACKNOWLEDGEMENTS

This research was supported by Energy R&D Management Center for the Ministry of Industry, Resources and Energy, Korea and by KOSEF.

REFERENCES

[1] R.K. Riley and M. R. Judd, Chem. Eng. Commun., 62 (1987) 151.

[2] B.H. Song, Y.T. Kim, and S.D. Kim, Chem. Eng. J., 68 (1997) 115.

[3] H.S. Ahn, W.J. Lee, S.D. Kim, and B.H. Song., Korean J. Chem. Eng., 16, (1999) 618.

[4] S.K. Foong, G. Cheng, and A.P. Watkinson, Can. J. Chem. Eng., 59 (1981) 625.

[5] W.J. Lee, Coal Gasification Characteristics in a Thermo-balance and Fluidized Bed Reactors, PhD Thesis, KAIST, Korea, 1995.

[6] Y.W. Jang, Steam Gasification of Bituminous Coal in a Bench-Scale Internally Circulating Fluidized Bed, MS Thesis, Kunsan National University, Korea, 2002.

[7] J.M. Lee, Non-catalytic and Catalytic Coal Gasification in an Internally Circulating Fluidized Bed Reactor, PhD Thesis, KAIST, Korea, 1998.

[8] S.K. Jeon, Coal Gasification Characteristics in a Circulating Fluidized Bed with Draft Tube, MS Thesis, KAIST, Korea, 1994.

Studies in Surface Science and Catalysis, volume 159
Hyun-Ku Rhee, In-Sik Nam and Jong Moon Park (Editors)

Steam gasification and combustion kinetics of gingko nut shell in a Thermobalance Reactor

Seon Ah Roh [†], Sung Real Son, Sang Done Kim[*]

Department of Chemical and Biomolecular Engineering and Energy & Environment Research Center, Korea Advanced Institute of Science and Technology, Daejeon, 305-701, Korea   (*e-mail:kimsd@kaist.ac.kr)
†Present address: Korea Institute of Machinery and Materials, 171 Jang-Dong, Yuseong-gu, Daejeon, Korea

1. Abstract

Pyrolysis kinetics of a gingko nut shell in a thermo gravimetric analyzer (TGA) and the steam gasification and combustion kinetics of chars from gingko nut shell in a thermobalance reactor have been determined. The effects of the reaction temperature (350 $^{\circ}$C – 950 $^{\circ}$C) and steam partial pressure (0.4 - 0.8 atm.) on the gasification kinetics and that of oxygen partial pressure (0.4 - 0.8 atm.) on the combustion kinetics have been determined in a thermobalance reactor. The activation energy and the pre-exponential factor were determined from the Arrhenius plot based on the shrinking core model. In the steam gasification reaction, the activation energy and the pre-exponential factor are found to be 40.8 kJ mol^{-1} at 1.7 s^{-1}atm.$^{-1}$, respectively. The reaction order is found to be 0.32 with respect to water partial pressure at 750 $^{\circ}$C. In the combustion reaction, the activation energies and pre-exponential factors are found to be 92.8 kJmol^{-1} at 136.8 s^{-1}atm^{-1} and 9.2 kJmol^{-1} at 0.012 s^{-1}atm^{-1} in the reaction control and the pore-diffusion control regimes, respectively.

2. Introduction

The gasification of biomass is one of the promising technologies for thermochemical conversion [1]. Bio-fuels are clear and important renewable alternative energy resources. A number of biomass resources such as agricultural waste, forestry waste and municipal solids and industrial wastes [2, 3] can be used as fuels. Gingko nuts are one of the important agricultural products in Korea. Thereby, gingko nut shells were used as a gasification material in this study. Gasification is a flexible process to produce fuel gases of variable compositions and energy contents depending on the gasification process, gasification agent and the operating conditions. In a gasifier, char conversion is usually the rate-limiting step and thus, the successful design and modeling of a gasifier require reliable kinetic data. Since the steam gasification reaction is endothermic, heat should be supplied to the gasification reaction. This can be done by partial combustion of biomass within a gasifier using a hypo-stoichiometric amount of air as a gasification agent. In this study, the pyrolysis kinetics of the gingko nut shell were determined in a TGA and the combustion and steam gasification kinetics of the gingko-nut shell chars were determined in a thermobalance reactor.

3. Experimental

The devolatilization characteristics of gingko-nut shell (mean size = 0.7 mm) were determined under N_2 atmosphere in a TGA (Setaram TGA 92). The N_2 flow rate was adjusted to 300 m l min^{-1} and then 25 mg of the gingko nut shell was heated to 900 °C, at a rate of 30 °C min^{-1}. The char particles were prepared by devolatizing the gingko nut shell that was heated from room temperature to 900 °C, at a rate of 20 °C min^{-1} and then maintained there for 30 min at 900 °C under N_2 atmosphere in a Lindberg furnace. To obtain the kinetic data, the prepared char samples were placed in a stainless steel wire mesh basket suspended from an electronic balance in a thermobalance (0.055 m-i.d. × 1.0 m-high) reactor with a 4 kW external heater. The reactor was heated to the desired temperature under N_2 flow. At the desired temperature, a mixture of air and N_2 (O_2 partial pressure of 0.06 - 0.25 atm) was admitted to the reactor at a flow rate of 2 l min^{-1} for the kinetics of combustion and a mixture of steam and N_2 (steam partial pressure of 0.4 - 0.8 atm) for the kinetics of steam gasification. During the reaction, weight variation of the sample was continuously monitored by a personal computer. Details of the thermobalance reactor and the experimental procedure can be found elsewhere [4].

The analyses of the gingko nut shell used in this study are shown in Table 1.

Table 1. Analyses of the gingko nut shell.

Proximate analysis	Wt%	Ultimate analysis	**wt%**
Volatile matter	**79.3**	C	**48.6**
Fixed carbon	**20.2**	H	**7**
Ash	**0.5**	O	**44**
Moisture	**6.8**	N	**0.37**
Heating value (cal/g)	**4700**	S	**0.03**

4. Results and discussion

4.1. TGA analysis

The thermo-gravimetric (TG) and differential thermo-gravimetric (DTG) curves of the gingko nut shell are shown in Fig. 2 where the moisture losses take place up to 200°C followed by the pyrolysis reaction. Then, the major weight loss due to the main degradation occurs at around 360°C. This zone is referred to as the active pyrolysis zone where the evolution of volatile compounds occurs during decomposition of the primary hemi-cellulose and cellulose [5].

4.2 Gasifiacation and combustion kinetics

The shrinking core and the volume-reaction models have been examined to interpret the conversion-time data of combustion and steam gasification of the gingko nut shell char [4]. The shrinking core model provides the better agreement with the experimental data. With the shrinking core model, the relationship between $[1-(1-X)^{1/3}]$ and the reaction time t at 350°C -

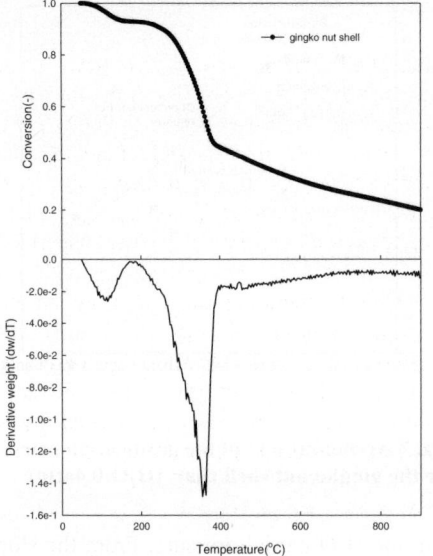

Fig. 2 Typical TG and DTA curves of the gingko nut shell.

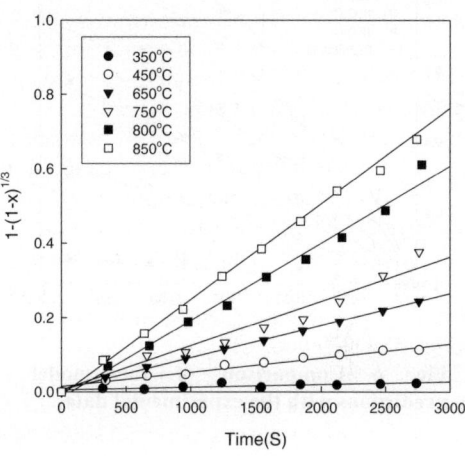

Fig. 3 Plot of the conversion fraction of the gingko nut shell char vs. the reaction time (s) in the gasification reaction. (H$_2$O:0.4 atm).

850°C for the steam gasification is shown in Fig. 3 where the shrinking core model predicts the experimental data very well.

From the Arrhenius plot of logarithm of the reactivity (*ln* k) *vs*. 1/T at a constant H$_2$O partial pressure, the activation energy and the pre-exponential factor are found to be 40.8 kJ mol^{-1} and 1.7 s^{-1}atm^{-1}, respectively (Fig. 4).

The effect of the H$_2$O partial pressure on the reactivity is shown in Fig. 5. As can be seen,

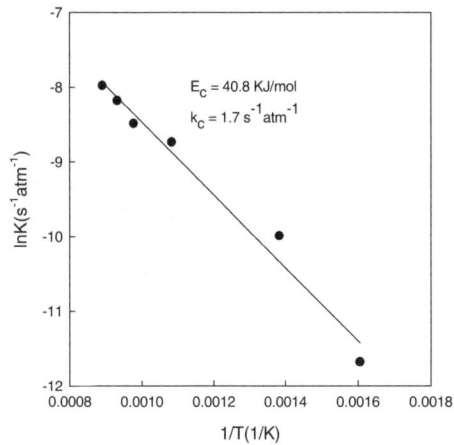

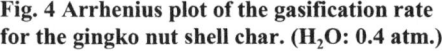

Fig. 4 Arrhenius plot of the gasification rate for the gingko nut shell char. (H$_2$O: 0.4 atm.)

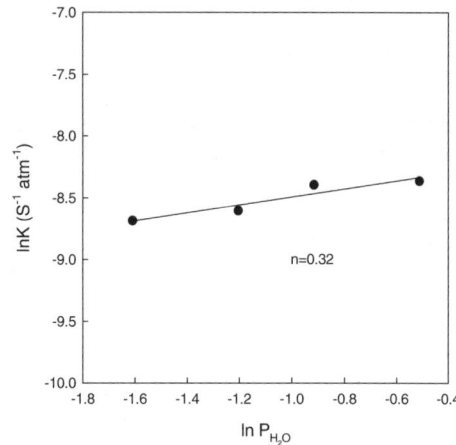

Fig. 5 Effect of the steam partial pressure on the reactivity at T=750 °C.

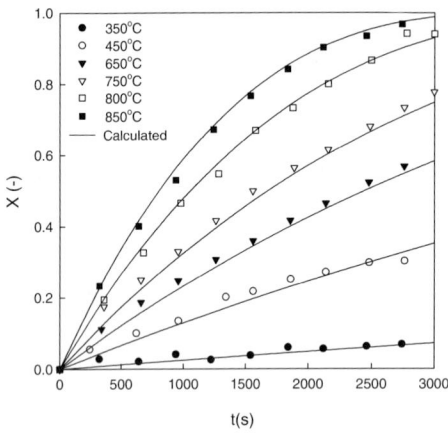

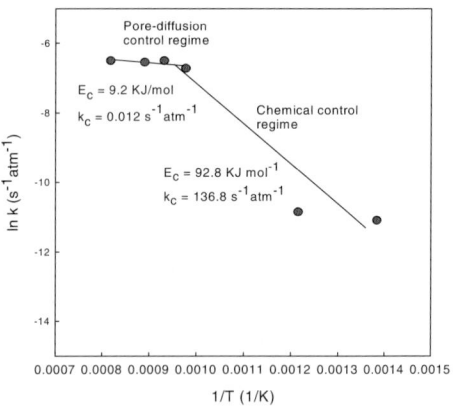

Fig. 6 Comparison of the model predictions with the experimental data.

Fig. 7 Arrhenius plot of the gasification rate for the gingko nut shell char. (H_2O:0.4atm)

the average reaction rate constant is proportional to the H_2O partial pressure. From the slop of the line (*ln* k vs *ln* H_2O partial pressure), the reaction order, n, is found to be 0.32.

Therefore, the steam gasification reaction rate of the gingko nut shell-char can be represented by the following kinetic equation as:

$$\frac{dX}{dt} = 1.7 P_{H_2O}^{0.32} \exp(-\frac{40.8}{RT})(1-X)^{\frac{2}{3}} \tag{1}$$

In Fig. 6, the calculated values from the kinetic equation are compared with the experimental data. As can be seen, the kinetic model predicts the experimental data very well.

In the same manner, the activation energy and the pre-exponential factor of the combustion are determined from an Arrhenius plot. As can be seen the kinetic equation of the combustion can be expressed as:

$$\frac{dX}{dt} = k_0 P_{O_2} \exp(-\frac{E}{RT})(1-X)^{\frac{2}{3}} \tag{2}$$

where E = 92.8 kJ mol^{-1} and k = 136.8 s^{-1}atm^{-1} in the chemical reaction control regime (450 – 700°C) and E = 9.2 kJ mol^{-1} and k = 0.012 s^{-1}atm^{-1} in the pore-diffusion control regime (700 – 850°C), respectively. The activation energy of the gingko nut shell combustion is lower than that of coal-char (128.93 kJ/mol) and waste tire-char (134.79 kJ/mol) since gingko nut shell has higher H/C ratio [6,7,8].

REFERENCES

[1] A.V. Bridgwater, Fuel, 74 (1995), 631
[2] D.P.C, Fung, S.D. Kim, Koean J. Chem.Eng, 7 (1990), 109.
[3] S.A. Roh., S.R. Son, S.D. Kim, Key Engineering Materials, 278 (2005), 637
[4] J.S. Lee, S.D. Kim, Energy, 21 (1996), 343.
[5] K. Raveendran, A. Ganesh, K.C. Khilar, Fuel, 75 (1996), 987.
[6] J.M. Lee, Y.J. Kim, W.J. Lee, S.D. Kim, Energy-The Int. J. 23, (1998), 475.
[7] J.R. Kim , S.D. Kim, Energy-The Int. J. 19 (1994), 845.
[8] C.D. Blasi, F. Buonanno, C. Branca, Carbon, 37 (1999), 1227

Studies in Surface Science and Catalysis, volume 159
Hyun-Ku Rhee, In-Sik Nam and Jong Moon Park (Editors)
© 2006 Elsevier B.V. All rights reserved

Release behavior of trace metals from coal during high temperature processing

Y.Sekine, K.Sakajiri, R.Inoue, E.Kikuchi, M.Matsukata

Department of Applied Chemistry, Waseda University, 55S-602, 3-4-1, Okubo, Shinjuku, Tokyo 169-8555, JAPAN(e-mail:ysekine@waseda.jp)

1. Introduction

Coal is regarded as an important energy source in the future, however, the pollutant emission from coal utilization is worried to damage for environment or human health. Recently, the emission trend of trace metals from coal utilization is an important issue[1]. During high temperature coal utilization process, trace metals are distributed to bottom ash, fly ash and flue gas[1]-[5]. Therefore, it is necessary to clarify the behavior of trace metals in coal during high temperature processing.

Based on thermodynamic property of each element, the behavior of trace metals in the coal combustion process was generally classified into 3 groups as follows[6]-[7]:

Group 1 elements which are emitted as gas even in low temperature flue gas (323 K); Hg

Group 2 elements which are partly vaporized in the boiler but which may condense on particles at electrostatic precipitator (410 K); Zn, Se, Sb

Group 3 elements which are not mostly vaporized in the boiler (1423 K); V, Cr, Mn, Co, Ni

Referring to the classification, we investigated the temperature dependency of release of trace metals in coal combustion. We already reported the behavior of these three types of elements during high temperature coal processing and reported elsewhere[8]. So in this paper, we investigated the effect of atmosphere for the emission behavior of trace elements.

2. Experimental

In this study, SS013 coal[8] whose particle size was <150 μm, provided from CCUJ (Center for Coal Utilization, Japan), was used. A mullite tube (42 mm i.d. and 1 m length) was used for the reactor. About 0.1 g of coal sample was set in a platinum basket, and inserted into the reactor heated at 573-1573 K beforehand. Combustion atmosphere was air; 5.6 L min^{-1} and 0 or 10 % of steam was supplied together. Pyrolysis atmosphere was N$_2$; 5.6 L min^{-1}. Steam gasification atmosphere was N$_2$; 5.6 L min^{-1} and 30 % of steam was supplied together. Hold time was 0-20 min. The residue remaining in the platinum basket after defined period for high temperature processing was collected carefully. As coal property is not uniform, above operation was repeated ten times, and 1 g of coal sample was treated as a total. The collected residue was dissolved into acid media using a microwave sample pretreatment system (Ethos Plus, Milestone General Co.; solvent HNO$_3$ 7 ml and HF 1.5 ml in the first step, and HNO$_3$

2ml, HF 0.5 ml and $HClO_4$ 3 ml in the second step). The contents of trace elements (Zn, Se, Sb, Hg) in the solution was measured by ICP-AES (CIROS-120, Rigaku Co.). The content of Se, Sb and Hg was measured with hydride generation method. And then the residual fraction of trace metals was defined as follows.

$$\text{Residual fraction }[\%] = \frac{\text{Content in ash }[\mu g\,(g-coal)^{-1}]}{\text{Content in coal }[\mu g\,(g-coal)^{-1}]}$$

3. Results and discussion

3.1. Temperature dependency of release of trace metals in coal combustion

Residual fractions of Zn, Sb, Se, Hg during SS013 coal combustion are shown in Fig.1. The residual fraction of Zn, Se, Sb and Hg was changed with the rise of temperature. The residual fraction of Zn decreased over 1273 K drastically. And the residual fraction of Se, Sb and Hg was decreased in temperature range of 373-773 K. These results were mostly coincided with the speculated volatility from thermodynamic property of each element mentioned above.

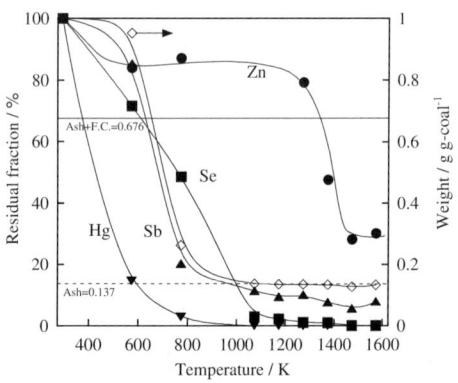

Fig. 1 Residual fraction of group 1 and 2 metals during combustion at various temperatures.

3.2. Effect of atmosphere

In our early study[8], it was found that trace metals in coal were only released in the initial stage (< 2min) of combustion. Trace metals were changed those chemical forms by some factors in initial stage of combustion. Therefore, we discussed about the effect of O_2 and steam in operating atmosphere on the release behavior of trace metals. The residual fraction of Hg, Se, Sb and Zn at various temperatures, when SS013 coal was treated under various conditions (Combustion with/without steam, Pyrolysis and Steam gasification) for 10 min, was shown in Fig.2.

The residual fraction of Hg at 573 K after combustion without steam and pyrolysis was about 15% and 25% respectively, meanwhile the residual fraction after combustion with steam (10% of steam) and steam gasification (steam 30%) were about 60% and 80%. These results showed that Hg release was suppressed under the steam existence. Hg was released under steam absence as $HgCl_2$ (m.p. 550 K) or Metal-Hg (m.p. 234 K) whose volatility was relatively higher than other compounds of Hg. Under steam existence, however, it was considered that Hg was released from coal via the reaction scheme as formula (1). At first Hg in coal reacted with steam, and $Hg(OH)_2$ generated immediately. Next the $Hg(OH)_2$ changed into HgO easily by heating, and HgO was decomposed to Hg $\uparrow$ and O_2 at 773 K. So the decomposition temperature of HgO was higher than the melting point of $HgCl_2$ or Metal-Hg.

$$Hg(OH)_2 \rightarrow HgO \rightarrow Hg\uparrow, O_2\uparrow \text{ (773 K)} \tag{1}$$

The order of the residual fraction of Se was: pyrolysis>steam gasification>combustion. This result shows that the release of Se was promoted under combustion comparably than the case of pyrolysis and gasification. It was considered that Se in coal was oxidized to SeO_2 which sublimed at 588 K. Under combustion (steam 10%), Se in coal was oxidized to SeO_2 once, and the SeO_2 which is deliquescent, reacted with steam, and H_2SeO_3 which had high volatility (m.p. 333 K) was generated [Formura (2)]. Additionally the residual fraction under steam gasification was almost same as the case of pyrolysis, because Se was not oxidized under both conditions.

$$SeO_2 + H_2O \rightarrow H_2SeO_3 \tag{2}$$

The residual fraction of Sb at 573 K under pyrolysis and steam gasification was both almost 0%, meanwhile that of Sb under combustion was about 60-80%. From this result, it was found that the release of Sb was suppressed against the case of Se. This was because that a part of Sb in coal, which would be released under inert atmosphere, was oxidized to Sb_2O_3 (m.p. 929 K) under combustion. Here, nevertheless the melting point of Sb_2O_3 was 929 K, the residual fraction at 773 K was decreased under combustion. It was considered that the releasing rate of Sb was faster than the oxidizing rate above 773 K. Next, under combustion the residual fraction under steam existence was larger than the case of steam absence. We assumed that Sb was released under steam existence via such as hydroxide or oxo-acids, because these compounds generally had high volatility.

The residual fraction of Zn after combustion without steam at 1273 K was about 80 %, meanwhile the residual fraction after pyrolysis was about 30 %. Also at 1423 and 1573 K, the residual fraction of Zn after combustion was greater than in the case of pyrolysis. This was considered that under N_2 atmosphere, Zn in coal was released as the compound that should be volatized at each temperature, however under air atmosphere, a part of Zn were oxidized into ZnO (mp 2521 K) which was very stable at high temperature. When the combustion was carried out with additional steam, the residual fraction of Zn was same as the cases in the absence of steam. However, the residual fraction of Zn after steam gasification was about 60 %, greater than the residual fraction after pyrolysis (about 30 %) and smaller than the residual fraction after combustion (about 80 %). This was considered that under steam gasification, a part of Zn, which should be emitted at 1273 K, was oxidized into ZnO by additional steam.

4. Conclusion

We compared and discussed about the release behavior of trace metals from coal under the condition of coal combustion, pyrolysis and steam gasification. Under the condition of combustion, the residual fraction of Zn and Sb was grater than in case of pyrolysis and the residual fraction of Se was smaller. Because a part of trace metals in coal was oxidized during combustion. Moreover the release of Se and Sb was promoted when steam existed in atmosphere, and the release of Hg was suppressed.

As a result, it was found that the release of trace metals was affected by temperature, and atmosphere. The volatility of chemical form of trace metals in raw coal and the chemical change of trace metals during high temperature heat processing were very important for the prediction of the trend for the emission of trace metals from coal.

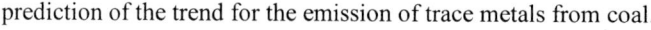

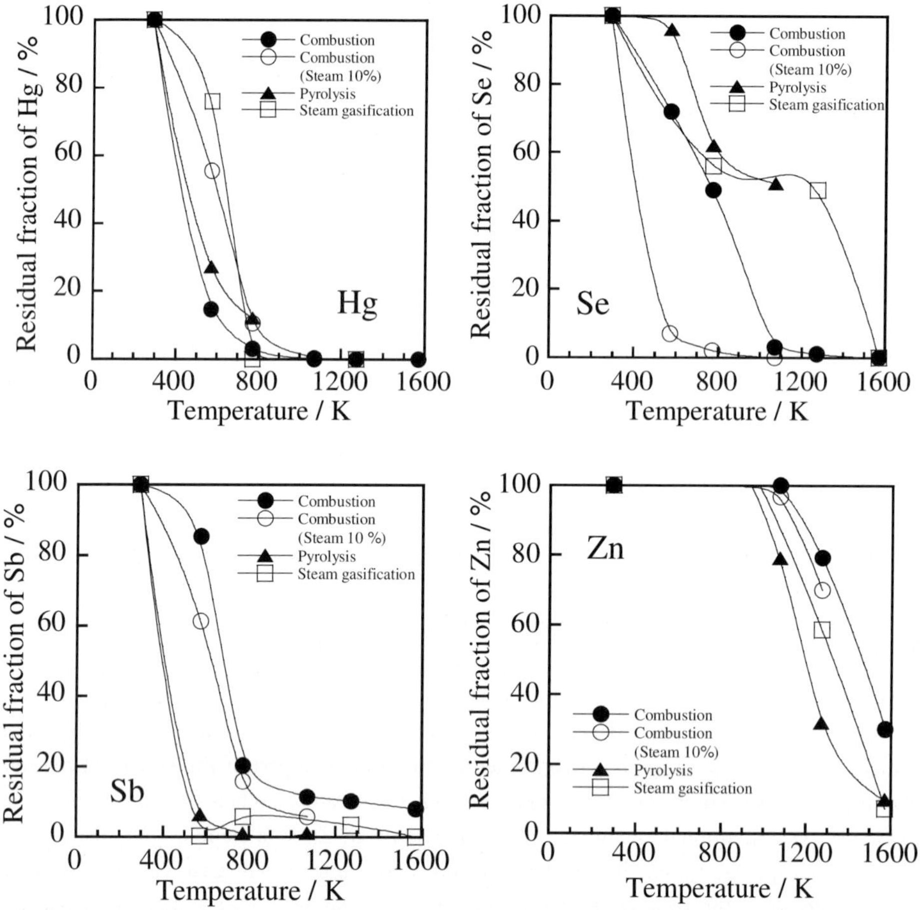

Fig. 2 Effect of atmosphere on residual fraction of Hg, Se, Sb, Zn.

References

[1] L.L.Sloss and I.M. Smith, *Trace element emission*, CCC/34, IEA Coal Research (2000)

[2] R.Yan, D.Gauthier and G.Flamant, *Fuel*, 80 (2001) 2217-2226

[3] J.J.Helble, W.Mojtahedi, J.Lyyranen, J.Jokiniemi and E.Kauppinen, *Fuel*, 75 (1996) 931-939

[4] C.L.Senior, J.J.Helble and A.F.Sarofim, *Fuel Processing Technology*, 65-66 (2000) 263-288

[5] K.C.Galbreath, D.L.Toman, C.J.Zygarlicke and J.H.Pavlish, *Energy & Fuels*, 14 (2000) 1265-1279

[6] L.B.Clarke and L. L.Sloss, *Trace Elements-emission from Coal Combustion and Gasification*, IEACR/49, IEA Coal Reserch (1992)

[7] T.Yokoyama, and K.Asakura, *Proc. Trace Element Workshop 2000*, Yokohama, Japan, Research group on trace element (2000) 69-80

[8] Y. Sekine, K.Sakajri, R.Inoue, E.Kikuchi and M.Matsukata, *Proc. Pittsburgh Conference* (2004)

Studies in Surface Science and Catalysis, volume 159
Hyun-Ku Rhee, In-Sik Nam and Jong Moon Park (Editors)
© 2006 Elsevier B.V. All rights reserved

Destruction of chlorinated organic solvents in a molten carbonate with transition metal oxides

H.-C. Yang[a], Y.-J. Cho[a], H.-C. Eun[b], J.-H. Kim[a], Y. Kang[c]

[a]Nuclear Fuel Cycle R&D Group, Korea Atomic Energy Research Institute
P.O. Box 105, Yuseong, Daejeon, 305-600, Korea (e-mail:nhcyang@kaeri.re.kr)

[b]Quantum Energy Chemical Engineering, University of Science and Technology

[c]Department of Chemical Engineering, Chungnam National University

ABSTRACT

This study investigated the destruction of chlorinated organic solvents in a molten carbonate reactor, in which powdered transition metal oxides are floating. The addition of powdered transition metal oxides could reduce the operating temperatures at a given oxidation efficiency. The collection of chlorines in the molten salt was not influenced by the tested operating temperatures of 750-950°C. Nearly all the acid from the destruction of the tested chlorinated solvent is collected in the tested molten salt oxidation system, indicating that no acid gas scrubbing system is necessary.

1. INTRODUCTION

Chlorinated organic solvents are widely used in the chemical industry. Wastes generated from their use include a significant quantity of chlorines. While an incineration is considered as an effective tool for the treatment of these chlorinated organic wastes, the environmental acceptability of the emissions of dioxins is a major criterion for the application of the incineration process [1]. There is therefore a developing need for an alternative oxidation process that can effectively destroy chlorinated organics without an emission of chlorinated organics. One of the promising alternatives is a molten salt oxidation. In a hot molten alkali carbonated salt, the hydrogen chloride or chlorine first released at the stage of a dehydrochlorination of the chlorinated organics are trapped in the form of an alkali chloride. Therefore the recombination of toxic chlorinated organics is prevented [2,3]. The purpose of this study is to establish the influence of a metal oxide addition as well as the operating temperature for the MSO reactor performance.

2. METHODS

The schematic diagram of a lab-scale MSO system with a designed capacity of 0.5 kg PVC plastics/h is shown in Fig. 1. The salt mixture consists of 50 mol% Li_2CO_3 and 50 mol% Na_2CO_3. The eutectic temperature of the used binary salt mixture is 505 °C. The selected

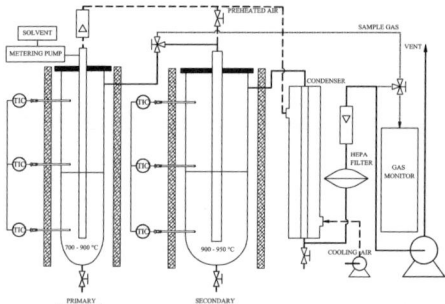

Fig. 1. Two-Stage Molten Salt Reactor System.

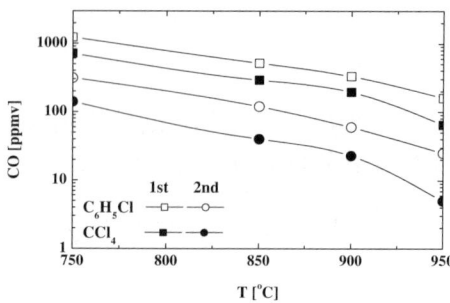

Fig. 2. The emissions of carbon monoxide from each MSO reactor

Table 1 Summary of primary MSO reactor test conditions

Test run No.	T1-1	T1-2	T1-3	T1-4	T2-1	T2-2	T2-3	T2-4
Temperature, °C	750	850	900	950	750	850	900	950
Solvent feed rate	0.15 kg C_6H_5Cl/h (1.2 vol.% in air)				1.2 kg CCl_4/h (5.4 vol.% in air)			

solvents tested in this study are carbon tetrachloride and chlorobenzene. The detailed description of the MSO system and the operating procedure are described in a previous work [4]. The feed rates of C_6H_5Cl and CCl_4 were fixed at 1.2 vol.% and 5.4 vol.% in air, respectively. Test conditions are summarized in Table 1. Four investigated molten salt temperatures of the primary reactor were 750, 850, 900 and 950°C. The temperature of the secondary MSO reactor is fixed at 900°C throughout all the test runs. During the test runs, the concentrations of O_2, CO_2, CO and NOx are measured using a combustion efficiency analyzer (TESTO-300). The speciation of the products of an incomplete combustion (PICs) is performed by the GC(HP5890)-MSD(HP5972) system.

3. RESULTS AND DISCUSSIONS

3.1. Organics Destruction

The overall chemical reaction of the chlorinated organics by Li_2CO_3-Na_2CO_3 eutectic salt is given by the following equation.

$$C_aH_bCl_c + \frac{c}{2}Na(or\ Li)_2CO_3 + (a + \frac{b-c}{4})O_2 \rightarrow (a + \frac{c}{2})CO_2 + \frac{b}{2}H_2O + Na(or\ Li)Cl \qquad (1)$$

As the products of an incomplete reaction, some hydrocarbons, carbon monoxide and hydrogen chlorides can be emitted from the MSO reactor. GC-MSD analysis of the off-gas samples indicated that the decomposition products of C_6H_5Cl include a trace amount of C_2H_2 and C_2H_4. But their emissions were in a range smaller than 100 ppb during the lowest temperature test for chlorobenzene (T1-1). During the other test runs, no gaseous organic species were detected in the off-gas sample. This indicated that the secondary MSO reactor had a good function for a destruction of a trace amount of hydrocarbons from the primary reactor. In addition, carbon-containing organic species were not found in the spent salt samples. During the course of each test run, the temperatures were not deviated by over 1°C from the set temperatures. This indicates that the MSO reactors have an excellent mixing

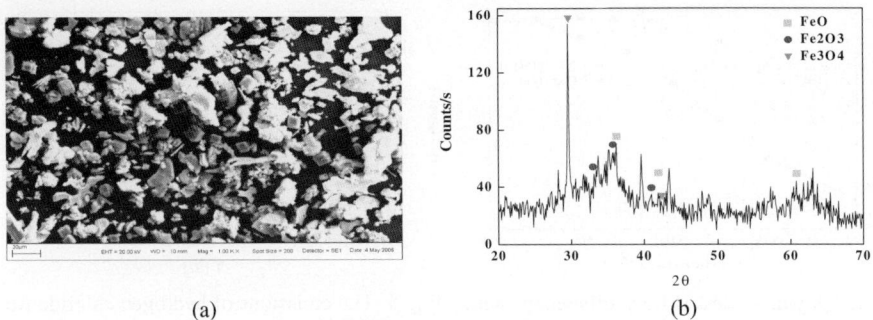

(a)   (b)

Fig. 3. SEM photograph (a) and XRD patterns (b) of iron oxide powders in the spent salt sample

capacity and the heat transfer is completely uniform within the molten salt bed. It can therefore be positively said that the incomplete reaction products of the above reaction (1) are continually mixed with the oxidizing air until their oxidation is completed.

3.2. Carbon Monoxide Emission

Carbon monoxide (CO) emissions from the primary and secondary MSO reactors are plotted in Fig. 2. For the C_6H_5Cl destruction, the CO emissions of the primary reactor were greatly influenced by the reactor temperature. The averaged CO emissions in the 750°C and 850°C test runs (T1-1 and T1-2) were 1219 ppm and 513 ppm, respectively. The averaged CO emissions from the primary reactor at the higher temperature test runs (T1-3 and T1-4) were about 330 ppm and 159 ppm, respectively. They showed a strong dependency of the CO emissions on the temperature. During these high temperature test conditions, a further oxidation in the secondary reactor was substantial and the final CO emissions were averaged to be about 60 ppm and 25 ppm, respectively. As shown in Fig. 2, the CO emissions from the primary reactor during the CCl_4 destruction were also greatly influenced by the temperature.

3.3. Effect of Metal Oxide Particles

As discussed, an effective oxidation in molten carbonate occurs at rather high temperatures over 900°C. But it is known that molten salts are very corrosive and vaporizable at these temperatures. Thus we attempted to find a catalytically active molten salt to decrease the reactor temperatures. We measured the emissions of CO and hydrocarbons from the primary MSO reactor by increasing the temperature from 900°C to 980°C. Two duplicates of same condition were performed with or without an addition of ferric oxides of 10 wt% of molten salt. The SEM photograph and the powdered XRD patterns of the ferric oxides in the spent salt samples, which are shown in Fig. 3, revealed that the ferric oxides (Fe_2O_3) were partially reduced to FeO and Fe_3O_4 by emitting free oxygen to enhance the oxidation of the organics or carbon monoxide and that they were agglomerated into about 20 μm. Based on the measured CO_2, CO and hydrocarbon emissions, the oxidation efficiency (OE) was determined as follows:

$$OE = \frac{M_{CO_2}}{M_{CO_2} + M_{CO} + M_{HCs}} \times 100\% \qquad (2)$$

Fig. 4 typically shows the effect of an addition of powdered Fe_2O_3 in reducing the reactor temperature with a given destruction efficiency. The temperature f a 99% oxidation of the total carbons (T_{99}) was achieved at a temperature around 950°C without Fe_2O_3. T_{99} was reduced to reduced to about 920°C by adding 10 wt% of the powdered Fe_2O_3.

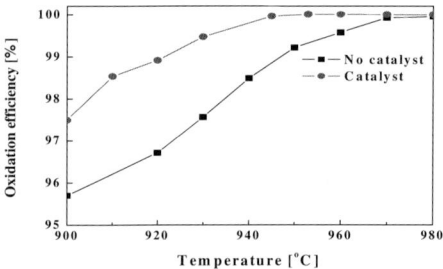

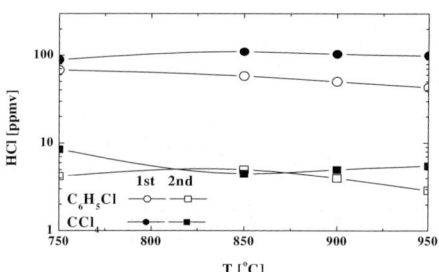

Fig. 4. Organics destruction efficiency with adding Fe₂O₃ powders

Fig. 5. The emissions of hydrogen chloride from each MSO reactor

Table 2 Chlorine collection efficiencies in the two stage MSO reactor system

Primary reactor temperature, °C	750	850	900	950
C_6H_5Cl	99.956	99.954	99.958	99.970
CCl_4	99.997	99.998	99.998	99.999

3.4. Chlorine Collection Efficiency

Regardless of the tested condition, hydrogen chloride was found to be the only chlorine-containing species in all the flue gas streams. The emissions of HCl from each MSO reactor are plotted in the Fig. 5 and the overall chlorine collection efficiencies are listed in Table 2. HCl emissions from the primary reactor were not varied very much with the reactor temperature. The averaged HCl emissions from the primary reactor were about 50 ppm during the C_6H_5Cl destruction and about 100 ppm for CCl_4. The final HCl emissions were maintained to be less than 10 ppm. The overall chlorine retention efficiencies were in the range between 99.954 and 99.999%. This indicates a significant advantage for the use of the molten salt oxidation reactor system. Acid gas scrubber system producing large amounts of waste water is not necessary since nearly all the acid from the destruction of highly chlorinated organics is reactively scavenged in the molten salt [5]

4. CONCLUSION

Destruction of the organics and the further oxidation of the destroyed products in a molten carbonate reactor are effective at the temperatures above 900°C. However, the addition of powdered transition metal oxides can reduce the operating temperatures in a given oxidation efficiency. The retention of chlorines in the molten salt was not influenced by the tested operating temperatures. Nearly all the acid from the destruction of tested chlorinated solvent is collected in the tested two-stage molten salt oxidation system, indicating that no acid gas scrubbing system is necessary.

REFERENCES

[1] D. Wang, X. Xu, M. Zheng and C.H. Chiu, *Chemosphere* 40 (2002) 857-863.
[2] B.H. Edwards, J.N. Paullina and C.J. Kathleen, *J. Hazard. Mater.* 12 (1985) 201-205.
[3] M. Alam and S. Kamath, *Environ. Sci. Technol.* 32 (1998) 3986-3992.
[4] H.C. Yang, Y.J. Cho, H.C. Eun, J.H. Yoo and J.H. Kim, *CJChE* 81 (2003) 713-718.
[5] H.K. Lee, B.R. Deshwal, and K.S. Yoo, *KJChE* 22 (2005) 208-213.

Studies in Surface Science and Catalysis, volume 159
Hyun-Ku Rhee, In-Sik Nam and Jong Moon Park (Editors)

Photodegradation of a volatile organic compound by fluidized bed reactor with TiO$_2$/SiO$_2$ and metal-TiO$_2$/SiO$_2$

Sang-Keun Lee, Jung-Sun Kim, Il-Kyu Kim and Jea-Keun Lee[*]

Department of Environmental Engineering, Pukyong National University, Busan, 608-737, KOREA
(e-mail:leejk@pknu.ac.kr)

ABSTRACT

The photodegradation of trichloroethylene (TCE) in a fluidized-bed reactor with a TiO$_2$/SiO$_2$ and a metal-TiO$_2$/SiO$_2$ photocatalyst is investigated. The best conditions of the superficial gas velocity (U$_g$) and the annulus gap for the TCE removal are found at 2.0 of U$_g$/U$_{mf}$ and 5 mm, respectively. TCE removal efficiency decreases by increasing the inlet TCE concentration and it increases by increasing the UV light intensity. A steady state operation with a 99% removal efficiency is maintained for 50 hours at a superficial gas velocity of 2.0 U$_{mf}$. Photodegradation of TCE with metal-TiO$_2$/SiO$_2$ is more efficient than that with TiO$_2$/SiO$_2$, except for Cu-TiO$_2$/SiO$_2$. The Pt-TiO$_2$/SiO$_2$ achieves the highest photodegradation efficiency in this study.

1. INTRODUCTION

Many chlorinated organic compounds such as trichloroethylene (TCE) have been widely used as industrial solvents for degreasing metals and a dry cleaning. A considerable amount of soil and groundwater has been contaminated by chlorinated organic compounds like TCE due to leaks from underground storage tanks and an improper disposal. TiO$_2$, as a photocatalyst, has been of considerable interest because it has various attractive properties such as a nontoxicity, high stabilization, low cost, and a high photoactivity at a low temperature for a chemical waste treatment.

In this study, we investigate the photocatalytic degradation of TCE in the gas-phase using a fluidized-bed reactor with TiO$_2$ photocatalyst. Effects of several important parameters such as the superficial gas velocity, annulus gap, UV light intensity, inlet concentration and different metal loadings of the TiO$_2$ photocatalyst were examined.

2. EXPERIMENTAL

2.1. Annulus fluidized-bed photoreactor

A small Pyrex grass tube (60 mm O.D. × 1000 mm-height) was located at the center of a larger Pyrex grass tube (70, 80, 90 mm O.D. × 1000 mm-height). A gas distributor was also used to provide a gas stream for an uniform fluidization of the TiO_2/SiO_2 photocatalysts. A black light fluorescent lamp (15W, Philips, F15TB.BLB) was installed inside the small pyrex tube for an effective ultraviolet light irradiation. The reactor used in this study consisted of a TCE feeding device, a photocatalytic fluidized bed reactor and a detector. TCE gas was prepared by mixing air and TCE through a grass saturator. Due to its high concentration, the concentration was adjusted to obtain 20 ± 5 ppmv by adding air in to the mixing chamber. The minimum fluidization velocity (U_{mf}) of the TiO_2/SiO_2 photocatalyst was determined by a pressure drop and the fluidized bed height in the reactor with a variation of the superficial gas velocity (U_g). The concentration of TCE was measured by a gas chromatography/ECD.

2.2. Preparation of photocatalysts

SiO_2 (particle diameter = 250 ~ 417 μm, surface area = 561 m^2/g), transparent to UV light, was used as a support to improve the fluidization quality of the TiO_2 since TiO_2 powder is classified into the Geldart C group having poor fluidization characteristics. A precursor solution for the TiO_2 coating onto the silica gel was prepared by using tetraisopropoxide, 2-propanol, HCl, and deionized water. After mixing this solution for 120 minutes at room temperature, silica gel was added to the aqueous colloidal suspensions of TiO_2 and then it was dried at 80 ℃ for 24 hours. It was calcined at 500 ℃ for 1 hour. Copper (II) nitrate, palladium (II) nitrate hydroate, and hydrogen hexachloroplatinate (IV) hexahydrate were used as precursors for Cu (II), Pd (II), and Pt (IV) for the TiO_2/SiO_2, respectively.

3. RESULTS AND DISCUSSION

3.1. Effects of the superficial gas velocity and the annulus gap

The effects of the superficial gas velocity (U_g) on the TCE removal are shown in Fig. 1. The TCE removal efficiency initially increases with the U_g/U_{mf}, reaches a maximum at U_g/U_{mf} = 2.0, and then decreases with a further increase of U_g/U_{mf}. This result shows a similar tendency to the a previously reported result of Yue et al. [1]. Therefore, it is found that the photocatalysis of the TCE decomposition requires a sufficient high residence time and a suitable TCE gas velocity to form a proper bubble size for the intimate contacts between the photocatalysts and the TCE molecules.

In addition, the effects of the annulus gap (5, 10, 15 mm) on the TCE removal efficiency with the TiO_2/SiO_2 photocatalyst are also shown in Fig. 1. As shown, the TCE removal efficiency exhibits a maximum value at a 5 mm annulus gap (U_g/U_{mf} = 2.0) and it decreases with a further increase in the annulus gap at the U_g/U_{mf} range of 2.5 to 4.0. Therefore, the annulus gap is an important parameter to

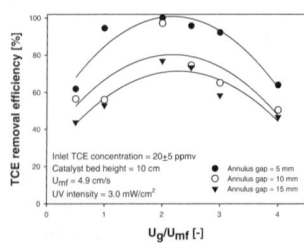

Fig. 1. Effects of annulus gap on TCE removal efficiency.

maintain a proper bubble size for a light transmission [2, 3].

3.2. Effect of UV light intensity

Fig. 2 shows the effects of the intensity on the reaction rate of TCE.   As shown, the TCE reaction rate increases with an increasing UV light intensity and it seems that more electron-hole pairs are produced by the UV light.   Thus, more photons can bring forth a greater degradation of the TCE. Obbe and Brown [4] reported that the dependency of the photoreaction rate on the ultraviolet light intensity followed a power law.   The reaction rate on the ultraviolet intensity follows a power law such as;

$$R = R_0 I^N$$

where the exponent value of the power law obtained from the least square method is found to be 0.96, as established in previous studies [5].

3.3. Effect of the inlet TCE concentration

The effects of the inlet TCE concentration on the photodegradation in the fluidized bed reactor are shown in Fig. 3.   As shown, the gas stream with the lower inlet concentration of TCE has a higher removal efficiency.   This means that limited and fixed amounts of active sites on the TiO_2/SiO_2 are present in the fluidized bed reactor system

3.4. Long-term photocatalytic activity

Variation of the TCE removal efficiency in the reactant gas stream as a function of the time in the fluidized-bed photoreactor is shown in Fig. 4.   At the optimal gas velocity ($U_g/U_{mf} = 2$) with an inlet TCE concentration (100~150 ppmv), a steady state operation with a 99% removal efficiency was maintained for 50 hours.   After a termination of the operation, a yellowish discoloration of the TiO_2/SiO_2 was observed due to an accumulation of the by-products which were produced during the photocatalytic degradation of the TCE.

3.5. Effect of the various transition metals doping on TiO_2/SiO_2

Fig. 5 shows the effects of various transition metal loadings for the TiO_2/SiO_2 photocatalyst on the TCE removal efficiency.   Even though several different Cu loading contents have been tried, a better TCE removal efficiency was not obtained.   But a better TCE removal conversion efficiency was

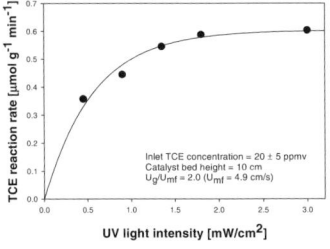

Fig. 2. Effects of the UV light intensity
on the TCE photoreaction rate.

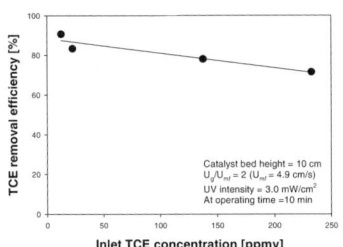

Fig. 3. Effects of the inlet TCE concentration
on the TCE removal efficiency.

584

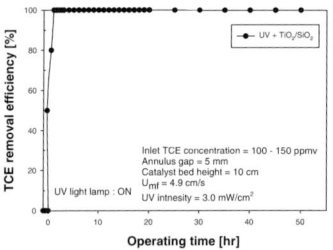

Fig. 4. Long-term photocatalyst activity.

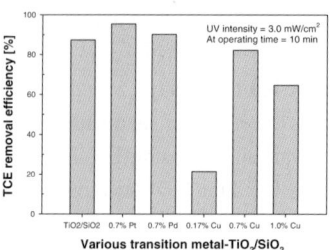

Fig. 5. TCE removal efficiency with various transition metals doped on TiO_2/SiO_2.

observed in the presence of the TiO_2/SiO_2 photocatalysts with a Pt or Pd loading. These transition metals appear to act as electron acceptors which cause a reduction of the recombination rate of the electron-hole pairs produced by the UV irradiation.

3.6. Intermediates and products

The intermediates and end products of the TCE were measured over a retention time by GC/MS (SHIMADZU, QP-5050A) in this experiment. According to the GC/MS chromatogram, dichloromethane (CH_2Cl_2), chloroform ($CHCl_3$), and carbon tetrachloride (CCl_4) were identified as intermediates of the TCE exposed to a photocatalytic reaction. Also an evolution of carbon dioxide (CO_2) and hydrochloric acid (HCl) was observed during the reaction.

4. CONCLUSION

The photodegradation characteristics of TCE have been studied in a fluidized bed photoreactor with TiO_2/SiO_2 and transition metal-TiO_2/SiO_2 catalysts. The best conditions of U_g and the annulus gap for TCE removal are found at 2.0 of U_{mf} and 5mm, respectively. The TCE removal efficiency decreased by increasing the inlet TCE concentration and it increased by increasing the UV light intensity. Photodegradation of the TCE with the transition metal-TiO_2/SiO_2 photocatalysts was more efficient than that with the TiO_2/SiO_2 photocatalyst except for Cu-TiO_2/SiO_2. The Pt-TiO_2/SiO_2 achieves the highest photodegradation efficiency in this study. The degradation of TCE to CO_2 and HCl by a photocatalystic reaction has been accomplished and the formation of reaction intermediates such as dichloromethane (CH_2Cl_2), chloroform ($CHCl_3$), and carbon tetrachloride (CCl_4) was observed.

ACKNOWLEDGEMENT
This work was supported by the Korea Basic Science Institute (2004).

REFERENCES
[1] P. L. Yue, F. Khan and L. Rizzuti, Chem. Eng. Sci., 38 (1983) 1893.
[2] D. Iatridis, P. Yue, L. Rizzuti and A. Brucano, Chem. Eng., 45 (1990) 1.
[3] T. H. Lim and S. D. Kim, Korean J. Chem. Eng., 21 (2004) 905.
[4] T. N. Obbe and R. Brown, Environ. Sci. Technol., 29 (1995) 1223.
[5] D. F. Ollis, E. Pelizzetti and N. Serpone, Environ. Sci. Technol., 25 (1991) 1523.

Studies in Surface Science and Catalysis, volume 159
Hyun-Ku Rhee, In-Sik Nam and Jong Moon Park (Editors)

585

Removal of benzene from benzene-laden air in a biotrickling filter

Sung-Ho Hong and Jea-Keun Lee[†]

Dept. of Environ. Eng., Pukyong National University 599-1 Daeyeon-dong, Nam-gu, Busan, 608-737, Korea (e-mail:leejk@pknu.ac.kr)

ABSTRACT

Experiments have been conducted to investigate the performance of a bench-scale biotrickling filter for removing benzene from benzene-laden air. The operating parameters are inlet loading rate of benzene, empty bed retention time (EBRT), liquid recirculation rate, temperature, and pH. The elimination capacity of benzene from benzene-laden air increases with increasing the EBRT and the inlet loading rate. Further, the removal efficiency increases as liquid recirculation rate decreases. However, it decreases as inlet loading rate increases. The optimum pH and temperature for the removal efficiencies are in the range of $7.0 \sim 7.5$ and $25 \sim 40\,^{\circ}\mathrm{C}$, respectively.

1. INTRODUCTION

Biological processes are often used for removing volatile organic compounds (VOCs) with low concentration (<1000 mg/m^3). These processes have advantages of simple configuration, low capital and operation costs and minimum secondary pollution production. Biofilter process which is one of biological processes has been used successfully for odor abatement and for VOCs control in industry for several decades [1]. However, this process should be satisfied with the several requirements such as adequate supplying of nutrients, humidification of induced gas and proper controlling of pH, humidity and pressure drop in bed.

The biotrickling filter process is a relatively new technology that employs inert media and receives liquid nutrients and buffer through a nozzle system positioned on top of the biotrickling filter. Due to better control of pressure drop across the bed, pH and nutrient feed, biotrickling filter can be operated more consistently as compared with natural media biofilters [2]. However, most inert and synthetic packing media are more expensive than natural organic packing media and have a longer startup. In the previous study [3], zeolite-contained polyethylene media were developed and used for removing toluene from contaminated air in a biotrickling filter. The media have the advantage of improving the cell immobilization, reducing the biofilm detachment, and preventing the clogging problem in biotrickling filter process.

Experimental results showed the packing media used in a bench-scale biotrickling filter was effective for treating gaseous toluene. The goal of this study is therefore to extend the previous study in examining the effects of empty bed retention time (EBRT), liquid recirculation rate, pH and temperature on the performance of a biotrickling filter for treating benzene from benzene-laden air.

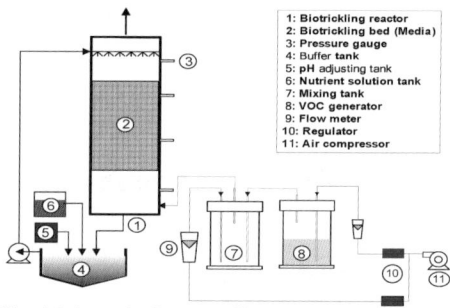

1: Biotrickling reactor
2: Biotrickling bed (Media)
3: Pressure gauge
4: Buffer tank
5: pH adjusting tank
6: Nutrient solution tank
7: Mixing tank
8: VOC generator
9: Flow meter
10: Regulator
11: Air compressor

2. EXPERIMENTAL

Fig. 1 Schematic diagram of biotrickling filter process.

2.1. Apparatus

A schematic of the experimental bench-scale biotrickling filter system for treating gaseous benzene is shown in Fig. 1. The biotrickling filter reactor is constructed of cylindrical stainless steel with 0.4 m in internal diameter and 1.2 m in height. The reactor is packed 0.6 m above the gas distributor with the zeolite-contained polyethylene media which was used in the previous study [3]. The media are originally inoculated with a defined microorganism consortium containing a *Bacillus cereus 1*. The buffer tank of 0.4 m ID and 0.5 m in height is installed to control nutrients and pH in the circulating liquid.

2.2 Procedure

The experimental conditions investigated in this study are; inlet loading rate of benzene $(20 - 130 \text{ g/m}^3 \cdot \text{hr})$, EBRT $(20 - 40 \text{ s})$, liquid circulation rate $(0.96 - 40 \text{ m}^3/\text{m}^3 \cdot \text{media} \cdot \text{hr})$, pH $(5.0 - 8.5)$, and operating temperature $(5 - 45 \text{°C})$. The pH of the circulating liquid in a buffer tank is controlled by adding NaOH and HCl. The compositions of nutrient solution are $0 - 3.5$ g/l K_2HPO_4, $0 - 4.5$ g/l KH_2PO_4, 18.1 g/l KNO_3, 1 mg/l $FeSO_4$, 500 mg/l $MgSO_4 \cdot 7H_2O$ and 20 mg/l $CaCl_2 \cdot 2H_2O$. Gas samples are collected from each of four sampling ports using 2L teflon bags, and analyzed immediately using a gas chromatograph (Shimadzu QP 5050A) equipped with a flame ionization detector.

3. RESULTS AND DISCUSSION

3.1 Elimination Capacity

Fig. 2 shows the variation of elimination capacities of benzene from benzene-laden air according to the inlet loading rates at different EBRTs. As shown in the figure, the elimination capacity increases as the EBRT increases at a constant inlet loading rate. Further, the elimination capacity linearly increases up to a value of inlet loading rate, and then the

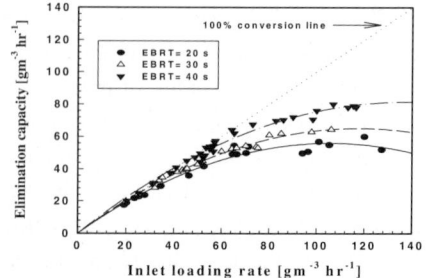

Fig. 2 Elimination capacity with regard to benzene loading rate.

increment of elimination capacity becomes lower with further increase of inlet loading rate. The maximum value for an elimination capacity is about 78 g/m^3·hr at an EBRT of 40 s.

3.2 Benzene Removal Efficiency

The variation of removal efficiencies of benzene according to the inlet loading rate and the liquid recirculation rate is shown in Fig. 3. As shown in the figure, the removal efficiency of benzene decreases with increasing the inlet loading rate and decreasing the liquid recirculation rate. At an inlet loading rate less than 40 g/m^3·hr, benzene removal efficiency shows a little variation as the liquid recirculation rate increases. However, the removal efficiency

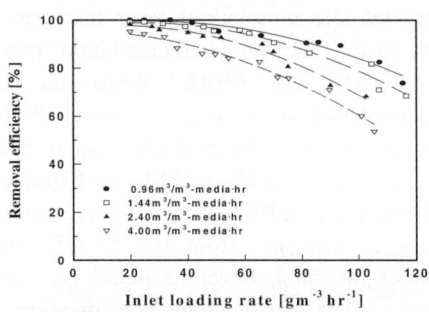

Fig. 3 Removal efficiency vs. inlet loading rate.

remarkably decreases when the inlet loading rate increases to 100 g/m^3·hr. Such a tendency can be explained as follows: The thickness of liquid film layer formed on the biofilm surface increases when the liquid recirculation rate increases, resulting in mass transfer limitation. As a result, the removal efficiency decreases. Yang and Allen [4] also reported that the reduction of H$_2$S removal efficiencies at shorter residence times was not necessarily caused by the insufficient reaction time between the H$_2$S molecules and the biomass, but could be caused by the slow step in the overall removal process of H$_2$S diffusion from the gas phase into the liquid phase where the microorganisms exist.

Microbial activity and biotrickling filter performance are strongly influenced by pH. Generally, each species of microorganisms used in biotrickling filters is most successful over a certain range of pH and will be inhibited or killed if conditions move outside this range. The ranges for different species may be narrow or broad. Some species do well at high pH, and some at low pH. Fig. 4 shows the variation of benzene removal efficiency as a function of pH in the circulating liquid. As shown, the removal efficiency tends to increase with pH to some extent and then begins to decrease, whose maximum occurs at pH between 7 and 7.5. This result shows a similar trend to that of Lu et al. [5]. They reported that most of bacteria for removing benzene, toluene, ethylbenzene, and xylene (BTEX) may have a high biological activity in a weak basic environment.

Fig. 5 shows the effect of operating temperature on benzene removal efficiency. In the temperature range of 25~40℃, high removal efficiencies are observed consistently with a little variation (93 - 94 %). However, the removal efficiencies decrease rapidly to 55 and 75 % at low temperature below 10℃.

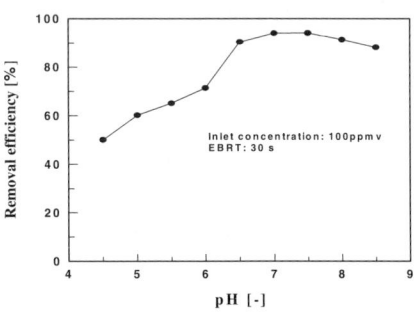

Fig. 4 Effect of pH on benzene removal efficiency.

588

3.3 Pressure Drop

Fig. 6 shows the variation of pressure drop as a function of liquid recirculation rate and EBRT during a steady-state operation in the biotrickling filter. The pressure drop through the biotrickling filter increases with increasing the liquid recirculation rate and decreasing the EBRT. When the liquid recirculation rate is increased from 0.99 to 4.4 m³/m³-media·hr, the pressure drop is changed from 0.03 to 0.15 mmH₂O at EBRT 40 s. However, at EBRT 20 s, the pressure drop sharply increases from 0.8 to 4.5 mmH₂O under the same range of liquid recirculation rate. The reason why the pressure drop sharply increases with the liquid recirculation rate at low EBRT is that the higher liquid recirculation rate in the biotrickling filter reduces the space available for gas flow through the media. Moreover, the gas flow rate increases with decreasing the EBRT. As a result, the higher the liquid recirculation and gas flow rates, the higher the pressure drop in the biotrickling filter.

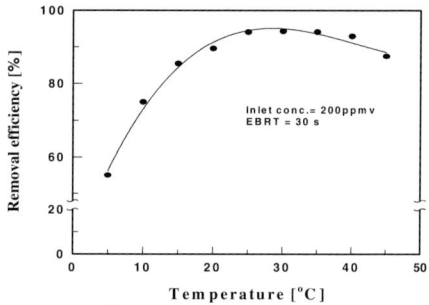

Fig. 5 Effect of temperature on benzene removal efficiency.

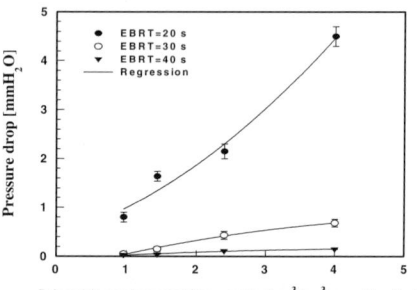

Fig. 6 Pressure drop vs. liquid recirculation rate.

CONCLUSION

Experimental results show that the elimination capacity of benzene from benzene-laden air increases with increasing the EBRT and the inlet loading rate. The removal efficiency of benzene decreases with increasing the inlet loading rate and decreasing the liquid recirculation rate. The optimum pH and temperature for the removal efficiencies are in the range of 7.0~7.5 and 25~40℃, respectively. The pressure drop through the biotrickling filter increases with increasing the liquid recirculation rate and decreasing the EBRT.

REFERENCES

[1] K.H. Lim and S.W. Park, Korean J. Chem. Eng., 21 (2004) 1161.
[2] G. A. Sorial, F. L. Smith, M.T. Suidan, A. Pandit, P. Biswas and R. C. Brenner, J. Environ. Eng., 123 (1997) 530.
[3] S. H. Hong, C. S. Lee, K. H. Kim, J. G. Jang and J. K. Lee, 9th ASCON on FB & TPR (2004) 439.
[4] Y. Yang and E. R. Allen, J. Air & Waste Manage. Assoc., 44 (1994) 863.
[5] C. Lu, M. R. Lin and C. Chu, Advances in Environ. Research, 6 (2002) 99.

FUEL CELLS AND ELECTROCHEMICAL
REACTION ENGINEERING

Studies in Surface Science and Catalysis, volume 159
Hyun-Ku Rhee, In-Sik Nam and Jong Moon Park (Editors)

Effects of Catalyst Loading and Oxidant on the Performance of Direct Formic Acid Fuel Cells

Hyo-Song Lee, Jae-Keun Yu, Jin-Yong Kim, Ki-Ho Kim, Young-Woo Rhee[*]

Department of Chemical engineering, Chungnam National University
220, Gung-dong, Yuseong-gu, Daejeon, 305-764, Korea

1. INTRODUCTION

Direct methanol fuel cell (DMFC) has been extensively studied as a promising power and energy supplier for portable devices. However, methanol, used in DMFC, has several problems such as fuel crossover and CO poisoning of the cathode catalyst [1,2]. As a solution for these problems, formic acid was proposed as a substitute fuel for miniature fuel cells [3]. The theoretical open circuit potential (OCP) of formic acid is 1.45V, which is higher than that of hydrogen (1.23V) and methanol (1.18V). In addition to these advantages, slow diffusion of formic acid through the Nafion membrane is expected to cause a low fuel crossover and is able to use high concentrations of formic acid.

In this study, we investigated the effects of anode catalyst loading and oxidant on the performance of DFAFC at various temperatures to better understand the significance of anode catalyst in a DFAFC system.

2. EXPERIMENTAL

The catalyst inks were prepared by dispersing the catalyst nanoparticles into an appropriate amount of Millipore water and 5wt% Nafion solution. Then, both the anode and cathode catalyst inks were directly painted using a 'direct painting' technique onto either side of a Nafion 117 membrane. A carbon cloth diffusion layer was placed on to top of both the anode and cathode catalyst layers [3-5]. The active cell area was 2.25cm^2.

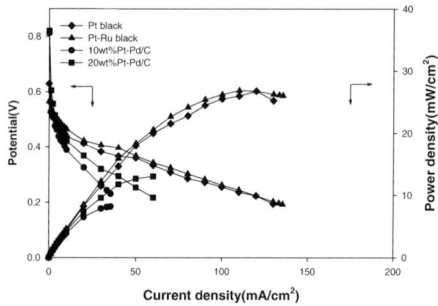

Fig. 1. Effect of catalyst types on the performance of fuel cell at 25 ℃ (9M HCOOH, Air).

In this study, commercial catalysts were used. As the cathode catalyst Pt black was used. Different amounts of Pt black, Pt-Ru black, 10wt% Pt-Pd/C and 20wt% Pt-Pd/C catalysts were used for the anode.

3. RESULTS AND DISCUSSION

Fig. 1 shows the effect of catalyst types on the DFAFC performance at 25 ℃. We used 4-types of anode catalyst such as Pt black, Pt-Ru black, 10wt% Pt-Pd/C and 20wt% Pt-Pd/C. The concentration of formic acid was 9M and air was used as oxidant. Among them Pt-Ru catalyst showed the best power density of 27 mW/cm^2. Pt-Pd catalysts showed the best OCP of 0.820V. However, their cell performance decreased rapidly as the current density increased. It is speculated that the large amount of carbon could inhibit the transfer of protons in the anode, causing poor performance of fuel cell. So, for the using of supported catalyst, we should investigate optimum content of catalyst and carbon.

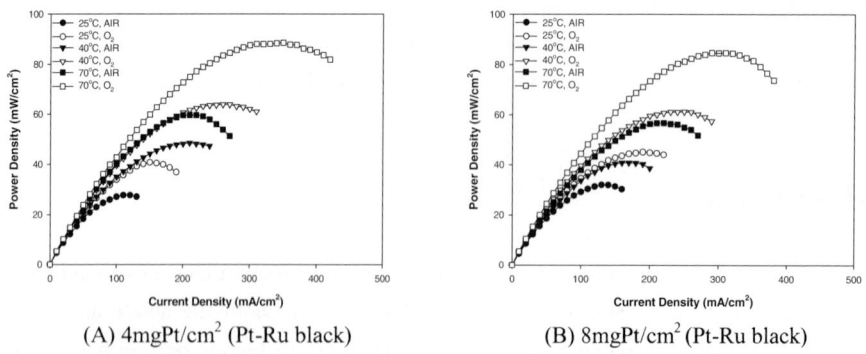

(A) 4mgPt/cm^2 (Pt-Ru black)   (B) 8mgPt/cm^2 (Pt-Ru black)

Fig. 2. Effects of various temperature and oxidant on the DFAFC power density.

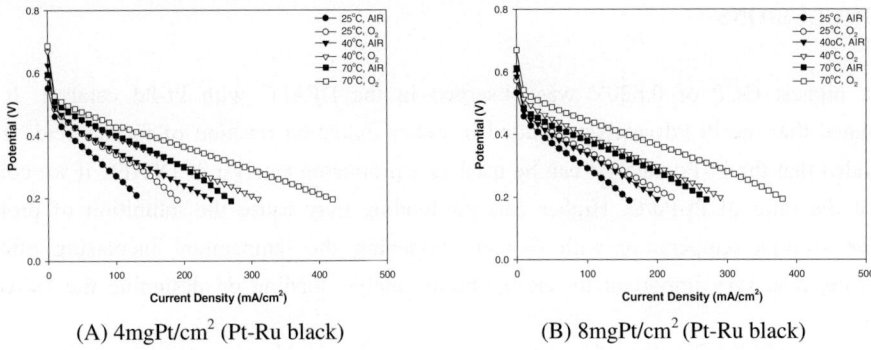

(A) 4mgPt/cm^2 (Pt-Ru black)   (B) 8mgPt/cm^2 (Pt-Ru black)

Fig. 3. Effects of various temperature and oxidant on the DFAFC potential.

Figs. 2 and 3 show the effects of oxidant and temperature on the DFAFC performance for the two catalyst loadings. The maximum power density of 89mW/cm^2 was observed at 70℃ with air. The cell performance increases with the temperature increase and the change of oxidant from air to oxygen significantly enhances the cell performance. In case of the fuel cell with air as an oxidant, the fuel cell with catalyst loading of 8mgPt/cm^2 has higher performance than that of 4mgPt/cm^2. In case of the fuel cell with oxygen, however, the fuel cell with catalyst loading of 4mgPt/cm^2 has higher performance. It is believed that the thick catalyst layer of higher catalyst loading inhibits the transfer of protons. For the catalyst loading of 4mgPt/cm^2 the power density increases by 114% when the temperature changes from 25 to 70℃ whereas only 78% power density increase is observed for 8mgPt/cm^2 with the same temperature change. Fig. 4 shows the effect of catalyst loading on the cell performance. The fuel cell with 4mg catalyst/cm^2 demonstrates lower performance than the others of high catalyst loading at 25℃. In this case, it is speculated that the catalyst loading is not enough to effectively catalyze the oxidation reaction of fuel at the anode. Similar results for DMFC have been reported in the literature [6].

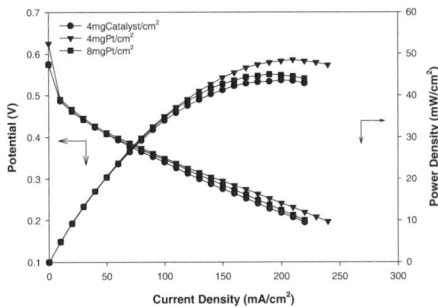

Fig. 4. Effect of anode catalyst loading on the DFAFC performance.

4. CONCLUSIONS

The highest OCP of 0.820V was observed in the DFAFC with Pt-Pd catalyst. It is speculated that the Pt-Pd catalyst resulted in better oxidation reaction of formic acid. It is concluded that the Pt-Pd catalyst can be used as a promising catalyst of DFAFC if we could control the ratio of Pt-Pd/C. Higher catalyst loading may cause the inhibition of proton transfer at high temperature with oxygen, lessening the temperature increasing effect. Therefore, it is very important to use optimum catalyst loading on designing the DFAFC system.

ACKNOWLEDGMENTS

This work was supported by Korea Energy Management Corporation (2003-N-FC03-04-03-0-000)).

REFERENCES

[1] T. H. Yang, G. G. Park, P. Pugazhendhi, W. Y. Lee, C. S. Kim, Korean J. Chem. Eng., 19 (2002) 417.
[2] C. H. Pak, S. J. Lee, S. A. Lee, H. Chang, Korean J. Chem. Eng., 22 (2005) 214.
[3] C. Rice, R. I. Masel, P. Waszczuk, A. Wieckowski, T. Barnard, J. Power Sources, 111 (2002) 83.
[4] S. Ha, C. Rice, R. I. Masel, A. Wieckowski, J. Power Sources, 112 (2002) 655.
[5] C. Rice, R. I. Masel, A. Wieckowski, J. Power Sources, 115 (2003) 229.
[6] N. Nakagawa, Y. Xiu, J. Power Sources, 118 (2003) 248.

Studies in Surface Science and Catalysis, volume 159
Hyun-Ku Rhee, In-Sik Nam and Jong Moon Park (Editors)

Steady-state and Dynamic Operations of 3W DMFC stack

Jung-Han Ryu and Sung Min Cho*

Department of Chemical Engineering, Sungkyunkwan University, Suwon 440-746, Korea

1. INTRODUCTION

The dynamic behavior of fuel cells is of importance to insure the stable operation of the fuel cells under various operating conditions. Among a few different fuel cell types, the direct methanol fuel cell (DMFC) has been known to have advantages especially for portable applications [1-2].

The transient response of DMFC is inherently slower and consequently the performance is worse than that of the hydrogen fuel cell, since the electrochemical oxidation kinetics of methanol are inherently slower due to intermediates formed during methanol oxidation [3]. Since the methanol solution should penetrate a diffusion layer toward the anode catalyst layer for oxidation, it is inevitable for the DMFC to experience the high mass transport resistance. The carbon dioxide produced as the result of the oxidation reaction of methanol could also partly block the narrow flow path to be more difficult for the methanol to diffuse toward the catalyst. All these resistances and limitations can alter the cell characteristics and the power output when the cell is operated under variable load conditions. Especially when the DMFC stack is considered, the fluid dynamics inside the fuel cell stack is more complicated and so the transient stack performance could be more dependent of the variable load conditions.

In this paper we report the effect of varying loads on a small size DMFC stack (10 cells with 9 cm^2 active-area each). The transient responses of the stack voltage have been investigated upon variable current load conditions to obtain the information on the dynamic characteristics of the stack. Also, the transient responses of the stack current upon changing fuel flow rates have been monitored to obtain the optimal operating conditions for the stack.

2. EXPERIMENTAL

The DMFC used in this study was a 10-cell stack with 3cm x 3cm active area (total active area for the stack of 90 cm^2). The 8 middle membrane electrode assemblies (MEAs) were sandwiched by two graphite bipolar plates and 2 side MEAs were sandwiched by a graphite bipolar plate and an end plate. In the stack the fuels of methanol solution and air were introduced through a port to be distributed evenly to 10 separate flow channels.

MEAs used in this study were prepared in the following procedure [5]. The diffusion backing layers for anode and cathode were a Teflon-treated (20 wt. %) carbon paper (Toray 090, E-Tek) of 0.29 mm thickness. A thin diffusion layer was formed on top of the backing layer by spreading Vulcan XC-72 (85 wt. %) with PTFE (15 wt. %) for both anode and cathode. After the diffusion layers were sintered at a temperature of 360°C for 15 min., the catalyst layer was then formed with Pt/Ru (4 mg/cm^2) and Nafion (1 mg/cm^2) for anode and with Pt (4 mg/cm^2) and Nafion (1 mg/cm^2) for cathode. The prepared electrodes were placed either side of a pretreated Nafion 115 membrane and the assembly was hot-pressed at 85 kg/cm^2 for 3 min. at 135°C.

3. RESULTS AND DISCUSSION

The maximum power obtained from the fabricated stack at a temperature of 50°C was 5W. Since the total active area of the stack is 90 cm^2, the average power density of the stack is found to be 55.6 mW/cm^2, which is close to that of a unit cell at the temperature. When a square pulse with a fixed current density is loaded to the stack, the open circuit voltage (O.C.V.) is expected to drop quickly to a voltage corresponding to the current load and then the voltage should recover back to the O.C.V. when the pulse ends. In experiment, upon the current loads of 0 ~ 1.6A applied to the stack as consecutive pulses, the voltage response is monitored by time as shown in Fig. 1. As the current load increases, the voltage drop increases correspondingly. In case of a load step change from zero current to a fixed current, the measured voltage drops instantaneously at first to a voltage lower than the steady state voltage corresponding to the fixed current, and returns relatively slowly to the steady state voltage. This behavior was observed identically for the current densities applied from low to high value. When a load step change occurs from a fixed current to zero current, however, the voltage response rather gradually approaches to the steady state voltage at relatively low current densities, but the response tends to show a sharp rise resulting in an overshoot and slow relaxation back to the steady state voltage at higher current densities.

Fig. 2 shows the dynamic response of stack voltage to the step changes of various applied current densities. Like the former case of applied current pulses, the response exhibits the overshooting and relaxation which is caused by the methanol oxidation kinetics on the catalyst surface. The steady state stack voltage was found to be the same for both pulse and step loads with the same current density.

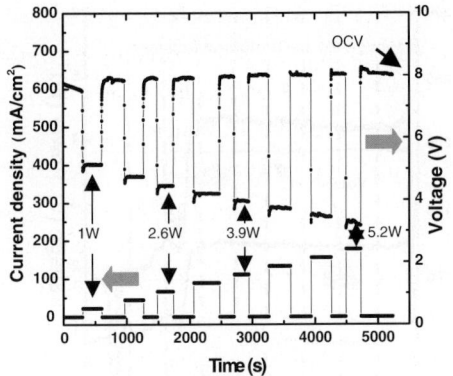

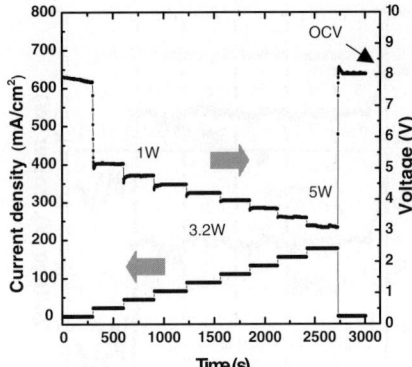

Fig.1. Dynamic response of output voltage with pulse-wise current loads.

Fig.2. Dynamic response of output voltage with Step-wise current loads.

When 2M methanol solution is fed to the stack at a flow rate of 2 ml/min and the stack is operated at a constant voltage output of 3.8V, the transient response of the stack current density is shown in Fig. 3 varying the flow rate of air to the cathode. The stack was maintained at a temperature of 50°C throughout the experiment. As shown in the figure, while the stack current is maintained at the air flow rates higher than 2 L/min, the stack current begins unstable at the slower flow rates. A similar result is shown in Fig. 4 for varying methanol flow rate at an air flow rate of 2 L/min. At a methanol flow rate of 8 ml/min, the current density reaches initially a current density value of about 130 mA/cm^2 and then starts to decrease probably due to methanol crossover. As the methanol flow rate decreases, the stack current density increases slowly until the methanol flow rate reaches 3 ml/min because of the reduced methanol crossover. The current density drops rapidly from the methanol flow rate of 2 ml/min.

It is worthwhile to note here that in Figs. 3 and 4 the current densities measured at a methanol and air flow rates of 2 ml/min and 2 L/min, respectively show the inconsistency even at the same operating conditions. The reason is expected to be due to the water produced at the cathode. Comparing Figs. 3 and 4, the measured current density response looks quite unstably scattered at low air flow rates in Fig. 3, while it seems to be stable even at the low methanol flow rates in Fig. 4. This instability of measured current density in Fig. 4 could have caused by the water formed at the cathode during the operation of the stack. In fact, it has been observed from the experiment that the water produced at the cathode looks accumulated in the cathode flow channels for a while and it bursts out intermittently to the cathode outlet. Based on the experimental observations, for the stable operation of the 5W

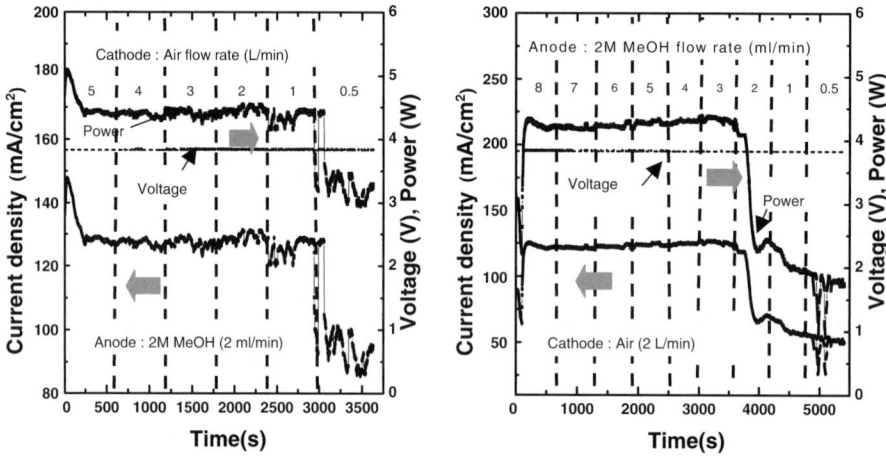

Fig. 3. Dynamic response of power output
with changing air flow rate.

Fig. 4. Dynamic response of power output
with changing methanol flow rate.

DMFC stack fabricated in this study, the minimal operating conditions are the flow rates of 3 ml/min and 2L/min for 2M methanol and air, respectively.

4. CONCLUSIONS

This study has shown the dynamic behavior of a 5W DMFC stack when the current loads have changed by pulses and steps. In order to determine the optimum operating conditions of the stack, the dynamic behavior of the stack current has been studied under a constant voltage output of 3.8V, varying the flow rate of 2M methanol solution and air. For the stable operation of the 5W stack, the minimal fuel flow rates are found to be 3 ml/min and 2L/min for 2M methanol and air, respectively.

REFERENCES

[1] C. Xie, J. Bostaph, J. Pavio, J. Power Sources, 136 (2004) 55.
[2] H.-Y. Cha, H.-G. Choi, J.-D. Nam, Y. Lee, S. M. Cho, E.-S. Lee, J.-K. Lee, C.-H. Chung, Electrochimica Acta, 50 (2004) 795.
[3] J. Kallo, J. Kamara, W. Lehnert, R. Helmolt, J. Power Sources, 127 (2004) 181.
[4] C. Lim, C. Y. Wang, J. Power Sources, 113 (2003) 145.

Studies in Surface Science and Catalysis, volume 159
Hyun-Ku Rhee, In-Sik Nam and Jong Moon Park (Editors)

Performance of an Anode-supported Solid Oxide Fuel Cell in a Mixed-gas Configuration

Nguyen Xuan Phuong Vo, Suk Woo Nam[*], Sung Pil Yoon[*], Jonghee Han, Tae-Hoon Lim and Seong-Ahn Hong

Fuel Cell Research Center, Korea Institute of Science and Technology (KIST)
P.O. Box 131, Cheongryang, Seoul, Korea

1. INTRODUCTION

It is well known that conventional solid oxide fuel cells (SOFCs) which, compared to other types of present electric generation devices, offer a number of distinctive features due to their primarily ceramic structure and high operating temperatures (800-1000°C) [1]. Unfortunately, although they are not subject to the problems of other fuel cell types, they have problems of their own that prevent SOFCs from commercialization. Materials problems associated with high operating temperature (1000°C) including electrode sintering, interfacial diffusion between electrodes and electrolyte, mechanical stress due to different thermal expansion coefficients of the cell components, and limited choice of expensive interconnects have given rise to recent attention of SOFCs operated at reduced temperatures (500-800°C), while still maintaining the power densities achieved at high temperatures. The key issues of overcoming excessive overpotentials of electrodes and ohmic loss of the electrolyte at such lower temperatures are the extension of the triple-phase boundary (TPB) to electrodes and the fabrication of thin film electrolytes as well as new design of one-chamber cell. As a method for improving electrode performance, Yoon et al. [2] demonstrated that thin films of electrolyte deposited within the pores of electrodes by a sol-gel coating technique can provide paths for oxide ions and expand the reaction zone into a new triple phase boundary (TPB). In addition, conventional Y_2O_3 stabilized ZrO_2 (YSZ) electrolyte has recently been effectively replaced by higher ionic conductive materials such as doped ceria. Thin samaria-doped ceria film, $Sm_{0.2}Ce_{0.8}O_{1.9}$ (SDC), has been widely studied because its high ionic conductivity is about 30-times higher when compared to that of YSZ at 800°C [3]. Even though CeO_2 is unstable at low oxygen partial pressures [4], this demerit can be overcome by operating at a lower temperature and a higher oxygen partial pressure when compared to the conventional operating conditions for an SOFC running on hydrogen fuel at 1000°C. As the thin electrolyte can no longer be the mechanically supporting component, one of the porous electrodes must take over this function and a thin electrolyte film may be deposited on the electrode layer followed by the deposition of the other porous electrode. Then a single cell is put into a single-chamber station and exposed to a uniform flow of a mixture of practical fuel and oxidant at reduced temperatures. Because of no need to separate the reactants, the imperfect layer of thin electrolyte with cracks or pinholes is of no concern, provided that there is no short-circuit between anode and cathode materials.

In the present work, we report results on the fabrication and performance of anode-supported, thin SDC electrolyte fuel cells operated in a single chamber configuration where methane and oxygen served as the gas mixture.

2. FUEL CELL FABRICATION

For the anode, NiO-SDC powder at a weight ratio of NiO:SDC = 70:30 from Praxair (surface area of 5.58 m^2/g, particle size < 4.4 μm) was mixed with 10 wt.% graphite and then pressed using a 25 mm stainless steel mold under a 2500 psi load. An electrolyte slurry consisting of SDC powder, α-terpineol, EtOH, dispersant, and plasticizer was screen-printed onto the pre-sintered anode at 1000°C and subsequently fired at 1300°C in air for 3 h. The dual layer cell was treated in flowing, diluted hydrogen at 600°C for 1 h to reduce the NiO in the anode to Ni prior to cathode deposition. The cathode slurry consisting of LSM powder and methylcellulose was also deposited on the electrolyte layer by screen-printing with 1 cm x 1cm area and then fired at 1100°C in N_2 for 2 h, a condition which was found to give good adhesion between the cathode and the electrolyte. To the complete trilayer structures, the sol-gel coating method using SDC sol [2] was applied. The cell was dipped in the SDC sol 10 times, each followed by calcination in N_2 at 600°C for 2 h. Shown in Fig. 1 are the optical image and microstructures of the cell as characterized using a scanning electron microscope. The SDC electrolyte film can be controlled to be less than 20 μm thick and bound tightly to the anode substrate. The electrolyte layer is not fully dense but enough to avoid short-circuit between porous anode and cathode layers and appropriately work in one-chamber fuel cell station.

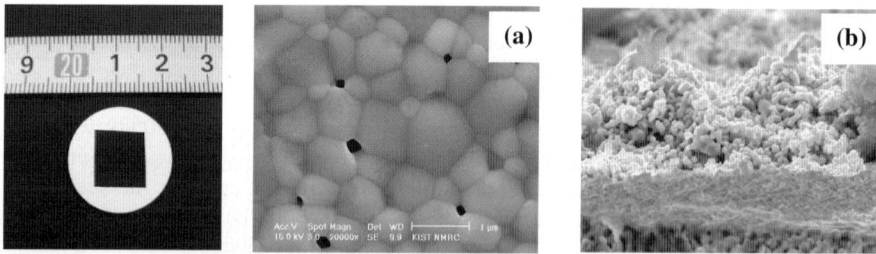

Fig. 1. Optical image and SEM micrographs of a single cell: (a) the SDC electrolyte surface view; (b) LSM-SDC-NiSDC cross view.

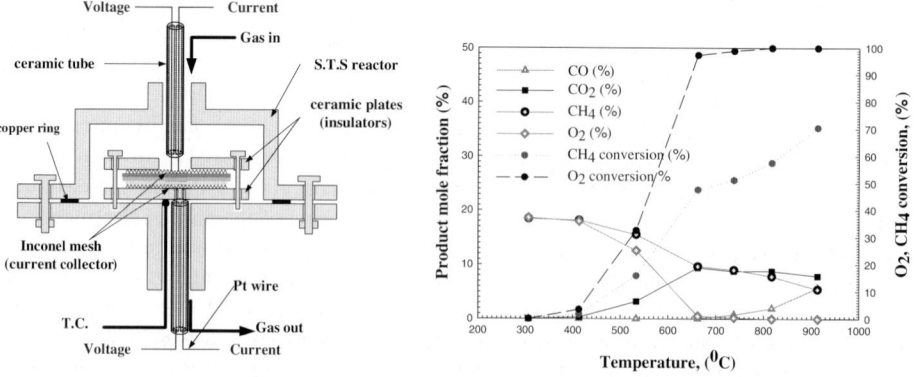

Fig. 2. A schematic illustration of the mixed-gas test cell.

Fig. 3. Catalytic activity of LSM (0.2 g, calcined at 1100°C), CH_4 flow rate of 20 ml min^{-1}, O_2/C=1.0

The single cell thus fabricated was placed in a single chamber station as illustrated in Fig. 2. A humidified mixture of methane and oxygen was supplied to the station so that both electrode compartments were exposed to the same composition of methane and oxygen. For the measurement of the cell temperature, a thermocouple (TC) was placed approximately 4 mm away from the cathode site. For the evaluation of the fuel-cell performance, Pt wires and Inconel gauzes were used as the output terminals and electrical collectors, respectively.

The cell tests were carried out by measuring the terminal voltage between the two electrodes during cell discharge by a Solartron 1287 electrochemical interface.

3. OPERATING CONDITIONS

For single chamber fuel cell operation, the operating temperature must be sufficiently high to ensure that all the oxygen that reaches the anode is totally consumed in the partial oxidation reaction, but low enough that cathode's activity towards complete oxidation of the fuel is minimal [5]. From a measurement of electrodes' catalytic activities, a suitable operating temperature range can be anticipated. Fig. 3 showed catalytic activity of the cathode material, $La_{0.85}Sr_{0.15}MnO_2$ (LSM), for partial oxidation of methane. An oxygen conversion of 32-97% was observed at 530 to 660°C, obtained from LSM powders calcined at 1100°C for 2 h. The oxygen was almost completely consumed at temperatures higher than 650°C, which is a result of the increase in catalytic activity of LSM for methane partial oxidation. Thus, the cathode material restricts the operation of the single chamber cell using methane as a fuel to temperatures lower than 700°C. Conversion of oxygen over the anode material, reduced Ni-SDC, was found to occur at much lower temperatures, by 350°C. Thus, the anode material restricts the operation of single chamber fuel cell using methane as a fuel to temperatures of approximately 350°C and above. It is clear that the expectant operating temperature range for the single cell comprised of Ni-SDC and LSM materials utilized here is from 350°C to 700°C. A mixture of methane and oxygen, with a fixed ratio of oxygen to methane R = 0.6, was supplied to the station at a flow rate of 200 mL min^{-1} from 500°C to 700°C. The R value of 0.6 is larger than the stoichiometry of 0.5 for methane partial oxidation.

4. FUEL CELL PERFORMANCE

All fuel cell experiments were carried out using the anode supported Ni SDC (SDC) / SDC / LSM (SDC) cells described above. Temperatures reported are those at the fuel cell reaction zone. Fig. 4 shows the voltage V and power density P versus the current density J for the anode-supported SOFC operated on the R = 0.6 fuel mixture at 200 mL.min^{-1}.

Under operating conditions, peak power density increased gradually from 500°C to 650°C, shifted to higher current density. It reached the maximum value of about 375 mW cm^{-2} at 650°C under a current load of 750 mA cm^{-2} when the cell was exposed to the mixed gas with a O_2:C ratio of 0.6.

The peak power density was enhanced when the operating temperature increased, but was limited to 650°C. The reason for the performance decrease in the cell operated over 650°C may result from increase in catalytic activity of LSM for methane partial oxidation.

We reported the dotted and dash performance curves illustrated in Fig. 5 for an anode-supported Ni5Al (SDC) / SDC / LSM (SDC) cell running on humidified methane and air mixtures in mixed-gas mode at 700°C with O_2:C ratio of 1.5 and 1.67, respectively [6]. In the previous paper, we suggested that the decrease of cell performance for mixed-gas SOFC results from an increase in the anode polarization and is due to the low catalytic activity of the

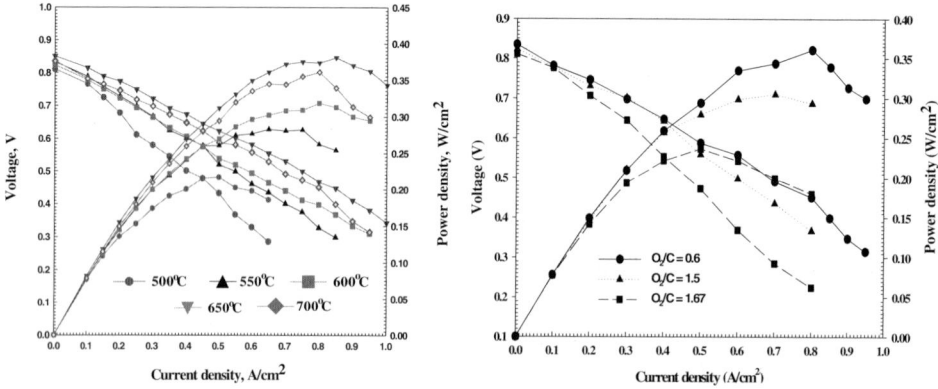

Fig. 4. *I-V-P* characteristics of the single cell in one-chamber station from 500°C to 700°C.

Fig. 5. Performance comparison for mixed-gas single cells at 700°C with different O_2/C ratio.

anode brought on by the excessive amount of oxygen, which easily consumes the hydrogen and carbon monoxide produced. In this work, we have developed the mixed-gas SOFC system having a higher catalytic activity for the partial oxidation of methane. Fig. 5 shows the increase of anode polarization according to the O_2:C ratios. The solid curves exhibit a higher performance measured from the anode supported Ni SDC (SDC) / SDC / LSM (SDC) cell exposed to gas flow conditions of 125 mL.min^{-1} methane + 75 mL.min^{-1} oxygen (R = 0.6) in the mixed-gas test cell shown in Fig. 2.

5. SUMMARY

Thin films of Samaria-Doped Ceria (SDC) were formed on porous substrates using screen printing and within the pores of electrodes using a sol-gel coating method. The SDC-electrolyte film, with a controllable thickness of less than 20 μm, was bound tightly to the anode substrate. The maximum power density of the single cells was 375 mW·cm^{-2} at 650°C with a 25% methane in mixed gas mixture. The long-term stability of the cell as well as the effects of flow rate, gas composition, fuel type, etc. on the cell performance will be continuously studied to get the optimization of operating conditions.

ACKNOWLEDGEMENTS
This work was financially supported by Korea Institute of Science and Technology.

REFERENCES

[1] V. Dusastre and J.A. Kilner, Solid State Ionics, 126 (1999) 163.
[2] S.P. Yoon, J. Han, S.W. Nam, T-H. Lim and S-A. Hong, J. Power Sources, 136 (2004) 31.
[3] K. Eguchi, T. Setoguchi, T. Inoue and M. Arai, Solid State Ionics, 52 (1992) 165.
[4] H. Uchida, H. Suzuki and M. Watanabe, J. Electrochem. Soc., 145 (1998) 615.
[5] Z. Shao, C. Kwak and S.M. Haile, Solid State Ionics, 175 (2004) p. 41.
[6] N.X.P. Vo, S.P. Yoon, S.W. Nam, J. Han, T-H. Lim and S-A. Hong, Key Eng. Mat., 277-279 (2005) 460.

Studies in Surface Science and Catalysis, volume 159
Hyun-Ku Rhee, In-Sik Nam and Jong Moon Park (Editors)
601

Study on Ceria Coating Effect on H$_2$S Tolerance in the Anode of Molten Carbonate Fuel Cell

Hary Devianto, Sung Pil Yoon*, Suk Woo Nam, Jonghee Han, Tae-Hoon Lim

Fuel Cell Research Center, Korea Institute of Science & Technology, Seoul 136-791, Korea

1. Introduction

It is well established that sulfur compounds even in low parts per million concentrations in fuel gas are detrimental to MCFCs. The principal sulfur compound that has an adverse effect on cell performance is H$_2$S. A nickel anode at anodic potentials reacts with H$_2$S to form nickel sulfide. Chemisorption on Ni surfaces occurs, which can block active electrochemical sites. The tolerance of MCFCs to sulfur compounds is strongly dependent on temperature, pressure, gas composition, cell components, and system operation (i.e., recycle, venting, and gas cleanup). Nickel anode at anodic potentials reacts with H$_2$S to form nickel sulfide. Moreover, oxidation of H$_2$S in a combustion reaction, when recycling system is used, causes subsequent reaction with carbonate ions in the electrolyte [1]. Some researchers have tried to overcome this problem with additional device such as sulfur removal reactor. If the anode itself has a high tolerance to sulfur, the additional device is not required, hence, cutting the capital cost for MCFC plant. To enhance the anode performance on sulfur tolerance, ceria coating on anode is proposed. The main reason is that ceria can react with H$_2$S [2,3] to protect Ni anode.

2. Experimental

2.1. Surface modification of anode

We prepared ceria on Ni substrate by sol-gel coating method. Ceria sol solution was prepared with ceria sol solution (Alfa, 20% in H$_2$O, colloidal dispersion) mixed with ethanol (99.9%, Hayman) with weight ratio (1:2) and stirred. Ceria was deposited on Ni substrate by dip coating method. The variation number of dipping was carried out to obtain different coating ratio. The anode was completely dipped into the ceria sol solution for several seconds and dried at a temperature of 50°C for 24 hours in air atmosphere followed by calcination at 700°C for 30 minutes in 5%H$_2$-N$_2$ atmosphere.

2.2. Verification of surface modification effect

In order to analyze the electrolyte filling contents in the prepared anode, we set up a single cell by 10x10 cm^2. It was made of LiAlO$_2$ matrix sheets, 56% porosity and 2 mm thickness, modified anodes coated by ceria in each coating ratio, Lithiated in situ oxidation Ni cathode, 70% of porosity, electrolyte (Li$_2$CO$_3$:K$_2$CO$_3$=62:38 mol ratio) eutectics and gases, H$_2$:CO$_2$:H$_2$O=72:18:10 for anode and Air:CO$_2$=67:33 for cathode. After the cell operation at 650 ℃, the electrolyte filling contents were analyzed by direct comparison of mass difference before and after the electrolyte extraction. In case of the extraction of electrolyte immersed in cathode and matrix, 10 vol% acetic acid aqueous solutions was used, however for anode case, distilled water was used to prevent metal dissolution. Single cell operation was carried out with pre-treatment condition from 25°C to 450°C for 3 days in air atmosphere and from 450°C to 650°C for 3 days in CO$_2$ atmosphere. After pretreatment the temperature was held at 650°C for 180 hours in MCFC gas condition followed by 80 ppm H$_2$S introduction for 180 hours and back to normal gas. The total operating time was more than 450 hours. The Open Circuit Voltage (OCV), performance, internal resistance and N$_2$ cross over were measured during cell operation

3. Results and discussion

3.1. Morphology of the modified anode

Ceria sol with small particle diameter (~10 nm) makes the ceria coating layer form a rough thin layer. It would give additional advantages such as increasing wettability due to surface modification. In our experiment, 1-5 wt% of anode coating condition was carried out due to maintaining well distributed CeO$_2$ particles. CeO$_2$ distributions along with the depth direction of the porous Ni-10wt%Cr anode were investigated by cerium mapping of EDAX. Cross section of 0.7 mm in thickness of porous coated anode with 5 different coating ratios were analyzed as can be seen from Fig 1. It was shown in the 150 times magnification that when the coating ratio is higher than 3 wt%, there were ceria particles accumulations on the Ni-10wt%Cr surface. Three samples, 2.83%, 3.84 wt% and 4.67 wt%, indicated porous anode surface (peripheral side) accumulated by ceria. While two samples, 1.15 wt% and 2.01 wt%, indicated well distributed ceria in the porous anode. During dip coating process, the samples were completely dipped into ceria sol for several seconds. The ceria sol filled porous anode from the surface to the center of porous anode. Drying process was used to remove all volatile matters from the ceria sol, thus ceria particle will remain in the pores. These sequencing process were repeated in order to achieve the requested value. With increasing the number of dipping times, the ceria particles impregnated inside the anode pores increased. Thus, the anode pore size decreased due to ceria layer covered the nickel surface. When the coating ratio reached 3 wt%, the ceria particles near the anode surface blocked the pores, which inhibited ceria particles journeyed from the surface to the center of the anode. As a result, ceria particles were accumulated on the surface.

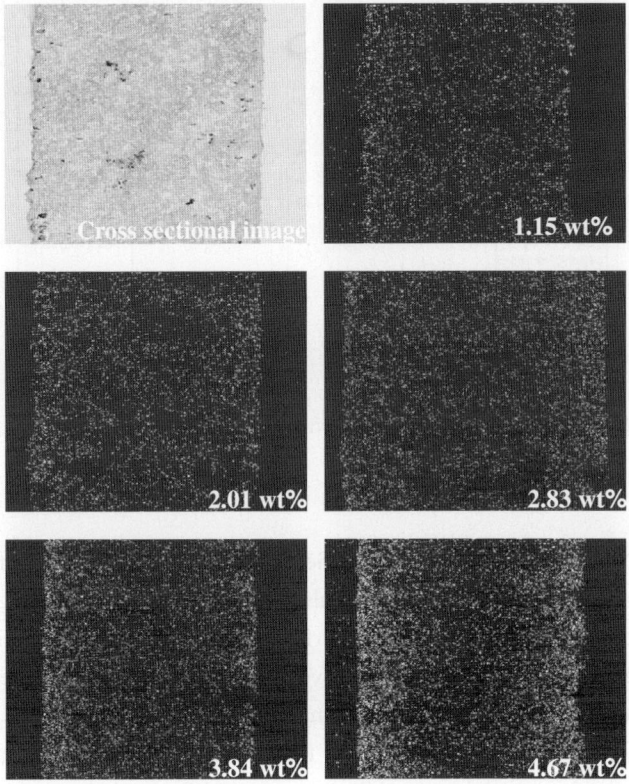

Fig. 1. Uniform distributions of cerium through ceria coated 10wt%Cr anode.

3.2. Sulfur tolerance

Adding 80 ppm of H_2S in H_2 fuel to the single cell during the cell operation caused the performances to decrease in both pure Ni-10wt%Cr and CeO_2 coated anode as shown in Fig. 2. However, the voltage loss can be suppressed to less than 0.1 V in the CeO_2 coated anode. It showed better performance rather than pure Ni-10wt%Cr which had voltage loss more than 0.4 V at 150 mA/cm². CeO_2 coating layer can reduce the effect of poisoning because of its ability to remove H_2S especially in a highly reducing gas at high temperature [2,3]. N_2 cross over and internal resistance (IR) were checked to evaluate whether the single cell worked normally. All data showed that the N_2 cross-over were below 2% at the anode outlet. All internal resistance (IR) data showed the value below 6 mΩ. It was also detected that nickel

604

reacted with sulfur to form nickel sulfide in the form of Ni_3S_2 in pure Ni-10wt%Cr anode, similar result showed by Nowak, et al [4] and CeO_2 coated anode. However, in the coated anode, besides nickel sulfide, ceria sulfide (Ce_2O_2S) was formed.

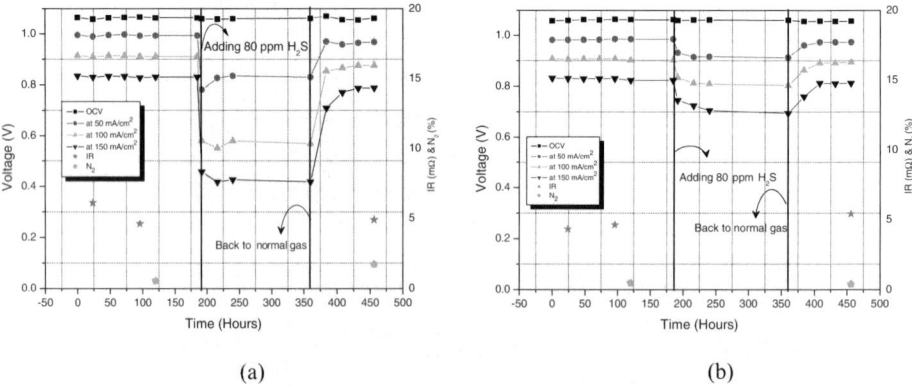

(a)   (b)

Fig. 2. Performance curve of (a) pure Ni-10Cr and (b) CeO_2 coated anode.

4. Conclusions

On the results of the cell tests with the modified anodes, the amount of electrolyte filling contents in the anode pores. Reaction site also increased in anode owing to uniform distribution of electrolyte on the anode pore surface. Moreover, ceria coating improved the anode sulfur tolerance. It had ability to suppress voltage drop caused by sulfur deposition. The ceria-coated anode showed 0.4 V higher performance than that of the uncoated anode, when 80 ppm of H_2S was introduced for 180 hours at 150 mA/cm^2. CeO_2 coating layer can reduce the effect of poisoning because it has the ability to remove H_2S especially in a highly reducing gas at high temperature. The Ni based anode was poisoned by H_2S to form Ni_3S_2, which caused a voltage drop. However, the ceria coated anode had ability to suppress this voltage drop caused by sulfur deposition. Ceria reacted with H_2S to form Ce_2O_2S.

REFERENCES

[1] A.J. Appleby, F.R. Foulkes, Fuel Cell Handbook, 6th ed., New York, Krieger Pub. Co, 1992.
[2] K.B. Yi, Ceria-Zirconia oxide high temperature desulfurization sorbent, Ph.D. Thesis, Louisiana State University, 2004.
[3] D. P. Harrison, F.R. Groves, W. N. Huang, A. L. Ortiz, J.D. White, Y. Zeng,, et al., High Efficiency Desulfurization on Synthesis Gas, Final Report, Louisiana State University, 1998.
[4] J.F. Nowak, M. Lambertin, J.C. Colson, A kinetic and morphological study of the sulphidation of Ni/23 Cr and Ni/33 Cr alloys in S$_2$ vapour and hydrogen sulphide, Corrosion Science, 18 (1978) 971.

Studies in Surface Science and Catalysis, volume 159
Hyun-Ku Rhee, In-Sik Nam and Jong Moon Park (Editors)

A study on surface treatment of Nafion membrane and its effects on cell performance

M. Prasanna, E. A. Cho[*], H.-J. Kim, T.-H. Lim and I.-H. Oh

Fuel Cell Research Center, Korea Institute of Science & Technology, Seoul 136-791, Korea

1. Introduction

One of the technical challenges in developing polymer electrolyte membrane fuel cells (PEMFCs) is reduction of the stack fabrication cost possibly by lowering Pt loadings for the electrodes. For that reason, control of the electrolyte/electrode interfacial structure can be a key technique in fabricating PEMFC stack since the electrochemical reactions producing electricity occur on the electrolyte/catalyst interfaces that are in contact with hydrogen or oxygen gas, so-called three phase boundaries (TPBs). In a previous study [1], Cho et al. reported that by roughening surface of Nafion® membrane by ion beam bombardment, maximum power density of the single cell operating on hydrogen and oxygen was almost doubled probably due to the enlarged interfacial area between the electrolyte membrane and the electrode catalyst layer. However, reduction of Pt loading is more critical in PEM fuel cells operating on hydrogen and air due to the significant kinetic loss for oxygen reduction reaction. In this work, to improve Pt utilization and hence to reduce Pt loading, effects of Pt loading and surface roughness of Nafion® membrane on the electrochemical characteristics of the PEM fuel cell operating on hydrogen and air was examined by measuring cell performance, cyclic voltammogram, and impedance spectroscopy.

2. Experimental

2.1. Preparation of ion beam treated membranes

As an electrolyte, Nafion® 112 (Du Pont, Inc) membrane was pretreated using H_2O_2, H_2SO_4 and deionized water before ion beam bombardment. The prepared membranes with a size of 8 × 8 cm^2 were mounted on a bombardment frame with a window size of 5 × 5 cm^2, equal to the active area of the test fuel cells, and dried up at 80 ℃ for 2 hr. Then, the mounted membrane was brought in a vacuum chamber equipped with a hollow cathode ion beam source as described in the previous study [1]. Ion dose was measured using a Faraday cup. Ion density bombarded onto the membrane surface was 10^{15}, 10^{16}, 5×10^{16}, 10^{17} ions/cm^2 at the ion energy of 1.0 keV. Vacuum level in the chamber was kept to be 0.133 ~5.3 $\times 10^{-5}$ kPa during the bombardments. Surface morphology of the membranes were analyzed using scanning electron microscopy (SEM, Hitachi S-4200).

2.2. Fuel cell tests

Catalyst ink was prepared by mixing 40 wt% Pt/C (Johnson Matthey, Inc.) with isopropyl alcohol (Baker Analyzed HPLC Reagent) and 5 wt% Nafion® solution (Du Pont, Inc.). Active electrode area was 25 cm^2 with catalyst loading of 0.3 mg-Pt/cm^2 for anode. Pt loading for cathode was in the range from 0.1 to 0.55 mg-Pt/cm^2. Hydrogen and air was fed to the anode and cathode, after passing through a bubble humidifier at a temperature of 80 and 65 °C, respectively, at a cell temperature of 80 °C. Performance of the single cells was evaluated by measuring *i-V* characteristics using an electronic load (Daegil Electronics Inc, EL 1000P).

3. Results and discussion

3.1. Characterization of the membranes

Fig. 1 shows SEM images for the surface of the untreated and the surface treated Nafion® 112 membranes with ion dose density from 10^{15} to 10^{17} ions/cm^2 at ion beam energy of 1 keV. With increasing the ion dose density, surface of the membrane was clearly roughened. Surface of the membrane seemed to have a nodule-like structure at ion dose densities of 10^{15} and 10^{16} ions/cm^2 and to have a whisker-like structure at 5×10^{16} and 10^{17} ions/cm^2. In the previous study [1], it was reported that with increasing ion dose density from 10^{15} to 10^{17} ions/cm^2, RMS roughness of the ion beam bombarded membrane increased from 21 to 204 nm without changing ionic conductivity of the membrane.

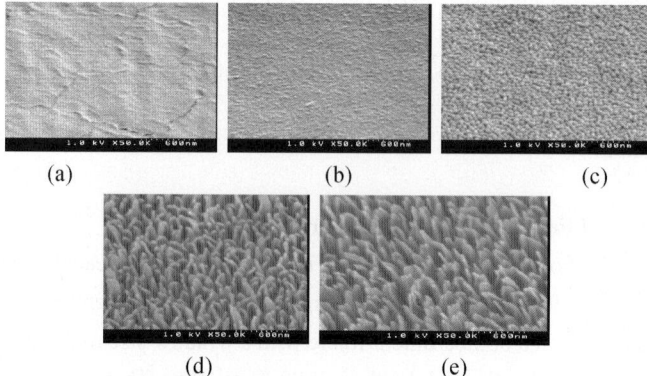

Fig. 1. Surface SEM images of the (a) untreated and untreated membrane at ion beam dose of (b) 10^{15}, (c) 10^{16}, (d) 5×10^{16} and (e) 10^{17} ions/cm^2. Ion beam energy was 1 keV.

3.2. Effects of ion dose density on cell performance

To investigate effects of ion dose density on cell performance, single cells were prepared using the untreated and the treated Nafion® 112 membranes. Fig. 2 shows i-V curves for the single cells measured at 80 °C. Open circuit voltage of the single cells was almost same to be 0.94 V. At current densities from 50 to 1,100 mA/cm^2, the single cell employing the membrane bombarded at an ion density of 5×10^{16} ions/cm^2 exhibited the highest voltage; at a current density of 800 mA/cm^2, cell voltage was measured to be 0.55, 0.54, 0.56, 0.59 and 0.55 V at ion dose densities of 0, 10^{15}, 10^{16}, 5×10^{16} and 10^{17} ions/cm^2, respectively.

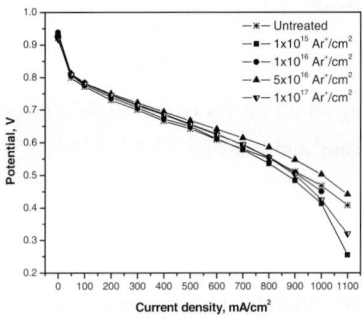

Fig. 2. Effects of ion dose density on *i-V* curves. Cathode catalyst loading = 0.2 mg-Pt/cm^2, H$_2$/air.

3.3. Effects of catalyst loading on the cell performance

Effects of the cathode catalyst loading on the cell performance was investigated for the untreated and the treated membrane at 5×10^{16} ions/cm^2 with changing cathode Pt loading from 0.1 to 0.55 mg-Pt/cm^2. Fig. 3(a) and (b) show i-V curves for the single cell using the untreated and the treated membrane, respectively, at various Pt loadings. With increasing Pt loading from 0.1 to 0.4 mg-Pt/cm^2, cell voltage at a given current density increased gradually. However, with Pt loading of 0.55 mg-Pt/cm^2, the cell voltage was drastically lowered at high current densities due to mass transport limitation caused by thick catalytic layer. At catalyst loading below 0.4 mg-Pt/cm^2, the single cell using the treated membrane exhibited higher cell voltage by 10~40 mV, implying that by using the treated membrane, cell performance could be improved and hence Pt loading could be reduced. The lower performance of the single cell using the treated membrane than that using the untreated membrane at 0.55 mg-Pt/cm^2 could be attributed to higher mass transport resistance from the gas channel to the catalyst particles on the deep valleys formed of the membrane surface by the bombardments.

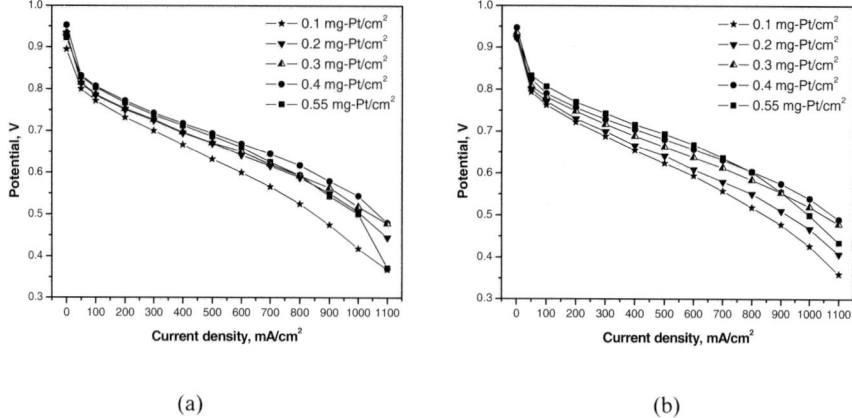

(a)   (b)

Fig. 3. Effects of catalyst loading on i-V curves of the single cells using (a) the untreated and (b) the treated membrane (5×10^{16} ions/cm^2 and 1 keV).

4. Conclusions

By ion beam bombardment onto Nafion® 112 membrane using Ar ions with ion dose densities from 10^{15} to 10^{17} ions/cm^2 at ion energy of 1 keV, surface of the membrane was roughened. With increasing the ion dose density up to 5×10^{16} ions/cm^2, cell performance and Pt utilization increased and charge transfer resistance decreased, reflecting that interfacial area between the electrolyte membrane and the electrode catalyst increased with the surface roughness. However, with ion dose density of 10^{17} ions/cm^2, a portion of catalyst probably could fall into the deep valleys of the roughened membrane and hence be isolated from the electrode, resulting in decreases in cell performance.

With increasing catalyst loading from 0.1 to 0.55 Pt-mg/cm^2, performance of single cells using the untreated and the treated membrane (5×10^{16} ions/cm^2, 1 keV) was improved up to Pt loading of 0.4 Pt-mg/cm^2 and then lowered at 0.55 Pt-mg/cm^2 due to mass transport limitation. Compared with the single cell using the untreated membrane, the single cell using the treated membrane exhibited higher performance, implying that by the ion beam bombardment on the membrane surface, cell performance could be improved and hence catalyst loading could be reduced.

REFERENCES

[1] S.A. Cho, E.A. Cho, I.-H. Oh, H.-J. Kim, H.Y. Ha, S.-A. Hong, J.B. Ju, J. Power Sources (2005) in press

Studies in Surface Science and Catalysis, volume 159
Hyun-Ku Rhee, In-Sik Nam and Jong Moon Park (Editors)

609

Preparation of Pt/mesoporous carbon catalysts and their application to the methanol electro-oxidation

Heesoo Kim[a], Pil Kim[a], Kyunghee Choi[b], Jongheop Yi[a]*

[a]School of Chemical and Biological Engineering, Institute of Chemical Processes, Seoul National University, Shinlim-dong, Kwanak-ku, Seoul 151-744, South Korea

[b]National Institute of Environmental Research, MOE, Inchon, 404-170, South Korea

*Corresponding author (Tel:+82-2-880-7438, E-mail: jyi@snu.ac.kr)

1. INTRODUCTION

There have been many studies for the application of porous carbons due to their excellent properties such as chemical inertness, conductivity, and resistance in harsh environments. One of the promising application areas of porous carbon is its use as a catalyst support for direct methanol fuel cell (DMFC). Although conventional carbon supports are good enough for obtaining highly dispersed catalysts in case of low metal loading, formation of large metal aggregates with increasing metal loading is inevitable, especially, in the DMFC system where high metal loadings (even up to 60 wt.%) are required for obtaining high power density [1]. In this respect, it is believed that excellent textural properties of mesoporous carbons (MCs) such as high surface area and uniform pore structure make them well suited as supports for highly dispersed metal catalysts in the DMFC system. In this work, MCs with different pore structure were prepared using ordered silica templates and they were used as supports for Pt catalysts with high loadings. In order to investigate the performance of the prepared catalysts in the methanol electro-oxidation, cyclic voltammograms were measured using conventional three-electrode system.

2. EXPERIMENTAL

MCM-48 and SBA-15 were synthesized for use as templating materials according to the method described in literatures [2-4]. The calcined mesoporous silica was impregnated with a solution of carbon precursor (sucrose) containing sulfuric acid. The resulting slurry was placed in an oven for 6 h at 373K and for 4 h at 433K, followed by carbonization at 1073K in a stream of nitrogen. The silica template was removed using HF or NaOH solution, to yield the ordered mesoporous carbon. The MCs derived from MCM-48 and SBA-15 were denoted as CMK-1 and CMK-3, respectively. The supported Pt catalysts were prepared by an impregnation method. Mesoporous carbons were pretreated for 2 days in HNO₃ solution (30wt.%). The supports were slowly added into an aqueous solution of Pt salt (H₂PtCl₆) with constant stirring. The resulting slurry was dried overnight at 353K, and then it was reduced at 523K for 3 h in a stream of hydrogen. For the purpose of comparison, the Pt catalyst supported on Vulcan carbon was also prepared by an impregnation method.

3. RESULTS AND DISCUSSION

N_2 adsorption-desorption isotherms revealed that MCs had high surface area (>1200 m^2/g) and large pore volume (>1.0 cm^3/g). From SAXS patterns of the prepared materials, it was confirmed that pores of SBA-15 and CMK-3 retained highly ordered 2-dimensional hexagonal type arrangement [5], while MCM-48 had 3-dimensional cubic type pore structure. It should be noted that a new scattering peak of (110) appeared in the CMK-1 after the removal of MCM-48 template. Furthermore, the pore size of CMK-1 and the wall thickness of MCM-48 were found to be 2.4 nm and 1.3 nm, respectively. This result demonstrates that a systematic transformation of pore structure occurred during the replication process from MCM-48 to CMK-1 [6].

Fig. 1 shows the TEM images of the reduced catalysts. The Pt/Vulcan catalyst showed irregular Pt particles with large aggregates, indicating poor dispersion of Pt particles on the Vulcan support. On the other hand, Pt particles supported on CMK-3 and CMK-1 were highly and uniformly dispersed. It is known that pores or channels in the mesostructured materials can serve as an individual nanoscale reactor or a barrier for preventing metal aggregation or metal growth [7]. The Pt/Vulcan catalyst had no such pores or channels to serve as a nanoscale reactor or a barrier, and therefore, metal growth easily occurred in the Pt/Vulcan catalyst during the reduction process. In addition, it was observed that pore structures of Pt/MC catalysts were unchanged even after the reduction. These features of MC supports are very important because pore structure is directly related to the catalytic activity and the deactivation behavior.

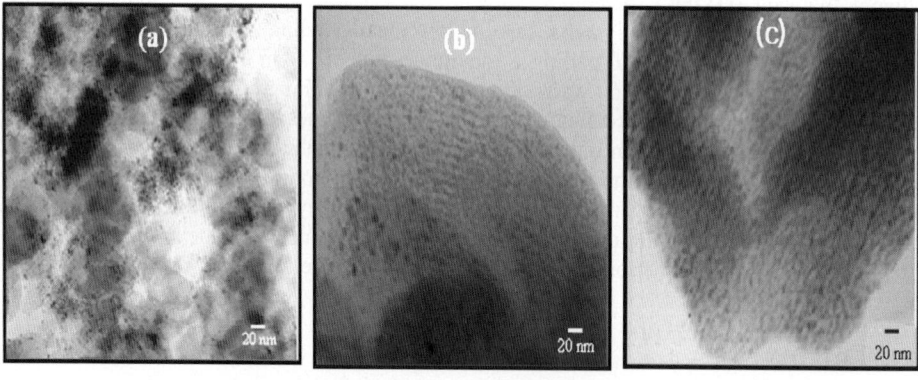

Fig. 1. TEM images of reduced catalysts: (a) Pt(40wt.%)/Vulcan, (b) Pt(40wt.%)/CMK-3, (C) Pt(40wt.%)/CMK-1.

Fig. 2 shows the cyclic voltammograms over the supported Pt catalysts obtained in the absence of methanol. Since the measurement conditions were the same in all cases, metal dispersion on the supports could be directly compared. Clearly, the integrated area of the hydrogen electro-oxidation peak for Pt/MCs was larger than that for Pt/Vulcan catalyst, indicating higher metal dispersion in the Pt/MC catalysts. The specific metal surface areas of the supported catalysts were calculated as a function of metal loading, as shown in Fig. 3. It was observed that line slope for Pt/Vulcan catalyst, compared to that for Pt/MC catalysts, was more steep, suggesting that metal dispersion of Pt/Vulcan catalyst was decreased more drastically with increasing metal loading. In addition, Pt/CMK-1 retained slightly higher metal dispersion than Pt/CMK-3, which was believed to be due to favorable pore structure of CMK-1 for metal dispersion, although fundamental reason for the different metal dispersion between Pt/CMK-1 and Pt/CMK-3 has not been well elucidated yet. It is believed that CMK-1 may serve as a promising support in the methanol electro-oxidation.

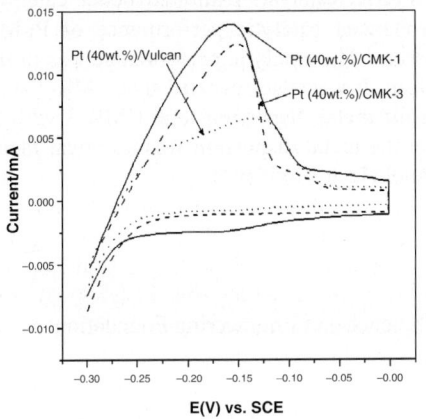

Fig. 2. Cyclic voltammograms over supported Pt (40wt.%) catalysts in H_2SO_4 solution at 298K (scan rate = 20mV/s).

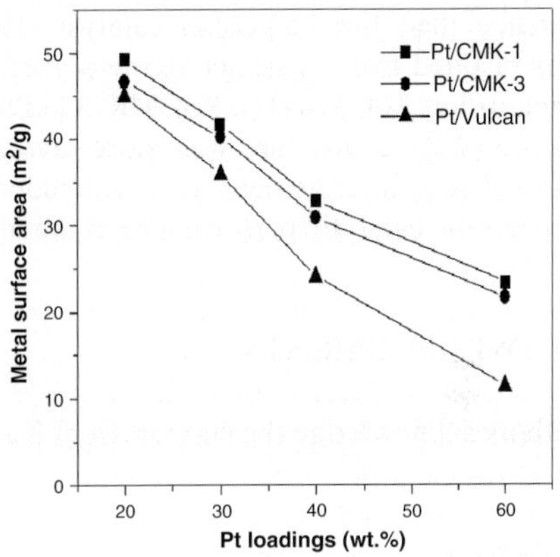

Fig. 3. Metal surface areas of supported Pt catalysts as a function of metal loading.

Fig. 4 shows the current density over the supported catalysts measured in 1 M methanol containing 0.5 M sulfuric acid. During forward sweep, the methanol electro-oxidation started to occur at 0.35 V for all catalysts, which is typical feature for monometallic Pt catalyst in methanol electro-oxidation [8]. The maximum current density was decreased in the order of Pt/CMK-1 > Pt/CMK-3 > Pt/Vulcan. It should be noted that the trend of maximum current density was identical to that of metal dispersion (Fig. 2 and Fig. 3). Therefore, it is concluded that the metal dispersion is a critical factor determining the catalytic performance in the methanol electro-oxidation.

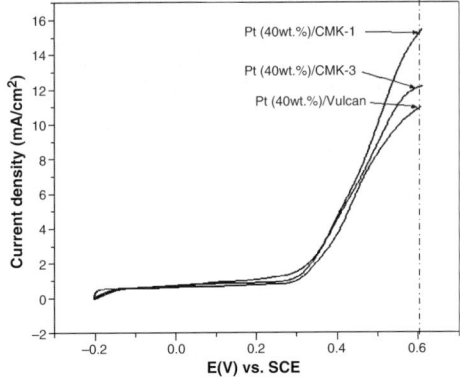

Fig. 4. Current density over supported Pt (40wt.%) catalysts in 1M CH_3OH containing 0.5M H_2SO_4 at 298K.

4. CONCLUSIONS

Mesoporous carbon materials were prepared using ordered silica templates. The Pt catalysts supported on mesoporous carbons were prepared by an impregnation method for use in the methanol electro-oxidation. The Pt/MC catalysts retained highly dispersed Pt particles on the supports. In the methanol electro-oxidation, the Pt/MC catalysts exhibited better catalytic performance than the Pt/Vulcan catalyst. The enhanced catalytic performance of Pt/MC catalysts resulted from large active metal surface areas. The catalytic performance was in the following order; Pt/CMK-1 > Pt/CMK-3 > Pt/Vulcan. It was also revealed that CMK-1 with 3-dimensional pore structure was more favorable for metal dispersion than CMK-3 with 2-dimensional pore arrangement. It is concluded that the metal dispersion was a critical factor determining the catalytic performance in the methanol electro-oxidation.

ACKNOWLEDGEMENTS

The authors acknowledge the support from Korea Science and Engineering Foundation.

REFERENCES

[1] C. Park, S.J. Lee, S.-A. Lee and H. Chang, Korean J. Chem. Eng., 22 (2005) 214.
[2] D. Zhao, J. Feng, Q. Huo, N. Melosh, G.H. Fredirckson, B.F. Chemlka and G. D. Stucky, Science, 279 (1998) 548.
[3] T. Kang, Y. Park, J.C. Park, Y.S. Cho and J. Yi, Korean J. Chem. Eng., 19 (2002) 685.
[4] R. Ryoo, S.H. Joo and J.M. Kim, J. Phys. Chem. B, 103 (1999) 7435.
[5] S. Jun, S.H. Joo, R. Ryoo, M. Kruk, M. Jaroniec, Z. Liu, T. Ohsuna and O. Terasaki, J. Am. Chem. Soc., 122 (2000) 10712.
[6] L.A. Solovyov, V.I. Zaikovskii, A.N. Shmakov, O.V. Belousov and R. Ryoo, J. Phys. Chem. B, 106 (2002) 12198.
[7] S.H. Joo, S.J. Choi, I. Oh, J. Kwak, Z. Liu, O. Terasaki and R. Ryoo, Nature, 412 (2001) 169.
[8] H.-P. Liang, H.-M. Zhang, J.-S. Hu, Y.-G. Guo, L.-J. Wan and C.-L. Bai, Angew. Chem. Int. Ed., 43 (2004) 1540.

Studies in Surface Science and Catalysis, volume 159
Hyun-Ku Rhee, In-Sik Nam and Jong Moon Park (Editors)

613

Development of Anode Catalyst for Internal Reforming of CH_4 by CO_2 in SOFC System

Byung Gwon Lee, Jung Shik Kang, Dae Hyun Kim, Sang Deuk Lee and Dong Ju Moon[*]

Hydrogen Energy Research Center, Korea Institute of Science and Technology,
39-1 Hawolgok-dong, Seongbuk-gu, Seoul 136-791, Korea (*djmoon@kist.re.kr)

1. INTRODUCTION

Considering environmental perspective, the reduction and sequestration of CO_2 have been received much attention due to its greenhouse gas and global warming effect. Therefore the catalytic reforming of CH_4 by CO_2 has been considered as an attractive technology. Above all, this reaction has an advantage such as the production of synthesis gas and the reduction of greenhouse gas, simultaneously. It is of special interest from an industrial perspective since it produces synthesis gas with a low H_2/CO ratio, which can be preferentially used for the Fischer-Tropsch synthesis. Furthermore, both CH_4 and CO_2 are the cheapest reactants and most abundant carbon-containing materials. However, the catalytic reforming of CH_4 by CO_2 has significant drawbacks. It is highly energy consuming and susceptible to coke formation. For these reasons, the various catalysts have been investigated for the reforming reaction. Because of the high cost and limited availability of noble metal catalysts such as Pt, Rh and Ru, many studies on the transition metal catalysts have been investigated to reduce the coke formation. However, there is no commercial catalyst for the catalytic reforming process. In order to overcome these problems, the authors suggested internal reforming of CH_4 by CO_2 as a new reforming system for co-production of syngas and electricity in a solid oxide fuel cell (SOFC) system [1]. In this system, the catalytic anode material is one of the most important factors affecting the electrocatalytic performance in SOFC system.

In this work, the catalytic reforming of CH_4 by CO_2 over Ni based catalysts was investigated to develop a high performance anode catalyst for application in an internal reforming SOFC system. The prepared catalysts were characterized by N_2 physisorption, X-ray diffraction (XRD) and temperature programmed reduction (TPR).

2. EXPERIMENTAL

2.1. Preparation of catalyst

The Ni based anode catalysts were prepared by a physical mixing method. NiO (99.99%, Sigma-Aldrich Co.), YSZ (TZ-8Y, TOSOH Co.), MgO (98%, Nakarai Chemical Co.) and CeO_2 (99.9%, Sigma-Aldrich Co.) were used as raw materials. The physically mixed catalyst was prepared by mixing of NiO, YSZ and MgO or CeO_2 powders and then pulverizing the mixture using a ball mill (Samhung Machinery Co., SH-BM-1). The resultant material was dried at 100 ℃ for 24 h in an air and calcined at 1350 ℃ for 2 h.

2.2. Characterization of catalyst

Physical properties of the prepared catalysts were measured by an adsorption analyzer [Quantachrome Co., Autosorb-1C]. The structure of prepared catalysts were investigated by XRD [Simmazdu Co., XRD-6000] with a Cu-Kα radiation source (λ = 1.54056 Å), voltage of 40.0 kV, current of 30.0 mA and scan speed of 5.0 deg/min. Also, temperature-programmed reduction (TPR) profiles of the samples were investigated by a sorption analyzer [Micromeritics Co., Autochem II] and obtained by heating the samples from room temperature to 1100 ℃ at a rate of 10 ℃/min in a 5 % H_2/Ar gas flow (50 ml/min).

2.3. Catalytic reforming system

The catalytic reforming of CH_4 by CO_2 was carried out in a conventional fixed bed reactor system. Flow rates of reactants were controlled by mass flow controllers [Bronkhorst HI-TEC Co.]. The reactor, with an inner diameter of 0.007 m, was heated in an electric furnace. The reaction temperature was controlled by a PID temperature controller and was monitored by a separated thermocouple placed in the catalyst bed. The effluent gases were analyzed by an on-line GC [Hewlett Packard Co., HP-6890 Series II] equipped with a thermal conductivity detector (TCD) and carbosphere column (0.0032 m O.D. and 2.5 m length, 80/100 meshes), and identified by a GC/MS [Hewlett Packard Co., 5890/5971] equipped with an HP-1 capillary column (0.0002 m O.D. and 50 m length).

3. RESULTS AND DISCUSSION

To reduce the formation of carbon deposited on the anode side [2], MgO and CeO_2 were selected as a modification agent of Ni-YSZ anodic catalyst for the co-generation of syngas and electricity in the SOFC system. It was considered that Ni provides the catalytic activity for the catalytic reforming and electronic conductivity for electrode, and YSZ provides ionic conductivity and a thermal expansion matched with the YSZ electrolyte.

Table 1. The physical properties of prepared catalysts

Catalyst	BET surface area (m^2/g)		Total pore volume (cc/g)		Active metal surface area (m^2/g)	
	Before	After	Before	After	Before	After
Ni-YSZ-MgO	8.9	13.4	0.003	0.005	0.081	0.059
Ni-YSZ-CeO_2	10.2	15.7	0.004	0.007	0.098	0.076

The characteristics of prepared catalysts are summarized in Table 1. BET surface area and total pore volume of Ni-YSZ-MgO catalyst were 8.9 m^2/g and 0.003 cc/g, respectively, while those of the Ni-YSZ-CeO$_2$ catalyst were 10.2 m^2/g and 0.004 cc/g. The BET surface area of the Ni-YSZ-CeO$_2$ and Ni-YSZ-MgO catalysts after the reaction increased 54 % and 75 % also the total pore volume increased 51 % and 67 %, respectively. It was found that the BET surface area of the catalysts increased by the formation of carbon on the catalysts [3, 4].

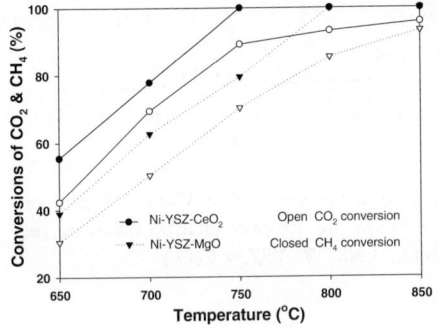 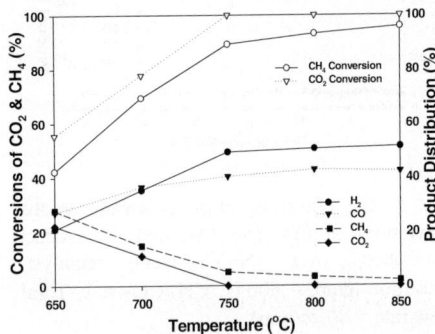

Fig. 1. The effects of temperature on the conversions of CO$_2$ and CH$_4$ over Ni-YSZ-CeO$_2$ and Ni-YSZ-MgO catalysts.

Fig. 2. The effects of temperature on the conversions of CO$_2$ and CH$_4$ and the selectivity over Ni-YSZ-CeO$_2$ catalyst.

Figure 1 shows the effects of reaction temperature on the conversions of CO$_2$ and CH$_4$ over Ni-YSZ-CeO$_2$ and Ni-YSZ-MgO catalysts. It was found that the Ni-YSZ-CeO$_2$ catalyst is showed higher catalytic activity than the Ni-YSZ-MgO catalyst at temperature range of 650 ~ 850 ℃ and the maximum activity was observed at above 800 ℃, the optimum temperature for internal reforming in SOFC system [5]. In our previous work, it was identified that Ni-YSZ-MgO catalyst was deactivated with reaction time, however Ni-YSZ-CeO$_2$ showed stable catalytic activity more than Ni-YSZ-MgO catalyst under the tested conditions [6].

The effects of temperature on the conversions of CO$_2$ and CH$_4$ and the product distribution over Ni-YSZ-CeO$_2$ catalyst are represented in Fig. 2. The concentrations of H$_2$ and CO were slowly increased with increasing reaction temperature but those of CO$_2$ and CH$_4$ were decreased. Moreover, the concentrations of H$_2$ and CO over Ni-YSZ-CeO$_2$ catalyst were slightly higher than those over Ni-YSZ-MgO [7].

Figure 3 represents the effects of time on stream on the conversions of CO$_2$ and CH$_4$ over Ni-YSZ-CeO$_2$ catalyst. The catalytic reforming reaction of a mixture of CO$_2$ (50 vol%) and CH$_4$ (50 vol%) was carried out at the reaction conditions of 800 ℃, atmospheric pressure and total flow rate of 20 cc/min for 30 h. It was found that the catalytic activity and selectivity of Ni-YSZ-CeO$_2$ catalyst were very stable during the tested conditions. The conversions of CO$_2$ and CH$_4$ over Ni-YSZ-CeO$_2$ catalyst were obtained 100 % and 94 %, respectively. X-ray diffraction (XRD) patterns of the Ni-YSZ-CeO$_2$ catalysts before and after the reaction are presented in Fig. 4. It was found that NiO phase was changed into NiC after the reaction for 30 h. It should be noted that even though a metal carbide and surface carbon were formed during the reaction, Ni-YSZ-CeO$_2$ exhibited high stability for 30 h.

616

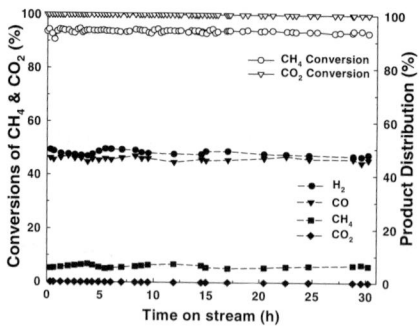

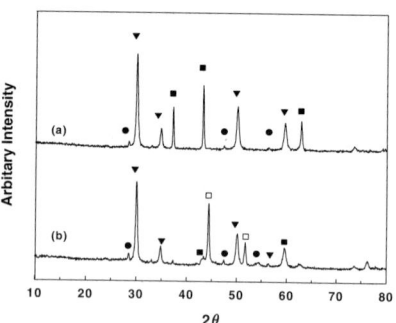

Fig. 3. The effects of time on stream on the conversions of CO_2 and CH_4 and the product distribution over Ni-YSZ-CeO$_2$ catalyst. (Reaction temp. = 800 ℃, CH_4/CO_2 = 1, Total flow rate = 40 cc/min)

Fig. 4. XRD patterns of Ni-YSZ-CeO$_2$ catalysts (a) before and (b) after catalytic reforming (■: NiO, □: NiC, ▼: YSZ, ●: CeO$_2$).

The results suggest that Ni-YSZ-CeO$_2$ is applied as an anode catalyst for internal reforming of CH_4 by CO_2 in the SOFC system. It is considered that CeO_2 played an important role of oxygen storage and supply, preventing carbon formation on catalyst surface. Furthermore, the produced syngas can be used to produce high valued chemicals such as methanol and dimethyl ether (DME).

4. CONCLUSIONS

The Ni-YSZ-CeO$_2$ catalyst in the catalytic reforming of CH_4 by CO_2 displayed higher activity than Ni-YSZ-MgO catalyst. It was found that Ni-YSZ-CeO$_2$ exhibited high stability for 30 h even though metal carbide and carbon were formed on the surface of catalyst. The conversions of CO_2 and CH_4 over Ni-YSZ-CeO$_2$ catalyst were obtained 100 % and 94 %, respectively. The results suggest that the Ni-YSZ-CeO$_2$ catalyst is applied as a catalyst anode material for internal reforming of CH_4 by CO_2 in the SOFC system.

REFERENCES
[1] D. J. Moon, J. W. Ryu, S. D. Lee, Stud. Surf. Sci. & Catal., (2004) 153.
[2] Freni, S., Cavallaro, S., Mondello, and Frusteri, F. , Catal. Commu., 4 (2003) 259.
[3] S. B. Tang and F. L. Qui, Catal. Today, 24 (1995) 253.
[4] S. Wang and G. Q. Lu, Energy & Fuels, 12 (1998) 248.
[5] D.J Moon, J.W. Ryu, T.Y. Kim, B.G. Lee and S.D. Lee, U.S. Patent 10830225 (2004).
[6] D.J Moon, J.W. Ryu, T.Y. Kim, D. M. Kang, J. M. Park, S. D. Lee and B. G. Lee, Korea Patent 2003-0074934 (2003).
[7] D. J. Moon, J. W. Ryu, Catal. Today, 87 (2003) 255.

Studies in Surface Science and Catalysis, volume 159
Hyun-Ku Rhee, In-Sik Nam and Jong Moon Park (Editors)

Studies on the Internal Reforming of CH_4 by CO_2 in SOFC System

Jong Woo Ryu[a], Jong Min Park[a], Eun Hyung Choi[a], Kye Sang Yoo[a], Suk In Hong[b] and Dong Ju Moon[a*]

[a]Hydrogen Energy Research Center, Korea Institute of Science & Technology, 39-1 Hawolgok-dong, Seongbuk-gu, Seoul 136-791, Korea (djmoon@kist.re.kr)

[b]Department of Chemical & Biological Engineering, Korea University, 5-ga, Anam-dong, Sungbukku, Seoul, Korea

1. INTRODUCTION

The CH_4 reforming by CO_2, yielding synthesis gas, has received much interests because of the feasibility of enhancing natural gas utilization and converting greenhouse gases into valuable feedstocks [1]. Furthermore, both CH_4 and CO_2 are the cheapest reactants and abundant carbon-containing materials. For these reasons, studies on the development of catalysts with high activity and resistance against coking have been reported during past decades. However, the CH_4 reforming by CO_2 has two major problems, endothermic reaction consuming much energy and carbon formation causing rapid deactivation of catalysts. Recently, Park et al. [2] reported studies on the direct oxidation of CH_4 in a solid oxide fuel cell (SOFC). In his work, the direct electrocatalytic oxidation of dry methane was verified to have a reasonable performance. Ishihara et al. [3] presented studies on the partial oxidation of methane over a fuel cell reactor for simultaneous generation of synthesis gas and electric power. Moon et al. [4] reported that electrocatalytic reforming of CH_4 by CO_2 in a solid oxide fuel cell system had some advantages over catalytic reforming. The synthesis gas generated by internal reforming can be used as fuels for power generation in SOFC and raw materials for the production of high valued chemical and fuels. In this work, the cogeneration of a syngas and electricity by the internal reforming of CH_4 by CO_2 was investigated over electrolyte supported cell (ESC) in SOFC system.

2. EXPERIMENTAL

2.1. Preparation of anode catalyst and ESC

NiO powder (99.99 %, Sigma-Aldrich Co.) and YSZ powder (TZ-8Y, TOSOH Co.) were first mixed by ball milling method and then CeO_2 (99.9 %, Sigma-Aldrich Co.) powder was added to the mixture. The sample was pulverized into a fine powder by a ball mill. The electrolyte supported cell (anode | YSZ | (LaSr)MnO$_3$)) was prepared by tape casting and screen printing methods.

2.2. Catalytic reforming system

The catalytic reforming of CH$_4$ by CO$_2$ was carried out in a fixed bed reactor system. The details can be found elsewhere [4]. The characteristics of the catalysts before and after the reforming reaction were investigated by N$_2$ physisorption, XRD and SEM.

2.3. Internal reforming over ESC in SOFC system

In the SOFC system, the flow rate of reactants was controlled by mass flow controllers (Bronkhorst HI-TEC Co.). The ceramic and inconel reactors with an inner diameter of 0.025 m were heated in an electric furnace. The Pt wires were connected to Pt meshes placed on both side of electrode surfaces as a current collector. A PID temperature controller was used to control the reaction temperature. The outlet gases were analyzed by an on-line GC equipped with a thermal conductivity detector (TCD) and using a carbosphere column (0.0032 m O.D. and 3.048 m length, 80/100 meshes). The electrochemical properties were measured by a Solatron 1287 Electrochemical Interface with a Solatron 1260 Impedance/Gain phase analyzer.

3. RESULTS AND DISCUSSION

In this work, CeO$_2$ was selected as a modification agent of NiO anodic catalyst for the cogeneration of syngas and electricity in the SOFC system, in order to reduce the formation of carbon deposited on the anode side. The characteristics of prepared NiO, CeO$_2$ and Ni-YSZ-CeO$_2$ catalyst were measured by N$_2$ physisorption. BET surface area of the NiO, CeO$_2$ and NiO-YSZ-CeO$_2$ catalyst were 1.34 m^2/g, 81 m^2/g and 10.2 m^2/g, and total pore volume of the NiO, CeO$_2$ and NiO-YSZ-CeO$_2$ catalyst were 0.0006 cc/g, 0.48 cc/g and 0.004 cc/g, respectively.

Figure 1 shows the effects of reaction temperature on the conversions of CO$_2$ and CH$_4$ and product distribution over Ni-YSZ-CeO$_2$ catalyst in the catalytic reforming of CH$_4$ by CO$_2$ in a fixed bed reactor system. It was found that the maximum activity was observed at above 800 ℃. Especially, CO$_2$ conversion over Ni-YSZ-CeO$_2$ catalyst was almost 100 % at 800 ℃. It was found that Ni-YSZ-CeO$_2$ catalyst showed higher catalytic activity than the others (not shown). The H$_2$ and CO produced in the catalytic reforming can be applied as fuel gases in the SOFC system.

The effects of reaction temperature on the conversions of CH$_4$ and CO$_2$ and the product distribution in the internal reforming of CH$_4$ by CO$_2$ over ESC (NiO-YSZ-CeO$_2$ | YSZ | (LaSr)MnO$_3$) of SOFC system are represented in Figure 2. It was found that the conversions of CO$_2$ and CH$_4$ and the concentrations of H$_2$ and CO increased with increasing reaction temperature. The conversions of CO$_2$ and CH$_4$ over ESC were 31.8 % and 32 % at 800 ℃, respectively. In our previous work [4], it was determined that the carbon produced in the catalytic reforming was destroyed by the reaction of oxide ion in the SOFC system. It was considered that internal reforming over ESC is a more desirable process than catalytic reforming, although the conversions of CO$_2$ and CH$_4$ in the internal reforming were low. In internal reforming system, the catalytic reforming must be proceeded to make electricity. It was considered that catalytic conversion of fuel gas over ASC is higher than that over ESC because of high surface area of the anode catalyst. Therefore the development of ASC and SOFC system is needed to improve conversions of CO_2 and CH_4 over ASC.

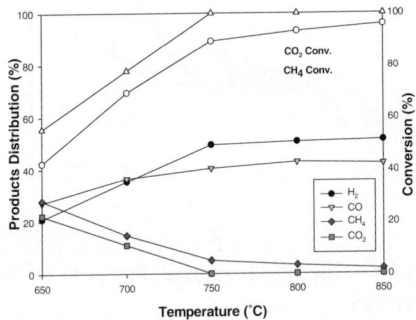

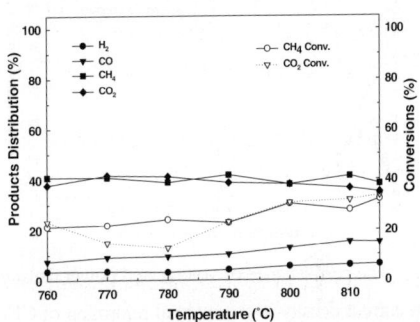

Fig. 1. Effects of temperature on the catalytic performance in the catalytic reforming of CH_4 by CO_2 over NiO-YSZ-CeO$_2$ catalyst in a fixed bed reactor system.

Fig. 2. Effects of temperature on the catalytic performance in the internal reforming of CH_4 by CO_2 over ESC of SOFC system.

Figure 3 and 4 show SEM images for the surface of anode catalyst and the cross section of ESC (NiO-YSZ-CeO$_2$ | YSZ | (LaSr)MnO$_3$), respectively. The micro structure of the catalyst electrode was characterized by SEM (Hitachi Co., S-4200). The morphology of particles over NiO-YSZ-CeO$_2$ was uniformly distributed.

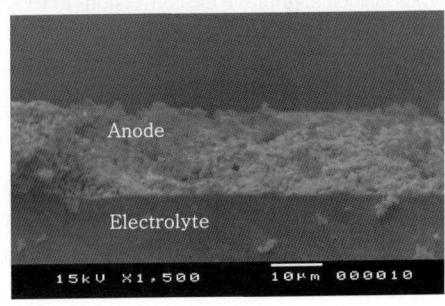

Fig. 3. SEM image for the surface of anode catalyst in ESC (NiO-YSZ-CeO$_2$ | YSZ | (LaSr)MnO$_3$).

Fig. 4. SEM images for the cross-section of ESC (NiO-YSZ- CeO$_2$ | YSZ | (LaSr)MnO$_3$).

The internal reforming of CH_4 by CO_2 in SOFC system was performed over an ESC (electrolyte supported cell) prepared with Ni based anode catalysts. Figure 5 shows the performance of voltage and power density with current density over various ESC (Ni based anodes | YSZ | (LaSr)MnO$_3$) at 800 °C when CH_4 and CO_2 were used as reactants. To improve the contact between single cell and collector, different types of SOFC reactor were used [5]. In the optimized reactor (C), it was found that the open-

620

circuit voltage over ESC (NiO-YSZ-CeO$_2$ | YSZ | (LaSr)MnO$_3$) was 1.02 V and the maximum power density of 52 mW/cm^2 was obtained at the current density of 100 mA/cm^2 at 800 ℃.

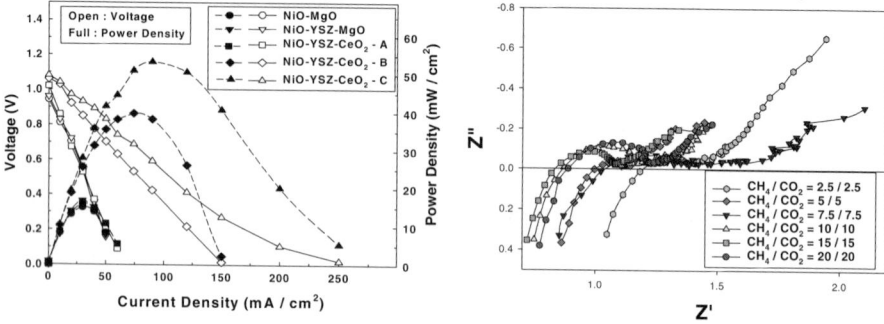

Fig. 5. The performance of voltage and power density with current density in the internal reforming of CH$_4$ by CO$_2$ over ESC (anodes | YSZ | (LaSr)MnO$_3$) of SOFC system.

Fig. 6. The effects of total flow rate of fuels (CH$_4$/CO$_2$ = 1) on the impedance in the internal reforming of CH$_4$ by CO$_2$ over ESC (NiO-YSZ-CeO$_2$ | YSZ | (LaSr)MnO$_3$) of SOFC system.

The effects of total flow rate of fuels (CO$_2$/CH$_4$ = 1) on the impedance in the internal reforming of CH$_4$ by CO$_2$ over ESC (NiO-YSZ- CeO$_2$ | YSZ | (LaSr)MnO$_3$) of SOFC system are represented in Figure 6. It was considered that the total resistance was dependent on the total flow rate because the conversions of CO$_2$ and CH$_4$ over ESC were affected by contact time in the internal reforming system.

The results suggest that the development of high performance catalyst and ASC (anode supported cell) is needed to improve the conversions of CO$_2$ and CH$_4$ and electrical performance.

4. CONCLUSIONS

The internal reforming of CH$_4$ by CO$_2$ over ESC (anode catalyst | YSZ | (LaSr)MnO$_3$) of SOFC system was successfully operated to coproduce synthesis gas and electricity, simultaneously. The results suggest that the electrocatalytic internal reforming of CH$_4$ by CO$_2$ over ESC of SOFC system is an attractive process more than the catalytic reforming, and studies on the internal reforming over ASC will be needed.

REFERENCES

[1] J.R. Rostrup-Neilsen, Catal. Today, 71 (2002) 243.

[2] S. D. Park, J. M. Vohs, R. J. Gorte, Nature, 404 (2002) 265.

[3] Ishihara T., Yamada T., Akbay T. and Takita Y. Chem. Eng. Sci., 54 (1999) 1535.

[4] D. J. Moon, J. W. Ryu, Catal. Today, 87 (2003) 255.

[5] J. M. Park, D. J. Moon, .J. S. Kang, K. S. Yoo, S. P. Yoon, S. W. Nam and S. D. Lee, Theo. & App. of Chem. Eng., 11 (2005) 497.

Studies in Surface Science and Catalysis, volume 159
Hyun-Ku Rhee, In-Sik Nam and Jong Moon Park (Editors)
© 2006 Elsevier B.V. All rights reserved

The Surface Fractal Investigation of Anode Electrode of Molten Carbonate Fuel Cell

Jung-Ho Wee, Chang-Sung Jun and Kwan-Young Lee[*]

Department of Chemical and Biological Engineering, Korea University,
5-1, Anam-dong, Sungbuk-ku, Seoul 136-701, Korea
[*]Corresponding author: kylee@korea.ac.kr

Abstract

In order to describe the geometrical and structural properties of several anode electrodes of the molten carbonate fuel cell (MCFC), a fractal analysis has been applied. Four kinds of the anode electrodes, such as Ni, Ni-Cr (10wt.%), Ni-Ni$_3$Al (7wt.%), Ni-Cr (5wt.%)-Ni$_3$Al(5wt.%) were prepared [1,2] and their fractal dimensions were evaluated by nitrogen adsorption (fractal FHH equation) and mercury porosimetry. These methods of fractal analysis and the resulting values are discussed and compared with other characteristic methods and the performances as anode of MCFC.

1. Introduction

The wetting ability of an anode electrode of a molten carbonate fuel cell is a primary factor in the electrochemical reaction occurring in the state of 3 phases interface (fuel gas-electrolyte-anode electrode). Therefore, the wetting ability of an anode electrode is one of the key factors of the cell performance [3-5]. The wetting ability of a porous solid is the function of the surface chemical property as well as physical properties such as irregularity and dimension of its inner surface structure. Fractal analysis has become a new and powerful method to describe the geometric and structural properties of porous solid, since their complex patterns are better described in terms of fractal geometry as long as the requirement of self-similarity is satisfied [6,7]. The parameter named the value of fractal dimension, D_S, is the number used to quantify these properties. It lies between 2 and 3. A regular and smooth surface possesses $D_S=2$ and a higher D_S value suggests a more irregular and space-filling surface. In our study, by use of nitrogen adsorption (fractal FHH equation) and mercury porosimetry, the values of fractal dimensions of 4 kinds of anode electrodes, which were prepared in the previous study [8,9], were evaluated to investigate their relative pore structure quantitatively. And then, we compared the values of fractal dimensions with the contact angle which is representative of the wetting ability.

2. Theory

The wetting ability of the anode electrode was evaluated as the contact angle measured by the capillary rise method. The value of fractal dimension of anode electrode of MCFC was calculated by use of the nitrogen adsorption (fractal FHH equation) and the mercury porosimetry.

2.1. Wetting ability of anode electrode

The wetting ability of an MCFC electrode is closely related to the performance of cell operation especially including electrochemical reaction, and can be expressed as contact angle between electrolyte and electrode. The surface energy of 3 phases, geometric structure of anode electrode and viscosity of electrolyte become the key factor of determining the contact angle. The capillary rise method for calculation of contact angle of irregular 3 dimensional pores was as follows [10]. When the porous anode electrode is in contact with electrolyte, the electrolyte rises up within the pore by capillary force and the differential height of electrolyte within the pore of radius r with differential time can be expressed simply by Washburn equation (1) [10].

$$\frac{dh}{dt} = \frac{\bar{r}\,\gamma_{lv}\cos\theta}{4\eta h}$$

(1)

Here, $\bar{r}$ is the average effective radius of pore, γ_{lv} is surface tension between liquid and vapor, θ is the contact angle, η is the dynamic viscosity of the electrolyte, and h is the height elevation of the electrolyte within pore at time t. In the experiment, the amount of electrolyte wetted within the anode electrode, m, expressed as $h = m/\rho AP$, was measured instead of the height, h. Integrated Eq. (1) for t becomes Eq. (2).

$$m = \sqrt{\cos\theta}\ \rho A P \sqrt{\frac{\bar{r}\,\gamma_{lv}}{2\eta}}\ \sqrt{t}$$

(2)

Here, A is the contacting surface area of anode electrode facing with electrolyte and P is the porosity of anode electrode. The average effective radius of pore, $\bar{r}$, could be calculated from the results of the capillary rise method using ethanol, which shows a contact angle of $0°$ with the anode electrode. And then, the contact angle θ could be acquired as the slope from the plot of m versus $\sqrt{t}$ in Eq. (2).

2.2. Nitrogen adsorption (fractal FHH equation) method

FHH (Frenkel-Halsey-Hill) theory is valid for multi molecules adsorption model of the flat surface material. When this model is applied for the surface fractal in the range of capillary condensation, in other words, in the state of interface which was controlled by the surface tension between liquid and gas, the modified FHH equation can be expressed as Eq. (3).

$$N \propto [\ -\ \ln X\ ]^{\,D_s\,-\,3}$$

(3)

Here, N is the amount of adsorption and X is the relative pressure (P/Po). D_s can be calculated through the slope of a log–log plot of Eq. (3) by a single nitrogen adsorption isotherm data.

2.3. Mercury porosimetry method

Considering each equation of Pfeifer [11], Washburn [12] and Rootare-Prenzlow [13], the surface area of porous solid, S_{RP}, measured by porosimeter can be expressed as Eq. (4).

$$S_{RP} \propto P^{\,Ds\,-\,2}$$

(4)

Here, P is the equilibrium pressure. Therefore, D_s of the anode electrode can be also calculated through the slope of a log–log plot of Eq. (4) by the data from the mercury porosimeter.

3. Experimental

The four kinds of anode electrodes, Ni, Ni-Cr (10wt.%), Ni-Ni$_3$Al (7wt.%), and Ni-Cr (5wt.%)-Ni$_3$Al, were prepared according to the procedure reported [8,9]. An experimental apparatus for measuring wetting ability was manufactured. It was composed of the wetting part of electrode and the measuring part of a mass-increased electrode by the wetted electrolyte. The electrolyte of the wetting part was bottled in an alumina crucible surrounded with quartz and the same composition gases of the anode part of real MCFC were introduced for the wetting ability measurement [9]. The data needed for calculation of D$_s$ were measured by BET (Micrometrics, ASAP 2010) and mercury porosimetry (Micrometrics, Autopore II 9420).

4. Results and Discussion

The wetting abilities of the four kinds of electrode are shown in Fig. 1. The wetted amount of electrolytes in Ni$_3$Al-added electrodes was larger than others. The contact angles, which were calculated by Eq. (2), of four electrodes, are listed in Table 1. The values ranged between 80.1 and 84.6°. The values of fractal dimension of the four kinds of anode electrode were obtained from a log–log plot of Eq. (3) (Fig. 2) and Eq. (4) (Fig. 3) by nitrogen adsorption method and mercury porosimetry method, respectively. The results are listed in Table 1. Their ranges of fractal dimensions were between 2.73 and 2.86 from the FHH equation and between 2.70 and 2.77 from the mercury porosimeter. A remarkable slope linearity of log-log plot of Eq. (4) was shown in the range of pore diameter between 1.2 and 2.8 μm and thus the electrode have self-similarity in this range. Due to their value of fractal dimension over 2.7, it was concluded that the four kinds of anode electrode had very irregular and rough surface structures.

Physically, the wetting ability increases (the contact angle decreases) as the values of the fractal dimension of the electrode increases if the electrode material is same. However, in this study, we could not obtain a good correlation between the fractal dimensions and the wetting abilities as shown in Table 1. It means that not only the physical properties such as the surface irregularity and roughness but also the chemical interaction between electrolyte and electrode were important in wetting ability.

With the experimental results about the wetting ability and the fractal dimension of four kinds of anode electrodes, we could conclude the following. The addition of Ni$_3$Al could make the electrolyte wet the electrode very well. The pore structures of all the electrodes prepared in this study were highly irregular and rough. Finally, the chemical properties of the surfaces were as important as the physical properties in determining the wetting ability of the electrodes in this study.

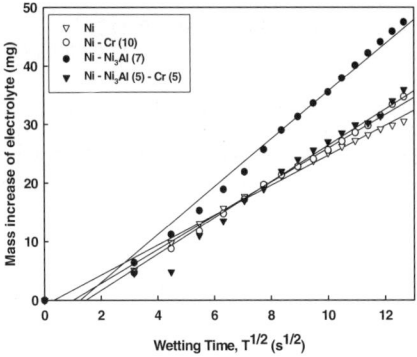

Fig. 1. Mass increases of anode electrodes with time by the contact with electrolytes.

624

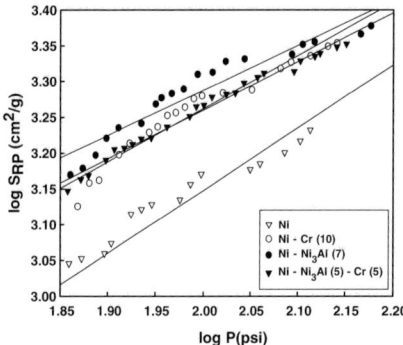

Fig. 2. Plot to calculate surface fractal dimension by mercury porosimetry data.

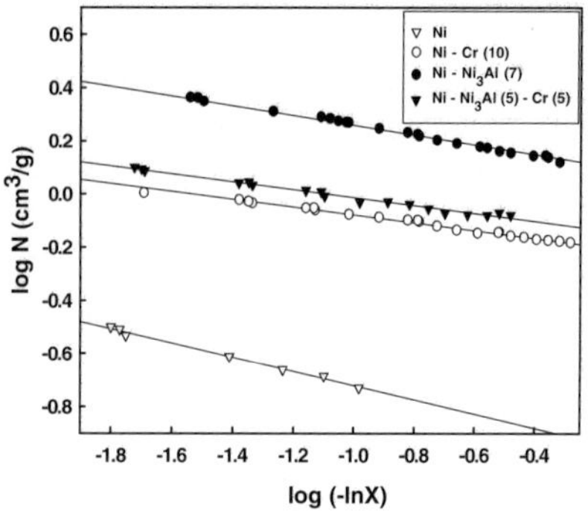

Fig. 3. Plot to calculate surface fractal dimension by nitrogen adsorption method.

Table 1 Surface fractal dimension determined by nitrogen adsorption and mercury porosimetry, and contact angle.

Anode electrode	Sintering temp. ($^\circ$C)	Initial porosity (%)	FHH equation (Nitrogen adsorption)	Mercury porosimetry	Average D_s.	Contact Angle with electrolyte θ ($^\circ$)
Ni	600	55.6	2.73	2.77	2.75	84.6
Ni-Cr (10wt.%)	700	57.4	2.86	2.74	2.80	83.2
Ni-Ni$_3$Al (7wt.%)	900	60.8	2.80	2.70	2.75	80.1
Ni-Cr (5wt.%)-Ni$_3$Al (5wt.%)	900	57.9	2.85	2.76	2.81	84.3

Acknowledgements
The authors appreciate the support by research grants from the Korea Science and Engineering Foundation (KOSEF) through the Applied Rheology Center (ARC) at Korea University.

References
[1] Y.-S. Kim, K.-Y. Lee and H.-S. Chun, *J. Power Sources*, **99** (2001) 26.
[2] H.-S. Choo, K.-Y. Lee, Y.-S. Kim and J.-H. Wee, *Intermetallics,* **13** (2005) 157.
[3] Y. Mugikura and J. R. Selman, *J. Electrochem. Soc.,* **143** (1996) 8.
[4] L. Suski, A. Godula-Jopek, and J. Oblakowski, *J. Electrochem. Soc.,* **146** (1999) 11.
[5] J.-H. Wee and K.-Y. Lee, *J. Appl. Electrochemistry,* **35** (2005) 521.
[6] S. Miller and R. Reifenberger. *J. Vac. Sci. Technol. B,* **10** (1992) 3.
[7] D. Vanderputten, J. T. Moonen, H. B. Brom, J. C. M. Brokkenzijp and M. A. I. Michels Maj. *Phys. Rev. Lett.,* **69** (1992) 3.
[8] J.-H. Wee, D.-J. Song, C.-S. Jun, T.-H. Lim, S.-A. Hong, H.-C. Lim and K.-Y. Lee, *J. Alloy Compd.,* **390** (2005) 155.
[9] J.-H. Wee, D.-J. Song, C.-S. Jun, T.-H. Lim, S.-A. Hong, H.-C. Lim and K.-Y. Lee, *J. Alloy Compd.,* **390** (2005) 161.
[10] A. Lundblad, B. Bergman, *J. Electrochem. Soc.* **144** (1997) 984.
[11] P. Pfeifer and D. Avnir, *J. Chem. Phys.,* **79** (1983) 7.
[12] E. W. Washburn, *Phys. Rev.,* **17** (1921) 374.
[13] H. M. Rootare, and C. F. Prenzlow, *J. Phys. Chem.,* **71** (1967) 8.

Studies in Surface Science and Catalysis, volume 159
Hyun-Ku Rhee, In-Sik Nam and Jong Moon Park (Editors)
© 2006 Elsevier B.V. All rights reserved

Development of Highly Compact PROX System for PEMFC Fuel Processor

**Seong Ho Lee[2], Heon Jung[3], Young-Seek Yoon[2], Byong-Sung Kwak[2]
and Kwan-Young Lee[1,*]**

[1] Department of Chemical and Biological Engineering, Korea University,
  5-1, Anam-dong, Sungbuk-ku, Seoul 136-701, Korea
[2] Energy & Environmental Research Team, R&D Center, SK Corporation,
  140-1, Wonchon-dong, Yusung-ku, Taejon 305-712, Korea
[3] Clean Energy Department, Korea Institute of Energy Research,
  71-2, Jang-dong, Yusung-ku, Taejeon 305-343, Korea
[*] Corresponding author: kylee@korea.ac.kr

Abstract

A preferential oxidation (PROX) reactor for a 10 and 25-kWe polymer electrolyte membrane fuel cell (PEMFC) systems is developed. Pt-Ru/Al_2O_3 catalyst powder with a size of 300–600 μm is used for the PROX reaction. To minimize pressure drop and to avoid hot spots in the catalyst bed, the reactor is designed as a multi-stage, multi-tube system. The steady and transient performances were investigated using gasoline-reformed gas. A newly designed heat exchanger type PROX system with non-pellet type coated catalyst was also designed and applied to a small-scale PROX system.

1. Introduction

A polymer Electrolyte Membrane Fuel Cell (PEMFC) is operated with hydrogen either from a storage tank or generated by the reforming of the fuels. Even though, in transportation applications of PEMFCs, the use of pure hydrogen is known to be the best method [1], a hydrogen infrastructure is not established at present and the storage of hydrogen on-board still has many technical problems. Therefore, an on-board reforming system is considered as a solution to the hydrogen supply problems for fuel cell powered vehicles [2]. An on-board reforming system consists of a hydrogen generating unit and a CO clean-up unit. Since CO acts as a poison to a PEMFC, it must be reduced to very low concentrations of about 10 ppm, depending on the anode material and operation conditions of the PEMFC [3,4]. In general, CO is reduced by a series process of the water gas shift (WGS) reaction and the preferential oxidation (PROX). It is known that the CO concentration in reformate after WGS is 1~2% [5,6]. The activities of Pt-Ru catalysts with various Pt/Ru ratios were investigated by us [7,8]. Considering the CO conversion and hydrogen consumption by methanation, composition of the catalyst was determined in this study. The determined catalyst was applied to compact 10 and 25kWe PROX systems successfully by a proper reactor design without a water cooling unit. A heat exchanger type, non-pellet catalyst was applied to a small-scale PROX system.

2. Experimental

2.1. Catalyst and activity test

A mixed solution of platinum and ruthenium precursors was prepared by adding H_2PtCl_6 and $RuCl_3$ at a certain ratio to de-ionized water. The solution was impregnated on γ-alumina of size 300–

600 μm. Impregnation was achieved by an incipient wetness method. The resulting paste was dried overnight until it became a powder.

The catalyst (0.15 g) was loaded into a quartz tube reactor (internal diameter = 4 mm). The catalyst was pretreated in nitrogen at 400°C. Simulated gasoline reformate was used for the activity test of the catalyst. The composition of the simulated reformate was 36 wt% H_2, 17 wt% CO_2, 28 wt% N_2, 17 wt% H_2O, 1 wt% CO, and air was added additionally as the oxidant. The total flow rate was maintained at 100 ml/min. The test was performed over the temperature range of 120–280°C at various flow rates of inlet air.

2.2. 10 and 25 kWe PROX systems

The 10-kWe PROX system is shown in Fig. 1. The reactor was designed as a dual-stage multi-tube system in order to minimize pressure drop and to allow efficient heat transfer. The 300–600 μm catalyst was used. Each stage consists of 128 parallel tubes with an internal diameter of 0.5 in. The system has an inter-stage cooling plate which is a thin disk fitted between the first and second stages. The catalyst volume is 1.2 L and the actual dimension of the reactor is 20 cm (Φ) x 20 cm (h). The gas distributor, which was specially designed for the homogeneous distribution of reactant to each parallel reactor, was installed in the inlet part of each stage. The design of the 10kWe PROX was also used for 25kWe PROX. The number of parallel tubes was increased and finally the size of the reactor was 30 cm (Φ) x 30 cm (h).

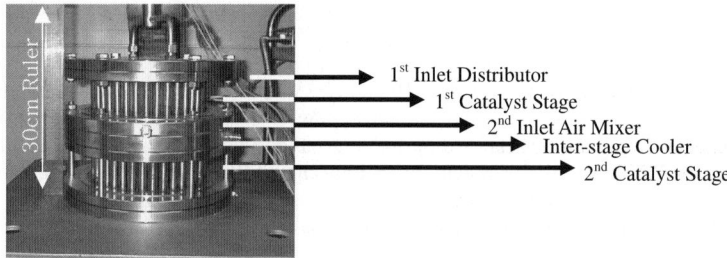

Fig. 1. 10-kWe dual-stage PROX system.

2.3. Heat-exchanger type PROX system with non-pellet catalyst

A small-scale PROX system was manufactured in a type of heat exchanger using non-pellet catalyst. Pt-Ru catalyst screened was impregnated on the support sheet. The support sheet was made by coating γ-Al_2O_3 on porous SUS-mesh plate (thickness 1.0 mm). The surface area of the catalyst sheet was 96 m^2/g. The catalyst sheet was applied to a heat exchanger type reactor of PROX as shown in Fig. 2. The PROX reactor was manufactured as a unit module and tested. Fig. 3 is the test-set of the PROX. Air was applied as the coolant.

3. Results and Discussion

3.1. Pt-Ru PROX catalyst

CO conversions over the catalysts with various ratios of Pt/Ru are shown in Fig. 4. The catalyst with composition E showed the highest conversion among the screened catalysts over the temperature range of 120-240°C. As the composition was changed from A to E, methane concentration in effluent gas was increased. Although the catalyst with composition E among the screened catalysts showed the best performance for the CO removal, the catalyst with composition D was chosen for the PROX, since, as shown in Fig. 5, hydrogen consumption by methanation was too high in case of the catalyst with composition E.

3.2. Development of 10kWe and 25kWe PROX system

Fig. 6 shows 10kWe single-stage (catalyst volume, 0.06 L/kWe) and dual-stage (catalyst volume, 0.12 L/kWe) performances. In single-stage performance, the outlet CO concentration decreased with the increase in the inlet air. However, when the air was fed over 1.4 of stoichiometric O_2/CO ratio, the outlet CO concentration increased on the contrary. This means that excess inlet air brought about the large extent of H_2 oxidation and hence CO selectivity became low at the elevated temperature. Dual stage performances were evaluated adding the first and secondary air to each stage. The O_2/CO ratio of the 1st stage was held constant and the amount of the secondary air was controlled so that the outlet CO concentration of the 2nd stage could be below 20ppm. The optimum steady state was determined at the condition that the total amount of inlet air was minimized and the outlet CO concentration of the 2nd stage maintained below 20ppm. When the 1st oxygen stoichiometry was in the range of 1.0-1.2, the amounts of total inlet air could be minimized as shown in Fig. 6. At the optimum condition, the outlet CO concentration was below 20ppm and the hydrogen loss was below 2%.

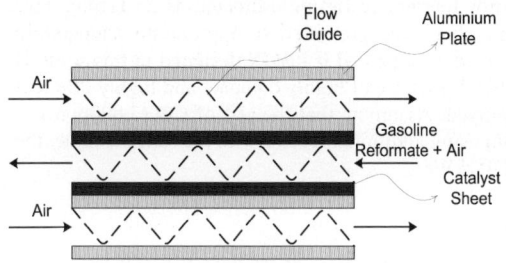

Fig. 2. Schematic diagram of heat exchanger type PROX.

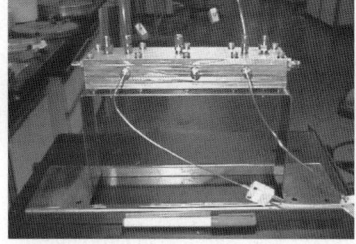

Fig. 3. Test-set of heat exchanger type PROX.

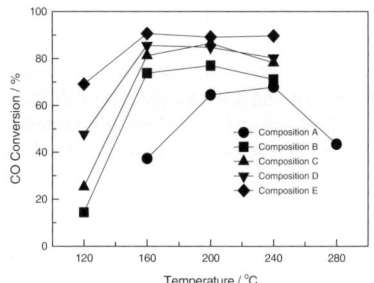

Fig. 4. CO conversions over the catalysts with various compositions of Pt/Ru. GHSV 80,000hr^{-1}, O_2/CO = 1.0 (stoichiometric ratio).

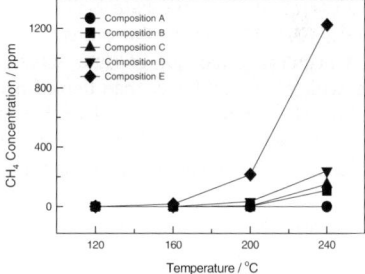

Fig. 5. Methane concentrations in the effluent gas. GHSV 80,000 hr^{-1}, O_2/CO = 1.0 (stoichiometric ratio).

The design of 10kWe PROX was also used for the 25kWe PROX. When the inlet CO concentration in the reformate gas was 10,000ppm, the outlet CO concentration was 1,500ppm at 1.2 of the oxygen stoichiometric ratio. Also, when the inlet CO concentration in the reformate gas was 1,500ppm for the second stage, the outlet CO concentration could be lowered less than 20ppm at 2.5 of the oxygen stoichiometric ratio. These results could be obtained by the dual-stage 25kWe PROX given in section 2.2.

3.3. Development of PROX with a heat exchanger type (Non-pellet catalyst)

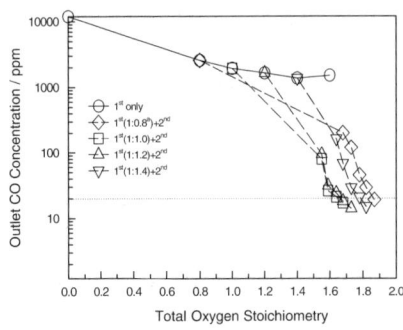

Fig. 6. 10kWe Steady Performance.
(^{a}:O_2/CO ratio for the first stage alone).

The heat exchanger type PROX was tested in a 0.4kWe scale using gasoline reformate containing 1% CO. The test was performed under adiabatic and non-adiabatic condition, respectively. Under adiabatic condition, the outlet CO concentration was 3,000ppm consuming 1.0 % hydrogen. On the other hand, under non-adiabatic condition using air coolant, the outlet CO concentration was 2,000ppm consuming 0.08 % of hydrogen. At this condition, the temperature of air (20 L/min) used as coolant increased from normal temperature to 93°C. Considering 80% of hydrogen utilization in 0.4kWe PEMFC stack, the amount of air flow to cathode can be estimated as 25 L/min. This air should be preheated to appropriate temperature (70-80°C) before entering stack. The required heat can be supplied if PROX designed in this study is used. Therefore, it can be concluded that a new PROX system, of highly compact and highly efficient thermal-integrated system, was successfully developed. Assuming that the rest of CO (2,000ppm) in the outlet reformate can be treated less than 20ppm using additional reactor with the same volume, the specific catalyst volume is estimated as low as 0.04L/kWe.

4. Conclusions

10kWe PROX was developed with highly compact volume (catalyst volume 0.12L/kWe) and it showed good performance. The outlet CO concentration was below 20ppm at steady state consuming 1.8% of hydrogen. A 25kWe PROX with compact volume was also developed and it showed steady performance. The outlet CO concentration was also controlled below 20ppm. To meet the challenge of non-pellet application, a small scale heat-exchanger type PROX was manufactured and evaluated. Using one stage of the PROX, the outlet CO concentration could be reduced from 10,000ppm to 2,000ppm. Considering the use of dual stage, the catalyst volume could be estimated as low as 0.04L/kWe, which is much lower than that of packed-bed design. The heat exchange type PROX has other benefits. The air heated by heat exchange could be used efficiently for other sub-system in PEMFC system since the air fed into the auto-thermal reformer or cathode of PEMFC should be heated to each operating temperature for the proper operation.

Acknowledgements

The authors appreciate the support by research grants from the Korea Energy Management Corporation (KEMCO), SK Corporation, and National RD&D Organization for Hydrogen & Fuel Cell.

REFERENCES

[1] M.J. Bradley, Future Wheels: Report Submitted to U.S. Department of Defense by Northeast Advanced Vehicle Consortium, Nov. 2000, p. 6.
[2] A. Docter, A. Lamm, *J. Power Sources* **84** (1999) 194.
[3] S. Chalk, FY 2000 Progress Report for Fuel Cell Power Systems, Energy Efficiency and Renewable Energy Office of Transportation Technologies, U.S. Department of Energy, Oct. 2000, p. 163.
[4] S. Gottesfeld, U.S. Patent 4,910,099
[5] M.M. Schubert, M.J. Kahlich, H.A. Gasteiger, R.J. Behm, *J. Power Sources* **84** (1999) 175
[6] M.J. Kahlich, H.A. Gasteiger, and R.J. Behm, *J. Catal.* **182** (1999) 430.
[7] Seong Ho Lee, Jaesung Han, Kwan-Young Lee, *Korean J. Chem. Eng.*, **19** (2002) 431.
[8] Seong Ho Lee, Jaesung Han, Kwan-Young Lee, *Journal of Power Sources*, **109** (2002) 394.

Studies in Surface Science and Catalysis, volume 159
Hyun-Ku Rhee, In-Sik Nam and Jong Moon Park (Editors)

Dynamic Simulation of Plate-Type Reformer and Combustor System for Molten Carbonate Fuel Cell

Jae Min Hong[a], Ju Seok Lee[a], Ahrim Yoo[a], Sang Deuk Lee[b] and Dae Ryook Yang[a*]

[a]Department of Chemical and Biological Engineering, Korea University,
1 Anamdong, Seonbukku, Seoul, 136-713, Korea

[b]Reaction Media Research Center, Korea Institute of Science and Technology
39-1 Hawolkokdong, Seongbukku, Seoul, 136-791, Korea

1. INTRODUCTION

Molten Carbonate Fuel Cell (MCFC) is one of the most promising clean energy sources and it has been a popular research subject recently. For MCFC, hydrogen is the feed material to generate electricity and can be produced by steam reforming, which converts the light hydrocarbons to hydrogen. Since the combined steam reforming and water gas shift reaction in the reformer is endothermic, a high temperature heat source is required to operate the reactor. Therefore, the reformers are combined with combustors in order to provide the required heat source [1, 2]. The most popular reformer types for MCFC with external reforming are concentric annular-type reformers and plate-type reformers. Plate-type reformers are more compact than annular-type reformers with catalyst beds. In the plate-type design, the reformer plates are arranged in a stack alternating the reformer and combustor plate as in Fig. 1. The difficulty of the plate-type reformer is to provide the generated heat to the reformer evenly in order to attain high conversion while preventing hot spots. In order to investigate whether the designs can overcome the difficulty, mathematical models can be utilized. Although various reformer models were suggested by previous researchers, the pressure in the reactor was assumed to be constant in their models. However, since pressurized operation of MCFC is preferred to improve the performance, the effect of pressure on the reaction should also be considered for modeling. In this paper, dynamic simulation of a plate-type reformer is perfomed considering the pressure drop using the momentum balances unlike previous models [3, 4]. From the simulation results, the interaction between the reformer and the combustor is analyzed, and the Model Predictive Control (MPC) system is applied to control the hydrogen production rate and reformer temperature.

---

* Authors to whom correspondence should be addressed. Email: dryang@korea.ac.kr

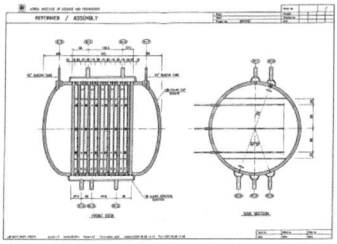

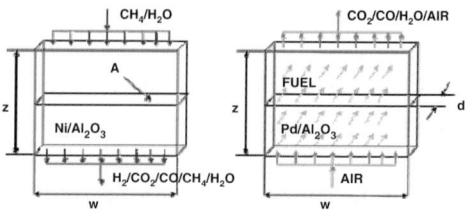

Fig.1 Plate type reformer/combustor system.　　　Fig.2 Unit model of reformer and combustor.

2. REFORMER/COMBUSTOR CONFIGURATION AND DYNAMIC MODEL

The reformer feeds and combustor air flow in a counter current manner as shown in Fig. 2. In order to transfer heat to the reformer evenly throughout the interface between reformer and combustor, the combustor is designed to feed the fuel through the holes distributed over the combustor. In this manner, the feed will mix with air incrementally and generate heat throughout the combustor plate evenly. The combustor plate is packed with a Pd catalyst and the reformer uses a Ni/Al_2O_3 catalyst.

Based on this configuration, the reformer and combustor are modeled with partial differential equations. Since the thickness of the plates is relatively small, only the flow direction is considered. Using the equation of continuity, the component mass balances are constructed and the energy balance considering with heat loss and momentum balance are established as follows.

Equation of continuity:

$$\frac{\delta C_{i,r}}{\delta t} = -\frac{\delta(v_r C_{i,r})}{\delta z} + \frac{(1-\varepsilon_r)}{\varepsilon_r}\rho_{cat,r}\eta_i \gamma_r \quad , \qquad \frac{\delta C_{i,c}}{\delta t} = -\frac{\delta(v_c C_{i,c})}{\delta z} + \frac{F_{fuel}}{\varepsilon_c LA}\eta_i a \qquad (1)$$

Equation of Energy:

$$\frac{\delta h_r}{\delta t} = -\frac{\delta(v_r h_r)}{\delta z} + \frac{W}{A}U(T_c - T_r) + (1-\varepsilon_r)\rho_{cat,r}\Delta H_r^T r_r \qquad (2)$$

$$\frac{\delta h_c}{\delta t} = -\frac{\delta(v_c h_c)}{\delta z} - \frac{W}{A}U(T_c - T_r) + \frac{F_{fuel}}{LA}\Delta H_c^T a \qquad (3)$$

Equation of motion: $\quad \dfrac{\delta P_r}{\delta z} = -1.75\dfrac{\rho_r v_r^2}{d_{p,r}}\dfrac{(1-\varepsilon_r)}{\varepsilon_r^3} \quad , \quad \dfrac{\delta P_c}{\delta z} = -1.75\dfrac{\rho_c v_c^2}{d_{p,c}}\dfrac{(1-\varepsilon_c)}{\varepsilon_c^3} \qquad (4)$

Equation of state: $\quad \rho_{m,r} = P_r / RT_r \quad , \quad \rho_{m,c} = P_c / RT_c \qquad (5)$

Equation of heat loss: $\quad M_b C_p (dT_b / dt) = U_{b,c}A_{b,c}(T_c - T_b) - U_{b,a}A_{b,a}(T_b - T_a) \qquad (6)$

In the model equations, A represents the cross sectional area of reactor, a is the mole fraction of combustor fuel gas, C_i is the molar concentration of component gas, C_p the heat capacity of insulation and F is the molar flow rate of feed. The ΔH denotes the heat of reaction, L is the reactor length, P is the reactor pressure, R is the gas constant, T represents the temperature of gas, U is the overall heat transfer coefficient, v represents velocity of gas, W is the reactor width, and z denotes the reactor distance from the inlet. The Greek letters, ε is the void fraction of catalyst bed, ρ the molar density of gas, and η is the stoichiometric coefficient of reaction. The subscript, c, cat, r, b and a represent the combustor, catalyst, reformer, the insulation, and ambient, respectively. The obtained PDE model is solved using Finite Difference Method (FDM).

3. RESULTS AND DISCUSSIONS

From the steady-state simulation result as in Fig. 3, the temperature of the reformer at the entrance showed a temperature drop due to the endothermic reforming reaction. In the middle of the reformer, the temperature is increased by the heat supplied from the combustor. On the other hand, the combustor temperature is lower at the entrance (exit section for reformer) compared to the rest of the part due to fresh air supply. At the exit, the reformer temperature has a tendency to fall slightly due to the counter-current flow pattern. The Fig. 4 shows the dynamic simulation results of the reformer outlet gas when the mole flow rate of reactant feed is increased by 10% at 1 sec. The hydrogen yield increases due to the plentiful reactant amount initially, and the yield dropped due to the decrease of the reformer temperature. Figure 5 shows the dynamic simulation results when the mole flow rate of combustor fuel is increased by 10% at 30 sec. From this simulation, it is found that the interactions between variables are significant. In order to provide a stable feed to a fuel cell which might require frequent load change, a reliable control system should be devised. In this study, standard Model Predictive Control (MPC) is applied to control the hydrogen production rate and reformer temperature by manipulating the reformer feed and combustor fuel flow rates. Figure 6 shows the simulation result of the control system when the set-point of the H_2 flow rate is increased by 10% at 15 sec. The MPC shows satisfactory performance while the performance of convention PID control is not acceptable due to interaction.

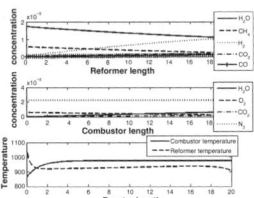

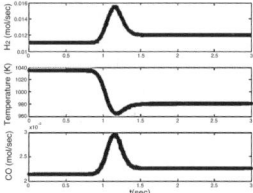

Fig. 3. Steady-state profile of reformer and combustor.

Fig. 4. Reformer outlet profile with reactant feed rate change by 10% at 1 sec.

632

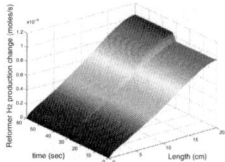
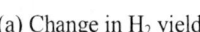
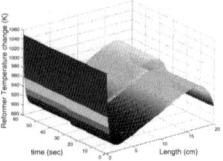

(a) Change in H₂ yield   (b) Change in reformer temperature

Fig. 5. Result of reformer dynamic simulation with combustor feed rate change by 10% at 30 sec

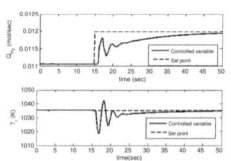

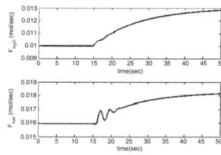

(a) Change in controlled variables   (b) Change in manipulated variables

Fig. 6. Performance of control system with set point change in H₂ production rate by 10% at 15 sec

4. CONCLUSIONS

In this study, a plate-type reformer for MCFC has been designed and analyzed. Also, a control system using MPC has been applied to provide a stable hydrogen source to fuel cell. The plate-type reformer is configured the reformer and combustor in alternating manner to maximize the heat transfer between the reformer and combustor. The models for plate-type reformers and combustors have been developed in a PDE form and the effects of the important manipulated variables on the crucial output variables have been investigated. From the simulations, it is found that the interactions between variables are significant and an efficient control system is required. Thus, the MPC is designed to control the H_2 production rate and the reformer temperature by manipulating the reformer feed and the combustor fuel flow rates.

REFERENCES

[1] A.J. Appleby and F.R. Foulkes, Fuel cell handbook, Van Nostrand Reinhold, 1989.

[2] M. Yamaguchi, T.Saito, M. Izumitani, S. Sugita and Y. Tsutsumi, Analysis of Control Characteristics Using Fuel Cell Plant Simulator. IEEE Transactions on Industrial Electronics, 37 (1990) 378.

[3] C.R.H. de Smet, M.H.J.M. de Croon, R.J. Berger, G.B. Marin and J.C. Schouten, Chemical Engineering Science, 56 (2001) 4849.

[4] W. He and Kas Hemmes, Operating characteristics of a reformer for molten carbonate fuel-cell power-generation systems. Fuel Processing Technology, 67 (2000) 61.

Studies in Surface Science and Catalysis, volume 159
Hyun-Ku Rhee, In-Sik Nam and Jong Moon Park (Editors)

Electrochemical properties of various transition metal oxides for energy storage

Hyun Wook Jung[a]* **and Yeon Uk Jeong**[b]

[a]Dept. of Chemical and Biological Engineering, Korea University, Seoul 136-713, Korea

[b]Dept. of Inorganic Materials Engineering, Kyungpook National University, Daegu 702-701, Korea

1. INTRODUCTION

Rapid growth in the markets of portable electronic devices and hybrid electric vehicles for clean air has created enormous interests and challenges for the improvement of electrochemical power sources. There are several types of electrochemical power sources such as batteries, fuel cells, and electrochemical supercapacitors, etc. While batteries and fuel cells offer high energy densities, electrochemical capacitors give high power density. There are two kinds of electrochemical capacitors: One type is electrical double layer capacitor (EDLC) which operate via an electrostatic reaction and the other is pseudo capacitor which use a redox reaction [1]. In the case of the EDLC, an activated carbon with high surface area $(2000\text{-}3000\text{m}^2/\text{g})$ supplies specific capacitance value of $\sim 300\text{F/g}$ in acid electrolytes. Pseudo capacitor using amorphous hydrous ruthenium dioxide provides $\sim 700\text{F/g}$ [2]. To develop new and inexpensive materials, various attempts have been pursued [3-7]. Transition metal oxides with nanocrystalline, amorphous, or metastable phases have been synthesized in this study by novel solution-based chemical synthesis procedures. Furthermore, their electrochemical reactions and performances have been investigated.

2. EXPERIMENTAL

2.1. Synthesis procedures

Amorphous $WO_3 \cdot xH_2O$ was prepared by adding dilute HCl to 100mL of a 0.1M aqueous solution of $Na_2WO_4 \cdot 2H_2O$. At pH < 1, $WO_3 \cdot xH_2O$ was obtained through filtering, washing with deionized water, and drying in air. Amorphous $WO_2 \cdot xH_2O$ and $Na_yWO_3 \cdot xH_2O$ were also synthesized by a reduction reaction. 50mL of 0.25M $Na_2WO_4 \cdot 2H_2O$ was reduced with 100mL of 0.25M $NaBH_4$ at various pH conditions. At pH < 2, $WO_2 \cdot xH_2O$ was formed, which was then soaked in methanol for 3 hours, filtered, washed with methanol, and dried in air. At pH=7, $Na_yWO_3 \cdot xH_2O$ was produced through filtering, washing with deionized water, and drying in air. In order to prepare amorphous $ReO_3 \cdot xH_2O$, 25mL of 0.1M $NaReO_4$ was reduced with 20mL of 0.25M $NaBH_4$ at pH=1. The product was then filtered, washed with deionized water and dried in air. A part of the product was also soaked in acetone and dried in vacuum at $100°C$.

2.2. Materials analysis and electrochemical characterization

X-ray powder diffraction was used to identify the phases of samples using a Philips Model APD 3520 powder diffractometer with CuKα radiation. A Hitachi S-4500 Field Emission Scanning Electron Microscope (SEM) was used to investigate the particle size and morphology. The thermal behaviors were observed using a Perkin-Elmer Series 7 Thermogravimetric Analyzer (TGA). A Cameca SX50 Wavelength Dispersive Spectroscopy (WDS) was employed for quantitative analysis of Na_xWO_3 samples. The surface areas of powder samples were measured by B.E.T. For the evaluation of electrical resistivity, four probe techniques were applied with the samples of pressed pellets. For the electrochemical measurement, electrodes were prepared by mixing the active material with 15-25wt% acetylene black and 5wt % PTFE. The electrodes were fixed on a tantalum current collector. The cyclic voltametric behaviors of electrochemical capacitors were predicted by an EG&G model 273 potentiostat.

3. RESULTS AND DISCUSSION

3.1. Tungsten oxides

As illustrated in Fig. 1, X-ray diffraction shows that the as-prepared $WO_3 \cdot xH_2O$, $WO_2 \cdot xH_2O$ and $Na_yWO_3 \cdot xH_2O$ are amorphous to X-ray. The amorphous phases transform to crystalline phases at higher temperatures. WDS analysis indicates the sodium to tungsten ratio is 0.37:1. Figs. 2 and 3 show SEM morphology of $Na_{0.37}WO_3 \cdot xH_2O$ sample and its TGA behavior, respectively. A weight loss of about 12 % below 200°C corresponds to the loss of water. Fig. 4 shows the cyclic voltammograms of amorphous $WO_3 \cdot xH_2O$, $WO_2 \cdot xH_2O$, and $Na_{0.37}WO_3 \cdot xH_2O$ collected with a scan rate of 2mV/s in H_2SO_4 electrolyte. WO_3 is a well known electrochromic material that undergoes a color change by accepting H^+ or Li^+.

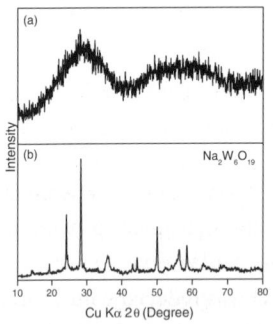

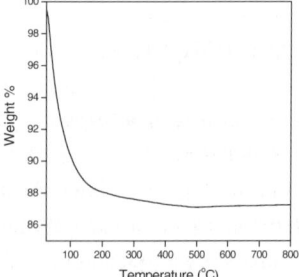

Fig. 1. X-ray powder diffraction patterns of $Na_{0.37}WO_3 \cdot xH_2O$: (a) as-prepared and (b) after heating in TGA in N_2 atmosphere to 800°C.

Fig. 2. SEM picture of $Na_{0.37}WO_3 \cdot xH_2O$.

Fig. 3. TGA plot of $Na_{0.37}WO_3 \cdot xH_2O$.

The sharp peak around 0V in Fig. 4 may be caused by the insertion of protons into the empty A sites of the poorly crystalline WO_3 with a distorted perovskite structure. Amorphous

$WO_2·xH_2O$ is expected to show better electrochemical capacitor behavior than amorphous $WO_3·xH_2O$ since the former has better conductivity and the electrochromic insertion reaction can be suppressed due to its different structure. However, $WO_2·xH_2O$ gives only a small improvement in the capacitor behavior (Fig. 4b) compared to $WO_3·xH_2O$ (Fig. 4a). Fig. 4(c) shows the cyclic voltammogram of $Na_{0.37}WO_3·xH_2O$. This material provides good capacitor property in the range of -0.1 to 0.3V, and it has a capacitance value of 50 F/g, corresponding to a participation of 0.06 H^+ per $Na_{0.37}WO_3·xH_2O$. The better capacitance behavior of $Na_{0.37}WO_3·xH_2O$ could be explained by its better electronic conductivity compared to $WO_3·xH_2O$ and the presence of Na^+ ions in the A sites of the $Na_{0.37}WO_3·xH_2O$. Fig. 5 compares the electronic conductivities of WO_3 and $Na_{0.37}WO_3$ samples, which were dried at 200°C in air to remove water contents in the samples.

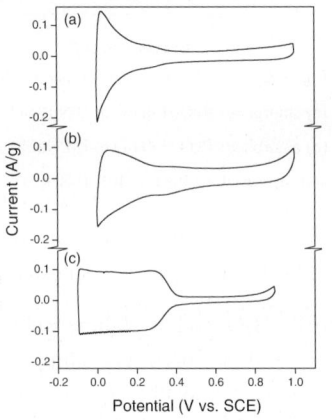 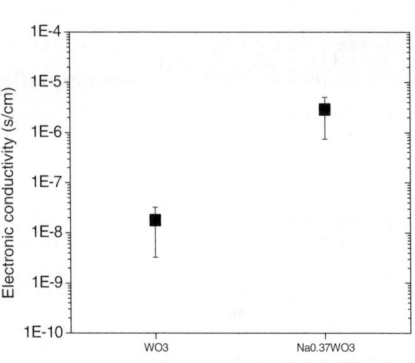

Fig. 4. Cyclic voltammograms of various tungsten oxides: (a) $WO_3·xH_2O$, (b) $WO_2·xH_2O$, and (c) $Na_{0.37}WO_3·xH_2O$.

Fig. 5. Comparisons of electronic conductivities of WO_3 and $Na_{0.37}WO_3$.

3.2. Rhenium oxides

Fig. 6 displays the X-ray powder diffraction patterns of rhenium oxide. While the as-prepared sample is amorphous to X-ray, the sample soaked in acetone and dried at 100°C clearly exhibits sharp reflections corresponding to ReO_3. The large difference between the two X-ray patterns suggests that the processing conditions play a key role in the crystallinity and surface characteristics. As shown in the TGA plot of the as-prepared sample (Fig. 7), the weight loss of about 10% below 100°C results from the loss of water.

Fig. 8 shows the cyclic voltammograms of the rhenium oxides collected with a scan rate of 2 mV/s. The as-prepared $ReO_3·1.4H_2O$ sample gives a capacitance value of 150F/g in 10N H_2SO_4, but over a narrow voltage range of -0.2 to 0.3V (Fig. 8a). The crystalline ReO_3 obtained by drying the acetone-soaked sample provides a much higher capacitance value of 516F/g (Fig. 8c), in accordance with the participation of 0.56 protons per Re. The cyclic voltammogram of the as-prepared sample in a neutral electrolyte (1M NaCl) is shown in Fig. 8b. Replacement of acid electrolyte by a neutral electrolyte leads to a widening of the voltage range, but the capacitance value is lowered from 150F/g to 50F/g. A larger and heavier Na^+ ion compared to H^+ ion and the lower ionic conductivity of the neutral electrolyte result in a lower capacitance.

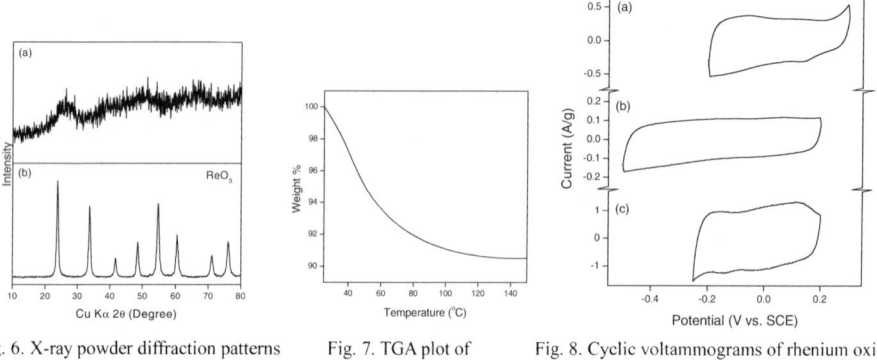

Fig. 6. X-ray powder diffraction patterns of $ReO_3 \cdot xH_2O$: (a) as-prepared sample and (b) sample soaked in acetone and dried at 100°C in vacuum.

Fig. 7. TGA plot of as-prepared $ReO_3 \cdot xH_2O$

Fig. 8. Cyclic voltammograms of rhenium oxides: (a) amorphous $ReO_3 \cdot 1.4H_2O$ in 10N H_2SO_4, (b) amorphous $ReO_3 \cdot 1.4H_2O$ in 1M NaCl, and (c) crystalline ReO_3 in 10N H_2SO_4.

4. CONCLUSIONS

Amorphous $Na_yWO_3 \cdot xH_2O$, $WO_3 \cdot xH_2O$, $WO_2 \cdot xH_2O$, $ReO_3 \cdot xH_2O$, and crystalline ReO_3 have been successfully synthesized by solution-based synthesis procedures, providing smaller particle size as well as metastable phases. It has been found that synthesis conditions and drying temperatures influence the surface area, water contents in the samples, electrochemical performances for capacitors. $Na_yWO_3 \cdot xH_2O$ shows better capacitor property than $WO_3 \cdot xH_2O$ and $WO_2 \cdot xH_2O$, but has a smaller capacitance value of 50F/g in a voltage range of -0.1 to 0.3V. While amorphous $ReO_3 \cdot xH_2O$ gives a smaller capacitance of 150F/g, the crystalline ReO_3 with high surface area supplies a capacitance of 516F/g. Further optimization of the synthesis conditions may lead to higher capacitance values than those by $RuO_2 \cdot xH_2O$.

REFERENCES

[1] B.E. Conway, Electrochemical Supercapacitors, Kluwer Academic/Plenum, New York, 1999.
[2] J.P. Zheng, P.J. Cygan, and T.R. Jow, J. Electrochem. Soc., 142 (1995) 2699.
[3] O.R. Camara and S. Trasatti, Electrochim. Acta, 41 (1996) 419.
[4] K.C. Liu and M.A. Anderson, J. Electrochem. Soc., 143 (1996) 124.
[5] Y. Takasu, S. Mizutani, M. Kumagai, S. Sawaguchi and Y. Murakami, Electrochem. Solid-State Lett., 2 (1999) 1.
[6] H.Y. Lee and J.B. Goodenough, J. Solid State Chem., 144 (1999) 220.
[7] Y.U. Jeong and A. Manthiram, J. Electrochem. Soc., 148 (2001) A189–A193.

Studies in Surface Science and Catalysis, volume 159
Hyun-Ku Rhee, In-Sik Nam and Jong Moon Park (Editors)

637

Principal design parameters of electro-catalysts for PEMFCs

Gu-Gon Park, Young-Jun Sohn, Sung-Dae Yim, Tae-Hyun Yang, Young-Gi Yoon, Won-Yong Lee, Koichi Eguchi[a], and Chang-Soo Kim[*]

Fuel Cell Research Center, Korea Institute of Energy Research

71-2, Jang-dong, Yuseong-ku, Daejeon, 305-343, Republic of Korea

[a]Department of Energy and Hydrocarbon Chemistry, Graduate School of Engineering,

Kyoto University, Nishikyo-ku, Kyoto 615-8510, JAPAN

Abstract

Design parameters of the anode catalyst for the polymer electrolyte membrane fuel cells(PEMFCs) were investigated in the aspect of active metal size and inter-metal distance. Pt-Ru catalysts which have various morphologies could be prepared by using systematically pretreated Ketjenblacks. The electro-catalysts were characterized to get their physicochemical properties by XRD, and Ar adsorption/desorption. And the performances of electro-catalysts were evaluated and compared as an electrode of PEMFCs by single cell test. Based on the relationship between the I-V performance and the morphology of catalysts, the basic parameters for the preparation of catalysts are suggested to be in the ranges of about 2.0 to 2.8nm and 5.0 to 14.2nm for the active metal size and inter-metal distance, respectively. In addition, as long as the structure of the electrode can be optimized for the each of new electro-catalysts, the active metal size is a more important design parameter rather than inter-metal distances.

1. INTRODUCTION

Electro-catalysts which have various metal contents have been applied to the polymer electrolyte membrane fuel cell(PEMFC). For the PEMFCs, Pt based noble metals have been widely used. In case the pure hydrogen is supplied as anode fuel, the platinum only electro-catalysts show the best activity in PEMFC. But the severe activity degradation can occur even by ppm level CO containing fuels, i.e. hydrocarbon reformates[1-3]. To enhance the resistivity to the CO poison of electro-catalysts, various kinds of alloy catalysts have been suggested. Among them, Pt-Ru alloy catalyst has been considered one of the best catalyst in the aspect of CO tolerance[1-3].

For the support material of electro-catalysts in PEMFC, Vulcan XC72(Cabot) has been widely used. This carbon black has been successfully employed for the fuel cell applications for its good electric conductivity and high chemical/physical stability. But higher amount of active metals in the electro-catalysts, compared to the general purpose catalysts, make it difficult to control the metal size and the degree of distribution. This is mainly because of the restricted surface area of Vulcan XC72 carbon black. Thus complex and careful processes are necessary to get well dispersed fine active metal particles[4,5].

In this work, we tried to suggest the principal design parameters of the electro-catalysts. Among various parameters, active metal size and inter-metal distance were intensively investigated in terms of electrode performance. To elucidate the effect of each parameters, several kinds of sample catalysts were prepared. But it was difficult to the desired sample variation using just Vulcan XC 72 as a support material. By adopting Ketjenblack, which has larger specific surface area, this obstacle could be overcome. In the present work, Ketjenblack, as received and chemically treated, was used as the support materials of catalysts. The morphologies and single cell performance of catalysts were examined.

2. EXPERIMENTAL

As received Ketjenblack EC 300J was treated with NaOH to change its physicochemical properties. The treated samples were dried at 120 °C for 12hrs and designated as S700, S800 and S900 followed by their treated temperatures. SRaw was used as abbreviation for the untreated Ketjenblack. The 30wt% Pt-Ru alloys catalysts were prepared by modified borohydride method[6]. The prepared catalysts were dried at 110 °C and designated as Pt-Ru/SRaw, Pt-Ru/S700, Pt-Ru/S800 and Pt-Ru/S900. By this process, the electro-catalysts which have the similar metal size but different inter-metal distances were successfully prepared.

The slurries of electro-catalysts were prepared by mixing together the catalysts and appropriate amount of 5wt % Nafion solution(Du Pont) including some kinds of dispersant[8]. The electrodes were made by spraying method with these well mixed inks. Two electrodes and Nafion 112 membrane were hot pressed with the condition of 50kgf/cm^2, 120°C for 3min to fabricate MEAs(Membrane Electrode Assembly).

3. RESULTS and DISCUSSIONS

Table 1 shows that the physicochemical properties of the support material were modified by the pre-treatment process. The particle sizes, Dp, which are summarized in the Table 1 were calculated from the X-ray diffraction patterns of prepared catalysts and a commercial catalyst(30 wt% Pt-Ru/C : E-TEK) by using Scherrer's equation. To avoid the interference from other peaks, (220) peak was used. All the prepared catalysts show the particle sizes of the range from 2.0 to 2.8nm. It can be thought that these values are in the acceptable range for the proper electrode performance[7]. For the prepared catalysts, notable differences are inter-metal distances(X[nm]) compared to commercial one. Due to their larger surface areas of support materials, active metals are apart from each other more than 2 ~ 3 times distance than commercial catalyst. Pt-Ru/SRaw has the longest inter-metal distances.

Fig. 1 shows I-V characteristics of MEAs fabricated with prepared catalysts. As anode electrodes, Pt-Ru/SRaw, Pt-Ru/S700, Pt-Ru/S800 and Pt-Ru/S900 catalysts were evaluated,

Table 1
Characteristics of support materials and prepared electro-catalysts

	SRaw	S700	S800	S900	Pt-Ru /SRaw	Pt-Ru /S700	Pt-Ru /S800	Pt-Ru /S900	Pt-Ru/C (E-TEK)
S_{BET}[m^2/g]	712	485	458	406	493	343	349	290	150
Dp^{XRD}[nm]	-	-	-	-	2.31	2.36	2.18	2.74	2.11
X^{XRD}[nm]	-	-	-	-	14.19	11.83	10.55	13.64	5.83

Dp^{XRD} : Crystallite size calculated by Scherr's equation with XRD data.
X^{XRD} : Inter-metal distance calculated by XRD data[6].

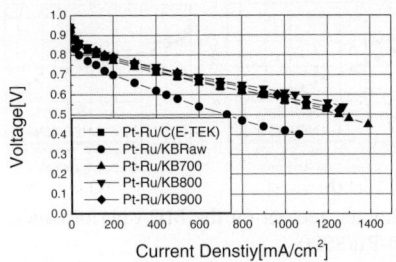

Fig. 1. Performance evaluation of prepared electro-catalysts as an electrode of PEMFC. Cell temperature:70°C, active area: 50cm², platinum loading: anode(0.3mgPt/cm²)/cathode(0.45mg Pt/cm²), fuel utilization: H_2/O_2 = 80%/50%, RH : 100% RH, pressure ; H_2/O_2 = 0 psig/0 psig.

and commercial E-TEK catalyst(30wt% Pt-Ru/C) was used as a reference. 40wt% Pt/C(Johnson-Matthey) catalyst was employed for the cathode electrode. In this single cell evaluation, 30% of Nafion loading(*NFP*), which are optimized for the commercial E-TEK catalysts, was applied for the all electrodes. In this test condition, Pt-Ru/S800 shows slightly better activity than commercial catalyst, but Pt-Ru/SRaw is inferior to the other catalysts. At least in these test conditions, the catalytic activity can be correlated to the relatively long inter-metal distances of evaluated electro-catalysts.

When considering the morphology of prepared electro-catalysts are different to each other especially to the commercial one, one can think that the structure of electrode which was optimized to the commercial catalyst may not be optimum. So, the for the better electrode structures was conducted by investigating the effect of *NFP*. Fig. 2 is a schematic of electrode which depicts the effect of Nafion content[9]. For the conventional electro-catalysts, the range of 30 ~ 35 % *NFP* is reported as optimum value[10].

Because the newly prepared electro-catalysts have larger specific surface areas and longer inter-metal distances, a higher *NFP* value was adopted. In Fig. 3, the electrodes which have different electro-catalysts are compared at the Nafion contents of 30 and 45 %. Aside from the commercial catalyst, all prepared catalysts shows enhanced performance when the *NFP* increases from 30 % to 45 %. The excess amount of Nafion ionomer can hinder the gas diffusion as well as increase the electric resistance in the electrode. As a result, the commercial catalyst shows decrease of performance as the *NFP* increases. Compared to other cases, case (b) shows biggest degree of performance enhancement. This can be understood by considering the morphological characteristics, such as larger surface areas, longer inter-metal distances and similar active metal sizes, of Pt-Ru/SRaw.

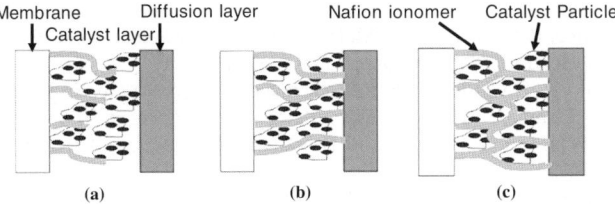

Fig. 2. Schematic representation of electrodes. (a) Content of Nafion too low: not enough catalysts with ionic connection to membrane. (b) Optimal Nafion content: electronic and ionic connections well balanced. (c) Content of Nafion too high: catalyst particles electronically isolated from diffusion layer. Reproduced from [9].

640

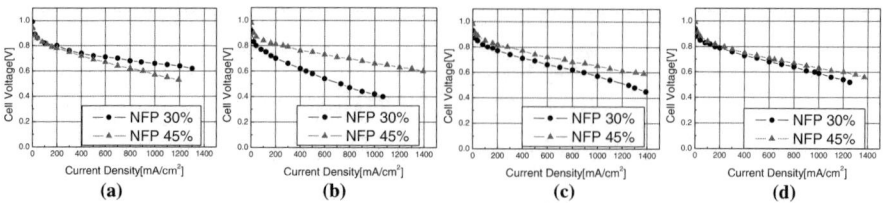

Fig. 3. The effect of Nafion ionomer content on the MEA performance. (a) E-TEK catalyst, (b) Pt-Ru/SRaw, (c) Pt-Ru/S700, (d) Pt-Ru/S900.

4. CONCLUSIONS

Design parameters of the anode catalyst for the polymer electrolyte membrane fuel cells were investigated in the aspect of active metal size and inter-metal distances. Various kinds of catalysts were prepared by using pretreated Ketjenblacks as support materials. The prepared electro-catalysts have the morphology such as the sizes of active metal are in the range from 2.0 to 2.8nm and the inter-metal distances are 5.0 to 14.2nm. The electro-catalysts were evaluated as an electrode of PEMFC. In Fig. 1, it looked as if there was a correlation between inter-metal distances and cell performance, i.e. the larger inter-metal distances are related to the inferior cell performance.

But when the contents of Nafion ionomer was increased from 30 to 45 % to find out the better electrode structures, the Pt-Ru/SRaw, which had showed the lowest single cell performance, became the best electro-catalyst. By this result one can conclude that as long as the structure of the electrode can be optimized for the each of new electro-catalysts, the active metal size is a more important design parameter rather than inter-metal distances. Furthermore, when the electro-catalysts are designed, the principal parameters should be determined in the consideration of the electrode structures which affect on the electron conduction, gas permeability, proton conductivity, and so on.

ACKNOWLEDGMENTS

This work was supported by Ministry of Science and Technology and by Ministry of Commerce, Industry and Energy, Republic of Korea.

REFERENCES

[1] S.J. Lee, S. Muerjee, E.A. Ticianelli, J. McBreen, *Electrochim. Acta,* 44 (1999) 3283.
[2] T.R. Ralph, M.P. Hogarth, *Platinum Metals Rev.,* 46 (2002) 117.
[3] E. Antolini, L. Giorgi, F. Cardellini, E. Passalacqua, *J. Solid State Electrochem.,* 5 (2001) 131.
[4] M. Watanabe, M. Uchida, S. Motoo, *J. Electroanal. Chem.,* 229 (1987) 395.
[5] H. Bonnemann, W. Brijoux, R. Brijoux, R. Brinkmann, E. Dinjus, T. Jonssen, B.K. Angew, *Chem. int. Ed.* 30 (1991) 1312.
[6] G.-G. Park, T.-H. Yang, Y.-G. Yoon, W.-Y. Lee, C.-S. Kim, *Int. J. Hydrogen Energy,* **28 (2003) 645.**
[7] H.A. Gasteiger, S.S. Kocha, B. Sompalli, F.T. Wagner, *Applied Cat. B : Environmental,* 56 (2005) 9.
[8] T.-H. Yang, G.-G. Park, P. Pugazhendhi, W.-Y. Lee, C.-S. Kim, *Korea J. Chem. Eng.,* **19 (2002) 417.**
[9] E. Passalacqua, F. Lufrano, G. Squadrito, A. Patti, L. Giorgi, *Electrochimica Acta,* **46 (2001) 799.**
[10] Z. Qi, A. Kaufman, *J. Power Sources,* 113 (2003) 37.

MICRO-REACTION TECHNOLOGY

Studies in Surface Science and Catalysis, volume 159
Hyun-Ku Rhee, In-Sik Nam and Jong Moon Park (Editors)

Improvement of product yield and selectivity in microreactors by combining fluid segments of different concentrations and sizes

Nobuaki Aoki and Kazuhiro Mae

Dept. of Chemical and Engineering, Graduate School of Engineering, Kyoto University, Kyoto-daigaku Katsura, Nishikyo-ku, Kyoto 615-8026, Japan

1. INTRODUCTION

Microreactors can serve as novel devices for improving yield and selectivity of desired products because of inherent advantages of small dimensions such as improved mass- and heat-transfer. For this purpose, mixing performance in microreactors is an essential issue. Mixing in microreactors is mainly driven by molecular diffusion, since the reactor miniaturization also leads to low Reynolds number in each reactor channel and thus to laminar flow. The mixing time scales with the square of diffusion length. Reduction of diffusion length is, therefore, required for fast mixing in microreactors. To reduce diffusion length, reactant fluids are split into many laminated fluid segments at the mixer section. The interdigital mixer [1], V-micro-jetmixer [2], and SuperFocus mixer [3] are examples of the micromixers using this mixing principle. Fluid segments in the mixers have the configuration of the same reactant concentration and segment size. However, we can also design configurations of fluid segments by combining fluid segments of different reactant concentrations and sizes. Such combinations give us a new reactor operation that enables us to precisely and flexibly design the mixing and the concentration profile in microreactors and a possibility to improve product yield and selectivity. From this viewpoint, we examine influences of these combinations on product yield and selectivity in microreactors using computational fluid dynamics (CFD) simulations.

2. SIMULATION METHODS

We first explain the setting of reactors for all CFD simulations. We used Fluent 6.2 as a CFD code. Each reactant fluid is split into laminated fluid segments at the reactor inlet. The flow in reactors was assumed to be laminar flow. Thus, the reactants mix only by molecular diffusion, and reactions take place from the interface between each reactant fluid. The reaction formulas and the rate equations of multiple reactions proceeding in reactors were as follows: A + B → R, $r_1 = k_1 C_A C_B$; B + R → S, $r_2 = k_2 C_B C_R$, where R was the desired product; and S was the by-product. The other assumptions were as follows: the diffusion coefficient of every component was 10^{-9} m^2/s; the reactants reacted isothermally, that is, k_1 was fixed at

1 m^3/(kmol·s) and $k_1/k_2 = 10$; and the mean residence time was 5 s. From the CFD simulations using these settings and the fluid segment configurations described in the subsequent sections, we obtain concentration profiles of the components in reactors, and then calculate relations between the yield of R, Y_R, and the conversion of A, x_A, from the mass-weighted average concentration of the components in the cross section of reactors at each axial position.

2.1. Four-segment configuration

We studied effects of combinations of the reactant concentrations and the segment widths for reactors where each reactant fluid is split into the two laminated fluid segments. The total number of fluid segments is thus four. Fig. 1 illustrates the configurations of fluid segments at the reactor inlet including the concentration of each reactant and the width. In the reactor shown in Fig. 1(a), the two fluid segments of the reactant j have the common concentration C_{j0} and size. In the other reactors, the fluid segments located in the middle of the reactor channel have the higher concentration (b), the lower concentration (c), the larger width (d), or the smaller width (e) than those located in the both sides of the reactor channel. The width of the channel was fixed at 200 μm.

2.2. Three-segment configuration

We examined effects of concentrations of the reactants and the segment widths for reactors where one reactant fluid is fed in the middle of a reactor channel and the other reactant is split into two laminated fluid segments and then fed in the both sides of the channel. The total number of fluid segments is three. Fig. 2 illustrates the configurations of fluid segments at the reactor inlet. The three widths of fluid segments fed in the middle of reactor channels were applied. We assumed that all the reactors take the same total mole flow of each reactant and that the width of the channel is fixed at 200 μm. Thus, the segment width located in the both sides of channels and the concentration of each reactant in fluid segments are changed with the width of fluid segments located in the middle of the channels as shown in Fig. 2. We also exchanged the location of fluid segments of the two reactants.

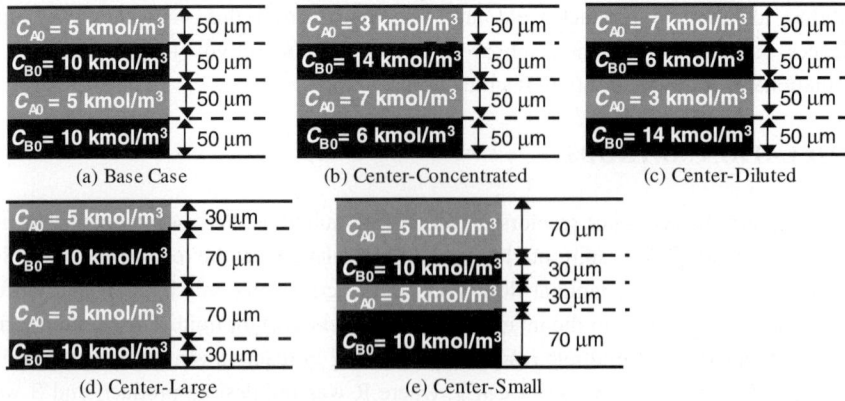

Fig. 1. Reactant concentrations and widths of fluid segments (four-segment configuration)

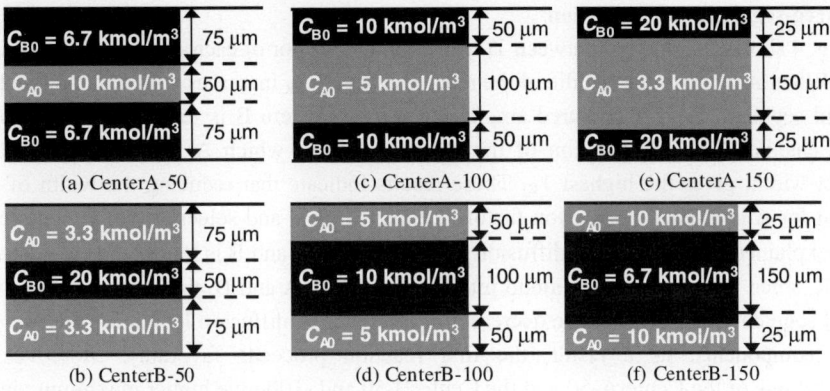

Fig. 2. Reactant concentrations and widths of fluid segments (three-segment configuration)

3. RESULTS AND DISCUSSION

3.1. Four-segment configuration

Fig. 3 shows the relation between Y_R and x_A in the reactor of each configuration shown in Fig. 1. Here, we refer the reactors as the name shown in the figures of the previous sections. Compared at the same x_A, the Center-Concentrated and the Center-Large show higher Y_R, namely, higher selectivity of R than those of the Base case. The reason for this tendency is speculated as follows: For the Base case, the fluid segments located in the both sides of the reactor channel have one interface between the two reactants, whereas the fluid segments located in the middle of the channel have the two interfaces; thus, the reactants in the fluid segments located in the both sides of the channel mix slower than that in the middle; therefore, the slow mixing causes low Y_R especially at the high x_A in the Base case as compared with those in the Center-Concentrated and the Center-Large. In contrast, for the Center-Diluted and the Center-Small, the effect of fluid segments in the both sides is dominant by the wide segment widths, leading to slower mixing and lower Y_R than those in the Base case.

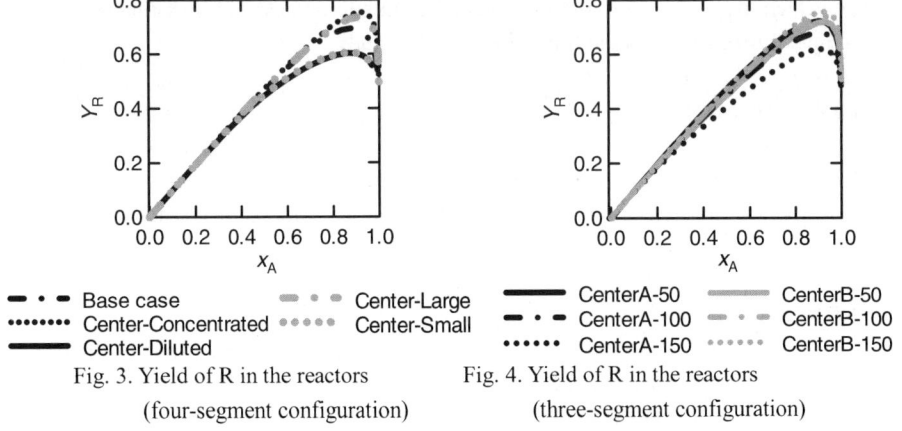

Fig. 3. Yield of R in the reactors
(four-segment configuration)

Fig. 4. Yield of R in the reactors
(three-segment configuration)

644

3.2. Three-segment configuration

Fig. 4 shows the relation between Y_R and x_A in the reactor of each configuration shown in Fig. 2. When A is fed in the middle of the reactor channel, Y_R increases with reducing width of the fluid segment for A. Compared among the reactors where B is fed in the middle of the reactor channel, the configuration of the CenterB-150, in which A is fed by the smallest segment width, gives the highest Y_R. These results indicate that reducing the width of fluid segment for A is a crucial operation factor to enhance yield and selectivity of R. The reason can be explained as follows: The diffusion length between A and B is longer than that between B and R, since R mainly exists around interfaces between A and B; thus, reducing the width of fluid segments for A causes the decrease in difference in diffusion length for the two pairs of the components; as a result, the first reaction proceeds favorably. Moreover, the configurations of the CenterA-50 and the CenterB-50 and -100 give higher maximum yield of R than that of the Base case, even though the widths of fluid segments of both reactants for the first three are equal to or larger than that for the Base case. Besides the merit in product distribution, the three-segment configuration has also an advantage in the operability of microreactors. The large segment sizes in these three-segment configurations lead to avoid excessively small channels for splitting reactant fluids into fluid segments and thus a high pressure drop in the channels.

4. CONCLUSION

We have studied effects of fluid segment configurations, that is, combinations of reactant concentrations and width of laminated fluid segments on product distribution of multiple reactions in microreactors using CFD simulations. We have applied two configurations of fluid segments: each reactant is split into two fluid segments; and one reactant fluid is fed in the middle of reactor channel and the other reactant is split into two fluid segments. The second configuration can give higher yield and selectivity of desired products than those of the first, even though the sizes of fluid segments are equal to or larger. This result shows that a proper design of fluid segment configurations is one of effective methods to enhance product yield and selectivity with improving the operability in microreactors. Experimental validation of the computational results is the future issue of this study.

ACKNOWLEDGMENT

We have conducted this research within the Project of Micro-Chemical Technology for Production, Analysis and Measurement Systems financially supported by the New Energy and industrial Development Organization (NEDO). We would appreciate the Micro Chemical Plant Technology Union (MCPT) for their support.

REFERENCES

[1] W. Ehrfeld, K. Golbig, V. Hessel, H. Löwe and T. Richter, Ind. Eng. Chem. Res., 38 (1999) 1075.
[2] St. Ehlers, K. Elgeti, T. Menzel and G. Wießmeier, Chem. Eng. Proc., 39 (2000) 291.
[3] V. Hessel, S. Hardt, H. Löwe and F. Schönfeld, AIChE J., 49 (2003) 566.

Studies in Surface Science and Catalysis, volume 159
Hyun-Ku Rhee, In-Sik Nam and Jong Moon Park (Editors)
© 2006 Elsevier B.V. All rights reserved

Computational study on the micro-channel fuel processors

S. Um[a] , H.-M. Jung[b], S.-D. Yim[a], W.-Y. Lee[a], and C.-S. Kim[a]

[a] Fuel Cell Research Center, Korea Institute of Energy Research,
P.O. Box 103, Yuseong-Gu, Daejeon, 305-343, Republic of Korea

[b] School of Aerospace and Mechanical Engineering, Hankuk Aviation University,
200-1 Hwajeon-Dong, Deokyang-Gu, Goyang, Gyeonggi-Do, 412-791, Republic of Korea

1. INTRODUCTION

Various miniaturized energy conversion systems have been developed to provide stable and long-lasting power, particularly for portable electronics for the past decades. One of the most promising alternative power technologies is based on polymer electrolyte fuel cells (PEFCs) using micro-channel reactor design, which have high energy density and easy-to-recharge advantages over other potential energy resources such as batteries. While the PEFC system would be best fit for the power electronics on ever-increasing energy demand, there are still significant technical barriers in developing miniaturized PEFCs for the limited-space applications: mainly system integration and fuel supplier. Particularly for supplying enough amount of fuel for the small scale power devices, the available high density hydrogen resources and the chemical generation of hydrogen-rich gas mixture are major challenging issues for the micro-scale systems. Recently, MeOH steam reforming processor has been emerged as a competitive fuel reformer for this miniaturized application.

Numerous kinetics modeling efforts have been performed for MeOH steam reforming over $Cu/ZnO/Al_2O_3$.[1-9] A comprehensive kinetic model for the MeOH fuel reforming processes based on surface reaction mechanism was presented by Jiang et al. [6] who considered irreversible Langmuir-Hinshelwood rate expression for the rate of disappearance of methanol. Peppley et al. [7-8] developed a kinetic model more extensively in that they included all three of the possible reactions; methanol-steam reforming reaction, water-gas shift reaction, and methanol decomposition reaction using the analysis of elementary surface reactions. This model has provided useful insight and reasonable predictions of the kinetics, however experimental data are needed to determine the partial pressure of each component in this model. Recently, Lee et al. [9] introduced a simple methanol decomposition model and presented various studies on the kinetics of methanol steam reforming by using a commercial catalyst Synetix 33-5. They reported Langmuir-Hinselwood rate expression for the MeOH steam reforming and power-law kinetics of the reforming. Also, they compared the kinetic models with a series of experimental data.

In this work, the MeOH kinetic model of Lee et al. [9] is adopted for the micro-channel fluid dynamics analysis. Pressure and concentration distributions are investigated and represented to provide the physico-chemical insight on the transport phenomena in the micro-scale flow chamber. The mass, momentum, and species equations were employed with kinetic equations that describe the chemical reaction characteristics to solve flow-field, methanol conversion rate, and species concentration variations along the micro-reformer channel.

2. MeOH STEAM REFORMING

The catalytic combustor provides heat for the endothermic reforming reaction and the vaporization of liquid fuel. The endothermic reforming reaction is carried out in a parallel flow-type micro-channel of the reformer unit. It is well known that the methanol steam reforming reaction for hydrogen production over the $Cu/ZnO/Al_2O_3$ catalyst involves the following reactions [10]. Eq. (1) is the algebraic summation of Eqs. (2) and (3).

$$CH_3OH + H_2O \Leftrightarrow 3H_2 + CO_2 \quad \text{(methanol steam reforming)} \quad \Delta H_{298}^{\circ} = +48.96 \ kJ/mol \quad (1)$$

$$CH_3OH \Leftrightarrow CO + 2H_2 \quad \text{(methanol decomposition)} \quad \Delta H_{298}^{\circ} = +90.13 \ kJ/mol \quad (2)$$

$$CO + H2O \Leftrightarrow CO2 + H2 \quad \text{(water-gas shift reaction)} \quad \Delta H_{298}^{\circ} = -41.17 \ kJ/mol \quad (3)$$

The catalytic steam-reforming process of methanol on $Cu/ZnO/Al_2O_3$ catalyst primarily produces hydrogen and carbon dioxide. In addition, the minor quantities of carbon monoxide are also produced. This mechanism is explained in terms of parallel reactions [11].

The micro-channel MeOH fuel processors consist of vaporizer, fuel reformer, and combustor. It requires two different types of fuel vaporizers: one is for fuel reforming and another for combustor. In the present work, the flow pattern in the reformers as shown in Fig. 1 is analyzed by the computational modeling. This mechanism-based model was validated against the experimental data set available in Lee et al. [9] and then applied to various parametric studies of the micro-channel fuel processors targeted for the optimal catalyst loading and fuel reforming purpose. This model can also be used to investigate the effects of various parameters, such as the operating temperature of the micro-channel, S/C ratio, and W/F ratio of reactant flow on the performance of the micro-channel fuel processors for the optimal design.

Fig. 1 Photograph of micro-channel plates.

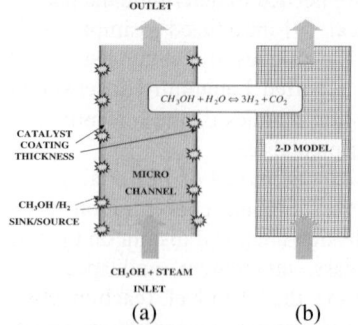

(a)   (b)

Fig. 2 Schematic diagram of a micro-channel of reformer (a) and meshes in 2-D model (b)

3. COMPUTATIONAL MICROCHANNEL FLUID DYNAMICS MODELING

Fig. 2 shows a schematic diagram of a micro-channel of reformer section to be examined in this study. A multi-physics computer-aided numerical model framework integrating kinetics, mass transport, and flow dynamics in micro-channel reactors has been established.

The present model assumes:
- The ideal gas mixtures due to highly operating temperature;
- incompressible and laminar flow;
- negligible pressure drop along the micro-channel;
- constant reformer temperature

The inlet methanol molar concentration was determined by the mass of catalyst, S/C ratio, and W/F ratio. Here, steam-to-carbon (S/C) ratio is defined as the ratio of steam molecules per carbon atom in the reactant feed and W/F ratio as the amount of catalyst loading into the channel divided by the amount of methanol molar flow rate. For more information on the design parameters, physical properties, and operating conditions, refer to Jung et al. [12].

4. RESULTS AND DISCUSSIONS

Since the W/F ratio depends on the mass of catalyst, the W/F ratio may be identified by the function of catalyst loading weight.

The simplified reformer model was validated against the experimental data available in the literature [9] as shown in Fig. 3 which shows good agreement between the experimental data and the numerical prediction over wide range of operating temperatures.

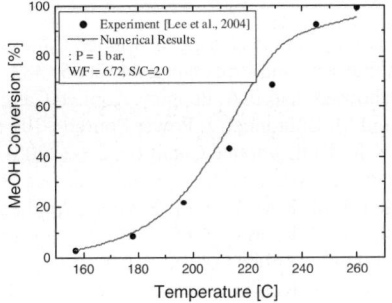

Fig. 3 Comparison of computational results against experimental data

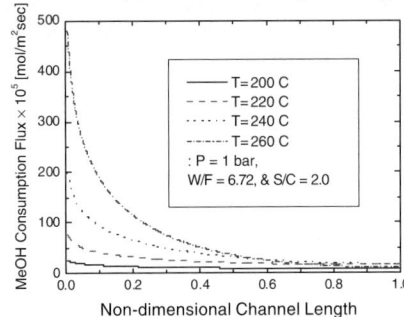

Fig. 4 Variations of methanol consumption flux along the channel length with increasing the reformer temperature (P_{in} = 1 bar, W/F = 6.72 kg·s/mol)

Fig. 4 features the effect of the reaction temperature on the methanol consumption flux. The methanol consumption flux is related to variation of methanol molar concentration along the channel. The methanol concentration decreased exponentially along the channel length. At high temperatures, the activation energy is increased, corresponding to reaction rate. Consequently, the hydrogen concentration increases along the channel as the reaction temperature is elevated. In this study, the hydrogen production rate is three times of the MeOH consumption rate according to Eq. (1).

5. CONCLUSIONS

Numerical models integrating surface kinetics and transport phenomena are introduced such that extensive computational studies are performed to obtain better design concepts on the micro-channel fuel processors. Comparisons are made to validate the numerical model against the reference experimental data with good agreement and then applied to various operating conditions of the micro-channel fuel processors. It was found that the resultant hydrogen concentration in the reaction chamber increased along the channel as increasing of the reactor temperature. The computer-aided models developed in this study can be greatly utilized for the design of advanced micro-channel fuel processors research.

ACKNOWLEDGMENTS
This work was supported by Ministry of Science and Technology (MOST) and by Ministry of Commerce, Industry and Energy (MOCIE), Republic of Korea.

REFERENCES
[1] P. Mizsey, E. Newson, T. Truong, and P. Hottinger, Applied Catalysis, 213 (2001) 233.
[2] L. Nowicki, Chemical Engineering and Processing, 44 (2005) 383.
[3] S.P. Asprey, B.W. Wojciechowski, and B.A. Peppley, Applied Catalysis, 179 (1999) 51.
[4] J. Agrell, H. Birgersson, and M. Boutonnet, J. Power Sources, 106 (2002) 249.
[5] V. Agarwal, S. Patel, and K.K. Pant, Applied Catalysis, 279 (2005) 155.
[6] C.J. Jiang, D.L. Trimm, and M.S. Wainwirght, Applied Catalysis, 97 (1993) 145.
[7] B.A. Peppley, J.C. Amphlett, L.M. Kearns, and R.F. Mann, Applied Catalysis, 179 (1999) 21.
[8] B.A. Peppley, J.C. Amphlett, L.M. Kearns, and R.F. Mann, Applied Catalysis, 179 (1999) 31.
[9] J.K. Lee, J.B. Ko, and D.H. Kim, Applied Catalysis, 278 (2004) 25.
[10] A.V. Pattekar, M.V. Kothare, S.V. Karnik, and M.K. Hatalis, IMRET5, Strasbourg, France, 2001.
[11] G.H. Charles, An Int. to Chemical Engin. Kinetics & Reactor Design, John-Wiley & Sons, 1997.
[12] H.-M. Jung, S.-D. Yim, S. Um, Y.-G. Yoon, G.-G. Park, Y.-J. Sohn, T.-H. Yang, W.-Y. Lee, and C.-S. Kim, 3rd Int. Conf. on Fuel Cell Science, Engin. and Tech., Michigan, USA, 2005.

Studies in Surface Science and Catalysis, volume 159
Hyun-Ku Rhee, In-Sik Nam and Jong Moon Park (Editors)
649

Microreaction technology in practice

Kwang Ho Song[a]**, Youngwoon Kwon**[b]**, and Jaehoon Choe**[b,*]

[a]Department of Chemical & Biological Engineering, Korea University, Seoul 136-713, Korea

[b]Corporate R&D, LG Chem Research Park, Daejeon 305-380, Korea

1. INTRODUCTION

Microreaction systems offer many exceptional technical advantages for a large number of applications in the chemical and pharmaceutical industry. The large surface-to-volume ratio of a microreactor allows for precise process control and heat management. In this work, the microreactor is used in the production of pharmaceutically active compounds such as quinolone carboxylic acid derivatives having antibacterial activity. The compound is a fluoronaphthyridone carboxylic acid with a novel pyrrolidine substituent (anhydrous (R,S)-7-(3-aminomethyl-4-methoxyiminopyrrolidin-1-yl)-1-cyclopropyl-6-fluoro-4-oxo-1,4-dihydro-1,8-naphthyridine-3-carboxylicacid) and it is the most active agent against gram-positive species including strains observed to be resistant to other fluoroquinolones [1-3]. A key step in the synthesis of this compound is the protection of the enamino pyrrolidinone with t-butoxycarbonyl anhydride [4]. This is a highly exothermic reaction and the reaction heat generates impurities. The impurities in the product also lengthened the time of the next filtration step by making the filtration cake sticky, which also resulted in lowering the overall yield. The calorimetric study shows an adiabatic temperature rise of 33 °C, which is sufficient to initiate the formation of byproducts [5].

2. EXPERIMENTAL

The protecting reaction of the enamino pyrrolidinone with t-butoxycarbonyl anhydride was carried out by mixing 4-(N-t-butoxycarbonyl)-4-aminomethylene-pyrrolidin-3-one (Boc-AMP) with 1.2 molar equivalents of t-butoxycarbonyl anhydride (t-Boc$_2$O) to make 1-(N-t-butoxycarbonyl)-4-(N-t-butoxycarbonyl)-aminomethylene-pyrrolidin-3-one (Boc$_2$-AMP). Several types of reactors, including batch reactors, a continuous stirred tank reactor (CSTR) ,

---

*Corresponding author, E-mail: choej@lgchem.com

a tubular reactor, in-line static mixer reactors in series, and a microreactor were investigated to achieve better selectivity and product yield.

The reaction was compared by addition of 4 molar equivalents of 22.5% aqueous KOH and 6 molar equivalents of 50% aqueous KOH in a 0.5 L batch reactor. Four molar equivalents of 50% aqueous KOH were used in both 5 L and 8000 L batch reactors. In a CSTR, the 8, 10, and 16 molar equivalents of 50% aqueous KOH were tested. In a tubular reactor, a 0.3175 cm (O.D) tube, 1 m in length, with a cooling jacket was used at 0 °C with 6 molar equivalents of 50% aqueous KOH. A Kenics type in-line static mixer reactor was also applied to the reaction. The Kenics type in-line mixer is composed of a series of twisted mixing elements arranged axially within the tube. As the fluids pass through each twisted element, the flow divided by half. Each Kenics mixer unit has 27 twisted elements and the reactor dimensions are 19 cm in length and 3.3 mm in inner diameter with 0.476 cm in outer diameter. One feed consisted of Boc-AMP and t-Boc$_2$O and the other feed was 50% aqueous KOH solution. The two feeds mixed in a union tee fitting and then flowed through an in-line static mixer reactor.

A microreactor was also applied to this reaction. The slit interdigital micromixer was purchased from IMM (Mainz, Germany). The width of the interdigital channels is 25 μm. HPLC pumps were used to feed the two reaction solutions. One is a mixture of Boc-AMP and 1.2 molar equivalents of t-Boc$_2$O. The other is a 50% aqueous KOH solution. The microreactor was immersed in a temperature controlled cooling bath at 15 °C. The product was quenched with an acid, and samples were taken for HPLC analysis.

A Waters 2690 HPLC system equipped with a Waters 996 photodiode array detector (Waters, WA) was used for the analysis. All measurements were taken with a 1.2 ml/min mobile phase flow rate at a wavelength of 300 nm on a Capcell-pak C18 column (4.6 mm i.d. × 250 mm, particle size 5 mm, Shiseido, Japan). Data integration was performed with Millennium software. For Boc$_2$-AMP: ^{1}H NMR (CDCl$_3$, δ, ppm) 10.10 (s,1H), 7.30 (s, 1H), 4.40 (d, 2h), 3.95 (d, 2H), 1.55 (m,18H). MS (FAB, m/e): 313 (M+H).

3. RESULTS AND DISCUSSION

In the batch reactor, Boc-AMP is reacted with t-Boc$_2$O to make Boc$_2$-AMP by dropping aqueous KOH solution at a jacket temperature of –20 °C since this reaction is highly exothermic. In the 0.5 L batch reactor, the product yields were 94% with 4 molar equivalents of 22.5% aqueous KOH and 96% with 6 molar equivalents of 50% aqueous KOH. In the batch reactor, the removal of heat generated by the reaction was not fast enough to maintain the temperature of the reaction mixture as low as to prevent the formation of impurities. The results in the plant scale experiments showed that the product yield was dropped from 96% at 0.5 L lab scale reactor to 40~75% in the 8000 L plant scale reactor. Table 1. shows that the batch reactor volume greatly affects the product yield of Boc$_2$-AMP.

The reaction was also tested with a CSTR. The KOH solution and the mixture of Boc-AMP and t-Boc were mixed in the CSTR and the residence time was 30 sec. The results showed that the product yield was 82% when the molar ratio of Boc-AMP to KOH was 10. The product yields were 81% and 76% when the molar ratios of Boc-AMP to KOH were 8 and 16, respectively. In the CSTR, KOH was continuously fed in the reactor and the product, Boc_2-AMP, could react with fresh fed KOH to make byproducts.

Table 1. Production of Boc_2-AMP with batch reactors

Reactor volume (L)	Product yield (%)	Jacket temperature (°C)
0.5	94-96	-20
5	89	-20
8000	40~75	0

Tubular reactor was, therefore, applied to minimize a back mixing with fresh KOH and Boc_2-AMP. However, only 27% of Boc-AMP was converted to Boc_2-AMP due to the phase separation between t-Boc_2O and aqueous KOH solution.

To improve the mixing quality in the tubular reactor, Kenics type in-line static mixer reactor was employed. The in-line static mixers were designed to mix two or more fluids efficiently since an improved transport process such as flow division, radial eddying, flow constriction, and shear reversal eliminated the gradients in concentration, velocity and temperature. However, only 70 % conversion was achieved with one Kenics mixer unit. As shown in Table 2, five mixer units were required to achieve the maximum conversion.

Table 2. Production of Boc_2-AMP with Kenics in-line static mixer reactor

Number of Kenics mixer	Conversion (%)
1	70.0
2	91.0
3	95.1
4	96.4

When the reaction was performed in the microreactor, the maximum conversion of 97.0 % was attained when the flow rate of Boc-AMP solution was 9 ml/min and the molar equivalents of KOH to Boc-AMP was 13 as shown in Fig. 1. Optimum operating conditions were obtained from a statistical method by using factorial design [6]. The yield decreased over the KOH equivalency of 13 in Fig. 1, since the phase separation between the t-Boc_2O and the aqueous phase was observed due to the increased water content with increasing KOH equivalency. As the heat transfer performance of the microreactor was greatly improved compared with conventional reactors, higher reaction temperature could be admissible.

In this experiment, the reaction temperature was isothermally controlled at 15 °C. The heat of reaction was completely removed using microreactor so that virtually no byproducts were produced during the reaction. It can be compared with other reactors described above, which should be operated at 0 °C or –20 °C to avoid side reactions.

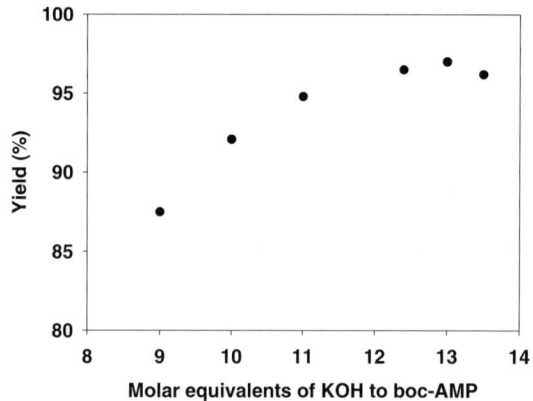

Fig. 1. Effect of KOH equivalency on the yield of Boc_2-AMP production using microreactor

4. CONCLUSION

Microreactor was successfully applied to a pharmaceutical intermediate at a fast exothermic Boc protecting reaction step among several types of reactors investigated above. The maximum conversion was achieved by avoiding side reactions since efficient heat removal was possible due to the extremely large surface to volume ratio and the effective mixing within a short residence time. The microreactor has very short residence time compared to batch and tubular reactors so that excess KOH is required for a maximum conversion of reactant. The results show that the maximum conversion requires KOH and Boc-AMP in a molar ratio of 13 and total flow rate of 13 ml/min using the microreactor. The annual production volume of Boc_2-AMP could be produced with only 25 microreactors, which would means a dramatic decrease in investment and operational costs compared to the operation with 8000 L sized batch reactor.

REFERENCES

[1] C.Y. Hong, Y.K. Kim, J.H. Chang, S.H. Kim, H. Choi, D.H. Nam, Y.Z. Kim, and J.H. Kwak, J. Med. Chem., 40 (1997) 3584.
[2] E.C. Taylor, S.R. Fletcher, and S. Fitzjohn, J. Org. Chem., 50 (1985) 1010.
[3] J.H. Chang, W.S. Kim, T.H. Lee, and K.Y. Moon, Process for preparing a protected 4-aminomethyl-pyrrolidi-3-one, US Patent No. 6 307 059 B1 (2001).
[4] T.W. Greene and P.G. Wuts, Protective Groups in Organic Synthesis, John Wiley & Sons, New York,1999
[5] J. Choe, Y. Kim, and K. H. Song, Org. Proc. Res. Dev., 7 (2003) 187.
[6] J. Choe, Y. Kwon, Y. Kim, H. Song and K.H. Song, Korean J. Chem. Eng., 20 (2003) 268.

Studies in Surface Science and Catalysis, volume 159
Hyun-Ku Rhee, In-Sik Nam and Jong Moon Park (Editors)

Development of a CO remover employing microchannel reactor for polymer electrolyte fuel cells

Gu-Gon Park[a], Sung-Dae Yim[a], Sang-Phil Yu[a], Won-Gyo Lee[b], Young-Gi Yoon[a], Sukkee Um[a], Chang-Soo Kim[a], Koichi Eguchi[c], Yong-Gun Shul[b]

[a]Fuel Cell Research Center, Korea Institute of Energy Research
71-2, Jang-dong, Yuseong-gu, Daejeon, 305-343, Korea

[b]Department of Chemical Engineering, Yonsei University, Seoul 120-149, Korea

[c]Department of Energy and Hydrocarbon Chemistry, Graduate School of Engineering,
Kyoto University, Nishikyo-ku, Kyoto 615-8510, JAPAN

1. INTRODUCTION

Small size PEMFCs (polymer electrolyte membrane fuel cells) would be an attractive power source for portable electronic devices. However, there are still significant technical barriers in developing small PEMFCs. One of the technical challenges for the systems is the development of a small size and light weight hydrogen supplying system. Among various hydrogen supplying systems for PEMFCs, methanol reformation processors employing microchannel reactor have been received a great attention because of their high energy density and instant recharge time of liquid fuel. The microchannel reactor has been employed to minimize the complicated chemical plants mainly due to its advantages for chemical reactions such as high surface to volume ratio revealing several orders of magnitude higher compared to traditional chemical reactors and low linear dimensions enhancing heat transfer and mass transfer in the reactor [1]. On the other hand, the methanol steam reformer produces significant amount of CO with hydrogen mainly due to thermodynamic constraints of the reaction. Thus, it is necessary to integrate CO removal reactor with the fuel processor to reduce CO concentration less than 10ppm to prevent the deactivation of Pt-based electrocatalyst for PEMFCs by CO. As an effective CO removal process, preferential oxidation (PrOx) of CO is regarded as the most promising and practical method.

Actually, various efforts have been made to develop the compact and efficient microchannel PrOx reactor for portable PEMFC applications. Goerke et al. [2] reported micro PrOx reactor employing stainless steel microchannel foil and Cu/CeO_2 catalyst. They showed more than 99% CO conversion at less than $150^{\circ}C$ and residence time of 14ms while CO selectivity was about 20%. Chen et al. [3] also developed microchannel reactor made of stainless steel for CO selective oxidation. The microchannel PrOx reactor used Rh-K/Al$_2$O$_3$ catalyst and revealed 99.5% CO conversion at CO=1%, GHSV=100,000h^{-1} and 157~200°C. CASIO [4] is also developing mluti-layered microchannel reactor being integrated with methanol reformer, CO PrOx reactor, vaporizer and catalytic combustor made of glass for 10W class PEMFC system for laptop PC. They controlled CO concentration less than 50ppm by using the CO PrOx reactor employing Pt/Al$_2$O$_3$ catalyst at 180°C. These previous studies show that the performance of microchannel reactor critically depends on the design of microchannel plate, reactor fabricating method and reactor integration.

In this study, we developed microchannel PrOx reactor to control CO outlet concentrations less than 10 ppm from methanol steam reformer for PEMFC applications. The reactor was developed based on our previous studies on methanol steam reformer [5] and the basic technologies on microchannel reactor including design of microchannel plate, fabrication process and catalyst coating method were applied to the present PrOx reactor. The fabricated PrOx reactor was tested and evaluated on its CO removal performance.

2. Experimental

2.1. Fabrication and integration of microchannel reactor

We prepared microchannel reactor employing stainless steel sheet 400μm thick patterned microchannel by a wet chemical etching. The microchannel shape and dimension were decided by computer simulation of flow distribution and pressure drop of the reactants in the microchannel sheet. Two different types of patterned plates with mirror image were prepared [5]. The plate has 21 straight microchannels which are 550μm wide, 230μm deep and 34mm long as revealed in Fig. 1(b).

 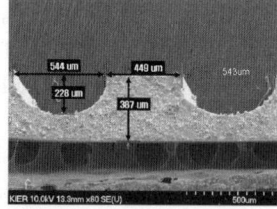 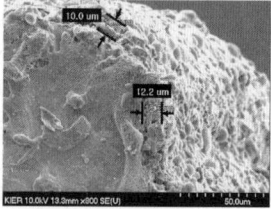

(a) Microchannel patterned plate   (b) Cross-sectional view   (c) Al$_2$O$_3$ coated channel

Fig. 1. SEM images of microchannel plate

CO preferential oxidation reactor was coated with Ru/Al$_2$O$_3$ catalyst (Sud Chemie). To enhance the catalyst coating strength on the stainless steel sheet, the microchannel sheets were precoated with alumina. The alumina was uniformly and strongly coated on the plate with average thickness of 10μm as shown in Fig. 1(c). The alumina coated plates were then coated by a Ru/Al$_2$O$_3$ catalyst by slurry coating method. The catalyst slurry was made of Ru/Al$_2$O$_3$ catalyst, 20wt% Alumina sol (NYACOL$^{®}$ AL20DW colloidal alumina, PQ Corporation), distilled water and 2-Propanol. After drying in air, catalyst coated metal structure was calcined at 300°C. About 50mg of the catalyst was coated on each microchannel plate.

The fabricated microchannel plates were stacked together and tightened by end-plates to prevent the leak and mixing of gases as shown in Fig. 2. The dimensions of the reactor excluding fittings were about 60 mm x 40 mm x 3 mm respectively.

Fig. 2. CO removal reactor integrated with microchannel plate

2.2. Experimental set-up

The experimental apparatus is consists of reformed gas feeding sections, CO PrOx reaction section in the reactor, and the analysis section with a gas chromatograph system. Simulated reformed gas composition was 75 vol.% H_2, 24 vol.% CO_2 and 1.0 vol.% CO. The dry reformed feed stream was fed with O_2 (λ=1) into the microchannel reactor by MFC (Brooks 5850E). Water vapor (10vol.% of reformed gas) was also fed into the reactor by a syringe pump.

The CO PrOx reaction was conducted in the temperature range of 100°C to 240°C and the reaction temperature was controlled by electrical heater.

The product stream was separated using a cold trap maintained at 5°C and the the composition of dry reformed gas was analyzed by a gas chromatograph (Agilent 6890N).

The catalyst activity is defined as the ratio of the reacted CO divided by input CO, and the catalyst selectivity towards CO in the H_2-rich stream is defined as the ratio of the O_2 consumed for CO oxidation over total O_2 consumed for both H_2 and CO oxidation.

3. RESULTS AND DISCUSSION

Fig. 3 shows the activity of CO preferential oxidation in the microchannel reactor with respect to reaction temperature at different reactor space velocity. At 50,000h^{-1} of reactor space velocity, all CO was removed and the outlet CO was controlled less than 10ppm in the temperature ranges from 130 to 200°C. However, at 100,000h^{-1}, the temperature windows for high CO conversion more than 99.99% decreased as 160 to 185°C. In addition, the CO conversion steady decreased as the reaction temperature increases above 210°C mainly due to the superior oxidation of H_2 compared to CO. With increasing reaction temperature CH_4 production was also evident. At 50,000h^{-1} of reactor space velocity, produced CH_4 was 300ppm and 1700ppm at 183°C and 230°C, respectively. The CO selectivity was about 50% at the temperature ranges showing high CO conversion more than 99%.

For the practical use of this CO removal reactor, the microchannel reactor should be operated carefully to maintain operating temperature ranges because the reaction temperature is critical for the microchannel reactor performance such as CO conversion, selectivity and methanation as disclosed in the above results. It also seems that the present microchannel reactor is promising as a compact and high efficient CO remover for PEMFC systems.

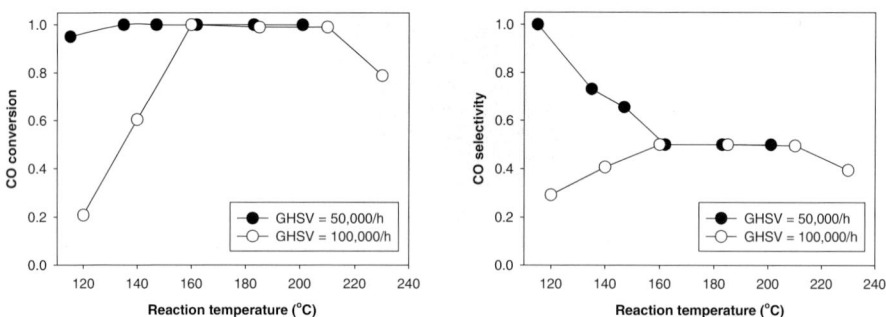

Fig. 3. CO conversion and selectivity with respect to reaction temperature in a microchannel reactor

4. CONCLUSION

A microchannel reactor for CO preferential oxidation was developed. The reactor was consisted of microchannel patterned stainless steel plates which were coated by Ru/Al_2O_3 catalyst. The reactor completely removed 1% CO contained in the H_2-rich reformed gas and controlled CO outlet concentration less than 1ppm at 130~200°C and 50,000h^{-1}. However, CH_4 was produced from 180°C and CO selectivity was about 50%. For high performance of present PrOx reactor, reaction temperature should be carefully and uniformly controlled to reach high CO conversion and selectivity, and low CH_4 production. It seems that the present microchannel reactor is promising as a CO removal reactor for PEMFC systems.

ACKNOWLEDGMENTS

This work was supported by Ministry of Science and Technology and by Ministry of Commerce, Industry and Energy, Republic of Korea.

REFERENCES

[1] W. Ehrfeld, V. Hessel, H. Lowe, Microreactors, WILEY-VCH, Weinheim, 2003, pp. 6-8.

[2] O. Goerke, P. Pfeifer, K. Schubert, Appl. Catal. A, 263 (2004) 11.

[3] G. Chen, Q. Yuan, H. Li, S. Li. Chem. Eng. J., 101 (2004) 101.

[4] K. Yamamoto, Y. Kawamura, N. Ogura, T. Yamamoto, T. Terazaki, Proc. 2004 Fuel Cell Seminar, San Antonio, Texas, USA, 2004.

[5] G.-G. Park, S.-D. Yim, Y.-G. Yoon, W.-Y. Lee, C.-S. Kim, D.-J. Seo, K. Eguchi, J. Power Sources, 145 (2005) 702.

Studies in Surface Science and Catalysis, volume 159
Hyun-Ku Rhee, In-Sik Nam and Jong Moon Park (Editors)

657

A miniature methanol steam reformer for polymer electrolyte fuel cell

Suk Woo Nam*, Myeong-Ju Ha, Jonghee Han, Sung Pil Yoon, Tae-Hoon Lim, and Seong-Ahn Hong

Fuel Cell Research Center, Korea Institute of Science and Technology
Seoul 136-791, Korea
*Corresponding author: swn@kist.re.kr

1. INTRODUCTION

A fuel cell is an electrochemical reactor that combines the oxidation of a fuel with oxygen reduction to produce electricity and heat. Since fuel cells convert chemical energy of reaction directly into electrical energy without an intermediate combustion stage, energy conversion efficiency from fuel to electricity is much higher than conventional methods of power generation that are restricted by Carnot limitations. Higher conversion efficiency without direct combustion of fuel yields fewer emissions, making fuel cells ideal for power generation in the future.

The fuel utilized in the fuel cell is mainly hydrogen since its electrochemical reaction rate is much faster than other fuels. Methanol and formic acid can directly participate in the electrochemical reaction, but their reaction rates are an order of magnitude lower than hydrogen. Therefore, hydrogen is usually produced from other fuels by using a separate fuel processor and subsequently supplied to the fuel cell.

Small fuel cells combined with miniature fuel processors have been under intensive development as main power sources or battery chargers to extend the operation time of portable electronic devices. Among a variety of fuel reforming reactions, much attention is focused on steam reforming of methanol to provide hydrogen to a polymer electrolyte fuel cell (PEMFC) since a relatively lower amount of CO (1% or less) is produced, significantly reducing the volume and weight of a subsequent CO clean-up reactor. Endothermic methanol steam reforming, however, must be closely coupled with fuel combustion reaction to enhance reforming efficiency. That is, efficiency of the reformer is governed mainly by rate of heat transfer from the combustion section of the reformer to the reforming catalyst. A number of reformers based on heat exchangers or micro-structured plates have been suggested to facilitate heat transfer between the combustion and steam reforming sections of the reformer [1-6]. Recent studies on micro-structured fuel processors for fuel cell are well summarized by Hessel et al. [1].

In this study, an integrated methanol reformer including an evaporator and a combustor was fabricated and tested. Previous tests of the reformer with a number of on-off cycles revealed that non-uniform temperature distribution caused hot spots within the combustion plate, resulting in cracking of the welded region of the reformer. Therefore, emphasis was made to achieve a uniform temperature distribution within the reformer. In addition, start-up characteristics of the complete reforming system were investigated.

2. EXPERIMENTAL

An integrated reformer was made from a number of stainless steel plates with a thickness of 0.6mm. In the evaporator and steam reforming section, gas channels of 1mm width and 1mm depth separated by 1mm fins were formed by photochemical etching. Fig. 1 shows the configuration of the reformer. An evaporator plate and a reformer plate were in close contact with a combustor plate. Fuel (hydrogen from anode outlet of a PEMFC) and air were introduced to the combustor through a central distributor to two combustion catalyst plates with a space velocity of 10^4 h^{-1}. The combustion catalyst was made by coating Pt/ceria catalysts in the pores of Ni plate. Porosity and pore size of the porous Ni plate were 50-60% and 3-10 micrometers, respectively. The combustion products were gathered at the outlet manifold and then flowed through the outer region of the evaporator and the reformer, respectively.

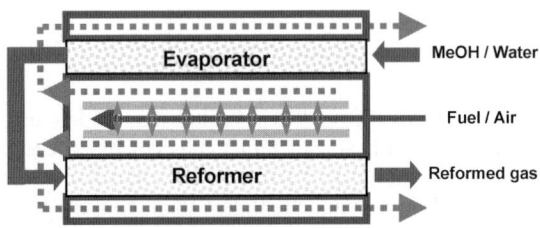

Fig. 1. Configuration of an integrated reformer.

A methanol/water mixture with a molar ratio of 1/1.5 was introduced to the evaporator where a porous Ni plate was inserted. Then reactant vapor was flowed into the reformer through an external manifold to the reforming catalyst with a space velocity of 7,000 h^{-1}. Cu/ZnO/Al$_2$O$_3$ catalyst from Süd-Chemie (MDC3) was used in the form of either crushed powders or catalyst layers coated on porous Ni plate. Fig. 2 shows pictures of the integrated reformer containing the evaporator and combustor. Micro-laser welding was used to join a number of reactor plates. The reformer was 22.5cc (24mm x 72mm x 13mm) and 106g without thermal insulation, but the volume and weight were significantly increased to 153cc (42mm x 104mm x 35mm) and 169g after insulation. The complete reforming system was prepared by connecting the reformer with a PROX(preferential CO oxidation) reactor. The PROX reactor was a packed-bed type reactor in which a Pt/Al$_2$O$_3$ catalyst doped with Co was used.

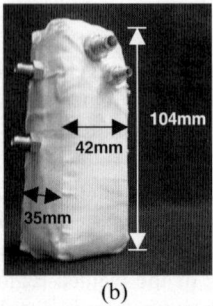

(a)   (b)

Fig. 2. Photographs of the integrated reformer : (a) without and (b) after the thermal insulation.

3. RESULTS AND DISCUSSION

3-1. Effect of introducing gas distributor in the combustion chamber

A reformer combined with a PEMFC must be capable of thermal cycling to meet the requirements of a portable electronic device which is frequently turned on and off. To ensure stable operation of the reformer, it is important to avoid hot spots inside the reformer. In our previous experiments, hydrogen and air were supplied to the combustion chamber without using a gas distributor. Fig. 3-(a) shows a thermal image of the combustion plate during the hydrogen oxidation, indicating that a hot spot (white area in Fig. 3) was formed near the gas entrance region. Thermal cycling caused cracks in the welding region of combustion plates, especially near the region where the hot spot was formed. When the gas outlet temperature was kept at about 300°C, the temperature difference between the gas inlet and outlet, measured by thermocouples, was about 200°C. By introducing a gas distributor in the combustion chamber, the temperature difference was reduced to less than 50°C. A thermal image of the combustion plate as shown in Fig. 3-(b) confirms that a relatively uniform temperature distribution can be obtained by introducing a gas distributor in the combustion chamber.

Fig. 4 shows the evolution of temperature in the methanol steam reformer combined with a combustion plate equipped with a gas distributor. In this case hydrogen was used as a fuel for start-up at room temperature. As the reformer temperature reached near 300°C in about 5 min, methanol/water vapor was introduced to the reformer. It can be clearly seen that temperature within the reformer became relatively uniform after 25 min of operation.

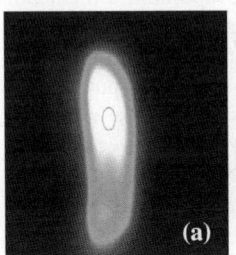 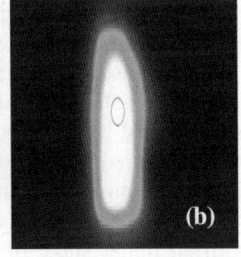 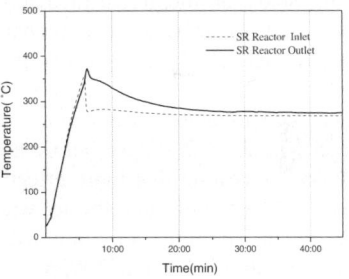

Fig. 3. Thermal images of combustion plate: (a) without and (b) with a gas distributor (Hydrogen and air were introduced to the combustion plate from the top).

Fig. 4. Evolution of temperature in the methanol steam reformer during the start-up and steady-state operation

3-2. Performance of an integrated methanol reformer combined with a PROX reactor

A complete methanol reforming system was constructed by connecting the integrated reformer with a PROX reactor. Fig. 5 shows the evolution of temperature at the gas outlet of the evaporator, reformer and PROX reactor during the start-up. Temperature of the reformer became stable in 5 min after introduction of the reactant. The reformer produced hydrogen up to 1.5L/min with methanol conversion higher than 95%, enough to run a 100W PEMFC.

Fig. 6 shows the evolution of the CO concentration after the introduction of the methanol/water mixture to the reformer, indicating a CO spike at the initial stage of operation. CO concentration of up to 2% (dry basis) was observed at the reformer outlet by gas chromatography, but it was reduced to 0.6% in 20 min as shown in Fig. 6. CO concentration at the outlet of the PROX reactor operating at an O_2:CO ratio of 1, increased rapidly to 1100ppm at the initial stage, but it decreased to near 200ppm immediately. The CO spike at the initial stage of the reformer's operation is believed to be due to the relatively high temperature of the reformer and PROX reactor. Further research is needed to remove the CO spike and to reduce the CO concentration down to 10ppm.

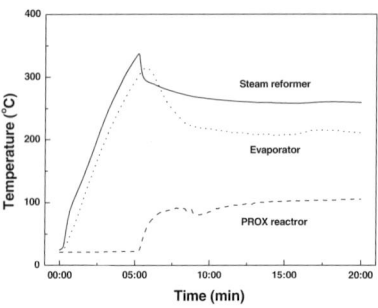

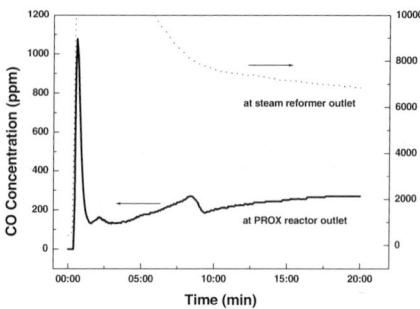

Fig. 5. Evolution of temperature in the reformer and PROX reactor during the start-up.

Fig. 6. Evolution of CO concentration after introduction of reactants into the reformer.

4. CONCLUSIONS

In this study, an integrated methanol steam reformer containing an evaporator, a methanol reformer and a combustor was prepared to produce hydrogen-rich gas for a small polymer electrolyte fuel cell. Emphasis was made to achieve a uniform temperature distribution inside the reformer. By adding a gas distributor to the combustion chamber, it was possible to decrease the temperature gradient to $50^{\circ}C$ in the reformer. The power density of the integrated reformer was about $2.3W_e/cc$ before thermal insulation, but it decreased to $0.43W_e/cc$ after the insulation, suggesting that more effective scheme is required to fabricate a compact reformer. The total efficiency of the reformer was about 55% based on lower heating values of methanol and hydrogen. The reformer combined with a PROX reactor produced a reformed gas with the CO concentration of about 200ppm, though with an initial spike of 1100ppm. Research is underway to reduce CO concentration and to increase the power density of the reformer.

ACKNOWLEDGEMENTS

This work was financially supported by Korea Institute of Science and Technology.

REFERENCES

[1] V. Hessel, H. Löwe, A. Müller, G. Kolb, Chemical Micro Process Engineering – Processing and Plants, Wiley-VCH, Weinheim (2005) 281-408.
[2] P.J. de Wild and M.J.F.M. Cerhaak, Catalysis Today, 60 (2000) 3.
[3] J.D. Holladay, E.O. Jones, M. Phelps, and J. Hu, J. Power Sources, 108 (2002) 21.
[4] J.D. Holladay, E.O. R.A. Dagle, G.G. Xia, C. Cao, and Y. Wang, J. Power Sources, 131 (2004) 69.
[5] V. Cominos, S. Hardt, V. Hessel, G. Kolb, H. Löwe, M. Wichert, R. Zapf, Proceedings of the 6th IMRET, New Orleans, USA, 2002, p. 113.
[6] P. Pfeifer, K. Schubert, M. Fichtner, M.A. Liauw, and G. Emig, Proceedings of the 6th IMRET, New Orleans, USA, 2002, p. 125.
[7] L. Pan and S. Wang, Int. J. Hydrogen Energy, 30 (2005) 973.

MODELING, SIMULATION AND CONTROL
OF CHEMICAL REACTORS

Studies in Surface Science and Catalysis, volume 159
Hyun-Ku Rhee, In-Sik Nam and Jong Moon Park (Editors)

Reactor Simulation for an Up-flow Anaerobic Sludge Blanket Process

Katsuhiko Muroyama[a], Toshiyuki Nakai[a], Yusuke Uehara[a], Yasunori Sumida[a] and Akihiko Sumi[b]

[a]Dept. of Chemical Engineering, Kansai University, 3-3-35 Yamate-cho, Suita, Osaka, 564-8680, Japan

[b]KOBELCO ECO-SOLUTIONS CO., Ltd., Kobe, Hyogo 651-0072, Japan

1. INTRODUCTION

The UASB process has found wide applications in treating wastewaters from various food processing industries such as sugar production from beets, bean blanching, dairy, potato starch production and vine breweries (Hulshoff Pol et al., 1986 [1]; Zeevalkink et al., 1986 [2]; Nanninga et al., 1986 [3]; Craverio et al., 1986 [4]; Yoda et al., 1989 [5]). This technology is simple for the facility unit and enables to treat organic wastewater at a high efficiency, but the available information on the design and scale-up of UASB reactors is rather scarce. In the previous study [6], to design a UASB granular bed reactor on a rational basis, the reaction kinetics for biodegradation of volatile acid components including acetic, propionic and butyric acids (respectively abbreviated as Ac, Pr and Bu) commonly contained in the wastewater from a beer brewery factory was investigated. Then, the reaction scheme was identified and their Monod kinetic parameters were evaluated for the degradation of each single component in a batch UASB reactor. In addition, the degradation sequences and their kinetics were also identified to evaluate their Monod kinetic parameters for the multiple substrate systems with combinations of the three volatile acid components with or without an inhibition mechanism.

In the present study, the UASB reactor was modeled in terms of the dispersed plug flow and the Monod type of rate equations to construct the differential mass balance equations for the anaerobic biodegradation of single and multiple substrates components of the volatile fatty acids.

2. ANAEROBIC DECOMPOSITION MECHANISM FOR VFA COMPONENTS

Figure 1 schematically shows the degradation schemes for the volatile acid components (Ac, Pr and Bu). The detailed stoichiometric relationships of the degradation sequences were given for the single and multiple components of the volatile acids in the previous study [6]. We assumed the substrate decomposition rates in the suspension of biocatalyst granules to be represented by the Monod kinetics on the basis of the amount of molecular carbon, and then the Monod kinetic parameters were evaluated for the decomposition of each single component, Ac, Pr, or Bu. The Monod equation for a single substrate is given by,

$$-r_S = \frac{-r_{S,max}C_S}{K_S + C_S} \qquad (1)$$

In the systems for decomposition of multiple organic acid substrates, the decomposition of acetic acid was inhibited by the presence of butyric acid through a competitive mechanism, and the decomposition rate was correlated well by a puely competitive inhibition model as given by the bellowing equation.

$$-r_{Ac}^{I} = \frac{-r_{Ac,max}C_{Ac}}{K_{Ac}\left(1+C_{Bu}/K_{I,Ac}\right)+C_{Ac}} \tag{2}$$

The decomposition of propionic acid was inhibited by the presence of acetic acid and the inhibition behavior was approximated by purely non-competitive inhibition model as given by following equation.

$$-r_{Pr}^{I} = \frac{-r_{Pr,max}C_{Pr}}{\left(K_{Pr}+C_{Pr}\right)\left(1+C_{Ac}/K_{I,Pr}\right)} \tag{3}$$

The Monod kinetic parameters were evaluated by least squares fitting procedures, for the single and multiple substrate systems with/without mutual inhibition, and were indicated in Table 1 [6]. The value of $-r_{S,max}$ indicates the linear decomposition rate.   It is clear that the decomposition rate for propionic acid is significantly lower than those for acetic acid and butyric acid.

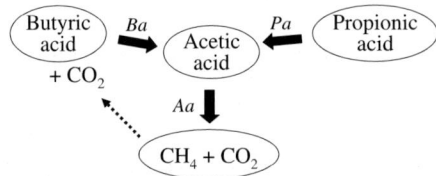

Fig. 1 Reaction sequence for anaerobic decomposition of organic fatty acids considered; *Aa*, *Ba* and *Pa* are yield coefficients for respective decomposition reactions.

Table 1 Monod parameters for various substrates with/without inhibition

Substrates to be decomposed	$-r_{S,max}$ [mol-C/g-dry cell/min]	K_S [mol-C/l]	$K_{I,j}$ [mol-C/l]
Acetic acid (Ac)	1.20×10^{-5}	5.76×10^{-3}	
Propionic acid (Pr)	6.80×10^{-7}	6.80×10^{-4}	
Butyric acid (Bu)	4.34×10^{-6}	8.70×10^{-3}	
Acetic acid inhibited by butyric acid	1.20×10^{-5}	5.76×10^{-3}	1.20×10^{-3}
Propionic acid inhibited by acetic acid	6.80×10^{-7}	6.80×10^{-4}	2.84×10^{-3}

3. REACTOR MODELING AND SIMULATION

The UASB reactor was modeled by the dispersed plug flow model, considering decomposition reactions for VFA components, axial dispersion of liquid and hydrodynamics.   The differential mass balance equations based on the dispersed plug flow model are described for multiple VFA substrate components considered.

$$\frac{1}{Pe}\frac{d^2Y_{Bu}}{dZ^2} - \frac{dY_{Bu}}{dZ} - \omega_{Bu}\frac{Y_{Bu}}{1+\beta_{Bu0}Y_{Bu}} = 0 \tag{4}$$

$$\frac{1}{Pe}\frac{d^2Y_{Pr}}{dZ^2} - \frac{dY_{Pr}}{dZ} - \omega_{Pr}\frac{Y_{Pr}}{\left(1 + \beta_{Pr0}Y_{Pr}\right)\left(1 + Ac/K_{I,Pr}\right)} = 0 \tag{5}$$

$$\frac{1}{Pe}\frac{d^2Y_{Ac}}{dZ^2} - \frac{dY_{Ac}}{dZ} + Ba\cdot\left(\frac{C_{Bu0}}{C_{Ac0}}\right)\cdot\omega_{Bu}\frac{Y_{Bu}}{1 + \beta_{Bu0}Y_{Bu}} + Pa\cdot\frac{2}{3}\left(\frac{C_{Pr0}}{C_{Ac0}}\right)\cdot\omega_{Pr}\frac{Y_{Pr}}{\left(1 + \beta_{Pr0}Y_{Pr}\right)\left(1 + Ac/K_{I,Pr}\right)}$$

$$- \omega_{Ac}\frac{Y_{Ac}}{1 + Bu/K_{I,Ac} + \beta_{Ac0}Y_{Ac}} = 0 \tag{6}$$

Boundary conditions at the inlet and outlet of the reactor are as follows (j=Bu, Pr, Ac).

$$Z = 0;\ \left.\frac{dY_j}{dZ}\right|_{Z=0^-} = Pe\left(\left.Y_j\right|_{Z=0^-} - 1\right)\ ,\ \text{and}\ Z = 1;\ \left.\frac{dY_j}{dz}\right|_{Z=1^-} = 0 \tag{7}$$

In the above equations, the dimensionless substrate concentration, Y_j, and the dimensionless axial distance Z are defined as follows.

$$Y_j = C_j/C_{j0},\quad Z = z/L \tag{8}$$

Here, L is the depth of granule bed [m], C_j is the substrate concentration and C_{j0} is the influent substrate concentration on the basis of carbon element [mol-C/l].

Pe, ω_j, and β_0 are dimensionless parameters relating to the operating conditions; Pe is Peclet number denoting the inverse of axial mixing intensity, ω_j denotes the inverse of volumetric loading rate per mass of granules, and β_0 denotes the dimensionless inlet substrate concentration as respectively defined as follows:

$$Pe = \frac{U_l L}{D_z},\quad \omega_j = \frac{(1-\varepsilon)}{U_l}\cdot\frac{\rho_p LA}{A}\cdot\left(\frac{-r_{s,max}}{K_s}\right) = \frac{W_p}{Ql}\cdot\left(\frac{-r_{s,max}}{K_s}\right)\ \text{and}\ \beta_0 = \frac{C_{j0}}{K_j}. \tag{9}$$

4. RESULTS AND DISCUSSION

In the following, the performance of a UASB reactor with the same size of a pilot plant [7] is evaluated according to the reactor simulation model incorporated with the Monod kinetic parameters for the hypothetical influent composition for the three VFA components as indicated in Table 2.

In Fig. 2, the axial distributions of the dimensionless concentration of the VFA components are shown. The concentration jumps for the substrate species at the inlet of the column inside indicate the effect of axial liquid mixing. The degree of decomposition is smallest for the propionic acid because its decomposition rate is quite slow compared with other substrate components. This means that the inclusion of large amount of such a slow degradable component as propionic acid inevitably leads to a significant reduction in the total substrate removal efficiency. The COD removal efficiencies of the VFA components are in order as, acetic acid: 0.765 > butyric acid: 0.705 > propionic acid: 0.138. The estimated value of the total COD removal efficiency is 0.561, which is significantly small compared to the experimental COD removal efficiency of 91% in a pilot plant UASB unit treating wastewater from a beer brewery [7].

Table 3 indicates a hypothetical operating condition treating wastewater containing only acetic acid with various flow rates. Figure 3 shows the axial variation of the relative ratio of the substrate (remaining acetic acid) with the change of the substrate loading rate (ω_{Ac}); the higher the substrate loading (smaller ω_{Ac}), the higher the remaining amount of the substrate, resulting in the lower substrate removal efficiency.

Table 2 Reactor dimension and influent condition

Column dimension		Influent constituent	
Height [m]	7.5	Butyric acid [ppm]	80
Diameter [m]	1.0	Propionic acid [ppm]	170
Flow rate [m/h]	5.5	Acetic acid [ppm]	400
Liquid hold-up [-]	0.549	COD_{cr} [ppm]	2200
Peclet number [-]	1.6	Granule density [kg/m^3]	110.8

Table 3 Operating conditions

Column height [m]	7.5
Column diameter [m]	1.0
Liquid flow rate [m/h]	4, 5, 7, 9, 11
Acetic acid conc.[ppm]	1000
Granule density [kg/m^3]	110.8
Liquid hold-up [-]	0.549
Axial Peclet number [-]	1.6

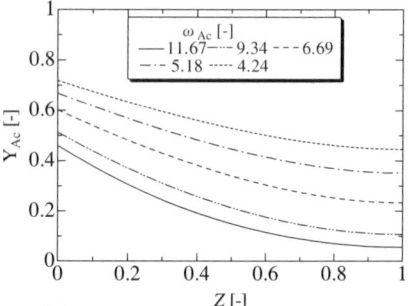

Fig. 2 Axial distributions of dimensionless substrate concentrations for multiple VFA components.

Fig. 3 Influence of flow rate on UASB reactor analysis

5. CONCLUSIONS

A pilot scale UASB reactor was simulated by the dispersed plug flow model with Monod kinetic parameters for the hypothetical influent composition for the three VFA components. As a result, the COD removal efficiency for the propionic acid is smallest because its decomposition rate is quite slow compared with other substrate components; their COD removal efficiencies are in order as, acetic acid: 0.765 > butyric acid: 0.705 > propionic acid: 0.138. And the estimated value of the total COD removal efficiency is 0.561. This means that the inclusion of large amount of propionic acid will lead to a significant reduction in the total VFA removal efficiency.

ACKNOWLEDGEMENT

This research was financially supported in part by the Kansai University Grant-in-Aid for Promotion of Advanced Research in Graduate Course, 2004.

REFERENCES

[1] L. Hulshoff Pol and G. Lettinga, Wat. Sci. Technol., 18 (1986) 41-53.
[2] J.A. Zeevalkink, and A.J.M. Jans, Starch, 38 (1986) 243-246.
[3] H.J. Nanninga and J.C. Gottschal, Wat. Res., 20 (1986) 97-103.
[4] A.M. Craverio, H.M. Soares, and W. Schmidell, Wat. Sci. Technol., 18 (1986) 123-134.
[5] M. Yoda, H. Yamauchi, M. Kitagawa, Yosui To Haisui, 31 (1989) 49-54.
[6] K. Muroyama, T. Nakai, Y. Uehara, Y. Sumida, and A.Sumi, J. Chem. Eng. Japan, 37 (2004) 1026-1034.
[7] M. Kaji and A. Sumi, Shinko Pantec Gifo (Shinko Pantec Co. Ltd.), 41 (1997) 9-17.

Studies in Surface Science and Catalysis, volume 159
Hyun-Ku Rhee, In-Sik Nam and Jong Moon Park (Editors)
© 2006 Elsevier B.V. All rights reserved

665

The Simulation and Control of the Reactive Distillation Process for Dimethylcarbonate Production

Yong Hee Jang[a], Ahrim Yoo[a], Byung Sung Ahn[b] and Dae Ryook Yang[a*]

[a]Department of Chemical and Biological Engineering, Korea University,
1 Anamdong, Seonbukku, Seoul, 136-713, Korea

[b]Reaction Media Research Center, Korea Institute of Science and Technology
39-1 Hawolkokdong, Seongbukku, Seoul, 136-791, Korea

1. INTRODUCTION

Dimethylcarbonate (DMC) is an important chemical, which is used as a solvent, an octane booster in gasoline to meet oxygenate specifications, and as a starting material for organic synthesis *via* carbonylation and methylation replacing the use of poisonous phosgene and dimethyl sulfate. It is also used as a precursor for polycarbonate resins instead of toxic phosgene [1]. Even though many kinds of methods for DMC production are reported, most of these methods have several drawbacks such as low yields, high cost of the starting materials and separation problems. One of the alternatives of DMC production is the use of Reactive Distillation (RD) for transesterification of Ethylene Carbonate (EC) and methanol. The RD is a combination process, where both separation and reaction are considered simultaneously in a single column. In this paper, an optimal process configuration was proposed to produce the DMC using RD efficiently.

It is a fairly new approach to produce DMC with RD system. One of the advantages of RD is less-demand of separation effort. If the reactant, EC is in excess in the reaction, almost pure DMC can be obtained from the top of the RD column, but the separation of unreacted EC and by-product ethylene glycol (EG) is necessary. If the methanol is in excess, an azeotropic mixture of methanol and DMC will be produced at the top and almost pure EG from the bottom. In this study, two alternatives of the possible configurations of RD system as shown in Fig. 1 are compared. Also, a dynamic model is developed to investigate the crucial issues such as feed ratio control, the effect of reactant recovery column and so on. Based on the dynamic simulation, control system configuration is designed using PI control and Model Predictive Control (MPC) for handling the operational objectives such as temperature constrains, productivity, and the energy saving.

---

* Authors to whom correspondence should be addressed. Email: dryang@korea.ac.kr

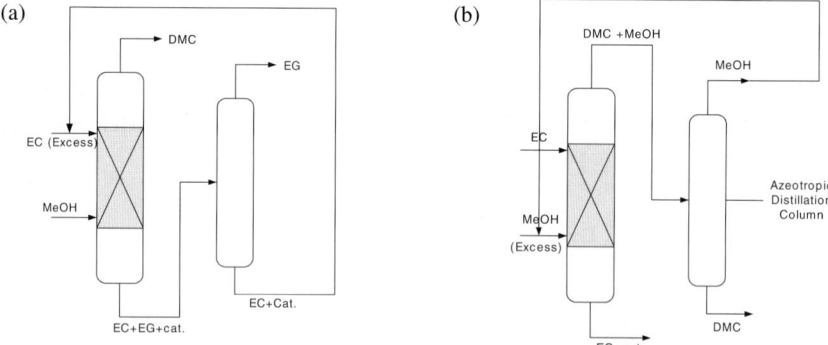

Fig. 1. RD Configurations for DMC production: (a) EC excess process, (b) MeOH excess process.

2. MODELING OF RD SYSTEM

2.1 Steady-state model

To simulate the RD system, the MESH model is used, which is assumed that each plate is in vapor-liquid equilibrium. The MESH equations are as follows:

(M) Material balance: $\quad L_{j-1}x_{i,j-1} + V_{j+1}y_{i,j+1} + F_j z_{i,j} - L_j x_{i,j} - V_j y_{i,j} + R_{i,j} = 0$ $\qquad$ (1)

(E) Phase Equilibrium relation: $\quad y_{i,j} - K_{i,j}x_{i,j} = 0$ $\qquad$ (2)

(S) Mole fraction Summations: $\quad \sum_{i=1}^{C} y_{i,j} - 1.0 = 0, \quad \sum_{i=1}^{C} x_{i,j} - 1.0 = 0$ $\qquad$ (3)

(H) Energy balance: $\quad L_{j-1}H_{L_{j-1}} + V_{j+1}H_{V_{j+1}} + F_j H_{F_j} - L_j H_{L_j} - V_j H_{V_j} - Q_j = 0$ $\qquad$ (4)

The detail information on reaction can be found in the literature [2]. These equations are quite nonlinear and the solution can be obtained by using Naphthali-Sandholm algorithm.

2.2 Dynamic model

A dynamic model should be consistent with the steady-state model. Thus, Eqs (1) and (4) should be extended to dynamic form. For the better convergence and computational efficiency, some assumption can be introduced: the total amounts of mass and enthalpy at each plate are maintained constant. Then, the internal flow can be determined by total mass balance and total energy balance and the number of differential equations is reduced. Therefore, the dynamic model can be established by replacing component material balance in Eq. (1) with the following equation.

$$\frac{dM_j x_{i,j}}{dt} = V_{j+1}y_{i,j+1} - L_{j-1}x_{i,j-1} - V_j y_{i,j} - L_j x_{i,j} + R_{i,j}$$ $\qquad$ (5)

3. CONTROLLER DESIGN

The main control objectives of this system are to maintain high conversion of EC and to minimize the energy consumption under given constraints. In this study, PI controller and Model predictive controller (MPC) were investigated to satisfy the objectives. For testing performance of the control systems, the dynamic model obtained in above section is used. Since the RD system is a multiple input and multiple output (MIMO) system, the paring of controlled variables (CV) and manipulated variables (MV) is needed for PI control. Using the Relative Gain Array (RGA) method, the top composition was controlled by the distillate flow rate, and the bottom composition was controlled by the amount of heat added to the reboiler. However, the minimization of the energy consumption cannot be satisfied with the PI controller and the performance of it was limited due to the significant interaction between variables. Therefore, the MPC is preferred for this RD system. The objective function of MPC was constructed to satisfy the set-point of top and bottom composition and at the same time to minimize the energy consumption.

4. RESULTS AND DISCUSSIONS

The steady-state simulation results of proposed configurations are compared. The top composition of DMC is 28.77% in the case of methanol excess process as in Fig. 2. On the other hand, that of DMC is 38.18% in EC excess process. However, the methanol excess process has a higher DMC conversion and productivity as shown in Table 1. In other aspects, the azeotropic separation of methanol and DMC is relatively easy at an elevated pressure such as 10 atm. Also, the unreacted EC which is high-cost reactant material will decompose at high temperature in the existence of base homogeneous catalyst. Therefore, the methanol excess configuration should be used for DMC production. Using the decided configuration, a dynamic model is established and the effects of important variables are investigated and the performances of control configurations are also studied. The Figures 3 and 4 show the results of control systems when feed flow rate was increased from 187.56 g/hr to 202.56 g/hr (Figs. 3d and 4d).

Table 1. Comparison of MeOH excess and EC excess processes.

Operating Condition	Methanol Excess Process	EC Excess process
Top product rate (g/Hr)	666.89	28.74
Bottom product rate (g/Hr)	145.00	783.15
Weight fraction of DMC in Top product (%)	28.77	38.18
Conversion (%)	99.50	47.32
DMC production rate (g/Hr)	191.86	10.97

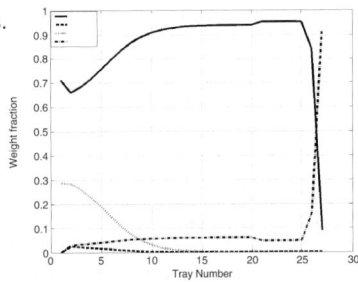

Fig. 2. Steady-state simulation result of MeOH excess process.

668

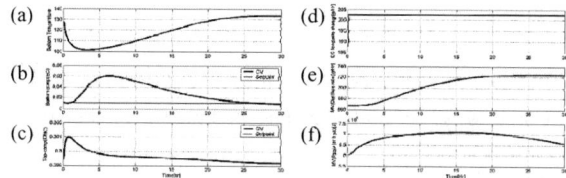

Fig. 3. PI control results when EC feed change form 187.56 g/hr to 202.56 g/hr

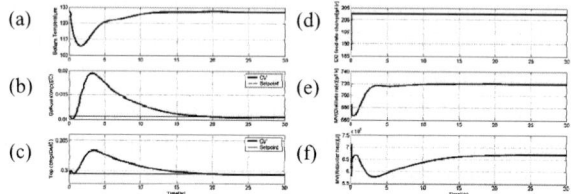

Fig. 4. MPC results when EC feed change form 187.56 g/hr to 202.56 g/hr

For the operating conditions, the set-points of EC and DMC compositions at the top and bottom are 0.01 and 0.2996, respectively, and the bottom temperature should not exceed 140℃ to prevent the decomposition of reactants. From these plots, it can be concluded that the MPC outperforms the PI controller in terms of response speed in disturbance rejection, maintaining the variables at set points, and optimization capability. Especially, the PI controller failed to maintain the DMC composition set-point due to the slow long-term dynamics caused by the interaction between the RD column and azeotropic recovery column.

5. CONCLUSIONS

In this study, a model of RD system for DMC production is developed using MESH equation. For the DMC production using RD, there are two possible configurations. One is EC excess process and the other is methanol excess process. From the comparison of steady-state simulation results, the methanol excess process configuration is superior to the EC excess process in the sense of productivity and ease of separation. Also, the effects of manipulated variables on process outputs are investigated to design the control strategy using the dynamic model. The PI control is not adequate for this system and the use of MPC is recommended.

REFERENCES

[1] Franco Rivetti, "The role of Dimethylcarbonate in the replacement of hazardous," *International Journal of Control*, 23(4) (1989) 123.
[2] Yun-Jin Fang and Wen-De Xiao, "Experimental and modeling studies on a homogeneous reactive distillation system for dimethyl carbonate synthesis by transesterification," *Separation and purification technology*, 34 (2004) 255.

Studies in Surface Science and Catalysis, volume 159
Hyun-Ku Rhee, In-Sik Nam and Jong Moon Park (Editors)

Hydrodynamic Analysis of a Novel Photocatalytic Reactor Using Computational Fluid Dynamics

Shane J. Cox and Adesoji A. Adesina[*]
Reactor Engineering & Technology Group, School of Chemical Engineering & Industrial Chemistry, University of New South Wales, Sydney, Australia
    [*] a.adesina@unsw.edu.au

ABSTRACT

This study investigates the hydrodynamic behaviour of an annular bubble column reactor with continuous liquid and gas flow using an Eulerian-Eulerian computational fluid dynamics approach. The residence time distribution is completed using a numerical scalar technique which compares favourably to the corresponding experimental data. It is shown that liquid mixing performance and residence time are strong functions of flowrate and direction.

INTRODUCTION

Many industrial processes which employ bubble column reactors (BCRs) operate on a continuous liquid flow basis. As a result these BCR's are a substantially more complicated than stationary flow systems. The design and operation of these systems is largely proprietary and there is, indeed a strong reliance upon scale up strategies [1]. With the implementation of Computational Fluid Dynamics (CFD), the associated complex flow phenomena may be analyzed to obtain a more comprehensive basis for reactor analysis and optimization. This study has examined the hydrodynamic characteristics of an annular 2-phase (liquid-gas) bubble column reactor operating co- and counter-current (with respect to the gas flow) continuous modes.

Resident Time Distribution (RTD) is widely employed in the chemical engineering industry, as an analytical tool for characterizing flow dynamics within reactor vessels. RTD provides a quantitative measure of the back-mixing with in a reactor system [2]. However the cost and time involved in building and operating a pilot- or full scale reactor for RTD analysis can be economically prohibitive. As such we have implemented a numerical RTD technique through the FLUENT (ver. 6.1) commercial CFD package.

EXPERIMENTAL SETUP

The general operation of the pilot scale reactor has be previously described by Pareek et. al. [3]. However, modifications were required to allow the injection of the gas and liquid tracers, and their subsequent detection at the outlets. The liquid tracer, 5mL Methyl blue solution ($10gL^{-1}$), was injected via a syringe inserted into the liquid feed line. Outlet samples were measured with a Shimadzu 1601 UV-Vis Spectrophotometer at a wavelength of 635nm. A pulse (20mL) of helium gas tracer was introduced using an automated control system, with the outlet concentration monitored in real-time with a thermal conductivity detector. Runs were carried out based on a two-level factorial design

CFD MODEL DESCRIPTION
An Eulerian-Eulerian (EE) approach was adopted to simulate the dispersed gas-liquid flow. The EE approach treats both the primary liquid phase and the dispersed gas phase as interpenetrating continua, and solves a set of Navier-Stokes equations for each phase. Velocity inlet and outlet boundary conditions were employed in the liquid phase, whilst the gas phase conditions consisted of a velocity inlet and pressure outlet. Turbulence within the system was accounted for with the Standard k-ε model, implemented on a per-phase basis, similar to the recent work of Bertola et. al.[4]. A more detailed description of the computational setup of the EE method can be found in Pareek et. al.[5].

NUMERICAL RESIDENCE TIME DISTRIBUTION PROCEDURE
Numerical Residence Time Distribution (RTD) information was obtained from the solution to;

$$\frac{\partial \alpha_i \phi_i}{\partial t} + \nabla \cdot \left(\alpha_i v_i \phi_i - \alpha_i \Gamma_i \nabla \phi_i \right) = 0 \qquad [1]$$

Via a passive scalar method [6] where α_i denotes the volume fraction of the i-th phase, while Γ_i represents the diffusivity coefficient of the tracer in the i-th phase. The transient form of the scalar transport equation was utilized to track the pulse of tracer through the computational domain. The exit age distribution was evaluated from the normalized concentration curve obtained via measurements at the reactor outlet at 1 second intervals. This was subsequently used to determine the mean residence time, t_m and Peclet number, Pe [7].

RESULTS AND DISCUSSION
To establish the validity of the numerical scalar technique for RTD analysis, the normalized exit age distribution curve of both counter-current (Figure 1 (a-b)) and co-current (Figure 1 (c-d)) flow modes were compared. Table 1 shows that a good agreement was obtained between CFD simulation and experimental data.

Table 1: Liquid RTD results summary

Flow Direction	Liquid Flow (Lmin^{-1})	Gas Flow (Lmin^{-1})	Simulation		Experimental	
			T_m	Pe	T_m	Pe
co-current	1.5	2	7.67	4.74	10.20	4.61
counter-current	1.5	2	4.90	5.16	8.45	4.87
co-current	3	2	5.83	4.76	5.56	4.45
counter-current	3	2	4.91	4.21	4.55	5.01
co-current	1.5	5	8.87	4.82	10.34	4.69
counter-current	1.5	5	6.66	5.54	8.06	5.24
co-current	3	5	5.22	4.30	5.18	4.23
counter-current	3	5	4.94	4.85	4.87	4.49

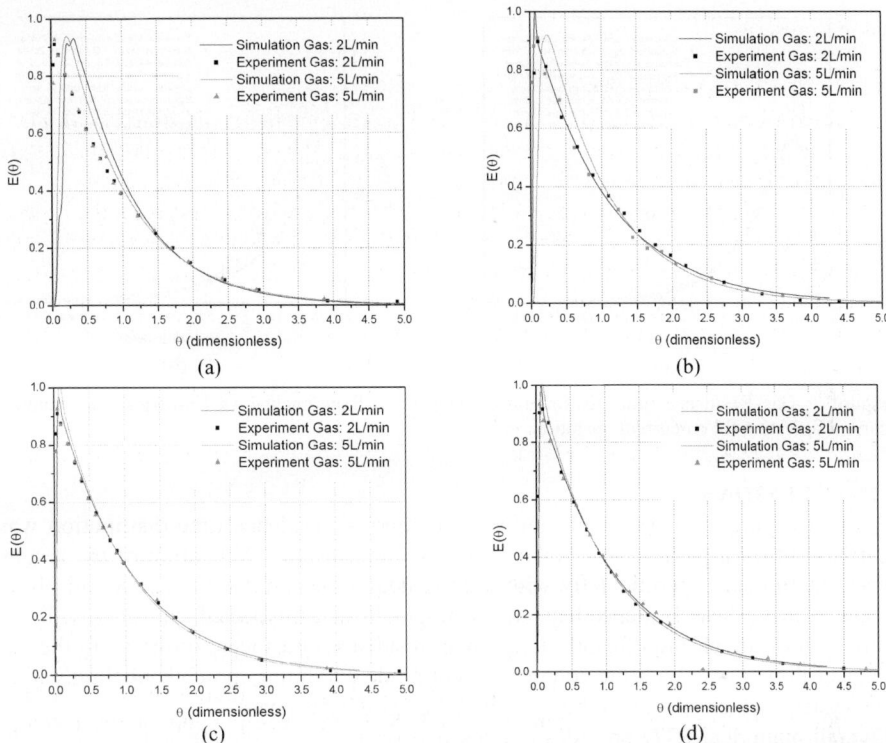

Figure 1: Liquid Residence time distribution comparison – Experimental vs Numerical: (a) counter-current operation Liquid flowrate: 1.5Lmin^{-1} (b) counter-current operation Liquid flowrate: 3Lmin^{-1} (c) co-current operation Liquid flowrate: 1.5Lmin^{-1} (d) co-current operation Liquid flowrate: 3Lmin^{-1}

The liquid mean residence time, t_m, in counter-current mode was significantly lower than in co-current mode. Statistical analysis (Yates method of ANOVA) indicates that along with the liquid flowrate, the direction of liquid travel is the most significant factor in the liquid mean residence time. This variation in the liquid phase mean residence time, suggests an increase in short-circuiting or channeling in the counter-current mode as a result of the gas-liquid interactions.

Figure 2 (a-b) shows the RTD profiles in the gas phase. In general gas phase mixing is somewhat more challenging to unravel due to wide fluctuations in gas phase concentration measurements with changing flow rate. [8]. However, as the results in Figure 2(a-b) show there is a reasonable degree of correlation between the numerical and experimental results. Significantly, less mixing is observed in the gas phase than in the slower moving liquid. The mean residence times (table not shown) for the gas phase, as with the liquid phase show a strong dependence upon the mode of operation (i.e. co- or counter-current.), furthermore it is the impact of the gas on liquid which contributed to the liquid phase channeling in counter-current mode. This dependence further highlights the intimate nature of mixing between the liquid and gas phases in bubble column reactors.

672

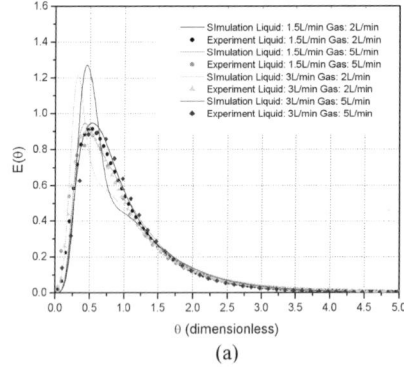
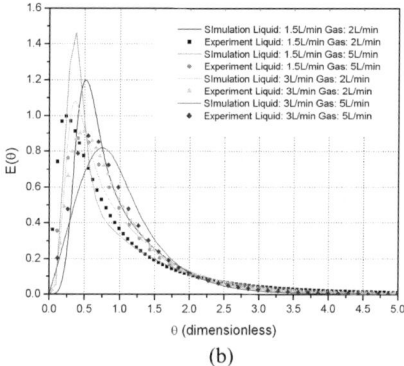

| (a) | (b) |

Figure 2: Gas Residence time distribution comparison – Experimental vs Numerical: (a) counter-current operation (b) co-current operation.

CONCLUSIONS

An investigation into the applicability of numerical residence time distribution was carried out on a pilot-scale annular bubble column reactor. Validation of the results was determined experimentally with a good degree of correlation. The liquid phase showed to be heavily dependent on the liquid flow, as expected, but also with the direction of travel. Significantly larger man residence times were observed in the co-current flow mode, with the counter-current mode exhibiting more channeling within the system, which appears to be contributed to by the gas phase.

Overall numerical RTD provides a quick and relatively simple method to accurately determining the flow characteristics within this type of reactor vessel.

ACKNOWLEDGEMENTS

The authors are grateful for the ARC funding for the research program on multiphase reactor system design and UNSW Engineering for a postgraduate scholarship to SJC.

REFERENCES

1. Ramachandran, P.A. and R.V. Chaudhari, Three Phase Catalystic Reactors. 1983: Gordon and Breach.
2. Gavrilescu, M. and R.Z. Tudose, Residence time distribution of the liquid phase in a concentric-tube airlift reactor. Chemical Engineering and Processing, 1999. 38(3): p. 225-238.
3. Pareek, V., M.P. Brungs, and A.A. Adesina, Photocausticization of Spent Bayer liquor: A Pilot-Scale Study. Advances in Environmental Research, 2003. 7(2): p. 411-420.
4. Bertola, F., M. Vanni, and G. Baldi, Application of Computational Fluid Dynamics to Multiphase Flow in Bubble Columns. International journal of Chemical Reactor Engineering, 2003. 1: p. A3.
5. Pareek, V.K., S.J. Cox, M.P. Brungs, B. Young, and A.A. Adesina, Computational fluid dynamic (CFD) Simulation of a Pilot-Scale Annular Bubble Column Photocatalytic Reactor. Chemical Engineering Science, 2003. 58(3-6): p. 859-865.
6. Brown, G.J. CFD Predictions of Particle and Fluid Residence Times in Industrial Vessels. in 6th world congress of chemical engineering. 2001. Melbourne.
7. Levenspiel, O., Chemical Reaction Engineering. 3rd ed. 1999.
8. Zahradnik, J. and M. Fialova, The effect of bubbling regime on gas and liquid phase mixing in bubble column reactors. Chemical Engineering Science, 1996. 51(10): p. 2491-2500.

Studies in Surface Science and Catalysis, volume 159
Hyun-Ku Rhee, In-Sik Nam and Jong Moon Park (Editors)
673

A Kinetic Study on the Solid State Grafting of Maleic Anhydride onto Isotactic Polypropylene in Supercritical CO_2

Gangsheng Tong[a], Tao Liu[a], Ling Zhao[a], Guo-Hua Hu[b, c] and Weikang Yuan[a]

[a]UNILAB Research Center of Chemical Engineering, State Key Laboratory of Chemical Reaction Engineering, East China University of Science and Technology, Shanghai 200237, Peoples Republic of China, [b]Laboratory of Chemical Engineering Sciences, CNRS-ENSIC-INPL, 1 rue Grandville, B. P. 451, 54001 Nancy, France (hu@ensic.inpl-nancy.fr), and [c]Institut Universitaire de France

1. INTRODUCTION

Solid state grafting of maleic anhydride (MA) onto isotactic polypropylene (iPP) in supercritical CO_2 is a useful post-polymerization method for functionalizing PP without significantly changing the molecular architecture of the PP backbone and can overcome drawbacks brought from conventional processes such as solution process [1], melt process [2-3] and solid state process [4]. Studies on the mechanism and kinetics of the maleation reaction in $scCO_2$ are of great importance for the development and application of this novel technology. In this work, we apply a maleation reaction mechanism involving the main elementary reactions to develop a general kinetic model on the basis of our experimental results. We then simulate the kinetics of the solid state PP maleation in $scCO_2$.

2. EXPERIMENTAL

Maleation of iPP was carried out in a high-pressure vessel. After a desired time interval, the vessel was quenched with running cold water to terminate the reaction. Samples were taken out and then purified for subsequent determination of the grafting degree [5].

3. MALEATION MECHANISM

Similar to inert solvents used in conventional solid state grafting processes, CO_2 can only impregnate into the amorphous region of PP [6-8], consisting of the monomer and initiator. Therefore, $scCO_2$ assisted solid state grafting of MA onto PP should be similar to the conventional solid state grafting process in many aspects, despite the fact that the former may follow some special mechanisms. Table 1 shows the mechanism of the free radical solid grafting proposed in this work comprises the classical initiation, propagation, chain transfer, and termination steps [9-10]. Meanwhile, the following hypotheses are made for the free-radical solid grafting mechanism: (1) The same type of hydrogen of the polypropylene is equally reactive; (2) The initiator efficiency and the rate constants of the radical reactions remain unchanged during the reaction; (3) Homopolymerization of MA is negligible; (4) Dismutation is the primary mode of radical termination; (5) All types of radicals are in a quasi-stationary state.

Based on the above hypotheses and further some mathematical manipulations, we obtained

Table 1. Mechanism of the free-radical graft of MA onto iPP with DCP as the initiator

Initiation	Chain Transfer
$I \xrightarrow{K_d} 2R^*$	$PM_nM^* + P \xrightarrow{K_{tr1}} PM_{n+1} + P^*$
$R^* + M \xrightarrow{K_{i1}} RM^*$	$RM_nM^* + P \xrightarrow{K_{tr1}} RM_{n+1} + P^*$
$R^* + P \xrightarrow{K_{i2}} RH + P^*$	$PM_nM^* + M \xrightarrow{K_{tr2}} PM_{n+1} + M^*$
$P^* + M \xrightarrow{K_{i3}} PM^*$	$RM_nM^* + M \xrightarrow{K_{tr2}} RM_{n+1} + M^*$

Propagation	Termination
$PM^* + nM \xrightarrow{K_P} PM_nM^*$	$PM_nM^* \xrightarrow{K_t} PM_{n+1}$
$RM^* + nM \xrightarrow{K_P} RM_nM^*$	$RM_nM^* \xrightarrow{K_t} RM_{n+1}$

I, M and P represent the initiator, the monomer and polypropylene, respectively.

the following equation for the grafting rate:

$$R_g = K_p[PM_n^*][M]$$

$$= 2K_p K_d[I][P][M] \times \left[\frac{K_{i2}}{(K_{i1}[M] + K_{i2}[P])(K_{tr1}[P] + K_{tr2}[M] + K_t)} + \frac{K_{tr1}}{K_t(K_{tr1}[P] + K_{tr2}[M] + K_t)} \right] \quad (1)$$

The concentration of the PP can be considered constant because it is much higher than those of all other reactants. Therefore, Eq. (1) can be reduced to:

$$R_g = K[I][M] \times \left[\frac{K_{i2}'}{(K_{i1}[M] + K_{i2}')(K_{tr1}' + K_{tr2}[M] + K_t)} + \frac{K_{tr1}'}{K_t(K_{tr1}' + K_{tr2}[M] + K_t)} \right] \quad (2)$$

where $K = 2K_p K_d[P]$, $K_{i2}' = K_{i2}[P]$, $K_{tr1}' = K_{tr1}[P]$.

Eq. (2) can be further simplified in form of $R_g = K_{app}[I][M]^\beta$. At the initial reaction time, $[I]$ and $[M]$ are close to $[I]_0$ and $[M]_0$. Under such conditions, the corresponding grafting rate R_{g0} can be expressed as $R_{g0} = K_{app}[I]_0[M]_0^\beta$. If K_{app} follows the Arrhenius law, the kinetic equation can be written as:

$$R_{g0} = A \exp(-\Delta E_a / RT)[I]_0[M]_0^\beta \quad (3)$$

A more general form of the above equation can be expressed as:

$$R_{g0} = A \exp(-\Delta E_a / RT)[I]_0^\alpha [M]_0^\beta \quad (4)$$

4. EXPERIMENTAL RESULTS

Fig. 1 shows the grafted anhydride content in PP-g-MA ($G_{MA}\%$) as a function of reaction time at different initial DCP concentrations ranging from 0.1515 to 0.4535 $mol \cdot L^{-1}$ under the following conditions: initial MA concentration = 1.424 $mol \cdot L^{-1}$, temperature = 413.15 K and CO_2 pressure = 18.0 MPa. Fig. 2 depicts $G_{MA}\%$ as a function of reaction time at different initial MA concentrations ranging from 0.712 to 2.136 $mol \cdot L^{-1}$ obtained with the following conditions: initial DCP concentration = 0.2267 $mol \cdot L^{-1}$, temperature = 413.15 K and CO_2 pressure = 18.0 MPa. Fig. 3 provides with $G_{MA}\%$ as a function of reaction time under 18.0

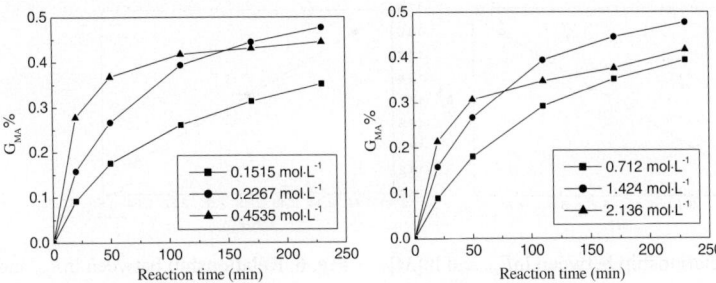

Fig.1. Anhydride content of PP-g-MA *vs* reaction
time at different initial DCP concentrations.

Fig.2. Anhydride content of PP-g-MA *vs* reaction
time at different initial MA concentrations.

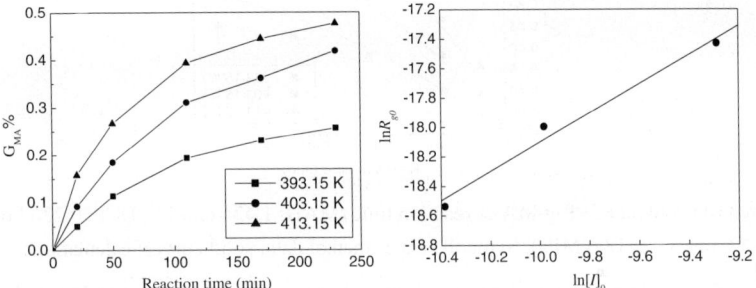

Fig.3. Anhydride content of PP-g-MA *vs* reaction
at different temperatures.

Fig.4. Curve of $\ln R_{g0}$ *vs* $\ln[M]_0$ time

MPa CO_2 atmosphere and at 393.15, 403.15 and 413.15 K, respectively. The initial MA and DCP concentrations are 1.424 mol·L^{-1} and 0.227 mol·L^{-1}, respectively.

5. DISCUSSION

From Figs.1-3, the initial grafting rates R_{g0} at different initial DCP concentrations, initial MA concentrations or temperatures, can be calculated from the grafted anhydride content of PP-g-MA obtained at 20 min of reaction. The power of the DCP concentration term α is correlated to be 0.99 from the linear relationship between $\ln R_{g0}$ and $\ln[I]_0$ in Fig. 4. That of the MA concentration β is correlated at 0.97 from the linearity between $\ln R_{g0}$ and $\ln[M]_0$ in Fig. 5. The relationship between $\ln R_{g0}$ and $1/T$ is linear, as shown in Fig.6, implying that the apparent rate constant K_{app} followed the Arrhenius law. The apparent active energy ΔE_a and the frequency factor A calculated from the straight line are 77.5 kJ·mol^{-1} and 1.36×10^6 L·mol^{-1}·s^{-1}, respectively. Based on the above results, the kinetic equation for solid state grafting of MA onto iPP in scCO$_2$ can be expressed as:

$$R_g = 1.36 \times 10^6 \times \exp(-77.5 \text{ kJ·mol}^{-1}/RT)[I]^{0.99}[M]^{0.97} \tag{5}$$

The authors' another study has obtained the following equation for the DCP decomposition kinetics: $K_d = 1.30 \times 10^{14} \exp(-139.2 \text{ kJ·mol}^{-1}/RT)$ [11]. Substitution of this equation to Eq. (5) allows calculating the kinetics of the solid state grafting of MA onto iPP in scCO$_2$. The results

676

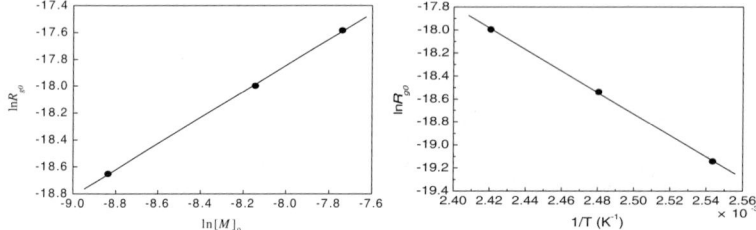

Fig. 5. Relationship between $\ln R_{g0}$ and $\ln[M]_0$ Fig. 6. Relationship between $\ln R_{g0}$ and $1/T$

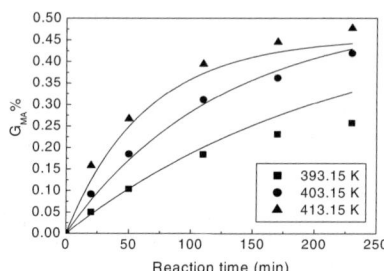

Fig. 7 Anhydride content of PP-g-MA vs reaction time (MAH: 1.424 mol·L^{-1}; DCP: 0.2267 mol·L^{-1}; CO$_2$ pressure: 18.0 MPa). Symbols: experimental data; solid curves: calculations.

are shown in Fig. 7. The average deviation between the calculated and experimental results is 9 %, indicating that Eq. (5) agrees with the experimental results satisfactorily.

6. ACKNOWLEDGEMENTS

The authors are grateful to the Ministry of Science and Technology of China for the support of a major project for international cooperation (Grant 2001CB711203), to the National Science Foundation of China and PetroChina for the support of a joint project on multiscale methodologies, and to the Association Franco-Chinoise pour la Recherche Scientifique et Technique (AFCRST) for the support of PRA Mx04-05.

7. REFERENCES

[1] J. Ma. García-Martínez, O. Laguna, E. P. Collar, J. Appl. Polym. Sci., 68 (1998) 483.

[2] G. Moad, Prog. Polym. Sci., 24 (1999) 81.

[3] D. Shi, J. Yang, Zh. Yao, Y. Wang, H. Huang, W. Jing, J. Yin, G. Costa, Polymer, 42 (2001) 5549.

[4] M. Rätzsch, M. Arnold, E. Borsig, H. Bucka, N. Reichelt, Prog. Polym. Sci., 27 (2002) 1195.

[5] T. Liu, G. H. Hu, G. Sh. Tong, L. Zhao, G. P. Cao, W. K. Yuan, Ind. Eng. Chem. Res., in press.

[6] A. I. Cooper, J. Mater. Chem., 10 (2000) 207.

[7] Ramesh R., Michael V., Sunggyu L, J. Appl. Polym. Sci., 39 (1990) 1873.

[8] Catoire B, Verney V, Hagege R, Michel A., Ploymer, 11 (1992) 2037.

[9] B. De Roover, M. Sclavons, V. Carlier, J. Devaux, R.Legras, and A. Momtaz., J. Polym. Sci. Part A Polym. Chem., 33 (1995) 829.

[10] X. L. Zhan, X. B. Yang, F. Q. Chen, J. Chem. Ind. & Eng. (China), 55 (2004) 110.

[11] G. Sh. Tong, T. Liu, L. Zhao, G. H. Hu, W. K. Yuan, in preparation.

Studies in Surface Science and Catalysis, volume 159
Hyun-Ku Rhee, In-Sik Nam and Jong Moon Park (Editors)
© 2006 Elsevier B.V. All rights reserved

Boundary conditions in TAP reactor modeling

Lei Gao, Xueliang Zhao and Dezheng Wang[*]

Department of Chemical Engineering, Tsinghua University
Beijing, 100084 China (E-mail: wangdz@flotu.org; Fax: ++86 10 62772051)

Abstract

The two boundary conditions (BC) in TAP reactor modeling: the use of the delta function to approximate the pulse input and the assumption of zero concentration (infinitely large pumping speed) at the reactor outlet, are discussed. When the delta function approximation is used, curve fitting should be limited to the latter ¾ part of the response curve for curves of width < 3 times pulse width, while for curves with widths > 4 times pulse width, it is a fair approximation. The infinitely large pumping speed BC is not a good assumption even for a 1,500 ls^{-1} pumping speed and broad outlet pulses with widths >1000 ms.

1. INTRODUCTION

A reason for using microkinetics in heterogeneous catalysis is to have comprehensive kinetics and a transparent reaction mechanism that would be useful for reactor design or catalyst development. Furthermore, in the long run, the experimental effort to develop a microkinetics scheme can be less than that for a Langmuir-Hinshelwood (LH) or power-law scheme because of the more fundamental nature of the reaction kinetics parameters.

An important need in the implementation of a microkinetics approach is the experimental apparatus to measure the elementary reaction rate parameters. The temporal-analysis-of-products (TAP) reactor developed by Gleaves and coworkers [1] is capable of meeting this need. It can be used to identify elementary step sequences and measure the rate parameters of these reaction steps. There are two ways to use the TAP technique: a qualitative method where gaseous products are time-resolved measured and a quantitative method where kinetic parameters are obtained by regression with experimental response curves. The TAP technique uses very sharp pulses and the times of elution of the pulses of different product species provide the data for deducing the sequence in which the elementary steps occurs. Good time resolution is important for good data and the structure of the reactor is important in determining the time resolution [2]. In the second quantitative method, experimental TAP curves are simulated by a reactor model and the kinetic parameters are obtained by regression and curve fitting. This paper discusses the second method, in particular, the idealizations made in the boundary conditions (BC) in the modeling of TAP reactors. It discusses some situations where the idealized BCs can lead to errors in parameter estimates and where more care is needed in formulating the BCs to derive more accurate kinetic parameters.

2. EXPERIMENTAL

Experiments were carried out in a pulse reactor system based on the TAP-2 reactor described by Gleaves et al. [3]. This consists of a small tubular reactor and a detector housed in a vacuum system pumped by a 1,500 ls^{-1} turbomolecular pump to a base pressure of

~2x10^{-7} torr. The reactor has four pulse valves for gas input. The pulse valves are solenoid valves held shut by mechanical springs and were constructed using a design by Otis and Johnson [4]. Each valve stem tip is filled with silicone rubber that makes a gastight seal when pressed by the springs against the polished surface of the orifice opening. MOSFET switches based on a design in Horowitz and Hill [5] controlled on-off times of the current passing through the solenoids to control the opening and closing of the valves. The four pulse valves are attached to a device that opens directly onto the mouth of a tubular reactor that contains the catalyst sample. Gases used were high purity grade from commercial sources.

The detector is a Stanford Research Systems RGA 200 mass spectrometer with the ionization cage aligned with the reactor outlet, and operated at the default resolution and sensitivity. The signal from its electron multiplier was measured with a SR 570 current-voltage amplifier and a SR 560 voltage amplifier in tandem, and a PC-controlled AD/DA board with <10 μs conversion times. Collected data were signal averaged over five to ten pulses after visual inspection of individual pulses to discard anomalous pulses, if any.

Numerical integration of the partial differential equations for simulations and parameter estimation were performed using the Crank-Nicholson scheme. Parameter optimisations were performed with the Nelder-Mead or simulated annealing algorithms. The accuracy of the computed curves was verified by halving time and space steps until the resulting plots of the response curves are visually non-distinguishable. It is estimated that the computed data points are precise to within 1%. Generally, a space step of 0.1 mm was used, and the time step was chosen to ensure stability and accuracy with the corresponding value of diffusivity.

3. RESULTS AND DISCUSSION

The two BCs of the TAP reactor model: (1) the reactor inlet BC of the idealization of the pulse input to the delta function and (2) the assumption of an infinitely large pumping speed at the reactor outlet BC, are discussed. Gleaves et al. [1] first gave a TAP reactor model for extracting rate parameters, which was extended by Zou et al. [6] and Constales et al. [7]. The reactor equation used here is an equivalent form from Wang et al. [8] that is written to be also applicable to reactors with a variable cross-sectional area and diffusivity. The reactor model is based on Knudsen flow in a tube, and the reactor equation is the diffusion equation:

$$\varepsilon_b(x)\frac{\partial C_A}{\partial t} = \frac{\partial}{\partial x}D_{e,A}(x)\frac{\partial C_A}{\partial x} \pm rxn/ads/desb\ terms \qquad (1)$$

Initial condition:

$$0 \leq x \leq L_1+L_2+L_3, \quad t=0, \quad C_A^{all\ zones} = 0 \qquad (2)$$

BCs at the entrance and exit of the reactor:

$$At\ x = 0\,(reactor\ inlet), \quad t \geq 0, \quad -D_{e,A}^{zone1}\frac{\partial C_A^{zone1}}{\partial x} = \delta(t-\tau)N_{pA}/A_r \qquad (\delta-function\ pulse) \quad (3)$$

$$At\ x = L_1+L_2+L_3\ (reactor\ outlet), \quad t \geq 0, \quad C_A = 0 \qquad (4)$$

The BCs have been previously discussed by Gleaves et al. [1], Zou et al. [3], Creten et al. [9] and others. Initial condition (2) can be accepted because its statement of an initially clean surface is an experimental statement. BCs (3, 4) are here further discussed with reference to the experimental apparatus. BC (3) states that the flux at the reactor inlet is a delta function and is the approximation that pulse injection occurs over an infinitely short time. This is discussed using experimental data on the speed of injection of the input pulse. BC (4) is the approximation that the gas concentration is zero outside the reactor tube. It implies that any gas eluting from the reactor tube is immediately removed, that is, the approximation is that the pumping speed of the vacuum system is infinitely large.

3.1. The use of the delta function to describe the pulse input

The motivation for discussing the delta function BC is actually experimental rather than theoretical because very sharp pulses of <0.1 ms full width at half maximum (FWHM) need sophisticated and expensive high power electronics, but if the requirement on the quality of the pulse can be relaxed to >0.5 ms FWHM, cheap homemade pulse valves can be used. Thus, the question is: when is one justified in using the delta function BC when one uses a pulse valve with >0.5 ms FWHM? Most workers assume that the delta function description is reasonable because the open-shut time of the pulse valve is much less than the residence time of the pulse in the reactor. Creten et al. [9] suggested the criterion that the input pulse width be two orders of magnitude less than the mean residence time of the pulse in the reactor, but this is a conservative estimate. There had been no quantitative assessment of the range of validity for using the delta function although this question is important because of the limit that the technology of solenoid valves imposes with its finite pulse widths.

This work uses numerical simulations to answer the question by solving equation (1) for pulse inputs of different widths. The curves in Fig. 1 show the response curves for pulse inputs of a delta function (curve A), and 0.5 ms (curve B) and 1 ms (curve C) rectangular steps into a 6 mm i.d., 41 mm long tube with a diffusivity of 525 mm^2/ms (in agreement with the Knudsen formula). Fig. 2 shows the measured widths of our solenoid valves. They are 0.4 to 0.6 ms, increasing to 0.8 ms for a very large pulse, and show that the pulse widths used in Fig. 1 are reasonable. In Fig. 1, curves B and C look different from curve A but this is only due to the finite width of the step function causing a time shift of the curve peak. When the curves are shown time-shifted to have the 0.5 ms (curve D) and 1 ms (curve E) step pulse curves have their peaks coincide with curve A, they show good fits to curve A over its latter ¾ part. The front ¼ part of curves D and E are different. Thus, for narrow response curves, e.g. those where the residence time is just 2 times the pulse width, one needs to ignore the front ¼ part of the curve in order for a delta function to be a good approximation to a finite width input pulse. The probably reason for the good fit to the latter part is that diffusion is a smoothing process and acts to smooth out differences with passing time. Fig. 3 shows the comparison with experimental data from a narrower and longer tube with a larger residence time of the pulse in the tube. In fig. 3, although the mean residence time of curve A is only 3 ms, the indication is that the delta function approximation is fair even with input pulse widths as wide as ¼ the mean residence time of the pulse in the reactor.

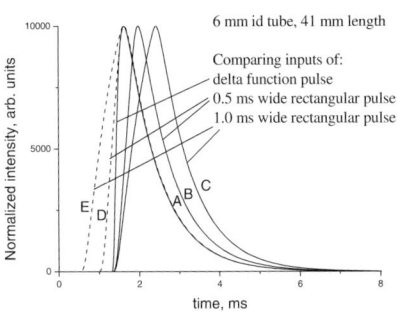

Fig. 1. TAP curves with different pulse widths.

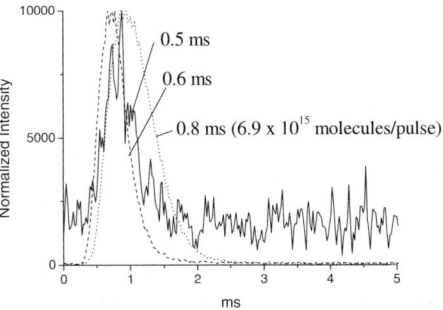

Fig. 2. Experimental input pulse FWHMs.

680

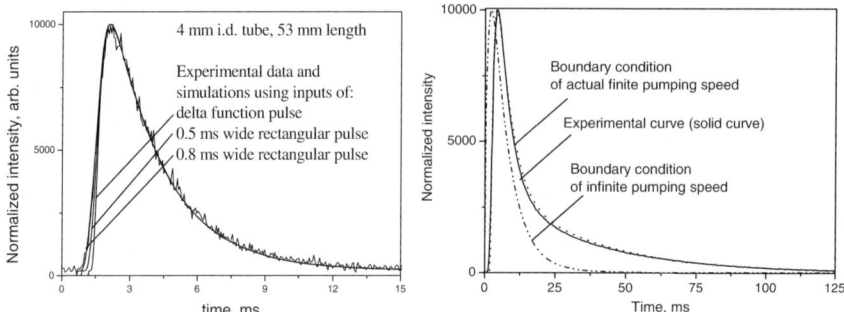

Fig. 3. TAP curves with different input pulse FWHMs.   Fig. 4. Curves with different outlet BCs.

3.2. The assumption of an infinitely large pumping speed at the reactor outlet

BC (4) is the approximation that the gas concentration (but not the flux) is zero outside the reactor tube. This implies the assumption that the pumping speed is infinitely large. Our experimental data shows that this can be a major source of error, even with a vacuum system with a pumping speed of 1500 ls^{-1}. A rough idea of the physical dimensions is that a pump with a pumping speed of 1500 ls^{-1} needs a mouth larger than 20 cm in diameter. Pumps for the average surface analysis vacuum system have pumping speeds of ca. 500 ls^{-1}. Fig. 4 compares experimental results from a 4 mm i.d., 145 mm long tube, a simulation that uses BC (4), and a simulation that uses an improved boundary condition that takes into account the actual finite pumping speed of the vacuum chamber. For the latter case, the finite pumping speed means that the concentration is non-zero in the vacuum chamber outside the reactor tube and thus the diffusion equation is extended to include the whole vacuum chamber, and the position of the BC is fixed at the location where the pump actually is.

4. CONCLUSION

Curve fitting using a delta function for the pulse input for a TAP reactor should be limited to the latter ¾ part of the response curve for curves of FWHM < 3 times pulse width, while for curves with FWHM > 4 times pulse width, it is a fair approximation for most of the curve. The assumption of a zero concentration at the reactor outlet is not good even for a pumping speed of 1,500 ls^{-1} and broad response curves with FWHM > 1000 ms.

Acknowledgements
We thank the National Natural Science Foundation (China) (grant 20273036) and the Ministry of Science and Technology of China (No. 2005CB221405) for supporting this work.

REFERENCES
[1] J.T. Gleaves, J.R. Ebner, T.C. Kuechler, Catal. Rev. – Sci. Engr., 30 (1988) 49.
[2] D.Z. Wang and Z.L. Li, Studies in Surf. Sci. and Catal., 133 (2001) 768.
[3] J.T. Gleaves, G.S. Yablonski, P. Phanawadee and Y. Schuurman, Appl. Catal. A 160 (1997) 55.
[4] C.E. Otis and P.M. Johnson, P.M., (1980), Rev. Sci. Instr., 51 (1980) 1128.
[5] P. Horowitz and W. Hill, The Art of Electronics, 2nd ed., Cambridge U. Press, Cambridge, 1989.
[6] B.S. Zou, M.P. Dudukovic, P.L. Mills, Chem. Eng. Sci., 48 (1993) 2345.
[7] D. Constales, G.S. Yablonsky, G.B. Marin, J.T. Gleaves, Chem. Eng. Sci., 59 (2004) 3725.
[8] D.Z. Wang, Z.L. Li, C.R. Luo, W.Z. Weng, H.L. Wan, Chem. Eng. Sci., 58 (2003) 887.
[9] G. Creten, D.S. Lafyatis and G.F. Froment, J. Catal., 154 (1995) 151.

Studies in Surface Science and Catalysis, volume 159
Hyun-Ku Rhee, In-Sik Nam and Jong Moon Park (Editors)
681

Mathematical modeling of atmospheric inorganic aerosols

Kee-Youn Yoo

Department of Chemical Engineering, Seoul National University of Technology,
172 Kongneung-2 dong, Nowon-gu, Seoul 139-743, Korea

1. INTRODUCTION

Atmospheric aerosols have a direct impact on earth's radiation balance, fog formation and cloud physics, and visibility degradation as well as human health effect[1]. Both natural and anthropogenic sources contribute to the formation of ambient aerosol, which are composed mostly of sulfates, nitrates and ammoniums in either pure or mixed forms[2]. These inorganic salt aerosols are hygroscopic by nature and exhibit the properties of deliquescence and efflorescence in humid air. That is, relative humidity(RH) history and chemical composition determine whether atmospheric aerosols are liquid or solid. Aerosol physical state affects climate and environmental phenomena such as radiative transfer, visibility, and heterogeneous chemistry. Here we present a mathematical model that considers the relative humidity history and chemical composition dependence of deliquescence and efflorescence for describing the dynamic and transport behavior of ambient aerosols[3].

2. MODELING OF ATMOSPHERIC INORGANIC AEROSOLS

One of the most challenging parts for the modeling of inorganic aerosols is the prediction of the partitioning of the inorganic components between aqueous and solid phases. A direct minimization of the Gibbs free energy implicitly predicts phase evolution without any *a priori* knowledge of the behavior of electrolyte solution. However, the direct minimization approach disregards the specific mathematical structure of solid-liquid equilibrium problem. As a result, it is computationally intensive to be used in large-scale applications. For an equation-based approach, provided the set of mass balance and equilibrium relations can be determined algorithmically to reflect the actual state of electrolyte system under the varying process condition, its solution corresponds to the minimum of the Gibbs free energy and thus predicts the physical state of electrolyte systems correctly. To attack this problem, we presented a canonical form of electrolyte solution systems and then give the mathematical interpretation of solid-liquid equilibrium stage[4]. The canonical framework reported for examples of inorganic aerosols is general and then of great useful to develop the efficient numerical algorithm for the prediction of the phase evolution of electrolyte solution systems. See the detailed discussion about the Karush-Kuhn-Tucker conditions and a primal-dual active-set/Newton method to compute the minimum and track solid salts at the equilibrium in [5-6].

The predicted solids based on thermodynamic measurements and models derived from them depend strongly on chemical composition. However, small aqueous aerosols remain meta-stable with decreasing RH until reaching a crystallization relative humidity(CRH). In contrast, solid aerosols take up water at the thermodynamically favored deliquescence relative humidity(DRH). This hysteresis causes a dependence of aerosol phase on RH history.

Transformation from a meta-stable phase, such as supersaturated solution, to a thermodynamically more favorable phase requires first the crystal nucleation of a germ of the new phase. According to the classical nucleation theory, the volume nucleation rate J (cm^{-3}sec^{-1}), describing the number of nuclei(i.e., a critical germ) formed per volume per time, is given by:

$$J = J_0 \exp\left(-\frac{\Delta G_{germ}}{kT}\right)$$

where J_0 is a pre-exponential factor that is related to the collision efficiency between supernatant ions and crystal interface and has a value of order 10^{24}-10^{36} (here we choose $J_0 = 10^{30}$) and ΔG_{germ} is the energy required for the formation of a critical germ given by:

$$\Delta G_{germ} = \frac{16\pi}{3} c_{germ} \frac{\sigma^3}{\left(\rho_0 RT \ln S\right)^2}$$

where c_{germ} is a geometric factor, σ_s is the surface tension of the germ in the medium, and S is the saturation ratio. Nucleation is a stochastic process, approximated by the Poisson distribution. The expectation time τ_{nucl} after which an aerosol particle of volume V_{pm} forms a single nucleus is given by:

$$\tau_{nucl} = \frac{1}{JV_{pm}}.$$

We assume, for our analysis on small particles, that the overall *crystallization time* is limited by the nucleation time for a single germ.

3. NUMERICAL ALGORITHM

Let x_a, a_a, k_a, and A_a stand for the concentration vector, the activity vector, the canonical equilibrium constant vector, and the component-based formula matrix for the species set $a = 1, s$ respectively. Let I_s be the index set of salts in the system and define the saturation ratio vector S by $\ln S = A_s^T \ln a_c - \ln k_s$. Then I_s has the *decomposition* $I_s = W + (M + N)$, where $W = \{i : x_{s,i} > 0, \ln S_i = 0\}$, $M = \{i : x_{s,i} = 0, \ln S_i > 0\}$, $N = \{i : x_{s,i} = 0, \ln S_i < 0\}$ with W, M and N denoting respectively the set of the *crystallized*, *supersaturated* and *subsaturated* salts.

In this work, we developed the safeguard active-set method by modifying the active-set method for thermodynamic equilibrium in order to include the classical nucleation theory. At t_n, assume that the partition $(W(t_n), M(t_n), N(t_n))$ and the crystallization time $t_{cryst}(t_n)$ for $M(t_n)$ are known. For a new feed vector and RH at t_{n+1}, compute $(W(t_{n+1}), M(t_{n+1}), N(t_{n+1}))$ and $t_{cryst}(t_{n+1})$ as follows:

1. **Update the crystallization set** $W(t_{n+1})$ Activate the supersaturated salts $\delta M(t_{n+1})$ that are expected to crystallize in the time interval (t_n, t_{n+1})

$$M(t_{n+1}) = M(t_n) - \delta M(t_{n+1}),$$
$$W(t_{n+1}) = W(t_n) + \delta M(t_{n+1}),$$

with $\delta M(t_{n+1}) = \{ i \in M(t_n) : t_{cryst,i}(t_n) \in (t_n, t_{n+1}) \}$.

2. **Newton Iteration** Compute x_l by Newton iterations on the reduced KKT system.

3. **Optimality Test** If x_l is a stationary point, then proceed as follows:

   a) If $\mathcal{W}(t_{n+1}) = \emptyset$, i.e., no constraints are active, then the current point is a local (unconstrained) stationary point and go to 4.

   b) If $\mathcal{W}(t_{n+1}) \neq \emptyset$, then compute the concentrations of the crystallized salts in the active set, $\bar{x}_s$. If $\bar{x}_s \geq 0$, then a local stationary point has been reached and go to 4; otherwise, drop constraints corresponding to the crystalline salts having *negative* concentration from the active set

$$\mathcal{W}(t_{n+1}) = \mathcal{W}(t_{n+1}) - \delta\mathcal{W}(t_{n+1}),$$
$$\mathcal{N}(t_{n+1}) = \mathcal{N}(t_{n+1}) + \delta\mathcal{W}(t_{n+1}),$$

   with

$$\delta\mathcal{W}(t_{n+1}) = \{\, i \in \mathcal{W}(t_{n+1}),\ x_{s,i} < 0 \,\},$$

   and go to 2.

4. **Update the Metastable set** Compute the saturation ratio vector S and update the metastable set:

$$\mathcal{N}(t_{n+1}) = \mathcal{N}(t_{n+1}) - \delta\mathcal{N}(t_{n+1}),$$
$$\mathcal{M}(t_{n+1}) = \mathcal{M}(t_{n+1}) + \delta\mathcal{N}(t_{n+1}),$$
$$\mathcal{M}(t_{n+1}) = \mathcal{M}(t_{n+1}) - \delta\mathcal{M}(t_{n+1}),$$
$$\mathcal{N}(t_{n+1}) = \mathcal{N}(t_{n+1}) + \delta\mathcal{M}(t_{n+1}),$$

   with

$$\delta\mathcal{N}(t_{n+1}) = \{\, i \in \mathcal{N}(t_{n+1}),\ \ln S_i > 0 \,\},$$
$$\delta\mathcal{M}(t_{n+1}) = \{\, i \in \mathcal{M}(t_{n+1}),\ \ln S_i \leq 0 \,\}.$$

Compute $\tau_{\text{nucl},i}$ for $i \in \mathcal{M}(t_{n+1})$ according to the classical nucleation theory $\tau_{\text{nucl},i} = 1/(J_{\text{nucl},i} V_{\text{pm}}(t_{n+1}))$, update the expected crystallization time $t_{\text{cryst},i}(t_{n+1})$ for the meta-stable phases $\mathcal{M}(t_{n+1})$

$$t_{\text{cryst},i}(t_{n+1}) = \min\left(t_{\text{cryst},i}(t_n), t_{n+1} + \tau_{\text{nucl},i}\right)$$

and go to 1.

4. NUMERICAL EXPERIMENTS and CONCLUSIONS

Here our interest is in the application of homogeneous nucleation theory to produce the comprehensive plots of meta-stable crystallization. Fig. 1 illustrates the meta-stable efflorescence paths(solid lines) of $(NH_4)_2SO_4$ and $(NH_4)_3H(SO_4)_2$ particles as a function of RH with the decreasing rate of $\Delta RH = 0.005$ min with the deliquescence paths($\Diamond$). Fig. 2 shows the expectation time of the aqueous particle composed of $(NH_4)_2SO_4$ and H_2SO_4

compared with Martin's experimental correlation(dotted lines)[3]. The numerical results show that the developed model is satisfactory to predict the deliquescence and efflorescence phenomena of atmospheric inorganic aerosols

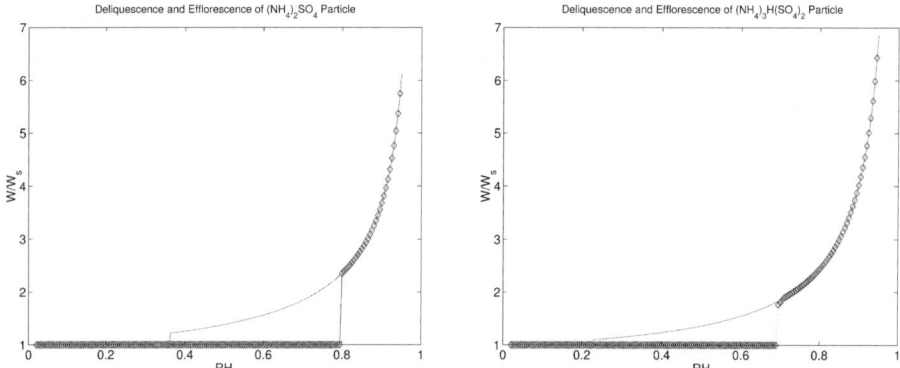

Fig.1. Deliquescence and efflorescence phenomena simulation of atmospheric $(NH_4)_2SO_4$ and $(NH_4)_3H(SO_4)_2$ particles at $D_{pm} = 1\mu m$ and 25°C.

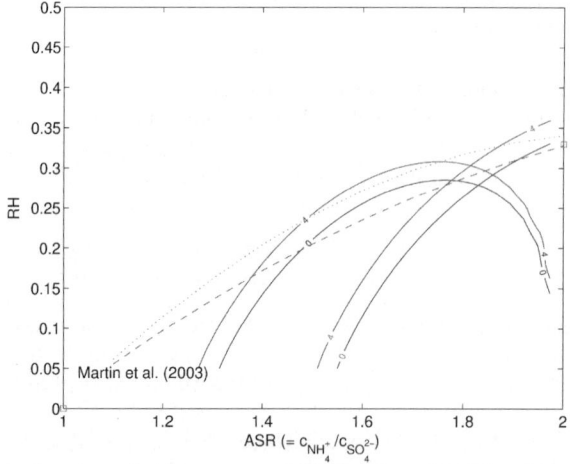

Fig.2. Expectation time of efflorescence of $(NH_4)_2SO_4$ and H_2SO_4 solution, $\log(\tau(min))$ according to ammonium-sulfate ratio (ASR) at $D_{pm} = 1\mu m$ and 25°C.

REFERENCES

[1] S. T. Martin, Chemical Reviews, 100 (2000), 3403.
[2] S. N. Pandis, A. S. Wexler and J. H. Seinfeld, J. Phys. Chem., 99 (1995), 9646.
[3] S. T. Martin, J. C. Schlenker, A. Malinowski, H. M. Hung, Y. Rudich, Geophysical Research Letters, 30 (2003)., 2102.
[4] K.-Y. Yoo, J. He, and N. R. Amundson, Korean J. of Chem. Eng., 21 (2004), 303.
[5] N. R. Amundson, A. Caboussat, J. He, J. H. Seinfeld, and K.-Y. Yoo, C.R. Acad. Sci. Paris, Ser. I 340 (2005), 683.
[6] N. R. Amundson, A. Caboussat, J. He, J. H. Seinfeld, and K.-Y. Yoo, Journal of Optimization Theory and Applications, in press.

Studies in Surface Science and Catalysis, volume 159
Hyun-Ku Rhee, In-Sik Nam and Jong Moon Park (Editors)
© 2006 Elsevier B.V. All rights reserved

Hydrogen production by methanol autothermal reforming with high conductivity honeycomb supports

Dong Hyun Kim* and Jitae Lee

Department of Chemical Engineering, Kyungpook National University,
Daegu, 702-701, Korea
dhkim@mail.knu.ac.kr

1. INTRODUCTION

Onboard production of hydrogen by reforming liquid fuels is an alternative to storing compressed hydrogen in fuel cell vehicles for the lack of the hydrogen infrastructure and safety problems. Among various liquid fuels, methanol is easy to reform at low temperatures (200 – 300 °C) and the products are almost exclusively hydrogen and carbon dioxide [1]. On the other hand, other hydrocarbon fuels such as gasoline and diesel are reformed at high temperatures (600 – 800 °C) and the reformate contains a high concentration of CO. The reformer for liquid fuels other than methanol is considerably more complicated than the methanol reformer, less suitable for onboard applications.

Among various methods of methanol reforming, the autothermal reforming, a combination of the exothermic methanol partial oxidation and the endothermic steam reforming with a net enthalpy change of zero, is simple to operate and requires a simple reactor configuration as the external heat exchange is not necessary [2]. A reaction equation of the autothermal reforming is

$$CH_3OH + 0.766\ H_2O + 0.117\ O_2 \rightarrow CO_2 + 2.766\ H_2 \quad (\Delta H\ at\ 473\ K = 0\ J/mol) \qquad (1)$$

Locally the reaction does not proceed according to Eq. (1). At the reactor inlet where oxygen is present, the exothermic partial oxidation predominantly occurs, and after the oxygen is completely consumed, the endothermic steam reforming alone proceeds. But the reaction stoichiometry and the heat effect shown in Eq. (1) are valid if the reaction proceeds to completion. Therefore heat supply or removal is not necessary if the reactor is designed to yield a complete methanol conversion. Due to local imbalance in heat generation and consumption, however, the temperature profile along the reactor normally exhibits a hot-spot near the reactor inlet. The hot-spot temperature can rise over 400 °C, causing rapid deactivation of the catalyst and increasing formation of CO by methanol decomposition and reverse water-gas-shift reaction [2]. Operation of the reactor within the safe temperature range (200 – 300 °C) is critical for stable catalyst activity and for minimal CO formation. To date, however, no satisfactory reactor design to solve the problems associated with the hot spot has been developed. The purpose of this study is to develop a reactor that can eliminate the hot-spot by rapidly transferring the heat generated in the inlet zone to the rest of the reactor, thereby flattening the temperature profile and suppressing the maximum temperature within the safe temperature range of the catalyst.

2. HIGH CONDUCTIVITY HONEYCOMB REACTOR

The reactor configuration we propose, shown in Figs. 1 and 2, allows rapid heat transfer along the axial direction of the reactor by conduction through the wall made of high conductivity metal such as copper or aluminum. The catalyst can be packed into the honeycomb cells or wash coated on the walls of the cells.

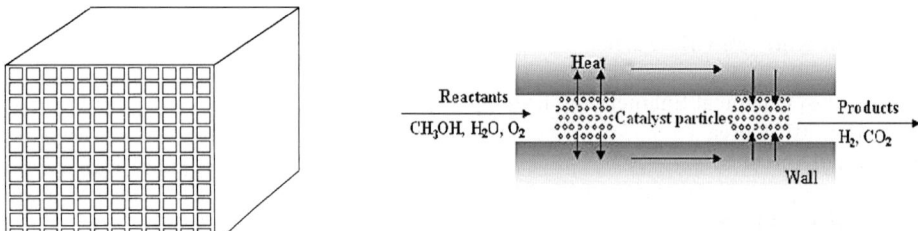

Fig. 1. Metal honeycomb   Fig. 2. A cell in the honeycomb

2.1. Reactor modeling

With decreasing cell size, the temperature difference between the wall of the cell and the catalyst particle in the cell would decrease to zero. For sufficiently small cell dimensions, we may assume the two temperatures are the same. In this case, the heat conduction through the wall becomes dominant and affects the axial temperature profile. As the external heat exchange is absent and the outside of the reactor is normally insulated, the temperature profile is flat along the direction transverse to the reactant flow, and the conditions in all channels are identical to each other. The energy balance is

$$\frac{A_w k_w}{A} \frac{d^2 T}{dZ^2} - \sum_i F_i Cp_i \frac{dT}{dZ} + (r_{POX}(-\Delta H_{POX}) + r_{SR}(-\Delta H_{SR})) = 0 \qquad (2)$$

where T is the temperature of the wall and the flow, A_W is the cross sectional area of the wall, A is the cross sectional area of the honeycomb including the frontal open area and A_W, k_W is the thermal conductivity of the wall, and Cp_i is the heat capacity of the component i, normally a function of temperature. Here we may define an effective thermal conductivity of the monolith $k_e = A_w k_w / A$ for the axial direction. The first term in Eq. (2) accounts for the heat transfer by wall conduction. In the absence of the first term, Eq. (2) becomes the energy equation of a conventional adiabatic reactor. At the entrance of the honeycomb, we assume that the heat transferred from the honeycomb by conduction is heating the feed, i.e.

$$-k_e \frac{dT}{dZ}\Big|_{Z=0} = \sum_i F_i^0 \int_{T_0}^{T(0)} Cp_i dT \qquad (3)$$

As a result, there is a jump discontinuity in the temperature at Z=0. The condition is analogous to the Danckwerts boundary condition for the inlet of an axially dispersed plug-flow reactor. At the exit of the honeycomb, the usual zero gradient is imposed, i.e.

$dT/dZ = 0$. In addition to the energy balance, the reactor model consists of two mass balance equations for oxygen and methanol and a pressure drop equation. The reaction rate of methanol partial oxidation is unclear in the literature and suggested to be faster than the steam reforming rate [3]. We observed little difference in the computed temperature profiles of the honeycomb reactor when the rate was varied from half to six times of the reforming rate. Hence we assume that the rate is equal to the steam reforming rate reported in [1].

3. RESULTS AND DISCUSSION

The reactor model was solved for the two cases: insulator wall and aluminum wall. In the case of insulator wall, heat transfer by conduction through the wall is absent, and the heat generated by the partial oxidation at the inlet of the reactor heats the reaction mixture. This case resembles a conventional packed-bed reactor where the heat conduction through the catalyst bed is poor and the contribution of the wall conduction is negligible. Simulation was carried out for the reactor and the operating conditions listed in Table 1. For metal honeycomb reactor, a shooting method was used to integrate the model equations. The simulation results are shown in Fig. 3. The temperature profile of the insulator wall shows a sharp peak reaching 565 °C at the reactor inlet, whereas that of the highly conductive aluminum wall reactor shows a relatively flat temperature profile with a maximum temperature of 270 °C at the inlet. The results clearly show that the heat conduction through the wall can destroy the hot-spot and keep the temperature profile below 300°C, the upper limit for stable activity with the Cu catalyst.

Table 1
Honeycomb reactor and operating conditions

Honeycomb reactor
Honeycomb with 200 cells per square inch (CPSI)
Hydraulic channel diameter: 1.45 mm, Length: 0.2 m
Open frontal area: 65 %, Catalyst: ICI 33-5 (Cu/ZnO/Al$_2$O$_3$), 0.2mm particles
Feed
Temperature: 200 °C
Flow rate per 1 cm^2 cross sectional area:
0.8 mol/h CH$_3$OH, 0.8 mol/h H$_2$O, 0.104 mol/h O$_2$, 0.39 mol/h N$_2$

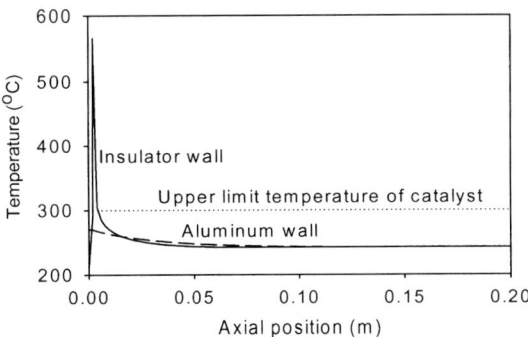

Fig.3. Temperature profiles of the honeycomb reactor. Insulator wall (——) and aluminum wall (--).

A thermal Peclet number, $Pe_h = L\sum F_i Cp_i / k_e$ can be defined for the reactor to represent the ratio of heat transport rates by fluid convection and by wall conduction. Fig. 4 shows the effect of the Peclet number determined at the feed condition, Pe_h^0, on the maximum reactor temperature. The Peclet number was varied by adjusting the feed flow rate. When the conversion of methanol is complete, the reactor exit temperature was 243 °C regardless of the Peclet number. The maximum reactor temperature decreased and approached the exit temperature with decreasing Peclet number or decreasing feed flow rate, showing that wall conduction took effect to remove the hot-spot and flattened the temperature profile. For increasing feed rate, the methanol conversion fell below 0.99 above $Pe_h^0 = 2$. At around $Pe_h^0 = 1.4$, the maximum temperature exceeded 300 °C, the upper limit temperature of the catalyst. In terms of methanol flow rate, $Pe_h^0 = 1.4$ corresponded to 1.87 mol $CH_3OH/cm^2/h$, producing 5.12 mol $H_2/cm^2/h$. In view of the reactor length of 20 cm (Table 1), the maximum production rate per liter of reactor volume was 256 mol H_2/h, which could power a 5.7 kW fuel cell, if a 1 kW fuel cell consumes hydrogen at a rate of 44.6 mol (1000 liter STP) H_2/h. In view of this high capacity, a compact reformer suitable for mounting on a fuel cell car can be developed with the honeycomb reactor.

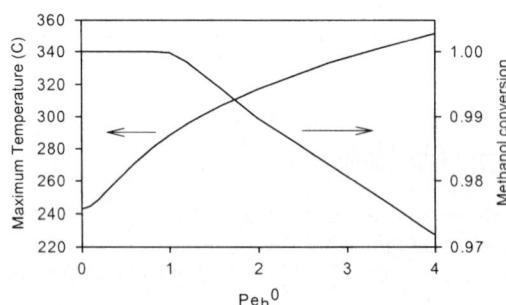

Fig.4. Effect of thermal Peclet number on maximum reactor temperature and methanol conversion.

4. CONCLUSION

Despite advantages of methanol autothermal reforming for onboard production of hydrogen, the hot-spot formed in the autothermal reactor has been a major problem to be solved in the development of an onboard reformer. This study has shown that reactors made of high-conductivity-honeycomb supports can destroy the hot-spot in the autothermal reforming and keep the reactor temperature within the safe temperature range of the catalyst. According to the simulation results, the honeycomb reactor has a high production capacity of hydrogen, exceeding 5000 liter H_2 per hour per liter of the reactor.

REFERENCES
[1] J. K. Lee, J. B. Ko and D. H. Kim, Appl. Catal. A 278 (2004) 25.
[2] N. Edwards, S. R. Ellis, J. C. Frost and J. G. Reinkingh, J. Power Sources 71 (1998) 123.
[3] S. Velu, K. Suzuki, M. P. Kapoor, F. Ohashi and T.Osaki, Appl. Catal. A 213 (2001) 47.

Studies in Surface Science and Catalysis, volume 159
Hyun-Ku Rhee, In-Sik Nam and Jong Moon Park (Editors)
689

Mechanochemical and chemical reaction engineering aspects in the break down of turbulent drag reduction of cationic surfactant solutions

Jin Hyun Kim[a], Jungrim Haw[b] and Chongyoup Kim[a*]

[a]Dept. of Chemical and Biological Engineering, Korea University
Anam-dong, Sungbuk-gu, Seoul 136-713, Republic of Korea

[b]Department of Materials Chemistry and Engineering, Konkuk University
Hwayang-dong 1, Gwangjin-goo, Seoul 143-701, Korea

*Corresponding author (cykim@grtrkr.korea.ac.kr)

1. Introduction

The pressure drop in a turbulent pipe flow is lowered (as much as 80% or sometimes even more) when a small amount of polymer, surfactant or fiber is added to the flowing Newtonian liquid. This phenomenon is called turbulent drag reduction. Since its first report researches have been focused on the utilization of turbulent drag reduction due to its concrete benefit in many systems such as long distance transport of crude oil, district heating and cooling, sewage, fire-fighting operation and irrigation.

In district heating systems (DHS), high temperature and high Reynolds number operations are involved. At these severe conditions, chemical additives such as polymers or surfactants are known to degrade thermally and/or mechanically. Recently Kim et al. [1] reported that Habon G solution showed break-down phenomena and the break-down could be caused by the chemical reaction with dissolved oxygen as well as by the association of micelles in the strong shear and extensional flow at the wall layer of turbulent pipe flow. In this report we reexamine the break-down of turbulent drag reduction with STAC $(CH_3(CH_2)_{17}N(CH_3)_3Cl)$, a cationic surfactant. We have investigated whether the same phenomenon is observed for differing surfactants and the same mechanism can be applied. The result shows that break-down is also observed in the STAC solutions and the chemical reaction of surfactant molecules with dissolved oxygen could be one of the mechanisms of degradation of surfactant solutions. Also, the slow transformation of shear induced state at the wall layer to cylindrical micelles appears to be responsible for the break-down. It is suggested that the fluid mechanical system with the break down phenomenon could be modeled from the viewpoint of chemical reaction engineering.

2. Experiment

For the experimental set-up we used the same equipment as used in Kim et al. [1]. The whole system consisted of a reservoir, a test section, a return tube and measuring equipment. The fluid comes out of the reservoir of $0.05m^3$ to pass through a 1/4-inch stainless steel tube in the test section. The actual inner diameter and length of the tubing are 4.25mm and 5.15m, respectively. Then the fluid is returned to the reservoir by flowing through 1/2-inch plastic tubing. For the surfactant Miconium STAC-65 supplied from Miwon Chem Co. was used as received. An equimolar amount of NaSal was added as the counterion to facilitate the formation of cylindrical micelles in the solution. The solvent was distilled water. To remove the dissolved oxygen, sodium sulfite was added. Surfactant solutions were prepared by diluting the master solution of 10,000ppm. 100ppm of sodium azide (NaN_3) was added to the master solution as a biocide. The master solution was diluted with preheated distilled water as desired.

Since the present study aims at carrying out the investigation of the break-down phenomenon and searching for the possible mechanism of the phenomenon, we have chosen the similar condition as in [1] for the wall shear stress to induce break-down: The reference temperature in the degradation studies was 60 °C. This value may be lower than the value used in a typical DHS. In a low-pressure system, however, it was necessary to use lower the temperature to avoid the formation of bubbles. For parametric studies, one of the variables was varied while the other variables were fixed at the reference condition (Temperature: 60 °C; Re: 8,000; Surfactant concentration: 200ppm; Volume of solution charged: 0.010 m³).

3. Results

The drag reduction experiments show that the characteristics of drag reduction are the same as Habon-G solution reported earlier: The amount of drag reduction stays almost the same for the first several hours. Then the amount of drag reduction falls off to zero abruptly as shown in Fig. 1. In the figure we plot the % friction reduction against sheared time, where the % friction reduction, % FR, is defined as follows:

$$\% \text{ FR} = \frac{f_N - f_S}{f_N} \times 100 \tag{1}$$

In the above equation, f_N and f_S are the friction factors for the Newtonian solvent and the surfactant solution, respectively, at the same flow rate. This was named 'break-down' in Kim et al. [1]. However the color change was not observed as in Kim et al. [1]. Even after the addition of excess amount of sodium sulfite it was impossible to remove dissolved oxygen completely. It appears that the dissolved oxygen attacked surfactant molecules, especially the head group. Hence the formation of rod-like micelles was hindered. The reaction of head group and dissolved oxygen could be confirmed by an IR-spectrum analysis as shown in Fig. 2 (One may confirm that there is a new peak around 1750 cm^{-1} indicating that there are oxygenated products). We also performed a UV analysis whether the amount of surfactants is diminished in some ways including the adsorption to the walls of vessels and pipes. But the amount of surfactants in solution remains the same. So the adsorption could not be a major mechanism for break-down even though it could affect the degradation partly. Therefore one of the major mechanisms for the break-down in drag reduction should be the chemical reaction of surfactants with dissolved oxygen.

4. Mechanism of Break-down

From the previous experimental results and reasoning, we propose the following two parallel mechanisms for the break-down of drag reduction of the cationic surfactant solution. Firstly, the cationic surfactant is attacked by dissolved oxygen, which results in fragmentation of the surfactant molecule. Considering the fact that thermal/oxidative degradation of polymers in solution progresses as chain reactions [2] this reaction may proceed according to a similar chain reaction. The fragments include insoluble hydrocarbons. Since the fragmented surfactant molecules cannot form cylindrical micelles the number of cylindrical micelles decreases as time goes on. This proceeds faster when the temperature is higher.

Secondly, at the tube wall there develops the shear induced state of cylindrical micelles [3]. The shear induced state has a very large viscosity, and hence at the wall there exists wall-slip. This shear induced state is in a dynamic state. Therefore when the shear induced state is returned to the almost quiescent reservoir, some of the shear induced state will be dissolved to become cylindrical micelles. If the shear induced state is given enough time before it reenters the turbulent flow, in other words, a large amount of solution is used, it will return to the equilibrium state of cylindrical micelles. This coalescence of micelles in turbulent flow is faster when the wall shear stress is larger, that is when Re is larger at a fixed tube diameter. When a sufficiently large amount of surfactant is present, the drag reduction will continue. When the minimum concentration for the maximum drag reduction is reached, as the effective micelles are diminished, drag reduction decreases. The rate of coalescence is sufficiently fast and hence the drag reduction breaks down when the concentration reaches to the critical value for break-down. When the initial concentration is high, the break-down commences later because the incipient concentration for break-down is the same at a given temperature. Due to this coalescence, break-down cannot be avoided even when oxygen scavenger is added to suppress the oxidative degradation.

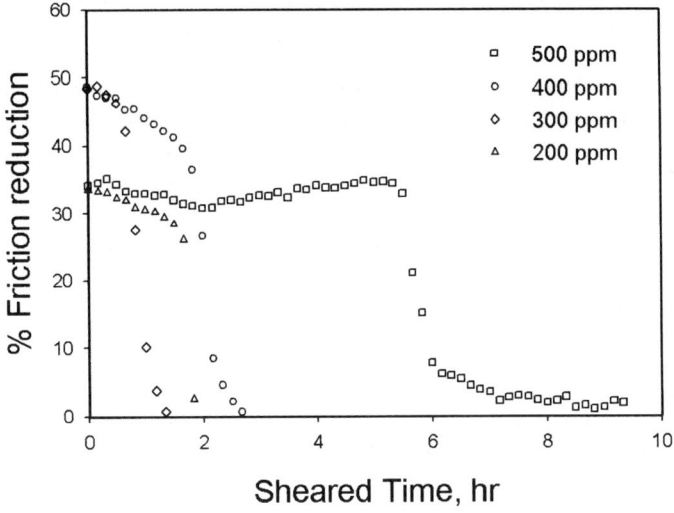

Fig. 1. % Friction reduction vs. shear time for STAC solutions.

692

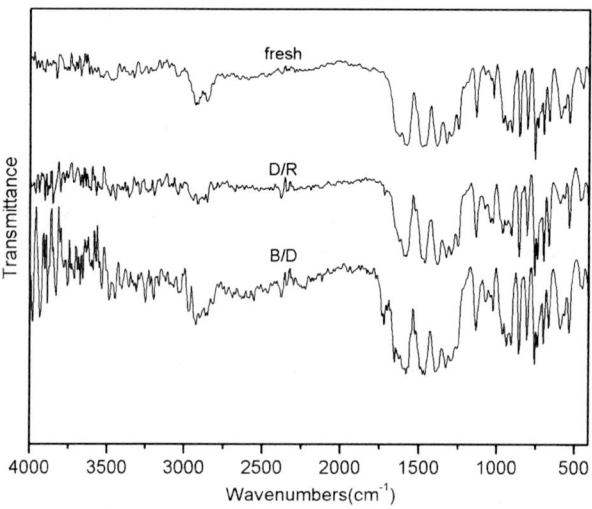

Fig. 2. FT-IR spectra of fresh, drag reducing and broken-down solutions.

5. Discussion

Until now, most of the degradation studies in turbulent drag reduction have been performed from the mechanochemical point of view [2, and the references therein]. But we propose a new concept for the analysis of the system from a chemical reaction engineering point of view. The whole system of district heating systems could be considered to be a series of chemical reactors in that the chain reaction occurs under the influence of high shear field during the pipe flow while a part of liquid remains in a reservoir while having chemical reactions without experiencing high shear field. If we model the system in this way we may be able to predict the degradation process more quantitatively. For this study, we need more kinetic studies for the reaction under the influence of high shear stress.

References

[1] C. Kim, S. R. Park, H. Yoon and J. R. Haw, *J. Chem. Eng. Japan*, **37**, 1326 (2004).
[2] T. Rho, J. Park, C. Kim, H. Yoon and H. Suh, *Polym. Degradation Stability*, **51**, 287 (1996).
[3] A. Gyr and H.-W. Bewersdorff, Drag reduction of turbulent flows by additives, Kluwer Academic Publishers, Dordrecht (1995).

Acknowledgement

The authors wish to acknowledge the financial support from the Applied Rheology Center (Project Number: R11-2000-088-01001-0), Korea University (ERC supported by KOSEF).

Studies in Surface Science and Catalysis, volume 159
Hyun-Ku Rhee, In-Sik Nam and Jong Moon Park (Editors)
693

Effect of various formulation on viscosity and melting point of natural ingredient based lipstick

Awang Bono, Ho Chong Mun and Mariani Rajin

Universiti Malaysia Sabah, 88999 Kota Kinabalu, Sabah, Malaysia.
e-mail: awang@pc.jaring.my

ABSTRACT

Lipstick is one of the decorative cosmetic products that command a unique market. Advanced technology has been used to manufacture modern lipsticks with vast functionality. In the last quarter of twentieth century, cosmetic industries exist with technology of their own. Every year, users were introduced with various new cosmetics products of the latest trend. The quality of lipstick is directly linked to the basic material used in the formulation. Varying the ratio of the ingredient used determines the final product characteristic. A common problem in pre-formulation of the cosmetic product is the optimisation of the mixture composition aimed to obtain a product with the required characteristic. In this work, various compositions of natural waxes, oils, and colorant were used to prepare the lipstick formulations. Mixture design was performed to obtain the optimum formulation. Twenty-five combination components were selected according to the D-optimal criterion. The viscosity and melting point of the lipsticks were studied. Contour graphics were formed to assess the change in the response surface in order to understand the effect of the mixture composition on lipstick characteristics.

Keywords : cosmetic, lipstick formulation, natural ingredient, mixture design, D-optimal

1.INTRODUCTION

Cosmetics certainly improve female attractiveness and this belief has led the cosmetic industry to be one of the most successful worldwide. The cosmetic industry in Malaysia is proven to be one of the most important economy source [1].In the last quarter of twentieth century, cosmetic industries exist with technology of their own. Every year, users were introduced with various new cosmetics products of the latest trend. The ingredients and basic material used in cosmetic formulations become the important criteria for consumers in choosing the cosmetic product, as their interest in health and safety issues grew. Natural-ingredient based product getting popular [2-3].

Lipstick is one of the decorative cosmetic products that command a unique market. Lipstick contains a variety of emollients, emulsifiers, preservatives, colorants and binders [4]. The quality of lipstick is directly linked to the basic material used in the formulation [5-6]. Varying the ratio of the ingredient used in formulation determines the final product characteristic such as texture, viscosity, hardness and melting point of the lipstick [7-9].

A common problem in pre-formulation of the cosmetic product including lipstick is the optimisation of the mixture composition aimed to obtain a product with the required characteristic. Mixture design represents an efficient approach for solving such optimisation problem [10]. It has been proved to be an effective tool to select the best lipstick formulation [11].

In order to understand the relationship between the mixture component, physical properties and consumer acceptance of the lipstick, various lipstick formulations have to be produced. The physical properties of each formulation should be studied. The consumer acceptance towards the product also should be investigated. However, only a part of this work will be discussed in this paper. Here, natural waxes, oils and solvent have been used to produce natural ingredient based lipstick formulations based on the formulation suggested by the statistical mixture design. Contour plot and response surface graph were formed in order to understand the relationship between the mixture component and physical characteristic of the lipstick.

2. METHODOLOGY

2.1 Statistical Mixture Design and Lipstick Formulation

The experimental settings were performed by mixture experimental design. The experimental design of five-components system was conducted by using Design Expert (version 6.10, Stat-Easy Inc., Minneapolis,USA). A set of candidate points in the design space was selected using the D-optimal criterion [10]. Finally, natural ingredient based lipstick formulation was prepared in laboratory scale according to the composition suggested by the mixture design. The essential ingredients used in lipstick formulation were natural waxes, oils, solvent and colorant. In this study, there are restrictions on the component proportions X_j that take the form of lower L_j and upper U_j constraint as follows:

$$L_j < X_j < U_j \tag{1}$$

The constraint of the component proportion as shown in Table 1 was adapted from the experimental results of a previous study [9].

Table 1
Constraint of the component proportion

Ingredient, Xi	Low Limit, L_j (%)	High Limit, U_j (%)
Castor oil, X_1	37	65
Beeswax, X_2	5	20
Candelilla wax, X_3	1	5
Carnauba wax, X_4	1	5
Solvent, X_5	5	20

2.2 Physical Properties characterization

The viscosity measurement was carried out using Brookfield Engineering Rheometer model HA DV-III equipped with small sample adapter, using spindle #28 and speed 10 rpm. Meanwhile, the melting point of the lipstick was observed with SMPI Melting Point Apparatus (Stuart Scientific).

3. RESULT AND DISCUSSION

Table 2 shows the design layout in terms of actual factor values and viscosity results from each experiment at room temperature. A set of candidate points in the design space is selected using the D-optimal criterion. In this work, 25 candidate points have been selected.

Table 2
Design points in the constrained region by the five blend components

Design Point	Design Factor					Response	
	Castor Oil X_1	Beeswax X_2	Candelilla X_3	Carnauba X_4	Solvent X_5	Viscosity (cP)	Melting Point (°C)
1	56.00	5.00	1.00	5.00	20.00	58000.00	61.80
2	39.00	20.00	3.00	5.00	20.00	59778.00	60.00
3	56.00	5.00	5.00	1.00	20.00	57000.00	62.10
4	65.00	5.00	5.00	3.00	9.00	59778.00	61.90
5	48.50	12.50	3.00	3.00	20.00	58444.00	64.40
6	56.00	5.00	5.00	1.00	20.00	57000.00	61.60
7	58.50	5.00	5.00	5.00	13.50	57889.00	61.30
8	56.60	12.20	3.00	3.00	12.20	59111.00	60.40
9	58.50	13.50	5.00	5.00	5.00	62000.00	60.60
10	65.00	5.00	1.00	1.00	15.00	56889.00	62.20
11	41.00	20.00	5.00	1.00	20.00	59556.00	58.90
12	65.00	5.00	5.00	3.00	9.00	59889.00	61.70
13	60.00	20.00	1.00	1.00	5.00	59889.00	61.80
14	56.00	5.00	1.00	5.00	20.00	57556.00	59.00
15	60.80	9.10	4.00	4.00	9.10	60222.00	62.80
16	45.00	20.00	1.00	1.00	20.00	59333.00	61.10
17	41.00	20.00	5.00	1.00	20.00	59111.00	59.90
18	65.00	11.00	1.00	5.00	5.00	60222.00	64.00
19	60.00	20.00	1.00	1.00	5.00	60111.00	58.70
20	48.50	20.00	1.00	5.00	12.50	59667.00	62.20
21	52.00	20.00	5.00	5.00	5.00	62444.00	61.10
22	65.00	11.00	5.00	1.00	5.00	59889.00	61.90
23	65.00	10.00	1.00	1.00	10.00	56667.00	59.00
24	65.00	5.00	3.00	5.00	9.00	58556.00	59.60
25	44.50	12.50	5.00	5.00	20.00	60778.00	61.90

3.1 Effect of mixture components on the viscosity and melting point of the lipstick

Since there are significant interaction effects between mixture composition and process factors, which is viscosity and melting point, the contour plot and response surface graphs are shown in Fig. 1 and Fig. 2. Both diagram describe the variation on the viscosity and melting point response as a function of the mixture composition. In order to represent the response evaluation in a bidimensional system, two of the variables were to be kept constant. In this case, castor oil and solvent composition were kept constant since these components have the less effect on viscosity and melting point. The figures show that the higher amounts of candelilla wax and carnauba wax give the higher value of viscosity and melting point. On the other hand, both viscosity and melting point of the lipstick reduced with the increases in beeswax composition.

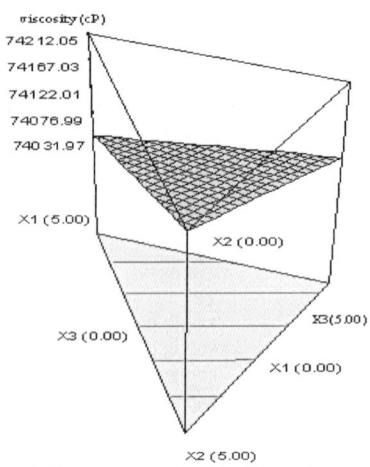
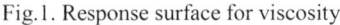

Fig.1. Response surface for viscosity

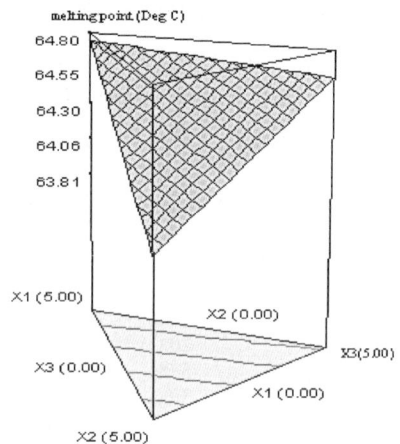

Fig.2. Response surface for melting point

4.Conclusion

Natural ingredients based lipstick formulations have been prepared. The effects of the natural waxes, oils and solvent compositions on the viscosity and melting point of the lipstick have been studied. The result indicates that the viscosity and melting point of the lipstick can be manipulated by changing the composition of natural candelilla wax, carnauba wax and beeswax in the formulation. Another important lipstick characteristic, which is hardness, will be studied. Consumer acceptance towards the product will be investigated. Finally, by relating the consumer data and instrumentation analysis, optimisation process will be conducted.

5. REFERENCES

[1]   A. Khan,.Perkembangan Industri Kosmetik. Majalah Wanita, (2000), 9.
[2]   P. Fridd, Natural Ingredient in Cosmetics II. Micelle Press, England.1996.
[3]   M.Grievson, , J.Barber, and A.L. Hunting, Natural Ingredient in Cosmetics, Micelle Press, New Jersey,1992.
[4]   G.I. Sackheim and D.D.Lehman, Chemistry For The Health Science, Prentice Hall, New Jersey, 1998.
[5]   B.Arifin, A.Bono & M.Rajin, Borneo Science Journal, Volume 12 (2002) 79.
[6]   D.F. Williams and W.H. Schmitt, Chemistry and Technology of the Cosmetic and Toiletries Industries, Blackie Academic & Profesional, Surrey, 1994.
[7]   A.Bono, B.Ariffin , M.Rajin & S.H. Keman, Proceeding of the International Conference On Chemical & Bioprocess Engineering, Volume 2 (2003) 1001.
[8]   M. Rajin, A.Tahir & A.Bono, Proceeding of 2nd Student Congress of Engineering & Technology, (2004) 144.
[9]   H. Butler Poucher's Perfumes, Cosmetics and Soap, 10th Edition, Kluwer Academic Publisher, Netherland , 2000.
[10]   R.H. Myers and D.C. Montgomery, Response Surface Methodology, John Wiley & Sons, Inc. New York, 2002.
[11]   F. Zanoti, S. Masiello, S. Bader, M.Guarneri, and D. Vojnovic, International Journal of Cosmetic Science, 20 (1998) 217

Studies in Surface Science and Catalysis, volume 159
Hyun-Ku Rhee, In-Sik Nam and Jong Moon Park (Editors)

697

An optimal model-based PID tuning method for the control of poly-butadiene latex reactor

Yeong-Koo Yeo[a], Tae-In Kwon[b] and Kwang Hee Lee[c]

[a]Department of Chemical Engineering, Hanyang University, Seoul 133-791, Korea
Fax: +82-2-291-6216, e-mail: ykyeo@hanyang.ac.kr
[b]Info Trol, Ltd., Yangcheon-Ku, Seoul 158-723, Korea
[c]LG Chemicals, Ltd., Yeochon, Chonnam, Korea

Abstract

The PBL (Poly-butadiene Latex) production process is a typical batch process. Changes of the reactor characteristics due to the accumulated scaling with the increase of batch cycles require adaptive tuning of the PID controller being used. In this work we propose a tuning method for PID controllers based on the closed-loop identification and the genetic algorithm (GA) and apply it to control the PBL process. An approximated process transfer function for the PBL reactor is obtained from the closed-loop data using a suitable closed-loop identification method. Tuning is performed by GA optimization in which the objective function is given by ITAE for the setpoint change. The proposed tuning method showed good control performance in actual operations.

1.   Introduction

Although many advanced control strategies have been developed including model-based control techniques [1,2], structurally simple proportional-integral-derivative (PID) controller is still widely used in industrial control systems [3]. The use of PID control algorithms in various application fields stems from the facts that the PI or PID controller structure is simple and its principle is easy to understand; the performance of the PID control is robust and acceptable in a wide range of applications. Tuning of PID controllers has attracted concerns of many researchers. If the target process approximates to the first or second order model, the tuning parameters can be obtained by the Ziegler-Nichols (Z-N), Cohen-Coon, ITAE (Integral of the Time-weighted Absolute Error) and IMC (Internal Model Control) methods [4]. An analytical derivative formula that enables to compute optimal tuning parameters for the anti-derivative-kick PID controller was derived based on the well-known Levenberg-Marquardt optimization method [5]. A design method for PID controllers was proposed based on the direct synthesis approach and specification of the desired closed-loop transfer function for disturbances [6]. So far tuning of PID controllers has relied mainly on open-loop analysis. But usually the open-loop test is prohibited in operating plants and disturbances and noises may cause unexpected control errors during closed-loop operation. For these reasons closed-loop identification has attracted much attention [7]. The transfer function using the closed-loop identification was calculated for the bioreactor controlled by a PID controller. The PBL(Poly-butadiene Latex) process considered in the present study is a typical nonlinear batch process and is controlled by PID controllers in cascade control structure. As operation batches proceed, dynamics of the process change and the control performance is getting worse. But PID controllers with fixed tuning parameters are used during the whole operation cycle. For this reason consistent product quality could not be achieved and the number of batches in one operation cycle was limited only to 44 ~ 47. Increase of the number of batches in one operation cycle while maintaining the product quality as desired is imperative to enhance the economics of the plant.

In the present study, we propose a tuning method for PID controllers and apply the method to control the PBL process in LG chemicals Co. located in Yeochun. In the tuning method proposed in the present work, we first find the approximated process model after each batch by a closed-loop identification method using operating data and then compute optimum tuning parameters of PID controllers based on GA (Genetic Algorithm) method.

2.   Closed-loop Identification of PBL reactor

PBL reaction begins with the injection of the reactant (Acrylonitrile butadiene styrene). The heat generated during the reaction is removed by the refrigerant (NH_3) flowing inside the internal tube. The reactor temperature is controlled by adjusting the level of the internal tube. As the operation batch is repeated, the polymer fouling is accumulated on the surface of the internal tube, causing decrease of cooling efficiency and poor control performance.

The identification of plant models has traditionally been done in the open-loop mode. The desire to minimize the production of the off-spec product during an open-loop identification test and to avoid the unstable open-loop dynamics of certain systems has increased the need to develop methodologies suitable for the system identification. Open-loop identification techniques are not directly applicable to closed-loop data due to correlation between process input (i.e., controller output) and unmeasured disturbances. Based on Prediction Error Method (PEM), several closed-loop identification methods have been presented: Direct, Indirect, Joint Input-Output, and Two-Step Methods.

The PBL reactor considered in the present study is a typical batch process and the open-loop test is inadequate to identify the process. We employed a closed-loop subspace identification method. This method identifies the linear state-space model using high order ARX model. To apply the linear system identification method to the PBL reactor, we first divide a single batch into several sections according to the injection time of initiators, changes of the reactant temperature and changes of the setpoint profile, etc. Each section is assumed to be linear. The initial state values for each section should be computed in advance. The linear state models obtained for each section were evaluated through numerical simulations.

3.   Results and Discussions

In the present work only the tuning of the parameters of the master controller is considered. The process model is identified based on the operation data of 35th batch for illustration. The operation data of any other batch can be used and identification and tuning after each batch would be most desirable. The computation time is 2 minutes on the platform based on the Pentium 5, which is quite acceptable for on-line application considering the cleaning and charging time of 20 minutes. On-line identification and tuning after each batch is planned in the plant. Fig. 1 shows results of operations of the reactor A. Fig. 1(a) shows the results of operation at 9th batch with the parameters without tuning, i.e., the parameters used in the 1st batch are still being used. Fig. 1(b) shows the results of operation at 10th batch with the parameters tuned by the closed-loop identification and GA optimization. As can be seen, oscillations are suppressed and the movement of the valve is more stabilized. From Fig. 1, we can see clear improvement of the control performance with the use of GA tuning method.

Fig. 2 shows results of operations of the reactor B. Fig. 2(a) shows the results of operation at 15th and 17th batches without tuning, i.e., the parameters used in the 1st batch are still being used. By tuning the parameters based on the closed-loop identification and GA method as before, we could achieve better control performance (Fig. 2(b)).

4.   Conclusions

The closed-loop identification and GA optimization were used to tune the parameters of the PID controller used in the PBL (Poly-butadiene Latex) reactor. The one cycle of operation consists of 44 – 47 batches. We first identify the model of the PBL reactor by the closed-loop identification followed by the determination of PID parameters using the GA optimization method. The process model is identified based on the single batch operation data for illustration. The operation data of any batch can be used and identification and tuning after each batch would be most desirable. The computation time is 2 minutes on the platform based on the Pentium 5, which is quite acceptable for on-line application considering the cleaning and charging time of 20 minutes. On-line identification and tuning after each batch is planned in the plant. The proposed tuning method showed good control performance in actual operations.

Nomenclature

A, B, C, D = n-dimensional system matrixs
e = white noise
G_C = transfer function of the controller
G_P = transfer function of the process
$\hat{G}_P$ = transfer function of the approximated process
K = matrix of kalman gain
K_C = controller gain
N4SID = numerical algorithms for subspace state space system identification
t = time [sec]
x = n-dimensional state vector
y, Y = process output

Literature Cited

[1] Chin, I. S., Chung, J. W. and Lee, K. S., *Korean J. Chem. Eng.*, **19**(2) (2002) 213.
[2] Arpornwichanop, A., Kittisupakorn, P. and Hussain, M. A., *Korean J. Chem. Eng.*, **19**(2), (2002) 221.
[3] Astrom, K. J. and Hagglund, T., PID controllers: theory, design & tuning, 2nd Ed., IAS Research Triangle Park, NC (1995).
[4] Seborg, D. E., Edgar, T. F. and Mellichamp, D. A., Process Dynamics and Control, John Wiley & Sons, New York, 272-309 (1989).
[5] Sung, S. W., Lee, T. and Park, S., *AIChE J.*, **48**(6), (2002) 1358.
[6] Chen, D. and Seborg, D. E., *Ind. Eng. Chem. Res.* **41**(19), (2002) 4807.
[7] van den Hof, P. M. J., Proceedings of the 11th IFAC Symposium on System Identification, Fukuoka, Japan, **4**, (1997) 1651.

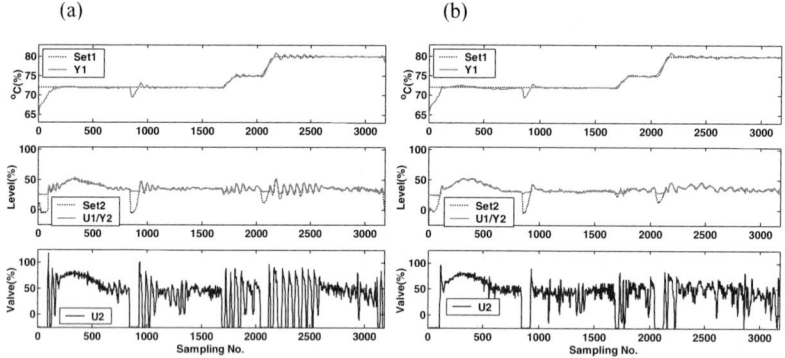

Figure 1. Results of closed-loop operations (reactor A):
(a) 9th batch: without tuning, (b) 10th batch: GA tuning

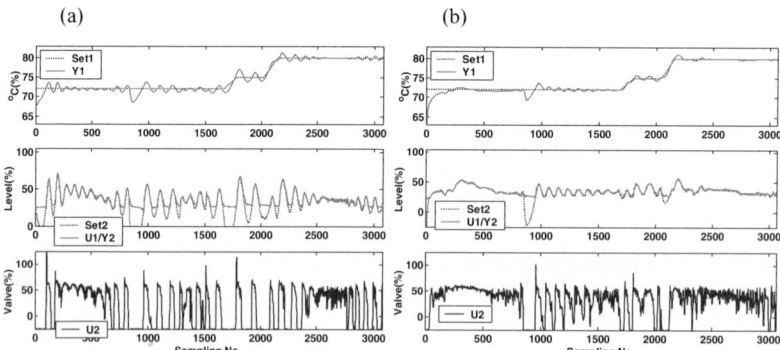

Figure 2. Results of closed-loop operations (reactor B):
(a) 15th batch: without tuning, (b) 16th batch: GA tuning

Studies in Surface Science and Catalysis, volume 159
Hyun-Ku Rhee, In-Sik Nam and Jong Moon Park (Editors)

Adsorption and Desorption Dynamics of Evaporative Fuel Gas in Canister of ORVR (On-Board Refueling Vapor Recovery) System

Dong Kyu Kim, Jeong Won Kang and Dae Ryook Yang[*]

Department of Chemical and Biological Engineering, Korea University,
1 Anam-dong, Seonbuk-gu, Seoul, 136-713, Korea

1. INTRODUCTION

The environmental restrictions have been much stronger recently and will be harder in the future, for the reason of a growing interest in the environment and need for preserving nature. The ORVR (On-Board Refueling Vapor Recovery) system is one of the efforts of environmental protections. An important objective of ORVR system is a significant reduction in leakage of evaporative fuel gas during fueling at the gas station. Nowadays, the ORVR system is mandatory in law for car production and the car makers have to prove the performance of their systems. The ORVR system is consisted of fuel tank, canister, valves (rollover valve, control valve) and connections. In this paper, the simulation model of canister in ORVR system is developed and the validity of the model is verified experimentally by the comparison of the adsorbed amounts of fuel during adsorption and desorption. The verified model can be used to design and manufacture the ORVR system.

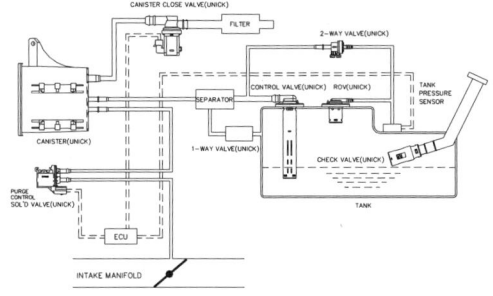

Fig. 1. The schematic diagram of ORVR system.

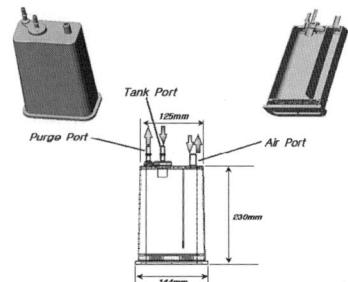

Fig. 2. The canister of ORVR system.

[*] Authors to whom correspondence should be addressed. Email: dryang@korea.ac.kr

2. ADSORPTION AND DESORPTION DYNAMICS IN CANISTER OF AN ORVR SYSTEM

The evaporative fuel gas from a fuel tank of a car during fueling must be completely adsorbed by rerouting the fuel gas through a canister (Fig. 1 and Fig. 2) of ORVR system, which is filled with activated carbon adsorbents. The adsorbed fuel gas in canister will be burned away in the engine by desorbing them with air while driving. When the evaporative fuel gas is adsorbed by an activated carbon, the considerable adsorption heat is generated and this heat hinders the adsorption of the evaporative gas in canister. Thus, to reflect the actual situation, the adsorption heat has to be considered in energy balance. The canister model can be simplified by considering only flow direction with cylindrical shape [1]. The balance equations are as follows [2, 3, 4]. The material and energy balances are considered with the adsorption equilibrium relationship by Harwell [3]. The PDE model is solved by Finite Difference Method (FDM) with ODE solver using MATLAB$^{®}$.

$$-u_f \frac{\partial C_{di}}{\partial z} - \left(\frac{1-\varepsilon}{\varepsilon}\right) r_{si} = \frac{\partial C_{di}}{\partial t} \quad (0 < z < z_T, t > 0) \quad \text{where} \quad r_{si} = \frac{\partial C_{si}}{\partial t} = K_{si} a_s (C_{si}^* - C_{si}) \tag{1}$$

$$u_f \rho_d c_{pd} \frac{\partial T_s}{\partial z} + \rho_d c_{pd} \frac{\partial T_d}{\partial t} = \left(\frac{1-\varepsilon}{\varepsilon}\right) h_s a_h (T_s - T_d) \quad \text{(Energy balance on the gas bulk phase)} \tag{2}$$

$$\rho_s c_{ps} \frac{\partial T_s}{\partial t} = h_s a_h (T_d - T_s) - \sum_{i=1}^{c} \Delta H_i K_{si} a_s (C_{si}^* - C_{si}) \quad \text{(Energy balance of the solid phase)} \tag{3}$$

$$C_{si}^* = \frac{C_{Ti} K_i C_{di}}{1 + \sum_{i=1}^{c} K_i C_{di}} \quad \text{where} \quad K_i = K_{0i} T_s^{-1/2} Pe^{-\Delta H_i / RT_s} \quad \text{(Equilibrium Relationship by Harwell)} \tag{4}$$

I.C. $(0 < z < z_T, t \le 0): C_{si}(z,0) = C_{si}^0(z), \ C_{di}(z,0) = C_{di}^0(z), \ T_s(z,0) = T_s^0(z), \ T_d(z,0) = T_d^0(z)$ (5)

B.C. $(z = 0, t > 0): \ C_d(0,t) = C_{di,in}, \ T_d(0,t) = T_{d,in}$ (6)

In model equations, u_f denotes the linear velocity in the positive direction of z, z is the distance in flow direction with total length z_T, C is concentration of fuel, ε represents the void volume per unit volume of canister, and t is time. In addition to that, K_{si} is the overall mass transfer coefficient, a_s denotes the interfacial area for mass transfer from the fluid to the solid phase, a_h denotes the interfacial area for heat transfer, ρ is density of each phase, c_p is heat capacity for a unit mass, h_s is heat transfer coefficient, T is temperature, P is pressure, and ΔH_i represents heat of adsorption. The subscript d refers bulk phase, s is solid phase of adsorbent, i is the component index. The superscript * represents the equilibrium concentration.

3. SIMULATION AND EXPERIMENT RESULTS FOR ADSORPTION AND DESORPTION IN CANISTER OF AN ORVR SYSTEM

3.1. Simulation conditions

The evaporated fuel gas generated from the fuel tank of a car during fueling enters the canister by manipulating the relevant valves in ORVR system. It is assumed that the evaporative fuel gas is mainly constituted of butane gas in this simulation because butane gas is the worldwide standard test gas for canister performance in ORVR system. The simulations are performed under following conditions. The fuel tank size of ORVR system is 45 liter, fuel temperature is 26.67°C (80°F), fueling rate is 10 GPM, and the ambient temperature is assumed to be 19.44°C (67°F). The fueling time is about 200sec while the leaked fuel gas is adsorbed in the canister with 30cm total length. The adsorbed fuel gas will be desorbed by air at the velocity of 0.05m/sec during operation of the vehicle in the opposite direction. The duration of desorption is about 450sec in our simulation. These conditions are same as the test conditions of UNICK Ltd. in Korea which manufactures the ORVR system.

3.2. Simulation and experiment results

A typical simulation result is shown in Fig. 3. Under the given conditions, the concentration of fuel gas in bulk phase at the exit (Fig. 3a) is zero and the concentration of evaporative fuel gas at solid phase (Fig. 3b) at the exit did not reach the equilibrium concentration of activated carbon during adsorption. These results indicate that the canister of ORVR system is properly designed to adsorb the evaporative fuel gas. The temperature changes in canister (Fig. 3c) during the operation remains in the acceptable range. The test results for different weather conditions showed that the canister design in this study can fulfill the required performance.

With this model and solver, the ORVR simulation system (ORVRSS) package is developed (Fig. 4a and 4b). Using this package, various operating conditions and designs can easily be tested. The ORVRSS can calculate the dynamic behavior of adsorption/desorption of a canister considering fuel tank design and refueling pattern. It can assist to find the optimal design factors of canister in ORVR system for various conditions, which are the length of canister, amount of activated carbon, and so on.

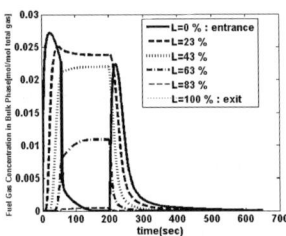 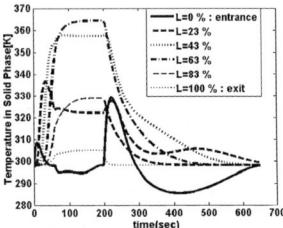

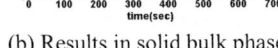

(a) Results in gas bulk phase   (b) Results in solid bulk phase   (c) Results in solid bulk phase

Fig. 3. The simulation results of concentration change (a and b) and temperature change (c) in canister.

704

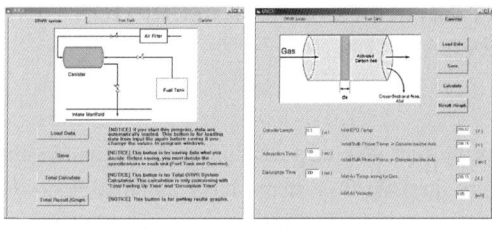

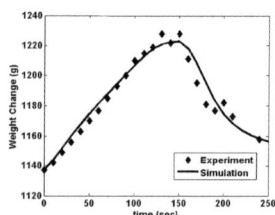

(a) ORVR tab   (b) Canister tab

Fig. 4. ORVR simulation system package.

Fig. 5. The comparison of weight change in canister between experiment and simulation during adsorption and desorption operation.

The validity of the model is tested against the experiment. A 1500cc canister, which is produced by UNICK Ltd. in Korea, is used for model validation experiment. In the case of adsorption, 2.4l/min butane and 2.4l/min N_2 as a carrier gas simultaneously enter the canister and 23l/min air flows into canister with a reverse direction during desorption. These are the same conditions as the products feasibility test of UNICK Ltd. The comparison between the simulation and experiment showed the validity of our model as in Fig. 5. The amount of fuel gas in the canister can be predicted with reasonable accuracy. Thus, the developed model is shown to be effective to simulate the behavior of adsorption/desorption of actual ORVR system.

4. CONCLUSIONS

The ORVR system is an important subsystem which reduces the contamination of evaporative fuel gas at gas station during the fueling. In this paper, a simulation model of adsorption and desorption of evaporative fuel gas in canister of ORVR system is developed. From the comparison between the simulations and experiments, the validity of the developed model is verified and the dynamics can be predicted. This PDE model can be used to design the canister of ORVR system effectively for diverse climate and operating conditions.

REFERENCES

[1] D. K. Kim and D. R. Yang, Proceedings of the 2003 KIChE Annual Meeting, Sunchon, Korea, Vol. 9, No. 1 (2003) 229.

[2] C. A. Silebi, W. E. Schiesser, Dynamic Modeling of Transport Process Systems, Academic Press, San Diego, 1992.

[3] Charles D. Holland, Athanasios I. Liapis, Computer Methods For Solving Dynamic Separation Problems, McGraw-Hill, New York, 1983.

[4] Bruce E. Poling, John M. Prausnitz, John P. O'connell, The Properties of Gases and Liquids 5th Ed., McGraw-Hill, Boston, 2001.

Studies in Surface Science and Catalysis, volume 159
Hyun-Ku Rhee, In-Sik Nam and Jong Moon Park (Editors)
© 2006 Elsevier B.V. All rights reserved

Effectiveness factor approximation by using a perturbation method

Jietae Lee and Dong Hyun Kim
Department of Chemical Engineering, Kyungpook National University
Taegu 702-701, KOREA, E-mail: jtlee@knu.ac.kr, dhkim@knu.ac.kr, Fax: +82-53-950-6615

1. INTRODUCTION

The ratio of the observed reaction rate to the rate in the absence of intraparticle mass and heat transfer resistance is defined as the effectiveness factor. When the effectiveness factor is ignored, simulation results for catalytic reactors can be inaccurate. Since it is used extensively for simulation of large reaction systems, its fast computation is required to accelerate the simulation time and enhance the simulation accuracy. This problem is to solve the dimensionless equation describing the mass transport of the key component in a porous catalyst[1,2]

$$\frac{d^2 y}{dx^2} + \frac{s}{x}\frac{dy}{dx} - \phi^2 f(y) = 0 \tag{1}$$

with boundary conditions of $y(1) = 1$, $dy(0)/dx = 0$. Here y is the dimensionless concentration of the key component, x is the dimensionless space variable in the catalyst, ϕ is the Thiele modulus and s is the shape factor of the catalyst ($s=0$ for the slab, $s=1$ for the cylinder and $s=2$ for the sphere). The dimensionless reaction rate function $f(y)$ is normalized to be $f(1)=1$. The effectiveness factor η is given as $\eta = \frac{s+1}{\phi^2}\frac{dy(1)}{dx}$

Various simple equations have been available. However, each equation has its own disadvantages. For example, the equation by Wijnggaarden et al.[3] showing excellent estimates of effectiveness factors for various problems fails to provides appropriate estimates for a large Thiele modulus when $f'(1)$ is negative[2]. It is well-known that

$\eta \approx 1 - \frac{f'(1)}{(s+1)(s+3)}\phi^2$ for a small ϕ and $\eta \approx \frac{(s+1)}{\phi}\sqrt{2\int_0^1 f(y)dy}$ for a large ϕ. Gottifredi

and Gonzo[4] obtained a simple matching equation for the above two asymptotes. However, their matching equation fails to provide accurate results unless $s=0$ (slab geometry). For some $f(y)$ (e.g. $f(y)=y^2$ and $s=2$), it does not follow the asymptote for a large ϕ.

Here, for better estimates, perturbation equations with terms up to ϕ^{10} are obtained for a small ϕ and the Euler transformation[4] is applied to accelerate convergence of this series. A different matching equation combining two asymptotes for the small and large ϕ is proposed for $\beta \equiv dy(1)/dx = \phi^2 \eta/(s+1)$. The proposed equation estimating the effectiveness factor is shown to have better accuracy for various reaction rate functions of $f(y)$.

2. SMALL THIELE MODULUS CASE

For a small ϕ, the regular perturbation method[5] can be applied with assuming $y(x)$ as $y(x)=y_{S0}(x)+ \phi^2 y_{S1}(x)+ \phi^4 y_{S2}(x)+....$ Substituting this $y(x)$ to Eq. (1), we have

$$\frac{d^2(y_{S0}+\phi^2 y_{S1}+\Lambda\ )}{dx^2}+\frac{s}{x}\frac{d(y_{S0}+\phi^2 y_{S1}+\Lambda\ )}{dx}-\phi^2 f(y_{S0}+\phi^2 y_{S1}+\Lambda\ )=0$$

$$\frac{dy(0)}{dx}=\frac{dy_{S0}(0)}{dx}+\phi^2\frac{dy_{S1}(0)}{dx}+\phi^4\frac{dy_{S2}(0)}{dx}+\Lambda\ =0 \tag{2}$$

$$y(1)=y_{S0}(1)+\phi^2 y_{S1}(1)+\phi^4 y_{S2}(1)+\Lambda\ =1$$

Since $f(y_{S0}+\phi^2 y_{S1}+\Lambda\ )=f(y_{S0})+f'(y_{S0})(\phi^2 y_{S1}+\phi^4 y_{S2}+\Lambda\ )+\Lambda\ $, terms of the order of

ϕ^0 in Eq. (2) are $\dfrac{d^2 y_{S0}}{dx^2}+\dfrac{s}{x}\dfrac{dy_{S0}}{dx}=0,\ \dfrac{dy_{S0}(0)}{dx}=0, y_{S0}(1)=1$ and its solution is $y_{S0}(x)=1$.

Terms of the order ϕ^2 are $\dfrac{d^2 y_{S1}}{dx^2}+\dfrac{s}{x}\dfrac{dy_{S1}}{dx}-f(1)=0,\ \dfrac{dy_{S1}(0)}{dx}=0, y_{S1}(1)=0$ and its

solution is $y_{S1}(x)=(x^2-1)/2(s+1)$. In this way, we can obtain higher order terms analytically. Symbolic programming language is used to obtain higher order solutions. From these equations, we can obtain the estimate of $\beta(\phi)$ as

$$\beta_S(\phi)=q_1\phi^2+q_2\phi^4+q_3\phi^6+q_4\phi^8+q_5\phi^{10}$$

$$=\frac{1}{3}\phi^2-\frac{f'(1)}{45}\phi^4+(\frac{f''(1)}{945}+\frac{2f'(1)^2}{945})\phi^6-(\frac{2f'(1)f''(1)}{6075}+\frac{f^{(3)}(1)}{25515}+\frac{f'(1)^3}{4725})\phi^8$$

$$+(\frac{32f'(1)^2 f''(1)}{467775}+\frac{71f'(1)f^{(3)}(1)}{4209975}+\frac{2f'(1)^4}{93555}+\frac{17f''(1)^2}{1403325}+\frac{f^{(4)}(1)}{841995})\phi^{10}$$

for $s=2$. Derivatives of $f(y)$ at $y=1$ can be computed numerically via the backward difference method[6]. The convergence of series can be accelerated with transformations[6]. Among various transformations, the Euler transformation, $\beta_S(\phi)=q_1\phi^2+q_2\phi^4+0.875q_3\phi^6+0.5q_4\phi^8+0.125q_5\phi^{10}$, is very effective when signs of terms are alternating and used here. Fig. 1 shows convergence of the series and the performance of the Euler transformation for $f(y)=y^2$ and $s=2$.

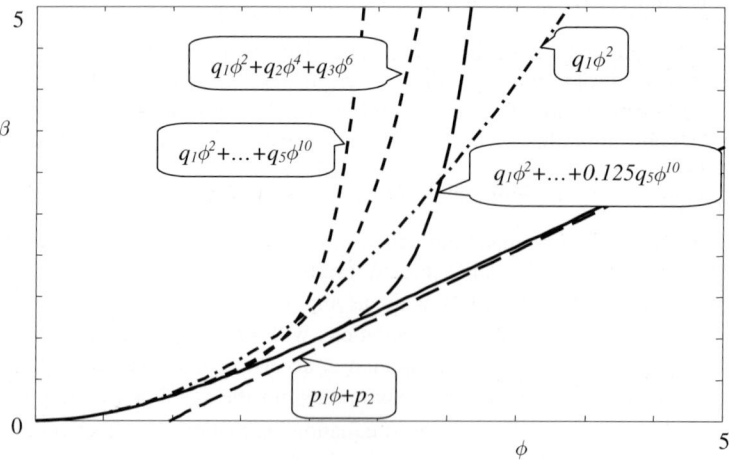

Fig. 1. Estimates of $\beta=dy(1)/dx$ for $f(y)=y^2$ and $s=2$ (solid: exact).

The valid region of $\beta_S(\phi)$ can be obtained by finding the point that both slopes of asymptotes for small and large ϕ, $2q_1\phi + 3q_2\phi^3 + \Lambda + 10q_5\phi^9 = \sqrt{2\int_0^1 f(y)dy}$. Let its real solution near $\phi=2$ be ξ (when such root does not exist, we use $\xi=2$). Simulations show that ξ is large enough and $\beta_S(\phi)$ is accurate for ϕ between 0 and ξ.

3. LARGE THIELE MODULUS CASE

For a large ϕ, the singular perturbation method should be used[5]. For this, we transform the independent variable x as $x=1-t$. Then Eq. (1) becomes

$$\frac{d^2y}{dt^2} - \frac{s}{1-t}\frac{dy}{dt} - \phi^2 f(y) = 0$$

$$y(t)\big|_{t=0} = 1, \quad \frac{dy}{dt}\bigg|_{t=1} = 0$$

(3)

We extend the variable t as $t = \tilde{t}/\phi$ and expand $y(\tilde{t})$ as $y(\tilde{t}) = y_{L0}(\tilde{t}) + \phi^{-1}y_{L1}(\tilde{t}) + \Lambda$. Substituting it to Eq. (3), we can obtain a differential equation for each order. For the order of ϕ^0, we have $\dfrac{d^2y_{L0}}{d\tilde{t}^2} - f(y_{L0}) = 0, y_{L0}(0) = 1, \dfrac{dy_{L0}(\phi)}{d\tilde{t}}\bigg|_{\phi\to\infty} = 0$ and $\dfrac{dy_{L0}(0)}{d\tilde{t}} = -\sqrt{2\int_0^1 f(y)dy}$.

Higher order terms require solving linear boundary value problems numerically. Here, as the estimate of $\beta(\phi)$ effective for a large ϕ, $\beta_L(\phi) = p_1\phi + p_2 = \sqrt{2\int_0^1 f(y)dy}\,\phi + p_2$ is used, where p_2 is estimated as

$$p_2 \approx \beta_S(\xi) - p_1\xi + \vartheta$$

$$\vartheta = \begin{cases} -\beta_S(\xi) + p_1\xi, & s = 0 \\ 0.52\sqrt{2\int_0^1 f(y)dy} - 0.41, & s = 1 \\ 0.43\sqrt{2\int_0^1 f(y)dy} - 0.46, & s = 2 \end{cases}$$

Here, ϑ is a correlation equation which minimizes errors of $\left|p_2 - (\beta_S(\xi) - p_1\xi)\right|^2$ for $f(y)=y^n$, where n is between -0.2 and 1.

4. MATCHING EQUATION

Now a matching equation combining $\beta_S(\phi)$, an asymptote for a small ϕ, and $\beta_L(\phi)$, an asymptote for a large ϕ, is given as

$$\hat{\beta}(\phi) = \begin{cases} \beta_S(\phi) = q_1\phi^2 + q_2\phi^4 + \Lambda + q_5\phi^{10} & \text{if } \phi \le \xi \\ p_1\phi + [\beta_S(\xi) - p_1\xi] + (1 - \dfrac{\xi}{\phi})\vartheta & \text{otherwise} \end{cases}$$

(4)

It is noted that $\hat{\beta}(\phi) = \beta_S(\phi)$ for a small ϕ below ξ, $\hat{\beta}(\phi) = p_1\phi + \hat{p}_2$ for a large ϕ and $\hat{\beta}(\phi)$ is continuous.

5. DISCUSSIONS

The proposed equations have been tested for the various reaction rate functions and compared with equations by Wijnggaarden et al.[3] and Gottifredi and Gonzo[4]. Results for the power law reaction rate function of $f(y) = y^n$ are given in Fig. 2. Except for $f(y)=y^{-0.5}$

having multiple steady-states, our estimates of β show relative errors below 2% for $s=2$.

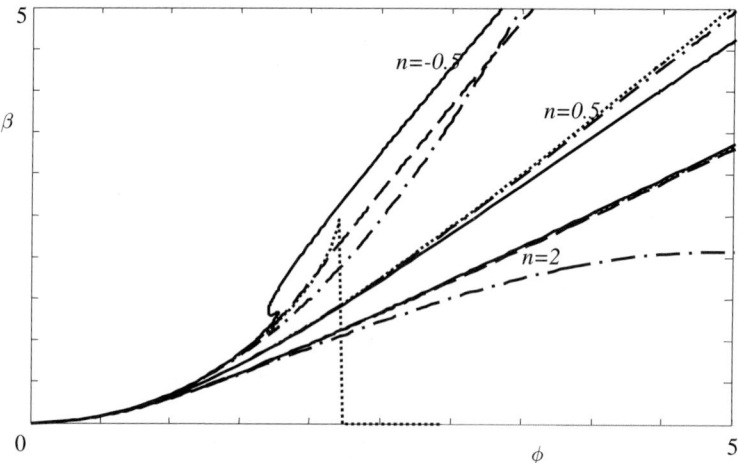

Fig. 2. Estimates of $\beta(\phi)$ for the power law reaction rate functions ($f(y)=y^n$) (solid: exact, dashed: proposed, dotted: Wijnggaarden et al.[3], dash-dotted: Gottifredi and Gonzo[4]).

Similar results have been obtained for exothermic first-order reaction rate functions and Langmuir reaction rate functions for $s=2$.

6. CONCLUSION

We proposed a simple analytic equation for effectiveness factor and compared it with those of Wijnggaarden et al.[3] and Gottifredi and Gonzo[4]. Simulations show that for problems with multiple solutions all three equations are inaccurate. Any single equation will not fit to estimate multiple solutions. Equation of Wijnggaarden et al. fails to provide accurate estimates for reaction rate functions with negative derivatives. It is found to be useful for $f'(y)>0.5$. Equation of Gottifredi and Gonzo is found to be useful for problems with $s=0$ and $f'(1)\int_0^1 f(y)dy < 0.75$. For problems without multiple solutions, the proposed equation shows below 2% relative errors for $s=2$, 6% relative errors for $s=1$ and 16% relative errors for $s=0$.

REFERENCES

[1] R. Aris, The Mathematical Theory of Diffusion and Reaction in Permeable Catalysts. Vol. I The Theory of the Steady State, Clarendon, Oxford, 1975.
[2] D.H. Kim and J. Lee, Chemical Engineering Science, 59 (2004) 2253.
[3] R.J. Wijnggaarden, A. Kronberg and K.R. Westerterp, Industrial Catalysis, Wesley-VCH, Weinheim, 1988.
[4] J.C. Gottifredi and E.E. Gonzo, Chem. Eng. Journal, in revision (2004).
[5] A. Nayfeh, Perturbation Methods, Wiley, New York, 1973.
[6] W.H. Press, S.A. Teukolsky, W.T. Vetterling and B.P. Flannery, Numerical Recipes in Fortran. 2nd Ed., Cambridge University Press, New York, 1992.

Studies in Surface Science and Catalysis, volume 159
Hyun-Ku Rhee, In-Sik Nam and Jong Moon Park (Editors)

Kinetics of norbornene synthesis and continuous reactor modeling study

Sang-Beom Lee[a], Sung-Hwan Cho[a], Young-Whan Park[a], and Huyn-Ku Rhee[b]

[a]LG Chem, Ltd., 104-1 Moonji-dong, Yuseong-gu, Daejeon 305-380, Republic of Korea

[b]School of Chemical and Biological Engineering, Seoul National University,
San 56-1 Sillim-dong, Gwanak-gu, Seoul 151-744, Korea

1. Introduction

With the growing interest for the polynorbornene, photoresist polymer, and cyclic olefin copolymer, the synthesis norbornene or bicyclo[2,2,1]-2-heptene (NBN) has drawn significant attention because it is one of the most important precursor for these materials. Norbornene is produced by the reaction between ethylene and cyclopentadiene (CPD) via the Diels-Alder condensation process at elevated temperature and pressure [1,2].

2. Experimental procedure

Synthesis of norbornene is examined in a well mixed semi-batch reactor to which ethylene is supplied continuously. CPD is obtained from thermal decomposition of dicyclopentadiene (DCPD) because it is always present as a dimer at ambient temperature. DCPD as a raw material, toluene as a solvent and cyclohexane as an internal standard are put into a 200ml autoclave equipped with a magnetic stirrer (stirring speed 400rpm) and thus the initial concentration of DCPD is 0.8mol/L. The reactor purged with helium is heated to the desired temperature and pressurized with ethylene gas. Samples are taken out periodically during the course of reaction and analyzed by gas chromatography on HP-1 column with FID detector.

3. Reaction kinetics

To develop the rate equations suitable for process modeling and reactor design, experimental data have been analyzed on the basis of the postulated reaction mechanism [2] given in Table 1. Here the formation of polymer is excluded because it is not detected under our experimental conditions. All of the reactions are equilibrium-limited and the net rates for the formation of each component with some assumptions [3] are given as follows:

$$r_1 = -k_1[DCPD] + k_{-1}[CPD]^2 \tag{1}$$

$$\begin{aligned} r_2 = {} & k_1[DCPD] - k_{-1}[CPD]^2 - k_2[CPD]^{n_{21}}[C_2H_4]^{n_{22}} \\ & + k_{-2}[NBN] - k_3[NBN][CPD] + k_{-3}[DMON] \end{aligned} \tag{2}$$

$$r_3 = k_2[CPD]^{n_{21}}[C_2H_4]^{n_{22}} - k_{-2}[NBN] - k_3[NBN][CPD] + k_{-3}[DMON] \tag{3}$$

$$r_4 = k_3[NBN][CPD] - k_{-3}[DMON] \tag{4}$$

where the subscripts 1, 2, 3 and 4 on r represent DCPD, CPD, NBN and DMON, respectively.

By using the method of Levenberg-Marquardt [4] the activation energies and frequency factors for individual rate constants are determined as given in Table 2 and the reaction orders with respect to CPD and ethylene are estimated to be $n_{21} = 0.96$ and $n_{22} = 0.94$, respectively. When these kinetic parameters are used, the numerical simulation results turn out to be in good agreement with the experimental data as shown in Fig. 1. This indicates that the estimated kinetic parameters may be utilized for the purpose of reactor design.

Table 1 Postulated reaction mechanism for the synthesis of norbornene

Thermal decomposition	$DCPD \underset{k_{-1}}{\overset{k_1}{\rightleftharpoons}} 2CPD$
Diels-Alder reaction of CPD and ethylene	$CPD + C_2H_4 \underset{k_{-2}}{\overset{k_2}{\rightleftharpoons}} NBN$
Diels-Alder reaction of NBN and CPD	$NBN + CPD \underset{k_{-3}}{\overset{k_3}{\rightleftharpoons}} DMON$

Table 2 Estimated values of kinetic parameters

Reaction rate constant k_i	k_1	k_{-1}	k_2	k_{-2}	k_3	k_{-3}
Frequency factor k_{oi}	2.656×10^{12}	2.295×10^3	9.978×10^7	1.820×10^{14}	9.555×10^{16}	1.194×10^{15}
Activation energy Ea_i (J mol^{-1} K^{-1})	1.401×10^5	5.450×10^4	9.195×10^4	1.862×10^5	1.967×10^5	1.749×10^5

4. Optimal operating conditions for continuous reactors

Based on the kinetic mechanism and using the parameter values, one can analyze the continuous stirred tank reactor (CSTR) as well as the dispersed plug flow reactor (PFR) in which the reaction between ethylene and cyclopentadiene takes place. The steady state mass balance equations may be expressed by using the usual notation as follows:

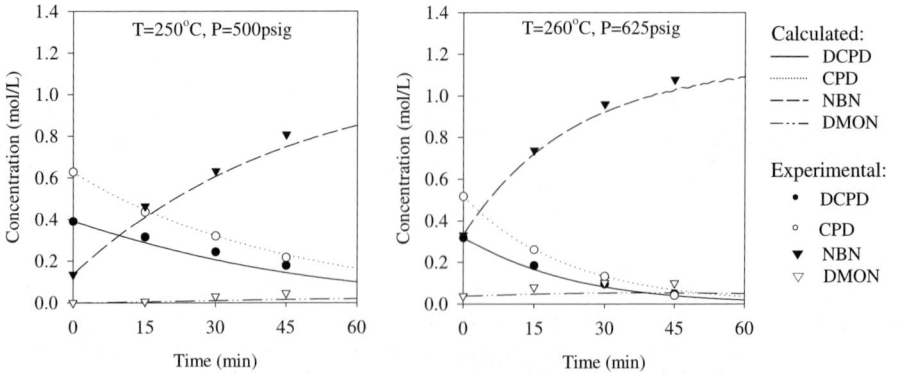

Fig.1. Comparison of numerical simulation results and experimental data.

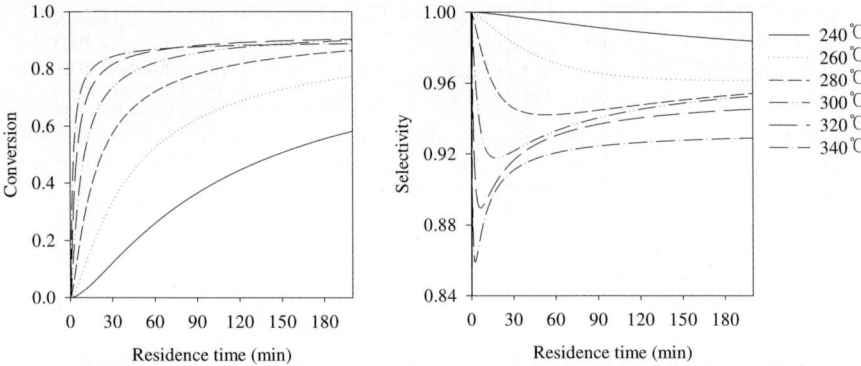

Fig.2. Effect of temperature on the conversion and selectivity in a CSTR at 500psig.

For CSTR,
$$C_{j0} - C_j = -\tau\, r_j \qquad (5)$$

For dispersed PFR,
$$-\frac{1}{Pe}\frac{d^2 C_j}{dz^2} + \frac{dC_j}{dz} = \tau\, r_j \qquad (6)$$

$$\text{at } z = 0, \quad -\frac{1}{Pe}\frac{dC_j}{dz} + C_j = C_{j0} \qquad (7)$$

$$\text{at } z = 1, \quad \frac{dC_j}{dz} = 0 \qquad (8)$$

for j = 1, 2, 3 and 4. For the dispersed PFR model we apply the Dankwerts boundary conditions and assume that the axial dispersion coefficient is the same for all the components.

4.1. CSTR

The effect of temperature on the reactor performance is investigated under various conditions by using Eq.(5). As shown in Fig. 2, the conversion increases with temperature due to thermal reaction of norbornene synthesis while the selectivity to norbornene decreasse owing to the formation of dimethanooctahydronaphthalene (DMON) by the subsequent reaction of NBN further with CPD. The reaction pressure is found to exercise almost the same influence on the reaction performance as the temperature. On the basis of the simulation results, the best reaction performance is achieved at around 320 ℃ and 1200 psig to yield 94% conversion and 92% selectivity with a residence time of 120min. The mole ratio of ethylene to DCPD in the feed has a similar effect on the reaction performance to the temperature and pressure as shown in Fig.3. However, the rate of change becomes insignificant above the mole ratio of ethylene/DCPD=4. Similar results are obtained for a plug flow reactor (PFR) except for the higher conversion and selectivity at a shorter residence time because of the high concentration near the reactor entrance.

4.2. Dispersed PFR and recycle reactor

Simulation studies are also conducted for a dispersed PFR and a recycle reactor at 260 ℃, 500 psig and feed with DCPD=0.32 mol/min, CPD=0.96mol/min and ethylene=3.2mol/min. Peclet number (Pe) or the recycle ratio is selected as a variable parameter for the dispersed PFR or for the recycle reactor, respectively. Conversion approaches to that of PFR over Pe=50 as can be seen in Fig.4. It is also worth mentioning that the reactor performance is improved with recycle if the residence time is low.

712

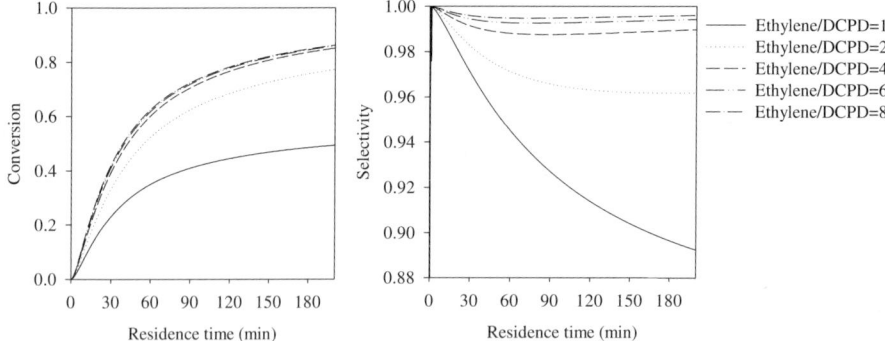

Fig.3. Effect of the mole ratio of ethylene to DCPD in a CSTR at 260 ℃ and 500 psig.

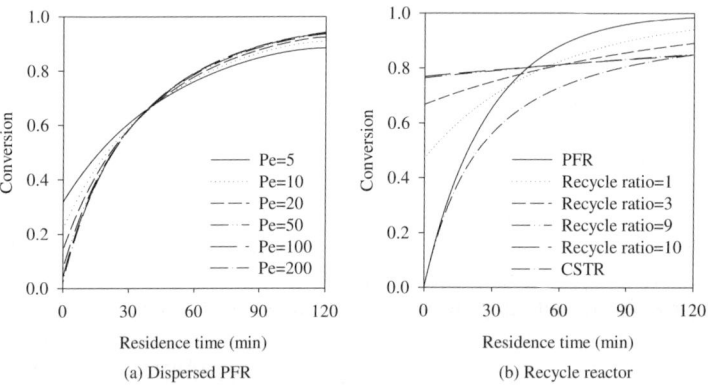

(a) Dispersed PFR   (b) Recycle reactor

Fig.4. Effect of Pe in a dispersed PFR and recycle ratio in a PFR with recycle stream at 260 ℃ and 500 psig.

5. Conclusions

The kinetic parameters associated with the synthesis of norbornene are determined by using the experimental data obtained at elevated temperatures and pressures. The reaction orders with respect to cyclopentadiene and ethylene are estimated to be 0.96 and 0.94, respectively. According to the simulation results, the conversion increases with both temperature and pressure but the selectivity to norbornene decreases due to the formation of DMON. Therefore, the optimal reaction conditions must be selected by considering these features. When a CSTR is used, the appropriate reaction conditions are found to be around 320°C and 1200 psig with 4:1 mole ratio of ethylene to DCPD in the feed stream. Also, it is desirable to have a Pe larger than 50 for a dispersed PFR and keep the residence time low for a PFR with recycle stream.

References
[1] N. B. Lorette and L. Jackson, US Patent No. 3 766 283 (1973).
[2] R. Kotwica and A. Marbach, US Patent No. 6 479 719 (2002).
[3] T. G. Lenz and J. D. Vaughan, J. Phys. Chem., No. 93 (1989) 1592.
[4] M. E. Davis, Numerical methods and modeling for chemical engineers, John Wiley & Sons, New York, 1984.

Studies in Surface Science and Catalysis, volume 159
Hyun-Ku Rhee, In-Sik Nam and Jong Moon Park (Editors)
© 2006 Elsevier B.V. All rights reserved

Variation of Reaction Stages And Mole Composition Effect on Melamine-Urea-Formaldehyde (MUF) Resin Properties

Awang Bono[a], Duduku Krishnaiah, Mariani Rajin and Nancy J. Siambun
Chemical Engineering, School of Engineering & Information Technology
Universiti Malaysia Sabah, 88999 Kota Kinabalu.
Tel: +6088 320600 (ext3135), Fax: +6088 32034, [a]Email: awang@pc.jaring.my

1. INTRODUCTION

MUF resin is widely used as an adhesive in wood industries, coating technology, paper industries and a main material in kitchenware production. In various applications, different resin properties are needed to suit its application. Important resin properties are for example higher resin solubility, low curing period with lower temperature and catalyst amount, good stability for longer shelf life, and lower free formaldehyde emission, as formaldehyde is very toxic, and can cause cancer [1]. One of the factors that affecting the MUF resin properties is the mole composition. The mole composition is a ratio of formaldehyde to amino compound i.e. melamine or urea, in each reaction stages [2]. Besides that addition of other chemical such as sorbitol, may be added to enhance the resin properties [3]. In Tutin (1998) [4] and Gapud et al. (1999) [5] researches, addition of sorbitol has improve the resin stability and solubility in water and also cured resin flexibility.

Using experimental design such as Surface Response Method optimises the product formulation. This method is more satisfactory and effective than other methods such as classical one-at-a-time or mathematical methods because it can study many variables simultaneously with a low number of observations, saving time and costs [6]. Hence in this research, statistical experimental design or mixture design is used in this work in order to optimise the MUF resin formulation.

2. MATERIALS AND METHODOLOGY

2.1. Statistical Mixture Design and MUF Resin Formulation

The statistical mixture design for 5–components was carried out by using Design Expert, D–Optimal criterion (Version 6.10, Stat–Easy Inc., Minneapolis USA). In this study, there are restriction on the component proportions X_j that take the form of lower L_j and upper U_j constraint as $L_j < X_j < U_j$. The constraint of the component proportions as shown in Table 1 was adapted from the experimental results of the previous study [2,5]

Table 1
Constrain of the component proportion

Ingredient, X_j	Low Limit, L_j (%)	High Limit, U_j (%)
Formalin, X_1	50	60
Melamine, X_2	25	30
Urea added initially, X_3	4	10
Urea added after post refluxing, X_4	6	15
Sorbitol, X_5	1	10

714

2.2. MUF Resin Synthesis

The process of MUF resin synthesis and physical properties was adopted from previous research [2, 5]. The MUF resin was produced with certain amount of formaldehyde to amino compound and sorbitol, and 40°C End Point polymerisation. By using Design Expert D-optimal software, 25 MUF resin formulations were generated. The resin synthesis was carried out in 3-reaction stages. In the first stage, MUF resin raw materials were mixed together and undergo additional process at alkaline condition and low temperature. The reaction rate was gradually increased by slowly increasing the resin temperature, while the pH gradually dropped. In the second stage, the resin was in acidic condition and high temperature and thus optimum polymerisation was occurred. The resin allowed to reflux sufficiently to avoid crystallisation of melamine. Finally in stage three of resin synthesis, the polymerisation rate was minimized, and then additional urea was added. The resin was then cooled down to ambient room temperature, and the physical properties were analysed.

2.3 Physical Characterization of MUF Resin

The MUF resin pH was determined using pH meter model pH 340-A/SET 1–MTM. The pH meter was calibrated before it was used to determine the pH of the resin. The viscosity was determined using the Cole-Parmer 98936-15 viscometer (R2 spindle, 100rpm speed). The storage life was a test of shelf life of the MUF resin under the ambient environment. Resin was first stored at ambient room temperature. Viscosity of the resin was checked for every three to four days. The ratio of water that can be added into resin before it turned turbid or precipitated is called resin solubility. The resin solubility was determined by divide the weight of resin and the weight of water added into resin before it turned turbid or precipitated. The curing period of a resin was defined as the time period for the resin to be hardened after application in a 30°C and 1.0% of NH_4Cl powder (as hardener).

3. RESULTS AND DISCUSSION

3.1 Analysis of MUF properties

3.1.1 MUF resin viscosity

Viscosity is a measure of resistance to flow. The viscosities of amino resin are substantially higher than water. Low viscosity adhesive have a tendency to flow through veneers, which are often highly porous with a consequent detrimental effects on finishing treatments. Fig.1 shows that the viscosity of resin is most affected by sorbitol, urea added after post refluxing and formalin. The viscosity is increasing with decreasing amount of formalin (60.06%–50.00%), increasing amount of sorbitol (1.00%–11.06%) and urea added after post refluxing (6.00%–16.06%).

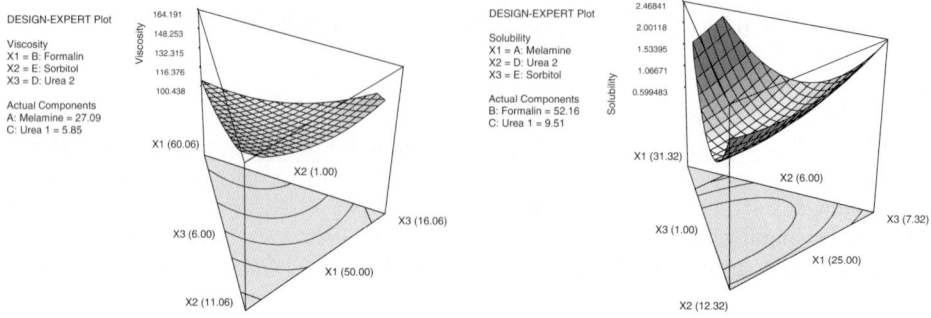

Fig.1. A 3D surface plot for MUF resin viscosity   Fig. 2. A 3-D surface plot for MUF resin solubility

3.1.2 MUF resin solubility

MUF resin with high solubility can dissolve a large amount of water without precipitating [8]. This property is very important to enable it to be mixed with large amount of water, or other waterborne material. Resin with high solubility can easily washed away from utensil, and will not cause blocking of pipelines during cleaning. A more advantageous use of solubility property would be to further dilute the resin in order to reduced the resin solid content to a certain level to reduced cost [9]. Fig. 2 shows that the solubility of resin is most affected by melamine, urea added after post-refluxing and sorbitol. The solubility property is increasing, with increasing amount of melamine (25.00%–31.32%), sorbitol (1.00%–7.32%) and urea added after post-refluxing (6.00%–12.32%).

3.1.3 MUF resin curing period

Curing period is a time needed for the resin to cure in the presence of hardener. Most industries such as in coating, plywood, or fibreboard, paper and plastic industries, a resin with shorter curing period is preferred, to help shorten the assembling period, and hence increase the production rate [10]. Fig. 3 shows that the curing period of resin is most affected by urea added initially, urea added after post-refluxing and sorbitol. The curing period is increasing with decreasing amount of sorbitol (9.39%–1.00%), increasing amount of urea added after post-refluxing (6.00%–14.39%) and constant amount of urea added initially (4.00%–12.39%).

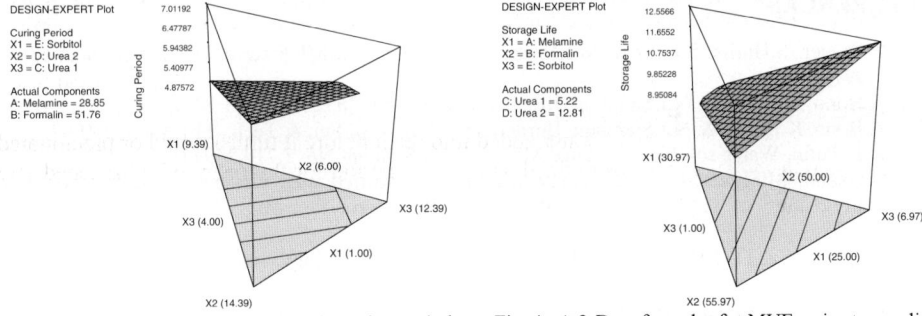

Fig. 3. A 3-D surface plot for MUF resin curing period   Fig. 4. A 3-D surface plot for MUF resin storage life

3.1.4 MUF resin shelf life

Resin with good shelf life or storage stability is very important to enable it to be kept for long-term usage. Besides that, resin with good stability can also enable import and export activities. Resin with short stability, must be used shortly after their preparation, and unexpected scheduling disruption can lead to losses of product [11]. Fig. 4 shows that the storage life of resin is most affected by sorbitol, melamine and formalin content. The storage life is 3-D surface plot shows that, the storage life is increasing with decreasing amount of melamine (30.97%– 25.00%), increasing amount of sorbitol (1.00%– 6.97%) and constant amount of formalin (55.97%– 50.00%).

3.2 Optimum Formulation for MUF Resin

The MUF resin formulation is built up from combination of certain amount of formalin, melamine and urea (in initial and post refluxing stages) and also sorbitol. Variation on the formulation gives different resin properties. The optimum resin properties give the optimum MUF resin formulation. From the properties analysis data, the optimum formulation is determined by using Mixture Experimental Design D-optimal criterion. The selective criteria

716

for optimum resin properties should have good viscosity, maximum solubility in water, minimum curing period at room temperature and maximum storage life. The optimum resin properties is obtained from the MUF resin with formulation consisting of 59% formalin, 30% of melamine, 4% of urea added initially, 6% of urea added after post-refluxing and 1% of sorbitol. This resin has 96cp viscosity, lower curing period and long storage life period.

4. CONCLUSION

By using Mixture Experimental Design D-optimal criterion 25 resin formulations with combination of various percentage of melamine, formalin, urea added initially and after post-refluxing and sorbitol. The data analysis shows that each resin property is predominantly affected by three main components. The viscosity of MUF resin is mostly affected by formalin, sorbitol and urea added after post-refluxing. The solubility property is mostly affected by melamine, sorbitol and urea added after post-refluxing. The curing period is mostly affected by sorbitol, and urea added initially and after post-refluxing. The storage life is mostly affected by melamine, sorbitol and formalin. Data analysis from Mixture Experimental Design also helps determined the optimum MUF resin formulation. MUF resin with optimum resin properties, gives the optimum MUF resin formulation.

REFERENCES

[1] R.Breyer, S.Hollis, S.G.Jural, & J.Joseph, Low mole ratio MUF resin, U.S Patent No.5,681,91 (1997).
[2] A.Bono, K.B.Yeo & N.J.Siambun, Jurnal Teknologi, No.38 (2003), p43-52.
[3] A.Bono, K.B.Yeo & N.J.Siambun, Borneo Sci.: A Jour. of Sci. & Tech., 11 (2002), p59-67.
[4] K.K.Tutin, Water-soluble sulfonated MF resins, U.S Patent No. 5,710,239 (1998).
[5] B.Gapud, M.Shoemake, & E.Searcy, MF resins modified with dicyandiamide & sorbitol for impregnation of substrates for post-formable decorative laminates, U.S Patent No.6,001,92 (1999).
[6] R.H.Myers & D.C.Montgomery, Response Surface Methodology, John Wiley & Sons, Inc., NY, 2002.
[7] N.J.Siambun, K.B. Yeo, M.Rajin and A.Bono, (2), International Conference on Chem.& Bioprocess Eng., Kota Kinabalu, 2003. p996-999
[8] L.Shih-Bin, G.Joel, P.Ball, G.Vaughn, International Waterborne, High solid, & powder coating Symp., New Orleans LA USA, 2000.
[9] I.Skeist, Handbook of Adhesives, Van Nostrand, New York, 1990.
[10] A.Pizzi, Wood Adhesives: Chem. & Tech., Marcel Dekker Inc., NY, 1989.
[11] S.H.Bernard, Adhesives Recent Development, Noyes Data Corporation, London, 1976.

NANO MATERIALS SYNTHESIS
AND APPLICATION

Studies in Surface Science and Catalysis, volume 159
Hyun-Ku Rhee, In-Sik Nam and Jong Moon Park (Editors)

Effect of the calcine condition on surface structure of titania nanocrystal photocatalyst

Kongkiat Suriye and Piyasan Praserthdam[*]

Center of Excellence on Catalysis and Catalytic Reaction Engineering,
Department of Chemical Engineering, Chulalongkorn University, Bangkok 10330 Thailand.

In this work, effects of atmosphere during calcination process on photoactivities for decomposition of ethylene were investigated. TiO_2 were prepared by sol-gel and then calcined under N_2 plus increasing amounts of O_2 at temperature 723 K. Conversion of ethylene increased with increasing surface defect. Increasing this defect occurred when amount of O_2 during calcination process increased. Surface defect of TiO_2 samples was determined by CO_2-TPD and ESR, while XRD, SEM, TEM and BET were used to characterize other physical properties of TiO_2 samples.

1. INTRODUCTION

Titania is a material of great interest. In case of photocatalysis, titania can be used in many applications such as water cleaning and delay of ripening of fruit because TiO_2 has an excellent chemical stability and the positive holes photogenerated on it have strong oxidation power [1]. Titania nanocrystal is usually prepared by sol-gel method [2]. It always exists structural defects on the surface and inside the titania particles [3]. Using electron spin resonance (ESR), photo generated electrons was found to be trapped at surface defect (oxygen vacancy site, Ti^{3+}) or in the bulk at Ti^{4+} sites and holes trap at lattice oxygen ions [4]. In this manner, surface defects are good for high photoactivity because they are used as an active site on which electron donor or acceptor is adsorbed. However, the bulk defect lowers the photoactivity because they provide sites for the recombination of the photogenerated electrons [5].

In this work, effect of atmosphere during calcination process on surface defect was investigated using photocatalytic activity for oxidation of ethylene. CO_2-TPD and ESR were used to determine this surface defect.

2. EXPERIMENTAL

Titania nanocrystals were synthesized via sol-gel process of titanium ethoxide according to the procedure described in [6]. The resulting materials were then dried and calcined at 723 K in flowing oxygen at concentration of 1%, 10%, 21%, 70% and 100% (v/v) in N_2, respectively, for 2 hours and the heating rate was at $10°C/min$. The resulting TiO_2 crystals were subjected to characterization with XRD, SEM, TEM and BET. The surface structure of titania such as Ti^{3+} and Ti^{4+} were monitored using CO_2-TPD and ESR. In former case, it was

---

[*]Piyasan.p@chula.ac.th

carried out using 1 g of a titania sample. Titania was dosed by 1% CO_2 in He for 1 h in liquid nitrogen and then desorped in a range of temperature from 123 to 253 K by level controlling. And the later, Electron spin resonance spectroscopy (ESR) was conducted in vacuum at temperature 77 K using JEOL, JES-RE2X Electron Spin Resonance Spectrometer.

The photocatalytic experiments were performed in a horizontal quartz tube which it have TiO_2. Illumination was provided by 500 W mercury lamps, located above the horizontal quartz tube. The reactant was 0.1% (v/v) ethylene in air. In case of Photo-Catalyst test, reactor effluent samples were taken at 30 min intervals and analyzed by GC. The composition of hydrocarbons in the feed and product stream was analyzed by a Shimadzu GC14B (VZ10) gas chromatograph equipped with a flame ionization detector. In all case, steady state was reached within 3 h.

3. CATALYST NOMENCLATURE

The nomenclature used for the catalyst samples in this study is following:
TiO_2(X): the calcined titania with X% of O_2 in feed during calcination.

4. RESULTS AND DISCUSSION

Nanocrystals titania was prepared by sol-gel method. X-ray diffraction result is shown in Figure 1, all samples were anatase phase. Based on Sherrer's equation, these samples had crystallite sizes about 7 nm. From XRD results, it indicated that titania samples showed the similar of crystallinity because the same ordering in the structure of titania particles make the same intensity of XRD peaks.

Table 1
Information of prepared titania.

| Sample | Surface area (m^2g^{-1}) | Crystallite size (nm) | |
		XRD[a]	TEM[b]
TiO_2(0)	116	7	~ 7
TiO_2(10)	113	7	~ 6
TiO_2(21)	101	7	~ 7
TiO_2(70)	99	7	~ 6
TiO_2(100)	99	7	~ 7

[a]Calculated from Sherrer's equation.
[b]Calculated from 50 particles of titania.

The measured BET surface areas of titania samples were in the range of 99-116 m^2/g. It was found that surface area of titania decreased (as shown in Table 1) with increasing %O_2 during calcinations process whereas the crystallite size was apparently constant.

Thermal desorption spectra of CO_2 from a titania surface are shown in figure 2. It revealed two desorption peaks at temperature ca. 175 and 200 K. As reported, surface of titania have two structures which is similar to the results found by Tracy et al. [7]. Based on their study, it was confirmed that one peak at ca. 170 K was attributed to CO_2 molecules bound to regular five-coordinate Ti^{4+} site considered as the perfected titania structure. The second peak at ca. 200 K considered as the CO_2 molecules bound to Ti^{3+} referred to the defected titania structure. Thus, based on TPD results, it was observed that CO_2 desorption peak areas at ca. 200 K apparently increased with increasing %O_2 during calcination process.

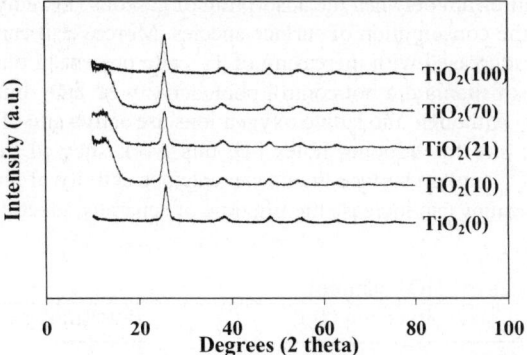

Fig. 1. XRD patterns of TiO_2 samples.

Based on ESR analysis of titania samples as shown in figure 3, all titania samples exhibited mainly one signal at g value of 1.996. According to the results of Nakaoka et al, this peak can be attributed to Ti^{3+} at surface. From figure 3, intensity of ESR spectra per surface area of all titania samples which is corresponding to the amounts of Ti^{3+} present in titania surface. The presence of Ti^{3+} on titania surface increased with increasing %O_2 during calcination process as also seen from CO_2-TPD. Although, oxygen vacancies can be easily formed by reduction of TiO_2 in high temperature [8]. However, this results showed that the increased of Ti^{3+} amount did not result from the reduction of Ti^{4+}. This result is in good agreement with that obtained by Zhao et al. [9]. They proposed that the removal of the surface hydroxyl account for the formation of the new oxygen vacancies site. During calcination process, organic and inorganic residues were oxidized and then it released energy. This energy increased with increasing % O_2 during calcination process. When this energy increased surface hydroxyl can remove easily and then increasing of oxygen vacancies can occur.

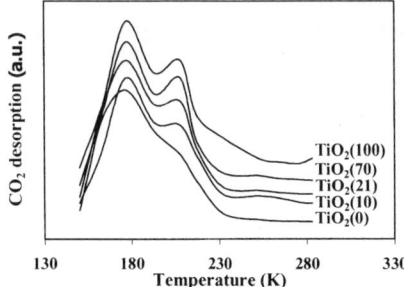

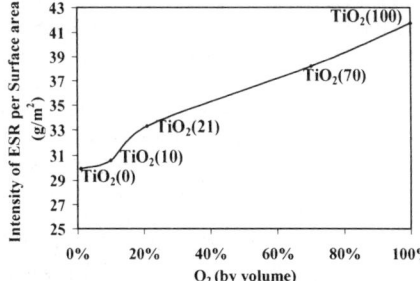

Fig. 2. Thermal desorption spectra for CO_2 Adsorbed on titania samples

Fig. 3. Intensity of ESR spectra per surface area at different %O_2 in N_2.

720

In our experiment, photocatalytic decomposition of ethylene was utilized to probe the surface defect. Photocatalytic properties of all titania samples are shown in table 2. From these results, conversions of ethylene at 5 min and 3 hr were apparently constant (not different in order) due to the equilibrium between the adsorption of gaseous (i.e. ethylene and/or O_2) on the titania surface and the consumption of surface species. Moreover it can be concluded that photoactivity of titania increased with increasing of Ti^{3+} site present in titania surface. It was found that surface area of titania did not control photoactivity of TiO_2, but it was the surface defect in titania surface. Although, the lattice oxygen ions are active site of this photocatalytic reaction since it is the site for trapping holes [4], this work showed that the presence of oxygen vacancy site (Ti^{3+} site) on surface titania can enhance activity of photocatalyst, too. It revealed that oxygen vacancy can increase the life time of separated electron-hole pairs.

Table 2
Photocatalytic properties of TiO_2 samples.

Sample	C_2H_4 conversion (%)[a]		Rate (mol/g cat.h)[b]	
	Initial[c]	SS[d]	Initial	SS
$TiO_2(0)$	41	40	75	74
$TiO_2(10)$	46	43	85	79
$TiO_2(21)$	47	48	86	88
$TiO_2(70)$	59	60	109	110
$TiO_2(100)$	62	60	114	110

[a]Photocatalytic reaction was carried out at 40-55 °C, 1 bar, and 0.1% ethylene in air.
[b]Error 5%.
[c]After 5 min of reaction.
[d]After 3 h of reaction.

ACKNOWLEDGMENT

The authors would like to thank the Thailand Research Fund (TRF) and TJTTP-JBIC for their financial support of this project.

REFERENCES
[1] F. Akira, H. Kazuhito and W. Toshiya, TiO_2 Photocatalysis fundamentals and applications, 46.
[2] S. Sivakumar, P. Krishna, P. Mukundan and K.G.K. Warrier, Mater. Lett., 57 (2002) 330.
[3] T. Torimoto, R.J. Fox and M.A. Fox, J. Electrochem. Soc., 143 (1996) 3712.
[4] Y. Nakaoka and Y. Nosaka, J. Photochem. Photobiol. A: Chem., 110 (1997) 299.
[5] K.Y. Jung and S.B. Park, J. Photochem. Photobiol. A: Chem., 127 (1999) 117.
[6] C.C. Wang and J.Y. Ying, Chem. Mater., 11 (1999) 3113.
[7] T.L. Thompson, O. Diwald and J.T. Yates, J. Phys. Chem., B 107 (2003) 11700.
[8] J. Weidmann, Th. Dittrich, E. Konstantinova, I. Laucrmann, I. Uhlendorf and F. Koch, Solar Ener. Mater. and Solar Cells, 56 (1999) 153.
[9] Q. Zhao and X. Wang, T. Cai, Appl. Surface Sci., 225 (2004) 7.

Studies in Surface Science and Catalysis, volume 159
Hyun-Ku Rhee, In-Sik Nam and Jong Moon Park (Editors)

Effect of oxidative treatments on catalytic property of carbon nanofiber composite

Ping Li*, Jie Wu, Tie-Jun Zhao, Jing-Hong Zhou, and Wei-Kang Yuan

UNILAB of Research Center of Chemical Reaction Engineering, East China University of Science and Technology, Shanghai 200237, People's Republic of China

1. INTRODUCTION

Carbon nanofibers (CNFs) have shown promise for many potential applications such as polymer reinforcements, energy storage materials, and catalysts or catalyst supports. Main advantages of CNFs used as catalytic materials are the possibilities of tailoring their microstructure by proper selection of a preparation method, and controlling their surface chemistry via surface modification. Furthermore, the mesoporous macrostructure of CNFs can be expected to reduce the resistance of inner pore diffusion a great deal during the reaction, and thus to relieve consecutive reactions such as coking. However, the CNFs as grown generally through catalytic decomposition of carbon containing gases are in the form of fine powders that are very difficult to be applied in catalytic reactors such as fixed bed reactors, thereby hindering their use for industrial purposes [1]. Nevertheless, only limited literature can be found in shaping of CNFs so far [2].

In this paper, a novel CNF composite of desired shape and size, as well as of sufficient mechanical strength, is proposed and synthesized deliberately by a shaping process. The shaping procedure involves steps of mixing the CNFs with a polymer binder, followed by press molding and thermal treating at high temperatures in an inert atmosphere. The polymer binder is carbonized during thermal treating, providing a carbon network to bind the isolated fibers altogether. The CNF composite is further activated through gas (air) or liquid (HNO3) phase oxidative treatment so as to modify its physicochemical properties. This paper focuses on the effect of oxidative treatments on the physicochemical properties of the CNF composite, especially its catalytic property. The catalytic property of the CNF composite is investigated for oxidative dehydrogenation of ethylbenzene to styrene (ODE) using O2 as the oxidant. The discussed ODE is supposed to be an alternative process for the conventional production of styrene that could be achieved with high efficiency over carbon catalysts [3]. The textural property and surface chemistry of the CNF composite are determined with N2 physisorption measurement and temperature-programmed desorption (TPD), respectively. The relationship

---
* Corresponding authors. Tel.: +86-21-64252169; fax: +86-21-64253528.
E-mail addresses: unilab605@ecust.edu.cn (P. Li), ychdai@ecust.edu.cn (Y.-Ch. Dai).

between the catalytic property of the CNF composite and its textural property and surface chemistry is discussed. A commercially available coal-based activated carbon catalyst is selected for comparison.

2. EXPERIMENTAL

CNFs were synthesized by catalytic decomposition of CO/H_2 mixture at 600 °C for 16 h over 20 wt% Fe/γ-Al_2O_3 catalyst which was placed inside a horizontal quartz reactor beforehand. Close observation using high resolution transmission electron microscopy of individual fibers of synthesized CNFs indicates a fishbone microstructure with a hollow core throughout the fiber. The CNF powders as grown were then mixed with the polyglycol-dissolved phenolic resin. The CNF mixture was molded using a pellet compressing machine to produce pellets of designed size and configuration. Followed by curing of the pellets at 160 °C, carbonization of the polymer binder in pellets, via which isolated CNFs were agglomerated, was carried out at 700 °C in a flowing Ar stream for 6 h. The prepared CNF composite was further subjected to gas or liquid phase activation using air or HNO_3 as the oxidant, i.e. heated in stagnant air at 400 °C for 10 h, or boiled in a concentrated HNO_3 solution for 0.5 h. Catalytic behaviors of various carbon materials for the ODE reaction were investigated in a fixed bed flow reactor at a reaction temperature of 380 °C. Gaseous reactants were composed of 2 vol% O_2 and 2 vol% ethylbenzene, balanced with Ar. Each catalyst weight was 1 g, with the corresponding feed gas flow kept at 100 mlmin-1. Reaction organic products were detected by a HP 4890 gas chromatograph, and CO and CO_2 in the gas were monitored by an ABB Questor GP process mass spectrometer.

3. RESULTS AND DISCUSSION

3.1. Textural characterization

The results of textural characterization of different carbon materials tested in the present study are listed in Table 1. The carbon materials include the CNF powder, the CNF composite, the composite activated with air or HNO_3, and the activated carbon. It is shown that the BET surface area and total pore volume of the CNFs are obviously decreased after shaping. Further boiling in concentrated HNO_3 solution only leads to a slight change in the specific surface area of the CNF composite. However, the activation effect using air on the CNF composite is pronounced, inducing an increment in the specific surface area and a renewal of the total pore volume. The activated carbon is full of micropores, leaving a high specific surface area.

3.2. Surface functional groups

The surface oxygen-containing groups on carbon materials can be decomposed to CO and CO_2 at different temperatures. Table 2 gives the total amounts of CO and CO_2 desorbed from the surface of each carbon material during temperature programmed desorption (TPD) terminated at 900 °C. It can be found that the amounts of the surface oxygen-containing groups on the composites after oxidative treatments are extremely larger than those on the CNFs and on the untreated composite. Liquid phase oxidation of the composite brings about the greatest amount of surface groups, regardless of the lowest CO/CO2 ratio. The CO-forming surface groups are mainly responsible for the basic character of carbon materials, while CO2-forming surface groups for acidic character [4]. It can be deduced that the HNO3-treated composite is more acidic than the untreated composite, however, the air-treated composite is more basic than the untreated one, consulting the CO/CO2 ratios. The activated carbon presents the least amount of surface oxygen-containing groups normalized per square meter.

Table 1

Textural properties of carbon materials

Sample	BET surface area (m2g-1)	Pore volume (cm3g-1)	Micropore volume (cm3g-1)	Average pore diameter (nm)	Pellet density (gcm-3)
CNFs	211.9	0.405	0.012	10.5	-
Composite	138.6	0.265	0.032	9.2	1.111
Composite-air	234.6	0.364	0.036	8.0	0.975
Composite-HNO3	154.6	0.250	0.038	8.9	1.060
Activated carbon	1067.5	0.515	0.356	2.5	0.875

Table 2

Amounts of CO and CO2 desorbed from carbon materials in TPD

Sample	Amount desorbed (μmolg-1)			Amount desorbed (μmolm-2)			CO/CO2
	CO	CO2	total O	CO	CO2	total O	
CNFs	270.5	35.1	340.7	1.28	0.17	1.62	7.71
Composite	285.4	106.6	498.6	2.06	0.77	3.60	2.68
Composite-air	1987.0	380.5	2748.0	8.47	1.62	11.71	5.22
Composite-HNO3	1597.4	1237.0	4071.4	10.33	8.00	26.33	1.29
Activated carbon	535.1	141.3	817.7	0.50	0.13	0.76	3.79

3.3. Catalytic performance of CNF composites for ODE reaction

As illustrated in Fig. 1, the activated carbon displays the highest conversion and selectivity among all the catalysts during the initial reaction period, however, its catalytic activity continues to decrease during the reaction, which is probably caused by coke deposition in the micropores. By contrast, the reaction over the CNF composites treated in air and HNO3 can reach a pseudo-steady state after about 200 min. Similiar transient state is also observed on the CNFs and the untreated composite. Table 3 collects the kinetic results after 300 min on stream over catalysts tested for the ODE, in which the activity is referred to the BET surface area. The air-treated composite gives the highest conversion and styrene selectivity at steady state.

It has been suggested that the active sites for the ODE reaction over carbon catalysts are the basic surface groups [4]. As the air-treated composite possesses the greatest amount of CO-forming surface groups based on the sample weight, while the HNO3-treated composite

724

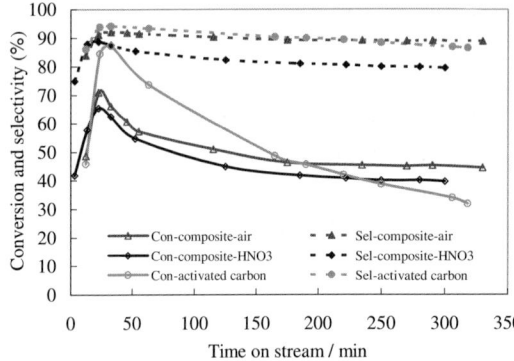

Fig. 1. Conversions and styrene selectivities of the ODE over CNF composites after air and HNO₃ treatments and activated carbon as a function of time

Table 3
Kinetic results obtained on different carbon materials after 300 min on stream lowest for the activated carbon.

Sample	Conversion (%)	Selectivity of styrene (%)	Selectivity of CO2 (%)	Selectivity of CO (%)	Activity *103(μmolm-2s-1)
CNFs	39.5	87.4	3.7	1.0	2.77
Composite	27.9	82.2	2.9	1.3	2.99
Composite-air	45.2	89.0	3.6	1.1	2.87
Composite-HNO3	39.7	79.5	6.1	1.8	3.82
Activated carbon	34.0	86.8	3.0	1.2	0.47

contains the most CO-forming groups based on the sample area and the activated carbon contains the least (see Table 2), the reaction results obtained on these catalysts are understandable. With respect to the by-product selectivities, it is notable that the absolutely high CO_2 selectivity is given on the HNO3-treated composite, implying a relationship between the CO_2 selectivity and the CO_2-forming surface groups. In conclusion, the catalytic property of the CNF composite could be adapted for the expected reactions by means of oxidative treatments.

ACKNOWLEDGEMENTS

The authors acknowledge the support from the NSFC (No. 20376021), the NSFC/China-Petro. major project (No.20490200), and the special fund of STCSM (No.036505010).

REFERENCES

[1] T.G. Ros, A.J. van Dillen, J.W. Geus and D.C. Koningsberger, Chem. Eur. J., 8 (2002) 1151.
[2] M.J. Ledoux, R. Vieira, C. Pham-Huu and N. Keller, J. Catal., 216 (2003) 333.
[3] N. Keller, N.I. Maksimova, V.V. Roddatis, M. Schur, G. Mestl, Y.V. Butenko, V.L. Kuznetsov and R. Schlogl, Angew. Chem. Int. Ed., 41 (2002) 1885.
[4] M.F.R. Pereira, J.L. Figueiredo, J.J.M. Órfão, P. Serp, P. Kalck and Y. Kihn, Carbon, 42 (2004) 2807.

Studies in Surface Science and Catalysis, volume 159
Hyun-Ku Rhee, In-Sik Nam and Jong Moon Park (Editors)

725

Screening of metal oxide catalysts for carbon nanotubes and hydrogen production via catalytic decomposition of methane

S.H.S. Zein[*], A.R. Mohamed and S.P. Chai

School of Chemical Engineering, Engineering Campus, Universiti Sains Malaysia, Seri Ampangan, 14300 Nibong Tebal, Seberang Perai Selatan, Pulau Pinang, Malaysia

A number of catalysts prepared from transition metals such as copper (Cu), iron (Fe), nickel (Ni), cobalt (Co) and manganese (Mn) on TiO_2 support were tested for the decomposition of methane into hydrogen and carbon. These catalysts were used in the experiments without any pretreatment. The experimental results show that the activities of the metal-TiO_2 catalysts decreased in the order of NiO/TiO_2 > CoO/TiO_2 > MnO_x/TiO_2 ≈ FeO/TiO_2 ≈ CuO/TiO_2. NiO/TiO_2 catalyst exhibited extremely high initial activity in the decomposition of methane. The optimum NiO doping on TiO_2 for the decomposition of methane were obtained at 20mol% NiO. The effective promoters for the catalyst was investigated using 15mol%M/20mol%NiO/TiO_2 catalysts (where M = MnO_x, FeO, CoO and CuO). 15mol%MnO_x/20mol%NiO/TiO_2 was found to be an effective bimetallic catalyst for the catalytic decomposition of methane into hydrogen and carbon, giving higher catalytic activity, attractive carbon nanotube formed as well as longer catalytic lifetime.

1. INTRODUCTION

Carbon nanotubes are one of the most innovative material technologies of the twenty first century, because of their many desirable material properties [1-5]. For the synthesis of carbon nanotubes, several methods have been developed (mainly arc discharge, laser ablation, and chemical vapor deposition). The development of a reliable source of large quantities of carbon nanotubes is dependent on better production methods. The abundance of natural gas, which contains primarily methane, can be better utilized by increasing its use as a source of chemicals in place of its predominant use as a fuel. The decomposition of methane to hydrogen and carbon nanotube is of current interest as it is an alternative route to the production of carbon nanotubes from natural gas. The decomposition of methane at higher temperature attracts considerable interest today because the conversion of methane at this condition is higher [6-8]. However, at higher temperature, the catalyst deactivates very fast due to the formation of encapsulating type carbon on the catalyst. Thus, in order to put this process into practice, a catalyst with high activity without any treatment prior to its use becomes necessary.

2. EXPERIMENTAL PROCEDURE

All the catalysts used in this work were prepared by conventional impregnation method. The selected dopant concentrations were actually relative to the molar quantity of the TiO_2 support.

---

[*] E-mail: chhussein@eng.usm.my

The desired amounts of the transition metal nitrates were dissolved in deionized water, and then the solution was impregnated in the TiO_2 powder. The resulting paste was dried in an oven and calcined in a ceramic crucible at 900 °C. The catalysts were then sieved to a size of 400-500 μm. The activity tests for the developed catalysts were carried out at atmospheric pressure in a stainless steel fixed-bed reactor (length: 600 mm and diameter: 10.92 mm) at 998 K and gas hourly space velocity of 2700 h^{-1}. High purity methane (99.999% purity) was mixed with argon (99.999% purity) before entering the reactor. The product gases were analyzed using an on-line gas chromatograph (Hewlett-Packard Series 6890, USA). The fresh catalysts were investigated from X-ray diffraction (XRD) patterns measured by Siemen D-5000 diffractometer, using Cu-KR radiation at room temperature. The deposited carbons were analyzed using transmission electron microscope (Philips TEM CM12).

3. RESULTS AND DISCUSSION

3.1. Screening of catalyst components

Table 1 shows the effect of catalyst supports in methane conversions and hydrogen yield for the decomposition of methane at 998 K and gas hourly space velocity (GHSV) of 2700 h^{-1}. The tested supports were TiO_2, SiO_2, Al_2O_3, and MgO. These supports were chosen because of their good activity towards methane activation. In Table 1, TiO_2 showed the highest methane conversion among the other tested supports. As a result, TiO_2 was chosen as a catalyst support in this study. Table 2 summarizes the catalytic activity of metal oxide-TiO_2 catalysts in the decomposition of methane and the results show that CoO/TiO_2 and NiO/TiO_2 catalysts were active, whereas MnO_x/TiO_2, FeO/TiO_2 and CuO/TiO_2 catalysts did not cause any significant in decomposition of methane. The NiO/TiO_2 catalysts exhibited high initial activity in the methane decomposition at 998 K with the methane conversion of 60%. Therefore, this study was focused on the direct decomposition of methane over NiO/TiO_2 catalyst.

Table 1
The effect of catalyst supports on methane conversions and hydrogen yield in the methane decomposition at 998 K and GHSV of 2700 h^{-1} at steady state.

Catalyst support	Conv. (%)	H_2 Yield (%)
MgO	1.2	1.2
TiO_2	7	7
Al_2O_3	4.3	4.3
SiO_2	0.4	0.4

Table 2
The effect of metal oxide-TiO_2 catalysts on methane conversions and hydrogen yield in the methane decomposition at 998 K and GHSV of 2700 h^{-1} at steady state.

Catalyst	Conv. (%)	H_2 Yield (%)
15 mol%MnO_x/TiO_2	<1	<1
15 mol%FeO/TiO_2	<1	<1
15 mol%CoO/TiO_2	11	11
15 mol%NiO/TiO_2	60	60
15 mol%CuO/TiO_2	<1	<1

3.2. The effect of NiO loading on TiO$_2$ support

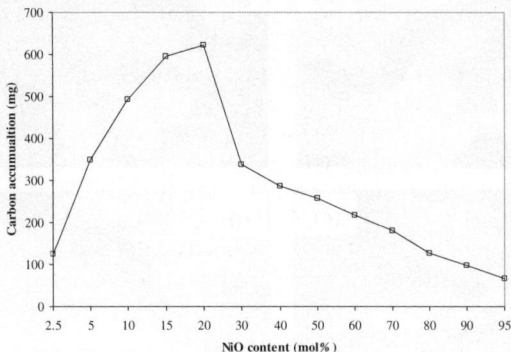

Fig. 1. The relationship between the carbon accumulation and the NiO doping on TiO$_2$ support in the methane decomposition at 998 K and GHSV of 2700 h^{-1} at steady state.

NiO concentration was varied from 2.5 to 95 mol% on TiO$_2$ support as shown in Fig. 1. The optimum NiO doping on TiO$_2$ for decomposition of methane obtained was at 20 mol%NiO. Further increase the NiO content on TiO$_2$ support leads to lower the carbon accumulation. NiO/TiO$_2$ catalysts with different NiO doping were investigated using XRD as to reveal the reason of lower carbon accumulation being observed over high-loaded NiO catalysts. The XRD results obtained indicated that as the amount of NiO loaded increased, the number of Ni$^{\circ}$ sites were also increased and sintered to form larger NiO particles which lead to a catalyst deactivation.

3.3 The effect of promoter on 20 mol% NiO/TiO$_2$ catalyst

Table 3 shows the performance of the promoted-catalysts for the decomposition of methane to hydrogen at 5, 60, 120 and 180 min of time on stream. The results in Table 3 revealed that the activity of the parent catalyst and MnO$_x$-doped catalyst remained almost constant until 120 min of time on stream. The activity of the other promoted-catalysts, on the other hand, decreased with an increase in the time on stream. The data for the CoO-doped catalyst and 20 mol%NiO/TiO$_2$ could not be recorded at 120 min and 180 min, respectively because of the pressure build-up in the reactor. This finding indicates that adding MnO$_x$ enhances the stability and the resistibility of the NiO/TiO$_2$ catalyst towards its deactivation.

Table 3
The performance of the catalysts doped with transition metals on 20 mol% NiO/TiO$_2$ catalyst for hydrogen production at 998 K and GHSV of 2700 h^{-1} at steady state.

Catalyst	H$_2$ Concentration (%)			
	5 min	60 min	120 min	180 min
20 mol% NiO/TiO$_2$	61	62	61	-
15 mol% CuO/20 mol% NiO/TiO$_2$	61	67	46	34
15 mol% MnO$_x$/20 mol% NiO/TiO$_2$	59	58	56	48
15 mol% FeO/20 mol% NiO/TiO$_2$	57	50	44	27
15 mol% CoO/20 mol% NiO/TiO$_2$	66	59	-	-

728

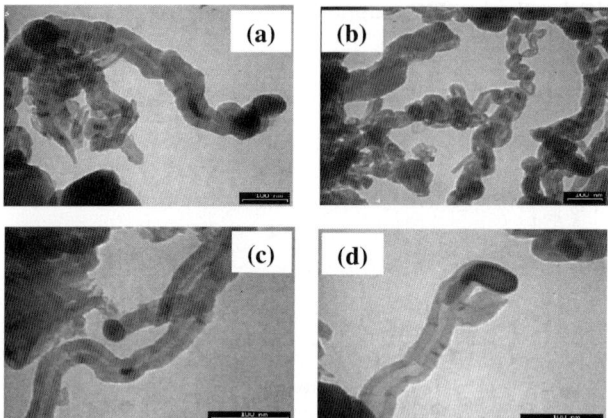

Fig. 2. TEM images of carbons deposited by methane decomposition at 998K and GHSV of 2700 h^{-1} on (a) 15 mol%CoO/20mol%NiO/TiO$_2$, (b) 15mol%FeO/20mol%NiO/TiO$_2$, (c) 15mol%CuO/20mol% NiO/TiO$_2$ and (d) 15mol%MnO$_x$/20mol%NiO/TiO$_2$.

The deposited carbons on promoted-20mol% NiO/TiO$_2$ catalysts were further studied using TEM. The result obtained elucidates that doping second metal influences the carbon morphology remarkably. Fig. 2a shows that the morphology of the carbon formed on 15mol%CoO/20mol%NiO/TiO$_2$ catalyst was an encapsulating type. Chain-like type carbon (Fig. 2b) was observed on 15 mol%FeO/ 20mol%/NiO/TiO$_2$ catalyst. The carbon formed on 15 mol%CuO/20mol%NiO/TiO$_2$ catalyst (Fig. 2c) was a mixed type carbon which exhibited poor-turbostratic wall. Well-crystalline carbon nanotube with a hollow structure was observed on 15 mol%MnO$_x$/20mol%NiO/TiO$_2$ catalyst (Fig. 2d). The formation of carbon nanotube with well-crystalline structure and its growth in a proper orientation prevent the active catalyst particle, found at the tip of the nanotube, from being encapsulated by graphitic layers. This was the reason why the 15 mol%MnO$_x$/20mol%NiO/TiO$_2$ catalyst maintained its activity in the decomposition of methane.

ACKNOWLEDGEMENT

The authors would like to acknowledge the financial support provided by the Academy of Sciences Malaysia under Scientific Advancement Grant Allocation (SAGA) (Project No. 6053001) and Universiti Sains Malaysia under USM short-term grant (Project No. 6035146).

REFERENCES

[1] L.B. Avdeeva, D.I. Kochubey and S.K. Shaikhutdinov, Appl. Catal. A, 177 (1999) 43.
[2] M.A. Ermakova, D.Y. Ermakov, G.G. Kuvshinov and L.M. Plyasova, J. Catal., 187 (1999) 77.
[3] M.A. Ermakova, D.Y. Ermakov and G.G. Kuvshinov, Appl. Catal. A, 201 (2000) 61.
[4] R. Aiello, J.E. Fiscus, H.-C. Zur Loye and M.D. Amiridis, Appl. Catal.A, 192 (2000) 227.
[5] K. Otsuka, H. Ogihara and S.Takenaka, Carbon, 41 (2003) 223.
[6] T.V. Reshetenko, L.B. Avdeeva, Z.R. Ismagilov, A.L. Chuvilin and V.A. Ushakov, Appl. Catal. A, 247 (2003) 51.
[7] S.H.S. Zein, A.R. Mohamed and P.S.T. Sai, P.S.T, Ind. Eng. Chem. Res., 43 (2004) 4864.
[8] S.H.S. Zein and A.R. Mohamed, Energ. Fuel., 18(5) (2004) 1336.

Studies in Surface Science and Catalysis, volume 159
Hyun-Ku Rhee, In-Sik Nam and Jong Moon Park (Editors)
© 2006 Elsevier B.V. All rights reserved

Synthesis of crystalline quantum dots in AOT-stabilized water-in-CO$_2$ microemulsions

Yutaka Ikushima[a], Juncheng Liua[a, b] and Poovathinthodiyil Raveendran[a]

[a]Research Center for Compact Chemical Process,
National Institute of Advanced Industrial Science and Technology, 4-2-1 Nigatake,
Miyagino-ku, Sendai 983-8551, Japan

[b]Department of Chemical Engineering, Auburn University, Auburn, Alabama 36849, USA

Utilization of supercritical carbon dioxide for the synthesis and processing of nanomaterials is an important issue in green nanotechnology. Although researchers have previously used water-in-CO$_2$ (w/c) microemulsions for the synthesis of metal and semiconductor quantum dots, the high cost of the specially designed fluorinated surfactants make this non-viable industrially. In this work, we demonstrate the synthesis of Ag, AgI, and Ag$_2$S nanoparticles with high crystallinity in the w/c microemulsions prepared using the commercially available AOT surfactant, in the presence of perfluoropentanol as co-solvent.

1. INTRODUCTION

Recent efforts to eliminate the environmental hazards associated with chemical processes (green chemistry) are of high relevance to the still emerging nanoscience. One of the important issues in green chemistry is the use of environmentally friendly alternatives to the conventional volatile organic solvents. Supercritical CO$_2$ (scCO$_2$) has been regarded as a "green" solvent due to its non-toxicity, abundance, low cost, and tunability of solvent parameters. Thus the synthesis and processing of nanomaterials in supercritical CO$_2$ is of importance. In particular, the use of water-in-CO$_2$ (w/c) reverse microemulsions (RMs) has attracted much interest in recent years as nanoscopic solution templates for compartmentally growing metal and semiconductor nanoparticles [1]. The reverse micellar organization provides spatially confined micro-aqueous templates for the synthesis of the nanoparticles from their water soluble precursor. Also, once the nanoparticles are formed, the surfactant interfacial monolayer helps to passivate the surface and quench the growth of the nanoparticles which would otherwise aggregate due to the high surface energy of these systems.

In addition to the environmentally benign attributes and the easily tunable solvent properties, other important characteristics such as low interfacial tension, excellent wetting behavior, and high diffusion coefficients also make scCO$_2$ a superior medium for the synthesis of nanoscale materials [2]. Previous works on w/c RMs showed that conventional hydrocarbon surfactants such as AOT do not form RMs in scCO$_2$ [3]; AOT is completely insoluble in CO$_2$ due to the poor miscibility of the alkyl chains with CO$_2$, restricting the utilization of this medium. Recently, we had demonstrated that the commonly used surfactant,

AOT, could form w/c RMs in the presence of the commercially available perfluoropentanol (F-pentanol) as a co-surfactant, and the RMs formed could provide polar micro-aqueous for highly ionic chemicals[4,5]. Herein, we present the synthesis of crystalline nanoparticles of Ag, AgI, and Ag_2S (which have potential application as photoelectric and thermoelectric devices) in the polar micro-aqueous domains of the w/c RMs stabilized by the AOT/F-pentanol (AOTF) surfactant/co-solvent combination, suggesting the possibility of the commercial utilization of $scCO_2$ in nanomaterials synthesis.

2. EXPERIMENT

SCF grade CO2 (99.999% purity) supplied by the Nippon Sanso Co. Ltd. AOT (99%, Sigma Ultra, MW 444.56), purchased from Sigma Chemical Co. Ltd. (USA), was vacuum dried at 60 °C for 24 hours prior to use. F-pentanol, dehydrated ethanol (Tokyo Kasei Kogyo Co. Ltd.), KI, $AgNO_3$, doubly distilled and de-ionized water prepared by ultra-filtration, reverse osmosis, deionization & distillation (Wako Pure Chem. Ind. Ltd.), $NaBH(OAc)_3$ (which was used as reducing agent, Aldrich) were used as purchased. The w/c RMs were prepared by adding 0.016M AOT and 0.24M F-pentanol to the UV cell, followed by an aliquot of 0.10M silver nitrate aqueous solution to achieve the desired amount of water. $ScCO_2$ was pumped into the UV cell by an HPLC pump and thermal equilibrium was attained. The CO_2 pressure was raised with continuous stirring until a single-phase, optically transparent w/c RMs were formed. For the synthesis of Ag nanoparticles, freshly prepared $NaBH(OAc)_3$ in ethanol solution was slowly injected into the UV cell until a final pressure of 34.50 MPa was achieved. Stirring was stopped after the addition of the reducing agent so as to allow spontaneous nucleation and growth of silver nanoparticles. The estimated concentration of the silver nitrate and NaBH (OAc) $_3$ in $scCO_2$ microemulsions was 30 mM and 15mM respectively, and the fluid phase contained 20 μ L ethanol. The high-pressure UV cell consists of a stainless-steel cell with two sapphire windows. It has a volume of 2.2 cm^3 and withstands a maximum pressure of 45 MPa. The mixtures in the cell were stirred by a magnetic stirrer. CO_2 pressure was controlled by a back-pressure regulator (880-81, JASCO Co.) to an accuracy of 0.01 MPa in the pressure range of 0~50 MPa. All the measurements were carried out at 38.0 (±0.1) °C. The procedures for synthesizing the silver iodide and silver sulfide nanoparticles are similar.

AOTF w/c RMs bearing the silver, silver iodide and silver sulfide nanoparticles were depressurized slowly and the nanoparticles in the cell were collected and re-dispersed in ethanol. Finally, the sample grids for the TEM (FEI TECNAI G^2) measurements were prepared by placing a drop of ethanolic dispersion of nanoparticles on the copper grid. The morphology and size distribution of the silver, silver iodide, and silver sulfide nanoparticles were determined by TEM at an operation voltage of 200kV. The crystallinity of the silver, silver iodide, and silver sulfide nanoparticles was studied by electron diffraction techniques.

3. RESULTS AND DISCUSSION

The Ag nanoparticles were synthesized by the reduction of Ag^+ ions in the water core of the AOT/F-pentanol (AOTF) w/c RMs using $NaBH(OAc)_3$. The acidic nature of the water core due to the formation of carbonic acid makes conventional reagents such as $NaBH_4$ and hydrazine unsuitable as they will react with CO_2 under the experimental conditions. The W values used in the experiment (corrected for the water dissolved in the $scCO_2$ continuous phase) for synthesizing silver, silver iodide, silver sulfide nanocrystals are 4.7, 5.6, and 5.9 nm, respectively. Although the water loading ($W= [H_2O]/[surfactant]$)) in the microemulsion system is not high under, these nanometer-sized micro-aqueous domains are particularly attractive for the synthesis of metal and semiconductor quantum dots having sizes below 10 nm.

The silver nanoparticles synthesized were redispersed in the ethanol (after decreasing the CO_2 pressure), resulting in single-phase yellowish-green solution. The morphology and size distribution of the silver nanoparticles were investigated by using transmission electron microscopy (TEM). Figure 1 shows the high-resolution TEM image of the silver nanoparticles. The characteristic spherical silver nanoparticles were observed with a relatively narrow particle size distribution (3~11 nm range). The TEM image of the particles indicates the interlinking between individual nanoparticles plausibly through the interdigitation of the surfactant tails of adjacent micelles. A histogram of silver particle size distribution corresponding to the micrograph of Figure 1 is presented in Figure 2. Particle sizes were measured using Scion Image software. The mean particle diameter observed is 6.0 nm (standard deviation, $\sigma = 1.3$ nm) and more than 95 % of the nanoparticles are in the size range from 4 to 8 nm, showing that the aqueous cores of the w/c RMs can act as effective templates for the synthesis of relatively monodisperse metal quantum dots and the surfactant interfacial monolayer act as passivation contacts for the stabilization of the silver nanoparticles formed inside these templates

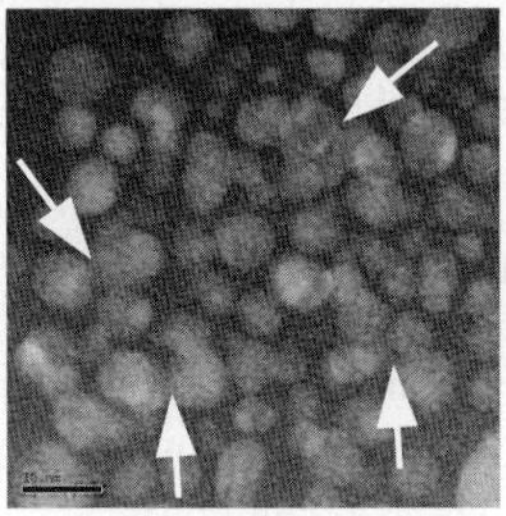

Fig. 1. Representative TEM images of Ag nanoparticles synthesized by the AOT w/c RMs with W = 4.7 at 38.0°C and 34.50MPa. (Scale bar: 10 nm). The arrows indicate the self-assembly of the adjacent nanoparticles through the surfactant tails.[6]

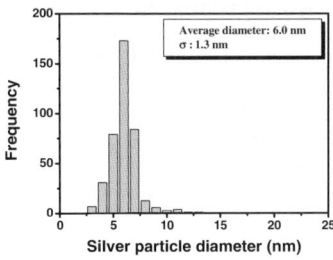

Fig. 2. Histogram[6] showing the size distribution of the silver nanoparticles corresponding to Fig. 1

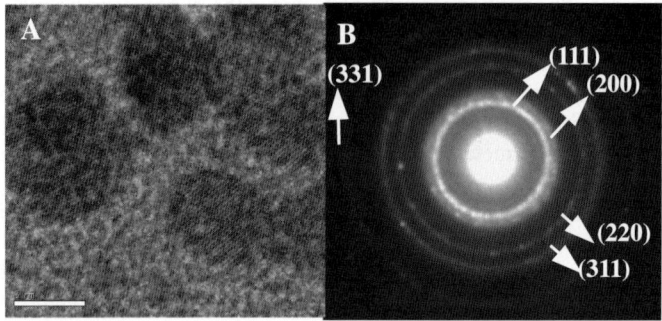

Fig. 3. (A) High resolution TEM image (scale bar: 5 nm) and (B) the ED pattern of the Ag (0) nanocrystals.[6]

Furthermore, the multiple lattice fringes in the high-resolution TEM image (Figure 3 A) suggest high degree of crystallinity of the particles. In order to clarify the crystal structure of the silver nanoparticles prepared, the electron diffraction (ED) technique was used in the study. Figure 3 B shows the ED pattern of the silver nanoparticles and the diffraction rings well correspond to the crystalline planes of face-centered-cubic (fcc) structured silver nanocrystals.

We have further demonstrated the synthesis and stabilization of relatively monodisperse silver iodide and silver sulfide nanocrystals by the chemical reactions in the water core of the AOT/F-pentanol (AOTF) w/c RMs at moderate temperature and CO_2 pressures. TEM results confirm the formation of relatively monodisperse silver iodide and silver sulfide nanocrystals with an average diameter of 5.7 nm ($\sigma = 1.4$ nm) and 5.9 nm ($\sigma = 1.65$nm), respectively. Although the current experiments are restricted to the synthesis of metal and semiconductor nanoparticles, the approach can be expanded for a range of inorganic reactions and organic, especially for some important catalytic reaction promoted by in situ prepared highly effective nano-sized metal catalyst, using $scCO_2$ as a bulk phase.

RERERENCES
[1] M. Ji, C.M. Wai and J.L. Fulton, J. Am. Chem. Soc., 121 (1999) 2361.
[2] K.P. Johnston and P.S. Shah, Science, 303 (2004) 482.
[3] T.A. Hoefling and E.J. Beckman, J. Phys. Chem., 95 (1991) 7127.
[4] J.C. Liu, Y. Ikushima and J. Supercrit. Fluids, 33 (2005) 121.
[5] J.C. Liu, P. Raveendran, Z. Shervani and Y. Ikushima, Chem. Commun., (2004) 2582.
[6] J.C. Liu, P. Raveendran, Z. Shervani and Y. Ikushima, Chem. Euro. J., 11 (2005) 1854.

Studies in Surface Science and Catalysis, volume 159
Hyun-Ku Rhee, In-Sik Nam and Jong Moon Park (Editors)
© 2006 Elsevier B.V. All rights reserved

Gas phase production of fine SiO$_2$ particles from tetramethoxysilane

T. Kojima[a], K. Tachi[a], S. Komiya[a], T. Uchiyama[a], S. Kato[a], H. Shibuya[b] and S. Uemiya[c]

[a]Department of Applied Chemistry, Seikei University, Musashino, Tokyo, 180-8633 Japan

[b]JGC Company, 2-3-1 Minatomirai, Nishi-ku, Yokohama, Kanagawa, 220-6001 Japan

[c]Department of Applied Chemistry, Gifu University, 1-1 Yanagido, Gifu, 501-1193 Japan

1. INTRODUCTION

Recently, fine silica powder with diameter of micron orders has been noted as functional material for filler of IC, raw material for silica glass and so on with the growth of semiconductor and communication industries [1].   The manufacturing processes of the fine silica powder are roughly classified into two methods; liquid phase and vapor phase syntheses.   The representative liquid phase method is the sol/gel method of hydrolysis of alkoxysilanes [2, 3], which enables the production of pure, fine, and uniform silica powder, however its continuous operation is difficult [4].   The vapor phase methods include productions of silica fines from silicon tetrachloride by combustion method [5] and from molten silica by spraying.   However, there has been reported little knowledge on the gas phase synthesis of silica fines.

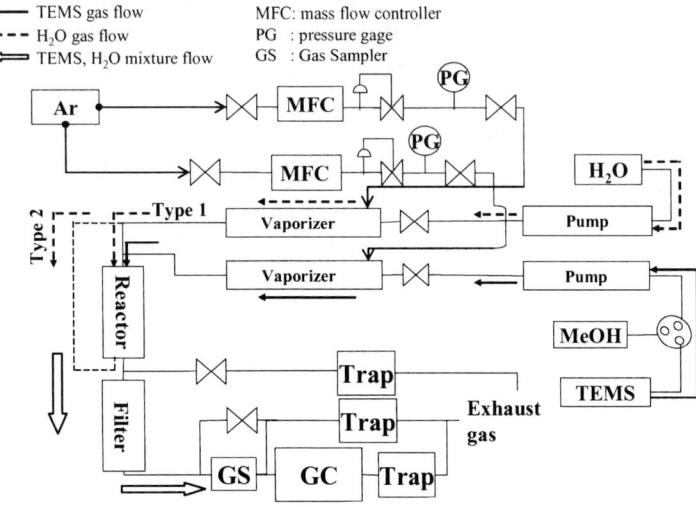

Fig. 1. Experimental flow sheet

In our previous paper [6], the authors have demonstrated that production of fine silica powder is possible by gas phase hydrolysis of tetramethoxysilane (TEMS). In this communication, we report the effect of the shape of reactor and operational condition, especially mixing condition on conversion of the reaction and properties especially diameter of produced silica fines.

2. EXPERIMENTAL

The experimental flow sheet is shown in Fig.1. Career gas (Ar or CO_2) was fed to the vaporizers through mass flow controllers. TEMS and water were separately fed to the vaporizers at predetermined flow rates through tube pumps. Then, the vaporized raw materials were sent to the reactors together with carrier gas. The silica fines produced by the hydrolysis reaction were captured with a tubular glass filter directly connected under the reactor. A part of the produced gas was collected with a gas sampler, and was analyzed by gas chromatograph. The other gas from the reactor than the analyzed was introduced into a bubbler filled with CH_3OH, and was exhausted into a flame. The silica particles collected were analyzed by the SEM, XRD, BET methods.

In the present study, two types of the reactors were use. The configuration of the Reactor Type 1 is shown in Fig. 2. By making structure of the upper part of the reactor dual, TEMS and H_2O were separately fed and mixed in the reactor.

Reactor type2 shown in Fig. 3, in which the steam vapor was first fed from the bottom of the reactor then transported through an internal thin tube in the heated part of the reactor, and then mixed with TEMS fed from the top of the reactor without heating. Thus the shape of the reactor was improved to generate smaller particles than in Type 1. Furthermore, CO_2 was also used instead of Ar to increase the heat capacity of carrier gas.

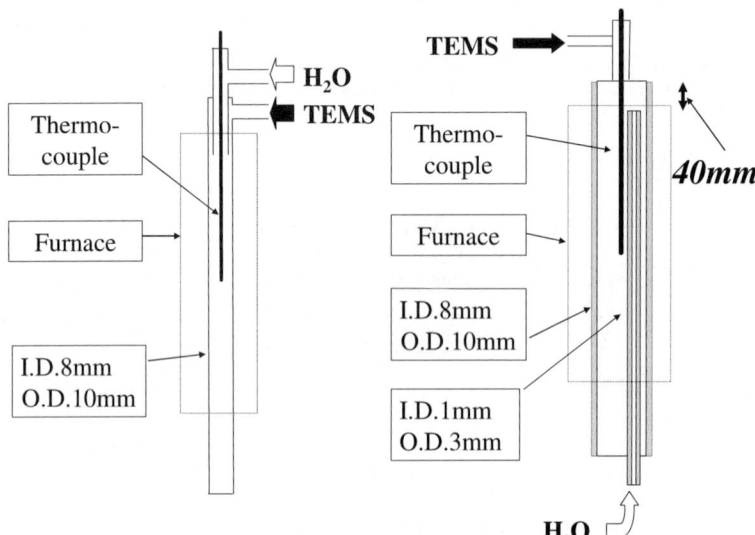

Fig. 2. Detailed structure of the reactor type 1   Fig.3. Detailed structure of the reactor type 2

3. RESULTS AND DISCUSSION

3.1. Influence of concentrations of reactant and temperature profiles on conversion of TEMS

The results on the TEMS conversions using Ar. as carrier gas, determined from the produced gas analysis are shown in Fig. 4. It was found that the TEMS conversion start to increase from around 700 °C, was accelerated with temperature and attained to almost 100% at around 950 °C, even when only TEMS was supplied. In case with high steam concentration, the conversion was drastically increased, which suggests that hydrolysis become dominant. It was also found that the Reactor Type 2 gives higher conversion than Type 1, which is explained by the rapid heating of TEMS by heated steam.

3.2. Comparison of average particle size between two reactor types

The SEM photographs of the produced silica fines produced in Reactor Type1 and Type2 (with carrier gases of Ar and CO_2) at 1223 K are shown in Fig.5. The average particle diameters of the produced particles were given as Green Diameter calculated from the SEM. As a result, the average particle diameters for Type 1 (Ar), Type 2 (Ar) and Type 2 (CO_2) were 155 nm, 145 nm, and 127 nm, respectively. It was demonstrated from the results that smaller fine particles were produced by rapid heating of TEMS under coexistence of water vapor. Under the condition, only nucleation reaction is expected to be accelerated, by suppressing particle growth through heterogeneous reaction. It is suggested that a lot of fine particles were able to be produced by rapid heating of TEMS especially with carrier gas with high heat capacity.

On the other hand, the particle size distribution of the particles prepared by the reactor type 2 looks more broad, which may be attributed by the wider gas residence time distribution in the reaction zone. In order to reduce the dispersion of particle size, i.e., residence time distribution, back mixing should be prevented.

It is expected that the optimum operation conditions of the gaseous phase reactions for the synthesis of fine silica particles will be determined by changing the reactor configuration and operational conditions leading to the change in gas mixing and heating conditions.

4. CONCLUSION

Influences of operational conditions such as temperature and reactor configuration on the conversion of TEMS hydrolyses reaction were investigated. It was found that the essential condition for 100% conversion to produce spherical silica particles without mutual coagulation is high water concentration with high temperature. The mixture condition of the gas leading to the heating condition of TEMS was changed by changing the shape of reactor and carrier gas, which resulted in the change in averaged diameter of the produced silica particles. It is suggested that a lot of fine particles are able to be produced by rapid heating of TEMS.

736

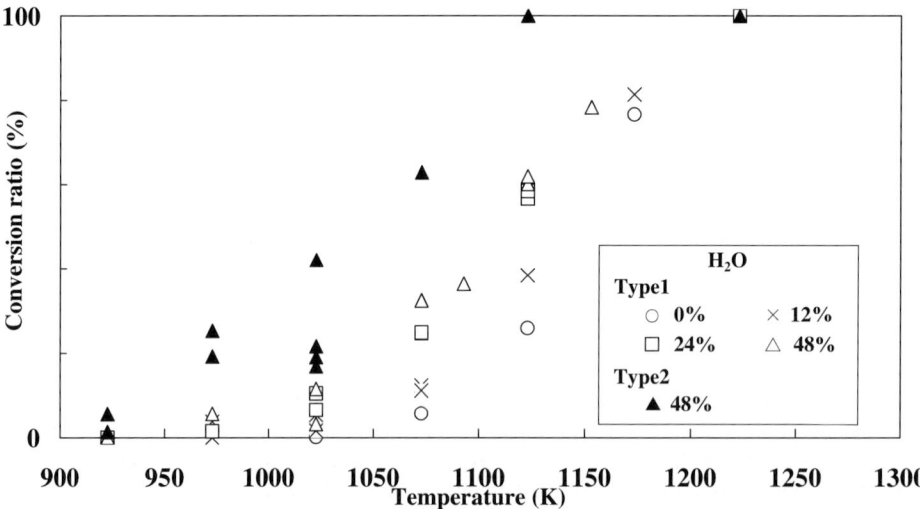

Fig. 4. TEMS conversion (TEMS 60cm^{3}min^{-1}, total 500cm^{3}min^{-1}, Ar balance, at 25℃)

Reactor type 1 Ar ⎯ 1μ m   Reactor type 2 Ar ⎯ 1μ m   Reactor type 2 CO$_2$ ⎯ 1μ m

Fig. 5. SEM photographs of silica product (950℃, TEMS 60cm^{3}min^{-1}, H$_2$O 240cm^{3}min^{-1}, total 500cm^{3}min^{-1}, at 25℃)

REFERENCES

[1] M. Ochiai, Product and Application of High-Purity Silica (in Japanese), CMC Press, Tokyo, 1991, Chap.5, 246-263.

[2] M. Kimata, Production of Mono-dispersed Particle by Series Conversion of TEOS (in Japanese), 31st Autumn Meeting of the Society of Chemical Engineers, Japan, p.227, Fukuoka, Japan (1998).

[3] M. Kimata, M. Koizumi and M. Hasegawa, Kagaku Kogaku Ronbunshu, 22 (1996) 1366.

[4] I. Shinohara and Y. Yamamoto, Kagaku Kogaku Ronbunshu, 25 (1999) 973.

[5] Y. Mitani, Kagakushochi, 37 (1995) 83.

[6] T.Kojima, T. Uchiyama, S. Kato, H. Shibuya and S. Uemiya, Kagaku Kogaku Ronbunshu, 30 (2004) 306.

Studies in Surface Science and Catalysis, volume 159
Hyun-Ku Rhee, In-Sik Nam and Jong Moon Park (Editors)

Growth of GaN nano-structures using Ga(mDTC)₃ precursor

Chinho Park*, **Jin-ho Kim, Deoksun Yoon, Woo-Sik Jung and Tae Jin Lee**

School of Chemical Engineering & Technology, Yeungnam University, Gyeongsan 712-749, Rep. of Korea

The hydride-organometallic vapor phase epitaxy (HOMVPE) method was applied to grow GaN nanorods on sapphire substrates. Particularly, a Ga(mDTC)₃ precursor solution coating technique has been investigated to realize the GaN nanorods at relatively lower growth temperature. It was found that the GaN nanorod growth on sapphire follows the Stranski-Krastanow mode. The GaN nanorod growth without using Ga(mDTC)₃ precursor was also investigated, and it was found the growth follows the Volmer-Weber mode. The size of GaN nanorods obtained in this study was 600~650 nm in diameter.

1. INTRODUCTION

The GaN nano-structures such as nanorods and nanotubes are expected to have great potential in the novel nanotechnological applications [1]. Because of the large bandgap and structural confinement of GaN nanorods, the fabrication of visible and UV optoelectronic devices having relatively low power consumption is potentially feasible [2]. The one-dimensional GaN nano-structures have been recently investigated by several research groups using various methods [3-8]. The methods include the Ga sublimation in NH_3 environment, organometallic vapor phase epitaxy (OMVPE) using trimethyl gallium (TMGa) and NH_3 with or without catalyst, and hydride vapor phase epitaxy (HVPE) using metallic Ga and NH_3.
Although random and irregular type GaN nanorods have been prepared by using transition metal nanoparticles, such as Ni, Co, and Fe as catalysts and carbon nanotubes as the template, the preparation of controllable regular array of straight GaN nanorods has not yet been reported. Fabrication of well-ordered nano-structures with high density is very important for the application of nano-structures to practical devices.

The hydride-organometallic vapor phase epitaxy (HOMVPE) is a technique of using TMGa, HCl, and NH_3 as precursors for the epitaxial growth of thick, high-quality GaN films [9]. The HOMVPE technique combines the benefits of two conventional growth techniques, i.e. OMVPE and hydride VPE, to accomplish high crystal quality with good process controllability as well as high growth rate. This technique is suitable for the growth of structures with high aspect ratio to realize the LED display device on sapphire substrate without going through the costly photolithographic process.

In this study, we report on the GaN nanorod growth by HOMVPE technique with or without using a new precursor, tris(N,N-dimethyldithiocarbamato)gallium(III) (Ga(mDTC)₃). The structural and optical properties of GaN nanorods were characterized by x-ray diffraction (XRD), scanning electron microscopy (SEM), and photoluminescence (PL).

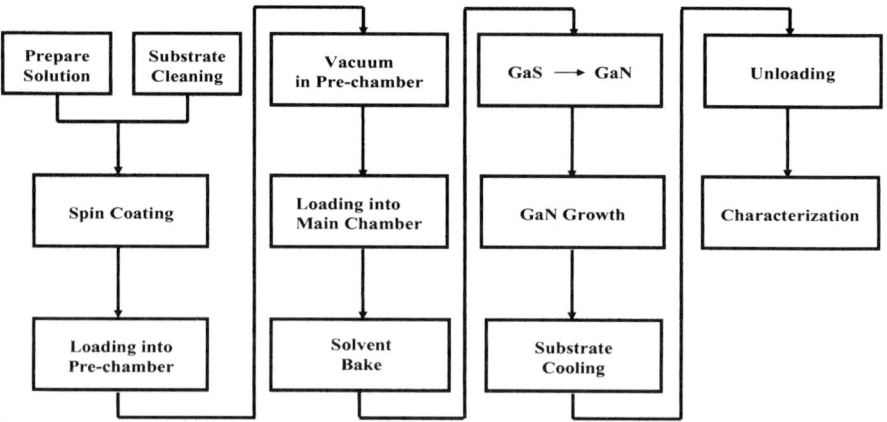

Fig. 1. GaN nano-rod growth procedure using Ga(mDTC)₃ precursor

2. EXPERIMENTAL

The (0001) Al_2O_3 substrate was first degreased, cleaned and rinsed in TCE, acetone and methanol in series with ultrasonification for 5 min each, and the cleaned substrate was blow-dried in a N_2 flow. The Ga(mDTC)₃ precursor solution was spin-coated onto the cleaned and dried sapphire substrate, when desired, and the substrate was loaded into the pre-chamber, where solvent was removed by heating. The substrate was then loaded into the growth chamber and placed in the downstream of the reactor under a NH_3/N_2 environment, and it was heated to the growth temperature (700 ~ 800 ℃). The typical HOMVPE growth sequence was then followed to grow GaN nanorods, which was shown in Fig. 1. The substrate heating procedure and the reactant switching sequence was found to be critical in visualizing the GaN nano-structures.

In other experiments, the GaN nanorods were grown on sapphire substrates without using Ga(mDTC)₃ precursor, and the structures were also characterized. The growths in those experiments were carried out with the V/III ratio of 250, the HCl/TMGa ratio of 4.04 at the growth temperature of 700 ℃. The growth conditions were chosen on the basis of preliminary runs to optimize the process parameters such as the V/III ratio, the HCl/TMGa ratio and the growth temperature. The grown GaN nanorod structures were characterized by the scanning electron microscopy (SEM), X-ray diffraction (XRD), and photoluminescence (PL) spectroscopy.

3. RESULTS AND DISCUSSION

Fig. 2 shows the SEM image of the GaN micro-pillars grown on (0001) Al_2O_3 substrate using Ga(mDTC)₃ precursor. The GaN pillar structures were found to grow preferentially near the edge of the substrate compared to near the center of the substrate, which were oriented normal to the substrate. Unexpectedly, the diameters of the GaN rod structures ranged from 5 μm to 20 μm. It was also found that the surface between the rods was covered by the GaN epitaxial film, which indicated that the growth followed the Stranski-Krastanow mode [10]. The preferential pillar structure growth near the edge of the substrate could be explained by the relatively inefficient surface diffusion of ad-species on the surface near the edge of the

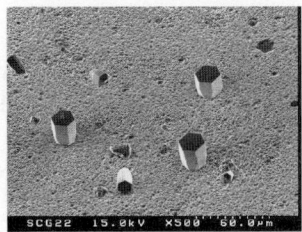

Fig. 2. The SEM Image of the GaN nano-structures grown on (0001) Al₂O₃ substrate.

substrate which promotes the 3-D type growth behavior at otherwise the same growth environment. Thin epitaxial films between the GaN pillars were believed to be formed by the 2-D type growth following the nucleation from the ad-species.

Fig. 3 shows the XRD diffraction patterns and PL spectrum of the GaN structures grown on (0001) Al₂O₃ substrate using Ga(mDTC)₃ precursor. It was observed that the three distinctive diffraction peaks indexed as (0002) and (0004) of the wurtzite GaN structure, and (0006) of the sapphire appeared, which indicates that the GaN structures formed in this study by HOMVPE technique are single crystalline in nature. In the PL spectrum, the sharp, near-band emission at 3.43 eV was observed, and a broad band was observed in the energy range of 2.1~2.6 eV. The sharp and intense near-band emission peak and a weak yellow band emission indicate that the GaN structure has good crystalline quality.

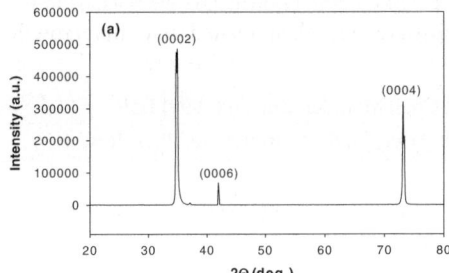
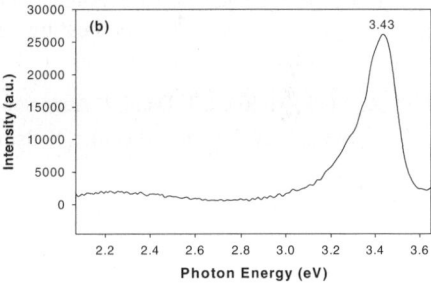

Fig. 3. (a) XRD diffraction patterns and (b) PL spectrum of the GaN nano-structures grown on (0001) Al₂O₃ substrate using Ga(mDTC)₃ precursor.

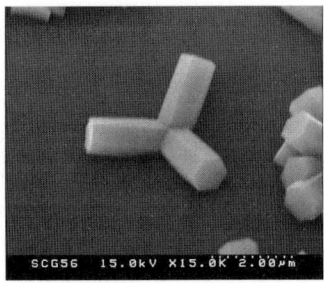

Fig. 4. SEM image of the GaN nanorods grown by HOMVPE technique without using Ga(mDTC)₃ precursor.

4. SUMMARY

The GaN nanorod growth was investigated in this study by using HOMVPE technique. Particularly, the use of $Ga(mDTC)_3$ precursor was investigated for the first time to realize the GaN nanostructures. It was found that the GaN growth on sapphire follows the Stranski-Krastanow mode, when the $Ga(mDTC)_3$ precursor was used, and the GaN growth follows Volmer-Weber mode, when the $Ga(mDTC)_3$ precursor was not used. The size of GaN nanorods obtained in this study was 600~650 nm in diameter.

ACKNOWLEDGEMENTS

This work was supported by the Korea Research Foundation Grant funded by Korean Government (MOEHRD: project #D00199).

REFERENCES

[1] S. Nakamura, Science, 281 (1998) 956.

[2] S. Nakamura, T. Mukai and M. Senoh, Appl. Phys. Lett., 64 (1994) 1687.

[3] W. Han, S. Fan, Q. Li and Y. Hu, Science, 277 (1997) 1287.

[4] C. C. Chen and C.C. Yeh, Adv. Matter., 12 (2000) 738.

[5] J. Y. Li, X.L. Chen, Z.Y. Qiao, T.G. Cao and Y.C. Lan, J. Cryst. Growth, 213 (2000) 408.

[6] M. Yoshizawa, A. Kikuchi, M. Mori, N. Fuhita and K. Kishino, Jpn. J. Appl. Phys., 36 (1997) L459

[7] D. Elwell, R.S. Feigelson, M.M. Simkins and W.A. Tiller, J. Cryst. Growth, 66 (1984) 45.

[8] H.-M. Kim, D.S. Kim, Y.S. Park, D.Y. Kim, T.W. Kang and K.S. Chung, Adv. Mater., 14 (2002) 991.

[9] O. Kryliouk, M. Reed, T. Dann, T. Anderson and B. Chai, Mat. Sci. Eng., 6 (1999) B59.

[10] B. Daudin, F. Widmann, G. Feuillet, Y. Samson, M. Arlery, and J.L. Rouvie're, Phys. Rev., B56 (1997) R7069.

Studies in Surface Science and Catalysis, volume 159
Hyun-Ku Rhee, In-Sik Nam and Jong Moon Park (Editors)

741

Dehydrogenation of ethylbenzene with carbon dioxide over carbon nanofiber supported iron oxide

Tie-Jun Zhao[a, b]*, Yi-jian Sun[a], Xiong-Yi Gu[a], Ping Li[a], De Chen[b], Ying-Chun Dai[a], Wei-Kang Yuan[a] and Anders Holmen[b]

[a]UNILAB, State Key Laboratory of Chemical Reaction Engineering, East China University of Science and Technology, Shanghai 200237, China
[b]Department of Chemical Engineering, Norwegian University of Science and Technology, N-7491 Trondheim, Norway

1. INTRODUCTION

Large amounts of styrene (ST) is commercially produced by dehydrogenation of ethylbenzene(EB) on potassium-promoted iron oxide catalysts in the presence of excess superheated stream at 650-570°C. The dehydrogenation of EB with carbon dioxide (DE-CO_2) to styrene is believed to be an energy-saving process and the equilibrium conversion can be significantly improved compared with the commercial process [1]. The current industrial catalysts for styrene production are not active for DE-CO_2, and various supported catalysts have been tested for DE-CO_2. It is found that active carbon or active alumina supported iron oxides or vanadium oxides are effective on DE-CO_2, and the best results are obtained on alkali- promoted iron oxides supported on active carbons [2, 3]. However, active carbons not only display complicated physico-chemical properties, but they are also sensitive to the oxidant at the elevated temperature, particularly on the metal-mediated condition. In fact, these catalysts suffer from deactivation [3], and a development of the new catalyst is of great interest. Recently, carbon nanofiber (CNF) has been employed as catalyst support due to their unique properties such as graphitic and fibrous structure, high external surface area et. al [4, 5]. Various methods have been employed to deposit the active metal on CNF and the resulted catalysts display better activity or selectivity in gas or liquid phase reactions [4]. However, to our best knowledge, no study has yet been carried out on the carbon nanofiber supported iron oxide catalyst for the EB dehydrogenation to ST. In this study, a series of carbon nanofiber supported iron oxide catalysts are prepared, characterized and tested in EB dehydrogenation. It aims at developing a new effective catalyst and at elucidating the role of carbon nanofiber and the nature of the iron oxide as well as the promoter in the reaction.

2. EXPERIMENTAL

2.1. Catalyst and carbon nanofiber preparation

Fish-bone carbon nanofiber is grown on 20wt%Ni/Al_2O_3 catalyst under methane/ hydrogen (80/10ml/min) mixed gas at 600°C as previously described in the literature [5]. The as-grown carbon nanofiber is treated in a mixture of H_2SO_4/ HNO_3 (1:1) at 90°C for 0.5 h to remove the catalysts used for synthesis, and to introduce the oxygen-containing groups on the surface simultaneously. Oxidized CNF supported iron oxide catalyst is prepared by the

---

* tiejun@chemeng.ntnu.no

deposition-precipitation method as follows: CNF (about 9.5g) is suspended in 500ml of water at 90°C under strongly mechanical stirring. Fe(NO₃)₃·9H₂O(3.64g), and CO(NH₂)₂ (4.05g) are added and the pH is adjusted to 3-4 using nitric acid. After 16 h reaction, the mixture is cooled to room temperature, filtered, washed for three times, followed by drying at 120°C overnight. Furthermore, Li, K et. al. are introduced by impregnation using the nitrite salt and then dried at 120 °C overnight. The dried catalyst is finely calcined in argon atmosphere for 1 h at 400°C.

2.2. Catalytic test

Catalytic tests of the catalysts are carried out in a fixed bed (12.0mm i.d., 50cm length) at 823 K and atmospheric pressure. 2.0g sample of the catalyst is held in the isothermal zone by quartz wool. Flows of Ar and CO_2 are controlled by mass flow controllers, and desired mixtures are fed into the ethylbenzene vaporizer. The temperature of the vaporizer is adjusted to obtain the desired volume ratio of EB/CO_2/Ar mixture (2:30:48ml/min). The corresponding partial pressure of EB is 2.5 kPa. The reaction products are analyzed by on-line GC (Agilent 4890D) equipped with a PEG20m/6210-5% packed column(2 m) to analyze aromatic product using FID and with a TDX-1 packed column to determine permanent gases using TCD.

2.3. Catalyst and carbon nanofiber characterization

The XRD study is performed in Rigaku, D/Max2550VB/PC instrument, using CuKα radiation (λ=1.5405 nm) and a graphite crystal monochromator. The microstructure of the catalysts is studied by a JEOL JEM 2010 TEM with an accelerated voltage of 200KV and a lattice resolution of 0.18nm. Transmission specimens are prepared by ultrasonic dispersion of the respective samples in ethanol, followed by dropping the suspension to a holey carbon-coated copper grid.

3. RESULTS

3.1. Dehydrogenation of ethylbenzene with CO_2

The results of catalytic activities in the dehydrogenation of ethylbenzene with various iron oxide based catalysts are shown in Fig. 1(a-b). The number in the parentheses of the catalyst codes indicates the weight fraction of metal per gram carbon. On oxidized CNF alone less than 20% conversion of EB is observed after 3 h on stream. The conversion of ethylbenzene

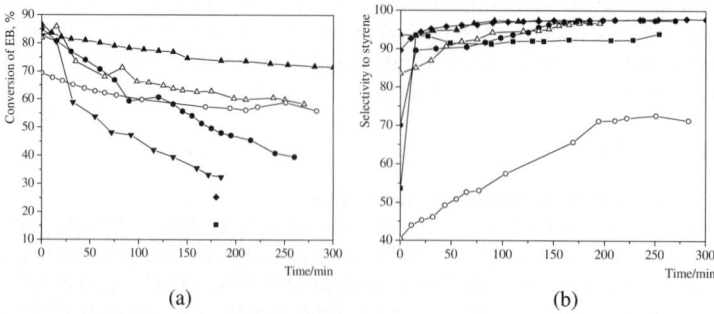

(a)   (b)

Fig. 1. Catalytic performance of various catalysts on DE-CO₂ CNF-ox(■); Fe(5)/CNF-Ar(◆); Fe(5)/CNF(▼); Ca(0.5)Fe(5)/CNF(●); Li(0.5)Fe(5)/CNF(△); K(0.5)Fe(5)/CNF(▲); K(0.5)Fe(5)/CNF-HCl(○)

in the presence of carbon dioxide was found to be much higher than in argon over Fe(5)/CNF catalyst. Since commercial iron catalyst is normally promoted with alkali or alkali earth metals, the effect of a small fraction of alkali or alkali earth metal salt addition are examined. From Fig. 1(a), addition of alkali metal salt not only enhances the conversion, but also significantly improves the stability of catalyst. The order of the effect of alkali or alkaline earth metal on the catalyst activity for styrene formation is as follows: K>Li>Ca. Among the tested catalysts, K(0.5) Fe(5)/CNF catalyst shows the best performance, where the conversion of ethylbenzene is maintained at more than 75% after 5 h on stream. The specific activity decreases gradually, but the decrease in activities of the Li and K doped Fe/CNF catalysts is not so remarkable after 4 hours of reaction.

Fig.1(b) represents the selectivity to styrene as a function of time for the above catalysts. It is observed that the selectivity to styrene is more than 95% over carbon nanofiber supported iron oxide catalyst compared with about 90% for the oxidized carbon nanofiber. It can be observed that there is an increase in selectivity to styrene and a decrease in selectivity to benzene with time on stream until 40 min. In particular, when the carbon nanofiber which has been treated in 4M HCl solution for three days is directly used as support to deposit the iron-precursor, the resulting catalyst shows a significantly lower selectivity to styrene, about 70%, in contrast to more than 95% on the similar catalyst using oxidized carbon nanofiber. The doping of the alkali or alkali metal on Fe/CNF did not improve the steady-state selectivity to styrene, but shortened the time to reach the steady-state selectivity.

3.2 Carbon nanofiber and catalyst Characterization

The effect of oxidation pretreatment and oxidative reaction on the graphitic structure of all CNF or CNF based catalysts has been studied by XRD and HRTEM. From the diffraction patterns as shown in Fig. 2(a), it can be observed the subsequent treatment do not affect the integrity of graphite-like structure. TEM examination on the tested K(0.5)-Fe(5)/CNF catalysts as presented in Fig.2(b), also indicates that the graphitic structure of CNF is still intact. The XRD and TEM results are in agreement with TGA profiles of fresh and tested catalyst: there is no obviously different stability in the carbon dioxide atmosphere (profiles are not shown). Moreover, TEM image as shown in Fig. 2(b) indicates that the iron oxide particles deposited on the surface of carbon nanofiber are mostly less than less than 10 nm.

An increasing intensity of the diffraction peaks of hematite is observed when comparing the dried and calcined catalyst as shown in Fig. 2(a), indicating that hematite forms at higher temperatures. No obvious diffraction peaks to lithium such as lithium iron oxide ($LiFe_5O_8$) could probably be ascribed to the small fraction of lithium or overlapped peaks between hematite and lithium iron oxide. The diffraction peak intensity of magnetite in tested catalysts increases significantly.

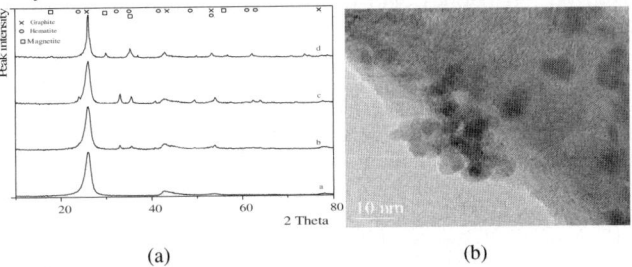

(a)   (b)

Fig. 2. XRD patterns of various catalysts (a) and TEM images of tested K(0.5)Fe(5)/CNF catalyst(b)
a: CNF-ox; b: Dried Li(0.5)Fe(5)/CNF; c: Calcined Li(0.5)Fe(5)/CNF; d: Tested Li(0.5)Fe(5)/CNF

4. DISCUSSION

From Fig.2 (a), A solid phase transformation from hematite, Fe_2O_3 to magnetite, Fe_3O_4, is observed, indicating that the active sites of the catalyst are related to Fe_3O_4. Suzuki et. al also found that Fe_3O_4 plays an important role in the formation of active centers by a redox mechanism [6]. It is also observed that the hematite itself relates to the formation of benzene at the initial periods, but no obvious iron carbide peaks are found on the tested Li-Fe/CNF, formation of which is considered as one of the reasons for catalyst deactivation [3, 6].

It is found that the addition of potassium is effective to improve the catalyst activities than that of lithium and calcium. According to the proposed redox mechanism, the role of carbon dioxide seems to be to keep the iron oxide in a high valence state such as Fe_3O_4 for styrene formation, doping the alkali metal can enhance the carbon dioxide adsorption. The higher stability of K-Fe/CNF than Fe/CNF can be explained by the stable magnetite phase, due to the enhanced adsorption of CO_2 by potassium doping.

Lower conversion and selectivity were observed on potassium promoted iron oxide supported non-oxidized carbon nanofibers. The inert and hydrophobic properties of non-oxidized carbon nanofiber are advantageous in suppressing the side reaction, but on the other side, making the deposition of metal precursor difficult owing to the lack of anchoring sites. The oxygen-containing groups probably act as nucleation and anchoring sites on the surface of carbon nanofiber [7]. Using non-oxidized carbon nanofiber as support, the resulted $Fe(OH)_3$ is apt to deposit in the solution, but not on the surface of carbon nanofiber, so the interaction between the carbon nanofiber and metal oxide nearly disappears, showing that the proper interaction is critical to obtain the higher selectivity to ST.

5. CONCLUSIONS

Dehydrogenation of ethylbenzene with carbon dioxide is carried out on various carbon nanofiber supported alkaline-promoted iron-oxide catalysts prepared by deposition-precipitation method. K-Fe/CNF shows the best performance among the above catalyst at 550°C, indicating that potassium enhances the activation of carbon dioxide. XRD analysis of the above catalysts indicates that the magnetite transformed from hematite plays the important role in formation of styrene. XRD and HRTEM examination on the tested catalysts presents the graphitic structure of carbon nanofiber is intact and the iron oxide particles with less than 10 nm are deposited on the surface of carbon nanofiber. Non-oxidized carbon nanofiber supported catalyst shows the significantly lower selectivity due to the lack of anchoring sites on the surface of carbon nanofiber.

ACKNOWLEDMENT

This work is done under the support of the National Science Foundation of China (NSFC 20376021, 20490200, 20506004), and the Special Foundation of State Key Laboratory in Shanghai (036505010).

REFERENCES

[1] T. Badstube, H. Papp, R. Dziembaj and P. Kustrowski, Appl. Catal. A: General, 204 (2000) 153.
[2] R. Dziembaj, P. Kustrowski, T. Badstube and H. Papp, Top.Catal., 11-12 (2000)317
[3] A. Sun, Z. Qin and J. Wang, Appl. Catal. A: General, 234 (2002) 179.
[4] K.P. De Jong and J.W. Geus, Catal. Rev. Sci. Eng., 42 (2000) 481.
[5] T. Zhao, D. Chen, Y. Dai, W. Yuan and A. Holmen, Ind. Eng. Chem. Res., 43 (2004) 4595.
[6] M. Sugino, H. Shimada, T. Turuda, H. Miura, N. Ikenaga and T. Suzuki, Appl. Catal. A: General, 121 (1995) 125.
[7] J.H. Bitter, M.K. Lee, A.G.T. Slotboom, A.J. Dillen and K.P. De Jong, Catal. Let., 89 (2003)139.

Studies in Surface Science and Catalysis, volume 159
Hyun-Ku Rhee, In-Sik Nam and Jong Moon Park (Editors)

Oxidative dehydrogenation of propane over carbon nanofibers

Zhi-jun Sui, Jing-hong Zhou and Ying-chun Dai

State Key Laboratory of Chemical Engineering, East China University of Science & Technology, Shanghai, 200237, P. R. China

1. INTRODUCTION

As a new kind of carbon materials, carbon nanofilaments (tubes and fibers) have been studied in different fields [1]. But, until now far less work has been devoted to the catalytic application of carbon nanofilaments [2] and most researches in this field are focused on using them as catalyst supports. When most of the problems related to the synthesis of large amount of these nanostructures are solved or almost solved, a large field of research is expected to open to these materials [3]. In this paper, CNF is tested as a catalyst for oxidative dehydrogenation of propane (ODP), which is an attractive method to improve propene productivity [4]. The role of surface oxygen complexes in catalyzing ODP is also addressed.

2. EXPERIMENTAL

The CNF was synthesized by catalytic deposition of 80 vol% CO/H_2 mixtures on a 20 wt% NiFe (molar ratio 1:1) alloy catalyst. To remove metallic inclusions, the CNF was repeated washed in 2 mol/L HCl over a period of 7 days. Then, the CNF sample was filtered, washed by a large amount of deionized water until the pH of filtrate was close to 7 and dried at 120 ℃ overnight. This sample was named CNF-R. Before testing as catalysts for ODP, CNF-R was calcined in air at 500℃ for 2 h and designated as CNF-RA hereafter. To improve the graphitization extent of CNF, the as-grown CNF was treated at 1700 °C for 12 h under protection of argon, which was referred to CNF-HT. Prior to test as catalyst for ODP, the CNF-HT needed to be further oxidized. Three methods are used: 1) calcined in air at 600℃ for 1 h (CNF-HA); 2) immersed in the mixture of concentrated HNO_3 and H_2SO_4 (1:1) for 1 day (CNF-HL); 3) CNF-HL was treated in Ar at 800 ℃ for 2 h; the sample temperature decreased to 60 ℃ under Ar protection; and then contact with air at 60 ℃ for 2 h (CNF-HB).

Catalytic experiments were performed at 1 atm pressure in a conventional fixed bed flow reactor made of stainless steel. The CNF catalysts were charged (0.8 g, particle size < 0.18 mm) without inert diluents. The catalysts were placed on quartz wool in the isothermal zone of the reactor. Free volume of the reactor was packed with silica and quartz wool. The blank run results showed that the homogeneous reaction could be ignored under the experimental conditions used in this work. The reaction gases were composed of 4 vol % propane, 8 vol% oxygen and balance Ar (total flow rate is 100 ml/min). Analyses of reactants and products were carried out by using two separate on-line gas chromatographs (Agilent 4890D) with

TCD detectors, using a HP-Plot Q capillary column (30m x 0.53mm) for hydrocarbons and a TDX packed column (2m) for permanent gases.

Structures of the CNF were characterized by HRTEM (JOEL JSM2010, Japan). The textural properties had been obtained from N_2 adsorption-desorption isotherms (ASAP 2010, Micromeritics, USA) at $-196\,℃$ after out-gassing the samples at 190 ℃ and 1 mmHg for 6 h. X-ray diffraction (XRD) was performed on a Rigaku D/Max2550VB/PC (Rigaku, Japan, Cu Kα radiation). Thermal gravimetry (TG), using air as carrier gas, was used to characterize the structure stability of the CNF and determine the ash content. Temperature-programmed surface reaction (TPSR) was performed to determine the role of surface oxygen complexes on catalyzing ODP. The samples were first temperature-programmed desorption in Ar to 650 ℃, cooled in Ar to 150℃, and then the sample temperature was raised at a constant ramping rate of 10 ℃/min in a gas mixture of 8 vol% C_3H_8/Ar (total flow rate, 50 ml/min). Concentrations of Ar, C_3H_8, C_3H_6, C_2H_4, CH_4, CO_2 and CO were calculated by the signal intensities of mass 40, 39, 41, 27, 16, 44 and 28 respectively. No C1~C2 hydrocarbon is detected by quadrupole mass spectrometer (Questor, ABB Extrel, USA).

3. RESULTS AND DISCUSSION

3.1. Catalyst Structure

HRTEM picture (Fig.1(a)) shows that CNF-R is estimated of a diameter of 30~40 nm and has a hollow core. The graphene layers are about 15~20 ° inclining to the axis. After heat treatment, CNF-HT, the structure remained, but graphene layers stack more regularly.

The BET surface area of CNF-R and CNF-HT are 152.5 and 141.6 m^2/g respectively. After oxidation, the surface areas increase a little (160~170 m^2/g). All the samples have small micropore volumes (<0.008 cm^3/g). XRD results imply a 0.339 nm d $_{002}$ spacing for CNF-HT, which is smaller than that of CNF-R (0.341 nm). The onset weight loss temperature (temperature needed to reach 5 % burn-off) of the CNF-HT is 660 ℃ while the CNF-R is only 540 ℃. These results indicate the graphitizaton extent of CNF increases after heat treatment. The ash content of CNF-HT is 0.06 wt%. So, the side effects of the metal on catalytic performances can be ruled out.

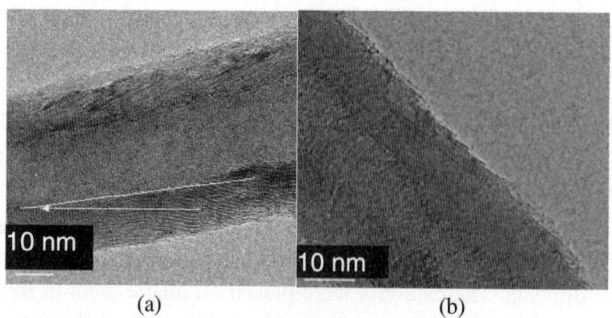

(a)   (b)

Fig. 1. HRTEM images of CNF-R(a) and CNF-HT(b)

3.2. Catalytic performances

The catalytic products for all the CNF catalysts include C_3H_6, COx and trace amount of

C_2H_4. Propene selectivity as a function of propane conversion is studied on CNF-RA under different reaction conditions. The results are shown in Fig. 2. The C_3H_6 selectivity decreases with the increase of C_3H_8 conversion, which implies consecutive reaction of C_3H_6 degradation. Apparently, a high reaction temperature and O_2/C_3H_8 ratio is favored for high propene yield. Extrapolating the results in Fig.2 to zero conversion shows that the theoretical propene selectivity of ODP on CNF-RA increases when the reaction temperature is raised. But the theoretical propene selectivity does not exceed 80% on this sample. Except that propane undergoes oxidative dehydrogenation to form propene and combustion of propane and propene to form COx, CNF gasification (when the temperature is above 450℃) should be included. As mentioned above, high propene yield and theoretic selectivity could be achieved at high reaction temperature and O_2/C_3H_8 ratio, which are also benefit for CNF gasification and restrain achieving higher propene yields on CNF-RA.

It is found that the CNF-HT has not catalytic activity for ODP. After oxidation, all the three samples show highly catalytic performances, which are shown in Fig.3. CNF-HL has the longest induction period among the three samples, and it has relatively low activity and propene selectivity at the beginning of the test. During the induction periods, the carbon balance exceeds 105% and then fall into 100±5%, which implies the CNF structure is stable and the surface chemistry of CNF reaches a dynamic equilibrium eventually. These results indicate that the catalytic activity of ODP can be attributed to the existence of surface oxygen complexes which are produced by oxidation. The highest propene yield(18.96%) is achieved on CNF-HL at a 52.97% propane conversion.

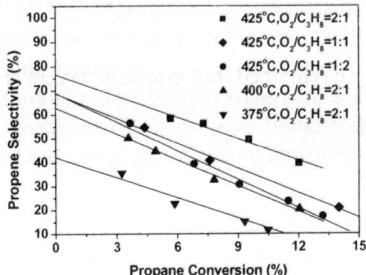

Fig.2. Propene selectivity as a function of propane conversion over CNF-RA

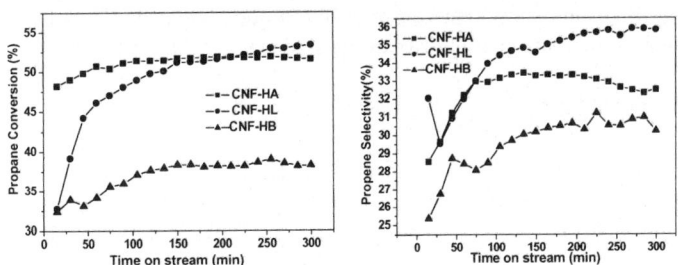

Fig. 3. Evolution of catalytic performances with time for CNF catalysts;
reaction temperature, 550 ℃; W/F=37.9 (mol C_3H_8)/g h

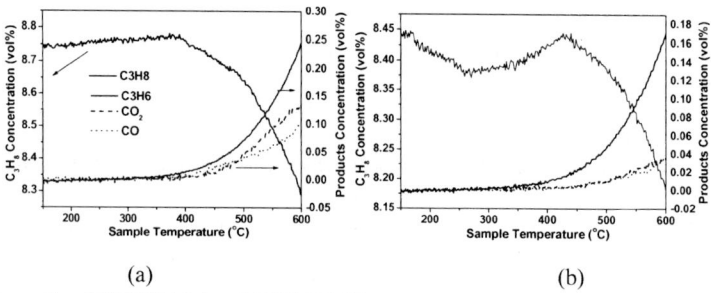

(a)   (b)

Fig. 4. TPSR results of CNF-RA(a) and CNF-HA(b)

3.3. TPSR results

TPSR results are presented in Fig. 4. Propene is produced when the sample temperature is above 350℃ on both samples, which means converting of propane over CNF catalysts could occur without oxygen. The desorption products amounts are 0.35 and 0.26 mmol/g for CNF-RA and CNF-HA respectively while the percentages of propene in the desorption substances over these two sample are 51.4% and 87.7%. These results imply that the propene selectivity may increase, at least partly, due to restriction of oxidation of propane to COx by heat treatment at the cost of catalytic activity.

It's worth pointing out that no products (propene and COx) is detected by QMS in TPSR runs under the following circumstances: 1) TPD of CNF-RA to 1000℃, then cooled it down to 150℃ in Ar before the TPSR runs; 2) CNF-HT without calcinations in air; 3) no CNF sample is presented. Based on these results we can deduce that converting of propane over CNF occurred on their surface oxygen complexes containing C=O bonds (carbonyl-like groups and basic oxides), which could exist on CNF surface after TPD to 650℃ [5]. These structures are thought to be responsible of catalyzing redox reactions on carbon surfaces [6].

4. CONCLUSIONS

CNF could be the effective catalyst for ODP, but the high propene yield can only be achieved at high reaction temperature and O_2/C_3H_8 ratio. Heat treatment of CNF at 1700℃ for 12 h could increase its graphitization extent and enables it to operate at 550℃ without apparent gasification. TPSR results show that carbonyl-like groups could be the active sites for ODP over CNF. Heat treatment also could restrain the side reaction of oxidation of propane to COx.

ACKNOWLEDGEMENT

The author would thank the National Science Foundation of China (NO. 20490200 and NO. 20376021) for the finical support

REFERENCES

[1] K.P.De Jong and J.W. Geus, Catal Rev.- Sci. Eng, 42 (2000) 48.
[2] M. J. Ledoux, R. Vieira, and C. Pham-Huu, J. Catal, 216 (2003) 333.
[3] P. Serp, M. Corrias and P. Kalack, Appl Catal, A, 253 (2003) 337.
[4] F. Cavani and F. Trifirò, Catal. Today, 24 (1995) 307.
[5] J. L. Figueiredo, F. R. Pereira, M.M.A. Freitas and J. J. M. Órfão, Carbon, 37 (1999) 1379.
[6] C.A. Leon y Leon D. and L.R. Radovic, Chem. Phys. Carbon, 23 (1993) 213.

Studies in Surface Science and Catalysis, volume 159
Hyun-Ku Rhee, In-Sik Nam and Jong Moon Park (Editors)

Synthesis and purification of carbon nanotubes by arc discharge

W.H. Hong[a], D.-W. Park[a]*, S.-M. Oh[a], W.S. Ahn[a], J.H. Oh[b], C.M. Lee[b] and W. P. Tai[b]

[a]Department of Chemical Engineering, Inha University, Nam-gu, Incheon 402-751, Korea

[b]School of Materials Science and Engineering, Inha University, Nam-gu, Incheon 402-751, Korea

1. INTRODUCTION

Carbon nanotubes (CNTs) are attracting many researchers in materials science since 1991 [1]. CNTs have applications such as field emission displays, nanoscale transistors, supercapacitors, secondary batteries and various composites. CNTs are currently being prepared using many kinds of methods including arc-discharge, chemical vapor deposition, laser ablation and catalytic decomposition of hydrocarbon [2]. Among them, the arc discharge has been known to be very simple and cheap, and to have merit in massive production. However, the method reveals lower yield of CNTs in the products containing fullerenes, amorphous carbon, and some graphitic sheets [3]. Highly purified CNTs are generally required for sophisticated measurements and practical applications, but the yield CNTs is very low although the various purification methods have been tried [4]. In this study, high purity MWNTs and SWNTs using arc discharge have been investigated.

2. EXPERIMENTAL

The schematic diagram of the arc discharge apparatus is shown in Fig. 1. Two graphite rods were used as the anode with the diameter of 10 mm and the cathode with the diameter of 6 mm. The anode was controlled until the distance between the anode and cathode was very small to approx. 1 mm.

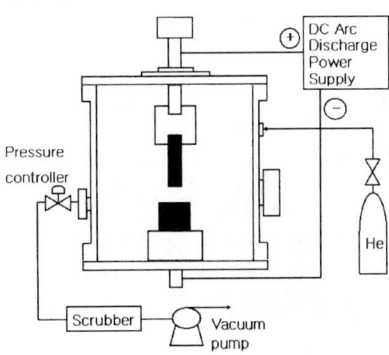

Fig. 1. Schematic diagram of DC arc discharge

The discharge current of 75 A was applied between two electrodes under helium at from 250 torr to 760 torr. The produced MWNTs were purified by the thermal annealing in air atmosphere. The purification procedure was as follows. The collected samples were grinded to small pieces and transferred into the furnace. The samples were heated at 600 ℃ for 30 min in air stream to remove carbonaceous particles.

For producing SWNTs, transition metals, such as Ni, Co, FeS were used as catalyst. We made a 3 mm-hole in the anode and filled a mixture of catalyst and graphite powder, which the total amount of catalyst in a graphite powder was adjusted to 5 wt%. The pressure of helium gas was in the range of 250~760 torr. To purify the products, synthesized powders was grinded and transferred to the heating machine, where the CNTs powder was heated at 470 ℃ for 90 min in air to remove carbonaceous particles. To remove the catalysts, the annealed powder was immersed and filtered in 10 % hydrochloric acid (HCl). The sample was washed out in deionized water.

Fourier Transform (FT) Raman spectroscopy (Model RFS 100/S, BRUKER Co.) using ND:YAG laser was used to analyze the products on their structure electronic and vibration properties. The morphology of CNTs was observed by scanning electron microscopy (SEM, Model S-4200, Hitach Co.) and transmission electron microscope (TEM, Model JEOL 2000FX-ASID/EDS, Philips Co.).

3. RESULTS AND DISCUSSION

Soot deposited on the chamber wall contained mostly carbonaceous particles, where no MWNTs were contained. The deposits on the cathode consist of two portions; the inside is black fragile core and the outside hard shell. The inside included MWNTs and polyhedral graphitic nanoparticles. The outer-shell part consisted of the crystal of graphite.

Fig. 2 shows the SEM and TEM images of MWNTs purified by the thermal oxidation. Fig. 2(a) is the SEM image of the ground raw sample. Many carbonaceous particles were observed. The samples were oxidized at 760 Torr under air ambient. Fig. 2(b) shows the improvement in the nanotube-to-nanoparticle ratio during oxidation. The basic idea of the selective etching is that amorphous carbons can be etched away more easily than MWNTs due to the faster oxidation reaction rates [2]. In Fig. 2(c), MWNTs were opened and thinned through oxidation. This results means that the significant difference in the oxidation reaction rates of the carbon nanotube caps as compared to the cylindrical surfaces results in the production of open nanotubes during oxidation [5]. The reaction with oxygen started from the edge of nanoparticles and then proceeded to their centers. Compared with nanoparticles, it took more time for MWNTs to be completely burnt out, since MWNTs were much longer than nanoparticle.

Fig. 2. The SEM images of (a) the raw sample, (b) the purified MWNTs and (c) the TEM image of the purified MWNTs

751

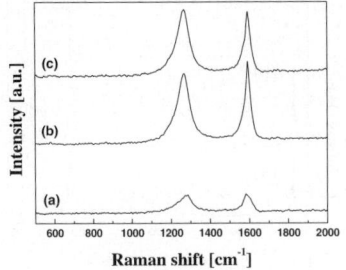

Fig. 3. Raman spectra of MWNTs synthesized under different pressure; (a) 250 Torr, (b) 500 Torr, (c) 760 Torr

Therefore the cap of the MWNTs was removed using the purification process.

Fig. 3 shows the Raman spectra of the MWNT samples as a function of helium pressure. The peaks around 1280 cm⁻¹, called the D-mode, are known to be attributed to amorphous carbons and defects of nanotubes, whereas the peaks around 1600 cm⁻¹, called the G-mode, are known to be due to the graphitic structure of carbon atoms. The G-mode of produced MWNTs was shifted to a lower wave number region (1595 cm⁻¹) by the strain of the forming tube [6]. The intensity of MWNTs synthesized under 250 Torr was lower than at other pressure. And the ratio of the G-mode to the D-mode was the highest at pressure of 500 Torr. The highest purity of MWNTs was obtained when the pressure of helium is 500 Torr.

Fig. 4 shows the SEM images of SWNTs purified by the thermal oxidation and acid-treated. Fig. 4(a) shows a SEM image of the raw soot. In addition to the bundle of SWNTs, carbonaceous particles are shown in the figure. These structural features might be caused by various in the arcing process because of an inhomogeneous distribution of catalysts in the anodes [7]. It can be seen that the appearance of SWNTs was curled and quite different from that of MWNTs. Fig. 4(b) shows a decrease of amorphous carbons after oxidation. The basic idea of the selective etching is that amorphous carbons can be etched away more easily than SWNTs due to the faster oxidation reaction rate [2]. Since the CNTs are etched away at the same time, the yield is usually low. The transition metals can be etched away by an acid treatment. Fig. 4(c) shows the SEM image of the acid-treated sample, where the annealed sample was immersed in 10 % HCl.

Fig. 5 shows typical Raman spectrum for SWNTs, the Raman spectra of SWNTs have fingerprint features, which is quite different from those of graphite, MWNTs and amorphous carbon.

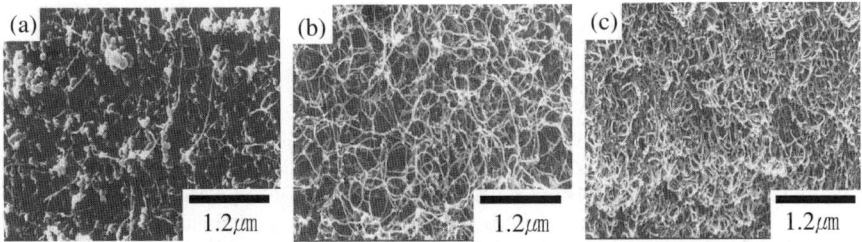

Fig. 4. SEM images of (a) the raw sample, (b) the thermally treated sample, (c) HCl-treated sample

752

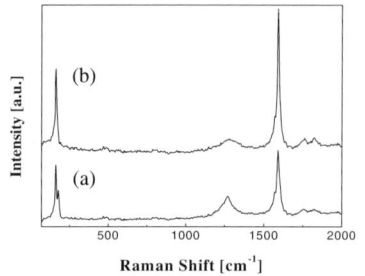

Fig. 5. FT-Raman spectra of (a) the raw SWNTs sample and (b) the purified sample (Ni, Co and FeS were used as a catalysts)

The radial-breathing mode (RBM) of SWNTs was clearly observed at 161 cm^{-1}. The frequency of the RBM is in inverse proportion to the diameter of the SWNTs [8], and the relation can be described by the equation, $w_r = 223.75/d$, where w_r is the RBM frequency in cm^{-1} and d is the SWNTs diameter in nm. The RBM at 161cm^{-1} corresponds to a diameter of 1.4 nm. The shoulder peak at 1572 cm^{-1} appearing on the left of the main peak (called the G-band) at 1592 cm^{-1} originates from the splitting of the E_{2g} mode of graphite and is one of the characteristic Raman scatterings of SWNTs [8]. The broad peak at 1274 cm^{-1} is known as the D-band. This band appears if the sample contains carbon with a disordered structure of the SWNTs themselves have incomplete wall structures [2]. The peak intensity ratio of the G-band to the D-band is a good measure of the SWNTs content, and this value increased for purification.

4. CONCLUSIONS

We have investigated the high yield and purity of MWNTs and SWNTs by arc discharge. MWNTs were synthesized under different pressure of helium. The SWNTs were synthesized using transition metal, which played as a catalyst during the formation process. High-yielded synthesis of MWNTs by arc discharge was achieved under high helium pressure of 500 Torr and high purity MWNTs and SWNTs were obtained through the thermal oxidation and acid-treated.

ACKNOWLEDGEMENT

This work was supported by Korea Research Foundation Grant (KRF 2003-005-D00011). (2004).

REFERENCES
[1] S. Iijima, Nature, 354 (1991) 56.
[2] J.M. Moon and Y.H. Lee, J. Phys. Chem. B., 105(5) (2001) 677.
[3] Y.H. Lee, D.J. Bae, Y.C. Choi, W.S. Kim and K.S. Kim, Carbon Science, 2 (2001) 120.
[4] Z. Shi, Y. Lian, F. Liao and X. Zhou, Solid State Comm., 112 (1999) 35.
[5] T.W. Ebbesen, P.M. Ajayan, H.Hiura and K. Tanigaki, Nature, 367 (1994) 519.
[6] K.S. Kim, Y.S. Park, H.J. Jeong, S.C. Lim, Y.S. Lee and Y.H. Lee, Carbon Science, 1 (2000) 53.
[7] B. Liu, E. Olssen, R. Yang and S. Zhang, Chem. Phys. Lett., 320 (2000) 365.
[8] H. Kajiura, S. Tsutsui, H. Huang and Y. Murakaami, Chem. Phys. Lett., 364 (2002) 586.

Studies in Surface Science and Catalysis, volume 159
Hyun-Ku Rhee, In-Sik Nam and Jong Moon Park (Editors)

753

A novel catalyst for PTA manufacture: Carbon nanofiber supported palladium

J.-H. Zhou*, Y. Cui, J. Zhu, P. Li, T.-J. Zhao, Y.-C. Dai and W.-K. Yuan

UNILAB, State Key Laboratory of Chemical Reaction Engineering, ECUST, Meilong Rd. 130, Shanghai, 200237, P. R. China *jhzhou@ecust.edu.

1. INTRODUCTION

Carbon supported palladium catalyst is widely used in hydrogenation reactions. However, to make it more effective for some specific application, e.g. terephthalic acid (TA) hydropurification, its preparation, generally from simple precursors, is complicated and laborious. Hydrogenation of crude terephthalic acid (CTA) over Pd/C is a common production process for purified terephthalic acid (PTA), a raw material for polyethylene terephthalate(PET) manufacture, during which the main process is hydrogenation of 4-carboxybenzaldehyde (4-CBA)[1]. Since the Pd/C catalyst was initially developed for CTA hydropurification, there have been numerous efforts on improving measures or looking for alternatives of the Pd/C catalyst, considering some disadvantages such as easy deactivation, easy contamination of the PTA. However, all these efforts have not been so successful hitherto due to the strong acidity of the reaction systems under industrial conditions [2]. Carbon nanofiber(CNF) has preferable properties, such as high surface area, inertness to strong acid and base, high electronic conductivity, and large fraction of graphite edge, etc., therefore, CNF is suggested to be a potential catalyst support. Actually, the use of CNF as catalyst support has been investigated by several authors and tested in various reactions. All the results indicate that CNF is a prospective novel catalytic material [3]. Herewith, we have developed a palladium on CNF catalyst and have applied to CTA hydropurification.

In this work, various conditions have been used for preparing catalysts to screening out a proper one mainly for CTA hydrogenation. The palladium precursor, impregnation time, calcination and reduction will be taken into account.

2. EXPERIMENTAL

2.1. Preparation of CNFs

CNFs were synthesized through the CCVD (Chemically Catalytically Vapor Deposition) by decomposing carbon containing gases on appropriate catalysts. Typically 1.0 g catalyst (ultrafine Fe_3O_4) was loaded in a quartz reactor mounted in a horizontal tubular furnace. After reduction of the catalyst in a 25% hydrogen-argon stream at 600 ℃ for 3 h, a mixture of carbon monoxide/hydrogen (4:1) was charged into the reactor. More than 10 g platelet CNF could be obtained after 16 h growth.

2.2. Preparation of Pd/CNFs

CNF supported palladium catalysts(Pd/CNFs) were prepared by wet impregnation. CNF was slurried in deionized water for about 15 min, and then the palladium precursor solution, which contained a desired amount of Pd (0.5 wt%), was titrated to the slurry. The latter was kept being agitated for a preset period of time, and then filtered (without washing) and finally dried at about 120 ℃ overnight.

2.3. Catalytic performance tests

The performance of the catalyst for the CTA hydropurification was evaluated in a batch autoclave reactor under conditions similar to those in the industry. 90g of CTA containing about 3000 ppm o f 4-CBA and 240 ml of water were charged to the reactor with 1g catalyst loaded. Hydropurification of the CTA was conducted at 280℃ in the reactor under stirring (800 rpm) and 0.7 MPa hydrogen pressure. Samples taken after 0.5 h of reaction were analyzed with HPLC [4]. The catalytic performance of the Pd/CNF catalyst was characterized by 4-CBA's conversion.

3. RESULTS AND DISCUSSION

CNFs, prepared by proper choice of the synthesis conditions, were supported palladium and used for CTA hydrogenation. Results indicated that Pd/CNF catalysts behave satisfactorily. The conversion of 4-CBA reached 98.3% with our novel Pd/CNF catalyst, while 90% with commercial Pd/C under similar evaluation conditions. This may attribute to the unique mesoporous structure of CNF support reducing the diffusion resistance.

3.1. Physical properties of the CNF support

Fig.1 compared a typical pore size distribution of an industrial Pd/C for CTA purification with that of the Pd/CNF prepared by the authors. The activated carbon support had mainly micropores of less than 2 nm. On the other hand, the pores of the CNF were larger and their average diameter was about 10 nm. Meanwhile, our BET analysis indicated that although the specific surface area of activated carbon was greater than 1000 m^2/g, more than 80% of the surface was micropore surface to which the reactants were difficult to access. By contrast, the CNF's specific surface area was only about 200m^2/g. However, it was almost totally external surface, to which the reactants were much more accessible. Thus, the CNF supported catalysts were expected to perform well in liquid reaction systems especially when large molecules were involved and presenting strong diffusion resistance. Inspection of the SEM pictures in Fig.2 would lead to similar conclusions. Skeins of fibers in CNF granules accumulated optionally and chaotically, formed much greater pores which were good for reactants access compared to that in activated carbon.

3.2. Effect of palladium precursor

Two most frequently applied palladium precursor, H_2PdCl_4 and $Pd(NH_3)_4Cl_2$, were used to prepare Pd/CNF by wet impregnation. Results of the CTA hydrogenation showed that under the same palladium loading ratio (0.5Pd%, see Table 2), the Pd/CNF from H_2PdCl_4

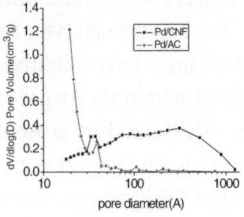

Fig.1. Pore distribution of Pd/C and
Pd/CNF

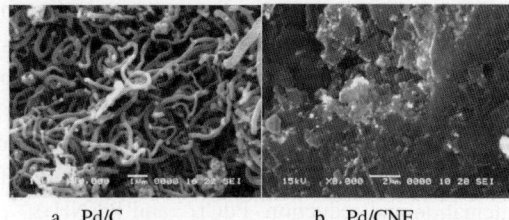

a Pd/C   b Pd/CNF

Fig.2. SEM pictures of Pd/C and Pd/CNF

exhibited a higher catalytic activity than that from $Pd(NH_3)_4Cl_2$. This is in accordance with the literature concerning application of palladium catalysts to other systems. Zeta potential of CNF aggregates, slurried in deionized water, was measured by a Malvern Zetasizer as about +50 mV. Thus it is believed that the interaction of the CNF surface with the negative $PdCl_4^{2-}$ ions was stronger than that with the positive $Pd(NH_3)_4^{2+}$ ions, retarding shifting and sintering of active metal on the support and resulting in a finer palladium dispersion. It was proved by our chemisorption results. Pd dispersion of catalysts after reduction were 15.3% and 13.5% when using $Pd(NH_3)_4^{2+}$ and $PdCl_4^{2-}$ as precursor separately.

3.3. Effect of impregnation time

Table 1 presents the effect of the impregnation time on the catalytic activity in CTA hydrogenation. The catalytic activity decreased with increasing impregnation time. This decrease in catalytic activity could not be related to the amount of the Pd deposited. In fact, the ratio of the loaded Pd amount to Pd precursor amount initially present in the impregnation solution was above 99.5%, thus, the deposited Pd at 0.5 h impregnation time was almost the same as at 20 h impregnation time. Hence, for short impregnation times, Pd would be mainly adsorbed in the outer shell of the CNF aggregates, while for long impregnation times, more metallic atoms seemed to be deposited inside the pores, thus inaccessible for large molecules

Table 1 Effect of impregnation time on the catalytic activity

No.	Impregnation time (h)	4-CBA Conversion (%)
1	0.5	96.2
2	3	96.5
3	6	89.7
4	15	89.0
5	20	90.4

Palladium precursor: $H_2Pd Cl_4$

such as 4-CBA. Therefore, the catalytic activity   decreased with impregnation time as it exceeded 3 h. It was noted that an appropriate impregnation time to prepare Pd/CNF with a high activity was about 0.5 to 1 h.

3.4. Effects of calcination and reduction

Calcination in air and reduction with hydrogen were found to have a profound influence on Pd/CNFs property (Table 2). After calcination in air at 250℃ for 2 h, the catalytic activity decreased evidently when using H_2PdCl_4 as the precursor, but change only slightly when using $Pd(NH_3)_4Cl_2$. However, when first calcinated and then reduced with a H_2/Ar gas stream at 300℃ for 1.5 h, both catalysts activity increased significantly. This could be attributed to changes in composition and crystalline structure of palladium species on support during the calcination and reduction. $PdCl_4^{2-}$ and $Pd(NH_3)_4^{2+}$, adsorbed on the CNF surface during the wet impregnation, were decomposed during the calcination to PdO. Compared to original precursor $PdCl_4^{2-}$ and $Pd(NH_3)_4^{2+}$, whose reduction potential were 0.591 and 0 V separately, PdO formed after calcination, with reduction potential 0.917V, was more difficult to reduce to effective active Pd atom during the liquid hydrogenation process. Hence, the activity after calcinations decreased. Yet, most of the PdO species were transformed to Pd atom when reduced in the H_2/Ar. Thus catalysts activity after reduction increased significantly when they were used for CTA hydrogenation. Moreover, after calcination and reduction, the difference in catalytic activity between catalysts from different palladium precursors diminished for that the calcination change the different reduction precursor $PdCl_4^{2-}$ and $Pd(NH_3)_4^{2+}$ to actual the same PdO, although the interaction between the support and metal was still slightly different for the reason mentioned in 3.2.

Table 2 Effects of calcination and reduction on the catalytic activity

Palladium	4-CBA Conversion (%)		
precursor	before calc.	after calc.	after calc. and redu.
H_2PdCl_4	85.8	45.5	98.3
$Pd(NH_3)_4Cl_2$	31.7	34.2	92.4

4. CONCLUSIONS

1) Preliminary research indicates that CNF, with special morphology and porosity, is a prospective support for palladium catalyst with high activity when applied to CTA hydropurification.

2) Preparation conditions of Pd/CNFs by wet impregnation method, such as palladium precursor, impregnation time, calcinations and reduction, are proved to have profound effect on the catalytic property. The catalyst prepared by impregnating H_2PdCl_4 precursor in an hour, then calcinated in air and reduced in 20%H_2/Ar is believed to perform better in CTA hydropurification than the industrial Pd/C under laboratory conditions.

REFERENCES

[1] I.P. Wheaton, S.A. Cerefice, US Patent No.4 476 242 (1984).
[2] S.H. Jhung, A.V. Romanenko, K.H. Lee et al., Appl. Catal. A: General, 225 (2002) 131.
[3] N.M. Rodriguez, M.S. Kim and R.T.K. Baker, J. Phys. Chem., 98 (1994) 13108.
[4] L.G. Chen and J.H. Zhou, J. of East China University of Sci. and Tech., 30 (2004) 719.

Studies in Surface Science and Catalysis, volume 159
Hyun-Ku Rhee, In-Sik Nam and Jong Moon Park (Editors)

Synthesis and characterization of Mn, Pr doped ZnS and CdS/ZnS nanoparticles

Kwan Hwi Park[a], Hyun Uk Kang[a], Jun Woo Lee[b], Sang Sig Kim[b], Sung Hyun Kim[a]*

[a]Department of Chemical & Biological Engineering, Korea University, 1,5-Ka, Anam-dong, Sungbuk-ku, Seoul 136-701, Republic of Korea

[b]Department of Electrical Engineering, Korea University, 1,5-Ka, Anam-dong, Sungbuk-ku, Seoul 136-701, Republic of Korea

Nanoparticles of Mn and Pr-doped ZnS and CdS-ZnS were synthesized by wet chemical method and inverse micelle method. Physical and fluorescent properties were characterized by X-ray diffraction (XRD) and photoluminescence (PL). ZnS nanoparticles annealed optically in air shows higher PL intensity than in vacuum. PL intensity of Mn and Pr-doped ZnS nanoparticles was enhanced by the photo-oxidation and the diffusion of luminescent ion. The prepared CdS nanoparticles show cubic or hexagonal phase, depending on synthesis conditions. Core-shell nanoparticles enhanced PL intensity by passivation. The interfacial state between CdS core and shell material was unchanged by different surface treatment.

1. INTRODUCTION

For nanoparticles doped with luminescent ions, an efficient energy transfer occur from host to luminescent ions [1]. ZnS doped with luminescent ions generates light with wavelengths in the visible range. Tm and Li-codoped, Mn-doped, and Sm-doped ZnS semiconductors emit blue, orange and red light respectively [2-3]. ZnS nanoparticles codoped with Mn and Pr may show stronger white emission than bulk ZnS powders codoped with these ions. Nevertheless, there is severe non-radiative recombination through the surface states of the nanoparticles. Optical annealing is one of methods to reduce the severe non-radiative recombination due to polymerization and photo-oxidation [4] and enhancement of the crystal quality of the nanoparticles [5].

Core/shell-type nanoparticles overcoated with higher band gap inorganic materials exhibit high PL quantum yield compared with uncoated dots due to elimination of surface non-radiative recombination defects. Such core/shell structures as CdSe/CdS [6] and CdSe/ZnS [7] have been prepared from organometallic precursors.

In this study, we investigated the rare-earth material doping, optical annealing, and core-shell structure for enhancing the luminescent intensity. The synthesis of nanoparticles was first described, and the luminescent properties of undoped, doped ZnS nanoparticles optically annealed in vacuum or air were analyzed. Then, the crystalline and PL properties of CdS nanoparticles were discussed.

758

2. EXPERIMENTAL

ZnS nanoparticles were synthesized by stirring the mixed solution which consisted of Na_2S and $Zn(NO_3)_2 \cdot H_2O$. To synthesize Mn and Pr-doped ZnS nanoparticles, $Pr(NO_3)_3 \cdot 6H_2O$ and $Mn(NO_3)_2 \cdot H_2O$ were added to $Zn(NO_3)_2 \cdot H_2O$ aqueous solution and thereafter Na_2S solution was injected into the mixed solutions. The UV light of 325nm from a He-Cd laser was used as the light source for optical annealing. CdS-ZnS core-shell nanoparticles were synthesized by water/AOT/ether inverse micro-emulsion system. $Cd(NO_3)_2 \cdot 4H_2O$, $Na_2S \cdot 9H_2O$, and $Zn(NO_3)_2 \cdot 4H_2O$ as precursors of CdS-ZnS core-shell nanoparticles and dodecanethiol as a capping material were used.

3. RESULTS AND DISCUSSION

3.1. Characterization of Mn and Pr-doped ZnS nanoparticles

PL spectra of the undoped ZnS nanoparticles optically annealed in air (a) and vacuum (b) are plotted in Fig. 1. PL intensity increases with optical annealing time. In the PL spectrum of the undoped ZnS nanoparticles, the emission band peaks appear at around 420nm. A comparison of Fig.1a and 1b reveals the more significant increase in PL intensity of the nanoparticles annealed in air than vacuum. Optical annealing in vacuum helps in enhancing the crystallinity. In addition to the enhancement of the crystallinity, the optical annealing for ZnS nanoparticles in air induces the photo-oxidation of nanoparticles. PL intensity is increased more by photo-oxidation than the enhancement of crystal quality of the nanoparticles by the annealing.

PL spectra of Mn-doped ZnS nanoparticles optically annealed in air (a) and in vacuum (b) are shown in Fig. 2. For Mn-doped ZnS nanoparticles, the PL band is seen at around 585nm. When Mn-doped ZnS nanoparticles were annealed in air, PL intensity is increased more significantly with UV irradiation time compared with ones annealed in vacuum. PL spectra of Pr-doped ZnS nanoparticles are shown in Fig. 3. The broad emission at 430 nm corresponds to the emission of the undoped ZnS nanoparticles. The other peak is related to the Pr-related complexes. The effect of the optical annealing in air is more notable than in vacuum on the enhancement of luminescent intensity. The increase of PL intensity for Pr-doped ZnS nanoparticles in air is more rapid than undoped or Mn-doped ZnS nanoparticles.

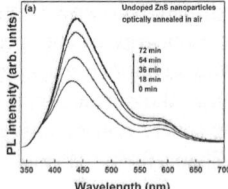 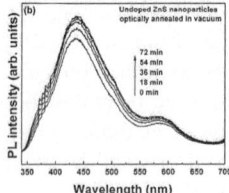

Fig. 1. PL spectra of undoped ZnS nanoparticles optically annealed in air (a) and in vacuum (b)

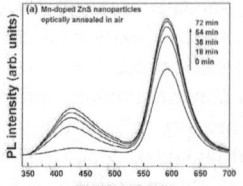 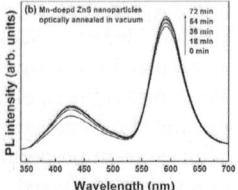

Fig. 2. PL spectra of Mn-doped ZnS nanoparticles optically annealed in air (a) and vacuum (b)

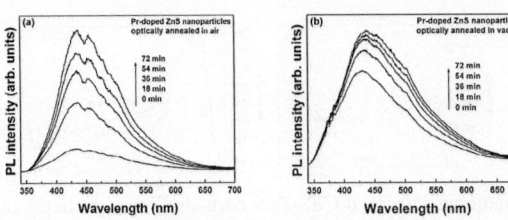

Fig. 3. PL spectra of Pr-doped ZnS nanoparticles optically annealed in air (a) and vacuum (b)

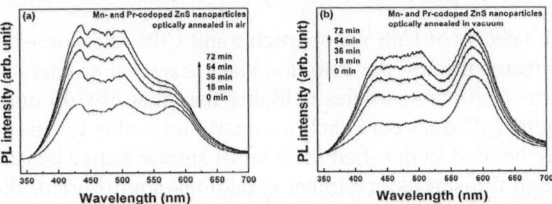

Fig. 4. PL spectra of ZnS: (Mn, Pr) nanoparticles optically annealed in air (a) and vacuum (b)

Fig. 4 shows PL spectra of Mn and Pr-codoped ZnS nanoparticles optically annealed in air and vacuum. Mn and Pr-codoped ZnS nanoparticles emit light of white color. The PL intensity of the Pr-related peaks increased more rapidly than that of Mn-related peak, for the codoped ZnS nanoparticles annealed in air. The different rates may be associated with the luminescent ions. Pr-related complexes are increased with the increasing UV irradiation time, but Mn ions are constant. In case of the annealing in vacuum, Pr-related peaks are initially weaker in intensity than Mn-related peaks due to small Pr-related complexes.

3.2. Characterization of the CdS-ZnS core-shell nanoparticles

Fig. 5 shows XRD patterns of CdS nanoparticles. For the synthesized nanopartices, a mean crystallite size (D) calculated by the Scherrer formula is between 3 and 5nm [8]. XRD pattern of hexagonal CdS in pattern of (a), (b) and (c) of Fig. 5 consists of two broad peak at 25~30° and 44~52°. The peak at 25~30° is due to the convolution of (002) and (101) peak of hexagonal phase. The peak at 44~52° may be due to the convolution of (110), (103), and (112) peak of hexagonal phase. XRD pattern of Fig. 5(d) shows convolution of cubic and hexagonal phase. The peak of Fig. 5(d) at 25~30° may be due to both (101) peak of hexagonal structure and (111) peak of cubic. The peak of Fig. 5(d) at 44~52° may be convolution of (103) peak of hexagonal structure and (220) and (311) peak of cubic structure. The structure of CdS changes to cubic phase with increasing concentration of cadmium nitride. The cubic structure of (111), (220), and (311) peak are shown in XRD pattern of CdS synthesized in ultrasonic condition

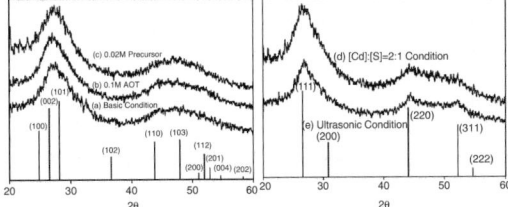

Fig. 5. X-ray diffraction pattern of the Cadmium Sulfide nanoparticles were synthesized in 0.05M AOT and 0.04M precursors condition (a), 0.1M AOT condition (b), 0.02M precursors condition (c), $(Cd(NO_3)_2 \cdot 4H_2O)$: $(Na_2S \cdot 9H_2O)$=2:1 condition (d), and ultrasonic condition(e)

760

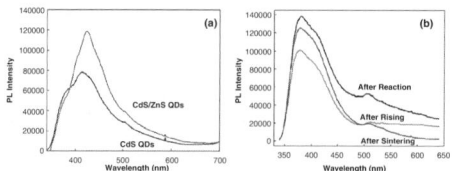

Fig. 6. PL spectra of CdS nanoparticles and CdS-ZnS core-shell nanoparticles (a) and CdS-ZnS core-shell nanoparticles by surface treatment (b)

Fig. 6 shows PL spectra of CdS nanoparticles and CdS-ZnS core-shell nanoparticles. In PL spectrum of CdS nanoparticles, the emission band is seen at around 400nm. The emission band of CdS-ZnS core-shell nanoparticles is higher than that of CdS ones at around 400nm. The PL enhancement of CdS-ZnS core-shell nanoparticles is due to passivation which means that surface atoms are bonded to the shell material of similar lattice constant and much larger band gap [9]. Although the surface treatment conditions are different, the emission band of CdS-ZnS core-shell nanoparticles is same in PL spectra of Fig. 6(b). This indicates that interfacial state between CdS core and shell material was unchanged by different surface treatment.

4. CONCLUSION

Undoped, Mn, and Pr-doped ZnS nanoparticles synthesized by wet chemical method were optically annealed in air or vacuum. PL emission increased with annealing time. This increase is attributed to the photo-oxidation, enhancement in the crystal quality, and diffusion of the luminescent ions. PL intensity of nanoparticles annealed in air increased more significantly due to the photo-oxidation compared with the nanoparticles annealed in vacuum. Mn and Pr-codoped ZnS nanoparticles emitted white light due to the effects of dopants. The optical annealing enhanced the emission intensity.

CdS and CdS-ZnS core-shell nanoparticles were synthesized by inverse micelle method. Crystallinity of CdS nanoparticles was hexagonal structure under the same molar ratio of Cd and S precursor. However it was changed easily to cubic structure under the condition of sonication or higher concentration of Cd than S precursor. The interfacial state between CdS core and shell material was unchanged by different surface treatment.

REFERENCES
[1] R.N. Bhargava and D. Gallagher, Phys. Rev. Lett., 72 (1994) 416.
[2] L.V. Zavyalova, A.K. Savin and G.S. Svechnikov, Displays, 18 (1997) 73.
[3] A.B. Stambouli, S. Hamzaoui and M. Bouderbala, Thin Solid Films, 283 (1996) 204.
[4] T. Tang, M. Yang, and K. Chen, Ceramics Int., 26 (2000) 153.
[5] A.L. Stepanov, and V.N. Popok, Surf. Coat. Tech., 185 (2004) 30.
[6] C. Choo, T. Sakamoto, M. Tohara, K. Tanaka, R. Nakata and N. Okuyama, Surf. Sci., 445 (2000) 480.
[7] X.G. Peng, M.C. Schlamp, A.V. Kadavanich, and A.P. Alivisatos, J. Am. Chem. Soc., 119 (1997) 7019.
[8] M. L. Curri, G. Leo, M. Alvisi, A. Agostiano, M. Della Monica and L. Vasanelliz, J. Coll. Inter. Sci., 243 (2001) 165.
[9] K.K. Song and S.H. Lee, Curr. Appl. Phys., 1 (2000) 167.

Studies in Surface Science and Catalysis, volume 159
Hyun-Ku Rhee, In-Sik Nam and Jong Moon Park (Editors)

Preparation of titania nanoparticles of anatase phase by using flame spray pyrolysis

D.J. Seo[*], M.Y. Cho and S.B. Park

Department of Chemical and Biomolecular Engineering, Center for Ultramicrochemical Process System, Korea Advanced Institute of Science and Technology, Daejeon, Korea, 305-701

ABSTRACT

Properties of titania particles prepared by the flame spray pyrolysis were investigated by XRD, TEM, SEM, and compared with those prepared by the conventional spray pyrolysis. Unlike the conventional spray pyrolysis, the ratio of anatase phase to rutile phase increases with the increase of flame temperature, and pure anatase titania particles were obtained when the mixing-cup temperature of the flame was higher than 1100℃. The pure anatase titania particles had 36nm of crystallite in diameter and 44m^2/g of surface area This is close to the maximum surface area obtained from spherical titania particles of 36 nm in diameter. Short residence time in the order of mili-second and high rate of heating and cooling in the flame and growth zone are responsible for the formation of anatase phase at much higher temperature than 900℃.

1. INTRODUCTION

Titania photocatalyst is used for air and water purification, photo-splitting of water to produce hydrogen, odor control and disinfectant. Crystal structure and crystallite size of titania particles are one of the most important factors that affect on the photoactivity. Photoactivity of anatase is higher than that of rutile, and increases with crystallite size [1]. Therefore, to increase photoactivity, it is desirable to find a route for the synthesis of the pure anatase titania with large crystallite size.

In general, the increase of preparation or calcination temperature helps to increase the crystallite size of anatase titania. However, anatase phase is thermally unstable and is easily converted to rutile phase. Moreover, reactive surface area decreases with increasing the preparation or calcination temperature.

Various methods are applied to the synthesis of titania particles including sol-gel method, hydrothermal method [2], citrate gel method, flame processing and spray pyrolysis [1]. To utilize titania as a photocatalyst, the formation of ultrafine anatase titania particles with large crystallite size and large surface area by various ways has been studied [4].

In this work, flame spray pyrolysis was applied to the synthesis of titania particles to control the crystal structure and crystallite size and compared with the particles prepared by the conventional spray pyrolysis

[*] Currently working for LG Chem, Ltd

2. EXPERIMENTAL PROCEDURE

As a starting material, TTIP (Titinium tetraisoproxide, Aldrich Chem. Co. ltd., 98%) was dissolved in distilled water by adding nitric acid. Total titanium ion concentration was fixed at 0.5M. The precursor solution was converted into droplets by ultrasonic nebulizer of 1.7MHz. These droplets were transported to the reaction region by carrier gas.

In flame spray pyrolysis, reaction region was diffusion flame formed by the combustion of fuel and oxidizer gas out of the concentric flame nozzle. The flow rate of propane as a fuel was varied from 1 to 5L/min, and that of oxygen as an oxidizer was fixed at 10 times of propane. The mixing-cup temperatures of flame were 900, 1100, 1400, 1600, 1900℃ in each case. The mixing-cup temperatures were estimated by measuring the temperature increase of water passing through stainless steel tube in the flame [5]. To compare with conventional spray pyrolysis, same precursor solution was atomized by the same ultrasonic nebulizer and precursor droplets were carried into electrical furnace which operated at various temperatures from 600℃ to 1100℃ by air. The system of conventional spray pyrolysis was shown in our previous work [6].

The properties of titania particles were investigated using X-ray diffraction (XRD, Model D/MAX-RB, Rigaku Ltd.), scanning electron microscopy (SEM, Model 535M, Philips Ltd.), transmission electron microscopy (TEM, Model 2000EX, JEOL Ltd.). The crystallite sizes were estimated by Scherrer's equation and the composition of rutile phase in titania were estimated from the respective integrated XRD peak intensities.

3. RESULTS AND DISCUSSION

In conventional spray pyrolysis, titania particles has rutile phase without calcination when preparation temperature is sufficiently high as shown in Fig.1. Rutile phase appeared with the increase of preparation temperature and surpassed anatase phase when the preparation temperature was above 1000℃. When the preparation temperature was below 800℃, rutile phase was not observed. However, the fraction of rutile phase increased and that of anatase phase decreased with the increase of preparation temperature. Fig. 2 shows the average sizes of anatase crystallite and the fractions of rutile phase as a function of preparation temperature. Rutile phase appeared at 900℃ and its fraction increased from 31% at 900℃ to 97% at 1100℃. The average size of anatase crystallite increased from 12nm to 47nm with increase of preparation temperature. However, pure anatase titania particles had small crystallite size of just 19nm.

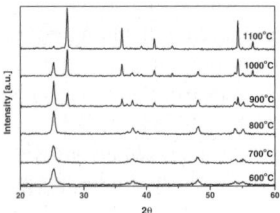
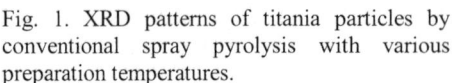
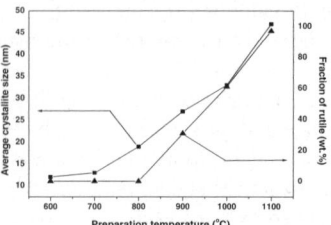

Fig. 1. XRD patterns of titania particles by conventional spray pyrolysis with various preparation temperatures.

Fig. 2. Average anatase crystallite sizes and fractions of rutile phase as a function of preparation temperature in conventional spray pyrolysis.

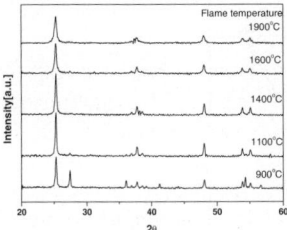

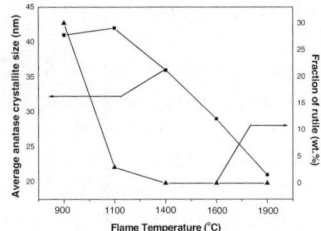

Fig. 3. XRD patterns of titania particles prepared by flame spray pyrolysis with various flame temperatures.

Fig. 4. Average anatase crystallite sizes and fractions of rutile phase as a function of flame temperatures.

Strikingly different results were observed in titania particles prepared by flame spray pyrolysis. Fig. 3 shows that the ratio of anatase to rutile was increased with the increase of flame temperature. This feature was also reported when titania particles were prepared by direct flame processing of TTIP [7]. Fig. 4 shows the average sizes of anatase crystallite and the fractions of rutile phase as a function of mixing-cup temperature. Pure anatase titania particles with average crystallite size of 36nm were formed. It is almost twice as large as the diameter of particles prepared by conventional spray pyrolysis.

The reason for the formation of anatase phase at such a high temperature might be explained as following. The as-prepared ultrafine titania particles are liquefied at sufficiently high temperature because melting point of nanoparticles are lower than that of bulk titania (1850℃). The liquid titania particles are supercooled and became metastable states. The residence time in the flame is only in the order of mili-second so that the metastable phase has no time to become thermodynamically stable phase, rutile.

Another distinguishing feature of titania prepared by flame spray pyrolysis is the decrease of anatase crystallite size with the increase of flame temperature. Generally, the increase of preparation temperature increases the crystallite size in other processes such as sol-gel method, hydrothermal method [2, 3], flame processing and conventional spray pyrolysis. The decrease of crystallite size was directly related to the decrease of particle size. Fig. 5 shows SEM and TEM images of titania particles prepared by flame spray pyrolysis.

At low temperature flame, as-prepared titania particles had bimodal particle size. However,

Fig. 5. SEM images of titania particles prepared by flame spray pyrolysis with various flame temperatures (a) 900℃ (b) 1100℃ (c) 1400℃ (d) 1600℃ (e) 1900℃ (f) TEM image of titania particles at 1600℃

764

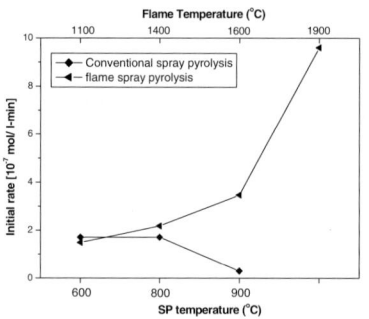

Fig. 6. Comparison of initial rate with titania particles prepared by flame spray pyrolysis and spray pyrolysis.

submicrometer-sized titania particles disappeared with the increase of flame temperature, and nanometer-sized titania particles appeared.

Fig. 6 also shows the striking difference in photoactivties of the particles prepared by conventional spray pyrolysis and flame spray pyrolysis. As the preparation temperature was increased, the rate of TCE decomposition in liquid phase was decreased in the conventional spray pyrolysis whereas the reaction rate kept increasing with the increase of flame temperature.

4. CONCLUSIONS

Compared with the conventional spray pyrolysis, flame spray pyrolysis produces titania particles that are strikingly different in crystal phase and surface area. The fraction of anatase phase increases with the increase of flame temperature while it decreases with the increase of preparation temperature in the conventional spray pyrolysis. The surface area and photoactivity also show the opposite trend with the preparation temperature. This is explained due to the shorter residence time in the flame than in the furnace used for conventional spray pyrolysis. The short residence time prevents the anatase phase titania from transforming to the thermodynamically stable phase which is rutile. High rate of heating and cooling is also responsible for the formation of small size crystallite anatase phase at high flame temperature. It is evident that the flame spray pyrolysis produces photocatalyst of high activity but further investigation is needed to explain the reason why photoactivity is increased with the preparation temperature.

Acknowledgements

The work was supported by BRAIN KOREA 21 program and CUPS (Center for Ultramicrochemical Process System)

REFERENCES
[1] K.Y. Jung and S.B. Park, Korean J. Chem. Eng., 18[6] (2001) 879.
[2] S.W. Nam and G.Y. Han, Korean J. Chem. Eng., 20[1] (2003) 180.
[3] C.H. Cho, D.K. Kim and D.H. Kim, J. Am. Ceram. Soc., 86[7] (2003) 1138.
[4] M. Anpo, T. Shima, S. Kodama and Y. Kubokawa, J. Phys. Chem., 91 (1987) 4305.
[5] D.J. Seo, S.B. Park, Y.C. Kang and K.L. Choy, J. Nanoparticle Res., 5 (2003) 199.
[6] S.B. Park, Y.C. Kang, J. Aerosol Science, 28[suppl.1] (1997) S473.
[7] G. Yang, H. Zhuang and P. Biswas, Nanostruct. Mater., 7[6] (1996) 675.

Studies in Surface Science and Catalysis, volume 159
Hyun-Ku Rhee, In-Sik Nam and Jong Moon Park (Editors)
© 2006 Elsevier B.V. All rights reserved

765

Synthesis of composite particulates with tailored morphologies using dry particle coating

J.-B. Jun, J.-H. Lee, J.-G. Park, J.-G. Koo and J.-I. Kim

Chemical R&D Center, Cheil Industries Inc.,
332-2 Gocheon-dong, Uiwang-si, Gyeonggi-do 437-711, Republic of Korea

1. INTRODUCTION

The modification of particle surface by coating process which alters the surface properties, morphologies and/or functionality of fine particles or powders is a method for preparing highly functional materials [1]. Therefore, it has gained much interest in many industrial applications that require improving and controlling the particle properties such as catalytic activity, electric, electrostatic, optical, magnetic, rheological properties, wettability, solubility, etc. In general, surface modification of particles is carried out by employing wet-based chemical processes such as coacervation, interfacial and/or *in-situ* polymerization. Although, wet-based processes have been used most commercially in pharmaceutical, food, and agricultural industries, it becomes less desirable due to its long process time and recent environmental issues over the resulting waste water, streams and the volatile organics. Dry coating process, as opposed to wet coating process, enables direct attachment of fine (guest) particles onto coarse (host) particles, excluding the use of any solvents or water. In this dry process, composite particles can be formed in either a discrete or continuous coating using a number of different devices such as some types of milling machine, mechanofusion, hybridizer, magnetically assisted impaction coater (MAIC), and theta composer [2-5].

The morphology of composite particulates as well as the functionality of particles usually determines the properties of composites. A model flow of the composite formation using hybridization system is shown in Fig. 1. When the two components are very different in size, the smaller particles tend to adhere onto the larger particles, which is usually referred as ordered mixing or structured mixing [6]. Then, the ordered mixture is converted to the composite particulates through the embedding and/or filming stages by the impaction forces.

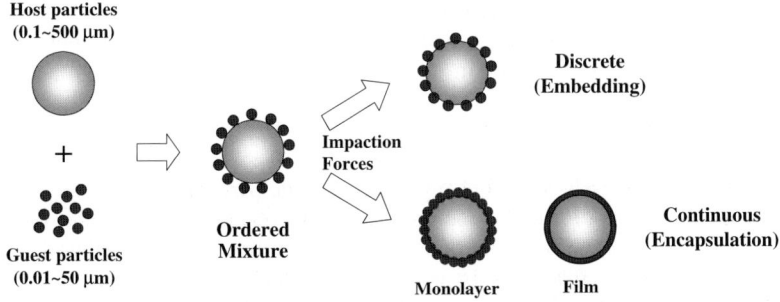

Fig. 1. Schematic model flow of hybridization system

So, not only the operating conditions but also the properties of particles such as size, specific gravity (or density), hardness, T_m and morphology can significantly affect the final morphology of the composite particulates. The objective of this work is to investigate the morphological changes of metallic/organic composite particulates in view of the operating conditions and the guest material design.

2. EXPERIMENTAL

In this study, a hybridizer (NHS-0) manufactured by Nara Machinery Co. of Japan was used as a dry coating device. The hybridizer is composed of a very high-speed rotating rotor with six blades, a stator and a powder recirculation circuit [3, 4]. The powder placed in the processing vessel is subjected to high impaction and dispersion due to the high rotating speed of the rotor. The rotor speed was controlled from 6,000 to 14,000 rpm and the processing time was fixed at 9 min. And the total charge of powders in the hybridizer was fixed to 20 g. Metal (nickel/gold)-plated conductive particles in the micron size (4~5 μm) were used as host particles. Guest particles used were divinylbenzene (DVB)-crosslinked polystyrene (PS) fine particles in the submicron size (0.1~0.4 μm) which were synthesized via an emulsion polymerization method. The characteristics of the materials used in the experiments are given in Table 1. The coating performance was examined with changing the mixing fraction and the properties of guest fine particles. A scanning electron microscope (SEM) was utilized to study the surface morphology and particle shape after dry coating.

Table 1
Physical properties of materials used

Sample	Average size [μm]	Specific gravity [g cm^{-3}]	DVB Content [wt%]	Remarks
Host	4.3	2.7	-	Ni/Au-coated polymer bead
Guest	0.12, 0.25, 0.35	1.2	0, 10, 20, 50	(crosslinked) PS

3. RESULTS AND DISCUSSION

Host particles of metallized conductive beads were coated with crosslinked PS fine particles with varying the operating conditions and the guest particle properties. The coating performance was investigated in view of the morphological change of composite particles. To evaluate the structure of coated composite particles, an image analysis for SEM photographs was carried out. Individual composite particles were classified into three states and graded as follows: embedding (-1), filming (+1), and intermediate (0) which contains both embedding and filming of guests at similar rate in one particulate. Total 100 composite particles were analyzed and numbered, and then the sum of individual points was utilized to decide morphological structure of composite particle whether the coating is close to embedding or encapsulation.

Fig. 2 shows the morphological change of composite particles with different guest sizes by varying the rotor speed. Here, the mixing fraction of guest/host particles was fixed at 0.08. As one can see, the rotor speed is very important factor in determining whether the coating of guest particles is formed discretely or continuously. But, the morphological difference of composites with the change of guest size was not distinct at the same rotor speed.

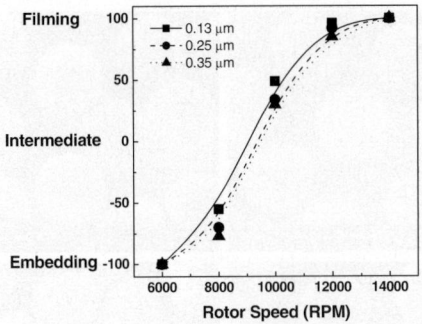

Fig. 2. Morphological changes of composite particles as a function of rotor speed at different sizes of guest particles crosslinked with 20% DVB

Honda *et al.* disclosed that the size of particles and the size ratio of guest to host are important factors for the formation of a monolayer coating [2]. However, though the size and size ratio significantly affect the adhesion of guest particles and so the formation of ordered mixture and/or close compaction, it does not guarantee whether the guest particles deform or embed during coating process.

Fig. 3 shows the effect of the crosslinking degree of guest particles on the coating morphology. The morphological change of composites showed almost same tendency at relatively low crosslinking degrees, DVB 10 and 20%, of guest particles. But, the coating with highly crosslinked guest particles (DVB 50%) was likely to produce composite particles of embedding structure at mild hybridization conditions. The coating with linear (non-crosslinked) PS fine particles mostly showed filming morphologies regardless of rotor speed and so the data was not listed. Typically, the hybridizer produces coated composite particles where the forces that guest and host particles are subjected to are very high. The mechanical energy, produced by the collision and the friction of particles, is applied to the fine particles and either deform the particles or cause the fine particles to be embedded on the host particle surface [3]. Therefore, the higher energy generated by faster rotating can cause increased deformation of guest particles and finally filming of coating. Also, the increased stiffness according to the crosslinking makes the guest particles resist the filming deformation.

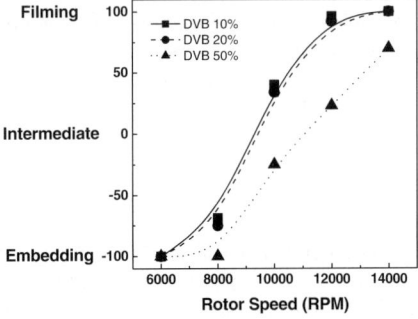

Fig. 3. Morphological changes of composite particles as a function of rotor speed at different DVB crosslinking degrees and same size (0.25 μm) of guest particles

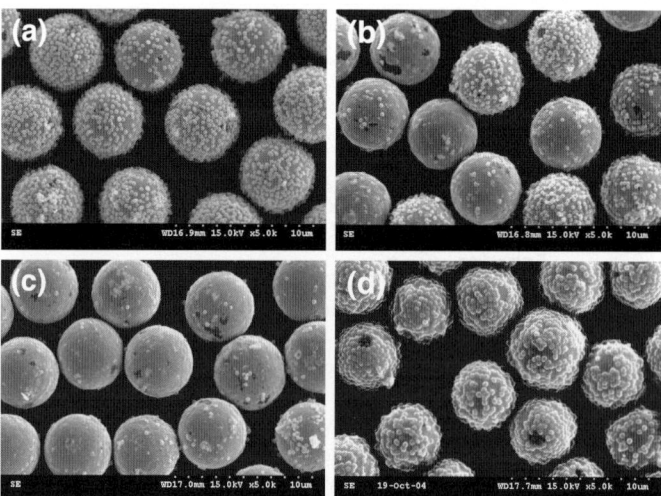

Fig. 4. SEM photographs of composite particulates by dry coating at different conditions: (a), (b) and (c) for composites with 0.25 μm PS particles crosslinked with 20% DVB at 6000, 10000, and 14000 rpm, respectively. (d) for composites prepared at 10000 rpm with 0.35 μm PS fine particles having different crosslinking degree (core of 0.25 μm with 50% DVB and shell with 10% DVB)

SEM images of differently coated composite particles are shown in Fig. 4. Typical morphologies for the discretely coated (bump-type) particles (Fig. 4(a)) and the encapsulated coating particles (Fig. 1(c)) were observed according to the rotor rotating speed. Fig. 4(b) showed intermediate state between embedding and filming of coating. From the above results, the rotor speed for producing the intermediate morphology was found to range 9000~11000 rpm depending on guest particle property. From this point, an interesting concept that core maintains the spherical shape and shell goes to filming by controlling the crosslinking degree of guest particles was originated. In practice, composite particles of encapsulated bump morphology were observed using the core-shell guest particles (Fig. 4(d)).

4. CONCLUSIONS

The encapsulation of host particles with a continuous film of guest materials was attained under which the impaction force (or energy) is enough to deform and conglutinate the guest particles. In conclusion, the morphology of composite particles using the hybridizer was strongly dominated by the rotating hybridization condition and the guest particle property.

REFERENCES

[1] K. Iinoya, K. Gotah and K. Higashitani, Powder Technology Handbook Second Edition, Marcel Dekker, 1997.
[2] H. Honda, M. Kimura, F. Honda, T. Matsuno and M. Koishi, Colloid Surface A, 82 (1994) 117.
[3] R. Pfeffer, R. N. Dave, D. Wei and M. Ramlakhan, Powder Technol., 117 (2001) 40.
[4] M. Koishi, Hybridization. In: S. Yuichi (eds.), Technologies and Applications of Polymeric Ultramicrospheres, CMC Publishing, Tokyo, 2001.
[5] M.R. Mohan, R.N. Dave and R. Pfeffer, AIChE Journal, 49 (2003) 604.
[6] J. Hersey, Powder Technol., 11 (1975) 41.

Studies in Surface Science and Catalysis, volume 159
Hyun-Ku Rhee, In-Sik Nam and Jong Moon Park (Editors)
© 2006 Elsevier B.V. All rights reserved

Preparation of nitrogen-doped TiO$_2$ powder by plasma jet

Yong-Tae Park and Dong-Wha Park

Department of Chemical Engineering and Regional Research Center for Environmental Technology of Thermal Plamsa, Inha University, 253 Yonghyun-Dong, Nam-Gu, Incheon 402-751, Korea

1. INTRODUCTION

Nanostructured titanium dioxide (TiO$_2$) has been of great interest for a variety of applications including photocatalysis[1], nanoceramics, photovoltaics, and gas sensors. Nano-sized TiO$_2$ has been fabricated by using sol-gel, sputtering, combustion flame, and thermal plasma[1-6]. The thermal plasma process has unique characteristics for preparing nanopowders, because high temperature and rapid cooling gave the short processing time.

The photochemical activity of pure TiO$_2$ has been investigated extensively for decades, and it has been revealed that the primary limitation is poor solar spectrum photon absorption because of its wide band gap. Recently, it has been reported that narrowing band gap can be achieved by doping TiO$_2$ with other elements such as nitrogen[7], sulfur, carbon, etc. For example, Ihara et al.[8] reported nitrogen doping shifts the absorption band as well as narrows the band gap.

In this work, we aimed to investigate the effect of doping nitrogen on the characteristics and demonstrate the feasibility of improvement of photocatalytic activity of TiO$_2$ under visible light.

2. EXPERIMENTAL

Pure TiO$_2$ and nitrogen doped TiO$_2$ nanopowders were synthesized from titanium tetrachloride (TiCl$_4$, 99.9%, Aldrich Co.) in a thermal plasma reactor, which is shown in Fig.1. The system mainly consists of a DC plasma torch (W cathode and Cu anode), a reaction tube, a quenching chamber and filter. The torch, reaction tube and chamber were properly cooled by water. Plasma was generated by Ar for pure TiO$_2$ or Ar + N$_2$ mixture gas for nitrogen doped TiO$_2$ with a total flow rate of 15 L min^{-1}. The reaction time was 10 minute in any case, and the process pressure was 750 torr. TiCl$_4$ was injected into plasma region by Ar or N$_2$ carrier gas. The synthesized powders were collected mainly at the reaction tube wall. The exhaust gases passed through a scrubber to remove chlorine compounds and other contaminants.

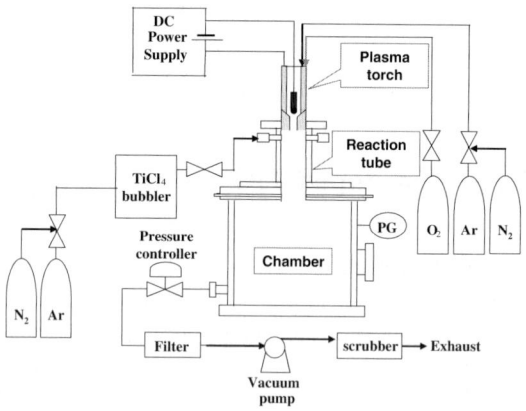

Fig.1. Experimental apparatus for preparation of nano sized nitrogen doped TiO_2

3. RESULTS AND DISCUSSION

Three kinds of samples were prepared by varying reactant gas compositions. When nitrogen was added, the color of synthesized powder was changed from white to yellow and with increase of nitrogen content, it became more yellowish. From the color of synthesized powder, the nitrogen content in the sample can be estimated.

Crystal structure of synthesized powder was analyzed by X-ray diffractometer (DMAX-2500, Rigaku Co.). Figure 2 compares XRD patterns of the powders synthesized at various conditions with those of P-25. In the case of sample prepared under pure O_2 flow (Fig. 2(b)), the main phase was confirmed to be anatase with a little rutile phase and no difference was observe compared with P-25. When small amount of nitrogen was added ($O_2:N_2 = 20:1$, Fig. 2(c)), pure anatase phase was synthesized without rutile phase. Interestingly, when excess N2 was used ($O_2:N_2 = 1:40$, Fig. 2(d)), the main peak of anatase shifted to lower angle, which is the characteristic peak of nitrogen-doped TiO_2 (TiOxNy)[6]. From the results of XRD analysis, it was concluded that for the synthesis of nitrogen doped TiO2, excess amount of nitrogen is needed.

Scanning electron microscopy (S-4200, Hitachi Co.) and particle size analyzer were used to investigate morphology and particle size. Figure 3 shows SEM image of nitrogen doped TiO_2 synthesized under excess nitrogen flowing ($O_2:N_2 = 1:40$). Average particle size was 30 nm, measured by particle size analysis. Generally, nano sized materials can be synthesized by thermal plasma method, taking advantage of rapid heating and cooling procedure, which is its unique characteristic. No change in morphology and particle size was observed with different reactant gas composition.

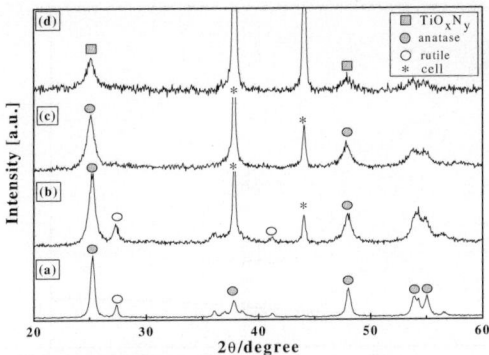

Fig.2. XRD patterns of commercial TiO_2 and synthesized powder at various conditions
(a) Commercial TiO_2 (b) Synthesized TiO_2 (c) O_2:N_2=20:1 (d) O_2:N_2=1:40

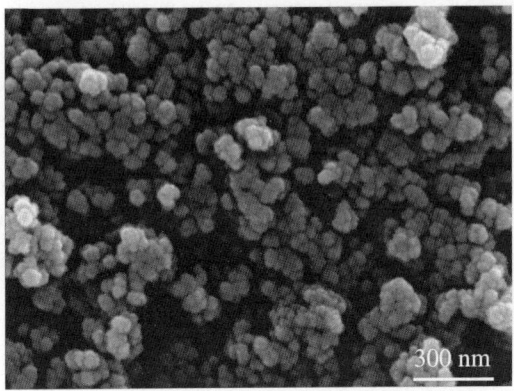

Fig. 3. SEM image of nitrogen doped TiO_2

The optical properties were investigated using UV/Vis. (Lambda 25, Perkin Elmer Co.) spectra with the range from 200 to 600 nm. The optical properties of photoelectrochemcal materials are very important because it means the number of photons absorbed on the materials. UV spectroscopy was performed on four samples to observe changes in optical properties as a function of reactant gas composition, as shown in Fig. 4. It was observed that the spectrum of synthesized TiO_2 under pure oxygen flow was similar to that of commercial TiO_2 powder (P-25). With nitrogen doping, the absorption peak was shifted to longer wave length (visible range) and with increase of nitrogen content, the absorbance in visible region increased, indicating a change in effective band gap energy. The band gap narrowing could be attributed to the interstitial nitrogen species in the TiO_2 lattice.

772

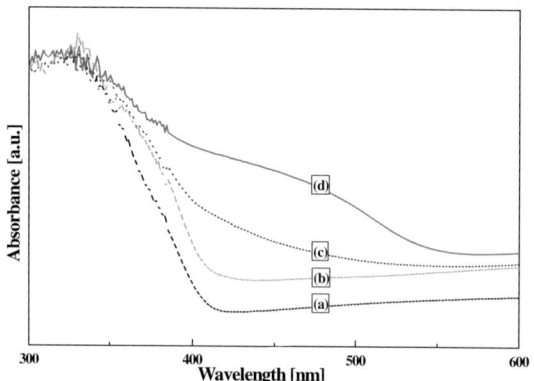

Fig.4. UV/Vis. spectra of commercial TiO_2 and synthesized powder at various conditions
(a) Commercial TiO_2 (b) Synthesized TiO_2 (c) O_2:N_2=20:1 (d) O_2:N_2=1:40

In summary, nitrogen doped TiO_2 has been successfully synthesized by thermal plasma method. Nitrogen doping density and crystal structure were found to be strongly dependent on the reactant gas composition. TiO_2 with anatase phase was synthesized under pure oxygen flow, while excess amount of nitrogen yielded nitrogen doped TiO_2. Compared with TiO_2, nitrogen doped TiO_2 showed more absorbance in visible region and lower band gap energy. Nitrogen doped TiO_2 prepared by thermal plasma method is very promising for photocatalysis applications, and they are under investigation in our group.

ACKNOWLEDGMENTS

This work was supported by INHA UNIVERSITY Research Grant.

REFERENCES

[1] S.M. Oh, S.S. Kim, J.E. Lee, T. ishigaki and D.W. Park, Thin Solid Films, 435 (2003) 252.
[2] Y.L. Li and T. Ishigaki, Thin Solid Films, 407 (2002) 79.
[3] Y.L. Li and T. Ishigaki, Chem. Mater., 13 (2001) 1577.
[4] S.M. Oh, J.G. Gong and D.W. Park, J. Chem. Eng. Jpn., 34 (2001) 283.
[5] J.E. Lee, S.M. Oh and D.W. Park, Thin Solid Films, 457 (2004) 230.
[6] S.M. Oh, J.G. Li and T. Ishigaki, J. Mater. Res., 20 (2005) 529.
[7] R. Ashi, T. Morikawa, T. Ohwaki, K. Aoki and Y. Taga, Science, 293 (2001) 269.
[8] T. Ihara, M. Miyoshi, Y. iriyama, O. Matsumoto and S. Sugihara, Appl. Catal. B: Environ., 42 (2003) 403.

Studies in Surface Science and Catalysis, volume 159
Hyun-Ku Rhee, In-Sik Nam and Jong Moon Park (Editors)

Preparation of nickel powders from nickel salt by ultrasonic chemical reduction method

S.Y. Jeon, J.S. Yun, D.H. Son, C.H. Hwang, S.S. Hong, and S.S. Park[*]

Division of Applied Chemical Engineering, Pukyong National University, Busan 608-739, South Korea

Submicron nickel powders have been synthesized successfully from aqueous $NiCl_2$ at various temperatures and times with ethanol-water solvent by using the conventional and ultrasonic chemical reduction method. The reductive condition was prepared by the dissolution of hydrazine hydrate into basic solution. The samples synthesized in various conditions were characterized by the means of an X-ray diffractometry (XRD), a scanning electron microscopy (SEM), a thermo-gravimetry (TG) and an X-ray photoelectron spectroscopy (XPS). It was found that the samples obtained by the ultrasonic method were more smoothly spherical in shape, smaller in size and narrower in particle size distribution, compared to the conventional one.

1. INTRODUCTION

Ultrasonic technology recently has attracted interest as an alternative to conventional processing. It has been reported that ultrasound may enhance chemical and physical changes in a liquid medium through the generation and subsequent destruction of cavitation bubbles [1, 2]. Over the past decade, fine nickel powder have been studied extensively due to its potential applications such as conducting paints, rechargeable batteries, chemical catalysts, optoelectronics, magnetic recording media, etc. [3]. Recently, it has attracted a great deal of attention as the inexpensive internal electrode of a multiplayer ceramic capacitor (MLCC) in the electronic industry. In the other hand, to prepare fine metal powders, ball milling, electrodeposition, thermal plasma, polyol process, chemical vapor deposition, wet chemical reduction in aqueous solution, and many other methods have been developed [4]. As the needs for the desired properties of fine metal powder and the economical aspects of process, one of candidate methods is the wet chemical reduction method [5]. In this method, the morphology of nickel powder, such as the shape, the size, and the size distribution of particles, can be easily controlled by reaction parameter, solvent composition, a nucleation agent, a reduction agent, etc. [6, 7].

In this study, the chemical reduction in aqueous solution using conventional and ultrasonic hydrothermal reduction method were conducted for the preparation of fine nickel powders from the aqueous solution of nickel salt by reducing with hydrazine. The differences in the reaction parameters and final product properties resulting from two methods were identified to find the effects of ultrasound.

---

[*] To whom correspondence should be addressed.
  E-mail: sspark@pknu.ac.kr

2. EXPERIMENTAL

A commercially available ultrasonic cleaner was used for the preparation of nickel powders from nickel salt in aqueous solution. This cleaner, Model 3210 (Branson Ultrasonic Corp., CT), is normally used as a cleaning apparatus, working at a frequency of 47 kHz with the power of 130 W that consists of a stainless-steel bath of 5.17 l capacity and has an ultrasonic transducer attached to the bottom of the bath. A liquid solution temperature in the bath can be varied from room temperature to maximum of 80 °C.

Starting material was nickel chloride hexahydrate ($NiCl_2 \cdot 6H_2O$). Doubly distilled water and reagent-grade ethanol were used for preparing alcohol-water solvent. Sodium hydroxide (NaOH) and sodium carboxyl methyl cellulose (Na-CMC) were used as a pH control agent and a dispersant, respectively. Hydrazine hydrate ($N_2H_4 \cdot H_2O$) was used as a reducing agent. All reagents were used in as-received form with no further purification. Starting solution was prepared by agitating for 15 min after adding 0.8 mol l^{-1} of $NiCl_2 \cdot 6H_2O$ and 4 g l^{-1} of Na-CMC into ethanol-water solvent (the volume ratio of ethanol = 0.4) at room temperature. The pH of starting solution was controlled to be about 12.0 by adding NaOH. Slurry was obtained by adding slowly 2.0 of molar ratio of $N_2H_4 \cdot H_2O$ to $NiCl_2 \cdot 6H_2O$ into the starting solution, agitated continuously, and then heated at various temperatures for 40 min by using a conventional and ultrasonic heating unit. The addition of $N_2H_4 \cdot H_2O$ turned the slurry's color to black typically within a few minutes, which indicates the nucleation of nickel particles. After mixing and heating, the slurry became supersaturated and precipitated. Precipitate was separated from mother liquor by vacuum filtration, and then washed repeatedly in distilled water until pH became 7. The washed precipitate was dried at 70 °C for 24 h in a vacuum dry oven.

3. RESULTS AND DISCUSSION

To find the effect of reaction temperature and ultrasound for the preparation of nickel powders, hydrothermal reductions were performed at 60 °C, 70 °C and 80 °C for various times by using the conventional and ultrasonic hydrothermal reduction method. Table 1 shows that the induction time, when starts turning the solution's color to black, decreases with increasing the reaction temperature in both the method. The induction time in the ultrasonic method was relatively shorter, compared to the conventional one. It assumes that hydrothermal reduction is faster in the ultrasonic method than the conventional one due to the cavitation effect of ultrasound.

Table 1
Comparison of induction time and the properties of samples prepared in the conventional and ultrasonic hydrothermal reduction method

Method	Reaction temp. (°C)	Induction time (min)	Particle size (μm)	Tap density (g cm^{-3})
Conventional	60	12.9	0.54	1.39
	70	8.3	0.34	1.35
	80	6.0	0.32	1.34
Ultrasound	60	10.0	0.30	1.45
	70	6.0	0.27	1.40
	80	3.5	0.23	1.38

From XRD analyses (Fig. 1), the crystalline peaks of nickel were only observed for all of the samples obtained for 40 min by both the method. From SEM analyses (Fig. 2), the particles in all of samples were almost spherical shape without agglomeration. The average particle sizes of the samples were ranged from 0.32 μm to 0.54 μm in the conventional method and from 0.23 μm to 0.30 μm in the ultrasonic method. The particle size distribution of the samples obtained by the ultrasonic method was much narrower than that of the samples obtained by the conventional one. Above results indicates that the average particle size of the samples obtained by the conventional method is relatively larger at the same condition than that obtained by the ultrasonic one because the spherical surface of nickel particles is much irregular and some large particles exists along with the small particles in the samples obtained by the conventional one. Also, it is known that the average particle size decreased with increasing reaction temperature in the conventional and ultrasonic method because the formation of many nuclei during the nucleation period suppresses the growth of particles.

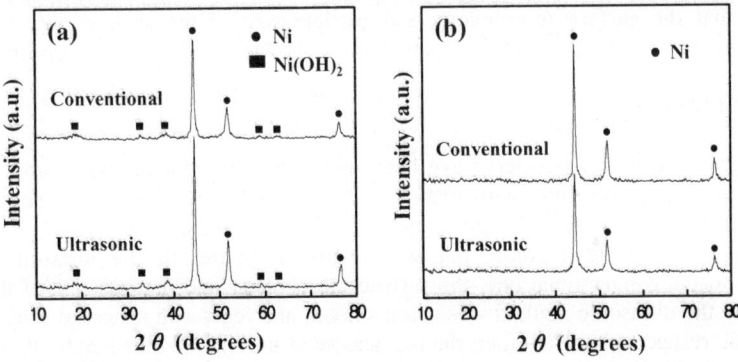

Fig. 1. XRD patterns of the samples prepared at 80 °C for (a) 10 min and (b) 40 min by the conventional and ultrasonic hydrothermal reduction method

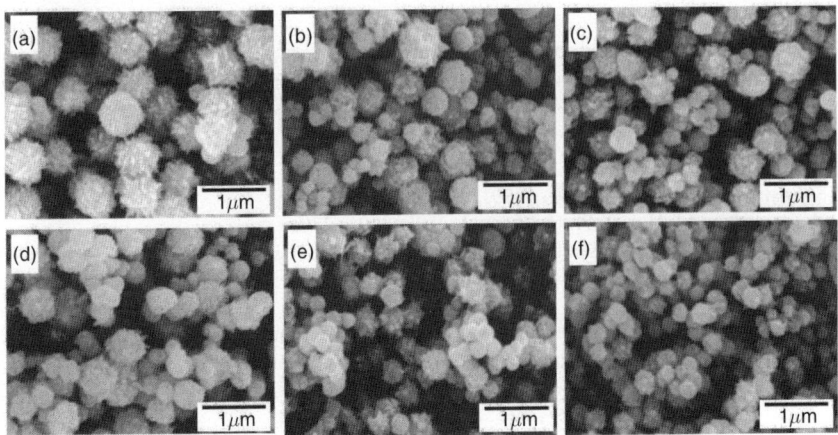

Fig. 2. SEM micrographs of the samples prepared at various temperatures for 40 min by the conventional and ultrasonic hydrothermal reduction method; (a) 60 °C, (b) 70 °C, and (c) 80 °C in the conventional method and (d) 60 °C, (e) 70 °C, and (f) 80 °C in the ultrasonic method

From TG results, it was found that the weight loss occurs at about 300 °C, and weight gain of about 20~24 % starts at about 320 °C and then stops at about 600 °C in all of the samples. The weight loss at about 300 °C may be due to the dehydration of $Ni(OH)_2 \cdot xH_2O$ or the decomposition of $Ni(OH)_2$ into NiO and the weight gain at above 320 °C may be due to the thermal oxidation of Ni into NiO [7]. Even though XRD results reveal that the single phase of crystalline nickel exists only in all of the samples, the possibility of existence of $Ni(OH)_2$ on the surface of samples can be checked. From XPS results, it was found that $Ni(OH)_2$ exists on the surface of all of the samples. It means that the weight loss at about 300 °C in TG results is caused by the dehydration of $Ni(OH)_2 \cdot xH_2O$ or the decomposition of $Ni(OH)_2$ existed on the surface of all of the samples. Therefore, it is known that the surface area of the samples obtained by using both the method is not reduced perfectly to form pure nickel powders due to the rapid and strong reduction of hydrazine hydrate used as a reduction agent.

As the previously shown in Table 1, the tap density of the sample obtained by using the ultrasonic method was relatively higher than that obtained by the conventional one. The reason is that the surface morphology and particle size of the sample obtained by the ultrasonic method are much smooth and small as the shown in SEM results, respectively.

4. CONCLUSIONS

The spherical fine nickel powders have been prepared from aqueous $NiCl_2$ and hydrazine hydrate at various temperatures with ethanol-water solvent by the conventional and ultrasonic hydrothermal reduction method. The induction time decreased with increasing the reaction temperature in both the method, but was relatively shorter in the ultrasonic method. Compared to the conventional one, the surface morphology and particle size of the sample obtained by the ultrasonic method was much smooth and regular in spherical shape and was much small, respectively. Therefore, the tap density of the sample obtained by the ultrasonic method was relatively higher than that obtained by the conventional one.

ACKNOWLEDGEMENTS

This research was supported by the Program for the Training of Graduate Students in Regional Innovation which was conducted by the Ministry of Commerce, Industry and Energy of the Korean Government and by Brain Busan 21 Project.

REFERENCES

[1] K.S. Suslick, Science, 247 (1990) 1439.
[2] M.A. Beckett and I. Hua, J. Phys. Chem. A., 105 (2001) 3796.
[3] S.H. Park, C.H. Kim, Y.C. Kang and Y.H. Kim, J. Mat. Sci. Lett., 22 (2003) 1537.
[4] K. Yurij, F. Asuncion, R.T. Cristina, C. Juan, P. Pilar, P. Ruslan and G. Aharon, Chem. Mater., 11 (1999) 1331.
[5] H.G. Zheng, J.H. Liang, J.H. Zeng and Y.T. Qian, Mat. Res. Bull., 36 (2001) 947.
[6] Y.T. Moon, H.K. Park, D.K. Kim and C.H. Kim, J. Am. Ceram. Soc., 78 (1995) 2690.
[7] K.H. Kim, Y.B. Lee, E.Y. Choi, H.C. Park and S.S. Park, Mat. Chem. Phys., 86 (2004) 420.

Studies in Surface Science and Catalysis, volume 159
Hyun-Ku Rhee, In-Sik Nam and Jong Moon Park (Editors)

777

Preparation of Mg(OH)₂/PMMA core-shell nanocomposite by emulsion polymerization

Eun Ju Park[a], Jin Ho Kim[a], Myung Jun Moon[b], Chan Park[c] and Kwon Taek Lim[a]*

[a]Division of Image Science and Engineering, Pukyong National University,
[b]Division of Chemical Engineering, Pukyong National University
[c]Division of Materials Science & Engineering, Pukyong National University, Pusan 608-739, Korea.

Nanosized magnesium hydroxide ($Mg(OH)_2$) was modified with 3-(trimethoxysilyl)propyl methacrylate (γ-MPS) as silane coupling agent, and then core-shell nanocomposites of inorganic/organic pair were synthesized by the emulsion polymerization using methyl methacrylate (MMA) as the shell monomer. The compatibility of $Mg(OH)_2$ surface with MMA was highly improved through the condensation of γ-MPS. The structure of core-shell composite was examined by transmission electron microscope (TEM).

1. INTRODUCTION

Encapsulated particles consisting of an inorganic core and polymer shell are of interest in various applications, such as cosmetics, inks, and paints because of their better physical properties [1-3]. Though many efforts have been devoted to the preparation core-shell type polymer/inorganic composites [4,5], few work was done for $Mg(OH)_2$ which is known to be an environmentally friendly halogen-free flame retardant additive. The lack of adhesion between an acrylic matrix and inorganic particles is sometimes responsible for a premature rupture. A solution for improving interface adhesion can be found through the grafting of a polymerizable organic molecules onto inorganic particle surface [6,7]. We report here the results on the preparation of new core-shell nanocomposite of $Mg(OH)_2$ and PMMA by emulsion polymerizations. This can be achieved through the preliminary treatment of the $Mg(OH)_2$ surface with γ-MPS leading to an organophilic coating and capable of later copolymerizing with MMA.

2. EXPERIMENTAL

2.1. Materials

Magnesium hydroxide (provided from Skynics), with average particle size of 50 nm, was dried at 110 ℃ under vacuum for 24 h prior to use. 3-(Trimethoxysilyl) propyl methacrylate (γ-MPS, from Sigma-Aldrich) and methyl methacrylate (MMA, from Junsei Chem.) were purified before use. Ammonium persulfate (APS, from Junsei Chem.), polyoxyethylene (50) nonyl phenyl ether (NP-1060), and ammonium (POE) alkyl arylether sulfate (Eu-S133D) were used without further purification.

2.2. Polymerization

The grafting of γ-MPS onto Mg(OH)$_2$ was carried out under argon atmosphere according to the literature procedure.[7] Modified Mg(OH)$_2$ was isolated by ultracentrifuge(Hanil, MF550) at 3000 rpm and washed with toluene for three times to remove unreacted γ-MPS. In a typical emulsion polymerization of MMA, NP-1060 as nonionic surfactant and Eu-S133D as anionic surfactant were introduced to deionized water solution in a 250ml three-neck round bottom flask equipped with an anchor-like stirrer. Then γ-MPS treated Mg(OH)$_2$ in MMA was added to the solution. The resulting mixture was stirred at 600 rpm for 30 min under argon stream. Then, the suspension was heated to 80℃ before addition of the initiator (ammonium persulfate, 0.05 wt% relative to MMA) to start polymerization. The polymerization was performed with a stirring speed of 700 rpm for 3 h. The recipe and the reaction parameters investigated in this study are shown in Table 2.

2.3. Characterization

IR characterizations of pristine and modified powders were performed using a BOMEM Hartman & Braun FTIR spectrometer. TEM experiments were performed with a JEOL JEM-2010 (accelerating voltage of 200kV) for inorganic particles and a HITACHI H-7500 microscope (accelerating voltage of 80kV) for inorganic particle/PMMA nanocomposite. The samples were prepared by dropping the dilute suspension solution of particles on a copper grid coated with a carbon membrane. BET analysis of pristine and modified powders were carried out on a MICRO MATRIX ASAP-2000. The particles were dried on the 300℃ for 4 hrs under vacuum before BET analysis.

3. RESULTS AND DISCUSSION

The reaction scheme for core-shell nanocomposite is illustrated in Fig.1. The grafting of γ-

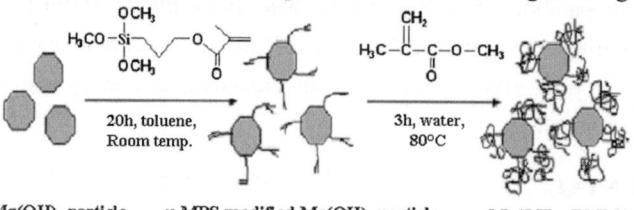

Mg(OH)$_2$ particle   γ-MPS modified Mg(OH)$_2$ particle   Mg(OH)$_2$ /PMMA

Fig. 1. Schematic representation of the process for the synthesis of Mg(OH)$_2$/PMMA nanocomposites.

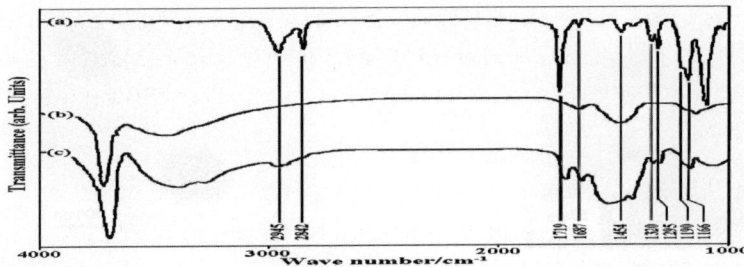

Fig. 2. IR spectra of: (a) γ-MPS reagent, (b) pristine magnesium hydroxide, (c) γ-MPS modified magnesium hydroxide

MPS occurred successfully on the nanosized $Mg(OH)_2$ particles. Fig. 2 shows IR spectra of original γ-MPS (a), $Mg(OH)_2$ before modifying with γ-MPS (b), and $Mg(OH)_2$ after modifying (c). The effective silanisation was evident from the FT-IR spectrum where strong α,β-unsaturated ester bands at 1687, 1320, and $1190 cm^{-1}$ and methyl bands at ca. $2900 cm^{-1}$ were observed. BET analysis of pristine and γ-MPS modified $Mg(OH)_2$ was carried out to determine grafting ratio of γ-MPS on the surface. The specific surface area of $Mg(OH)_2$ particles were decreased from 213.3 $m^2 g^{-1}$ to 113.3 $m^2 g^{-1}$ upon modifying with γ-MPS. (Table 1) The grafted density of γ-MPS was roughly calculated to be 29.4 $\mu mol/m^2$ by dividing reacted γ-MPS by the specific surface area of $Mg(OH)_2$.

The emulsion polymerization of MMA was attempted with different ratio of modified $Mg(OH)_2$, MMA, surfactant, and APS. (Table 2) Among the recipes, most stable core-shell nanocomposite latex solution was resulted from Exp. 3. TEM micrograph of pristine $Mg(OH)_2$ and prepared core-shell composite are depicted in Fig 3 and Fig 4, respectively. The pristine $Mg(OH)_2$ particles are aggregated form with mean diameter of 50 nm. The morphology of the core-shell composite is seen to be grapelike composites by PMMA nodules with diameter ca 100 nm. Light PMMA shells coat the dark grafted cores, and most of these microspheres have multiple cores. This core aggregation is likely attributed to the pre-aggregation of pristine $Mg(OH)_2$ particles.

From this observation, it is suggested that MMA could copolymerize with double bond on $Mg(OH)_2$ surface pretreated with γ-MPS and grow like nodule and eventually produce core-shell type nanocomposites.

4. CONCLUSIONS

Core-shell nanocomposite of $Mg(OH)_2/PMMA$ with an average particle size of ca. 500nm where $Mg(OH)_2$ is the core and PMMA is the shell was successfully prepared by the emulsion polymerization of MMA in the presence of surface modified $Mg(OH)_2$. The grapelike core-shell microspheres with PMMA nodules could be obtained as stable latex.

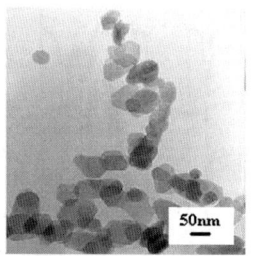

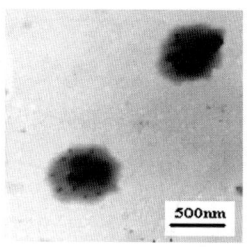

Fig. 3. TEM micrograph of Mg(OH)₂ particle of non-spherical shape with a mean diameter around 50nm.

Fig. 4. TEM micrograph of core-shell nanocomposite of γ-MPS modified Mg(OH)₂ with PMMA.

Table 1
Brunauer, Emmett and Teller (BET) analysis of pristine magnesium and γ-MPS modified magnesium hydroxide

Inorganic powder	Particle size (nm)[a]	Specific surface area (m^2 g^{-1})	γ- MPS surface density (μmol m^{-2})[b]
Pristine Mg(OH)₂	50	213.3	-
γ-MPS modified Mg(OH)₂	50	113.3	29.4

[a] determined by dynamic laser scattering (DLS)
[b] calculated from reacted γ-MPS and specific surface area of pristine Mg(OH)₂

Table 2
Recipes for emulsion polymerization of Mg(OH)₂/PMMA nanocomposite

Exp. No.	Mg(OH)₂ (g)	Monomer MMA (g)	Surfactant NP-1060 (g)	Surfactant EU-S133D (g)	APS (g)	DI water (g)
1		10	0.26	0.05	0.05	15
2	0.1	10	0.26	0.05	0.05	15
3	1	5	0.13	0.025	0.025	15

REFERENCES

[1] Y. Haga, S. Inoue, T. Sato and R. Yosomiya, Angew. Makromol. Chem., 49 (1986) 139.
[2] C.H.M. Caris, A.M. Herk, P.M. Loussia and A.L. German, Br. Polym. J., 21 (1989) 133.
[3] M. Hasegawa, K. Arai and S. Saito, J. Polym. Sci., Part A: Polym. Chem., 25 (1987) 3231.
[4] R. Stephane, P.L. Celine, R. Serge, M. Christophe, D. Etienne and B.L. Elodie, Chem. Mater., 14 (2002) 2354.
[5] S.R. Lee and S.D. Seul, J. Korean Ind. Eng. Chem., 13 (2002) 125.
[6] D. Etienne, A. Maher, M. Fabrice, M. Pierre and F. Michel, Macromol. Symp., 151 (2000) 365.
[7] R. Stephane, M. Christophe, B. L. Elodie, K. Etienne and R. Serge, Nano Lett., 4 (2004) 1677.
[8] N. Tsubokawa, K. Maruyama, Y. Sone and M. Shimomura, Polym. J., 21 (1989) 475.

Studies in Surface Science and Catalysis, volume 159
Hyun-Ku Rhee, In-Sik Nam and Jong Moon Park (Editors)

Carbothermal synthesis of nano-sized tungsten carbide catalyst

Tuan Huy Nguyen, Thanh Vinh Nguyen, Cyrus G. Cooper and Adesoji A. Adesina*

Reactor Engineering & Technology Group, School of Chemical Engineering & Industrial Chemistry, The University of New South Wales, Sydney NSW 2052 Australia
*Corresponding Author: Phone: +61-2-9385-5268; Fax: +61-2-9385-5966
Email: a.adesina@unsw.edu.au

Highly dispersed tungsten carbide particles have been obtained via low temperature – programmed propane carburization of supported tungsten sulphide. Thermogravimetric analysis showed that irrespective of support type, the solid – state conversion was a 2-step process involving the substitutionary elimination of sulphur atom to produce a tungsten carbide. The relatively low activation energy obtained ($50 - 75$ kJ mol^{-1}) also suggested that the solid carburization was diffusion – controlled. Moreover, gas phase kinetic analysis implicated a 1st order dependency on $H_2:C_3H_8$ ratio.

1. INTRODUCTION

The application of early transition metal carbides as effective substitutes for the more expensive noble metals in a variety of reactions has been demonstrated in several studies [1-2]. Conventional preparation route via high temperature (>1200K) oxide carburization using methane is, however, poorly understood. This study deals with the synthesis of supported tungsten carbide nanoparticles via the relatively low-temperature propane carburization of the precursor metal sulphide. In order to optimize the carbide catalyst properties at the molecular level, we have undertaken a detailed examination of both solid-state carburization conditions and gas phase kinetics so as to understand the connectivity between phase kinetic parameters and catalytically-important intrinsic attributes of the nanoparticle catalyst system.

2. EXPERIMENAL

Calculated amount of tungstic acid (H_2WO_4) were dissolved in dilute HNO_3, thioacetamide (CH_3CSNH_2), urea ($CO(NH_2)_2$) and equivalent weight of the support (Al_2O_3, SiO_2, TiO_2 and ZrO_2) to achieve a catalyst with 12wt%W. The conical flask was kept in a 363 K shaker-bath for 4 hours. The resulting tungsten sulphide slurry is then filtered and oven-dried at 393 K for 14 hours. Temperature-programmed carburization studies using different $H_2:C_3H_8$ ratios (1:1-5:1), heating rate (1-20 K min^{-1}) and carburization temperature (773-973 K) were performed on a ThermoCahn TherMax200 TGA to determine the kinetics of the propane carburization reaction:

$$WS_2 + \left(\frac{3+4x}{3}\right)H_2 + \left(\frac{3-x}{3}\right)C_3H_8 \rightarrow WC_{1-x} + 2H_2S + C_2H_6 \tag{1}$$

Carburization was carried out in a fixed-bed stainless steel reactor (ID = 6 mm). Typically about 0.5 g of metal sulphide was sandwiched between 2 layers of quartz wool placed centrally in a temperature-programmed furnace. The sample was heated at 10 $Kmin^{-1}$ to the desired temperature in the presence 100 ml min^{-1} flow rate of H$_2$ and C$_3$H$_8$. After carburization, the sample was cooled down to room temperature under N$_2$ blanket. For each TGA run, 100 mg of metal sulphide catalyst was placed inside the quartz sample boat and subjected to total flow of 55 mlmin^{-1} of the carburizing gas. Weight changes were continuously monitored every 5 seconds via the Cahn Win TGA software.

3. RESULTS AND DISCUSSION

3.1 Thermogravimetric Analysis

Figs. 1(a to d) show the individual thermograms at different heating rates for tungsten carbide formation over each of the four supports employed using a carburizing gas containing H$_2$:C$_3$H$_8$ = 5. The appearance of, at least, 2 distinct peaks suggests that the metal sulphide conversion involved the production of an intermediate phase – possibly, an organometallic sulphide (similar to the oxycarbide species reported when metal oxide substrate was used [3]) – at a relatively low temperature followed by final formation of the metal carbide, WC$_{1-x}$ (higher temperature peak). The appearance of the temperature peak also depended on the heating rate implicating the occurrence of solid intraphase changes during carburization. This behaviour has also been previously observed in bimetallic carbides [4].

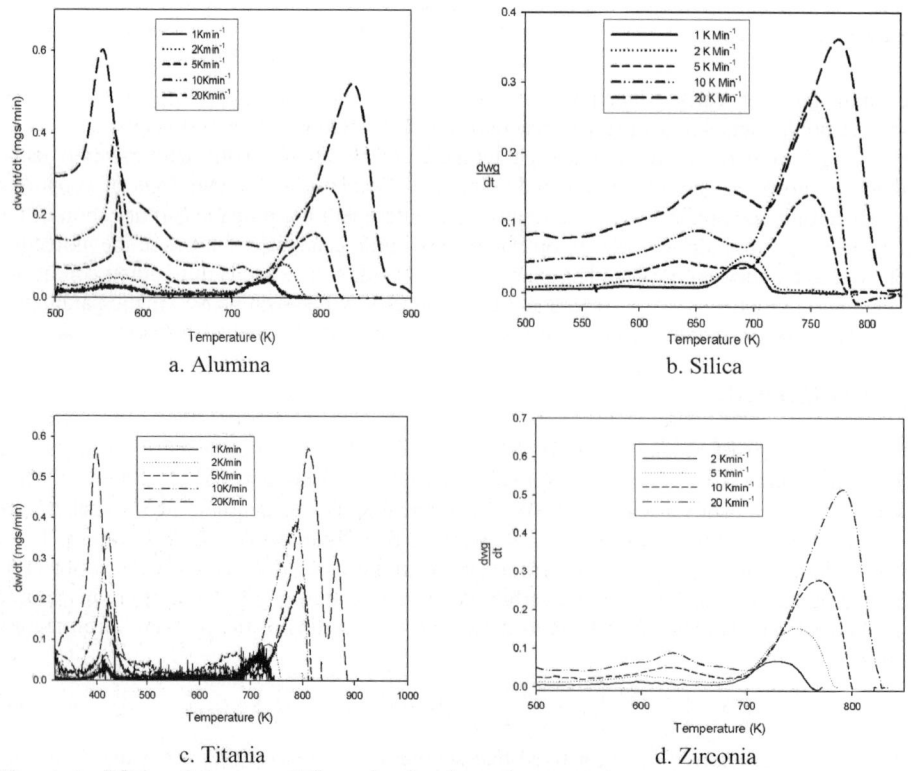

a. Alumina   b. Silica

c. Titania   d. Zirconia

Figs. 1. (a-d) Rate of change at different heating for various supports

As a result of these temperature – programmed runs, additional experiments were conducted isothermally at temperatures between 723 – 873 K. Based on the second phase peak (peak 2), the conversion – time profiles at different temperatures exhibited a characteristic Sigmoid shape and may therefore be used to interpret the solid carburization kinetics. As detailed in Brown [5], the transient conversion data may be described by

$$\left[-\ln(1-\alpha) \right]^{\frac{1}{n}} = kt \qquad\qquad (2)$$

where α = solid conversion, t = time, k = rate constant and $n \leq 2 \leq 4$. Although n appeared to increase from 2 to 4 with increasing temperature (an indication that thermally – stable phases of the tungsten carbide were formed at different temperature via intraphase mechanism), estimate of k at various temperatures provided the activation energy values, E_A displayed in Table 1. The relatively low E_A values (50 – 75 kJ mol^{-1}) suggest that the solid conversion is probably governed by a diffusion – controlled mechanism. This is consistent with the diffusion of carbon atoms to the metal lattice to replace sulphur atoms in the substitutionary reaction. We believe that propane carburization proceeded via the interaction of deposited carbon (from propane dehydrogenation) with the metal sulphide. This proposition was also supported by additional runs in which the feed $H_2:C_3H_8$ ratio was varied. As seen from Figs. 2 (a to d), a linear relationship exists between the peak height (a relative measure of reaction rate) and the $H_2:C_3H_8$ ratio, H_2 appeared to play a dual role in the carburization reaction – the removal of evicted sulphur atom (as H_2S) and excess carbon. The similarity in the gas phase kinetic plots for all supports is consistent with the view that identical mechanism is involved in each case.

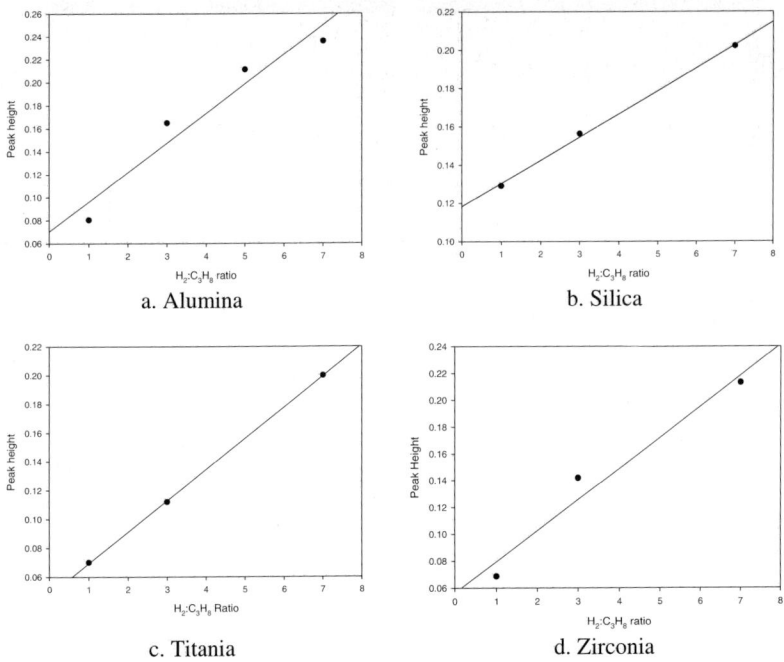

Fig. 2. (a to d) Peak height versus $H_2:C_3H_8$ ratio

784

Electron micrographs (scanning and transmission) showed that tungsten carbide is well dispersed on the surface of each support as nanosized particles (20 – 50 nm) as typified by the images in Figs. 3 (a & b). However, BET surface area decreased in the order alumina > silica > titania > zirconia. With highest surface area obtained for each support being 240, 133, 18 and 9 m^2g^{-1} respectively.

Table 1
Kinetic parameters for all four supports

Arrhenius parameters	Al_2O_3	SiO_2	TiO_2	ZrO_2
E_A (kJ mol^{-1})	75	65	50	50
Frequency factor, k_o (min^{-1})	9500	5000	350	400

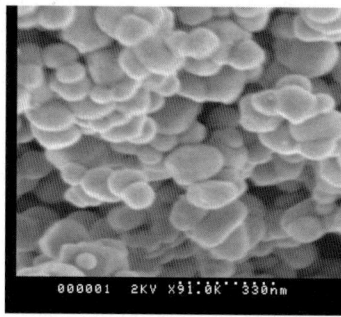 

Fig. 3. (a & b) SEM and TEM image of TiO_2 supported catalyst

ACKNOWLEDGMENTS

This work was produced as part of the activities of the ARC Centre for Functional Nanomaterials funded by the Australian Research Council under the ARC Centres for Excellence Program.

REFERENCES

[1] S. Monteverdi, M.M. Bettahar, D. Begin and F. Mareche, Fuel Proc. Tech., 119 (2002) 77.
[2] T.H. Nguyen, E.M.T. Yue, Y.J. Lee, A. Khodakov, A.A. Adesina and M. P. Brungs, Catal. Comm., 4 (2003) 353.
[3] T-H. Nguyen, A.A. Adesina, E.M.T. Yue, Y-J. Lee, A. Khodakov and M. P. Brungs, J. Chem. Technol. Biotechnol., 79 (2004) 286.
[4] T.H. Nguyen, T.V. Nguyen, Y.J. Lee, T. Safinski and A.A. Adesina, Materials Research Bulletin, 40 (2005) 149.
[5] M.E. Brown, Introduction to Thermal Analysis: Techniques and Applications, Kluwer Academic Publishers, Dordrecht, 2001.

Studies in Surface Science and Catalysis, volume 159
Hyun-Ku Rhee, In-Sik Nam and Jong Moon Park (Editors)

New acid catalyst comprising Keggin-type heteropoly acid supported on mesoporous silica for dehydration of acetic acid

Chang-Soo Woo, Lianhai Lu and Ho-In Lee

School of Chemical and Biological Engineering & Research Center for Energy Conversion and Storage, Seoul National University, Seoul 151-744, Korea

The first attempt to synthesize and characterize Keggin-type heteropoly acid supported on various mesoporous silicas and its application to acid catalysis in the formation of acetic anhydride via dehydration of acetic acid were described in this study. A variety of characterization techniques such as N_2 adsorption, TEM and XRD were applied.

1. INTRODUCTION

Keggin-type heteropoly acids have attracted much attention due to their stronger acidities than those of conventional mineral acids such as H_2SO_4, HCl, HNO_3 and so forth [1]. Tungstophosphoric acid (PW; $H_3PW_{12}O_{40}$) has the strongest acidic strength among the heteropoly acid family, but its surface area is too small to be used as an acid catalyst by itself. In addition, the molecular size of PW is very large (~1.2 nm), and thus it is impossible to introduce PW into the pores of conventional zeolites (<0.7 nm). Therefore, it is important to select a suitable substrate for PW loading. So far, porous silica has been mainly used to support PW for various acid catalyzed reactions. Normal porous silica has a wide range of pore size distribution with a large portion of big pore. However, the random pore distribution is not preferred for the point of improving reaction selectivity. Pure mesoporous slica SBA-15, which was first developed by Zhao et al. in 1998, seems to be a good candidate for supporting PW, because it has very large surface area, large pore volume and big uniform pore size [2]. Typically, its hexagonally arrayed mesopore is about 6 nm in size which can be tuned finely by controlling the synthetic conditions. Taking the advantage of structural characteristics of SBA-15, we selected it as a fundamental material. A series of SBA-15 materials were synthesized and applied for supporting PW species. Although there are some reports about PW supported on silica [3, 4], to the best of our knowledge, no attempts to support PW on SBA-15 have been reported. Dehydration of acetic acid has been known as an acidic catalyzed reaction related with hydroxyl group on silica surface [5]. In this study, we tried to investigate the catalytic activity of PW on SBA-15 for the reaction.

2. EXPERIMENTAL

Various SBA-15 materials were prepared by changing the synthetic conditions. For example, catalysts S1 and S2 were synthesized by applying different aging temperature of 353 K and 373 K, respectively [2]. Aging at 373 K after adding trimethylbenzene (TMB) as a swelling agent with the weight ratio of TMB/surfactant such as 0.2 and 0.4 gave catalysts TS1 and TS2, respectively [2, 6]. And then desired amounts of PW were impregnated on these various SBA-15 materials. Textural properties and BET surface areas of all the prepared catalysts were investigated by nitrogen adsorption at 77 K utilizing a Micromeritics ASAP 2010. Transmission electron microscopic (TEM) images were observed with a JEM-2000 EXII electron microscope operating at 200 kV. X-ray diffraction (XRD) spectra were obtained by using an MQC Science M18XHF22-SRA. Dehydration of acetic acid was performed using a quartz tube reactor at 823 K for 1 h, and the products were analyzed by a GC equipped with a flame ionization detector.

3. RESULTS AND DISCUSSION

3.1. Characterization

Table 1 summarizes the nitrogen adsorption data of all the catalysts. The pore sizes of S1 and S2 remained same with increasing the loading amounts of PW, meanwhile the pore volumes and BET surface areas were decreased, suggesting the PW species was highly dispersed on S1 and S2 with monodisperse pores [7]. However, TS1 and TS2 exhibited different behavior. According to Schmidt-Winkel et al. [6], the pore structure of TMB-added SBA-15 is mesostructured cellular foam (MCF) with ink-bottle pores in which large cells are connected by narrower windows even though pure SBA-15 possesses one dimensionally arrayed hexagonal mesopores. The mesoporous structures of various SBA-15 materials could be inferred from TEM images as illustrated in Fig. 1. Due to these characteristic mesoporous structures, the average cell sizes of TS1 and TS2 were slightly decreased while the window sizes were maintained with PW loading. It suggests that PW species was not uniformly distributed in the pores of substrates as in the case of S1 and S2, but exclusively dispersed on the cell parts. As shown in Fig. 2, adsorption pore distribution corresponding mainly to the cell part clearly demonstrated that main peak at ~11 nm was decreased together with the broad bands in small pore ranges below ~10 nm due to the introduction of PW, but desorption pore distribution originated mainly from the window part didn't change regardless of PW loading. Therefore, X-ray diffraction peaks from Keggin structure of PW were detected over PW/TS1 and PW/TS2, but no peaks were observed over PW/S1 and PW/S2 at the same loadings as shown in Fig. 3.

787

Table 1

Textural properties and catalytic activities of PW supported on various SBA-15 materials

Catalyst	Avg. pore size / nm	Cell size / nm	Window size / nm	Pore volume / cm^{3}g^{-1}	BET s.a. / m^{2}g^{-1}	Selectivity / %
S1	6.4	-	-	1.05	933	29
10% PW/S1	6.4	-	-	0.77	672	55
20% PW/S1	6.4	-	-	0.71	622	25
S2	9.2	-	-	0.99	824	65
10% PW/S2	9.2	-	-	0.92	724	82
20% PW/S2	8.9	-	-	0.83	669	19
TS1	10.2	11.4	8.6	1.59	627	70
10% PW/TS1	10.6	11.7	8.6	1.53	577	63
20% PW/TS1	10.8	12.0	8.5	1.32	486	45
TS2	12.3	13.6	11.1	1.74	565	92
10% PW/TS2	12.7	13.8	11.2	1.67	525	89
20% PW/TS2	13.0	13.9	11.1	1.46	451	83

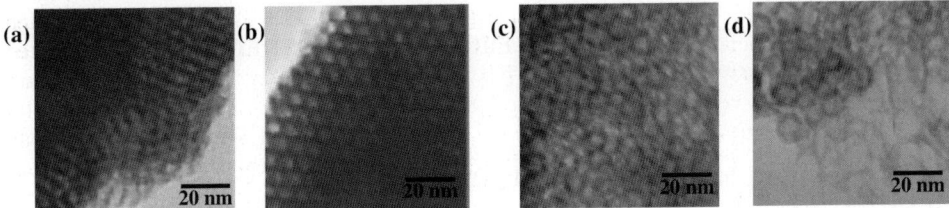

Fig. 1. TEM images of various SBA-15 materials; (a) S1, (b) S2, (c) TS1 and (d) TS2.

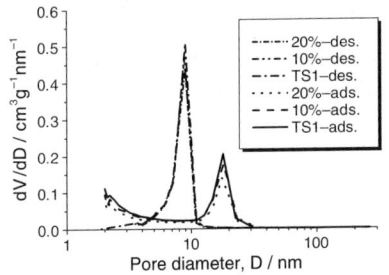

Fig. 2. Pore size distributions of TS1 after PW loading.

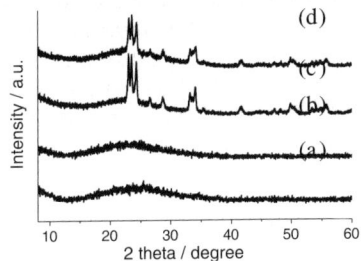

Fig. 3. XRD patterns of (a) 20% PW/S1, (b) 20% PW/S2, (c) 20% PW/TS1 and (d) 20% PW/TS2.

3.2. Catalysis

Conversions of all the catalysts were similar as ~50%, but selectivity for acetic anhydride via dehydration of acetic acid increased with increasing the average pore size in case of pure mesoporous silicas as depicted in Table 1. According to Martinez *et al.* [8], high diffusion rate of the reactant favors the dehydration of acetic acid. Therefore, it was reasonable that TS2 with the largest pore size presented the best activity among the pure mesoporous silicas. However, after PW was loaded on S1 and S2, the selectivity increased at 10% of PW loading, and then decreased at 20%. The enhanced selectivity of S1 and S2 upon PW loading was believed to be caused by high dispersion of PW on the internal surface of SBA-15 with uniform pore size [7, 9]. It was noticeable that the selectivity over TS1 and TS2 did not change significantly but just slightly diminished after PW loading. These phenomena were caused by the fact that PW species was not uniformly distributed but locally dispersed on the cell parts due to the typical MCF structure, resulting in low dispersion of PW at the same loadings.

4. CONCLUSIONS

The selectivity for acetic anhydride in the catalytic dehydration of acetic acid could be controlled by the pore size of pure mesoporous silica SBA-15. New acid catalyst comprising Keggin-type heteropoly acid supported on SBA-15 enhanced the activity effectively when tungstophosphoric acid was highly dispersed on the silica substrate.

ACKNOWLEDGEMENT

This work was financially supported by the ERC program of MOST/KOSEF (Grant No. R11-2002-102-00000-0).

REFERENCES

[1] I.V. Kozhevnikov, Chem. Rev., 98 (1998) 171.
[2] D. Zhao, Q. Huo, J. Feng, B.F. Chmelka and G.D. Stucky, J. Am. Chem. Soc., 120 (1998) 6024.
[3] V.M. Mastikhin, S.M. Kulikov, A.V. Nosov, I.V. Kozhenikov, I.L. Mudrakovsky and M.N. Timofeeva, J. Molec. Catal. A: Chemical, 60 (1990) 65.
[4] F. Marme, G. Coudurier and J.C. Védrine, Micropor. Mesopor. Mater., 22 (1998) 151.
[5] M.C. Libby, P.C. Watson and M.A. Barteau, Ind. Eng. Chem. Res., 33 (1994) 2904.
[6] P. Schmidt-Winkel, W.W. Lukens, Jr., D. Zhao, P. Yang, B.F. Chmelka and G.D. Stucky, J. Am. Chem. Soc., 121 (1999) 254.
[7] N.–Y. He, C.–S. Woo, H.–G. Kim and H.–I. Lee, Appl. Catal. A: General, 281 (2005) 167.
[8] R. Martinez, M.C. Huff and M.A. Barteau, Appl. Catal. A: General, 200 (2000) 79.
[9] C.-S. Woo, N.-Y. He and H.-I. Lee, J. Ind. Eng. Chem., *submitted*.

Studies in Surface Science and Catalysis, volume 159
Hyun-Ku Rhee, In-Sik Nam and Jong Moon Park (Editors)
© 2006 Elsevier B.V. All rights reserved

Synthesis, characterization and catalytic activity of titanium containing mesoporous materials with TS-1 wall structure

Kyoung-Ku Kang[a, b] and Hyun-Ku Rhee[a]*

[a]School of Chemical and Biological Engineering and Institute of Chemical Processes, Seoul National University, Seoul 151-744, Korea

[b]Research Institute of Chemical & Electronic Materials, Samsung Cheil Industries, Inc., Uiwang-si Gyeonggi-do 437-711, Korea.

TS-1/MCM-41 catalysts synthesized by the dry gel conversion method are shown to have hexagonal mesopores. The catalytic activity of synthesized TS-1/MCM-41 catalysts was tested with epoxidation reaction of olefins to reveal that both the conversion of olefins and selectivity to epoxide are higher than those of Ti-MCM-41.

1. INTRODUCTION

The synthesis of titanium-silicates having zeolitic properties was first described in 1967 [1]. More than ten years later, synthesis of titanium containing MFI zeolite was reported [2]. Several other synthesis methods and applications were also reported to demonstrate the possibility of preparing titanium-containing zeolites, but clear evidences for the presence of titanium in framework locations were not reported until 1986 when the synthesis and characterization of the TS-1 was reported [3]. The presence of titanium in the silicalite-1 structure gave rise to high liquid phase oxidation activity with H_2O_2 [4]. Although titanium containing zeolites are useful as catalysts in chemical industry, they have the major limitation arising from their micro pore size, which is too small for the synthesis of large organic compounds.

After the discovery of M41S mesoporous molecular sieves, a significant effort has been made to engineer their structural and surface properties [5]. Due to their mesoporous natures (20–100Å), the titanium-substituted mesoporous materials showed a good potential as epoxidation catalysts for larger molecules [6]. However, Ti-MCM-41 was much less active than TS-1 for liquid phase oxidation and it turns out that the hydrophilic/hydrophobic property of titanium-containing mesoporous materials is an important factor in regard to the catalytic activity.

In this work, highly active epoxidation catalysts, which have hydrophobic surface of TS-1, were synthesized by the dry gel conversion (DGC) method. Ti-MCM-41 was synthesized first by a modifed method and the TS-1/MCM-41 catalysts were subsequently synthesized by the DGC method. The catalysts were characterized by the XRD, BET, FT-IR, and UV-VIS spectroscopy. TS-1/MCM-41 catalysts were applied to the epoxidation of 1-hexene and cyclohexene with aqueous H_2O_2 to evaluate their activities for the epoxidation reaction.*

---

* To whom correspondence should be addressed:
hkrhee@snu.ac.kr, Fax: +82-2-880-1560, Tel: +82-2-880-7405.

2. EXPERIMENTAL

A synthesis gel for Ti-MCM-41 was prepared by the following procedure. Template solution was prepared with distilled water, TMAOH (tetramethylammonium hydroxide) and CTMABr (cetyltrimethylammonium bromide) at 303K. The inorganic sources, which are silica source (TEOS: tetraethyl orthosilicate) and titanium source (TEOT: titanium ethoxide), were prepared by mixing under dry nitrogen. The reactant gel was autoclaved at 383 K for 24 h. The reaction mixture was then treated by hydrothermal restructuring process with strong acidic solution (1 M HCl) and TMAOH solution [7]. Finally, the solid product was hot-filtered, washed, dried and calcined.

The TS-1/MCM-41 catalysts were synthesized in two steps [8]. The first step was involved with the preparation of TPAOH impregnated mesoporous materials and the second step was the DGC process. The TPAOH impregnated Ti-MCM-41 was prepared with calcined Ti-MCM-41, TPAOH (1 M solution of water) and ethanol under stirring by impregnation method. The parent gels were prepared with a TPAOH/Ti-MCM-41 ratio of 1/3 by weight. After 4 h, ethanol and water were removed in a rotary evaporator at room temperature and solid products were dried in a convection oven at 373 K for 48 h. The DGC process was carried out at 448 K for 3 h to obtain TS-1/MCM-41-A and for 6 h to obtain TS-1/MCM-41-B. However, the mesoporosity of Ti-MCM-41 was lost when the DGC process was carried out for 9 h.

The samples were dried at 373 K for 24 h and calcined at 823 K for 5 h. All the synthesized catalysts were characterized by the various analysis techniques such as XRD, BET and FT-IR and UV-VIS.

The catalytic activities of synthesized catalysts were measured for epoxidation of olefins (1-hexene and cyclohexene, respectively).

3. RESULTS AND DISCUSSION

The XRD pattern of Ti-MCM-41 synthesized by the modified synthesis method is presented in Fig. 1. Here we observe more than three distinguishable peaks, which can be indexed to different (*hkl*) reflections of hexagonal structure. These are the (100), (110), (200), and (210) peaks [5]. The highest intensity of (100) peak suggests that this material has a highly ordered hexagonal structure.

Fig. 2 shows wide angle XRD patterns of TS-1/MCM-41-A, TS-1/MCM-41-B and TS-1.

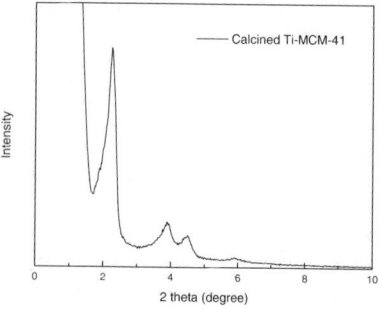

Fig. 1. XRD pattern of Ti-MCM-41

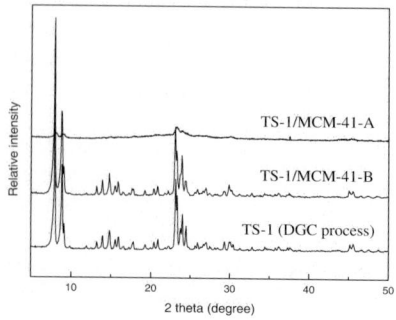

Fig. 2. XRD patterns of TS-1/MCM-41-A, TS-1/MCM-41-B and TS-1.

All the samples were synthesized by using the DGC process. The synthesized samples show typical XRD patterns of crystalline MFI zeolite [1-3], confirming that all the samples have the TS-1 structure. In particular, TS-1/MCM-41-B shows fully developed XRD patterns of TS-1. On the other hand, the XRD intensity of TS-1/MCM-41-A is less than that of TS-1/MCM-41-B, and this indicates that the growth of TS-1 structure depends on the DGC process time.

The nitrogen physisorption isotherm and pore size distributions for the synthesized catalysts are shown in Figs. 3 and 4. The Type IV isotherm, typical of mesoporous materials, for each sample exhibits a sharp inflection, characteristic of capillary condensation within the regular mesopores [5, 6]. These features indicate that both TS-1/MCM-41-A and TS-1/MCM-41-B possess mesopores and a narrow pore size distribution.

The variation in the lattice vibration of the solid products was examined by utilizing the FT-IR technique at successive DGC process times and the results are presented in Fig. 5. The absorption bands at 550 cm^{-1} and 450 cm^{-1} are assigned to the vibration of the MFI-type zeolite and the internal vibration of tetrahedral inorganic atoms. The band 960 cm^{-1} has been assigned to the O-Si stretching vibration associated with the incorporation of titanium species into silica lattice [4]. This indicates that the amorphous wall of Ti-MCM-41 was transformed into the TS-1 structure.

Fig. 6 presents the UV-VIS spectra of the catalysts. UV-VIS spectra of the catalysts show

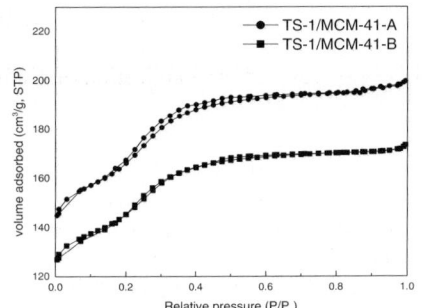

Fig. 3. Nitrogen physisorption isotherms of TS-1/MCM-41-A and TS-1/MCM-41-B.

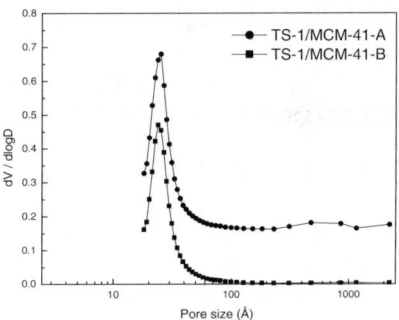

Fig. 4. Pore size distributions of TS-1/MCM-41-A and TS-1/MCM-41-B.

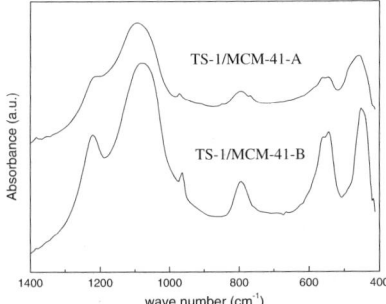

Fig. 5. FT-IR spectra of TS-1/MCM-41-A and TS-1/MCM-41-B.

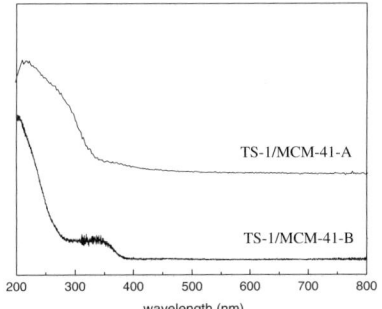

Fig. 6. UV-VIS spectra of TS-1/MCM-41-A and TS-1/MCM-41-B.

absorption band at ca. 220 nm, indicating titanium in tetrahedral coordination [9]. UV-VIS spectral data suggest that much favorable environment for titanium to take tetrahedral position in the subsequent DGC process is established in the titanium containing mesoporous materials, in which the mesoporous materials are the mother supports.

The catalytic activities of synthesized catalysts are given in Table 1. The TS-1 catalyst exhibited the highest epoxide yield and the best catalytic performance for the epoxidation of 1-hexene. The conversion of cyclohexene, however, is the lowest over TS-1. In case of TS-1/MCM-41-A and TS-1/MCM-41-B, the selectivity to epoxide is much higher than that of Ti-MCM-41. Moreover, the conversion of 1-hexene as well as cyclohexene is found larger on the TS-1/MCM-41-A and TS-1/MCM-41-B than on other catalysts. While the epoxide yield from 1-hexene is nearly equivalent to that of TS-1, the yield from cyclohexene is much larger than those of the other two catalysts. These results of olefins epoxidation demonstrate that the TS-1/MCM-41-A and TS-1/MCM-41-B possess the surface properties of TS-1 and mesoporosity of a typical mesoporous material, which were evidently brought in by the DGC process.

Table 1
Experimental results of olefins epoxidation with H_2O_2 over Ti-containg catalysts.

Catalysts	Conversion (%) (1-hexene/cyclohexene)	Yield (%) (1-hexene/cyclohexene)	Selectivity (%) (1-hexene/cyclohexene)		
			Epoxide	1,2-diol	1-ol+1-one
TS-1	38/3	24/3	62/98	8/-	30/-
Ti-MCM-41	26/10	5/1	18/13	21/17	61/69
TS-1/MCM-41-A	48/25	21/16	44/63	26/14	30/23
TS-1/MCM-41-B	41/19	21/14	52/74	27/8	21/19

4. CONCLUSIONS

Titanium containing hexagonal mesoporous materials were synthesized by the modified hydrothermal synthesis method. The synthesized Ti-MCM-41 has highly ordered hexagonal structure. Ti-MCM-41 was transformed into TS-1/MCM-41 by using the dry gel conversion process. For the synthesis of Ti-MCM-41 with TS-1(TS-1/MCM-41) structure TPAOH was used as the template. The synthesized TS-1/MCM-41 has hexagonal mesopores when the DGC process was carried out for less than 3 ~ 6 h. The catalytic activity of synthesized TS-1/MCM-41 catalysts was measured by the epoxidation of 1-hexene and cyclohexene. For the comparison of the catalytic activity, TS-1 and Ti-MCM-41 samples were also applied to the epoxidation reaction under the same reaction conditions. Both the conversion of olefins and selectivity to epoxide over TS-1/MCM-41 are found higher than those of other catalysts.

REFERENCES
[1] D.A. Young, US Patenet No. 3 329 (1967) 481.
[2] M. Taramasso, G. Perego and B. Notari, US Patent No 4 410 (1983) 501.
[3] G. Pergo, G. Bellussi, C. Corno, M. Taramasso and F. Buonomo, Stud. Surf. Sci. Catal., 28 (1986) 129.
[4] T. Tatsumi, M. Nakamura, S. Negishi and H. Tominaga, J. Chem. Soc., Chem. Comm., (1990) 476.
[5] C. T. Kresge, M. E. Leonowicz, W. J. Roth, J. C. Vartuli and J. S. Beck, Nature, 359 (1992) 710.
[6] A. Corma, M. T. Navarro, and J. Perez Pariente, J. Chem. Soc., Chem. Commun., (1994) 147.
[7] K-K. Kang and H-K. Rhee, Stud. Surf. Sci. Catal., 141 (2002) 101.
[8] K-K. Kang and H-K Rhee, Stud. Surf. Sci. Catal., 154 (2004) 497.
[9] A. J. H. P. van der Pol, A. J. Verduyn, and J. H. C. ban Hooff, Appl. Catal., 92 (1992) 113.

NOVEL REACTORS AND PROCESSES

Studies in Surface Science and Catalysis, volume 159
Hyun-Ku Rhee, In-Sik Nam and Jong Moon Park (Editors)

Reactor Engineering Studies on the BiodeNOx absorption process

F. Gambardella, J.G.M. Winkelman and H. J. Heeres

Department of Chemical Engineering, Stratingh Institute, RijksUniversiteit Groningen, Nijenborgh 4, 9747 AG Groningen, The Netherlands.

1. INTRODUCTION

The BiodeNOx process is a novel process concept to reduce NO emissions from flue gases of stationary sources like power plants and other industrial activities [1]. The concept combines a wet chemical absorption process with a novel biotechnological regeneration method. In the wet chemical absorption step, flue gas components are absorbed into an aqueous solution of $Fe^{II}(EDTA)^{2-}$ (EDTA= ethylene-diamino-tetraacetic acid). The following reactions take place:

$$NO_{(aq)} + Fe^{II}(EDTA)^{2-} \leftrightarrow Fe^{II}(EDTA)^{2-}(NO) \tag{1}$$

$$O_{2(aq)} + 4Fe^{II}(EDTA)^{2-} + 2H_2O \rightarrow 4Fe^{III}(EDTA)^{-} + 4OH^{-} \tag{2}$$

The second reaction needs to be suppressed as much as possible as the product $Fe^{III}(EDTA)^{-}$ is not capable of reacting with NO. In the biochemical regeneration step, the loaded iron-chelate solution containing $Fe^{II}(EDTA)^{2-}(NO)$ and $Fe^{III}(EDTA)^{-}$, is contacted with specific micro-organisms that convert the nitrosyl complex back to the original $Fe^{II}(EDTA)^{2-}$ complex and N_2 gas. In addition, $Fe^{III}(EDTA)^{-}$ is reduced to the original $Fe^{II}(EDTA)^{2-}$ compound [2]:

$$6Fe^{II}(EDTA)(NO)^{2-} + C_2H_5OH \rightarrow 6Fe^{II}(EDTA)^{2-} + 3N_2 + 2CO_2 + 3H_2O \tag{3}$$

$$12Fe^{III}(EDTA)^{-} + C_2H_5OH + 3H_2O \longrightarrow 12Fe^{II}(EDTA) + 2CO_2 + 12H^{+} \tag{4}$$

A multidisciplinary research project involving chemical engineering groups from three different Dutch Universities and input from industry was initiated in the year 2000 to demonstrate the feasibility of the BiodeNOx process and to gain insights in the fundamentals of both the wet chemical absorption unit and the bio-reactor. For simplicity, we have focused on a BiodeNOx process configuration with a separate absorber and bioreactor. Our activities in the project involved the design of the absorber unit. For this, the kinetics of the main- and side reaction (eq's 1 and 2) need to be known. We have performed an experimental study to determine the kinetics. Subsequently, a reactor model for a counter current packed column absorber was developed. The results will be summarized in this contribution.

2. RESULT AND DICUSSION

2.1 Kinetic studies

The intrinsic kinetics of the reactions taking place in the scrubber, i.e. the reaction of NO with the iron chelate forming an iron nitrosyl complex (eq. 1) and the undesired oxidation reaction of the iron chelate complex (eq. 2) were determined in dedicated stirred cell contactors. Typical process conditions were: T = 25-55 °C; $[Fe^{II}(EDTA)^{2-}]$ = 1-100 mol/m^3; [NO] = 1-1000 ppm; pH = 5-8 and an oxygen level ranging between 1 and 20 vol%.

The experimental results imply that the main reaction (eq. 1) is an equilibrium reaction and first order in nitrogen monoxide and iron chelate. The equilibrium constants at various temperatures were determined by modeling the experimental NO absorption profiles using the penetration theory for mass transfer. Parameter estimation using well established numerical methods (Newton-Raphson) allowed determination of the equilibrium constant (Fig. 1) as well as the ratio of the diffusion coefficients of $Fe^{II}(EDTA)^{2-}$ and NO [3].

The oxidation reaction (eq. 2) is an irreversible reaction with first order in oxygen and second order in iron chelate. The kinetic constants at various temperatures were determined using a penetration theory based expression derived by De Coursey [4] and the results are presented in Fig. 2.

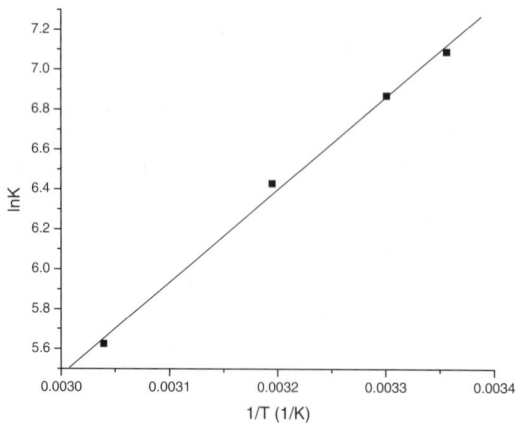

Fig. 1. Equilibrium constant as a function of the temperature for the reaction between NO and $Fe^{II}(EDTA)^{2-}$

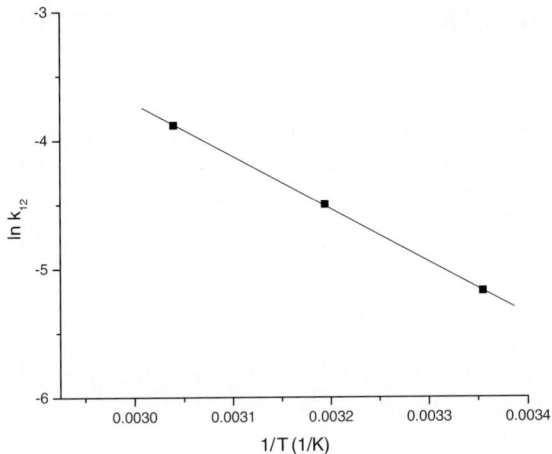

Fig. 2. Kinetic constant of the oxidation reaction (eq. 2) as a function of the temperature

The temperature dependence of the kinetic constant may be expressed as (pH = 7):

$$k_{12} = 5.3 \cdot 10^3 \cdot e^{-\frac{4098}{T}} \quad (m^6/mol^2\ s).$$

2.2. Absorber modeling

All experimental results obtained in this study have been integrated in a rate based reactor model describing the simultaneous absorption of NO and oxygen in an aqueous $Fe^{II}(EDTA)$ solutions in a counter current packed column reactor operated at isothermal and steady state conditions. Under a typical BiodeNOx condition ($C_{FeII(EDTA)}$ = 30 mol/m³, T = 323 K, C_{NOin} = 250 vppm, C_{O2in} = 5 % vol) combined with a superficial liquid velocity of 0.01 m/s, a superficial gas velocity of 1 m/s, a volumetric gas flow rate of 556 m³/s and using 1" metal Pall rings packing material, the height of the scrubber to achieve 90% NO removal efficiency was calculated 0.91 m by the model. A remarkable improvement in absorber performance may be expected when operating the column at lower temperatures (T < 323 K). The required column height for 90% removal efficiency is reduced considerably, see Fig. 3 for details. In addition, the rate of oxidation is also significantly reduced at low temperatures.

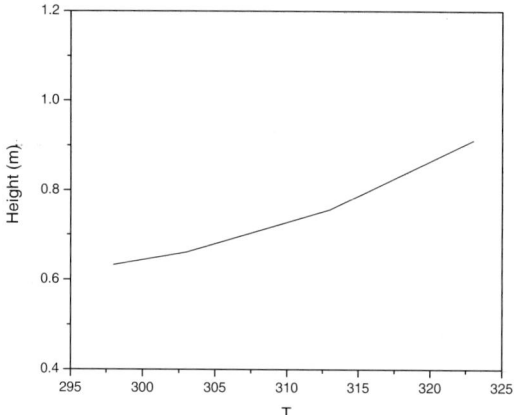

Fig. 3. Calculated column height for 90% NO removal efficiency as a function of the temperature

3. CONCLUSIONS

The kinetics of the reactions taking place when contacting flue gas containing NO and O_2 with aqueous $Fe^{II}(EDTA)^{2-}$ solutions were determined. With this information, a steady state, rate based BiodeNOx absorber model was developed for a counter current packed column absorber unit. The model has been applied to determine the optimum reaction conditions for the absorber. Absorber performance is improved considerably when operating at the low end of the temperature range.

REFERENCES

[1]   C.J.N. Buisman, H. Dijksman, P.C. Verbaak, A.J. den Hartog, WO Patent No. 96/24434.
[2]   P. van der Maas, P. van den Bosch, B. Klapwijk and P. Lens, Biotechnology and Bioengineering, 90 (2005), 433-441.
[3]   F. Gambardella, M.S. Alberts, J.G.M. Winkelman and H.J. Heeres, Ind. Eng. Chem. Res., 44 (2005) 4234-4242.
[4]   W.J. de Coursey and R.W. Thring, Chem.Eng.Sc., 44 (1989) 1715.

Studies in Surface Science and Catalysis, volume 159
Hyun-Ku Rhee, In-Sik Nam and Jong Moon Park (Editors)

Polymerization of methyl methacrylate in supercritical carbon dioxide with PDMS based stabilizers: A study on the effect of stabilizer anchor groups

H. S. Ganapathy, H. S. Hwang, Y. T. Jeong and K. T. Lim*

Division of Image Science and Engineering, Pukyong National University, Pusan 608-739, Korea.

The effect of different type of polydimethysiloxane (PDMS) based stabilizers on the dispersion polymerization of MMA in supercritical carbon dioxide (scCO$_2$) has been investigated in terms of their anchor group architecture. PDMS chain was used in each case as a CO$_2$-philic portion of the stabilizer and five different PMMA-philic endgroups were investigated: an alcohol, an acetate group, a methacrylate unit, and two block copolymers containing acrylate groups. A trifunctional block copolymer, PDMS-b-(PMMA(1.1K)-co-PMA(0.5K) was found to be very effective stabilizer due to the efficient stabilizer anchor-soluble balance (ASB) which produced excellent yield, high molecular weight, and uniform particles of micron sized PMMA.

1. INTRODUCTION

Supercritical carbon dioxide (scCO$_2$), essentially a 'green' solvent and environmentally benign, is rapidly becoming a viable alternative solvent for polymer synthesis and materials processing [1,2]. However, scCO$_2$ is a poor solvent for polymerization, an effective stabilizer is necessary to disperse the growing polymer chains in the reaction medium [3]. Though, a number of stabilizers based on block and graft copolymers of fluoro and siloxane polymers have been described, there have been a limited number of studies into the effect of surfactant architecture on stabilizing ability. Canelas et al. investigated the anchor-soluble balance (ASB, the ratio of polymer-philic to CO$_2$-philic portions in a surfactant) of PS-b-PDMS which was used to stabilize dispersion polymerizations of styrene in scCO$_2$. [4] It was demonstrated that the ASB of the stabilizer has a dramatic effect on both the progress of the reaction and the morphology of the resulting polystyrene colloid.

In this study, PDMS stabilizers with different anchoring groups were examined for use in dispersion polymerizations of MMA in scCO$_2$. Five PMMA-philic groups were investigated as anchoring units: an alcohol group, an acetate group, a methacrylate group, and block copolymers with poly (methacrylic acid) (PMA) block and (PMMA-co-PMA) blocks (Fig. 1). These anchor groups were combined with an identical CO$_2$-philic PDMS chain (M$_n$ ~5000) allowing their individual anchoring ability to be investigated.

Table 1

Polymerization results for different end capped PDMS stabilizers

Entry	Stabilizer	Wt % of Stab.	Mn[a]	PDI[a]	Yield (%)
1	PDMS-OH (5K)	5	24,800	1.64	60
2	PDMS-OAc (5K)	5	35,000	1.68	64
3	PDMS-mA (5K)	5	105,900	1.64	76
4	PDMS-mA (5K)	10	125,600	1.65	77
5	PDMS-b-P(MMA(1.1K)-co-MA(0.5K))	5	183,000	1.67	89
6	PDMS-b-P(MMA(1.1K)-co-MA(0.5K))	10	119,000	1.64	91
7	PDMS-b-PMA(1K)	5	124,900	1.66	84
8	PDMS-b-PMA(1K)	10	119,500	1.57	88

[a]Determined by GPC with PMMA standard

2. EXPERIMENTAL

2.1. Materials and synthesis

Poly(dimethyl siloxane) monomethacrylate (PDMS-mA) (5K), monohydroxy terminated PDMS (PDMS-OH) (5K) (Aldrich) were used as received. Ester capped Polydimethyl-siloxane monoacetate (PDMS-Ac) (5K) was prepared by the acetylation of PDMS-OH. PDMS block copolymers were synthesized as reported previously [5,6]. In a typical polymerization for MMA in scCO$_2$, 0.5 g of MMA, 0.005g (1wt% of monomer) of AIBN, 5 wt % (to monomer) of a PDMS stabilizer, and a teflon-coated stir bar were introduced into a stainless steel reactor (4ml). The reactor was pressurized by ISCO syringe pump (Model 260D) containing compressed CO$_2$. Following pressurization, the reactor was heated to 65 °C by immersing in a water bath. The polymerization was conducted at 345 bar for 15 hrs. After

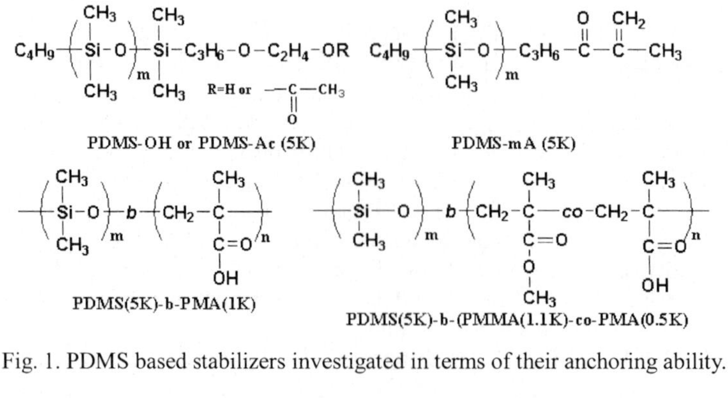

Fig. 1. PDMS based stabilizers investigated in terms of their anchoring ability.

polymerization, any unreacted MMA was extracted with liquid CO$_2$ at 70 bar at ambient temperature and the product was collected and weighed.

3. RESULTS AND DISCUSSION

The results of the polymerization of MMA are summarized in Table 1. The effect of the surfactant anchor group on stabilizing activity was investigated in terms of PMMA yield, molecular weight, and morphology. The unmodified PDMS-OH (5k) led to lower in yield and molecular weight (Table 1, entry 1) with highly flocculated morphologies (see Fig. 2. A), possibly due to short anchoring group. Moreover, this stabilizer is likely to undergo a certain degree of self-association as a result of the hydroxyl end group, which may prevent it from sufficiently anchoring to the PMMA [7]. When the same reaction was repeated in the presence of PDMS-acetate, there was negligible improvement observed in yield and molecular weight with a flocculated morphology of PMMA particles, which were similar to that of particles produced by PDMS-OH. The methacrylate terminated PDMS led to considerably high PMMA yield and molecular weight. (Table 1, entry 3 and 4). In this case, the stabilizer is covalently grafted into the growing PMMA chains via the methacrylate end group during polymerization. This would significantly enhance the ability of the PDMS to anchor to the growing polymer and may explain the observed improvement in the polymer product. Similar trend of results were observed recently, for a fluorinated stabilizer, perfluoropolyether (PFPE) with a range of varying anchor group functionality for the dispersion polymerization of MMA in scCO$_2$ [8]. In the case of block copolymers, though PDMS-b-PMA resulted in good yield and high molecular weight, particles were not uniform sized. On the other hand, a trifunctional block copolymer, PDMS-b-P(MMA(1.1K)-co-MA(0.5K)) was found to be the most successful stabilizer, which produced fine white powder with excellent yield, high molecular weight with uniform particles (Table 2, entry 5 and Fig. 2 C). This indicates that excellent CO$_2$-soluble/anchoring balance is achieved in the stabilizer.

According to theory and simulation studies by Johnston [7] and coworkers, colloidal suspensions will be stable above the upper critical solution pressure (UCSP) if the graft

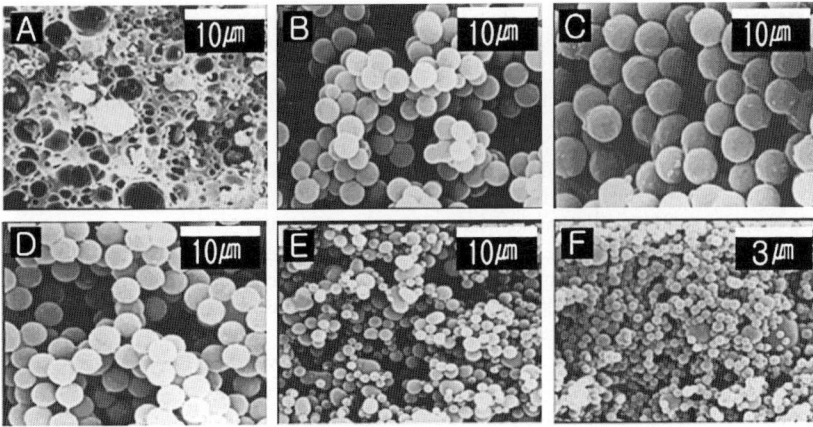

Fig. 2. Scanning electron micrographs (JNM-ECP400-JEOL) of PMMA particles produced with different surfactants: (A) PDMS- OH (5.0 wt %), (B) PDMS-mA (5.0 wt %), (C) PDMS-b-P(MMA(1.1K)-co-MA(0.5K)) (5.0 wt %), (D) PDMS-b-P(MMA(1.1K)-co-MA(0.5K)) (10.0 wt %), (E) PDMS-co-PMA (5.0wt %), and (F) PDMS-co-PMA (10.0 wt %). Parameters in the parenthesis are wt % of surfactant with respect to monomer.

density of stabilizer is high enough to sufficiently cover the growing polymer surface and if the grafted chains do not adsorb too strongly on the surface. It has been shown from NMR studies that the trifunctional surfactant, PDMS-*b*-P(MMA-*co*-MA), adsorbed more than 80% over the polymer surface compared to PDMS-*b*-PMA, which has absorbed only 60% over the polymer surface [6]. This suggests that the additional PMMA anchor group in the trifunctional surfactant enhanced the adsorption and thereby stabilizes the growing polymer particles effectively. Furthermore, it has been shown that a careful balance between the size of the anchor group (PMMA-philic) and the amount of the soluble component (CO_2-philic: PDMS) is also shown to play a major role in stabilization [9]. If the balance leans toward the soluble component, the dispersant will be more soluble in the CO_2 continuous phase and show a poorer adsorption onto the particle surfaces, leading to larger and irregular particles. Due to the short anchoring chain in PDMS-OH and PDMS-OAc, the balance shifted towards the PDMS thereby resulted in poor stabilization. On the other hand, by incorporating PMMA block into PDMS-PMA copolymers, the excellent CO_2-soluble/anchoring balance can be achieved. Thus of all stabilizers investigated in this study, the trifunctional PDMS stabilizer, was found to be an effective stabilizer for PMMA polymerization in CO_2.

4. CONCLUSION

Different type of PDMS based stabilizers were investigated on the dispersion polymerization of MMA in supercritical carbon dioxide (scCO_2) in terms of their anchor group architecture. While anchoring group consists of an alcohol or acetate did not work well, the methacrylate end group, methacrylic acid group gave better stabilizing effect. The trifunctional block copolymer, PDMS-b-(PMMA(1.1K)-*co*-PMA(0.5K) was found to be very effective stabilizer due to the efficient stabilizer anchor-soluble balance (ASB) which produced excellent yield, high molecular weight, and uniform micron sized PMMA particles.

REFERENCES

[1] H. M. Woods, M. M. C. G. Silva, C. Nouvel, K.M. Shakesheff, and S. M. Howdle, J. Mater. Chem., 14 (2004) 1663.

[2] K. Ryu and S. Kim, Korean J. of chem. Eng., 13 (1996) 415.

[3] J. L. Kendall, D. A. Canelas, J. L. Young, and J. M. DeSimone, Chem. Rev., 99 (1999) 543.

[4] D. A. Canelas, and J. M. DeSimone, Macromolecules, 30 (1997) 5673.

[5] K. T. Lim, S. E. Webber, and K. P. Johnston, Macromolecules, 32 (1999) 2811.

[6] G. Li, M. Z. Yates, K. P. Johnston, K. T. Lim, and S. E. Webber, Macromolecules, 33 (2000) 1606.

[7] M. L. O'Neill, Q. Cao, R. Fang, K. P. Johnston, S.P. Wilkinson, C. D. Smith, J. L. Kerschner, and S. H. Jureller, Ind. Eng. Chem. Res., 37 (1998) 3067.

[8] H. M. Woods, C. Nouvel, P. Licence, D. J. Irvine, and Steven M. Howdle, Macromolecules, 38 (2005) 3271.

[9] C. Lepilleur, and E. J. Beckman, Macromolecules, 30 (1997) 745.

Studies in Surface Science and Catalysis, volume 159
Hyun-Ku Rhee, In-Sik Nam and Jong Moon Park (Editors)

Microwave synthesis of oxovanadium phthalocyanine used as charge generation material

J.H. Park, J.U. Im, D.H. Son, S.D. Lee, G.D. Lee, and S.S. Park[*]

Division of Applied Chemical Engineering, Pukyong National University, Busan 608-739, South Korea

The feasibility of synthesizing oxovanadium phthalocyanine (VOPc) from vanadium oxide, dicyanobenzene, and ethylene glycol using the microwave synthesis was investigated by comparing reaction temperatures under the microwave irradiations with the same factors of conventional synthesis. The efficiency of microwave synthesis over the conventional synthesis was illustrated by the yield of crude VOPc. Polymorph of VOPc was obtained through the acid-treatment and recrystallization step. The VOPcs synthesized in various conditions were characterized by the means of an X-ray diffractometry (XRD), a scanning electron microscopy (SEM), and a transmission electron Microscopy (TEM).

1. INTRODUCTION

Over the past decade, microwave technology has been applied in such varied fields as drying, food processing, organic synthesis, ceramic processing, composite joining, decomposition processing, and waste treatment [1]. The reason was that microwave processing has attracted potential as an alternative to thermal heating because of the inherent advantages of microwave heating, which is selective, direct, rapid, internal, and controllable [2]. The heating effect associated with microwaves in chemical reaction is mainly due to dielectric polarization such as dipolar and interfacial polarization, although superheating can also be important at rapid heating [3]. Polar solvents such as water, methanol, DMF, ethyl acetate, acetone, chloroform, acetic acid, and dichloromethane are all heated when irradiated with microwaves. Non-polar solvents such as hexane, toluene, diethyl ether, and CCl_4 do not couple and therefore do not heat with microwave irradiation. Phthalocyanines have continuously been the subject of research due to their wide application fields, such as in organic pigment, chemical sensor, electro-chromic display devices, photovoltaic cells, xerography, optical disk, catalysis, and nonlinear optics [4]. These versatile features have stimulated attempts on the synthesis of various metal phthalocyanines or new phthalocyanine derivatives with objective of developing new materials which may show improved or more functional characteristics.

The introduction of microwave presents an excellent new option for the synthesis of VOPc from vanadium oxide, dicyanobenzene, and ethylene glycol. In the present study, the effectiveness of synthesizing crude VOPc from vanadium oxide and dicyanobenzene under the two synthetic methods was investigated by comparing reaction temperatures. Also, the preparation of fine crystal VOPc was investigated from the crude VOPc synthesized at

---

[*] To whom correspondence should be addressed.
E-mail: sspark@pknu.ac.kr

optimum condition through the acid-treatment and recrystallization step.

2. EXPERIMENTAL

Ethylene glycol (EG) was used as solvent. 50 ml of EG, 0.15 mol of dicyanobenzene and 0.0125 mol of vanadium oxide were charged into a 250 ml reaction flask, which was fitted with a modified thermocouple, a reflux condenser, and a motor-driven stirrer. The reactant was stirred uniformly at 100 rpm. It was gradually heated up to 120 °C with heating rate of 2 °C/min, then heated to between 130 °C and 190 °C with heating rate of 0.25 °C min^{-1}. It was maintained in that temperature ranges within the reaction flask for 4 h, using a heating mantle as a conventional heating source and a microwave unit as a microwave heating source. From reaction product thus obtained, liquor was filtered off with a reduced pressure. Then, cake was washed with methanol. After filtration, the cake was acid-treated for 1 h by 100 ml of 0.02 M H$_2$SO$_4$ solution, alkali-treated for 1 h by 100 ml of 0.02 M NaOH solution, and then washed with hot distilled water until washing solution became neutral. After filtration, sample was dried at 70 °C over 24 h in a dry oven, whereupon the yield of crude VOPc was obtained [5].

To produce amorphous VOPc, 5.0 g of crude VOPc was added into 250 ml of concentric H$_2$SO$_4$ solution, and then the mixture was stirred slowly at 5 °C for 2 h. After acid-treatment, cake was collected by filtration, washed with distilled water until washing solution became neutral, then dried at 70 °C over 24 h in a dry oven. To produce fine crystal VOPc, 5 g of amorphous VOPc was added into 90 ml of NMP/H$_2$O solution, and then stirred slowly at 80 °C for 1 h. After recrystallization, cake was collected by filtration, washed with methanol, and then dried. All polymorphs were assayed by XRD analysis.

3. RESULTS AND DISCUSSION

Fig. 1 shows the yield of crude VOPc synthesized, as a function of reaction temperature varying from 130 °C to 190 °C for 4 h in the conventional and microwave synthesis. It indicates that the yield of crude VOPc increased with increasing the reaction temperature in both the synthesis. The high yield exhibited at lower temperature range in the microwave synthesis can be attributed to a significant increase in the reaction rate of reactive species. It means that the microwave synthesis can be expected to produce more-efficient VOPc synthesis by effecting enhanced diffusion between the reactive species due to the selective and internal heating, together with the differential polarization effect [3].

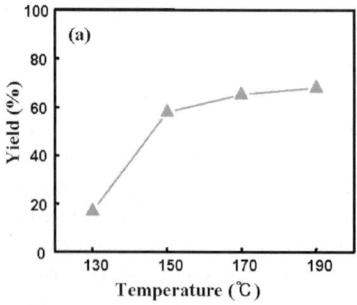
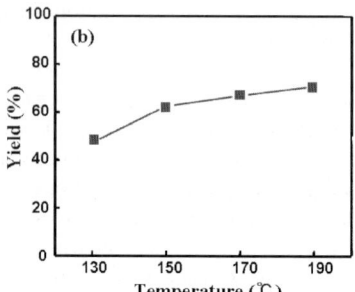

Fig. 1. The yield of the crude VOPcs obtained at various temperatures for 4 h by (a) conventional and (b) microwave synthesis.

Figs. 2 and 3 shows typical SEM pictures and XRD patterns of crude VOPcs obtained at 150 °C for 4 h in the conventional and microwave synthesis. As shown in Fig. 2, the smaller particle size and narrower size distribution are obtained in the microwave synthesis, compared to conventional one. From XRD results in Fig. 3, it can be calculated that the crystallite sizes of crude VOPcs obtained by the conventional and microwave synthesis are about 44 nm and 48 nm, respectively. Thus, the fact that particle size is smaller and crystallite size is larger in microwave sample, compared to conventional sample is probably caused by the microwave non-thermal effect [3].

Fig. 2. SEM pictures of the crude VOPcs obtained at 150 °C for 4 h by (a) conventional and (b) microwave synthesis.

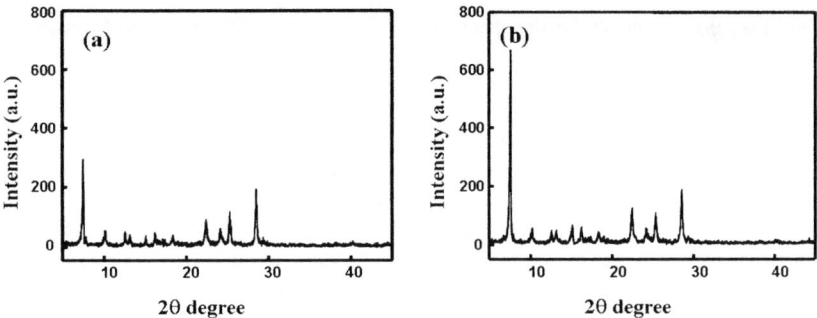

Fig. 3. XRD patterns of the crude VOPcs obtained at 150 °C for 4 h by (a) conventional and (b) microwave synthesis.

To prepare the charge generation material of photoreceptor used in xerography, the crude VOPc synthesized at 150 °C for 4 h in the microwave synthesis was acid-treated, and then recrystallized. As shown in Fig. 4, the amorphous VOPc can be obtained from crude VOPc by acid-treatment and the fine crystal VOPc can be obtained from amorphous VOPc by recrystallization. From XRD results, it can be calculated that the crystallite size of fine crystal VOPc is about 18 nm. As shown in Fig. 5, the fine crystal VOPc is well dispersed with uniform size. It indicates that this fine crystal VOPC can be probably used as the charge generation material of photoreceptor. Thus, further research will be required to measure the electrophotographic properties of fine crystal VOPc.

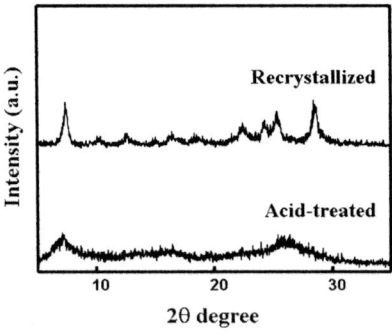

Fig. 4. XRD patterns of the VOPc samples obtained from crude VOPc by acid-treatment and recrystallization.

 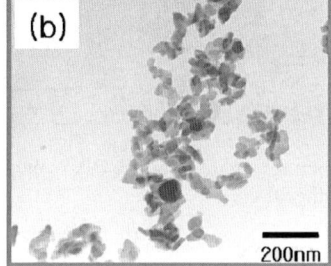

Fig. 5. TEM pictures of the VOPc samples obtained from crude VOPc by (a) acid-treatment and (b) recrystallization.

4. CONCLUSIONS

The microwave synthesis described in the present paper has proven to be quite effective due to its intense internal heating, compared to conventional synthesis. The yield of crude VOPc increased with increasing the reaction temperature under both synthetic methods. Fine crystal VOPc was prepared successfully from the crude VOPc obtained by microwave synthesis through the acid-treatment and recrystallization step.

ACKNOWLEDGEMENTS

This research was supported by the Program for the Training of Graduate Students in Regional Innovation which was conducted by the Ministry of Commerce, Industry and Energy of the Korean Government and by Brain Busan 21 Project.

REFERENCES

[1] J. An, L. Bagnell, T. Cablewski, C.R. Strauss and R.W. Trainor, J. Org. Chem., 62 (1997) 2505.
[2] W.H. Sutton, Ceram. Trans., 59 (1995) 3.
[3] A. Burczyk, A. Loupy, D. Bogdal and A. Petit, Tertahedron, 61 (2005) 179.
[4] Y. Wu, H. Tian, K Chen, Y. Liu and D. Zhu, Dyes and Pigments, 37 (1998) 317.
[5] European Patent EP 0 443 107 A2 (1991).

Studies in Surface Science and Catalysis, volume 159
Hyun-Ku Rhee, In-Sik Nam and Jong Moon Park (Editors)

A simulated countercurrent moving bed reactor for oxidation of CO at low concentration over Pt/Al$_2$O$_3$

Duangkamol Na-Ranong[a], Yuichi Saito[b], Takanori Yotsumoto[b], Mohammad Kazemeini[c] and Takashi Aida[*,b]

[a] Department of Chemical Engineering, Faculty of Engineering,
King Mongkut's Institute of Technology Ladkrabang, Bangkok 10520, Thailand

[b] Department of Chemical Engineering, Tokyo Institute of Technology,
O-okayama 2-12-1-S1-37, Meguro-ku, Tokyo 152-8552, Japan

[c] Department of Chemical and Petrochemical Engineering, Sharif University of Technology,
Azadi Ave., P.O. Box 11365-9465, Tehran, I.R. Iran

[*] taida@chemeng.titech.ac.jp

1. INTRODUCTION

A polymer electrolyte membrane fuel cell (PEMFC) has high potential for various mobile applications but it has a serious limitation on CO content in its hydrogen feed [1]. CO contamination has to be removed to the level lower than 10 ppm. Hydrogen produced from a steam reforming of hydrocarbons generally contains CO about 1%. Recently, many catalysts have been investigated for selective oxidation of CO in hydrogen-rich gas. Several investigators have shown that some noble metals can totally remove this trace amount of CO. However, a significant amount of H$_2$ is also simultaneously consumed [2]. In our research, we are focusing on enhancement of the selectivity towards CO oxidation of reformate gas based on the design of a reactor instead of the selection of a catalyst.

When a mixture of two gases having strong but different affinities to an adsorbent is fed countercurrent to the moving bed of the adsorbent, the gases are adsorbed onto the bed in different zones separately due to the chromatographic effect. Under the situation, the stronger adsorbate is carried by the adsorbent toward the inlet of the sweep gas and the other moves toward the outlet part of the bed. As the result, the two adsorbates are separated. The more different the adsorption affinities, the sharper the separation. For the case of a reactive adsorbent, a chemical reaction and the separation occur simultaneously and this unit is so-called a countercurrent moving bed reactor (CMBR).

In this study, Pt/Al$_2$O$_3$ having high activity for CO oxidation and different affinities for the adsorption of CO and H$_2$ was selected as a catalyst/adsorbent. In a conventional packed bed reactor (PBR), the surface of the catalyst is dominantly covered by CO$_{ads}$ with small amount of O$_{ads}$, the CO conversion is therefore low. Several investigations on periodic operation have illustrated that the reaction front with comparable amount of the two adsorbed species leads to enhancement of the CO conversion. Conceptually, this type of the reaction front should be generated by application of a CMBR, as well. Figure 1 illustrates an image of

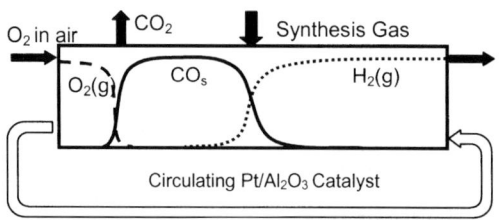

Fig. 1 Concentration profiles of CO, O_2 and H_2 inside a countercurrent moving bed reactor.

a net concentration profile in gas and solid phase for CO and H_2 inside a CMBR when a mixture of CO and H_2 is introduced at the meet with CO at the inlet part to form a reaction front and excess oxygen can be drawn out from the CO_2 outlet. Consequently, the consumption of H_2 due to the oxidation decreases.

In this study, a simulated countercurrent moving bed reactor (SCMBR) with four parts by switching the inlets and outlets of the parts cyclically is employed in order to avoid abrasion occurring from the movement of a solid catalyst. Based on the above concepts, we focused on the performance of a SCMBR for the oxidation of CO at low concentration in absence of H_2 over Pt/Al_2O_3 catalyst/adsorbent. For the first step of the overall reactor design, the performance of a SCMBR is experimentally investigated and compared with that of a PBR for the reaction.

2. EXPERIMENTAL

Figure 2 illustrates the rotating disk type SCMBR used in this study. The reactor consisted of four Pyrex tubes so that the reactor was divided into 4 parts (1-4). Pt/Al_2O_3 was packed with the same amount in each part. This reactor had two inlet ports and two outlet ports so that it was divided into two zones; *CO zone* and *O_2 zone*. Each zone consisted of two Pyrex tubes connected in series. At the inlet port of *CO zone*, a mixture of CO and N_2 was introduced so that this zone functioned for adsorption of CO, in other words trapping of CO. At the inlet port of *O_2 zone*, a mixture of O_2 and N_2 was introduced so that this zone functioned for oxidation of adsorbed CO. At the top of the Pyrex tubes, there were two metal disks. The upper disk was fixed to the outside frame, whereas the lower disk was designed to rotate counterclockwise cyclically so that the reactor tubes moved countercurrent to the gas flow. Since the rotation was 90 degree for each switching, a complete cycle needs four switchings. Cyclic Period (τ) is the time to complete one cycle. The reactor had an electrical furnace, so it can be operated at elevated temperature. The temperature of the reactor was controlled using a PID controller.

The catalyst/adsorbent was a 60/80 mesh pellet of 1 wt% Pt/Al_2O_3 prepared by impregnation. It was packed in each tube for 1.0g (1.0g×4 tubes). For this arrangement, the dead volume of each tube was only 3 cm^3. The feed to each zone was $C_{CO, CO\ zone}$ = 5,000 ppm, $C_{O2, O2\ zone}$ = 10,000 ppm (balanced by N_2). Total flow rate for each zone was 100 mlNTP·min^{-1}. The temperature of the catalyst bed was 373K.

CO concentration at the outlet of each zone was continuously measured using a CO analyzer (Shimadzu CGT-7000). To evaluate the performance of the reactors, the conversion of CO for the PBR (X_{CO}) with 4g of catalyst and the time-average conversion of CO for the SCMBR ($\overline{X}_{CO}$) with 2g of catalyst in each zone were calculated and compared. It should be noted that the CO concentration wave used for Eq. (1) was obtained when the system is at cyclic steady state (after 30 min of operation).

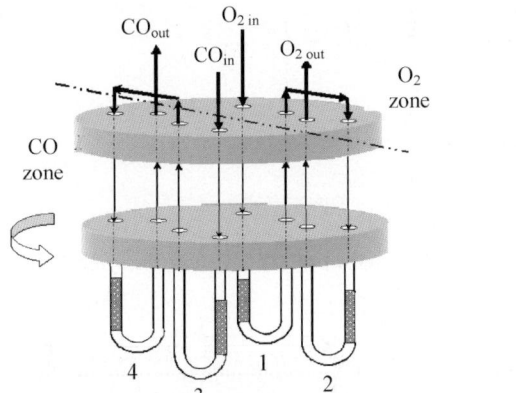 

Fig. 2 A schematic diagram and a photo of the SCMBR used in this study.

$$\overline{X}_{CO} = 1 - \frac{\int_{t}^{t+\tau} (C_{CO,outlet}\big|_{COzone} + C_{CO,outlet}\big|_{O_2 zone}) dt}{\tau \, C_{CO,\,inlet}} \tag{1}$$

3. RESULTS AND DISCUSSION

Figure 3 shows that $\overline{X}_{CO}$ was higher lined than X_{CO} for all over the range of $\tau = 2 - 15$ min. $\overline{X}_{CO}$ increased significantly for the short periods and reached the maximum around 4 min. Beyond the maximum it gradually declined with the period. Figure 4 shows the concentration waves of CO observed at the outlet of each zone for the periods of 4 and 8 min. At the outlet of *CO zone*, CO was not observed in gas phase just after the switching. From several seconds after the switching, CO appeared in the elute and reach the maximum. CO disappeared by the next switching. Large amount of CO was observed with the longer period. At the outlet of *O₂ zone*, CO was observed even right after the switching. Its concentration increased to the maximum and decreased to zero again before the next switching. These results imply that, in *CO zone*, CO is consumed by CO adsorption and CO oxidation over the catalyst/adsorbent at the beginning after the rotation. After the surface was completely covered by the adsorbed species, then CO breaks through to the outlet of *CO zone*. In *O₂ zone*, desorption of the adsorbed CO from the surface or flushing of gaseous CO in the dead volume take place first, and then adsorption of O onto the surface occurs and after several seconds passed the surface was completely covered by O atoms. Comparison of the results for the period of 4 and 8 min shows that CO concentration waves for *CO zone* depended upon the period whereas the one for *O₂ zone* was independent of the period. Therefore, $\overline{X}_{CO}$ was improved when the reactor was switched before CO adsorption saturated in the *CO zone*.

Our preliminary experiment using a PBR showed that the order of the rate of the CO oxidation over Pt/Al₂O₃ was positive with respect to O₂ concentration and was negative with respect to CO concentration. This result indicated that the reaction occurs via L-H mechanism with strong adsorption of CO. For this type of reaction, the surface coverage of O is suppressed by CO adsorption, and therefore the rate of CO oxidation is slow in a PBR. On the other hand, adsorption of O onto the surface occurs in the *O₂ zone*. In other words, in the countercurrent flow of CO_{ads} and O_2 there exists a reaction front where the net concentration

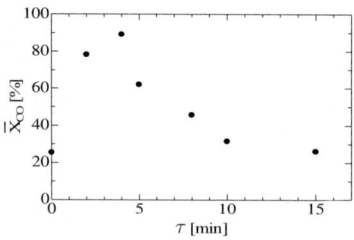

Fig. 3 Dependence of time averaged conversion of CO on the period of rotation; the symbol plotted at $\tau = 0$ min represent the result obtained from the PBR.

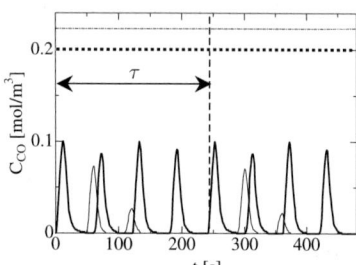 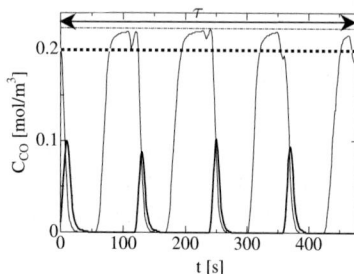

Fig. 4 Concentration waves of CO observed at the reactor outlet after reaching cyclic steady state for (——) *CO zone* and (——) *O_2 zone* and (······) at the inlet of *CO zone* with a period of (a) $\tau = 4$ min, (b) $\tau = 8$ min.

of CO and O_2 in gas and solid phase is comparable (see Fig. 1), and hence so that the rate is accelerated. The reaction front pattern cannot achieve in a conventional cocurrent PBR. Note that in the SCMCR the average contact time is only the half of that of PBR, and therefore, we can clearly say that the reaction rate was enhanced by the countercurrent operation. As a result, the amount of the catalyst can be reduced a SCMBR to achieve the same performance of PBR.

4. CONCLUSION

Performance of a simulated countercurrent moving bed reactor, SCMCR is experimentally investigated for oxidation of CO at low concentration in the absence of hydrogen over Pt/Al_2O_3 catalyst/adsorbent. The time-average conversion of CO obtained in the SCMBR was higher than the conversion of CO obtained from a conventional PBR for all over the tested range (period = 2–15 min). For the next step, the effects of operating variables on its performance are planed for both CO oxidation in the absence of H_2 and H_2-rich gas system.

Acknowledgement

One of the authors (D.N.-R.) would like to gratefully acknowledge The Hitachi Scholarship Foundation for sponsoring Hitachi Research Fellowship HSF-04082.

REFERENCES

[1] J. J. Baschuk and X. Li, *Int. J. Energy Res.*, **25**, 695 (2001).
[2] S. H. Oh and R. M. Sinkevitch, *J. Catal.*, **142**, 254 (1993).

Studies in Surface Science and Catalysis, volume 159
Hyun-Ku Rhee, In-Sik Nam and Jong Moon Park (Editors)

Continuous cyclopentenone synthesis with static mixer reactor

Jaehoon Choe[a], Youngwoon Kwon[a], Jong-Ku Lee[a], and Kwang Ho Song[b,*]

[a]Corporate R&D, LG Chem Research Park, Daejeon 305-380, Korea

[b]Department of Chemical & Biological Engineering, Korea University, Seoul 136-713, Korea

1. INTRODUCTION

Cyclopentenone structures are present in a wide variety of natural products. Among these cyclopentenone-containing compounds, 3-methyl-2-cyclopentenone is a key intermediate for preparing therapeutically important natural products such as cortisone, precapnelladiene and trichothecenes [1]. It is also a useful compound for preparing various metallocene catalysts for olefin polymerization. For example, 3-methyl-2-cyclopentenone can be an important starting material for ansa-ligand of ansa-zirconocene compound, which is active toward copolymerization of ethylene and norbornene as a metallocene catalyst [2]. Enol structure in 3-methyl-2-cyclopentenone can be synthesized by base catalyzed intermolecular aldol condensation. High temperature is required when the aldol releases water to form 3-methyl-2-cyclopentenone. The conjugated structure of the enone products is, however, unstable especially at a higher temperature since the conjugate nature of enone allows reacting further to produce higher molecular weight species. 3-methyl-2-cyclopentenone was synthesized in a batch mode according to previously reported procedures. However, the procedures required long reaction times or showed low yields with low selectivity or low conversion. In this work, we studied a simple and practical procedure for synthesizing 3-methyl-2-cyclopentenone with high yield.

2. EXPERIMENTAL

2.1. Preparation of 3-methyl-2-cyclopentenone at atmospheric pressure

The two phase mixture of toluene and aqueous NaOH solution was boiled under reflux at 80 °C in a batch reactor. Acetonylacetone was then slowly added to the mixture and the reaction was maintained at 80 °C for 6~24 hours. After the reaction, the resulting solution was cooled down in an ice bath.

2.2. Preparation of 3-methyl-2-cyclopentenone at continuous high pressure system

---

*Corresponding author, E-mail: khsong@korea.ac.kr

The high pressure continuous reactor consists of five Kenics type in-line static mixers, that were connected in series [3]. Each reactor unit has 27 Kenics elements and dimensions of 19 cm tube length and 3.3 mm inner diameter. Acetonylacetone and 1 % NaOH aqueous solution were pumped into the in-line static mixer reactor using two independent HPLC pumps. The in-line static mixer reactors were immersed in a constant temperature controlled oil bath at 200 °C so that the reaction mixture was heated to the reaction temperature. When the reaction was completed, the fluid was cooled down rapidly in a constant temperature cold bath at 0 °C. At the end of the cooling line, a backpressure regulator was placed to allow experiments to be run at 34 bar.

2.3. Separation of 3-methyl-2-cyclopentenone from the reaction mixture

The resulting solutions from above experiments were saturated with sodium chloride and then diethyl ether was added to extract 3-methyl-2-cyclopentenone from aqueous NaOH solution. The diethyl ether in the organic phase was then removed by a rotary vacuum evaporator. The heavy cut residue was separated by a simple distillation. The collected 3-methyl-2-cyclopentenone was dried over solid anhydrous $MgSO_4$ and filtered.

2.4. Analytical methods

Quantitative analysis of 3-methyl-2-cyclopentenone was performed by a 6890N gas chromatograph (Agilent Technologies) equipped with a 7863 series automatic injector, a flame ionization detector and HP ChemStation data system. An AT-1000 capillary column (15 m ×0.53 mm ID×1.2 μm film thickness, Alltech) was used. As 3-methyl-2-cyclopentenone is a thermally labile compound and breaks down in a standard hot flash injection before reaching the column, a cool on-column inlet with electronic pressure control was used instead of a hot flash injection. The oven temperature was programmed to be maintained at 50 °C for the first 4 min, then increased to 90 °C at 10 °C /min and held for 7 min, then increased to 300 °C at 30 °C /min, and the cool on-column inlet tracked the oven temperature. The internal standard method was employed for quantitative analysis of mixtures, and toluene was used as an internal standard.

3. RESULTS AND DISCUSSION

3-methyl-2-cyclopentenone was synthesized under atmospheric and high pressure conditions. When the reaction took place at the two phase batch reaction system under atmospheric condition, the tar formation was greatly reduced. This is because 3-methyl-2-cyclopentenone extracted in the toluene phase was separated from the aqueous base solution so that byproducts produced by the further contacting with the aqueous base were reduced [4, 5]. This indicates an improvement of reaction selectivity up to 95.8 %. The reaction selectivity can be defined as a ratio of synthesized 3-methyl-2-cyclopentenone to consumed acetonylacetone. In the two phase batch reaction, an overall yield as high as 68.9 % could be obtained. However, the conversion was only 72 % as shown in Table 1. As the differences in boiling point temperatures between acetonylacetone (b.p. 80 °C) and 3-methyl-2-cyclopentenone (b.p. 75 °C) are only 5 °C at 16 Torr, a separation of close boiling point compounds using distillation is very difficult and costly. 3-methyl-2-cyclopentenone is usually used as an intermediate chemical so that a high degree of purity is required to prevent impurity generation from subsequent reaction steps. The two phase batch reaction system showed some selectivity improvement since it was operated at a low-temperature. The conversion was however, too low for commercial applications. The reaction conversion could be increased by increasing the base concentration, but increasing the base concentration in a two phase batch mode causes the significant decrease in 3-methyl-2-cyclopentenone selectivity with increasing acetonylacetone conversion.

Table 1
Results of the two phase batch reaction system

Aqueous NaOH (% w/w)	Reaction time (h)	Conversion (%)	Selectivity (%)	Yield (%)
0.1	24	72	95.8	68.9
1	12	77	53.2	40.8
10	6	92	52.2	48.0
20	6	95	47.3	44.9

The aldol condensation is usually pushed to completion by dehydration at an increased temperature. When 3-methyl-2-cyclopentenone was synthesized under high pressure and high temperature, fast cooling of the product was essential to minimize further tar forming side reactions. In a batch operation, fast cooling of product is not possible especially with a large volume reactor. However, the reaction temperature can be controlled easily with the in-line static mixer reactor in a continuous mode, and therefore the selectivity of 3-methyl-2-cyclopentenone increases.

Table 2
Results of the continuous high pressure reaction system

Temperature (°C)	Residence time (min)	Conversion (%)	Selectivity (%)	Yield (%)
170	20	99.9	61.3	61.2
170	15	99.8	62.2	62.1
190	3.3	99.8	67.1	67.0
200	4	99.9	55.8	55.7
200	3.3	99.9	58.2	58.2
200	2	99.3	70.1	69.6

The two phase batch operation in Table 1 showed that the acetonylacetone conversion

812

increased from 72 to 95 % and it is accompanied by the decrease in selectivity of 3-methyl-2-cyclopentenone from 95.8 to 47.3 %. However, the selectivity of 3-methyl-2-cyclopentenone increased from 55.8 to 70.1 %, in the continuous mode, which was accompanied by slight decrease of conversion from 99.9 to 99.3 % as shown in Table 2. It is important to keep the conversion of acetonylacetone over 99.0 % since the unreacted acetonylacetone in the feed and 3-methyl-2-cyclopentenone are very difficult to separate from each other. The other benefit of using the continuous reactor is the reduction of reaction time. Since the reaction was carried out under high temperature and high pressure, the rate of the reaction was also increased and the reaction time was therefore considerably reduced from several hours to two minutes as shown in Tables 1 and 2.

4. CONCLUSION

Our experimental results demonstrated the advantages of the continuous flow process using a high temperature, pressurized in-line static mixer reactor for direct synthesis of 3-methyl-2-cyclopentenone. The application of a continuous flow process and in-line mixing techniques for this reaction step could result in a significant difference between a conventional batch reactor and an in-line static mixer reactor. Unlike two phase batch mode, the continuous process for this intermolecular aldol condensation reaction could avoid scale-up difficulties. We found that this reaction required tight control over temperature at a quenching step. The precise control of the reaction time was found to increase acetonylacetone conversion and 3-methyl-2-cyclopentenone selectivity. This could also eliminate major large scale equipment in a batch process. 3-methyl-2-cyclopentenone could be produced at a 0.5 kg/hr through this continuous process and the production rate could be improved easily with increasing feed and attaching more in-line static mixer reactors in parallel.

ACKNOWLEDGMENTS

This study was supported by research grants from the Korea Science and Engineering Foundation (Project No. R11-2000-088-02009-0) through the Applied Rheology Center.

REFERENCES

[1] A.Rossi, P. Kapahi, G. Natoli, T. Takahashi, Y. Chen, M Karin, and M.G. Santoro, Nature, 403 (2000) 103.
[2] Y.W. Park, Fulvene, metallocene catalysts and preparation method thereof, and preparation of polyolefines copolymer using the same, US Patent No. 20050004385 A1 (2005).
[3] N. Harnby, M.F. Edwards, and A.W. Nienow (eds.), Mixing in the Process Industries, Butterworth-Heinemann, Oxford, 1992.
[4] L. Bagnell, M. Bliesse, T. Cablewski, C.R. Strauss, and J. Tsanaktsidis, Aust. J. Chem., 50 (1997) 921.
[5] J. An, L. Bagnell, T. Cablewski, C.R. Strauss, and R.W. Trainor, J. Org. Chem., 62 (1997) 2505.

Studies in Surface Science and Catalysis, volume 159
Hyun-Ku Rhee, In-Sik Nam and Jong Moon Park (Editors)

813

Hydrogen production from biomass-ethanol at ambient temperature with novel diaphragm reactor

Y. Sekine[a*], S. Asai[a], E. Kikuchi[a], M. Matsukata[a], F. Haga[b]

[a] Department of Applied Chemistry, Waseda University, 55S602, 3-4-1, Okubo, Shinjuku, Tokyo 169-8555, JAPAN

[b] Central Research Center, Nissan Motor Co. Ltd., 1, Natsushima, Yokosuka, Kanagawa 237-8523 Japan

1. INTRODUCTION

Recently, fuel cells have commanded attention to establish high-efficiency hydrogen production processes. Some catalytic processes have been considered, but they have typically entailed numerous problems (high temperatures, catalyst deactivation, and coking).

Low energy pulsed (LEP) discharge is a simple hydrogen production process. This novel technique requires neither high temperature nor pressure: the reaction takes place at room temperature and atmospheric pressure. We have successfully reformed hydrocarbons using this LEP discharge [1–6].

Additionally, ethanol is an alternative fuel that might replace fossil fuels because ethanol is producible from various renewable sources (biomass, etc.). It offers many advantages such as transportation, storage, and a low environmental burden compared to methanol [7]. Recently, we developed a new reforming process from an ethanol-water mixture into synthesis gas in the vapor phase using carbon fiber electrodes [8]. In the present study, we undertook direct reforming of an ethanol-water mixture using LEP discharge. Results demonstrate the feasibility of this novel liquid fuel reforming process.

2. EXPERIMENTAL

Diaphragm discharge is a liquid phase discharge at ambient temperature and atmospheric pressure [9]. The reactor configuration, as shown in Fig. 1 [10], consists of Pyrex tube and a pair of electrodes set in the liquid fuel. The insulated teflon-membrane (diaphragm membrane), with its single pinhole, is placed in the gap of electrodes.

This membrane has a very important role in this discharge. The pinhole centralizes the charge, thereby generating a discharge in the liquid phase; liquid phase reforming can take place. We chose an ethanol-water mixture of 50/50 mol% as a reforming fuel. All outlet gas products were analyzed using a gas chromatograph equipped with FID and TCD (GC14-B; Shimadzu Corp.). The

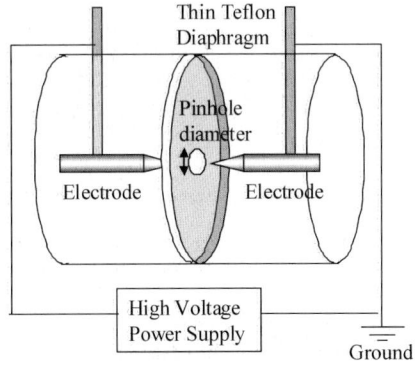

Fig. 1. Reactor configuration

liquid fuel was analyzed using GC-MS (GCMS-QP2010; Shimadzu Corp.) after the reaction. The non-equilibrium pulsed discharge was generated using a DC power supply. Waveforms of current and voltage were observed using probes (Tektronix Inc.) and a digital phosphor oscilloscope (Tektronix Inc.).

3. RESULTS AND DISCUSSION

3.1 Effect of the electrode shape

Figure 2 shows the effect of the shape of electrodes. The discharge showed the highest energy efficiency when using a couple of coaxial needle type of electrodes (Type A). It was considered that the input power for dielectric breakdown was very low in type A. On the other hand, the diaphragm discharge with a couple of flat-plate electrodes (types B and C) was difficult to control and the discharge was unstable.

Table 1 shows energy efficiency calculations based on lower heating value (LHV). Energy efficiency (E_{eff}) based on LHV is defined as Eqn. 1.

$$E_{eff} = (\Sigma E_{output} / \Sigma E_{input}) \times 100 \tag{1}$$

Furthermore, energy efficiency based on higher heating value (HHV) is defined as Eqn. 2.

$$E_{eff} = ((\Sigma E_{output} + E_{lh}) / (\Sigma E_{input} + E_{lh})) \times 100 \tag{2}$$

$$(E_{lh}: \text{the value of latent heat of } C_2H_5OH)$$

Input power is not the energy consumption at the discharge-gap, but the presented value on the high-voltage power supply. Results show that type A electrodes offer the highest energy efficiency (LHV, 30.2%; HHV, 30.8%). For that reason, the pair of coaxial needle type of electrodes was selected for subsequent experiments.

Table 1
Energy efficiency calculations based on LHV

Formula	Formation / Consumption rate mmol min^{-1}	Δ Hc MJ / mol	Efficiency KJ / min
H_2O	0.069	0.00	0.00
C_2H_5OH	1.303	1.37	1.78
Electricity	82.5 W		4.95
E_{input}			6.73
H_2	2.076	0.242	0.502
CO	0.778	0.283	0.220
CH_4	0.340	0.890	0.303
CO_2	0.013	0.000	0.000
C_2H_6	0.044	1.56	0.068
C_2H_4	0.362	1.41	0.510
C_2H_2	0.332	1.30	0.431
E_{output}			2.035
Efficiency			30.2 %

Conditions: gap distance, 6.0 mm; C_2H_5OH conc., 50 mol%; pinhole diameter, 1.0 mm; diaphragm thickness, 1.0 mm; shapes of electrodes, pin-to-pin type (Type A).

3.2. Effect of input power

Next, we investigated the effect of input power on gas formation rates and on carbon-selectivity. As shown in Fig. 3, the amount of product increased as the input power increased. The selectivity to carbon monoxide and C2 hydrocarbons was constant despite the change of the input power. These results showed an identical trend to that of the gas phase

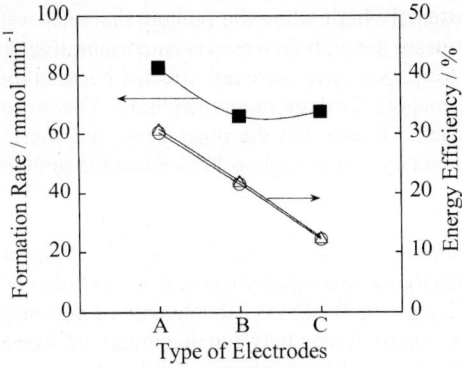

Fig. 2. Effect of shapes of electrodes on energy efficiency in diaphragm discharge: O, energy efficiency based on LHV; △, energy efficiency based on HHV; ■, input power; gap distance, 6.0 mm; C_2H_5OH conc., 50 mol%; pinhole diameter, 1.0 mm; diaphragm thickness, 1.0 mm.

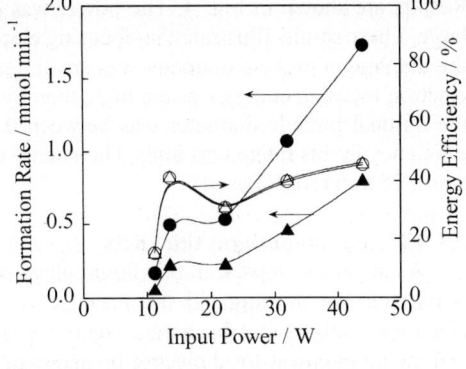

Fig. 3. Effect of input power on energy efficiency in diaphragm discharge: O, energy efficiency based on LHV; △, energy efficiency based on HHV; ●, H_2 formation rate; ▲, CO formation rate; gap distance, 6.0 mm; C_2H_5OH conc., 50 mol%; pinhole diameter, 0.5 mm; diaphragm thickness, 1.0 mm.

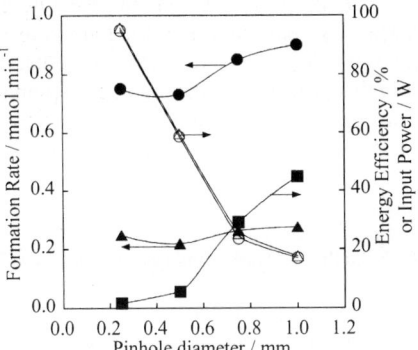

Fig. 4. Effect of pinhole diameter on gaseous formation rates and on carbon selectivity: O, energy efficiency based on LHV; △, energy efficiency based on HHV; ■, input power; ●, H_2 formation rate; ▲, CO formation rate; gap distance: 6.0 mm; C_2H_5OH conc., 50 mol%; diaphragm thickness, 1.0 mm.

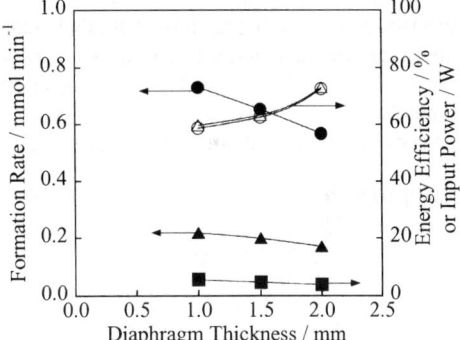

Fig. 5. Effect of diaphragm thickness on products formation rates/efficiency; O, energy efficiency based on LHV; △, energy efficiency based on HHV; ■, input power; ●, H_2 formation rate; ▲, CO formation rate; gap distance, 6.0 mm; C_2H_5OH conc., 50 mol%; pinhole diameter, 0.5 mm.

reaction using LEP discharge. Demonstrably, the number of electrons in the gap of electrodes serves as an important factor in this liquid phase discharge. The energy efficiency also increased as the input power increased.

3.3. Effect of the diaphragm pinhole diameter

The influence of diaphragm pinhole diameter was investigated. The correlation between input power for dielectric breakdown and the pinhole diameter (0.25–2.0 mm) was examined.

816

Results are shown in Fig. 4. The power was extremely high when the pinhole diameter was large. These results illustrate that focusing electrons on the pinhole was very important. Even if the diaphragm pinhole diameter was small, gas formation rates were not affected because the electron focusing energy was too high, thereby damaging the diaphragm membrane. Therefore, the optimal pinhole diameter was between 0.5 and 1.0 mm. On the other hand, the energy efficiency in this range was high. The energy efficiency was as high as 92% when the pinhole was 0.25 mm (HHV based).

3.4. Effect of diaphragm thickness

As noted previously, in this liquid phase discharge, focusing electrons on the pinhole was important. We investigated the influence of diaphragm thickness (the length of focusing electrons' path) on this discharge. Figure 5 shows the correlation between diaphragm thickness and the input power for dielectric breakdown. This figure shows no marked difference among these data. The important element in this discharge was not the volume of focused electrons, but the density of focused electrons.

4. CONCLUSIONS

Using the diaphragm membrane, liquid phase discharge was achieved at ambient temperature and atmospheric pressure. The main gas in products was H_2. In addition, CO, CH_4, C2 compounds and small amounts of CO_2 were produced. A couple of coaxial needle-type electrodes showed the highest yield and stable discharge. The gaseous formation rates showed an increasing trend according to the increase of input power. This result was identical to that for the vapor phase discharge. The most important factor in this process was the diaphragm pinhole diameter. It was important to focus electrons on the microscopic pinhole. This system is extremely simple and small, making it a good candidate for hydrogen production.

REFERENCES

[1] Y. Sekine, K. Urasaki, S. Asai, M. Matsukata, E. Kikuchi, S. Kado, *Energy & Fuels*, 18(2), 455-459 (2004)
[2] S. Kado, Y. Sekine, T. Nozaki, K. Okazaki, *Cat. Today*, 89(1-2) 47-55 (2004)
[3] S. Kado, Y. Sekine, K. Urasaki, K. Okazaki, T. Nozaki, *Stud. Surf. Sci. Catal.*, 147, 577-582 (2004)
[4] S. Kado, K. Urasaki, Y. Sekine, K. Fujimoto, T. Nozaki, K. Okazaki, *Fuel*, 82, 2291-2297 (2003)
[5] S. Kado, K. Urasaki, H. Nakagawa, K. Miura, Y. Sekine, *ACS Books Utilization of Green House Gas*, 852, 303-313 (2003)
[6] S. Kado, K. Urasaki, Y. Sekine, K. Fujimoto, *Chem. Commun.*, 415-416 (2001)
[7] J. Llorca, P. Ramirez de la Piscina, J. Sales, N. Homs, *Chem. Commun.*, 641 (2001)
[8] Y. Sekine, K. Urasaki, S. Asai, M. Matsukata, E. Kikuchi, S. Kado, *Chem. Commun.*, 78-79 (2005)
[9] Z. Stara, F. Krcma, Z. Raskova, *abst. 16th Int. Symp. Plasma Chemistry* 296 (2003)
[10] Y. Sekine, S. Asai, K. Urasaki, M. Matsukata, E. Kikuchi, S. Kado, F. Haga, *Chem. Lett.*, 5, 658, (2005)

Studies in Surface Science and Catalysis, volume 159
Hyun-Ku Rhee, In-Sik Nam and Jong Moon Park (Editors)
© 2006 Elsevier B.V. All rights reserved

Autothermal reaction of ethanol through Pd-Ag membrane reactor prepared by sequential electroless deposition

Ying-Chi Liu[a]   Wen-Hsiung Lin[b]   Wen-Hui Wang[a]   Hsin-Fu Chang[a*]

[a]Department of Chemical Engineering, Feng Chia University Taichung, Taiwan 407, R.O.C.

[b]General Education Center, Chienkuo Technology University Changhua, Taiwan 500, R.O.C.

1. ABSTRACT

In this communication we have studied the autothermal reaction of ethanol in the membrane reactor, Pd-Ag/PSS membrane reactors were prepared by sequential electroless plating of palladium and silver, with a film thickness of 20 μm and Pd/Ag weight ratio of 78/22, on the porous stainless-steel tube. Ethanol-water mixture (n_{H2O}/n_{EtOH}=1 or 3) and oxygen (n_{O2}/n_{EtOH}=0.2, 0.776 or 1.035) were fed concurrently into the membrane reactor packed with Zn-Cu/Al$_2$O$_3$ industrial catalyst (MDC-3). The reaction temperatures were set at 593~723 K and the pressures 3~10 atm. The selectivity of CO$_2$ increased with increasing flow rate of oxygen, while the selectivity of CO remained almost the same. The hydrogen flux to the permeation side increased proportionately with increasing pressure; however, it reduced slightly with increasing oxygen input. This is probably due to the fast oxidation reaction that consumes hydrogen before the onset of the steam reforming reaction. The effect of oxygen plays a vital role on the ethanol auto-thermal reaction, especially for a Pd-Ag membrane reactor in which a higher flux of hydrogen is required.

2. INTRODUCTION

People are interested in renewable energy worldwide and increased in the last decade [1]. Hydrogen, as a clean energy, can be obtained from renewable sources or from conventional hydrocarbons through thermal or catalytic processes. In this study, ethanol was fed with water and oxygen to produce hydrogen by auto-thermal reforming reaction in a Pd/Ag membrane reactor. The combustion reaction of mixed ethanol-air is highly exothermic and may produce acetaldehyde and ethylene, and also coke and soot, which are severe poisons to fuel cell electrodes. The steam reforming reaction (C$_2$H$_5$OH + H$_2$O → 2CO + 4H$_2$, ΔH_R=+260 kJ/mol) is strongly endothermic, requiring an external energy to heat the reactor to the 1073 K, which is necessary to achieve high conversion at residence times of ~1 s [2]. The water-gas shift reaction (CO + H$_2$O → CO$_2$ + H$_2$, ΔH_R= −41 kJ/mol) has a favorable H$_2$ equilibrium only at a lower temperature and with large amounts of water added [3]. Ideally, to perform the partial oxidation with stream reforming of ethanol combined with the water-gas shift reaction [4] in a single Pd-Ag membrane reactor will take advantage of the heat generated by the oxidation reaction required for the energy consumption of the steam reforming reaction and of a thermodynamically unlimited H$_2$ production in a palladium-silver membrane reactor.

In this study, a porous stainless-steel tube (PSS , with a diameter ca. 3/8 inch, pore size

818

ca. 4 μm and effect surface area 20 cm^2) was used as the support for Pd-Ag coating. Sequential electroless plating of palladium and silver was conducted on the PSS tube and reached a final film thickness of 20 μm and Pd/Ag weight ratio of 78/22. The temperature effect associated with heat of reaction that influences on reaction paths is thoroughly examined. The hydrogen production from auto-thermal reaction is judged under different oxygen feed rates.

3. EXPERIMENTAL

3.1. Preparation of Pd-Ag/PSS composite membranes

Pd-Ag/PSS membranes were prepared using the conventional electroless plating method. The activation procedure using palladium chloride which was reduced by tin chloride to form Pd nuclei, which triggered the subsequent electroless Pd plating. The plating procedure was repeated until the desired thickness of coating was reached. After each deposition, the membrane was sintered and annealed in the hydrogen atmosphere at 723 K for more than 18h [5].

3.2. Membrane characterization analysis

Permeation measurements were conducted on the Pd and Pd-Ag/PSS membranes at elevated temperature (623 K to 873 K) and pressures (up to 1 MPa). Surface morphology of the deposited layer was observed with a scanning electron microscope (SEM, S3000N, HITACHI Co.) equipped with an energy dispersive spectrometer (EDS, HORIBA Co.).

3.3. Auto-thermal reaction of ethanol through Pd-Ag membrane reactor

The auto-thermal reaction of ethanol occurred in the shell side of a palladium membrane reactor in which a Zn-Cu/Al$_2$O$_3$ industrial catalyst (MDC-3) was packed with silica powder. Ethanol-water mixture (n$_{H2O}$/n$_{EtOH}$=1 or 3) and oxygen (n$_{O2}$/n$_{EtOH}$=0.2, 0.776 or 1.035) are fed concurrently to the shell side. The reaction temperatures were set at 593~723 K and the pressures were 3~10 atm.

Catalytic activity was measured in terms of ethanol conversion, defined as:

$$X = \frac{F_{EtOH}^{in} - F_{EtOH}^{out}}{F_{EtOH}^{in}} \tag{2}$$

Products distributions were evaluated through their selectivity and the selectivity of specie i was defined as:

$$S_i = \frac{0.5F_i^{out}}{F_{EtOH}^{in} - F_{EtOH}^{out}} \tag{3}$$

where F$_i$ is the molar flow rate of species i.

4. RESULTS AND DISCUSSION

4.1. EDS analysis of Pd-Ag/PSS membrane reactor

Figure 1(A) shows the Pd-Ag deposit over and inside of porous stainless-steel substrate analyzed by EDS. The penetration depth is about 5μm from the surface, after that the amount of diffusion is too small to be measured. The amounts of Pd and Ag diffuse into the substrate were depicted in Fig. 1(B). The composition of Pd-Ag alloy is nearly identical to that on the outer surface and remains the same no matter how deep the alloy penetrated into the pore. This proves that the electroless plating process appears to be quite promising due to the possibility of uniform deposition on complex shapes.

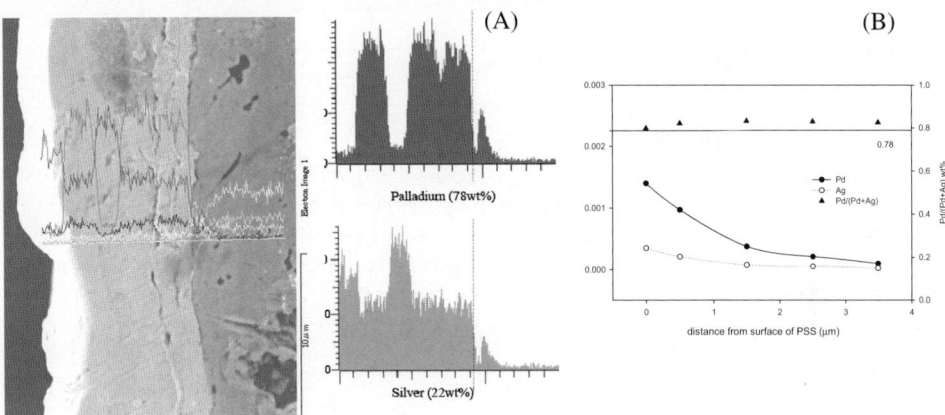

Fig.1. Energy dispersive spectroscopic analysis (A) deposition distribution of cross section (B) distribution of Pd and Ag inside of PSS substrate

4.2. Autothermal reaction of ethanol

At 623K and at WHSV=5h^{-1}, ethanol conversion in steam reforming reaction (n_{EtOH}:n_{H2O}:n_{O2}=1:1:0) is well below that when low oxygen is fed (n_{EtOH}:n_{H2O}:n_{O2}=1:1:0.2) into the reactor, indicating that the steam reforming reaction is not triggered at low temperature. Figs. 2(A) and 2(B) illustrate when we raise the temperature from 623 K to 723K, at n_{H2O}/n_{EtOH}=1, the ethanol conversion increases from 38~40 % to 77%~81%, irrespective of oxygen addition. As the temperature increases, the oxidation reaction rate also increases and more heat evolved to supply for steaming reforming, the ethanol conversion is much higher at higher temperatures than that at lower temperatures, especially at lower pressures, as shown in Fig. 2(C). Evidently, the auto-thermal reaction is prevailing with increasing temperature and decreasing pressure than the steaming reforming reaction. Regarding to hydrogen production, at low oxygen feed and normal water-ethanol feed (n_{EtOH}:n_{H2O}:n_{O2}=1:1:0.2), the hydrogen flux increases both with increasing temperature and pressure, especially at 9 atm, as shown in Fig. 2(D). Figure 2(E) shows that the hydrogen yield increases with increasing pressure, at different WHSVs and temperatures. In the condition of WHSV =5 and 8, the hydrogen yield of the former is larger than the latter. Figure 2(E) also shows that at the low temperature the hydrogen yield is affected by the space velocity, while at the high temperature is the WHSV effect becomes less moderate. When the oxygen in feed is raised (n_{O2}/n_{EtOH}= 0.776, 1.035), Fig. 2(F) shows that at different pressures the selectivities of CO and CO_2 vary with the amount of oxygen added to the reactor. The pressure does not retain a substantial effect on the production of CO and CO_2. The selectivity of CO_2 increases with the addition of oxygen, while the selectivity of CO remains almost the same. Although the hydrogen flux to the permeation side increases proportionately with increasing pressure, it reduces slightly with increasing oxygen input. This is probably due to the fast oxidation reaction that consumes hydrogen before the onset of the steam reforming reaction.

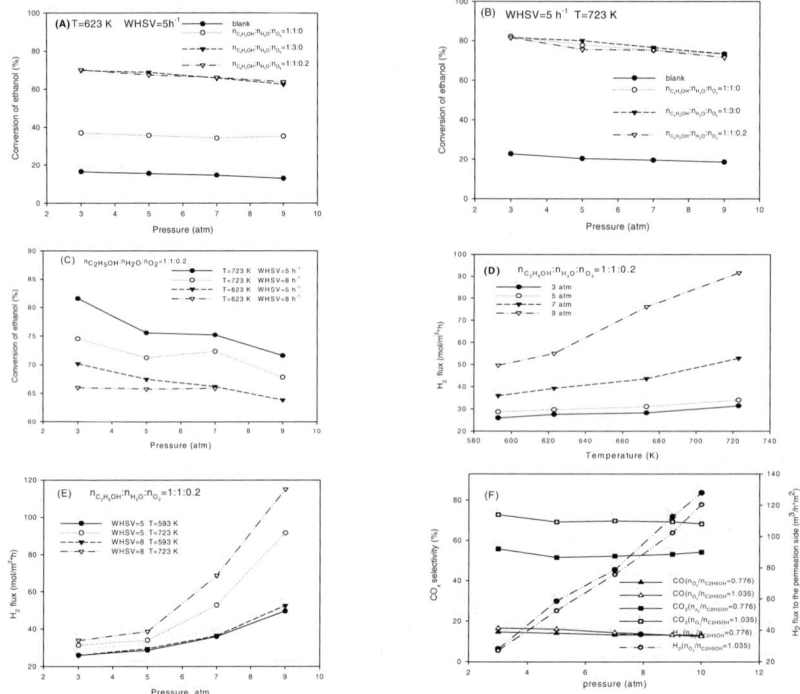

Fig.2. Ethanol autotherml reaction in the Pd/Ag membrane reactor at various temperatures and pressures

5. CONCLUSIONS

Autothermal reaction gains the merit that the heat evolved from the exothermic partial oxidation can supply for the need of endothermic steam reforming reaction. The amount oxygen in the feed is remarkably significant on the ethanol autothermal reaction, especially for a Pd-Ag membrane reactor from which a higher flux of hydrogen is obtained. The hydrogen flux, increasing both with temperature and pressures, is independent of WHSV at lower temperatures. If oxygen in the feed is not sufficient, it would be likely the steam reforming reaction prevailing. Inversely, high O_2 flow will shift the reaction scenario to be partial oxidation dominating, and selectivity of CO_2 increases with increasing oxygen feed.

ACKNOWLEDGEMENTS

This study was performed under the support of the National Science Council of the Republic of China, with the contract number NSC93-2214-E-035-004 and NSC92-2214-E- 027-001, to which the authors wish to express their thanks.

REFERENCES

[1] Breen, J.P., Burch, R. and Coleman, H.M., 2002., *Appl. Catal. B.*, 39: 65.
[2] Cavallro, S. and Freni, S., 1996., *Int. J. Hydrogen Energ.*, 21: 465.
[3] Fierro, V., Klouz, O., Akdim, C. and Mirodatos, C., 2002., *Cataly. Today*, 75: 141.
[4] Fierro, V., Akdim, O. and Mirodatos, C., 2003., *Green Chemistry*, 5: 20.
[5] Lin, W.-H. and Chang, H.-F., 2005. C, *Surf. Coat. T.*

Studies in Surface Science and Catalysis, volume 159
Hyun-Ku Rhee, In-Sik Nam and Jong Moon Park (Editors)

Decomposition of biomass by microwave plasma process

M. Kobayashi, K. Konno, H. Nagazoe, T. Yamaguchi, K. Onoe

Department of Life and Environmental Sciences, Faculty of Engineering, Chiba Institute of Technology, 2-17-1 Tsudanuma, Narashino, Chiba, 275-0016, JAPAN

1. INTRODUCTION

Biomass is a renewable natural resource unlike coal and oil, which are non-renewable resources. Biomass is easily accessible from various industries including agriculture, forestry, fishery and livestock [1]. The plant-based biomass consists mainly of cellulose and lignin (the content of cellulose is greater than that of lignin), and the ratio of cellulose and lignin varies depending on the type of plant-biomass [2]. In this study, to elucidate decomposition mechanism of biomass induced by microwave plasma, argon was chosen as a microwave plasma source because argon does not get consumed during reaction. As biomass base compounds, cellulose and lignin were used. As biomass, sugarcane bagasse and Alaska cedar sawdust were chosen. Since a cellulose/lignin ratio of bagasse is different from that of sawdust, we could also study the effect of the ratio on the decomposition rate and the product distribution.

2. EXPERIMENTAL

2.1. Materials

As a plasma source, we used commercially available argon gas. The physical properties of biomass samples are shown in Table 1. As base compounds for biomass, we used cellulose (less than 200 mesh from Merck Co.) and lignin (less than 200 mesh from Kanto Kagaku Co.). As plant-based biomass, we used Alaska cedar sawdust (less than 16 mesh, denoted as sawdust hereafter) and sugarcane bagasse (less than 16 mesh, from Okinawa, Japan, denoted as bagasse hereafter). Quantitative analysis of holocellulose was based on a modified chlorite method, whereas that of lignin was based on the sulfonic acid method. As pretreatment, all the materials were dried at 380 K for 3 hours in argon.

Table 1
Physical properties of each sample

	Elemental analysis [wt%] (d.a.f)				Component [wt%] (d.a.f)			Proximate analysis [wt%](d.b)		
	C	H (H/C)	N	O (diff.) (O/C)	oily components**	cellulose components	lignin	VM	ash	FC
cellulose	42.6	6.2 (1.75)	0.0	51.2 (0.90)	0.7	99.3	0.0	95.5	0.0	4.5
sugarcane bagasse	47.7	6.1 (1.54)	0.3	45.9 (0.72)	3.0	73.8	23.2	82.5	4.5	13.0
Alaska cedar sawdust	48.5	6.4 (1.58)	0.0	45.1 (0.70)	2.6	70.0	27.4	79.7	1.2	19.1
lignin	63.6	5.9 (1.11)	0.1	30.4 (0.36)	1.5	0.0	98.5	46.6	19.9*	33.5

* ignition residue,   ** benzene/ethanol soluble

822

2.2. Experimental apparatus and procedure

We used a low-pressure continuous reactor with microwave plasma [3]. Quartz wool supported a 1.5-gram dried biomass sample in the center of a cylindrical quartz tube reactor (25 mm inside diameter, 2 mm thickness, and 500 mm length). The reactor tube was set up so that the upper part of the sample could be situated facing the center of plasma. The sample depth for cellulose and lignin was 5 mm, whereas that for sawdust and bagasse was 20 mm because of a high bulk density of the biomass (0.15 g cm^{-3}). Pure argon was passed downwardly through the reactor at 4.0 kPa. The microwaves of 2.45 GHz were irradiated at 300 J/s, and the microwave irradiation time was varied from 1 to 10 min. The gas yield was determined from the weight difference of the reactor tube before and after reaction. The oil was collected from the reactor tube as well as from the ice-cooled trap and then extracted in benzene for 48 hours at 298 K, subsequently dried under vacuum for 60 min at 333 K, and finally dried under an atmospheric pressure for 60 min at 380 K. The oily product yield was determined from the weight difference before and after extraction. The reaction residue from cellulose, bagasse and sawdust was found to contain water-solubles. The elemental analysis was performed for residue. Gaseous products were analyzed by FID (SHINCARBON S column) and TCD (Shimadzu GC-14B, SHINCARBON T) gas chromatograph.

3. RESULT AND DISCUSSION

3.1. Effect of microwave irradiation time on conversion and product yield during cellulose or lignin decomposition

Figure 1 shows the conversion and product yields as a function of microwave irradiation time. After 1 min, the conversion of cellulose reached 40.1 %, comparable to that of lignin (37.2 %). After 5 min, the conversion of cellulose further increased and reached 90 %, whereas that of lignin reached 50 % after 2 min and stayed constant thereafter. To investigate the cause of the different conversion between cellulose and lignin, molar element analysis was conduced. As results, we can see that cellulose showed a decrease in moles of all the elements until 5-min microwave irradiation, whereas lignin showed a decrease in moles of carbon and oxygen until 2-min microwave irradiation and a decrease in moles of hydrogen until 3-min microwave irradiation. The molar change of carbon and oxygen was

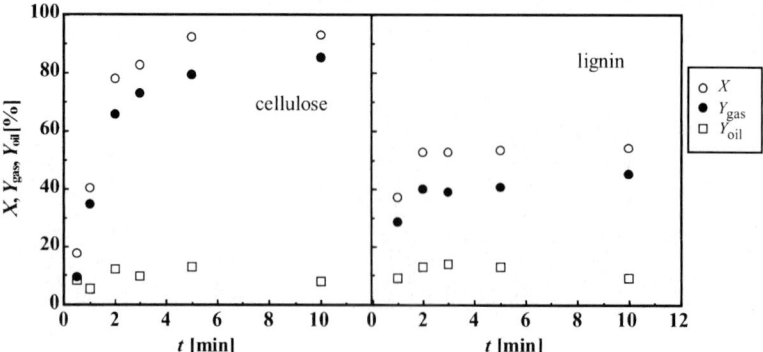

Fig. 1. Time change of conversion and product yield (cellulose, lignin)

the same for both cellulose and lignin. After 10 min irradiation, cellulose gave carbon residue of 5.8 mmol/g, and lignin gave that of 34.0 mmol/g. These values correspond to the difference in the moles of carbon and oxygen contained before decomposition (cellulose: 5.8 mmol/g; lignin: 34.0 mmol/g). This indicates that conversion proceeded with the reaction between carbon and oxygen and that it stopped when all oxygen atoms were consumed, suggesting that the reason for the higher conversion of cellulose is that cellulose contains more oxygen atoms (relative to carbon atoms) than lignin does.

3.2. Characteristics of gaseous product during cellulose or lignin decomposition

Figure 2 showed the cumulative moles of gases Q_m produced from 1g of cellulose or lignin. Both cellulose and lignin yielded hydrogen and carbon monoxide as major products throughout the microwave plasma irradiation. The ratio of hydrogen/carbon monoxide was roughly 0.70 for cellulose and 1.35 for lignin. Only small amounts of carbon dioxide and hydrocarbons and a very small amount of water were detected. Bassilakis, $et.al.$ reported that for thermal process of wheat straw, TG-FTIR detected large amounts of carbon dioxide and carbon monoxide (at 1073K, 6 wt.% and 10 wt.%, respectively) and 23 wt.% of water [4]. However, in this study, the ratio of carbon dioxide/carbon monoxide for cellulose and lignin was only 0.05 and 0.07, respectively. Graef, $et.al.$ also reported that for the microwave helium plasma process with pelletized lignin, the ratio of carbon dioxide/carbon monoxide was very small (0.05) [5]. Kamei $et\ al.$ [6] obtained H_2 and CO during Yalluron coal decomposition by microwave plasma technique, and Yalluron coal also yielded a very small amount of carbon dioxide. In addition, for plasma process in gas phase, carbon dioxide is known to decompose into carbon monoxide under hydrogen atmosphere, and for microwave plasma process, water vapor is known to decompose into hydrogen and carbon monoxide under methane gas atmosphere.

3.3. Characteristics of bagasse or sawdust decomposition

Figure 3 shows cumulative moles of gases produced from bagasse and sawdust. Hydrogen and carbon monoxide were produced as main gases throughout the conversion from both bagasse and lignin. In case of bagasse, the experimental values for hydrogen and carbon monoxide were greater than the calculated values throughout the conversion from both cellulose and lignin. On the other hand, in case

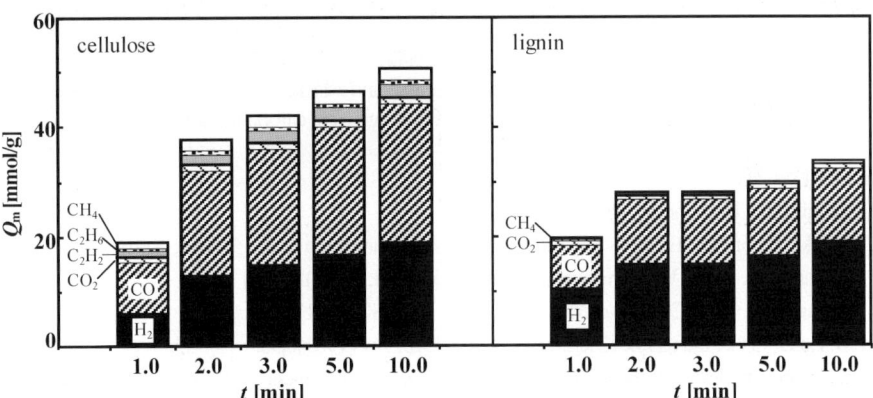

Fig. 2. Accumulated mol numbers of gaseous products against time (cellulose, lignin)

824

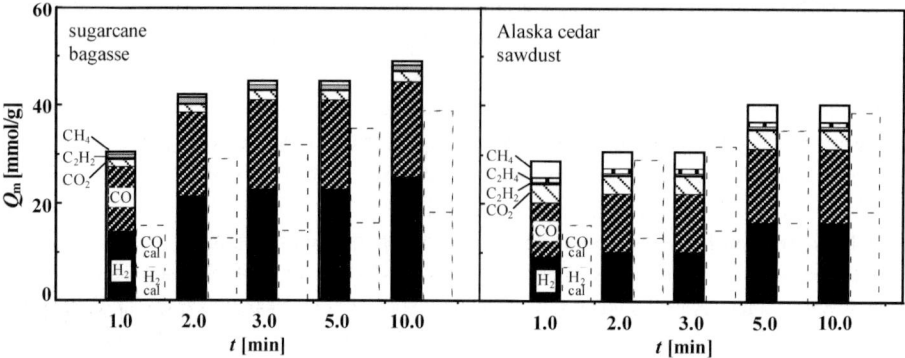

Fig. 3. Accumulated mol numbers of gaseous products against time (sugarcane bagasse, Alaska cedar sawdust)

of sawdust, within the initial 1-min irradiation, the experimental values of hydrogen and carbon monoxide were greater than the calculated values, but after the 1-min irradiation, the experimental values became smaller than the calculated values.   This can be explained by the high contents of methane and carbon dioxide produced within the initial 1-min irradiation.   By the fact that bagasse had the experimental value of hydrogen produced greater than the calculated value and the fact that sawdust had the experimental value of hydrogen produced lower than the calculated value, but had an increase in methane, it was suggested that for the microwave plasma reaction, the decomposition of lignin was influenced by the increase in the production of gases, and not by the consumption of hydrogen produced from cellulose.

4. CONCLUSION

Decomposition of plant-based biomass (sugarcane bagasse, Alaska cedar sawdust) and base compounds (cellulose and lignin) by the microwave plasma process was developed, and the following results were obtained.
1) Cellulose showed a greater conversion and a higher yield of gases than lignin.
2) Both cellulose and lignin produced hydrogen and carbon monoxide as main gaseous products.
3) Compared to cellulose and lignin, sugarcane bagasse and Alaska cedar sawdust initially showed a higher conversion and produced high amounts of hydrogen and carbon monoxide.

REFERENCES
[1] T. Ogi, J.Jpn. Inst. Energy, 78(4) (1999) 232.
[2] S. Saka, Baiomass · Energy · Environment, Industrial Publishing & Inc., Tokyo, 2001.
[3] O. Kamei, K. Onoe, W. Marushima, T. Yamaguchi, Fuel, 77(13) (1998) 1503.
[4] R. Bassilakis, R.M. Carangelo, M.A. Wójtowicz, Fuel, 80 (2001) 1765.
[5] M. Graef, G.G. Allen, B.B. Krieger, ACS Symp. Ser., 144 (1981) 293.
[6] K. Kamei, W. Marushima, M. Kobayashi, K. Onoe, T. Yamaguchi, S. Kawai, Y. Itoh, J. Jpn. Inst. Energy, 78(8) (1999) 664.

Studies in Surface Science and Catalysis, volume 159
Hyun-Ku Rhee, In-Sik Nam and Jong Moon Park (Editors)

825

Development of novel process for γ-butyrolactone production

H. Jeong[a,b], K. Kim[b], T. Kim[b], S. Cho[b] and I.K. Song[a]*

[a]School of Chemical and Biological Engineering, Institute of Chemical Processes,
 Seoul National University, Shinlim-dong, Kwanak-ku, Seoul 151-744, South Korea

[b]Korea Institute of Energy Research, Jang-dong, Yuseong-ku, Daejeon 305-343, South Korea

*Corresponding author (Tel:+82-2-880-9227, E-mail: inksong@snu.ac.kr)

1. INTRODUCTION

γ-Butyrolactone (GBL), tetrahydrofuran, and 1,4-butanediol are small volume commodities of considerable industrial interest, as indicated by both present market and growth prospects [1,2]. Several processes have been developed for the production of GBL, tetrahydrofuran, and 1,4-butanediol [3]. Among the commercial processes, hydrogenation of maleic anhydride is the most efficient and direct method to produce GBL. Therefore, much attention has been made on the development of efficient catalysts, which enable GBL to be produced by the hydrogenation of maleic anhydride with high conversion and selectivity. However, production of GBL from maleic anhydride has not been entirely successful from the practical point of view, because of low GBL yield and severe operating conditions [4]. Therefore, it is necessary to develop a new catalyst showing better activity under mild operating conditions. The present study includes the formulation of hydrogenation catalyst and the activity test toward the hydrogenation of maleic anhydride into GBL under mild operating conditions. The effects of hydrogen pressure, temperature, and catalyst loading on the catalytic performance were examined in order to optimize the reaction parameters for the maximum yield of GBL. On the basis of these fundamental studies, we finally designed a pilot-scale reactor that can produce 50 kg of GBL per one batch.

2. EXPERIMENTAL

Pd-Mo-Ni/SiO$_2$ catalyst was prepared by co-impregnation of silica support (Deggusa AROSIL 300 having BET surface area of 300 m^2/g) with aqueous solutions of palladium chloride, nickel nitrate, and ammonium molybdate tetrahydrate. The calculated loadings of Pd, Ni, and Mo were 2.3 wt.%, 4.5 wt.%, and 25 wt.% respectively. The impregnated silica catalyst was then calcined under the air stream (30 ml/min) at 450°C for 3 h. Prior to the catalytic reaction, the supported catalyst was reduced at 300°C for 3 h with a mixed stream of hydrogen (10 ml/min) and helium (30 ml/min). Catalytic reaction was carried out in a stainless steel (SS 316) autoclave reactor (1L). For the catalytic reaction, known amounts of maleic anhydride, catalyst, and solvent (tetrahydrofuran) were charged into the reactor. After charging the reactant mixture, the reactor was flushed with nitrogen for three times and then hydrogen was charged into the reactor to the pressure of 60-80 atm. The reactant mixture was heated up to 220-260°C and constantly stirred for 300 min. To make up for consumed hydrogen, that is, to maintain constant hydrogen pressure during the reaction, hydrogen was continuously added to the reactor. At the end of the reaction, the autoclave was cooled down to ambient temperature and slowly depressurized. The reaction products were analyzed by GC (HP 6890) and HPLC (Waters 2690).

3. RESULTS AND DISCUSSION

3.1. Catalytic activity and reaction kinetics

In order to investigate the effect of reaction temperature and pressure, hydrogenation of maleic anhydride was carried out by varying reaction temperature (220-260°C) and hydrogen pressure (60-80 atm). In the catalytic reaction, a mixture of 10 g of catalyst (Pd-Mo-Ni/SiO$_2$), 60 g of maleic anhydride, and 250 ml of tetrahydrofuran (solvent) was constantly stirred with an impeller speed of 600 rpm. Each catalytic reaction was conducted for 300 min. The typical catalytic performance plotted as a function of reaction temperature and hydrogen pressure is shown in Fig. 1. At constant hydrogen pressure of 70 atm, the GBL yield was increased with increasing reaction temperature. At the reaction temperature of 250°C, the GBL yield was over 90%. However, the GBL yields were nearly constant beyond 250°C and decreased with increasing reaction temperature at above 260°C due to the formation of by-products such as butyric acid and propionic acid. When considering GBL yield and energy consumption required for the process, it is concluded that the optimum reaction temperature would be around 250°C. The effect of hydrogen pressure on the GBL yield was somewhat complicated. The GBL yield was not increased in a monotonic fashion with increasing hydrogen pressure. The GBL yield was the maximum at the hydrogenation pressure of 70 atm. The GBL yield at 250°C and 70 atm was more than 90%. At hydrogenation pressure higher than 70 atm, the GBL yield decreased with increasing hydrogen pressure. In conclusion, the Pd-Mo-Ni/SiO$_2$ catalyst exhibited an excellent catalytic performance at mild reaction conditions.

Catalyst loading is also one of the major factors in determining reaction kinetics in a three-phase batch reactor like our catalytic reaction system. The effect of catalyst loading on the reaction rate was elucidated by conducting catalytic reaction under standard conditions as followings; maleic anhydride = 120g, reaction temperature = 250°C, hydrogen pressure = 70 atm, and catalyst loading = 5-20g. The effect of catalyst loading on the reaction rate in a slurry batch reactor can be described by the following equation [5,6].

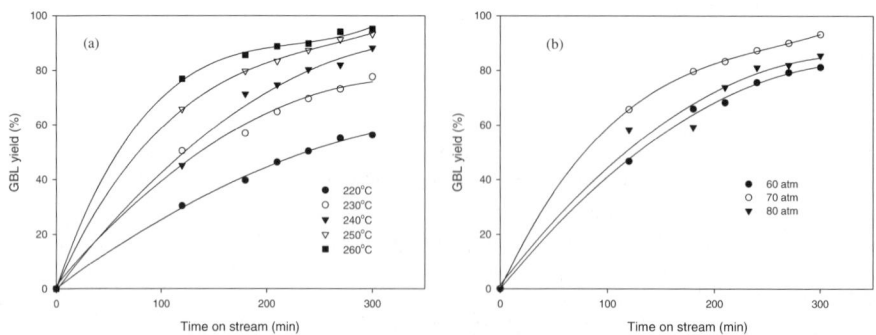

Fig. 1. GBL yield as a function of reaction temperature and hydrogen pressure:
(a) Hydrogen pressure=70 atm, and (b) Reaction temperature=250°C.

$$\frac{C_i}{-r} = -\frac{1}{K_L a} - \frac{1/K_s a + 1/\eta k}{m} \tag{1}$$

Equation (1) consists of various resistance terms. $1/K_L a$ is the gas absorption resistance, while $1/K_s a$ corresponds to the maleic anhydride diffusion resistance and $1/\eta k$ represents the chemical reaction resistance. The reaction rate data obtained under the reaction conditions of 250°C and 70 atm were plotted according to equation (1). Although catalytic reaction data with respect to time on stream were not shown here, a linear correlation between reaction rate data and catalyst loading was observed as shown in Fig. 2. The gas absorption resistance ($1/K_L a$) was -1.26 h, while the combined reaction-diffusion resistance ($1/K_s a + 1/\eta k$) was determined to be 5.57 h. The small negative value of gas absorption resistance indicates that the gas-liquid diffusion resistance was very small and had several orders of magnitude less than the chemical reaction resistance, as similarly observed for the isobutene hydration over Amberlyst-15 in a slurry reactor [6]. This indicates that absorption of maleic anhydride in solvent was a rapid process compared to the reaction rate on the catalyst surface.

Under the present reaction conditions, we observed the formation of succinic anhydride almost simultaneously together with the formation of GBL. The hydrogenation of maleic anhydride yields succinic anhydride, and the subsequent hydrogenation of succinic anhydride produces GBL. The rate of hydrogenation of maleic anhydride to succinic anhydride was very fast compare to that of succinic anhydride to GBL. When the reaction was carried out without solvent, tetrahydrofuran was not produced. The above results indicate that the Pd-Mo-Ni/SiO₂ catalyst under our experimental conditions played an important role for the selective formation of GBL. Therefore, it is inferred that the catalyst composition may influence the route by which tetrahydrofuran was formed, probably due to the different absorption mechanism of maleic anhydride, succinic anhydride, and GBL.

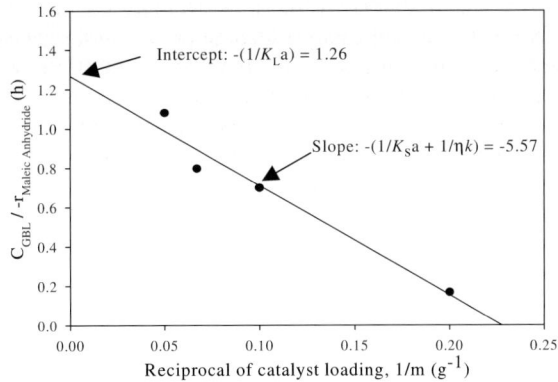

Fig. 2. Linear plot of equation (1) showing the influence of catalyst loading on the reaction rate.

3.2. Reactor scale-up

A pilot-scale reactor that can produce 50 kg of GBL per one batch was designed. The reactor size was calculated on the basis of maleic anhydride conversion of 100% and GBL yield of 50%. By assuming the saturated liquid density of reactant and product at the reaction conditions of 250°C and 70 atm under the condition of no solvent, the density and specific volume of total mixture were calculated. Critical properties of reactant and product, and liquid density of total reaction mixture could be described by the critical property equation [7], as summarized in Table 1.

We come to conclude that a pilot-scale reactor with capacity of ca. 120 L was necessary to produce 50 kg of GBL per one batch. By assuming the compressed liquid conditions, specific volume of total mixture was also calculated in order to compare with the result calculated on the basis of saturated liquid conditions. However, total volume of mixture was not greatly changed. At the compressed liquid conditions, total volume of mixture was decreased only 6% compared to that calculated at the saturated liquid conditions.

Table 1
Calculation of reactor volume for the pilot-scale GBL production (50 kg GBL/batch)

	T_c(K)	P_c(atm)	V_c(cm^3/mole)	Z_c
Succinic anhydride	806.9	66.8	237.0	0.2390
γ-Butyrolactone	740.4	58.6	231.5	0.2234
Maleic anhydride	715.2	71.9	222.0	0.2722

	Liquid density ρ^{sat} (250℃, g/cm^3)	Weight fraction	Density of mixture (g/cm^3)	Specific volume (cm^3/g)	Total volume of mixture (liter)
Succinic anhydride	1.22854	0.4525			
γ-Butyrolactone	1.06764	0.4525	1.098	0.9111	119.3
Maleic anhydride	0.7978	0.095			

4. CONCLUSIONS

Effects of pressure, temperature, and catalyst loading on the performance of Pd-Mo-Ni/SiO$_2$ catalyst were examined in a batch reactor for the production of GBL from maleic anhydride in order to determine the optimum reaction conditions for the maximum GBL yield, with an aim of designing a large scale reactor. The Pd-Mo-Ni/SiO$_2$ catalyst showed an excellent catalytic performance in the hydrogenation of maleic anhydride into GBL under the mild condition of 250°C and 70 atm. The Pd-Mo-Ni/SiO$_2$ catalyst exhibited more than 90% GBL yield. From a series of investigation, we came to conclude that a pilot-scale reactor with capacity of 120 L was necessary to produce 50 kg of GBL per one batch at 250°C and 70 atm. The pilot-scale reactor is now in successful operation.

REFERENCES

[1] S. Minoda and M. Miyajima, Hydrocarbon Process., 49 (1970) 176.
[2] Z.M. Zong, Y.L. Peng, Z.G. Liu, S.L. Zhou, L. Wu, X.H. Wang, X.Y. Wei and C.W. Lee, Korean J. Chem. Eng., 20 (2003) 235.
[3] A.M. Brownstein, Chemtech., 21 (1991) 506.
[4] S.J. Conway and J.H. Lunsford, J. Catal., 131 (1991) 512.
[5] S.T. Warna, Comput. Chem. Eng., 20 (1996) 39.
[6] C.M. Zhang, A.A. Adesina and M.S. Wainwright, Chem. Eng. Process., 42 (2003) 985.
[7] R.H. Perry, Perry's Chemical Engineers' Handbook, McGraw-Hill, New York, 1984.

Studies in Surface Science and Catalysis, volume 159
Hyun-Ku Rhee, In-Sik Nam and Jong Moon Park (Editors)

829

Kinetic study on alkaline decomposition of organophosphorus

Yasuhiko Takuma, Shigeru Kato and Toshinori Kojima

Department of Applied Chemistry, Faculty of Engineering, Seikei University, 3-3-1 Kitijoji Kitamati, Musashino-shi, Tokyo 180-8633, Japan

1. INTRODUCTION

Organophosphorus insecticides with high toxicity used from 1950's to 1970's, such as parathion, TEPP and so on, have been banned in Japan. However, unused quantities are still stored in many agricultural sites. Indeed, there is a significant likelihood that even the current permissible organophosphorus pesticides will ultimately be prohibited from use in the near future due to their endocrine-disrupting effects, so destruction methods for these compounds should be developed.

The main purpose of this work is development of small-scale and mobile decomposition system of these chemicals. A number of studies on decomposition of organophosphorus insecticides have been conducted [1-3]. It is well known that organophosphorus insecticides are decomposed by hydrolysis under alkaline condition, and its mechanisms have been studied [4]. Even so, relatively few papers have addressed the development of kinetic equations for reactor design. In this study, we aim to get kinetic equations for their decomposition under alkaline condition. As organophosphorus, we used parathion, fenitrothion, diazinon, malathion and phenthoate.

2. METHODOLOGY

2.1. Experiment: Relations between decomposition rate and alkaline concentration

At first, relations between decomposition rate and alkaline concentration were evaluated. 0.2mol/l of ethanol solution of organophosphorus insecticides and 1.0, 2.0 and 4.0mol/l of ethanol solution of sodium hydroxide were prepared. The insecticide and the sodium hydroxide solutions were kept at reaction temperature (35 °C) in a water bath, and equi-volumes of two solutions were mixed to get 10ml. Thus, the concentrations of agrochemicals and NaOH are the half of those of premixed solutions, before the reaction. After a reaction time, 50ml of distilled water was added, and the residual insecticide and others were extracted with 15ml of chloroform three times. Then the aqueous phase was changed to be acidic by adding hydrochloric acid. Weakly acidic compounds in the aqueous phase were extracted with 15ml of chloroform by three times. Sodium sulfate was added to these chloroform solutions for dehydration. Then these solutions were filled up with chloroform to 50ml and were quantitatively analyzed with GC-FID.

2.2. Experiment: Relations between decomposition rate and temperature

Dependences of reaction rate constants on temperature were evaluated. Experiments were carried out at 20, 35 and 50°C using 0.2mol/l of organophosphorus insecticide and 2.0mol/l of sodium hydroxide solutions. Experimental methods except the above conditions were same as those in 2.1.

2.3. Prediction of frequency factor

Theoretical frequency factor for the three compounds were evaluated with Eq.(1) [5] assuming the rate limiting elementary reaction steps are organophosphorus and ethoxide,

$$A = \sigma \left(\frac{8kT}{\pi\mu} \right)^{1/2} N_A \qquad (1)$$

where, A: frequency factor [$m^3 \cdot (mol \cdot s)^{-1}$], σ: collision cross-section [m^2], k: Boltzmann constant [$J \cdot K^{-1}$], T: temperature [K], μ: reduced mass [kg], N_A: Avogadro number [mol^{-1}]. Collision cross-section and reduced mass were calculated by Eqs. (2) and (3) respectively. In these equations, R is molecular radius [m] and m is molecular mass [kg].

$$\sigma = \pi(R_A + R_B)^2 \qquad (2)$$

$$\frac{1}{\mu} = \frac{1}{m_A} + \frac{1}{m_B} \qquad (3)$$

Table 1
Molecular radius and molecular mass

	Radius [10^{-8} m]	Mass [10^{-25} kg]
Parathion	5.372	4.836
Fenitrothion	5.255	4.604
Diazinon	5.709	5.054
$CH_3CH_2O^-$	2.557	0.7482

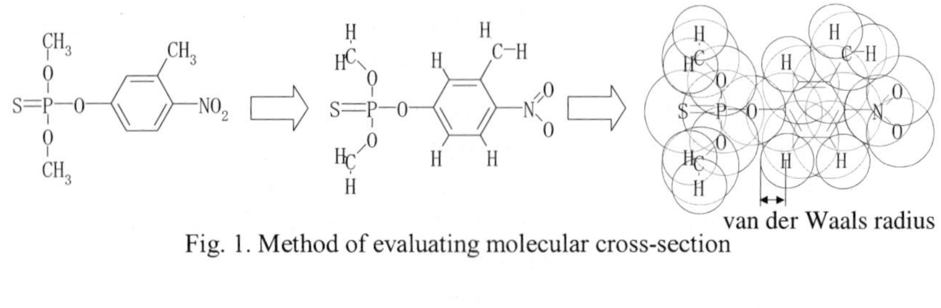

Fig. 1. Method of evaluating molecular cross-section

Molecular radius, R was estimated as the radius of equivalent circle to the projection area of the right configuration in Fig.1.

3. RESULTS AND DISCUSSION

3.1. Relations between decomposition rate and alkaline concentration

From the results of experiment 2.1, we confirmed decomposition reaction is pseudo first-order, and calculated pseudo first-order decomposition rate constants. Then from relationship between each first-order reaction rate constant and sodium hydroxide concentration, we confirmed that the reaction is expressed by second-order with expression first-orders for both of sodium hydroxide and fenitrothion.

Almost same relationships were also found for other two organophosphorus insecticides, parathion and diazinon. So second-order decomposition rate constants were used to evaluate all of the organophosphorus insecticides, where the second-order decomposition rate constants were calculated by dividing the first-order decomposition rate constants by the sodium hydroxide concentration. Rate constants of malathion and phenthoate could not be obtained, because these reaction rates were too fast to analyze.

3.2. Relations between decomposition rate and temperature

Second-order reaction rate constants for the three compounds at 20, 35 and 50°C were evaluated as in the methodology section of 2.2. Also, theoretical frequency factors are evaluated by Eq.(1). To calculate the frequency factors, we used the value shown in Table 1.

Second-order decomposition rate constants experimentally determined and frequency factors theoretically predicted are plotted on Fig.2. Good linear expressions were demonstrated, which suggest the reasonability of assumption of limiting elementary steps. From this figure, activation energies for three compounds, parathion, fenitrothion and diazinon are calculated. The obtained activation energies are shown in Table 2 together with theoretical frequency factors.

Table 2
Frequency factors and activation energies

		Parathion	Fenitrothion	Diazinon
Second-order reaction rate constant at 20°C	$[l \cdot (mol \cdot s)^{-1}]$	8.28×10^{-5}	2.26×10^{-4}	1.52×10^{-4}
at 35°C	$[l \cdot (mol \cdot s)^{-1}]$	4.34×10^{-4}	1.03×10^{-3}	4.17×10^{-4}
at 50°C	$[l \cdot (mol \cdot s)^{-1}]$	1.43×10^{-4}	5.20×10^{-3}	1.67×10^{-3}
Frequency factor	$[l \cdot (mol \cdot s)^{-1}]$	4.74×10^{9}	4.60×10^{9}	5.14×10^{9}
Activation energy	$[kJ \cdot mol^{-1}]$	77.2	74.5	76.8

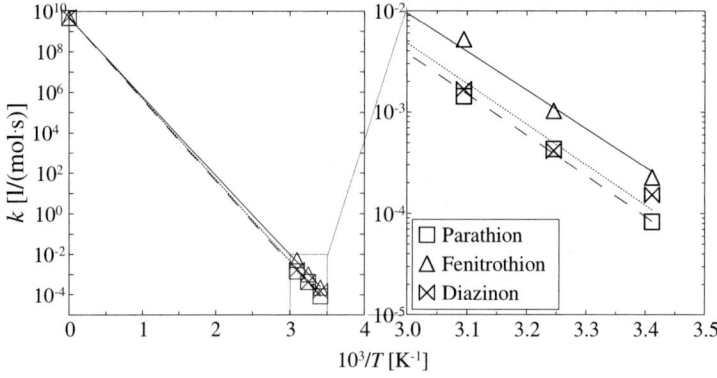

Fig. 2. Arrhenius plot

From the difference among the reaction rates, organophosphorus insecticides used in this work were divided into two groups. One group consists of the compounds with high decomposition rates such as phenthoate and malathion, the other with low rates such as parathion, fenitrothion and diazinon. The difference of reaction rate comes from the difference of these structures. The former compounds with high rates have a structure with central sulfur atom between the phosphoric acid functional group and the rest part and the latter, central oxygen. The mean bond enthalpy between carbon and sulfur is 259kJ/mol, and between carbon and oxygen is 360kJ/mol [5]. The former indicate the weaker bond than the latter, which explains the easier decomposition of malathion and phenthoate.

4. CONCLUSION

Decomposition kinetics of five compounds by alkaline hydrolysis were measured. For three compounds, second-order reaction rate constants and activation energies were given. These five compounds could be divided two groups with high and low decomposition rates.

References
[1] R. D. O'Brien and I. Yamamoto, Biochemical Toxicology of Insecticides, Academic Press, New York, 1970.
[2] F. Hong, S. O. Pehkonen and E. Books, J Agric. Food Chem., 48 (2000) 3013.
[3] T. Okajima and K. Maekawa, Environmental Chem., 11 (2001) 491.
[4] A. Dannenberg, S. O. Pehkonen, J. Agric. Food Chem., 46 (1998) 325.
[5] P. W. Atkins, Physical Chemistry Forth Edition, Oxford University Press, Oxford, 1990.
[6] The Chemical Society of Japan, Kagaku-Binnrann Kisohenn Kaitei 5 Hann (Chemical Handbook, Basic, Fifth Edition), Maruzen, Tokyo, 2004.

Studies in Surface Science and Catalysis, volume 159
Hyun-Ku Rhee, In-Sik Nam and Jong Moon Park (Editors)
© 2006 Elsevier B.V. All rights reserved

Effects of pH and mixing temperature on the continuous precipitation of lanthanum phosphate

Motoaki Kawase, Tomomitsu Suzuki, Hiroshika Goshima, Tunekata Kobata, and Kouichi Miura

Department of Chemical Engineering, Kyoto University

Kyotodaigaku-Katsura, Nishikyo-ku, Kyoto 615-8510, Japan

1. INTRODUCTION

The LAP phosphor, lanthanum phosphate doped with cerium and terbium ($LaPO_4$: Ce^{3+}, Tb^{3+}), is used for fluorescent lamps and computer CRTs due to its high efficiency and high color fidelity. Monodisperse spherical particles of 1–10 μm diameter are required for these purposes. When lanthanum phosphate particles are synthesized by using a stirred tank reactor, the productivity is low and and the particle size distribution is broad. These problems result from inhomogeneous mixing and nucleation due to the mechanical contact by stirring. The authors have successfully improved the monodispersity by applying a tubular reactor to lanthanum phosphate precipitation. In the tubular reactor, the mixing problems were prevented and uniform reaction time was attained. Monodispersed spherical particles were successfully prepared at a high yield. In this study, the effects of the important factors of precipitation, i.e. pH, reactant feed concentration, and temperature, were investigated.

2. EXPERIMENTAL

Fig. 1 shows the reactor setup used for continuous precipitation of lanthanum phosphate. A tube made of silicone rubber was used as a reactor. The inner diameter of the reactor tube

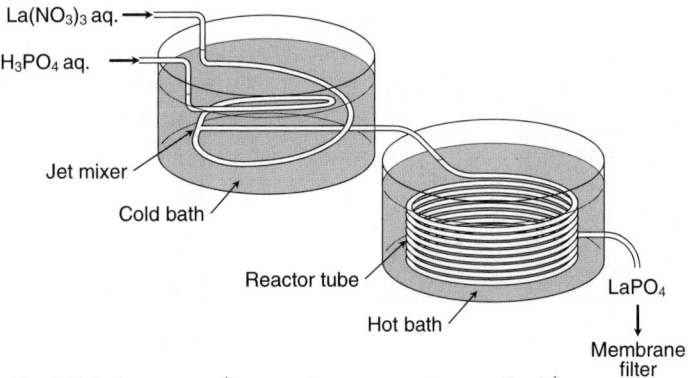

Fig. 1 Tubular reactor （in case of two-temperature synthesis）

was $d_t = 3.0$ mm and the length was $L_t = 0.1–5.3$ m. The first 15 cm of the reactor tube was submerged in a constant temperature bath at 0°C while the rest part was in another bath at 20-80°C. Only the hot bath was used for carrying out some isothermal experiments. The mixing part at the inlet of the reactor was in a T shape made of polypropylene, the inner diameter of which was 3.0 mm. Reactant solutions were 100 mol/m^3 La(NO)$_3$ and H$_3$PO$_4$ solutions. The pH of the solutions was conditioned by adding a HNO$_3$ or NH$_3$ solution. The residence time of the solution was 0.1–120 s. The particles were filtrated with a mixed cellulose ester membrane filter of 0.1 μm in pore size, and dried in vacuum, and then total weight of the particles was measured to calculate the particle yield. The diameter of the particles was measured and the size distribution was determined, and then the particle number density was estimated. Scanning and transmission electron microscopes (SEM & TEM) were used to observe the obtained particles.

3. RESULTS AND DISCUSSION

3.1. Effects of pH

Figs. 2a, 2b, and 2c show the SEM images of the particles obtained at pH = 1.3, 1.4, and 1.6 in the isothermal precipitation at 30°C. Ribbon-like particles were obtained at a low pH of 1.3. No spherical particles were obtained at pH lower than 1.2. At a high pH of 1.6, the mean diameter became as small as $D_{50} = 1.01$ μm, the geometric standard deviation was $\sigma_g = 1.46$, and the agglomeration of the particles was observed, although the particle yield increased to $Y_C = 27\%$. Monodisperse spherical particles ($D_{50} = 1.68$ μm, $\sigma_g = 1.39$) were able to be synthesized at a high yield, $Y_C = 20\%$, at an optimal pH of 1.4.

When a weak acid was used as reactant, the electrolytic dissociation was necessary to supply the precursor of precipitation. The concentration changes of phosphoric species, H$_3$PO$_4$, H$_2$PO$_4^-$, HPO$_4^{2-}$, and PO$_4^{3-}$, depending on pH were estimated from dissociation equilibrium. The observed particle yield was proportional to the H$_2$PO$_4^-$ concentration. Concentration changes of the other phosphoric species did not match with the yield change. Fig. 3 shows the particle yield, size, and number density against the estimated H$_2$PO$_4^-$ ion concentration. When the pH increased, the conversion increased due to enhanced electrolytic dissociation of phosphoric acid, and consequently the particle yield increased. These facts suggest that H$_2$PO$_4^-$ was the actual precursor of the precipitation. The number density was proportional to the square of the precursor concentration as shown in Fig. 3.

By increasing the pH, increasing the precursor concentration, the yield and the particle

Fig. 2.   Effect of pH on particle shape and size.   (a) pH = 1.3, (b) 1.4, (c) pH=1.6.

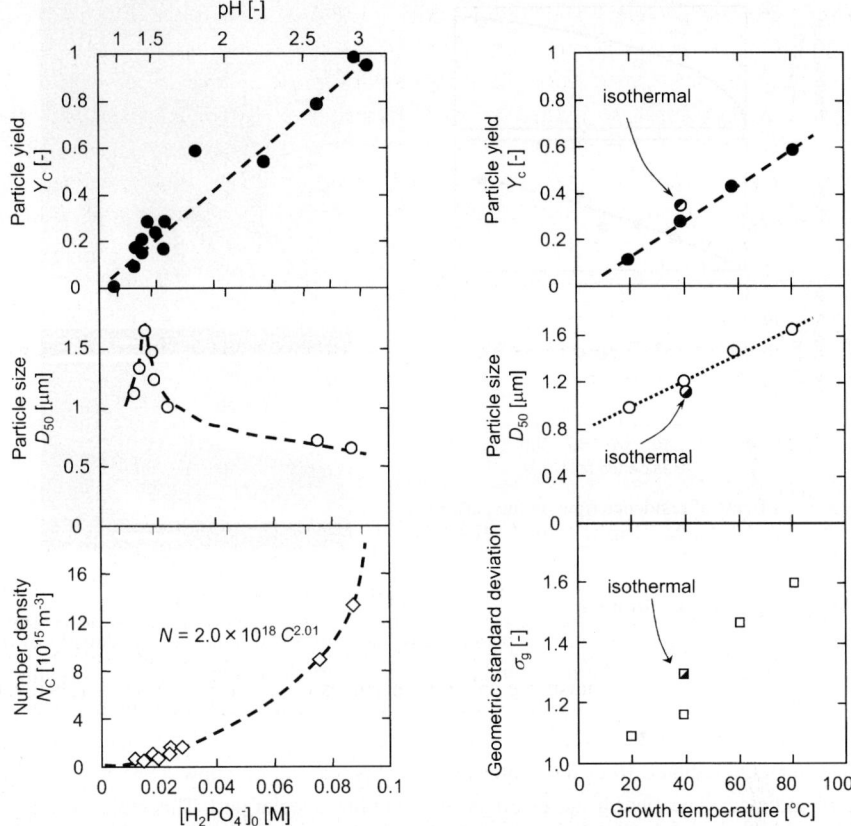

Fig. 3 Effects of pH on the particle yield, size, and number density. (30°C isothermal)

Fig. 4 Effects of temperature on the particle yield, size, and number density.

size increased. At too high precursor concentration, however, the nucleation was too rapid and a lot of small particles were obtained. That is why an optimum pH existed for attaining the maximum particle size.

3.2. Effects of the mixing temperature

In order to investigate the effects of the mixing temperature, two-temperature synthesis experiments were carried out with only mixing part cooled to 0°C. Fig. 4 shows the effects of growth temperature on the yield, size, and geometric standard deviation of the particles obtained by the isothermal and two-temperature precipitation. By comparing the results of isothermal and two-temperature synthesis at a growth temperature of 40°C, it is found that the monodispersity of the particle size was improved in the latter case while the yield and size were not greatly different. Lowering the mixing temperature, the particle growth near the reactor inlet was reduced, only the nucleation took place and the uniform crystal growth was attained in the following hot region. As shown in Fig. 4, the growth after the nucleation was able to be enhanced by increasing the growth temperature.

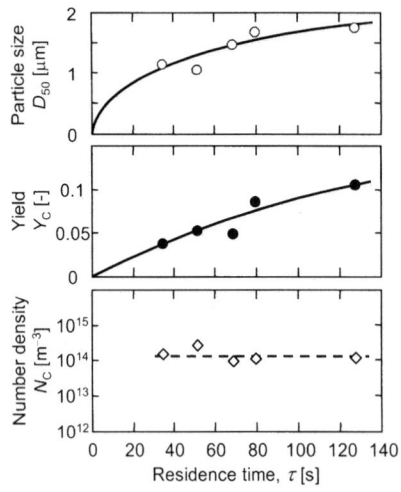

Fig. 5. Effects of residence time on the particle size, yield, and number density. (0°C, pH = 1.7)

Fig. 6. TEM images of lanthanum phosphate particle. (0°C, pH = 1.7)

3.3. Particle growth mechanism

Fig. 5 shows the residence time dependence of the size, yield, and number density of the particles obtained under isothermal conditions. The particle number density was almost constant independently of the residence time. This suggested that lanthanum phosphate particles did not grow by agglomeration. In the spherical particles of lanthanum phosphate, acicular elements radiated from the center [1,2]. Figs. 6a and 6b show TEM images of an obtained particle. The clear fringes were observed in the acicular part. This indicates that the particle growth resulted from the growth of the acicular crystallites.

4. CONCLUSIONS

By investigating the effects of pH, it was shown that $H_2PO_4^-$ was the actual precursor of lanthanum phosphate precipitation. The particle yield was proportional to the $H_2PO_4^-$ concentration. The number density was proportional to the square of the $H_2PO_4^-$ concentration. A maximum particle size was obtained at an optimum precursor concentration as a result of the tradeoff of the nucleation and growth.

Temperature determined the competition between the nucleation and particle growth. At low temperature, only nucleation took place. This allowed uniform particle growth, leading to monodisperse particle growth. Separation of the stages of nucleation and growth was effective for preparing the monodisperse particles.

REFERENCES

[1] Kawase, M., Masuda, T., Nakanishi, A., Kijima, N., Miura, K.; *Proc. 17th ISCRE (Hong Kong, Aug., 2002)*, MS#0168.
[2] Kawase, M., Miura, K., *et al.*; *Proc. 10th APCChE Congress (Kitakyushu, Sep., 2004)*, 1C-10.

POLYMER REACTION ENGINEERING

Studies in Surface Science and Catalysis, volume 159
Hyun-Ku Rhee, In-Sik Nam and Jong Moon Park (Editors)

837

Simulation and application of the ideas for productivity enhancement in an LDPE tubular reactor

Sang Seop Na and Jinsuk Lee

Samsung Total Petrochemicals Co., Ltd.,
411-1, Dokgod-Ri, Daesan-Up, Seosan-Si, Chungnam, Korea 356-711

1. INTRODUCTION

Reflecting the situation that the worldwide LDPE (low density polyethylene) demand is still slightly increasing, it becomes very important to enhance the productivity with existing facilities by minimum investment. The productivity enhancement is mainly done by changes of hardware, operating conditions and chemicals used such as initiators and chain transfer agents. Some examples of hardware change are the changes of initiator and monomer injection points, cooling zone division change, adopting different reactor spools, and so on.
Prediction of new operating conditions by which the same product qualities can be reproduced is one of the big problems encountered in the productivity enhancement operation. Without proper simulation model, one has to rely only on the trial and error based on the operation experiences, which usually costs a great deal.

Samsung Total has developed an LDPE reactor simulation model which can calculate not only conversion, concentration of each species, temperature, pressure, long chain branching, short chain branching but also MWD(molecular weight distribution) along the reactor axis. The MWD calculation based on kinetics which is used in this study is first applied to LDPE tubular reactor to the best of authors' knowledge. Here we briefly present the model we developed and some examples how the ideas of productivity enhancement were realized which are rarely reported in public domain so far.

2. MODEL DEVELOPMENT

LDPE polymerization reaction consists of various elementary reactions such as initiation, propagation, termination, chain transfer to polymer and monomer, β-scission and so forth [1-3]. By using the rate expression of each elementary reaction in our previous work [4], we can construct the equations for the rate of formation of each component.

Monomer, Initiators, CTAs, Polymers:

$$r_M = \frac{dM}{dt} = -k_p M \lambda_0 - k_{trm} M \lambda_0$$

$$r_{A^j} = \frac{dA^j}{dt} = -k_d^j A^j - k_x^j A^j$$

$$r_{S^k} = \frac{dS^k}{dt} = -k_{trs}^k S^k \lambda_0$$

$$r_{R_1} = \frac{dR_1}{dt} = \sum_j 2k_d^j A^j - k_p MR_1 - k_t R_1 \lambda_0 - \sum_k k_{trs}^k R_1 S^k + \sum_k k_{trs}^k \lambda_0 S^k - k_{trm} R_1 M$$

$$+ k_{trm} \lambda_0 M - k_{trp} R_1 \mu_1 + k_\beta \lambda_0 - k_\beta R_1 + k_{\beta*} \lambda_0 - k_{\beta*} R_1$$

$$r_{R_n} = \frac{dR_n}{dt} = k_p MR_{n-1} - k_p MR_n - k_t R_n \lambda_0 - \sum_k k_{trs}^k R_n S^k - k_{trm} R_n M$$

$$- k_{trp} R_n \mu_1 + nk_{trp} \lambda_0 Q_n - k_\beta R_n - k_{\beta*} R_n$$

$$r_{Q_n} = \frac{dQ_n}{dt} = k_t R_n \lambda_0 + \sum_k k_{trs}^k R_n S^k + k_{trm} R_n M + k_{trp} R_n \mu_1 - nk_{trp} \lambda_0 Q_n + k_\beta R_n + k_{\beta*} R_n$$

Moments of Living Polymers:

$$r_{\lambda_0} = \frac{d\lambda_0}{dt} = \sum_j 2k_d^j A^j - k_t \lambda_0^2$$

$$r_{\lambda_1} = \frac{d\lambda_1}{dt} = \sum_j 2k_d^j A^j + k_p M\lambda_0 - k_t \lambda_0 \lambda_1 + \left(\sum_k k_{trs}^k S^k + k_{trm} M + k_\beta + k_{\beta*} \right)(\lambda_0 - \lambda_1) + k_{trp}(\lambda_0 \mu_2 - \lambda_1 \mu_1)$$

$$r_{\lambda_2} = \frac{d\lambda_2}{dt} = \sum_j 2k_d^j A^j + k_p M(2\lambda_1 + \lambda_0) - k_t \lambda_0 \lambda_2 + \left(\sum_k k_{trs}^k S^k + k_{trm} M + k_\beta + k_{\beta*} \right)(\lambda_0 - \lambda_2) + k_{trp}(\lambda_0 \mu_3 - \lambda_1 \mu_1)$$

Moments of Dead Polymers:

$$r_{\mu_0} = \frac{d\mu_0}{dt} = k_t \lambda_0 \lambda_0 + \sum_k k_{trs}^k S^k \lambda_0 + k_{trm} M\lambda_0 + k_\beta \lambda_0 + k_{\beta*} \lambda_0$$

$$r_{\mu_1} = \frac{d\mu_1}{dt} = k_t \lambda_0 \lambda_1 + \sum_k k_{trs}^k S^k \lambda_1 + k_{trm} M\lambda_1 + k_\beta \lambda_1 + k_{\beta*} \lambda_1 + k_{trp}(\lambda_1 \mu_1 - \lambda_0 \mu_2)$$

$$r_{\mu_2} = \frac{d\mu_2}{dt} = k_t \lambda_0 \lambda_2 + \sum_k k_{trs}^k S^k \lambda_2 + k_{trm} M\lambda_2 + k_\beta \lambda_2 + k_{\beta*} \lambda_2 + k_{trp}(\lambda_2 \mu_1 - \lambda_0 \mu_3)$$

LDPE tubular reactor is divided into several reaction zones according to the feed injection points. Here we apply mixing cell model for tubular reactor which considers the reactor axis as series of cells which is conceptually the same as CSTRs in series. In this study 40 cells are used for each reactor spool of 10 m long. The mass balance equation of a single cell at steady state can be written as follows.

$$\frac{X_{feed}}{\tau_{feed}} - \frac{X}{\tau} + r_X = 0$$

Here X and r_x denote the concentration of a species and its reaction rate respectively. τ denotes mean residence time. In order to have deep understanding of the end user properties the prediction of MWD (molecular weight distribution) is essential. The numerical techniques of calculating MWD so far developed include approximation [5-6], Monte-Carlo simulation [7], numerical fractionation [8-10] and population balance method [11]. The population balance method used here is one of the most reliable methods since it is based on the kinetic equations without any manipulation or simplifying assumption. In order to calculate MWD one has to define the concentration of living and dead polymer with chain length n as follows.

$$R_1 = \frac{\sum_j 2k_d^j A^j + \sum_k k_{trs} S^k \lambda_0 + k_{trm} \lambda_0 M + k_\beta \lambda_0 + k_{\beta*} \lambda_0 + \dfrac{R_{1,feed}}{\tau_{feed}}}{k_p M + k_t \lambda_0 + \sum_k k_{trs} S^k + k_{trm} M + k_{trp} \mu_1 + k_\beta + k_{\beta*} + \dfrac{1}{\tau}}$$

$$R_n = \frac{\dfrac{k_p M R_{n-1}}{A} + \dfrac{R_{n,feed}}{A\tau_{feed}} + \dfrac{nk_{trp}\lambda_0}{A\left(nk_{trp}\lambda_0 + \dfrac{1}{\tau}\right)}\dfrac{Q_{n,feed}}{\tau_{feed}}}{1 - \dfrac{nk_{trp}\lambda_0}{A\left(nk_{trp}\lambda_0 + \dfrac{1}{\tau}\right)}\left(k_t \lambda_0 + \sum_k k_{trs} S^k + k_{trm} M + k_{trp}\mu_1 + k_\beta + k_{\beta*}\right)}$$

$$Q_n = \frac{k_t R_n \lambda_0 + \sum_k k_{trs}^k S^k R_n + k_{trm} R_n M + k_{trp} R_n \mu_1 + k_\beta R_n + k_{\beta*} R_n + \dfrac{Q_{n,feed}}{\tau_{feed}}}{nk_{trp}\lambda_0 + \dfrac{1}{\tau}}$$

$$where,\ A = k_p M + k_t \lambda_0 + \sum_k k_{trs}^k S^k + k_{trm} M + k_{trp}\mu_1 + k_\beta + k_{\beta*} + \dfrac{1}{\tau}$$

When the concentration of living and dead polymer with chain length n are calculated as above, the mass fraction of n-mer (polymer with chain length n) can be defined as follows.

$$f_n = \frac{weight\ of\ n-mer}{weight\ of\ total\ polymer} = \frac{mV(nR_n + nQ_n)}{mV\left(\sum_{i=1}^{\infty} iR_i + \sum_{i=2}^{\infty} iQ_i\right)} = \frac{nR_n + nQ_n}{\lambda_1 + \mu_1}$$

3. RESULTS AND DISCUSSION

First we tuned the simulation model using existing operation conditions. Product properties as well as conversion and temperature profile along the reactor axis closely coincided with the actual data after properly choosing the kinetic constants and other operation parameters.

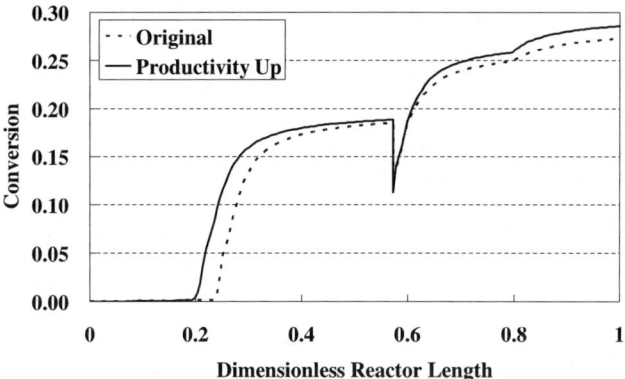

Fig. 1. Conversion profiles under the operation conditions of original and enhanced productivity.

840

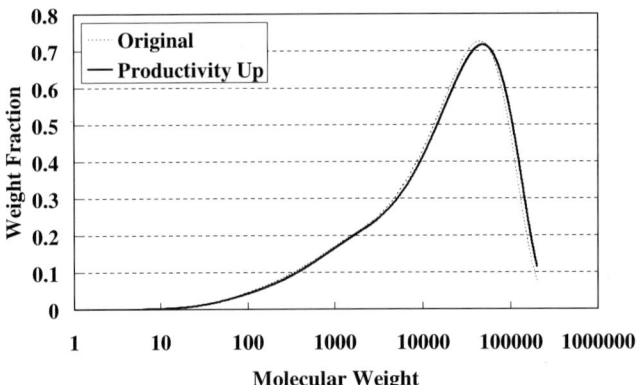

Fig. 2. MWDs under the operation conditions of original and enhanced productivity.

This is a further deepened work of what Samsung Total accomplished[12-14] several years ago. Several operation conditions including hardware modification which may enhance the productivity were deduced and simulated using the simulation model. Some ideas were already applied to commercial plant when they were concluded practically reasonable while some are on the waiting list. One of the examples of productivity enhancement is shown in Fig. 1 and Fig. 2 which compare the conversion profiles and MWDs under original and revised operation conditions. As shown in these two figures the productivity was enhanced while MWD does not change much.

4. CONCLUSION

By using the simulation model developed in Samsung Total we applied the ideas of productivity enhancement successfully to LDPE plant and accomplished considerable productivity increase. The MWD as well as the melt index and density calculated by the simulation model convinced us of applying the ideas to commercial plant. The end user property prediction capabilities of the model will be refined further by integration of physicochemical and statistical approaches and be one of the next potential research items.

REFERENCES
[1] B. J. Yoon and H.-K. Rhee, Chem. Eng. Commun., 34 (1985) 253.
[2] S. Goto, K. Yamamoto, S. Furui and M. Sugimoto, J. Appl. Pol. Sci., 36 (1981) 21.
[3] P. Feucht, B. Tilger and G. Luft, Chem. Eng. Sci., 40 (1985) 1935.
[4] J. Lee, J. Y. Ham, K. S. Chang, J. Y. Kim and H.-K. Rhee, Pol. Eng. Sci., 39 (1999) 1279.
[5] M. Wulkow, Macromol. Theory Simul., 5 (1996) 393.
[6] P. Pladis and C. Kiparissides, Chem. Eng. Sci., 53 (1998) 3315.
[7] H. Tobita and S. Saito, Macromol. Theory Simul., 8 (1999) 513.
[8] F. Teymour and J. D. Campbell, Macromolecules, 27 (1994) 2460.
[9] T. J. Crowley and K. Y. Choi, Ind. Eng. Chem. Res., 36 (1997) 1419.
[10] W. J. Yoon, J. H. Ryu, C. Cheong, and K. Y. Choi, Macromol. Theory Simul., 7 (1998) 327.
[11] H. Nordhus, O. Moen and P. Singstad, Polymer Reaction Engineering III, Palm Coast, U.S.A., 1997.
[12] J. H. Ryu and J. Lee, KIChE Autumn National Meeting, Pohang, Korea, 2000.
[13] M. S. Kim, J. H. Ryu and J. Lee, Proc. SML'01, Lucca, Italy, 2001.
[14] J. Lee, B. Y. Chung, M. J. Lee and K. Lo, Korean Patent Issued, 10-2004-0095084 (2004).

Studies in Surface Science and Catalysis, volume 159
Hyun-Ku Rhee, In-Sik Nam and Jong Moon Park (Editors)
© 2006 Elsevier B.V. All rights reserved

841

A comparative study ethylene/1-hexene copolymerization with [*t*-BuNSiMe₂Flu]TiMe₂ catalyst via various activators

Nawaporn Intaragamjon[a,b,*], Takeshi Shiono[c], and Piyasan Praserthdam[a,*]

[a] Center of Excellence on Catalysis and Catalytic Reaction Engineering
Department of Chemical Engineering, Faculty of Engineering,
Chulalongkorn University Bangkok, 10330 Thailand

[b] Ikeda-Shiono Laboratory, Chemical Resource Laboratory,
Faculty of Chemical engineering, Tokyo Institute of Technology
4259 Nagatsuta-Cho, Midori-ku, Yokohama-Shi, Kanagawa-Ken, 226-8503 Japan

[c] Polymer Chemistry Group, Department of Applied Chemistry
Division of Materials Chemistry and Chemical Engineering
Graduate School of Engineering, Hiroshima University
Kagamiyama 1-4-1, Higashi-Hiroshima 739-8527 Japan

1. INTRODUCTION

Due to the catalyst abilities and co-catalyst, we can control the amounts of alpha-olefin insertion, short chain branching distribution and the triad distribution of copolymers. The different microstructure of copolymer will affect the polymer properties such as melting and glass transition temperatures, melt viscosity, and mechanical and optical properties, all of which define the type and useful range of application of these materials [1-4]. Marks et.al [5] has investigated alpha-olefin comonomer insertion enchantment concentrate on catalyst and cocatalyst nuclearity using many kinds of catalyst. They found that in single site polymerization system, the sequence of polymer insertion strongly depended on the bimetallic complexes of catalyst and cocatalyst that were employed in the system. In this report, we reported the copolymerization behaviors between the [*t*-BuNSiMe₂Flu]TiMe₂ catalyst (denoted as catalyst 1 hereafter) with the different activators, such as MAO series, MMAO series, B(C₆F₅)₄ and Ph₃CB(C₆F₅)₄. The main objective of this present study was also to determine the influence of cocatalyst employed on the obtained polymer microstructure toward ethylene/1-hexene copolymerization.

2. EXPERIMENTAL

Polymerization was performed in a 100 mL glass reactor equipped with a magnetic stirrer and carried out as semi-batch method. First, the reactor was charged with MAO and then 1-hexene in the certain amount was added. Consequently the system was changed to ethylene atmosphere system. When reaction medium was saturated with ethylene monomer,

---

* e-mail to piyasan.p@chula.ac.th, Tel : +66-2-218-6883

polymerizations were started by adding the 1 ml (20 µmol) of catalyst solution. For the borane and borate system, polymerization was started by the successive additions of alkylaluminum, the borane or borate, and the complex to the monomer solution. Polymerization was conducted on the certain time and terminated with acidic methanol and precipitated in acidic methanol, filtered, adequately washed with methanol, and finally dried under vacuum at 60 °C for 6 h. The polymer obtained was characterized using GPC and ^{13}C NMR to investigate the molecular weight and microstructure respectively.

3. RESULT AND DISCUSSION

Results of ethylene/1-hexene copolymerization with various activators such as MAO, d-MAO, MMAO, d-MMAO, and borate are shown in Table 1. It indicated that activities of every type of polymerization with d-MAO and d-MMAO exhibited the highest activities among any other activators. Enhanced activities arising from d-MAO and d-MMAO were attributed to less TMA and TIBA in the dried MAO and MMAO, respectively. Considering the results when MAO and d-MAO were used, it was reported that the presence of TMA resulted in chain transfer reaction and also the excess of TMA in catalyst system would reduce the Ti(IV) species to inactive lower valance state [6]. Thus we can propose that activities of ethylene/1-hexene copolymerization with catalyst 1/aluminoxane system depended on the amounts of TMA present in the aluminoxane. Moreover, when compared the results between MAO – d-MAO systems and MMAO – d-MMAO systems, it revealed that TMA had stronger effect more than TIBA indicating significantly decreased activities in MAO systems compare to the MMAO system.

Table 1. Results of polymerization of ethylene/1-hexene polymerization with catalyst 1[a]

Entry	Co-Catalyst	Time (min)	Yield (g)	Activity[b]	M_n[c]	MWD[c]	N[d]
1	MAO	15	0.27	53	0.15	1.39	180
2	d-MAO	15	2.69	538	2.81	1.44	96
3	MMAO	15	2.02	405	3.93	1.99	51
4	d-MMAO	15	2.64	527	6.34	1.84	42
5	Borate[e]	7	0.29	123	9.49	2.98	3

[a]Polymerization conditions: Ti = 20 µmol, Al/Ti = 400, Solvent = toluene 30 ml, Temperature = 40°C. atmospheric pressure.

[b]Activity = kg(polymer) mol^{-1}(Ti) hr^{-1}.

[c]Number of average molecular weight and molecular weight distributions were measured by GPC analysis using poly styrene as reference. (x 10^{-4})

[d]Number of polymer chain calculated from yield and M_n

[e]Borate compound (Ph$_3$CB(C$_6$F$_5$)$_4$) 20 µmol and using Oct$_3$Al 20 µmol as a scavenger.

According to the ethylene consumption rate profile for ethylene/1-hexene copolymerization in figure 1, it indicated that deactivation of active species occurred in the borate system. On the other hand, MAOs and MMAOs systems exhibited the consistence in ethylene consumption rate when polymerization was conducted for several minutes. The consumption rate of ethylene when d-MAO, MMAO, and d-MMAO were employed in ethylene polymerization and ethylene/1-hexene copolymerization are relatively higher than that of when MAO was used as a cocatalyst.

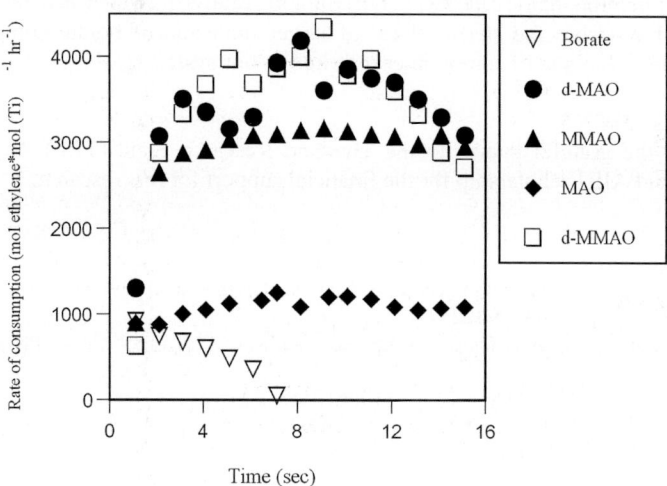

Fig. 1. Ethylene rate consumption profile of ethylene/1-hexene copolymerization with activators

In the view of molecular weight and molecular weight distribution we found that, the molecular weight of polymer obtained from MMAO system gave the larger number of molecular weight than the MAO system. This phenomenon was attributed to the present of TMA and TIBA in MAO and MMAO respectively. The molecular weight distribution was independent with any aluminoxane type. However, it essentially depended on types of cocatalyst (aluminoxane and borate)

Table 2. Triad distribution of ethylene/1-hexene copolymer

Entry	Co-Catalyst	EEE	HEE	HEH	EHE	EHH	HHH	%E	%H
1	MAO	0.07	0.17	0.16	0.12	0.43	0.04	40	60
2	d-MAO	0.11	0.12	0.13	0.08	0.56	0.00	36	64
3	MMAO	0.105	0.15	0.17	0.11	0.155	0.31	42.5	57.5
4	d-MMAO	0.14	0.14	0.12	0.11	0.39	0.1	40	60
5	Borate	0.00	0.08	0.18	0.07	0.37	0.30	26	74

The characteristics of ^{13}C NMR spectra for all copolymers were similar. The triad distributions for all copolymer from ^{13}C NMR monomer insertion are shown in Table 2. Based on the triad distribution of ethylene/1-hexene copolymers in Table 2, we found that microstructure of copolymer obtained from aluminoxane system was slightly different in monomer incorporation, but found significantly when borated system was applied. We suspected that this difference was arising from the differences in bimetallic complex active species between [aluminoxane]⁻[catalyst]$^{+}$ and [Borate]⁻[catalyst]$^{+}$ which had the electronic and geometric effects from the sterric effect of larger molecule of borate compare to the aluminoxane on the behaviors of comonomer insertion in our systems.

ACKNOWLEDGEMENT
We have to give the grateful thanks to the Thailand Research Fund (TRF) Royal Golden Jubilee program and AIEJ scholarship for the financial support for this research.

REFERENCES
[1] A. G. Simanke, G. B. Galland, R. Quijada, R. S. Mauler, Polymer 40 (1999) 5489.
[2] W. Kaminsky, Cat. Today 20 (1994) 257.
[3] V. K. Gupta, S. Satish, I. S. Bhardwag, J. Macromol. Sci, Rev., Macromol. Chem. Phys., C34 (1994) 439.
[4] M. Bochmann, J. Chem. Soc. Dalton Trans., (1996) 255.
[5] T. J. Marks, L. Liable-Sands, A. L. Rheingold, J. Am. Chem. Soc., 124 (2003) 12725.
[6] E. O. Dare, G. A. Olatunji, D. S. Ogunniyi, Eur. Polym. J., 40 (2004) 2333.

Studies in Surface Science and Catalysis, volume 159
Hyun-Ku Rhee, In-Sik Nam and Jong Moon Park (Editors)
© 2006 Elsevier B.V. All rights reserved

Development of a reduced-order model for metallocene-catalyzed ethylene-norbornene copolymerization reaction

Seung Young Park[a], Jongku Lee[a], Kyu Yong Choi[b], and Kwang Ho Song[c]

[a]LG Chem, Ltd/ Research Park
104-1 Moonji-Dong, Yuseong-Gu, Daejeon 305-380, Korea

[b]Department of Chemical and Biomolecular Engineering
University of Maryland, College Park, MD 20742, USA

[c]Department of Chemical & Biological Engineering
Korea University, Seoul 136-713, Korea

1. INTRODUCTION

In ethylene-norbornene solution/bulk copolymerization with rac-Et(1-indenyl)$_2$ZrCl$_2$/ methylaluminoxane catalyst system, penultimate model (Markov 2nd order model) provides improved predictions of polymer yield and copolymer properties than terminal model (Markov 1st order model) [1, 2]. The penultimate model takes into account the effect of penultimate monomer unit adjacent to an active monomer site in a growing polymer chain on the monomer insertion rates. It is believed that a bulky norbornene molecule exhibits a steric hindrance particularly on the incoming norbornene molecule [3]. Although the penultimate model yields better prediction of polymer yield than the terminal model, it tends to underestimate the norbornene composition in the copolymer at high bulk phase norbornene concentrations. Our experimental data shows that norbornene content in the copolymer increases up to 75 % ~ 80 % as norbornene concentration in the bulk phase is increased. It is possible that not only the secondary norbornene but also the tertiary norbornene molecule from the catalytic active site might also affect the copolymerization reaction. It has been observed by ^{13}C-NMR spectroscopy analysis that norbornene triads can be formed with certain metallocene catalyst systems [4], but norbornene microblocks longer than three unit length (tetrad, pentad, etc) are hardly seen [5]. This suggests that additional norbornene insertion into a norbornene diad to form a norbornene triad might be quite faster than the norbornene homopolymerization rate, while in the penultimate model the reaction rate for norbornene triad is as slow as the norbornene homopolymerization rate.

The Markov 3rd order or higher model can be used to account for the effect of a tertiary norbornene in the polymer chain on the reaction rate and copolymer composition. Higher order models, however, require an increased number of reaction parameters to be determined. For example, in penpenultimate model (Markov 3rd order model), 16 propagation rate constants should be determined, whereas 8 rate constants are needed in the penultimate model. In this work, we propose a reduced-order Markov model (ROMM) to effectively reduce the number of reaction parameters.

2. REDUCED-ORDER MODELS

2.1. Kinetic modeling

In the proposed model, the penpenultimate effect of norbornene in a growing polymer chain is considered. The kinetic scheme for propagation reaction of ethylene (M_1) and norbornene (M_2) is shown in Eqs. (1) ~ (6):

$$M_1^* + M_1 \xrightarrow{k_{11}} M_1^* \tag{1}$$

$$M_1^* + M_2 \xrightarrow{k_{12}} M_{2.1}^* \tag{2}$$

$$M_{2.i}^* + M_1 \xrightarrow{k_{21,i}} M_1^* \quad (i = 1, 2, \cdots, k) \tag{3}$$

$$M_{2.i}^* + M_1 \xrightarrow{k_{21,k}} M_1^* \quad (i > k) \tag{4}$$

$$M_{2.j}^* + M_2 \xrightarrow{k_{22,j}} M_{2.j+1}^* \quad (j = 1, 2, \cdots, k) \tag{5}$$

$$M_{2.j}^* + M_2 \xrightarrow{k_{22,k}} M_{2.j+1}^* \quad (j > k) \tag{6}$$

where $M_{2.i}^*$ represents the growing polymer chain of i monomer units of M_2 before M_1 monomer block, and k is the critical block length. For $i > k$, we assume that there is no induction effect of M_2 monomer unit. Note that no penultimate effect of ethylene molecule is assumed in the above kinetic model. Here, a small ethylene molecule can be viewed as a spacer between the bulky norbornene molecules [6]. The reactivity ratios are defined as

$$r_1 = \frac{k_{11}}{k_{12}}, \quad r_{2i} = \frac{k_{22,i}}{k_{21,i}} \quad (i = 1, 2, \cdots, k) \tag{7}$$

We determined the reaction parameters using the optimal parameter estimation technique with the experimentally obtained copolymer yield and norbornene composition data. Based on the literature report, we assume that $k = 3$ [5]. Fig. 1 shows that the estimated rate constant values depend on the norbornene block length. Note that the reaction rate constant

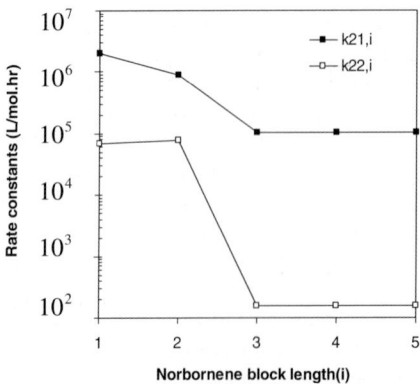

Fig. 1. Profiles of $k_{21,i}$ and $k_{22,i}$ rate constants against norbornene block length attached to catalyst active sites.

for norbornene triad ($k_{22,2}$) is about 500 times larger than the rate constant for norbornene tetrad ($k_{22,3}$ or norbornene homopolymerization rate). Fig. 2 shows the ethylene and norbornene polymerization rates predicted by the proposed model. The copolymer composition predicted by the predicted model shows an excellent fit with the experimental data (Fig. 3).

2.2 Monomer sequence length analysis

Tritto and coworkers report the detailed ^{13}C NMR spectroscopy analysis of sequence distributions of monomer units in ethylene-norbornene copolymers synthesized over various types of metallocene catalysts [2, 4]. From the analysis of the copolymer microstructure at tetrad level, they conclude that the penultimate model yields better results in estimating the monomer sequence in ethylene-norbornene copolymers at low to medium bulk phase norbornene concentrations ([NB]/[E] (mol/mol) < 10). They suggest that a 3rd order or more complex model might be needed at higher norbornene concentrations or when a metallocene catalyst of strong steric hindrance effect is used [2]. Table 1 shows that with *rac*-Me$_2$Si(2MeBInd)$_2$ZrCl$_2$ catalyst that might cuase a strong steric hindrance due to indene substitution, the reduced-order model gives a better estimation of monomer sequence distribution than the terminal or the penultimate model.

Fig. 4 shows the comparison of the three models in predicting the average norbornene block length in the copolymer with the experimental data. It illustrates that the reduced-order

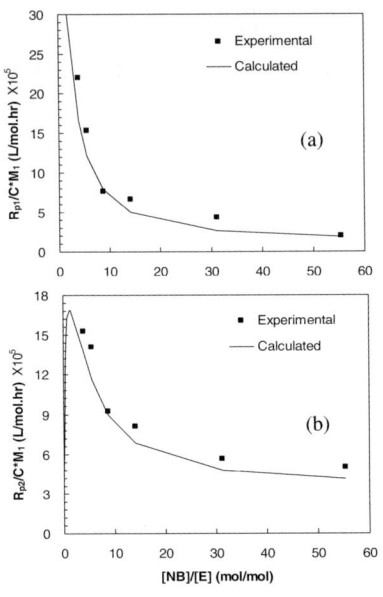

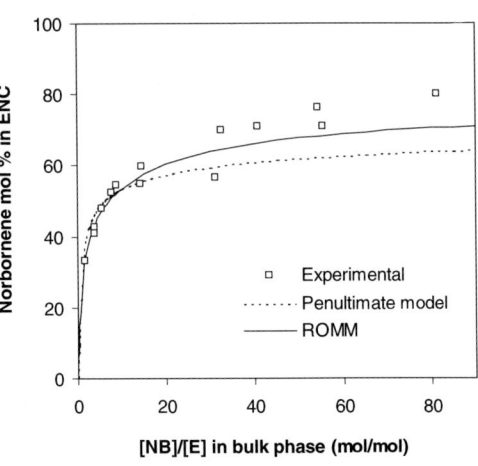

Fig. 2. Calculation of (a) ethylene and (b) norbornene reaction rates by ROMM.

Fig. 3. Copolymer composition diagrams calculated by penultimate model and ROMM.

848

Table 1
Sequence analysis of ENC with *rac*-Me$_2$Si(2MeBInd)$_2$ZrCl$_2$ catalyst

Tetrad level	Experimental [2]	Terminal	Penultimate	ROMM
EEEE	0.1609	0.1422	0.1360	0.1405
NEEE	0.1718	0.2005	0.1935	0.1992
NEEN	0.0826	0.0707	0.0688	0.0706
ENEE	0.3325	0.3372	0.3264	0.3316
NNEE	0.0045	0.0046	0.0047	0.0045
NENE	0.2366	0.2377	0.2271	0.2351
NNEN	0.0032	0.0032	0.0033	0.0032
ENNE	0.0000	0.0039	0.0004	0.0037
NNNE	0.0079	0.0001	0.0072	0.0079
NNNN	0.0000	0.0000	0.0326	0.0000
Sum of relative error squared		0.3221	0.0816	0.0627

model proposed in this study yields longer norbornene block lengths than the penultimate
model at higher norbornene concentrations in the bulk phase.

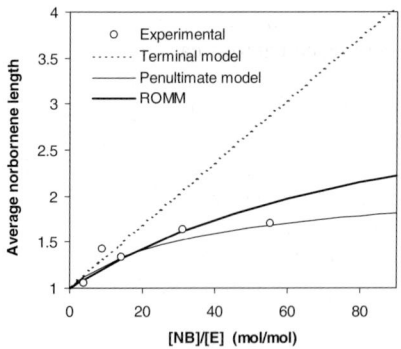

Fig. 4. Average norbornene block length calculation with models and
comparison with experimental data.

REFERENCES
[1] S. Y. Park, K. Y. Choi, K. H. Song, and B. G. Jeong, Macromolecules, 36 (2003) 4216-4225.
[2] I. Tritto, L. Boggioni, J. C. Jansen, K. Thorshaug, M. C. Sacchi, and D. R. Ferro, Macromolecules, 35 (2002) 616-623.
[3] D. Ruchatz and G. Fink, Macromolecules, 31 (1998) 4674-4680.
[4] I. Tritto, L. Boggioni, M. C. Sacchi, P. Locatelli, D. R. Ferro, A. Provasoli, Macromol. Rapid Commun., 20 (1999) 279-283.
[5] C. H. Bergström, T. L. J. Väänänen, J. V. Seppälä, J. App. Polym. Sci., 63 (1997) 1071-1076.
[6] D. Ruchatz and G. Fink, Macromolecules, 31 (1998) 4669-4673.

Studies in Surface Science and Catalysis, volume 159
Hyun-Ku Rhee, In-Sik Nam and Jong Moon Park (Editors)

Polymerization of ethylene with embedded metallocene catalysts

Jin Suk Chung[a]**, Dong Min Shin**[a]**, Gi Bae Moon**[a]**, Eun Woo Shin**[a]**, Kyu Yong Choi**[b]

[a]School of Chemical Engineering and Bioengineering,
University of Ulsan, Mugeodong, Nam-gu, Ulsan 680-749, Republic of Korea

[b]Department of Chemical and Biomolecular Engineering, University of Maryland, College Park, MD 20742, U.S.A.

1. INTRODUCTION

Metallocene catalysts have received considerable attention because they exhibit high activities for olefin polymerization. The polymers produced by these catalysts exhibit the properties that are different from the polymers produced by Ziegler-Natta type catalysts [1-2]. However, to use metallocene catalysts in a slurry or gas phase processes, it is desirable to heterogenize the them such that they can be employed as a "drop in" catalyst in the existing process. A homogeneous metallocene catalyst can be modified to a heterogeneous system in many ways. One popular method is to impregnate the metallocene catalyst onto an inorganic support such as silica or alumina [3-5]. In this work, we investigate the ethylene polymerization to high-density polyethylene by a liquid slurry process using $Et[Ind]_2ZrCl_2$ /MAO catalyst embedded into a polystyrene prepolymer. Here, the active metallocene compound is first embedded into a polystyrene prepolymer by polymerizing styrene at low reaction rate. Then, the solid embedded catalyst is used to polymerize ethylene in the subsequent polymerization reaction. The use of polystyrene as a carrier for the catalyst has an advantage in that there is no inorganic material such as silica or alumina will be present in final polymer.

2. EXPERIMENTAL

2.1. Materials

Styrene monomer was purified by vacuum distillation over CaH_2. Inhibitor in styrene was removed using activated alumina. N-heptane was purified by distillation over sodium to remove the trace of residual moisture. The purified styrene and n-heptane were stored over activated alumina under nitrogen blanket. $Et[Ind]_2ZrCl_2$ (Strem Chem.), MAO (modified methylaluminoxane, type 3A, Akzo Novel) were used without further purification.

2.2. Preparation of embedded catalyst

To prepare embedded catalyst, polymerization was first carried out in a small agitated glass reactor at room temperature for 1 hr in n-heptane at very low styrene concentration using $Et[Ind]_2ZrCl_2$/MAO catalyst with Al/Ti mole ratio of 20-200. After the reaction, a small part of the solid fraction (embedded catalyst) was isolated from the liquid phase for the characterization and the other fraction was transferred to glass reactor for the ethylene polymerization. More detailed description of experimental works can be found elsewhere [6].

2.3. Polymerization

The ethylene polymerization was carried out using a 12 OZ glass reactor equipped with a two blade impeller under a constant ethylene pressure of 20 psi. A predetermined amount of solvent (n-heptane), monomer, MAO and embedded catalyst were charged in series into the reactor. Polymerization was carried out at 70°C with agitation speed of 800 rpm. The polymer obtained was washed with excess amount of methanol containing hydrochloric acid solution and dried in vacuo for 24 hrs. The polymerization rate was determined from the amount of consumed ethylene, measured using a mass flow meter. DSC analyses (Dupont V4.0B) was carried out at a rate of 10 °C /min, and the results were obtained in the second scan.

3. RESULTS AND DISCUSSION

We analyzed the embedded particles with differential scanning calorimetry to identify the property of polystyrene. As shown in Fig. 1, the embedded particles show a small peak around 100 °C, which is typical in atactic polystyrene [7]. It is desirable that embedding polymer has a similar melting temperature as the final polymer (polyethylene) because a big difference in the melting temperatures between the two polymers may cause a gel problem and poor mechanical properties.

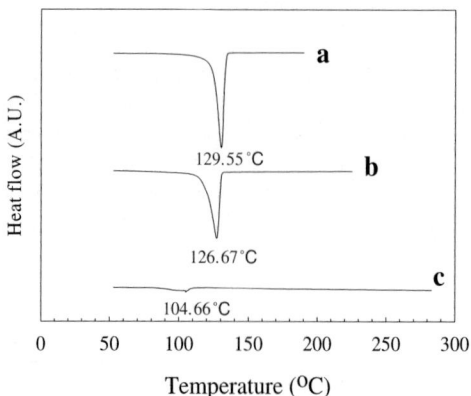

Fig. 1. DSC thermogram of polymers: (a) polyethylene with homogeneous catalyst;
(b) polyethylene with embedded catalyst; (c) catalyst embedded polystyrene.

On the other hand, the polymer prepared by the embedded catalyst shows T_m around 130 °C, which is a typical melting temperature of high density polyethylene. There was little activity difference between the polyethylene produced by embedded particles and those by homogeneous catalysts. The results of ethylene polymerization using embedded catalyst and homogeneous catalyst are summarized in Table 1 and Fig. 2.

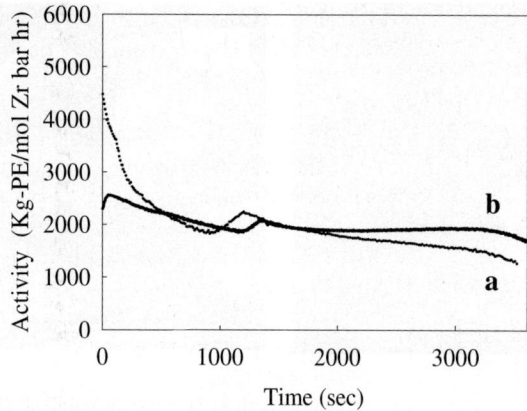

Fig. 2. Activity profiles of ethylene polymerization: (a) with homogenous catalyst; (b) with embedded catalyst.

The catalyst activity of the embedded catalyst is slightly higher or similar with that of homogeneous catalysts. It is thought that the nature of active site of metallocene catalyst is not affected by the embedding process. The catalytic activities are maintained up to 1 hr and there is no serious deactivation. It is notable that the bulk density of the produced polyethylene is 30% higher when the embedded catalyst is used. The bulk density of polymer is a very important feature in olefin polymerization processes because a poor bulk density can lower the reactor throughput [8]. The scanning electron microscope images of the polyethylene with homogeneous and embedded catalyst are shown as Fig. 3. We found not only the morphology of the produced polyethylene particles with embedded catalyst was remarkably improved, but also the fine particle content of product polymer was reduced. Fine particles in the polymer product can cause some troubles such as abnormal pressure drops, heat transfer problems in process operation. Our experimental results indicate that it is feasible to improve the polyethylene particle morphology by slurry phase polymerization using the embedded catalyst without troublesome immobilization of homogeneous metallocene onto support material. Some applicable methods for the gas phase polymerization are under investigation.

Table 1. Results of ethylene polymerization with embedded catalyst and homogeneous catalyst.

Catalyst Type	Styrene/Cat. (mol/mol)	Specific Activity (g PE/mol·Zr hr)	Bulk density (g/cc)	Morphology
Embedded catalyst	100	2.3×10^6	0.203	
	200	2.0×10^6	0.206	Regular particle
	400	1.7×10^6	0.261	
Homogeneous	-	2.1×10^6	0.194	Irregular particle

Reaction temperature = 70 °C, Al/Zr (mol/mol) = 1000

852

Fig. 3. SEM image of polyethylene particles: (a) with homogeneous catalyst; (b) with embedded catalyst.

4. CONCLUSIONS

The simple catalyst embedding technique has been applied to ethylene polymerization in slurry In this technique, active catalytic components are embedded into styrene polymer matrix. The resulting polyethylene shows better morphology and higher bulk density than those produced by homogeneous catalyst. No activity loss was also observed with the embedded catalyst. It is believed that the embedded catalyst technique is very useful because it does not change the kinetic characteristics of the catalyst but it improves the catalyst performance and process abilities.

ACKNOWLEDGEMENT
The authors acknowledge the financial support from University of Ulsan (2003-0086) for this work.

REFERENCES
[1] W. Kaminsky and H. Sinn, Adv. Organomet. Chem., 18 (1980) 99.
[2] H. S. Cho, J. S. Chung, J. H. Han, Y. G.. Ko, W. Y. Lee, J. Appl. Polym. Sci., 70 (1998) 1707.
[3] K. Soga and M. Kaminaka, Macromol. Chem. Rapid Commun.., 13 (1992) 221.
[4] K. Soga and M. Kaminaka, Macromol. Chem. Phys., 195 (1994) 1369.
[5] G. G. Hlatky, Chem. Rev., 100 (2000) 1347.
[6] J. S. Chung, B. G. Woo, K. Y. Choi, Macromol. Symp., 206 (2004) 375.
[7] F. M. Rabagliati, R. A. Cancino, A. M. Ilarduya, S. Muñoz-Guerra, Eur. Polym. J., 41 (2005) 1013.
[8] F. J. Karol, Catal. Rev. –Sci. Eng., 26 (1984) 557.

Studies in Surface Science and Catalysis, volume 159
Hyun-Ku Rhee, In-Sik Nam and Jong Moon Park (Editors)

Synthesis of active Ni(II) α-diimine catalysts for ethylene polymerization: study on substituent and cocatalyst effects

Bijal K B, Gi Wan Son, Chang-Sik Ha and Il Kim*

Department of Polymer Science and Engineering, Pusan National University, Jangjeon-dong, Geumjeong-gu, Busan 609-735, Korea

1. INTRODUCTION

Recently, intense efforts were put forth in the development of new single-site olefin polymerization catalysts based on late transition metals. [1]. The Ni(II) α-diimine catalysts exhibit high activities which are comparable to those of metallocene catalysts [2]. But the sensitivity of metallocenes to polar substituents is largely responsible for an increase interest in the late-transition metal complexes as olefin polymerization catalysts. They have the potential to yield polymers with different microstructures and are less oxophilic and therefore more tolerant towards functionalized monomers [3]. The microstructure of polyethylene produced by Ni(II) α-diimine catalysts varies from strictly linear to highly branched. The degree of branching in the polyethylene and polymer architecture can depend on a number of factors; the metal, the steric bulk of the diimine ligand and the reaction condition.

We synthesized a series of Ni(II) complexes bearing bulky diimine ligands, which serve as good catalysts for the polymerization of ethylene (Scheme 1). Polymerizations with these catalysts have been explored under a variety of reaction conditions and the substituent effects on the catalytic activity were investigated. Generally a threshold amount of cocatalyst is needed to effectively activate the precatalyst, presumably to scavenge impurities that may poison the active catalyst. The highly active catalysts produced high molecular weight (MW) polyethylene with different microstructure.

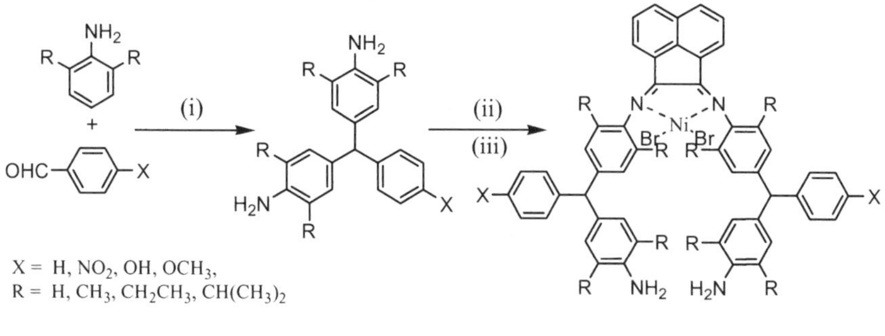

Scheme 1. Synthesis of Catalysts: (i) HCl; (ii) Acenaphthenequinone, MeOH; (iii) (DME)NiBr2, CH₂Cl₂.

2. EXPERIMENTAL

2.1 Materials and general procedure for the synthesis of catalysts

All experiments were performed using standard high vacuum or schlenk techniques. Polymerization grade of ethylene (SK Co., Korea) was purified by passing it through columns of Fisher RIDOX catalyst and molecular sieve $5\text{Å}/13\text{X}$. All solvents were purified under nitrogen and stored over molecular sieves. The reagents were purchased from Aldrich Chemical Co. and used without further purification.

The bidentate ligands were prepared by the Schiff-base condensation of two equivalents of the desired 2,6-dialkyl substituted anilines with acenaphthenequinone as in the scheme 1. The pre-catalysts, formed by addition of the ligand to $(DME)NiBr_2$ are isolated and purified. The products were characterized by ^1H, ^{13}C NMR, GPC, DSC and Elemental Analysis.

2.2 Polymerization procedure

Ethylene was polymerized with 2.5 µmol catalysts in toluene / CH_2Cl_2 for 30 min. at different temperatures and 5 psig ethylene pressure. Polymerization rate was determined from the rate of consumption, measured by a hotwire flow meter (model 5850 D from Brooks Instrument Div.) connected to a personal computer through an A/D converter. Polymerization was quenched by the addition of methanol containing HCl (5 v/v %). The polymer was washed with an excess amount of methanol and dried under vacuum at $50\,°C$.

3. RESULTS AND DISCUSSION

Hydroxyl and amino functionalized Ni diimine complexes were synthesized without the need of any protecting groups. These complexes were used to polymerize ethylene with high activity using alkyl aluminum halides as inexpensive cocatalysts even at very low Al to Ni ratio (23eq.). Generally high pressures are required to ensure rapid initiation, but at normal pressure itself rapid initiation has showed in our case. Even though our focus were mainly on the effect of remote substituents (Fig.1), we observed a striking effect of the ortho substituents

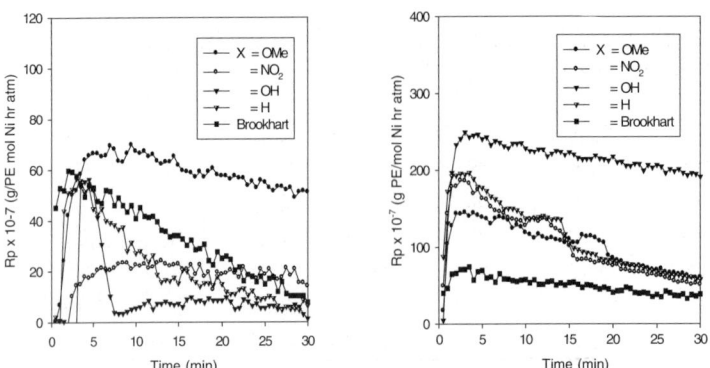

Figure1. Activity curves showing substituent effect with MAO and EAS, in turn.

in polymerization activity that is quite contrary to the reports on diimine systems so far [2,5]. It was noted that the activity of the complexes having high steric bulkiness (isopropyl) adjacent to the metal center was found to be decreased by 2-4 fold when compared to less steric bulky methyl group. A possible explanation may be due to high overcrowding effect of the triaryl system. The polymerization has been performed at temperature range from 10 to 50 ℃. As the temperature decrease, the MW and catalytic activity increase, but the branching degree of the resulting polymer decreases [3,6]. Even though the solubility of hydroxyl substituted catalyst in CH_2Cl_2-EAS system was very low, it showed very high activity and gave a high MW polymer. The rates of associative displacement and chain transfer are greatly related by the extreme steric bulkiness of the diimine ligands. Toluene was found to be the better solvent than CH_2Cl_2.

When the ortho substituent effects followed the general trend, the effects of remote substituents dramatically alter the catalytic activity and polymer microstructure (Table 1). The bulky substituents block the axial approach of olefins enhancing the ratio of chain propagation and thus permit the formation of polymers with high MW in the order of 10^5 [3]. The MW value decreases with the order of adjacent substituents as isopropyl > ethyl > methyl while with remote substituents the order follows as, -OMe > -NO2 > -OH. Methyl substituted catalyst gave less branched polymer while isopropyl produced highly branched ones. Branching of the carbon atom was found to increases with (i) increase in steric bulkiness and (ii) decrease in temperature. The catalyst degradation also affects the MW [4]. The MWD values were found to be broader in the range of 3.5 - 5.7 and this may be attributable to the presence of any foreign active species due to the interaction of cocatalyst with free amino group or with the remote substituents. In addition the presence of a great number of branches is one of the factors broadening the MWD values.

Comparing the reactivity of the catalysts (Fig.1 and Table 1), the -OH substituted catalyst showed highest activity with EAS and the least to –OMe substituted catalyst. But with MAO the trend was reversed. Even though it is difficult to establish a direct influence for these push-pull substituents on the metal center, their roles have been try to investigate by cyclic

Table 1
The effect of remote substituents on the ethylene polymerization.

	X	Cocatalyst[a]	Activity[b]	Mn	Mw	Mw/Mn[c]	Tm[d]	Branches[e]
1	H	MAO	22.06	59024	154659	2.62	115.3	109
2	H	EAS	115.06	56906	155313	2.73	-	112
3	NO2	MAO	18.25	54638	253170	4.63	117.3	81
4	NO2	EAS	98.36	52237	186559	3.57	-	130
5	OMe	EAS	78.68	49353	201209	4.08	-	99
6	OMe	MAO	50.07	48352	275586	5.70	115.6	100
7	OH	MAO	10.33	29888	129026	4.32	127	28
8	OH	EAS	213.81	33526	146951	4.38	124	39

[a] Al/Ni = 300, [b] R_p x 10^{-7}(g-PE/ mol Ni h atm), [c] Determined by GPC, [d] Determined by DSC (℃), [e] Branches per 1000 carbon atoms determined by ^1H NMR

voltammetric measurements. The half wave potentials of the redox couple for the Ni complexes having -H, -OH, -OCH$_3$, and -NO$_2$ subsitutent groups were determined to be 0.755 V, 0.781 V. 0.714 V. and 0.713 V vs. Ag/AgCl reference, respectively. The substituents may change the orientation of the catalyst structure and hence the reduction potential also. The electronic effects of the push-pull substituents on the immediate *para* position of the phenyl ring have dramatically accelerated the reaction rate, was reported to be the formation of Lewis acid-base complexes of the methoxy group with excess aluminium cocatalyst that will reduce electron density on the ligand and therefore lead to a more electrophilic metal center than would normally be expected [7]. The same effect was nourished with our system having extremely remote substituents. Even though the methoxy group in catalyst, entry 6 (Table 1) is in the remote *para* position to the metal center, the similar trend follows is quite tracing. However, the similar tendency was not observed with -OH catalysts and the effects distinctly depend on the cocatalysts. The choice of aluminium cocatalyst has a significant effect on the ethylene polymerization reactions. The anomalous behavior observed with hydroxyl-substituted catalysts is still indecipherable. The hydroxyl-substituted catalyst produced comparatively low-branched polyethylene as 39 branches/1000 carbons with EAS and 28 with MAO at 30°C.

4. CONCLUSIONS

The effects of newly synthesized diimine ligands on the catalytic structure were studied. The polymerizations by the catalysts bearing bulky substituted α-diimine ligands yielded high molecular weight polyethylene, which varied with the structure of the catalyst. High activity was maintained over a wide range of cocatalyst ratio. EAS was found to be an efficient cocatalyst for the present catalyst systems under almost all conditions. Even though the measured activities of these catalysts were affected by several variables and are often difficult to interpret, the experimental evidences substantiate the remote substituent effect.

ACKNOWLEDGEMENTS
This work was supported by the Basic Research Program of the Korea Science & Engineering Foundation, the Center for Ultramicrochemical Process Systems, and the Korea Institute of Industrial Technology Evaluation and Planning. The authors are also grateful to the BK 21 Project, the Center for Ultramicrochemical Process Systems (ERC) and the National Research Laboratory Program.

REFERENCES
[1] S. D. Ittel, L. K Johnson, and M. Brookhart, Chem. Rev., 100 (2000) 1169.
[2] L. K Johnson, C. M. Killian, and M. Brookhart, J. Am. Chem. Soc., 117 (1995) 6414.
[3] D. P. Gates, S. A. Svejda, E. Onate, C. M. Killian, L. K. Johnson, P. S. White, and M. Brookhart, Macromolecules, 33 (2000) 2320.
[4] D. G. Musaev and K. Morokuma, Topics in Cata., 7 (1999) 107.
[5] G. J. P. Britovsek, S. P. D. Baugh, O. Hoarau, V. C. Gibson, D. F. Wass, A. J. P White and D. J. Williams, Inorganica Chimica Acta, 345 (2003) 279.
[6] F. Alobaidi, Z. Ye and S. Zhu, Polymer, 45 (2004) 6823.
[7] S. A. Svejda and M Brookhart, Organometallics, 18 (1999) 65.

Studies in Surface Science and Catalysis, volume 159
Hyun-Ku Rhee, In-Sik Nam and Jong Moon Park (Editors)

Newly designed nickel(II)-based catalysts for the polymerization of ethylene

Gi Wan Son, Bijal K B, Chang-Sik Ha and Il Kim*

Department of Polymer Science and Engineering, Pusan National University, Jangjeon-dong, Geumjeong-gu, Busan 609-735, Korea

1. INTRODUCTION

Olefin polymerization and oligomerization catalyzed by transition metal complexes have attracted great attentions in both academic and industrial research [1,2]. In the past late transition metal catalysts were generally accepted that they produced oligomers because of the accelerated β-hydride elimination. An industrial process, Shell Higher Olefin Process (SHOP), is an example to produce oligomers by using homogeneous chelate Ni(II) complexes. However, Brookhart in 1995, in which they produced high molecular weight (MW) polyethylene (PE) for the first time with cationic Ni(II) α-diimine complexes, has revived the investigation of late transition metal complexes as potential catalyst precursors [3]. Considering these previous findings, we were interested in exploring and synthesizing some new ligands in order to develop new late-transition metal complexes for olefin polymerization catalysis. We report here the synthesis of novel nickel complexes bearing ligands with aryl rings of different *ortho* substituents, resulting in highly active catalysts for the polymerization of ethylene.

2. EXPERIMENTAL

2.1 Materials

All reaction were performed under a purified nitrogen atmosphere using standard glove box and schlenk technique. Polymerization grade of ethylene (SK Co., Korea) was purified by passing it through columns of Fisher RIDOX catalyst and molecular sieve 5 Å/13X. Organic solvents were distilled from Na/benzophenone and stored over molecular sieves (4 Å). Used all reagents were purchased from Aldrich Chemical Co. except MAO and used without purification. MAO was purchased from Akzo Chemical as 8.4 wt.-% total Al in toluene. α,α−Bis(4-amino-3,5-dimethylphenyl)toluene was synthesized according to the literatrure [4].

2.2 Synthesis of ligands and metal complexes

General procedures for the synthesis of ligands and metal complexes are shown in Scheme 1. For the synthesis of **2a** acenaphthenequinone (0.38 g, 2.1 mmol) and α,α−bis(4-amino-3,5-dimethylphenyl)-toluene(excess) were dissolved in 50 mL of CH_3OH in a round-bottom flask. Five drops of formic acid were added, and the sealed solution was stirred at 50 °C overnight. After filtration, the red solid was washed with hot methanol and dried to give 1.2 g of red powder in 50 % yield.

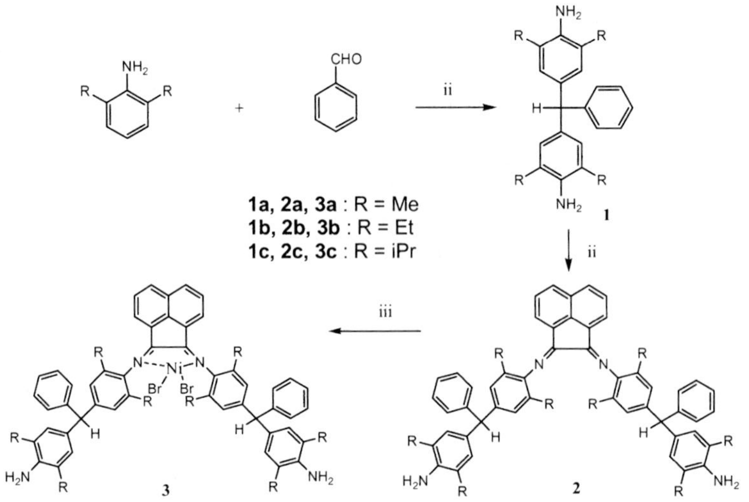

Scheme 1. synthesis of ligand and catalysts : (i) HCl; (ii) Acenaphthenequinone, MeOH; (iii) (DME)NiBr$_2$, CH$_2$Cl$_2$.

For the synthesis of **3a**, 1.1 equivalent of the ligand **2a** (0.25 g 0.31 mmol) and (DME)NiBr$_2$ (0.068 g 0.28 mmol) were stirred in 20 ml of CH$_2$Cl$_2$ for overnight. After removing of the solvent, the brown powder was washed three times with diethyl ether (yield = 0.28 g, brown powder).

2.4 Characterization

^{1}H-NMR, ^{13}C-NMR spectra were recorded on a Varian Gemimi 200 (200MHz). All chemical shifts are reported in parts per million relative to tetramethylsilane. The intrinsic viscosity was measured in decalin. At 135 ℃ using an Ubbelohed viscometer and the average of MW was calculated by following equation[5]: $[\eta] = 6.2 \times 10^{-4} \, \overline{M}_v^{0.7}$.

3. RESULTS AND DISCUSSION

The general structure of bidentate α-diimine ligands used in this study was shown in Scheme 1. They are prepared by the conventional Schiff-base condensation of two equivalents of the desired 2,6-dialkyl substituted anilines with acenaphthenequinone. The pre-catalysts, formed by addition of the ligands to (DME)NiBr$_2$. The reactions were complete in all cases in less than six hours, and the products are isolated by filtration or evaporation and washed with ether to yield desired catalyst precursors in quantitative yields.

The catalyst precursors were tested in solution polymerization runs at 1.3 bar of ethylene pressure in toluene at temperature (T$_p$) between 10 and 50 °C and the results are summarized in Table 1. The active catalysts are generated in situ in toluene by the addition of MAO to the catalyst precursor in the presence of ethylene. Methyl substituted catalyst (**3a**/MAO) showed the highest activity while isopropyl homologue (**3c**/MAO) the lowest activity. The **3a**/MAO catalyst showed higher activity than **3c**/MAO by 2-fold at low T$_p$ (say 10 °C). However, as T$_p$

Table 1. Results of ethylene polymerization [a]

Entry	Cat..	Loading (μmol)	Temp. ($^\circ C$)	Cocatalyst $(Al/Ni)^b$	$R_p^c \ x \ 10^{-7}$ (g-PE/mol-Ni h bar)	$M_v^d \ x \ 10^{-3}$
1	3a	7.5	10	500	93.39	204.5
2			30	500	65.13	160.0
3			50	500	21.05	77.9
4	3b	7.5	10	500	57.20	
5			30	500	61.97	179.7
6			50	500	18.71	
7	3c	7.5	10	500	47.76	
8			30	500	43.28	191.3
9			50	500	51.28	
10	3a	7.5	30	25	18.30	
11			30	50	58.96	67.8
12			30	100	40.61	73.2
13			30	300	100.06	170.5
14			30	500	65.13	160.0

[a] 250 mL reactor, 80 mL of toluene, 30 min runs, atmospheric pressure, [MAO]/[M] = 700 equiv.
[b] Al/ni, the molar ratio of MAO and Ni catalyst
[c] Average rate of polymerization for the period of polymerization.
[d] Viscosity average molecular weight.

increases to 50 °C, the reverse order was observed between them. Obviously these results come from the difference in steric bulk of the bidentate α-diimine ligands.

In ethylene polymerizations by Ni(II)-based α-diimine catalysts, the aryl groups are roughly perpendicular to the coordination plane so the bulky substituents on the aryls are positioned at the axial directions to retard associative chain transfer reactions [6,7]. At elevated temperatures, the aryl groups may freely rotate away from the perpendicular orientation, resulting in increased associative chain transfers and a resulting decrease in MW of the PE. In addition such free rotation makes the structure of the cationic active species more unstable, resulting in fast decrease of activity.

As expected from above explanation, the polymer MW (as M_v) vary dramatically with the modification of ligand structure. Reducing the steric bulk of the diimine ligands by substituting of methyl groups for isopropyl groups resulted in decreased MW (Table 1). Similar results have been reported for Ni(II) and Pd(II) systems and Fe(II) and Co(II) systems [1,3,8]. These results demonstrate that steric bulk around the active metal centers is a key to retarding chain transfer in order to obtain high MW polymer. The M_v value was also decreased monotonously as T_p increased.

In order to investigate the effect of MAO on the kinetic behavior of polymerization, we have run ethylene polymerizations at 30 °C by using a wide range of MAO concentration ([Al]/[Ni] = 25 to 500), even though this range is still narrow considering the MAO concentration needed to activate metallocene complexes. Catalytic activity increases with increasing in the concentration of MAO and the highest activity was found around [Al]/[Ni] = 300 (Table 1). This value to achieve the maximum activity is much lower than that needed in conventional metallocene catalyst systems, in which really excess amounts (> 10,000) of MAO are frequently needed. In addition **3a** catalyst affords PEs with high MW (M_v ranging from 67800 to 170500). It is interesting to note that the M_v value does not decrease as MAO concentration increases, which is different from conventional metallocene and Ziegler-Natta catalysts. It means that, within our experimental range, transfer to MAO cocatalyst is not a major factor controlling MW.

On the while β-elimination from the growing polymer chain seems to be a main termination reaction in this catalyst system. The resulting oligomer and polymer chains bearing unsaturated bond at their chain ends lead to chain branching or chain transfer. The formation of high MW polymers is possible because the steric protection of the vacant axial coordination sites reduces the rate of associative displacement from β-eliminated olefin-hydride complexes and thus reduces chain transfer rates. A range of polyethylene materials with high MW and degrees of branching from linear to over 100 branches per 1000 carbon atoms was accessible by simple variation of temperature and ligand architecture as shown in Table 1. At higher temperatures, unsaturated oligomer and polymer chains resulting from activated β-elimination are polymerized to yield highly branched materials as expected because of β-elimination and activated reinsertion mechanisms leading to chain branching.

4. CONCLUSIONS

A series of newly designed Ni(II) complexes bearing α-diimine ligands with different steric bulk were demonstrated to be very active catalyst systems in ethylene polymerizations combined with MAO. The activity, polymer MW and polymer microstructure could be tuned by modifying the catalyst architecture and by controlling polymerization parameters such as temperature and MAO concentration. In general both polymer MW and number of branches increased as steric bulk of the catalyst and polymerization temperature increased. Relatively small amount of MAO cocatalyst (say [Al]/[Ni] = 300) was needed to fully activate the catalyst and the chain transfer to MAO was a minor termination.

ACKNOWLEDGEMENTS
This work was supported by the Center for Ultramicrochemical Process Systems and the Korea Institute of Industrial Technology Evaluation and Planning. The authors are also grateful to R01-2003-000-10020-0 from the Basic Research Program of KOSEF and to the Brain Korea 21 Project and the National Research Laboratory Program.

REFERENCES
[1] S.D. Ittel, L.K Johnson, M. Brookhart, Chem. Rev., 100, 1169 (2000), and reference cited therein.
[2] G.J.P. Britovsek, V.C. Gibson, D.F Wass, Angew. Chem. Int. Ed. Engl, 38, 428 (1999).
[3] L.K Johnson, C.M. Killian, M. Brookhart, J. Am. Chem. Soc. 117, 6414 (1995).
[4] Hunter, C. A. J. Am. Chem. Soc. 120, 5303 (1992).
[5] R. Chang, J. Polym. Sci., 8, 35 (1957).
[6] D. J. Gates, S. A. Svejda, E. Onate, C.M. Killian, L.K. Johnson, P.S. White, M. Brookhart, Macromolecules 33 (2000) 2320.
[7] D. H. Camacho, E. V. Salo, J. W. Ziller, Z. Guan, Angew. Chem. Int. Ed. 43 (2004) 1821.
[8] Small, B. L.; Brookhart, M.; Bennett, A. M. A. J. Am. Chem. Soc. 117, 4049 (1998).

Studies in Surface Science and Catalysis, volume 159
Hyun-Ku Rhee, In-Sik Nam and Jong Moon Park (Editors)

861

Application of Wiener type predictive controller to the continuous solution polymerization reactor

In-Hyoup Song[a,b] and Hyun-Ku Rhee[a]

[a]School of Chemical and Biological Engineering, Seoul National University, Sillim-dong, Kwanak-gu, Seoul 151-744, Korea

[b]Institute of Process Engineering, Swiss Federal Institute of Technology Zürich, CH-8092 Zürich, Switzerland

1. INTRODUCTION

Inherent nonlinear characteristics of polymerization processes demand a nonlinear control scheme in general. Being started from PID controller, advanced nonlinear control techniques have been successfully applied to the control of polymerization processes. Among these, Wiener model predictive control (WMPC) has been highlighted because of its simplicity in the design and implementation [1]. The Wiener model is a special kind of nonlinear one which consists of a static nonlinear block and a dynamic linear part. Because of this cascade structure, any kind of automatic controller based on Wiener model can guarantee the global optimum of objective function. We focus on the WMPC based on subspace identification, which is able to deliver reliable LTI models directly from input/output data requiring only modest computational complexity without the need of (nonlinear) iterative optimization procedure. The main computational tools employed are QR and SVD decompositions [2].

I/O data-based prediction model can be obtained in one step from collected past input and output data. However, there still exists a problem to be resolved. This prediction model does not require any stochastic observer to calculate the predicted output over one prediction horizon. This feature can provide simplicity for control designer but in the presence of significant process or measurement noise, it can bring about too noise sensitive controller, i.e., the control input is also supposed to oscillate due to the noise of measured output.

In this work, therefore, we aim to combine the stochastic observer to input/output prediction model so that it can be robust against the influence of noise. We employ the modified I/O data-based prediction model [3] as a linear part of Wiener model to design the WMPC and these controllers are applied to a continuous methyl methacrylate (MMA) solution polymerization reactor to examine the performance of controller.

2. MATHEMATICAL MODEL FOR POLYMERIZATION PROCESS

For simulation study, the mathematical model is used as the virtual plant. To describe a continuous solution MMA polymerization reactor, ordinary differential equations are set with due regard to the mass and energy conservation laws and white noise is added to simulate noisy system. The reactor temperature is regulated by the coolant in the jacket and the flow rate of feed mixture consisting of solvent and monomer is adjusted by feed pump. The jacket inlet temperature and feed flow rate are regulated to control the weight averaged molecular weight (M_w) and the monomer conversion. Especially, the glass transition effect is considered in modeling so that the mathematical model can reflect the nonlinear characteristics more distinctly. The reference operating condition and the configuration of reactor are given in Table 1. For the details on the mathematical model, one may refer to the previous work [4].

3. WIENER MODEL PREDICTIVE CONTROL

As mentioned above, the backbone of the controller is the identified LTI part of Wiener model and the inverse of static nonlinear part just plays the role of converting the original output and reference of process to their linear counterpart. By doing so, the designed controller will try to make the linear counterpart of output follow that of reference. What should be advanced is, therefore, to obtain the linear input/output data-based prediction model, which is obtained by subspace identification. Let us consider the following state space model that can describe a general linear time invariant system:

$$x(k+1) = \mathbf{A}x(k) + \mathbf{B}u(k) + e(k), \quad y(k+1) = \mathbf{C}x(k+1) + \mathbf{D}u(k+1) + \mathbf{K}e(k) \quad (1)$$

where $x(k) \in R^n$ is the state variable, $u(k) \in R^m$ is the input vector, and $y(k) \in R^l$ is the output vector, while $\mathbf{K}$ and $e(k)$ denote Kalman filter gain and an unknown innovation sequence whose covariance matrix is given as $E\{e(k)e(k)^T\}=\mathbf{S}$. One can derive the following input and output prediction model:

$$\hat{\mathbf{y}}_f = \mathbf{L}_w \mathbf{w}_p + \mathbf{L}_u \mathbf{u}_f \quad (2)$$

in which $\mathbf{w}_p$ is defined as $(\mathbf{y}_p^T \ \mathbf{u}_p^T)^T$ with the last p known values of inputs $\mathbf{u}_p \in R^{pm}$ and outputs $\mathbf{y}_p \in R^{pl}$, respectively, where

$$\mathbf{y}_p = \left(y(k-p+1)^T \quad \Lambda \quad y(k-1)^T \quad y(k)^T \right)^T ,$$
$$\hat{\mathbf{y}}_f = \left(\hat{y}(k+1)^T \quad \Lambda \quad \hat{y}(k+p-1)^T \quad \hat{y}(k+p)^T \right) \quad (3)$$

and $\mathbf{u}_p$ and $\mathbf{u}_f$ are defined in a similar manner. To ensure the integral action in the controller, one can include an internal disturbance model of typical load disturbances and this approach was implemented by introducing the integrated noise model [5]. Consider the noise input $e(k)$ as an integrated noise,

$$e(k+1) = e(k) + a(k) \quad (4)$$

where $a(k)$ is a white noise signal. Substituting Eq.(4) into Eq.(1), one obtains

$$z(k+1) = \mathbf{A}z(k) + \mathbf{B}\Delta u(k) + a(k), \quad \Delta y(k+1) = \mathbf{C}z(k+1) + \mathbf{D}\Delta u(k+1) + \mathbf{K}a(k) \quad (5)$$

where $z(k)=x(k)-x(k-1)$. By successively substituting the first equation into the second one of Eq. (5), one obtains the input and output vector form

$$\Delta \hat{\mathbf{y}}_f = \Gamma_N \mathbf{z}(\mathbf{k}) + \mathbf{H}_N \Delta \mathbf{u}_f = \mathbf{L}_w \left(\Delta \mathbf{y}_p \ \Delta \mathbf{u}_p \right) + \mathbf{L}_u \Delta \mathbf{u}_f \quad (6)$$

This incremental prediction model should be converted into the output prediction form so that it can be used for the design of controller: i.e.,

$$\hat{\mathbf{y}}_f = \mathbf{y}_t + \mathbf{S}_w \mathbf{L}_w \Delta \mathbf{w}_p + \mathbf{L}_u \mathbf{S}_u \Delta \mathbf{u}_f \quad (7)$$

in which $\mathbf{y}_t$ denotes the vector $[y(k) \ \ y(k) \ \ \Lambda \ \ \ y(k)]^T$ where $y(k)$ is augmented p-times. Even though this prediction model could be identified from the incremental values of input and output, it is difficult to design the input signal. In practice, therefore, it is more convenient to identify $\mathbf{L}_w$ and $\mathbf{L}_u$ first and then use these matrices to form the prediction model.

To install the stochastic observer to Eq. (7), the linear prediction model is modified and this can give compensation for the currently calculated predicted output by incorporating the difference between the real output $\mathbf{y}_t$ at present time and the predicted output obtained one sampling time before Y_{f-1}. The modified prediction model is expressed as follows:

$$\hat{\mathbf{y}}_f = \mathbf{y}_t + \mathbf{S}_w \mathbf{L}_w \Delta \mathbf{w}_p + \mathbf{L}_u \mathbf{S}_u \Delta \mathbf{u}_f + \mathbf{K}_f (\mathbf{y}_t - \mathbf{y}_{f-1}) \quad (8)$$

where $\mathbf{K}_f = \mathbf{diag}(\mathbf{k}_k, \Lambda, \mathbf{k}_{k+p})$, $\mathbf{k}_k = \mathbf{P}_k \mathbf{C}^T \mathbf{R}^{-1}$ and $\mathbf{P}_{k+1} = \left[\left(\mathbf{A}\mathbf{p}_k \mathbf{A}^T \right)^{-1} + \mathbf{C}^T \mathbf{R}^{-1} \mathbf{C} \right]^{-1}$.

in which $\mathbf{K}_f$ denotes the gain matrix determined similarly to that in Kalman filter approach. It is worth noting that $\mathbf{L}_w$ and $\mathbf{L}_u$ are obtained directly not from state space model but from block Hankel matrices consisting of collected output and input data. So, there is no explicit calculation of state space model. For the calculation of Kalman gain, however, it is necessary to calculate the state space model. Here, we note that $\mathbf{Y}_{pres}$ is a vector which has as many elements as the prediction horizon and each element is given by the current output y_k. Now Eq. (8) can be utilized to predict the output and to design a predictive controller. For detailed information, one may refer to the previous work [5].

With this prediction model one can set up a suitable quadratic objective function as follows:

$$\min_{\mathbf{u}_f} \mathbf{J}(\mathbf{u}_f) = \{\hat{\mathbf{y}}_f - \mathbf{r}_f\}^T \mathbf{Q} \{\hat{\mathbf{y}}_f - \mathbf{r}_f\} + \mathbf{u}_f^T \mathbf{R} \mathbf{u}_f + \Delta \mathbf{u}_f^T \mathbf{R}_\Delta \Delta \mathbf{u}_f$$

$$(9)$$

subject to input constraints $\mathbf{C}_u \cdot \mathbf{u}_f \geq \mathbf{C}_f$

4. IDENTIFICATION AND CONTROL PERFORMANCES

Fig. 1 illustrates the identification result, i.e., validation of identified model. The 4-level pseudo random signal is introduced to obtain the excited output signal which contains the sufficient information on process dynamics. With these exciting and excited data, $\mathbf{L}_w$ and $\mathbf{L}_u$ as well as state space model are calculated and on the basis of these matrices the modified output prediction model is constructed according to Eq. (8). To both mathematical model assumed as plant and identified model another 4-level pseudo random signal is introduced and then the corresponding outputs from both are compared as shown in Fig. 1. Based on the identified model, we design the controller and investigate its performance under the demand on changes in the set-points for the conversion and M_w. The sampling time, prediction and

864

Table 1. Reference conditions.

Initial charge	
Monomer	460 mL
Solvent	340 mL
Initiator	4.6 g
Feed concentration	
Monomer	5.42 mol/L
Solvent	4.35 mol/L
Initiator	0.02 mol/L
Operating conditions	
Jacket inlet temp.	70 °C (55~85 °C)
Feed flow rate	15 mL(0~30 mL)

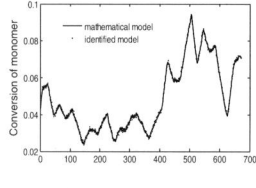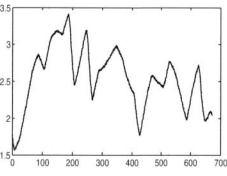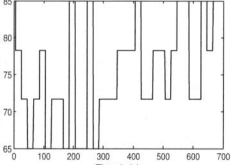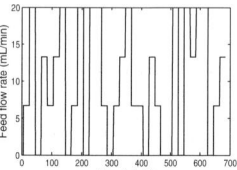

Fig. 1. Validation of identified model.

control horizons are 1, 15 and 5 min, respectively. At 400 min, the set-point for the conversion is changed from 0.16 to 0.12, while the set-point for M_w is increased from 120,000 to 200,000. Because of the interactive dynamics of the reaction system, the controller predictively decreases the jacket inlet temperature and the feed flow rate to decrease the conversion and increase M_w.

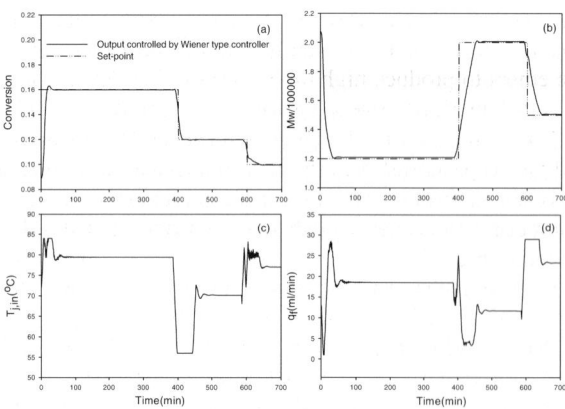

When the set-points for M_w and conversion are changed again at 600 min the controller predictively increases both the feed flow rate and jacket inlet temperature. Conversion decreases due to the increase of feed flow rate but the feed flow rate reaches its upper bound very quickly. Therefore, both inputs are decreased and these bring the conversion and M_w to their respective set-points through interactive dynamics. When compared with the other prediction model like Eq. (7), the present prediction model is found to exhibit an improved prediction performance and the designed controller shows a good set-point tracking performance under hard input constraints.

Fig. 2. Control performance for set-point tracking.

REFERENCES

[1] Henson, M. A. and Seborg D. E., *Nonlinear Process Control,* Prentice Hall: New Jersey, 1997.
[2] Van Overschee, P. and De Moor, B, *Subspace Identification for Linear Systems: Theory, Implementation, Applications.* Kluwer Academic Publishers: Dordrecht, 1996.
[3] Song, I. H. and H.-K. Rhee, *Ind. Eng. Chem. Res.,* **43**, 7261, 2004.
[4] Ahn, S. M., Park, M. J. and Rhee, H. K., *Ind. Eng. Chem. Res.,* **38(10)**, 3942, 1999.
[5] R. Kadali, B. Huang, A. Rossiter, *Contr. Eng. Prac.* **11**, 261, 2003.

Studies in Surface Science and Catalysis, volume 159
Hyun-Ku Rhee, In-Sik Nam and Jong Moon Park (Editors)

Synthesis of polycarbonates by copolymerization of carbon dioxide and glycidyl methacrylate using ionic liquids

Dae-Won Park*, Na-Young Mun, Hye-Young Ju, Youngson Choe and Sang-Wook Park

Division of Chemical Engineering, Pusan National University, Busan 609-735, Korea
E-mail: dwpark@pusan.ac.kr (D.-W. Park)

1. INTRODUCTION

As a kind of potential approach, one of the most promising areas of CO_2 utilization is its application as a direct material for polymer synthesis. To date, many excellent reviews in different period have made good description of this topic [1-3], where the importance of catalyst was never overestimated. Among catalysts reported, zinc dicarboxylates (for example, zinc glutarate) are known to afford the most active catalysts for the copolymerization of CO_2 and cyclic ethers to produce high molecular weight polycarbonates [4]. However, the conversion in the copolymerization was reported to be far below 100 %. In addition to the organozinc catalyzed copolymerization, the discovery of polymerization using aluminum and chromium complexes has expanded the utility of CO_2 as a comonomer in the production of polycarbonates [3,5,6]. Recently, the use of ionic liquids as environmentally benign media for catalytic processes or chemical extraction has become widely recognized and accepted. Ionic liquids have low vapor pressure, excellent thermal stability, and special characteristics in comparison with conventional organic and inorganic solvents. Many reactions catalyzed with ionic liquids and showing high performance have been reported [7].

In our previous work [8], we reported the synthesis of (2-oxo-1,3-dioxolan-4-yl)methacrylate (DOMA) from carbon dioxide and glycidyl methacrylate (GMA) using quaternary salt catalysts. In the present work, we studied the catalytic performance of alkylmethyl imidazolium salt ionic liquid in the synthesis of polycarbonate from the copolymerization of CO_2 with GMA. The influences of copolymerization variables like catalyst structure and reaction temperature on the conversion of GMA and the yield of the polycarbonate have been discussed.

2. EXPERIMENTAL

Glycidyl methacrylate (purity; 98 %) was purchased from Aldrich. Ionic liquids based on 1-n-ethyl–3-methylimidazolium (EMIm), 1-n-butyl–3-methylimidazolium (BMIm), 1-n-hexyl–3-methylimidazolium (HMIm) with different anions such as Cl⁻, BF_4^-, PF_6^- were prepared according to the procedures reported previously. Copolymerization of glycidyl methacrylate (GMA) and CO_2 were carried out in a 50 mL stainless steel autoclave equipped with a magnetic stirrer. The reactor was charged with 40 mmol of GMA, 2 mmol of catalyst and then purged several times with CO_2. The mixture was heated to a desired temperature. The reactor was then pressurized with CO_2 at 140 psig (9.6 atm) and the reaction started. After a certain period of reaction time, the pressure was reduced to atmosphere to terminate the copolymerization. The polymer was separated by precipitation followed by filtration. The polymerization yield was determined by gravimetry.

3. RESULTS AND DISCUSSION

The synthesis of polycarbonate from GMA and CO_2 was carried out using ionic liquids. The characteristic peaks from ^{1}H-NMR analysis ($CDCl_3$) in Fig. 1 are as follows: 4.8~5.2 ppm (peak a: poly carbonate), 3.15~4.0 ppm (peak b: ether). The IR spectra shown in Fig. 2 also confirmed the synthesis of carbonate. The spectra showed absorption at 1785cm^{-1} (C=O of polycarbonate group) and the disappearance of absorption due to epoxide ring at 910cm^{-1}. The conversions of GMA and molecular weight of polycarbonates with imidazolium ionic liquids of different cations such as EMIm$^+$, BMIm$^+$, HMIm$^+$ and those of different anions like Cl$^-$, BF$_4^-$, and PF$_6^-$ are summarized in Table 1. Imidazolium salt ionic liquids bearing Cl$^-$ anion showed good reactivity for the copolymerization CO_2 and GMA. The conversion of GMA was affected by the structure of imidazolium salts; the conversion increased from EMImCl (Run 1) to BMImCl (Run 2) probably due to a higher anion activating ability of BMIm$^+$ compared to EMIm$^+$.. However, when the alkyl group of the imidazolium salt increased to HMImCl (Run 3), the GMA conversion decreased because HMIm$^+$ is too bulky for the GMA to form a complex with it more easily. Separate experiments with different alkyl chain length in alkyl methyl imidazolium salts of tetrafluoroborate anion (BF$_4^-$) showed the same dependence of GMA conversion on the alkyl group as those of Cl$^-$.

The effects of anions in the addition of CO_2 to GMA are compared in Run 5, 8 and 9. The conversion of GMA increased in the order of PF$_6^-$ < BF$_4^-$ < Cl$^-$, which is consistent with the order of nucleophilicity of anions. More nucleophilic anion will be easier to attack the epoxide ring to form reaction intermediate. EMImBF$_4$ (Run 8) produced only a corresponding monomeric five-membered cyclic carbonate DOMA. There was no reaction when EMImPF$_6$ (Run 9) was used as catalyst. The conversion of GMA increased as the temperature increased from 60 ℃ to 140 ℃. Even though the GMA conversion from 100 to 140 ℃ did not change so much, TON increased significantly. It means that the portion of copolymer is higher at higher temperature. Longer reaction time also increased the conversion of GMA and TON (Run 1, 10, 11). However, molecular weights of the polycarbonates did not vary significantly and polydispersity was near to one for all the experiments in Table 1.

Based on the above results and previous works [3,9] on the reaction of epoxides and CO_2, we tentatively propose the plausible mechanism for the copolymerization of GMA and CO_2 (scheme 1). Alkylmethyl imidazolim salt (QX) and epoxide (GMA) reacted to synthesize an active species followed by chain propagation involving a concerted insertion of the epoxide. However, more detailed mechanistic studies are needed to clearly understand the polymerization steps.

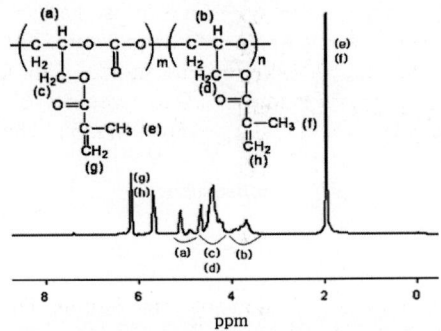
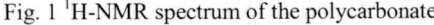

Fig. 1 ¹H-NMR spectrum of the polycarbonate

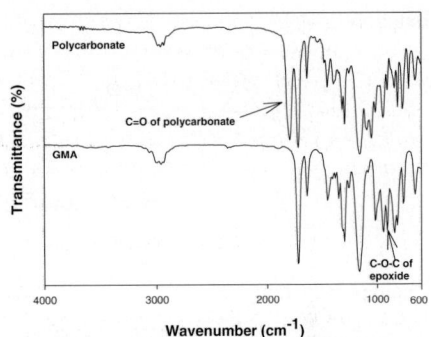

Fig. 2 FT-IR spectra of polycarbonate

Table 1
Synthesis of polycarbonates from glycidyl methacrylate and carbon dioxide with different ionic liquid catalysts

Run	Ionic liquid	Temp.	Time (h)	Conversion (%)[a]	TON[b]	TOF[c]	Mn[d]
1	EMImCl	60	6	50.7	1.9	0.32	3731
2	BMImCl	60	6	60.0	3.9	0.65	3830
3	HMImCl	60	6	44.8	1.9	0.31	3707
4	EMImCl	80	6	80.3	6.2	1.03	3808
5	EMImCl	100	6	93.0	6.4	1.07	3800
6	EMImCl	120	6	94.9	9.1	1.52	3764
7	EMImCl	140	6	100	15.2	2.53	3725
8	EMImBF₄	100	6	47.0	0	0	-
9	EMImPF₆	100	6	0	0	0	-
10	EMImCl	60	24	72.2	4.0	0.17	3756
11	EMImCl	60	48	78.9	5.9	0.12	3727

Polymerization conditions: GMA 40 mmol, catalyst 2 mmol, Pco₂ 140 psig
[a] Conversion is defined on the basis of GMA
[b] Turnover number: gram of polymer/gram of ionic liquid
[c] Turnover frequency: gram of polymer/(gram of ionic liquid)/h
[d] Data from GPC

868

Initation

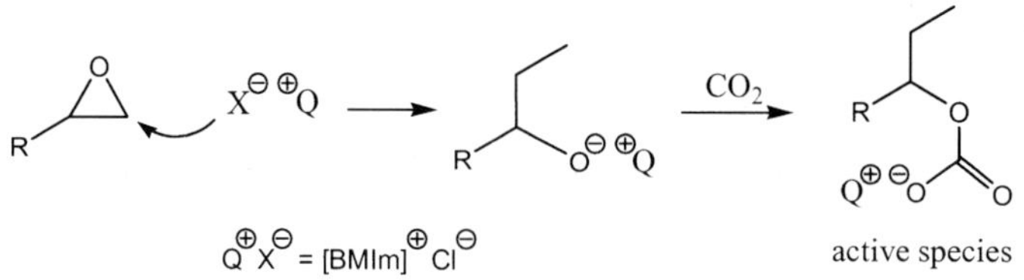

$$Q^{\oplus}X^{\ominus} = [BMIm]^{\oplus}Cl^{\ominus}$$

active species

Chain propagation

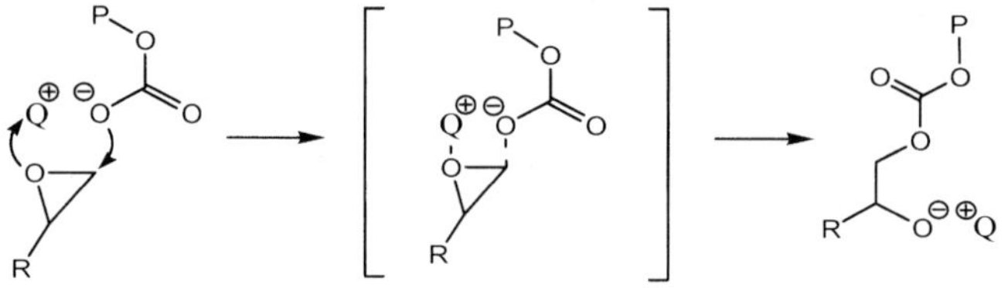

Scheme 1. Proposed mechanism for the copolymerization of GMA and CO_2

4. CONCLUSIONS

Ionic liquids based on 1-alkyl-3-methylimidazolium salt showed good catalytic performance for the synthesis of polycarbonates from the copolymerization of GMA and carbon dioxide. The ionic liquid of more bulkier alkyl chain length and of more nucleophilic anion exhibited higher reactivity. However, the TON of the polycarbonate decreased when hexyl containing ionic liquids was used in place of butyl group. Higher temperature and longer reaction time favored the yield of polycarbonate.

ACKNOWLEDGEMENT

This work was supported by Korea Science and Engineering Foundation (R-01-2005-000-10005-0), and Brain Korea 21 and Brain Busan 21 program.

REFERENCES

[1] S. Inoue, H. Koinuma, T. Tsuruta, J. Polym. Sci. Polym. Lett., B7 (1969) 287.
[2] A. Rokicki, W. Kuran, J. Macromol. Sci. Rev. Macromol. Chem., C21 (1981) 135.
[3] D. J. Darensbourg, J. C. Yarbrough, C. Ortiz, C. C. Fang, J. Am. Chem. Soc., 125 (2003) 7586.
[4] A. Rokicki, US Patent 4, 943 (1990) 677.
[5] D. J. Darensbourg, M. W. Holtcamp, Coord. Chem. Rev., 153 (1996) 155.
[6] T. Aida, M. Ishikawa, S. Inoue, Macromolecules, 19 (1986) 8.
[7] C. E. Song, W. H. Shim, E. J. Roh, J. H. Chio, Chem. Commun., (2000) 1695.
[8] J. Y. Moon, J. G. Yang, D. W. Park, J. K. Lee, Korean J. Chem. Eng., 14 (1997) 507.
[9] F. Li, L. Xiao, C. Xia, B. Hu, Tetrahedron Letters, 45 (2004) 8307.

Studies in Surface Science and Catalysis, volume 159
Hyun-Ku Rhee, In-Sik Nam and Jong Moon Park (Editors)

A Statistical study of phenol-formaldehyde resol resin synthesis reaction

Young Woo Nam[a], Tae Uk Park[a] and Dong Kwon Kim[b]

[a]Dept. of Chemical and Environmental Engineering, Soongsil University, 1-1 Sangdo-5dong, Dongjak-gu, Seoul 156-743, Republic of Korea

[b]Division of Interface Engineering Laboratory, KRICT, P. O. Box 107, Yusong, Taejeon 305-600, Republic of Korea

1. INTRODUCTION

Phenol-formaldehyde (PF) resins were synthesized to manufacture non-flammable insulating foam. When alkali catalyst, for example, barium hydroxide $(Ba(OH)_2)$, was present, resol resins are produced[1]. In the analysis of molecular species of resol, capillary GC-MS had been used to separate hemiformal-type compounds(acetylated hydroxybenzylhemiformals) [2]. The final structure of resins produced depends on the reaction condition. Formaldehyde to phenol (F/P) and hydroxyl to phenol (OH/P) molar ratios as well as reaction temperature were the most important parameters in synthesis of resols. In this study, the effect of F/P and OH/P wt%, and reaction temperature on the chemical structure (mono-, di- and tri-substitution of methyrol group, methylene bridge, phenolic hemiformals, etc.) was studied utilizing a two-level full factorial experimental design. The result obtained may be applied to control the physical and chemical properties of pre-polymer.

2. EXPERIMENTAL

A two levels of full factorial experimental design with three independent variables were generated with one center point, which was repeated[3]. In this design, F/P molar ratio, Oh/P wt%, and reaction temperature were defined as independent variables, all receiving two values, a high and a low value. A cube like model was formed, with eight corners. One center point (repeated twice) was added to improve accuracy of the design. Every analysis results were treated as a dependent result in the statistical study.

All 10 PF resins were produced with $Ba(OH)_2$ catalyst for 300 min with varying F/P molar ratio, OH/P wt %, and reaction temperature(Table 1). The resins were stored frozen at -18℃ until analysis. Molecular species in resol were analyzed by GC after trimethylsilylation of sample with N,O-bis(trimethylsilyl)trifluoracetamide in pyridine[2]. A glass column (3 m × 2 mm I.D.) packed with 3% OV-1 on 100-120 mesh Chromosorb W HP was applied. Injection port and detector temperature were maintained at 280℃, and the oven was programmed from 120 to 300℃ at 8℃/min.

Table 1
Properties of the 10 PF resins studied

PF Resin	F/P molar ratio	OH/P wt %	Temp.(℃)
1	1.5	0.5	80
2	1.5	0.5	90
3	1.5	1.5	80
4	1.5	1.5	90
5	2.5	0.5	80
6	2.5	0.5	90
7	2.5	1.5	80
8	2.5	1.5	90
9	2.0	1.0	85
10	2.0	1.0	85

Table 2
Analyzed results of 10 PF resins

PF resin	Substituted monomer			Methylol groups		Methylene bridges		Condensed dimers	Phenolic hemiformals	Free phenol
	mono-	di-	tri-	ortho	para	o-p'	p-p'			
1	10.27	4.85	2.37	6.07	4.20	4.54	0.61	6.65	6.63	8.89
2	8.26	4.62	2.29	5.35	2.91	4.98	0.61	6.87	5.81	6.18
3	7.61	3.92	2.41	3.82	3.79	3.95	0.45	5.31	4.99	6.21
4	7.91	4.18	2.10	4.48	3.43	6.12	0.77	7.96	4.50	5.77
5	10.34	5.65	2.36	6.72	3.61	5.34	0.69	8.13	8.03	6.00
6	10.04	5.53	2.26	6.63	3.41	5.41	0.64	7.97	7.24	3.04
7	12.56	5.62	4.25	4.98	7.58	5.43	0.81	7.72	11.19	4.20
8	7.72	5.03	2.68	4.21	3.51	5.82	0.72	8.05	6.88	4.00
9	8.97	4.47	3.22	4.98	3.99	5.42	0.74	7.63	6.33	5.56
10	9.12	4.85	3.20	5.16	3.96	5.41	0.71	7.62	6.67	5.64

All values are wt% of full mixture.

3. RESULTS AND DISCUSSION

From GC chromatogram, 16 molecular species in resol were identified. They were categorized as substituted monomers, condensed dimers, and phenolic hemiformals. Results of the 10 analyzed PF resins are shown in Table 2. Calculated results demonstrated o-p'/p-p' bridges ratio, methylene bridges ratios to methylols, and o-methylols ratios to p-methylols(Table 3). All depependent variables were analyzed by utilizing ANOVA and multiple regression models.

The F-test indicates the dependence of the dependent variables with the independent variables, P level indicates the statistical significance of the correlation(Table 4). The F-test results for the relation of the amount of *ortho* methylol phenols with F/P molar ratio and the reaction temperature were low, however, for the OH/P wt %, the F-test result was very significant, indicating a clear dependence of *ortho* methylol phenols on the OH/P wt %. It can also be seen that P level values for the relation between the amount of *ortho* methylol phenols and both F/P molar ratio and reaction temperature are above the set P value of 0.05, while for the OH/P wt%, the P value is under the set value. This data indicated that the relations of dependent variables *ortho* methylol phenols with independent variable OH/P wt% is statistically significant at the 0.05 significance level, while the relation of dependent variables *ortho* methylol phenols with F/P molar ratio and reaction temperature are not statistically significant.

F/P molar ratio showed an increasing effect, while reaction temperature demonstrated a decreasing effect on mono-substitution of methylol group. Three independent variables demonstrated no statistically significant effect on di-substitution of methylol groups. Tri-substitution of methylol groups, *para* methylol phenols, *o-p'* methylene bridges, dimer formation, the amount of free phenols, and the fraction of *o-p'/p-p'* were dependent on all of the three independent variables.

From the P level value, the relations of dependent variable phenolic hemiformals with independent variable F/P molar ratio was statistically significant at the 0.05 significance level, while the other two independent variables were not statistically significant. It was found that the F/P molar ratio and the OH/P wt% shows an increasing effect on the *p-p'* bridges, while reaction temperature demonstrated no statistically significant effect. The OH/P wt% and reaction temperature were found to have effect on *ortho/para* methylol fraction. The reaction temperature always had increasing effect, whereas the OH/P wt% had a decreasing effect.

Table 3
Calculated results of 10 PF resins

PF resin	Methylols *ortho/para*	Methylene bridges *o-p'/p-p'*	Methylene bridges/methylols
1	1.44	2.15	0.65
2	1.84	2.64	0.83
3	1.01	2.90	0.70
4	1.31	3.34	1.00
5	1.86	1.92	0.79
6	1.95	2.11	0.79
7	0.66	2.38	0.61
8	1.20	2.61	1.04
9	1.25	2.45	0.85
10	1.30	2.43	0.84

Table 4
F-test results and P level values for all variables of 10 PF resins.

	Substituted monomer			Methylol group		Methylene bridges	
	mono-	di-	tri-	ortho	para	o-p'	p-p'
F-test results							
F/P molar ratio	485.5	31.42	3540	61.36	3969	3630	17690
OH/P wt%	101.5	6.250	2916	408.9	4853	689.1	6561
Reaction temp.	521.4	0.801	2652	6.531	9735	5890	1.000
P level values							
F/P molar ratio	0.029	0.112	0.011	0.081	0.010	0.011	0.005
OH/P wt%	0.061	0.242	0.012	0.031	0.009	0.024	0.008
Reaction temp.	0.028	0.535	0.012	0.237	0.006	0.008	0.500

	Condensed dimers	Phenolic hemiformals	Free phenols	Methylen bridges o-p/p-p'	Methylols ortho/para
F-test results					
F/P molar ratio	64520	281.5	2970	2525	0.049
OH/P wt%	841.0	0.049	476.7	3630	846.8
Reaction temp.	23100	88.86	1229	1139	176.9
P level values					
F/P molar ratio	0.003	0.038	0.012	0.013	0.611
OH/P wt%	0.022	0.862	0.029	0.011	0.022
Reaction temp.	0.022	0.067	0.018	0.019	0.048

4. SUMMARY

A two level full factorial experimental design with three variables, F/P molar ratio, OH/P wt %, and reaction temperature was implemented to analyses the effect of variables on the synthesis reaction of PF resol resin. Based on the composition of 16 components of 10 samples, the effect of three independent variables on the chemical structure was analyzed by using 3 way ANOVA of SPSS. The present study provides that experimental design is a very valuable and capable tool for evaluating multiple variables in resin production.

ACKNOWLEDGEMENT
This work was supported by ECO-Technopia-21.

REFERENCES
[1] A. Gardziella, L. A. Pilato, and A. Knob, Phenolic resins, chemistry, applications, , safety and ecology, Springer-Verlag, Berlin, 2000.
[2] L. Prokai, J. Chromatography, 333(1985) 161.
[3] T. Holopainen, L. Alvila, P. Savolainen, and T. T. Pakkanen, J. Appl. Polym. Sci., 91 (2004)2492.

Studies in Surface Science and Catalysis, volume 159
Hyun-Ku Rhee, In-Sik Nam and Jong Moon Park (Editors)

Highly *cis*-selectivity and low molecular weight distribution polymerization of 1,3-butadiene with cobalt(II) pyridyl bis(imine) complexes in the presence of ethylaluminum sesquischloride: effect of methyl position in the ligand

Bu Ung Kim, Jae Sung Kim, Kyoung Ju Lee, Chang-Sik Ha and Il Kim[*]

Department of Chemical Engineering, Pusan National University, Jangjeon-dong, Geumjeong-gu, Busan 609-735, Korea

1. INTRODUCTION

Selective 1,4-*cis* polymerization of 1,3-butadinene (BD) is of much importance, since the resulting high *cis*-poly(BD)s can find wide applications as synthetic rubbers [1]. Industrially, catalyst systems such as $TiI_4/I_2/Al(i\text{-}Bu)_3$, $CoCl_2 \cdot Py/AlEt_2Cl/H_2O$, $Ni(OCOR)_2/AlEt_3/BF_3OEt_2$ and $NdCl_3/EtOH/Al(i\text{-}Bu)_3$ have been used for the production of poly(BD)s which have high 1,4-cis contents (94-98%), but the molecular weight and molecular weight distribution (MWD = 3-4) of the resulting polymers in these systems cannot be well controlled [2]. As homogeneous catalysts which can better control molecular weight, d-block transition metal metallocene/MAO catalyst systems such as $CpTiCl_3/MAO$, Cp_2TiCl/MAI, Cp_2VCl/MAO, and Cp_2Ni/MAO have been reported. However, all these systems showed considerably lower 1,4-*cis* selectivity (79-91%) [3]. We report herein an extremely active, cobalt-based single-site catalyst system, which can effect both high 1,4-*cis* selectivity and control on the molecular weight of the polymer products.

Recently new cationic tridentate Fe(II)- and Co(II)-based pyridyl bis(imine) catalysts were synthesized by Brookhart et al. and were widely used for the polymerization of olefins [4]. Some distinguished features of the diimine ligand catalysts are their high electrophilicities, cationic metal center and the use of sterically bulky pyridyl bis(imine) ligands. This type of catalysts is characterized by reduced oxophilicity and thus a higher tolerance for functional groups. In this sense, they might be used as catalysts for the polymerization of various monomers. However, there are no detailed reports on the polymerization of 1,3-butadiene by using pyridyl bis(imine) Co(II) complexes. In order to get deep insight into the polymerization of various monomers by late transition metal catalysts, we decided to undertake an investigation into late transitionmetal-catalyzed 1,3-butadiene polymerization. The cobalt(II) precatalysts employed were prepared by changing the position of dimethyl substituents in the aryl ring (Scheme 1). It is well known that protective bulk of the ortho substituents above and below the metal center is critical to achieve high activity and high molecular weight polymer in olefin polymerizations [4]. Relative steric hindrance imposed on the metal center can be easily changed by simply changing the position of the two methyl substituents on the aryl rings, i.e. 2,3-, 2,4-, 2,5-, 2,6- and 3,5-Me$_2$.

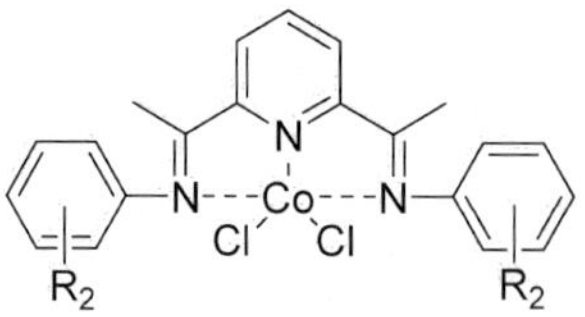

R_2	Catayst
3,5-Me_2	**cat(3,5-Me)**
2,3-Me_2	**cat(2,3-Me)**
2,4-Me_2	**cat(2,4-Me)**
2,5-Me_2	**cat(2,5-Me)**
2,6-Me_2	**cat(2,6-Me)**

Figure 1. The structure of bis(imino)pyridyl cobalt(II) precatalysts utilized in this study.

2. EXPERIMENTAL

2.1 General Methods and Materials

All reactions were performed under a purified nitrogen atmosphere using standard glove box and Schlenk techniques. Polymerization grade of 1,3-butadiene (SK Co., Korea) was purified by passing it through columns of Fisher RIDOX catalyst and molecular sieve 5 Å/13X. Used all reagents were purchased from Aldrich and used without purification. Ethylaluminum sesquischloride was obtained as a 0.9 M total Al solution in toluene. Literature procedures [5] were used to synthesize Co(II)-based pyridyl bis(imine) complexes shown in Figure 1.

2.2 Characterization

The molecular weights of polybutadiene were determined by gel permeation chromatography (GPC) using a Waters M515 series system in tetrahhydrofuran (THF) at 25 ℃ as calibrated with polystyrene standards. The microstructure of polybutadiene was determined by using ^{1}H NMR and ^{13}C NMR spectra recorded on a Varian Unity Plus 300 (300 MHz) spectrometer in $CDCl_3$ at 25 ℃ using tetramethylsilane as an internal reference. ^{1}H NMR: δ = 4.8–5.2 (=CH2 of 1,2-butadiene unit), 5.2–5.8 (–CH of 1,4-butadiene unit and –CH of 1,2-butadiene unit). ^{13}C NMR: δ = 27.4 (1,4-*cis*-butadiene unit), 32.7 (1,4-*trans*-butadiene unit), 127.7–131.8 (1,4-butadiene unit), 113.8–114.8 and 143.3–144.7 (1,2-butadiene unit).

2.3 Polymerization procedure

Solution polymerizations of 1,3-butadiene were carried out in a high-pressure glass reactor (40 mL) connected with a vacuum system. In a typical procedure, 4 μmol of precatalyst (EAS/precatalyst = 100 mol/mol) was dissolved in 20 mL of toluene. The polymerization started by adding 1.08 g of 1,3-butadiene and EAS to the solution in this order. The reaction mixture was stirred at a specific temperature (30 to 70 °C) for 40 min. The resulting solution was poured into acidified methanol (100 mL of a 5% v/v solution of HCl). The polymer was then isolated by filtration and washed with methanol before drying overnight at 40 °C. Polymer yield was determined by gravimetry.

3. RESULTS AND DISCUSSION

Table 1 summarizes the results of the solution polymerizations of 1,3-butadiene by using various catalysts in combination with EAS. The **cat(2,3-Me)** containing 2,3-methyl

Table 1. Solution polymerization results for butadiene using cobalt(II) pyridyl bis(imine) complexes. Polymerization conditions: [1,3-butadiene]= 1 mol/L; [Cat.] = 2.00 × 10^{-4} mol/L; [EAS]/[Co] = 100; toluene = 20 mL; polymerization temperature = 30 ℃.

Entry No.	Catalyst	Time (min)	Yield (%) [a]	M_n (× 10^{-5}) [b]	M_w/M_n [b]	Triad fractions c (%) [c]		
						1,4-*cis*	1,4-*trans*	1,2
1	**cat(3,5-Me)**	40	10.71	1.57	1.21	97.73	1.28	0.99
2	**cat(2,3-Me)**	40	25.70	1.34	1.24	96.64	2.43	0.93
3	**cat(2,4-Me)**	40	21.83	1.28	1.17	98.19	1.17	0.64
3, d	**cat(2,4-Me)**	48h	16.1	1.07	1.37	95.78	2.4	1.82
4	**cat(2,5-Me)**	40	16.52	1.45	1.29	98.62	0.92	0.46
5	**cat(2,6-Me)**	40	5.65	1.22	1.24	97.29	2.18	0.53
6	**cat(2,4-Me)**	10	0.99	0.57	1.29	98.63	0.65	0.72
7	**cat(2,4-Me)**	15	1.86	1.20	1.11	97.99	1.23	0.78
8	**cat(2,4-Me)**	30	17.10	1.24	1.19	97.30	1.97	0.73
9	**cat(2,4-Me)**	90	59.34	1.62	1.24	96.48	2.63	0.89

[a] Yield defined a mass of dry polymer recovered/mass of monomer used.
[b] Determined by GPC.
[c] Measured by 1H-NMR spectra and 13C-NMR.
[d] Polymerization conditions are the same as Entry No. 3 except [MAO]/[Co] = 1000 and polymerization time = 48 h.

substituents in the aryl rings showed the highest activity in 40 min of polymerization. The activity decreases in order of **cat(2,3-Me)** > **cat(2,4-Me)** > **cat(2,5-Me)** > **cat(3,5-Me)** > **cat(2,6-Me)**. It is surprising to note that **cat(2,6-Me)**, which shows very high activity in ethylene polymerizations up to 11 kg $mmol^{-1}h^{-1}bar^{-1}$, shows the lowest activity. It demonstrates that the propagation rate of active species-1,3-butadiene coordinated complexes is different from that of active species-ethylene coordinated species. The 1,3-butadiene monomer is differentiated from ethylene monomer in that it contains conjugated diene with relatively bulkier structure than ethylene. In this sense the electronic and steric environments of the active species imposed by growing polymer chain should be quite different each other. These made the molecular weights resulting from butadiene polymerizations and from ethylene polymerizations with the same catalyst systems completely different.

The data in Table 1 show that all catalysts produce polymers with narrow molecular weight distribution (MWD < 1.3). These MWD values are narrower compared to other late transition metal catalysts such as $Ni(acac)_2$/MAO [6], and $Ni(oct)_2/AlF_3/AlEt_3$. In order to check whether the narrow MWD values are induced by the living character of active species, a series of polymerizations were performed by using **cat(2,4-Me)**/EAS system by changing the time from 10 to 90 min (see Entry No. 3 and from No. 6 to 9 in Table 1). Both activity and molecular weight were not linearly increased according to time, demonstrating that the present catalyst systems do not have living character, while they show polymerization behaviors which can be generally found in single-site coordination catalyst systems.

Polymerization temperature is one of the most important parameters in both scientific and technological senses. A series of polymerizations were carried out with **cat(2,4-Me)**/EAS system over a temperature range between 10 and 70 °C and the results are summarized in Table 1. The catalyst system showed only negligible activity at 10 °C and the highest activity at 50 °C. As the temperature increase, the molecular weight decreases monotonously, remaining MWD narrow. In addition *1,4-cis* addition decreases as the temperature increases, but still highly stereospecific. It is also interesting to note the difference between 1,3-butadiene and ethylene polymerizations. In the latter case with the same catalyst system, it showed high activity at 10 °C but negligible activity at 70 °C. In addition the **cat(2,4-Me)** catalyst showed very high activity when it is associated with MAO in ethylene polymerizations; however, the same system showed very low activity in butadiene polymerization (see Entry no. 3' in Table 1). These results indicate that the bulkiness of monomer and its resulting coordinating species and cocatalyst structure influence the propagation rate.

4. CONCLUSIONS

A series of prototypical Co(II)-based pyridyl bis(imine) complexes studied as precatalysts for polymerization of 1,3-butadiene showed very high activities in the presence of EAS as a cocatalyst, producing high molecular weight polymers with narrow MWD (< 1.3). The Co(II) catalyst bearing two *o*-aryl methyl substituents (**cat(2,6-Me)**/EAS) showed the lowest activities among the catalysts of this study to give a high MW polymers with narrow MWD (= 1.22). All catalysts yielded *cis*-1,4-polybutadiene up to 98.6% of *cis*-1,4 regular structure. The protective bulk of the *ortho* substituents above and below the cobalt center is not critical to achieve high molecular weight polymer in butadiene polymerizations, which differentiated from ethylene polymerizations with the same catalyst systems.

ACKNOWLEDGEMENTS

This work was supported by the Center for Ultramicrochemical Process Systems, Pusan National University Grant and the Korea Institute of Industrial Technology Evaluation and Planning. The authors are also grateful to grant No. R01-2003-000-10020-0 from the Basic Research Program of KOSEF and to the Brain Korea 21 Project and the National Research Laboratory Program.

REFERENCES

[1] G. C. Eastmond, A. Ledwith, S. Russo and P. Sigwalt, Pergamon Press: Oxford. U.K, Vol. 4, pp53-108 (1989).

[2] For examples, see: (a) W.M. Saltman and Link, T. H. Ind. Eng.Chem. Prod. Res. Dev., 3, 199 (1964) (b) M. Gippin, Ind. Eng.Chem. Prod. Res. Dev., 1, 32 (1962) (c) K. Ueda, T. Yoshimoto, J. Hosono, K. Maeda and T. Matsumoto Kogyo KagakuZasshi 66, 1103 (1963) (d) J. Yang, J. Hu, S. Feng, E. Pan, D. Xie, C. Zhong and Ouyang, J. Sci. Sin. (Engl. Ed.), Z3, 734 (1980) (e) H. Hsieh and H. C. Yeh, Rubber Chem. Technol., 58, 117 (1984)

[3] (a) L. oliva, O. Longo, A. Grassi and P. Ammendola, Macromol. Chem. Rapid Commun., 11, 519 (1990). (b) G. Ricci, S. Italia, A. Giarrusso and L. Porri, J. Organomet. Chem., 451, 67 (1993). (c) G. ricci, C. Bosisio and L. Porri, Macromol. Chem. Rapid Commun., 17, 781 (1996). (d) G. Ricci, A. Panagia and L. Porri, Polymer 37, 363 (1996). (e) K. Endo, Y. Matuda and S. Aoki, Prepr, Jpn, 42, 282 (1993).

[4] S. D. Ittel, L. K. Johnson and M. Brookhart, Chem. Rev., 100, 1169 (2000).

[5] B. L. Small, M. Brookhart, A. M. A. Bennet, J. Am. Chem, Soc., 120, 4049 (1998).

Studies in Surface Science and Catalysis, volume 159
Hyun-Ku Rhee, In-Sik Nam and Jong Moon Park (Editors)

Co/Zn double metal cyanide catalyzed ring-opening polymerization of propylene oxide: effect of cocatalysts on polymerization behavior

Anas K., Seung Tae Baek, Chang-Sik Ha and Il Kim*

Department of Polymer Science and Engineering, Pusan National University, Jangjeon-dong, Geumjeong-gu, Busan 609-735, Korea

1. INTRODUCTION

For many years, polyether polyols have been made using conventional base catalysis in industry. The catalysts generally used are alkali metal hydroxide such as potassium hydroxide. Double metal cyanide (DMC) complexes catalysts discovered in early 1960s by researchers at General Tire and Rubber Co. are well known catalysts for epoxide polymerization. The catalysts are highly active, and give polyether polyols that have low level of unsaturation and narrow molecular weight distribution (MWD) compared with similar polyols made using basic (KOH) catalysts [1]. While DMC catalysts offer significant advantages, unlike KOH, DMC catalysts must normally be activated before the epoxide can be added continuously to the reactor. Usually, a polyol initiator (or starter) and a DMC catalyst are combined and heated under vacuum prior to the addition of small proportion of monomer [2]. Long induction period, say several hours, increases cycles time, which undercuts the economic advantages of faster polymerizations. It was reported that by controlling the type and amount of the co-complexing agent during preparation of the catalyst the catalytic activity, initiation time and the unsaturation level in polyether polyols can be tuned [3].

In order to develop DMC catalysts having higher activity and shorter induction periods in epoxide polymerizations, study of the role of each catalyst component is firstly needed. However, scientific studies of the DMC catalyst are scarce considering the commercial importance of the DMC catalysts [3-5]. In this study we have prepared a DMC catalysts by using zinc chloride ($ZnCl_2$) and hexacyanocobaltate(III) ($K_3Co(CN)_6$) in the presence of tertiary butyl alcohol (^tBuOH) and poly(tetramethylene ether) glycol (M_n = 1800; PTMEG) as complexing agents, and investigated the effect of the addition of various ionic complexes such as quaternary ammonium salts and imidazolium salts. This study is a way to get insight on the factor influencing the activity of DMC catalyst and on the understanding of the mechanistic pathway of DMC catalyzed polymerization.

2. EXPERIMENTAL

2.1 Preparation of catalyst and polymerization

The DMC catalysts of the general formula, $Zn_3[Co(CN)_6]_2 \cdot xZnX_2 \cdot yH_2O \cdot$ complexing agents were prepared by using $ZnCl_2$ and $K_3Co(CN)_6$ in the presence of ^tBuOH and PTMEG as complexing agents [3, 6].

Polymerizations of propylene oxide were carried out by using 1 L autoclave (Parr) at various temperatures [3,6]. 70 g of PPG (functionality = 2) was used as an initiator (or starter). The reactor was charged with the initiator, the DMC catalyst (0.1 g) and a prescribed amount of quaternary ammonium salt or imidazolium salts, and then purged with several times with nitrogen. The mixture was heated to 90 °C and evacuated for over 2 hours during stirring in order to remove traces of water contained in the initiator. Then 10 g of PO monomer was introduced into the reactor at a desired polymerization temperature (T_p). Additional monomer was started to be introduced gradually when an initial pressure drop, indicating activation of the catalyst, occurred in the reactor. The polymerization was stopped when the total amount of added monomer reached 400 g for the effectiveness of stirring.

2.2 Measurements

X-ray photoelectron spectroscopy (XPS) analysis of the catalysts was performed on an ESCALAB 250 induced electron emission spectrometer with AlKα1 (1486.6 eV, 12 mA, 20 kV) X-ray sources. Infrared (IR) spectra were recorded on a Shimadzu IRPrestige-21 spectrophotometer and X-ray diffraction (XRD) patterns of the catalysts were obtained with a RINT2000 wide angle goniometer 185 using Cu Kα radiation at 40 kV and 30 mA. The hydroxyl value (OHV) is defined as the equivalent amount of KOH corresponding to the hydroxyl groups in 1 g of polymer and analyzed according to ASTM D-4274 D. The total degree of unsaturation of polyols was measured by titration method according to ASTM D2847. Molecular weight distribution (MWD) was measured using a Waters 150 instrument operated at 25 °C, with a set at 10^4, 10^3, and 500 angstroms columns in tetrahydrofuran solvent. A Brookfield viscometer model DV III (Brookfield Instruments), with a small scale sample adapter and spindle no. 21, was used to measure the viscosity of the polymer samples.

3. RESULTS AND DISCUSSION

Eight quaternary ammonium salts and seven imidazolium salts were chosen in this study. The quaternary ammonium salts (QAS) used are (1) tetrapropylammonium chloride (TPACl) (2) tetrabutylammonium chloride (TBACl) (3) tetrahexylammonium chloride (THACl) (4) tetraoctaylammonium chloride (TOACl) (5) tetradodecylammonium chloride (TdodecylACl) (6) tetrabutylammonium bromide (TBAB) (7) tetramethylammonium bromide (TMAB) and (8) tetrabutylammonium iodide (TBAI). And the imidazolium salts used are (1) 1-ethyl-3-methylimidazolium chloride (EMImCl), (2) 1-butyl-3-methylimidazolium chloride (BMImCl), (3) 1-hexyl-3-methylimidazolium chloride (HMImCl), (4) 1-ethyl-3-methylimidazolium tetrafluoroborate (EMImBF$_4$), (5)1-Butyl-3-methylimidazolium tetrafluoroborate (BMImBF$_4$), (6) 1-Octyl-3-methylimidazolium tetrafluoroborate (OMImBF$_4$), and (7) 1-ethyl-3-methylimidazolium hexafluorophosphate (EMImPF$_6$). Figures 1 and 2 show the plots of propylene oxide consumption versus reaction time obtained by adding 1 mmol of QASs and imidazolium salts, respectively.

The polymerization reactions exhibit very short induction periods (several minutes) in comparison with that obtained by DMC catalyst alone under the same conditions. In the case of DMC catalyst alone it can be seen that the reaction rate decreased gradually with increased monomer addition. No such deactivation is observed if QASs are added during polymerizations. The induction period was found to be decreased with an increase in the number of the carbon atoms in the quaternary ammonium salt. These results show that the total carbon number of the quaternary ammonium salts is an important factor influencing the catalytic behavior of DMC compounds. From Fig. 1, it is clear that the activity of the QAS containing chlorine counterion is decreasing in the order of TdodecylACl > TOACl > THACl

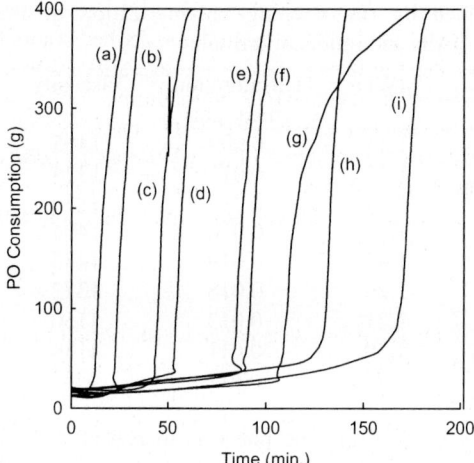

Fig. 1. PO consumption rate obtained by adding various QAS cocatalysts at 115 °C: (a) TDACl, (b) TOACl, (c) TMABr, (d) THACl, (e) TBAI, (f) TBACl, (g) cat. only, (h) TPACl, and (i) TBABr. Polymerization conditions: catalyst = 0.1 g, initiator (PPG-750) = 70 g, and QAS = 1 mmol.

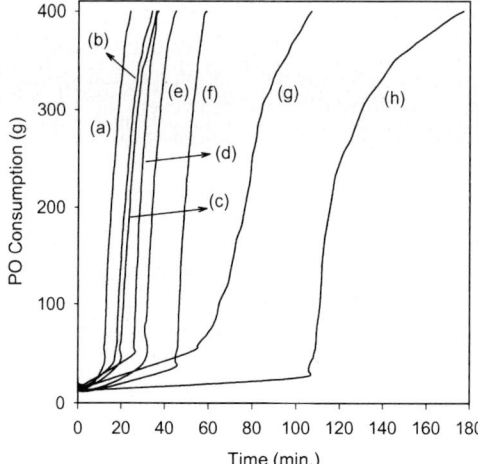

Fig. 2. PO consumption rate obtained by adding various imidazolium salt cocatalysts at 115 °C: (a) OMImBF4, (b) EMImPF4, (c) BMImBF4, (d) EMImCl, (e) EMImBF4, (f) BMImCl, (g) HMImCl, and (h) cat. only. Polymerization conditions: catalyst = 0.1 g, initiator (PPG-750) = 70 g, and imidazolium salt = 1 mmol.

> TBACl > TPACl, while in the case of bromine as anion the reverse was taking place according to the alkyl chain length; i.e. TMAB > TBAB. Both the high activity and the short induction period obtained by simply adding QASs are the best level comparing to reported results so far.

The similar favorable results were also obtained by adding imidazolium salts (Fig. 2). The induction period was increased when chloride was an anion with the increase of the alky chain length. When tetrafluroborate is an anion, the induction period was decreased with the increase of alkyl chain length.

It is not clear why QAS and imidazolium salt yield favorable results in DMC catalyzed polymerizations. One possible reason is that since the organic phase reaction is the rate controlling step, the total carbon number of the quaternary salt which participates in the

Table 1 Characterization of polymers obtained by ring-opening polymerization of PO catalyzed by DMC catalysts in the presence of QASs and imidazolium salts

Compound	M_n	M_w	MWD	Unsaturation (meq/ gm)	Viscosity
DMC only	6043	7430	1.23	0.009	1465
TPACl	6553	8821	1.35	0.011	1324
TBACl	6057	7795	1.28	0.015	1256
THACl	6247	8560	1.37	0.016	1404
TOACl	6129	9005	1.47	0.019	1485
TdodecylACl	6176	7794	1.26	0.018	1372
TBABr	5909	7219	1.21	0.013	1150
TBAI	5871	7638	1.30	0.011	1180
TMABr	6333	7805	1.23	0.011	1462

formation of the active intermediate of catalyst in the organic phase is increased with an increase in the total carbon atom of the quaternary ammonium salt. In addition for anions with high polarizabilities more organophilic QASs are favorable. However, more detailed mechanistic study seems to be needed to explain these effects more clearly.

All polymers, regardless of the use of ionic complexes, have high molecular weight and narrow molecular weight distribution (1.21-1.47) (Table 1). It demonstrates that the addition of QASs and imidazolium salts (not shown) do not influence the termination reaction so much. The narrow MWD of polyols keeps the viscosity molecular low. However, the level of unsaturation of polymer chain became increased slightly by the addition of QAS and imidazolium salt, demonstrating isomerizations are activated together with the accelerated propagation reaction. Considering induction period depends on the exchange equilibrium between the dormant site and active sites during the initial period of polymerization [3], the QAS and imidazolium salt additives may prevent the strong coordination of complexing species to the dormant sites, hence increase the propagation rate and also accelerate various isomerization reactions resulting in unsaturated chains. The detailed mechanistic discussion will be made elsewhere. Above results demonstrate that fine-tunings of DMC catalyst such as activity, induction period, and level of unsaturation are possible only by using simple additives.

ACKNOWLEDGEMENTS
This work was supported by the Center for Ultramicrochemical Process Systems and the Korea Institute of Industrial Technology Evaluation and Planning. The authors are also grateful to grant No. R01-2003-000-10020-0 from the Basic Research Program of KOSEF and to the Brain Korea 21 Project and the National Research Laboratory Program.

REFERENCES
[1] I. Kim, S. Lee, WO 2004045764 (2004); Chem. Abstr., 141 (2004) 7649.
[2] J. Kuyper and G. J. Boxhoorn, J. Catal., 105 (1987) 163.
[3] I. Kim, J.-T. Ahn, C.-S. Ha, C.S. Yang and I. Park, Polymer, 44 (2003) 3417.
[4] I. Kim, J. –T. Ahn, D. –W. Park, S. –H. Lee, I. Park, Stud. Surf. Sci. Catal. 145 (2003) 529.
[5] L. Wu, A. Yu, M. Zhang, B. –H. Liu, L. Chen, J. Appl. Polym. Sci. 92 (2004) 1302.
[6] I. Kim, S. H. Byun, C.-S. Ha, J. Polym. Sci.: Part A: Polym. Chem. 43 (2005) 4393.

Author Index

Kim Dong Kwon	869	Kobata T.	833
Kim Dong Kyu	701	Kobayashi M.	821
Kim E.Y.	121	Koh J.-Y.	445
Kim G.-J.	205, 313	Kojima T.	115, 733, 829
Kim H.	297	Komai S.	257
Kim H.B.	429	Komiya S.	733
Kim H.-J.	605	Konno K.	821
Kim Heesoo	265, 609	Koo J.-G.	765
Kim Honggon	301	Kook Y.H.	385
Kim I.	853, 857, 873, 877	Koyama M.	9
Kim I.-K.	581	Krishnaiah D.	713
Kim J.	233	Kubo M.	9
Kim J.H.	237	Kwak B.-S.	625
Kim J.-H.	473	Kwon T.-I.	697
Kim J.-H.	577	Kwon Y.	649, 809
Kim J.-I.	765	Laxmi Narasimhan C.S.	55
Kim J.-J.	509	Lee B.G.	613
Kim J.M.	317, 437	Lee C.M.	749
Kim J.S.	873	Lee C.W.	469
Kim J.-Y.	589	Lee D.H.	557
Kim Jae-Sung	509	Lee D.I.	61
Kim Jin Ho	777	Lee D.-K.	145, 393, 545
Kim Jin Hyun	689	Lee D.S.	149, 165, 401, 477
Kim Jin-ho	737	Lee E.-M.	509
Kim Joo-Sik	317, 437, 553	Lee E.Y.	549
Kim Jung-Sun	581	Lee G.-D.	237, 253, 261, 801
Kim K.	189, 825	Lee H.-C.	313
Kim K.H.	61	Lee H.-I.	353, 785
Kim K.-J.	457	Lee H.-S.	589
Kim K.-S.	273	Lee H.W.	129
Kim Ki-Ho	589	Lee J.B.	501
Kim Kyung-Hoon	225, 329, 433	Lee J.-B.	445
Kim M.H.	305	Lee J.D.	249, 425
Kim M.-I.	225	Lee J.H.	473
Kim P.	265, 297, 609	Lee J.-H.	765
Kim S.	437	Lee J.K.	385
Kim S.-C.	145, 393, 545	Lee J.-M.	509
Kim S.D.	103, 557, 565, 569	Lee J.W.	329, 361
Kim S.G.	253	Lee J.-W.	345
Kim S.H.	757	Lee Jae Sung	201
Kim S.-H.	313	Lee Jang Woo	361
Kim S.J.	141, 429, 513	Lee Jea-Keun	581, 585
Kim S.-J.	205	Lee Jeewon	125, 137
Kim S.S.	757	Lee Jinsuk	837
Kim S.W.	137	Lee Jitae	685, 845
Kim T.	825	Lee Jong-Hwan	441
Kim T.-H.	273, 277	Lee Jongku	705
Kim T.Y.	141, 429, 513	Lee Jong-Ku	809
Kim Y.	149	Lee Ju Seok	629
Kim Y.C.	301	Lee Jun Woo	757
Kim Y.-H.	125, 137	Lee K.H.	697
Kim Y.M.	461	Lee K.J.	873
Kim Y.-M.	477	Lee K.T.	449
Ko Y.-J.	301	Lee K.-Y.	297, 621, 625

STUDIES IN SURFACE SCIENCE AND CATALYSIS

Advisory Editors:
B. Delmon, Université Catholique de Louvain, Louvain-la-Neuve, Belgium
J.T. Yates, University of Pittsburgh, Pittsburgh, PA, U.S.A.

893

Volume 106 **Hydrotreatment and Hydrocracking of Oil Fractions**
Proceedings of the 1st International Symposium / 6th European Workshop,
Oostende, Belgium, February 17-19, 1997
edited by **G.F. Froment,B. Delmon and P. Grange**

Volume 107 **Natural Gas Conversion IV**
Proceedings of the 4th International Natural Gas Conversion Symposium,
Kruger Park, South Africa, November 19-23, 1995
edited by **M. de Pontes, R.L. Espinoza, C.P. Nicolaides, J.H. Scholtz and M.S. Scurrell**

Volume 108 **Heterogeneous Catalysis and Fine Chemicals IV**
Proceedings of the 4th International Symposium on Heterogeneous Catalysis and
Fine Chemicals, Basel, Switzerland, September 8-12, 1996
edited by **H.U. Blaser, A. Baiker and R. Prins**

Volume 109 **Dynamics of Surfaces and Reaction Kinetics in Heterogeneous Catalysis.**
Proceedings of the International Symposium, Antwerp, Belgium,
September 15-17, 1997
edited by **G.F. Froment and K.C.Waugh**

Volume 110 **Third World Congress on Oxidation Catalysis.**
Proceedings of the Third World Congress on Oxidation Catalysis, San Diego, CA,
U.S.A., 21-26 September 1997
edited by **R.K. Grasselli,S.T.Oyama, A.M. Gaffney and J.E. Lyons**

Volume 111 **Catalyst Deactivation 1997.**
Proceedings of the 7th International Symposium, Cancun, Mexico, October 5-8,
1997
edited by **C.H. Bartholomew and G.A. Fuentes**

Volume 112 **Spillover and Migration of Surface Species on Catalysts.**
Proceedings of the 4th International Conference on Spillover, Dalian, China,
September 15-18, 1997
edited by **Can Li and Qin Xin**

Volume 113 **Recent Advances in Basic and Applied Aspects of Industrial Catalysis.**
Proceedings of the 13th National Symposium and Silver Jubilee Symposium of
Catalysis of India, Dehradun, India, April 2-4, 1997
edited by **T.S.R. Prasada Rao and G.Murali Dhar**

Volume 114 **Advances in Chemical Conversions for Mitigating Carbon Dioxide.**
Proceedings of the 4th International Conference on Carbon Dioxide Utilization,
Kyoto, Japan, September 7-11, 1997
edited by **T. Inui, M.Anpo,K. Izui,S.Yanagida and T.Yamaguchi**

Volume 115 **Methods for Monitoring and Diagnosing the Efficiency of Catalytic Converters.**
A patent-oriented survey
by **M. Sideris**

Volume 116 **Catalysis and Automotive Pollution Control IV.**
Proceedings of the 4th International Symposium (CAPoC4), Brussels, Belgium,
April 9-11, 1997
edited by **N. Kruse, A. Frennet and J.-M. Bastin**

Volume 117 **Mesoporous Molecular Sieves 1998**
Proceedings of the 1st International Symposium, Baltimore, MD, U.S.A.,
July 10-12, 1998
edited by **L.Bonneviot, F. Béland, C.Danumah, S. Giasson and S. Kaliaguine**

Volume 118 **Preparation of Catalysts VII**
Proceedings of the 7th International Symposium on Scientific Bases for the
Preparation of Heterogeneous Catalysts, Louvain-la-Neuve, Belgium,
September 1-4, 1998
edited by **B. Delmon, P.A. Jacobs, R. Maggi, J.A.Martens, P. Grange and G. Poncelet**

Volume 119 **Natural Gas Conversion V**
Proceedings of the 5th International Gas Conversion Symposium, Giardini-Naxos,
Taormina, Italy, September 20-25, 1998
edited by **A. Parmaliana, D. Sanfilippo, F. Frusteri, A.Vaccari and F.Arena**

Volume 120A **Adsorption and its Applications in Industry and Environmental Protection.**
Vol I: Applications in Industry
edited by **A. Dąbrowski**

894